MW01257565

THE AMBONESE CURIOSITY CABINET

EFFIGIES
GEORGII EVERHARDI RUMPHII, HANOVIENSIS ÆTAT^S: LXVIII.

Cæcus habens oculos tam gnavæ mentis acutos,
Ut nemo melius detegat aut videat:
RUMPHIUS hic vultu est Germanus origine, totus
Belga fide et calamo: cætera dicet opus.

ex tempore posuit — N. S. Gub: Am b:

THE AMBONESE CURIOSITY CABINET

GEORGIUS EVERHARDUS RUMPHIUS

Translated, edited, annotated,
and with an introduction by

E. M. BEEKMAN

YALE UNIVERSITY PRESS
NEW HAVEN AND LONDON

The lines from "The Blind Seer of Ambon" used in the epigraph
are from *Travels* by W. S. Merwin, © 1992 by W. S. Merwin,
reprinted by permission of Alfred A. Knopf, Inc.

Publication of this book was made possible by the generous support of the
Prins Bernhard Fonds (Prince Bernhard Foundation)
and the Nederlands Literair Produktie- en Vertalingenfonds
(Foundation for the production and translation of Dutch literature).

Frontispiece: This is the only known portrait drawn from life. Rumphius' son, Paulus Augustus, drew the likeness sometime between October 1695 and July of 1696 in Kota Ambon, about six years before his father's death. The print states that Rumphius was sixty-eight when he posed for this portrait, and that the governor then in office wrote the Latin encomium. This was Nicolaes Schaghen, who was Ambon's governor from 1691 to 1696. The Latin verse reads in translation: "Though he be blind, his mental eyes are so sharp that no one can best him at inquiry or discernment. Rumphius is a German by birth but his loyalty and pen are completely Dutch. Let the work say the rest."

Designed by Sally Harris/Summer Hill Books.
Set in Janson type by Tseng Information Systems, Inc., Durham, North Carolina.
Printed in the United States of America by Edwards Brothers, Inc., Ann Arbor, Michigan.

Library of Congress Cataloging-in-Publication Data
Rumpf, Georg Eberhard, 1627–1702.
[D'Amboinsche rariteitkamer. English]
The Ambonese curiosity cabinet / Georgius Everhardus Rumphius ; translated, edited, annotated, and with an introduction by E. M. Beekman.
p. cm.
Includes bibliographical references (p.) and index.
ISBN 0-300-07534-0 (alk. paper)
1. Crustacea—Indonesia—Maluku—Early works to 1800. 2. Mollusks—Indonesia—Maluku—Early works to 1800. 3. Mineralogy—Indonesia—Maluku—Early works to 1800. I. Beekman, E. M., 1939– . II. Title.
QL441.5.I5R8513 1999 595.3'09598'5—dc21 98-40046

A catalogue record for this book is available from the British Library.

The paper in this book meets the guidelines for permanence and durability
of the Committee on Production Guidelines for Book Longevity
of the Council on Library Resources.

2 4 6 8 10 9 7 5 3 1

For Faith: my Susanna

Examine how your *Humour* is inclin'd,
And which the *Ruling Passion* of your Mind;
Then, seek a *Poet* who *your* way do's bend,
And chuse an *Author* as you chuse a *Friend.*
United by this *Sympathetick Bond*,
You grow *Familiar*, *Intimate* and *Fond;*
Your *Thoughts*, your *Words*, your *Stiles*, your *Souls* agree,
No Longer his *Interpreter*, but *He.*
Earl of Roscommon

The old men studied magic in the flowers,
And human fortunes in astronomy,
And an omnipotence in chemistry,
Preferring things to names, for these were men.
R. W. Emerson

Nestas fábulas vãs, tão bem sonhadas,
A verdade, que eu conto nua e pura,
Vence toda grandíloca escriptura.
Camões

Verhaalt hem, dat u heeft een droeve lange nagt
Met een geleende pen en oogen voortgebragt.
G. E. Rumphius

I take a shell in my hand
new to itself and to me
I feel the thinness the warmth and the cold
I listen to the water
which is the story welling up
I remember the colors and their lives
everything takes me by surprise
it is all awake in the darkness
W. S. Merwin

CONTENTS

ILLUSTRATIONS

The original sixty plates, reproduced from the first edition of *D'Amboinsche Rariteitkamer* (1705), were inserted in the present volume at the point of first reference. This principle did not eliminate the confusion of the original; the reader will find textual descriptions of certain items far removed from their appearance on a plate. This cannot be prevented unless one performs drastic surgery on the illustrations, since the subject matter was determined by the availability of space on the etcher's plate and not by the reader's convenience.

Listed here are the modern scientific names of the shellfish, stones, and other curiosities illustrated in the plates. Each plate number appears in the upper right corner of the plate. The original "captions," at the ends of some chapters, were supplied by Simon Schynvoet for the first edition (see the Introduction, p. xc). Where those captions appear in the current book is indicated by page number.

		Book 2	Page of Plate	Page of Original Caption
	5.	*Mancinella alouina* (Röding)?		
	6.	Perhaps *Thais aculeata* (Röding) or *Thais bituberculata* (Lamarck)		
XXV	A.	*Phalium glaucum* (L.)	120	121
	B.	*Phalium areola* (L.)		
	C.	*Phalium bisulcatum* (Schubert & Wagner)		
	D.	*Casmaria ponderosa* (Gmelin)		
	E.	*Casmaria erinaceus* (L.)		
	1.	*Phalium bandatum* (Perry)		
	2.	*Phalium flammiferum* (Röding)		
	3.	*Cassis tessellata* (Gmelin), from West Africa		
	4.	*Phalium exaratum* (Reeve)?		
	5.	*Phalium saburon* (Bruguière), from the Mediterranean and West Africa		
	6.–9.	Specimens of either *Casmaria erinaceus* (L.) or *Casmaria ponderosa* (Gmelin)		
XXVI	A.	*Chicoreus ramosus* (L.)	122	124
	B.	*Cymatium lotorium* (L.)		
	C.	*Chicoreus brunneus* (Link)		
	D.	*Homalocantha scorpio* (L.)		
	E.	*Cymatium pyrum* (L.)		
	F.	*Haustellum haustellum* (L.)		
	G.	*Murex tribulus* L.		
	1.	*Chicoreus axicornis* (Lamarck)		
	2.	*Hexaplex cichoreum* (Gmelin)		
	3.	*Murex pecten* Lightfoot		
	4.	*Bolinus brandaris* (L.), from the Mediterranean		
	5.	*Bolinus cornutus* (L.), from West Africa		
XXVII	A.	*Tonna tessellata* (Lamarck)	129	132
	B.	*Malea pomum* (L.)		
	C.	*Tonna perdix* (L.)		
	D.	*Tonna cepa* (Röding)		
	E.	*Purpura persica* (L.)		
	F.	*Rapa rapa* (L.)		
	G.	*Bula ampulla* (L.)		
	H.	*Atys naucum* (L.)		
	I.	The land snail *Pythia scarabeus* L.		
	K.	*Ficus subintermedia* (d'Orbigny)		
	L.	*Architectonica perspectiva* (L.)		
	M.	*Nassarius arcularius* (L.)		
	N.	*Nassarius pullus* (L.)		
	O.	The land snail *Planispira zonaria* (L.)		
	P.	The land snail *Nanina citrina* (L.)		

		Book 2	*Page of Plate*	*Page of Original Caption*
	3.	*Conus aulicus* L.		
	4.	Perhaps *Conus pennaceus* Born		
XXXIV	A.	*Conus aurisiacus* L.	155	154
	B.–D.	*Conus ammiralis* L.		
	E.	*Conus pulcher* Lightfoot, from southwest Africa		
	F.	*Conus acuminatus* Hwass., from the Red Sea		
	G.	*Conus genuanus* L., from West Africa		
	H.&I.	*Conus imperialis* L.		
	K.	*Conus ermineus* Born, from the Caribbean and West Africa		
	L.	*Conus achatinus* Gmelin		
	M.	Perhaps *Conus spurius* Gmelin, from the West Indies		
XXXV	A.	*Lambis chiragra* (L.)	157	162
	B.–D.	represent younger stages of *Lambis chiragra* (L.)		
	E.	*Lambis lambis* (L.)		
	F.	*Lambis lambis* (L.)		
	H.	*Lambis truncata sebae* (Kiener)		
XXXVI	G.	Younger stage of *Lambis lambis* (L.)	159	162
	I.	*Lambis millepeda* (L.)		
	K.	*Lambis scorpius scorpius* (L.)		
	L.	*Strombus latissimus* L.		
	M.	*Strombus epidromis* L.		
	N.	*Strombus canarium* L.		
	O.	*Strombus vittatus vittatus* L.		
	P.	*Strombus minimus* L.		
	6.	Perhaps a juvenile specimen of a species of the Turbinellidae family		
	7.	A juvenile specimen of *Turbinella pyrum* (L.)		
XXXVII	Q.	*Strombus lentiginosus* L.	161	162
	R.	*Strombus aurisdianae* L.		
	S.	*Strombus luhuanus* L.		
	T.	*Strombus gibberulus* L.		
	V.	*Strombus labiatus labiatus* (Röding)		
	W.	*Strombus mutabilis mutabilis* Swainson		
	X.	*Strombus marginatus succinctus* L.		
	Y.	*Strombus dentatus* L.		
	1.	Immature *Lambis chiragra* (L.)		
	2.	*Harpulina arausiaca* (Lightfoot), from Ceylon (Sri Lanka)		
	3.	*Harpulina lapponica* (L.), from Sri Lanka and southern India		
	4.	Cannot be identified; a species of the family Volutidae		

		Book 3	*Page of Plate*	*Page of Original Caption*
	5.–13.	cannot be identified, but they are European minerals added by Schynvoet		260, 263
LII	A.&B.	are Mamacur armlets	264	278
	1.–10.	are European crystals		266, 268
	2.	[*not* 12] might be twinned crystals		
	3.–6.	might be quartz crystals		
LIII		Piece of ambergris, viewed lengthwise	297	
LIV		Same piece of ambergris, viewed from one end	298	
LV		Dendrites; none from the Indies	322	323
LVI	A.–D.	Agates	325	324
	E.	A geode		
	F.	An agate		
	G.–I.	Pyrolusite dendrites, but not from the Indies		323
	1.–4.	European "achates"		
LVII		None of these objects can be identified with certainty, but with the possible exception of A., they are not from the Indies	332	333, 344, 347, 369, 372
LVIII		Fossil shells from Europe	351	352
LIX	A.–H.	European fossils	354	353,372
	D.–F.	Fossil *Echinoidea*		
	G.–H.	Fossil fishes		
LX		Fossil shells from Europe	356	355, 372
	Nos. 1–3	are fossil crustaceans (Brachyura)		

ACKNOWLEDGMENTS

As was the case when Rumphius was alive, an author today has to secure financial guarantees. In the present instance, aid was underwritten by the Prins Bernhard Fonds (Prince Bernhard Foundation), and the Nederlands Literair Produktie- en Vertalingenfonds (Foundation for the production and translation of Dutch literature), both located in Amsterdam, the Netherlands. Even more gratifying were individual gestures of devotion to Rumphius and Maluku. Mr. Bart Eaton quietly backed up his interest and love for the project with concrete inducements, while Mr. W. Buijze supported my endeavor as well with a grant in memory of his father, Adriaan J. Buijze. The present effort was also underwritten by the Tropisch Museum.

There is another, dearer coin. A book such as this, once the work of one man, would now sustain a thousand-and-one specialists. I do not embody that many and needed a number of experts to supply me with information or to correct my assumptions. I am very grateful to the following scholarly support groups without whom my annotations would have been more tentative or nonexistent. Despite their guarantees, the usual caveat applies: whatever errors remain are my responsibility. I feel privileged to have met and worked with Professor L. B. Holthuis, emeritus carcinologist in Leiden, who reviewed Book 1 and provided the identifications. A superior era of graceful scholarship retired with him. Book 2 was expertly examined and corrected by Mr. H. L. Strack, from Dordrecht, who supplied the numerous identifications that this book on mollusks required. In 1990 Mr. Strack organized a scientific expedition to Ambon in order to locate and verify Rumphius' invertebrates. The detailed expertise which he acquired from that venture he graciously put at my disposal in order to complete the first integral English translation of a major work by Rumphius.

I also needed to call on humanists. For Latin and Greek I relied mostly on my colleague, Professor Marios Philippedes, and Professor C. Heesakkers helped out on some things while I was in Leiden. The very difficult problem of Chinese in Rumphius' work—conveyed in wondrous concoctions because twice garbled—I had to relegate entirely to my other university colleague, Professor Alvin P. Cohen. Not every puzzle could be solved, but we got pretty close. Portuguese problems were passed by Professor F. C. Fagundes. Seventeenth-century Dutch has its own peculiarities, worsened by the haphazard printing practices of those days. In order to verify surmises and educated guesses, I had a fruitful night's session with René van Stipriaan in his pungent environment, and Dr. Hans Heestermans was an efficient and valuable fax-fellow after I returned to the States. Dr. Heestermans shares my conviction that Rumphius was a poet as well as a scientist. Professor A. Teeuw lent a hand with Malay, and Mr. J. P. Puype, of the Armamentarium in Delft, reviewed seventeenth-century weapons for me. Roelof van Gelder located some essential illustrative materials, and Nick Burningham of the Western Australian Maritime Museum managed to rig me an *orembaai* with India ink.

Strange denizens live in a text such as this one. Exotic items casually stroll by and do not introduce themselves. They have to be tracked down, and in order to do so, it was mandatory that I spend some length of time in the libraries and archives of Holland. Lacking American funds, I was lucky to obtain support from the Foundation for the Production and Translation of Dutch Literature, which sponsored my stay at the Translator's House in Amsterdam during the first part of the summer of 1996, and the Netherlands Organization for Scholarly Research granted me a Visiting Scholar Grant for the next three months, enabling me to live in Leiden and do research there and in other locations in Holland. I am grateful for the support from Professor D. J. van de Kaa and Professor Theo D'haen.

While wandering through the labyrinth of European archives and libraries, one soon becomes persuaded of the need of a cicerone. They are in short supply, but I want to express my thanks to Ms. F. Pieters and Ms. J. C. D. de Sommaville of the Plantage Library of the University of Amsterdam and Mr. J. van Rosmalen at the KITLV in Leiden. Such a guide is even more needed in Germany, and I am grateful to Christel Lentz in Idstein and Gerhard Steinl in Hungen for their help in trying to unravel Rumphius' youth. Despite my crucial stay in Europe, the writing of this book constantly demanded tracking down new leads or checking old ones. That is difficult to do transatlantically, and I am very grateful to Mr. W. Buijze and Dr. E. M. Joon for their unflagging support whenever I called upon them. *Terima kasih.*

I could not have done the present project without the generous, untiring, and ceaseless support of Dr. E. M. Joon in Usquert. In fact, I would not have been able to start this project if he had not offered the initial impetus that enabled me to contemplate this quixotic venture. He never hesitated or remonstrated and was always ready to jump into the breech when the occasion arose. His help and support have been invaluable, and his dedication to bring Rumphius to the world may have been tried, but it was never vanquished.

Finally, I am glad to have had the opportunity of working with Ms. Jean E. Thomson Black at Yale University Press. Nor should I neglect to mention the patience and efficiency of Alice Izer, who was my proxy at the computer, for I am allergic to things that glow in the night.

Illustration Credits

The frontispiece portrait of Rumphius, the illustrations on the opening pages of books 1, 2, and 3, and the sixty plates are reproduced from the first edition of *D'Amboinsche Rariteitkamer* (1705); they were photographed from the edition located at the Beinecke Rare Book and Manuscript Library, Yale University. The map of Indonesia was taken from Albert S. Bickmore, *Travels in the East Indian Archipelago* (New York: D. Appleton and Company, 1869). The detailed map of Ambon and surrounding islands was provided by Wim Buijze, who also supplied the map of the Banda fortress used in the notes. The signature of Rumphius when he was a young man is printed with permission of the Hessiches Hauptstaatsarchiv, while the signature from his old age is printed with permission of the Algemeen Rijksarchief in The Hague, The Netherlands. Permission to reproduce Andries Beeckman's painting of seventeenth-century Batavia was granted by the Koninklijke Instituut voor de Tropen in Amsterdam. Reproductions of localities where Rumphius lived in Larike, Hila, Kota Ambon, as well as illustrations in the notes of a sero, bubut, Makassarese standards, and a saloacco shield are courtesy of the Royal Institute of Linguistics and Anthropology (KITLV) in Leiden, The Netherlands. The *orembaai* was drawn by Nick Burningham of the Western Australian Maritime Museum, while the drawing of the Argonauta from the *Miscellanea Curiosa* is reproduced courtesy of Leiden University Library. Vinckboon's drawing of a contemporary funeral, used in the notes, comes courtesy of Collection Frits Lugt, Institut Néerlandais in Paris.

PRINCIPAL SOURCES

Abbreviations by Title

FSBA	Fürst zu Solms-Braunfels'ches Archiv, Braunfels.
G	*Rumphius Gedenkboek, 1702–1902*. Ed. M. Greshoff. Haarlem: Koloniaal Museum, 1902.
HHStAW	Hessisches Hauptstaatsarchiv Wiesbaden.
M	*Rumphius Memorial Volume*. Ed. H. C. D. de Wit. Baarn: Hollandia, 1959.
MC	*Miscellanea Curiosa sive Ephemeridum Medico-Physicarum Academiae Naturae Curiosorum*. Norimbergae: Wolfgangi Mauritii Endteri, 1683–1687.
MCG	*Miscellanea Curiosa sive Ephemeridum Medico-Physicarum Germanicarum Academiae Imperialis Leopoldinae Naturae Curiosorum*. Norimbergae: Wolfgangi Mauritii Endteri, 1689.
OED	*Oxford English Dictionary*. 17 vols. Oxford: Oxford University Press, 1933.
WNT	*Woordenboek der Nederlandsche Taal*. 29 vols. Eds. M. de Vries, H. Heestermans et al. 's-Gravenhage: Martinus Nijhoff & SDU Uitgeverij, 1882–1998. (unfinished)

Abbreviations by Author

De Clercq	F. S. A. De Clercq, *Nieuw Plantkundig Woordenboek voor Nederlandsch Indië*, 2d. rev. ed. Amsterdam: J. H. de Bussy, 1927.
De Haan	F. de Haan, *Priangan. De Preanger-Regentschappen onder het Nederlandsch Bestuur tot 1811*. 4 vols. Batavia: Bataviaasch Genootschap van Kunsten en Wetenschappen, 1910–1912.
De Wit	H. C. D. de Wit, "A Checklist to Rumphius's Herbarium Amboinense" in *M*, pp. 339–460.
Engel	H. Engel, "The Echinoderms of Rumphius," in *M*, pp. 209–223.
Heyne	K. Heyne, *De Nuttige Planten van Nederlandsch Indië*, 2d. rev. ed., 3 vols. Batavia: Departement van Landbouw, Nijverheid & Handel in Nederlandsch-Indië, 1927.
Holthuis	L. B. Holthuis, "Notes on Pre-Linnean Carcinology (Including the Study of Xiphosura) of the Malay Archipelago," in *M*, pp. 63–125.
Leupe	P. A. Leupe, "Georgius Everardus Rumphius, Ambonsch Natuurkundige der zeventiende eeuw," *Verhandelingen der Koninklijke Akademie van Wetenschappen*, vol. XII (Amsterdam: C. G. van der Post, 1871), pp. 1–63.

Lewis & Short — Charlton T. Lewis and Charles Short, *A Latin Dictionary.* Oxford: Clarendon Press, 1879.

Martens — E. von Martens, "Die Mollusken (Conchylien) und die übrigen wirbellosen Thiere im Rumpf's Rariteitkamer," in *G*, pp. 109–136.

Pliny — Pliny, *Natural History.* 10 vols. Trans. H. Rackham, W. H. S. Jones, and D. E. Eichholz. Loeb Classical Library. Cambridge: Harvard University Press, 1938–1962.

Rouffaer — G. P. Rouffaer and W. C. Muller, "Eerste proeve van een Rumphius-bibliographie," in *G*, pp. 165–219.

Sewell — William Sewell, *English and Low-Dutch Dictionary.* Amsterdam, 1691.

Strack — H. L. Strack, *Results of the Rumphius Biohistorical Expedition to Ambon (1990). Part I. General Account and List of Stations.* Monograph series Zoologische Verhandelingen, 289. Leiden: Nationaal Natuurhistorisch Museum, 1993.

Valentijn — François Valentijn, *Oud en Nieuw Oost-Indiën.* 5 books in 8 volumes. Dordrecht and Amsterdam, 1724–1726.

Wichmann — Arthur Wichmann, "Het aandeel van Rumphius in het mineralogisch en geologisch onderzoek van den Indischen Archipel," in *G*, pp. 137–164.

Wilkinson — R. J. Wilkinson, *A Malay-English Dictionary (Romanised).* 2 vols. Reprint ed. London: Macmillan, 1959.

Yule — Henry Yule & A. C. Burnell, *Hobson-Jobson: being a Glossary of Anglo-Indian colloquial words and phrases.* 1886; reprint ed. New Delhi: Munshiram Manoharlal, 1979.

MEASURES AND WEIGHTS

Length

voet [foot]	between 10 and 12 *duim* or inches.
duim [inch]	between 1/10th and 1/12th of a foot; about 2.5 centimeters or 0.98 inches.
vinger [finger]	both in terms of length of a finger or its width; if width of a finger, it was the equivalent of one quarter of a palm (= 3 centimeters) or 0.30 inches.
hand [hand]	in English usage this was equivalent to four inches.
span [span]	presumably equal to the distance between the tip of the little finger and that of the thumb of an outstretched hand; often said to be equal to two palms, i.e., 6 centimeters or 2.36 inches.
vadem [fathom, also spelled "*fadom*"]	as a measurement of length on land equal to 1.698 meters or 66.85 inches; as a measurement of depth at sea, the VOC's fathom was equivalent to 1.88 meters or 74 inches.
el or *elle* [ell]	about 0.70 meters or 27.56 inches; also known in India as *asta.*
schrede [pace]	either two-and-a-half *voeten* or 25 inches, or a surveyor's pace (and Rumphius had been a surveyor), which was 5 *voeten* or 50 inches.
houtvoet	same as "*Groninger voet,*" or 0.292 meters, i.e., 11.50 inches; an "*Amsterdamsche houtvoet*" was 0.283 meters or 11.14 inches.
Duitsche mijl [German mile]	equal to 7,536 meters or 4.68 English land miles; in the seventeenth century an English mile was 1,760 yards or 5,280 feet.
a *musket shot*	calculated to be approximately 200 meters or 218 yards.

Weights

ons or *once* [ounce]	ordinarily 1/16th of a *pond* (pound)—which varied from 430 to 494 grammes—or about 31 grams, just over an ounce.
drachma [quint or dragme]	1/8 of an *ons* or about 4 grams; 8 dragmes equaled an ounce.
grein [graine]	as an apothecary weight 1/5760 of a medicinal *pond* or 1/20 of a scruple since 20 graines equaled a scruple.
scrupel [scruple]	an apothecary weight equivalent to 1/528 of a *pond* or 20 *grein* or graines.
pistol ball	its maximum size was 11 millimeters or 0.43 inches.
musket ball	its maximum size was 18 millimeters or 0.71 inches.

cannonball [stone]	a one-pounder (a pedrero ball, for instance) would have a diameter of nearly 2 inches, a two-pounder almost 2½ inches, and a cannonball weighing three pounds would have a diameter of approximately 3 inches.

Coins (also used as weights)

stuiver [stiver]	was 1/20 of a *gulden* [guilder] or the equivalent of 8 *duiten;* between 17 and 20 millimeters in diameter.
dubbeltje [dime]	coin worth 2 *stuivers* [stivers].
schelling [shilling]	silver coin worth 6 *stuivers*, about 25 millimeters in diameter.
sesje or *zesje*	a small silver coin worth 6 *stuivers* or the equivalent of a *schelling*.
reaal [real]	Rumphius probably refers only to the silver real, worth anywhere from 3 to 8 *stuivers*. It weighed approximately 27 grams or just under an ounce.
rijksdaalder [rixdollar or dollar]	a silver coin worth 50 *stuivers* or in British money at the time, the equivalent of 4 shillings and 9 pence; about 40 millimeters across, and weighing almost 30 grams.
dukaat or *ducaton* [ducat]	if of silver, it had the same value as a *rijksdaalder;* if of gold, it was the most important gold coin at the time, weighing 3.5 grams.
pagoda	small gold coin, minted in India, about 5 millimeters diameter.
kupang	a measure of weight for gold, about 3.4 grams.
tael or *tail*	a Japanese coin, worth three and a half guilders at the time and weighing about 3.5 ounces.

Volume

mutsje [quartern]	less than a quarter of a pint.
kan [can]	between 1.4 and 2 liters, or between 3 and 3-½ pints.

INTRODUCTION: RUMPHIUS' LIFE AND WORK

The seventeenth century was not a sentimental age and accordingly was not prone to personal confessions. Autobiographical writings were scarce, and personal information about anyone not of the ruling class is very difficult to find and mostly a matter of luck if one does. In Rumphius' case we have, except for scattered references to himself in his posthumous works, some letters, a brief autobiographical statement that is part of a letter from 1680,[1] and a long narrative poem about his youth, written in Latin and printed in his *Herbal*.[2] Unfortunately, the poem breaks off at the most crucial point of disclosure. Yet there is enough other evidence to make it possible, by inference or by conjecture, to arrive at a plausible degree of veracity, even if absolute objective truth cannot always be achieved. Rumphius also lived in complicated times, but one cannot adequately address their complexity in a relatively brief introduction. I hope, therefore, that readers will grant my cursory remarks the benefit of the doubt and avail themselves of the additional information in the notes.

LIFE IN EUROPE

Georg Everhard Rumphius[3] was born in Hesse, in central Germany, at a time when that country was still a contentious chaos of aristocratic competitors. The first third of Rumphius' life, for instance, was entwined with the fortunes of the Solms family.[4] The House of Solms had existed since the twelfth century but split into various branches during the next two hundred years. In 1384 the town and castle of Braunfels (west of Giessen) became the principal seat of the Solms counts. Between 1420 and 1436, the domain was divided again, with Count Bernhard II of Solms-Braunfels securing Braunfels, Hungen, and Wölfersheim (the last two towns are southeast of Giessen) and Count Johann V controlling holdings such as Lich and Hohensolm, towns which are not pertinent here. In 1550 Count Philipp (1494–1581), of the Bernhardine line, introduced Protestantism into his domain, and his son, Conrad (1540–1592), established the Reformed faith in both Braunfels and the hilly region called Wetterau (northeast of Frankfurt). Conrad's religious persuasion had been influenced by his brother-in-law, Count Johann VI of Nassau-Dillenburg, a descendent of an important Protestant family. William of Orange, the leader of the Dutch revolt against Spain and the founder of Holland's House of Orange, was a Nassau-Dillenburg. Count Conrad had married William of Orange's sister, Elisabeth of Nassau-Dillenburg. This is just one of many instances of the pervasive influence of the Netherlands in the fortunes of this German region and of the man posterity has celebrated as the supreme chronicler of tropical nature.

Count Conrad of Solms-Braunfels had five sons. This plethora of siblings caused a fraternal division in 1602. Only Count Johann Albrecht I (1563–1623) and Wilhelm I (1570–1635) are important to our story, although it should be mentioned that another Solms-Braunfels, Countess Amalie (1602–1675), became one of the most important figures in Holland after she married William of Orange's third son, Frederik Hendrik, in 1625, thereby becoming the mother of Stadtholder Willem II and the grandmother of Willem III, later the king of England. The 1602 division granted Greifenstein and Wölfersheim to Count Wilhelm I, who, given his talents as a fortifications expert, turned the castle at Greifenstein into the region's most formidable stronghold. Rumphius' father, August Rumpf, worked for Count Wilhelm and accompanied him to Prague and Hungary in 1623 and 1624, when the count was in the service of the Holy Roman Emperor, Ferdinand II (1578–1637), as a fortifications engineer. The count's brother, Johann Albrecht I, was of a different persuasion. He had been assigned Braunfels and, not surprisingly for a nephew of William of Orange, was a staunch supporter of the Protestant cause. He was an intimate of the elector palatine Frederick V (1596–1632), the unfortunate "Winter King," whose defeat in 1620 at the Battle of White Mountain, outside Prague, inaugurated the brutal hostilities of what became known as the Thirty Years War (1618–1648). Johann Albrecht followed his dethroned employer (whose mother was Louise, one of William of Orange's daughters) into Dutch exile and died in The Hague in 1623. His daughter was the aforementioned Amalie von Solms, and his son, Johann Albrecht II, may well have hindered August Rumpf's son. The third branch was that of Solms-Hungen in the Wetterau region, headed by Count Otto II.

With the Thirty Years War,[5] considered by most historians the worst disaster for Germany prior to the twentieth century, we have introduced the other important force in Rumphius' youth. He was a child of war, born when Wallenstein, the imperial general who inspired Schiller to write his dramatic masterpiece at the end of the next century, drove the Protestant King Christian IV back into Denmark. During the initial phase of this succession of brutal campaigns, the imperial, that is to say, Catholic and Habsburg, cause of the Holy Roman Emperor Ferdinand II was far more successful than that of the Protestant Union. But there was no victory on either side decisive enough to end the slaughter. The political and religious complexities of this chaotic conflict are far too numerous and difficult to be presented in this brief summary; suffice it to say that what began as a religious struggle between Protestant and Catholic soon became mired in local political strife. After 1624, the erstwhile "German war" was usurped by foreign powers, and Germany became a battlefield where the fate of the Habsburg dynasty was decided. Whatever the European perspective may have been, several aspects of the war are relevant to Rumphius' life. One was that most of the fighting was done by armies of mercenaries that lived off the land, plundering and pillaging their way through a region until they left it a wasteland. The horrors of such devastations, deliberately planned by military commanders, inspired what is unanimously considered the greatest German novel of the seventeenth century, *Der Abentheurliche Simplicissimus Teutsch* (1669), by Hans Jakob Christoffel von Grimmelshausen (?1622–1676), and the series of engravings known as the "Miseries of War" (1632–1633) by the French engraver Jacques Callot (1592–1635).

Grimmelshausen and his masterpiece[6] represent an instructive corollary to Rumphius' youth in Germany. Only six years older, Grimmelshausen came from the same region that Rumphius called home. Both men became soldiers at the age of eighteen, and both had witnessed the devastation of Protestant Hesse by Catholic imperial troops. Both found solace in the written word after the middle of the century, and Grimmelshausen's novel even contains a possible clue to Rumphius' immigration to the tropics. At the end of chapter 20 in book 3, Simplicissimus implies that the then-current expression for leaving home—vanishing, disappearing—was "going to where the Pepper grows,"[7] that is to say, the East Indies. Until he became blind, Grimmelshausen's younger compatriot never regretted that he made the metaphor a reality.

The first part of *Simplicissimus* gives an account of the war in Rumphius' native region. A large portion of the first book takes place in Hanau,[8] the city Rumphius would proudly claim as his own on the title pages of his major works, and his father most likely experienced the siege that Grimmelshausen describes, because it is documented that he was in Hanau in February of 1638.[9] In chapter 29 of book 1, the 1635 taking of Braunfels by Protestant troops is mentioned—poor Braunfels was captured five times by the two sides—while a little later refugees from what had been the fertile Wetterau region are described as "looking cross-eyed from hunger" and "croaking at Hanau's doorsteps" from starvation.[10] Rumphius' birthplace, Wölfersheim, was located in the Wetterau, a hilly region east of Frankfurt, between the Taunus Mountains and Vogelsberg.

In book 3 Grimmelshausen states that to act decently and honestly was known as "to be Dutch."[11] At this time Holland was a seventeenth-century incarnation of the Promised Land. Golo Mann corroborates this in his biography of Wallenstein when he writes that the German Protestants "pinned their hopes on Holland. Those who were of the Protestant faith looked to Holland."[12] Whereas Grimmelshausen writes about Switzerland as "an earthly paradise" because it is a country of peace, which makes it as strange "as Brazil or China,"[13] Rumphius pays comparable homage to Holland in his autobiographical poem, in Latin, about his youth and early manhood. He apostrophizes the country that would be his employer for the greater part of his life.

> Vah! quae deliciae! quae gens in pace locata!
> Perpetua hic requies, & bona vita viget.
> [lines 75–76]
>
> (Oh, what delights, what peace these people enjoy!
> Quiet reigns eternal here, and the good life thrives.)

Another thing that Grimmelshausen's novel teaches us is that those who survived the Thirty Years War might not have lost their faith in God, but they did cease to believe in religion. Very soon after the onset of the wars, religion could no longer claim to have a moral imperative. Time and again, Grimmelshausen shows how irrelevant "sides" were and that an intelligent person must become cynical after years of religious fanaticism. At one point Simplicissimus tells a Calvinist minister that he follows "neither St. Peter [= Catholic] nor St. Paul [= Protestant]. I simply believe what is stated in the twelve articles of the common Christian faith, and I will not decide to join either one or the other, until one of them has succeeded in persuading me with conclusive proof that it is the only true and saving faith."[14] Lack of tolerance had cost Germany at least 20 percent of its population, while in areas like Hesse 40 percent of the rural population had been destroyed by violence, famine, and the plague.[15]

A concomitant disgust with aristocracy was inevitable. In a celebrated baroque allegory, Grimmelshausen described how the societal tree, nourished at its roots by such "inferior people" as artisans and peasants, was top-heavy with aristocrats. The nobility lived off the peasants and commoners, who

> nevertheless gave the tree its strength and renewed its vigor when this had been depleted. . . . And they had to endure much from those who sat in the tree, and for good reasons, for the tree's entire weight rested on them and this pressed them so hard that all their money was squeezed out of their purses and from their strongboxes even if they had been secured with seven locks. And if there was no money forthcoming, certain commissioners curried them with combs, called military raids, which squeezed lamentations from their hearts, tears from their eyes, blood from under their nails, and marrow from their bones.[16]

This lopsided situation did not change significantly after the war. Rumphius, for instance, discovered that by the middle of the century a ruler such as Idstein's Count Johannes could still legally invoke the power of the Augsburg Peace of 1555, which decreed that local rulers could impose their creed on their subjects, under the motto *cujus regio, ejus religio*, or "he who rules the land, dictates its religion." Such a dictum was mocked by the expedience of circumstance. This is where politics and religion intersected and where hypocrisy flourished. France's Henry IV set the example when he converted to Catholicism for the sake of his throne. Even as exemplary a figure as William of Orange began life as a Lutheran, converted to Catholicism for an inheritance, and died a Calvinist. During the Thirty Years War combatants of each persuasion switched religions with cynical ease, making a mockery of a nobleman's honor or ethical obligations. And if one disobeyed the religious imperative, this immediately implied the possibility that one would disobey secular and civil law as well. Rumphius confronted such persistent intolerance in Idstein even when the war was over. After so much bloodshed one had to turn away from patriotism and one's native realm if there was the opportunity to go where the pepper grew.

One has to keep these negative influences in mind when one considers the first third of Rumphius' life, those years which made it possible for him to remove himself to what was, quite literally, the other side of the earth. He fled from chaos, from a ruined country that had been pillaged, raped, and murdered into exhaustion. He escaped from fraudulent authority, hypocritical religion, and social inequality, a state of affairs which did not warrant allegiance to anything human. And yet given these conditions, one must say that Rumphius was fortunate.

His father, August Rumpf (?1591–1666),[17] was a much-sought-after engineer and *Baumeister*. Although the last term included the responsibilities of an architect, it also implied the more practical duties of supervising the actual building of a structure or repair works. Grimm's dictionary states that in the seventeenth century, *Baumeister* was the equivalent of *Bauherr*, what the Romans called *aedilis*. The latter can be described as a superintendent of public works. August passed on this combination of art and practicality to his elder son, who was to teach geometry and engineering to a nobleman's sons, as his father did, and redesign strongholds in the Indies in the service of the Dutch East Indies Company. It was a profession that trained one to be attentive to detail.

August Rumpf was employed by German nobility during and after the Thirty Years War. His employers were local nobles within an area, roughly speaking, between the city of Giessen in the north and Frankfurt-am-Main in the south and bounded by the Aar River in the west and the Salz River in the east. The right side of this irregular parallelogram in eastern Hesse—an area that includes the region called Wetterau (from the Wetter River) to the west, Vogelsberg to the east, and Gelnhausen (Grimmelshausen's birthplace) and Hanau in the south—is the site of Rumphius' youth. The region bears a remarkable resemblance to New England and upstate New York. Large valleys alternate with rolling hills, and, like New England, it is densely wooded with both conifers and deciduous trees. Although the father freelanced for any number of aristocrats from this region, the most important ones for our purposes were the three branches of the Solms family (August worked at one time or another for each one of them), branches of the Nassau family, and the counts of Hanau. One of his earliest masters was Count Wilhelm I of Solms-Greifenstein, with whom he signed a contract in Regensburg in April of 1623[18] and whom, as mentioned before, he accompanied to Prague in 1623–1624 when the count served the Catholic emperor as a fortifications expert. August Rumpf was working for the same count in 1625, in Birstein, a town near the Salz River, south of Vogelsberg. We know this from a letter he wrote to Wilhelm I,[19] dated 13 August 1625. This most submissive letter—August's son would have avoided such a servile tone and been satisfied with politeness—tells us some interesting facts. For one thing, Rumpf admits to being afraid of roving bands of soldiers. Such "dangerous times" make him want to stay close to loved ones; hence he would like to find housing in Wölfersheim, a small town thirty-seven miles from Birstein, a distance a horseman could cover in a day. The problem, however, is that his employer had not paid him, in either goods or cash, for such services as teaching the count's sons. The disclosure that they do not seem to have been very educable is less important than the fact that education was a private affair, usually—but not entirely—restricted to the upper classes.

Two years later, August Rumpf petitions the same count to be excused from accompanying him on a journey because his wife will give birth to a child in less than two months. The letter, dated 29 August 1627, was written and posted in Wölfersheim.[20] It follows that, if everything went according to nature, a child was born in late October or early November of 1627. We know that Rumphius was August Rumpf's first child,[21] so we can say that Georg Eberhard Rumpf, better known to posterity as Georgius Everhardus Rumphius, was born in the fall of 1627 in Wölfersheim (in the Wetterau region of eastern Hesse), the son of August Rumpf (*Baumeister* of Wilhelm I, Count of Solms-Greifenstein) and Anna Elisabeth Keller (?1600–1651).

We know about Rumphius' mother from a document that details a protracted dispute about a pew in Wölfersheim's Reformed Church.[22] It states that August Rumpf was the son-in-law[23] of Carl Keller, a prominent citizen, and when one considers the Keller family, one has discovered the link which made it possible for Rumphius in later life to write such flawless Dutch. Anna Elisabeth's brother, Johann Eberhard Keller, was a high functionary in Emmerich and Kleve (or Cleve), towns and regions very close to the Dutch border and only twelve miles from the Dutch city of Nijmegen. At the time these places were hostage to a succession conflict between Johan Sigismund of Brandenburg, a Protestant nobleman, and Wolfgang Wilhelm von der Paltz-Neuberg, the Catholic challenger. The Netherlands and France joined the fray, until it was by treaty concluded in 1614 that Kleve and Emmerich would go to the elector of Brandenburg. Hence Rumphius' uncle was a governor in the Kleve region for the elector of Brandenburg. Dutch was spoken there,[24] and it was customary at the time for a younger sister to live with her married brother to learn how to run a household. Johann Eberhard's son, called Johann Wilhelm Keller, had an even closer connection to the Netherlands. This cousin of Rumphius fought with Frederik Hendrik of Orange, acquired a lucrative sinecure in Utrecht, and was even appointed Holland's diplomatic representative in Moscow. He lived in Middelburg, once an important town in the province of Zeeland. Johann Wilhelm's brother-in-law, and Rumphius' other cousin, by the name of Hendrik Becker, also became a high officer in Frederik Hendrik's army.[25] I consider it likely that Rumphius' mother knew Dutch, while at a later date her son lived in a Dutch-speaking neighborhood of Hanau.

There were other children, but one cannot be certain of the number. We know of at least two siblings. One was a sister, who was baptized Anna Catherina Rumpf in the Reformed Church in Hanau, 2 February 1645.[26] This is important information, for it proves that the Rumpf family, including Rumphius, was Reformed.[27] This also fits with our knowledge that the Braunfels-Greifenstein line of the Solms family was Reformed. In fact, the towns of Wölfersheim and Hungen are still predominantly Reformed to this very day. The other child was a younger brother, Johann Conrad Rumpf (1639–1661), who is said to have killed someone in Hanau in 1661. One source states that the victim was a scrivener; another version alleges that at around ten o'clock one night, Rumpf's son, without provocation, drove his sword through the man's back up to the hilt. He was executed for his crime in the marketplace of Hanau's New Town.[28]

Rumphius spent his early childhood—at least up to the age of eight, at the most ten—in Wölfersheim. There he also received his early education, most likely from Johann Georg Venator, who was Wölfersheim's teacher from 1621 to 1635, whereafter he became its Protestant pastor until 1664.[29] A locally venerated scholar, Venator (more simply: Jäger) taught his young charges Latin, Greek, and Hebrew. One must remember that in the seventeenth century children began to learn very early in life and that it was a privilege, not a compulsory right. Most education was private and conferred distinction, marking its recipient as of a better social class. Rumphius also received instruction from his father, since part of August's duties was tutoring his employer's sons. In his autobiographical poem, Rumphius states unequivocally that his father taught him mathematics (line 21), but a later source indicates that August also taught his son how to draw, probably together with the count's sons.[30] We may take for granted that the *Baumeister* instructed his son in the engineering skills required of a fortifications expert. This was not a subject taught in any school at the time and could only be acquired by means of private instruction and apprenticeship.

The collateral horrors of the war did not spare Wölfersheim. The plague devastated the region in 1635, killing many people. Wölfersheim's church register has the melancholy entry that in 1635 the population of Rumphius' birthplace totaled "54 souls, 38 elderly: old men and women, 10 girls, and 6 schoolboys,"[31] of whom Rumphius was most likely one. This was what was left of the town's normal population of 900! Among the plague victims was August's employer, who died in February of 1635. The count's son and heir, Wilhelm II (1609–1676), would not honor August's contract with Wilhelm I, so the elder Rumpf had to look elsewhere for work.

Survival was not certain in those perilous times, so August Rumpf was very prudent to hedge his bets and recruit commissions outside the Wölfersheim area while his employer was still alive. He was also forced to do so for economic reasons since, as mentioned, aristocratic employers often did not pay their bills. This was not an exception but rather the rule. Apart from the very great princes, German nobility had none of the luster with which popular imagination would like to endow aristocrats from the distant past. Regardless of local encomiums to the ruling class, most members of the first estate led a far from enviable existence. Golo Mann summed it up concisely: "Dynasties with too many descendants that did not have the wherewithal to support them abounded. . . . They lived close together in castles where stuffed bruins and boars bared their teeth while a few weather-beaten ancestral portraits constituted the rest of the decor, cold corridors smelled of spiders' webs and rot, and they had no wish to stay for the rest of their lives. Constantly revised family compacts meant that they had to divide among themselves the paltry incomes squeezed from their estates. . . . They sought their fortune in war."[32] Martial fortune, however, was usually very cruel. The year before August's employer died, his relation, Count Conrad Ludwig, was forced to leave his Braunfels domain after it had been captured by imperial troops, and did so with nothing more than two horses and a pack mule.[33] When his younger brother, Johann Albrecht II, came to reclaim Braunfels after the Peace of Münster in 1648 (which ended the Thirty Years War), his property was so impoverished that the family could neither buy nor borrow bed linen. The count's councillor wrote to a colleague that "whoever wants to come here, should pick up his bed and bring it along."[34] It was therefore not unusual that, despite constant employment, Rumphius' father was often impecunious. This explains his son's later request to the Dutch East Indies Company (VOC) to give the money they owed him to his aging parent.[35]

Owing to these practical demands of survival, we find various sources stating that August Rumpf was already on the books as *Baumeister* in Hanau as early as 1630,[36] a position that became permanent in 1637, because in that year he is listed as Hanau's official *Baumeister* in the employ of the count of Hanau-Münzenberg.[37] But the lack of documentary evidence makes it impossible to know for sure when Rumpf moved to Hanau, though probably not before 1650. If Rumphius lived in Hanau before that time, he either boarded at the *Gymnasium* or lived with relatives. On the other hand, plague and famine were also visited upon Wölfersheim and the Wetterau region, which did not recommend them either. But whatever the case may be, August Rumpf had to work, and he did not sever his ties with the Solms family. It is known that in 1642 he worked for Johann Albrecht II, Count of Solms-Braunfels, and later for Reinhard, Count of Solms-Hungen, and his successor, Count Moritz, as well. War leaves a great deal of work for a man of August Rumpf's profession and when the local aristocracy began to restore their properties after the peace of 1648, Rumphius' father is documented to have worked on the restoration or completion of any number of castles (Hadamar, Wächtersbach, Büdinger, Weilburger, and Heidelberg castles, among others), of churches (the one in Birstein, for instance), even his son's *Gymnasium*, which had been left unfinished because of the war.[38]

The city which Rumphius would always associate himself with prospered only after refugees from the Netherlands and Wallonia came to settle there. Prior to the end of the sixteenth century, Hanau was an insignificant little town,[39] but after Philipp Ludwig II, Count of Hanau-Münzenberg (1576–1612), introduced the Reformed Church into his domain, things started to change for the better. His widowed mother married Count Johann VII of Nassau-Siegen (1561–1623), a great champion of the Reformed Church.

Philipp Ludwig was raised in the Reformed atmosphere of Dillenburg and educated at Herborn and Heidelberg, two prominent Protestant educational institutions. In keeping with the *cujus regio, ejus religio* principle, Philipp Ludwig established the Reformed Church in Hanau after he became its count in 1595. From that time on, the city of Hanau was a staunch supporter of the Protestant cause. After the Catholic victory at Nördlingen in 1634, for instance, Hanau was the only significant Protestant bastion left in that part of Germany. In 1597 Philipp Ludwig welcomed Flemish and Protestant Walloon refugees from the Spanish Netherlands. They had fled because of their Protestant faith and thought they would find safety in Frankfurt. Instead, Frankfurt's Lutheran city council pressured them to abjure their interpretations of Protestantism, and they were forced to turn to the ruler of the adjacent town of Hanau. Philipp Ludwig's recruitment of these mostly Dutch-speaking Protestants turned out to be a boon for Hanau. The Netherlandic Protestants were skilled craftsmen and good businessmen, and, aware of their economic prowess, Philipp Ludwig granted them extensive privileges, such as guaranteed freedom of religion, tax-free trade, and a charter that permitted them to build a new community, which became known as Hanau's "Neustadt" (New Town).

The New Town prospered and soon outstripped the old one ("Altstadt") in both size and economic prosperity. In character it was also physically different from the medieval old town, since the Netherlanders constructed their city with streets that crossed one another at right angles, a feature also noteworthy a couple of decades later when the Dutch built New Amsterdam on the Hudson. Neustadt Hanau had a large central market square that was lined with the houses of the more prominent citizens. It was in this square that Rumphius' younger brother was executed in 1661 and where, two centuries later, a monument was erected to the Grimm brothers, who were born in Hanau in 1785 and 1786. In 1600, three years after their city charter, the Netherlandic immigrants laid the first stone of what turned out to be a unique church. It consisted of two connected structures, a dodecagonal building for the Walloons and an octagonal one for the Dutch-speaking community. The young Rumphius must have worshiped in the latter.

Skilled silver- and goldsmiths, jewelers, metalworkers, and, later, manufacturers of earthenware, these immigrants brought prosperity to Hanau. In 1661, two Netherlanders, Daniel Behaghel and Jac. van de Walle, established the manufacture of majolica, or faience, a very profitable kind of earthenware. The influence of the foreigners was so pervasive that one commentator alleges that they were responsible for the "Hanau type": a person who was enterprising, freedom loving, and equipped with a healthy dose of horse sense.[40] Their other major contribution was in education. The original inhabitants of the Altstadt had little interest in pedagogy, preferring to teach their children a trade or sending them off to Hanau's New Town as unskilled laborers. But education was very important to the Protestants, and it was Philipp Ludwig's stepfather, Count Johann of Nassau-Siegen, who insisted that he establish a school of higher learning in Hanau. The count of Hanau did so in 1607, but he died in 1612 before actual construction had begun. It was his widow, Katherina Belgica,[41] William of Orange's daughter, who persisted and had the school built and a faculty assembled and who improved its program, until in 1623 it could call itself a "Gymnasium Illustre," a *Hohe Schule*, as it was also called at the time, similar to the established institution in Herborn, near Giessen.[42] This was the *Gymnasium* Rumphius attended and which he praised in his autobiographical Latin poem (lines 25–26):

> Claro Gymnasio fluvialis Hanovia nota est,
> Hic colui Musas Palladiumque chorum.
> (Hanau by the river is known for its famous *Gymnasium*,
> Here I studied the Muses and the Palladian Chorus [= intellectual subjects].)

The curriculum would have comprised the traditional liberal arts, that is to say, grammar, rhetoric, logic, arithmetic, geometry, music, and astronomy. The poem makes it clear that Rumphius was more interested in the humanities than in the sciences. He notes that he "did not have a great talent" for mathematics and that his father had to tutor him in that field. His preference was for "the shining art of the Aönides" (line 20), that is to say, the muses, the arts, and for "the secrets of NATURE," which he wanted to learn "from childhood" (lines 23–24). One can safely say, therefore, that Rumphius was well educated and that his interests were different from his father's. Rumphius did not want to stay in Hesse and follow his father's career of *Baumeister.*

According to his poem (line 27),[43] Rumphius left Hanau at eighteen, which would have been in 1645. He left because, as he states in his letter to Mentzel, he was "burning with an insatiable desire to know foreign lands," a wish echoed in his verse (line 28). The poem also speaks of his eagerness to see Italy because of its "ancient fame" (lines 29–30), an understandable desire for a young man steeped in the classics. But what actually happened to Rumphius in the next three years is still a mystery. The strange events of 1645 or 1646 can only be surmised from his autobiographical poem, which, unfortunately, stops just when it should tell us how and why its subject got to Portugal. The bizarre story of how Rumphius left Germany can be reconstructed from the poem, but we lack objective confirmation, and the subsequent drama in Portugal can only be guessed at from the scattered references in his posthumous work.

The opportunity to leave Germany was provided by a man whom Rumphius identifies in his poem as "the son of the generous and renowned Solms family, the youngest of three brothers, the Count from Holland who had returned to his native Germany, where his brother ruled in Greifenstein" (lines 31–34). From this little information we can deduce that this count, who, according to Rumphius, was "born from the blood of a crocodile" (line 42), was Ernst Casimir of Solms-Greifenstein (1620–1648), one of four sons of Count Wilhelm I of Solms-Greifenstein, August Rumpf's former employer. The poem speaks of either 1645 or 1646, and the nobleman and his eldest son had died in the same year, 1635. Hence the next oldest son, named after his father, became Count Wilhelm II and was in office at the time of Rumphius' departure from Germany. Of the two remaining brothers, Ernst Casimir was the younger. The poem states that the Rumpfs, both father and son, succumbed to the Solms' blandishments, which is not surprising if one realizes that August had worked for many years for Ernst Casimir's father and had more than likely taught his son. Rumphius may well have been a fellow pupil.

Ernst Casimir's short life was devoted to the military. Unmarried, constantly in debt, he lived off his mother. The two main geographical locations of the poem, Venice and Holland, fit this man, because Ernst Casimir served as an officer in the army of the Dutch States-General. His last tour of duty in Holland was in Maastricht under command of his nephew, Johann Albrecht II of Solms-Braunfels. The latter had spent most of his life in the Netherlands. He had been the governor of Utrecht, where his only son (Heinrich Trajectinus) was born, and then became governor of the strategically important city of Maastricht in the southern Netherlands, which is where his younger relative served under him until 1645. Both men, coincidentally, died in the same year.[44]

The young German nobleman was an "*Obrist*," or lieutenant-colonel, when he died. He seems awfully young to hold such a rank; today we would also question the possibility of a young man of twenty-five being in a position to lead mercenaries. It becomes perfectly plausible, however, when we remember that in that age of patronage, his nephew was a high officer in the Dutch government, his grandmother was William of Orange's sister, and his niece's husband, Frederik Hendrik, was the *Stadtholder!* With connections like that, a successful career is almost a matter of course.

The connection with Venice becomes understandable as well if one knows that Ernst Casimir died in 1648 at Kandia, in Crete.[45] The German soldier fell victim to Venice's seemingly endless conflict with the Ottoman Empire. Europe was too busy with a host of other conflicts and left the Republic of Venice to its own devices during its desperate struggle to retain its territory. Constantinople slowly but surely dispossessed Venice of domain after domain, including Crete, which had been subject to Venice for over four centuries, since 1204. During the Fifth Turkish War (1645–1648) Venice lost the island to the Turks. In 1645 the Turks landed an army of 50,000 men and captured the town of Canea, in Crete's west. Venice was obviously desperate for manpower; hence there seems little doubt that Ernst Casimir was not lying about his employer. His death proves it, because in 1648 the Turks laid siege to the capital of Crete, Kandia, but failed to capture it. The siege would go on for more than twenty years: Kandia did not surrender to the Turks until September of 1669.

Given the political and geographical circumstances of that time, it also made sense to convey recruits to a Dutch harbor for transport to the Mediterranean and the Greek archipelago. The poem indicates that Ernst Casimir went along, and there is no reason why he shouldn't have. But somewhere in Holland the scenario was rewritten and the plot changed. We will probably never know what happened exactly because, as in a dream, the details in the poem are quite clear, but the narrative transitions are vague.

The recruits were first transported all the way north to Wesel, "where the river Lippe joins the Rhine" (line 50). Loaded onto "six ships," they went down the Rhine toward Holland. Again Rumphius is quite specific: they went down the Waal, and in Rotterdam they were transferred to six other ships (lines 56–60). By this time the German recruits were becoming suspicious about their destination and began to complain, but an incipient mutiny was prevented by more promises and by the threatening attitude of the sailors (lines 61–70). The hirelings seem to have believed that they would be paid "what they were owed" (line 72) in Amsterdam, but when, after sailing across the Zuiderzee, they disembarked in that city, the ships suddenly left, and they were stranded in a strange town in a foreign country. This "lost crowd of Germans" (line 96), no longer commanded by their countryman, were at a loss what to do, but somehow they were later ("*postquam*"—line 103) transported to Texel, the island off the coast of northern Holland where the fleets that went to either the East or the West Indies rendezvoused while waiting for a favorable wind. In Texel the hoodwinked young Rumphius encountered a crowd of Frenchmen (line 106) who had also been recruited "for the Venetian Doge" (line 104), and after waiting for twenty days, two thousand men (line 117) were crammed into three ships and the fleet set sail for an unknown destination.

The first ship flew the admiral's flag, the second was called the *Prince of Orange* ("Auriacus Princeps"—line 114), and the third, which held August's unlucky son, bore the name of *Black Raven* ("Niger corvus"—line 123). Only when they were at sea and under full sail did they discover that their ship was going to Brazil.[46] The rest of the poem describes them sailing south through the Channel, past England, through the Gulf of Biscay, their encounter with bad weather, and the torment of strange diseases. Both foul weather and the distemper that killed several men are described in what can only be called rhetorical clichés. The entire passage seems melodramatic, but it most likely tells the truth.

Rumphius was probably part of an ill-fated attempt to relieve Dutch colonials in Brazil. This expedition left the Netherlands in the spring of 1646 in the hope of reinforcing what was otherwise a precarious situation in South America, largely the fault of the Dutch West Indies Company (WIC).[47] It had been founded in 1621 in imitation of Rumphius' later employer, the Dutch East Indies Company, which was better known by its Dutch initials, VOC. Holland's merchantile oligarchy created the WIC in order to harass the Spanish enemy in the Americas and dislodge the Portuguese from the western shore of Africa. In practical terms this meant that the WIC was primarily a privateering outfit during the first half of its existence—this period, from 1621 to 1647, is the only one of interest here; the WIC lasted until 1791—and a slave trader for the second. Little about the WIC could be called inspiring. It required more money than the far more successful VOC to get started; in fact it did not have enough capital to become operational until two years after its incorporation, and it was bankrupt for most of its existence. Its greatest public and financial success was Piet Hein's capture of one of Spain's Mexican silver fleets off Cuba in 1628, a celebrated event that took place when Rumphius was still an infant.

On the whole, the WIC can only be called a failure, though it made various individuals very rich, but some of its nonachievements had considerable historical significance. A case in point is the WIC's attempt to colonize a region that is now New York and Connecticut. But what concerns us here is the WIC's strange Brazilian adventure. In 1623 the WIC's governing body decided to dislodge the Portuguese, that is to say, Iberian power, from the northern coast of South America. In 1624 it launched an attack on Bahia, achieved a Pyrrhic victory, and a pattern was set for the next three decades. The only time that Dutch Brazil prospered was during the tenure of Johan Maurits van Nassau (1604–1679), who was governor of that northern region of Brazil from 1637 to 1644. He was recalled to Holland in the latter year owing to budgetary restraints. Though the Brazilian venture cost a great deal of money, the WIC's directorate never provided sufficient funds to make the attempt a lasting success—an economic policy typical of Dutch colonialism—and by the 1650s Holland's influence in the South American continent was marginal. It was during that earlier time of crisis that Rumphius was snared.

Johan Maurits went home to The Hague and to his mansion (which is now better known as the Mauritshuis Museum). The WIC had replaced him with Holland's favorite form of government: the committee. It consisted of five men, one of whom was Hendrik Haecxs, who kept a journal which describes the same journey as the one narrated in Rumphius' poem. When in 1645 and 1646 the local *moradores*, or Portuguese Brazilians, expelled the Dutch all across the region, the WIC lacked the funds for a large rescue operation and had to ask the States-General, the Dutch Republic's governing body, for aid in men and money.[48] The States-General finally conceded a subsidy by the end of 1645, and in February and May of 1646, a fleet of ships with a considerable number of troops left for Brazil.[49]

There are all kinds of factual problems that cannot be dealt with here. Sources differ on the number of ships, number of troops, exact date of departure, and who was included, but it is clear that ships left from both Flushing (Vlissingen) in Zeeland and the island of Texel, north of Holland, and that they rendezvoused in the English Channel. It is also clear from contemporary witnesses such as Pierre Moreau, Michel van Goch's French secretary, that no one left before February of 1646, because they were experiencing one of the worst winters of that era.[50] During the six months it took them to get to Brazil,[51] they had nothing but contrary winds, storms, gales, and enormous seas.[52] This severe weather was due to "the most inopportune season of the year for sailing"[53] and produced huge problems, conditions which tally with Rumphius' poem. The same can be said of the putrid water, spoiled food, and the strange diseases Rumphius reports (lines 153–218). Moreau also discusses the bad relationship between the soldiers and sailors, even an abortive mutiny by the "German troops."[54] It also becomes understandable that a ship could be fairly easily captured or shipwrecked on the Lusitanian coast. Moreau notes that after rounding Cape Finisterre, they sailed "along the coast of Portugal," including a stretch of between "forty and fifty miles over against the city of Lisbon."[55] There are unfortunately no exact matches, but there seems little doubt that the journey described in the second half of Rumphius' poem (lines 103–218) is the same as the 1646 expedition to relieve the siege of Pernambuco as described by Moreau and Haecxs. It is just a difference of perspective—the very real disparity between a common hireling below deck and the officials and commanders who quartered on the poop-royal.

Casualty rates in Brazil were staggering, and Rumphius was spared certain death when he landed, instead, in Portugal. We do not know how he got there. Was he shipwrecked or was the *Black Raven* captured? Nor do we know whether he was pressed into military service by the Portuguese or lived there as some kind of hostage. All one can say for certain, on the basis of scattered references in his works, is that he knew Portuguese well[56] and that he refers to himself as being part of a company of German troops. We also have a fair idea of how long he stayed in Portugal. In the letter to Mentzel, written in 1680, he says that when he felt the urge to "know foreign lands," he "went first to Portugal," and after "three years" he "traveled to the East Indies." He adds that this was "28 years ago," which leads us to conclude that the year he went to the tropics was 1652, since the letter was dated 1680.[57] We know conclusively that Rumphius left for Java the day after Christmas in 1652, and we also have evidence that he was working back in Hesse in the summer of 1649; hence it is reasonable to assume that he was in Portugal between 1646 and 1649.

It becomes clear from his letter to Andreas Cleyer, dated 18 August 1682,[58] as well as from passages scattered throughout his *Herbal*, that Rumphius was some kind of military man during those lost years in Portugal. The Catholic government had no scruples about hiring outsiders, but, as some of Rumphius' passages reveal, the local population did not take kindly to the foreign heretics. Rumphius might well have been part of frontier troops, because when he does mention specific locations, they are usually border regions.[59] Be that as it may, there is no doubt that he was a member of a German contingent. For instance, in chapter 45 of book 6 of his *Herbal* he introduces a particular variety of tough tropical shrub that could be made to grow into impenetrable fences. After describing the Indonesian plant, Rumphius goes on to say, "I saw [the same] in the South of Portugal, which is called *Algarbia*, and I understood the Portuguese to say that they had learned [to grow such natural fences] from the Brazilians."[60] Such hedge fences grew to a height of four feet and could not be breached with bare hands, "but we German soldiers knew how to cut a door in it with our cutlasses, hacking through it as if it were butter."[61] In a completely different text that describes algae, Rumphius informs the reader, "I have also found this grass a great deal in Portugal, in the river Minius which separates Portugal from Gallicia. It is a very placid stream wherein it is so dangerous to bathe that several of our soldiers drowned in it if they so much as got that grass around their legs, for it pulled them under, especially where the bottom was like Quick-sand."[62] One final instance of a reference to Portugal is in the 1682 letter to Cleyer, wherein Rumphius states that he once saw a variety of parsley in Portugal that was a great favorite of the local population. They would guard the plants jealously and prevent anyone from harvesting them, but "we Soldiers sometimes cut some off to use as a Salad or a vegetable, despite being pelted with slingshots."[63]

Apart from such scattered references, primarily in his *Herbal*, we have no way of knowing what Rumphius did during this Portuguese interim. He must have experienced things worthy of Simplicissimus, but whatever they may have been, surviving them only confirmed the fundamental grit of his character. The man who returned to Germany sometime in 1649 was still young in years but not in experience. I have the feeling that he had turned into something of a recalcitrant loner who was no longer content to submit to a restricted career at home. But the seventeenth century was an age of patronage, and in Germany too one did not fare well without an aristocratic protector.

Rumphius was back in Hesse by the summer of 1649, as mention is made of a visit by both father and son to Idstein on 16 July of that year.[64] Idstein is a town in the Taunus Mountains, northwest of Frankfurt and Hanau. It was ruled by Count Johannes von Nassau-Idstein (1603–1677), to whom Idstein and Wiesbaden were assigned after yet another fraternal division of an aristocratic domain. A staunch adherent of the Protestant cause, Count Johannes lost his property to the emperor after the Protestant defeat at the battle of Nördlingen in 1634. The count lived in exile for over a decade, first in Metz, then in Strasbourg, and could not return to his home until 1646. Upon his return, he began to repair his war-damaged property. He recruited Flemish craftsmen,[65] installed a French formal garden, and started various building projects to repair and improve his castle. He was a cultured man who commissioned a "Florilegium" to record the plants from his garden, acquired an art collection, and even wrote a moral text for his numerous offspring. It is strange that such a man, who also had suffered from the privations of the Thirty Years War, would show intolerance toward his subjects, yet it is documented that this educated Protestant man felt compelled to persecute individuals as witches with the full connivance of his subjects and officials.[66] The same intolerance made the count, who, after all, was the absolute secular and spiritual ruler of his domain (*cujus regio, ejus religio*), condone an anti-Calvinist campaign that disturbed his youthful *Bauschreiber*, who had just returned from Portugal after the Peace of Münster.

It was August Rumpf who secured the job as Idstein's *Bauschreiber* for his son. The Count of Idstein negotiated Rumphius' contract (*Bestallung*) with his *Baumeister*, who continued to live in Hanau and only came to Idstein if the building project demanded his presence. Rumphius' contract, signed in September of 1650, appointed him to the position of something like a project controller.[67] What the job of *Bauschreiber* entailed, at least in Idstein, gives some idea of the complexity of the position, and it also indicates the importance of August Rumpf's profession. No wonder Rumphius was such a valuable asset to the VOC during the next few years. This particular *Bauschreiber* was contractually obliged to oversee the work crews and be their liaison with the castle hierarchy. One can't help wondering whether one of the reasons that Rumphius got the job was that he spoke Dutch and could communicate with the Flemish workmen. The *Bauschreiber* also had to supply the drawings (what we now would call blueprints) for whatever structure was going to be built. It is known, for instance, that Rumphius drew the sketches for the ceiling decorations of shells and snails for one of the formal garden's grottoes.[68] He also had to teach several of his employer's children, using a curriculum that covered "Arithmetic, Geometry, Architecture, and related Arts." The last probably referred to drawing and painting,[69] and this particular *Bauschreiber* had learned all five skills from his *Baumeister* father. In return, the count agreed to pay Rumphius fifty *Gulden* per year; provide him with a horse for official business; supply him with clothes, a pair of boots, and a cloak; and arrange housing,[70] board, and a daily ration of wine. On paper this sounds like a lucrative position, but the son encountered the same problem his father had contended with in Wölfersheim twenty-five years earlier. Less than six months after signing the contract, August Rumpf is asking the count to pay what he owed his *Bauschreiber*. In other words, Count Johannes was yet another petty nobleman scraping by.

His domain's poor soil (consisting of pulverized slate) made commercial farming impossible, nor was there any industry. The timber he needed for his building projects had to come from elsewhere, as is documented by a text dated 28 July 1651. It describes the arrival of rafts of logs down the river, an event attended by Rumphius, the count, and various other notables.[71] Yet Count Johannes, unlike many of his peers, did not mind spending available funds on culture. He had a library, for instance, which, not surprisingly, was well stocked

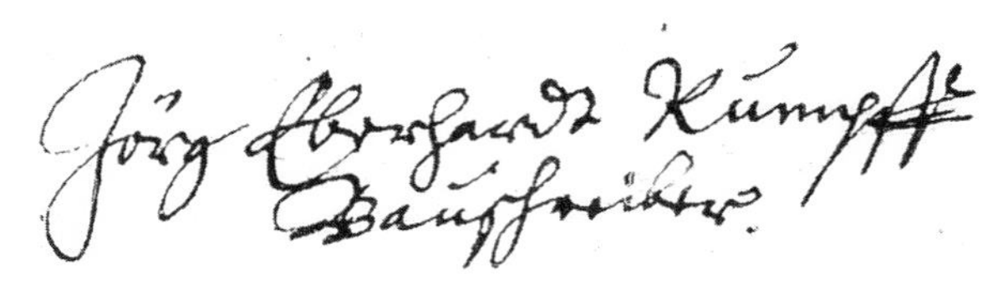

Signature of the young Bauschreiber *Rumphius from 1651, when he still signed his name as Jörg Eberhardt Rumpf.*

with works on architecture, but it also contained a copy of Pliny's *Historia Naturalis*, a work that later became almost as important to Rumphius as the Bible. But things did not turn out well for Rumphius, even though at first it must have been a relief to lead a normal existence again after the Portuguese detention. Rumphius lasted in Idstein only a year and a half.

Among the various reasons one can suggest for Rumphius' departure from Idstein is an ethical one. The Rumpf family was Reformed and devoted to Hanau. This implies their firm belief in religious tolerance and, as mentioned before, a love of freedom. After all the horrors of the Thirty Years War, it must have been offensive to watch a ruler who was cultured and who himself had suffered the privations of a religious war reinstitute intolerance in the form of witch hunting and the denunciation of Calvinism. The count might have done both for political and ideological ends, but that would only emphasize how noisome Idstein was to August Rumpf's elder son. In any case, the young Rumphius remonstrated against the denunciation of the Calvinists in public, dissenting behavior that reached the ears of his employer.[72] The difference between the two men can be judged from Rumphius' attitude toward the lore he learned from Indonesian herbalists. He never mocked them but treated their knowledge with respect. As he states in the preface to his *Herbal*, he recorded their information "because there always lies hidden among those fables a modicum of truth or latent attribute, as is true of the fables of *Ovid* and other *Poets*." The same distaste for peremptory behavior also made him refuse the position of *fiscaal* in 1661. It was a prestigious position in the colonial Indies, comparable to our office of public prosecutor, implying a higher income and social position. But for Rumphius the idea of hounding people must have been particularly abhorrent.

A major personal tragedy might also have provided the impetus to strike out on his own. Rumphius' mother died in December of 1651. She had been ill for a long time, and the domestic crisis forced August to recall his son to Hanau.[73] It is clear, however, that the younger Rumpf never again returned to Idstein.[74] Rumphius' mother, Anna Elisabeth Keller, was buried on the twentieth of December 1651,[75] leaving the widower with serious financial problems. It could well be that the son did not want to be an additional burden on his father, though I would also think that after his mother's death he no longer felt any obligation to remain in Germany. By the end of December 1651, Rumphius was gone. We know this from his father's letter from Hanau to the count in Idstein (dated 26 December 1651): he does not know "whether my son is still at his place or if he traveled somewhere without my knowledge, he could also be sick or in some other trouble, as is prone to happen to youth."[76]

We know that exactly one year later Rumphius sets sail for the Indies as an *adelborst*, translated in Sewel's contemporary dictionary as "Gentleman souldier." This was not a rank of great distinction, yet it was a notch up the social ladder, clearly separating Rumphius from more common personnel. Most likely, young Rumpf was helped by his mother's family, particularly by his influential cousin and his mother's nephew, Johan Wilhelm Keller. Keller lived in Middelburg, an important town in Zeeland, a province that had a strong presence on the board of directors of both the VOC and WIC. It is also known that the Kellers returned to their native region every so often. For instance, Johann Wilhelm was back in Wölfersheim in either 1650 or 1651, and his father, Johann Eberhard, was back in 1652. This very real possibility of personal intervention by his mother's family—something his father could not have done, nor would the elder Rumpf's employers have had any influence with the VOC—explains why the young *Bauschreiber* signed on with the Dutch company, why he could enjoy the modest status of *adelborst*, and why he rose rather rapidly in the company's hierarchy during the next couple of decades.

Georg Eberhard Rumpf left his family, Germany, and the European continent the day after Christmas of 1652. He would never see the continent, his country of origin, or his kin again. Father and son did correspond with each other from time to time, and in 1663 Rumphius instructed his employer, "out of filial duty," to send some back pay to "my old father, by the name of Augustus Rumph, Baumeister in Hanau."[77] By that time August was no longer a widower. He had married Anna Margarethe in 1660. She was the widow of the erstwhile driver of the count of Hanau-Münzenberg.[78] He died six years later and was buried in Hanau's New Town on 11 April 1666.[79]

Rumphius signed on with the VOC, or the Dutch East Indies Company, under the name of "Jeuriaen Rumph"[80] and sailed on 26 December 1652. He left Holland on the *Muyden*,[81] a 400-ton yacht that sailed from Texel, the same point of departure as for the disastrous Brazil journey seven years before. This time, all went well. On the eighteenth of April 1653, the ship had already reached the Cape of Good Hope, and it arrived safely in Java on the first of July. Captain Evert Theunisz. Harnay had lost only nine people on this rapid voyage of only six months.[82]

LIFE IN THE INDIES

"When one approaches the Isles of the Indies from the Sea, it is the *Palma Indica* or *Cocos-tree* that first comes into view, its crown exceeding all others." Rumphius never forgot the sight and with a fine sense of imaginative design "appointed it the Captain of [his] *Ambonese Herbal*," beginning his description of the coconut palm on the first page of the first chapter of the first book of his magnum opus.[1] Coconut palms lined the canals of Batavia, headquarters of the Dutch East Indies Company. The VOC, or Company (so indicated hereafter), had been in existence for about half a century when Rumphius set foot on Java, and Batavia had been its capital over the three decades since Coen, Holland's equivalent of Albuquerque, had founded it in 1619. Rumphius aptly called it the "Water-Indies" and distinguished this collection of more than thirteen thousand large and small islands from the "Old Indies," that is, the subcontinent of India and what used to be known as Indochina. As he told the readers of his *Herbal:* "The Water-Indies stretch from Sumatra, then Java, and the other large Islands, in a long chain to the East, once nicely called the eyelids of the world by *Julius Scaliger*, then with the South-easter Islands they curve to the North and North-west, through the Moluccos, to end in the North with the *Phillipine* or *Manilhase* Islands. In the most remote corner of these Water-Indies to the East, one will find the three governments, of Amboina, the Moluccas, and Banda; these are enclosed in the East and separated from the South Sea by the land which one is wont to call *Nova Guinea*."[2] It turned out that this remote corner of the island realm was to be home to Rumphius for most of his life, but in July 1653 Java itself must have seemed the most distant shore imaginable.

Batavia was an impressive and welcome sight after half a year at sea. At first Rumphius

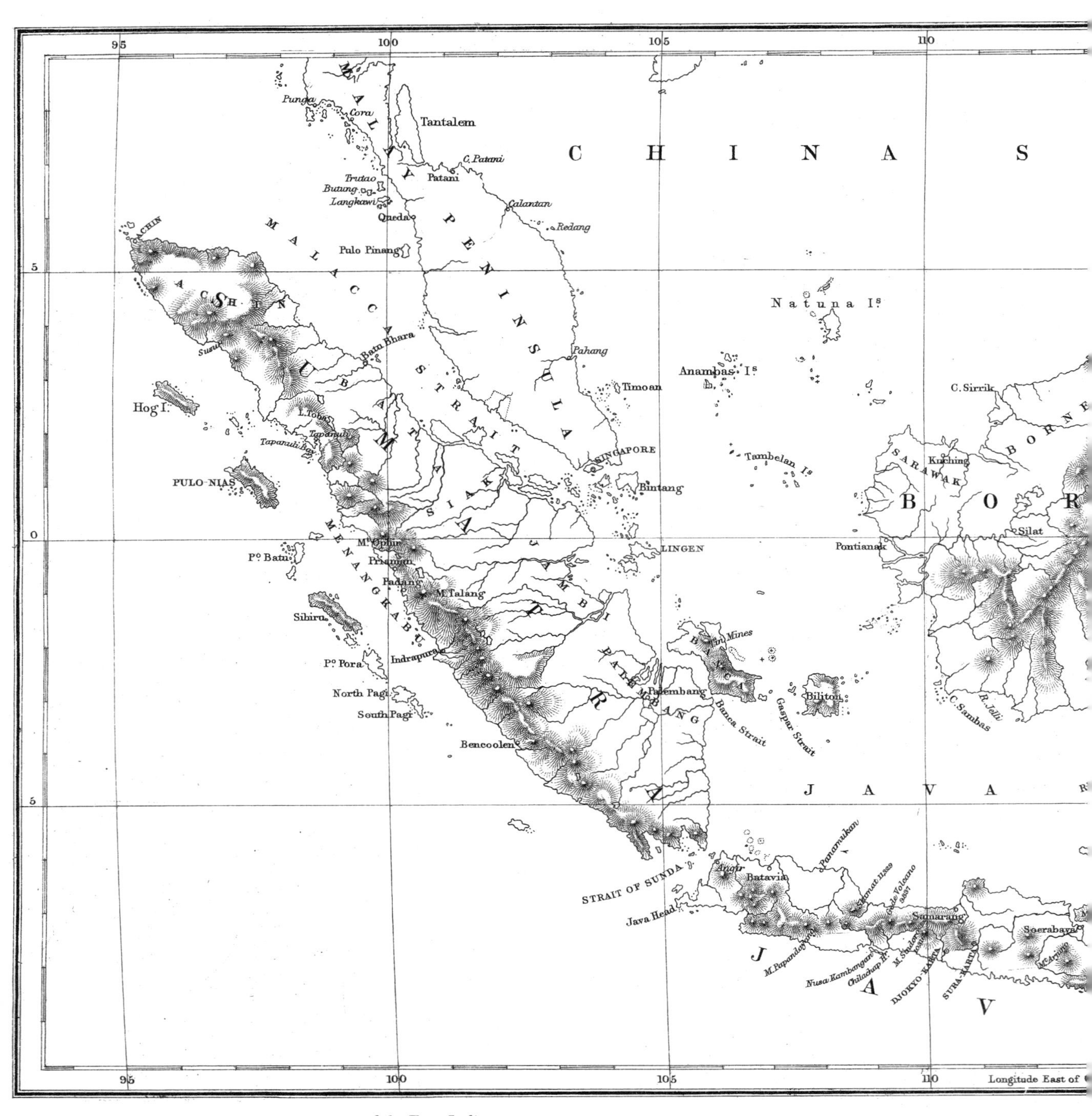

Nineteenth-century map of the East Indies.

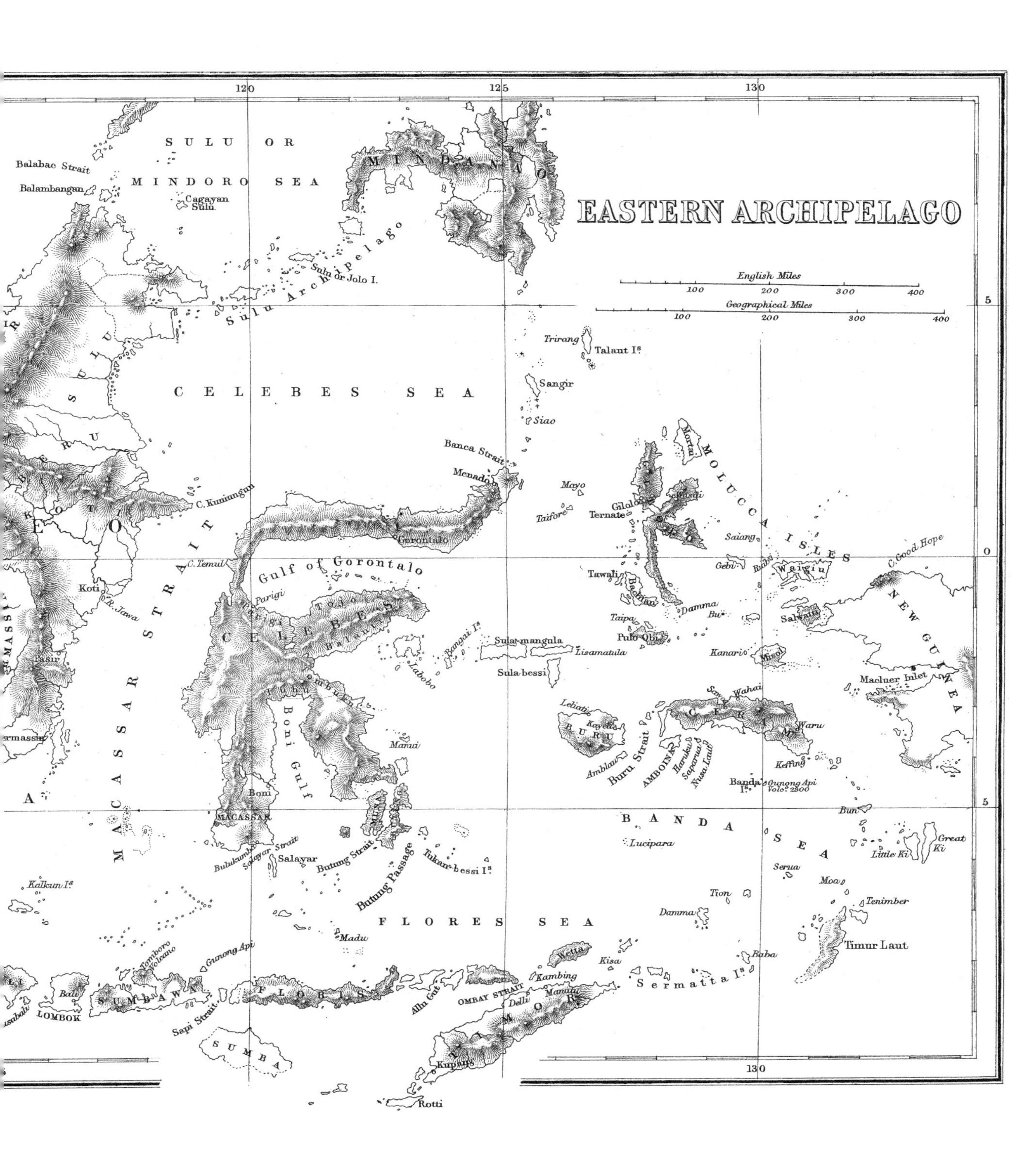

EASTERN ARCHIPELAGO
English Miles
100
200
300
400
Geographical Miles
100
200
300
400
120
125
130
SULU OR MINDORO SEA
CELEBES SEA
MACASSAR STRAIT
FLORES SEA
BANDA SEA
MINDANAO
Sulu Archipelago
Sulu or Jolo I.
Balabac Strait
Balambangan
Cagayan Sulu
Trirang
Talaut Is
Sangir
Siao
Banca Strait
Menado
Gorontalo
Gulf of Gorontalo
C. Kuniungan
C. Temul
Koti
R. Jawa
Pasir
Parigi
CELEBES
Boni Gulf
Boni
MACASSAR
Bulukumba
Salayar Strait
Salayar
Butung Strait
Butung Passage
Tukan-bessi Is
Madu
Kalkun Is
Sula-mangula
Sula-bessi
Lisamatula
Labobo
Manui
MOLUCCA ISLES
Morty
Gilolo
Ternate
Mayo
Taifore
Tawali
Batchian
Damma
Taipa
Pulo Obi
Saiang
Gebi
Waigiu
Salwatti
Kanari
Misol
C. Good Hope
NEW GUINEA
Macluer Inlet
Leliati
Kayeli
BURU
Amblau
Buru Strait
AMBOINA
Haruku
Saparua
Nusa Laut
Wahai
CERAM
Waru
Keffing
Banda Is
Gunong Api Volc. 2300
Lucipara
Bun
Great Ki
Little Ki
Serua
Moa
Tion
Damma
Tenimber
Timur Laut
Baba
Kisa
Wetta
Kambing
Sermatta Is
Delli
OMBAY STRAIT
Allu Gat
TIMOR
Kupang
Rotti
FLORES
SUMBA
Sapi Strait
Gunong Api
Tomboro Volcano
SUMBAWA
LOMBOK
Bali

must have been blinded by the dazzling white walls of the fort (known as the Castle) where he was quartered. Like the immured city itself, these walls had been built from white coral-stone, which glistened in the tropical sun. The Dutch used the coral not only because it was cheap and easily available but also because this material absorbed seventeenth-century bullets, rendering them harmless.[3] Rumphius had arrived during Java's best season, the dry monsoon, which lasts from April to September. This is the exact opposite of Ambon's; there the preferred season is between December and March and the wet monsoon is from May to September.[4]

It was customary to give the new arrivals three days off so that they could recover from the rigors of the ocean voyage.[5] Most of the soldiers disappeared into taverns and whorehouses on the infamous Lepel- and Zand-streets,[6] but I would think that Rumphius' inquisitive disposition urged him to investigate the Castle first, which contained a zoo that housed such unusual animals as rhinos, tigers, and crocodiles, and an elephant house where the Company's pachyderms were kept.[7] And when Rumphius toured the town of Batavia itself, he must have been amazed by the strange phenomenon of a Dutch town under a hot tropical sun. Stone houses, fronting straight tree-lined canals, had been built in square blocks as orderly as those in Hanau's New Town. They were just as occluded as residences in Amsterdam or Delft, a serious error in a climate where ventilation is a fundamental health requirement. The houses lacked gardens but were adorned with gables and stoops, and the streets, paved with the same coral used for the city walls, were blessed with sidewalks. But if his eyes were momentarily deceived, his nose would have told him that he was now in a country that was in every sense as far removed from Hesse as one could possibly imagine. The air was piquant, markets were redolent, local people smelled of spices, but the Europeans stank. The Dutch considered bathing an abomination and ridiculed the Javanese, who washed themselves every day in the *kali*, or river. Batavia's inhabitants must also have convinced him that the strange attempt to replicate Holland in a Javanese swamp had failed. Besides Chinese and Japanese immigrants, greater Batavia was host to an ethnic miscellany from all over the archipelago. They spoke their own languages and dressed according to their local traditions. For a man as linguistically gifted as Rumphius, it must have been fascinating to wander around this Asian Babel. This first encounter with the Indies must have been both a shock and a delight.

Company contracts were for five years. We know very little about Rumphius' first tour of duty between 1652 and 1657, yet it does not seem likely the Company would let a man of his talents go to waste on guard duty.[8] The VOC needed men of quality, because they got all too few. The Dutch themselves were not inclined to move, nor did Holland have a large surplus population such as England did. The result was a payroll of social misfits from the Netherlands and an unending stream of foreigners, particularly Germans. Under the circumstances, a man of Rumphius' manifold abilities could not be ignored. He might have had introductory letters to a local patron, but since each recruit was interviewed upon arrival and asked about his civilian occupation,[9] competent authorities must have discovered that this man possessed many qualities beyond the two basic requirements of being able to speak Dutch and being a member of the Reformed Church. He also spoke German, which, as just mentioned, was a language much in demand in the VOC, not to mention Portuguese. The latter was still pervasive in the Indies, especially in the eastern part of the archipelago. More important, Rumphius could write. But he would have passed up the chance of becoming a colonial Bartleby, because in the seventeenth century scriveners wrote from six in the morning till six at night, and if there was a lot to copy—and there usually was, since the VOC produced a prodigious amount of red tape—they continued to labor at night by

"De Vischmarkt te Batavia," painted by Andries Beeckman in 1656.

candlelight.[10] His education also distinguished him, though his knowledge of the classics was probably less important than his command of Latin. But Rumphius' most marketable skills were his military experience in Portugal and the engineering training he had received from his father, especially in the construction of redoubts. He had practiced these skills in Idstein, and they must have been put to good use by the VOC during the next few years, though we do not know what he did at the beginning of his VOC career. All we know for sure is that by the time the novice *adelborst* was twenty-six, he was posted to the eastern end of the archipelago; in 1666 Pieter Marville, governor of Ambon at the time, stated in an official letter to his superiors in Batavia that Rumphius "has resided in these parts for some 13 years already."[11] Rumphius later said that "fate" had brought him to the "remotest islands in the East,"[12] but in reality he was very likely on board the fleet for the Moluccas which left Batavia on 8 November 1653 under the command of the controversial Arnold de Vlamingh van Oudshoorn.

When Rumphius arrived in the Indies, Java was far less important to the VOC than the Moluccas or Ceylon or their trading posts in India.[13] The Moluccas had been the original target of the Dutch, and they established their first trading post in the Indies on Ambon, the island Rumphius was never to leave. The lure had been the spice in the shape of an antique handcrafted nail: the clove. As Rumphius expressed it: "This Noble fruit was in former ages only found in the Moluccan Islands, principally on *Mackian*, which quite rightly is considered the Mother of Cloves; this is a place beneath the Equator's burning heat, in the furthest reaches of the Eastern Ocean, and its name was barely known even to the neighboring peoples, so that in Europe reports of this Noble fruit caused the Kings of *Castile* and Portugal to vie with each other and dispatch fleets to find these hidden Islands, so much akin to a second Colchis."[14] The Lusitanian Argonauts found this tropical Colchis first, only to be dispossessed by the Dutch, who came three-quarters of a century later. The local populations were not overly fond of the Portuguese, who, according to St. Francis Xavier, showed remarkable skill in conjugating the verb "to steal . . . inventing new tenses and participles."[15] Hence the Dutch were welcomed as allies. One must remember that the Dutch trading company had at first no inclination to obtain merchandise at gunpoint. It had only commercial objectives in mind and wanted to achieve these by treaty and diplomacy rather than by conquest.

There were three main reasons—none of which was quite as straightforward as is alleged here—for the subsequent bellicosity of the Dutch. The first was profit. In every treaty they negotiated, they obtained promises of exclusivity. From the point of view of any European nation in the seventeenth century, this was of primary importance and, once obtained, worth defending at any cost. The Portuguese had tried it first, Spain practiced it in the New World, and the British were ready to vie with the Dutch anywhere at all. Holland's prosperity had been born from the restrictions imposed on that country by Philip II of Spain; hence political motives as well as commercial initiative were the causes of the burgeoning colonial enterprise.

One must also remember that this large insular realm was not a homogeneous entity. Only once had it been forcibly united. In the fourteenth century the prime minister of the Javanese Madjapahit Empire, Gajah Mada, established a large realm by conquest. Not until after the Second World War, when the free nation of Indonesia was born (equally under Javanese control) can the same be said to have happened again. Most of the time the archipelago consisted of various principalities which rose to power only to be replaced by others. Within each island several factions could be contesting one another's sovereignty, and the competitors would look outside their borders for aid and comfort. Except for Java and southern Sumatra, most islands were really littoral civilizations, established by enterprising Javanese or the seagoing Malay. Allegiances and friendships blossomed and withered with dismaying regularity until an outside force entered that was strong enough to unite the countless factions.

Such a force was Islam. As has been true of so much in the history of the Malay archipelago, this religious force first came to those shores via trade routes. First came the Hindu culture from India, then Buddhism; the Chinese Empire also exerted its influence; and by the turn of the thirteenth century the most lasting non-Western authority appeared. The new faith spread rapidly during the fifteenth century, energized by the fervent zeal typical of recent converts. It has been pointed out that one of Islam's attractions to many chieftains and overlords was its sanction of territorial conquest as a holy task. Islam provided a convenient justification for enlarging commercial realms.[16]

An equally fanatical piety sent the Portuguese in the fifteenth century to sail uncharted oceans in order to extend their crusades against the hated Moors. Those "pagans" considered not only the Portuguese but also anyone else—the disdain was quite democratic—who did not profess adherence to the Prophet to be *kafirs*, or infidels. In the archipelago, Islam's strongholds were Atjeh in northern Sumatra, Malacca on the Malay Peninsula, and the Kingdom of Ternate in the Moluccas. Therefore, when the Portuguese traveled to the legendary Spice Islands in the Moluccas, they intended not only to monopolize one of the most lucrative commodities in the world but also to strike a blow at Islam.

Not until the twelfth century did the Moluccas become a power to reckon with. The original inhabitants had little use for their clove and nutmeg trees but learned quickly enough from Chinese and Arab traders. After some time the Kingdom of Ternate became the dominant power in these islands. By the seventeenth century, this sultanate controlled the northern part of Halmahera, the southern portion of Mindanao (the southernmost of the larger Philippine islands), the eastern third of the northern peninsula of Celebes, the Sulu Islands, and Seram, with the adjacent islands of Ambon, Haruku, and Saparua. Its nearest rival was the Sultanate of Tidore, which ruled over the "bird's head" of New Guinea, the southern half of Halmahera, and the islands of Batjan and Makian. To the west was Ternate's most powerful neighbor, the Kingdom of Macassar, or Gowa, which ruled over the major portions of south and central Celebes, part of the eastern coast of Borneo, and Sumbawa, one of the major Sunda Islands.

Just as tiny Holland had become a major European power, so did the small island of Ternate exercise an influence disproportionate to its size. The subjects of Ternate's sultan were devout Muslims, and the kingdom could look to Gowa (that is, Macassar) for a staunch ally in combating the *kafirs*, be they the Dutch, the Portuguese, or native tribes. By the first quarter of the seventeenth century the Dutch had trouble maintaining their monopoly of the spice trade in Ambon and in Seram, as well as elsewhere. Native populations were pressed into service on the cora-cora fleets, which were used for policing the Banda Sea to ensure contractual adherence to Dutch policy, allowing no one else access to the clove harvest. To make sure that their supply remained constant and that no surplus could tempt their rivals, the Dutch resorted to the infamous policy of either cutting down the clove trees or peeling their bark and leaving them to rot and die. As Rumphius put it in an expressive phrase, the Dutch seemed to "wage war on trees rather than men."[17] Extirpating the spice trees also meant that the livelihood of the people was destroyed, and with it went any desire for cooperation.

As is true of any guerilla war—and the Dutch should have remembered this from their own war of liberation—an oppressed population nursing hatred of its overseers will readily allow an ally to infiltrate their land while providing it with all the assistance it can muster. Ambonese chiefs made common cause with the sultan of Ternate, who, although legally bound by treaties with the Dutch, did his best to undermine their authority and secretly aided the smuggling of the Buginese and the people of Macassar. Any overt act by the European power to enforce its trade agreement only provoked recalcitrance, and thus caught in a vicious circle, both the Dutch and their native opponents resorted to brutal vengeance in a seemingly endless vendetta.

Treaties and reconciliations alternated with warfare. Time and again the governor of Ambon, who also ruled over the neighboring islands, asked for assistance from Batavia. Such military aid was usually successful, but as soon as the Dutch squadron had sailed out of the Banda Sea, unrest and uprisings continued. The Kingdom of Macassar not only provided an outlet for the illegal spices from the Ambon region but also provided the rebellious forces with ships and manpower.

In 1648 the sultan of Ternate, Hamdja, died. He had signed a treaty with Van Diemen which confirmed him in his suzerainty over Seram, Buru, and the Mohammedans of Ambon, but he had also guaranteed that the VOC would be the exclusive purchaser of the cloves. He was succeeded by Mandar-sjah, an ally of the Dutch, but a man much disdained in his own realm as a *makan-babi* (eater of pork), because it was thought that he had converted to Christianity and European ways. The situation was complicated by religious quarrels, with the devout Madjira, Mandar-sjah's viceroy, revealing himself as the rival of his king, and in the ensuing palace revolution, Mandar-sjah was forced to seek refuge with the Dutch. In 1650, Batavia sent De Vlamingh van Oudshoorn to the Moluccas to set things right. He did so, but at a cost which exceeded the desired result. Rumphius, who was surprisingly militant in his ardor for what was after all his adopted country, summed up the situation as follows:

> Up to now we have shown how sundry Surgeons sought to heal the old wound of the Company's Clove trade in Amboina with balms of Persuasion, with Clysters of punishment and strong swaddlings of contracts, agreements, and the like, but to no avail. And now we will be directed by the rules of medicine (Hippocrates, section 8, aphorism 6) and dictate that what art cannot heal needs to be cured with iron, and what the iron cannot heal must be cured by fire. Therefore we will now present two renowned Camp-Barbers, to wit Gerrit Demmer and Arnold de Vlamingh, of whom the first brought to the old and stinking wound the iron and sharp corrosives, while the latter took to burning and cutting off entire limbs, and in this fashion restored Ambon's body back to health.[18]

Arnold de Vlamingh van Oudshoorn (1608–1661) was governor of Malacca in 1645, and of Ambon in 1647, and gained lasting infamy for the manner in which he dealt with the uprisings in Ternate, Ambon, and Seram in 1650. At the time, De Vlamingh had his supporters, although he was also denounced by his contemporaries, even by the Reformed Church.[19] He was a man of what seem now contradictory impulses, but in his day they smoothly dovetailed. A devout Christian, he labored for the promulgation of the Holy Writ, though at the same time it was under his command that the punitive fleets of cora-coras struck terror in the native populace whenever they were discovered off the coasts. These naval expeditions became known as the infamous *hongi* raids. His health was poor, yet De Vlamingh must have had an indomitable will. Often ill and bedridden, he led an attack up a mountain while leaning on a cane.[20] Before he gave the order to attack the stronghold of Assahudi in 1655, De Vlamingh sank to his knees on the beach and said a prayer, yet Dassen reports that when the general had captured Ternate's admiral, Saidi, and was presented with the bleeding captive, De Vlamingh knocked out his teeth, stabbed him through the throat and tongue, and smashed the roof of his mouth.[21]

De Vlamingh's nemesis was Madjira, who also became legendary. When he was surprised in his fortification on the Seramese peninsula, called Howamohel, it was reported that though an entire file of Dutch soldiers fired their weapons at him, Madjira escaped with mere scratches and nicks. The Dutch never managed to capture him, either dead or alive.

There is little point in rehearsing the numerous battles between the VOC and Ternate and its allies. The pattern seldom varied. Madjira would foment an uprising, bring support from Macassar in the form of both ships and manpower, punish the local population if it had submitted to the Dutch, and brace himself for the inevitable attack. De Vlamingh would land his troops, punish the local population if it had submitted to the viceroy from Ternate, devastate the countryside, and attack the *benteng* of his opponent. This was a redoubt that was usually constructed, in time-honored fashion, on top of a mountain, with seven fortifications ascending one after the other, and that had to be taken before the final assault could be ordered. After the conclusion of a campaign De Vlamingh would fall back and wait for reinforcements from Batavia while Madjira would sail to Macassar and obtain aid from his

Muslim ally. Obviously, such a contest could go on indefinitely, and this was one of the reasons that De Vlamingh practiced his version of a scorched-earth policy while at the same time ensuring, in the most literal sense, that the monopoly of the VOC was maintained by not just excluding any other trading power but also destroying the commodity itself.

The major theater of war turned out to be Howamohel, the long and narrow peninsula which pinches off Seram's west and which Rumphius called that island's "bread basket." It was relatively close to Macassar, had numerous sago palms that ensured ready provisions, and had a mountainous terrain that provided tactical advantages for the insurgents. The decisive battle was waged in July 1655 near the harbor Assahudi on Howamohel. The Dutch were victorious, but Madjira managed to escape to Macassar. Fierce fighting also subdued the islands of Manipa, Buru, and Batjan; the Banda Islands had met a similar fate several years earlier.

Since world consumption of cloves had fallen off and could be supplied by a lower yield, the Company ordered that only the islands of Ambon and its three smaller neighbors, the Uliasser Islands, be allowed to cultivate the spice. Second, the inhabitants were told to live on the coastal areas, which made for better control, and, third, depopulation programs were ordered. This meant that the entire peninsula of Howamohel was turned "into an eternal wilderness" and its population transported to Hitu on Ambon on pain of death; if anyone objected to this forced removal from ancestral ground, "his head would be placed in front of his feet."

De Vlamingh was replaced by Jacob Hustaerdt as the fifteenth governor of Ambon in 1656 and died in 1661 in a storm off Madagascar on his way back to the Netherlands. The legend goes that his superior, Governor-General Maetsuycker, dreamt that he heard De Vlamingh call for help and saw him go down with his ship; he noted the date of his dream, and it turned out to be the same day of De Vlamingh's demise.

All accounts of the wars just described mention the constant support of Macassar for Mohammedans. It was certainly clear to Rumphius and others who were living at the time, and evident from the treaties, that religion was a key factor in these campaigns and may have been the major reason for the undue ferocity of the fighting on both sides.

Support from Macassar was substantial. In 1652 the forts near Assahudi and La-ala in Howamohel each counted a garrison of 300 Macassarese. Rumphius reports that in 1642 thirty ships from Macassar sailed to Seram to aid the forces of Madjira.[22] The king of Macassar encouraged what the VOC deemed the smuggling of spices and never stopped selling them to other European nations, such as Portugal or Britain. The Portuguese had been trading with Macassar since 1528, and the city had been their major trade center since 1641. In 1666, the British offered Sultan Hassan Udin their support in case of war with the Dutch. Time and again Madjira escaped to Celebes only to return to do battle in the Moluccas with fresh troops and ships from Macassar. Deliberate provocations such as the killing of Dutch crews and the taking of hostages only intensified the mood of mutual distrust, and the government in Batavia also realized that the Moluccas would never be truly pacified if Macassar remained the dominant power in the region.

The inevitable war with Macassar began in December 1666 and did not end until December 1669, after nearly three years of ferocious battle. The people from Macassar had earned a reputation as outstanding soldiers, implacable foes of Christianity, and skilled mariners. Though the Dutch controlled the seas, Macassar's lighter ships usually managed to escape them and supply the insurgents with food, ammunition, and manpower.

The Dutch commander who earned his fame in this war was Cornelis Speelman. He led 600 Europeans, had support from the Buginese, who were traditional enemies of Macassar, and, ironically, also had troops from Ambon under the command of the legendary Captain Jonker. Speelman was successful despite tough resistance and the decimation of his own ranks by tropical diseases. He finally laid siege to the city of Macassar itself and forced Sultan Hassan Udin to sign what became famous as the Treaty of Bongaja (18 November 1667).[23] The Dutch received reparations both for the shipwrecked crews which had been murdered and for the cost of the war; they furthermore obtained a trade monopoly, excluding the British and Portuguese from any further dealings with Macassar. As Vlekke notes: "Of still greater consequence was the delimitation of Macassar's sphere of interest, which was definitely reduced to the city itself and its near surroundings. Even in this small area the Company exercised its authority and was permitted the fortress Rotterdam, while Dutch coinage was declared to have legal value under the control of the Company, even if for practical reasons it was brought under the nominal authority of native princes."[24]

In March 1668 war flared up again, and Speelman was proven right in placing little faith in the execution of the letter of the treaty. Rumphius had noted that the forces from Macassar used poisoned darts during the Ambonese wars,[25] and now Speelman had occasion to find that some twenty-five years later they used poisoned bullets,[26] with the result that even superficial wounds were incapacitating. A lengthy siege was needed to take the "castle of the king." Trenches were dug to reach its bulwarks and to provide access for the sappers. Casualties due to tropical diseases were staggering among the Dutch troops; more than half either were seriously ill or had died. Within the sultan's surrounded fort, conditions were not much better, yet the war persisted. In June Speelman found sufficient encouragement to order a final assault. The first phase of battle took twenty-four hours of unceasing combat. Four days later the Dutch troops finally breached the twelve-foot-thick main wall and found themselves confronting reinforced dwellings, which had to be taken in house-to-house fighting. Not until nine days after the assault had begun did Speelman finally vanquish the king's redoubt and capture its symbol, the holy canon called *anak-Macassar*. Macassar ceased to be the formidable power it had once been, and the eastern regions of the archipelago were pacified into progressive decline.

Historically, one can see what happened: the VOC had turned from a commercial enterprise into a colonial power. In defending and expanding its mercantile interests, it found itself faced with the decision not merely to delimit these but also to ensure with military might that no more interference be tolerated, either by Western rivals who were always ready to occupy its place if the VOC failed, or by local powers interested in furthering their own cause or in simply defending what they held dear. Vlekke summarizes this fundamental transition in the following manner:

> The war against Macassar is more than a mere episode in the endless series of warlike expeditions undertaken by the Company in the Indies. It marked the beginning of a new period. In the three decades between 1650 and 1680, all the major Indonesian states disintegrated. Ternate had been the first to lose its independence, but Macassar, Mataram, Bantam [both were kingdoms in Java], and even Atjeh followed within a few years. There is undoubtedly a connection between the fact that about 1650 the Dutch Company had definitely established its naval supremacy and the sudden collapse of the Indonesian states. The very nature of these states made a rapid succession of ups and downs in political life unavoidable, but now, for the first time in the history of the archipelago, an outside power, ready to intervene at any moment, stood watching the developments; and this outside power was of a character quite different from that of the Indonesian

> states. It was not subjected to all the vicissitudes through which a sultanate with its never-failing harem intrigues had to go. On the contrary, its government and the succession of its rulers were regulated and stable. It was slow to act, but it never loosened its grip. Too often the Indonesian princes made the mistake of calling upon this foreign power to intervene in their common quarrels, and in the civil wars that broke out in each sultanate because of the absence of fixed laws of succession, and although the Company heartily disliked to hazard troops and money in inland wars, it was resolved, once it had become involved in such a war, to see it through and to have its expenses repaid.[27]

The events here, merely outlined, were part of Rumphius' life and are constantly referred to in his work, including *The Ambonese Curiosity Cabinet.* Of the three Moluccan Wars, which he dubbed the "Madjerasensian Wars," after Madjira, the VOC's implacable enemy from Ternate, Rumphius had no personal knowledge of the first one (1635–1638), but it seems likely that the second one (1650–1656) was the reason he was ordered to the eastern archipelago, and some of his texts suggest that he could well have seen combat during that time.[28] The third he heard about, but by that time he was physically incapable of participating.

Soon after his arrival in the Moluccas, in 1654,[29] the VOC's military authorities were beginning to use Rumphius for purposes other than combat. He made a sketch of the village Cambello (or Combello) on Howamohel, the long and narrow peninsula off western Seram. Tradition said that the village had purloined the clove trees from Makian and had brought them to Cambello, where three huge and ancient trees were considered to be "the mother of all Ambonese Clove Trees."[30] Rumphius also drew a plan of the fort there in the same year, 1654.[31] It had been built of stone in 1646 at the order of Governor Demmer, De Vlamingh's equally tough predecessor, and was called "Hardenberg." The reason for the fort's existence as well as Rumphius' presence there is that Howamohel—frequently mentioned in his work—had become one of the major theaters of war during the VOC's long and arduous campaign to eradicate local rebels, Macassar's power, and the continuing machinations of the Portuguese.[32] Around this time he was already referred to as a mathematician,[33] and before his first tour of duty was finished, he was back doing what he had practiced in Idstein: overseeing building projects. His title was *fabryck,*[34] which a contemporary source explains as being a kind of director of public works.[35] Rumphius was now also an officer, even if of the lowest grade, since he had been promoted to *vaandrig* in 1656.[36] The equivalent of what the British called an ensign, *vaandrig* was the lowest grade of officer in the infantry, a position Rumphius held from 1656 to 1657.[37] Just before the end of his first tour of duty, Rumphius' technical expertise was called upon again when in 1657 he was ordered to inspect the location for a defensive structure on Banda, the largest of the "Nutmeg Islands." He was referred to officially as "Engineer and Ensign" and would return to the same locale five years later.[38] For a moment it appeared as if Rumphius had come full circle.

When his contract expired and he was faced with signing up for a second tour, Rumphius took matters into his own hands. He informed Jacob Hustaerdt, then governor of Ambon, that he "was not well disposed to military duty"[39] and requested a transfer to the civilian branch of the VOC, specifically to replace Evert van Hoorn at his post on the coast of Hitu, northern Ambon. Hustaerdt granted his request. In his missive to his superiors in Batavia, dated 9 September 1657, the governor made the shrewd observation that Rumphius would "undoubtedly be of greater service to the Noble Company" as a factor than if he remained a soldier.[40] Hustaerdt hired Rumphius at the rank of junior merchant (*onderkoopman*) and put him in charge of the Company's interests in Larike.

The Dutch East Indies Company, Rumphius' employer for the rest of his life, was in many ways the most successful business enterprise of the seventeenth century. Its trade settlements, or "factories" (after the Portuguese *fundação*), were found from Persia to New Guinea and from the Cape of Good Hope to Japan. This huge and complicated enterprise has been thoroughly dissected by modern scholars, resulting in a large number of investigations.[41] For our present purposes some general remarks will suffice. Toward the end of the sixteenth century Dutch merchants realized that Portuguese expansionism had lost its momentum. The price of pepper had skyrocketed, and when they realized that Linschoten had shown them the way to Asia, a number of enterprising individuals set out to realize a hostile takeover of the spice trade.[42] At first they squandered their resources among a number of competing companies, but persuaded of the wisdom to consolidate, the rivals acceded to the government's desire to form a single company and in a rare instance of Dutch cooperation, the VOC was created in 1602. The erstwhile rivals became six "Chambers" from different cities in Holland, Amsterdam being the dominant one, and the directors of the different ventures became the executive officers of the new company, a board known as de Heren XVII ("the Seventeen Lords"). The Company's headquarters were in Amsterdam, and, after 1619, its Asian administrative center was in Batavia. The colonial capital housed a governor-general (selected in Amsterdam) and a ruling council (the Council of the Indies). All the Company's far-flung settlements had to report in writing to Batavia, which in turn was responsible to the members of the board in Amsterdam. Many factories had different organizational structures, but those need not be detailed here. Suffice it to say that in Ambon the executive officer was the governor, assisted by various councils, and the other important echelons were *opperhoofd*, or "head" of a region, senior merchant (*opperkoopman*), merchant (*koopman*), and, finally, junior merchant (*onderkoopman*).

The Company was a nation unto itself, with an army, a navy, and a judicial system. Once a person became a VOC employee, he was for all intents and purposes the Company's indentured slave. The Company decided how and when to pay its employees; travel to and from one's post was at the mercy of the Company's navy; a strict chain of command had to be adhered to; and, as with modern company policies concerning patents, the Company owned outright (copyright did not exist) whatever its employees created, be it a book or a basket. Communication was at the VOC's behest, and it has been noted that it took twenty months or longer for Amsterdam to receive an answer from Batavia, an effort which one scholar compared to "communicating with the nearest star in the Milky Way."[43] That is important to remember when reviewing the sad history of Rumphius' manuscripts.

The VOC was a *business* enterprise, not an instrument of ideology. It shared the bottom-line mentality of modern corporations: profit was its primary motivation. Violence was not a political tool but a corporate tactic, and most of the VOC's military campaigns were fought in order to establish or maintain a monopoly. The essentially commercial character of the VOC is emphasized by the fact that the authority of civilian businessmen superseded that of the military. Nevertheless, Mercury and Mars were closely allied.

It was very common that someone arrived in the Indies as a soldier and ended up doing something entirely different. Six governors-general began their careers as soldiers. One of those, Anthony van Diemen (1593–1645), had attained the highest office just before Rumphius came; he was in power from 1636 to 1645. The same holds for the man who was responsible for the only botanical text that rivals Rumphius' extraordinary *Ambonese Herbal*. Hendrik Adriaan van Reede tot Drakenstein (1636–1691), author of the *Hortus Malabaricus* (published in twelve volumes. in Amsterdam between 1678 and 1693), arrived in Batavia as a

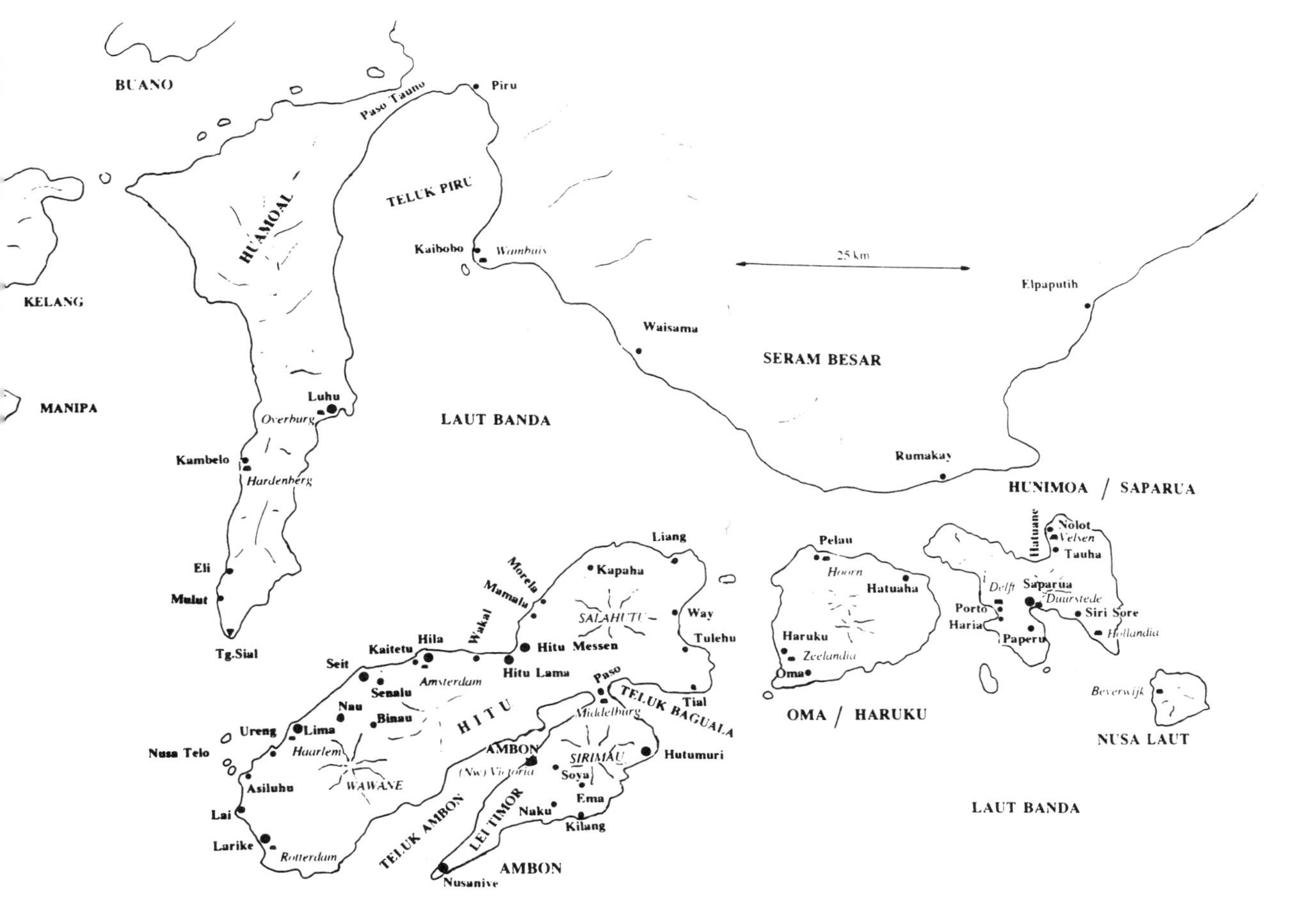

Map of Ambon and neighboring islands. The island of Ambon, at the bottom of the map, is actually two peninsulas, called Hitu and Leitimor, joined by the very narrow isthmus, known as Paso, which connects Ambon Bay (Teluk Ambon) with Baguala Bay (Teluk Baguala). To the east of Ambon are the Uliasser Islands: Oma (or Haruku), Saparua (or Hunimoa), and Nusa Laut. Above them hovers the large island called Seram (or Ceram). *Seram besar* means "Great Seram." Off the west coast of Seram and north of Ambon is the narrow peninsula known as "Little Seram," frequently mentioned by Rumphius. On this map it is called Huamoal, but in Rumphius' day it was also known as "Hoamohel" or "Hoeamohel." *Teluk* = bay; *laut* = sea; *tg.* = *tanjung* = cape; *nusa* = island; *wai* (or *way*) = river; *paso* = isthmus. Italicized names, such as *Rotterdam* or *Delft*, indicate remnants of colonial redoubts.

common soldier in 1657, even though he came from a noble family.[44] There was, therefore, no stigma to using the martial profession as a means to an irenic end.

A major problem for the VOC was the high level of vacancies in the tropics. There were frequent shortages of personnel due to illness and death, and settlements quite often had to make do with interim functionaries. Rumphius' long life—he reached the age of seventy-four—is an exception, not the rule. Officials came and died with alarming frequency; hence it is a remarkable coincidence that the governor-general who came to Java in the same year Rumphius did holds the record for the longest term in the office of any colonial governor-general: a quarter of a century, from 1653 to 1678. Johan Maetsuycker (1606–1678) was a superior man in an age that seemingly had an endless supply of them, but it is particularly his humane response when disaster overwhelmed Rumphius that recommends him to posterity. Johan Maetsuycker notwithstanding, death had a greater dominion than the VOC. Rumphius was under the authority of four different superiors in Batavia after Maetsuycker's death, and he had to deal with nineteen different superiors in Ambon.

Yet Ambon was a far more salubrious post than Java; it was five times healthier than Batavia.[45] Ambon (its name means "dew" or "vapor" [*embun*] in Malay) is a small island, thirty-two miles long and ten miles wide, in the northern regions of the Banda Sea, 04° 25′ north and 128° 05′ east. Its 386 square miles is divided into two peninsulas: the larger, northern one is called Hitu and the smaller, southern one Leitimor (Rumphius spells it "Leytimor"), which means "eastern side." The two are connected by a narrow alluvial isthmus which, in Rumphius' day, was called Passo Baguala. Here they used to haul ships on rollers across the isthmus, which is only 198 yards wide.[46]

Ambon resembles a lobster claw, with the Hitu peninsula as the large upper pincer, and Leitimor as the much smaller lower one. The terrain is mostly mountainous, though the soil is fertile and the island covered with vegetation. There is some level land along the coasts, but it is never wider than just under a mile. Innumerable small, unnavigable rivers become torrents during the rainy season, or wet monsoon, when the winds blow from the southwest from May to September. Ambon has a wet, tropical climate, with an annual rainfall of over 140 inches, but steady sea breezes make the humidity bearable. Rumphius' home was not a furnace; the average temperature is 79 degrees, with temperatures in the eighties during the dry season (December to March) and in the low seventies during the rainy season. The island has one major liability: it is subject to tectonic earthquakes.

Rumphius lived thirteen years on the coast of Hitu, undoubtedly the happiest period of his life. His first post was Larike, a small settlement in the extreme western part of Hitu, a word that means "seven" in Ambonese.[47] Larike was situated below a mountain on the beach near a tall rock, which Rumphius called "Sugar Loaf" and which he mentions in his *Curiosity Cabinet*.[48] When Rumphius lived there, the town had 831 "souls," who, despite handsome profits from the clove trade, remained poor because, as Rumphius reports, "they squander their pennies quite readily and are always in debt."[49] The Dutch built a stone reduit, or blockhouse, at the foot of the mountain, where a fairly broad river disembogued into the Banda Sea. Called "Rotterdam," it covered about fifty feet square and had been constructed in 1633 at the orders of Aert Gijsels, who was Ambon's governor from 1631 to 1634. Although Larike was considered the least of the VOC's settlements in Hitu, it was, according to Rumphius, not all that insignificant either. When Rumphius resided there, Rotterdam had a garrison of twenty-four and was armed with ten pieces of ordnance.[50] The resident

Fort Rotterdam at Larike, Ambon.

junior merchant had an assistant and lived in a modest dwelling outside the reduit.[51] Coincidentally, Larike was the birthplace of Willem van Outshoorn, the first governor-general to be born in the Indies. He was in office when Rumphius died.

Rumphius lived in Larike for three years, and during his tenure there he began to devote himself to the serious study of his tropical environment. He now had time and opportunity, and nature encouraged him to enjoy the world again. He had survived a devastating war, service in the foreign military, excruciating sea journeys, and had left his familiar world behind. During the first few years in the Indies he was still under close supervision, but now, in this remote settlement, he was his own master and could heed his own counsel. Rather than consider Rumphius lonely in Larike, one should imagine him happy there. Rumphius strikes me as a fine example of Emerson's self-reliant man, someone to whom self-reliance is not only mental fortitude but also self-respect. For Rumphius, what Thoreau called the "seed-time of character" was in Germany and Portugal; now came the time of the "harvest of thought."

This most unusual of VOC employees was finally able to lose and restore himself in nature's "everlasting Now."[52] But his investigations were hardly systematic at first. Like the coral reefs of this tropical region, separate observations were recorded according to their own individual necessity, and slowly, over time, an aggregate form began to insinuate itself. Rumphius' nature writings were part of a dynamic process, not a methodological fulfillment of a prior assignment. Rumphius belongs with Isaiah Berlin's foxes: those people "who pursue many ends, often unrelated and even contradictory, connected, if at all, only in some *de facto* way, for some psychological or physiological cause, related by no moral or aesthetic principle. . . . their thought seizes upon the essence of a vast variety of experiences and objects for what they are in themselves, without, consciously or unconsciously, seeking to fit them into, or exclude them from, any one unchanging, all-embracing . . . unitary inner vision."[53] As Rumphius himself phrased it in the present text, he would "rather be amazed at the possibilities which nature affords . . . and refrain from trying to be too wise or from examining things which are beyond our understanding, and be content if we can only guess at them."[54] I do not share the notion that Rumphius went to the Indies with a clear objective in mind and that he started to execute a master plan as early as 1654.[55] Nor does it seem plausible that isolated events or a youthful interest in nature caused such a major decision as to remove oneself to the other side of the world. An alleged reading of Orta in Portugal[56] or a chance encounter with Pliny's work in Idstein did not inspire a young man to program the next fifty years of his life, nor was a formal garden in Hesse sufficient catalyst to pledge one's existence to the production of 1,661 pages of botanical descriptions. The impulse to flee to the Indies was more likely economic and social; even adventure is a more acceptable inspiration than the duty to adhere to a calculated strategy, because Rumphius had more pressing matters to attend to, such as survival. This is not to deny that his early enthusiasm for nature prompted all sort of suggestions, but these fleeting awarenesses remained on the margin. Over time they multiplied and repeated themselves until they converged and, after the confrontation with the tropics, grew into a conscious love of nature that converted Rumphius to the task of writing a comprehensive natural history of eastern Indonesia. I would suggest this took place in the 1660s.

Rumphius himself contributed to this illusion of a programmed and orderly life. In the 1680 letter to Mentzel he stated that "as soon as I came to the Indies I began methodically writing a History of the trees, animals, fish and minerals of these Islands."[57] But other pronouncements argue against this official and orderly version intended for a scientific academy. In his preface to the completed *Ambonese Herbal* he calls himself a "lover"[58] of natural sci-

Fort Amsterdam at Hila, Ambon.

ence who had produced an "imperfect Chaos" of texts that had been written in the order "the plants [had] daily presented themselves" to him,[59] which hardly argues for an a priori plan. Some of his correspondence[60] also echoes the notion of a chaotic mass of papers which recorded his impressions in the haphazard sequence chance dictated every day. The structure of the finished product was imposed at a later date, and the realization that he could emulate Pliny's grand achievement was most likely of no consideration while in Larike. In Hila he could devote even more time to the extracurricular activities which Rumphius, borrowing an expressive phrase from Pliny, characterized as "*lucubrationes.*" In Roman times this wonderful phrase meant "study by lamplight," "nightwork," "nocturnal meditations," while later English usage reduced it to the general meaning of "study." Rumphius had Pliny's sense in mind, and it must have been in Hila that he was seduced by the possibility of composing a natural history of the Moluccan islands in imitation of his Roman avatar.

The promotion to Hila in 1660 was an official acknowledgment of Rumphius' professional talents. Hila was the capital of the Hitu coast, halfway across the length of the island, to the east, on the northwestern coast of the Hitu peninsula. It was located on a level piece of land not far from the beach and had in Rumphius' day 871 inhabitants.[61] The settlement and the nearby stronghold, called Fort Amsterdam, enjoyed a beautiful location which made Rumphius' alleged friend, François Valentijn,[62] wax lyrical with enthusiasm.

> This is the most delightful country in all of Amboina, not only for the beautiful plains, and the glorious river,[63] wherein one can bathe, but also for the surrounding pleasant hills, where one is wont to go Deer-hunting on horseback. Now there are also some Moorish villages, not to mention a Christian village, right next to each other, and the fenced-in house of the Head [of the Hitu coast][64] is nearby. One also derives great pleasure from a large and gorgeous forest of Mangga trees,[65] just outside these villages, and very near the Fort, which supplies an abundance of fine Manggas to the people from Hila and to many [inhabitants] of Amboina. I have experienced the time when, in the middle of the day, a large company of Lords and Ladies would be sitting under the fresh

> foliage of these trees, where they were not only delivered from the sun's heat, but were also treated to the astonishing sight of deer and wild pigs, which we had carefully driven down from the nearby mountains and uncommonly delightful hills, running past them through this Mangga forest to the Shore, where the same were caught and shot . . . a recreation no pen can describe.[66]

Since Hila was the second most important post after Victoria Castle in the city of Ambon, the Dutch constructed a fortification there, called "Amsterdam" by De Vlamingh. Rumphius' former commander enlarged the stone reduit that was already there until it was fifty feet square, and then surrounded it with a ten-foot-high wall and two opposing bastions. A sergeant was responsible for supervising the garrison of thirty soldiers and sixteen cannon.[67] From the watchtower one had a magnificent view across Piru Bay, with Seram's coast in the distance.

Rumphius lived here, according to Valentijn's opinion, "like a Prince, and with greater repose than many a King, since the greater burden of his office is borne by his Assistant and the Sergeant."[68] This is an exaggeration, because the duties of the merchant were demanding. The supposed free time, which for Valentijn was merely a perk, was essential to Rumphius. One would suppose that during these years in Hila he completed the greater portion of his research. But from a personal point of view as well, it must have been a dream come true for this expatriate Hessian. Every morning a freshly killed deer was placed on his doorstep, and Valentijn testifies to the fact that this venison was "incredibly delicious" and surpassed the best beef from Holland. There were plenty of fish, and the merchant also kept cows, sheep, geese, ducks, and chickens, while his stables sheltered horses equipped with beautiful "French and English saddles, with superb bridles, some of which are of pure silver." He had a large armed vessel (an "Orembaay") at his disposal, manned by forty rowers plus a gunner. He lived in what Valentijn calls an "elegant" house, large enough to accommodate a number of guests. This was necessary, because it is clear that his post was often visited by the upper echelon from the city of Ambon, less than thirty miles away. "Beautiful cabbages, endives, superb lettuce that is much better than some Dutch ones, beautiful Parsley, Chinese Radishes, and other useful Potherbs" were raised by a gardener in the residence's adjoining garden. There was even a small zoo, called a "vivarium" in that era. Valentijn's account, which fairly reeks of envy, makes it perfectly understandable that Rumphius took delight in his good fortune. He boasted to his father about it in 1663; the following year August wrote, not without a hint of glee, to Count Johannes in Idstein that "this past Fair I got a Letter from my Son Georg Everhard from the Island Amboina from Hila Castle where he is the Commander and everything is fine with him. He is also a Merchant and a Lord in [that] Country [and] has no desire to return to Hanau."[69]

Professionally speaking, Rumphius had succeeded. His abilities, not connections, had secured for him the plum appointment in Hila. What had recommended him for the second most lucrative position in Ambon was the fact that, as Governor Hustaerdt phrased it, this was a man of "solid abilities" who, most important, "knows how to suit himself to the temper of the Ambonese" and who "was versed in reading and writing the Arabic script." In other words, he could get along with the local population and could speak and read and write their language.[70] This underscores once again Rumphius' linguistic prowess, a talent he shared with his maternal cousin, Johann Wilhelm Keller. He could read, write, and speak German, Dutch, Malay, and Latin. He also spoke Portuguese and Ambonese and might have known some Chinese and Macassarese, in addition to a limited knowledge of Hebrew and Greek. One could even argue for a more extended menu.[71] Nor was his expertise in "all kinds of mathematics" ignored. In 1663 he accompanied Simon Cos, Hustaerdt's successor, on an inspection tour of the fortifications on the Banda Islands.[72] What Rumphius' extraordinary memory registered about that trip was the natural phenomenon of "white water," or "milk sea," and sailing for an entire day through a sea of jellyfish.[73] But the most important development in the early sixties was the metamorphosis of the Company man into the learned "Rumphius."

In Hustaerdt's official missive from 1657 Rumphius is still referred to as "Juriaen Rumph," but three years later the same governor speaks of him as "Juriaen Rumphius." Thereafter "Juriaen" vanishes as well and only "Georgius Everardus" remains. Someone in the seventeenth century Latinized his name only if he had intellectual pretensions; money and social position were not pertinent to this distinction. Rumphius makes his mission quite clear in a letter to the directors in Amsterdam in 1663, one year after signing a new five-year contract, which included the confirmation of his rank as merchant, at a salary of sixty guilders per month.[74] He is asking the VOC's permission to buy books and instruments in Amsterdam and have them transported on the company's ships, because he has

> begun a work that describes in Latin such plants, field products, animals, etc. which I have seen or which have been reported to me during the time of my residence in the Indies, as well as others in the future. I have sorted out their proper names, both from ancient Greek, Arabic and Latin authors, as well as from recent ones, compared them with and differentiated them from one another, and drawn fitting pictures of them from life (to the best of my artistic abilities); I supplied all of them with their characteristics and powers, derived from the aforementioned ancient authors but particularly from carefully wrought personal experience, in part also investigated and gathered by myself, in addition to other investigations Scientiam Physicam et Mathematicam; and all of this done with more effort, diligence and methodical order than one can trust to have happened up till now.[75]

Batavia granted his request, as did Amsterdam, but not until May of 1665. Given the length of time required for communication between Europe and Asia, it is perfectly possible that Rumphius did not know of the positive response until 1666.

By that time Pieter Marville, the new governor of Ambon from 1666 to 1667, had distinguished Rumphius with the highest honor his VOC career would know: he appointed him interim senior merchant (*opperkoopman*) and "*Secunde*," that is to say, the region's second in command. In his recommendation to Batavia asking that Rumphius be confirmed in this promotion, he left us with a rare contemporary impression of the man: "He has the reputation and the physical bearing of a man who has right knowledge and experience of these lands, particularly the coast of Hitu. He leads a good and modest life, he has an immoderate though humble and civil manner, and most of all possesses a just and scrupulous mind, [and is] not greedy or covetous."[76] This is curiously phrased, and some[77] have stumbled over the one negative word—"immoderate," for *onbillyck*—which seems to pertain to Rumphius' social intercourse. We will probably never know exactly what this man was like, but I would surmise from the few remarks about his behavior and from the tone of some of his more personal letters that Rumphius had a direct manner, could be brusque, perhaps peremptory, and may have had a volatile temperament.[78] Rumphius has hitherto been canonized as a long-suffering saint for science, but this ignores the flint of his character and the iron in his soul. It also discounts his military experience. He was, if anything, not servile, feckless, or fawning, but rather a man who did not suffer fools gladly. A sentimental view also ignores

Rumphius' sense of humor, which is in evidence throughout his work. It includes the bawdy, an aspect of his character that does not suit a saintly sentimentalist and that earned him the censure of less generous minds as late as 1959,[79] if not more recently.

In any event, Rumphius held the position of what the British would have called a "supercargo" for less than a year. Batavia disagreed with the appointment for reasons that had nothing to do with Rumphius. In March of 1667 he returned to his post as the resident merchant in Hila. Either to ease the disappointment or to offer a token of gratitude, Rumphius' superiors in Ambon gave him a small parcel of land, which he had selected himself when he was in Victoria Castle. It was southeast of the town hall near the castle's square and was to be his property free and clear and could be passed on to his heirs. The generosity is undercut by typical Dutch parsimoniousness: the land was given because "if it were sold it is judged not to fetch more than about 100 Rixdollars."[80]

And there were heirs now, even though his mother had been dead for fifteen years, his father had died about a year before, and his younger brother had been executed. He shared his life in Hila with a woman called Susanna. We know next to nothing about her and then only because Rumphius mentioned her in his *Herbal.* He named a rare terrestrial orchid that only flourished during the wet monsoon after her: "Flos Susannae," or the Susanna Flower. He had done so because there was no native name for this orchid and "to commemorate the person who, when alive, was my first Companion and Helpmate in the gathering of herbs and plants, and who also was the first to show me this flower."[81] It is a tender tribute but tantalizing in what it does not state. We do not know whether Susanna was Rumphius' legal wife. In other texts she is referred to as his "housewife," but there is nothing to indicate a legal status. Nor is she ever connected to a last name, a deficiency that suggests she was not Dutch. After Susanna's death Rumphius married a Dutch widow, and, since she was European, we know her name in full, but it was common practice at the time to give a native woman a Christian first name only. Susanna could, therefore, have been either an Ambonese or, more likely, a woman of mixed blood. That Susanna could not have been European is indicated by her knowing the local flora; no Dutch woman would have had such intimate knowledge of her tropical environment. Rumphius and Susanna had at least three children, perhaps more: a son, Paulus Augustus, and at least two or three daughters.[82]

His third contract had expired, and it was time to negotiate a new one. Rumphius was perfectly content to stay in Hila and had no desire to get trapped in the bureaucratic waste of endless meetings if he lived in the capital's Victoria Castle. From 1667 on, he no longer hides the fact that his "curious studies," and not the VOC, were paramount in his life. In July 1669 he informs his immediate superior, Jacob Cops, governor of Ambon from 1669 to 1672, that his "private" studies will contribute just as much to the "general good" as his labors as a merchant,[83] openly admitting that his VOC career had only been a "masque, which I had to wear all this time in order to earn the daily bread for me and mine."[84] And so he tries to negotiate a leave of absence of about eight to ten months so he can devote himself to the "quiet advancement of his curious studies,"[85] whereafter he wants to move to Batavia with the same merchant rank. The leave of absence was not granted, but Batavia did allow Rumphius to cut back on his duties as long as it did not harm the interests of the Company. After some wrangling he was also allowed to come to Batavia, by an order of February 1669. A hassle with sailing accommodations delayed his departure, and by the time things had straightened out it was too late. The halcyon years on Hitu were over, and the lucubrations had turned into a permanent night. On 9 May 1670, Governor Cops reported to Batavia that "the Merchant Rumphius became blind several weeks ago."[86] He was forty-two years old.

It was a major calamity that pushed Rumphius' sanity to the very limit. Milton became blind at about the same age, and both men have been praised for the pacific acceptance of their tragedies. But that might well be hagiography, for neither had a submissive character, though both may well have found solace in their faith. I think the rueful elegy to light in *Paradise Lost*, book 3, or the fury at its loss in *Samson Agonistes* ("O dark, dark, dark amid the blaze of noon") is much closer to the private truth. Rumphius expressed his sorrow at every opportunity.[87] In the preface of his *Herbal* he states that God saw fit to "visit his sight with a sad accident" which forced him into "a sad long night" where he was compelled to labor "with borrowed pen and eyes."[88] As late as 1680 he writes in his open letter to Mentzel that a "terrible misfortune . . . suddenly took away from me the entire world and all its creatures . . . compelling me to sit in sad darkness."[89] I even think that Rumphius unwittingly embodied his private horror in the symbolic correlative of the "Chama," "a Beast . . . dreadful to behold" not only because it is nothing but a skin but also because it is blind.[90] Hence I think that a man who had so lovingly and diligently consulted the native genius of tropic nature would entirely concur with Milton's moving lines from *Paradise Lost*, book 3:

> Thus with the year
> Seasons return, but not to me returns
> Day, or the sweet approach of Ev'n or Morn,
> Or sight of vernal bloom, or Summers Rose,
> Or flocks, or herds, or human face divine;
> But cloud in stead, and ever-during dark
> Surrounds me, from the chearful waies of men
> Cut off, and for the Book of knowledge fair
> Presented with a Universal blanc
> Of Natures works to mee expung'd and ras'd,
> And wisdome at one entrance quite shut out.

Not accustomed to a sightless existence from birth, Rumphius had a daily struggle not to give in to despair, to learn to cope, and, when the initial shock had eased, to find an interior illumination that would help him to diffuse his senses throughout his body until it could "look at will through every pore."[91] His work proves that Rumphius succeeded.

What he himself called, with a term from Pliny, *suffusio oculorum*[92] (an "opacity of the cornea") and *cataracta nigra*[93] ("black cataracts") would now be diagnosed as glaucoma, a progressive debility which in the seventeenth century irrevocably terminated in total blindness even if, as was true for Rumphius, it was preceded by a deceptive period of merely diminishing sensitivity to light.[94] Rumphius never mentioned the endless hours of study by candlelight but chose to blame his affliction on the tropical sun, saying that "precisely in the year I thought to leave Ambon, [I] searched all the beaches and hills for the last time, perhaps all too hastily and carelessly, not heeding any discomfort nor heat of the sun, which is quite fierce in these regions, in order to obtain a complete knowledge of whatever herbs remained. From such walking in the heat of the Sun my sight was struck to such an extent by a *Suffusio* or *Cataracta Nigra*, that I had lost most of it within three months."[95] That the sun can destroy one's sight is proven by those poor Renaissance mariners who used the backstaff for measuring "the height of the sun above the horizon by looking directly into its glare, with only scant eye protection afforded by the darkened bits of glass on the instruments' sighting holes. A few years of such observations were enough to destroy anyone's eyesight."[96] Whether the blindness was caused by the sun or by the strain of nocturnal studies, or both, Rumphius' life was forever changed.

Official reactions to his plight illustrate the two different sides of the VOC in the seventeenth century. Rumphius' immediate supervisor in Ambon, the same Jacob Cops[97] he had wrangled with about his contract demands, was only concerned about the effect the merchant's affliction would have on Hila's business performance. Not sanguine about the manager's future, Cops ordered Rumphius to relinquish his post and move his entire household to the city of Ambon. Despite great misgivings and physical pain, Rumphius had to comply, and in June of 1670 he was precisely in the place he had tried so long to avoid: the bureaucratic world of Victoria Castle. Rumphius objected to this treatment and filed a complaint with the governor-general in Batavia. This was still Johan Maetsuycker, and, as was consistent with his reputation, he ordered Cops to continue Rumphius' salary "until further notice" not only because there might still be hope of a cure but "even more so due to his lasting, good, and irreproachable services, and that, furthermore, more regard for his person could have been shown while he was still in Hitu." He also instructed Cops that the blind merchant remain on the Governor's Council and be permitted to attend any other meeting "where he was wont to appear"—for instance the Political Council, of which, as merchant in Hila, Rumphius was automatically a member—and retain "his old seat and rank, without any diminution in that respect."[98]

It must have been a bitter irony that disaster afforded him the leisure to pursue his true vocation. Work made it possible for Rumphius to endure his misfortune, and work he did. Besides attending a full agenda of meetings, Rumphius performed what we today would call consulting work, and most astonishingly, if we remember his circumstances, he, like Milton, produced his masterpieces after he was struck blind. In fact, all of Rumphius' known writings were written or completed after 1670. And he still had a household. The Rumphius family now lived in what was once considered one of the finest cities in the eastern archipelago.

Kota Ambon—*kota* means "town" and will henceforth be added in order to distinguish the town from the island—was built on a beautiful bay that provided a safe haven for VOC ships. Situated on a fine beach and flat land, it had expanded inland over the years away from its first structure, Victoria Castle, which had been constructed to repulse seaborne attacks. Gently rolling hills rose imperceptibly toward Mount Soya and the rugged mountains of Leitimor beyond. Pieter Marville's decree from 1666 ordered a rational design of the town, so that in Rumphius' day its orderly divisions and straight streets resembled Hanau's Neustadt or Manhattan. The streets, wrote Valentijn, "are quite straight, moderately broad, with firm ground, but not paved. When it rains heavily, the water is quickly gone, soaked up by the ground, which is spongy."[99] These streets divided the city into twenty or thirty blocks, with more than one thousand houses, not counting public buildings. Three rivers flowed through the city: Way Tomo to the east, along Victoria Castle; Way Alat, even further east, near Red Mountain, or Batu Merah; and the Elephant River in the west. The last, originating on Mount Soya, had been named after a rock in the shape of an elephant, locally called Batu Gadjah. One will find all three rivers frequently mentioned by Rumphius.

Rumphius' modest house was on what in the 1920s was called the Olifantstraat ("Elephant Street"). Built of wood on top of a stone foundation, it had a low, sloping roof, covered with *atap* (a roofing material made from palm leaves), and a floor of red tiles. A distinctive feature of the VOC era was its paneled windows, with small panes of glass. Like

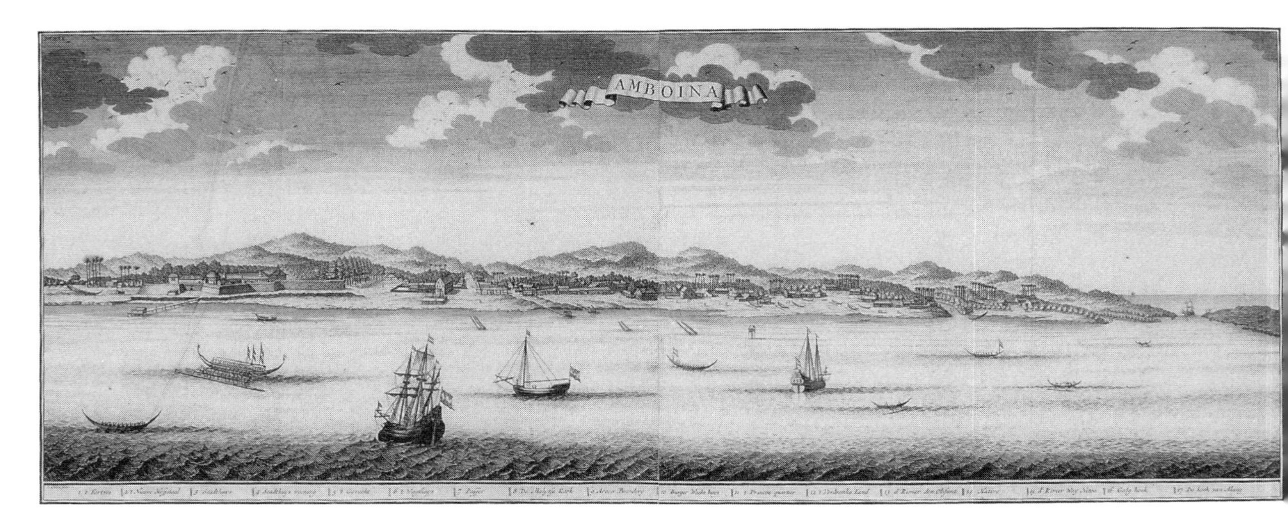

Contemporary view of the city of Ambon.

most houses in the Indies, it was surrounded by a porch (also called a gallery) shielded from the sun by the continuous roof that was supported by wooden pillars.[100] The house had no second story and had been constructed of timber and wooden beams, joined like a ship so that it could withstand the frequent earthquakes.[101] This was not the case with the stone houses (or *petak*) of the Chinese.

Kota Ambon had a considerable Chinese population, which, after 1625, concentrated along Chinese Street. This street ran west of Victoria Castle to Elephant River and was close to the shore and the open market (*pasar*) which Valentijn so admired.[102] It was there that on Saturday, the seventeenth of February 1674, Rumphius, Susanna, and one or two daughters were enjoying the spectacle of Chinese New Year. It was a quiet, tropical evening lit by a bright moon. At half past seven the bells in Victoria Castle suddenly began to toll, and people who stood talking to each other lost their footing and were tossed to the ground. Ambon had been struck by one of the worst earthquakes in its history. Seventy-five stone *petak* houses collapsed on Chinese Street, killing 79 people.[103] A contemporary eyewitness account reports that

> the housewife of Merchant G. E. Rumphius, with her youngest daughter,[104] as well as the widow of the former Secretary of the Justice Council, Joannes Bastinck, were called over by a certain Chinese woman and invited in, now when they felt the first shaking they wanted to walk away, but instead they got the wall of the house on their backs, and thus lamentably suffocated beneath it, notwithstanding that they were dug up as soon as was possible. Just before that time the aforementioned Rumphius had gone for an evening stroll past the very same house and had even been called over by his wife and daughter, but he demurred; which shows that God was disposed to keep him from harm, for if he had sat down, there is no doubt that there would have been no escape for him, due to his blindness. It was a piteous sight to see that man sit next to his corpses, and also to hear his lamentations concerning both this accident and his blindness.

The victims were buried the next day with full honors.

Altogether 2,322 people died, a huge number considering that, for instance, in 1696 the

Rumphius' house in Kota Ambon after 1670. Photo from the 1930s.

entire European population on Ambon was 1,013.[105] More than half, or 1,461 people, perished in Hila, Rumphius' former residence. Hitu's coast appears to have been hit by a tsunami. As Rumphius described it: from the deep "arose three dreadful waves coming from a subterranean whirlpool of sulphur, and they stood tall like walls, fiery and white on top and black below, and of a murky and stinking water which went three different ways. The greatest one set course along the coast of Hitu, and mowed down villages, people, cattle, and houses, as if they had been swept away."[106]

The Moirai were still not through with Rumphius. Misfortune closely attended him for the three lightless decades he lived in Kota Ambon, frozen permanently in the rank of merchant, living by the good graces of the Company, and at the mercy of substitute eyes and pen. First his vision and then Susanna had been taken from him; now his writing became a target. It was the only thing that had sustained him, but it was not allowed to comfort him. One has to realize that during a (for his era) long life, Rumphius published only his brief account of the earthquake and thirteen short texts in an arcane scholarly publication. Everything else remained in manuscript, victim of all sorts of mishaps, and was not published until after his death, sometimes centuries later.

He was still working on his *Herbal* in 1670. After he became blind, he had to start all over again, because there was no one to whom he could dictate Latin. He started the text from scratch in Dutch,[107] but at least he still had all his own illustrations, in color, augmented with work done by others after 1670. All these original drawings were reduced to ashes in a major fire that burned down the entire European quarter in Kota Ambon on 11 January 1687. The newly dictated text was barely salvaged.[108] The fire destroyed not only Rumphius' original illustrations but also his manuscripts, his botanical and zoological collections, and his library.[109] We can imagine the magnitude of this loss if we recall how difficult it was to acquire books from Europe. On top of that, in 1682 he was forced to sell his collection of curiosities (mostly shells), which he had gathered over a period of twenty-eight years, to Cosimo III (1642–1723), Grand Duke of Tuscany,[110] which left him with very few specimens, if any at all. And in 1695 someone stole sixty-one of the new illustrations from his office, which prompted Rumphius to admit to Isaac de Saint Martin that "the loss of the figures is mostly irreparable, given my weak old age."[111] Before the fire, Rumphius had completed a history of Ambon, as well as a geographical description of the island which was to serve as the history's introduction, and together they were to form a single unit. But in 1680 Batavia declared it a "secret document" because it was so "useful," and the work remained under lock and key, only to be consulted with special permission.[112] The only reason the description of Ambon survived the fire is that a copy happened to be in the house of a colleague, so that Governor Padtbrugge could order a scrivener to duplicate the document by hand.[113]

Rumphius could not do anything, of course, without amanuenses. Who helped him in the 1670s is not known, perhaps a clerk from Victoria Castle, but thereafter he received assistance from Batavia on a fairly regular basis. We have already noted how Governor-General Maetsuycker helped and protected Rumphius. Maetsuycker's successor, Rijklof van Goens (in office from 1679 to 1681), sent him a clerk in 1679 and the "temporary assistant," Daniël Crul, in 1680.[114] The first illustrator to assist Rumphius seems to have been J. Hogeboom, since there is a botanical drawing by him for Rumphius' *Herbal*, dated 1685.[115] It is difficult to say when Rumphius' son, Paulus Augustus, began to help his father, but he was probably working for him in 1686, as both draughtsman and secretary. But Paulus could not keep up with this extra work, because, as Rumphius wrote to Cleyer on 15 May 1688, his job demanded too much of his son's time.[116] But before Paulus quit he was obliged to teach drawing to a "Sailor," as Rumphius called him.[117] His name, ironically, was Philip van Eyck and he transferred from Batavia to Ambon on orders from Johannes Camphuys, governor-general from 1684 to 1691, one of Rumphius' staunchest supporters. Van Eyck drew illustrations for Rumphius from 1688 to 1696. Before he left for Holland in the latter year, he had taught his craft to Pieter de Ruyter, who worked for Rumphius from about 1694 to 1699. There is no doubt that Rumphius taught his son to draw, so that, indirectly, his own artistic gift kept on serving him after his blindness. The fifth and final artist who helped Rumphius was Cornelis Abramsen, a professional who worked for the ubiquitous Andreas Cleyer, for Governor-General Camphuys, and later for Cornelis Chastelein, who oversaw the final stages of completing the *Herbal.* Abramsen had worked earlier for Rumphius, and his original drawings from 1696 are the only surviving illustrations in color.[118] Johan Philip Sipman (1666–1725) was Rumphius' most helpful assistant, from 1692 to 1696.[119]

There is no doubt that Rumphius had help and support for his large enterprise, but the final responsibility remained his. It is astonishing how much work this aging man was able to do despite the major catastrophes of 1670, 1674, and 1687. The history of the *Herbal* manuscript is rather complicated (for more details, see the "Works" section below), but to put it briefly, Rumphius had finished the first six books of his magnum opus in 1690 and sent them to Batavia to be shipped to Holland. Governor-General Camphuys, an ardent naturalist, had them copied so the manuscript did not leave the Indies until 1692. They never reached Europe. On 12 September 1692, a French naval squadron attacked and sank the *Waterland*, the ship that was to bring this rare cargo to Holland.[120] This reversal did not stop Rumphius either; in fact, it seems he took the opportunity to make additions and corrections in Camphuys' copy of his text, sending "33 sheets of writing and 102 new figures" to Batavia in 1694.[121] Books 7–9 were put on board the retour-fleet in February of 1696, and the last three went in January of 1697. The entire *Herbal* was safely in Holland by August of 1697.

But Rumphius did not rest. He had written other works during the long labor on his *Herbal*, including the present volume, *The Ambonese Curiosity Cabinet*, which remains his best-known effort and which was finished in 1698 or thereafter (see the discussion below), but he continued to add to the *Herbal*. By September of 1701 he had finished an *Auctuarium*, or "augmentation," which was copied in Batavia (they had learned their lesson) and dispatched to Holland in May of 1701. The *Ambonese Herbal* was not printed in its entirety until nearly half a century after Rumphius' death. The cause for the outrageous delay—there were already interested parties in 1700 who wanted to print the entire work—was the VOC's paranoid monopolistic policy. They first stated a flat rejection in 1700: printing these books would be considered "inexpedient and cannot be permitted in any way." [122] Two years later they reversed themselves and grudgingly granted their permission, "but without the Company incurring any costs; therefore, when the aforementioned books are to be surrendered for printing, they shall be examined by the Amsterdam Chamber in order to determine which passages could be judged detrimental to the Company and are to be lifted from the same." [123] No one took them up on this dubious offer until Joannes Burman, professor of botany at Amsterdam University, submitted his proposal in 1739. One should mention that the VOC headquarters in Amsterdam did express their admiration and esteem in 1697 in a typical business manner. To "recompense Rumphius somewhat for his many labors," they resolved to promote his son, Paulus Augustus, from junior merchant to merchant, at a salary of sixty guilders per month, identical to what his father had made. Given the difficulty of communication, father and son did not know about this largesse until the beginning of 1699, and they jointly expressed their gratitude in a letter dated 25 September 1699.[124] But there can be no doubt that this was cold comfort to an author who had been thwarted by fate at every critical juncture of his life.

The only substantive recognition Rumphius received while still alive was his election in 1681 to a prestigious scientific association in Germany. The Academia Naturae Curiosorum had been founded by four physicians in Schweinfurt (now in Bavaria but at the time a free imperial city) in 1652, only four years after the end of the Thirty Years War, with the intention of promoting medicine and natural history. Its motto—*nunquam otiosus*, or "never idle"—perfectly matched Rumphius' life. Of the four founding fathers only Johann Michael Fehr (1610–1688) is of importance to Rumphius' life. He was the academy's second president (1666–1686) and was in office when Rumphius was admitted; Fehr was also the man to whom he sent the nautilus shell which had so miraculously dropped from the Ambonese sky.[125] The Academia Naturae Curiosorum was the first scientific organization of its sort, established a decade before the comparable Royal Society in London. Seeing the need for its own house organ, the academy started publishing its yearly *Miscellanae Curiosa Medico-Physica* in 1670. Not only was this effort the first of its kind in the world but it also assured Rumphius the only scholarly publications during his lifetime: thirteen notices, or "*observationes*," published between 1683 and 1698.

The academy existed in this form until 1687, when the Holy Roman Emperor Leopold I granted it imperial privileges, whereafter it was known as Sacri Romani Imperii Academia Caesareo-Leopoldina Naturae Curiosorum until 1870. For a long time the organization was located in a bewildering series of German cities, until in 1878 it found a permanent haven in Halle. It has been there ever since, and after a continuous existence of nearly three and a half centuries, the Deutsche Akademie der Naturforscher Leopoldina (as it is now called) can boast of a most impressive list of members. Some names of importance for the present context are Mentzel (matriculated as member no. 57), Cleyer (no. 81), Valentini (no. 118), Seba (no. 398), Linnaeus (no. 464), Burman (no. 497), not to mention subsequent luminaries such as Goethe, Louis Agassiz, Liebig, Alexander von Humboldt, and Baron van der Capellen, governor-general of the Dutch East Indies from 1818 to 1826 (matriculated in November 1826 as no. 1303), who, as will appear, commemorated Rumphius in the nineteenth century.

Rumphius matriculated as member no. 98 and received the most felicitous cognomen of "Plinius," a name, along with his membership, he proudly added to the title pages of his major works.[126] Christian Mentzel, court physician of the Kurfürst of Brandenburg, Friedrich Wilhelm, was primarily responsible for ensuring that the academy recruited members from distant regions and that the *Miscellanea* printed reports about "Curiosities from the Far East."[127] It is due to such contacts that Rumphius and his work entered the European learned world. Serendipity is, no doubt, partly responsible, but one should not forget that the learned community in the seventeenth century was a small one and that there was an easy interchange between scholars and officialdom.

Rumphius was fortunate to live in the Indies during the seventeenth century, because that century knew more distinguished colonial administrators than perhaps any other era. Quite apart from military genius, there were such contemporaries as Maetsuycker, Speelman, Camphuys, and Isaac de Saint Martin, who were cultured and educated individuals with more than a passing interest in the various branches of knowledge. Colonial society during the first century of Dutch rule in the Indies has been often portrayed as an intellectual and cultural wasteland inhabited by ferocious scavengers bent on enriching themselves. It may come as a surprise that, for instance, one will find a far more delicate appreciation of nature in the journal of Admiral Van Eck than in the writings of the historian Valentijn. A junior merchant in the Company's hierarchy, Herbert de Jager, was a linguist of genius, fluent in several Oriental languages, who taught Arabic, Persian, and Malay and corresponded with Rumphius about botany. During a most trying existence in the tropics, Jacob Bontius managed to write a pioneering work on tropical medicine, and Ten Rhijne, a physician in Batavia, wrote a monograph on leprosy wherein he correctly argued that the drinking water was contaminated by disease-carrying "little beasts." Van Reede tot Drakenstein, governor of Malabar, recruited Brahmin sages to act as a sort of college and pass judgment on the tropical plants he collected on official journeys and which were described in his celebrated *Hortus Malabaricus*. Camphuys, the governor-general who helped Rumphius so much, wrote a historical survey of the foundation of Batavia, and Speelman, the conqueror of Macassar, delighted in enlivening his correspondence with puns, while he also displayed a genuine interest and extensive knowledge of those parts of the archipelago he came to know personally. And what is certainly very odd, one will find seventeenth-century colonial reports and administrative journals interspersed with verse. Bontius, Rumphius, Valentijn, and Governor Camphuys inserted poems in their work; that the Muse seems to have wilted in the tropics is less important than the fact that she stirred anyone at all.

Mentzel connected Batavia to Europe. Admiral (and subsequent Governor-General)

Cornelis Speelman (who died in 1684) corresponded with Mentzel about poisonous darts used in Macassar, a subject Rumphius discussed repeatedly, particularly in his description of the Upas tree.[128] One of Speelman's protégés was Andreas Cleyer, who corresponded with Mentzel. Rumphius was friendly with Cleyer, and Governor-General Camphuys (who succeeded Speelman) knew and protected them both. It was Cleyer who, with Mentzel's support, proposed Rumphius for membership in the Academia. Cleyer knew Van Reede tot Drakenstein in Batavia in the late 1670s,[129] and both knew the physician Ten Rhijne, though Ten Rhijne and Cleyer became enemies. Robert Padtbrugge, governor of Ambon between 1683 and 1687, had corresponded with Cleyer in 1669 and probably stayed in contact sporadically thereafter because both were ardent botanists. Padtbrugge knew Rumphius very well and secured a parcel of land for his burial plot. Cleyer had a friend, Sebastian Scheffer, in common with Michel Bernhard Valentini, which was the reason that the German scholar from the University of Giessen established personal contact with the educated community in Batavia. The result was that Valentini's *Natur- und Materialien-Kammer*, printed in 1704, contained the first translated selections of Rumphius' work after his death. This is just one example of how tightly connected the existing world of privilege and knowledge was and how vital to one's career.

The last decade or so of Rumphius' life seems to have passed in relative peace, though troubled by the constantly deferred hope of ever knowing his work would appear in print. He married for a second time, but we do not know when. His second wife was Isabella Ras, widow of Abraham Wittekam. Rumphius survived her as well; she died in 1689.[130] He filled his days with writing almost to the very end. The *Auctuarium* to the *Herbal* was finished only nine months before his death. Professional duties also helped to pass the time, but by the middle of 1700 Rumphius was said to be "subject to many debilities due to his increasing old age" and he resigned from the various governing bodies to which he still belonged.[131]

It is difficult to say what his last years were like. His son was stationed in Larike, but we do not know whether a daughter lived nearby. Perhaps he found some companionship in a small Bengali slave called Cour, whom Rumphius inherited from his stepdaughter Giertje Wittekam,[132] and in a Javanese mongoose, of which he was rather fond. Rumphius described this Indonesian version of Kipling's Rikki-tikki-tavi with affection in the very last chapter of his *Herbal*, that is to say, chapter 83 of the *Auctuarium*—perhaps an incongruous ending to a book about plants, but it befits Rumphius' character. He clearly liked the little animal and wanted to present it to his readers, but he "doubted that [his] weakness and old age would allow him to finish the work on Native Animals [he was] presently working on,"[133] so he added this final chapter as an insurance against oblivion. Rumphius' description corresponds to Kipling's, with the emphasis on the animal's cheerful nature:

> He can stand up, walk, and sit on his two hind legs, like a Monkey; with his two front paws he grabs whatever one gives him to eat, and if he has nothing to do, he puts them on his chest, like a poor sinner. . . . It is even more wondrous how tame this Animal becomes, and he gets quite used to people he sees daily. You can't grab hold of him too well by his back, for he will immediately slip out of your hands if you hold him like that; but he will let you pick him up with your hands from the front, and he will presently take hold of them, and play with them, and do the same with the legs of whoever walks by. In short, he becomes so tame, that he even sleeps with people.[134]

Rumphius died on 15 June 1702, at the age of seventy-four. He was buried in a plot outside town, near Batu Gadjah, granted him by Governor Padtbrugge. Someone erected a marble tomb in Rumphius' honor—perhaps it was Adriaan van der Stel, who was governor of Ambon from 1706 to 1720[135]—and the spot acquired a certain sanctity for the Ambonese population, who took care of it out of respect.[136] But during the British interregnum on Ambon, from 1796 to 1802,[137] the British governor sold Rumphius' burial plot and tomb to a private operator, who demolished it, desecrating the grave in the hope of finding treasure. Failing to do so, he sold the marble and limestone for profit.[138] Such a crime made the place *sial*, or unlucky, and the local people began to avoid it. In 1823 the French naturalist Lesson visited Kota Ambon, and from where he was staying he could see a luxuriant stand of banana trees that found "nourishment in Rumphius' grave, while their verdant canopy shaded the sepulcher of this celebrated man"[139]—a floral tribute Rumphius might well have appreciated.

Lesson informed the governor of the Moluccas, P. Merkus, of this scandal, and he in turn relayed the information to the reigning governor-general, Baron van der Capellen, when he came to Ambon during an inspection tour in 1824. Van der Capellen was a fellow member of the erstwhile Academia Naturae Curiosorum, and out of respect for Rumphius, he bought the plot, had the grave restored, and erected a simple monument, which was unveiled on 8 April 1824.[140] This tumulus was destroyed by a bomb during the Second World War and finally replaced by a *tugu*, as a monument is called in Indonesian, which was dedicated on 22 April 1996 in Kota Ambon.[141]

Rumphius' signature from his old age.

WORKS (EXCEPT *THE AMBONESE CURIOSITY CABINET*)

It is difficult, sometimes impossible, to know when Rumphius wrote his texts, so in the following survey I list them chronologically according to the date of their publication, even though that covers nearly two centuries. This way there is some order based on known facts, and it will also reveal how little was ever reprinted in the modern era.

Texts

1675

Waerachtigh Verhael, Van de Schrickelijke Aerdbevinge / Nu onlanghs eenigen tyd herwaerts, ende voornaementlijck op den 17. February des Jaers 1674. voorgevallen, In/ en ontrent de Eilanden van Amboina (Batavia, 1675). [True Relation, of the Terrible Earthquake / That took place some time ago, to wit on 17 February of the Year 1674, In and around the Islands of Amboina]

Although his name is not printed on the title page, it is almost a certainty that Rumphius wrote this text in 1674. He states in his *Ambonsche Landbeschrijving* (see below), p. 6, that "the most terrible" earthquake happened in "Februario 1674, of which we rendered a particular description."[1] The report has been transcribed and translated by W. Buijze and published as G. E. Rumphius, *Waerachtigh Verhael Van de Schrickelijcke Aerdbevinghe In en Ontrent de Eylanden van Amboina*, ed. and transl. W. Buijze (Den Haag, 1997).

1683

"De Caryophyllis aromaticis," in *Miscellanea Curiosa* [hereafter *MC*] *sive Ephemeridum Medico-Physicarum Academiae Naturae Curiosorum. Decuriae II. Annus Primus. Anni 1682.* (Norimbergae: Wolfgangi Mauritii Endteri, 1683), pp. 50–53.

This is the first half of a letter to Mentzel, dated 20 September 1680 in the journal's text, of which the first section presents a rare autobiographical statement.

"De Corticibus Massoy & Culilawan: mari Indico noctiluco, & aliis, nomine saltem insignitis," in *MC* (1683), pp. 55–57.

The second part of the letter to Mentzel mentioned above. The entire text was translated into German and published by M. B. Valentini (1704; see below).

1685

"Observatio XXI: De Corallis in genere. Androsace, Fucis marinis, unguibus odoratis, lignis olentibus, pyritibus, silicum generibus &c. praemissa ad Cl. Dn. G. Everhardi Rumphii observationes sequentes, de rebus variis & raris ex Indiis ad me transmissis," in *MC. Annus Tertius. Anni 1684* (1685), pp. 70–74.

Although listed under Mentzel's name, the title, as well as the last sentence, makes it clear the material is Rumphius': it ends with "Jam enumerandae sunt Observationes Rumphianae, rerum exoticarum à se transmissarum, juxta contenta in ejus literis" ("We must enumerate Rumphius' observations about exotic things, which he conveyed himself, together with the content of his letter").

1686

The next four "observationes," nos. XXII–XXVI, are under Rumphius' name. Written in 1680 and 1683, as mentioned in the printed texts.

"De Agallochi generibus & Arbore excoecante," in *MC* (1686), pp. 74–75.
"De Ligno Lacca dicto, Funibus Sylvaticis, Resina Dammar Selan, & Damar Battu," in *MC* (1686) pp. 76–77.
"De Coralliorum quibusdam speciebus & lithodendris," in *MC* (1686), pp. 77–79.
"De Unguibus odoratis ex Murice & Pseudopurpura, Argyrodamante, Androdamante, Schoenantho Ambonico, Pisi Americani specie Ambonica; Herba sentiente Ambonica," in *MC* (1686), pp. 79–81.

The first section of the last text is an extract of chapter 17, book 2, of the present *Ambonese Curiosity Cabinet;* the next section is part of chapter 14 in book 3; while the third section became part of chapter 19 in book 3.

"De Ceraunia spathula metallica," in *MC. Annus Quartus. Anni 1685* (1686), p. 212.

Original from 1683, as dated in text. This short piece became part of chapter 9 in book 3.

1687

"Observatio XXXIX. D. Christiani Mentzelii, De Radice Chinensium Gîn-Sên," in *MC. Annus Quintus. Anni 1686* (1687), pp. 73–79.

This too was mostly written by Rumphius, as Mentzel's text indicates. Original from 1684, as dated in text.

"De Ovo marino, Porcellanis, seu Conchis venereis," in *MC* (1687), p. 222.

This has bits from chapter 23 of book 2.

1689

"De Ceraunia Metallica," in *MCG. Annus septimus. Anni 1688* (1689), pp. 5–8.

Original from 1687, as dated in text. This text combines passages from chapters 7, 8, and 9 in book 3.

"De Nautilo velificante & remigante," in *MCG*, pp. 8–9.

Original from 1687, as dated in text. This is part of the opening section of chapter 3 in book 2.

1697–1698

"De Caryophyllis Regiis Ambonicis," in *MCG. Decuriae III. Annus quintus et sextus. Anni 1697 & 1698* (Francofurti & Lipsiae: Johannem Michaelem Rüdiger & Engelbertum Streck, 1697), p. 308.

The foregoing fourteen texts were the only ones Rumphius saw in print while he was still alive. He died in 1702.

1704

Six botanical texts and fourteen letters in:

Michael Bernhard Valentini. *Museum Museorum Oder Vollständige Schau-Bühne Aller Materialien und Specereijen / Nebst deren Natürlichen Beschreibung / Election, Nußen und Gebrauch/Aus andern Material-Kunst- und Naturalien-Kammern / Oost- und West Indischen Reisz-Beschreibungen / Curiosen Zeit- und Tag-Registern / Natur- und Arßney-Kündigern/wie auch selbst-eigenen Erfahrung / zum Vorschub Der Studirenden Jugend / Materialisten/ Apothecker und deren Visitatoren / Wie auch anderer Künstler / als Jubelirer/Mahler/Färber/ u.s.w. also verfasset / und Mit etlich hundert sauberen Kupfferstücken Unter Augen geleget von D. Michael Bernhard Valentini* (Franckfurt am Mäyn: Johann David Zunners, 1704).

This book comprises three volumes in one, of which the third is the most important in terms of Rumphius. It is entitled:

Oost-Indianische Send-Schreiben / von Allerhand raren Gewächsen / Bäumen / Jubelen/Auch anderen zu der Naturkündigung und Arßney-Kunst gehörigen Raritäten / Durch Die Gelehrteste und Berühmteste Europäer / So vormahlen in Oost-Indien gestanden / Als D. Cleyern / Rumphen / Herbert de Jager, ten Rhyne &c. Alda gewechselt/Und aus Deroselben in Holländischer Sprachgeschiebenen / Originalien in die Deutsche Mutter-Sprache überseßet von D. Michel Bernhard Valentini.

This section contains twelve letters from and to Rumphius (pp. 3–60), followed by a selection of descriptions of tropical flora. These include six botanical texts which we can be sure were written by Rumphius,[2] printed thirty-seven years before the first volume of his *Herbal* was published (see below). The letters are important not only because so few are extant but also because they contain some rare personal observations. Several were quoted in the biographical introduction.

1705

D'Amboinsche Rariteitkamer (Amsterdam: François Halma, 1705).

Discussed separately below.

1724

Advys, en consideratien, ingegeven by den Koopman, Georgius Everhardus Rumphius, aangaande de Fortificatie van 't Kasteel Victoria [Advice, and reflections concerning the Fortification of Victoria Castle, submitted by the Merchant Georgius Everhardus Rumphius] (Dated

23 August 1686 in text), in François Valentyn, *Oud en Nieuw Oost-Indiën, vervattende Een Naaukeurige en Uitvoerige Verhandelinge van Nederlands Mogentheyd In die Gewesten*, 5 books in 8 vols. (Dordrecht & Amsterdam: Joannes van Braam & Gerard onder de Linden, 1724–1726), 2.2:238–239.

This vast—nearly 5,000 pages—and confusing work[3] devotes a considerable portion to Ambon and things Ambonese: parts 1 and 2 of volume 2, part 1 of volume 3, and a portion of part 2 of the same volume, which contains a disquisition on Ambonese shells. Valentijn purloined a great deal from Rumphius, but he also published this one particular text under Rumphius' name: a feasibility report on moving Fort Victoria to another location.

1741

Het Amboinsche Kruid-boek. Dat is, Beschryving van de meest bekende Boomen, Heesters, Kruiden, Land- en Water-Planten, die men in Amboina, en de omleggende eylanden vind, Na haare gedaante, verscheide benamingen, aanqueking, en gebruik: mitsgaders van eenige insecten en gediertens, Voor 't meeste deel met de Figuren daar toe behoorende, Allen met veel moeite en vleit in veele jaaren vergadert, en beschreven in twaalf boeken, door Georgius Everhardus Rumphius, Med. Doct. van Hanau, Oud Koopman en Raadspersoon in Amboina, mitsgaders onder de naam van Plinius Indicus, Lid van de Illustre Academia Naturae Curiosorum, in 't Duitsche en Roomsche Ryk opgerigt. Nagezien en uitgegeven door Joannes Burmannus, Med. Doct. en Botanices Professor in den Hortus Medicus te Amsterdam, Medelidt van het Keyzerlyke Queekschool der onderzoekers van de Natuurkunde; Die daar verscheide Benamingen, en zyne Aanmerkingen heeft bygevoegt. Volumes 1 & 2 (Amsterdam, By François Changuion, Jan Catuffe, Hermanus Uytwerf; In 's Hage, By Pieter Gosse, Jean Neaulme, Adriaan Moetjens, Antony van Dole; Te Utrecht, By Steven Neaulme, 1741). [The Ambonese Herbal, Being a Description of the most noteworthy Trees, Shrubs, Herbs, Land- and Water-Plants, which are found in Amboina, and the surrounding islands, According to their shape, various names, cultivations, and use: together with several insects and animals. For the most part with the Figures pertaining to them, All gathered with much trouble and diligence over many years, and described in twelve books by Georgius Everhardus Rumphius, Med. Doct. from Hanau, Former Merchant and Counselor in Amboina, also known under the name of Plinius Indicus, Member of the Illustre Academia Naturae Curiosorum, founded in the Holy Roman Empire. Corrected and published by Joannes Burmannes, Med. Doct. and Professor of Botany in the Hortus Medicus in Amsterdam, Fellow Member of the Imperial Academy of Inquiry into Nature; Who has added various Names and his Observations.]

One should mention that within this herbal the main title is variously printed as "*Amboinsch Kruid-boek*," also "*Kruyd-boek*," "*Kruydboek*" and "*Kruidboek*." I use "*Amboinsche Kruidboek*," just "*Kruidboek*," or "*Herbal*." As far as we know, Rumphius did not earn a degree in medicine, nor is this distinction mentioned on the title page of *D'Amboinsche Rariteitkamer*.

1743

Het Amboinsche Kruid-boek. Volumes 3 and 4.

Same as above.

1747

Het Amboinsche Kruid-boek. Volume 5.

Same as above, except for the publishers: (Amsterdam, By François Changuion, Hermanus Uytwerf; In 's Hage, By Pieter Gosse, Jan Neaulme, Adriaan Moetjens, Antony van Dole, 1747).

1750

Het Amboinsche Kruid-boek. Volume 6.

Same as above.

1755

Het Auctuarium, ofte Vermeerdering, op het Amboinsch [sic] *Kruyd-boek. Dat is, Beschryving van de overige Boomen, Heesters, en Planten, die men in Amboina, en de omleggende eilanden vind, Allen zeer accuraat beschreven en afgebeeldt na der zelver gedaantes, met de verscheide Indische benamingen, aanqueking, en gebruik, door Georgius Everhardus Rumphius, Med. Doct. van Hanau, Oud Koopman en Raadspersoon in Amboina, mitsgaders onder de naam van Plinius Indicus, Lid van de Illustre Academia Naturae Curiosorum, in 't Duitsche en Roomsche Ryk opgerigt. Nu voor 't eerst uitgegeven, en in het Latyn overgezet, door Joannes Burmannus, Med. Doctor, en Botanices Professor in het Illustre Athenaeum, en de Hortus Medicus te Amsteldam, Medelidt van het Keizerlyke Queekschool der onderzoekers van de Natuurkunde; Die daar verscheide Benamingen, en zyn Aanmerkingen heeft bygevoegt* (Amsterdam, By Mynard Uytwerf, en de Wed. S. Schouten en Zoon, 1755). [The Auctuarium, or Addition, to the Ambonese Herbal. That is, Description of the remaining Trees, Shrubs, and Plants, which are found in Amboina, and the surrounding islands, All very accurately described and depicted according to their proper shapes, with their various Indian names, cultivation, and use, by Georgius Everhardus Rumphius, Med. Doct. from Hanau, Former Merchant and Counselor in Amboina, also known by the name of Plinius Indicus, Member of the Illustre Academia Naturae Curiosorum, founded in the Holy Roman Empire. Now published for the first time, and translated into Latin, by Johannes Burmannus, Med. Doctor, and Professor of Botany at the Illustre Athenaeum, and the Hortus Medicus in Amsterdam, Fellow Member of the Imperial Academy of inquirers into Nature; Who has added various Names, and his Comments.]

This *Auctuarium* is, for convenience' sake, often referred to as the seventh volume, as it is in the present book.

This *Herbal* is Rumphius major work. It contains 1,661 folio pages, divided into 876 chapters, and illustrated with 695 plates. It describes about 1,200 species of plants. The scope of the book was defined with characteristic modesty in Rumphius' preface: "It bears the title of being an *Ambonese Herbal* because it cherishes no exalted opinion of itself; though with the understanding that it exhibits such plants which one will not only find in the Ambonese Isles, but also in the neighboring Moluccas, Banda, and others, and in Java itself. The work has called itself Ambonese because it was wrought in Amboina, and according to the shapes the plants exhibit there."

The framework of the *Herbal* conformed to the established pattern (after Aristotle), which divided botanical descriptions into three main groups: trees, shrubs and herbs. Each group could be divided again according to a more particular design of the author. Rumphius modeled his work on similar principles and may have found certain modifications in the work of his predecessor Carolus Clusius, who in his *Rariorum plantarum Historia* (1576) added such categories as "fragrant flowers," "odorless flowers," and "poisonous, narcotic, or pungent plants." The first five books of Rumphius' *Herbal* dealt with "edible, fruit-bearing, aromatic, and wild trees"; book 6 concerned itself with "all sort of shrubs which stand upright"; book 7 included "such Shrubs which cannot stand upright on their own, but which creep along with a long and slender trunk, or wind around other trees, the like we call in the Indies *Forest-ropes*, or *Taly-Outang* in Malay." Books 8–11 described herbs, while the twelfth presented plants "which do not wax on land but in the Sea, and which have a mixed nature of wood and stone, and which are called Sea-trees or *Coral-plants*."

Each entry described the plant or tree in detail, followed by its name in various languages, always including its Dutch, Latin, Malay, and Ambonese names, and Rumphius often provided Javanese, Hindostani, Portuguese or Chinese nomenclature. These names were not derived from Linnaeus but were either invented by Rumphius himself or transcribed from native appellations. After this a plant's habitat was given and its various usages described. In terms of the latter, primary attention was paid to a plant's medicinal applications, for Rumphius not only meant his work to be "diverting to Lovers of Nature" but also to be "of use and service to those who live in the East-Indies."

The original manuscript and illustrations have not survived. The manuscripts which at present reside in the Netherlands are copies. As already mentioned, it has been alleged that Rumphius began writing his *Herbal* as early as the end of 1653, but I consider that doubtful. All we know for sure is that in the 1663 letter to the VOC directors in Amsterdam, he says that he has started such a work. Between 1663 and 1670 he also produced his own illustrations. After he became blind in 1670, Rumphius had to switch to dictation in Dutch (his original text had been in Latin) and had other people produce the illustrations. It was said by Dirck de Haas, governor of Ambon, to be mostly finished by January 1687—720 chapters and half of the illustrations—when fire destroyed the artwork. By 1690 he had the first six books finished once again and sent them to Batavia. Governor-General Camphuys took the time to have the manuscript and illustrations copied for his own purposes, a job that lasted until 1692. That year they were finally conveyed to Amsterdam on the ship *Waterland* but they never reached their destination, because the ship was sunk by a French fleet on 12 September 1692. After he was informed of this disaster, Camphuys ordered his own copy copied. In 1694, Rumphius finished his corrections of the first six books and sent his emendations to Batavia so they could be incorporated into the text that was in the process of being copied. In January 1696, Rumphius sent books 7–9 to Batavia, and in January 1697 the final three. In 1696 the first nine books were dispatched to Holland and arrived safely in August of 1697. In January of 1697 the final three books also went to sea, accompanied by Sipman and Van Eyck, and arrived safely in Amsterdam. Rumphius finished the *Auctuarium* (or volume 7) by September 1701 and sent it to Batavia, where it was copied and dispatched to Holland in May 1702.

Joan Burman (1707–1778), a pupil of the internationally renowned Boerhaave, announced in 1736 that he would translate Rumphius' text into Latin and publish it. This was done by a consortium of publishers between 1741 and 1750, followed by the *Auctuarium* in 1755. Burman printed his Latin translation alongside Rumphius' original Dutch text in two parallel columns per page. Stylistically the Dutch original is more expressive than Burman's rather perfunctory Latin. I wonder whether Boerhaave's edition of Swammerdam's *Bybel der Natuure* was Burman's model. That book appeared, with a Latin translation by H. D. Gaubius, in 1737. The Latin translation was on the left and the original Dutch on the right, printed on the same page in two columns. The first volume of Burman's edition of Rumphius' *Kruidboek* did not appear until 1741. The *Herbal*'s first edition was probably 500 copies.[4] In 1750 appeared a second edition of six volumes by M. Uytwerf in Amsterdam. There are no major differences. The *Herbal* was never reprinted, though it was subjected to much examination in the years to come.

One strange fact is Linnaeus' scant mention of Rumphius' *Kruidboek* in his *Species Plantarum* of 1753. There are only twenty references.[5] Carolus Linnaeus (1707–1778) lived in Holland from 1735 to 1738, and stayed with Burman in 1737; hence he must have known the *Rariteitkamer*, published in 1705, as well as the *Kruidboek*, which Burman had in his possession in manuscript form. By 1750 Rumphius' *Herbal* was published except for the *Auctuarium*.

After Linnaeus, however, there have been at least eleven epitomes, indices, or taxonomic keys of Rumphius' *Kruidboek:* Olaf Stickman (1754), Burman (1755 and 1769), Radermacher (1781), Buchanan-Hamilton (1828 and 1832), Henschel (1833), Hasskarl (1866), J. J. Smith (1905; on orchids), E. D. Merrill (1917), K. Heyne (1927), and H. C. D. de Wit (1959).

The *Amboinsche Kruidboek* is a continent of a book, and this is not the place to discuss it at length. I merely want to add one final observation. Most scholars know that the VOC censored texts, but it is not easy to find proof. Because Valentini published six botanical texts by Rumphius which had not passed through the hands of the VOC officials, one can compare them with what was permitted to be printed as the final version. One text describes the Company's main cash crop, nutmeg, and one will find that (2:20) they excised a passage saying that when the harvest was too abundant, the Company burned the nuts in huge heaps or made the owner burn them (Valentini, p. 84). A subsequent passage is left out (2:20) because it stated that on Lontar many slaves died while harvesting the nutmeg (Valentini, p. 86). Rumphius knew full well that the Company would read such remarks, so they indicate that he possessed a social conscience and was not afraid to mention abuses.

The reputation of the *Amboinsche Kruidboek* has remained untarnished, a claim echoed in the following remarks by a Dutch botanist, M. Greshoff, who once wrote that Rumphius' influence is still "so great and immediate that it seems at times as if he were still among us, learning about the plants and animals of the Indies and inspiring us to ever more detailed investigations of the lands of the East Indies Company. He laid the foundation in the *Ambonese Herbal* for collecting native names of plants. . . ."[6]

1797

A description of the Alfura people on Ceram. Quoted from Rumphius' *Ambonsche Landbeschrijving* (see below), in J. S. Stavorinus, *Reize van Zeeland over Kaap de Goede Hoop en Batavia, naar Samarang, Macasser, Amboina, Suratte, enz. gedaan in de jaaren 1774 tot 1778*, 2 volumes (Leyden: A. en J. Honkoop, 1797), 1:268–271.

For readers of English this book is available in an excellent contemporary translation by S. H. Wilcocke, who also revised the text and added numerous notes: *Voyages to the East-Indies: By the Late John Splinter Stavorinus, Esq. Rear Admiral in the Service of the States-General*, transl. Samuel Hull Wilcocke, 3 volumes (London: G. G. and J. Robison, 1798); reprint ed., 2 volumes (London: Dawsons of Pall Mall, 1969), 2:357–361.

Wilcocke unwittingly demonstrates the unwarranted precedence Valentijn was given over Rumphius for a long period of time. People simply were not aware that most of what they read in Valentijn had been written by others and incorporated without due credit. The greater portion of Valentijn's text on Ambon and things Ambonese he stole from Rumphius. So it makes one cringe when Stavorinus ends his direct quote from Rumphius' *Landbeschrijving* with: "Thus far the relation of Mr. Rumphius, who being a man of some experience and much reputation, deserves credit in some instances" (2:361). It is also curious that Stavorinus knew about Rumphius' unpublished manuscript and had easy access: "I met with the following account of them [= the Alfuras], in the description of *Amboyna* composed by Rumphius, which, having been prohibited by the government at *Batavia*, has never been printed, but of which a manuscript copy is preserved in the secretary's office at *Amboyna*" (2:357). See also Wilcocke's fulsome praise of Valentijn (2:354 n.) when quoting from his description of Ambonese animals, knowledge the reverend had stolen from his "Bosom-friend" Rumphius.

1856

"Antwoord en rapport op eenige pointen uit name van zeker heer in 't vaderland, voorgesteld door den Edelen Heer Anthony Hurt, Directeur-Generaal over Nederlandsch Indië en beantwoord door Georgius Everardus Rumphius, oud Koopman en Raadspersoon in Amboina," ed. B. W. A. E. Sloet tot Oldhuis. [Report containing answers to some points raised by a certain gentleman in the fatherland, presented by His Honor Anthony Hurt, Director-General of the Dutch East Indies and answered by Georgius Everardus Rumphius, former Merchant and Counselor in Amboina], in *Tijdschrift voor Staathuishoudkunde en Statistiek* (1856), 13:125–141.

Text probably from 1684. This fascinating text, which contains a number of things, answers queries by Gisbert Cuper (1644–1716), mayor of Deventer and close friend of Nicolaes Witsen. Most likely it was Witsen who sent these questions on to Batavia, whence to the VOC's resident sage in Ambon. Rumphius describes New Guinea at some length, lore about Loro Kidul and the "Kalappa laut" (which found its way into the mistaken but wonderful chapter 8 in the *Herbal*'s book 12), "white water" (which is also discussed in chapter 28 in book 3 of the present volume), and China and, to my knowledge, makes the only mention of a member of his mother's family.

1871

"Advys over den Ambonschen landtbouw, ingegeven by den Oud-Coopman Georgius Everardus Rumphius" [Advice on Ambonese agriculture, delivered by the Former Merchant Georgius Everardus Rumphius], printed in appendix 2 by P. A. Leupe in "Georgius Everardus Rumphius, Ambonsch Natuurkundige der zeventiende eeuw," in *Verhandelingen der Koninklijke Akademie van Wetenschappen* (Amsterdam: C. G. van der Post, 1871), 12:42–44. Dated 23 August 1686 in text.

The work is a shrewd and straightforward recommendation, displaying Rumphius' knowledge of the Ambonese and the terrain where they had to make a living. He notes that rice did not flourish on Ambon and that the people subsisted on a diet of pulse and rhizomes. He astutely recommends that the Company urge the Ambonese to plant bananas and coconut palms. It is also interesting that he suggests that the Company not force new cultivation of cloves on the local population.

1910

De Ambonese Historie. Behelsende Een kort Verhaal Der Gedenkwaardigste Geschiedenissen zo in Vreede als oorlog voorgevallen sedert dat de Nederlandsche Oost Indische Comp: Het Besit in Amboina Gehadt Heeft [The History of Ambon. Containing a short Relation of the most Memorable Events that took place both in Peace and war since the Dutch East Indies Company took Possession of Amboina], published in *Bijdragen tot de Taal-, Land- en Volkenkunde van Nederlandsch-Indië*, zevende volgreeks: tiende deel [vol. 64] ('s-Gravenhage: Nijhoff, 1910).

The text was completed in 1678, but the VOC would not permit its publication.

1983

De generale land-beschrijving van het Ambonsche gouvernement, en wat daeronder begreepen is, mitsgrs. een summaris verhaal van de Ternataansche en Portugeese regeringe, en hoe de Nederlanders eerstmaal daerin gecomen sijn [General Geography of the Ambonese government, and what that entails, together with a summary narration of the Ternatan and Portuguese governments, and how the Dutch first came there], printed with the title: G. E. Rumphius, *Ambonsche Landbeschrijving*, ed. Suntingan Dr. Z. J. Manusama (Jakarta: Arsip Nasional Republik Indonesia, 1983).

Rumphius originally intended this to be the introduction to his *Ambonese Historie;* it was probably completed between 1674 and 1678. No original manuscript survived, and it only exists in copies. Rumphius revised and emended his text in 1693 and 1697. The VOC prohibited its publication.

1997

G. E. Rumphius, *Waerachtigh Verhael van de Schrickelijke Aerdbevinge*, ed. and transl. W. Buijze (The Hague: W. Buijze, 1997).

Letters

Listed are letters known to have been written by Rumphius, or significant fragments of his letters quoted by someone else (but only if they are printed, hence accessible). An asterisk indicates a letter of particular interest.

Year	Date	Addressee	Source
1663		To his father	Quoted in August's letter, dated 25 October 1664.
*1664	20 August	To the XVII in Amsterdam	Leupe, pp. 40–42.
*1664	25 October	August Rumpf to Count Johannes	See p. lxv of introduction.
*1680	20 September	To Christian Mentzel	*MC. Decuriae II. Annus Primus. Anni 1682*, pp. 50–51.
1682	15 August	To Cosimo III	Van Benthem Jutting in *M*, pp. 198–199.
*1682	18 August	To Andreas Cleyer	Valentini, pp. 51–54.
	18 August	To Jacob de Vica (i.e., Vicq)	Valentini, pp. 54–55.
1683	20 May	To Herbert De Jager	Valentini, pp. 3–4.
	20 May	To Willem Ten Rhijne	Valentini, pp. 56–57.
1684	6 May	To Herbert De Jager	Valentini, pp. 27–30.
1687	20 August	To Willem Ten Rhijne	Valentini, pp. 49–51.
*1688	15 May	To Andreas Cleyer	Valentini, pp. 57–60.
*1689	14 September	To Herbert De Jager	Valentini, pp. 44–48.
1691	15 May	To Isaac de Saint-Martin	Blume, 1:58.
	16 May	To D. De Haas	Blume, 1:57.
1692	15 September	To Isaac de Saint-Martin	Leupe, pp. 45–46.
1694	8 June	To Isaac de Saint-Martin	Leupe, p. 47.
	6 July	To Willem Ten Rhijne	Quoted at length in the following item
*1694	9 July	Ten Rhijne to Camphuys	Leupe, pp. 51–52.
*1695	2 March	Camphuys to Rumphius	Leupe, pp. 48–51.
*1695	29 June	To Camphuys	Leupe, pp. 53–55.
1696	8 or 9 July	To Isaac de Saint-Martin	Leupe, pp. 56–57.
1699	19 September	To the XVII in Amsterdam	Leupe, p. 59.

Lost Manuscripts

There were at least three other texts, but the manuscripts have vanished. Anything by Rumphius is significant, but these items are particularly so, especially in view of what I think that Rumphius was finally contemplating.

First there was a projected volume of Ambonese fauna. A missive from the governor and council on Ambon to Batavia, dated 20 May 1697, tells the colonial government that Rumphius is finished with his *Herbal* but that he still has a manuscript called the "Ambonese Curiosity Cabinet" in "three books," as well as "three other books concerning Land-,

Air- and Sea-animals of these islands, although he is without hope that he will be able to perfect them because of his age and encroaching debilities. Nevertheless, he requests Yr. Highnesses most humbly, that he may retain a scrivener and a draughtsman, for what most likely are only a few remaining years of his life, in order to give his mind some work, without which he fears to end his days in melancholy" (original in Leupe, p. 26). He also mentions the same texts twice in the unpaginated preface to his *Herbal* (*Amboinsch Kruidboek*, book 1, folios 5 and 6 of the "*voorreden*"), and does so in a manner that indicates that he already had a good deal finished.

This lost enterprise is corroborated by Rumphius' own remark on why he included a description of the "moncus" in his *Herbal.* He did so, he writes, because he "doubted that weakness and old age would allow him to finish *the work on Native Animals I am presently working on*" (6:69; emphasis added).

Second, Rumphius had been working on a Malay dictionary and had finished over half of it before he became blind. We know this firsthand from his letter to former Governor-General Camphuys (29 June 1695), wherein he states: "I understand that Master Christiaan Geraerts bought a Malay Dictionary on Batavia's Thieves' Market, and that it now resides with you, this is, given the circumstances, my Malay Lexicon, and I would like to have it back, since it was not written for the public, but for my own contemplation. I wrote it myself 25 years ago, from the beginning up to the word Pandas under the letter P, the rest is by someone else" (original in Leupe, pp. 54–55).

Finally, Rumphius also mentioned a "Stone-book" (for instance, 3:17 of the present text) several times, as early as the third chapter of the *Herbal*'s first book (1:21–22). This was rescued from oblivion and became the third book of *The Ambonese Curiosity Cabinet.*

Perhaps one understands now why it is important to include Rumphius' lost manuscripts in a bibliographical survey of his achievement. My surmise is that Rumphius meant his readers to conceive of his total output as a single work, similar to Pliny's *Historia Naturalis* (Natural History). I think that this grand design gradually evolved in Rumphius' mind and that it was not, as said before, an a priori scheme. A structure effortlessly suggested itself to Rumphius during the last few decades of his life. The model was Pliny's only surviving work, a book that ranked with the greatest texts from antiquity and that was once so popular that there had been 222 editions by the beginning of the twentieth century.

Rumphius identified with the literary equestrian from Como (who lived from A.D. 22 or 23 to A.D. 79), a man who combined the pursuit of knowledge with a military and official career. Rumphius usually paid scant attention to his martial past or to his military duties for the VOC, but when he advised his readers in the *Herbal*'s preface that the book had not been written by a specialist, he described himself as "merely a lover of natural science, though one who ascribes to himself a moderate knowledge of the botanical arts, and who devoted his idle hours to it amidst his *military and publick duties*" (emphasis added). This is a direct echo—as is the entire preface—of Pliny's *prefatio.*[7] Rumphius was intimately acquainted with Pliny's *Historia Naturalis* and, like any educated man of his time, was thoroughly familiar with the scant details of the procurator's life as they had been narrated by his nephew. It had not escaped him that Pliny served three tours of duty in Germany, including one stint in southern Germany as the commander ("prefect") of a cavalry unit, or that Pliny died indirectly from the eruption of Vesuvius, the same calamity that buried Herculaneum and Pompeii in August of A.D. 79. The account by Pliny's nephew of that event emphasized earth tremors and the fearful spectacle of the sea being sucked away from shore, a description that matches Rumphius' of the 1674 quake in Ambon. Nor should we forget that Rumphius originally wrote in Pliny's language and that he constantly refers to him throughout his printed works. But more important, the tropical naturalist and the intimate of Vespasian share a number of important themes and inclinations, so that it is far more than a conceit when Rumphius compares himself to Pliny: "For if Plinius, while commanding the Roman fleet, could still find time to complete his laudable work, of which the world still profits to this very day, I surely had more time and opportunity in my tranquil Prefectura, to describe the plants of these Islands."

Since this passage is from the preface (folio 2) to his *Herbal*, people have customarily associated Pliny with Rumphius the botanist. But I think there was a larger similarity. Rumphius surpassed his father and had little in common with him. In voluntary exile on a distant tropical island, I think that Pliny became a "father," an eminence who would ease the loneliness of his intellectual isolation. The tangible representation of the ancient Roman was his *Natural History*, and I think that it slowly occurred to Rumphius that he was unwittingly composing a "Historia Naturalis Tropicae." And that is how we ideally should view Rumphius' *entire* achievement.

I cannot embark here on a detailed comparison of the Roman and the Indian Pliny but will have to restrict myself to a few characteristic parallels, if we accept that Pliny's *Natural History* is more than an encyclopedic epitome. If that were all, and if Rumphius had followed suit, we would be only comparing him to a taxonomist like Linnaeus. I concur with Mary Beagon that Pliny's work was an art of living: "The whole subject and purpose of Pliny's work is 'life.' The structure of his inquiries is dictated by that of the natural world as viewed by man, starting with the cosmos as a whole, and progressing through all its subdivisions, animal, vegetable, and finally mineral. It is within this unique structure that theory and scholarship are united with the practical arts of agriculture and seafaring, of government and warfare. Philosophy is united with practical philanthropy."[8] This passage already contains a number of correspondences. Tropical life was Rumphius' main subject, and his entire effort was guided by Pliny's dictum *rerum natura, hoc est vita, narratur* ("I tell the story of nature, that is to say, life").[9]

To contemplate nature means to be alive, hence Pliny's phrase, which would apply to both authors, *vita vigilia est* ("life is vigilance").[10] This caring and attentive observation of the natural world proceeds, in Pliny's case, with the inquiries into the heavens (book 2), geography and ethnography (books 3–6), zoology (that is, man, land animals, aquatic animals, birds, and some insects) (books 7–11), botany (books 12–19), medical botany (books 20–27), *materia medica* from other sources (books 28–32), and, finally, mineralogy (books 33–37). Rumphius' modesty prevented him from inquiring after heaven, but otherwise the achievements match. The major difference is that Rumphius improved on Pliny's model by incorporating ethnographic and medical information organically, that is, when the subject inspired it. This makes for a more facile and more natural ingestion of knowledge since, as with fiction, Rumphius' writing presents items where they would normally occur within the artful flow of a narrative. That being said, Rumphius covers Pliny's geography and ethnography in his *Ambonsche Landbeschrijving* and *Ambonese Historie*, zoology in the lost volumes concerning "Land-, Air- and Sea-animals" as well as the first and second books of *D'Amboinsche Rariteitkamer*, botany (including medical botany) in his *Amboinsche Kruidboek*, and mineralogy (including *materia medica* from other sources) in the third and longest book of *D'Amboinsche Rariteitkamer*. In other words, he wrote what I would call a "Historia Naturalis Tropicae."

Like Pliny, Rumphius preferred empirical evidence to theory and also clearly condemned man for trying to control nature. Nature is too immense (*improbus*) for our understanding, and it is vain to claim otherwise.[11] Nature is for both authors a sensual reality,[12] and both insisted that nature was benevolent and rational and prone to adapt herself to man's needs.[13] Rumphius also echoes one of Pliny's main themes: the evils of luxury and the attendant curses of greed and avarice.[14] And, finally, among many other conscious affinities, the two authors share a love for the small. As Beagon notes, Romans normally admired size and grandeur; hence it was unusual for Pliny to say that real "Nature is to be found in her entirety nowhere more than in her smallest creations. I consequently beg my readers not to let their contempt for many of these creatures lead them also to condemn to scorn what I relate about them, since in the contemplation of Nature nothing can possibly be deemed superfluous."[15] This is the belief that put faith in Swammerdam's *Bible of Nature or The History of Insects* (1737) and prompted Leeuwenhoek to examine his bodily fluids. It is also a credo that engendered Rumphius' similar love for such lowly denizens of nature as jellyfish[16] or for a small and insignificant plant he called the "Ambonese Mouse-ear," which he included in his vast *Herbal* so that "it will be remembered and since I did not want to pass it by because of its comely shape."[17]

THE AMBONESE CURIOSITY CABINET

Publishing History

The complete original title is as follows: *D'Amboinsche Rariteitkamer, Behelzende eene Beschryvinge van allerhande zoo weeke als harde Schaalvisschen, te weeten raare Krabben, Kreeften, en diergelyke Zeedieren, alsmede allerhande Hoorntjes en Schulpen, die men in d'Amboinsche Zee vindt: Daar beneven zommige Mineraalen, Gesteenten, en soorten van Aarde, die in d'Amboinsche, en zommige omleggende Eilanden gevonden worden. Verdeelt in drie Boeken, En met nodige Printverbeeldingen, alle naar 't leven getekent, voorzien. Beschreven door GEORGIUS EVERHARDUS RUMPHIUS, van Hanauw, Koopman en Raad in Amboina, mitsgaders Lid in d'Academiae Curiosorum Naturae, in 't Duitsche Roomsche Ryk opgerecht, onder den naam van PLINIUS INDICUS* (Amsterdam: François Halma, 1705). [The Ambonese Curiosity Cabinet, Containing a Description of all sorts of both soft as well as hard Shellfish, to wit rare Crabs, Crayfish, and suchlike Sea Creatures, as well as all sorts of Cockles and Shells, which one will find in the Ambonese Sea: Together with some Minerals, Stones, and kinds of Soil, that are found on the Ambonese and on some of the adjacent Islands. Divided into three Books, And supplied with the requisite Prints, drawn from life. Described by GEORGIUS EVERHARDUS RUMPHIUS, from Hanau, Merchant and Counselor on Amboina, also member of the Academiae Curiosorum Naturae, founded in the Holy Roman Empire, under the name of PLINIUS INDICUS.]

Its contents are: French title; frontispiece (printed on the bottom of the folio page is: " 'T Amsterdam, Gedrukt by François Halma Boekverkoper. 1705" [At Amsterdam, Printed by François Halma Bookseller. 1705]; title as given above, with a vignette representing Constantine the Great; then Rumphius' dedication to Hendrik D'Acquet, Mayor and Counselor of the City of Delft, dated 1 September 1699; followed by the dedication to D'Acquet by François Halma, dated September 1705; then the lengthy "Printer's Preface" by F. Halma, dated 1 November 1704; a table of contents; and, finally, Paulus Augustus' portrait of his father. Then the work itself commences with a short title first and a short "Introduction," whereafter the text proper begins immediately on the same page. Book 1 describes the "soft Shellfish" in 44 chapters illustrated with 16 plates; Book 2 describes the "hard Shellfish" in 39 chapters and 33 plates; this is followed by the superfluous epitome of the preceding second book by Sipman, comprising 27 pages, which end with the following statement: "This description and division of the Ambonese Cockles and Shells was compiled by Mr. Sipman, one of the greatest collectors and Lovers of rare things, who was a companion of Mr. Rumphius on Amboina, and he sent the same over here to his especial good friend, M. J. de Jong, who, considering it useful to this work, handed it over to us, wherefore we added it here, for the satisfaction of the scrupulous Lovers, and for a closer understanding of this Description" (folio 194). I have omitted it from the present translation. The third and final book describes "Minerals, Stones, and other Rare things" in 87 chapters, illustrated with 11 plates. Between chapters 35 and 36 is a "Description of a Piece of Gray Amber" by an anonymous author, "N.N.," who proves to be Nicolas Chevalier. The book ends with an index. Altogether, there are 340 folio pages, with 60 plates, and 5 vignettes.

The Ambonese Curiosity Cabinet was printed twice more during the eighteenth century, but after 1741 it was never reprinted or translated in its entirety until the present edition. The text of those two earlier editions was not altered; only some of the preliminary material changed.

D'Amboinsche Rariteitkamer [1740]

The only change in the complete title, given above, is that "*weeten*" is printed "*weete*," "*beneven*" becomes "*benevens*," while "Lid in d'Academiae Curiosorum Naturae, in 't Duitsche Roomsche Ryk opgerecht, onder de naam van PLINIUS INDICUS" was changed to "Lid van het Kyzerlyke kweekschool der onderzoekers van de Natuurkunde in 't Duitsche Roomsche Ryk opgerecht onder de naam van PLINIUS INDICUS" [Member of the Imperial academy of naturalists in the Holy Roman Empire founded under the name of PLINIUS INDICUS]. This last change is interesting in that the publisher uses the vernacular to inform his readers that Rumphius was a member of a prestigious organization. The Dutch is also more accurate and reflects the change in the academy's status after it received imperial privileges in 1687. Ironically, Burman's edition of the *Herbal*, which began to be published only a year later, kept the outdated Latin phrase. It is, of course, unfortunate that the printer eliminated the comma after "*opgerecht*" because the omission makes the title page of the 1740 edition read as if the Imperial Academy went by the name of "Plinius Indicus."

The vignette on the title page was also changed and now shows the Bay of Ambon; the most obvious difference is that the new publisher, Jan Roman de Jonge, dropped Rumphius' dedication to D'Acquet as well as Halma's. The prefatory material that remains is: Dutch title; frontispiece, at the bottom of which is printed " 'T Amsterdam Gedrukt by Jan Roman de Jonge, Boekverkoper 1740" [At Amsterdam Printed by Jan Roman de Jonge, Bookseller 1740]; title page; De Jonge's undated dedication to Burman; Paulus Augustus' portrait of his father; and then Halma's "Preface of the Printer to the Reader."

D'Amboinsche Rariteitkamer [1741]

The 1741 edition is virtually the same as the 1740 edition except for two different vignettes, but the text is substantially the same as that of the 1705 edition. This one was also published in Amsterdam by Jan Roman de Jonge, but in 1741. It is the working edition I used for translating the text, but not for the prefatory material.

The complicated history of the *Rariteitkamer*'s illustrations will not be examined here. Suffice it to say that Rumphius must have drawn some of the illustrations himself before 1670. For instance, he says that a drawing was made of the *living* nautilus (book 2, chapter 2, folio 60). In the only instance I can recall of a direct interpolation, Schynvoet or Halma added parenthetically that "this figure was lost" (folio 61) and substituted drawings of the dead animal (see Plate XVII). But, for instance, the drawing of the Argonauta which accompanies Rumphius' description in *Miscellanea Curiosa* is very much that of a live specimen.[1] We also know that some specimens were drawn by Rumphius' draughtsman, Pieter de Ruyter, because there are two drawings of crustaceans with his name on them.

Herman Strack is convinced that the collection of drawings of which De Ruyter's pictures are a part was the one used for the plates in the *Rariteitkamer.* This collection of forty-five sheets was drawn partly in Ambon and partly in Holland. They reside at present in the Royal Library in The Hague.[2] Strack is also certain that the fifty-four watercolors which Maria Sibylla Merian (1647–1717) allegedly produced[3] for the plates in the *Rariteitkamer* were never used for that purpose.

Because of the high incidence of printing errors in the 1740 and 1741 editions, I collated them with four copies of the first edition of 1705: one in the Beinecke Library at Yale University, the second in the library of the University of Amsterdam, the third in the "Lexicologisch Instituut" in Leiden, and the fourth in the library of the University of Leiden. I verified and compared eighty-five major errors in the later editions with the first one and found that, though Rouffaer is correct that there is no major discrepancy in the text, there is a difference in the number of mistakes. The later editions have 54% more errors than the 1705 edition; hence they cannot be *exact* reprints of the 1705 edition. I also noticed that the rate of error increases from book 1 to book 3, with the last book having the dubious distinction of including the most faults, sometimes egregious ones. This gradual deterioration is also evident in the style. In the third and final book it sometimes resembles shorthand or a telegraphic style, as if Rumphius had wearied of crafting polished prose. I think this is attributable to the fact that Rumphius was in a hurry to finish the work in his race against mortality, and that the quality of assistance changed from educated amanuenses such as Sipman or his own son to mere scriveners who did not always understand what he was dictating. Another plausible explanation is that he suddenly decided to publish his still unfinished book in order to profit from the rage for curiosity cabinets (see below).

It would be instructive to compare the printed version with the manuscript, even if it is only a dictation, but the manuscript has never been found. If one considers the publisher's strange dedication to D'Acquet in the first edition, which De Jonge eliminated from the 1740 and 1741 editions, one gets the suspicion that it was probably François Halma who caused it to disappear. He might even have sold it, given the then prevalent rage for curiosity cabinets, be they real or printed. Lacking hard evidence, one is left with just a few printed dates. If we momentarily assume them to be true, then Rumphius, according to his own dedication to D'Acquet in the first edition, had completed the manuscript by the first of September of 1699. This seems plausible, because the earliest date mentioned in the *Rariteitkamer* is 1651 and the most recent 1698. Now, we know that it took a long time to convey anything from the eastern archipelago to Europe; thus Halma's statement that he did not receive the manuscript until 1701 might not be a lie, but why it took four years to get the book in print is a good question.

Rumphius' bad luck continued after his death, for Halma places the blame for the delay completely on the author, despite his very strange confession that "rumor" said the delay was due either to his own negligence or to his pursuit of more profitable projects. His rather strained claims of innocence and condescending tone suggest that Halma was protesting too much. One can only conclude that the blame for the lengthy delay, which caused this book to appear posthumously, lies with Halma and that his attribution of incompetence to Rumphius is a red herring.

In his preface, here printed in translation, Halma praises Schynvoet far more than Rumphius. The main reason, of course, was commercial: Schynvoet was a living patron, not a deceased author. So he extols Schynvoet's editorial prowess, almost crediting him with having saved Rumphius' book. This is disingenuous. Simon Schynvoet (1652–1727) was not only a private businessman but also an official for the city of Amsterdam. His greatest claim to fame was advising Czar Peter the Great on how to construct formal gardens. He was an assiduous collector of curiosities, but he did not edit Rumphius' text in the modern sense of the word. He merely added his remarks at the end of each chapter, telling the reader where an illustration could be found on the plates and telling what specimens he, Schynvoet, had added and what his thoughts were about certain items. These comments have been preserved in the present edition. Furthermore, if Rumphius' text was in such sorry shape as Halma alleges, one can only submit that Schynvoet was an incompetent editor, because there is no indication, and we are grateful for it, that he ever interfered with the text of his superior. All Schynvoet did was *add* his commentary, which, in most cases, has little value and often seems an excuse for advertising his intimacy with fellow collectors, yet Schynvoet patronizes Rumphius as well. Halma boasts of Schynvoet's strenuous exertions to supply the "missing" illustrative material. One will gladly grant the saddlemaker and landscape architect an energetic disposition, but experts have concluded that most of his specimens bear no relation to Rumphius' text, often are European specimens, and, if tropical, are not from Indonesia. One other instance of the publisher's sly lobbying for the live patron at the expense of the dead genius is Halma's lengthy list of shell names in his "Preface to the Reader." This mostly repeats Schynvoet's local labels rather than Rumphius' far more expressive and poetic metonymies. One must be grateful that Burman and the later consortium of publishers took care of Rumphius' *Herbal*, but one should also note that, compared with Rumphius, both Halma and Schynvoet were execrable writers.

Schynvoet devoted most of his attention to book 2, which repeatedly is billed as the "Real Ambonese Curiosity Cabinet." The popular partiality for shells can be illustrated by the fact that, to my knowledge, book 1 and book 3 were never translated until the present effort and that book 3 was scrutinized only once in three centuries (by Arthur Wichmann in 1902), and book 1 three times (by Henschel in 1833, De Man in 1902, and especially by Holthuis in 1952), but that book 2 has been translated once and epitomized or indexed at least ten times. Those *clavi*, or indices, are the following, but, first, one should be reminded that bits and pieces of chapters 3, 17, and 23 of book 2, and chapters 7–9, 13, 17, and 19 of book 3 were originally printed in *Miscellanea Curiosa* (see the previous "Works" section).

1705

"Beschryving en Verdeling der Amboinsche Hoornen en Schulpen, Door den Heer Sipman, Doctor in de Medicynen, enz, enz." [Description and Division of the Ambonese Cockles and Shells, by Mr. Sipman, Doctor of Medicine, etc., etc.], in *D'Amboinsche Rariteitkamer* (Amsterdam: François Halma, 1705), pp. 167–194.

The 1740 and 1741 editions include this as well. There is no evidence that Sipman ever earned a medical degree.

1711

Thesaurus Imaginum piscium testaceorum & cochlearum quibus accedunt conchylia conchae univalviae & bivalviae, Denique Mineralia, quorum omnium maximam partem Georgius

Everardus Rumphius Med. Doct. collegit; jam vero Naturae Amator & Curiosus quidam in hunc ordinem digessit, & nitidissimè aeri incidi curavit (Lugduni Batavorum [= Leiden]: Petrum vander Aa, 1711).

This is an epitome of the text and pertains to the illustrations only; hence one cannot call this a new edition of the entire book. The condensation is only fifteen folio pages long and is called "Denominationes Figurarum, quae in hisce tabulis continentur." This is followed by copies of the sixty plates of the *Rariteitkamer*, followed by indices in Latin, Malay, and Dutch. Rouffaer[4] contends the perpetrator was Simon Schynvoet, but one could make a case for Sipman as well. An almost identical edition was printed in 1739.[5]

1726

"Verhandeling der Zee-Horenkens en Schelpen, ofte Dubbletten van Amboina" [Discourse concerning Sea-Whelks and Shells or Doublets from Amboina], in François Valentijn, *Oud en Nieuw Oost-Indiën*, volume 3, part 2 (Dordrecht & Amsterdam, 1726), pp. 517–586.

This lists what is contained in the second book of Rumphius' *Rariteitkamer*, in addition to which Valentijn adds those shells which Rumphius "and others either skipped or did not know about, and which I own and can display" (p. 518). As usual, the important words are those indicating what *he* has, because, as Valentijn readily admits, this will also give him the chance "for the first time to describe the various kinds of *Ambonese Shells* in my little *Cabinet* and add only those which I do not own as yet, only because I don't have them" (p. 518). This text also gives an overview of contemporary shell collectors in both the Indies and in Holland, as well as a list of the most precious shells in their collections.

1754

Valentijn's text was issued as a separate publication in 1754 by J. van Keulen in Amsterdam (and was translated into German by Ph. L. St. Müller, published in Vienna in 1773).

1764

Jacobi Petiveri, *Opera, Historiam Naturalem Spectantia; or Gazophylacium. Containing several 1000 Figures of Birds, Beasts, Reptiles, Insects, Fish, Beetles, Moths, Flies, Shells, Corals, Fossils, Minerals, Stones, Fungusses, Mosses, Herbs, Plants, with Latin and English Names. The Shells, &c. have English, Latin, and Native Names*, 2 volumes (London: John Millan, 1764).

Hidden in this work is a rare and, to my knowledge, first epitome of book 2 in English. It is to be found in the first volume, after plate 156 (which shows "Cochleae Pernambuc. &c Brasiliae & other Brasil Shells") and is entitled: "Aquatilium Animalium Amboinae, &c. Icones & Nomina. *Containing near* 400 Figures, *engraven* on Copper Plates *of* Aquatick *Crustaceous and Testaceous* Animals, as Lobsters, Crawfish, Prawns, Shrimps, Sea-Urchins, Eggs, Buttons, Stars, Couries, Concks, Perywinkles, Whelks, Oysters, Muscles, Cockles, Frills, Purrs, Scallops, *with divers others Sorts of* Sea *and* River Shell-fish; *all found about* Amboina, *and the* Neighbouring INDIAN Shores, *with their* Latin, English, Dutch, *and* Native Names. By James Petiver, *Apothecary;* F. R. S. London."

The epitome consists of a printed list of names on four pages, at the end of which is stated: "London: Printed for Mr. *Christopher Bateman* in *Paternoster-Row*, 1713." The list is followed by twenty-two plates, which are not exact duplicates of Rumphius' prints but engraved in imitation by a "Sutton Nicholls" (stated at the lower right-hand corner of plate 3). The twenty-two plates display a total of 387 items. They are followed by a supplementary list of the names of the shells on plates 21 and 22.[6]

1766

Georg Eberhard Rumphs (Welcher ehemals Medicinae Doctor, ältester Kaufmann und Ratsherr zu Amboina, imgleichen ein Ehren-Mitgleid der Kayserlichen Academie der Naturforscher unter dem Beynamen PLINIUS INDICUS gewesen) Amboinische Raritäten-Kammer oder Abhandlung von den steinschaalichen Thieren welche man Schnecken und Muscheln nennet, aus dem Holländischen übersetzt von Philipp Ludwig Statius Müller, öffentlichen ordentlichen Lehrer der Weltweiszheit zu Erlangen und mit Zusätzen aus den besten Schriftstellern der Conchyliologie vermehret von Johann Hieronymus Chemnitz, Königlich Dänischen Gesandschafts Prediger in Wien und Mitglied der Kayserlichen Academie der Naturforscher (Wien: Krauszischen Buchhandlung, 1766). [Georg Eberhard Rumph's (Who formerly was Doctor of Medicine, senior Merchant and Counselor on Amboina, also an Honorary Member of the Imperial Academy of Naturalists with the Cognomen of PLINIUS INDICUS) *Ambonese Curiosity Cabinet* or Treatise of the hard-shelled Animals which are called Snails and Mussels, translated from the Dutch by Philipp Ludwig Statius Müller, Professor of Universal knowledge at Erlangen (University), augmented with additions from the best Authors on Conchology by Johann Hieronymus Chemnitz, Pastor of the Royal Danish Legation in Vienna and Member of the Imperial Academy of Naturalists].

This work has three parts. The first is "Vorläufige Einleitung und Kleine Zusäße zu dem Werke des Rumphs dabey die bekantesten und gebrauchlichsten Benennungen einer jeden Schnecke und Muschel angezeigt, die Urtheile der besten und bewerthtesten Schriftsteller von Conchylien gesammelt, und noch hin und wieder eigene Anmerkungen und nähere Erläuterungen hinzugethan worden von Johann Hieronymus Chemnitz" [Provisional Introduction and Modest Additions to Rumphius' Work, together with the best known and most common Names shown for each Snail and Mussel, a collection of judgments by the best and most significant Writers on Conchology, augmented with occasional personal Remarks and further Commentary by Johann Hieronymus Chemnitz], pp. i-cxxviii.

The second part is Müller's actual translation of the second book only of Rumphius' *Rariteitkamer*, followed by Sipman's epitome. Chemnitz states in his preface that this Müller was a "born Dutchman" and "famed teacher of universal knowledge"—obviously, a man qualified for the job. The third part consists of the plates. They are presented in a separate booklet, are smaller than those by Rumphius, and were edited, that is, corrected.[7]

1833

Aug. Guil. Ed. Th. Henschel, *Clavis Rumphiana Botanica et Zoologica. Accedunt Vita G. E. Rumphii, Plinii Indici, specimenque Materiae Medicae Amboinensis* (Vratislaviae: Apud Schulzium et socios, 1833).

The "Clavis zoologica Thesauri Amboinensis" is on pp. 203-215 and is an index to the first and second books. Henschel took the trouble to indicate which animals Rumphius was the first to discover; he cites fifteen for book 1 and one hundred eleven for book 2, for a total of 126 animals. One should add five mollusks which, Henschel notes, were extremely rare and which "were never observed again by either his contemporaries or anyone else."

1902

E. von Martens, "Die Mollusken (Conchylien) und die übrigen Wirbellosen Thiere in Rumpf's Rariteitkamer," in *G*, pp. 109-136.

This is the first modern key to the second book of the *Rariteitkamer*. I used it as the basis for the identifications in book 2, whereafter this information was updated by H. L. Strack.

1959

W. S. S. van Benthem Jutting, "Rumphius and Malacology" in *M*, pp. 181–207.

1996

Since the return of the "Rumphius Biohistorical Expedition to Ambon" in 1990, H. L. Strack and associates have been preparing new identifications of the animals mentioned in the first two books of the *Rariteitkamer.* Detailed scientific discussions of the following groups have been accomplished.

C. Massin, "Results of the Rumphius Biohistorical Expedition to Ambon (1990). Part 4. The Holothurioidea (Echinodermata) collected at Ambon during the Rumphius Biohistorical Expedition," *Zoologische Verhandelingen, Leiden*, 307 (23 December 1996), pp. 1–53.

R. Houart, "Results of the Rumphius Biohistorical Expedition to Ambon (1990). Part 5. Mollusca, Gastropoda, Muricidae," *Zoologische Mededelingen, Leiden*, 70 (20 December 1996), 26:377–397.

1998

H. H. Dijkstra, "Results of the Rumphius Biohistorical Expedition to Ambon (1990). Part 6. Mollusca, Bivalvia, Pectinidae," *Zoologische Mededelingen, Leiden*, 71 (1997), pp. 313–343.

A. R. Kabat, "Results of the Rumphius Biohistorical Expedition to Ambon (1990). Mollusca, Gastropoda, Naticidae" [in press].

H. H. Kool and H. L. Strack, "Results of the Rumphius Biohistorical Expedition to Ambon (1990). Mollusca, Gastropoda, Naticidae" [in press].

Secondary Literature in English

Rumphius' conchological work was cited by many eighteenth-century authors, as well as later ones, but they are too numerous to list.[8] Petiver (1764) has already been mentioned, but he does not say anything about Rumphius himself. Stavorinus (1798) could be listed, but that was a translation of a Dutch original. Then there was Francis Hamilton's commentary on the first two books of the *Herbal* (1826, 1832), and toward the end of the nineteenth century Yule's frequent mention of Rumphius' *Herbal* in his unique and fascinating storehouse of Anglo-Indian idiom, *Hobson-Jobson*, an etymological dictionary on historical principles that never ceases to inform as well as delight (first published in 1886). Rumphius is frequently cited as an authority.[9] Hence it seems that John Crawfurd was most likely the first Englishman to write about Rumphius with any insight, in his entry "Rumpf (Georg Everard); Latinized Rumphius" in his *Descriptive Dictionary of the Indian Islands and Adjacent Countries* (1856).[10] There are mistakes in his biographical sketch, but that is inconsequential. Crawfurd must have read Rumphius attentively, or he was well informed, for he mentions the important fact of Rumphius' intimacy with the native peoples, a topic to which I shall return, but which was once rarely mentioned. He notes: "Rumphius was evidently a man of talents, sound sense, and indefatigable industry. Much of his information was obtained through the natives of the country, and his work affords ample evidence of his familiarity with their language. It was he that taught the natives of Amboyna the improved process for preparing sago, which is still followed by them, and for which his name is still remembered."[11] The last is true and is described in the first volume of his *Herbal.*

After Crawfurd, any real enthusiasm derives from the United States. There is, first of all, the wonderful example of A. S. Bickmore, who, as a young zoologist, traveled in 1865 all the way to the East Indies, lured solely by "the first collection of shells from the East that was ever described and figured with sufficient accuracy to be of any scientific value . . .

made by Rumphius." Bickmore was a New Englander (born in Tenant's Harbor in Maine in 1839, died in Nonquitt, a town south of New Bedford, Massachusetts, in 1914) and founded the American Museum of Natural History in New York in 1869, becoming its first director, until 1884. Bickmore described his travels in a book still attractive for its enthusiasm: *Travels in the East Indian Archipelago.*[12] He seems to have traveled with a copy of the *Rariteitkamer*, because he showed its plates to the native population so they would know what to look for.[13] Bickmore's copy seems to have been a first edition;[14] hence he traveled with quite an expensive Baedeker. He carried the book around with him on Ambon, because, he wrote, "It was my desire not only to obtain the same shells that Rumphius figures, but to procure them from the same points and bays, so that there could be no doubt about the identity of my specimens with his drawings. I therefore proposed to travel along all the shores of Amboina and the neighboring islands, and trade with the natives of every village, so as to be sure of the localities myself, and, moreover, get specimens of all the species alive, and thus have ample material for studying their anatomy."[15] The same thing was accomplished with far greater accuracy by a team of Dutch scientists more than a century later. In 1990, the "Rumphius Biohistorical Expedition," led by H. L. Strack, went to Ambon and proved, among other things, "the great accuracy and reliability of Rumphius' work."[16]

Bickmore, who had studied under Louis Agassiz at Harvard and who fought as a Union soldier in the Civil War,[17] was probably inspired by Crawfurd's account of Rumphius' life. He repeats Crawfurd's biographical sketch in all its (often wrong) particulars, even using some of the Englishman's vocabulary.[18] But his respect and admiration for Rumphius were genuine:

> I had been at Amboina a long time before I could ascertain where the grave of Rumphius is located, and even then I found it only by chance—so rarely is this great man spoken of at the present time. From the common, back of the fort, a beautifully-shaded street leads up to the east; and the stranger, while walking in this quiet retreat, has his attention drawn to a small, square pillar in a garden. A thick group of coffee-trees almost embrace it in their drooping branches, as if trying to protect it from wind and rain and the consuming hand of Time. Under that plain monument rest the mortal remains of the great naturalist.[19]

It was Alfred Russel Wallace (1823–1913) who inspired another American admirer of Rumphius to go to "those islands of the Great East—the Malay Archipelago."[20] David Grandison Fairchild (1869–1954), a botanist from Lansing, Michigan, devoted a long life to collecting tropical plants. He introduced more than 20,000 species to the United States and was one of the founders in 1938 of the tropical garden, twelve miles south of Miami, which is named after him. In 1939–1940, Fairchild went on a botanical collecting cruise in a modern junk called the *Chêng-Ho*, after the legendary Chinese admiral Rumphius wrote about in the present volume. The cruise was specifically intended to explore the eastern regions of the archipelago. In his wistful account of that journey—*Garden Islands of the Great East* (1943)—which reads like a rueful lament for a world that was forever disappearing, like the wake of the ship, Fairchild showed his empathy for the Hessian who came to Amboina "and fell in love, as a naturalist can, with the hundreds of fascinating plants which he found around him." Rumphius recorded his botanical finds in a book that "took rank immediately as the most remarkable work of its time, filled as it is with descriptions of hundreds of new species of plants, descriptions which stand today as models of accuracy and care."[21] Fairchild's understanding of Rumphius' work is contained in the word "care," which connotes the solicitude and benevolent esteem Rumphius felt for nature and for life.

Fairchild had E. D. Merrill's *An Interpretation of Rumphius's Herbarium Amboinense* (1917) on board. This work by the Harvard botanist is considered by most experts the definitive examination of Rumphius' *Herbal.* It discussed specimens located and gathered by a young scientist, C. B. Robinson (1871–1913), who paid with his life for his devotion to Rumphius' plants.[22] Elmer Drew Merrill (1876–1956), nicknamed the "American Linnaeus," published more than 500 books and papers, mostly on East Asian botany. On the basis of his familiarity with the region where Rumphius lived most of his life, Merrill concluded unequivocally that Rumphius was "one of the outstanding naturalists of all time."[23]

Neither Darwin nor Wallace ever mentions Rumphius' work. Darwin shared Rumphius' love for nature and for tropical nature in particular.[24] Darwin's only mention of Ambon is by way of Lesson. In his notebook from 1838 he records his careful reading of Lesson's travel account and his findings in the eastern archipelago on board *La Coquille.* We know that the French naturalist expressed his outrage at the desecration of Rumphius' tomb, but it failed to arrest Darwin's attention.[25] More is the pity, because Darwin would have admired Rumphius' descriptions as much as he admired Humboldt's. But his mentor had recommended the tropics of South America, and Darwin never set foot in Asia. The ignorance of Darwin's co-founder of the theory of natural selection is more puzzling.

Alfred Russel Wallace (1823–1913) spent eight years (1854–1862) in the East Indies, mostly in the eastern archipelago. Wallace was not anti-Dutch, as so many of his countrymen were—Raffles, for instance. Wallace preferred Dutch colonialism to the British variety, stating that he "believe[d] the Dutch system the very best that can be adopted";[26] he even went so far as to dismiss that sacred canon of Dutch anticolonialism, Multatuli's *Max Havelaar* (1860).[27] In his classic travelogue, *The Malay Archipelago* (1869), he records that he stayed on Ambon in 1857, 1859, and 1860. He visited many of the places Rumphius mentioned both on Ambon and on such neighboring islands as Ceram, Buru, and adjacent areas. Wallace became fervent in his admiration of Ambon Bay:

> Passing up the harbour, in appearance like a fine river, the clearness of the water afforded me one of the most astonishing and beautiful sights I have ever beheld. The bottom was absolutely hidden by a continuous series of corals, sponges, actiniae, and other marine productions, of magnificent dimensions, varied forms, and brilliant colours. The depth varied from about twenty to fifty feet, and the bottom was very uneven, rocks and chasms, and little hills and valleys, offering a variety of stations for the growth of these animal forests. In and out among them moved numbers of blue and red and yellow fishes, spotted and banded and striped in the most striking manner, while great orange or rosy transparent medusae floated along near the surface. It was a sight to gaze at for hours, and no description can do justice to its surpassing beauty and interest. For once, the reality exceeded the most glowing accounts I had ever read of the wonders of a coral sea. There is perhaps no spot in the world richer in marine productions, corals, shells and fishes, than the harbour of Amboyna.[28]

Rumphius had frequented these same marine forests nearly two centuries before.

It is a fitting coincidence that Wallace's host on Ambon was the chief medical officer, "a German and a naturalist."[29] "Amiable and well-educated," this compatriot of Rumphius was an "enthusiastic" scientist who relied for his specimens "almost entirely on native collectors."[30] One might call this Dr. Mohnike spiritual kin except that Rumphius' genius remained unique. Wallace was totally unaware of his great predecessor in "the Moluccas [that] classic ground for the naturalist."[31] He writes, for instance, that "in 1674 an eruption [on the west side of the island] destroyed a village,"[32] not realizing that there is no volcanic activity on Ambon, nor that it was one of the worst earthquakes in recorded history there, killing 2,322 people, including Rumphius' wife and one or more of his children.

Summary and Discussion of the Three Books

The first book contains sixteen plates and forty-four chapters, most of which describe crabs, lobsters, and the like. These crustaceans are described in chapters 1–27 (with one exception), of which chapter 26 more particularly discusses Decapod larvae and Mysidacea (small crustaceans also known as Opossum shrimp). Chapter 21 describes the horseshoe crab (*Tachypleus gigas*), which belongs to the Chelicerata, a phylum which includes scorpions and spiders. Echinoderms (such as sea urchins and sea cucumbers) are to be found in chapters 28–32, as well as in chapters 34, 35, and 37. The missing chapters in this sequence are chapter 33, which describes a naked mollusk (a sea hare [*Dolabella* spec.]), and chapter 36, which presents sea pens (of the order Pennatulacea). The seven remaining chapters describe a variety of animals. Chapter 38, for instance, discusses a sea snake. Chapters 39 and 40 introduce Tunicates, and in chapters 41 and 43 are Coelenterata (such animals as jellyfishes, sea anemones, and corals). Because he considered them closer to plants, Rumphius discussed corals in his *Herbal* (book 12 in volume 6). Chapter 42 describes a Siphonophore (genus *Physalia*), better known as the Portuguese man-of-war. The final chapter, 44, discusses "Wawo," masses of small seaworms (Eunicidea) which appear near beaches at specific times of the year. Rumphius' detailed description of their periodicity is still of more than casual interest.

Eight of Rumphius' names were retained by Linnaeus, who, altogether, based the names of twenty-three crustacea on Rumphius' work. More interesting, L. B. Holthuis noted that Rumphius is one of the earliest authorities, if not the first, to identify poisonous crabs and to speculate on the provenance of their toxicity. It seems that most crabs which are alleged to be toxic are Xanthids. Rumphius warns specifically against any crab that has black pincers (chapter 18), and as Holthuis explains, "Black- or dark-brown-fingered crabs are especially numerous among the Xanthidae and rarely found among other brachyuran families. The fact that the Ambonese evidently distrusted all black-fingered crabs coincides curiously with the unusual preponderance of species of Xanthidae over those of other families, among the forms that have been cited as poisonous, also outside the Moluccas."[33] The dangerous crabs with black pincers mentioned by Rumphius are in chapter 12 ("Cancer Vilosus," that is, *Pilumnus vespertilio* [Fabricius]), chapter 15 ("Cancer Floridus," that is, *Atergatis floridus* [L.]), chapter 17 (no. II, called "Cancer Aneus," that is, *Lophozozymus pictor* [Fabricius]), and chapter 18 ("Cancer Nigris Chelis," that is, *Zosimus aeneus* [L.]). Other crabs (of the Xanthidae) Rumphius considered very dangerous were what he called "Cancer Noxius" (chapter 16) and which Holthuis identifies as *Eriphia sebana* (Shaw and Nodder), as well as the "Cancer Ruber," or *Carpilius maculatus* (L.) (chapter 17). Other ones, not Xanthidae, which Rumphius put in this noxious classification are described in chapters 14 and 19 and perhaps include the "Cancer Raniformis" in chapter 11.

Crustacea are also present in the other two books. In book 2 one will find Pinnotheridae, small crabs that are often commensals of mollusks, in chapter 28 (under no. I, folio 128), chapter 30 (under no. I, folio 138), chapter 36 (under no. I, folio 153), and chapter 37 (under no. VII, folio 156). Then there are the Balanidae (barnacles and such) in chapters

26 (no. III, folios 121–122) and 37 (nos. XIV and XV, folios 158–159). In book 3, the "Cancri Lapidescentes" of chapter 84 are in fact fossil crabs from China.

Pre-Linnean literature on tropical carcinology was limited, so that, despite the enormous difficulties Rumphius encountered when he tried to assemble a library of authoritative texts, his writings were largely original contributions. In the preface to his *Amboinsche Kruidboek*, Rumphius makes a statement that applies equally to the *Rariteitkamer*: "I cannot deny that a great deal of writing about Indian plants saw the light before mine, for instance by the Portuguese for over a century now, or recently by learned men from the Dutch nation: but I must reserve the honor for myself of having broken the ice where it concerns the description of plants from these regions, since, as I did as well, the earlier Writers were only concerned with those plants that were found in the countries where they lived, but no one, to my knowledge, ever rambled through the forests of these Eastern regions, nor did anyone ever scrutinize them."

Rumphius' bedrock of authorities was composed of Aristotle (384–322 B.C.) and Pliny the Elder (A.D. 22 or 23–79). They represent a solid foundation to build on, and modern hubris should not dismiss them with uninformed condescension. Aristotle, for instance, has still a great deal to teach. Rumphius simply did not suffer from modernism's curse of neophilia. At various times he cautions his readers not to belittle ancient or, for that matter, native sources. In chapter 17 of the second book, he admonishes "that one should not immediately slight the writings of the Ancients, or accuse them of lying if you do not instantly find something the way they described it; for every day there appear things, which have been unknown for a long time, but which one can find in Ancient writings, which are often not understood correctly."

Rumphius followed Aristotle's emphasis on the priority of function (as opposed to the modern priority on form) as he found it applied in the Greek biologist's *Historia Animalium* (third century B.C.). In his very brief introduction to the entire book, Rumphius uses three general terms from Aristotle to aid him in organizing his material: "Malacostrea," "Ostracoderma," and "Zoophyta." Aristotle discussed most of the animals in books 1 and 2 in a general sense, unaware, of course, of the tropical species Rumphius was to discover. One should also remember that Aristotle's work as a whole and *Historia Animalium* in particular, were far more "contemporary" in Rumphius' day than we realize. Aristotle had been rescued from oblivion by Arab culture in the eighth and ninth centuries, but the first complete European edition of his work was not in print until 1498. For the next two centuries, Aristotle's authority was paramount.

Pliny had remained an authority throughout Roman civilization, and through every century of the Middle Ages, and was a principal source on natural history during the Renaissance. During the sixteenth century scientists began to question the veracity of Pliny's information, a debate that could only lead to a closer scrutiny of real nature. The specimens discussed in books 1 and 2 are relevant to book 9 of Pliny's *Natural History*, but both qualitatively and quantitatively there is no comparison.

This is also the case with authors who are closer to the time Rumphius was working; L. B. Holthuis' discussion of the literature that is relevant to book 1 clearly demonstrates Rumphius' superiority. Rumphius mentions about a dozen authors in book 1 that have some significance for his subject.[34] Of these twelve, Holthuis' list has only two that merit attention (a third, Seba, lived after Rumphius), and even so, Clusius (*Exoticorum Libri decem*, 1605) only mentions two tropical crustaceans, while Rumphius' contemporary in the Indies, Jacobus Bontius (*Historiae Naturalis & Medica Indiae* [in Piso, *De Indiae*], 1658) lists a small number of Indonesian crustaceans rather perfunctorily. The only work that Rumphius mentions several times is Rochefort's wonderful *Histoire naturelle et morale des Iles Antilles de l'Amerique* (1658). The reason is obvious: Rochefort described *tropical* animals, even if from a different hemisphere. But despite my admiration for his writing (some passages read like prose poems), one must admit that Rochefort lacks Rumphius' precision and care.

The second book is twice as long as the first, though it has fewer chapters (thirty-nine), yet twice as many plates (thirty-three). The reason for this is longer chapters that discuss a greater number of specimens. Book 2 presents the mollusks. Univalve Gastropods are described in chapters 5–17 and 19–26. Chapter 18 includes land, sea, and freshwater snails, and chapter 27 discusses marine polychaete worms. Bivalves, such as clams, scallops, cockles, mussels, and oysters, are featured in chapters 28–37.[35] Cephalopods (the *Nautilus pompilius* [L.], two *Argonauta* species, and the small cephalopod called a *Spirula spirula* [L.]) occupy chapters 2, 3, and 4. In chapter 38, Rumphius attempts a concordance of his work and that of Pliny, and chapter 39 tells collectors how, where, and when to collect tropical shells.

Martens' research disclosed that book 2 describes 339 species and 151 genera of mollusks, of which 157 species and 20 genera had never been mentioned before. The book contains other original contributions. Rumphius was the first malacologist of the tropics to describe the living *animals* in their habitats. In other words, his book was never intended to be just an index of shells. For instance, his description of the *living* nautilus in chapter 2 was the first and remained the only one for more than a century,[36] and he was the first to note that the *Argonauta* has the habit of latching on to a piece of passing debris and using it as a cover (chapter 3). Although Schynvoet dismisses it, Rumphius was the first observer to warn about what he called the "little bone" of certain mollusks, a radular tooth that injects a deadly venom into its prey (for instance, nos. X and XI of chapter 19; no. I of chapter 20; or no. X of chapter 21), an observation that was not duplicated, according to Martens, until the middle of the nineteenth century.[37] A related instance of Rumphius' perspicacity is his recommendation in chapter 23, based on native knowledge, that one should never eat the meat from smooth and shiny snails but only from coarse, unsightly ones. This corresponds with findings from biochemical research in snail toxins, conducted toward the end of the twentieth century. The most venomous marine snails are the cone snails, which, after they have been cooked, are lustrous and smooth, although Rumphius was probably referring to olives and cowries.[38]

Strack's 1990 expedition to Ambon proved the astonishing accuracy of Rumphius' information. After the passage of nearly three centuries, his matching of specimens and locations turned out to be nearly infallible, and Strack makes the important observation that "of all [the] vernacular names recorded by the expedition, about one third appeared to be identical to those recorded by Rumphius,"[39] yet another validation of this remarkable man, who was as great an ethnographer as he was a botanist and zoologist.

Such innovative techniques as field observations, precise notation of habitats and sites, and a superior talent for classification set Rumphius' work apart from his pre-Linnean peers. Rumphius gave most of the mollusks a binomial label, which in the case of his Latin nomenclature is generally a noun for the genus and an adjectival form or a genitive case for the species. Linnaeus transferred thirty-two unaltered names of mollusks from Rumphius' *Rariteitkamer* to the tenth edition of his *Systema Naturae* (1758). Martens called this prerequisite for modern approval "prophetically Linnean,"[40] a judgment that becomes only more interesting when we note that the same thing can be said about the Malay names. The Dutch vernacular ones are usually singular attributive nouns which highlight the beauty, not the science, of the animal, poetic metonymies that capture a particular vitality of its form. It comes as no surprise that, just as with the first book, Rumphius' treatise on mollusks had no serious rivals until the publication in 1757 of *Histoire naturelle du Sénégal: Coquillages*, by Michel Adanson (1727–1806), and it remained the most important source of information on mollusks from the Malayan region throughout the nineteenth century.

Chemnitz wrote in 1766 that "the number of classic writers on conchology is uncommonly small."[41] I have already mentioned the most significant epitomes of Rumphius' work that were printed prior to Linnaeus' 1758 edition of his *Systema Naturae*, which inaugurated modern zoological taxonomy. Suffice it to say that in the case of malacology, Aristotle's *Historia Animalium* and Pliny's *Natural History* (book 9) remained the basic foundation. One should remember that Rumphius' innovative exploration of applied or economic malacology was something he had learned from Pliny, who does not discuss anything in nature unless it has some human use. There are, of course, a large number of books which include shells in some form or fashion, but it is not my intention to compile a comprehensive list. Many have been discussed as subjects of the history of science, and that history has been examined many times.[42] The most important predecessors and contemporaries that are pertinent to Rumphius are the following.

Guillaume Rondolet (1507–1566), the French anatomist and ichthyologist, friend of Rabelais and author of *Libri de piscibus marinis* (1554–1555), of which there was a French edition, *Histoire entière des poissons*, published in 1558. Conrad Gesner (1516–1565), the Swiss naturalist whom Cuvier dubbed "the German Pliny," probably on the basis of his voluminous *Historia animalium*, published between 1551 and 1587, which contains more than 4,500 pages. The important edition for the present context is that work's third volume: *Historiae Animalium. Liber III. qui est de Piscium & Aquatilium animantium natura* (1558). Another author Rumphius mentions several times is "Bellonius," or Pierre Belon (1517–1564), a French naturalist. Belon's relevant work is *Histoire naturelle des estranges poissons marins* (1551). The reference works by the Polish naturalist Jan Jonston (1603–1675) were present in the library of almost any serious Dutch collector in the seventeenth century. His most important volume was *Historia naturalis*, published in Amsterdam in 1657. Then there is the Italian conchologist Filipo Buonanni (1638–1725)—also mentioned by Rumphius under the name of "Philippus Bonannus"—whom Rumphius needlessly considered a serious competitor, probably because Buoanni was very popular. His classic work was first published in Rome in Italian, *Ricreatione dell'occhio e della mente* (1681), and three years later in Latin, *Recreatio mentis et oculi de testaceis*. Finally, although his work was not contemporary, one should mention Albert Seba (1665–1736), if only because his thesaurus of the specimens he owned has been highly praised and because he did not employ Linneaus' binomial nomenclature. Seba's main work, and the one relevant to Rumphius is: *Locupletissimi rerum naturalium thesauri accurata descriptio* (1734–1765).

The third book never received the attention the second or the first one did. Yet it is the longest section in the *Rariteitkamer* and has more chapters than the first two combined (eighty-three), but only eleven plates. That last fact probably explains the relative lack of attention the final book has been accorded: it deals with items that are visually not very exciting or that, at the time, were practically unknown. Yet this also means that book 3 contains more firsts than its two predecessors and that, with the exception of the four chapters on amber, most of the other chapters present Indonesian or, more generally, Asian subjects for the first time in a printed European text.

Book 3 most certainly contains Rumphius' "Book of Stones," but it covers more than that, and one should also be aware that "stone" is a very flexible concept here, one that includes just about anything that is a concretion. In order to sort this inorganic salmagundi to some degree, I divide it into several categories, none of which is exclusive of what Rumphius considered to be even remotely associated. He opens with seven chapters on *metals*, of which chapters 1–3 deal with gold and silver, chapter 4 describes the Indonesian alloy "*suassa*," and chapter 7 the Asian alloy "*gans*" (*ganza*), while chapter 6 deals with the presence of iron in the archipelago and how it is forged (including the *pamor* technique—see book 3, chapter 4). Chapter 5 discusses the effect of hydrochloric acid on certain metals.

The second largest number of chapters is devoted to minerals and gemstones. Chapters 10–22 list a variety of *minerals:* chapter 10 discusses orpiment, chapter 11 alum, chapter 12 something called "misy" (which might be a ferrous sulphate), chapter 13 marl, and chapter 14 briefly lists some unknown stones. Chapter 15 explains the uses of realgar, chapter 16 ilmenite, which was used as "sand" to blot the ink of written documents and which still rustles out of VOC documents that were never consulted before. Chapter 17 describes some unknown "bubbling ashes" from Indochina, while chapter 18 is on firmer ground with Indonesian flintstones, or a kind of pyrites, which were used for firearms. Chapter 19 discusses another type of pyrites known as "marcasite," and chapter 20 talks about a kind of gypsum. Chapters 21 and 22 discuss quartz, and one should add here chapter 31, which briefly refers to "amarilite," a kind of hydrous ferric sulfate. The following five chapters also discuss minerals: chapter 79 gives a fairly lengthy presentation of Chinese alabaster, chapter 80 talks about the once legendary "amianthus" or "Salamander's Hair," which is a variety of asbestos, chapter 81 presents some "stone from Malacca" which might be a ferric oxide, chapter 82 a serpentine, while chapter 83 concludes with touchstones from Ambon.

Obviously related to minerals are *gemstones*, which one will find in chapters 39–45. Chapter 39 describes the better known gems such as diamonds, rubies, emeralds, sapphires, and so on. Chapter 40 discusses a chrysoberyl known as "cat's-eye" and the opal, two gems which were far more important to Asia than, say, diamonds. I think that the "Joc" of chapter 41 is jade, chapter 42 has a disquisition on agates, more particularly on what we call "dendrites," a term Rumphius assigns to something entirely different (see chapter 69), and chapter 44 might also be about some form of agate. Chapter 43 is a short text on three kinds of chalcedony, and chapter 45 tells one how to test whether a gem is genuine or not.

Rumphius devotes the most attention to what loosely may be called concretions in plants and animals, known as either *mestikas* or *guligas* in Indonesia. They are discussed, with three exceptions, in chapters 46–75. They are the same thing to most people, but Rumphius makes a clear distinction. Though he does not always adhere to his own demarcations, chapter 46 prescribes that a *mestika* (or *mustika*) is "any little stone that one finds in a plant, in wood or in other stones, in a way that is not nature's wont, and that was produced by the same, not from a corruption that would afflict an animal." A *guliga*, for instance the bezoar (chapter 55), is a calculus exclusively found in an animal, never in a plant, while *mestikas* come in both floral and faunal varieties. Perhaps the easiest way to distinguish them is that Rumphius considers *mestikas* magical fetishes, while *guligas* are for medicinal purposes, part of the *dukun*'s apothecary. On the basis of that distinction, I divide this large number of chapters as follows, though some overlap. The preponderance describes *mestikas* derived from a variety of sources: humans (chapters 47 and 48), pigs (chapter 50), deer (chapter 53), cats (chapter 54), millipedes and lizards (chapter 59), birds (chapters 60–62), fishes (chapter 63), and mollusks (chapter 64); while floral *mestikas* come from coconut palms (chapters 68 and 69), from the areca nut, from rice, and from garlic (all three discussed in chapter 70), from various trees (chapter 71), from fruits (chapters 72–74), and, finally, from the jasmine flower (chapter 75). *Guligas* appear to have been far less common. One variety is produced by porcupines (chapter 52); the other comes from goats and represents the well-known "bezoar," which, Rumphius states, can also come from monkeys (chapter 55). The once equally renowned "snake-stone," both Ceylonese and Indonesian varieties, is discussed in chapters 57 and 58. The *mestikas* from millipedes and geckos (chapter 59) and the one produced by the coconut (chapter 68 and 69) double as *guligas*. For good measure, Rumphius throws in a discussion of hairballs from wild pigs (chapter 51), which, though definitely a "curiosity," might also have some medicinal powers.

The remaining chapters could be combined under the heading "curiosities," though this medley of items hides some surprises. The most unclassifiable are "stones" that are found in trees and fruits but that are not *mestikas*. They appear to be plain pebbles which were enveloped by the waxing plant at some early stage of its growth (chapter 76). Then there are "stones which happen to have an unusual shape" (chapter 77), including large pieces of coral rock that appear to have human shapes. The chapter on "melitites" (chapter 78) elicits a good story, but it is not easily identified. Other odd items are a brief description of stone files used on Java (chapter 33) and "small pieces of iron" found inside trees (chapter 32). Rumphius calls them "*dendrites metallica*," where the first word has nothing to do with its modern meaning of a ramate pattern in a mineral or stone produced by a foreign mineral. Rumphius uses it in Pliny's sense of a "stone" found in a tree. Other disparate items are a chapter on porcelain (chapter 23) and one on dishes that will detect the presence of poison (chapter 24), native arm bracelets made of glass, called *mamakurs* (chapter 25), a story about a magic bracelet (chapter 26, repeated in chapter 76), beads from Timor (chapter 27), the phenomenon of the sea aglow at night (chapter 28), a mineral resin called "*mor*" (chapter 30), and directions on how the Ambonese clean gold jewelry (chapter 29). We are also told how the Ambonese obtain a crude form of salt (chapter 85), that there is a particular resin found in China (chapter 87), and that the Portuguese swear by something called "Lapis Cordialis" which will cure just about any infliction to which man is heir (chapter 86).

Reflecting the importance accorded it at the time, there are also lengthy texts on amber. Chapter 35 discusses ambergris, chapter 37 tells us about spermaceti, while the "black amber" mentioned in chapter 36 could be some kind of bitumen, but I have no idea what is meant by the "sealard" described in chapter 38.

Last but not least are the chapters devoted to fossils and to prehistoric tools and weapons. The latter are discussed in chapters 8 and 9 and represent the first European discussion of prehistoric objects from Indonesia.[43] Rumphius was also the first European to write about Indonesian fossils. The strange "stones" he describes in chapter 34 are Belemnites, some containing ammonites. One will find fossil shells and Echinidae (such as sea urchins) in chapter 65, fossilized gastropods in chapter 66, and fossilized lamellibranchia (such as oysters and clams) in chapter 67. In chapter 84 we are told about fossilized crabs from China, while earlier, in book 2, chapter 29, Rumphius has discoursed at length on fossilized Tridacna shells.

The chapters on prehistoric tools and on fossils are not the only original contributions of book 3. Rumphius' discussion of "*suassa*" is the first in Western literature, as is that chapter's mention of its use for samurai swords. This is also true of the remarks on the forging technique called "*pamor*" outside Java, while the report in chapter 7 is the first mention in the archaeological record of Bali's famous "Pejeng Moon," and the first discussion in world literature of any Indonesian antiquity. Our only knowledge of Indonesian "snake-stones" is contained in chapter 58, and the twenty-four chapters on *mestikas* remain the single most important source on the history of these fetishes.

One can tell from this summary that most of book 3 is original, hence uncharted in the relevant literature. Only mineralogy and gemstones had a written record, but that was almost exclusively European. Rumphius relied on Pliny, of course, who discussed minerals and gems in books 36 and 37 of his *Natural History*. Although he mentions a few other names, there are only two works that would have been of any significance to Rumphius. One was by "the father of mineralogy," his compatriot Georg Agricola (1490–1555), especially his *De re metallica* (1556), and the other was *Gemmarum et Lapidum Historia*, by Anselmus Boetius de Boodt, which was published in Hanau in 1609 (a later, revised edition appeared in Leiden in 1636). De Boodt, for instance, treats the bezoar (chapters 192–194), the "eagle-stone," or "Aëtites" (chapters 196–198), and the "thunder-stone," or "Chelonitis" (chapter 264). But even if there had been more technical literature, Rumphius was pretty much left to his own devices. There is no doubt that Wichmann is correct when he states that we have to thank Rumphius for "the first descriptions of Indonesian minerals. And they are precise enough to enable us to identify the various kinds in terms of contemporary knowledge. In this respect, Rumphius had no precursors." Book 3 provides us with a "topographical mineralogy of the Indonesian archipelago."[44]

Arthur Wichmann's discourse on the mineralogical and geological aspects of book 3 is, as far as I know, the only extensive discussion of anything in that book. Nobody paid any attention to it until the end of the nineteenth century, so that we have

> the rare case here of a work that today is almost as important and innovative as [it was] 197 years ago.[45] . . . This honor is due to Rumphius' precise observations and the reliability of his information, two aspects in which he remains unsurpassed. It allowed him to be of far greater use than if he had wasted his time developing theories whose time had not come yet, and which were not yet capable of finding a solution. This also spared him the fate of many of his contemporaries who, though "famous" at the time, will only elicit a smile today, that is, if their work is read at all. Hence today's naturalist can honor Rumphius as the outstanding guide and predecessor whose work can rightly be called: monumentum aere perennius.[46]

The sentiment is right, but the statement is not entirely correct. Besides the wealth of anthropological information, which has considerable importance to ethnographers, this final book also grants us a rare glimpse of Rumphius' intellectual convictions. What this information discloses is that he was a complicated man, with seemingly contradictory impulses. He clearly remained a devout Protestant all his life, one who believed in the literal truth of the Bible, while at the same time he trusted only the evidence of direct observation and argued for the need of data, that most modern of scientific requirements. He confesses to disliking the "new Philosophers" (chapter 28) with their "new philosophy" (chapter 42) and testifies to his faith in "Astral influences" (chapters 8 and 61), sounding very much like Paracelsus and Athanasius Kircher. He fulminates against *tapa*, an Indonesian variety of religious asceticism, as well as against the practice of magic (chapter 21), but he makes no bones about the value of native "superstitions" (chapter 46). There is no doubt that he believed in Paracelsus' doctrine of signatures (chapter 42), but one should remember that this was also fundamental to native medical practices, which, according to both Bontius and Rumphius,[47] were far more efficacious than European conventions. In chapter 9

Rumphius makes the moving admission that he "would rather be astounded by the unfathomable powers of nature, than to lapse into some kind of error because of too punctilious a scrutiny." This is a statement worthy of a poet, a Keats or Wordsworth,[48] but it would not fare well in the modern, post-Darwinian world of science, which prefers information to understanding. Yet his work certainly supports the contention that he too was what he called a medical practitioner from Malabar: an "Empiricus" (chapter 68). This word dates from classic Rome and referred to a physician who derived his knowledge from experience, not from theory. Hence in the monumental debate of the seventeenth century, Rumphius would have rejected the Cartesians and maintained his support of Pliny or, for that matter, Sir Thomas Browne, whom, rather surprisingly, he quotes at length in chapter 23. Perhaps even more important, the *Rariteitkamer* reveals that, as is the case with his *Herbal*, Rumphius had not only become completely acclimatized to the archipelago, but he also had developed a profound sympathy for Indonesian culture.

One will find evidence of Rumphius' respect for the Indonesians in his work and in comments from his superiors. It was already mentioned that Pieter Marville praised Rumphius' "knowledge and experience of these lands" in 1666, and Hustaerdt in 1660 specifically recommended Rumphius as a man "who knows how to suit himself to the temper of the Ambonese," which clearly indicates that this was a rare gift. It is also obvious that he was fluent in the language and knew the customs, religion, lore, and skills of the local peoples. He knew their food and medicine, their weapons and dress, their superstitions and stories. He was without a doubt the first (and for a long time the only) significant ethnographer of Indonesia. Not until the nineteenth century will one find a comparable description of an Indonesian region, though I doubt there was ever one as detailed.

Rumphius would have fulfilled Lévi-Strauss' wish "that every ethnologist were also a mineralogist, a botanist, a zoologist and even an astronomer."[49] Such knowledge was rare, and the VOC was quite happy to exploit it, but there is the intimation, particularly from comments made by Padtbrugge in 1683, that this same talent was also considered suspect. Rumphius often took the side of the local people against his powerful employer, nor did he scruple to lecture and criticize his European superiors. The following written comment by Governor Padtbrugge proves this: "I would have readily ordered the Council [to do this] according to the rules, but the merchant Rumphius is not on it. I respectfully hope to hear if he was not assigned to it on purpose or not, since we still have to talk to him, because he is in everything hostile and prejudicious to the Company's methods. Because in his pursuit of curiosities he is ruled by an uncommon desire, one that is often quite contrary to the Company's interests, as for instance in locating land for the Company, and other instances, as we experienced often enough already."[50] One gets the feeling that he often was more comfortable with the indigenous population than with his compatriots. For instance, Rumphius confessed with commendable honesty and grace that local experts taught him about plants and their uses. In a text on a certain tree he describes a procedure to extract oil from its wood. He wants to do it his own way, "but my Master, who taught me this art of distilling, by the name of Iman Reti, a Moorish Priest from Buro, admonished me that such was necessary, otherwise the wood would catch on fire."[51] Earlier he stated that "this wood was shown to me for the first time . . . by *Patti Cuhu*, a Regent or Orangkay of the Hitu village called Ely, a Man experienced in the knowledge of plants, who has helped me a great deal in this work, wherefore I deem it only decent, to commemorate him here."[52] Here we have this remarkable man in a nutshell. This European colonial does not condescend. Quite the contrary, he has no hesitation in calling a native informant his "Master"[53] and telling his readers that he could not have achieved what he did without the help of people he could easily have left unmentioned. Quite apart from the fact that they vouch for the authenticity of his work, these passages tell us why the people of Hitu called a hill after Rumphius, why he could so easily converse with Radja Salomon, the chief from Timor, why he obtained select information that no other European could ever have elicited, and why, after his death, the Ambonese regarded his tomb as a sacred place. From this perspective, one might say that when Rumphius went to the Indies, he came home.

Curiosity Cabinets

One reason that *D'Amboinsche Rariteitkamer* exists in its present form was Rumphius' desperate attempt, relying on the then current rage for curiosity cabinets, to get at least one of his books in print before he died. The material was otherwise slated to be included in the larger "natural history of the Indies," with the first two sections part of Rumphius' "Animal Book" and the final third the "Book of Stones," which had been mentioned by him in various places. At one point he must have suddenly realized that what for him were quotidian delights were, by virtue of their geographical provenance, exotic *mirabilia* in Europe, in a word, "curiosities." This was an ancillary decision, for a careful reading of *The Ambonese Curiosity Cabinet* reveals that its program and its tone are those of a natural history, not those of a catalogue of lifeless oddities. Like the *Herbal*, the present work depicts an Indies that is alive and sensual, with the author wanting to convince his readers of the tropical abundance he clearly loves. Rumphius' text is not an inventory of economic pride. Consider the second book. Rumphius' own preface states that the descriptions of shells represent "the true Curiosity Cabinet," an opinion repeated several times. The drawers of such a cabinet would have contained a neat array of shells, cleaned of any semblance of life. Yet the emphasis of the text is on the animal that lives within these handsome shelters: where one expects a conchology, one is surprised by a malacology. In short, these pages were conceived with a different intention than the one which caused it to be marketed.

The impetus for this sudden rush into print might have been the forced sale of Rumphius' collection of natural objects in 1682. I have no doubt that Rumphius was coerced into selling what for him must have been souvenirs in the most literal sense: memorials of the life that was extinguished in 1670. Though natural objects, these shells and stones were for their owner also palpable guides to his most private habitat. Whenever he refers to this sale, the tone is rueful or angry. Social etiquette counseled constraint when he wrote to their buyer, Cosimo III de Medici, Grand Duke of Tuscany (1642–1723): "May the Lord . . . favor the passage of these objects and may they arrive safely . . . so that no damage will be done to a treasure which I gathered over many years with much cost and labor and which, in the future, it will be impossible to acquire again, especially since I am now old and blind."[54] Even this sentence hints at scruple and loss, but Rumphius' recalcitrance comes through a little clearer when in the last chapter of the "true Curiosity Cabinet" he declares, for no pressing contextual reason, that he had a collection of 360 curiosities, gathered "over a period of 28 years," which was dear to him because it contained many unique items, and that he had to send this collection to "the Grand Duke of Tuscany at the insistence of several friends, to whom I was obliged."[55] This was not a deliberate sale beneficial to both commercial parties, as, for instance, when Seba or Ruysch sold his collection to Peter the Great in 1716 and 1717.

One commentator wrote that Rumphius mentioned this transaction only three times in the *Rariteitkamer*,[56] but this is incorrect. It preyed on his mind for the final two decades of his life, for one will find that the year 1682 is mentioned at least eight times, twice as often as what should be the far more traumatic date of 1674, and four times more often than 1670, the year he became blind. Nor should his reluctance give the impression that Rumphius was not generous. Quite the opposite; he gladly shared the beauty of the tropics. We have evidence that he gave natural objects to Camphuys, D'Acquet, Mentzel, Cleyer, Fehr, Witsen, and probably others we don't know about. And if one wants to protest that this was for ulterior motives, then one should be able to account for the wit and good humor which accompanied his gifts to D'Acquet or Camphuys.[57] Giving things away was not the problem. What vexed Rumphius was the compulsion, backed by some kind of high-pressure method, to relinquish part of his life to a stranger for money. And given that context, some passages in the present volume take on a different significance.

Rumphius subscribed to Pliny's denunciation of greed. There is his sly censure of rich and covetous people that opens the third book, or the very Plinian statement, in chapter 39 of the same book, that diamonds no longer get a chance to rest and grow old because "today's covetousness and pomp wants them in such profusion." But I think that beyond this age-old criticism, Rumphius had yielded to the Indonesian ethic of liberality. He mentions it in his work time after time and always with approbation. Fetishes, for instance, *mestikas*, lose their power if they are sold, while "curiosities" are deprived of this special dispensation as "semiophores" (as Pomian called them)[58] when they are treated as commercial inventory. The two are combined in the following statement, which concerns a magic bracelet that fell out of the sky (chapter 26 in book 3). Rumphius does not necessarily believe that a spirit gave this object to a particular man from Hila, but he "completely agreed with him" that after the man had sold these objects to him "they would no longer have any powers with me, which they say of all Curiosities which one has not found oneself or that were not given as a gift, but were bought with money." Similarly: "They also ascribe the *Mesticae* the general attribute, that they will be only lucky for the one who found it, or the person who received one as a gift, but not for someone who buys them" (chapter 46 in book 3), and in chapter 71 he states that the Chinese felt the same way. This is the Indonesian notion of *pusaka*, which bestows rarity on an object not by material but by metaphysical fiat, ruled by an Asian sense of what Barthes aptly phrased the "apprehension of the thing as event not as substance."[59] The Cosimo affair, therefore, represents a clash of Western capitalistic imperatives and a stubborn Asian metaphysic, with the Hessian deferring to the Indonesian. But Rumphius had not, of course, divested himself entirely of Europe. This same event also showed him a way to get his words in print, or perhaps someone else suggested it to him. Whatever the truth may be, he now collected the existing pages and presented them to the world as an example of what Descartes denigrated as the "curious sciences."

By the end of the seventeenth century, Holland was the commercial center of the traffic in curiosities. The activity of collecting had been indulged from the second half of the fourteenth century, with the emphasis on history, that is to say, antiquities and coins. The interest in natural objects was a relatively late development in Europe, but it became characteristic of Dutch collectors during Rumphius' lifetime,[60] while it is also important to realize that though collecting had once been an aristocratic or royal pastime, it became the passion of private citizens in the Dutch Republic. In Holland, natural curiosities usually meant items that were non-European, more often than not exotic things from the tropics. This was only possible because of the VOC's busy commercial intercourse with Asia. Most collectors were part of the urban elite, the upper middle class from cities and towns with offices of the VOC and WIC.[61] Although this inevitably constituted a relatively small, closely knit group of enthusiasts—most of whom were mentioned by Schynvoet or were part of Rumphius' existence—their enthusiasm made Dutch collections part of an international activity. For instance, the first major Dutch collector was a physician who lived in Enkhuizen. The house of Bernardus Paludanus (1550–1633), the Latinized version of Berent ten Broecke, became a tourist attraction for the traveling upper echelon of European aristocracy and intelligentsia. His cabinet, or collection, what in German was called a *Wunderkammer*, was visited by a large number of travelers, including most of the people who were connected to Rumphius' life in some way or another. In 1593 Paludanus opened his door to a young German student from Leiden. This was the subsequent founder of Rumphius' *Gymnasium*, Philipp Ludwig II, Count of Hanau-Münzenberg (1576–1612). Other German nobility followed, including, in 1611, Otto of Hesse-Kassel, and the Elector Palatine, Frederick V (1596–1632), better known as the "Winter King," who, while he was living in exile in The Hague, came to Enkhuizen in 1625. Prince Maurits came once, as did his brother, Frederik Hendrik, and there was always a steady stream of scholarly visitors. To restrict oneself to those of significance to Rumphius, one can mention calls by Clusius, Scaliger, Ole Worm, Jan Jonston, and Jan Huyghen van Linschoten.

During his working visit to Holland in 1697, Czar Peter the Great personally examined the curiosity cabinets of Nicolaes Witsen (1641–1717), Jacob de Wilde (1645–1721), Levinus Vincent (1658–1727), Nicolaas Chevalier (1661–1720), and Simon Schynvoet (1652–1727), all familiar to readers of the present text.[62] The Russian czar also conversed at length with Antoni van Leeuwenhoek (1632–1723), Herman Boerhaave (1668–1738), and Frederik Ruysch (1638–1731) and during his second visit to Holland, in 1717, saw the collection of Albert Seba (1665–1736). Seba had persuaded the czar to buy a previous collection in 1716.

Cosimo III de Medici (1642–1723) stayed in Holland twice: from December 1667 to February 1668 and during the summer of 1669. The grand duke of Tuscany spent a great deal of time with VOC management and with Pieter Blaeu (1637–1706), grandson of Willem Blaeu (1571–1638), the mapmaker and publisher who became the VOC's official cartographer in 1633. Pieter Blaeu, who was then in charge of a famous map shop on the Bloemgracht, guided Cosimo around Amsterdam almost every day,[63] bringing him to many private collections of curiosities, mostly objects from the Indies, but also more inclusive ones such as the famous one owned by the apothecary Jan Jacobsz. Swammerdam (1606–1678), father of the troubled genius Jan Swammerdam (1637–1680). The son's collection of insects so impressed Cosimo that he offered to buy it and suggested to the recently graduated physician that he would give him a job in Italy. The younger Swammerdam refused[64] and went on to write his classic study of insects, *Bybel der natuure* (Bible of nature), a book which Rumphius would have appreciated and which, like Rumphius' work, was only published posthumously, in 1737–1738. Cosimo bought a great number of things through Blaeu, thereby establishing the commercial obligation which resulted in the sale of the Ambonese collection some fifteen years later.

Cosimo was only able to see certain places and people because the VOC encouraged such visits,[65] just as thirty years later Peter the Great could tour the VOC's wharf in Zaandam only because Nicolaes Witsen invited him to do so. Witsen had been a director of the VOC since 1693 and was several times mayor of Amsterdam after 1682. Witsen's collection, large enough to warrant the publication of its own "atlas," or album of prints, prospered owing to his position and contacts. Neither his "cabinet" nor those owned by Vincent, Seba, De Wilde, or De Jongh would have been so distinguished if the VOC had not existed and these men unable to exploit their positions of power within the commercial hierarchy. To curry the favor of the man who decided the captaincies of the VOC's fleet, master mariners brought curiosities to Jacob de Wilde as presents. De Wilde was the secretary of the VOC, and Joan de Jongh was its accountant, a position which enabled him to obtain fresh specimens from the tropics.[66] Vincent had a large network of commercial contacts as well, but Witsen outdid them all. Not only was he a director of the Company, but he was related to two governors-general in the Indies. Witsen could influence appointments and was powerful enough to send personal emissaries to distant places in order to obtain objects he coveted.[67]

One can well imagine that the favor of these influential men was often solicited by powerful foreigners, which in turn only strengthened the international ties which had been initially knotted by commerce. Capitalism subsidized curiosity in a reciprocal relationship that enhanced and dignified both, though to be sure it was the social and commercial center of Amsterdam that decided which previously "useless" objects were now items of commercial value. Commerce and curiosity were closely allied during the seventeenth century and the so-called curiosity cabinets—which ranged from what were indeed cupboards of drawers to large, specially built display halls[68]—are a perfect symbol of that peculiar alliance. Studying curiosity cabinets means discovering a large part of the intellectual life of the seventeenth century, which in turn means disclosure of that era's economic vitality. This is too large a subject here, but if one examines it to any degree, one is struck by what amounts to an ecology of common interests, an internet of deliberately elected affinities which, in the final analysis, could only bring distinction to all participants. Collectors visited one another, their private catalogues circulated within this select group, and they corresponded with one another and with like-minded enthusiasts abroad. Most of the older works of scholarship that are mentioned in the present volume were part of private libraries all over Europe and, one might add, Asia. For instance, Valentini's *Schaubühne oder Natur- und Materialienkammer* was found in the libraries of two Dutch collectors,[69] while it is clear that Rumphius consulted Buonanni's manual on shells in Ambon. Ruysch, Seba, Vincent, and Leeuwenhoek became members of the Royal Society in London; Ruysch and Seba were members of the same Academia Naturae Curiosorum that Rumphius and Cleyer belonged to; and Ruysch was also a member of the Académie des Sciences in Paris.[70] Witsen's close friend Gisbert Cuper, who was a professor and subsequently mayor of Deventer, became a corresponding member of the French Académie des Inscriptions. Cuper and Witsen produced a copious correspondence with each other over the years, and it was Cuper who, through Witsen, sent a questionnaire to Batavia which was answered by Rumphius at Hurdt's request.

Witsen pulled strings for Herbert de Jager and owned some of that troubled man's drawings,[71] and he was also in close contact with Cleyer. The draughtsman Hendrick Claudius worked for both Cleyer and Witsen,[72] and Rumphius corresponded with Cleyer, Witsen[73] and De Jager. Witsen was also very "familiar"[74] with Boerhaave, the most famous medical man of his age. Boerhaave was responsible for the publication of Swammerdam's lifework, admired Seba's cabinet, and persuaded Seba to become a subscriber of Swammerdam's book.[75] Samuel Johnson wrote a life of Boerhaave after his biography of Sir Thomas Browne, the same author Rumphius quotes at length in the *Rariteitkamer*. Peter the Great spoke with Witsen about Siberia and with Simon Schynvoet about topiary art, and when the czar bought Seba's collection in 1716, he had not only thousands of curiosities transported to St. Petersburg but also the works of Aldrovandi, Jonston, Merian, and Rumphius.[76] Seba was on friendly terms with Witsen, Vincent, Ruysch, and Boerhaave and corresponded with many people all over Europe. He became a member of the Academia Naturae Curiosorum almost half a century after Rumphius, and Hans Sloane procured him membership in the Royal Society in London. Rumphius' dubious friend Valentijn knew Seba and visited his house twice; Linnaeus stopped by to admire Seba's acquisitions in 1735, although he did not approve of the apothecary's nonscientific, though artful, displays.[77]

The same passion was indulged in the colonial Indies as well. Valentijn reports that "there were many collectors [in the seventeenth century] both in the *Indies* as well as in *Holland*, and their numbers have notably increased in this [i.e., the eighteenth] century. The oldest *liefhebber* I personally knew in the *Indies* was the old gentleman called *Rumphius*, and thereafter his Son, Mr. *Paulus Augustus Rumphius;* the old gentleman began doing it as soon as he came to the Indies, to wit Anno 1655."[78] To be sure, Valentijn is restricting himself to collectors of shells only, but, as we saw, hardly anyone specialized to that degree in those days. Including father and son Rumphius, as well as himself, Valentijn lists nineteen collectors in Ambon and Batavia, including Padtbrugge's wife, Johannes Sipman, Schaghen's widow (Zara Alette van Genege), and Van der Stel, Ambon's governor from 1706 to 1720, who might have erected the memorial on Rumphius' grave. He then proceeds to give a detailed accounting of shell collectors in Holland. Ordering them by city, Valentijn's catalogue includes all the people mentioned by Schynvoet: he lists thirty for Amsterdam, including Schynvoet, whose collection Valentijn judges "incomparable";[79] four for Delft; three in Haarlem; six in The Hague; six in Rotterdam; two in Hoorn; six in Dordrecht, including himself; six in the province of Zeeland; and two in Overijsel.[80]

A great deal more could be said about this network, but it is clear that owing to its existence and to Holland's connection with Asia, Rumphius was not completely isolated in the far reaches of the eastern archipelago. Although he was on the very periphery of this web, any activity on his part would cause a reverberation throughout the invisible structure until it would be felt somewhere in Europe. The unusual relationship was only possible because scientific specialization had not yet tamed and channeled curiosity.

Wunderkammer and science could not co-exist, nor could curiosity and objective knowledge. We know what happened in the course of time, but during Rumphius' life the issue had not been settled yet. Rumphius' stated skepticism about the "new philosophers" is a natural distrust of Cartesian reason. Descartes mocked "*les sciences curieuses*" in his dialogue from 1701, entitled *La recherche de la vérité par la lumière naturelle.* Descartes' derisive spokesman, Eudoxe, makes it clear that those "curious sciences" include not only what would now be dismissed as the occult ("all those wondrous effects which one is wont to attribute to magic")[81] but also the humanities. Eudoxe separates the sciences from those "simple forms of knowledge which are acquired without any recourse to reason, such as languages, history, geography, or generally anything that depends merely on experience."[82] Science is governed by reason, while that other kind of knowledge is energized by curiosity. He then proceeds to eliminate the marvelous altogether by legislating that a man of science should not waste his time studying "experiences that are extraordinary or recondite" because this would require that he first examine "all the plants and stones that come from the Indies, that [he] had seen the Phoenix, in short, that he could not be ignorant of what is most strange in nature."[83] This is diametrically opposed to Rumphius' confession that he would rather be astounded by nature than dissect it (chapter 9 of book 3).

This tacit disapproval of anything not deduced by reason was also present, as Pomian pointed out, in the strange censure of the word *curieux* in eighteenth-century France.[84] While being an *amateur* was admirable because it meant that one "loved" something, being "curious" about something was negative because it implied immoderate desire to know. English did not know the use of the word "amateur"—someone who has a nonprofessional interest in something—until the early nineteenth century. During Rumphius' lifetime it had to make do with the Dutch noun *liefhebber*, the same word employed by Schynvoet in the present text. But the negative connotations, though not as strong, were part of English usage as well. Whereas its earlier meanings connoted a positive quality of attentive care, the word began to acquire during the eighteenth century the present sense of undue inquisitiveness. The age-old worry about the inherent potential for abuse when one wants to acquire knowledge is evident in the original meanings of *curiosus* in classical Latin: its positive meaning of "applying oneself with assiduous care to something" was corrupted in the post-Augustan era, until it meant "prying," even becoming synonymous with Rome's secret police. Hence what is, after all, a fundamental attribute of Homo sapiens was thought to be in need of control. Descartes accomplished this by denying primacy to experience and granting legitimacy only to reason and a priori mathematics. Rumphius never reached that threshold of the modern era. He remained loyal to Aristotle and Pliny, believed in concrete existence, and delighted in experiencing it firsthand: "experior, ergo sum" rather than "cogito, ergo sum."

Rumphius died just before (what Pomian called) "exuberant curiosity" was moderated and constrained. During the eighteenth century nature became order rather than abundance, its incoherent vitality trained by discipline and method: "In scholarly culture, the quest for miracles became the search for laws."[85] The exceptional gave way to the commonplace, the rare to the ordinary. The exotic was no longer a lure; scholars turned instead to the mundane. The "*sciences curieuses*" were once again, as they had been in the early Middle Ages, excluded from official culture, but this time by *both* church and science.[86] The *Wunderkammer* no longer profited from the aura of wonder but were first regularized as natural-history collections and then absorbed by public institutions to be of service to science.[87]

A great deal more needs to be said about Rumphius and his achievement, but this cannot be accomplished in the present circumstances. Perhaps it is enough to recapitulate some of the major points.

Rumphius was very much a man of his age and yet saw beyond it. He was a complicated man, as most superior people are. A devout Protestant, he was, like Linnaeus and the young Darwin, a creationist, yet his conception of nature was much closer to that of such "pagans" as Aristotle and Pliny than to the orthodox Christian view. A further irony is that, since he was premodern, his antique set of values made him far more sympathetic to Indonesian culture and magic than a more "modern" intellect would have been. Perhaps aided by the relative isolation and lack of intellectual intercourse with Europe, Rumphius had a greater incentive to become intimate with what in so many ways was an alien world. Yet this interest also resulted from his character and personality, which in general can be said to be a combination of steel and grace, delicacy and vigor, forbearance and impatience, a boundless curiosity about the world and nature and a stubborn indifference to the corporate lure of material acquisition.

Rumphius' writings are neither science in the modern sense nor religion. Perhaps one might characterize them as a third voice, a mode that can partake of both but that, as science, is ready to impart information yet is more interested in understanding, while as religion, it aspires to a sense of rapture but does not want to impose orthodoxy or ideology. This voice, as Schopenhauer knew, is art, the voice of the poet. As mentioned before, he will never be dismissed as a naturalist, yet if it were to happen, Rumphius would still survive as a great ethnographer. But he can be forever read as a poet. Rumphius had a generous supply of *caritas*, in both the classical Latin sense of "regard," "esteem," "affection," and the Christian notion of "charity." A Latin derivative of "caritas" was *caritates*, "loved ones." For Rumphius the denizens of tropical nature were his *caritates*. For though his stated objective was partly utilitarian—to give people practical medical knowledge—he reserved his most poetic texts for such "useless" items as jellyfish. They are good for nothing, but their very existence is enough of a reward. A Physalia, or Portuguese man-of-war, he describes as having a body that "is of a transparent color, as if it were a crystal bottle filled with that green and blue *Aqua Fort*, otherwise called *Aqua Regis*. The little sails are as white as crystal, and the upper seams show some purple or violet, beautiful to see, as if the entire Animal were a precious jewel." What he called a jellyfish, but is now considered a marine snail called the "Purple Sailor," he presents to us as "[appearing] to sail with a faint breeze, wondrous to see, such a fleet of easily a thousand little ships, that could sail so agreeably together. When one took them from the sea and put them in a dish with water, the little Sea-Gellies did manage to stand upright for a day or so, emitting a wondrously beautiful repercussion of light, as if the dish were filled with Precious stones." Rumphius' work is everywhere adorned with such splendid images. An ingredient in incense is compared to "a Basse in Musick which, when heard alone has no comeliness, but which when mixed with other voices, makes for a sweet accord"; a certain people in Celebes use shells to decorate their proas and "their champions do the same by hanging them on their houses, so that, when the wind blows, people will be warned by their clattering, that here lives a brave man, who won't be trifled with; so that these Shells are like the Trumpets of Fame." Consider the beautiful precision of the following simile: "Scales" on a shell which "when dried in the sun [are as] crumbly as parchment that was wet and then dried up again." Or "Night Shells" that are "black, with wide, light-red and whitish rays, that narrow towards the back, resembling a painting that depicts a dark night that is pierced by a light shining through." Consider also the astonishing visual memory of a blind man who remembers a ring he once owned whose gem reflected light: "I had such a ring on my finger while we were sailing along a shore, and I could see in it the green trees that stood on shore, also the movement of the boatman behind me, as well as the faces of those who were sitting next to me." Then there are tiny crabs that "walk on the sand like a foam" or the troublesome hermit crabs, which "carry off [their] beautiful shells, leaving me their old coats to peep at." One can augment these examples with a host of others. A poetic index of Rumphius' work would be as long as a scientific one.

Claude Lévi-Strauss called this a "*connaissance concrète*" (concrete knowledge),[88] which demonstrates that "theoretical knowledge is not incompatible with sentiment and that knowledge can be both objective and subjective at the same time, in order that the concrete relations between man and other living creatures sometimes color the entire universe of scientific knowledge with their own emotional shadings . . . especially in those civilizations which consider science entirely 'natural.' "[89] This is Rumphius' science, but "concrete knowledge" is also the curriculum of the poet. Rumphius' work is suffused with E. O. Wilson's "biophilia," which Wilson defines as "the innate urge to affiliate with other forms of life."[90] In Rumphius' work "burns a quiet passion"[91] not for control but for a constant understanding. His work is a perfect example of Wilson's contention that poetry and science are linked. Rumphius' descriptions in both the *Herbal* and the *Curiosity Cabinet*—for everything I have mentioned so far is, mutatis mutandis, equally true of his botanical writings—are distinguished by a Wilsonian "elegance," the "right mix of simplicity and latent power." Both science and poetry are "enterprises of discovery" which rely on metaphor and analogy in order to find "power through elegance." The "binding force" that connects the two and that Rumphius' work so beautifully demonstrates is "our biology, our relationship to other organisms";[92] that is why he can rejoice in "useless" jellyfish and pesky hermit crabs. In Rumphius, art and the protoscientific method are fruitfully conjoined, demonstrating Wilson's contention that "in principle, nothing can be denied to the humanities, nothing to science."[93]

ABOUT THIS TRANSLATION

The preceding paragraph also suggests the reason I embarked on this enterprise. I wanted to deliver this unusual achievement, which has been hidden in a language few will take the trouble to learn, to a larger audience. Seventeenth-century Dutch is a difficult medium to translate but is also beautiful because blessed by a relatively low tolerance for abstractions. I also wanted to narrate the life of this "wader, picking curiosities from the shadows" (to quote a verse from Ted Hughes) because that too is a tale that needs to be told.

The translation is as close as I can get to the original without sacrificing comprehension. There is a school of translating which condones substituting modern idiom for an earlier one, but to do so violates vocabulary and kills a specific analogical system. I refuse to do this. A horse is not a combustion engine, a wooden skittle-pin is not a Coke bottle, nor does one cast a cybernet upon the waters. A translator should want an audience to savor what was and still is unique, not degrade it to the lowest common denominator of communication. If that were all, then, in the case of Rumphius' writings, one should be satisfied with one of the epitomes. A reader should be an active participator, not a passive receptacle.

I therefore did not use any locutions that were not current prior to 1700, or, conversely, I diligently guarded against "modern" usage because it might be more convenient, that is to say, more readily available. Hence it is a blessing that I am not a scientific specialist devoted to expedience and therefore tempted to tamper with the text by substituting modern scientific nomenclature that simply did not exist in Rumphius' day. For instance, the noun *zenuw* did not mean "nerve," as it does today in modern Dutch, but "sinew"; hence the substantive *zenuwachtig*, which means "nervous" today, meant "sinewy" in Rumphius' day. *Nering*, which today means "trade" or "merchandise," was used by Rumphius in its oldest meaning of "food." Conversely, one might expect that "univalve" and "bivalve" are modern compounds, but they were already part of Aristotle's vocabulary. By trying to stick to this principle of matching linguistic environments, I found myself writing a more vivid prose, more Anglo-Saxon (or more Germanic, if you will) than Latinate: a body's "hollow," not "cavity." Given his time and his poetic talents, Rumphius wrote a clear, supple prose that was precise and natural, enlivened with surprising metaphors and analogies. "The speech I love is a simple, natural speech, the same on paper as in the mouth; a speech succulent and sinewy, brief and compressed, not so much dainty and well-combed as vehement and brusque . . . rather difficult than boring, remote from affectation, irregular, disconnected and bold; each bit making a body in itself; not pedantic, not monkish, not lawyer-like, but rather soldierly."[1] Montaigne's description captures Rumphius' style perfectly. One should not forget that we are listening to his *voice*, since the greater part of these texts was dictated, that is to say, *spoken* by him. One can imagine that the printed rhythms are the cadences of Rumphius' speech. I generally kept, therefore, the peculiar punctuation of the original, because, though it is for the most part done by the eighteenth-century compositor, it also was at one time dictated by Rumphius' speech patterns, and at times a modern rewriting would get lost in a perplexity of antecedents. I also kept the original spelling of various words. The spelling is always idiosyncratic and often bizarre. To apply modern standardization would be a mistake, however, since this would engender a totally different text. More familiar orthography is provided in the notes.

If one embarks upon a project such as this, one must confront the problem of explication. A rich and varied text such as this one, dealing with so many things that are still exotic as well as esoteric, demands a great deal of elucidation. The negative result is a sizable corpus of notes, while on the positive side it forces one to produce what may be described as a preliminary attempt at an authoritative edition. The qualifiers are in order because I have not systematically collated the three known editions of *D'Amboinsche Rariteitkamer.* It would be a daunting task, yet one that would pale beside the one of doing the same thing with the *Herbal.* On the other hand I have indicated problem areas and, more important to a present reader, I (with the help of others) have managed to explicate about 95 percent of what was in need of commentary. One must remember that Rumphius' books have never been reprinted in any form since the eighteenth century; hence the present edition is a pioneering effort, with all the virtues and flaws that entails.

This English version is a translation of the integral text of *D'Amboinsche Rariteitkamer,* except for two omissions. I have not included Sipman's epitome of the second book, because it is superfluous, nor have I translated the marginalia, because they are too expensive to reproduce and because they are no more than another layer of redundancy. Yet I have included all the commentaries by Schynvoet, no matter how irritating or irrelevant, because they have insinuated themselves over time as a permanent accretion, and Chevalier's unnecessary contribution has been kept because it too has become part of the text by default. Schynvoet has at least some value for the history of Dutch curiosity cabinets, but the best excuse for Chevalier is a built-in opportunity to compare Rumphius' prose with that of a contemporary. It is instructive at times to experience piffle in order to appreciate what is superior. It should also be mentioned that all translations are mine, unless otherwise noted.

I have kept the older Malay spelling in the notes, because most of the texts mentioned use it and because all secondary literature from before the Second World War was printed in the older spelling. It will make it easier for the nonspecialist to find the sources. For those who wish to know equivalents in modern Bahasa Indonesia, the following changes should be noted: the former spelling tj (tjemar) is now c (cemar); dj (djoeroek) is now j (jeruk); ch (chas) is now kh (khas); nj (njai) is now ny (nyai); sj (sjak) is now sy (syak), and oe (soedah) is now u (sudah). Of these, only the last change has been adopted because the orthography for that diphthong is not familiar to English readers.

D'AMBOINSCHE RARITEITKAMER,

Behelzende eene BESCHRYVINGE van allerhande zoo weeke als harde

SCHAALVISSCHEN,

te weeten raare

KRABBEN, KREEFTEN,

en diergelyke Zeedieren,

als mede allerhande

HOORNTJES en SCHULPEN,

die men in d'Amboinſche Zee vindt:

Daar beneven zommige

MINERAALEN, GESTEENTEN,

en ſoorten van AARDE, die in d'Amboinſche, en zommige omleggende Eilanden gevonden worden.

Verdeelt in drie Boeken,

En met nodige PRINTVERBEELDINGEN, alle naar 't leven getekent, voorzien.

Beſchreven door

GEORGIUS EVERHARDUS RUMPHIUS,

van Hanauw, Koopman en Raad in Amboina, mitsgaders Lid in d' *Academiæ Curioſorum Naturæ*, in 't Duitſche Roomſche Ryk opgerecht, onder den naam van

PLINIUS INDICUS.

T'AMSTERDAM,

Gedrukt by FRANÇOIS HALMA, Boekverkoper in Konſtantijn den Grooten.

1705.

The Ambonese
Curiosity Cabinet
Containing a Description of all sorts
both soft as well as hard
Shellfish
to wit rare
Crabs, Crayfish,
and suchlike Sea creatures
as well as all sorts of
Cockles and Shells
Which one will find in the Ambonese Sea:
Together with some
Minerals, Stones,
and kinds of Soil, that are found
on the Ambonese and on some of the Adjacent Islands.

Divided into three Books,
And supplied with the requisite Prints, drawn from life,
Described by

GEORGIUS EVERHARDUS RUMPHIUS,

From Hanau, Merchant and Counselor on Amboina, also Member
of the *Academiae Curiosorum Naturae*, founded in the Holy Roman Empire
under the name of

PLINIUS INDICUS

At Amsterdam
Printed by François Halma, Bookseller,
in Konstantijn den Grooten
1705

To his Noble Worship

HENDRIK D'ACQUET

Burgomaster and Councillor of the City of Delft;
Doctor, and famed practitioner of medicine, great
protector of the arts and sciences, and lover of
all handsome rarities, etc., etc.

YOUR NOBLE WORSHIP.

Some years ago I had the distinct pleasure of maintaining a written intercourse with Yr. Worship concerning the love of Curiosities, in particular *Conchylia*, whereof I sent most kinds, which I could obtain here (and which I call a *Synodus Marina*) to Yr. Worship, to nourish a true and tried Friendship; and since I was certain that such small gifts, from distant lands, would always be pleasing to Yr. Worship, because of your great knowledge of, and love for, these matters.

But since these things, Yr. Noble Worship, have very different names both here and in the Indies, as well as in Europe, which causes even Collectors not to understand one another at times, and since I had already described the majority of them, before I obtained any kind of writing about them from Europe, in particular by the Italian Gentleman Philippus Bonanus, I deemed it prudent to gather my Writings, such as they were, and sent them to Yr. Noble Worship, so that, if Yr. Worship deemed them worthy, You could deliver them to a printing press, so that the same might appear under the protection of Yr. Noble Worship's name, to whom I am devoted by means of my pen, and even more with my heart. Notwithstanding the aforementioned reasons, which more than suffice, I dedicate this work to Yr. Noble Worship, in order to further strengthen our friendship, and so I could present visible proof to all the world, at least to the eyes of Asia and Europe, of the love and esteem which I harbor for you.

I call this work, Yr. Noble Worship, the *Ambonese Curiosity Cabinet*, since it deals mostly with such Rarities as are found in the Ambonese sea, or on the beaches of the neighboring Islands, and which I, during my lengthy stay on Amboina, carefully collected and preserved, with much effort and expense; whereof Yr. Worship owns the majority in your outstanding Cabinet; not to mention so many other natural and artificial adornments which will not so much satisfy an audience as provoke them to wonder.

We divided the work into three books, Yr. Worship, according to the material itself; whereof the first contains a description of all kinds of soft Shellfish, such as Crayfish, Crabs, and suchlike Sea creatures, which I was able to track down in these waters and lands. The second book, which is the true Curiosity Cabinet, displays all kinds of Whelks, Shells, and rare things which occur and can be found in the Eastern Seas. The third book deals with some Minerals, Soils, and strange Stones which these Islands produce. We adorned the work throughout with artful Prints which, we hope, will be agreeable to the Lovers of Curiosities, and especially to Yr. Worship's taste. Meanwhile we durst promise you that, taking these Descriptions as a whole, they will not be distasteful to the Europeans; since it shows a number of Natural rarities which are not commonly known, so that their careful contemplation (which offers no small evidence of God's power and wisdom) might be particularly pleasing to those who are expert in these matters; whereof Yr. Noble Worship has more experience than most.

Herewith I conclude this Dedication to Yr. Worship, and after wishing Yr. Worship, as well as Your Wife, and your blessed Children, and all those you hold dear, the greatest welfare of soul and body, I remain:

Your Noble Worship.

Your Noble Worship's humble servant
and obliging Friend,

GEORGIUS EVERHARDUS RUMPHIUS.

Merchant and Councillor at Amboina.

At Victoria Castle
Amboina,
1 September
1699.

To the Same Gentleman.

Your Noble Worship.

It has been five years since Mr. EVERHARD RUMPHIUS wrote the aforegoing Dedication to Your Noble Worship on Amboina, at least as far as the actual content is concerned. The work of this Gentleman came into my hands because of Yr. Worship's help and kindness, but not until the year seventeen hundred and one, whereafter I proceeded to bring it to press, though we labored much longer over it than we had imagined, due to change, as is so often the case in human affairs: we even began to despair at times if we were ever to complete it, and did not know how we would satisfy the Collectors, who were eagerly awaiting it.

Therefore, Yr. Noble Worship, I must apologize to Yr. Worship, as I must do to the world at large, since we have prevailed too much upon its patience; it was particularly Yr. Noble Worship who urged me several times, in all honesty and fairness, to bring this job to a rapid end, but this was impossible, no matter how important it was to me, for two or three major reasons; once those are mentioned, Yr. Worship, as well as all Reasonable people, will doubtlessly quit me of the rumor, that I neglected this work either because of careless negligence, or because I was hampered by too much other work, for I have been accused of this several times already, and also of much worse.

The work, which had been rather nicely contrived, seemed at first, Yr. Worship, far more complete and ready for print than it subsequently proved; as far as the numerous drawings of Cockles and Shells were concerned, which the Author had promised us, or rather Yr. Worship, they never reached us, nor do I know why, but I never laid hands on them; so that, as one is wont to say, we had to make a virtue of necessity, and had to make do with what we had; that is, we had to supply those items which we had not received from the Indies with specimens from local Curiosity Cabinets, which we did at great cost and trouble; Yr. Worship helped us greatly in this matter, wherefore we are indebted to You, since without You it would have taken even longer. One should add, Yr. Noble Worship, that the Author, as people say, had barely touched on certain things, which therefore needed to be enlarged upon; while other items were entirely lacking and which, without harming the order and design of the work, could not be left out. This required time, scrutiny, and labor; and we do not want to hide from Yr. Worship that without the help and diligence of Mr. SIMON SCHYNVOET, who is a great arbiter and collector of these items, and a honored friend of both of us, nothing would have been accomplished; for he, because he knows everything completely, enriched the entire work not only with copies of the missing drawings, but also with his own careful observations, thereby enhancing the same with particular luster; as can be ascertained from his Additions, which were printed in a different letter.

All of this, Yr. Noble Worship, thwarted the rapid progress of the work, as anyone who is not a stranger to this business, can attest to. But finally, with God's help, we surmounted all these problems; meanwhile Yr. Noble Worship received the news from the East, that Mr. RUMPHIUS, the Author of this work, shared the common fate of mortality, and had passed on to the other life, so that his Honor was not able to see the harvesting of the fruit, which he had cultivated with so much labor and care, before he submitted it to your protection; wherefore I, since I conceive it to be my duty, and since I am also the nearest one to have such a right, at least in these parts, now replace his Honor, and submit this work to Yr. Noble Worship, on behalf of the deceased Gentleman as well, with all due respect, as the fruit of my press, in full assurance that not only the work itself, but also my debt of honor will please Yr. Noble Worship, and incur Your favor.

Meanwhile, Yr. Noble Worship, I recommend myself to Your continuing friendship, and wish that I were able to give Yr. Worship greater and more visible proof of my respect for your Person and accomplishments, regarding both public Governance, and the felicitous art of healing, wherein Yr. Noble Worship is, for a twofold reason, like unto a Father of the Fatherland, and particularly a Father of the City of Delft and its citizenry; which is jubilantly acknowledged by the entire community. But since this is presently not the case, I want to finish with the blessed wish that God will shower both Yr. Noble Worship and your Wife, that dear Companion of your bed, and her Children, with heavenly blessings, and will keep Yr. Noble Worship, notwithstanding your great age, which has caused some physical infirmities, safeguarding You for the governance of the Republic which has bestowed so much luster on Yr. Noble Worship and on Members of your Family for so many years; and I will always consider it a distinct honor if I can contribute something to Yr. Noble Worship's service and satisfaction; I hereby sincerely bear witness that I am and remain:

YOUR NOBLE WORSHIP

Amsterdam
1 September
1704

Yr. Noble Worship's most humble and obliging servant,

François Halma

FOREWORD
from the Printer
to the
Reader.

God is miraculous in all his works which, being the eternal Cause of everything, were known unto Him, from the beginning of the world. There is nothing in All of creation, including subterranean holes and caves, that does not manifest the visible evidence of the Creator's eternal wisdom, power, and goodness that granted it life, from the beginning of time, through his Divine will, and kept it to this day in all its variety; notwithstanding that God, being the Essence of all Creation, is not beholden to anyone to give a reason for his acts, which are therefore irrefutable and only to be praised, since He, who is the eternal essence, thought it good, and from whom only that can issue forth that is perfect. This Moses sang in the Israel of old: *The Lord is my rock, and His works are perfect;* and the edifice of Heaven and Earth with all that it contains, is ever regarded as the great stage, whereon God's virtues and perfections are most illustriously displayed; so that *his eternal power and Divinity is seen and understood in the creatures from the very beginning of the world,* as Paul says, *so that no one is to be excused.* How glorious, how exalted, how eloquent did Job and his Friends speak thereof, in many chapters of the Book of his long suffering, that one should be amazed at the excellence of the material, and the superiority of the expressions, for there one will find disclosed the greatest eloquence of the East, which we today cannot rival. And what do they always conclude, after it has been debated from both sides, or when these intellectual Orientals, proposed something which often did not favor their distressed friend? Nothing else but God's everlasting and unlimited power, contrasted to our nothingness and worthlessness as creatures: and one only needs to listen to God Himself at the end of said Book, where he recollects some examples and effects of his power, in order to be transported (unless one lacks either the use of or feeling for reason), not only by the Majesty of the words, for when one will confess along with Job: *To repent in dust and ashes;* that is, if one has thought or spoken recklessly and unseemly of God and his works and, though knowing Him as Lord from his illustrious creations, did not serve and worship Him as the Lord of all creation: for the greatest evil of mankind is, when man's heart has lapsed to sin from iniquity, and turned boastful, evil, and blind, so that it reviles itself by submitting to the lowliest of idols, and instead of praising God for all eternity, dost venerate the creatures, even unto the most despised ones; as the Apostle, in his letter to the Church at Rome, in the second chapter, instructs us at great length.

Now God's perfections and virtues display themselves not only on the surface of the earth, which makes a fine show with an endless variety of plants and animals in its richly embroidered carpet that no art can imitate, with man above all, the masterpiece, so to speak, of God's hand, wherein he displayed his Divine wisdom and omnipotence, and who he equipped, as the only creature of reason, with powers and skills, in order to know, with the greatest humility, and to acknowledge and pay homage to God, his Creator, and only Cause, as is fit. Nor did the bright beams of God's omnipotence shine only upon the heavens, where the great lights of Sun and Moon, go their appointed rounds, and where the multitude of stars preen with their beauty, and whereof the earth receives so much service and use, yea, without which we would almost see the earth changed back into its first confused clump: but so his light did shine upon the Sea and its currents the veins of that immeasurable whirlpool, wherein the Eternal words of the Creator, *the Sea bringeth forth*, etc. caused innumerable fishes, writhing creatures, and other animals to be, each according to its own kind.

And the Divine Harpist did not only praise Heaven and its sparkling lights, for he also sang: *Those that go down to the sea in ships, see the works of the Lord, and his wonders in the deep.* Which was proven and affirmed by certain experiences of all nations, in the four corners of the earth.

This granted the scrutinizers of Nature at all times the reason to whet their intellect, and render their pens to the service of discovering the masterpieces which the sea engenders in her womb; and so too that erstwhile gentleman, EVERHARD RUMPHIUS, a man of great learning and perspicacity, was captured by the desire, to produce an exact Description of the Animals and Plants from the Sea, that can be found near Amboina and other Moluccan Islands, and which he sent here from those Regions, to satisfy the curiosity of our Fellows, and which we now, o inquisitive Reader, make public through our printing press, at some cost, and, as we hope, to your satisfaction (though somewhat later than we had anticipated in the beginning). So here you have the long awaited *Ambonese Curiosity Cabinet*, wherewith you can satisfy your desire for these things, which seem to increase every day, both by reading the material, as by contemplating the wondrous prints, that were contrived without sparing money or effort. For it is certain that with this Work the experienced Author has left immortal proof of his knowledge, diligence and tireless zeal in his search for these natural curiosities; and who would deny that he did so handsomely and that he deserves to be honored by those who will come after? And so we can contemplate in his first Book the large and soft Shellfish; in his second, the wondrous Whelks and Shells; and in his third, the unusual treasures and effects of Nature, from those Eastern regions; and we are, so to speak, ravished at every step by his discourse concerning so many wondrous things. Hence we delight in following in the Author's footsteps, and to compass briefly, for the Reader's convenience, what the Work states at length.

He begins his first chapter, of the first book, with the *Locusta marina*, or Sea Crayfish, of which he particularly notes its thorny shield, and the strength of its tail, wherewith it defends itself so marvelously. Then we see the *Ursa Cancer*, a very strange Crayfish, though it tastes most pleasant to the tongue. Next appears the Pincer, with a description of the amazing power of its shears, or pincers, with which it performs incredible feats. And just as every animal lives in its own stuff, one will also find here a Mudman, which resorts to the slimy holes of muddy streams and which, being of a vile shape and color, is likewise unfit to eat. The Purse Crab seems hideous, and appears to be partly a Crab and partly a Crayfish, while it is also, contrary to the nature of all the others, an inhabitant of the land, and never of the water. The Stone Crab also gets its turn here, and is described as one of the best, not only because of its shape, but also to eat. The true Sea Crab is also touched upon, while the Spiny Crab is briefly depicted. The Moon and Dog Crabs are likewise sketched according to their particular characteristics, and shown in the Prints in all their variety. The rarity of the *Cancer Raniformis* occurs next, and it acquires quite a retinue of various Landcrabs in the next Chapter. The ugliness of two kinds of Thorny Crabs, inspires revulsion and dread and is quite rightly booked as monstrous. The Flower Crab, with the truly dangerous or poisonous ones, follow one after another; whereafter the red, and several black kinds are brought onto the stage. The Grass- or Moss Crab is also branded as harmful, and is therefore rejected by the Fishermen and thrown away. The *Calappoides* Crab also deserves its place, though consisting of more shell than meat, it is unfit for the mouth. But what is that monster that soon follows after, called the Perverse Crab, because everything about it seems to be perverse? And to be sure, its Description would warrant such a name. Now we proceed to the Wanderers; Crabs that, grown out of their own shell, look for a strange house that fits their body, and who often come to vicious blows about this; nor do they scruple at night, to exchange their old and dirty houses, like public robbers, for the finest ones of their Fellow Crabs, wherever they find them. And what do we think of the Pinne Guard, that tiny shrimp, that stands guard for its host as if it were a sentry, and that remains loyal to it until death do them part? The bearded Crab, and several kinds of Duck Crabs follow, while some of their peculiarities are mentioned. The Bloody and Fiery Sea Sheets, as well as the Sea Louse also find their places; followed by all kinds of Sea Apples, that receive a careful Description, which brings wonder to one's mind. There are still Sea Snails to come, while various Sea Stars also appear, all artfully depicted. And is there anyone who would not be afraid of the Medusa head, which, as the poets say, changed everything into stone, even when it had been decapitated; and who would not be petrified at the sight of such a monster, and not be subjected to a deadly fear? The Sea Arrows or Sea Darts are also given mention here, and they too deserve our admiration. Nor should one ignore the Seaplants, nor the Snakes, big and small, which live in the Indian sea. And what are the Sea Gellies or Sealungs, and what is their origin, since that is not certain at all? But the Mizzen deserves its own particular description, about which some things are noted that, though they seem incredible, are true nevertheless. And the Writer concludes his first book with a disquisition on the little Seaworms, called *Wawo*, of which he describes in detail their strange nature and the rare manner of catching them.

And there, scrupulous Examiner of Nature's secrets, you have a brief of what is contained in the first book of the Ambonese Curiosity Cabinet, and be amazed along with me by these wondrous creatures, though more so by their Creator, who caused them to be in his infinite wisdom. But do not tarry here; but please enter into the inner chambers of this Eastern Nature Cabinet, where your eyes will be dazzled by many and varied kinds of Whelks and Shells, painted with a motley of the most beautiful colors, so you will have reason to be grateful and amazed. Here we find the hard Shellfish most carefully anatomized, and thereafter described in a pleasing order. So feast your eyes first on that beautiful Mother-of-Pearl Shell, called the great Skipper, of a shiny appearance, and with several uses. But then go on to consider what is said about the smaller kind, which we really call the small Skipper. Who is not astonished by this small animal, which can navigate its slight and tender little boat according to the best rules of seamanship, so that it need not yield to even the most careful of Skippers? The Little Posthorn, which looks like those which grow on the head of Jupiter Hammon; the Sea Gellies' Boats, and the other Winkles are described here as well; also the Moon Eyes, Ruffs and Spurs; and many more of this kind? What the Collectors call Tops, because of their shape, are shown here in all their variety, most pleasing to the eye. The various Snail Whelks are carefully described and pictured here. They are followed by the Casks, so called after the protective headgear, that was once common in ancient times, and which these Whelks resemble to some degree. As to the Knobbles, these the Commentator could not bring together under one name, so that he called one branched Swiss pants, another the small branched Mulberry, also the Toadwhelk, Bed Ticking, and Hightails; and much more, all artfully displayed in our Prints. Various kinds of Gray Casques, or Bezoar Whelks, are also wondrously brought into print, followed by Curlwhelks, Footwhelks, Brandaris, Snipe's Head, and many more, which got their names from things they resemble, though they are still nameless in many parts of the World. The Author goes on to speak in detail of the Hornlids of Whelks, called Blatta Byzantia, and of the Onyx from Holy Writ, about which he disagrees with the famous and learned *Samuel Bochartus.* The Bell Whelks have various names and shapes, and the use of its fishes for Medicine, is revealed here as well. And then there comes the Sea Trumpet, used by the Tritons, or Neptune's train, as the Poets say, though they are used as real Trumpets by the Alphorese from Keram, whereof we will find here some precise descriptions, and wondrously beautiful images. They bear strange and wondrous names, too many to recount here; these are followed by the Needles, or Pens, long and narrow Whelks, which also have various names and shapes. But behold the beautiful crowned Beakers, and what are called the Toots; for instance the nicely spotted Music Whelks, Lace Whelks, and an almost infinite number of others, displayed here so artfully one would think they were alive. Next the Crabs or Flap Whelks are most handsomely presented, they have many names, according to the choice of the Collectors, each one mentioned in turn. Clack Dishes, or Rock Whelks, Turtle Whelks, the latter called after the handsome shape of the Shields of these crawling Land animals; they make a fine show on paper, as well as in their Descriptions, so that one's eyes will not be easily satiated. The engraver's needle provided such faithful copies of the Rolls or Dates, that not a single spot is missing. The Rock Clingers, Sea Ears and the Pox, and all sorts of this kind, are also powerfully portrayed here. And what strange likenesses have been described and pictured here of the Solens, Sea Pipes, or Sea Snakes? The Father Noah Shells are the memorable remnants of the general Flood, with which God removed the spots of evil from the first world. There is a good reason for the names of the Japanese Game doublets, as well as the toothed Venus doublet, just like many other Shells that are named here. The Furred Cloaks are beautiful to see, and highly prized, also the double Venus heart, that strikes the eye as being so graceful. One will find the large and the small Duckbeak, otherwise known as the external Gapers, described here, along with many others, and artfully rendered by the etcher's needle. The Duck Mussel, the winged little Bird, the Holsters, or Ham doublets, are all Mussels, of different shapes, which can be found among the curiosities in the Cabinets of the Collectors, and which you will find passing before your eyes here, drawn while they were alive. The art of etching has marvelously presented the many beautiful and rare Oysters, with different names, including the Polish Saddles, Cox Combs, the Lazarus Clappers and Rock doublets. And what can one say of the Venus shell, with hair all around, as natural as life? Nature certainly offers us marvelous images of things, which were unexpected; for instance the Agate stones of the next book. The names of other excellent and precious Whelks are printed in the short Appendix, and matchlessly rendered in copper as well as on paper, so that even the most exacting audience will be unquestionably satisfied.

We proceed, however, according to the Work's design, to the third Book, which discloses many treasures from the earth, and displays many wonders from Nature. Gold is the most precious thing on earth which, both in gravity and durability, outweighs all other minerals. And so it is only right that our learned Author begins with that omnipotent earthling, for one is certainly justified to call it such, since it was begotten inside the earth, nurtured, and carried to term, whereafter human wisdom gave birth to it, so to speak, from the lap of people, and finally, in order to put it into circulation, it was stamped with the image of the High Magistracy. But what does our Naturalist inform us? His Title, borne out by the Description, speaks of how gold is falsified in the Indies, discovered in two cunning ways, which appear to surpass the Europeans' own craftiness, but which the Author discloses here as a warning. Which is quickly followed by a water test for gold and silver, for those who are skilled in proving these metals, and who can profit from the several preparations and calculations. Then go on to see various ways of testing gold, what it will fix on, or when it will not prove. What kind of metal *Suassa* is, is told in great detail; then how some Minerals are tested by salt spirits; followed by a description of rare irons in the Indies; though mostly about the false Miracles perpetrated by a Priest king, on Java's East Coast, with rings and bracelets. Mention is also made of the metal *Gans*, which includes the Author's guess concerning a piece of a metal Wheel, which they like to believe fell out of the sky. But then the Thunder stones are brought to the fore, in all their various shapes and kinds, and one will encounter here so many assured reasons for them, both from the Author and the Commentator, that one will at least be persuaded of the truth, as opposed to the superstitions some people have contrived. The Thunder Shovels, which might also be called Thunder Chisels here, are dealt with specifically, revealing many wondrous things.

But why should we continue to name and quote these Wonders from Nature, when they are best seen and known in the Book itself, and to which we direct the attentive Reader for all other things; and, if so desired, the Reader can consult the list of Chapters. We hope meanwhile that this Work, though it did not achieve utter perfection, which no human endeavor ever does, will attract the approbation of the Collectors, and be granted a place in their libraries, which it deserves twice over.

We wish them all well, and that this Work may redound to their singular profit and delight.

Amsterdam the
1 of November
1704.

F. HALMA

THE AMBONESE CURIOSITY CABINET
Containing a
DESCRIPTION
of all sorts, both soft as well as hard
SHELLFISH

Introduction

The present Work is separate from the Description of Animals, and has acquired the title of *The Ambonese Curiosity Cabinet*, because in it are described those things, be they from living or lifeless Creatures, which, because of their rare shape, or because they are seldom encountered, are wont to be kept as *curiosities* by their Admirers, and is divided into 3 Books. The first contains those which are called soft Shellfish, in Latin *Pisches Crustaceos*, in Greek *Malacostrea*, and which have hard though breakable Shells; such as Crayfish, Crabs, Shrimp, Sea Apples and Sea Stars: whereof the last two are also called *Zoophyta* or *Plantanimalia*.

This is followed in the second Book by the second main Group, which by rights are called *Ostracoderma* or *Sclerostrea*, that is hard Shellfish, and which have a shell as hard as a stone or bone wherewith they cover their entire body, except for the mouth; such as Whelks, Bivalves, and Oysters. This is the true Cabinet of the Fanciers, or a gathering of all the *curious* Whelks and Bivalves extant, which I have been able to find in the *Sea of Ambon*, and wherefrom *Amboina* is known throughout Europe.

In the third Book you will find various *Minerals*, rare Stones, Soils and Saps, which are to be found in these Eastern parts.

The singular Attributes of the Animals in the first Book are, that they do not have blood, and that they do not shew their parts as clearly as do other animals. The singular aspects of the first main group is that they have eight feet, not counting the pincers, and some small legs near their mouths; that the head, the chest and back are fastened one to another, that they have eyes that jut out and move, among them are counted Crayfish and Shrimp, which have an elongated body, and a long tail, and go backwards.

Among the Crabs are those which have a roundish or pinched body, and go forth sideways. All the other ones are *Zoophyta*, which have more the shape of a fruit or plant, than that of an animal, without express parts, except for the legs, inside however a living flesh and veins, wherewith they move.

THE FIRST BOOK
of the
AMBONESE CURIOSITY CABINET
Dealing with Soft
SHELLFISH

by
Georg. Everhard. Rumphius

THE FIRST BOOK
of the
AMBONESE CURIOSITY CABINET
dealing with soft
SHELLFISH

CHAPTER 1
Of the Locusta Marina. Sea Crayfish. Udang Laut.

This Sea Crayfish for the most part resembles those that are found in *Italy* and in the Mediterranean Sea, but the Indian is far thornier, because the entire back, the part that is called the shield (*'t horax*)[1] for all these kinds, is so thorny, that one could scarcely grab it if all those thorns were not bent forward. And it has two especially large thorns above the eyes, and these are bent forward, with four smaller ones beneath the same, wherefore it cannot be taken from the front.[2] The hinder part of the Shield has small short little hairs or brushes, so that it appears to be woolly. It is in general 14, and 15 inches long, except for the two long Beards which are 18, and 20 inches long, in back nearly a finger thick, and round, after that gradually coming to a point, each one divided into three parts, and completely covered with thorns: between these are two other, shorter and thinner Beards, also divided into three parts, of which the foremost one is split into two horns, and whereof one part is always shorter than the other. Instead of pincers it has two feet, at the very end split in two like a pair of Tongs, otherwise similar to the others, except that they are larger. The other feet are eight in number, are not thorny, and divided in front into blunt claws. The tail consists of eight rings, ends in a point on one side, and has five fins towards the back, making for a broad tail.

The color of the raw Crayfish is a deep blue, with some whitish or faded spots here and there, somewhat red on the tail and on the horns. The legs are blue with white stripes. When cooked it is entirely red. The Male is always spinier than the Female; and if this is the same kind of Crayfish that *Suetonius*[3] writes about, when a poor Fisherman startled *Tiberius* on the cliffs of the Island *Capreae*, and the Emperor had them rub the man's Beard with [the beast], it is a wonder that it did not rip his entire face to pieces. It has a lot of meat, white, rather hard, and richly sweet; it is therefore no great dainty, although one can fill a whole plate with one, and that would be just from the legs and tail; for one does not eat anything from the body's hollow. The little Crayfishes look a great deal like *Scorpions*, and are partly transparent.

One calls this Crayfish in Latin *Locusta Marina Indica*.[4] In Malay *Udang Laut*,[5] *Udang* being a general name for everything resembling Crayfish and Shrimp. In Amboinese *Mitta Soa*, as if one said *Squilla retrogada*,[6] or in Malay *Udang-Ondor*, that is backwards-walking Crayfish: on Leytimor *Ulehi* and *Ulehir*.

It lives in the open Sea, yet near the shore, it goes into the *Seris*,[7] crawls into the Weels,[8] and is caught in nets. It cannot be taken hold of when alive, but needs to be stabbed with a harpoon: it has such a great strength in its tail that, if it wraps it around a stone while it is being pulled out or seized, one can scarcely dislodge it therefrom. While in the water it goes

Fol. 2.
N°. I.
A
B
A.

sideways, bearing its horns to the side, when it is searching for prey, for nothing is so bold that dares to meet it: but perceiving something it would rather shun, or feeling that one will seize it, it will crawl backwards and extends its horns to the front. It is also clever enough that, when surrounded with a net, and not seeing a way out, it will climb to the upper edge, and if one does not stop it, it will try to jump over it. Fishermen do not like to see it in their Weels because, feeling oppressed, it will greatly injure the other fishes.

The entire Crayfish is cooked in salt water, whereafter the tail and the legs are knocked into pieces, and all the white meat removed, and one should make a special kind of sauce for it. It will almost fill you, because it is so cloyingly sweet, and because it is so hard to digest.

Such Crayfish are called *Homars* by the *French* in the *Caribbean* Isles for which see *Histor. Antill. Rochefortii,*[9] Cap. 19., and which the Caribees[10] catch at night on shelves of sand and sandbanks, with the help of lit torches, to which these Crawfish are wont to crawl, whereafter they stab them with a harpoon, or otherwise observe them by Moonlight. For more about the *Locusta Marina* see *Gesnerum de Aquatilibus lib.* IV.[11]

Since Mr. Rumphius *did not provide an image of this* Locusta Marina, *Dr.* D'Acquet[12] *sent us two from his Curiosity Cabinet, which we, since they are so rare, and since they correspond with the description, judged meet to display herewith: see plate No.* I. A. *and* B.[13]

CHAPTER 2

Of the Ursa Cancer. Udang Laut Leber.

This is a strange kind of Shrimp or Crayfish, wide, not tall; resembling one of those lumpish gloves, which one calls wool mittens, a span[1] long and the width of a hand,[2] its entire body covered with a gray and coarse woolly stuff. The back is slightly bossed, with some spines; the tail is somewhat round, with a ridge down the middle, has five or six joints, and has wide fins at the end; it does not have pincers, but has five legs on each side, which end in black and pointed Bird claws, which are singly on the four frontal feet, but double on the hind ones, or split in two, like a shear, and bend backwards. It furthermore has two small feet near its mouth, which serve as its hands. It is widest in front, to wit five or six inches wide, and is sawed there on the sides, but the front of the Head has on each side two wide and thin flaps, made of the same shell, though not fixed to it, but moveable, toothed on the sides, which serve as fins. Between these flaps it has two short horns or beards in front, and behind the same, it has eyes on the sides. It mostly crawls on the bottom, with a slow gait, and which is where the Fishermen stab it with a light harpoon or barbed spear. It has a very white, hard and sweet meat, that tastes better than the Sea Crayfish, but it is rarely found.

Its name in Latin is, *Ursa Cancer* and *Squilla lata.*[3] In Malay: *Udang laut leber.* Ambonese *Uhut*, on Leytimor *Miju uhut*, or *Cattam gonosso*, of the hairy shield, like the Husk of a Coconut. In Javanese: *Udang Bladook.*

Ursa Cancer. *The Author provided the picture* C., *being the Male: and we added* D., *the female, sent to us by Doctor* D'Acquet; *see Plate No.* II.

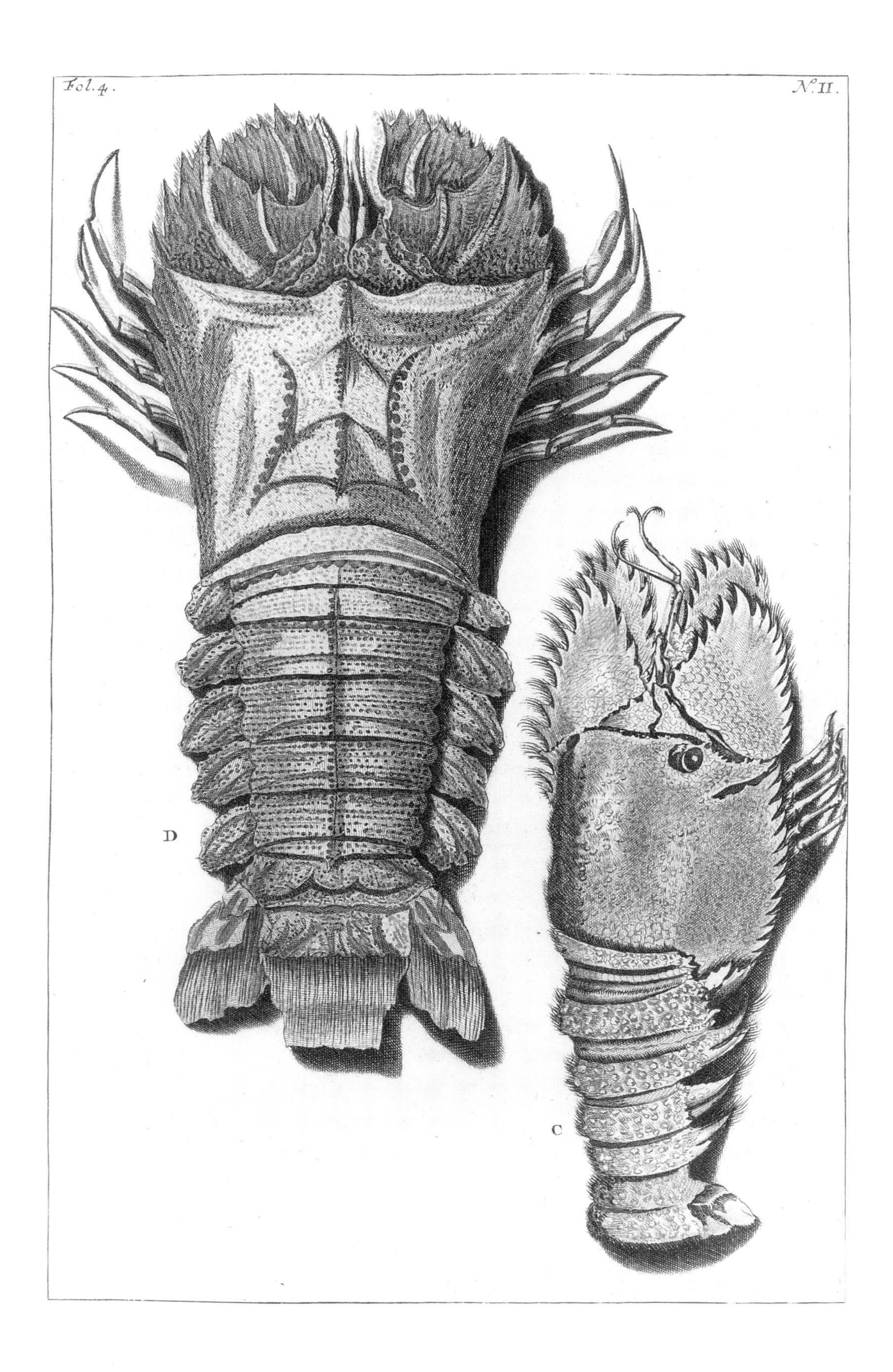

Vol. 4.
N.º II.
D
C

CHAPTER 3

Of the Squilla Arenaria. Pincher. Locky.

This Squilla also has the shape of a long Shrimp, is usually a hand long, two fingers wide, and is divided into two kinds. I, *Locusta* or *Squilla Arenaria Terrestris*. II, *Squilla Arenaria Marina*.

I. The *Squilla Arenaria Terrestris*, is the largest, usually a hand long, and easily two fingers wide, consisting entirely of joints, to wit, five broad, and three narrower joints, like so many wide bands, joined together like a half moon, and representing the body; it has a wide, spiny tail behind, made from sharp points, as if a crown, and it has four fins on either side. Under the five wide joints, that make up the body, it has five double, broad, scaly feet, resembling the blade of an oar, with which it can move at a good pace. It also has four pairs of smaller feet, the same shape as the former ones, on the four narrower bands of its body. It also has on its head two long, round fins, which are hairy or bearded on the sides, and between the same, four other shorter beards. The shears or pincers, which are the strangest thing of this Crayfish, have a completely different shape from those of other Crabs, because they are fashioned from three large parts, of which the back one is thick and curved, as with others; but the middle one is thin, wide, and the longest, with a deep notch in the back edge. The front one is narrow, divided into five or more curved teeth, like sickles, or bird's claws, which it can conceal in the aforementioned notch of the Center joint, as if in a sheath, like the French Spring-knives.[1] Below these shears, on either side of its mouth, it has another six small, finned (*Pinnatos*)[2] legs, that end in small shears. It has great power in the aforementioned Shears, called Pincers, not only for drilling into the bottom, tossing stones and sand aside with them, but also to smite small fishes with them until they die, whereafter it cuts them up into pieces, and brings them to its mouth with its six little legs; it can also hurt someone quite badly with them, if one tries to grab it, whilst it will also lash out with its spiny tail; wherefore one cannot handle it when alive, but [it] needs to be caught with a snare. It has big eyes, and white meat, like other Shrimp, also of the same flavor, good to eat, if it is found in sandy places, but they are not so good, if from muddy grounds, and are then seldom eaten. The Color of the raw one is a light russet, also whitish mixed with brown, but it is a pale red after it is cooked, but the pincers are completely white, some have speckles that are more or less the color of smoke.

This Crayfish is called in Latin *Squilla Arenaria Terrestris*,[3] because, though it does sojourn on Beaches, it only frequents those where the seawater does not reach, or on such Sandbars that remain mostly dry. Its common name in Malay is *Udang laut*. In Ambonese *Lokki* and *Loe*, but a smaller kind, no longer than a finger, they call *Miterna* on Hitu. Our Nation calls it Pincher; on Leytimor *Hehei-Lain*.

It has its dwelling around the beach, on level sandy shores, or near the mouths of some rivers, where one will see many little hillocks that were tossed up, like mole hills; here is where it digs itself down, three or four feet deep into the sand, until it finds hard stone ground. At night, or when the tide is going out, it emerges from there to look for food, which it then drags back to its hole. And that is how it is caught; one clears away as much of the Soil it has heaped up, until one sees its hole, then one places a snare there, made from some tough material, like horse hair, and fastened around a stick. One puts some Bait to one side, where it cannot reach it unless it crawls through the noose, and when it grabs it, it also pulls the noose tight and remains hanging from that stick, which it cannot bring into

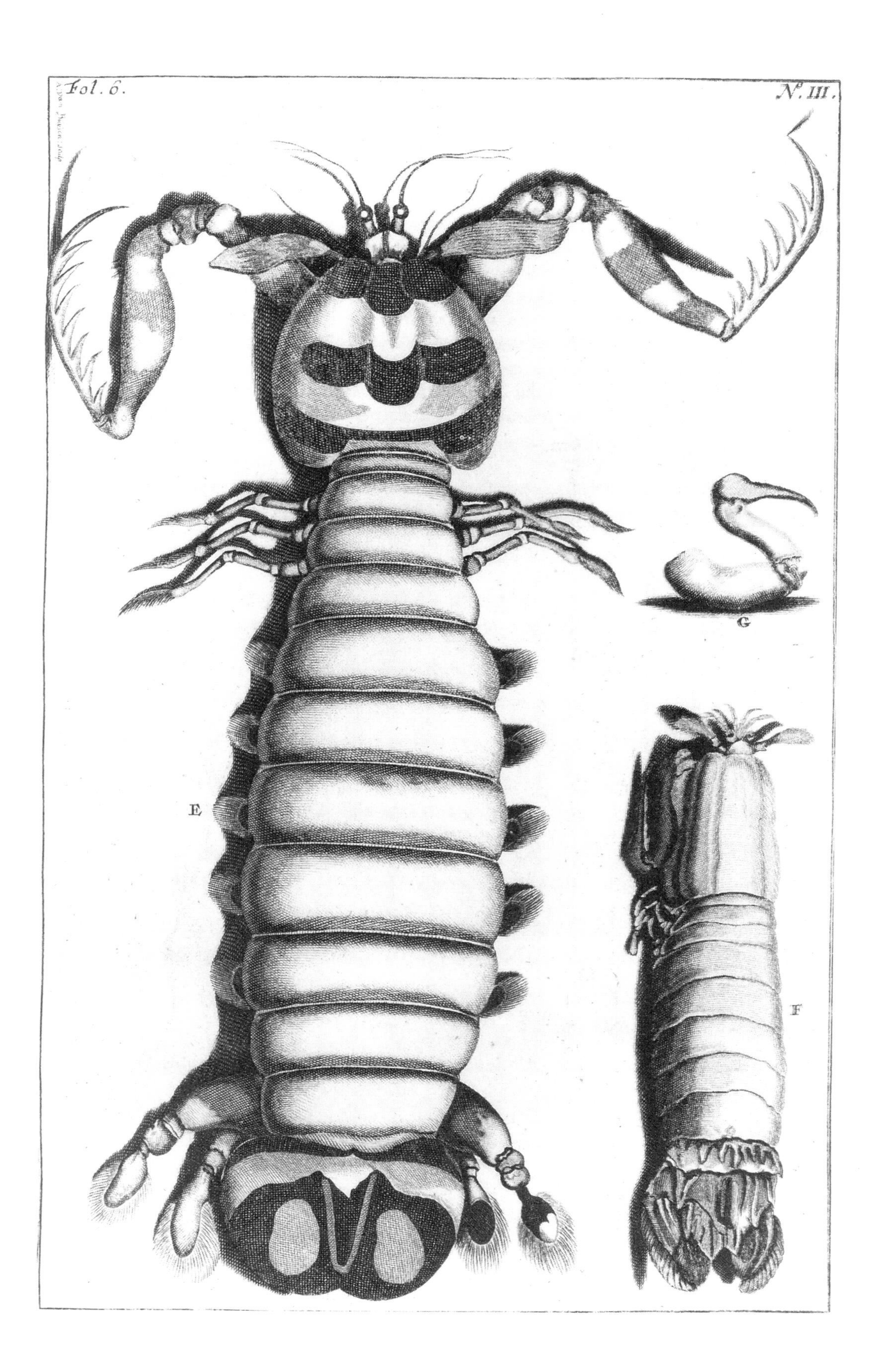

Vol. 6.
N.° III.
E
F
G

its hole, because it is lodged across the opening; but if one does not get to it quickly, it is able to cut the rope with its sharp pincers. The Natives usually eat it roasted, but when it has more mud than meat, they throw it away. The shears, or pincers, are kept as curiosities, because they are so quaint. The Inhabitants of *Buru* hold that its meat is better than that of other Shrimp and Crabs, and good for the stomach.

II. The *Squilla Arenaria Marina* has the same shape, but is smaller, and much more beautifully decorated, to wit with dark green, white and blue speckles over its entire body, and red at the ends of its legs. Their pincers are shaped differently, since the two back joints are somewhat round, green or speckled; the front joint is straight, without sickles, and red; which is why these Pincers, when broken off, almost resemble a little Swan with a red beak, altogether about half a finger long, and most of those who do not know the Animal, think it is the neck and beak of some animal. When cooked, it does not turn red, but a deathly green. The meat is better and whiter than the foregoing one; it dwells in similar holes; and on such Beaches, where large stones and sand lie below each other, and the water from the sea comes and goes by turns. It is also caught in the same manner with snares.

In Latin this Crayfish is called *Squilla Arenaria Marina*,[4] because it has its home in places where the Sea runs in and out, specifically *Squilla Cygnellorum*.[5] In other languages it has the same name as the one above. We call them Swan Crabs, due to the beautiful little Swans they make from the Pincers, and it is most often caught for that reason, when the pincers are torn from it while it is still alive; the rest of it is cooked and roasted, like Shrimp. It is hard to find, and even harder to catch, because one needs to watch the ebbing water very carefully.

A smaller kind of these is four fingers long, whereof the head or Shield makes up a fourth; the head comes to a sharp point, like that of a Shrimp, while the two in the center are split in two, and they have two little hairy leaves or fins on either side, crosswise. The rest of the body is the Tail, made from 9 circles or bands, of which the front one is narrow, the hind ones gradually wider; the actual Tail is the widest of the entire body, with many spines, fashioned from four stiff thorns, and with two hairy fins on the outer sides. Amidst these two thorns are yet two other sharp claws, over against each other, covered with a thorny-like Nail. Below the five widest circles that are in the back, it has 10 hairy little flaps, like legs. The front and narrowest circle has nothing, the next three center flaps have six legs beneath them, short, thin, and with the front joints turned up; in front of the same, near the shield, are the Shears or Pincers, being the most singular part of the Animal, differing from the two aforementioned Shrimp, because they are not pincers or tongs divided in two, but end in a straight thorn, that looks forwards, and that fits into a notch in the center joint, wherein it always hides this point, until it wants to hit or push something away with it; the beak is not red in this smaller kind, nor the body green, but yellow and grayish. The cooked one sometimes turns brown, sometimes a pale yellow. It has another six legs between the Pincers at the bottom of its mouth, which serve as its hands.

When it rains, it stays beneath the rocks; at other times it hides in the sand. The Swan-carrying Lokky,[6] is a cunning Killer of Fishes, if it gets into a *Seri* or Weel;[7] for though it be small, and its Pincers apparently insignificant, it can still wound large Fishes, and cut small ones in half, as if they were cut in twain with a knife; wherefore the Fishermen often find a goodly number of such dead Fish, and surprise[8] the Killer among them, who will not be taken alive, but one should stab it immediately in the Neck, which is its weakest part, whereby it is rendered harmless.

Squila Arenaria E. *and* Squila Arenaria Marina F. *The Little Sea swan* G. *is a Shear or Pincer of* F. *See plate no.* III.

CHAPTER 4

Of the Squilla Lutaria. The Mudman. Udang Petsje.

This Mudman has the same shape as the foregoing *Squilla Arenaria*, but its body is that of a Crayfish, consisting of two parts, of which the front one is the head; the shell is thin and soft, the length of seven inches, with two short beards on its head: it has four short legs on either side, which end in straight and pointed claws, and the center joint is bearded, or edged, with a beard-like little skin; the pincers are like those of other Crabs, and their middle joint is somewhat grainy, and supplied with beard-like little skins on the sides of the edge, the right usually larger than the left. The front joint is a sharp pincer, of which the lower, or rigid half, is only half the length of the upper one. The Tail consists of 6 circles or bands, except for the extreme joint, which resembles a nail, armed on either side with two thorns. The back of the tail is almost square, because of two edges that stick out, and every circle has a couple of bearded legs on the bottom. The raw one's Color, is an ugly, pale red mixed with gray, and when cooked it does not become truly red either. It has little, nay, hardly any meat, because its entire body and tail is full of greenish mud, and one will find some white and brittle meat only in the pincers, of no particular taste. It dwells in holes, like the foregoing ones, on the banks of muddy streams, where it appears at low tide, and crawls very slowly along the ground, as if it knew that it is not in great demand.

This one is called *Squilla Lutaria*,[1] that is Mud Crayfish, in Latin. The Natives from Celebes eat the meat from the pincers, but when I wanted to follow their example, I began to feel ill; wherefore I hold it to be a useless Crayfish, unless it is better in other Countries.

There is no picture of the Squilla Lautaria, D'Akquet *has contributed two kinds of the same for this purpose; See plate No.* V., *letters* K. *and* L. *both from Amboina.*

CHAPTER 5

Of the Cancer Crumenatus. Catattut. Purse Crab.

These monstrous Crabs have a mixed shape, that of a Crab and a Crayfish; moreover, they are inhabitants of the land, and never go into the water. Its Shield, or *Thorax*, is made up of four pieces, of which the three biggest ones are fastened to one another. The one furthest in front and also the smallest makes for the head, ending in a point, as does a Crayfish, below it are two eyes quite close to each other, and on the sides of the same are two beards, made up of two joints, whereof the lowest and shortest is rather wide and divided into branches, and the one in front is thin, and is nearly perpendicular with the lower one. Below this, and at either side of the mouth, are still two other long beards, which serve at the same time as hands, each one fashioned from four joints, all of them at right angles; but in such a way that it can put them together. The underside of the back of the body is covered with hair. They seldom appear during the day, but hide in the hollows of large rocks, but at night, most often when the Moon is dark, they come forth in order to look for food. The middle part is the real body cover, the other two are two round flaps, hanging over the side, which make the hinder part one-and-a-half hands wide, and of about the same length; this is followed by a thick, round Tail, made up of five parts or rings with broad fins; beneath this Tail is a thick, wrinkled Bag, like a bulging purse, whence the name. It has two very power-

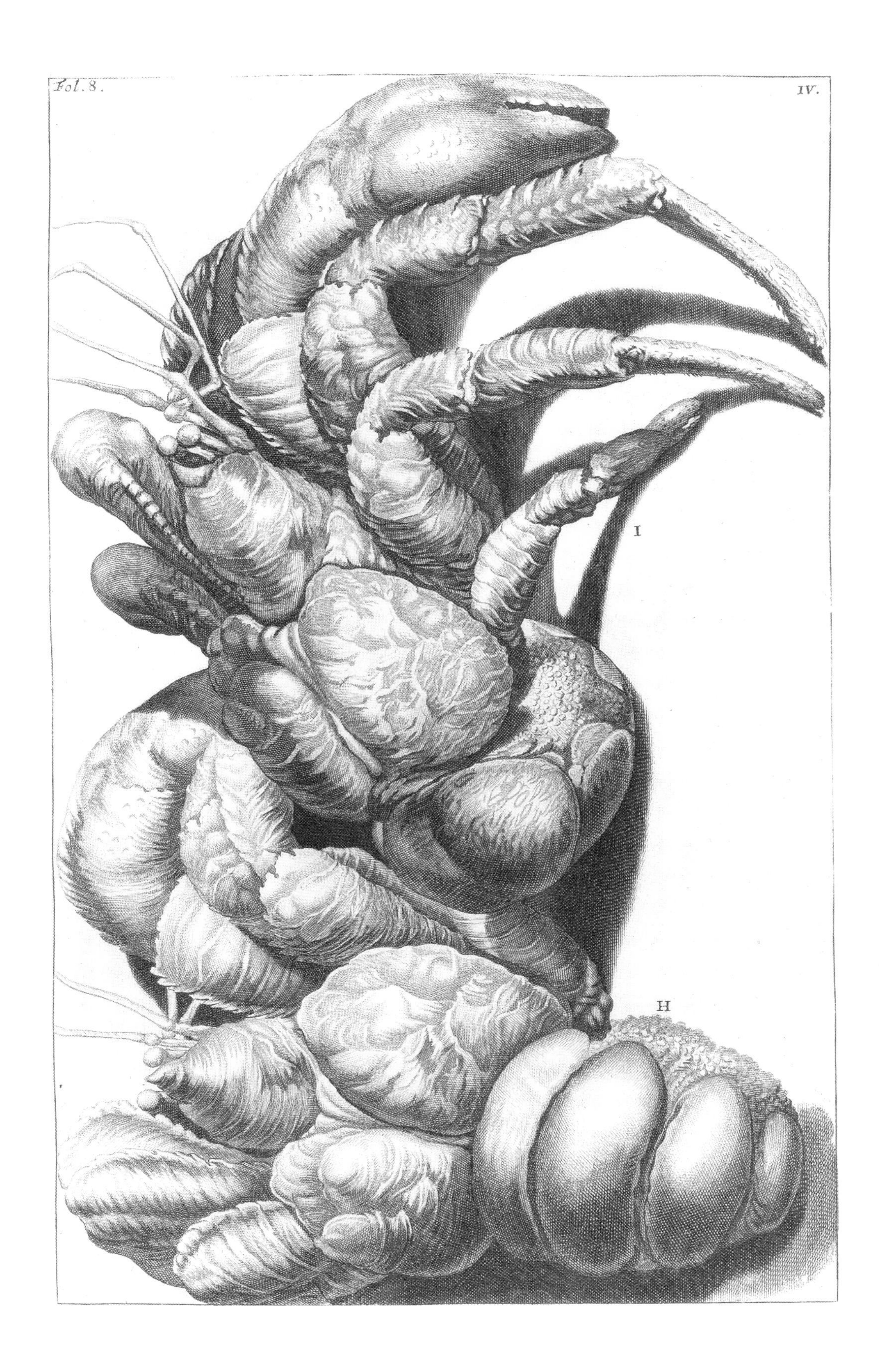
Fol. 8.
IV.
I
H

Tol.10.
N.°V.
K
M
L

ful Pincers, of which one, usually the right one, is smaller than the other, ribbed athwart as if with seawaves, and one will see short bristles on the ribs; it also has eight legs, shaped like the ones of Crabs, but much larger, and the two back ones are divided into pincers; the two flaps that hang down are scaly underneath, and the shield has some pits and furrows on top. The color of the living ones is a bright blue, with whitish stripes on the legs, and spots on the back. The flesh in the leg pincers is white and hard, in the body like unto that of other Crabs, and in the overhanging flaps there is a white soft marrow like fat, but the Purse is filled with a greasy substance, like butter, which is the best morsel of this Crab, and is the reason one catches it. A blackish vein goes all the way from the head, through the entire body, down to the tail, and this is the gut, ending in the extreme end of the tail. Next to this gut, is another, thin vein, like a white thread, which one should carefully remove along with the aforementioned gut, especially from the tail, if one wants to eat it, so that not even a little bit of it stays in there, because there is some hurt therein, and, once the same has been removed, one can without scruple eat the remaining buttery substance. This Crab has an enormous strength in its pincers, so that, if it has grabbed onto something, one would sooner rip a pincer from its body, than that it will let go of what it is holding, and the shells of the pincers are the thickest. However, if one tickles it under its tail, it sometimes lets go of what it was holding, and it can endure that fiddling so little, that, if one tickles it for a long time, its anger will be kindled so fiercely, that it will foam at the mouth, and will pinch its own tail with its pincers so much, that it will die. It can easily crack a Canary Nut[1] with its pincers, while we usually have a hard enough time knocking it open with a stone. I was once in an Orangbay,[2] where we had hung a Purse Crab on the mast, and under it happened to be a middle-sized Goat, and the Crab got the goat by the ear, and lifted it so high, that it was already off the ground, before we could come to its aid, but the pincer had to be smashed into pieces, before the Goat got loose.

This Crayfish or Crab, is called *Cancer Grumenatus*[3] in Latin, *Purse Crab* in Dutch, from the substantial Purse it has under its Tail; in Malay *Cattam Calappa*,[4] since it sojourns in the Coconut Trees, although another Crab, described later on, has the same name; also *Cattam Canary*, or *Cattam Mulana*, from the small Island *Mulana*, that is on the south side of the *Uliasser* Islands,[5] rocky and uninhabited, where it is caught often; in Ambonese *Catattut* and *Atattut*.

One will find it on such beaches that have steep cliffs, uninhabited, full of holes, with Coconut Trees about, although one will also find it where there are no Coconut Trees; as on the aforementioned small Island of Mulana, on the rocky southern corner of Oma, *Nusatello*,[6] or the Three Brothers, and on the South side of *Leytimor*,[7] on *Ema*'s rocky promontory; further also on the uninhabited Islands of *Lussapinje*[8] and some Islands near *Banda*[9] and Ternate;[10] furthermore on the uninhabited small Islands of *Tafuri*, *Bliau* and *Hiri*,[11] wherefrom it is called *Cattam Hiris*. It lives in hollow rocks, always on land, without ever getting into the water; it climbs Coconut Trees, and pinches off the nuts, and then searches under the tree for the ones that were thrown down, and knows how to bite them open with its pincers, and get the meat out; this is also the stuff with which to catch it. If one ties a piece of hard coconut meat to a small stick, and dangles it in the holes where it hides, then it will clamp onto it so hard that one can pull it out of there; which, as was said, one should do by a dark Moon, with lit torches.[12] One should not grab hold of it with one's hands, but with a stick that has been split like a pair of tongs, and toss a noose around its body. Nor should one put them or hang them together, unless one ties up the pincers first, otherwise they will pinch each other to death. One can keep it alive for some time with Calappus meat, even bring it to Batavia. They are cooked in one piece, thereafter the tail is opened separately, and the two aforementioned Veins carefully removed; the remaining marrow, like melted butter, along with the fat that is concealed under the overhanging flaps, is made into a thick sauce with pepper and vinegar or lemon juice, wherein one then mixes the white meat from the legs and pincers, and then one eats it all together. This is thought to be a dainty fit for the table of a Lord; the *Chinese* especially love it, so that one must pay as much as a quarter of a Rixdollar[13] for a fine Crab, and a schelling[14] for a medium one. Then there are those, however, both among ours as well as other Nations, who are fearful to eat this Crab, and not without reason, because they lack the knowledge how to remove the aforementioned white thread, which, when carelessly eaten, will cause a great distress and dizziness, though, as far as I know, it has not hurt anyone; and if it were to happen, that someone would feel ill from eating these Crabs, then there are ready remedies for this: for if one takes the roots of the *Papaja*-tree, or those of the *Siriboppar*,[15] one of these ground with water, together with black *Calbahaar*,[16] and drinks it down, this will cause all that harmful stuff to be vomited forth, if it is still in the stomach, and furthermore deprive it of its hurtfulness.

There are different opinions about the origin, and the generation of this Crab; many contend that they derive from the *Cancellis*, which are the little Crabs, called *Cuman* in Malay, that live in snail shells, about them later.[17] And to be sure, one will find among the *Cancellis* those that bear a great resemblance to young Purse Crabs; but I cannot approve this opinion, since the *Cancelli* are found on all kinds of beaches in great multitude, but the Purse Crabs rarely, and in few places. Then I have also seen the Purse Crab's own Eggs: and I have had little Purse Crabs, which were smaller than the biggest *Cancelli*, and espied a notable difference between the two: wherefore I judge these Crabs to have their own form of procreation, and I am strengthened [in my opinion] by what the Natives inform me, who find these Crabs in and on such rocks, where there are no *Cancelli* about.

Rochefort Hist. Antill. Lib. I. Cap. 21.,[18] in the third section, describes the Crabs called *Boursires*, which also want to be Purse Crabs, and which occur in the *Antillis* in the Mountains, and also have fearful pincers, wherewith they clash against each other as if they were armed men, and that once a year they descend from the mountains to the Sea in full Battle Array, in order to lay their eggs, etc. But the description does not tell us, whether they correspond to our Purse Crabs in either shape or color, for even among many Colors, no blue was mentioned, which ours are for the most part.

These Purse Crabs should not be put in salt, nor in fresh water for they will soon die. Also, one seldom finds them with their eggs. If one turns the Crab over, so that the purse is on top, and looks at it from behind, than it quite resembles an armed Man, wherefrom it, turned over like that in the picture, carries the name of *Don Diego* in full armor.

Cancer Crumenatus *etc. or Don Diego in Full Armor letter* H. *is the same from up above, and letter* I. *from another side; so that one can see its bodily shape and purse on Plate No. IV.*

CHAPTER 6

Of the Cancer Saxatilis. Cattem Batu.

True Crabs differ in shape from Crayfish, either because their bodies are round, or oblong from the side, and because they walk sideways. It is neither my intention nor my ability to describe all the kinds of this genus, because they are numerous, only the best known of those that became available to me; and which I divide into edible, and harmful ones. The best of the edible ones is the so-called *Cancer Saxatalis*, or Stone Crab; they have a common and familiar shape, with a round body, about half a foot wide, more or less, the front part of the shield has on each side nine teeth, like a saw, next two large holes that hold the eyes, and is thin and brittle, on the back somewhat bossed like a breastplate, otherwise smooth. It has two large, strong pincers, of a thick mass and smooth, of which the one is larger than the other, and on the two pincers blunt teeth like our grinders, further eight legs on either side; they conceal the tail under their belly, but there is a difference between the male and the female, because the male's tail is small, narrow, and lies within a notch in the belly, though one can lift it up; but the one of the female is larger, loose from the belly, and in due time one will see eggs hanging from it. The body's hollow is swollen with a watery brown moistness, like blood, while hanging from the belly flap is a hairy meager meat not fit to eat, but it has a white and yellow fat at the edges which is fit for food. The best meat is in the big pincers and in the legs; nothing suitable in the tail: the living one is a dark gray or blackish, but the cooked one becomes a uniform light red; it is not filled with goodly meat all of the time, but conforms to the Moon; for as long as the Moon waxes one will find nothing but slime and water in it; but when the Moon is full it has lots of meat and fat, which thereafter gradually diminishes again, until the day before the new Moon, when it is full up again.

Their name in Latin is *Cancer Saxatilis:*[1] in Malay *Cattam Batu*, because they have a dark gray color, resembling a stone.

They dwell in swampy places, that have a mixture of small stones and coarse sand, particularly around the roots of the Mangi trees,[2] near the exits of Rivers, both in and out of the water: but there is a noticeable difference in the taste, which is not as good as the ones that live only in the mud, but as good as those that live in rocky places, near some running water. These are generally held to be the best for eating. One can catch them with one's hands, or stab them with a harpoon, but most can be caught with a square scoop net, about a fathom wide, called *Tejang*, which has the four corners tied to sticks, so that one can lift it. One places this *Tejang* flat on the ground, at low tide, weighing it down at the edges with some small stones, and puts some bait of chicken guts in the center, or *Calappus*[3] meat. When the water rises, the Crab will crawl to the bait, and remains hanging there with its legs entangled in the net, which one notices from the buoy made from a light wood, which one ties to the net and lets float on the water.

Cancer Saxatiles. *The illustration is lacking, but Doctor* D'Acquet *sent us a picture. See Plate No.* V. *letter* M. K. *and* L. *belong to chapter IV.*[4]

CHAPTER 7

Cancer Marinus. Cattam Aijam.

This *Cancer Marinus*, or really Sea Crab, does not differ much from the foregoing *Cattam Batu;* hence the common man confuses one with the other, but I wanted to give each its particular Chapter. It is divided into two kinds, to wit: 1. *Cancer Marinus Laevis*, or smooth one, 2. *Cancer Marinus Sulcatus*, or furrowed one.

I. *Cancer Marinus Laevis*, or smooth one,[1] is smaller than the foregoing, has only six teeth or points on the sides, whereas the former has nine, and which are sharper as well, though shorter. The tongs of the pincers are longer, narrower, and finer toothed on the inside. The shield of most of them is smooth, a grayish dark-green. The rest is like the foregoing.

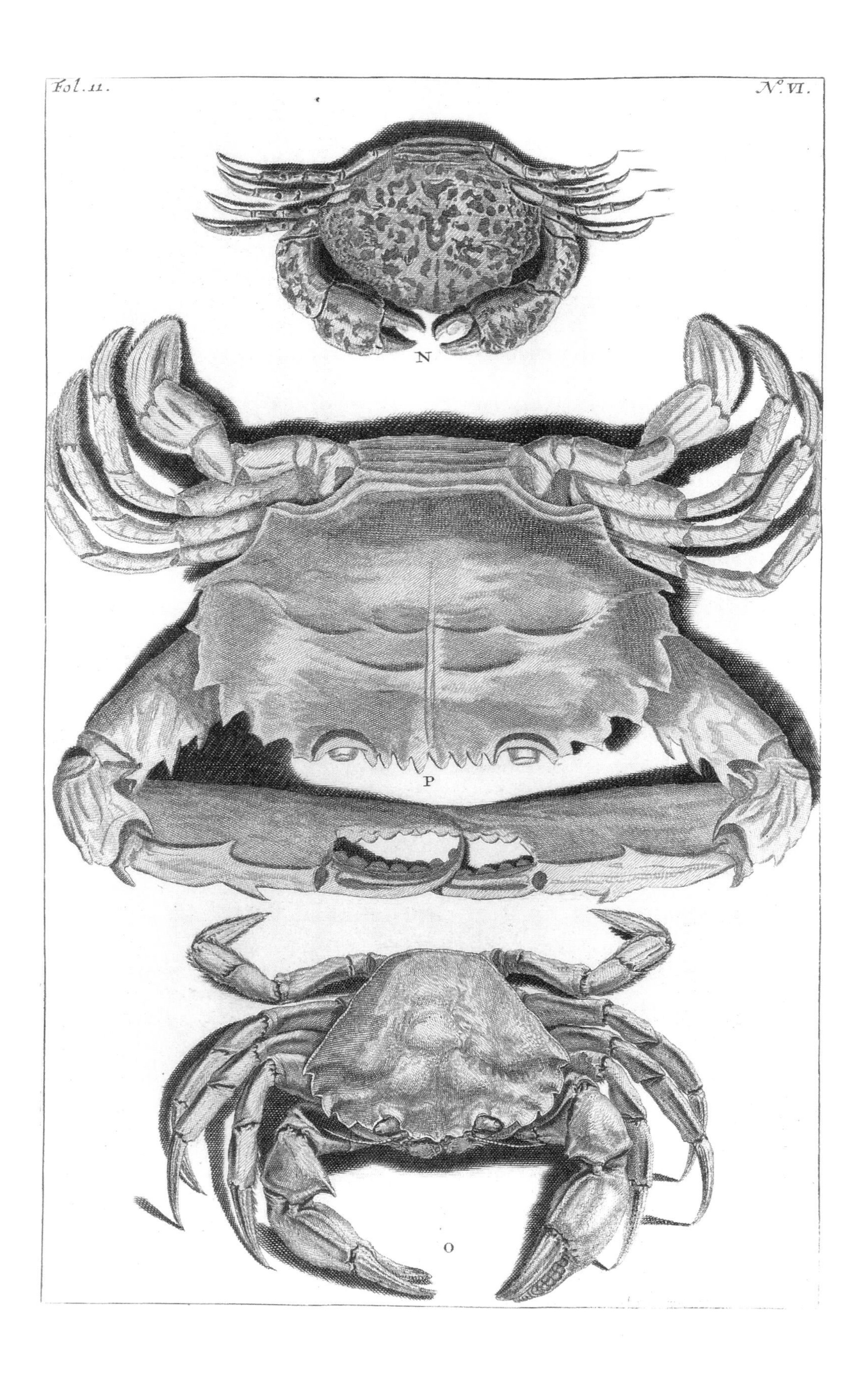
Vol. II.
N°. VI.
N
P
O

II. *Cancer Marinus Sulcatus,*[2] has a Shield that is a hand wide across and long, but always wider than it is long, has furrows across the shield, which stick out like ribs, but not too high, some also carry Shells and Oysters on their backs that have grown fast there. Raw, they are also a dark green, cooked they are red, of a uniform color, on the sides they have five teeth like a saw, up to the eyes, with small, short beards; they have eight legs, of which the six front ones end in pointed claws, and the hind ones in oblong little plates. The pincers are furrowed and like thorns, white at the tip, behind that a black band, the rest is like the body; the male has a narrow tail fastened to its belly; it never comes on Land but remains in the Sea, but it is either surprised on the beach when the tide is going out, in holes and stones, or caught with a draw-net, and is fit to eat. But it is true, however, that Experience has taught us that among all these edible Crabs, there are some that have made those who eat them ill. But one should not attribute this failing to the entire sort, but rather to some particular Crab that happened to have eaten some harmful wood or fruit, as, for instance, one suspects the fruits of the *Arbor excoecars,*[3] which look very much like the kernels of *Cataputia;*[4] at least, I know that when *Sardines* and other Beach Fishes have eaten these kernels, they were found to have a bitter taste, and upset one's stomach.

This Crab is called *Cancer marinus* in Latin. In Dutch Sea Crab. In Malay *Cattam aijam,* and *Uccu manu,* that is Chicken Crab, because its shield is furrowed and mossy; In Ambonese *Yhu hatan,* that is *Cattam Caju,* or Log Crab, because it is often found on old rotten logs, or trees that float in the Sea.

Cancer Marinus Laevis. N. *and* Cancer Marinus Sulcatus O. *The pictures were sent to us by Doctor* D'Acquet, *and another Ambonese one to complete plate No.* VI; *see letter* P.[5] *This is another kind of* Cancer Marinus Laevis; *but very rare.*

CHAPTER 8

Of the Pagurus Reidjungan.

The most common of the edible Crabs is the Prickly Crab, so called because of the two long thorns it has on its sides. The width of the shield is oblong, from head to tail, the length of four fingers long but six fingers wide, with sharp edges at the sides, each with a protruding thorn: from there up to the eyes, it has seven or eight teeth. The shield is smooth on some, grainy on others, like Shagreen leather,[1] while some have also three large spots on their backs, like eyes; the fresh one is brown, the cooked bright red. The Pincers are quite long, made from two joints, of which the front one is furrowed, and thorny, divided into two long teeth, which have many sharp teeth on the inside. The back joint is also thorny on the upper edges. The other eight legs are like ordinary Crabs.

These Crabs are called *Pagurus*[2] in Latin. In Malay *Reidjungan, Reidjucan,* and *Rindu Rindu,* in general also *Cattam bulan;*[3] in Ambonese on *Leytimor, Yatallan,* from the resemblance to the outstretched wings of the *Tallan* bird, that is, Frigate Bird.[4] Some also call it *Hyu manu,* which really belongs to the foregoing Crab.

Pagurus Reidjungan. *See Plate* VII. *letter* R.

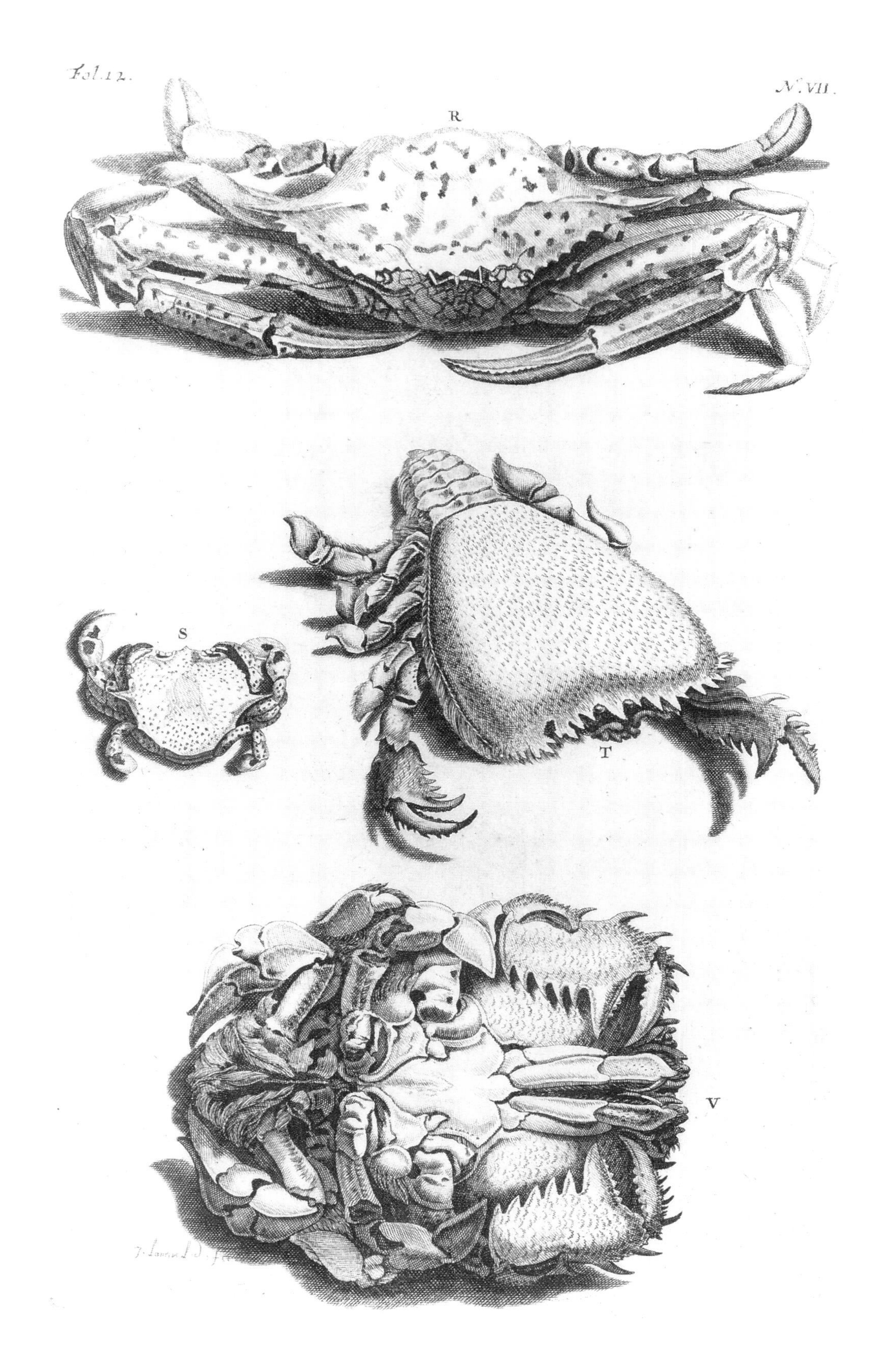
Tab. 12.
N°. VII.
R
S
T
V

CHAPTER 9

Of the Cancer Lunaris. Cattam Bulan.

The *Cattam Bulan* belongs to the foregoing genus as well, though it is smaller, the length of two fingers, and a little broader. Its Shield has two straight thorns at the sides, from there up to the eyes the jaws are finely toothed. The Pincers are short, thick and strong, thorny on top. The shield has some warts, or protruding wheals, and, moreover, is speckled with small red spots, like measles. The body is reasonably thick and swollen, which is why it is widely hunted for food. One will find it on all kinds of beaches where it hides under rocks, and is caught when the tide is receding. Otherwise it hides itself in the sand, but readily betrays where it is; since one usually becomes aware of some toil underfoot, when one steps on such a spot, which is where it is dug up.

Its name in Latin is *Cancer Lunaris*, in Malay *Cattam Bulan*, or better *Cattam Bulan Trang*,[1] and so in Ambonese, *Yu Hulam Rita*, or, shorter, *Yulan Rita*, that is, full-Moon Crab, because it is most often caught by the light of the full Moon when one puts a *Dammar*[2] on the beach, whereto they will crawl in troops, even beneath the sand, where one can feel them toiling.

Cancer Lunaris. *See Plate* VII. *letter* S.

CHAPTER 10

Of the Cancer Caninus. Cattam Andjin.

The Dog Crab is two inches long, somewhat less wide, somewhat square and tapering to the rear, with a thick body and smooth sides; alive, a blackish brown; cooked, a reddish purple. It has eight ordinary legs, of which the two front joints are hairy, and finely sawed at the edges. The Pincers are short, thick, and strong, smooth on top, and the tongs are white; with the female the left one is larger than the right one. The head is also smooth, with some barely perceptible beards, and the mouth is covered on each side with two rather broad legs; the stomach is also smooth, and conceals a fair amount of meat inside, particularly much fat or egg-stuff. It sojourns both on land and in the water, but most often beneath rocks; and if one happens upon it, it will hide in the sand. It walks quite fast, and has strong Pincers; they are edible, but with a difference, because those that stay too much on land are considered not as good as those which are caught under the rocks, when the sea recedes. The eyes are red in front, surrounded with white, and it is also whitish on the side of the shield.

Its name in Latin is *Cancer Caninus*.[1] In Malay *Cattam Andjin*, that is Dog Crab, because it walks so fast; In Ambonese, on *Leytimor: Hyu Sarissa Poëti*, to distinguish it from *Hyu Sarissa*, which is red when it comes out of the Sea. Also *Yu Mattacau*, for its red eyes, and *Hyu Asso:* on Hitu, *Lilu Maolo Yal*, that is Kanary Cracker.[2]

The second kind[3] is smaller, but square, an inch long and two wide, with smooth edges and high sides. The eyes are strangely shaped, because they are narrow below, with a thick little knob on top, being the right eye, and on top of that a straight little horn, without noticeable beards. The right Pincer is much smaller than the left one, which is short, broad, grainy on its back, and sawed on the sides, with a small black dot on the outermost tips. The eight legs are very long, thin, and as if clawed: On its back it has a Character like an

H and beneath the H are two more dots, red, as if incised, a narrow, small tail in a groove, and below that two small horns like a claw; digs itself into the sand, is edible, and has a lot of fat; the color of the raw one is a grayish yellow, the cooked one a deathly pallor. It walks even faster than the foregoing ones, so that one can count it among the *Hippos* of *Aristoteles*, which are small crabs, that walk so rapidly on the beach, that they seem like horses.[4] It carries the same name as the former, but one could distinguish it by calling it the *Hansze* crab,[5] because of the letter H.

The first kind has eight hairy legs, and a spot on the chest that has tiny hairs, like velvet; it has such strength in its Pincers, that it can open a Kanary nut, which it has in common with the aforementioned Purse Crab, the wild Boar, and the *Cacatoca*[6] Bird. It also knows how to climb Kalappus trees, throws the Nuts down, which it then proceeds to drill, looking for the meat. It extends its hollow passages under Houses, and crawls out of them at night, making a lot of noise. It also knows how to creep up on Chickens, grab one by the feet, and haul it to its hole, which causes that nocturnal noise, that one hears sometimes coming from the Chicken coops. If you pour hot water in their holes, they have to come out.

These Chicken Crabs are very large on uninhabited Islands, particularly on *Lussapinju*,[7] with a thick, bossed shield, and Pincers that are coarse and the length of a hand, the front joint round and grainy, dwelling mostly among *Pandang* shrubs,[8] and is inedible.

Cancer Caninus. *No picture provided.*

CHAPTER 11

Of the Cancer Raniformis.

The *Cancer Raniformis*,[1] is a rare Crab, about four inches long, and near the head three inches wide, consisting mostly of a globular shell on its back, covered all over with coarse and spiny little points, and near the Pincers it suddenly turns inward, and becomes narrower. The tail is short and suddenly comes to an end, consisting mostly of two noticeable joints, and ends in a narrow little snout, which it hides under its stomach, and that is why it seems to lack a tail, just like a toad. One also notes nothing more on the head than a small mouth covered with short little beards, and above that the shell is hairy, with small eyes on either side. The aforementioned shell is a dirty white, sharply toothed on the sides, or sawed, and like a thick leaf, ending in two short pincers. Behind that follow, on both sides, three legs of a singular shape, hairy on the sides, and ending in an oblong heart, that also resembles a small leaf. The two in back are also flat and hairy, and its little heart is slightly curved, protruding like a sickle, and protecting the tail. Under the shell, on either side, it has a channel or groove, wherein it can hide its six legs. It can draw in its pincers under its body in the same way, so that one will hardly notice them; and then it really looks like a toad. It is not much known on *Amboina*, and one finds it on the flat and stony beach of *Luhu:* whether it is fit to eat, I do not know; the Natives at least are fearful to eat it.

Its name in Latin is *Cancer Raniformis*. In Malay *Cattam Codoc*.[2] In Ambonese *Hyu Allaac*, the same as the others.

Cancer Raniformis. *See Plate* VII. *letter* T. *where one sees it from above, and letter* V. *from below: since it is so rare we thought it necessary to show it in two ways. This was also sent to us by Doctor* D'Acquet.

CHAPTER 12

Of the Cancer Terrestris Tenui Testa. Cattam Darat.[1]

There are different kinds of Land Crabs as well, of which we will describe the ones with thin shells in this Chapter, though some also go into the water.

Cancer Rubris Oculis: more oblong than round, not sawed, painted with small, pale-yellow flowers on its back. The eyes in the live ones are as red as Rubies; it has two short, thick, roundish, and smooth Pincers, with short tongs, at the ends slightly black, black dots on the Pincers, with eight short and thin legs, all with claws; lives also among the rocks.

Its name in Latin is *Cancer Rubris Oculis,*[2] in Ambonese, on *Leytimor, Hiu matta cou,* like the foregoing *Cancer Caninus,* though this name really belongs to this Crab, because its eyes are much redder than the other one. Like the foregoing, they are thought of as Land Crabs, and if they are found around the beach, and have a smooth shield, and are hairless, they are sometimes eaten, though this often goes wrong; because it happened in my day, in *Hitulamma,* that a Woman and her little girl ate nothing more than the Pincers of these Crabs (one judged them to have been the foregoing Dog Crabs, others said it was the *Cancer norius*),[3] yet were found dead by the fire, with the pieces of the Crabs still lying next to them. Otherwise, the general antidote for all such Crabs, is scrapings of black *Accarbahar,* the root of *Siriboppar,* or *Pissang Swangi,* and of *Lalan* grass, either drunk, or chewed with *Pinang,*[4] and the juice swallowed, which will expel this hurtful food by vomiting.

The *Cancer Villosus* is an entirely rough Crab, with an oblong body, two or three fingers wide, an inch long, the entire shield covered with coarse bristles, as it if were a Sea Apple,[5] round and with a smooth belly. The Pincers are also rough, short and thick; it has black tongs and eight short legs. One holds it to be similar to the large harmful one; it is not edible, lives in the sand under rocks, all of them round bellied, with the tail attached to it.

Its name in Latin is *Cancer Villosus,*[6] in Malay *Cattam Bulu,* in Ambonese *Yu Huta,* that is, *Cattam Rompot,* because of its rough, coarse body.

Cancer terrestris, tenui testa. *There are many kinds, but since the Author did not send us any Pictures, and since we were unable to get any, we did not depict them here.*

CHAPTER 13

Of the Cancer Vocans. Cattam Pangel.

These little Crabs also belong to the *Hippos,* but they are barely an inch long, almost square, wide in front, tapering to the rear, with smooth sides. The cooked ones do not become red, but stay a bluish green. The eyes stick out quite far, and are on thin stems; they have eight normal, narrow and smooth legs, two very dissimilar Pincers, of which the right one is much larger than the entire body, the edges and the frontal joint are sharp, grainy on the back, and with sharp little teeth inside. The shield is a little bossed, but smooth; the claws of the Pincers always have another color, and are either yellow or red; and the upper claw is also lighter, or whiter, than the lower one. The left Pincer is smaller than any of its legs, with two subtle claws. It lives on smooth, sandy beaches, and when the tide is receding one will see it continually waving the bigger Pincer above its head, as if it wants to call people, but when one gets near, it hides in the sand; it has a thick body and is edible.

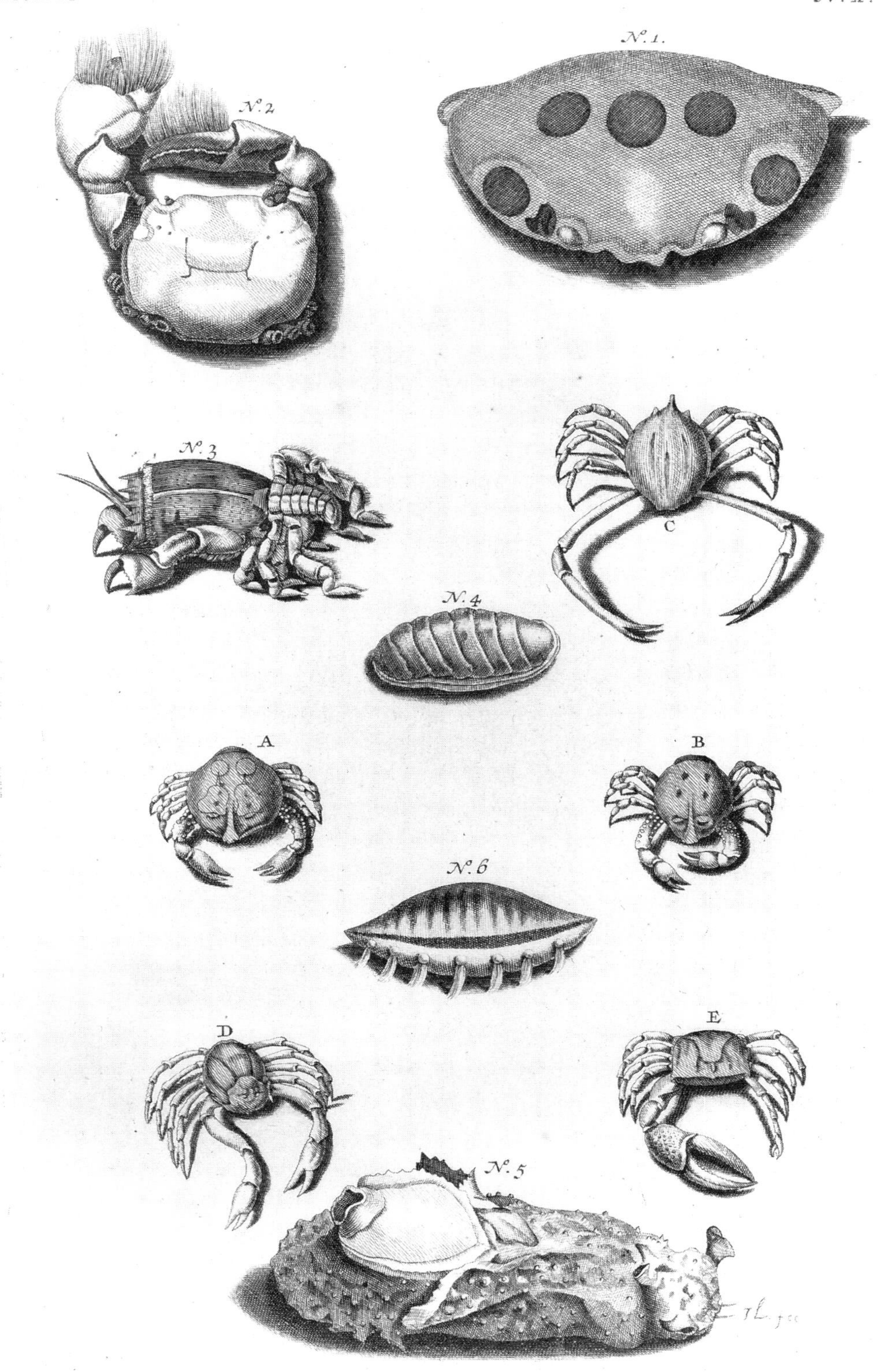
N°. 1.
N°. 2
N°. 3
C
N°. 4
A
B
N°. 6
D
E
N°. 5

Its name in Latin is *Cancer Vocans.*[1] In Malay *Cattam Pangel,* that is, the Caller, for the above reason.

The shield on the raw one is usually blackish, with white dots like Characters, and it can stow the long eyes in certain notches on the sides. It can walk so fast that a person can barely overtake it, and if one encounters one, it will dig itself so rapidly down into the sand, that one can hardly trap it while digging it up. People despise them as food, because they are so small, but not so Ducks, who will chase, dig up, and swallow even the smallest ones. The *Mippi* [*sic*] are described by *Plinius* lib. 9 cap. 31.[2] *Aristoteles* lib. 4. *Animalium*[3] calls them *Hippeas,* that is Horsemen, they both say that these are found on the beaches of *Phoenicia,* but *Bellonius*[4] found them on sandy *Syrtibus.*[5]

Cancer vocans. *See Plate* X. *letter* E.

CHAPTER 14

Of the Cancer Spinosus. Cattam Baduri.

These are ugly kinds of Crabs, of which we will describe two sorts in this Chapter. The first kind has a roundish body like a Spider, a shield that is two or three fingers wide, its back is wide and round, the front somewhat narrower, where a little flap hangs from its head, divided into two horns, like the head of a Grasshopper, with some smaller thorns on the sides; above this small flap are two other protruding thorns or horns, and behind them, all over the shield, many other smaller ones, so that it is thorny all over, a dirty gray and always covered with moss. The legs and the Pincers are also thorny; the body is mostly filled with a dirty brown liquid, scarce meat, which is why they are not eaten. One will find them on the beach when the tide recedes, under rocks, but never on dry land.

Their name in Latin is *Cancer Spinosus,*[1] in Malay *Cattam Baduri,*[2] in Ambonese, on *Leytimor: yhu* or *hyu, makeku huta,* that is *Pickol rompot,*[3] or Moss Bearer, while others call it *Hyu hatu talae,* because it lives in the hollows of Coral stone, especially when it blows hard at sea.

The second kind of the *Cancer Spinosus,* is the *Longimanus,*[4] an ugly and hideous animal, with a body that resembles a Spider as well, about two inches wide, narrow both in front and behind, also with an overhanging little flap or snout on its head, coming to a blunt point; the shield has three bosses in back, while it is also covered with short thorns and warts, which, on the living ones, are always covered with moss or other filth, so that one would hardly know it. It has a deep notch in its belly, wherein it hides its tail, a narrow chest, with 8 legs. Besides these, it also has two uncommonly long Pincers, so that the entire Crab looks to be nothing but Pincers, made from two long and one short joint, every joint five, six, or more inches long, the thickness of a finger, and almost triangular, of which the inside is grainy, the third or outer one is usually covered with blunt thorns, some large, some small, and the large ones are also divided into other smaller ones, also covered with moss. The front joint ends in a short pincer, of which the lower tong is curved downward; the color of both kinds is a dirty gray, or a smoky color, nor does it ever become red when cooked. The latter lives mostly on the bottom of the Sea, wherefore it is seldom caught; although it does sometimes crawl into the Weels, while it is also hauled up with nets at times. One sometimes finds them so large, that the Pincers, stretched out, cover more than an ell,[5] and the Fishermen

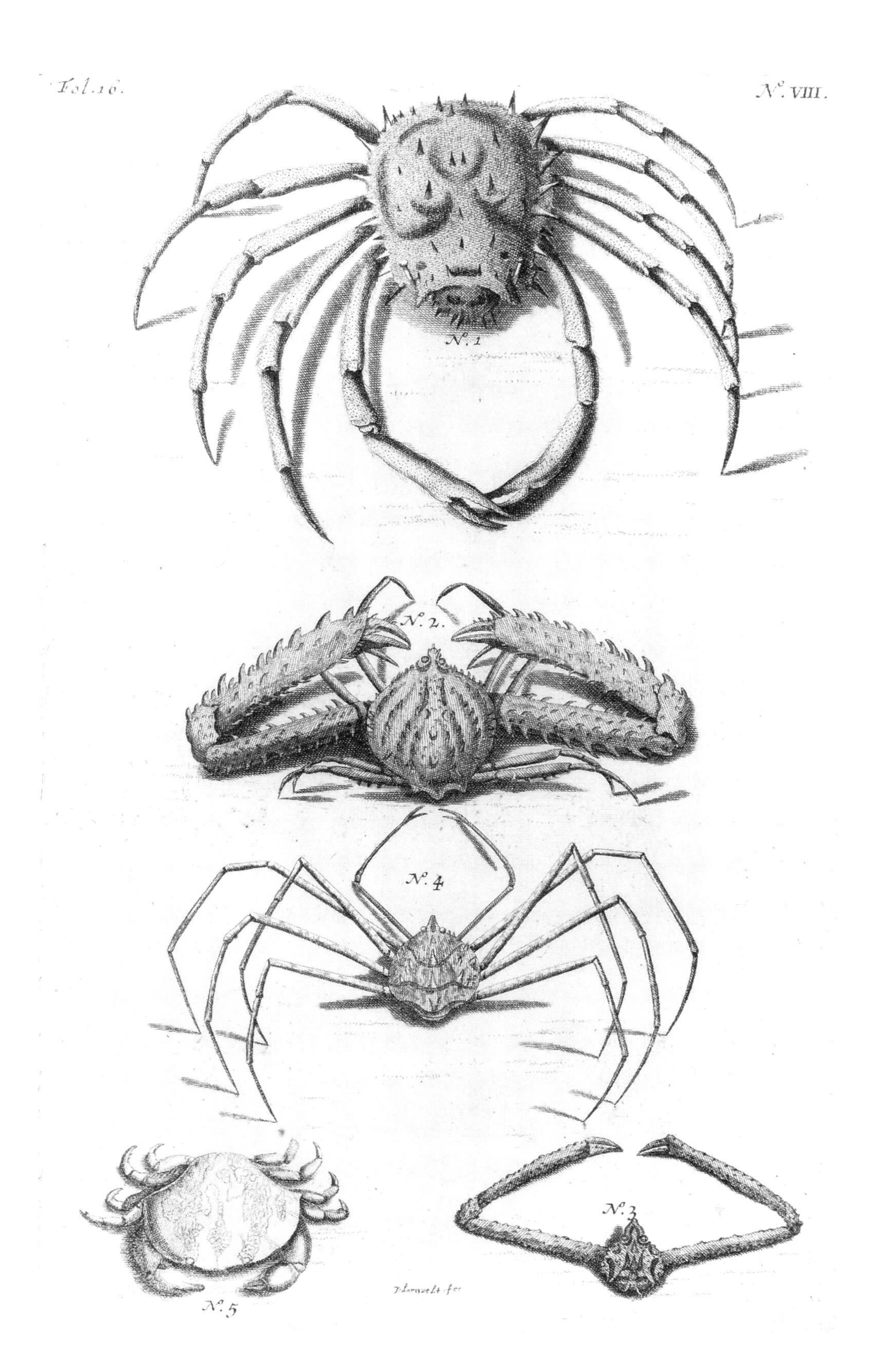
Tab. 16.
N°. VIII.
N°. 1
N°. 2.
N°. 4
N°. 5
N°. 3

abhor it, as if it were an ugly Sea monster, wherefore they usually throw it overboard again, regarding it as useless, nay, even harmful if eaten.

To distinguish it from the first one, one could call it *Cancer Spinosus longimanus;* in Ambonese it should have the name of *Yhu makeku huta* far more than the foregoing, that is Moss Bearer, because it is always covered with it. Also *Yu maccar,* that is, *Cateam*[6] *duri Sagu.*

The first kind is widely found on Amboina, especially near *Weynitu* on flat beaches, that are sand with small stones, and is considered edible; but the second one, or Longhand, is generally unknown in Ambon Bay, and is considered nasty.

The *Cancer Spinosus* also has a common beach kind,[7] that is 2 inches wide, slightly longer, with a dangling little shield in front, as if it had a rag over its mouth, the hind part of the body is somewhat wider, and comes to a point, after the fashion of Spiders. The legs are rather long, but the Pincers are much shorter than those of the *Longimanus,* otherwise spiny and knobbly all over, and hairy on its chest and belly.

Another larger and rarer Crab[8] seldom appears, because it dwells deep down in the Sea. Its body is 4 inches wide, narrowing in front, where one will see nothing but a knobbly nose, and two small eyes right next to each other. The entire back is covered with knobs, with clefts in between, containing sea moss, and small coral shrubs. It has uncommonly long Pincers, each one 9 inches long, covered with similar knobs, and, which is something unique, the two long joints have a shorter one between them. The other 4 feet on either side, have a normal appearance, except that in back they are covered with a host of sharp twigs on three sides. The color is speckled red and white; it has tiny, little beards which it can hide in hollow slits under its nose. The Ambonese are so afraid of it that, when they haul it up with the weel or fishing line, they immediately throw it back into the sea, because they consider it bad to eat, even though they cannot give one example, when it has harmed anyone. On Celebes' East Coast one will find them just as large, and they too are considered to be hurtful. They usually carry some kind of Sea plant on their backs or Pincers, most often little white coral trees,[9] (*Lithodendrum calcarium*) which is used to make lime, which is why it is often tossed up on the beach by a rough Sea, and dashed into pieces.

Cancer Spinosus, *see Plate* VIII. *No.* 1 Cancer Spinosus Longimanus. *This picture was lacking, but Doctor* d'Acquet, *provided two depictions for this purpose, to wit the* Cancer Spinosus Longimanus, *Major No.* 2. *and Minor No.* 3. *and one of the rarest of Crabs as well. The* Cancer Aragnoides, *see on the same Plate No.* 4., *and No.* 5. *is the* Cancer Floridus. *Mister* d'Acquet *also sent us an uncommonly large Crab, known to the Author, to represent the fourth kind of the* Cancer Spinosus, *which we call* Rock Crab, *because it looks like a Rock or Coral stone, see Plate No.* IX.

CHAPTER 15

Of the Cancer Floridus. Cattam Bonga.

This has the shape of an ordinary Crab, with an oblong shield; scarcely two inches long, two and a half wide, with a rather thick body; the shield is decorated with a large number of puckers, of a pale color, furthermore painted with yellow spots, of which some resemble lines, others spots, and flowers. The shield has sharp edges, on every side with a small corner, or tooth. The Pincers are, like the body, thick and round, but short, covered with puckers and kernels, like the body, and attached to them are sharp Pincers or tongs, chestnut brown, also 8 normal legs that end in sharp claws, which are also brown and coarsely covered with short little hairs. When the tide is ebbing, they live on the beach, and at high tide they sometimes crawl onto dry rocks, where they sometimes remove their old skin, which one will then find in its entirety, with legs, pincers and all, nicely painted and thin, that one will hardly know, where the animal crawled out, but which one will become aware of on the bottom, between the belly and the tail, where these empty little houses are open.

Its name in Latin is *Cancer Floridus*,[1] in Malay *Cattam Bonga*, that is Flower Crab, because it has the most beautiful shield of them all, as if it were strewn with flowers: The *Ambonese* also call it *Yu Nikimetten*, because the ends of the Pincers are somewhat black, which is why they are counted among those, which Nature has marked not for eating, wherefore it is thought to be a smaller kind of the big Black Tooth, described hereafter in Chapter 18. They are not eaten, because the Natives consider it a common sign, that all Crabs, that have Pincers of which the claws are black or brown, are not fit to eat, as if they had been marked by Nature.[2]

Cancer Floridus. *See plate* VIII. *No.* 5.

CHAPTER 16

Of the Cancer Noxius. Cattam Pamali.

These Crabs are considered truly harmful, and they do not have a uniform shape. The most common one does not differ much from the edible Sea crab, but has a slightly larger and thicker shield, which is pale or whitish, speckled with red spots and Characters that resemble flowers, but different from the *Cancer Floridus*, which has a shield decorated with many puckers, though it resembles the *Cancer ruber*.[1] Its eyes are quite far outside its head, and are red, [and it has] big and strong Pincers; it sometimes crawls in the Weels, but once hauled up, it is immediately thrown away, because it is held to be harmful, even deadly: It happened in my day at least, that on *Hitulamma*, a certain Woman had eaten only one claw of such a Crab, that she had roasted on coals, because she did not know what kind of Crab this was, and had fallen asleep while sitting by the fire, and was found dead on that spot,[2] as was her little daughter, who had also eaten of it, and did die shortly thereafter, but not before she indicated the Crab they had eaten from; which can also be looked up in Chapter 12 above. And so Nature has wisely ordered, that this harmful Beast dwells by itself on the bottom of the Sea, and will not come onto the beach, so that it cannot harm many of the people who look for their food on the beach during ebb tide.

Its name in Latin is *Cancer Noxius:*[3] in Malay *Cattam Pamali:* in Ambonese, on Leytimor, *Yu-Umali:* on Hitu *Lilu-Umali*, that is *Cancer Infaustus*, or harmful Crab; also *Cattam Bisa*, poisonous Crab, like the foregoing.

CHAPTER 17

Of the Cancer Ruber. Cattam Salissa.

This is a rare Crab, that is seldom found, somewhat larger than a normal Sea crab, 4 inches long, five wide, it has the thickest shield of all, which is smooth and even, the thickness of a knife, it has a blunt corner or tooth on either side, another one by each eye, and another four on its forehead, blunt and sort of round. The eyes are small, and hide

Fol. 16.

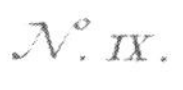

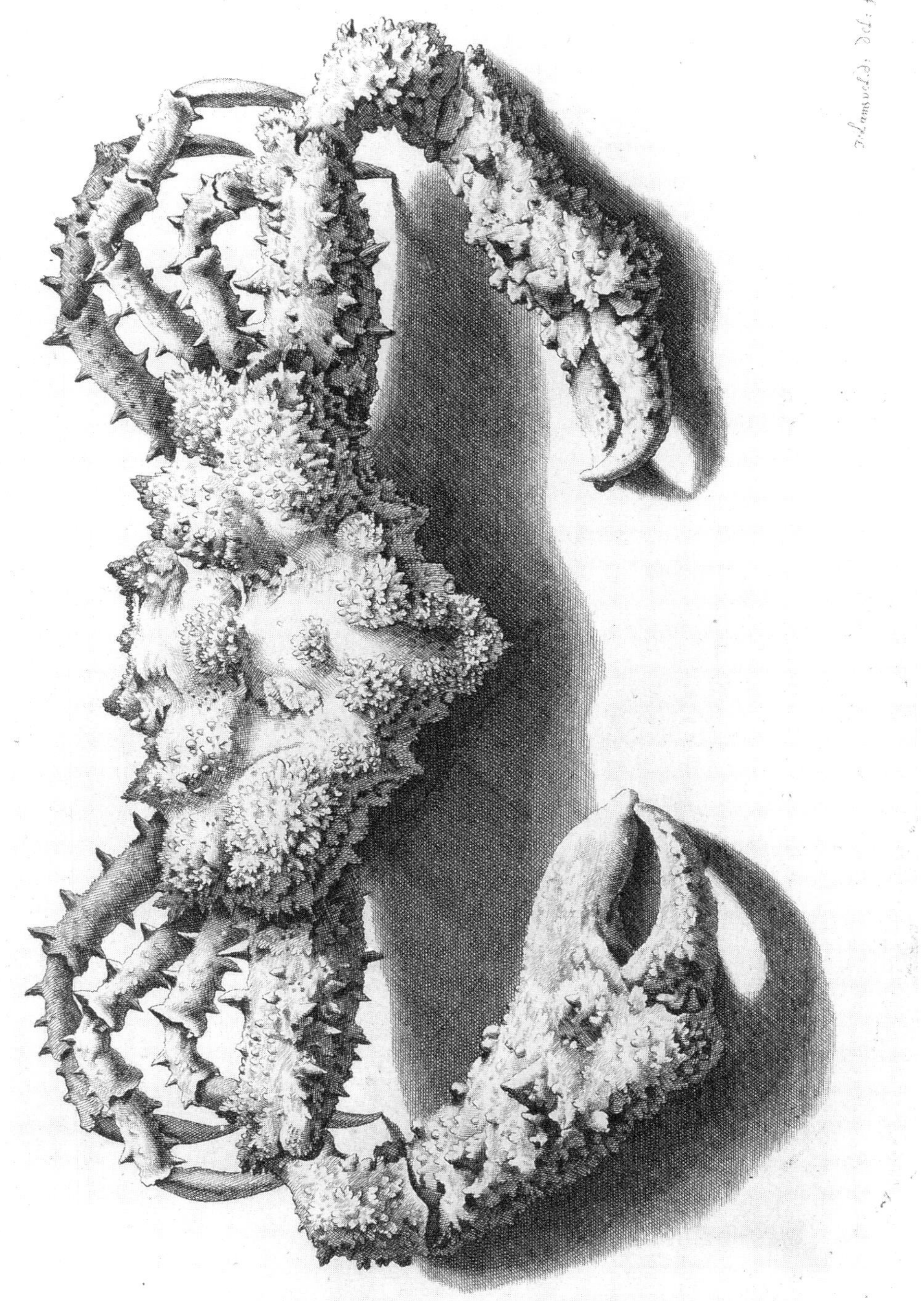

Fol. 20.
N°. 1
N°. XI.
N°. 2
N°. 3
N°. 4

in small holes. The shield has two flaps like half Moons hanging from the stomach, which seem to have been added to it, and which drop off in time;[1] the Pincers and legs are still not known, since I was always brought only bare shields. The shield's color is red, even of the living ones, when they come from the Sea, but not commonly, because the shield itself is pale, but one will see large red spots on it, the color of the fallen leaves of the Tree called *Salissa*, or *Catappan*,[2] which are red as a cooked Crayfish, and the three largest spots are lined up on its back like dubbeltjes, 3 smaller ones by the tail, sometimes four, and still another 4 by the eyes. These markings have convinced the Natives, that they should suspect these as well, and will not allow them as food. At least, they showed me in the village of *Hucconalo*, that they had harmed some People, who had caught them in *Ambon* Bay, and had imprudently eaten them; on the other hand, there are Natives who say that they do eat them. This Crab has 11 or 13 red eyes, in 3 rows on the shield, the largest the size of a stiver. Others have it, that those from *Hucconalo*, are unjust to call them harmful, because on *Caybobbo*[3] and *Bonoa*[4] they are considered to be the best, being full of yellow fat, and full of meat, and are called *Mule pipi* on *Bonoa*, that is Painted Cheeks.

Their name in Latin is *Cancer Ruber;*[5] in Malay *Cattam Salissa*, in Ambonese, on Leytimor: *Yu Sarissa*, on Oma *Uca Sarissa;* because what they call *Sarissa* on Leytimor, is what the people on Hitu pronounce as *Salissa*, in Malay the *Catappan* Tree.

Among the aforementioned *Cancer Aeneus*, as well as among the foregoing *Cancer Noxius*, belongs the following, called *Cattam Tambaga*, that is, *Cancer Aeneus*,[6] or Copper Crab, since it is red or the color of Copper, when it emerges from the Sea, not much larger than an ordinary Crab, nor does it differ much from them as to shape, though it does not have any hairs on its entire body, except for some soft little hairs at the end of its legs. The shield is closely marbled with little brown eyes, but it has 3 truly red spots around the tail, of which the center one is square; it also has some bulges which resemble a closed Purse, with its mouth in back; it has two small eyes, which are deeply hidden, between the notches in the head, and it has no beards at all. The Pincers are as black as coal, but not sharply toothed; the hindmost and broadest legs have their rear and broadest joint on the back part of the body, covering the same up to the neck of the Purse, and seem to protect the hind part of the body. The legs end in short, blunt little thorns. The belly is also marbled with red and white. One swears that this Copper Crab is very harmful, and therefore rarely appears in the light of day. It is smooth over its entire body, and shiny like porcelain.

I was brought a Crab from *Assahudi*,[7] under the name of *Yu Sarissa*, or *Cattam Catappan*, not much different from the above, except that its shell was not as red; the back was a pale yellow in the center, brown towards the sides, with a large red spot in the center of its back, and on either side five lesser ones, all red, with a few short blunt thorns on the sides of the shell. The belly is high off the ground, and has four smooth legs on either side, with long black claws; the right Pincer is twice as large as the left one, both with black tongs; in back a pale yellow and also smooth, the eyes do not stick out too far, and are whitish or a pale yellow, nor does it have any Beards.

Cancer Ruber. *See Plate* X. *No.* 1. *It came to us from the Cabinet of Mr.* de Jong.[8] *The* Cancer Aenus *was also lacking, and was supplied by Doctor* d'Acquet: *see Plate* XI. *No.* 4.

CHAPTER 18

Of the Cancer Nigris Chelis. Cattam Gigi Itam.

This Crab is dark brown on its back, without spots, but sharply sawed on the sides, also without beards; one sees two Pincers near its belly, both the same size, the back part a pale yellow, in front black; the legs are entirely coarse and spiny, usually with two rows of short thorns; all a pale yellow and covered with hairs; the two hind legs are very small, and curved over backwards, it has whitish and clear eyes, which shine at Night like Jewels if one goes into the Sea with a burning *Dammar*, and it usually has a smaller one nearby.

The smaller one is also entirely brown, has a small shield that has yellow lines like Characters, full of bumps, and so curled as if it had been carved, with a few corners on the sides, and without a beard; the Pincers are grainy and a dark yellow in front, with black claws, the legs are smooth on top, hairy below. One will find it on the Islands before *Assahudi*,[1] lying on the Hook of *Nussanive*,[2] where they crawl into the Weels, and are dragged up by the Fishermen, who immediately throw them away.

The largest is unknown to me, except for the Pincers, which are big and strong; the first and second joints have large protruding thorns on the sides, the front part is entirely grainy, with some furrows, the tongs or claws are black, with sharp teeth inside; it is a violet-brown when cooked, but the rest is a pale red. It too is reckoned to be harmful, so that the warning of the Poet applies to it:

Hic niger est hunc tu Romane caveto.[3]

Its name in Latin is *Cancer Nigris Chelis:*[4] In Malay *Cattam Gigi itam*, and so too in Ambonese, *Yu niki metten*, that is Black Tooth, because the Malay call *Gigi* or Tooth, what we call Pincers.

CHAPTER 19

Of the Cancer Lanosus. Cattam Bisa.

This rare Crab is also seldom encountered, which is a wise disposition of Nature, that one finds it only rarely, and that it dwells in the Deep, because this one and the foregoing Black Tooth are considered truly harmful Crabs. Its body is about the size of two fists, half a foot long and wide, its back greatly bossed, with 4 or 5 blunt teeth on the sides; the head comes to a point, where small eyes lie close together, or hide in the holes, without beards, but near the mouth there are 4 pointed claws on either side, which are its hands. It has two noticeably large and long Pincers, ending in strong claws, which are bare and white, shaped like the beak of a Parrot. The rear joint of this Pincer is, as usual, nearly triangular, of which one side is toothed; it also has four legs on each side, which are uneven, because the two front ones are the longest, and they are somewhat bent in the center joint, with a short sharp claw like a bird's foot at the front end; the third foot is only half that length, and the fourth one is even smaller, and is almost on its back, both armed with small round claws, looking quite like the sting of a Scorpion, but with this difference, that the sting of the third foot faces inwards, and the one on the fourth faces out. Over against these little claws one will see another one that is much shorter, the first joint looking like a Pincer, with which it can also pinch, and which one can also see in a way on the two front legs. The entire shell, legs, and Pincers, are covered with a dirty gray stuff that resembles moss; it feels like woollen cloth, but it is coarser on the sides, Pincers, and legs, like bristles, and this wool is so firmly attached, that one can barely scrape off a little, beneath it one will see a shell that is usually of a deathly pallor, quite thick on the back and on the Pincers; it has no exceptional meat inside, but is filled with a brown-black liquid, which colors the water purple, if one cooks it. It is an ugly beast to behold, and is held to be injurious, wherefore the Fishermen, if they get one in their Weels, toss it immediately back into the water.

Its name in Latin is *Cancer Lanosus:*[1] in Malay *Cattam Bisa*, that is *Cancer Venenatus.*[2] In Ambonese *Yu Teku hutta*, that is, a Crab that carries grass or moss.

Our Ambonese, both Christians and Moors, regard it as harmful, and throw it away, if they catch one, but it seems to me that this revulsion comes more from its ugly shape, than from experience. Because there are People who will not hesitate to roast it on coals and eat it, separating the meat from the black blood. This is done by the Inhabitants of *Bonoa*[3] and of *Serua.*[4] Similarly, there is enough experience that the Puffer Fish,[5] when eaten, is lethal, yet there are many who eat it without injury, as long as they know how to isolate the meat from the slimy veins, which contain the harm. It should be noted that, if it is mentioned in this Book that some Crabs are poisonous, that this is said in the usual, erroneous way because strictly speaking, they are not poisonous, but have a power that can make you dizzy or choke, and which sometimes can be expelled with a simple Sugar Syrup, or something that is greasy.

In the year 1692, they caught a small one of this kind that had gotten hold of a Sea plant, *Basta Laut*,[6] with its 4 hind legs, crosswise on one end, and in such a way, that the plant lay there as if it had grown there, and the crab covered itself with it as if it were a shield. It had been caught in a Weel, and when it was dry for a long time, it began to foam at the mouth, casting up many bright little bubbles that had the color of clear soap suds.

Another similar Crab had its entire back covered with a Sea plant, that had a spongy substance, and was divided into many blunt points or flaps, like a coral-stone, while it also had a twig of black *Accarbaar* among it. It seems that the rough backs of these Crabs can catch all sorts of things that stick to it, and they sometimes put their hind legs into these spongy plants.

Cancer Lanosus. *See Plate* XI. *No.* 1.

CHAPTER 20

Of the Cancer Calappoides. Cattam Calappa.

C*attam Calappa*[1] has the shape of a broken *Calappus*, though it is wider than it is long, to wit, 4 and 5 inches wide, and scarcely two long, resembling a little boat, or spoon made from the shell of a *Calappus* nut, on its back slightly furrowed in an oblique fashion, bossed and of a pale color, both when it is raw and when it is cooked. The shield has large flaps overhanging on either side, so that it is usually hollow below, with a narrow chest, which has 8 short and narrow little legs, but the Pincers are a great deal larger, and of a particular shape, resembling a child's small hand, broad and thin at the front, with a jagged corner that sticks out, like a Cock's Comb, with two small but sharp little claws. It never stretches the Pincers out in a straight line, but it is able to hide them, along with all its legs, under its shield in such a way, that one cannot see any of them, when it sits quietly on the beach. It has so little meat that it almost seems as if it consists entirely of shell, and is therefore held unfit for food. When it crawls on the beach under a full Moon, its shield reflects the light, which is how one recognizes it, otherwise one would have trouble distinguishing it from the sand, because it has the same color; one finds them often on the beach by the village *Hucconalo*, and also along the entire inner bay of the Bight of Ambon, where they leave them unmolested, because no one deigns them fit to eat; yet some are caught, cooked, dried, and kept as curiosities, most often because of the strange Pincers and the handsome shield.

Its name in Latin is *Cancer Calappoides:*[2] in Malay *Cattam Calappa*, after the shape of the *Calappus* nut. In Ambonese on *Leytimor Hyu hulan Pute*, that is, *Cattam Bulan*, *Cancer Lunaris*,[3] because of the way it reflects the light of a full Moon, while this is otherwise the name of another Crab, that was described previously. The true Malay call it *Cattam Salwacco*, after the shape of an oblong Ambonese shield.[4] This Crab resembles, at least in terms of its markings, the *Cancer Heracleoticus*[5] as described by *Bellonius*, and depicted in *Jonst. Histor. Animal. Aquitil. part 2. cap. 2 Sect. 3*,[6] where it is called *Gallus Marinus*, that is Sea Cock, because its shears resemble a Cock's Comb, though it differs from the true *Cancer Heracleoticus*, as described by *Gesner: Libri IV.* the *Cancris*.[7]

Cancer Calappoides. *See Plate* XI. *No.* 2. *which is the same from above, and No.* 3. *seen from below.*

CHAPTER 21

Of the Cancer Perversus. Balancas.

Jacob Bontius, *Hist. Indic. Libr. 5. cap. 31.*[1] describes this Crab as having a light-green, round shield, and a long tail, coming to as sharp a point as an Awl, and which, if it wounds someone, causes a great pain, as if one were stung by a Scorpion. He says that the meat is eaten, though it is not as good as that of other Crabs.

It is a monstrous Crab, with almost everything looking different from other Crabs, while it also belongs among the largest. Its upper body consists of two parts, of which the first and larger part represents a shield, the front edge round, but cut out like a half Moon in back, wherefrom hangs the hinder and smaller part with a skin; at its end it has a long, triangular tail, and it has sharp thorns on the sides. In other Crabs this hind part would be considered the head, while the first-named round part would be the tail, as many of our Europeans do. The big shield is usually the width of a span, sometimes more, its color is a smooth olive, and its top is covered with short little stings, under which one can see two stubs, whitish on top, which are the eyes. The back part of the body is, as was said, thorny on the edges, and at its end it has another small curve like a half Moon, in the center of which one will find the aforementioned long tail, about a hand long, barely a finger wide, and the side on top is also like a thorn, coming to a sharp point like an Awl. If one turns it over, it looks like a saucer, in the center of which one will discern the head, like a little clump, barely noticeable, except for some hairy flaps which appear to make up the mouth, with another two short Pincers, which close off the mouth. One will see five slender legs on each side, which they can pull in, in such a way, that one will see nothing but a bare shell whilst looking at it from above. Behind the legs is a little sack, that has some edible meat, the rest of the back part of the body is filled with a muddy substance, which is expelled from under the tail. It is also curious where the eggs are in this Crab; because if one looks inside the shell, one will not notice a single egg, and someone who is inexperienced will look for a long time before finding any, although it is filled with them. One has to know that the big

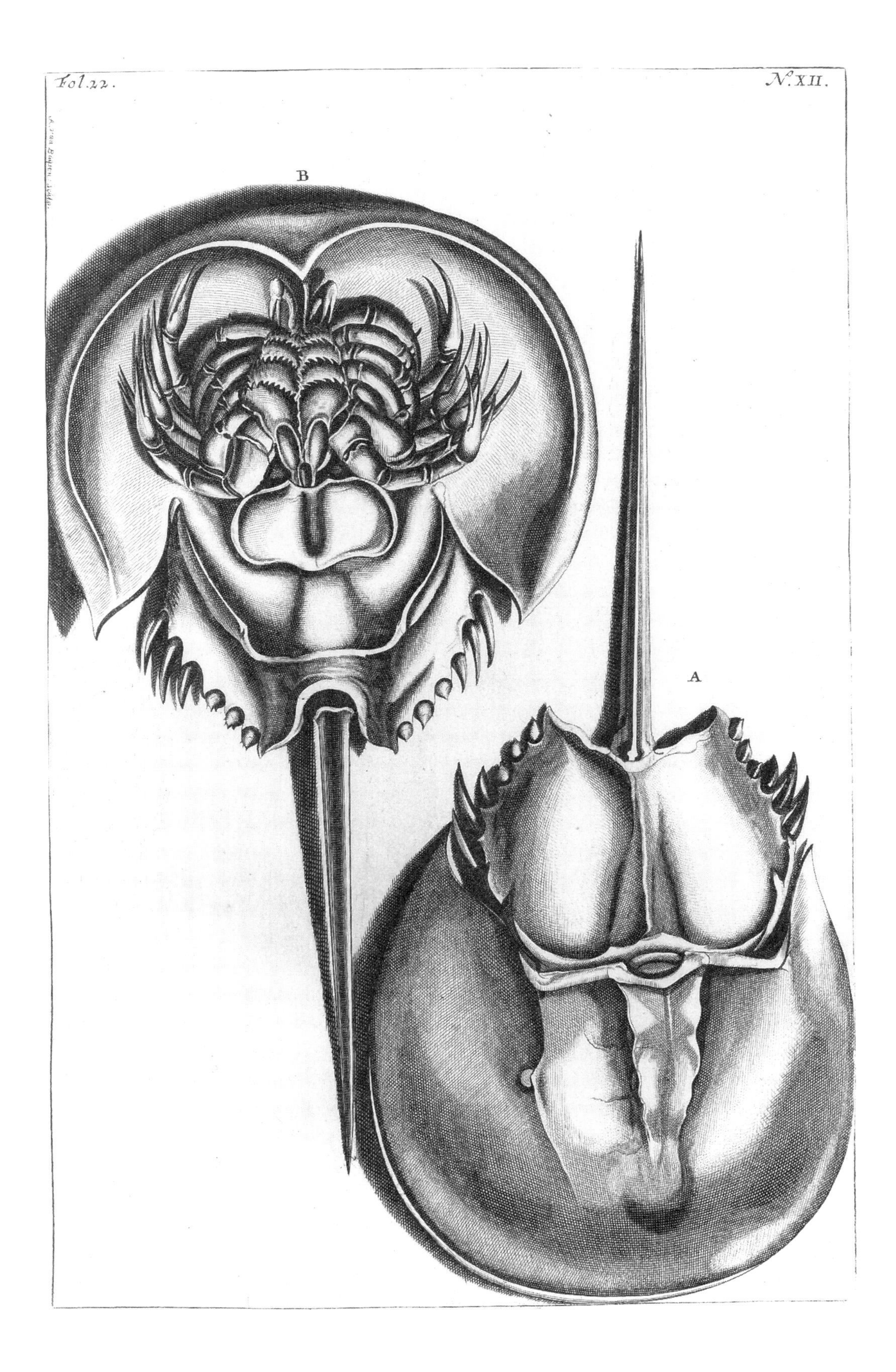
Fol.22.
N°.XII.
B
A

shell is lined inside, with a thin but stiff fleece, and the Eggs are hidden in profusion between it and the outer shell, having the color and size of a Javanese *Catjang*,[2] and they are the best part of this Crab; because one can use them to make a tasty *Bacassan*.[3]

It is mostly found on the northern coast of *Java*,[4] where there are swampy and flat beaches, always two by two, to wit, the male and the female, and, since the male is smaller, the female always carries the male on her back. They can go at a rapid pace, and stick their tails up in the air, with which they defend themselves.

The Javanese will never eat one that was caught by itself, saying that this would be detrimental, and cause dizziness. The shells are used as medicine, mostly against the *Sawan*[5] who plagues children.

Its name in Latin is *Cancer Perversus*,[6] Perverse Crab, for the reasons given above; in Malay *Balancas*.[7] We Dutch call them Sea Lice, because they look a little like Wall-Lice.[8] On Java *Mime* and *Mimi*.

Cancer Moluccanus, is also a strange Crab, and was described and depicted by Mr. *Clusius, Exotic. lib. 6. cap. 14.*,[9] under the name of *Cancer Moluccanus*, and was supposed to have been brought in the year of Our Lord 1603, from the Moluccan Islands [to Holland], and one can note there that in the Fatherland, they mistook the smallest half for the head of the Crab, that part that holds the sharp tail, and he [Clusius] even admits, that they could not find any sign of head or tail on the entire shell.

I am well aware that they find them these days in the Moluccan Islands; but they also sent me one that was caught in *Manado*, being the foreland[10] of *Celebes*.

The Cancer Perversus, *here also called Sea Spider, or Arrow Tail. See Plate* XII. *At* A. *one sees it from above, at* B. *from below.*

CHAPTER 22

Of the Cancelli. Cuman.

The front of these Wanderers[1] is shaped like a little Crab, while hindermost they look like a whelk,[2] and they live in the houses of strangers; which is why their origin is uncertain and, likewise, their shape. They are generally half a finger long, some much larger and some smaller; the front part of their body is a small Crab, of various shapes, some elongated, some short and wide, according to the snail's house that formed them. Their hinder part ends in something like a curl, fashioned like a Shrimp's tail but without a shell, and at the very tip of the tail they have a little fin like Crayfish do, wherewith they hold fast to the innermost whorl of the whelk. One of the Pincers, usually the right one, is always larger than the entire body, [and] has been shaped by its house in such a way that it can close the selfsame door with it as if with a shield, and cover its entire body. The inner sides of the Pincers are sharply toothed and can pinch hard and hold what has been grasped. Most are of a smoky color or the color of soil, with a little purple mixed in, with smooth Pincers of uniform color, like those that live inside periwinkles.[3] Others, that reside in elongated whelks or in the *Turbinatis*,[4] have a longer body and longer Pincers, although some of these are smooth, while others are hairy or covered with brushes, and with blue spots on the Pincers, ugly to behold. Then there are some that resemble Purse crabs[5] so much that many believe that the Purse crab springs from them; which is amiss, as I pointed out before in the appropriate chapter.

The flesh-like tail of some is partially transparent, and they also carry their eggs under there; if one cuts it off and places it in the sun, it will sweat out an oyl. Their dwellings are in all manner of shells which have the form of Snails or Winkles, but most often in those, which we call *Cochleas Margariticas & Semilunares* in Book Two. Each one lives in its own shell for a length of time until it has grown large and can no longer hide its body in its old house, and then it has to find another one of the same kind or, at least, it should not differ too much, and be somewhat larger; then it leaves its old house and by twisting its tail knows how to get into the new one in such a way, and live in it, and go forth with it, as if it had grown in there.

When it happens that many are bickering over the same little house, and who will live in it, they become quarrelsome and fight until the weak one has yielded to the stronger and left it the spoils. Such a fight happens mostly among the so-called *Cumans*, which live in Winkles or a similar kind of Snail, because those who have been shaped in elongated houses have to stay in their old houses until they find the like, for since they do not have a large Pincer on the right side they cannot close off the round mouth of Winkles. These little quarrelsome creatures have caused me much grief, because when I laid out all kinds of handsome whelks to bleach, even on a high bench, they knew how to climb up there at night, and carried off the beautiful shells, leaving me their old coats to peep at. But some of these thieves I caught again, those that by chance had gotten a house too large for them, or had a long tail, or thorns which did not allow them to move on. If one wants to chase them out of their houses, hold a glowing coal against the tail end of the shell, when they feel the heat they jump out, though some would rather burn, to wit those which in my opinion have grown fast in there and have not loosened their tails as yet; otherwise, you can also hold them in your hand and wait until they stick their heads out on their own accord, then grasp them quickly with your fingers, hold them lightly and turn them gently, but do not get them out by force because they will be pulled from their tail which will then be left in the shell, and rot all over again.

Their name in Latin is *Cancelli*, in Malay *Cuman*, in Ambonese *Uman*. As one can read in *Hist. Antill. Rochefort. lib. I. cap. 4.* the French inhabitants of the *Caribbean* Islands have the nice name of *Soldiers* for them, because they do not have their own home but command one wherever they go.[6] In *Languedocq Testa Rondolet*[7] they are also called *Bernard l'Eremite*, because they wander like Hermits, and live in hired Cells.

One finds them on all kinds of flat beaches where the Sea casts up a variety of seawrack,[8] especially if there is some foodstuff or dregs among it. When the water rises they retreat to the nearest rocks, and when they see a man coming they jump down with a purling sound and hide in the sand so quickly that one can barely catch one out of a hundred. One cannot carry the captives in either hand or pocket, because when they feel warmth they soon creep out of their little house and pinch with their Pincers, holding on so stubbornly to what they did grab that they will often be rather torn to pieces than let go. When after many fine and dry days there is some rain, they will crawl at night into the houses and rooms of people, and make such a din, that if one is asleep one is forced to wake up thinking that there are billy goats in the room. The inhabitants [of Ambon] use them for other tricks as well, when they put them under somebody's pillow, or in a covered pot under the bed. They are unfit to eat, at least in these Islands, but in the aforesaid *Hist. Antill.* one can read that the Caribbeans eat them and also use them as a physic; because once pulled from their shell, killed, and placed in the Sun, they excrete an oil which can be profitably smeared on *Guttam Frigidam* or cold sores. One can use them also favorably to expel the hardness and wens of the body. In Amboina they have also tried to use the tail of these *Cumans*, ground and wrapped up like a plaster, to cure the painful stings of the fish *Ican Swangi*[9] and other poisonous fishes. Its origin is as much debated as was the aforementioned Purse Crab; some judge them to grow from the remains of the Winkles when they have been cooked and not very well scraped out, so that its hindermost curl is tossed out along with its house, and that therefore the hind parts of these *Cumans* match the shape of Snails. Others hold, with better reason, that they procreate themselves, since they have under their tail as many eggs as other Crabs. It may be that both opinions agree with the truth; to wit, that the first ones grow in the relics of the cast-off Snails, and then spawn other ones which have to look for houses similar to the ones their Parents lived in. I will remain of this opinion until Experience has taught me differently since I hold entirely that every shell brings forth its particular *Cuman*, as one can perceive from their various shapes.

Then again there are those who hold that certain kinds of *Cuman* become so large that they can no longer find fitting houses, and from then on go naked, and finally turn into Purse Crabs, which I do not consider possible at all, and therefore countered in the previous Chapter. Nor do they dwell only in Winkles. For, if they cannot get whelks, they will also crawl into the empty husks of the *Parrang* fruit, which is called *Faba* and *Castanea Marina;*[10] also in the empty claws of large Crabs, and the like, that they find suitable for their body. *Arist. Lib. 4. Hist. Animal* and *Aelianus Lib. 7. Cap. 31.*[11] call them *Carcinades*, and the latter says that they are procreated naked, without a shell, but once grown larger they seek the empty houses of the *Purpura* and *Turbine. Plin. Lib. 9. Cap. 31.*[12] seems to call them *Pinnotheres,*[13] as if to say Holster or Shell-hunter, and which are to be distinguished from the *Pinnoteres* which he describes in the later Chapter 42.

Cancelli e turbine verrucoso[14] is a scant finger long, the shell is a light brown with red stripes and white little dots; on the left side it has a short, fat, crooked Pincer, two long smooth feet, and behind them another smaller and shorter one; on the right side it has a small Pincer, and feet like the former ones; at the very end of the tail, between two little fins, it has a small, somewhat elongated Pincer resembling a bird's claw, with which it appears to hold on to the inside of the whelk; it cannot live long out of the water.

Rochefort, Hist. Antill Lib. I. Cap. 24., part three, has found a precious use for the oil and water of these *Cumans*, to wit, when someone has sat under the poisonous *Mancenilie Tree*[15] which causes his body to swell and raises wheals on the skin, or that he otherwise has been hurt by its pernicious milk, he then will rub the spot with the clear water that is found in the shells of these Soldiers, or with the oyl of selfsame drawn out by the Sun, and these aids will at once cast out the poison of this fiery liquid, and place the Sufferer out of danger.

The Writer did not provide an illustration of these Cancelli Cuman, *wherefore they were not included here: we could have added some here, but since there are very few differences, such as in size, we thought it best not to do so, and to refer the Reader to Plate* V. *letters* K. *and* L., *which are also like them, while we also do show them in their houses in the next Volume, concerning Shells; it should be said that the ones in* Jonston *and* Gesnerus *seem like Scorpions to us, and we are doubtful of those pictures.*

CHAPTER 23

Of the Pinnoteres, or Pinna Guard.

The *Pinnoteres*[1] is a small Shrimp, at most the length of one's little finger, but generally as long as two finger joints, soft and with a thin shell, usually a light or fiery red, speckled with small white dots, sometimes also of a light blue and partially transparent, like a dark crystal, or ice. It has three little legs on either side, and two other small ones by the Pincers near the mouth. The Pincers are very pointed and sharp in front, and curved like claws, with which it can pinch very nastily, and would rather have the Pincer pulled off than let go. It usually carries its tail under the belly, curved, like all Shrimp, and on it one will see subtle little pins, with which it covers its eggs. One will never see this little animal out in the open, wandering all by itself, but it always sojourns in two kinds of shells, to wit, in the *Pinna* or Holster Shells, and in the *Chama Squamata* or Nail Shells, both of which are described in the next book;[2] when either one of these shells have attained their full size, it will have only one of these little Shrimp living with it, especially the *Chama*, because I have not found them in the average *Pinnis.* And so this little Shrimp lives therein, never leaving the Master of its house as long as it lives, but once dead, the little Shrimp jumps out. Therefore, one will not eat any of these if its Companion cannot be found inside, and if it happens to get out, and gets lost, then the animal inside the shell will die as well. Its duty to these shells is that it has to be on guard inside them like a doorkeeper, and warn them when there is any prey or danger at hand. Because, speaking with *Plinius lib. 9 cap. 42.*, the *Pinna*, which is called *Chama* in this country, being a large lumpish beast, cannot see, and opens its shell, luring the little fishes inside, which will swim and play there without fear. Now the little Pinna Guard, seeing its house full of strange Guests, pinches its *Hospes* in the flesh with its sharp Pincers, and that one will immediately close its shell, killing the enclosed fishes, and use them for his food, giving a portion of it to his Comrade, wherefrom *Plin. loc. cit.*[3] wants to prove, that one can find some passions of mutual love and friendship even among the Water animals; and it is surely a wonder, that this devouring Shellfish maintains such a faithful friendship with this little animal, which can live there without harm; except for this provision, that every Shell will not tolerate more than one little Shrimp, from which one can conclude, that the young Shrimp, begotten by the older ones, will need to look for other *Pinnas* or *Chamas*, wherein they may be permitted to live.

Their name in Latin is *Pinnoteres*, that is, Pinna Guard, and *Pinnophylax*, and is to be distinguished from *Pinnotheres*, which was described in the previous chapter.

I found two kinds of these little Guards in the Letter Shells,[4] during the month of August in 1638,[5] the first was a small Shrimp as long as a fingernail, of a bright orange, yellow and half transparent, with thin, white little legs. The other was a little Crab, of the *Cancellis Anatum* kind,[6] barely as large as a fingernail, with a bossed and raised shield, pointed up front, gray and soft; already had Eggs under its tail: they were both quite lively, and were walking up and down. It seemed to us that the first one was the young of a Crayfish, and the other a young little Duck Crab, which, being so small, crawls into this shell, because one will not find all of them.

We also lack a picture of these Pinnoteres *or Pinna Guards, but they are like the ones from the previous chapter; see what we have noted there.*

CHAPTER 24

Of the Cancer Barbatus.

This is a small Crab, not as large as a Rixdollar, shaped not much different from the ordinary *Cancer Caninus*,[1] with a reasonably thick body, large Pincers, which have a tuft of black hair or a little brush up there by the claws, that looks like the hair of the *Papuas*, and which distinguishes it from the other Crabs, and gives them a singular appearance; they are mostly males, because they have their tail stuck to their bellies; they sojourn in freshwater Rivers, but only in some; because one will find them on Amboina in the Elephant River,[2] with small, short beards; but one will find the biggest and rarest on *Ceram*, in the *Hatihau* River, some distance in the mountains, and which have fairly large, and somewhat curled brushes on their Pincers. The *Alphorese*[3] of that Country use them for food.

Their name in Latin is *Cancer Barbatus;* in Malay, provisionally, *Cattam Gigi Bulu*,[4] since I do not know their native name.

These bearded Crabs come floating down the Elephant River every year in large troops, shortly after the *Wawo*[5] is passed, for only two or three days; thereafter one will not find them again for the entire year: but I have noticed, that they do not come down every Year. They are good to eat and the biggest Pincers are kept as curiosities, since they are scarcely less bearded than the ones from *Ceram.*

The Cancer Barbatus, *or Beard Crab, is shown on Plate No.* X *and No.* 2.[6]

CHAPTER 25

Of the Cancelli Anatum. Cattam Bebec.

There are three or four other kinds of little Crabs, which are called Duck Crabs, because Ducks seek them out and like to eat them.

I. *Cancellus Anatum Primus*,[1] is the size of a dubbeltje, with a roundish and bossed shield, smooth and of a light gray, shining like any small stone, round in back, and coming to a point up front by the head, with little legs and Pincers, which in back are notched and grainy.

II. *Cancellus Anatum Secundus*,[2] is even smaller, narrow and with a long body, to wit, only half the width of a finger long from head to tail, the width of a finger across, with thin legs and Pincers, which you can barely see. They walk very rapidly sideways, and as soon as they perceive anyone, they know how to hide themselves in the sand with a wondrous speed. Therefore, even when you see a large troop of them together, when you go up to them, they will all hide in the sand, where the Ducks, who are allowed to roam on the beach here, readily know how to get them out of there, and swallow them; which they can do all the easier, because neither kind of little Crabs has special Pincers, the way the *Cancer Vocans*[3] has, which is also small, but is not pursued as often, because it knows how to pinch.

III. *Cancellus Anatum Tertius*,[4] is a little larger than the smooth one, with a round bossed body, which comes to a point behind, and which is rough all over its body, as if it were covered with sand, the little legs are delicate and meager, but the claws are long, and thin, and are divided into sharp pincers.

IV. *Cancellus Anatum Quartus*,[5] is about as large as a finger nail, with an uneven and bossed back, as gray as ashes, it has on either side 4 narrow and slender little legs, easily a finger joint long, and of which it can raise the two back ones over its body, and defends itself from behind; the claws are somewhat shorter on the legs, and end in subtle Pincers; they walk on the sand like a foam, and if one comes too close, they will be gone in a thrice; they can raise their body on the legs, like a spider.

Their name in Latin is *Cancelli Anatum*, in Malay *Cattam Bebec.*

The first kind, that which is the real *Cattam Bebec*, comes in troops onto the sand when it is neap-tide, and bask themselves in the Sun, where they glisten quite fairly with their smooth shields, and light-red little legs; now in order to surprise them, one should approach them stealthily, scoop them up with a little net, or turn them head over heels with a broom, so they will not be able to crawl so rapidly into the sand, and then pick them up quickly with your hands; but you can also dig for them, and get them out of the sand; these are tossed to the Ducks, both young and old ones, who like to eat them, and lay many Eggs because of them, because they are harmless in their Stomachs, since the Cancelli close their little legs, and roll up together like balls, not harming or stabbing the Stomachs of the Birds, the way Shrimp do, whereof the Ducks often die, as they do also from the sharp little Crystals, which they find everywhere on the beaches of the small Rivers and which they then swallow.

Cancelli Anatum. *There are several kinds, see Plate No.* X., *letters* A., B., C., D.

CHAPTER 26

Of the Foetus Cancrorum, or Bloody and Fiery Sea-Red Sheets.

These are small creatures similar to Crabs and Shrimp which during the first rain-months when the Sun is in *Taurus*,[1] come floating down rivers in huge numbers, and stay in the Sea near these deltas, and are caught there with stretched-out pieces of cloth, whereafter they are crushed and pickled, which turns them into a thick, brown paste which is thinned with Lemon-juice and then used for dipping; this is called with a Chinese word *Kitsjap*, and the aforesaid little Shrimps are called *Oulewan* in Hitu.[2]

There is also another kind of small Shrimp which at certain times of the Year leave the Sea in such multitudes to be thrown onto the beach, that they cover it and make it look purple-red; one finds them the most on *Java*'s beaches, but I remember that during the West monsoon I saw the beach of *Laricque*[3] also red in like fashion, consisting of similar little Shrimp, the majority scarcely bigger than lice. These could not have come from rivers because at the place where I saw them there were none, so they must have come from the Sea; it also happens often that one encounters such floes in the Java Sea, when it seems that one is sailing through blood. The *Javanese* and *Chinese* living in *Grissek*,[4] know how to catch these little shrimp in great numbers, to crush and pickle them, and make a similar kind of brown paste, which they call *Bolatsjang*. From this they fashion dry cakes which they bring to other Countries to be made into Sauces; but if something pure is desired one must order it specially from these folks, because the common *Bolatsjang* is prepared somewhat slovenly and dirtily. In Sebalt de Weert's Journey[5] through the Straits of *Magellan*, p. 15. An. 1599 on the 12th of January, it is told how upon rounding the point of the *Rio de Platas*[6] he saw the Sea as red as blood, and when he drew water from it found that it was filled with little red worms which, if put in one's hand, jumped away like fleas. There are some who maintain that the Whales excrete them at certain times of the Year, but this is not certain.

In the year 1670 a ship sailing from Holland to the Indies[7] fell away further West than the Skipper had intended, round about the rocks called *Penedos de St. Paulo* situated by the *Brazilian* coast at 2 degrees latitude North: wandering around there they saw one night how the sea behind them was entirely white and fiery in the shape of a half moon some 3 miles long and wide, and whereto, according to them, the Ship was drifting, and that alarmed them greatly because they feared that it was the beforenamed dry salvages or rocks. They hastily launched a boat to examine the bottom, but it was not found, and the white and fiery field came readily toward the Ship, so that they found themselves in the midst of it. They scooped up some of the water, which during the day did not differ even slightly from other Seawater; but dragging some linen clothes behind the Ship they found them the next day to have beautiful blue spots, which could not be washed out. Since one of these items of clothing was a nightshirt, a certain young Lady put it on and received a spot on the ball of her thumb that turned into an *Escharam*,[8] but with no great pain, and she had a scar until the day she died. In the year 1675, a ship sailing in the neighborhood of *Manado*,[9] found the Sea at night equally fiery, becoming as bright as clear Moonlight. They sailed through it because they knew that they had nothing to fear on account of the bottom, yet they did not examine the water at all.

In the Indian Ocean between *Persia* and *Mallabar*[10] one frequently encounters large areas and fields which are entirely incarnadine, and so intense and vividly colored that one would think to paint Carmine and Purple with it; if one scoops up some of this water one finds it to be full of little bubbles, as if they were merely Sea gellies, but altogether clinging to each other with a sliminess that does not burn; and if one handles them but a little they readily dissolve into water: our Sailors call these the Red Sheets.

Their color of red differs from that of the Red Sea as it appears at certain times of the Year, whereof a certain Skipper told me that, when he was moored in the Red Sea by the city *Mocha*,[11] the Sea first became as white as milk and then turned a brownish red, but in such a manner as if it were only a reflection; and this lasted for 10 days. The Natives explained to him that this happens yearly around September, but they could not tell him the cause.

CHAPTER 27

Of the Pediculus Marinus. Fotok. Sea Louse.

This has the mixed shape of a Shrimp and a Louse, because it does not stretch its legs out very far, which is why it resembles a louse. It is half a finger long, and an inch wide. The body consists of a shell, which is a brownish yellow, with small white spots or eyes. It has 5 legs on either side, of which the front one is the longest, instead of a Pincer; the 4 other ones end in a small little flap, which just barely peeps out of the shell. The tail is narrow and comes to a point, and it is so long, that when it holds the same under its body, it almost reaches its head; on the bottom hollow is a channel, wherein it hides its eggs. One sees nothing more on its head than two short beards.

These little animals crawl on the sand, with their tails stretched out behind; but if one wants to catch them, they hide instantaneously in the sand, where one can easily dig them up. The ones in our Ambon Bay are small, scarcely the length of a finger joint, but they are larger in *Banda*, so they make more of them; because there they are cooked and eaten like shrimp.

I also only know them by the name, *Bandanese Fotok*. I called them *Pediculus Marinus*[1] in Latin, that is, Sea Louse.

Pediculus Marinus, *or the Sea Louse. See Plate No.* X. *No.* 3.

CHAPTER 28

Of the Eschinus Marinus Esculentus. Sea Apple. Seruakki.

The Sea Apple is just as common in the East Indies as in the Mediterranean Sea, is divided into various kinds, shapes and colors, of which the common ones are the most like those from Europe. *Aristoteles lib. 4. Hist. Animal. cap. 5.* proposes 5 kinds of *Echinus*

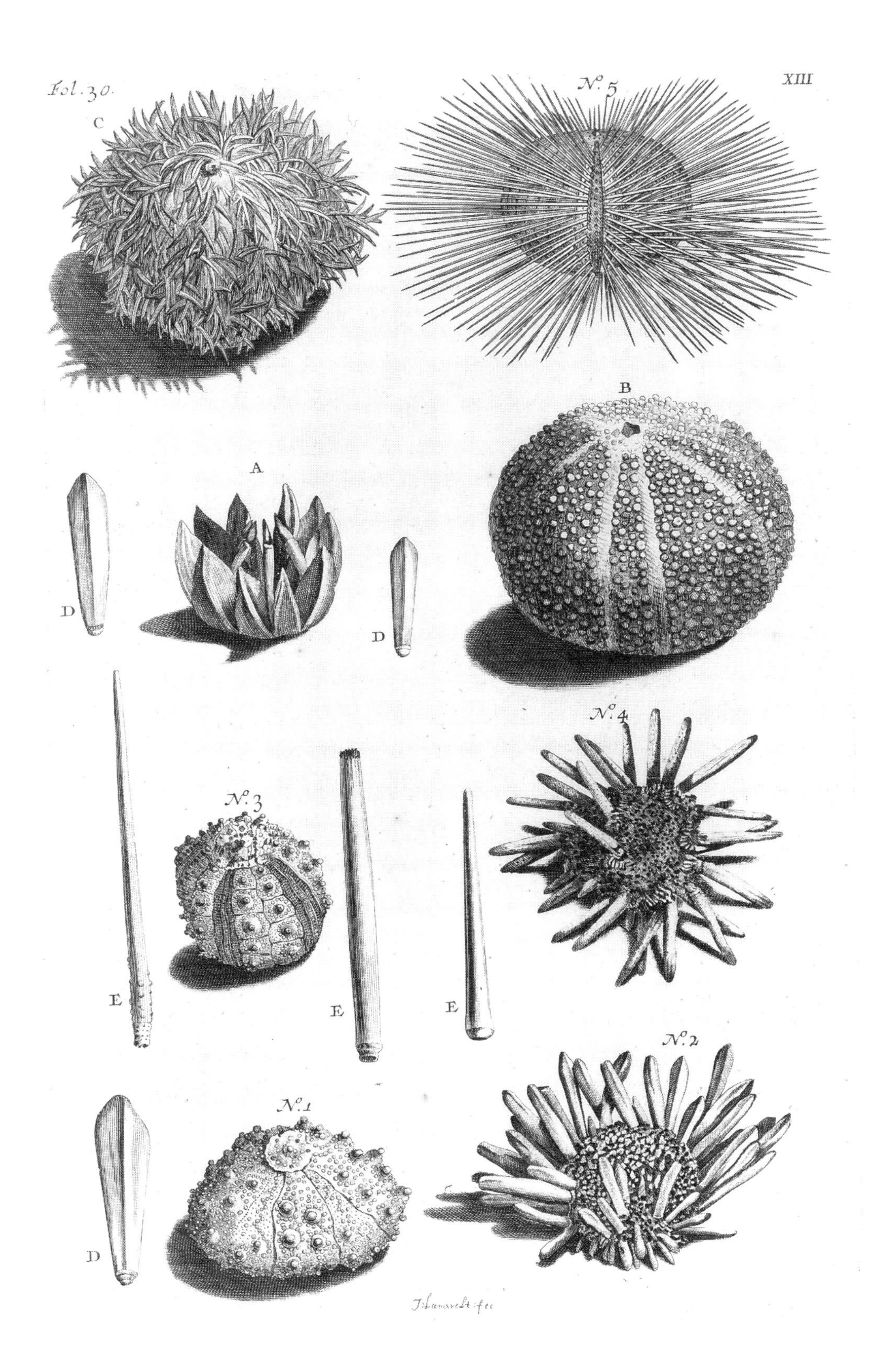
C
N°. 5
B
A
D
D
N°. 4
N°. 3
E
E
E
N°. 2
N°. 1
D
J. Lamsvelt fec

Marinus, to wit, 1. *Echinus vulgaris esculentus*, 2. *Spatagus*, 3. *Brissus*, 4. *Echinometra*, 5. *Genus quoddam pelagium longis ac praeduris spinis.*[1] *Dioscor. lib. 2. cap. 1.* does not give any kinds, and only speaks in general about their powers. I have found many other kinds in these Eastern Islands, not to mention the common ones, so that I am not sure where to place the ones mentioned above; however, I will also divide them into 5 genera, and give each one its particular Chapter. As 1. *Echinum vulgarem*, 2. *Echinometram digitatem*, 3. *Echinometram setosam*, 4. *Echinum sulcatum*, 5. *Echinum planum*,[2] all divided into various lesser kinds. But I deem it right to put their general description first, and to do so briefly, since they are known well enough from European books. Hence, all *Echini* or Sea Apples have a semi-round, vaulted, and breakable shell, in the shape of a Baker's oven, some round and even, some oblong, some pressed in and furrowed, others flat. One will see four wide shafts on the outside of the shell, coming together in the uppermost hole as if it were a *Centrum*. The shell itself is fashioned from five parts, which lock together with toothed or sawed seams, as in the cranium, but more regularly. The selfsame is also pierced with thousands of little holes, that are all lined up in rows, as if it were an artful piece of embroidery, some of them so small that one can barely see through them. Furthermore, the outer shell is covered with countless large and small warts, of which the large ones are regularly lined up within the aforementioned shafts, and on each wart a little foot, which in some resembles a blunt thorn, in others a finger, [and which] is attached to it with a fragile little fleece, and is in addition fastened to the flesh inside with a thin sinew[3] that goes through the little hole in the wart, and so everything, therefore, can move when these little animals need to be on the go, nor do they roll, as some would have it,[4] but they go straight ahead, unless they are bowled over by Sea waves, whereafter they right themselves again with their longest feet, which are at their backs. This perverse animal has its mouth towards the ground, and its arse up to the sky, but it is the shape of its body that demands this; because, since it is a flat globe, and since it must look for its food on the ground, its mouth must from necessity be down below, and the arse or the exit of the excrements over against it. The mouth is a round hole, at the bottom of the shell; in it 5 teeth stand over against each other, forming a cone, of a particular shape, with the tip pointed downwards, and flat on top. Every tooth is made from two striped and brittle little bones, loosely joined together, hollow inside with a little bone that runs through it, like a thorn, and which sticks out a little beyond the shell; these 5 points are really its teeth, which form the point of the cone, and behind it one sees some other curved little bones, which is where this crop (*ingluvies*)[5] divides into 5 stomachs, which lie against the shell, and which only contain some filthy brown water mixed with sand, and which close together again at the top, near the exit; these teeth hang there rather loosely, connected with thin and soft little skins, and if one takes the entire clump of little skins away, which becomes clean and dry, it looks altogether a great deal like an Emperor's Crown, provided one wraps it round with a thread, so they will not fall away from one another. One will find five other oblong little clumps between the aforementioned 5 stomachs, and they are of a yellow or whitish dry substance, and also lie against the shell, and these are called the Eggs, and look a great deal like roe, and this, in some kinds, is edible. There are also 5 ears attached to the shell near the edge of the Mouth, and through them run the bands, which join the crop to the five stomachs.

Their general name is: *Echinus Marinus*. In Dutch Sea Apple. In Ambonese *Seruakki*, on Hitu *Anay*, on Luhu *Sepalakke*, on Leytimor *Ulen huaa*, or Little Pot; but *Anay* is the name of the throat of a Weel, where the fishes dart into, and a Sea Apple, turned upside down, also looks a little like this. *Manuhoeloe est Echinus esculentus.*[6]

Now coming to the normal kind, which is described in this Chapter, we will find the following varieties.

I. *Echinus Esculentus,*[7] is the most common one, and has the largest globe, round all about, 5 and 6 fingers across, and 2 or 3 high, like a flat Baker's-oven, with a very thin and brittle shell, covered all over with short thorns, which are as long as a fingernail, as thick as a needle, the biggest ones always on top of the biggest warts, and all mobile; the color adjusts to the color of sand or ground where they are, to wit, usually a dirty white, sometimes a uniform color, other times covered all around with gray spots, which form 5 Circles; but the pieces of each Circle do not hang onto each other. These have the best Eggs, fit to eat, and they taste all right, like fish roe. One boils them in water, after carefully pressing the shell into pieces, and removing the five stomachs full of that brown blood, so it cannot corrupt the Eggs, which one then can eat separately. It also depends on what beaches one finds them, because one deems those the best, that dwell on clean sand, which has small stones mixed in with it, and is as white as possible. But any of those found on beaches that have black sand, mixed with mud, have a browner shell, and are more swollen around the Eggs, while they also have a more or less muddy taste; in a word, all those we include in this first kind, have a different shape, color and size, according to what the beach is like.

II. *Echinus Saxatilis,*[8] might be the *Brisius* of *Aristot.*, with a small and oblong shell, the size of a walnut, with a thicker and harder shell than the foregoing, a light-red or white color, with larger warts; the feet are longer and thicker than those of the foregoing, almost the length of a finger joint, stiff and sharp, so that one can barely grab it. They dwell in the dents and holes of coral-stone, where they sometimes grow so well that one cannot remove them, which is noteworthy of this animal; because if one wants to pull it away, it stiffens its thorns even more, so that they are everywhere in contact with the stone. They are not searched out for food, however, because the Eggs are somewhat bitter, but the shells are handsome and can be kept as curiosities, especially since they do not break easily.

III. *Echininus Niger,*[9] are also small, quite round, like a large Doublet button, of a blackish or earthen color, but their ten shafts are lighter, covered with very short and fine bristles, as if they were pieces of brushes. The shell would warrant keeping because of its beautiful shape, but it is so brittle, that one can hardly handle it. One finds them a lot around Ambon Bay, near *Ruma Tiga,*[10] where the beaches are rather stony and muddy; these too are not gathered for food. If one wants to keep the shells of these Sea Apples, one should put them on a board in the rain, until the feet fall off, then brush them with a soft brush, and let them dry in the Sun, after all the guts have fallen out. Then the right speculative thing to do is to look into these shells from the bottom, and one will notice how Nature has placed all the little holes in a perfect order, and in such a way that someone, who had never seen these animals alive, would otherwise only believe that they were artificial works of art.

IV. *Oculis Poliphemi,*[11] is a rare kind, that seldom sees the light of day since it dwells in the deep. It is round and bossed like half a cannon-ball, the size of two fists, without thorns or feet; covered with a dirty green slime, and has a strong smell of the sea, so that it seems it is rolled around, and is sometimes thrown onto the shore, and is caught in the dragnets. There is nothing useful about it, so it is immediately tossed back in again; one will find them in the Bight of Ambon. The Ambonese do not always cook the *Echinos,* but place them upside down on coals, and roast them like that; which, though it makes the meat harder, makes it also easier to get out. They make such a dainty from the Eggs, that they prefer it to chicken meat, a judgment with which a European would hardly agree. Yet *Dioscorid. lib. 2. cap. 1.* agrees with them, saying that they are good for the stomach, and that they expel water. He also says that it is good to smear the ashes made from their shells on all sorts of scabbiness, and for cleaning foul sores. He also tells us that they weigh themselves down with small stones when there is bad weather in the offing, so that they will stand more firmly, and

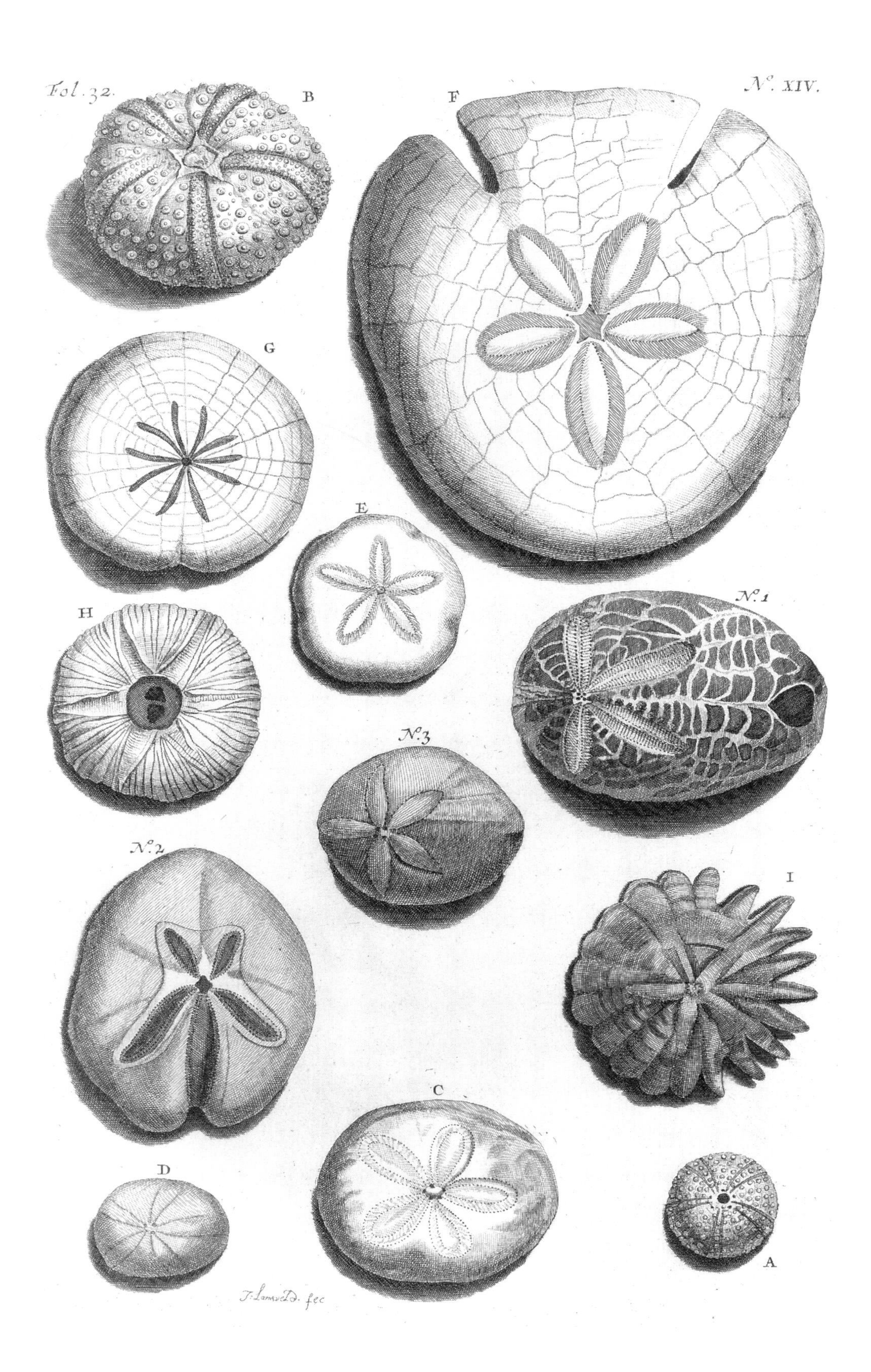

Fol. 32.
Nº. XIV.
B
F
G
E
H
Nº. 1
Nº. 3
Nº. 2
I
C
D
A

will not be tossed around by the waves;[12] but the Natives have nothing to say about this, nor have they ever seen such stones between those little feet, except that they are draped with moss, weeds, and other such plants that grow on the bottom of the Sea, and through which they might have to crawl. The Natives prepare them in this manner: They take a load of raw Sea Apples, break them into pieces, take the Eggs out; then put them in a green piece of *Bamboo* with *Sajor Songa* and *Sajor Pacu*,[13] making layers, and add a little salt and pepper; thus prepared, they cook all of it together.

Mr. Rumphius *provides us in this Chapter with a detailed story of the* Echinus Marinus Esculentus, *or Sea Apples, but no pictures, wherefore we deemed it necessary to add the same, as much as was possible. By the letter* A. *on Plate* No. XIII.[14] *we depicted the mouth of a Sea Apple, but opened, in order to see all its parts fittingly, because they are of a most unusual shape; letter* B. *shows the Apple, and letter* C. *the same with its Spines.*

The Echinus Saxatiles, *is on Plate* No. XIV. *letter* A. *but somewhat smaller, and we added a Sea Apple, which we call the Turkish Turban, which is very rare, and all of them were sent to us by Doctor* D'Acquet.

CHAPTER 29

Of the Echinometra Digitata Prema. Djari Laut.

I call *Echinometras*[1] all such Sea Apples, which are covered with long feet like fingers or arrows, and whereof we will describe the two most beautiful kinds in this chapter.

I. *Echinometra Digitata Prima Loblonga*,[2] may have a smaller shell than the common ones, but it is much thicker, harder, and hangs more firmly together, oblong, somewhat flat both on top and at the bottom, covered along the 10 shafts or rows with big warts as if with pearls, which are rather tenuous,[3] and innumerable smaller ones between them. Every big wart has a small hole in the top, and a circle around the foot. The mouth is round, also a little oblong, and also has five ears on the inner edge, pierced in the center; the teeth are like the common ones, but make for a blunter cone, also with the point looking down. The color of the shell is whitish, veering towards a light red, and the warts, once cleaned, shine like pearls, which lends this shell a great beauty. There is a foot on every wart, the shape and length of a finger, but not as thick, and there are smaller feet on the smaller [warts]. The big fingers are on the back and on the sides, narrow and round on the underside, where one will see a little hole, that fits the warts, fastened all around to the shell with a little skin; and that is joined to the other flesh with a sinew, that goes through the little hole of the wart, and which aids the animal in moving its feet. These fingers are triangular on top, and somewhat pointed, but the old ones become roundish. The color is light brown, which pales a little when they are dried, with two or three white bands in the center. These shells do resemble stones somewhat, and when struck against one another, sound like glass,[4] but they are dry and brittle inside, and can be scraped like hard chalk. The other smaller fingers are the length of one or two finger joints, wide on top, and somewhat flat like a spatula;[5] usually light brown, and of a denser substance, which is why they sink like a stone in water. These fingers have a special and wondrous attribute, to wit, that, when thrown into water, some, that is most of them, will sink, and lie flat, but some stand up straight as if dancing in the water, and others stand at an angle as when you are couching a pike.[6] I presumed for a long time that they assume the same stance, as does the living animal, but closer examination showed that such is not the case with all fingers. I cannot determine the true cause of this, but I judge that, if we were supplied with magnifying glasses, we could examine it more closely, and would undoubtedly find that the ones that lie down are of a thicker and denser substance, but the upright ones, that is, the fingers that stand up straight and float, are more tenuous and spongier in their upper parts; they have, therefore, more air within them, and this causes the lightest portion to rise up, and the heaviest part will sink to the bottom. When these fingers float for a long time on the beach, or between the rocks, they will eventually become heavy, and change entirely into stone, wherefore many suspect (and not entirely unlikely), that it is a special kind of Stone sea growth.

Their name in Latin is *Echinometra Digitata Prima*, and the fingers chiefly *Digiti Marini*,[7] in Malay *Djari Laut;* it has no special name for the Ambonese, except on Hitu, *Sulopay;* on Hila, *Hatu Opay*, and *Hatoyhu;* on Luhu, *Sapalakke.*

They occur the most on *Hitu*'s shore, on such beaches that have large stones as well as sand; the Natives use them for food like the earlier ones, to wit, the ovaries, which one will find inside, after carefully cleaning them of the meat and slime that clings to them. They grind the stone fingers a little at the tip, so that the sharp edges are gone, and hang them around their children's necks, instead of teethers,[8] saying, that they make for easier teething. Some of our Surgeons have experienced that, when they are pulverized and ingested, they have the same effect as *Oculi Canceri*,[9] to wit, they strongly expel Water, Bladder Stones, and Women's menses. I learned from the people of *Ternate*, and from the Inhabitants of the *Xulasse* Islands,[10] that they mix these fingers with those things, which they take against any poison they might have swallowed, and I often found the very same in the little Bundles which they carry with them everywhere at Sea or when going to War, and which usually consist of 7 pieces. As number 1 a black, 2 a white knotted, 3 a gray, 4 a red Calbahaar, 5 a *denticuli Elephantis*, described hereafter under *Solenes*, 6 these stone fingers, 7 an unknown little root, but in place of the same some other bundles had a small piece of *Solen arenarius*, or Sandpipe.[11] Others felt that the Secret consisted of their using these 7 things not only against ingested poison, but also against a sorcerer's enchantment, of which they are most afraid, especially such roguish tricks, as to deprive each other of their manly powers.

II. *Echinometra Digitata Secunda l. notunda*,[12] is a little larger than the foregoing, usually the size of a fist, truly round, and pressed down somewhat like a small cheese; [it is] the most artful and most decorative of Sea Apples, because its shell is covered with beautiful, large warts, shiny like pearls, each one with a little hole passing through it, and a small circle all around: these warts are in 10 rows of 5 pairs, and in each row, from the upper to the lower hole or mouth, 5 or 6, and in between another two or three half rows, each one of three or four warts; each wart has a finger on top, easily the length of one's middle finger, and the thickness of a Goose Quill, round and striped lengthwise, a little coarse, dark gray, with a flat little star on top that has many rays, its substance harder and stonier than the foregoing, wherefore they all sink in water. Between the rows are other, smaller ones, but they are somewhat tenuous, and mostly round on top. There is also a special kind of these, which have fingers that are shorter and thinner, also with round heads. The third kind has notched chaplets round the fingers, which are not more than 1-½ inch long. The fourth kind has long and slender fingers, coming to a blunt point at the top, white, light, and of a chalky substance. The five seams that hold the shell together, do not run straight, but are rather twisted and like a snake, and they have narrow notches, with which they hang onto each other, and the dried ones, therefore, fall easily apart, unless one ties them together with thread. The upper hole, or the exit for the Excrement, has 5 lesser holes around it, and in that part the shell is made up of many small bits, which easily fall apart, wherefore the entire *Tholus*,[13] or Keystone of this vault, usually collapses entirely, leaving a large hole, and which is scarcely to be prevented. In the mouth one will not see more than five ears at the edges either, but not with very many holes, that hold the bands which hold onto the Crown; the latter is like the foregoing; see below.

Their name in Latin is *Echinometra Digitata Secunda.* In particular *Cidaris Mauri;*[14] in Dutch Moorish Turbant because the cleaned shell looks a lot like an Arabian or Mongolian Turbant, that is covered with pearls or Precious Stones; the Ambonese do not have a special name for it.

If in a perfect state, it has 60 fingers in general, in 10 rows, among them 28 or 30 large ones, and 10 very small ones. Each large finger has 20 small pins around it, standing closely around the large one, in order to keep it upright. As was said already, it has 5 seams, which lock together in pairs, finely sawed like a *sutura oranii*,[15] but each part can be divided in two again, which stick a little better to each other.

One finds it along with the first one, on the coast of Hitu, and also in the Bay of Ambon, where there are beaches of coarse stones, but neither of these is as numerous anywhere as those from the foregoing Chapter. They are seldom eaten, nor do the fingers have a known use, except that the women use them to press star-shapes in their gingerbread; but a cleaned Turbant is from every point of view worthy to be kept among the most beautiful of Curiosities, and if it were to fall apart, one can put it back together again with some thin glue, which will then hold it together more firmly than it did before.

Siamese Sea Apples are one of these two kinds, but larger and round, of thicker substance, and the grains are [arrayed] more neatly. They know how to cut them out near the mouth, to lacquer them on the inside, or to cover them with little, thin silver plates, from which they then fashion very nice boxes or drinking cups, that are quite durable.

The Sea Styles[16] already described are also used against cholera,[17] if one ingests the same after it has been ground in water, and with a little ginger, just to give it a little taste.

The second or round kind, has several smaller kinds, of which the first is also round and in the shape of a cheese, covered with long and thin darts, easily the length of a finger, and as thick as a sail needle, the hinder part is coarse and covered with little knobs, the front half rough or sharp, ending in a blunt point like a Syringa.[18] Other little darts are smoother, thinner, and end in a sharp point like a needle. Another kind is also round, flatter than the foregoing, straight like a Spindle-whirle,[19] its feet as long as a finger joint, and jagged, the small crowns set in circles one on top of the other, a pleasure to behold. One of the first kinds, caught during one of the rainy months, August, had about 60 fingers, both large and small ones, which all sank in the water and laid down flat; this needs to be examined, in order to find out if it was the abundant Rain that caused this, since afterwards many of the dried and burst ones[20] began to straighten up again; and the same thing happened with the fingers of the first and second kind, which were found in the stomachs of a large Pufferfish.[21] The little crowns in the mouth of a Sea Apple, consist of 25 to 30 little bones, which can separate, of which the 10 biggest ones make up the five teeth.

In the year 1689, on the 19th of February, a Northwestern storm that blew for 8 days, tossed a multitude of Moorish Turbants on shore, which otherwise are seldom seen in the Bay of Ambon. When a finger is broken, the remaining little stump grows fast onto its wart, and does not move any longer. Some fingers are worn down and become chalky, and one can use them to write on a slate, like a style.

The Echinometra Digitata Prima, *is depicted on Plate No. XIII. Nos.* 1. *and* 2. *is the same, with its pins or feet, their various kinds shown by the letters* D,D,D.
The Echinometra Digitata Secunda, *or the second kind, which is even more wondrous than the foregoing, is depicted on the same Plate with Nos.* 3. *and* 4. *is the same with their pins or feet, whereof three kinds are indicated with the letters* E,E,E.

CHAPTER 30

Of the Echinometra Setosa. Bulu Babi.

This Sea Apple has a smaller shell than the two previous ones, otherwise it has the same shape as *Echinometra Secunda*, thinner and flatter in terms of its shell, and with small warts, among which are many that are even smaller. The biggest shells are the size of a fist or a small cheese,[1] covered all over with long darts, from 6 to 7 inches long, the thickness of a Sewing needle, but at the tip as pointed as a bristle. Others among these are shorter, and as thin as little hairs. They are a black brown, some a dirty black, straight, stiff, but so brittle, that, as soon as one bumps against them, they will break off: they are notched transversely on the sides from top to bottom like a thin screw, and feel rough to the touch. It does not have any special Eggs, and is considered unfit to eat.

Their name in Latin is *Echinometra Setosa*,[2] or *Echinus Setosus*, in Malay *Bulu Babi*,[3] in Ambonese *Hahuru* and *Hahulu*, all because of the resemblance to Pig's bristles: on Leytimor it is also called *Maccariwan*, that is, something that makes one laugh if it touches you; but more so for the spectators, than for the victim.

They can be found on flat sandy beaches, with few stones, sometimes together in troops, particularly when, at ebb tide, some shallow puddles are left behind. This villainous vermin endangers the beaches everywhere, especially for those who look for food, at ebb tide, and walk about one foot deep in the water, mostly at night, and then bump into this animal unawares, or step on it, thinking it is a stone. For as soon as you touch these sharp darts, or even when you begin to place your foot on it, you will get the points of the darts in your skin, which then break off, causing great pain. It happened to me as well, when, strolling by Moonlight in the Sea, looking for Curiosities among the rocks, and even though I was rather suspicious of this vermin, and consequently felt around very carefully under the rocks, I still got two or three of these darts in my hand, almost before I felt that I was touching something, because they are stretched out all around the animal like a Hedgehog or a Porcupine after it has raised its bristles. One is cured from this stinging by gently tapping on the injured member, so that it is partially deadened,[4] and the pieces of the dart [can then be] completely crushed; then you warm the member over a fire, as hot as you can stand it; or, if you have some, smear a poultice of gray *Crusta Accarbary*[5] on it, which numbs the pain, yet one still needs to tap the member so you can crush the dart. The shells are not saved, because they are too brittle: one sometimes keeps the thickest darts, that have the thickness of a Sail Needle, and which are nicely jagged at the bottom with chaplets, but the longest point one should break off, and these are the Sea Needles. During the day it can lay its darts down, under water when the Sun shines; it also does this when it is approached by a hand, that has been smeared with crushed Ginger, so that one can handle it without detriment, just as Ginger will also heal the pain which one gets from the stings.

The second kind[6] has a large and flat shell, supplied with short quills, that do not sting as fiercely as the foregoing, and their quills are much shorter, and can be kept as Curiosities.

Echinometra Setosa. *See Plate* No. XIII. No. 5., *the quills of which are very thin and fine, wherefrom we also call them Sea Needles.*

CHAPTER 31

Of the Echinus Sulcatus. Skulls.

This kind of Sea Apple differs somewhat from the normal shape, because, though this *Echinus* is also roundish and oval, it has the mixed shape of a Sea Star. At least, on its back one will see the division of a star, with 5 uneven rays, which are full of holes, and on them many short little feet [that are] much spinier, like small bristles. The shell is whitish, but gray around the Star, thin and very breakable, so that one can hardly handle them; on one corner of the sides it has a round hole, that would fit one's little finger, and which is its mouth. Near the other corner, on the lower belly, they have another hole, sideways, of which one of the lips curves up to the belly somewhat, and this is the passage for the excrements; and the three longest rays of the star on the back are made by three furrows. It is seldom found, does not have any particular name, so that I call it provisionally *Echinum Sulcatum;* the Dutch call them Skulls.[1] After its shell has been cleansed in the rain and then dried, it is kept among the Curiosities.

A second variety of this has a somewhat smaller shell, roundish, also oval, and not more than the thickness of a finger thick, without a star or furrows on its back, but with a normal mouth on its belly below. The shell is covered with many tiny little grains, and on them small, firm little spines like pieces of bristles, gray and shiny like *Amianthus*,[2] which it can raise when in the water, but which it lowers when it is pulled out of the sea. If one steps on it or grabs hold of it, it will cause a slight burning or itching in the hand, and make the member red without any pain. The shell has a round hole on one side, which is its mouth, so that the foregoing must be its exit. One finds them a great deal in the Bight of Ambon, in the *Weipitu* River.[3]

Echinus Sulcatus, *called* Skull *by us. See Plate No.* XIV. *letter* C., *a second kind at the letter* D. *which, because of their shape, are also called* Snake Eggs. *We added a marbled one, indicated by* No. 1. No. 2 *is a white one, though slightly grayish. All of these are from Doctor* D'Acquet's *Cabinet, and* No. 3. *was added by Mr.* De Jong, *and all of them we deemed necessary to add here, because of their variety.*[4]

CHAPTER 32

Of the Echinus Planus. Pancakes and Sea Reales.

This *Echinus* also cannot make up its mind whether it is a Sea Apple or a Sea Star, so that it can be put among either, and I have noted three varieties of it. 1. The first one,[1] which one calls Sea Reales,[2] is round, about a hand wide in *Diameter*, thin on the edge, as thick as a quill in the center, whitish or a light gray. It shows a Star of five rays on its back, which is collapsed a little, and marked with many notches: in the middle or *Centrum* there is a barely perceptible little hole, and another little round hole on its belly, which is the mouth, and from this there are also five furrows that spread out in a circle. It is covered with few short little spines, most of them on the ray of the star, and with even smaller ones on its stomach, so that it does not move in a particular way in the Sea, but lets the waves carry it from one place to the next.

There is a large and rare variety of this kind,[3] very thin and flat, 5 and 6 inches long and wide, about a quill thick in the center, and on its back one will see a Star with 5 rays, about the length of a joint, of which the front one is the longest, while the two back ones run into two notches or inlets, that are as much as two finger joints deep, as if they had cut two oars out of them, and they are open in the old ones, but closed in the young ones;[4] the young are purple, the old are an ashen gray mixed with dark purple. Their little feet are short little hairs or little spines, wherewith they can move quite rapidly under water, but once out of the water, they lie still in the same place. One will find them on the flat sand beaches, near the *Weynitu* River[5] on Leytimor, in the months of November and December. One might also, though rarely, find some that have five slits cut out of them,[6] the way the foregoing has two; but these notches have grown together at the edges.

The second kind has a thicker shell,[7] round though angular, with two longer rays, almost resembling a small buckler.[8] It also has a Star on its back, and a round mouth on its belly, and is covered all over with short, greenish little spines. The shell is whiter and firmer than the foregoing; it is pressed down a bit on the belly with five furrows, while it is slightly raised on the back, where the rays are full of little holes.

The third kind is the Sea Schelling,[9] well-nigh made the same way, the size of a Dutch Schelling,[10] the thickness of a quill, usually round or slightly angular, as if it also wanted to resemble a buckler. The belly is pressed in, and it is always thicker on the edges, with a star on its back, and covered all over with similar short little spines. One finds them in large numbers on the flat beaches of the *Uliasser* Islands, they have no use, except that one keeps the most beautiful ones as Curiosities, after one has first put them in the rain,[11] until all the little spines have fallen off; the last two kinds have a harder and more durable shell, but the first one is just as beautiful and just as rare.

Here also belong some small Stones,[12] which are found on the *Uliasser* Islands among the rocks in the Sea, and they are the size of a stiver, but with a small corner sticking out on one side, that is pierced by a hole, as if they were going to be used for [bead] curtains.[13] The under side is smooth and even, but the upper part is hollow, full of sharp rays that form a Star, like the buds[14] of some coral trees. It has no other sign of life, except for a covering of slimy flesh, that resembles that of a Sea gelly, which soon rots away if it is put in the air.

The first kind, the true Pancakes, are found to be 6 or 7 fingers wide, most of them round, and they have on one side two bights, as if little pieces had been cut from them, resembling somewhat the upper leather of a shoe, after it has been cut out by the Cobbler; I also found some with 5 of these bights, but one sees those rarely.

Echinus Planus, *called* Pancakes *by us, of which the Author provides us with these three drawings, which are depicted on Plate No.* XIV. *by the letter* E. *which are also called* Sea Reales. *The letter* F. *shows what is really the big* Pancake, *of which one will also find several smaller varieties, and an example can be found at the letter* G. *We added two more from Doctor* D'Acquet's *Cabinet, both very rare, of which one is shown by the letter* H., *being stony and hard; and the other by the letter* I., *which has protruding points around half of its circle, wherefore it is called the* Sunbeamed Pancake.[15]

CHAPTER 33

Of the Limax Marina. Sea Snail.

I found the following varieties of *Limax Marina.*[1] I. The first[2] one is like a true Road snail, roundishly oblong or oval, four finger widths long, and two wide, flat on the bottom, and half round on top. The back is covered with blackish and soft thorns, while under them lie 7 or 8 half circles like nails, or like the lames of a Coat of Mail,[3] black, of a horny substance, joined closely together, but in such a way, that the animal can bend them at will. From below it is completely like the Shell *Lopas*, or Stone-sticker,[4] to wit, it has a tough, yellow, slimy flesh, surrounded by a wrinkled edge, that is of a lighter color. It uses this to cling to rocks that are found in the Sea on uninhabited islands, especially the dry rice cape called *Siël*,[5] being the most southern point of little *Ceram.* One has to use force to pull them off those rocks, and then they curl up like a Hedgehog. It is called in Ambonese *Kokohot*, on Luhu *Talluul*, and are cooked and eaten by the Natives.

II. *Limax Marina Verrucosa*,[6] also has the shape of a crawling Road snail, but it is thicker and more bossed, its back covered with big warts, without a shield or thorn. One also finds these on rocks in the Sea; around the aforementioned cape, or on lonesome beaches that have large stones. It stays in one place for a long time, like the *Lopades*, which is why the spot where they cling to the rocks, will be bare and smooth; one will also find them hanging from the underside of rocks; even where they are covered with sand, like the little Donkeys (*Azelli*)[7] in Wine Cellars below the stones. Two long horns stick out from the front of its head, with which it seeks its way when it is crawling, as do all Road snails. In Ambonese it is called *Ulayl* and *Ulael. Tylos*,[8] in *Plinius*, is a kind of *Azellus*, and is called such because of the loathsome, tough Flesh (*Calleuze*),[9] and he might well be referring to one of the *Limaces*, or Sea Snails.

III. *Limax Tertia*,[10] is only a clump, more round than oblong, with a yellowish tough flesh, having a greasy smell, like the foregoing, and has nothing on its back but a small white bone, somewhat larger than a dubbeltje, with a protruding curved hook, looking almost like a ham.

It grows on the flat beaches in front of Victoria Castle, and is good to eat, but it is more esteemed by the *Panegeijers*,[11] than by our Ambonese.

There are two other varieties, mostly consisting of a clump of flesh, of which the first one carries a small round shield, like a flat saucer.

The other has an oblong little shield, almost square, like an Ambonese shield, half a finger long, and a small finger wide. Both are also kept as Curiosities.

Limax Marina, *or the* Sea Louse, *is depicted on Plate No.* X. *No.* 4. *and the* Limax Tertia, *No.* 5. *We have no pictures of the other ones; but No.* 6. *was added here, because it also belongs to that genus.*

CHAPTER 34

Of the Stella Marina. Bintang Laut.

Of the *Stella Marina*[1] I noted the following kinds.

I. The usual kind[2] has five feet, in the shape of a Star with five rays, because one usually calls the rays the feet, but one should say branches, because each branch is four or five inches long, round, and as thick as a thumb, smooth and bright blue on top, otherwise covered with low warts, whitish on the bottom, where one will also note that every branch

Fol. 40.
N°. XV.
F
B
D
C
E
A

has an opening or slit lengthwise, but closed tight, having otherwise no noticeable mouth. If one turns it over, these slits display countless tiny little feet, which the animal can stretch out or pull in. If it sticks all of them out at the same time, it almost looks like a millepede. One will also find that there are those with 6 branches, to wit, two small and three big ones [*sic*]; but these are smaller and grayer. They move slowly under water, where one can see them lie, a beautiful blue, on the bottom, and sometimes one suddenly becomes aware that they are hiding under the big rocks, which they do not accomplish with crawling, but by floating in the sea. Inside is a kind of hard, though watery flesh; they do not have different limbs, and smell strongly of the sea; and if one handles them roughly, such will cause some itching in one's hand.

Its name in Latin is generally *Stella Marina.* In Dutch Five-foot and Sea-star; in Malay *Bintang Laut;*[3] in Ambonese, on Hitu, *Sasanna:* its general name is *Sannawaru.*

II. *Stella Marina Minor,*[4] is just as large as a flat hand; the branches are flat, granulated on the back, light gray, or whitish, otherwise fashioned the same way as the foregoing; it is found in quiet bays and on flat beaches, such as in the Bight of Ambon, near cape *Martyn Fonso.*[5]

III. *Stella Marina Quindecim Radiorum,*[6] is found very rarely; it is 4 and 5 inches wide, divided all around into 12 or 14 branches, which are not all of the same length, each one has a joint in the center that it can bend, about a finger-joint long, with a russet shell, and covered with sharp spines, about the length of a finger nail, also russet as if it were a Sea Apple; on the bottom of the branches one will see a slit, and in it exceptionally tiny little feet. It dwells in the very depth of the Sea, where it is full of rough stones, such as the beach at *Laricque,*[7] where it is called *Hulupana.* If one is hurt by its spines, it will cause a very bad burning and great pain, which is why it is left unmolested. One can, however, dry it quickly and keep it, without its spines falling off.

IV. *Stella Marina Quarta,*[8] is the largest of all, 8 and 9 inches wide, also divided into 5 branches, which are flat on the bottom; on top they have a raised and pointed back, of a russet color, that is covered with tall black warts, hard and pointed like blunt thorns; on the bottom of the branches one can distinctly see the long slits, which close with sawed edges as does the [human] skull, where it sticks out innumerable little legs, soft, pliable, and with a little knob in front like a Snail, with which it not only moves, but which it also uses to look for food, so that these five tears serve as its five mouths. If one looks at this animal from above, it looks a great deal like a hard, baked Pasty[9] with black burned knobbles and edges. One will detect no life in this animal, except for its outstretched little feet, because the branches are stiff, and one will not see it make any movement on the ground, even if one touches it. If one wants to dry it, one should first place it in fresh and warm water, until the saltishness has been drawn from it, and dry it in the hot Sun or in smoke, and one can keep it then for a long time. But with rainy weather, or when moist air gets to it, it will easily fall apart.

Its name in Latin is *Stella Marina Quarta & Artocreas Marinum:*[10] In Dutch: Sea Pasties; in Ambonese *Sasanna,* like above.

One will find them often on *Ceram*'s northern coast by *Assahudi,* and on *Bonoa;* also in the inner Bay of Ambon.

V. *Stella Marina Quinta & Scolopendroides,*[11] is the smallest but liveliest of all the Sea stars, with a hideous shape, since its Body is small and flat like a Spider's, on top an ashen gray, smooth, and divided by five furrows, on top only covered by a little skin, but like a shell on the bottom, roundish but pentagonal. From this come five long and narrow branches, made up of nothing but vertebrae, which can bend to all sides, as if they were five long worms, furnished on the edges with two rows of little tiny feet, which make it look entirely like a millepede, so that one would think it were one, if one did not see the body. One does not see slits on the underside of the branches, as with the foregoing; only five short slits start from the mouth. The animal can move these branches quickly to all sides, as the Sea Cat can do with its arms, while it can also move rapidly with them, throwing the same with many curves forward, and then pulling the body after it. If one touches a branch, the animal will crawl away, and leave a piece of it in your hand; it is also wondrous to behold, how these branches, when cut into many pieces, will all move for a long time afterwards, like the chopped-off tails of Lizards. But if the entire animal begins to die, it will curl its branches over its head in a ball, and will die that way, all rolled up; there is nothing fleshy in the body, only a black blood, and is therefore not fit to eat.

Its name is *Stella Marina Scolopendroides;* in Ambonese, on Hitu, *Sanna Waru manuhulu;* something that resembles a Bird's Plumage, wherewith they mean the thin little legs on the branches.

One will find them on stony beaches, where one can hardly lift a stone without finding some under it, though they know how to hide very quickly and crawl away; one will see almost nothing but rays or branches, so that you would suppose it was a bunch of millepedes, which scares off anyone from touching them.

None of the Sea stars are used for food, except that I saw some Natives from the South-Easter Islands, using the first and fourth variety as food, but I did not experience how they prepared it; otherwise they make Bait or *Ompan* from it, which they put in the Weels in order to lure the Fish inside; to do this they take all sorts of Sea stars, roast them on coals, grind them up, and then mix them with other strong smelling things, and tie that in the belly of the Weels. One can read in *Hist. Antill.* that people noticed how the Sea stars in the *Caribbean* Islands, when they see thunder storms approaching, grab hold of many small stones with their little legs, looking to weigh themselves down, or hold themselves down as if with anchors, so they will not be tossed back and forth by the Waves.[12] The people on *Huamohel*[13] prepare them for food as follows: they take some of the first or fourth variety, cut them up into pieces, and squeeze the black blood out, then they cook them in water with some sour leaves, like *Condong* or *Tamaryn,*[14] and leave them like that for a couple of days; then they scrape off the outer, thick or coarse skin, cut them into smaller pieces, and cook them once again with *Santang Calappa;*[15] whereafter these people eat them.

Stella Marina, *or* Sea stars, *has many varieties, of which we did not obtain a picture, except for those on plate No.* XV. *depicted by the letter* A. *which, because of their shape, are also called* Sea Pasties, *the* Stella Marina Scolopendroides, *is indicated by the letter* B., *and is set round about with thorns, and letter* C. *is another kind, but entirely smooth. Because of its rarity, we added one with four rays, drawn at letter* D. *and one with five rays by the letter* E. *followed by one with nine points at letter* F., *and which were all sent to us by Doctor* D'Acquet, *and which all are 3 to 4 times larger than shown here.*

CHAPTER 35

Of the Caput Medusa. Bulu Aijam.

This monstrous Sea animal also belongs to the Sea Stars, dreadful to behold, and not easily touched by those who have never seen one; whereof I will describe two kinds.

I. The first one[1] has the Body or the head of a Spider, somewhat flatly round, and somewhat like a pentagon, about an inch wide; the shell is harder than that of the Sea Apples,

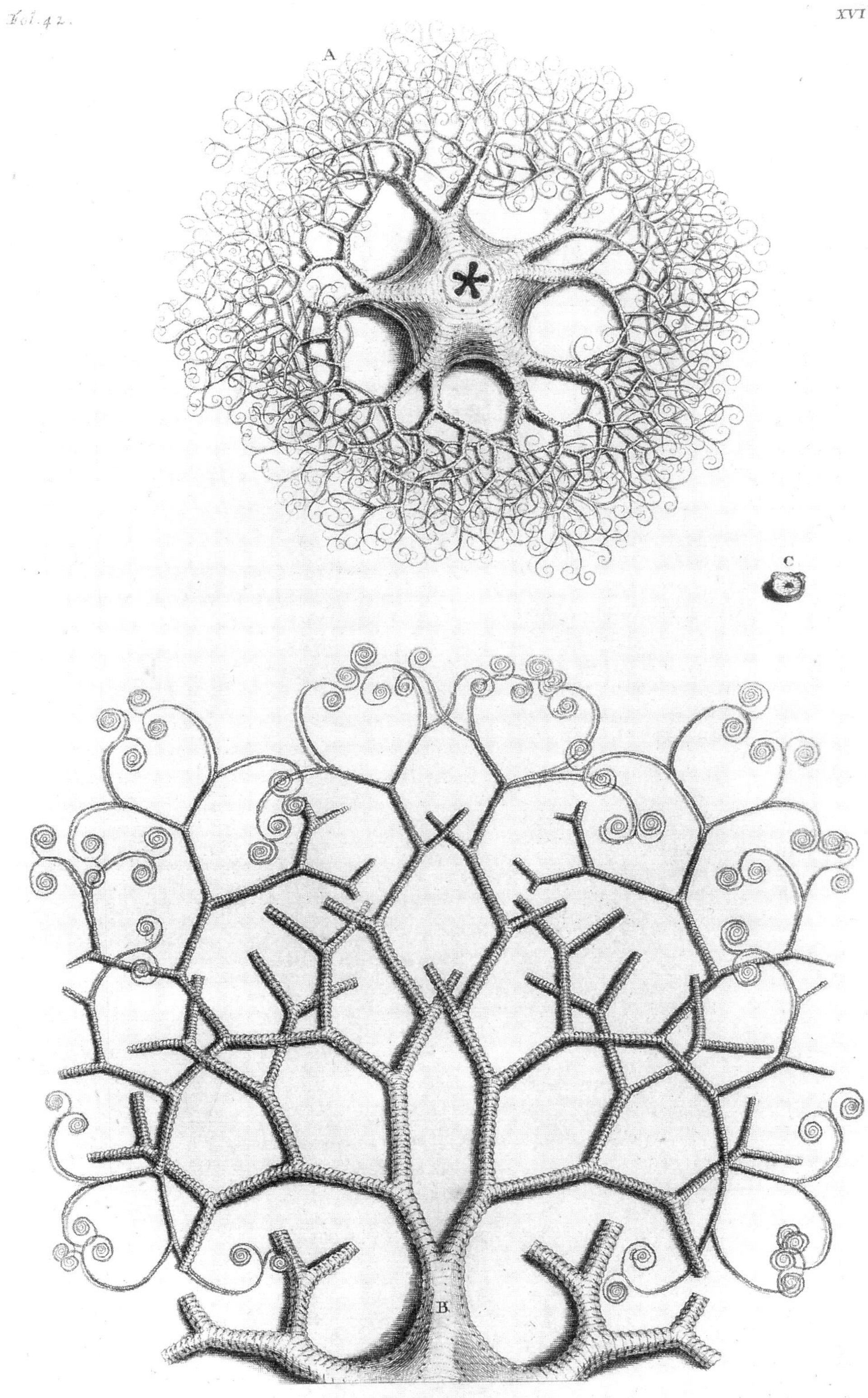
Vol. 42.
XVI.
A
C
B

covered with a soft skin on top, and on the underside it has an oblong slit, which is the mouth, furnished with many feet. This Head divides itself first into 5 pairs of 10 single Branches, of which every two have the same origin, and which all conjoined make up the aforementioned head, so that it really does not have a particular Body. Every branch divides into 3 others, and each one of those into 4 or 5 thinner ones, the entire branch being easily the length of a hand. All these branches or rays have been put together from innumerable stony Vertebrae, capable of movement to all sides, like the beards of a Sea Cat. On the underside one sees, in two rows, countless tiny little feet, finely notched, with a little yellow bud up front, as if they were the small threads of a flower, wherewith this Beast moves, and is of such a ghastly appearance, that one would hold it to be a clump of millepedes or tiny little Snakes. The normal color of both animals is russet, but one will also find coal-black, green, gray and yellow ones; what is inside cannot be easily recognized as parts of the gut, for it is all entangled, and mostly resembles the yellow meat of Mussels. Under water it resembles a stretched-out flower, and if one lifts it, it lets its branched feet hang down in a jumble, but if pulled out of the water, it closes its branches upwards, slinging them around the hand of whoever grabs it by its head, and who will be quite startled, thinking he took hold of some dead Sea plant, which then suddenly wraps itself around his hands. It also dies in that fashion, with all the branches closing upwards into a ball, looking quite like a head of cabbage (*Brassica*)[2] that wants to close up; but it can not be kept for very long, because it is so brittle, and is just some Vertebrae hanging loosely together. It dwells in the deep, where there are many coral stones, and one will usually find it wrapped around the little Sea Trees, but it is not found very often.

II. *Caput Medusae Secundum*,[3] is even rarer and stranger than the foregoing, but differs the most in the number of branches because the central head, which also resembles a Spider's body, is first divided into 5 pairs of 10 main branches, that are no longer than an oblong hand. Each one of them divides into two others, which are a foot or 1-½ long, but one is always longer than its neighbor. Each one of these divides again into 20 or 24 smaller branches, which face each other alternately, and these divide finally into yet more cross-threads, all put together from stony, but brittle vertebrae, and they can move to all sides. All the lesser branches are rough on the underside, which is how it crawls along, for it does not have feet, but it knows how to spread and advance its branches in such a way, that they do not get in each other's way. The head has a round hole below, which is its mouth, around which are a host of tiny little teeth; besides this, it has another 5 slits where the main branches commence, which also serve it as mouths. Its body is covered with a smooth skin, and the shell is much softer than that of the foregoing. Inside one will see 5 bloodsacks and eggs similar to those of the Sea Apples, but much smaller; the eggs are yellower or almost reddish, harder than in the Sea Apples, and one should look for the stomachs under the 5 main branches. It also dwells in the deep, beneath and on large Coral stones, where it spreads its branches all around, occupying a place that is easily four feet in *Diameter;* the body usually hangs from the rocks, and it slings its branches around the nearest Sea Tree, which it sometimes covers, especially the kind which we in *Lib. 12.* of our *Herbarium*, called *Pseudo Corallium rubrum.* It seldom comes into the light of day, and the Seamen sometimes haul it up with their anchors, but no one is so bold as to take this fearful animal away, and they are startled when there is an old *Orang Lamma*[4] among them, who dares to touch it with his hands, and watch it sling its branches around his hands, and yet not hurt him. Once it has been pulled from the sea, it curls all its branches and foliage together into a ball over its head, like a pomed cabbage,[5] and dies like that, but it will not last longer than to the next Rainfall, when it falls apart into pieces, unless one has hung it for a long time in smoke.

The general Latin name for both is *Caput Medusae*,[6] because I know of nothing else that resembles that Snakehead as much as this animal does. In Malay *Bulu Aijam*, after the Ambonese *Manuhulu*, that is, Chicken feathers, with which they compare this multitude of branches, though this name was in the previous Chapter also given to the *Stella Marina Quinta* or *Scolopendroides*, which is the same sort of animal.

The Inhabitants of *Hitu* and *Huamohel*, sometimes use it for food, taking the second or largest kind, throw away all the thin branches, and cook only its central body, with the nearest main branches, wherein the stomachs are hidden, from which they only eat the Eggs.

If one finds some small ones of the first kind, which close their little branches upwards quite orderly, and which have withstood thorough drying, they look a great deal like the so-called Rose of Jericho,[7] so that one can sell them to someone who is not in the know.

Caput Medusae. *Among all the Creatures I have encountered, this must be the most amazing; it is an animal that should amaze the careful observer and make him say: Lord, how wondrous is your creation! It is almost impossible to imagine this Sea star, because the profusion of limbs which, when dried, lie there so entangled, that one can only make an imaginary picture of it; which is why the Author could not have it properly done. Doctor* D'Aquet *had sent us one of these from his Cabinet, which we, as much as feasible, have depicted here, see Plate No.* XVI. *Letter* A. *and* B. *is one of its five main branches, which immediately divides into two, and each of those split into two, and so on, up to a tenfold doubling. For as far as we have been able to perceive, a ray or branch makes for about* 512 *ends, of which we by the letter* B. *could show only a few, in order not to be confusing, and if we had also drawn the others, which are now cut off, it would have seemed nothing but a tangled clump. So if we multiply this single one by five, (for this Star has five of such points), that makes* 2560. *ends altogether, which are all as fine as a curled thread, and entangled because of their number, are lying all through and over one another; now this being the case, we find that each point or branch, always going up by two, makes for* 512 *ends, which comprise* 1023 *joints, and multiplying the same by five, we arrive at* 5115 *joints. Every joint, in turn, exists of several vertebrae, which are all sort of rounded on the bottom, but flat on top, and furnished with a little thorn on every corner, as shown by the letter* C. *I have investigated this myself, and found that every joint (which are not all alike, but which are longer, the closer they get to the end), have from* 10 *to* 20, *indeed, in some I counted* 24 *of such vertebra bones: if we suppose that the same has an average of* 16 *of them, we come to an amount of* 81840. *Now this is only as far as the little* Caput Medusae *is concerned, as described here by the Author; the one I investigated, was, lying there all tangled up, about* 5 *or* 6 *inches long, and I think,that if the ends were stretched out, the diameter of the distance between them, would be about a foot, so that one can conclude from this, what kind the Author calls the second one, must be like. I have seen another one like that, which, if I imagine it lying straight, must have been three feet across, though its joints were no bigger than the foregoing. From this one can get an idea, that if the same was* 3, *or four joints longer, how many more ends, joints, and vertebrae it would have had.*

I have learned from the mouth of his Majesty the Tsar of Muscovy,[8] *when he graced me with his presence, in order to see my collections, that one will find many of these in the Caspian Sea, and that the manner of catching them, corresponds with that of the Author, to wit, that they poke the animal with a stick in the water, and it will sling its branches around the same, and since it holds on tight, it is easily pulled out of the water like that: and thereafter as well, at the house of Mr.* De Jong, *when in conversation with the Secretary of his Majesty the Tsar, this gentleman confirmed such, adding, that they find many, and very large ones in the same. The Writer says that these Sea stars cannot be kept very well, but I think that the reason for that is the hot Air over there. The one I encountered, was firm, and in one piece, but it may well be that it came from another region, (where the Air does not cause as much decay). It was very white, and the one belonging to Doctor* D'Aquet *a grayish yellow, which makes me believe, that each of these had a particular form.*

CHAPTER 36

Of the Sagitta Marina. Sasappo Laut.

Sea Arrows or Sea Darts are thin little sticks, which one finds inside certain Worms on some beaches, divided into two kinds: white and black.

I. The White Sea Darts[1] are true sticks, about 1-½ foot long, their ends barely the thickness of an Oaten pipe, and gradually narrowing, until they are as fine as a thread. Contrary to what is normal of heavy things, the thickest part of this little stick is at the top, and the thinnest at the bottom, to wit, in the body of a thick Worm, that hangs from it at the bottom, another half a foot longer than the arrow, in the shape of a belly-worm, (*lumbricus intestinorum*)[2] becoming gradually thicker going down, until the thickness of a little finger, and suddenly ends in a blunt point. The arrow is usually covered with a soft little skin, as if it were a sheath that was fastened to the worm, and has on either side a continual little edge, of a red, yellow and white color, as if it were a fringe, like that on the long, narrow Standards from Maccassar.[3] The Worm is white or flesh-colored, without circles, but flat, a little striped, of a hard, sinewy substance, filled on the inside with a dirty water, mixed with some sand; these Worms grow on flat sandy beaches, and the arrows always stand straight up, at high tide usually with the arrows above the ground, but covered with Sea-water, and when the water recedes, they also draw down into the soil, and do not remain sticking out above ground more than 3 or 4 fingers high. Therefore, if one wants them, one should observe them at high tide, then grab the upper part of the Worm, where the arrow sticks out, and tear it out with one motion; because if you start wiggling it, the worm will withdraw further and further down. When hauled out of the water, the fringe that hangs from the sides will presently wither away, until one does not recognize it anymore; if one puts it in fresh water or in the rain, for several days, then the worm will rot away, and one will be left with a white arrow, which is one to 1-½ foot long, round, and not entirely smooth, and which breaks very easily; one could think they were little broomsticks.

II. The second kind[4] is 2 and 2-½ feet long, of which the smallest half, being the true Worm, is smooth, and sticks in the sand, the rest towards the top; it is on one side overgrown with two rows of little combs, like cock's combs, but much thinner, and spread out by means of many thin rays, in a circle, like gills, also with subtle points sticking out, right on top of each other, and sideways on the worm, but in such a way, that one can get a finger between them. When they are in the water, they spread out beautifully like a flower, and move, but once out of the water, they close themselves up and over their heads, where they become gradually smaller. If one is careless enough to touch one of these spines, one will feel a burning, and the hand will turn red, followed by an unpleasant itching, followed by little blisters, as if one had touched a Nettle, and it lasts for three days, but if one grabs them from below and up, one will only feel the burning in one's hand, which is why one should grab them in this manner, after one has roughened up one's hands with some sand. It is not flesh-colored like the foregoing, but a pale blue and gray; by night they give off a fiery and greenish slime; the arrow of this one, is not white, as is the foregoing one from Hitu, but bluish. They only withdraw in the sand up till their combs, and can be found on the beach in front of Victoria Castle, close to the lowest waterline, where the shore begins to drop off.

Its name in Latin is *Sagitta Marina*,[5] in Dutch Sea Arrows and Sea Darts, in Malay *Sasappo Laut*, that is Sea broomsticks. In Maccassarese *Panadokki*.

They are found on beaches that are not covered for more than an ell at high tide, and that at low tide will still have some stagnant water. One will find the first kind on Hitu, on the beach of *Kaytetto*, the inner coast of Huamohel by *Lokki* and *Laala*, as well as near *Macassar;* the second kind I saw nowhere else except in Ambon Bay, in front of Victoria Castle. Nor did I experience any particular burning or itching in my hand when I pulled out the first kind, the one from Hitu, although I did so by the hundreds; nor did I notice if at night they gave off that fiery or bluish slime like crunched Millipedes, which the Ambonese kind does. The latter, if bleached in the sun and rain, does loose its bluishness, though they never become as white as the ones from Hitu, for which the cause might well be that the one grows in whiter and purer sand than the other. The cleaned arrows are only used as Curiosities, although they have to be locked into a piece of Bamboo, so they will not break. Howbeit, a man from *Macassar* imparted a great secret to me, to wit, that these arrows serve to draw the horrible *Maccassarese* poison from the wounds, which they inflict with their poisonous darts; one sticks a little piece of such an arrow into the wound, which it will then penetrate on its own accord, and suck out the poison, and then one should pull it out again with a thread that has been tied to it: so that it seems that Nature designed these Sea Arrows to counteract this poison. For in thickness and length they do resemble the *Maccassarese* darts a great deal, except that the poisonous little head has not been attached to them yet, and the pieces draw down into the wound, according to their former nature, when they were still in the sand and pulled downwards. This is how one cures the burning and the blisters caused by the last kind, if they have been handled roughly, or when the fine little spines from the combs stick in one's hand. Take some warm ashes mixed with sour Lemon sap, rub it on your hands two times or so, until the painful itching passes, then mash some *Crusta Cinerea Accarbary Nigri*,[6] after it has been rubbed on a stone with water, and smear that on the skin, which will draw out the burning and the venom.

CHAPTER 37

Of the Phallus Marinus. Buto Kling.

The following animals are really called *Zophista*[1] or *Plantanimalia*, which have the shape of a Plant or other lifeless thing, yet live under water; and, if one cares to observe them carefully, one will note three kinds, or divide them into three steps or rungs. The first and topmost rung includes those that are the most like animals, and wherein one can recognize some distinct limbs, among them are some that were described in the previous Chapters, such as the Sea Apples and the Sea Stars. On the second rung are those, that have equal parts of both, like the things that follow hereafter. The third and lowest rung are those, that come very close to plants and stones, and hardly show anything that looks alive, of which we described a number in the 12th Book of the Ambonese herbs,[2] but Nature is so confounded by the Element of water, that one finds things in it, which can hardly be assigned to any of these rungs, as if they were remnants of the original Chaos; because one will find here living, growing and mineral things all jumbled together, making for plants that live, stars that grow, and animals that counterfeit plants.

We assign the *Phallum Marinum* to the second rung of the *Zoophitis*, of which we will describe two kinds. The first and most common is the black or dark-gray one, very like the

Male member, or resembling the barrel of a Cannon, 6 or 7 fingers long; more than an inch thick, round, at the one or hindmost end thicker, just like the Curtal,[3] pointed in front, though tapering off roundly, where they pull a round little head in and out, from which they squirt a water that looks like whey; its body is bare, though it is striped or wrinkled lengthwise and sideways with some small veins. Most of them have a dark gray or blackish color, others are flesh-colored. Its substance is sort of hard or callous, while it shrinks down, and then stretches out, which makes it move, though slowly. Inside they have nothing but some wheyish water mixed with sand, and is otherwise surrounded by thick Calloused flesh. One finds them on flat beaches where grows all kinds of seaweed, especially *Acorus Marinus*.[4]

II. *Phallus Marinus Verrucosus*, is larger than the foregoing, 6 and 7 inches long, two inches thick in back, roundish and flat, its back covered with many warts or knobbles, flesh-colored or with a reddish color, pulling in and out, and looks inside the same as the foregoing. Both make their way without feet, with the movement of Snakes, on the ground, crawling slowly along. One will also find them on flat beaches that do not entirely dry up, although they have no seaweed or stones. The first one is often found near *Hitulamma*. The second kind in the Bay of Ambon.

Its name in Latin is *Phallus Marinus*,[5] in Malay *Buto Kling*,[6] in Dutch Kaffir Pricks,[7] on Hitu *Suheru*.

Plain folks use them as food, to wit, they cut them open at the belly, press all the water and sand out of them, then cook them in water, or roast them on coals, adding a little Lemon juice or *Bocassan*.

CHAPTER 38

Of the Anguis Marini. Ular Laut.

Next to the large Sea Snakes[1] which one will find in many places in Indies Waters,[2] especially in the great bay of *Nova Guinea*, that turns inland north of Wezel's Island, thereby separating *Onyn* from *Nova Guinea* or *Onyn Cubicy*[3] (called Ryklof van Goen's Bay[4] by some of our people who visited it in the year 1675); we also have a smaller variety, that greatly resembles the Blind-worm[5] Cecilia, about an ell long, and a finger thick, without an apparent head or tail, round, smooth and black, crawling along flat beaches at a slow pace. One will find it often in quiet bays where *Acorus Marinus*[6] grows, as, for instance, in the Bay of Ambon.

Its name in Latin is *Anguis Marinus*,[7] *sive Cecilia Marina*, in Malay *Ular Laut*.[8]

They are not particularly bad, the dread being greater than the damage; if one steps on it, it will sling itself around one's foot; if one leaves it be, it stays where it is; but if one handles it roughly, it will bite the hands until there is blood, but with no other damage.

CHAPTER 39

Of the Tethyis.

These are fleshy outcroppings, which one will find clinging to those rocks which are constantly covered with Sea Water; incarnadine or the color of flesh, though some are red, tough and sinewy of substance, with several shapes; some are like a piece of meat, with many notches and wrinkles; some are like nipples or fingers, wherefrom they got their

Greek name, on the outside smooth and slippery, and if one touches them, one will notice that they move a little, but if one grabs hold of them too hard, they will provoke a burning or itching in one's hand. Inside one will see nothing more than pipes that look like veins, and which are filled with water, which one can press out of them, as with nipples; they fasten onto the rocks, and one can only pry them off with force, but one usually leaves them in peace, since this is a nasty brood. I have cut up some of them, and found them to be a light red inside, like flesh, and they kept moving for some time after. When dried in the Sun they shrink together, and become like tough leather.

Their name in Latin is *Tethya,*[1] that is, Teats. *Euseb. Nierenb.*[2] *Hist. Nat. Lib. 6. Cap. 19. vocat eas Carnuculas molles*, and *Glandulas ex Athenaeo.*[3]

CHAPTER 40

Of the Sanguis Belille. Dara Belilli.

The plant or growth that produces the renowned *Dara Belilli*, appears to be of the genus *Tethyorum.*[1] The plant itself is not familiar to me, because it is said that it only grows in the Sea of *Solor*[2] and the surrounding Islands; but I will tell all that credible Men know about it. The first information came to me from Mr. Jacob van Wykersloot,[3] former Chief of Timor, in the year 1681, and states the following: True *Dara Bilelli* is not fishblood, but grows on rocky bottoms, and it is found around *Larentuque*, and near Mount *Lamaha*, on the West coast,[4] in water that is five or six fathoms deep; when it is brought to the surface, it turns out to be slimy stuff, almost like the Duck Weed in ditches, but browner and slimier, so that they have to use knives under water to cut it off the rocks or bottom in small pieces. Then they cook it in Sea Water, scumming it constantly, until it becomes as thick as porridge, and does not have any scum or filth anymore; then they put it in ordinary pieces of Bamboo, which they place upright in the Sun until it begins to stiffen, whereafter they usually put the pieces of Bamboo in smoke, so it will not spoil, or so no small Worms can get to it (which is prone to happen when it is freshly made), for it can never be too dry.

This agrees mostly with what *Radja Salomon*, King of *Apimantutu*,[5] told me, except that he said it can only be found on *Solor*; just as the *Solorese* themselves confess, that there are only a few families among them, that know how to prepare it. They are like pieces of meat, that grow on the rocks, sort of round below, with a long nipple or finger on top; their color is red, just as the aforementioned common ones that grow in Ambon, are white or flesh-colored. There are two kinds [of preparation]. The first and best is to tear off entire pieces, which sometimes have the rocky root still attached, cook them, cut them in long pieces or strips, an inch wide, as if osiered with glue:[6] black on the outside, inside like a dark glue, that is a combination of Sea-green and black, tough, with a very bitter and brackish taste. The other kind comes in pieces of Bamboo like clods of dried blood, brinier than the foregoing, not as bitter or as vile. They make this by grinding or cooking up big and small pieces together, then pouring it in the Bamboos as above.

I conclude from this tale that these *Solorese* pieces of meat, are merely a variety of the previously described *Tethyis*, and they differ from the Ambonese kind only in that they are red, bloody on the inside, and taste bitter.

The bitterness sounds so much like Snake wood,[7] that one would say that they mixed it with some Snake wood scrapings or sap, but eyewitnesses assured me, that this bitterness is natural to this meat.

Its name in Latin is *Sanguis Belille*, in Malay *Dara Belille*,[8] to wit, boiled blood that they sell in pieces of Bamboo, about a hand long and an inch thick, but the dried meat is called *Dagin Belilli*, which they sell, cut into short strips, as long and thick as a finger.

One finds it nowhere else but in the aforementioned places, and it is mostly sold to other Nations by the *Portuguese* who live on *Liffau* and *Larentuque*. The *Solorese* are also accused of greatly falsifying this *Dara Belille*, but this is usually nothing more than that they do not sufficiently skim it while cooking, and that they then pour it, scum and all, into the pieces of Bamboo, when it has not thickened enough yet, which often causes the pieces of Bamboo to be half empty, and the blood to be sandy, very brackish, but almost without bitterness. It is a Physick that is highly esteemed by the Portuguese, and the right kind is quite dear when it is sold; one cuts a small slice, the thickness of a knife, and the size of a dubbeltje, of the dry blood, that sticks in the piece of Bamboo, and this represents a dose. One grinds this in weak Arack,[9] or for fevers, in plain water, and drinks it down like that. At first it tastes very briny, but not unpleasant, almost like the Chinese *Bolatsjang*,[10] which is used to make sauces, then one notices a clear bitterness, that tastes like Snake wood. Its greatest power is against raging fevers, which it casts out through powerful sweating, but one should not drink anything for 3 or 4 hours after it, because that would hinder the sweating, even though one suffers a great thirst, which usually follows upon taking this remedy. It is also praised as a remedy for pleurisy, or stitches in one's sides, griping in the guts, Cholera and Mordexi,[11] that is caused by an overflow of gall: also against any poison that one has swallowed, smelling and tasting quite a lot like *Theriac*.[12] One also praises it as a remedy against colic, but I do not know that from experience, though I do about the foregoing; the only thing left to note is that, after taking this, one should not have any Lemons or any sour things with one's next meal, since this will be deadly. When the Portuguese use it against poison,[13] they add the Root *Lussa Radja*:[14] when it was first brought to Japan, it was so highly regarded, that one could get 100 Tail[15] Silver, or 125 Rixdollars, for a katti,[16] or 1 and 1 quarter pound of this blood, but then when greed came to encroach too much, the market was spoiled, and the Physick despised.

CHAPTER 41

Of the Pulmo Marinus. Papeda Laut.

This *Zoophitum*[1] belongs with those that are called *Holothuria* in Latin, and in Dutch Sea Gellies. When I saw them in Europe on Portugal's beaches they had almost the same shape, except that the Portuguese ones are larger, and have within them curls most fair. The Ambonese Pulmo is the size of a Trencher, and is on top bossed or rounded, like a lung, while on the bottom or inner side it resembles a dish, wherein one can see a pentacle of handsome branches or curls, somewhat resembling a flower. The body of the dish is partly transparent and the color of flesh, even as the Gelatines, which are made from boiled calves' feet, very delicate and trembling. The branches or rose is quite the color of ice, to wit, transparent and bluish; floating in the Sea it displays some life, because it moves, pulls itself together, and then opens up again. If one grabs hold of it firmly, it will burn a little in one's hand; as soon as it is removed from the water or cast-up on shore, one can no longer discern signs of life in it, and if it lays for only half a day in the Sun it will melt completely, leaving some thin little fleeces, and it is a pity that such a fair Creature neither does last, nor can be kept. One finds them often on clear nights, when the Sky is full of Stars, and the Wind blows a little, and finds them in the morning cast up on the beach. I have found small

Sealungs the size of a dubbeltje, clear, transparent as ice, with a red spot in the center that divided into 5 rays which contracted and expanded as if they were arteries that had their origin in the heart, and did not burn at all in one's hand.

Its name in Latin is *Pulmo Marinus*,[2] and wrongly *Patella Marina*,[3] a name which rightly belongs to a certain Shell, called *Lopas* in the next Book; in Dutch Sea Gellies and Sealungs, in Malay *Papeda Laut*,[4] after its resemblance to a porridge called *Papeda;* in Ambonese on Hitu, *Metuae* and *Methuay*, and called by others *Mothe* and *Lappia Mutin*, that is to say, Cold Papeda.

Many hold that these Sealungs spring from the snuffing of Shooting Stars, since one finds the Sealungs most often after some clear nights with many Shooting Stars have gone by. To be sure, when the snuff of such a *Meteorum* has fallen to earth, it is found to be a burnt-out, delicate, and light substance, resembling the Sealungs somewhat, but if one touches it, it falls away into ashes which, when shot into the Sea, is believed to turn into Sea Gellies. But what would seem to argue against this is that, as I said before, one finds such small Creatures among them that one could conclude they grew from a small beginning, but one can say against this that the aforementioned *Excrementum* of the Shooting Stars breaks up into many small parts when it falls into the Sea, and that each of those bits forms such a Creature. The Natives have no use for it, but the Chinese know how to prepare it into some kind of fare, and they say that they also make some sort of Physick from it. But our Nation feels that, since the Chinese are a deceitful and greedy people, they use these Sea Gellies only to add it to the mixture they use to stoke *Arak*,[5] and thereby making it proportionally strong and fiery. And this is indeed the case, though it is also not healthy for the head, and damaging to the Nerves, and therefore they cheat like this secretly, and if it does see the light of day, they are punished for it.

CHAPTER 42

Holothuria. Mizzens.

This strange kind of *Holothuria*[1] can be counted among the *Urticas Marinas;*[2] but I wanted to describe it by itself. One cannot grant it a real shape, neither that of an animal nor that of a plant. Its body is an oblong bladder, about a finger long and an inch thick, the whole body resembling a full bladder, and on its back are many little fleeces sideways, broad on the bottom, fastened onto the back, on top forming a point, similar to a half-sail, called a mizzen. All these little sails are at the top joined by a seam running across, which allows it to lower and raise these sails when it feels a Wind and wants to sail. The body is of a transparent color, as if it were a crystal bottle filled with that green and blue *Aqua Fort*, otherwise called *Aqua Regis*.[3] The little sails are as white as crystal, and the upper seam shows some purple or violet, beautiful to see, as if the entire Animal were a precious jewel. When the sails are set,[4] the body becomes almost triangular, whereof the head curves up and the backside is extended from its belly, which is bluer than the upper body, as if the *Aqua Fortis* were stored there, and the upper body were made of crystal. This sea-animal does not have a mouth apparently, and when it strikes its sails it nevertheless keeps on going in the water by means of the body's movements and that of the beards[5] suspended from it. On one side, if I remember it is the right side, and all around in the back hang a multitude of long, thin beards, of which the two largest ones must be two ells long and float behind; the others are small and large all intertwined; many are broken off and are growing out again. The large beards hang by a narrow neck from the body, followed by an oblong bladder and the rest of the beard the thickness of a quill or an oaten pipe, gradually becoming thinner; the other beards are the thickness of sail-twine and thinner, all covered with a host of little knobs, as if they were divided in so many joints. Their color is a beautiful blue, but always with a hint of green. They are very fragile and easily break off and remain hanging from whatever touched them, causing a dire burning and pain; wherever they touch, the skin turns red and becomes covered with blains; indeed, the entire side of the body will be filled with pain where it has been touched by them. The harm lurks mostly in the beards, because I have touched the body of Sea Gellies without injury. Fishermen are often troubled by these pests, since they remain hanging from the nets or fishing poles and lines, and are sorely hurt by them. But the good thing is that one does not encounter them the entire Year, but mostly with the outgoing East monsoon,[6] or in the month of August, or when the white water ends,[7] and it is spreading Westward. I remember that, Anno 1662, when I sailed in a Shallop[8] from *Pulo Ay* in *Banda*[9] to *Ambon* we went for an entire day through a sea covered with these Mizzens, which had come from the East in large schools or troops. A certain gentleman, our Governor[10] at the time, saw this, and said: there you can see that much venom is hidden on the bottom of the Banda Sea or in its white water, wherefrom comes forth such a multitude of poisonous Sea Gellies.

Their name is *Holuthuria Urticae specie & Epidromides Marinae;*[11] in Dutch: *Mizzens;* in Ambonese: *Hurun; Si Ambonitae Latinam callerent linguam, crederem ab urendi facultate, eis nomen dedisse.*[12]

Ordinarily one should shy away from touching them, if one does not want the hands to be burned; and I know nothing more injurious in the Sea that causes such a fierce burning; it is, therefore, even more amazing that there are Natives so bold that they dare use these villainous Sea Gellies as food, even making a Delicacy from them, cooking them only in salt water in a piece of Bamboo with *Sajor Songa*,[13] adding Spanish peppers and Lemon-juice; for they say that cooking deprives them of all harm, howbeit they do not dare eat the aforementioned Sea Lungs, but leave those to the Chinese. The ones cast on shore do not burn as much.

CHAPTER 43

Urtica Marina. Culat Laut.

Urtica Marina has almost the same shape as what the Italians call *Potto Marina*,[1] in the Mediterranean Sea. It resembles an outspread Flower about a foot wide, with notched and wrinkled leaves spread out on the bottom, of a greenish color, mingled with some red and white, of a fleshy Substance; inside are other, smaller leaves, curled in upon themselves as if they wanted to form a head; below it has a wide piece of flesh, reddish, that is its root, with which it clings to the sand, or stones, which lie under the sand; if one grabs hold of the central head, or touches it, it closes up; it also burns a little in one's hand, but it means little, and is quickly gone, because one has to grab them by this head if one wants to pull it out; they occur on such beaches where the lowest tide is still 4 or 5 feet deep; one considers the hollow head its Stomach, since, when little fishes get into it, it encloses them and digests them.

Its name in Latin is *Urtica Marina*.[2] In Malay *Culat Laut*.[3] In Ambonese *Ulat*, a common name for all Campernoyles,[4] because they consider this *Zoophitum* to be of that kind.

They may seem as distasteful as possible, but they make for a fine dish that might even appeal to a Lickerish fellow.[5] One tears them out, root and all, cooks them up in some salt water, pours the same out and wrings them out a little, then they cut them up into little pieces, and stew them once more with a spicy sauce. The Ambonese cook it their way, in pieces of green Bamboo, with *Sajor Songa*, layer upon layer.

For other kinds and a longer description see *Jonston*,[6] the Bloodless water animals *Lib. 4. Cap. 1.*

CHAPTER 44

Vermiculi Marini. Wawo.

These are small little Worms or rather worm creatures,[1] scarcely a foot long, some as thick as sail twine, but most like silken Floss,[2] all entangled in small clumps, which always has one that is larger, thicker and longer than the others, and which is considered to be the Mother. They have different Colors. Most are dark green, but among it also plays some dirty white or yellow, red, brown, and some blue. It is difficult to know its true shape, because they all cling together like an entangled hank of thread, which quickly breaks into pieces if one touches them; which is why one should very quickly observe them by the light of a candle as soon as they are out of the water; but if one lets them be in salt water until the next day, and then looks at them through a magnifying glass, one can discern the following shape; as was said, every clump has one that is larger than the others, which one calls the Mother, and it has the thickness of the coarsest grade of sail twine, sometimes like a thin quill, of a pale yellow or whitish color, sticking its little head just above the water, which shows only two little horns, like Snails have, while it has on either side four clear little feet, like the caterpillars. The others are very fine, like tiny hairs, not to be counted, which they stretch out and pull in. The children hang all over this Mother, of the thickness of fine sail twine or silken Floss, greenish, some as long as a hand, others a foot and a half; they are ribbed athwart their Bodies, as if they consisted of many joints, but they are so fragile that they presently break into pieces, if one handles them or lifts them up, and one can note the joints better in the ones that have been cooked. If one looks at them in the evening when they have just been hauled out of the water, one can clearly discern signs of life, but one cannot keep them alive until day dawns, even if one leaves them unmolested in salt water; signs of sight, hearing, and smell can also be detected, because it seems that sight makes them dart towards a burning torch or a light, but when the Moon has risen, they hide themselves again. One has to ascribe to them a sense of hearing, because they can be chased away if one makes a big noise, and it must be their smell that makes them swim so eagerly toward pregnant Women, or to Legs that have sores on them. One does not see these worms all year round, but only on the 2nd, 3rd, and 4th night after a full Moon, which takes place when the Sun is in Pisces, in February and March; one should look for them at that time after the Sun has set by the light of Torches, on such beaches, that have large rocks standing in water, and which are full of large cracks, but not sharp or spiny, but smooth, because it is around that kind of rock that one sees swarms of these little worms, floating on top of the water, and they come to those who have a burning torch in their proa, and one can scoop them out of the water with stretched pieces of cloth or fine sieves. The first two nights one finds them floating around the rocks, but thereafter further out at Sea. The Natives want you to be quiet while they are catching them, you cannot make any noise, neither while scooping nor with talking, and they let the proa drift gently all by itself. By the next full Moon these little worms have grown already, about the thickness of an oaten pipe, quite like young Millipedes, of a mixed green, brown, and white, and look indeed somewhat disgusting, but these have a special name, and are not considered the true *Wawo.* They disappear after the fifth night that follows upon the aforementioned full Moon, and one will not see them again for the entire Year, except, as stated, during the subsequent full Moon, when they have a different shape and name. The aforementioned full Moon, when the Moon is in Virgo and the Sun in Pisces, is the usual time for *Wawo,* though it might happen that one will find them earlier, to wit, with a full Moon, when the Sun has not yet entered Pisces. Nor does *Wawo* appear each year in the same quantity: When it is preceded by much warm rain, it will be plentiful, and one can scoop them up three nights in a row; but if there are many dry and hot days before, there will be little and only for one night, because one will see black splotches in the water of the Sea: One has also experienced that there is a very high tide every year, at least higher than the daily tide, before the *Wawo* arrives.

The common notion is, that these worms are the excrement of the aforementioned rocks, both those that are bare, as well as others that are hidden in the sand. At least where such rocks are not to be found, one will not find *Wawo* either: The biggest catches are in Ambon Bay, around Red Mountain, in the 3 *Liasser* Islands, on *Latuhaloij,* and in *Banda,* as well as in the *Molukkos.*[3]

One could give it the provisional Latin name of *Vermiculi Marini;*[4] the Malay name is unknown, perhaps because it does not occur in those parts: In common Ambonese *Wawo*[5] and *Wau,* which is also considered to be the word on Ternate. On Hitu, *Melatten,* on Leytimor, *Laur,* in the Uliasser Islands, *Melattonno,* on Banda, *Ule:* Since it is of two sorts, after two full Moons, they call the first and real ones *Wawo Kitsjil:* the other, larger one, after the subsequent full Moon, *Wawo Bezaar.*[6] Those from Hitu make even finer distinctions, and come up with three kinds. The first one is called *Malatten Salanay;* these are small tiny worms that hang together like a bunch of threads, appearing by full Moon in January, when the Sun enters Aquarius, but since this does not happen every year, besides being tiny and few, they are not harvested and are left as food for the fishes. The second and true kind is called *Melatten Yan,* that is *Fish Melatten,* as if this had already turned into living fishes or recognizable creatures. The third is called *Melatten Lalian,* that is Millipede *Wawo,* because it then has the shape of a Millipede; this one appears the next full Moon in April, and is deemed unfit to eat.

There is much ado about this *Wawo* in Amboina and Banda, and those people who are used to it, consider it a great delicacy, although when you look at it, it seems quite nasty; it is prepared in three ways: The first and tastiest way is to cook it fresh in green pieces of Bamboo, for which they use the fresh *Wawo* of the first kind; they let the same float in salt water, then remove with small sticks any filth that might have gotten stuck there while catching or scooping it up; then they thoroughly rinse it with fresh water, then they take the green leaves of *Sajor Songa,*[7] cut them into small pieces as you cut Tobacco, and add grated Coconut, Salt, Pepper, and, if you like, onion and garlic, mix all of this together with the *Wawo,* and put it into the Bamboos, put them near the fire and let them cook. It is not eaten by itself, but along with Fish and other food, after one has first dipped it into *Bocassan*[8] or some other sauce, and it has then a nice, rich taste, but only if it has been thoroughly cleaned beforehand of pieces of wood, coral, and shell, that float among the *Wawo.* One can mix the aforementioned *Sajor Songa,* or use it by itself as well, with *Sajor Pacu* or *Filix Esculenta;*[9] also the leaves of the little *Lignum Aquosum*[10] tree, called *Aijwaijl* in Ambonese, and called *Daun Laur* because it is used like this. The second way is to pickle it; and this is done, after the washed and cleaned *Wawo* has been thoroughly drained so that it is completely dry, by putting it in brine, with the juice of Lemon *Papeda*, and if it is kept carefully stoppered, one can keep it easily for a month or so, and make a sauce with it, wherein one dips the food in order to arouse an Appetite, as one otherwise does with *Bocassan;* when it is pickled like that, you will not see the shapes of the Worms anymore, because they completely melt away into a slime, especially if one does not drain the water off properly. Another way of making *Bocassan*, is that you do not handle the newly caught *Wawo* as long as they live, but pour them gently into baskets, and bring them on land while shaking them as little as possible; you first cook a sop[11] that is half water and half of the aforementioned Lemon-juice, with Vinegar, Ginger, *Lanquas*,[12] Peppers, and, for those who like it, also Garlic, and pour this over it; then you take Ambonese salt,[13] smoke it completely dry over a fire, rub it fine and sprinkle it over it, keep it in tight little pots so that no air can get to it, and you will have a fine *Bocassan* that can be used for dipping, and which one can keep for more than a Year, although the worms will melt entirely away. The third way to prepare it is by smoking it, when one takes the cleaned *Wawo*, that has been mixed with the aforementioned chopped herbs, put it on a Banana leaf, spread it thinly, cover with another Banana leaf, and then tie it between two thinly split Bamboos, or *Gabba Gabba*,[14] roast it over a fire, whereafter you hang it in the smoke; when you want to eat it, you remove it from between the leaves, and it will be as thin as a Pancake; break it into pieces and dip it in some sauce. All *Wawo*, especially the pickled kind, retains its rocky smell and taste, which does not bother those people who love it, though most of it is taken away by the aforementioned Spices. One has also found that it expels water, but it is forbidden for those who have any kind of sore on their bodies; people who have their legs covered with sores, should not let the *Wawo* get to them when they are in the Sea, since that will cause them to rot more and will prevent healing.

The second kind that appears during the following full Moon, and which already has the shape of Millipedes, is not or is rarely eaten because it looks so hideous, and though one will not experience any great harm from it, it will make many people dizzy and distressed; hence it is better to avoid it entirely. This last kind also illumes at night, giving off a clear light, which makes people avoid them even more, since they share this attribute with the Millipedes. In order to provide additional information about this Creature, I will give here the notes of several years, concerning [the times] when the true *Wawo* was caught.

Anno 1684 it came on the 3d, 4th, 5th of March, the full Moon having been on the first of March, and appeared in reasonable numbers, under the rocks, from red Mountain to *Hative Kitsjil;* one scoops them up either sitting in a proa, or going into the water up to one's waist, when only one among all of them holds a burning Torch in the hand, wherefore the Natives choose a Woman who has conceived, even if she remains seated with the Torch in the Proa.

Anno 1685 the full Moon was on 20 March, when the Sun enters Aries, just before and after the full Moon the weather was very hot and still, which is why they did not see much *Wawo* that year, because at night on the 22nd of March, that small stuff showed up, called *Wawo Ican*, since the same is left for the Fish. On the 23d following they caught the good ones, but in small number.

Anno 1686 it was full Moon on the 8th of March, and they were catching *Wawo* on the 11th of that Month, but still not much due to the continuing dry spell, although there was some heavy rain on those two evenings.

Anno 1687 it was full Moon on 27 February, and the following first of March should have been the time of the *Wawo*, but nothing could be seen, except for some slimy red threads, which did not have any shape or form, the foregoing drought and stillness being again the reason.

Anno 1688 it was full Moon on the 17th of March, and again it was hot and dry weather, which is why little *Wawo* was caught on the 19th and 20th, most of it having been engendered by the rain, which occurred two nights earlier.

Anno 1690, the 27th of March, at the full Moon in Aries, when it was quiet weather with little rain, one came upon a reasonable amount of *Wawo* both that evening and the next two.

Anno 1693, the 24th of March, being the third night after the full Moon, they began to catch *Wawo* 4 evenings in a row, in reasonably large amounts, because the weather was calm around this time, which is when the aforementioned worms are most likely to appear.

Anno 1694, the 11th of March, being two days after the full Moon in Pisces, they caught some *Wawo*, being fine and calm weather, but there was not much of it.

The Author did not provide us with any pictures for the foregoing Chapters that is from 36 *to this one, and, I believe, such was not done, because they dealt with things, which would be too much subject to spoiling, and would not be able to be preserved, such as the* Pulma Marinus, *the* Sea Gellies, *the* Anguis Marinus, *the* Sanguis belilla, *and others, which only consist of jelly and things that look like flesh, wherefore one has to be content with the descriptions. But those who desire to do so, can check* Jonston, *or even better,* Gesnerus, *since* Jonston *derived most of it from him anyway, who shows several of the foregoing although I do not know if they correspond to the ones* Mr. Rumphius *has described; because I have noted that, quite often, the shapes and names differ among the Authors and do not match.*

END OF BOOK I

The Ambonese

CURIOSITY CABINET

The Second Book; Dealing with the Hard SHELLFISH

by

Georg. Everhard. Rumphius

The Second Book,
About the Hard
SHELLFISH
Being the True
Ambonese
CURIOSITY CABINET

CHAPTER 1
About Hard Shellfish in General.

In this second Book we will describe the Hard Shellfish, called *Ostracoderma* and *Sclerostrea* in Greek and Latin, which have covered their entire body with a Shell as hard as Stone, except for the mouth, or, if they consist of two Shells, can enclose themselves so tightly within them, that one can hardly see anything of the Animal. Their parts and general Attributes we will not discuss extensively, but refer the Reader to the European Writers. We will divide them into three Main genera or orders (*Classes*), and then mention something specific about the Properties of each of them. The first order includes those which we call whorled univalves (*Quae constant una eaque contorta aut anfractuosa testa:*)[1] and which one properly calls *Cochleas*, or Snails and Whelks. The second includes the Single-shelled ones, those that have a Shell only on one side, *Univalvia*, commonly a single unturned Shell, and cling with the other side to rocks. In the third order are those one calls Two-shelled (*Bivalvia*), *Conchas* in Latin, in Dutch Shells, Mussels and Oysters. Those of the first order are so different that one can hardly arrange them in Main Groups, indeed, there are some among them of a mixed nature, of Fish and Whelk. But in order to aid Memory, we will divide them into twelve Main genera, while immediately warning the Reader that he will find some in each genus that will differ a great deal from one another, and that it does not matter in which Chapter they are discussed. We refer to them with the general name of Whelks, Sea Whelks and Sea Snails, without distinction, because we mostly get our curiosities from the Sea, or from such places where rivers discharge into the sea. The Malay generally call this animal *Bia*. The Javanese and other Malay call it *Crang*.[2] The Ambonese call it *Kima* and *Huri*. The second order is divided into two main genera, whereof the first comprises those that have only one bossed or vaulted shell, and cling with their other side to the rocks, and the second, those which have a single, unturned shell that has two openings. The third order is divided into five kinds, which will be discussed at the appropriate place.

One should note about the first order that all those that are formed like a snail, or are long and twisted, that these are closed off at the mouth by a lid, which is fastened to the flesh, and which they carry on their heads while crawling; but when they contract, they close their mouths so tightly with it that nothing of the flesh can be seen. The general name that the Malay have for this lid is *Matta*, that is, Eye; the point, or that part of the little whelk that comes to a point, and which our Nation holds to be the head, is for the Malay the tail, just as for them the widest part is the head, which we hold to be the undermost, where the mouth is. Besides the two little horns which most of them have on their head, and which stick out while they crawl, some of them have one other or several more snouts, which they stretch out by means of one or more branches, which are halfway hollow on the bottom, like a channel, and some have a sharp little Bone therein, which can hurt you, and wherewith they apparently are able to pierce something. The flesh from the head is a hard Gristle, bad to eat, and in its center lies the stomach, with the gut attached to it, and that which one calls the *Papaver*,[3] being a thin little sack, wherein one finds a greenish or blackish mud, which is not good to eat. The guts wrap themselves around the head below the neck, where they have their exit, and where they cast out their filth. The other flesh that is in the twists of the whelk, is not a gut, but more like fat, and which is also the best to eat. The other kind, that has an elongated body, with a long, narrow mouth, does not have a lid at its mouth, but covers itself either with a hard callous flesh (*callosa*),[4] wherein one sometimes can still see a little bone, or they crawl entirely inside, so that one cannot see any of it at all. Other Attributes, that some hang bare from the rocks, that some crawl on the ground, or that some hide themselves in the sand, we will describe in detail for each kind.

How Shellfish Grow.

Here one should note three Ways, which all Shellfish maintain while they grow: The first is that some Shellfish grow by adding matter to the mouth of the shell, (*per appositionem novae materiae*)[5] such as the *Murices* and *Turbinata*. This can be seen clearly in the *Murex*, because the front part of its shell is always thinner and clearer than its hinder parts. This material is generated from a tough spittle or slime, which the animal discharges, and which is formed into a thin shell, at the opening of the old mouth, and continues like that for about one-fourth of the circumvolution,[6] where it will rest and form a new toothed or branched mouth. The shell of the *Murex* appears to be triangular, so that each addition always forms one-third of the shell. But the undermost side, where the mouth is, is always wider than the two sloping ones from the back, so that it takes up almost half of the circumvolution. From this follows that this kind must change the mouth of its shell every so often, or advance it in accordance with the continuation of the twists. The second way is by stretching the entire body to all sides, (*per extensionem totius corporis testacei*)[7] since this kind always keeps the same mouth, but the entire body becomes larger. All *Porcellanae* and *Cauris* grow this way.[8] The third way comes about by adding and stretching at the same time, (*per appositionem & extensionem simul*) like all Shells, Mussels and Oysters, which not only add on to the frontal edge, but to the shell itself as well, or rather its plates (*Lamellae*),[9] thicken over its entire body, obtaining, so it seems, its food from thin little veins that are fastened to the *Spondylo*.[10] But one can note about all of them, that the old parts of the shells, especially the outermost skin and the protruding horns, do not get as much food, or no longer get as much from the inflow of the animal as does the youngest addition; because all the corners, thorns or nails which were once perfect but have broken off, do not grow back again.

CHAPTER 2

Nautilus Major Sive Crassus. Bia Papeda.

We propose to include in the first genus of univalve and twisted whelks, those which have a Mother-of-Pearl interior, or which at least shine so much that it seems Mother-of-Pearl. They have various shapes, and the one that excels is the *Nautilus*, having a mixed nature of Fish and Shellfish, whereby it differs from all other kinds of Shellfish: because the Fish that lives therein or that shapes this Shell, is a kind of *Polypus* or Plurifoot,[1] divided into two kinds, coarse and fine, to which we will devote their own particular chapters.

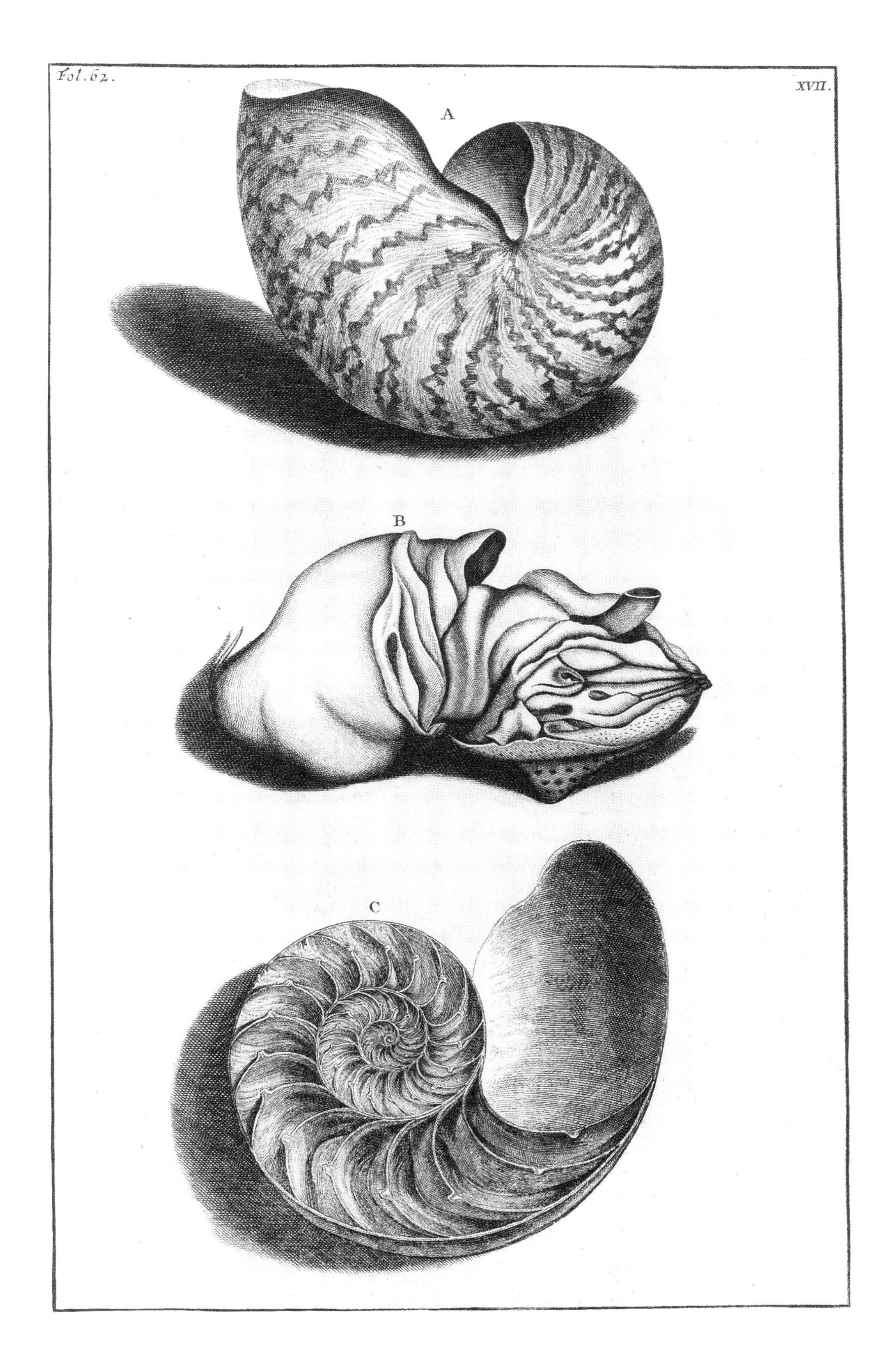
Vol. 62.
XVII.
A
B
C

Nautilus Major sive crassus, is the one usually called Mother-of-Pearl Whelk, of which we will describe the two most important parts, each one separately, to wit: the Shell or the House, and the Fish that lives therein. The shell has the appearance of an ordinary Snail, or of that feigned shell (*Cornu Ammonis*),[2] also something of a round little boat; for the keel is truly round, about 6 and 7 inches from front to back, the front part makes the little Boat, which is open on top, 4 and 5 fingers across. The stern rises above this flat part with a rounded curl, which is turned and wound in upon itself, with no opening anywhere, but completely vaulted. Inside, it is separated from the hold of the Boat by means of a partition, (*Parietem intergerinum*)[3] that has a small round hole in its center, through which one can pass a thick needle, wide on the outside, and ending inside in a short little pipe. Inside, such Partitions divide the curl into innumerable little Chambers, which all have that aforementioned hole in their centers. The shell has two crusts that lie firmly on top of each other, though they are barely the thickness of a knife. The outer crust is coarse, like other shells, and has fine little cracks, is somewhat uneven, and of a dirty or pale white; it is of one color at the keel, towards the back, where the curl commences, and is marked athwart with many brown and wide bands, which gradually become narrower until the top of the curl: Looking inside at the curl's other half, [one will find it to be] black and the lower part the color of silver. The inner crust is of two kinds, because the one belonging to the little Boat has the beautiful color of Mother-of-Pearl, though showing more green and red, giving off a reflection like a rainbow, though it is not transparent once it has been cleaned, whereafter it remains no more than the thickness of half a knife. The partitions or screens, are also of a Substance that is like Mother-of-Pearl, but much more beautiful, smoother and whiter, shining like silver, so that this Whelk surpasses the usual Mother-of-Pearl in the beauty of its color. The Inhabitant, or animal, needs be kin to the *Polypus*, but it has a particular shape, fashioned by the hollow of the shell, which it does not entirely fill, when it contracts. The hinder part of its body is somewhat hollowed out, and it is over against the curl; the upper part that faces us (although it really is the lower part of the animal, when it is crawling along on its own) is mostly flat or slightly bossed, gristly and rumpled, veering towards russet or a light brown, with some black spots, which discolor, just like the *Polypus;* the lower part, that lies against the keel, and which becomes the top when it is moving, is also gristly, but softer than the former, and covered with many warts. Between these, on the front part, one will see a clump of innumerable little feet with several flaps, placed one on top of the other, and covering the mouth on both sides, and each of these flaps is shaped like a child's hand, of which the outermost and largest is divided into 20 fingers or little feet, each half a finger long, and the thickness of a straw, round, smooth, without those little warts one will find on the beards of the *Polypus*, but rather wide in front, like the blade of an oar; below that is the second and third little flap, or small hand, also divided into 16 fingers, and beneath that yet more small ones with short fingers, up to the mouth. It can project or contract these fingers as it pleases, and they serve it not only as feet while crawling, but also as hands in order to bring prey to its mouth. The mouth is like the Beak or Bill of the *Caccatuwe* bird,[4] or like some Seacat, to wit, the upper part is large and curved, a little notched or toothed on the edges, the lower part small and concealed beneath the upper one. Both are somewhat hollow in back or are divided in half, and that is where they are attached to the flesh. The entire mouth is as hard as bone, black, veering towards a blue, like Indigo, surrounded at the back with a circle of white and tough flesh, and below that yet another little skin, which almost covers the entire beak, and the same is also entirely hidden by a slime and by the aforesaid little feet, so that one cannot see anything thereof, unless one pulls them apart. The Eyes are close to the sides, but more towards the keel, quite large like navels, fashioned without an apparent pupil,[5] but they have a hole in the same place, and are filled inside with a black-brown blood. From the hinder part of the body, which lies against the first partition, a long vein goes through all the holes of the partitions, and through all the little chambers up to the inner center, and this is the only thing with which the fish hangs on to the shell, and which also easily breaks off if one pulls the fish out, whereafter the chambers are empty. Below its snout it has a half round pipe, which is rolled up, however, made from a whitish flesh, like that of other Seacats, and in it lies concealed something that looks like a tongue; but it does not have any noticeable openings in its belly, although it is undoubtedly the same conduit pipe, through which the Seacat shoots forth its black blood.

In order to understand this animal better, we will explain the adjoining figure, the way it was taken from its shell and drawn,[6] according to the position it would have had when floating on top of the water, to wit, with the boat's opening towards the top; B,C,D,E, represent the upper gristly part, near C, lying against the black curl A, B being, a white skin that goes all around the sides. The eyes far more resemble an ear or a navel, and are pierced in the center. F, is the aforementioned conduit pipe. The entire belly G,H,I,K,L, is fashioned from a thin skin, which is the thickest on the sides by D; the innermost hollow is filled with entangled entrails and a dark-brown liquid.[7] Next to M, are two round quoits, looking like eggs, filled inside with a yellow slime. I, shows a wide band, that goes around the belly. K, indicates the aforementioned vein, that goes through all the little chambers. By N, are two egg-shaped yellow glands with little bladders on top, being, so it seems, the stomach of the animal; because from there, up to the black bill, is a wide throat, filled with sand, small stones and slivers[8] of Oysters and Mussel Shells. (Since this drawing is lost, another was added here in its place, with its own particular letters and particular explanation.) So when it floats on top of the water, it will stick out its head with all its beards, and spread them on the water, with the stern or curl always above the water, but on the bottom it crawls around turned about, with the little boat on top, and the head and beards on the ground, yet it progresses quite rapidly. It usually stays on the bottom, but it does sometimes crawl into the weels, but after a storm, when things quiet down again, one can see troops of them float on top of the water, because they were undoubtedly lifted up by the tempestuous waves, from which one can infer, that they stay in troops on the bottom as well. Such floating does not last long, however, because, pulling all its beards in, it turns its little boat over and goes to ground. On the other hand, one will often find the empty shells floating along, or tossed onto the beach, for this defenseless animal has no lid and easily falls prey to Crabs, Sharks and Caymans, wherefore one will find that the shell is most often gnawed at the edges, and whilst it does not hang firmly inside its shell, they can easily pull it out, and leave the empty shell to drift away. The young creatures of this *Nautilus*, no larger than a schelling,[9] have the beautiful Mother-of-Pearl color both inside and out, so that the rough shell only grows over it at a later time, and which commences at the front part or little boat.

Their name in Latin is *Nautilus major sive crassus.*[10] In Dutch Mother-of-Pearl Whelk. In Malay *Bia papeda,*[11] *Bia Coijn*, though this really refers to the shell. In High Malay, *Krang modang.* In Ambonese *Kika;* and since they include the next *Nautilus* as well, they call this kind *Kika lapia*, that is *Papeda Whelk.*

One will find them in all the seas of the Moluccan Islands, as well as around the thousand Islands before *Batavia* and *Java,*[12] but usually only the empty shell, because the animal is seldom found, unless it crawls into the weels.

The animal is eaten like other Seacats, but its meat is much tougher and is difficult to digest. The shell is used most often to fashion beautiful drinking bowls, such as those known in Europe;[13] to do this one should choose the biggest and smoothest ones, and look very carefully, that they do not have little holes, that one can see through anon, and which were made by certain hollow warts (a kind of *Balanis*),[14] a slimy worm that has a sharp little tooth, with which it can drill through this hard shell, when it comes to grow on it, and which then makes these shells unfit for this kind of work. One should place a complete one in something sour for 10 or 12 days, such as spoiled rice, vinegar, or water that had grape leaves rotting in it, and then the outer shell will come away, which one should rub away by means of hard scouring, beginning at the place where it is the thickest, and if it is not entirely gone yet, one should put it back in there again, until the Mother-of-Pearl has come through everywhere, which one then rubs with a weak aqua-fortis, until it has acquired its perfect luster, and finally rinse it with soapy water. The clean ones are cut through by the chambers, so that the four or five back ones become transparent; the next three or four chambers are cut out entirely, and one cuts a small open helmet in the innermost curl, and one can carve all kinds of figures on every side of the little boat, rubbing them with crushed coals mixed with wax or oil, until they stand out in black. The Natives, who do not want to take the trouble of cleaning them, carve the bottom in such a way, that it becomes a large spoon or boat, with which, for instance, one can eat *Papeda*, wherefrom this whelk got its name among these people; but since one serves a sauce of sour lemons and vinegar with *Papeda*, the Mother-of-Pearl on the inside gets a pale film from this, which one therefore should always rinse out with soap or water mixed with ashes.

One will find only one kind of this fish in *Plinius Lib. 9. Cap. 29–30.*,[15] and he calls it *Nautilum Pompilon* and *Nauplium*, whilst its description fits the next, thin *Nautilus:* but today's Authors from the Mediterranean Sea have remarked upon two kinds, of which *Bellonius* calls this first and thicker kind *Cochleam margaritiseram*,[16] and attributes to it many chambers inside a shell that shines like Mother-of-Pearl, so that it should be the same one which today's Greeks, according to what *Robertus Constantinus*[17] writes, call *Talamen tucta podiu*,[18] that is to say, *polythalamum*, or a whelk with many chambers. *Cardanus*[19] must have had a similar whelk, which he called *Cochleam Indicam*, shaped like a galley and fit for fashioning into precious and beautiful drinking cups.

On *Buro* they found a small stone in such a fish (perhaps in the *Glandulis*[20] which represents the fat), the size of a small bean, white like a piece of alabaster, of irregular shape, angular, and pitted as if it were made up of many small pieces; nonetheless hard and shiny. A Chinese woman found it and put it in a box where she kept it by itself for some time, certain that no one had gotten to it, but when she opened the same again she found that it had begotten a small little stone, the size of a lentil, but rounder and thicker, white and smooth as well, and two other little stones some time thereafter, which were the size of a mustard seed, but one of them was so brittle, that one could rub it to pieces with one's finger. From that time on, the bigger stone birthed no more, and one could not tell where the young stones had come from. But my guess is, that it must have been small pieces that fell off, and the places where they had been before were most likely an even white, and smooth, so that one could not notice that they had fallen off. She kept it, believing that they would bring good luck when looking for mussels, to make a fine *Bocassan*.[21] *Plinius* also described such birthing stones, *Lib. 37*, where he calls them *Peantides* and *Gemonides*.[22]

These whelks have the Nature, which they share with some other *Turbinatis*, that they sweat when it is damp weather, and that they do this so much that there will be drops on them, even if one has kept them in the house for years, and wipes them often, which is caused by the salt moisture, which is in the little chambers and which dries only gradually, and in order to get rid of this, one should wash them out often, and put them out in the sun again to dry.

In the *Hist. Antill. Cap. 19. Art. 4. Lib. 1.*[23] one can tell much better from the picture than from the description, that *Bourgou* is meant to be the *Nautilus Major*, because the Fish's shape is not stated nor is the outer shell described.

One should also be warned, that one should not be offended, when the new Authors call this whelk *Cochleam margaritiseram*, and we in Dutch Mother-of-Pearl Whelk, as if one really held them to be the mother of true pearls, howbeit this whelk is only called such because of the beautiful Pearly luster of its shells. For this see the new Curiosity-book by *Philip Bonannus, Part. 2. Class. I. N. 1 and 2.*[24]

The Nautilus Major, *is known to us as* Coquilie, *although the French call all whelks that, but for us it usually passes by that name, and we do not mean any other by it but that one. It is shown on Plate* No. XVII *by the letter* A. *but with more perfect spots, than usual; and in order to show which are the best, I present this one, instead of the one by the Author. Letter* B. *is the fish that the Author described, but since that drawing was lacking the letters, we give the Reader the description, the way it was written down before, in order to see for oneself, which parts of the Fish are meant by that.*

In order to satisfy the collectors even more, I had one of these Whelks cut through, the better to display the inner chambers, which, if not from the description, can be comprehended by the eye. It has been wondrously put together, and it is unimaginable, that the same are always filled by the fish, since its communion from one chamber to the next is in front an opening the thickness of a needle and in back no more than that of a fine little hair. I counted about 50 chambers in the same, and believe that there were more of them in the end; but the partitions are so feeble, that they, though they were cut with a very fine tool, still crumbled into pieces. See the figure, as well as the engraver could get it, on the same plate by the letter C.

CHAPTER 3

Nautilus Tenuis. Ruma Gorita[1]

Now this is the true *Nautilus* which we, to distinguish it from the foregoing ones, call the thin or fine one, and which seems to have been the only one known to the Ancients, because it has the right shape of a galley, with a narrow keel and two high sides, a small curl at the end, whereas the foregoing suddenly comes to a point without too much winding [see plate XVIII, letter A]. I received two kinds of these, big and small ones: The large *Nautilus* is a whelk of 5 to 7 inches long, and four inches high; but one also finds them the length of a span,[2] fashioned from a single thin shell that has the thickness of parchment, partially transparent, exceedingly white, like paper, though some are also yellowish. The keel is scarcely half a finger wide, half round like a circle, divided into many edges or notches, from which many folds go up the sides and which nearly all gather in the back of the head, where the curl begins to curve inward, but many leave off on the way, so that the entire whelk is covered with folds. The keel is blackish up to and over the curl, as if it were smoked, and this cannot be washed away. The sides are straight, and at an angle, and leave on top an opening that is 2 or 3 fingers wide, and in the back by the curl they have a protruding corner, like an ear; the Fish that lives there is notably similar to a *Polypus*[3] or Many-Feet, to wit of the sort that *Aristot.* calls *Bolitoena*,[4] its flesh entirely soft, supplied with 8 beards,[5] of which the 6 front ones are short, white and covered with warts like all Seacats,[6] and which it

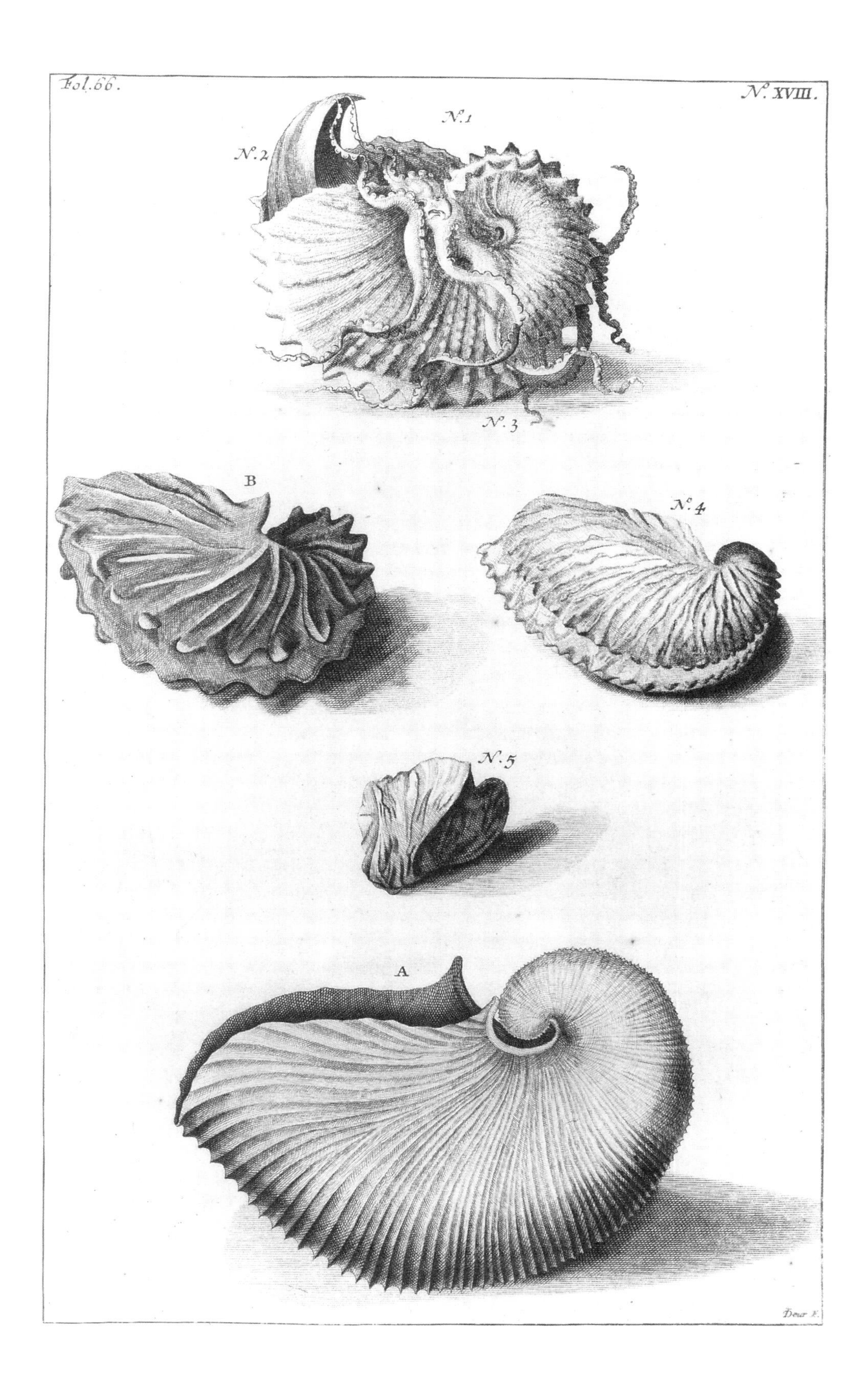
Fol. 66.
N.° XVIII.
N.° 1
N.° 2
N.° 3
B
N.° 4
N.° 5
A

spreads out like a rose when swimming. The two back ones are the same length, which it stretches forth along the curl in the back, lets them hang down in the water and steers its little boat therewith; for they are round and smooth, but at the end as wide as the blade of an oar, the color of silver, and it too has warts on the sides, so that this little boat is steered like a kind of Javanese sloop, called *Tingang*, commonly known as *Tinan*,[7] which is steered with 2 rudders. I did not discern a thin little fleece between the frontal beards, as Writers mention about the *Polypus* in the Mediterranean Sea; but it is true that the same beards are inside a sack made of skin, hanging from the head, which it expands a little with the two top beards, but that would not help much with sailing. On the contrary, I am of the opinion that it does not so much go asailing with this little skin, as it does with the hollow of the little boat, of which it raises the prow so that the wind can blow in it. For one will find that when this fish goes sailing it stows most of its body towards the rear, keeping only the two aforementioned rudders on the outside for steering; but when there is no wind, it will bring all its beards out, lowers the prow again, and rows. When it notices any kind of danger or ambush, it will pull all its flesh on board, turn the stern up, so that the little boat will take on water, and in this manner it goes to ground, so that, if one wants the shell complete with fish, one has to approach it very quietly in the wind and skillfully scoop it up; but this happens so rarely, that Fishermen consider it a great boon if they catch one, which is only likely to happen when there is a great stillness after a thunderstorm. With its little paws it grabs hold of bits of wood, which it finds in the sea, and with which it goes afloating. In the hollow of the little boat one can see the wide loose stomach, as in other Seacats, and under the beards in this stomach a wide pipe, with which it pumps out the water when it surfaces; on the flat part of the stomach and back, forming together a sack, one sees the same dark-brown little stars and spots like unto those on the *Polypus*, which change color. It lies loose in its shell, and it does not have a vein fastened inside the curl as the foregoing did; wherefore it can easily be flung from its shell, and it will come to float on top of the water, and, if fortunate, the fishermen will catch it before it is dashed to pieces against the rocks. On the bottom of the sea it will turn the little boat over, its head upside down, the way it is also seen with the keel up when it first comes to the surface, but it deftly turns the little boat over and bails the water out so it can float; its beards, as told, spread out like a rose. One has also seen them, with their beards clinging to the underside of a large tree leaf, and, while hidden under it, begin to float.[8] The eyes are not hollow like the foregoing, but full and clear.

In terms of the fish and the structure of the shell, the small *Nautilus* resembles the previous one the most, but the boat is much smaller and wider, three and four inches long, two and three fingers across, with a broad keel that is divided in fewer and duller notches, and therefore has fewer folds, but they also twist and turn and go obliquely towards a midpoint near the curl. Its shell is neither as pure nor as white as the foregoing, but a dirty horn color and as if smoked, the sides of the keel are also black, though one can get it to be somewhat whiter with a lot of washing in soapy water and bleaching. The fish is like the foregoing, and it uses its beards in the same manner as well, but it is more frequently found rowing than sailing, hiding especially under leaves and driftwood; otherwise it stays mostly on the bottom, and also winds up at times in weels,[9] which is the reason it is seen more often than the one mentioned before.

It is uncertain if both these fishes can live on their own in the sea when they fall out of their shells: at least, I had some brought to me fresh from the sea, which died anon, even though I put them in water. I have also found eggs in their bellies, which are round white grains bunched together in a clump, each of them with a small black speckle on top like an eye. Both of them have similar black beaks like the bill of a *Kakatu*,[10] which is hidden under the flesh. On the bottom of the *carina*[11] or shell one will find a small clump of eggs or roe that has the shape and color of other fishroe, and the entire little clump is wrapped in a thin little fleece; even though the entire house is not much bigger than a finger, they still have an ovary; lying on the keel, like a cushion.

Their name in Latin is *Nautilus tenuis & legitimus* and its little Boat, *Carina Nautili.* In Dutch *Doekhuyven*,[12] because of the many folds. In Malay, *Ruma gorita*, that is, *Domuncula Polypi*, just as the Greeks called it *Ovum Polypi.* In Ambonese *Kika wawutia.*

This whelk is found so rarely that it is priced very highly, even in the Indies. The Natives consider it a sign of good fortune if they find it, and keep it among their treasures, and they seldom display it except on holidays and times of public mirth, when the women bring it forth, when they perform the round dance, *Lego Lego;*[13] when the lead dancer carries this shell raised up high in her right hand; but they are not finical with them, even if they are somewhat smoked, torn, and full of holes; which is why one should get them forthwith out of their hands by means of money or handsome speech, if one wants them to be undamaged; for the ordinary kind that can hold 4 or 5 quarterns[14] in its boat, one will pay 1 Rixdollar[15] without bargaining, but one cannot get them for that anymore, and there have been those of 8 quarterns, or a Zeeland can,[16] which fetched 9 Rixdollars a piece, because they are uncommon, and I have not seen any larger than that. Those of the second kind are simple and of no value. I should mention a rare event here. A Sea Eagle (*Haliaetos*),[17] being a bird that constantly hunts at sea, took such a *Nautilus*, while it was floating in the Sea, and bore it aloft, but while his business was with the fish, and since he did not care about it as a curiosity, he struck his claws mostly into the fish, wherefore the shell came to fall out of his claws and, by rare fortune, it fell on a small spot of sand between rocks in such a way, that nothing was broken off, except for a small corner of the foremost edge; and a fisherman who was wandering thereabouts, quickly picked it up and brought it to me; and since this fell out of the sky like another *Palladium*,[18] as if a small likeness of the fabled ship Argos, I sent[19] it in the year 1683 as a memento to Mr. *Johan Michael Fehr, Philosophiae & Medicinae Doctor, Physicus Suinfurtensis Ordinarius & Academiae Curisorum Naturae per S. Romanum Imperium stabilitae electus Praeses dictus Argonauta:* the Collegium in which I was accepted as a Member in the same year under the surname of *Plinius*, upon the favorable recommendation of Mr. *Christianus Mentzelius, Med. Doctor, Archiater & Consiliarius* of the illustrious Kurfürst of Brandenburg, *Apollo* of the same Collegium. The Ancient Greeks called this fish *Nautilus; Suidas*,[20] *Nautes* and *Ovum Polypi; Athenaeus*,[21] *Naiplium* some also call them *Piscem Nauticum; Aristot. Hist. animal. Lib. 4. Cap. 2.* already noted two kinds; the first has a striped shell like the Jacob's shell, hollow inside, and not attached to the fish, this kind is small, related to *Bolitaena*, a kind of *Polypus*, [to be found] on all the beaches, and is often tossed out of its shell onto dry land, where it is either eaten or it spoils; the second kind is firmly attached to its shell like a road snail (*Limax*),[22] never leaves its shell though it does at times spread its beards on the water: the first is, without a doubt, our fine *Nautilus*, and the second the foregoing Mother-of-Pearl whelk. *Plinius*[23] describes it as a singular marvel of the sea, but, as was his wont, rather obscurely and with two names, which, so it seems, he considered to be two different things or fishes: the first he called, in *Lib. 9. Cap. 29. Nautilum* and *Pompilum*, which he describes in his own words as follows, and which I, because they are so concise, translated into Dutch, and offer them herewith. "Among the foremost marvels belongs the one, called *Nautilum* or *Pompilum*, and it, lying on its back, that is, turning up the open side of the shell, comes to the surface of the sea, and after it has bailed out all the water through its pipes, it rows readily away; it then curves the two foremost beards backwards (*Athenaeus* says that it raises them), spreads a thin skin between the same in such a way, that it can sail with it, and rows with the other beards, and steers its ship with the middle tail as if it were a rudder, and it sails there on the high seas, imitating the yachts [*Liburnicas*], and if it encounters anything frightful, it takes in water and goes to the bottom." Thereafter, in the following Chapter 30. he describes a fish which he thinks is different, from the story by *Mutianus*,[24] calling it *Nauplius*, with these words:

"Mutianus *has stated, that he saw another kind of fish* in Porpontide, *which lived in a little boat, or carried a little ship; it is a shell, fashioned like a little boat, with a round and raised stern, but with a sharp head*[25] *in front, wherein lives the Nauplius which, like unto a Seacat, only wants to please itself and play, which happens in two ways; for when the sea is still, the one that sits on top stretches its feet out, and rows with them in the water, but if a breeze entices it, then it will steer with them, and turns the hollow of the little boat into the wind, whereby it sails very easily. The delight of the one is to carry, and for the other to steer, and sometimes both together founder, which is sometimes held to be a sad omen*[26] *of a human tragedy, for as many contend, when they see this* Nauplius *suddenly founder, it is a sign of an impending shipwreck.*"

Both these descriptions fit our fine *Nautilus*, which *Jonston*,[27] after *Bellonius*, describes somewhat clearer; *Hist. Natur. de Exang. aquatilibus Titul. 3. de Turbinatis* where we will begin with *Jonston's* words. "The *Nautilus*, which is called by *Suidas*, *Nautes;* by others *Nauticus*, and generally *Nautilus*, and which appears to be the same as the *Nauplius* of *Athenaeus*, is called by today's Greeks *Thalamis Toukdapodius*,[28] as *Robertus Constantinus*[29] writes. We find two kinds of it in *Aristotle*. There is the one whose shell resembles the *Pectunculus*[30] but which is hollow and is not naturally attached to it; it often wanders on shore because it was tossed on dry land by the waves, where it falls out of the shell, or is caught, or dies; this kind is small, and is like the genus of *Polypus* called *Bolitoena*. The other hangs in the shell like a snail, and this one never goes outside, but sometimes it sticks its arm out."

Bellonius described it precisely. *He says that the shell consists of three parts (namely a keel and two rims, although they are one) and its sides appear to be fastened onto the keel, and often so large, that it can be grasped around by two hands; all of them are not thicker than parchment, with long stripes, at the edges jagged with notches and rounded off; but the hole, through which the* Nautilus *is fed, and crawls out of the shell, is large; this shell is brittle, white as milk, clear and smooth, and of the shape of a round ship, because it sails on top of the sea and brings itself up with its hollow shell inverted, in order to surface better, and once it has arisen it turns the shell over again, but between the arms of the* Nautilus *is a thin fleece, similar to that between the claws of flat-footed fowl, but much thinner, like unto a spider's web, though it is strong, wherein it lets the wind blow; the curls, which it has in abundance on either side, it uses as a rudder, and when it is frightened, it fills the shell with water and sinks to the bottom.*[31]

And I want to say this once again, that most of the time I did find that thin fleece between the two foremost Beards of the East Indian *Nautilus*, which the *Authors* ascribe to this fish, except for one kind, which comes hereafter; also that the name *Thalamis Toukdapodius* cannot be applied to the true or thin *Nautilus*, which does not have little rooms, but belongs to the Mother-of-Pearl whelk that was described in the previous chapter. *Bontius*[32] *Lib. 5. Cap. 27*, places the *Nautilus* among the *Polypos*, as *Plinius* does, and the shape of the shell does come quite close to resembling ours, but I cannot guess, what kind of fish he is dealing with, which burned his hands so badly that, after rubbing them with garlic in water, he had to leave them be so they would heal; for no one has heard that a *Polypus* burns, wherefore I would guess that some words were left out, and that one should understand him to be talking about the Sea Gellies.

Anno 1693, in January, they caught a *Nautilus* on *Hitu* that had a boat the length of 7 inches, 6 inches high, and differed somewhat in appearance from the ones mentioned above, especially as to the two back or upper beards. For the six beards were between 12 and 14 inches long, the front half very thin and narrow, the back covered with sucker warts (*Acetabulis*).[33] The two back or upper beards (which others hold to be the two front ones) were much thicker, at the back easily as thick as a finger and covered with sucker warts, alternately across from one another;[34] the front half had a thin and wide skin or cloth, like a Mizzen[35] turned upside down, to wit smaller in back, but broad in front, wherein it is different from the foregoing: and it seems that it sails as well as rows with these flaps, for with the sucker warts it grabs the sides of its boat, and it rows with the broader part, and the same with the other beards, which then are still lying overboard; and now when it wants to sail, it raises both flaps. But I have never experienced with any *Nautilus*, that the two mentioned beards have grown together by means of a thin little fleece that was between them, as you read in the European writers; wherefore I stick to my former opinion, that it accomplishes sailing with the foremost and hollow part of its boat. It was so clever that, when the fishermen pursued it in their small proa, it turned over first the right then the left side of its boat; wherefore, when the fishermen noticed it wanted to scoop up water and sink to the bottom, one of them had to jump overboard and overtake it while swimming. *Philippus Bonannus*, who is renowned nowadays, in his Book called *Recreatio mentis & oculi N. 13. Class. 1. part. 2.*, calls it in Italian *Polpo moscardino* and *moscarolo*, and says, that it is often caught in the Adriatic Sea, or in the Gulf of Venice on the Italian side, sometimes going to the beach amongst other Seacats in order to feed there, which no one in the East has ever noticed, because it is always seen only on the open sea and by itself. Our *Nautilus* has under the beards in the open stomach yet another pipe, through which it spews forth water, and does so quite high up, and quite often in the fishermen's faces; at the back of the pipe hangs a little bladder, similar to the one of the *Polypus* and *Sepia*, but herein one will not find a red, but a brown purple blood. Furthermore, it remains hidden from man's intelligence how this fish fashions its shell, to which it is nowhere fastened, and yet they grow together, never jumping out of it except when in danger, and one is left with the empty shell. The fish goes to the bottom but no one knows if it then proceeds to fashion itself another house; it is, after all, a very tender fish, and it will soon die if one handles it even a little; it cannot be kept alive for long even in sea water. One can somewhat whiten the black edges of the freshly caught ones by rubbing it with fine white sand, but they will not come off the top part of the stern. Its brown blood changes color throughout its body, and leaves spots, but it blanches in the dead ones. I also found in their stomachs pieces of other beards, and its eggs were lying outside the body in the hollow of the shell, though fastened to the fish.

We call the Nautilus tenuis, Nautilus, *while mariners call it the* Little Skipper *(because they see it often sail across the sea), but old devotees call it the* Linen Coif, *because of its shape and brittleness, because it far more resembles a cloth coif as people were wont to wear, than a whelk from such a wild element as is the sea. I saw such a* Nautilus, *which was the length of an Amsterdam* houtvoet,[36] *and which now, as I was told, is kept in the Cabinet of the Grand Duke of Florence,*[37] *who is a great fancier of excellent seagrowths. One should believe the Writer's remark about the raising of the little mizzen or small sail, with which this animal sails, since the Gentleman, quite apart from his careful inquiry, was his own eye-witness, while all other Writers usually have this word by mouth, just as I was made to understand from many seafarers, who assured me that the animal had in front a small raised sail; or it could be that it was another kind which is not known to the Writer. Mr.* Volkertsz,[38] *now deceased, one of the greatest Collectors and Experts of his time, had a similar one drawn from life, and had it, along with many other excellent whelks, etched by* Savry,[39] *and which are to this day in the excellent Cabinet of the Honorable Mr.* LaFaille,[40] *Chief Officer of the City of Delft. I have a picture of it, which I thought I would include here, and one can see it depicted on plate* XVIII. *by* No. 1.; No. 2. *is the so-called Mizzen, and* No. 3. *are the oars with which it steers or rows, the shell is entirely different, since it had knobs on its ribs. The British fancier,* M. Lister,[41] *gives it a much larger sail in his* Historia Conchyliorum, lib. IV. sect. IV., *most likely because he was guessing; but* Volkertsz' *one agrees quite a lot with the sail that* Gesnerus[42] *had depicted: although the shells of* Gesnerus, M. Lister, Bonanus, *and the present Writer resemble each other much better. Besides the two kinds which Mr.* Rumphius *presents to us, there are several more. The one depicted as* No. 4. *can be seen in several Cabinets; it is much finer than the foregoing, with a broader keel, wider stomach, made of a whiter substance, with knobs on its weals.*[43] *There is yet another kind that resembles this one a great deal, but it has a narrower keel, has finer knobs, and is without weals; some of these can be seen in the excellent Cabinets of Mr.* d'Aquet,[44] Feytemaas;[45] *and the biggest and finest known to us, is owned by Miss* Oortmans,[46] *a great fancier of such beautiful things. There is only one other kind which, as far as I know, is only in my possession, and is depicted on the same plate as* No. 5. *It has the shape of a cap of a farmer's wife from Waterland,*[47] *is quite broad, flat on top, has something of a keel, but has twisted weals and knobs, and is not as evenly sided as the other ones, which makes me think it is a miscarriage by a* Nautilus.

CHAPTER 4

Cornu Ammonis. Little Posthorn.

This is a small whelk in the shape of a ram's horn, or like the way they paint the God Hammon's ears,[1] the size of a *dubbeltje*. One might look upon it as an imitation of the *Nautilus major*, because it resembles the hindermost curl, but this is not the case, for it is a genus all by itself, since, though it does curl in upon itself, like a little ram's horn, [one will find that] the whorls do not hold fast to one another, but are asunder all the way to the very tip on the inside, where one will see a tiny knob like a small pearl, or like the nit of a louse. The outermost whorl has the thickness of an average quill, on the outside round and truly white, broken off in front, inside usually divided with many partitions, into many little chambers, shining like white Mother-of-Pearl, and every partition has a tiny little hole, with a piece of a little pipe going in, just like the *Nautilus major;* and this is why so many believe that these are tiny copies of the same, though I have found these Posthorns to be bigger than the smallest of the aforementioned *Nautilus*, which is also a beautiful Mother-of-Pearl both inside and out. These Posthorns, however, have a slimy animal in their foremost chamber, which clings to rocks with a thin and narrow thorn, which runs through the animal and the first number of holes, and is fastened onto the rocks. Now when the North wind blows, and troubles the sea, these little whelks are thrown off the rocks, which is why it always seems as if they were broken off at the mouth. The tips remain stuck to the rocks, and they are stiff enough to hurt anyone's feet if one steps on them. The Ambonese believe that they are generated from white Gullshit; this does not seem very likely at first, since one will not find them all the time, but only in certain months, when the North wind blows, and they are tossed in large numbers onto the beaches among other seawrack.

Their name in Latin is *Cornu Hammonis.*[2] In Dutch *Little Posthorn.* In Ambonese *Tay manusamal,*[3] that is Gullshit, for reasons given above.

We call the Cornu Ammonis *also* Little Posthorn, *like the Author, and though it be very small, it is nevertheless so beautifully fashioned, that it needs yield to nothing else. It is white and dull (that is without a sheen) on the outside, but inside like mother-of-pearl and transparently thin. It has innumerable little chambers, which are all connected, by means of a very small little hole. I have tried to cut one through the middle, and display it like that, as we did with the* Nautilus major, *but I have never succeeded because it is so fragile. It is depicted on Plate* XX. No. 1.

CHAPTER 5

Carina Holothuriorum. Sea Gellies' Boats.

These are also related to the *Nautili,* although they have the shape of a Winkle, just like the small kind which we will describe in the next chapter. They are nigh an inch wide, round, somewhat angular on the outermost whorl, somewhat flat on the bottom, and it has few whorls, just like the Land Snails. It has a single shell, thin and transparent, very light, on the bottom a light violet blue, becoming the color of lead on top; the mouth is wide, round, but it has a small little corner at the bottom that sticks out like a Flew or hanging lip, and they are white inside. The animal that lives in there, does not have the shape of a snail at all, but is like a kind of Sea Gelly that stands up straight like the joint of a finger, when it drifts out on the open sea; this little Sea Gelly is perfectly clear, like a small crystal, with a blue sheen, and it consists of nothing but slime, that is surrounded by a little skin, and if one keeps it but for a single day, it will surely pass away. These little boats are seldom seen, at least at Amboina we had them for the first time in August and September, Anno 1682, at the end of the Harvest [season],[1] when they came drifting up in large schools from the East, on the open sea, past *Manipa* and *Buro,* where the Natives scooped them up when the weather was dry. The little boat laid there with the opening up, and the little Sea Gelly stood upright like a small pillar, appearing to sail with a faint breeze, wondrous to see that such a fleet of easily a thousand little ships, could sail so agreeably together. When one took them from the sea and put them in a dish with water, the little Sea Gellies did manage to stand upright for a day, emitting a wondrously beautiful repercussion of light,[2] as if the dish were filled with Precious stones; but one could detect little life there, and they dried up gradually, so that I could still know their shape after four days. The Natives declared, that they had never seen such before.

Their name in Latin is *Carina Holothuri;*[3] one can count them among the *Neritas,*[4] which are small little whelks in the Mediterranean Sea, which also float on top of the water, and are so called in Greek from [their] swimming; we call them Sea Gellies' Boats. The Natives have no name for them.

The Author did not give us a picture of the Carina Holothuriorum, *or the* Sea Gellies' Boat, *but we had one drawn after one in our own collection; see plate* XX. No. 2.

CHAPTER 6

Cochlea Lunaris Major. Matta Bulan Besaar, or Matta Lembu.

The second part of the Main Genus, includes all those that have the shape of an ordinary Winkle, but have an interior that looks like Mother-of-Pearl; of these we will describe two kinds. The first kind, called Giants' Ears, are the largest of this genus, usually the size of two fists or more, shaped like a Winkle, with a large round mouth, which has a small growth[1] at its lowest corner, and therefore looks like a little earflap. The outer and largest whorl has two sides, one up and one down, and between these yet another ridge, all three covered with knobs. The shell is made from two thick layers, of which the outer one is grayish, the color of a corpse, somewhat coarse and full of cracks, but here and there spotted like snakes, with black, brown, and sometimes even Spanish green;[2] the smaller the horns, the more beautifully spotted they are, and also smoother; and one also finds those that are entirely gray without spots, but these are considered ordinary. The inner layer is a beautiful Mother-of-Pearl, not white, but showing all the colors of the rainbow, to wit, green, red, and blue. If one smashes them, there will be slivers that all keep the same color. The animal within has proportionally the same size, in front [it is] of a hard white flesh. The hindermost curls are pure fat, and in between is a large *Papaver* or Sand sack. It has a large round shield on its head, with which it locks its door tight, the width of a hand and of the thickness of a finger, hard as stone, as if it were a white pebble, on the outside slightly bossed like a lentil, white, and sometimes with a number of shallow dents. It is flat on the inside, although a few ridges do protrude slightly, and they suddenly come together to form a navel, which is covered with a russet or brownish little skin, that holds on very tightly, and this side is [connected to] the flesh of the animal; when it is crawling, it carries this shield on its head, but it does not venture far from home, keeping at least half of the shield inside the shell; it is so strong, that there is not a man who can pull it out, even if he grabbed it under the shell, for it is more likely that it would pull his hand inside. This shield belongs to those one calls *Umbilicos marinos*, or Sea Navels, of which this one is the largest. It can be found on such beaches that have steep rocks, against which the sea dashes with great force, and they are therefore hard to dive for, but wherever one does find them there are many of them together, as if Companies or Troops, like the *Tsjankor* on the coasts of *Coromandel* and *Ceylon*.[3]

Its name in Latin is *Cochlea Lunaris Major.*[4] In Dutch Giants' Ears. In Malay *Matta bulan besaar;* for one should know, that the Malay call *Matta Bulan*, that is, Moon's eyes, all those whelks, which carry a round, thickish and stony-hard shield, the outer sides of which resemble a full moon; it is more reasonable to call this one *Cochlea Margaritica*, than when *Cardanus* did so for the foregoing *Nautilo major;* because it has more expressly the shape of a snail, than does the same *Nautilus.*

The natives make much of it, that is as food, boiling the same in hot water, until the shield opens up; the best part of it is the tail, because the front flesh is rather hard, and all that green or black *Papaver* one should throw away, because it is bitter and sandy. The Kings of *Buton*[5] reserve this food for themselves, wherefore their subjects must bring these whelks only to them. We are more interested in the shell which, if it is nicely spotted, like a Leopard, without cracks and in one piece, is kept, else one stops up the mouth with clay and puts wax on top, so that the tartness does not get to the Mother-of-Pearl, and puts it for several weeks in vinegar or spoiled rice, but one should change the vinegar 2 or 3 times, and also scour it often, until the rough inner layer gives way, and the entire whelk becomes

Fol.70.
N.º XIX.
N.º 1
A
B
F
N.º 4
C
E
N.º 6
D
N.º 3
N.º 7
N.º 2
N.º 5
Deur F.

Mother-of-Pearl; which one then rubs all over with aqua fortis, and then rinses again with soapy water. The Japanese smash this Mother-of-Pearl shell, and put it on big Trunks or Cabinets,[6] in order to shape flowers and stars with them, for their black lacquer work, which renders it most handsome, displaying more beautiful colors than real Mother-of-Pearl.

One can use the stone shield, or Moon's eye, in place of a sleeker,[7] and I was also assured by some *Surgeons*, that it has all the powers of the *Oculi cancri*,[8] after one has crushed it, dissolved it in vinegar, and administered it, it removes clotted blood, incites the menses of Women, and expels bladder stones.

Hereto also belongs a small kind, not much bigger than the nail of a finger or thumb, shaped like an ordinary Winkle, with a wide round mouth, closed off with a small, round lid or stone, which makes a ring on the sides [that is] the color of lead, but which is white on top. The shell itself is on the outside the color of pale soil, sometimes with a few white speckles; veering on the inside a little toward Mother-of-Pearl. They are not very handsome, and therefore not worth keeping, but the Natives like to cook and eat them, especially with *papeda*, because they have a sweet, though scant amount of meat, which is why some call them *Bia papeda kitsjil*.[9] The Natives place the lids of the Giants' Ears in water, wherein they wash their young children, against the *Sawan*,[10] that is, the evil spirit, which seizes the children and makes them restless.

The Cochlea Lunaris major, *called* Giants' Ears *by the Author, are known to us as the pyed* Knobs Whelk: *one will find some that are a beautiful green, and then are called the green* Knobs Whelk; *see their pictures on plate* XIX. *Letter* A. *shows it from the top, and the same at letter* B. *from below. In its opening or mouth, witness the shield or lid,* Umbilicus Veneris, *to us* Venus navel, *indicated with the letter* F.

CHAPTER 7

Cochlea Lunaris Minor. Bia Matta Bulan.

All of the following share the shape of ordinary Winkles, covered with a thick double shell, and have their mouth closed by a round Moon's eye, which we call *Umbilicus Marinus*, or Sea Navel, of which we will describe five kinds.

I. *Cochlea Sulcata nigra.*[1] In Malay, *Krang sussu*,[2] because it somewhat resembles a small teat, has a thick and black shell outside, deeply furrowed sideways or along the whorls, furthermore somewhat coarse as if scaly, without any luster, but it sometimes has a few Spanish-green or white spots; it is a pale white Mother-of-Pearl inside; the lid is a dirty white on the bossed or outermost side, with some grains, at times somewhat black in the center; the animal is good to eat, but the shell is not handsome.

II. *Cochlea petholata.*[3] In Malay *Bia pethola*, is well-nigh the same shape as the foregoing, perhaps a little smaller, and the whorls are a bit angular in front; some are also quite round, on the outside completely smooth and decorated in several colors, like *Pethola* cloths, or the large *Ular pethola* Snake: the most common color is brown with black and white spots, but if it also has green or bright brown, they are considered the best; some of the ones that are completely round, have black stripes along the whorls instead of spots; inside they are yellowish, veering towards mother-of-pearl; its lid is one of the handsomest of *Matta bulangs*,[4] truly round, flat on the bottom, and, as usual, marked with a navel, on top it is almost a half circle or the shape of a lentil, mostly black, green and russet on the edges, smooth and shiny like the eye of an Ox. There is also a smaller kind,[5] blackish, with a green reflection; the animal has firmer flesh than the foregoing, tough and slimy, and therefore not fit to eat. These are rarely found, and therefore counted among the foremost curiosities, due to the beautiful color of its shell.

III. *Cochlea Lunaris aspera*,[6] has the shape of the first kind, also furrowed along the whorls and entirely coarse, with scales that stick out, which makes them quite spiny. The Moon's eye is like the foregoing, but plainer, with less of a sheen, while it is also mostly black, or a little russet on the edges; its biggest variety is the size of a Chicken egg, more spiny on the outside, while inside it is a yellow Mother-of-Pearl, which makes it resemble a glowing oven, and is called such, therefore; the smaller kind[7] is less spiny, and is white inside or a silvery Mother-of-Pearl, which is more common and plainer, and its Moon's eye is a little grainy; both are blackish on the outside and do not shine, and they sometimes have Spanish-green[8] spots. The best way to eat them is to throw the green *papaver*[9] away, and to cook them until the lids open up.

IV. *Cochlea Lunaris minima*,[10] is barely as big as a thumbnail, also furrowed on the outside, a light chestnut-brown, lusterless and without spines, yellow on the inside with some Mother-of-Pearl; the lid is also a round little shield, blackish and does not shine.

V. Big silver ovens[11] resemble this kind, except that they have an edge on the topmost circumvolution, which has some scales on it, also painted on the outside with black and white, though the white is the least, [and it is] silvery inside; they are seldom found. All of these are called in general *Bia matta bulan*,[12] and are brought to market to be sold as food, except for the second kind, which is rare, and occurs mostly in the Uliasser Islands.

The Cochlea Lunaris minor: *The first kind is shown on Plate* No. XIX. *letter* C. *The second kind by the letter* D., *which we call a* Nassauer, *besides which we show another four varieties from our Cabinet, which are all of an excellent color and design, see them on the same plate* Nos. 5., 6., 7., *and* No. 1. *is another kind for this Author than [that by] the letter* D. *We could add many more of these, but think it sufficient to show the best; it is a virtue if they are a beautiful silver at their mouths. The Author's third kind is illustrated by the letter* E. *these have brown-yellow spots on the outside, the blacker the better; we call them* Gold Mouths, *owing to their fine golden Mother-of-Pearl color; which is also why the Writer calls them Glowing Ovens. A second kind, see the same plate* No. 2. *have gray-green and brown spots; we call them* Silver Mouths *because of the white Mother-of-Pearl. We also add the spotted or* Pyed Silver Mouth, No. 3. *as well as the* Green Silver Mouth No. 4. *which is also very rare; the Author did not provide us with an illustration of the 4th kind, perhaps because it is of little value; the same thing for the 5th kind, wherefore his Honor refers to the first kind.*

CHAPTER 8

Umbilicus Marinus. Matta Bulan.

We will now tell about the lids of all the foregoing whelks together, called *Matta bulan*, that is Moon's Eyes by the Malay but European books call them *Umbilicus marinus*, after the shape of a navel that they have on the inner or flat side, with which they were fastened to the animal's flesh, and are wrongly held to be particular whelks.

I. The first and largest one is the lid of *Auris Gigantum*, Cap. 6.,[1] a hand's width in *Diameter*, and looks much more like a sleeker[2] of marble or a flintstone, than part of a whelk; it is flat inside, but the whorls give for some elevation because of their edges, covered with a

N°. 4
N°. 5
N°. 6
G
I
H
N°. 1
A
K
N°. 2
C
E
D
F
N°. 3
B

brown skin, that is firmly fastened thereto; the outer side is, as mentioned, white and bossed like a lentil, smooth on some, but on most uneven with some shallow dents. If one smashes the shield, one will find several thick skins lying on top of one another, which makes it apparent that it grows in layers over a period of time. If it stays on the beach, and is rolled around for a long time, the brown skin will be scoured off, and it will become so smooth, that it resembles a single flintstone.

The little shields of the small kinds, that are described in the same chapter, are not esteemed, and therefore not numbered.

II. *Umbilicus marinus niger;*[3] is the lid of the second and third kind of the foregoing chapter, the most beautiful of them all, the most shiny as well, and therefore to be considered the true *Umbilicus marinus;* but it clearly differs from the *Umbilicus,* which one finds in the Mediterranean Sea, since that one is indeed flat on the inside, and marked with a navel, but it is pressed down on the bossed side, with a noticeable little dent, russet or of an orange color: ours is, on the contrary, on the same side, half round, flat and smooth, usually black, with some green and russet mixed in, very smooth and shiny like the eye of an Ox, while it has the size of a ducat or less.

III. *Umbilicus granulatus,*[4] is the lid of the first and third small kinds, far less beautiful than the foregoing, and therefore not considered among the curiosities. The first one is not really black on the inner side either, but it is crooked and full of holes, on the outside a dirty white, without any sheen or graininess. The other lesser kinds are also part of them.

One will find these Moon's eyes on Buru and elsewhere, scattered on the beach without the whelks, with its brown skin mostly worn off, which causes it to be mistaken for a particular whelk; they have the characteristic that, if one places them on a flat little saucer, that has some vinegar or sour Lemon, and if one rubs them around the bottom three times or so, they will move by themselves, and some do this without rubbing, but will soon lie still again. The ones that were just taken from the whelks, will not do this or barely; from which it seems, that if they lie on the beach in the sun, they become spongy; because all such things which rot by themselves in vinegar, must be spongy [*Porosae*],[5] and have some air that is closed up within, and which seeks to drive the vinegar away, and this struggle makes the body move. The Natives hold that these Moon's eyes of the second kind are good eye stones, with which they rub inflamed eyes, especially when they get those little sores (*Hordeola*),[6] in order to expel the same; our Surgeons have also experienced that, when these stones have been pounded, and dissolved in vinegar, they have the same powers as the *Oculis cancri;* the *Umbilicus marinus* from the Mediterranean Sea, is used by the Turks, particularly in *Candia,*[7] against cholera or burning in the stomach, and this is when they place the same in the little hollow of the heart,[8] turning it a little until it is sticking; then they take it away and weigh it, and when they find it heavier than it was before, they hold this to be a sign, that it has drawn the moist vapors into itself.

The Umbilicus marinus, *or* Moon's eyes, *are known to us as* Venus navels, *of which there are many kinds. And though the Writer did not give us any pictures of them, we do show them, since we have many that agree with the description. The first kind, (see plate* XX. *letter* A.) *is the biggest one. The second kind by letter* B. *is exceedingly shiny and beautiful. The third is pictured by the letter* D. *and is on top completely covered with pearls. The fourth kind is the one from the Mediterranean Sea, indicated by the letter* C., *it has a reddish flesh color, and truly resembles a Navel. Letter* E. *is the same from below. Letter* F. *is another kind, but roundish and with beautiful designs. Letter* G. *is yet another kind: it looks like a greenish gray and has on the bottom a feebler design.*

CHAPTER 9

Cochlea Laciniata. Little Ruffs and Spurs.

These are flat Winkles, their inner whorls only slightly raised, and of two kinds.

I. *Cochlea Laciniata.*[1] In Dutch Little Ruffs; [it] is flat with few whorls, entirely rough on the outside, it has crooked flaps that stick out from the whorl's edge, and these are striped or furrowed as well, between the flaps it has little spines that are not as thin though blunt, and which cover the entire shell; there are usually 5 large flaps per circumvolution, and each one is divided in front by a cleft; there are some that, instead of flaps, have blunt spines all over their bodies; normally, their color is gray, without a sheen, a little red mixed in, seldom clean, but covered with white sea grit;[2] they are a true Mother-of-Pearl inside, and their lid is a round little slice, thin, dark brown, hollow on the outside with a dent, protuberant inside; the biggest ones are the size of a rixdollar, the normal ones that of a schelling. Their shape looks like the old collars, called Ruffs.[3]

II. *Calcar*[4] or spurs, are barely the size of a schelling, even flatter and thinner, with a noticeable edge along the whorls, that all around shows many protruding corners, hence resembling the shape of a spur, a dirty gray on the outside, rough and jagged, inside Mother-of-Pearl; its lid is a small round little shield, a little pressed down on the outside, with a dent, shiny like a reddish pearl. Both are fit to eat. They occur on such beaches that are flat, and have a mixture of small stones and coarse sand, for instance by the red mountain[5] in the Bight of Ambon. Those one wants to keep, need to be cleaned of the sea grit with aqua fortis, and carefully scraped with a small knife, taking care not to break off the protruding corners. The spurs have two shapes; some have the uppermost whorls raised like Winkles, though they are decked out with protruding teeth all the way to the inside; others are flat, straight like a spur.

The Cochlea Laciniata, *also* Little Ruffs, *are called* Little Dolphins *or* Bearded Men *by us; see their picture on plate* XX., *indicated by the letter* H. *They are a brownish gray, sometimes mixed with some red; though Mr.* Vincent[6] *shows one, that is entirely incarnadine: I never saw another one like it.*

The other one, called Calcar *or* Spurs *by the Author, we call little* Sunshell: *see it on the same plate, by the letter* I.; *to which we added the big* Sunshell, *indicated by the letter* K.[7] *this one is a grayish white, also quite rare and unusual.*

CHAPTER 10

Trochus. Bia Cucussan.

I call all such whelks *Trochus,*[1] that have the shape of a top, with which Boys play, or of a funnel turned upside down, to wit, having a wide foot, coming to a sharp point, like a short pyramid or cone; of which four kinds are noted.

I. *Trochus primus sive maculosus,*[2] is at the bottom three or four fingers wide, and not much higher, with protruding little corners on the sides of the uppermost whorls, somewhat toothed, but the side of the lower whorl is smooth; the outer skin has black and reddish spots, without a sheen; but the inner one is of a base Mother-of-Pearl; the mouth is narrow, and elongated; but the little shield, which covers the same, is round, very thin like a tin, flexible, the color of honey and has a large number of rings that come to a point.

J. d Later: fe.

II. *Trochus secundus,*[3] has a narrower *basis* or foot, but a higher top, granulated on the outside or generally covered with little kernels, a pale red, not speckled, and of a pale color.

These two are what one really calls *Trochus.* In Malay *Bia cucussan,* after a certain kitchen implement, which resembles a funnel turned upside down.[4] In Dutch *Tops.* In Bandanese *Tombor.*

III. *Trochus tertius sive papuanus, ut & Trochus longaevus.*[5] In Malay *Cucussan papouan,* is as large as the first kind, but is very coarse and wrinkled over its entire body, and what is more, the edges of the whorls are covered with knobs, thick and with a shell as hard as stone, on the outside a pallid gray, with some green in it, yet clear on the inside, without Mother-of-Pearl; its shield is also like a thin tin, and the animal can pull it back very far; the animal itself is very hard and has a tough meat, not fit to be eaten. This *Trochus* does not live in the sea, but clings to steep rocks that are splashed by seawater. On Amboina one finds them no bigger than a small schelling;[6] but in the Papuan Islands, *Manipa* and *Kelang,*[7] [one will find them] larger than a rixdollar. This animal is so tough, that someone who had not examined it would not readily believe it; at least, I would not have dared to write this on someone else's say-so, if I had not experienced it myself. The Papuas say that it can be kept for an entire year without food or drink, wherefore it is their custom, to put these little whelks with their clothes in their Tomtoms or straw trunks, as a keeper of their household stuff; and if the animal were to die before its appointed time, they think that something was stolen from their trunk. In the year 1675 I was sent about 12 of the largest kind from the Papuan Island *Messoal,*[8] and I put them in a large earthenware dish in my room, and kept them alive more than two months, and this after they had been underway for a month already; then, from a wrong kind of compassion, I put some water in the dish, and placed a rough sea rock in it, so that the little animals would not die from hunger; but they crawled immediately out of the water onto the dry side of the dish, while those who stayed in the water gradually died, so that I had lost half of them by the fourth month already; thereafter I let the rest crawl into the dish, and I found that the last one died in the ninth month; from which I conclude, that these animals suck their food from the briny moisture on the rocks that they cling to, and that they could not survive in water; though it is to be marveled that, nevertheless, they have the hardest shell and flesh of all whelks; wherefore they quite rightly are called *Trochus longevus,* or long-liv'd *Tops.* Since that time I have also found them on Nussanive's steep hook,[9] which did become covered with water, with the flowing tide, but they gradually crawled up higher; I kept those for 7 months, and after that even sent them to Batavia while they were still alive, so that there is a chance of sending these little whelks to Holland while they are alive. One could compare them to Bears, who, as is told, can live for half a year by sucking on their paws, and so sustain life;[10] because it is certain, that this animal must get its food from the tough slime, that it brings with it. In the year 1693, such a *Trochus* still lived after it had been locked up for an entire year.[11]

IV. *Trochus quartus,*[12] is smaller than the foregoing, and tends towards the shape of Winkles, because the tip is somewhat elevated, and the mouth, unlike the foregoing, is not below but more to the side; the animal has a thick, as if double, lip on the same, so that the entrance remains narrow; it is the size of a small schelling, with round whorls, and is covered with coarse grains, a pale red, mixed with green and gray, inside a little Mother-of-Pearl, and a thin, longish little shield. We call this kind, Fat Lip, that is *Labeonem.*[13]

Except for the third or *Papuan* one, all these kinds are fit to eat, but one should cook them for a long time; in order to get to them, one has to smash their houses, because they pull themselves so far inside, that one cannot get at them otherwise.

These are the kinds of the first main genus, which have Mother-of-Pearl inside. One will also find some *Quisquilas*,[14] that is riffraff,[15] or small fry that belongs to this genus, which I do not deem worthy to be described; howbeit the Collectors, who want to supply their Cabinets with all kinds, should have them. That should be understood of all the upcoming kinds, which also have their own *Quisquilias*, that is not mentioned by me.

The Trochus, *or the Author's* Tops, *are called by us* Pyramids, *and also* Begyne's Turds.[16] *He only supplies us with five pictures, which are indicated on plate* XXI., *with the letters* A.B.C.D. *and* E., *the others he did not deem necessary to describe; but since His Honor only shows us Ambonese ones, we thought it necessary to show some, especially the best ones (herewith), from other places;*[17] *see on the same plate* XXI. No. 1. *is an exceptional one of a light-brown color, with a thin white band all around it, made from fine white little pearls.* No. 2. *is another one, with a higher top and sharp points; is banded blue, orange and brown.* No. 3. *resembles letter* A. *though its joints are knobbly.* No. 4. *has the same shape, but instead of having colored flames, it has green, red and white spots.* No. 5. *has raised knobby rings, and has red and white spots.* No. 6. *is flatter, but has a sharp top, and has red and white spots.* No. 7. *has a higher top, and has notches or waves on its circumvolution, its color is white and brown.* N. 8. *is similar, but not notched as much.* Nos. 9. and 10. *have the same color, but differ in shapes and knobbles.* No. 11. *is a very fine striped one, with brown and gray spots, but once this layer has been removed, it is Mother-of-Pearl, beautiful blue, red and green, reflecting. Of all those that are pyramidal, only No. 12. excels. It is Chestnut-brown, all around with fine and even stripes, is very rare, and is called, from its shape, the Sea Barrel.*

CHAPTER 11

Cochlea Valvatae. Bia Tsjonckil.[1]

The third Main Genus of the univalve whelks includes the *valvatas sive semilunares*,[2] which have the shape of a small Winkle, and a little shield, the shape of a half moon, as their lid; the foregoing have a round shield, which does not contract when they are cooked, but turns sideways; the Ambonese generally call them, *Matta tahettul*, [*sponte aperientes*] [3] because it is easy to remove the cooked meat along with the lid. There are two kinds, smooth and striped ones; whereof we will describe the smooth ones in this chapter, and it has the following varieties.

I. *Valvata laevis prima sive vitellus*,[4] in Dutch: *Yolk*, is a round Winkle, its tip not sticking out, smooth and yellow like a yolk, so that one could think it was an egg yolk, if it were not adorned with some white little eyes arrayed on the lowest whorl; it is smooth and of a beautiful white inside; the lid is a half moon, which is so smooth and white on the outside, that one might hold it to be a little piece of white porcelain, but the edges are somewhat jagged and seamed; the inner side has a faded color, where it fastens onto the animal, and it closes level with the mouth, so that the animal cannot contract it. These are rarely found, mostly on beaches that have a mixture of sand and pebbles, on *Hitu*'s coast.

II. *Vitellus pallidus*,[5] is somewhat larger than the foregoing, of a pale white color, the mouth has a protruding corner at the bottom; the shield is also white, but far less beautiful than the first one, lined on top with deep furrows: it is called *the Jew*.

III. *Vitellus compressus*,[6] resembles a yolk that is spread out, because it is flatter than the foregoing, smooth, the color of liver, without any speckles; the mouth rather long but narrow, with a thin dark-brown little shield.

Fol. 78.
N°. XXII.
N°. 1
L
N°. 2
N°. 3
F
N°. 4
I
M
K
N
N°. 5
D
A
N°. 8
B
C
H
N°. 6
G
N°. 7
E
O

IV. *Valvata quarta*,[7] has the shape of the first one, but smaller, except that the mouth sticks out a little at the lower corner; its entire body is white, but along the center of the whorls one will find one and sometimes two or four rows of speckles that look like black drops; the shield is a grayish white, and is grainy along the edges. A smaller variety has 4 or 5 rows of black or brown drops across its white body, and has the name *Canrena lima* because of it.

V. *Valvata quinta*,[8] has the same shape, but it has a dark-gray or liver-colored body with white bands along its whorls, but most have only a white band; the little shield closely resembles the foregoing.

VI. *Valvata sexta*,[9] is smaller than all the foregoing ones, it has a dark liver color, without any designs, and is black around the mouth, which is why they are called Black Mouths; the little shield is also as hard as stone, a dirty white, and grainy.

VII. *Valvata septima sive albula*,[10] has few whorls; its tip is raised somewhat, with a wide mouth, that has a dark-brown or honey-colored lid over it, thin as a plate made of horn; most are exceedingly white. Another kind[11] has a black spot only near its mouth. The third kind[12] is rounder, an orange yellow over its entire body, with similar lids. These *Albulae* are specifically called *Issii palessu* by the Ambonese, that is, those that have more flesh than they can stow, because they crawl along with so much flesh outside, that one would think it were impossible that they could stow the same in their house, yet they are able to pull it in, and can also close their mouth with that dark-brown lid.

VIII. *Valvata octava sive tenuis*,[13] is smaller and has a thinner shell than the foregoing *albula*, it also has a rounder body, some are black, some have a faded, light-brown color, both [kinds are] decorated with little white snakes.

IX. *Valvata nona sive Gothica*,[14] is small, round, shaped like the little Black Mouth; somewhat violet or purplish at the sides of the mouth, its body is white, but covered with faded Characters, as if they were old Gothic letters; it closes off its mouth with a small shield that is like porcelain; one also calls them *Bia sarassa Kitsjil*, to distinguish it from the large one, which is the harp.[15] All these *Valvatae* have a hard, tough meat, and are not searched out for food, because they have the power to make you choke.

X. *Valvata decima fluviatilis sive rubella*.[16] In Malay, *Bia mattacou*,[17] that is, red eye; this Winkle lives in fresh-water rivers, where they empty into the sea, and have smooth rocks at the mouth; or on such stony beaches, that have fresh water welling up, as often happens near the roots of *Mangi Mangis*.[18] It is shaped like an ordinary Winkle, with a thin shell, on the outside blackish and somber, with a wide mouth, and a russet color on the edges, which gives them their name; the little shield is also in the shape of a half moon, smooth and shiny, on top it has a protruding little corner, like a tooth; it has veins that are black or a russet dirty-yellow, which run parallel to the twisted or crooked side, with the shape of an agate, smooth and shiny. One will find them sticking to hard, red stones, like the *Patellae*,[19] though it is easy to pull them off; it often has so many dirty-white warts, like grains, on its back, that one can hardly know the shell, these are its Young and they stay there, and will faithfully die there; and if one squeezes them, one already will find a tiny slimy creature inside. They are fit to eat, with a sweet taste, and they are brought to market as such, one will find them in relative abundance at the mouths of large rivers, but with this difference, that the true Red Mouths grow on hard and russet rocks, as they do on *Hitu*, which is why the entire beach is called *Mattacou*, because of similar stones and Winkles. But those that are found in swampy rivers, even though they also cling to hard stones, have little or sometimes no red at all around the mouth, though they have a sweeter taste. When the aforementioned Young have grown a bit, they leave the Mother's shell, and crawl onto the rocks. One will also find those with their warts rubbed off, leaving many yellow little rings, which adorn it. One will also find them in holes in hard, reddish earth on the banks of rivers.

Cochlea valvata, *which we call* Snail whelks; *the Author has provided us with some illustrations, depicted on plate* XXII. *and indicated with letters.* A. *is the first kind, there is no illustration for the* 2d, *but the* 3d *kind can be seen by the letter* B., *the* 4th *by the letter* C. *the* 5th *by* D. *the* 6th *by* E. *the* 7th *by* F. *the* 8th *by* G. *the* 9th *is lacking, but the* 10th *is shown by letter* H.; *while we add two outstanding ones as* No. 1. *and* No. 2.[20] *of which* No. 1. *is the only one known.*

CHAPTER 12

Valvata Striata. Bia Tsjonkil.

These *Valvatae* also have a little shield like a half moon, but they are more or less ribbed or striped all over their shell, and have the following varieties.

I. *Valvata striata prima sive alpina*,[1] is a very handsome Winkle, without a raised tip, but with a wide mouth and a thick shell, which is only striped on top, with a nice black design, as if it were a tangle of faggots, or a wild mountain range, the way the Alps, or other large mountains, are depicted on Maps; the mouth has a thick lip, light yellow on the sides; the shield is a light gray, shiny, hard as stone, and grainy, with a protruding little tooth in the upper corner. There are 3 different types of these; the first is the aforementioned true *Alpine;* the second[2] is rounder and more bossed, with deeper stripes and covered with black little stipples, which one cannot contrive to cohere into a design, which makes it look quite black; the third has pointed black mountains, which is why they are called little Spitsbergens,[3] and has hardly any stripes.

II. *Valvata secunda sive fasciata*,[4] is like the former, except that it has one, two, or three red bands along the whorls, otherwise decorated like the others. It is rarely found, mostly on *Puloron*[5] or other Banda Islands, which is why they are also called *Pulorons.*

III. *Valvata tertia undulata*,[6] is somewhat rounder and more bossed than the foregoing, and clearly striped or furrowed along the whorls, and somewhat toothed on the sides of the mouth; the shell is watered obliquely in black, which is why they are called Camelots;[7] some are watered with a yellowish color instead of black; the mouth and shield are like the foregoing, though the tip sticks out more, and ends in a short point; nor is there any yellow around the mouth, only a dirty white, and it has a broad lip, which slopes down at an angle towards the inside. All the foregoing kinds are really called *Bia tsjonckil*[8] in Malay, as if they said Picks,[9] because one has to get the meat out of the cooked ones by picking at it with a pin; we call them *Valvatas striatas* in Latin, in Ambonese on Hitu, *Matta Cahettul*, that is, those that open their *Matta* or shield by themselves, when they are cooked; one will find them on such beaches, where large rocks stick out of the sand, since at low tide they hide in the sand, but when the tide comes in, they crawl out and cling to the hollow rocks; they are the best ones to eat, because they have a sweet meat, and also yield a goodly sop. The best and the most occur in the Banda Islands; one will also find them on the hook of Nussanive and on *Outumurij.*[10] Those one wants to keep as curiosities should not be cooked, because that will bleach them, but they should be kept somewhere where they will not touch each other; nor should one permit fresh water or rain to get on them, but let them rot while they are dry.

IV. *Valvata granulata*,[11] this one has a white shell, usually covered with rough warts or kernels, arranged in rows, with furrows between them along the whorls; the mouth is very wide, with a grayish, grainy lid; the fresh ones will show some short bristles among the kernels or warts, which makes it seem as if they have black specks. A second kind[12] is smaller, a dirty white that is almost gray, with even coarser warts than the foregoing, though also deeply furrowed and without any sheen; the first is seldom found; the second can be found on Amboina near the Weynitu River, around the roots of Wakkat trees.[13] This kind also includes the small, dirty-white beach kind, common all over the Bay of Ambon.

V. *Valvata sulcata nigra*,[14] this one has deep furrows with protruding round ribs, a black shell that is mossy, but if one cleans it, there will appear small white stripes on the ribs, which makes them look speckled, though the black is always the commanding color; these furrows also make for jagged edges of the mouth; the shield is blackish, grainy on the outside, with a tooth such as with the former ones, and its tip does not stick out at all. They are called *Kima Ahussen* in Ambonese, because if one eats too many of them, they will be hurtful to those people who have a cough, and cause a slight itching in the throat.

VI. *Valvata sulcata alba*,[15] is smaller and rounder than the foregoing, with a protruding tip, a white shell, or with some red mixed in, usually without any design, though some might have a few black stipples on the back; the lip is very thick, and has some protruding teeth in front, which makes the mouth very narrow, and a small shield, like a piece of a nail.

VII. *Valvata compressa*,[16] is almost nothing but mouth, wide in front and with narrow whorls that suddenly come to a point, though without a tip, pressed flat and round; the shell is on the outside somewhat furrowed, with broad flat ribs in between, painted with black and white stripes and spots; the mouth, which encompasses and encloses it, is narrow owing to the same thick lips.

VIII. *Valvata minor*,[17] the *Valvatae minores* belong to the *Quisquilias* of this genus, which one will find everywhere near the edge of the water, and in various shapes, all not much larger than a finger nail. The nicest ones have the shape of the first one mentioned, though they have a thinner shell, and proportionally a wider mouth, smooth on the back, and decorated with black faggots. The second is flatter, deeper furrowed and without any luster. The third has a smooth shell, with a single color, which is usually a light red or yellow. The fourth kind also has a smooth shell, but is decorated with several bands: All of these are common on *Hitu*, where there are stony beaches, but they are unknown in the Bay of Ambon.

IX. *Valvata spinosa:*[18] In Dutch *River thorns.* These are thorny Winkles, which one will not find in the sea, but in rivers, and there are two kinds: the biggest ones are like a thumbnail, an earthy color or a very pale one, the top of the whorls are furnished with blunt thorns; one will find them by river exits, ugly and unsightly, but good to eat: the other ones are much smaller, like the nail of the little finger, black, with a thin shell, and also covered with sharp stiff little thorns; they are common in all rivers, where they sit on stones, and do quite vex people that are passing by because, if one steps on such stones, they will stick to one's foot with those thorns. They are most noticeable when there has been a sudden rainfall during dry weather. They are called *Hehul* in Ambonese.

Cochlea striata, *are for the most part like the former ones, though they differ, in that they have raised stripes. There are many kinds of these as well as of the former ones, of which some are called* Snailwhelks with White Mouths, *and others* Moonwhelks, *because they have on the inside around their mouths half of a white circle, wherein one detects the shape of a half Moon. On plate* XXII., *by the letter* I. *one will see the first kind, and by the letter* K. *the* 2nd. *by letter* L. *the* 3d. *the* 4th *by the letter* M. *the* 5th *by the letter* N. *the* 6th 7th *and* 8th *kinds are not illustrated, but the* 9th *is by the letter* O., *thus far the ones sent to us by the Author. The ones that follow were added by us because they are so rare.* No. 3.[19] *is the true* Moonwhelk, *of which very few are found.* No. 4. *is a second kind:*[20] *both of these have black and white spots.* No. 5. *is a* 3d *kind,*[21] *white and black with Orange bands.* No. 6. *is a* 4th *kind,*[22] *with brownish-white and black spots.* No. 7. *is a* 5th *kind,*[23] *a blackish gray with white bands and* No. 8. *is a true wide-mouth, with black and white spots. One could add more kinds, but be content with seeing the finest.*

CHAPTER 13

Cassides Tuberosa. Bia Cabesette.

The fourth Main Genus of the Univalves, includes those that have the shape of Casks[1] and ordinary snails; first the Casks, which we divide into, 1. *Cassides tuberosas,* 2. *Verrucosas,* 3. *Laeves,* 4. *Murices.*[2] *Cassis tuberosa,* or bossed cask, has the mixed shape of a snail, and a *Voluta,* because it has a broad head with many circles or whorls that come together, like the head of a *Voluta,*[3] but the body is big and bossed, like a *Cochlea;* all of them have a long narrow door or mouth, of which the outside is curled over, or thickly edged; they cannot close their mouth too tightly, since they only have an oval little bone, or a dark-brown, thin little shield, which they can pull quite far inside, so that one cannot see it. They consist of the following four kinds.

I. *Cassis tuberosa prima sive Cornuta.*[4] In Dutch *Horned Casks* or *Oxheads,* these are the largest of this genus, and have two shapes, which become one over time; because as long as they are not larger than one or two fists, they have many blunt horns on the side of the uppermost whorl, to wit, 11 and 12 on one half of a circumvolution, because the rest is covered with the mouth; the front, which we call the head, and where it is the widest, represents the *Voluta* or curl, where one sees the whorls come together, and that ends in the center with a short, sharp point; from this broad head the bossed ridge gradually narrows towards the front, which curves upwards like a tail; this end we call, with respect to the whelk, its tail, but with respect to the animal, it is its head; because the animal sticks its tongue out through this tail, which is hollow inside and open to the front; the back has 2 more rows with dull bosses, and the entire shell is engraved with short grooves, usually a dirty white, with some blackish or dark-brown spots here and there; as mentioned already, the mouth is long and narrow, and has a wide lip on the outside; which is furnished with round teeth on the inside, but is otherwise curled out and over; the opposing lip, part of the whelk itself, is very wide, and covers almost the entire stomach, is very smooth and shiny, as is the outer lip and the entire inner shell, veering from white to a dull yellow; the rest of the outer lip can be seen at various places on the *Voluta*'s door, which indicates, that this whelk grows by adding half a whorl at a time, which then passes over the previous lips, and this means that the animal can either strike down or remove all that stands in its way, by means of some wondrous innate quality. One sees this clearly, when one smashes the whelk, and finds only small remnants of the old lips on the inner whorls, though they are clearly perceptible on the outer *voluta;* and one should know that this is the case with all the subsequent whelks of this genus, especially the *Murex aculeatus.*[5] Now when this whelk gets to be the size of a man's head, it will not have as many small bosses anymore, but only four or five blunt horns that stick out on the uppermost whorl, just like the stubs of a goat's first horns, the earlier and lesser bumps on the upper circumvolution stay apparently on the inner convolutions of the *voluta;* the outer lip is then very thick and broadly curved outwards and

B
N°. 3.
A
N°. 2.
C
N°. 4.
N°. 1.
D

up, and behind it are broad black stripes; the outer shell has no sheen, but is cleanly and handsomely speckled; but they do have this one shortcoming, which is that they are often covered with sea grit and are entirely faded, that is, as far as the back sticks out of the sand, for the greater part of them is buried beneath it; this sea grit can sometimes eat away so deeply, that it makes little holes and dents in the shell, while the horns are often more than half eaten away. The animal has a thin, tough flesh, covered in front with a little thin oval bone, the color of honey and somewhat toothed, resembling the claw of a large bird.

Their name in Latin is *Cassis Cornuta*. In Dutch *Horned Casks* or *Oxheads*. In Malay *Bia Cabesette*[6] and *krang buku*.[7] In Ambonese *Hubussuta*, which means the same thing. In Butonese *Tandaca*. They are not easily assigned to any known genus among the old names, but they almost resemble the *Murex*. I have noticed that some have described or drawn *Aristotle*'s *Aporrhais*[8] in this manner; but since the *Aporrhais*, as its name indicates, is a large snail that, like a lump of stone, hangs from rocks, as if it were dripping from them, it can not be our cask, because one will only find it buried in sand, at times entirely, at other times with only a small part of its back sticking out. One will find them on flat, sandy beaches, in the Liasser Islands, especially on *Oma* by *Haruko*.[9] The Natives place them on coals in one piece, roast them, and smash the skins into pieces, to get the meat out. Those one wants to keep as curiosities, have to be dug from the sand in their entirety, and be of average size, because those have a clear shell, and the brown-black spots will appear after some scrubbing; otherwise, if there still is a little sea grit on it, one should place it in the rain for a few days, then scour it with sand, and rub it with some aqua fortis; but it should not have any flesh inside, because if the dead meat's slime seeps out, it will spoil the beautiful shine on the underside, which one then calls dead,[10] and which no art can restore, and this goes for all smooth whelks.

II. *Cassis rubra*.[11] *Red Cask*. This is a rare kind that is seldom found, smaller than the foregoing, about the size of two fists; the *Voluta* with its point protrudes a bit more, and the back is covered with round, and not very elevated knobs, while between them the shell is decorated with short grooves and ribs, a brown-black and also like a russet color, like chicken feathers; the belly or mouth is red like raw and bloody meat, as is the inner surface, on the one side of the mouth, which is also curled over, and the animal's flesh also veers towards red. One will find these buried in sand as well, like the foregoing, with their backs sticking out at times, and which will then be so overgrown, that they cannot be cleaned anymore; but those one finds beneath the sand, are usually clean, smooth and shiny; one rarely finds them, and then only on the Islands of Manipa and Bonoa, which is why they are counted among the rarest of curiosities; one finds them a lot on Buton, and are much sought after by the Malay for making motley arm bracelets.

III. *Cassis pennata*,[12] is another kind of the foregoing; it does not have knobs on its back, except for a few short ones on the upper whorl; the mouth does not have a curled lip, but becomes a single shell; the belly is not as red, but paler; but the back is more beautifully painted in black, brown and white, like chicken feathers, or like the marbled paper, which is called Turkish.[13] They are so rarely found, that in all my years I obtained only one, which was brought to me from *Ceram*'s northern coast.

IV. *Cassis aspera*.[14] In Dutch *Thorny Casks*. This one is not much larger than an egg, dark-gray over its entire body, and covered with many sharp little knobs that resemble blunt thorns, between them many wrinkles and grooves, and the thorny teeth can also be seen on the whorls of the *Voluta*, which bulges out a little; they do not have a curled tail in the back, like the foregoing *Casks*, at least nothing more than a tiny beginning of one; nor does the mouth have a lip that curls over, though it is somewhat toothed on the outside, while the side across from it is smooth, even, and of a light yellow. The animal closes the mouth with a long, narrow nail, which one could assume to be a *Unguis odoratus*,[15] but it cannot be used as such. This kind is seldom found on Amboina; the most beautiful ones come from the *Tukabessi* Islands,[16] which are next to *Buton*. A plainer sort occurs on *Ceram*'s southern coast, around *Kellimuri*. One will also find them on *Samatera*'s[17] western coast.

Cassides tuberosae, *we call* Casks. *The Author provided us with some pictures, which we indicate on plate* XXIII.; *although the one by the letter* C. *should really be with the* Voluta, *described further on: but we will not depart from the way the Author divided them. The first kind is pictured by the letter* A. *The second by the letter* B. *The third by the letter* C. *And the last by the letter* D. No. 1. *is the one we call a Cask: It is also called the knitted Cask or Whelk, because it has a host of small notches all over.*[18] No. 2. *is another kind, with brown flames and slightly knobbed; it is very uncommon.*[19] No. 3. *is another kind with fine and sharp notches;*[20] *and* No. 4. *has round bands all over, which have some brown spots.*[21]

CHAPTER 14

Cassides Verrucosae. Knobbles.

This kind of whelk consists mostly of knobs, like certain small, knobbly glasses, called Pimpeltjes,[1] for drinking brandywine; it is difficult to class them with particular kinds, because nearly every Island knows of a special sort; which is why I will only present the following ones, since they are the best known.

I. *Verrucosa prima sive Ceramica*,[2] is the largest of this genus, with a protruding head, like a *Turbo;* the rest of the shell is shaped like a *Murex.* The whorls have 7 knobs that stick out, resembling blunt thorns, which follow along the whorls until the very tip, and there are smaller ones on the back, with folds in between; the tail does not have a curl, nor a curved lip; the shell is faded on the outside, the color of soil, and black on the knobs, also very thick and hard, as if it were a knobbly stone; they are a dirty white around the mouth, and the animal closes the same with an oval *Unguis*,[3] the color of dark honey, and which, in a pinch, can be used as incense. One will find them on *Ceram*'s southern coast, where the beaches are covered with small, black stones.

II. *Verrucosa secunda*[4] is the size of a chicken egg, its head not sticking out as much as that of the foregoing, though it resembles the shape of a *Voluta* more; each whorl has 8 knobs per circumvolution, broader and more blunt than the foregoing, with a notch on one side, as if they had been forced together; the rest of the back is covered with short thorns, with ribs in between; the thorns get bigger around the snout, and the snout does not have a curl; the lip is single and thick and notched; the color is faded, a pale white, but the thorns are black, and for the most part so much covered with sea grit that they are difficult to clean; others have a thinner shell and have semi-circular scales near the mouth, which are the Young and partially grown knobs; a third kind has a flatter head, almost like a true *Voluta*, and with a whiter shell; a fourth kind has a tip that sticks out further, while the thorns are smaller though more pointed, black around the mouth, as is the rest of the shell. These are all considered ordinary Knobbles, that are well known and which occur on all kinds of beaches, though mostly on those that have a lot of coral stones, especially on the small Island called *Nussaanan*,[5] or Dove Island, that lies between Amboina and Ona;[6] its *Unguis* can be used for incense. They are called *Bia papuwa*[7] in Malay.

III. The third kind has a wider head, on the outside white and chalky, with blunt knobs,

F
N°. 1
B
C
G
A
N°. 4
E
N°. 3
N°. 2
I
D
H
N°. 5
N°. 6

wrinkly in between; white around the mouth, with a little yellow or violet mixed in. They are called Banda Knobbles,[8] because they occur the most in those Islands.

IV. Wide-mouthed Knobs,[9] have the form of *valvata,* to wit, a wide mouth, and a thin edge; but the shell is thick and usually covered with blunt knobbles, which are not as black as the former ones. They occur the most on *Leytimor*'s southern coast, and, like the foregoing, are eaten, or sometimes not, because their meat is slightly bitter.

V. Little Yellow-mouths,[10] are no bigger than a hazelnut, with a very narrow and wrinkled mouth, whereon one will see little yellow spots; the shell is chalky, with black knobbles sticking out, like blunt thorns. Then there is another kind, even smaller, but with a wider mouth; the shell has a tawny color, with a few dull knobbles. One calls these brown Knobbles; both of these have even more varieties, depending on the different Islands.

VI. Hairy ears,[11] these come closer to being *Buccina,*[12] because they have a protruding head or tip, while the rest of the body is high and bossed, furthermore ribbed, and covered with round knobs, and everywhere, but mostly on the back, covered with blunt bristles, which remain stuck to it even when they are dried; but the old ones lose them gradually; the mouth is narrow and covered with strange twists, somewhat resembling the shape of an ear, with wide, outstretched lips, that are smooth and shiny, while the rest of the whelk is coarse and pallid; they have an upturned snout or tail at the end; in front open and hollow. A second kind has a flatter belly, and the tail is straighter; the back is less knobbly, but is covered more closely with soft little bristles, like wool, which also stick to it. They are not often found, mostly on *Hitu*'s coast.

VII. Little Toads,[13] also have the shape of a *Buccinum,* rounded and flattened somewhat, with an edge on one side, striped or furrowed on the back, covered with short though prickly weals like the back of a Toad, of an ashen color; the tail does not have a curl, the mouth is wide, with a thick, notched lip. One finds these around the Bay of Ambon, around Victoria Castle, in muddy sand that is mixed with stones; called *Bia Codoc*[14] in Malay.

VIII. *Ranulae,* Frogs,[15] have almost the same shape, though they have a shorter and squatter shell, with a tail in back, that curves up a little; the ridges are also knobbed and wrinkled, but are not prickly, also of a gray and pallid color; the mouth has a thick and smooth lip, notched on the outside. One will find them along with the foregoing ones, though they are rarer, and they resemble a Frog with a blunt tail; they have also a variation, which is very small, coarse, wrinkly, and with a chalky shell, with a truly raised tail, the way the Meerkats[16] hold theirs.

IX. Burs,[17] are also a mixed form of *Buccinum* and *Cochlea,* short and squat, with a raised shell, with round little knobbles that stick out; of a grayish color and without a tail; the mouth's lip comes to a sharp point and is jagged; they resemble the burs that hang from sheep.

All of these, as well as the *Knobbles,* have a dark-brown or blackish nail, wherewith they can close the mouth, and which, for want of real incense, can be used as such. All have a coarse shell, are lusterless, and covered with sea grit and chalk, so that it takes a great deal of trouble to clean them with much scraping and aqua fortis; all of these can be considered *Murices.*

Cassides verrucosa, *we could not bring these under one name, since we have a separate name for each one, and which will be indicated at its proper place. The Author contributed nine kinds, on plate* XXIV., *indicated with letters. The one by* A. *is the branched* Swiss Pants, *to which we add* No. 1.[18] *which we call* Swiss Pants. *The* 2d *kind is depicted by the letter* B. *The* 3rd *by the letter* C., *and which we call the* large branched Mulberry. *The* 4th *kind by the letter* D. *The* 5th *kind by the letter* E., *and which we call the* small branched Mulberry. *The* 6th *kind by the letter* F., *called* Hairy Ears *by the Writer; because its outer skin is covered with small hairs, but we get them mostly clean and hairless, which is why they are called* Oorreliezen[19] or Earwhelks. *The* 7th *kind is the letter* G. *which we call* Toadwhelk, *because of its shape. The* 8th *kind is depicted by the letter* H. *which we call* Hightails. *The* 9th *kind is by the letter* I. *and with these belong also the* Turtle tails, or Bed-Ticking;[20] *the first name because of its small, pointy tail, the other because of its stripes or bands; there are several kinds of these, but the best ones are the* doubly branched *[ones], indicated with* No. 2.,[21] *while* No. 3. *is the smooth one, and we own one that has Orange bands.* No. 4.[22] *is a branched bastard kind.* Nos. 5.[23] *and* 6.[24] *are knobbled, but rare.*

CHAPTER 15

Cassides Laeves Sive Cinereae. Gray Casques.[1]

These are round and smooth, shaped like common snails, but always with a protruding tip, and include the following kinds.

I. *Cassis Cinerea laevis,*[2] this one attains the size of a fist, though it is usually smaller, round and with a smooth back, except that the uppermost whorl has some dark traces of knobbles; the color is an ashen gray, veering at times to a grayish, other times to a drab color; it has a raised and curled tail in back, which is wide open; the mouth has on the outside a thick, round, and arched lip, that has 3 or 4 sharp teeth at its lower corner; the lip across from this is fat and spread out like a wing, very smooth and shiny, and behind it one will see a thick seam, the length of the shell, which is the remnant of an old mouth. These Gray Casques also grow by means of adding, and in such a way, that the old lip stays, but below it a new thin shell comes creeping along which, when it encounters the corners and edges on the other side becomes soft again through the strength of the living animal, and lays itself down, wherefore one can often feel the old seams in the mouth where the new lip overcame it already; and sometimes one can see this old lip running along the back; the inner surface is pallid and partially transparent in the Young animal, but violet and smooth in the Old ones; they close off their mouths with a long, thin, and dark-brown little shield, which is not fit for incense. There is yet another variety,[3] that has a more wrinkled shell on the back, and noticeable teeth on the whorls, that are rather sharp. They gather in September, 20 to 30 in a troop, and lay their eggs, one after the other, but together, on the stones, at a depth of two fathoms. These eggs are short, branched, and brittle, like *Alga Coralloides,*[4] about an inch long, crammed close together in the center, on top with a blunt point, soft, slimy, and a light brown, about the thickness of sail twine; when they have become older, one will find the shapes of small snails inside them, wherefrom grow the aforesaid Casques. I found another form of the gray Casque's eggs, in the year 1694 in October, being a clump like a Duck's egg, a dirty yellow outside, like dirty wax, coarse, fashioned from many thin layers on top of each other, between them small partitions that made for many cells; when cut through, it proved to be denser inside, the color of flesh, a red mixed with white, of a spongy substance, though one could not find any shape of little whelks inside, on top sat such a Casque, while it clung below onto a little stone, and each side had another 3 or 4 little whelks. They occur on *Ruma tiga.*[5]

One calls them *Cassis Cinerea laevis* in Latin. In Dutch *Gray Casques.* In Malay *Bia bewang,*[6] because if one eats its cooked meat, it smells a little like garlic, and those who eat them a lot, will get a strong-smelling sweat from it; but others also call them *Bia Cabesette*

Fol. 84.
N°. XXV.
A
E
C
B
N°. 6
N°. 7
N°. 2
N°. 1
N°. 5
N°. 4
N°. 8
N°. 3
D
N°. 9

Kitsjil. One finds them in abundance in the Bight of Ambon, around the village called *Hucconalo.*[7]

II. *Areola.*[8] In Dutch, *Little Beds,* is also a Casque, with a round and smooth shell, with large square spots on its back, a tawny color, lined up in an orderly fashion like beds in a garden; they do not have the three sharp teeth on the mouth, as do the foregoing, but instead have a row of tiny little teeth: They are seldom found.

III. Bellies,[9] are also entirely round and with a short tip, with a wide mouth and a thin shell; one kind corresponds to the little beds, whereof they carry dark traces on their backs, moreover, it has narrow furrows along the whorls; the other one is somewhat larger and rounder, clearly furrowed without beds, but with a uniform color that is a pale yellow or grayish.

IV. *Fimbriata striata,*[10] striped seams, are small oval little Casques, veering towards the shape of a *buccina,* and are thus called, because they have a wide seam on one side of their mouth, which has short black bands on it, as well as some sharp little teeth; they are plaited along the back of the whelk, smooth, light-brown or tawny; they close off their mouth with a small, yellow and thin little shield, which can be pulled in quite far. One will also find some that are entirely white, or mixed with some russet, but these are very rare.

V. *Fimbriata laevis,*[11] smooth seams, are shaped like the foregoing ones, with a thinner and smoother shell, not plaited, a light-brown color, decorated with little brown snakes; the seam is narrow, though it also has small black bands and sharp teeth. One will also find smooth seams that do not have the little snakes, but instead rows of brown little dots.

A smaller kind[12] belongs to them as well, no bigger than a thumbnail, with a smooth shell, a dirty green, or a little speckled, with some weals along the whorls, and a narrow little seam along the mouth, with sharp little teeth.

Cassidis Laevis, &c. *the Author places these too among the foregoing genera, and not without reason, because they are also Casques and Casks, but we have different names for them, to wit,* Bezoar[13] Whelks, *because of the resemblance in terms of color and sheen; yet they are very different from one another. The first, which is shown on plate* XXV., *near the letter* A. *is the common* Bezoar Whelk. *The* 2d *kind is by the letter* B. *and is called by us the* small spotted Bezoar;[14] *there are two more kinds of these, one somewhat smaller and with slighter spots and not as jagged, while the other is much larger, and is present at* No. 1.[15] *Add to this the* striped Bezoar, *depicted by* No. 2.[16] *of which few are known.* No. 3. *is a* wild Bezoar,[17] *much lighter and thinner, than the other one: also very rare.* No. 4. *is an* obliquely striped Bezoar.[18] *The Author's* 3d *kind is by the letter* C. No. 5. *is another kind of this, but with deep furrows.*[19] *The* 4th *is by the letter* D. *and the* 5th *by the letter* E. *of which we show 4 other kinds, indicated by* Nos. 6., 7., 8. and 9.,[20] *of which the last is very uncommon, because it has not only a spotted seam near its mouth, but has another one running along its back.*

CHAPTER 16

Murices. Bia Unam.

The fourth kind, of the fourth Main Genus, is the branched and thorny Casques, which we hold to be East-Indian *Murices,* and which differ in shape, but not in quality, from those that are found in the Mediterranean Sea, having the mixed shape of a *Cassis* and a *Buccinum;* so that some of this genus may be considered a *Buccinum;* on the other hand, some among the *Buccina* resemble these Casques. There are the following kinds.

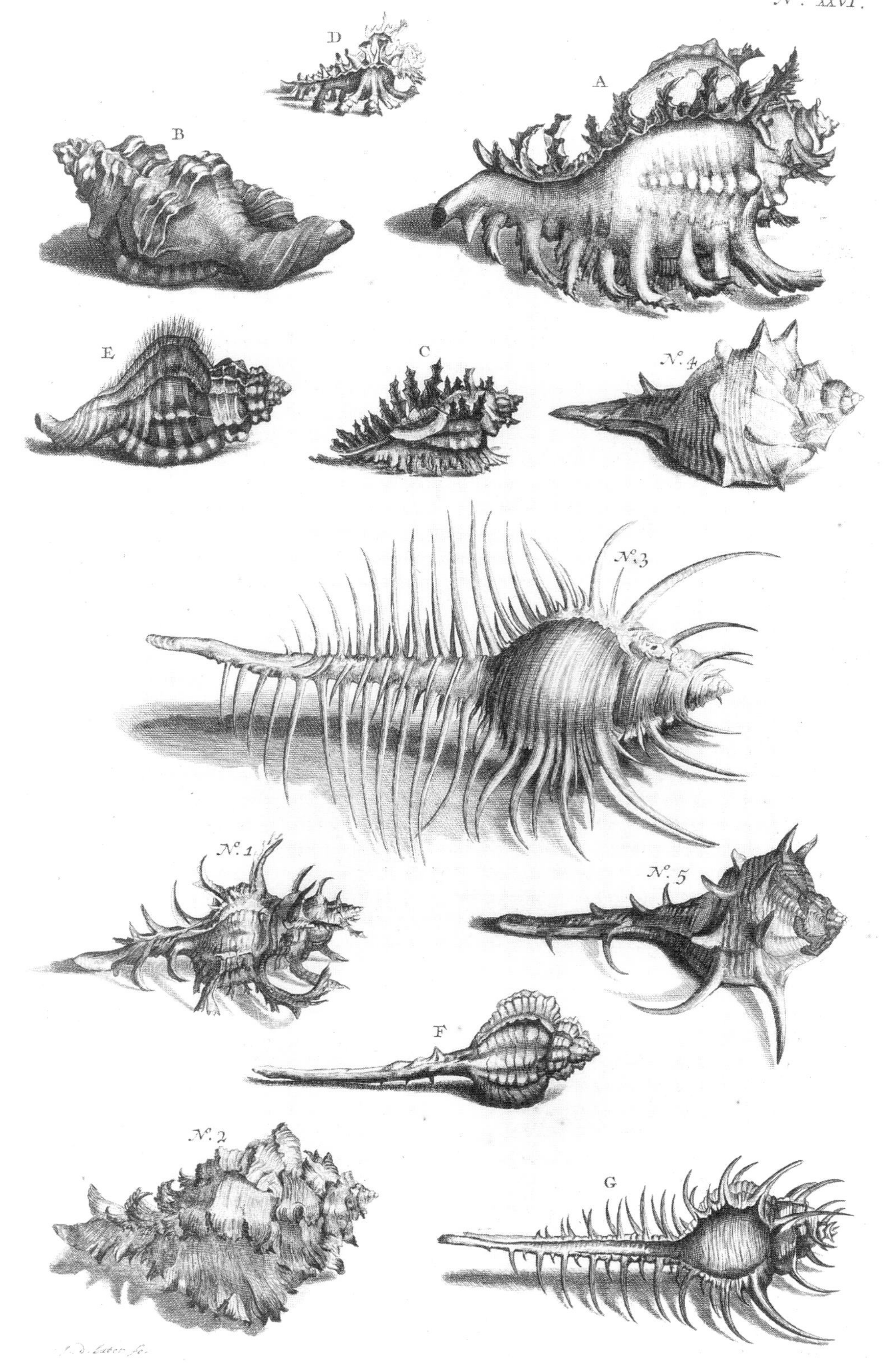
D
A
B
E
C
N°. 4
N°. 3
N°. 1
N°. 5
F
N°. 2
G

I. *Murex ramosus.*[1] In Dutch, Branched Casques, in Malay *Bia unam.*[2] In Ambonese *Lay noa* and *Palay noa,* but *Sasalon* on Leytimor. This one is the largest and most important of this genus, with a short shell that is pressed together and raised, and which seems to be triangular, of which the two most oblique sides form the back, and the third one the belly; every side has an edge furnished with crooked, thorny branches, to wit, 5 bigger ones on the body, and 3 on the tail or snout, which is broad and somewhat raised; each branch is divided again into other, smaller ones, wherefore it is entirely spiny, all of them bend backwards a little, hollow inside, and opened a little in front, as is the tail or snout; furthermore, the shell is coarse, ribbed and covered with blunt knobs along the whorls, the front part is clear, the back half chalky, which happens, because the front half is always the youngest part; one can clearly tell with this whelk, how they grow by addition, to wit, a third of a circumvolution each time, whereby all the opposing branches, as we already have mentioned several times, are softened and lowered by means of that remarkable attribute of this animal; the remains of the branched rows can still be seen on the head, which is thorny up to its very tip, but one will always see three perfect rows, as well as three snouts; this whelk is white on the inside, smooth and shiny like porcelain, a light red or incarnadine around the mouth, as are the teeth, which one sees on the branched rows; the mouth is closed off with a roundish shield, with a protruding little corner on top, and somewhat curved, like the claw of an animal, the size of a schelling, sometimes that of a rixdollar, as thick as an oaten straw, or a little more, but one side is always thicker and rounder; the outside is the color of horn, or a dirty gray, divided into several raised rings, the way they grew together, which usually corresponds to the number of additions, or thorny rows; the inside also has many rings running into one another, wrinkly and a dark brown, like resin,[3] but the one side is round and smooth. It is this shield that is the true *Unguis Odoratus,* to wit, the broad kind, called *Blatta byzantia*[4] in Apothecaries, about which more later.

This one also has a smaller kind,[5] with a whiter shell, and with shorter branches, which stand in five rows, the way the former one has three; its *Unguis* is also brown-black, not much bigger than a finger nail, also crooked, and is considered the best of all; the first kind stays at sea at considerable depth, and is found a lot in Ambon Bay by Cape *Martyn Alfonso,*[6] though usually not much larger than two fists, but on the *Aru* Islands,[7] and on *Nova guinea,* they can be as big as a head, and its *Unguis* bigger than a small hand; the smaller kinds one will find on the *Tukkabessi* Islands.[8]

II. *Murex Saxatilis,*[9] has the size of the foregoing, but its head or *turbo,* sticks out further, so that it looks more like a *Buccinum;* it does not have the three thorny rows, but has protruding knobs and wrinkles over its entire body; the mouth is simple, with a single lip, inside russet or yellowish, also without a snout; its *Unguis* is thin, flat, and plain. It is rarely found on very stony beaches, such as the one by old *Hative,* and so on.

III. *Murex Minor,*[10] has the following four varieties: 1. The gray one, which is as oval as a *Buccinum,* with a coarse body, wrinkly and gray, with 3 rows of blunt branches. 2. The black one,[11] commonly called the Burned Whelk, or Caltrop, in Malay *Bia papua;* this one is no longer than one's little finger, furnished with three rows of black and sharp branches, just as the entire whelk is black, as if it were burned, but in such a way, that one can see white furrows in between. 3. The brown one, which has the same size but has longer branches, and has a dark brown color. 4. The pale one is a pale yellow or a dirty white, with a broad lip at the mouth, a broad tail as well, that has blunt thorns on it. All four have a small *Unguis,* which, in a pinch, can serve as incense. The second kind is commonly encountered on all beaches that are rocky, but the other three are rarely found.

IV. The Little Scorpion[12] is a small *Murex*, with a round, blunt head, and a long straight tail, that has blunt branches on it, that have the shape of Scorpion legs, sometimes a dark gray, sometimes a dirty white; this one is rarely found.

V. The Dried Pear,[13] this is a raised, bossed, little whelk, with a twisted tail, its body covered with knobs and wrinkles, and with 6 or 7 rows of bristles as well, which are stuck to each other, as if they were little skins, though they are readily removed when scrubbed, if one does not want them there; its color is russet or pale brown, with black spots here and there; the mouth has a thick lip, with some furrows on the inside, and is notched.

VI. *Haustellum.*[14] In Dutch, Little Scoop. In Malay *Bia Sibor*, also has a round head, wrinkled and with three seams, without thorns: the tail is quite narrow, and easily 3 times longer than the whelk, hollow inside, through which the animal sticks out its tongue, and on top one will see 2 or 3 short little thorns; they are reddish in the mouth, and its *Unguis* is quite wrinkly and furrowed on the outside, and the same can be used for incense.

VII. *Tribulus.*[15] In Dutch *Spider.* In Malay *Bia Duri Lemon,*[16] this one is shaped like the foregoing one, but it has long sharp thorns on its three seams like Lemon thorns, though a little curved, and which also run along the tail, a uniform gray over its entire body. There is a rare variety[17] of this, with long thorns, that stand much closer together, like the teeth of a comb, and which have other smaller thorns where they originate; its *Unguis* can also be used for incense. One will find the first kind abundantly on flat sandy beaches, in the Bight of Ambon, and are a real plague to those who pull the Sean nets,[18] because the thorns hurt their feet. But the other kind is rare and seldom found, and is therefore counted among the rarest of curiosities. One calls them Little Combs or Lice Combs, in Malay *Bia Sissia.*[19]

The Murices, *we will, as we have so often done before, ignore the Author's names, and call them by the names which are known to us. The* First *kind, is depicted on plate* XXVI., *by the letter* A. *and is called the* Curl whelk *by us, because of its curled branches. I added* No. 1.[20] *because it is so uncommon; it is also called the* Heart whelk. *The* 2d *kind is shown by the letter* B., *which we call the* Foot whelk, *though also the* dried Pear. *The* 3d *kind, by the letter* C., *is called by us* Brandaris[21] *or sometimes* Burned whelk *as well, because it seems burned black.* No. 2. *is a double one of this kind,*[22] *and is very rare; the background is white and the branches or curls are black; there is yet another kind with black bands and curls. The* 4th *is by the letter* D., *of which I know a black, and a white one. The* 5th *kind, is by the letter* E. *and the* 6th *by the letter* F. *which we call the* Snipe's Head,[23] *and not unjustly so, because it looks like one; the* Feitemaas[24] *gentlemen display two, that are entirely blue,*[25] *which is highly unusual.* No. 4. *is a branched* Snipe's Head;[26] *and* No. 5. *a thorny one,*[27] *of which only one is known to me. The* 7th *kind, indicated by the letter* G. *is called the* Spider, *of which we depict a second, double one, by* No. 3.[28] *a very rare piece, worthy to behold, sent to us by Dr.* D'Aquet:[29] *Miss* Oortmans,[30] *has a similar one in her outstanding collection, as does Mr.* Van der Burg,[31] *another one is at Mr.* Vincent's, *and the last one resides with Mr.* De Jong,[32] *and besides these I know of no others like them.*

CHAPTER 17

Unguis Odoratus. Sche Chelet. Unam.

Now we will describe in particular what kind of things are known as *Unguis Odoratus,*[1] and from what kind of whelks they were taken.

Unguis Odoratus, is a cover or lid, in the shape of a horn or a claw, without any beauty, painted a color of dark honey, or a dark resin,[2] with which some whelks close off their mouths. The Ancients only got it from Purple Snails; but we will show here several whelks which bring forth the same, as do all the foregoing *Murices*, and several of the *Turbines*[3] and *Buccina*[4] that follow. This *Unguis* is also called *Onyx Marina*, and is a well-known incense throughout the Indies, being the *Basis* or the Chiefest among all of them, (*Thymiamata*)[5] to wit, the one that is censed on coals, or with which one *perfumes*, like *Aloe*[6] among pills; by itself it does not have an agreeable scent, but when broken into large chunks and laid on coals, it will first smell like roasted Shrimp, but then will immediately veer towards Amber, or, as *Dioscorides*[7] would have it, to *Castoreum;*[8] so that, if censed by itself it is not very pleasant, but when mixed with other incense, the same gives, so to speak, a manly power, and durability; for since most incenses consist of woods, resins, and saps, that have a sweet, flowery or cloying odor, one should mix the Sea Nail among them, in order to make them strong and durable. One can therefore compare this *Unguis* with a Basse in Musick which, when heard alone has no comeliness, but which when mixed with other voices, makes for a sweet accord, and maintains the same: of these I will describe the following.

I. *Onyx major sive taurina*,[9] is the largest and best-known of all, being the lid of *Murex major*, described in the foregoing chapter, as large as a Rixdollar more or less, also of the same thickness, somewhat round at the two corners, then coming to a point; outside a dirty gray, with some ribs or furrows; the interior is on one side thick, round and smooth; the rest is lower, thinner, and curled with many circles running through one another, of a color like dark resin.

The ones that come from *Onin*[10] and *Nova Guinea*,[11] are the biggest and fattest, but many among them have a burned smell, which comes about because the Savages roast the whelks over coals, so they can eat the meat from them. This *Onyx* is quite similar to what one calls in Apothecaries *Blatta Byzantia*, in new Greek *Blattion Byzantion*, which is to say, a leaf from *Byza*,[12] an erstwhile city in Africa, wherefrom one might have brought this incense, and not from *Byzantium*, which is *Constantinople*. Our *Unam* is also bent or curved inward somewhat, like the claw of a large animal; it is the most common among varieties of incense; one does not pound it into a fine powder, but merely into coarse pieces or slivers; in order to spread its fine smell, it should melt slowly over coals, and not be hastily burned; one mixes it with the dry incense *Astangi*,[13] or especially with Macassarese *Dupa*, described in its proper place; it also has some medicinal use other than what one finds in European books; because the Native women who are Healing Masters rub a little of it on a stone, and give this to drink against colic and stomach ache; they also use its smoke against mother sickness,[14] but if so used one has to roast it thoroughly.

II. *Onyx secunda*, is the lid of the next *Purpura*,[15] about half a finger long, a finger's width across, thicker and more black-brown than the previous one, also looking much more like a claw, bent and crooked; gray on the outside, dark-brown inside, and one side is also raised, round and flat, the rest lower and wrinkled, where it was fastened to the flesh. The Malay call this one chiefly *Unam Casturi*, that is, *Onyx moschata*, because, according to them, it has a smell sweeter than the previous one; which is why it is used more for fragrant salves, than as incense on coals.

III. *Onyx minima & moschata*.[16] Malay: *Unam Casturi*, it is as large as the nail of a finger, and is again bent and curved; derives from a smaller variety of the aforesaid *Murex ramosus*,[17] which occur in the Islands of *Aru*[18] and *Tuccabessii;*[19] this one is thought to be the best for incense.

IV. *Onyx quarta*,[20] this one also belongs with the *Blattas*, and is the lid of the subsequent *Buccina Tritonis*, it resembles a broad leaf, 5 inches long, 2-½ wide, the thickness of a straw, much less tortuous, and mostly flat; the outside is gray and chalky, the inside the color of honey, lighter than the previous one, and the one side does not stick out as much; as to smell, it yields to the previous kind, and is only used when there is a lack of the other one.

V. *Onyx quinta*,[21] is also a *Blatta* 4 inches long, and 2 wide, pointed at both ends, dark brown, and clear, with a notch in the center, scarcely as thick as an oaten pipe; this one is the lid of the *Buccina Aruana* and is preferred to the one just mentioned; but it is rarely found, because it only occurs on *Aru* and *Nova Guinea*, wherefrom the Malay and the Macassarese obtain it.

VI. *Onyx sexta*,[22] is thin and flat, and is the size of a thumbnail; on the outside it is ribbed, with protruding circles, it comes from the two previous *Murices*, *Haustellum*[23] and *Tribulus*,[24] common in Amboina, and is only used when there is a lack of others.

VII. *Onyx septima*,[25] this one has the shape and size of the *Onyx tertia*, but is thinner and flatter; they derive from the wide-mouthed, little Knobbles,[26] and they smell good as well.

VIII. *Onyx octava*,[27] is smaller and rounder than the foregoing, but thicker, and rough on the outside; it derives from the *Murex minor*, and is of slight value.

IX. *Onyx nona*,[28] is a completely different kind, that comes from a snail that thrives not in the sea but in swampy rice fields, thin and flat, scarcely the length of a finger joint; on the outside gray and chalky, and shining like silver on the inside. One finds these snails mostly in the rice fields of *Macassar*;[29] around *Marus*,[30] where they are called *Sisse*, and are described in the next chapter; but this *Unguis* smells bad and is not used very often. This kind makes what *Dioscorides*, *Lib. 2 Cap 8.*, writes more probable, in that the *Conchilia*, wherefrom he gets his *Unguis odoratus*, are supposed to grow in swamps in India, where also the *Nardus*[31] grows, which these snails are said to graze upon and from which the *Onyx* would beget its goodly smell; howbeit that till now I have not heard anything about such swamps where the *Nardus* is supposed to grow, on the contrary, it is generally agreed that it grows deep in the heart of the country, and in the mountains: but this can at least serve to say that one should not immediately slight the writings of the Ancients, or accuse them of lying if you do not instantly find something the way they described it; for every day there appear things, which have been unknown for a long time, but which one can find in Ancient writings, which are often not understood correctly: and so one can defend *Dioscorides*' entire text, if one explains him properly; for instance, when he says that the *Onyx* is a cover of a *Conchilium*, and is like unto the one that covers the Purple Snail, one should not understand that to mean the shell or the house, but the lid or little shield wherewith many whelks, especially all the *Murices*, close their mouths, the same little shield they have on their forehead above the nose: which is also maintained by *Nicolaus Mirepsicus*,[32] whose judgment it is that the *Onyx* is a small bone from the Purple Snail's nose: mention of the *Nardus*-rich marshes can be understood two ways; either that there are indeed mountains upstream of the Ganges, which produce the *Nardus*, and where at the feet of them there are indeed such swampy fields with Snails or Whelks, that do have an *Onyx*, which we just proved with the example of Maccassar, and whereof the best kind is called *Nardus Gangitis* by *Dioscorides;* or *Dioscorides* followed the mistake of the common man who, at that time, called *Malabathrum* also *Nardus*, which, as he wrote, wants to grow in marshy places, being unknown in Europe today, although the recent herbalists contend that it is the same as the leaves of wild cinnamon. One should not cavil too much about the smell, whether it should be translated as fragrant or as pungent; for we said above that, of itself, and smelled alone, it is not very pleasant, and that the same smell is judged according to different noses: for it will seem heavy to one, veering towards *Castoreum*, *Bitumen*[33] or roasted Shrimp; while others (as most of our Natives do) find it agreeable, and compare it to *Casturi*, that is, Musk.

The names of these 5 fragrant Nails are, in Latin *Unguis odoratus*, *Onyx marina*, and *Conchula Indica*, the flat thin kinds known specifically, with a new Greek word, as *Blattion Byzantion*, and *Blatta Byzantia* in the Apothecaries: in Arabic *Adfaro tibi*, or commonly pronounced as *Adfar altibi*, that is, savory Claw. In *Plinius*[34] *Lib. 32. Cap. 10.* one will also find *Ostracion*, because it resembles a Pot-sherd: and with the Hebrews *Schechelet:*[35] in Malay *Unam* and *Unam Carbou:*[36] in Macassarese, which is also Malay, *Ambelau:* in Butonese *Lacca nuga:* in Ambonese *Laynoa matta:* in Chinese *Lepi*[37] and *Hiole;* others call it specifically *Tanka nuda*, the little *Onyx*, and the big one *Ambelau* or *Unam Carbou. Schecheleth, Exod.* XXX. verse 34., is one of the four things which together comprised the holy incense, and is translated by most Translators as *Onyx;* which to me seems also most likely although the scholar *Samuel Bochartus*,[38] was pleased to translate it in his *Hierozoico Lib. V. Cap.* 20, *p. 808.* as *Bdellium*, apparently in relation to the fact that young Rabbis who have lived mostly in Europe, have held the same *Schechelet* to be a root, or something that derives from a plant, the way the High-German Jews today still call it Nailroots while referring to the herb *Caryophyllata;* he also read in *Galenus*,[39] that the black *Bdellium* has some white spots, which, because they resemble human nails, he calls *Onyx;* from which *Bochartus* concludes that *Schechelet* must be *Bdellium;* but since one will neither read nor hear anywhere that *Bdellium* is used as incense, and while, on the other hand, all Eastern nations use *Onyx marina*, one would do better, therefore, to adhere to the earlier translation. Nor does one have to split hairs over where in the Desert the Israelites obtained this *Onyx*, which comes from such distant seas, because they could readily get it from neighboring Araby: for it is already apparent in the aforesaid place in *Dioscorides*, that the best *Onyx* came from the Red Sea, and the small black kind from Babylon, whereto it was carried from *Bassora*[40] and the Persian sea; and so they could have gotten it from either side. Proof for this can be found in *Alcasnino*,[41] in his Tractatus on the Water Animals, in the Chapter on *Itar*, as he calls this *Unguis*, which is very similar to the text by *Dioscorides. Bochart.* p. 806. *Avicenna*[42] in the Chapter on *Aphdar* or *Unguis odoratus*, after he has repeated *Dioscorides*' words, adds this on his own: that the best *Ungues* comes from the Red Sea near *Mecha*[43] and the harbor *Judda*,[44] and that it has a pleasant smell, and is normally used in Healing. Thereafter he mentions even more places, where good *Onyx* occurs, to wit, all of *Jemen*,[45] or Happy Araby, *Bacharyn*,[46] and the castle *Abadan*[47] near the *Tigris*, *Bassora*, and other harbors on the Persian Gulf.

Byzantium or *Byzancium*, from which the *Blatta* got their name, was according to *Plinius*, *Lib.* 4. *Cap.* 5, a small region in Africa, near the lesser *Syrtes*, inhabited by the Libyphoenicians, where once stood the cities of Leptis *Major* and *Minor*, *Adrumetum*, *Ruspina* and *Tapsus*, and where one finds today the city of *Mahumeta;* so it appears that the *Blatta Byzantium* were once brought from these aforesaid places. The Chinese are more apt to use the large, round *Vugnis Lepi*,[48] that derives from the *Murex ramosus*, for medicine rather than incense; it is pulverized along with other medicinal herbs, then cooked in an Oyl called *Maju*, from which they make a salve for wounds. The aforesaid *Murex ramosus* is called *Tsohil Le*[49] in Chinese, that is, thornlike whelk, and occurs in the sea near *Quantung*,[50] but larger than in Amboina; only this *Unguis* is known to those people.

We call Unguis odoratus, Blatta Byzantia, *being lids or covers of Winkles. The writer did not give an illustration of this, but we had drawings made for the Fanciers, who may not know the same, of four particular ones that are the most important, and which can be found on plate* XX, *Nos.* 3. 4.[51] 5. *and* 6.[52]

CHAPTER 18

Cochlea Globosae.

The fifth Main genus of the Univalve Whelks includes those, that have a round form or shape, closely resembling the common Snails, as well as some others, which depart a little from the round shape, and which are all reckoned to be *Cochleas Globosas;*[1] consisting of the following familiar kinds.

I. *Cochlea striata sive olearia.*[2] In Malay *Bia minjac,*[3] this one is usually truly round, with quite a wide mouth, protruding ribs along the whorls, and notched at the side of the mouth; it has a light violet color, mixed with gray, though it has many brown speckles on the ribs; usually the size of a fist, but most have a uniform color without speckles: this Snail is used a lot by the Ambonese in order to scoop Oil out of Coconuts, when the same is cooked, and from which it derives its name: for this they chose those [snails] that have a thin edge or shell. The Animal within does not have a lid, but lies exposed like other Snails, and its eggs (*Favus*)[4] are a clump of entangled, thick and white threads, which one sometimes sees dangling from its mouth, but which does not produce any Young.

I found such a Whelk without an Animal inside, in the year 1667, which was covered by a slimy skin, completely filled with a *Melicera,*[5] or Ovary, consisting of innumerable little white branches, loosely hanging onto each other, and covered with white, transparent kernels, the size of Barley, but more oval, while nearly all of them had 2 little black dots, as if they would be the eyes of an Animal; the branches resembled *Lithodendrum calcarium*[6] from which lime is slaked. Today's Greeks call this Whelk *Cocholi batar,*[7] that is, *Cochlea pelagia.*

II. *Cochlea striata altera,*[8] is smaller and has a thicker shell; the thick and round ribs are separated by very small furrows, brown-yellow, and with white eyes; the mouth is narrow, with a thick and heavily notched edge; they are called thick-lipped Oil whelks.

III. *Cochlea pennata.*[9] In Dutch Partridges. In Malay *Bia Culit bawang,*[10] is a wide-mouthed Snail with few whorls, and a thin shell, painted white and light brown, like the feathers of a Fowl or Partridge.

IV. *Cochlea pennata altera,*[11] is rounder than the foregoing, with a thinner shell, almost like parchment, which is why the Natives compare it to an Onion-skin, and is therefore really called *Culit bawang;* it is very light, when the beast is out of it, and the animal, like the ones before and the ones to follow, is also bare and without any lid.

V. *Cochlea patula.*[12] In Dutch Wide-mouth, is almost nothing but a gaping mouth, with few whorls that suddenly come to a point; the shell is very thick, hard as stone, but thin and jagged near the mouth, furthermore wrinkly, gray on its back with black and white spots; the mouth is closed off with a thin black-brown lid. One kind is bigger than a Duck's egg, and is rarely found. Another kind is much smaller and more bossed, gray and chalky.

VI. *Rapa.*[13] In Dutch Turnip, has a round body with a flat head, at the back a short, curled tail, like a pig's tail, with a thin and light shell, somewhat coarse and wrinkled, of a light lemon yellow, and hard to find.

VII. *Bulla.*[14] In Dutch Bubbles, which is a special form of Snails, all rolled up, with few whorls, and a wide oval mouth, with an outside that is longer than the rest of its body, divided into three kinds: The first,[15] is the biggest and has the thickest shell, as large as a round Plum, and speckled all over its body, brown, and black, in the manner of Plovers' eggs, and covered with some kind of sliminess. The second one,[16] has a thin shell and a wider mouth, pale white, and with very fine stripes or furrows. The third kind,[17] is the smallest and the

Fol. 92.
XXVII
Q
N
C
I
N°. 1
B
A
F
E
D
L
O
P
K
R
M
G
H

thinnest, almost like a water bubble, also white, painted sideways with many blackish and brown stripes. The first and second kinds are common; but the third one is rare; when the last one is painted with black and red stripes, they are called Prince's Flags.[18]

VIII. *Cochlea Imbrium.*[19] In Malay *Bia Ribut*,[20] is a fairly flat Snail with a pointed head, and a narrow mouth, deeply notched inside and turned over at the edges; a drab color with dark brown, but with broad, blackish stripes at the sides, as if they were feet; the opposite side of the mouth is also somewhat sharp, so that it looks like a flat Frog, it does not have a beautiful shape or color, though some are kept, because one can perceive the shape of an animal with legs in it. One finds them by the sea shore, under rotted shrubbery, leaves and pieces of wood, both on the beach as well as further inland; often even in the mountains, where no people roam, nor is it likely that they suddenly crawled up there from the beach; wherefore it is generally believed that a strong wind, when it was raining, picked them up down below, and tossed them up there: but I think it more likely that they were generated up there by the rain, because one will find them both small and large.

IX. *Ficus.*[21] In Dutch a Fig, while others call it a Lute, is also a Snail with an exceptional shape; because the body is round, with a flat head, while in back it narrows, like a Pear; wide and with an oval mouth, a thin shell, dark gray or the color of earth, without a sheen, well-nigh wrinkly and coarse; it is kept, however, as a Curiosity.

X. *Umbilicata*,[22] is a Snail that is pressed flat, on the bottom entirely flat, on top bulging a little, with many narrow whorls that run into one another, sharp at the edges; on the bottom it is shaped like a Navel, but around the center one will see an open, small dent, that is notched at the edges; the upper surface is somewhat wrinkled, on top with brownish and white circles and with short, fine furrows between them; the mouth is narrow with a thin little lid: There are also two smaller kinds of this, no larger than a finger nail; of which the one is raised more, gray, wrinkled, and without a sheen; the second is flatter, not wrinkled, but smooth, with small light-brown flames: one would think they were Land Snails, but they are not, since they live on the beach in sea water.

XI. *Arcularia major*;[23] is a small Snail, as large as a thumbnail, with a pointed head like a *turbo*, plaited on its back, and notched on the edges of the circumvolution, with a thick shell and of a dirty-white or yellowish color; the mouth is narrow with a thick lip, furnished with sharp little teeth, and closed off with a thin, small yellow lid.

XII. *Arcularia minor*;[24] is as large as the nail of one's little finger, shaped like the foregoing, but with a raised boss, and not notched, of a dark-gray color, but smooth and shiny, the mouth is very narrow and has a thick lip. The Malay call them *Bia Totombo*,[25] that is, *Arcularias*, because they are used for the little straw trunks, called *Totombo* in Malay, *Totomme* by the people of Ternate, and which are square little trunks, artfully woven by the Inhabitants of East *Ceram* and *Goram*[26] from certain tree leaves, with rows of these little whelks sewn on top with straw yarn, after the little boss has been ground off; when new, it is quite comely, but it will not last a hundred years, because the straw yarn rots all too easily.

XIII. *Serpentuli*,[27] Little Snakes, these are flat snails, rolled up like a coiled snake, that have two shapes, both with thin shells, with a curled lip at the mouth. The first and larger kind, is the Elephant's snout, of a uniform brown color. The second is smaller and flatter with some blackish or dark-gray little flames. There is yet a third kind, smaller than these two, mostly brown and smooth, which does not sojourn in sea water, like the foregoing, but on land, in shrubbery, and among the roots of Trees. Such as these, and even smaller ones with a thicker shell and smoother, are found on the beach of the Persian Gulf, around *Gamaron*,[28] and which we have also found around *Hatuwe* on *Ceram*'s northern coast.

XIV. *Cochlea terrestris*,[29] is also counted among them, because it bears a close resemblance to the foregoing two kinds, and, because of its beauty, is considered a curiosity; its shape is like a common snail, with a thin and light shell, of various colors; most are a light yellow, with one or more white bands; others with brown bands; some are entirely brown, with and without bands, some are brown only for the top portion, while the lower half is white or a light yellow, but they are not found very often.

XV. *Cochlea lutaria*,[30] or Mud Snails, have two kinds, large and small. The large ones have the shape of a common snail, or like the foregoing *Vitellus*,[31] with a thin shell, dark green mixed with brown, like the dried leaves of *Cotihomera*,[32] with some thin, yellowish little veins that run athwart; the mouth is large and round, closed off with a rather thick lid or oval little shield; this is the *Onyx* No. IX. of the previous chapter. They will stick out two horns when they crawl, like other Land snails, while carrying the shield on their back; they have a round mouth below, with which they suck up mud and water. The second kind is smaller, and has the same shape, but has a more pointed tip. In Makkassar they call the first kind *Sisso Capong:* The second sort *Sisso Potir:* in Tombuko,[33] *Wonko* or *Wonke:* on Bali, *Kakol.* In *Makkassar* both are dug up from the mud of the rice fields, while in *Tombocko* also from the muddy banks of rivers, where they grow as big as a small fist; the meat is good to eat, they are cooked in water, and the meat is removed with a Lemon thorn, although one can also suck it out, because their tips are usually broken off; their lids are used there for incense, although both are inconsiderable, and are only used when other *Unam* are lacking; but the smaller ones are considered better. In Tombocko the small one is thorny, and is called *Senipa*, one can keep them alive in water troughs and convey them across the sea.

There are three kinds of the Makkassarese *Sisso.* The first and biggest one is called *Sisso salombe*, which has the size of a small fist, is smooth and black, but if held against the light one will see that two or three blacker stripes run through it; its shield has the shape of an ear, [is] thick, and is as hard as bone, a pale brown on the outside, inside like a dark mother-of-pearl, and not fit for incense. 2. *Sisso capong* is the middle-sized one; and *Sisso potir* is the smallest, with a pointed snout, while its lid is a plain *Unam.*

The lid of the large *Cochlea Lutaria*, is oblong like a blunt half moon, the thickness of a knife, a dirty gray on the outside, but inside whitish and shining like unpolished silver; in muddy rivers they will sink so deeply into the mud, that they will reach firm ground, although one will also see them on rocks, around such muddy places. One could give them the name *Pomatias*,[34] from the Greek, that is, Lid Snails, since they are the only ones among land snails that have a thick lid; one will find them on *Celebes*, *Java*, *Baly*, and *Sumatra*, wherever are muddy rice fields, and where they will be the size of a fist; when the rice fields dry up, they hide in the dried mud, where they will remain, until the rain harvest returns. They are good to eat, if one cooks them in water, or roasts them on coals, after smashing the front tip or tail, one can easily suck them out, or empty them with a Lemon thorn; if one puts them in water troughs, one can keep them alive, convey them across the sea, and transplant them into other pools, like the Romans used to do with similar kinds, which they got from Africa; they are thought to be particularly healthy for people suffering from fever, or those who are getting consumption. They have brought me such snails from *Cuchin*,[35] and I would guess that they also occur on *Ceilon*, anywhere where there are swampy rice fields, and are much sought after for food.

XVI. This genus also has some *Quisquilias*,[36] various small snails and whelks, which do not warrant a particular description; the most important of these, are what are called Little Stones [*lapillos*], nary as big as a finger nail, with a painted head, like the *buccina*, a narrow mouth with a thick lip; some are brown with little white eyes;[37] others are wrinkled and nicely decorated with stripes; there are also other small snails, like unto the *serpentuli*, but with a raised little snout.[38]

Cochlea Globosa, *are the kinds, which we call* Bell Whelks, *of which each one has its own particular name. The first is depicted on plate* XXVII. *by the letter* A. *which is known under the name of* spotted Bell whelk. *The* 2d *kind is by the letter* B., *to which we add another one,* No. 1.,[39] *being the* knobbly Bell whelk, *which is rarely found. The* 3d *kind is by the letter* C. *and is called the* Partridge whelk, *also the* Plover egg. *The* 4th *kind by the letter* D. *is called the* Onion-skin *by us. The* 5th *kind is depicted by* E. *The* 6th *kind with the letter* F. *is the* Turnip bell *or* Turnip whelk. *The* 7th *kind, by the letter* G., *is called the* Agate bowl, *or the* Plover egg; *the letter* H. *shows one from the mouth side; it has many kinds, among them one that is entirely white. The* 8th *kind with the letter* I. *we call the* Magician, *also the* Magic-snail. *The* 9th *kind by the letter* K. *which is known to us as the* Pear whelk; *it is also called the* Spanish Fig, *because of its shape, and there are several kinds. The* 10th *is depicted by the letter* L. *which we call the* Whorl whelk, *while it is also called the* Perspective whelk, *because if one looks into it from below, one will note that it is very deep while gradually narrowing. The* 11th *kind with the letter* M. *The* 12th *by* N. *The* 13th *by the letter* O. *which we call the* Bell snail. *The* 14th *by* P. *is called the* Post whelk *by us. The* 15th *by the letter* Q. *is a* banded Onion-skin. *The* 16th *by the letter* R. *is another kind of* Post whelk, *of which many different kinds are known.*

CHAPTER 19

Buccinum. Bia Trompet.

The sixth and seventh Main genera, are the *Turbinata*, which resemble each other so closely in form, that one could mistake many kinds of one genus for that of the other; but to be understood better, we will call *Buccina* such whelks, whose tutel[1] or *turbo* is smaller than the rest of its body, or, at least, not much longer; on the outside mostly rough, knobbly and ribbed, and these are the ones we will deal with in this chapter. Their general name in Latin is *Buccinum*.[2] In Malay *Bia trompetta*.[3] In Ambonese *Kima Tahuri:* howbeit only the largest ones are to be understood by this, the ones that can be used as Trumpets.

I. *Buccina Aruana*,[4] named after the Island *Aru*, from where they are brought, although they also occur in neighboring *Nova Guinea*, otherwise unknown in the Ambonese region: this is the biggest and most lumpish of all the ones I have seen, heavy and with a thick shell, more than a foot and a half long, a span high, with few whorls that make for a short tutel, with a short tail or snout in the back; scaly and cracked over its entire body, white and without a sheen; the lid is a thin *Blatta*,[5] oblong and black-brown, and is counted among the best *Onyx*, mentioned above in Chapter 17, the Fifth Kind.

II. *Buccinum Tritonis*,[6] is next to the foregoing the biggest and most beautiful of the *Turbinata*, and to this one really belongs the name of *Bia Trompetta*, *Krang Seroney* and *Kima Tahuri*, the way they paint the *Tritones*,[7] or Watermen: we Dutch call them Whelks [Kinkhoorens]; other Writers *Turbo magnus;* it resembles a *Turbo* more than a *Buccinum;* it is round and has a thick body, ending in a long tutel, without a tail, smooth on the outside, but the upper edges of the whorls are notched, as if they were hemmed with strands of *Paternosters;*[8] they are handsomely painted, like the feathers of Hens, or more like Turkish paper,[9] and also shine quite beautifully; the mouth is wide and round, its edge notched on

A.
B
D
N°. 1
C

the right side, and trimmed with teeth, they are as bright as daylight[10] inside, or fiery red, smooth and shiny like porcelain; the Animal has a coarse tough flesh, the thickness of an arm, wrinkled like the neck of a Turtle, with brown and russet speckles, to which an oblong shield is fastened, 5 inches long, 3 wide, the thickness of a knife, which is the *Onyx*, that was described earlier in Chapter XVII, Fourth Kind.

The largest ones are more than 1-½ feet long, 6 or 7 inches high, with the point usually somewhat broken, and its body covered with coarse white and red stipples, which one has to scrape off with a penknife, after they first have been softened with aqua fortis. These are considered to be among the most excellent of Curiosities, and if they are perfect, they can fetch a Rixdollar in these Islands as well. They are seldom found here in *Amboina*, but come mostly from the Southeastern Islands; they stay deep down in the sea, and sometimes crawl into the Weels, such as the one I got here on *Hitu*, of the largest kind, with an outstretched neck as thick as a leg; and when I wanted to cut off the shield, the Beast pulled both my hand and the knife inside, so that I had a hard enough time, just cutting through that tough meat. The *Alphorese* on *Ceram* use these Whelks for their Trumpets, making a hole in the center ring, where they blow into it, and this sound can be heard over a great distance and tells the neighboring Villages to come together, and such shell Trumpets are also used by the *Tartars* in their Armies, as one will find in Father *Martinus*[11] in his Tartar war, and which the Chinese also told me.

The animal has a reddish meat in back, or rather, fat, fit for food, but the meat in front is too tough; one will find pieces of small shells, coral and pebbles in its stomach.

They got the name *Kinkhooren* [Whelks], because they cause a *kinken*[12] or rustling or soughing sound, when you hold their mouth against your ear, and the common people persuade each other, that this is a sure sign of it being genuine, because you hear therein the soughing of the open sea: But there is something wrong here, because one does not become aware of this soughing on every day or at all hours, but only during the day, when the air is moved by wind, rain, or the voices of people; whilst during silent nights, one will not notice any rustling, even though one has the genuine ones. In Europe they also persuade each other, that it is healthy for feverish people, to drink water from these Shells. And they have another rare attribute, which they share with other Whelks, and that is when we have constant rainy weather they sweat so much that there will be drops on them, which I even noticed on some, which I had for more than 16 years already; and this happens again and again, when the rains come, even though one wipes them dry every day.

In order to maintain their living luster, one should sometimes, at least every two years, place them in salt water for several hours, which is called: *to give the Shells to drink;* whereafter one should rinse them with fresh water, and dry them in the sun, and rub them with a woolen rag very softly and for a long time, until they become warm, which makes them glisten like a mirror. These *Buccina Tritonis* are called *Tsjanku*, or *Tsjaku* in Chinese; on the Islands *Liukieuw*, or *Lequeos;*[13] wherefrom the Tartars get them to turn them into Trumpets.

Hist. Antill. Lib. Cap. 19. Art. 8. calls all Whelks *Cornets de Mer*, whereby I understand the white ones, which look like ivory, to be the *Subulas*, or Marlin spike;[14] but the shape of the *Trompette Marine*, or sea Trumpet, is our *Buccina Tritonis.*

III. *Buccina tuberosa*,[15] or knobby Whelk, is much smaller than the foregoing, usually the length of a hand, its entire body wrinkled, and covered with knobbles, with neither luster nor beauty, a dirty white color, yet, because it is so rare, considered a Curiosity, but inside, it is as beautifully white as Porcelain; it is called *Hector.*

IV. *Buccina tuberosa rufa*,[16] or the red knobby Conch, is even shorter than the foregoing, also entirely knobbly and wrinkled, very much covered with sea grit or chalk, and therefore difficult to clean; of a russet color, with black spots on the knobbles; the mouth is round, very wrinkled on the sides, and closed with a thin dark-brown lid, just like the foregoing; neither is usable for incense; one rarely finds them, and if one does, it usually will be on beaches with boulders. On *Tombucco* or *Celebes*' East Coast,[17] this kind is called *Honka*, and is much sought after by their champions, when they want to go to war, not without superstition: because they have to be brown-red on the outside and a beautiful fire-red on the inside; the knobbles around the mouth must also correspond with each other in a certain way; and then they tuck Ginger[18] along with various other small roots in it, also little notes with characters on them, and then they tie them into their belts which they fasten around their loins, believing, that they will henceforth be lucky and invulnerable in Battle; and this makes them as bold, as the Greek hero Ajax before Troy, which is why we gave this Whelk the surname of *Ajax;* just as the foregoing white knobbled[19] one got the surname of *Hector.*

V. *Pseudo Purpura.*[20] Bastard Purple Snail. I call it that, because it somewhat resembles the shape of the Purple Snails from the Mediterranean, and it has red, bloody flesh inside, which one will not see in any other snail; but whether one can get a red blood or sap from it, fit for painting, has not yet been examined; it has a mixed form, of both a *Murex* and a *Buccinum*, so that one could place it with either genus; it is 5 and 6 inches long, pointed in front and with many whorls, with a remarkably thin snout in back; the uppermost whorls of the edges are covered with knobbles, at times as sharp as dull thorns; the rest of the shell is a uniform dark gray, or the color of soil, without a sheen; and if one scrapes off the upper fleece (which is not easy to do), the shell is white, with blackish threads across; the mouth's edge is thin, sawed with sharp teeth; across from it, it is smooth and of the color purple; the Whelk's inner space is furrowed with thin little ribs: the Animal therein, is hard, and as red as raw meat, good to eat; carries on its head an oblong little shield, the length of a finger joint, and crooked like the claw of a beast, just closing off the mouth, a dirty gray and smooth on the outside, dark brown on the inside and somewhat wrinkled. This lid is of the best *Onyx Marina*, called *Unam Casturi* by the Malay, that is, *Onyx moschata*,[21] because it is generally thought, that this is the most fragrant of all *Unams*, but I disagree; because it does not smell like *Moschus* at all, but rather like burned Shrimp mixed with amber and I would prefer the *Unguis* of the first kind, unless it is better in lands other than Amboina. It is described in Chapter XVII, Second Kind.

One will find a reasonable number of these Whelks in the Bay of Ambon, on the North shore, and they appear together in troops, but only in certain months, and then leave again for the ocean deep. Those one wants to keep for Curiosities, should not be robbed of its little top fleece, because otherwise it will look like it has the mange, since one can rarely remove all of it.

VI. *Fusus*, a Spindle;[22] this is an oblong little Conch, of which the top half is a *Turbo* or Cone, whereon the whorls are marked off with deep clefts and, furthermore, furrowed, with a protruding furrow in the center of each whorl, like a seam; the other half is a long and narrow snout, inside hollow like a pipe, and a little curled at the end; it has a grayish color over its entire body, without a sheen, and is covered with a thin fleece, which cannot be removed; but the point, and the end of the snout, are blackish; the mouth is small and round, covered with a roundish little shield, that is of a blackish brown, but not suitable for incense: When just taken from the sea, they are covered with something woolly, that can be easily rubbed off.

VII. *Fusus brevis*, Blunt Spindle,[23] does not differ much from the foregoing, except that it has a thicker body, and a short, thick snout, the whorls covered with knobs, dark gray and

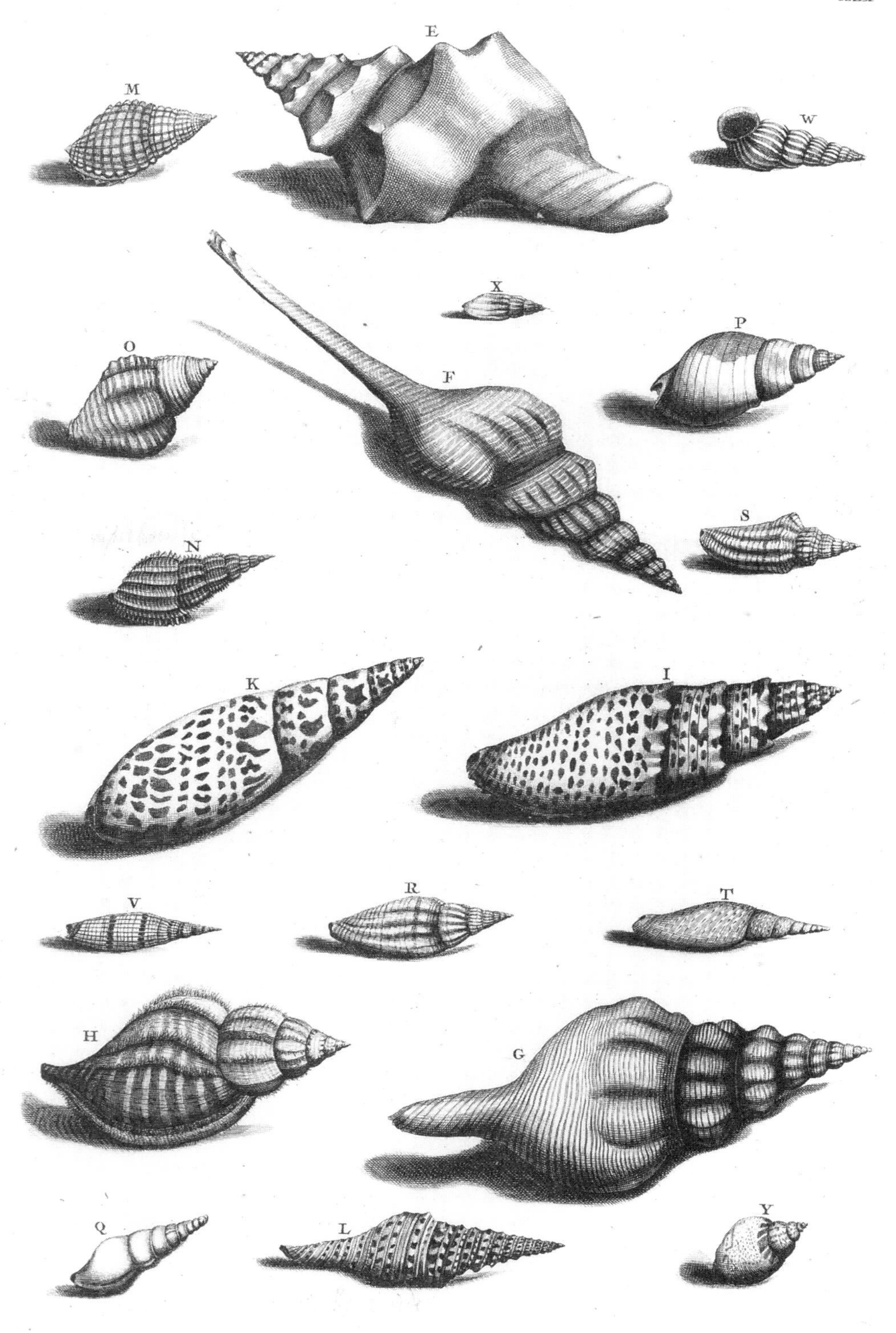
E
M
W
X
P
O
F
S
N
K
I
V
R
T
H
G
Q
L
Y

furrowed like the foregoing. It has a second kind, that has round whorls, without knobs: both their shields are rough on the outside and divided into rings, and can be used to some degree for incense; although it is not mentioned in the Chapter on *Unguis Odoratus.*

The true Spindle is 9 and 10 inches, the blunt Spindle 4 and 5 inches long; barely an inch thick in the center, and therefore named after the shape of a spindle, and is seldom found.

There is yet another, even rarer kind, that does not come into the light of day, unless one of them crawls by accident into a weel, because it sojourns in the ocean deep. It is larger than an ordinary Spindle, also furrowed and entirely white, and it has knobbles or bumps along the whorls, which are somewhat sharp towards the tips.

VIII. *Buccinum pilosum primum sive crassum*,[24] in Dutch, Hairy Fatlips, or Bearded Mannikins; these have various shapes, all with a thick tail, and a broad, jagged lip, furrowed all over its body, and covered with several rows of bristles, rough and without any sheen: One kind has roundish whorls, on the side that is over against the mouth, with a thick, protruding seam: Another kind is not only furrowed, but also covered with knobbles and bumps; the mouth is small, russet on the inside, and closed with a lid like an *Unguis.*

IX. *Buccinum pilosum tenue*, is no more than two inches long, with a thin shell, also furrowed, and is covered with long, soft hairs, which stand together in rows, and are easily rubbed off, whereas those of the foregoing held firm: one thinks they are the females of the first one.[25]

X. *Mitra papalis*,[26] the Pope's Crown, is an oblong smooth little Whelk, of an unusual shape; because the first whorl is as long as the rest of its body, oblong as a roll, entirely without a tail, its body white, but densely covered with red drops, that are in rows of sorts, and whereof most are square, as if they were precious stones on a Pope's Crown: the upper edges of the whorls are nicely notched or toothed, as is the mouth's lip. This Whelk has a Harmful Animal as Inhabitant, with a hard, tough, white and slimy flesh; concealed in its snout, that sometimes hangs out like a long tongue, is a small bone, like a thorn, with which it can deliver a poisonous sting, when you hold it in your hand, and has killed some people: when the Animal itself is cooked and eaten, it will bring on a lethal choking, and is therefore rejected as food; although the ordinary folks of *Kaybobbo*, a village on Big *Ceram*, where there are a lot of them, sometimes roast them on coals, and eat them without harm: who could tell, that such a holy Pope's Crown would conceal such deadly venom?

The second kind[27] is smaller, either little or finely notched on the sides, while the drops are a brownish red, which are called brown Pope's Crowns. A third kind[28] stays small, about half a finger long, red mostly, with white flames, and deeply notched or toothed along the outline of the whorls: the first two kinds are rare; but the third one is common.

XI. *Mitra Episcopi*,[29] a Bishop's Mitre, has the shape of the foregoing, but is narrower and more elongated, without teeth on the whorls, except for the mouth, where it is sharply toothed; smooth all over its body, painted with big red spots, so that there is as much red as white: The flesh conceals the aforementioned little bone, wherefore one should shun them as food: It is 4 or 5 inches long, and scarcely an inch thick; when the Animal's flesh rots, it changes into a black slime, like ink. It is common on the beach of *Kaybobbo.* One finds there an even smaller kind,[30] scarcely the length of one's little finger, somewhat furrowed on its body, the color of fire, hence called fiery Pope's Crowns.

XII. *Turris Babylonica*, Tower of Babel:[31] This one has many whorls coming to a pointed cone, like a *Strombus*, with a tail or snout in back, deeply furrowed along the whorls, of which one sticks out along the back; the background is white, with many black little dots, which resemble doors and windows, the way they have been painted on the Tower of Babel.

This Whelk also has the particular sign of its genus, to wit, it has a cut in the upper corner as if a piece had been broken out of it; the lid on the mouth is a black-brown *Unguis:* It also has two smaller kinds; of which one[32] has small pale little dots; the other[33] is mostly black; they are similar in shape to the foregoing Spindle, except that they are smaller and stippled.

XIII. *Buccinum granulatum planum*,[34] is scarcely the length of one's little finger, the lowest or first whorl is wide, where it is roundish and pressed flat a little, with two protruding seams; one at the mouth the other opposite: furthermore, its body is covered with round grains, that stand in rows, and sometimes look like *Paternosters*, and which are mostly white; the center ones, forming a band, are red and black. Fresh ones have short and stiff little hairs on the length between the grains.

XIV. *Buccinum granulatum rotundum*,[35] is a single *Turbo*, with a round body, and without seams; usually covered with round grains as well; a uniform pale white, without bands.

XV. *Buccinum aculeatum*,[36] is about half a finger long, and prickly all over; it has coarse ribs athwart the whorls, while it is also furrowed, and of a light-brown color.

XVI. *Buccinum undosum*,[37] is a short, bossed little Whelk, the length of a finger joint, with a thick shell, with 5 ridges right across the whorls, resembling sea waves, and deeply furrowed along the whorls; it has sharp little teeth at the mouth: The ridges are black or the color of enamel, the rest is a faded color. A second kind[38] is mostly round, without waves, but also furrowed, and covered with something woolly, that sticks to it. A third kind[39] consists mostly of a large round whorl, with a short round little tail in back, a thin shell, and is surrounded with many ribs, that are tied around like thread; this one is seldom found; but the first two are common.

XVII. *Buccinum lineatum*,[40] is also a single and smooth *Turbo*, scarcely the length of one's little finger, surrounded with very fine or small black stripes, as if it were wound with black thread: the shell is light, with small little teeth at the mouth: It has the same venomous little thorn in its flesh, like the Pope's Crowns, for which one should be on one's guard, so it will not hurt anyone's hand.

XVIII. *Digitellus*,[41] the Little Finger, is also but half a finger long, with a thick shell, and lip, with a short little tail in back, grainy over its entire body, of a pale reddish color, and with some bands at times: The front *turbo* is always crooked, and its point is dull, so that it resembles a small finger, that points to something. A second, smaller kind,[42] is smooth, white, and shiny like porcelain: this one is rarely found.

XIX. *Turricula*, the Little Tower.[43] In Malay *Bia bidji gnemon:*[44] This is a special kind of *Buccina*, oblong, pointed above as well as below, of which the lower whorl is very long, several times longer than the remaining *turbo* or point; the shell is thick, pleated a little at the edge of the upper whorl, and covered across with very fine little furrows: some are russet or yellow all over its body: some have red and black bands: some are somewhat pleated, with or without bands. The smallest kind[45] is once again smooth, or a little pleated at the edges, and it has a blackish color; the mouth is oblong, narrow as a slit, and the Animal has neither bone nor lid.

XX. *Turricula plicata*,[46] pleated Little Tower, this one has a thicker body and shell, mostly gray, with protruding pleats as ridges, and furrowed across, so that it is entirely rimpled and rough: but there is yet another kind,[47] which is plainer, but the same ridges are sharper, and they usually have bands: the Malay call this second kind, *Bia bidji gnemon*, because of the resemblance to *Gnemon* Kernels, which are like our Oilstones.[48] There are many different ones, but we put all of them under these two kinds.

XXI. *Turricula filis cincta*,[49] this one has a rounder body, enwrapped with black or brownish and protruding little ribs, as if with iron wire.

XXII. *Turricula granulata*,[50] this one is even smaller, and the first whorl constitutes most of the body, all around thickly covered with tiny kernels: The one kind is a light gray: The other has a few bands of red, black, and blue kernels, as if they were *Paternosters*, and they sometimes are called such.

XXIII. *Buccinum angulosum*,[51] also seems to be pleated, but the ridges are narrower and sharper; it has a thin and light shell, and is a uniform dark gray.

XXIV. *Buccinum scalare*,[52] the Wentletrap, is a rare, small, white little Whelk, on the outside surrounded with many spots, which are athwart the whorls, like small scales, and which rise orderly like a spiral staircase.

XXV. *Buccinum spirale*,[53] being the tiniest of this kind, hardly the length of a finger joint, gray or brown, coarse, and wrinkled. They are called *Bozios*[54] in Portugal (although this is the common name for all *Buccina*), and they are often sold to cloisters, where the Nuns, in order to pass the time, use them to make headbands and chaplets.

XXVI. *Buccinum foliorum*,[55] are also among the smallest Conches, with thin shells, one round whorl coming suddenly to a point, as sharp as a needle; it is finely ribbed along the whorls, of a green-gray color, with blackish stipples; it closes its mouth with a thin round little lid. One will find hosts of them on the leaves and branches of small trees and shrubs, that grow on the beach, particularly on *Mangium fruticans*.[56] The Natives take the biggest ones, cook them, and eat them.

This same genus also has its *Quisquiliae*, or small riffraff, in various shapes, which one will find here and there on stony beaches, and their shape demonstrates, that they belong among the *Buccina*.

Tsjanko[57] brought from the coast, does not resemble any of the Ambonese Whelks; except perhaps the *Pseudo-purpura*, although it has no bumps, nor thorns, but is round and comes to a blunt point, with a thick snout: this Whelk is scarcely one span long, very heavy and with a thick shell, to wit, easily the thickness of a quill; the outside skin is an unsightly, dirty gray, mixed with yellow, cracked here and there, and with slivers that have come off; it is orange-yellow inside its mouth, but inside it becomes gradually whiter, but not like Mother-of-Pearl; one will see three protruding ribs on the left side of the mouth, which go up inside like a spiral staircase; it is used for making arm rings, when it is sawed athwart into pieces. Ordinary people believe it has a King, but upon closer examination one has found, that it is a female or Queen, not much different from the ordinary troop; except that her whorl goes the wrong way, to the right, when one holds the mouth straight above one, while it turns to the left in all other Whelks, from the point of view of the observer. According to what the Divers tell us, one finds the *Tsjanko* on the bottom of the sea, at a certain time of the Year, gathered together by the hundreds in a troop, from which one knows, that they have hidden this Queen among themselves, and, as one guesses, are impregnating her: for shortly thereafter one will find in the same place, a strange *Melicera*,[58] or Ovary, that has many crystal-like grains, hanging, like grapes, around a stem that stands straight in the sand, from which the Young *Tsjanki* come forth. This King, as they call it, is very valuable to the Natives, who give as much as a hundred *Pagoden*[59] a piece for it, because it is found so rarely: and ordinary people may not hide it, but have to bring it to their Kings.

The Buccinum, *like the ones before, have different names for us. See the first kind on plate* XXVIII. *indicated with the letter* A. *The second kind by the letter* B. *are the ones called* Trumpets, *also* Triton-Horns. *One will find very large ones of this kind, where a hole has been made in the third curl or circumvolution, through which someone, blowing powerfully, can make a big noise. Mr.* Van der Burch[60] *has put a smaller one at our disposal, indicated with* No. 1. *which still contains the entire animal, and resembles what was indicated with* K. *and* L. *on the* Fifth *plate, in Book one; but with this difference, that the legs and pinchers are somewhat sharper and thinner. The* 3d *and* 4th *kinds, indicated by the letters* C. *and* D., *are called* Oil Cakes *by us, because they have the same shape and color; indeed, if they are polished really well, they glisten just as if they were smeared with oil; the one by the letter* D. *is a single one; and the one by the letter* C. *a double, and is also called the bossed* Oil Cake; *there are several other kinds of these, with two that excel, but are rarely found; to wit, the* wrinkled one, *and the* Oylcake with raisins. *The* 5th *kind is depicted on plate* XXIX. *with the letter* E. *The* 6th *kind, with the letter* F. *which are called* Spindles *but also* Tobacco Pipes. *The* 7th *kind, with the letter* G. *And the* 8th *with the letter* H. *The* 9th *does not have a picture. But the* 10th *kind is indicated by the letter* I., *and is the* Pope's Crown, *a Whelk that was once of the first order: I seem to remember, that they paid two hundred guilders for it; and even if their value has gone down a bit, they are still good, provided they have very bright red spots, which is then called* Premier Color, *also if their color is the same from below up to the tip of the point, without lessening anywhere, and is then deemed* very beautiful,[61] *and in particular they should not have any seams or furrows. The venomous little sting, which the Author noted in this fish, should be ignored by the Roman Catholic collectors, and the same for the one in the* Bishop's Mitre, *indicated by the letter* K. *being the* 11th *kind. We call them* Orange pens, *because their spots are a beautiful clear Orange: The* Feitemaas Gentlemen[62] *own one of these with beautiful Lemon-yellow spots: except for these, I do not know of any others. The* 12th *kind, indicated with the letter* L., *is called* Pyramid *by us, of which there are again several kinds. The* 13th *kind has no picture; But the* 14th *kind is by the letter* M., *which we, because of its knobbliness call* Ricepudding Whelk. *The* 15th *kind, with the letter* N., *is called the little* Thistle Whelk. *The* 16th *is depicted by the letter* O. *The* 17th *by the letter* P. *The* 18th *by the letter* Q. *The* 19th *with* R. *which are usually called* Band-shells *and of which there are many kinds. The* 20th *by the letter* S. *is one kind of the foregoing. The* 21st *by the letter* T. *The* 22nd *with the letter* V., *this one, because of its beautiful color of the bands, is called the* States-Flag Shell.[63] *The* 23rd *kind does not have a picture. But the* 24th *by the letter* W. *is presented to us by the Author as a* Wentletrap Whelk, *although it is only a secondary kind: Many of these are found on our beaches, and differ only slightly and are wrongfully called* Whelks *by our Fishermen: it would have been convenient to provide the correct* Wentletrap *here, but since the picture was lacking when we made this plate, we will depict it at the first opportunity. The* 25th *kind is indicated with the letter* X., *and the* 26th *with the letter* Y.

The Author did not give us a picture of the Tsjanko, *but we call the same the* Offering Whelk, *because the Heathens use the same in their offerings, wherewith they pour fragrant oil or balm (as they say). They usually arrive here, with crude decorations on the outside, and carved out deeply inside, from which one must conclude, that they were made like that for some purpose. I have never encountered any of these in one piece, but Mr.* De Jong[64] *has a few right now, from Ceilon, but too late, because the plate was finished already; which is why I could not depict it: for that matter, it is coarse, lumpish, without color or any design, and therefore not worth making a new plate.*

CHAPTER 20

Strombus. Needles. Sipot.

The seventh genus includes the *Strombi*, which are narrow, long Whelks, with many whorls that come to a long point, resembling a wooden nail. They are really called *Strombi* or *Turbines*,[1] in Dutch Needles, or Pegs;[2] in Malay *Bia Krang*, *Djarong* or *Sipot*.[3] There are two kinds, simple ones, and angular or knobbly ones. The simple ones have a

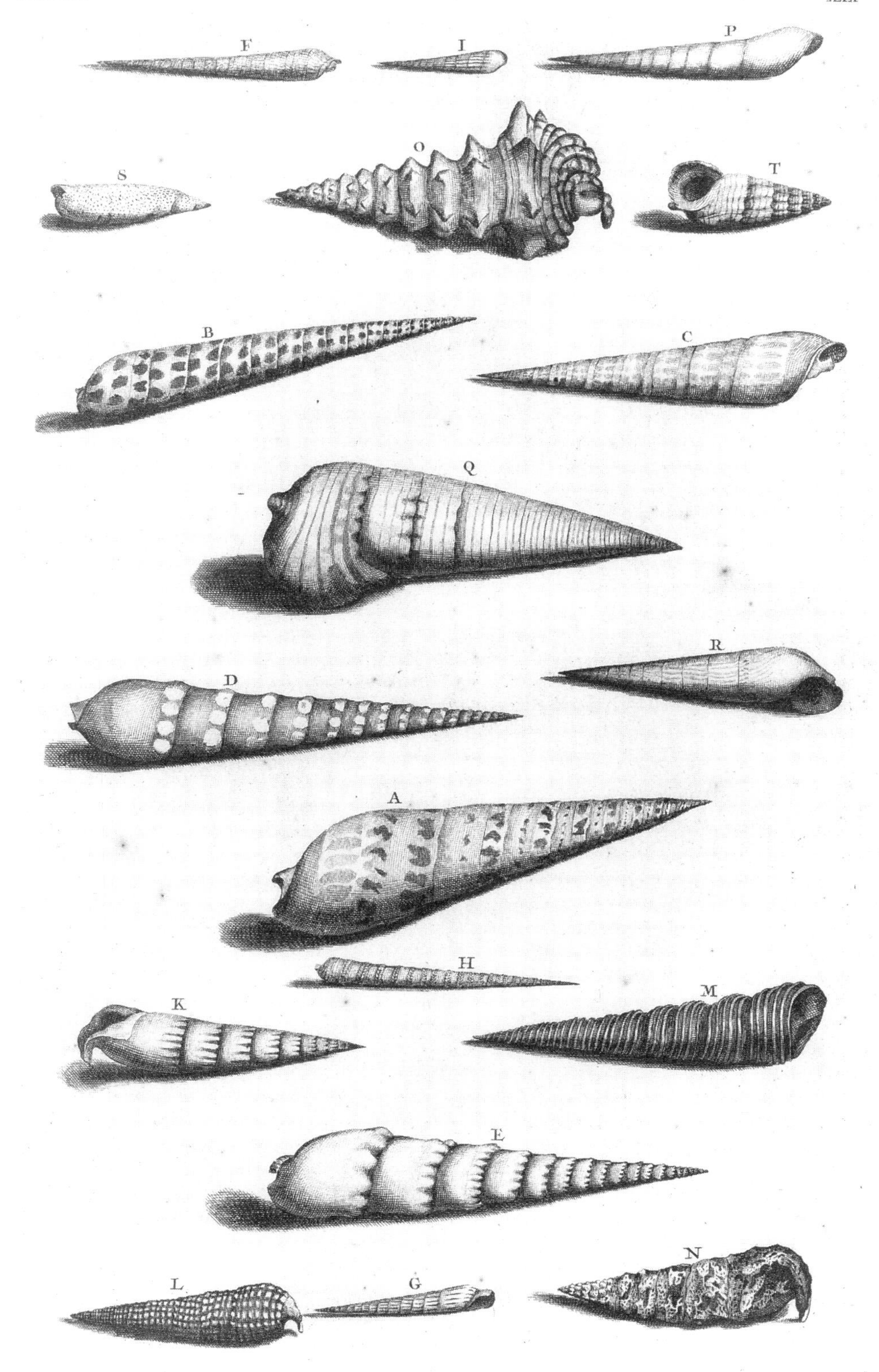
F
I
P
S
O
T
B
C
Q
R
D
A
H
K
M
E
L
G
N

smooth shell. The whorls have no notches, or very small ones, between them, or they are somewhat ribbed, as follows.

I. *Strombus primus sive subula*,[4] an Awl or Marlin Spike, is the largest of this genus, at most the length of a hand, but usually shorter, with a smooth shell, a dirty white like ivory, with black-blue spots or stripes on the side of the whorls: the Animal is white, very hard, with tough flesh, not fit to eat; it closes off its mouth with a thin, little lid, which hardly covers the Animal: there is a venomous little bone in its flesh, which is supposed to be lethal, and it also causes a great deal of pain, if one is wounded by the sharp shell, or its point.

II. *Strombus secundus*,[5] is of the former length, but narrower, no thicker than a finger, its whorls are furnished with big black drops and also bulge somewhat in the center.

III. *Strombus tertius*, is the common kind; of which some are whitish, with lead-colored spots and stripes;[6] some are white, with black little stipples along the edge of the whorls;[7] others have a blunt point, like a spigot which one sticks into a vat; then there are others that have russet and crooked stripes, as if there was some entangled thread on top of it.[8]

IV. *Strombus quartus*,[9] is light brown, with large, white eyes, and here too the whorls bulge somewhat in the center; these are rarely found.

V. *Strombus quintus*,[10] has a thin shell, a light red or the color of fire, with little white snakes, and is also rarely found.

VI. *Strombus dentatus*,[11] notched Needle, is also light brown, and adorned along its whorls with blunt little teeth, as well as with thin stripes.

From the *Lussapinjo* Islands[12] comes yet another strange Marlin Spike, toothed or notched along the whorls, like the foregoing, but of a pale yellow color, like dirty ivory, with tiny stripes, and therefore called the ivory Marlin Spike.

VII. *Strombus septimus*,[13] is the length of a little finger, and the thickness of a quill, while it has fine grooves across the whorls, resembling a Unicorn.

VIII. *Strombus octavus sive Lanceatus*,[14] the Pikeman, is a small and narrow little Needle, white and smooth, with many black upright stripes along its whorls, as if one saw a stack of raised pikes.

IX. *Strombus nonus sive granulatus*,[15] grainy Needle, is small stuff, the size of Sewing needles, not as thick as a straw, covered with many small grains; some are also ribbed and notched, of various shapes.

X. *Strombus chalybus*,[16] Sail needles, are the smallest of this genus, somewhat angular, but furrowed; some are quite blue, like blued steel: some have white and black little stipples.

XI. *Strombus caudatus albus*,[17] white Spouts, are like the other Needles, but have a curled tail in back; the white ones have furrows that go sideways across the whorls, and the greater the number of furrows, the rarer they are, and they close off their mouth with a thin yellow little lid; some are entirely white; some are encompassed by fine, black, or brown stripes,[18] and these are the rarest; a third kind[19] is entirely coarse, and notched, indeed, almost thorny.

XII. *Strombus caudatus granulatus*,[20] is entirely grainy; some are completely gray; of others the largest kernels are white, and those are more handsome. Both of these kinds are called Tutels, because they resemble the tutel of a Flagon.

XIII. *Strombus tympanorum*, or *Tympanotonos*.[21] In Dutch Drum screws, because of their resemblance to the little piece of wood with which the Drummers tighten their drums; its whorls are distinguished by notches, and they are ribbed as well, of a uniform tawny color, and without a tutel; a rarer kind is entirely white; these are the rarest ones of this genus, at least on *Amboina;* but they are also found on the south side of Ceram in the region of *Kellimuri:* larger, and more beautiful ones can be found on *Java*, and *Sumatra*.

XIV. *Strombus tuberosus*,[22] knobbed Tutels, is the first of the angular Needles, covered with large knobs, and nicely painted with black spots and stripes.

XV. *Strombus angulosus*,[23] rough Drum screws, these have a very rough shape, not only are they knobbled and angular, but they also have deep furrows, are coarse, lusterless, and chalky, therefore hard to clean; they do not have a noticeable tail in back, but the mouth there is slightly crooked.

XVI. *Strombus fluviatilis*,[24] River needles, *Sessu* in Ambonese, *Sipot aijer*[25] in Malay, or simply *Sipot.* These are long narrow Needles, with a thin and light shell, grayish green, or the color of mud, without any sheen or beauty, 4 and 5 inches long, scarcely the thickness of a finger: another kind is smaller, with a blunt point, and has blackish stripes: some are also slightly angular along the whorls. One will find them in the mouths of all kinds of Rivers, where it is muddy, where they lie hidden in the mud, and where they are found in large numbers, and brought to market, because they are a good food; even the juice is used for eating *Papeda:* one should, however, let them lie in fresh water for half a day or night, so they will spew forth some of that sand and mud; they have a sweet taste, but if one wants to eat them, one should break a large piece off the tip, whereafter one can suck them out, or empty them with a pin; their mouth is closed with a thin, blackish little lid.

XVII. *Strombus palustris*,[26] in Ambonese *Sipot kitsjil*, in Makkassarese and Malay *Borongan*, this one is shaped like the Marlin Spikes, but shorter, uglier, the color of mud, usually with the tip broken off, inside white and smooth, and the mouth is closed with a lid; it lives in marshy Sagu forests, is unknown in Amboina, but all the more so on *Ceram*, *Buro*,[27] or *Celebes;* this one is also fit to eat, and is therefore much sought after and is prepared like the aforementioned ones.

XVIII. *Strombus palustris laevis*,[28] has a thick shell, and the shape of the common Needles, with a small notch in the corner of the mouth, smooth, black or a dark brown, this one too lives near swampy Rivers among the roots of Trees, and they are also good to eat.

XIX. *Terrebellum*,[29] the Cooper's auger, this one I also count among the *Strombi*, because of its long and narrow shape, although it has a particular form, almost like that of the Rolls;[30] because it has only one large whorl, a narrow and long mouth, and comes suddenly to a point, like a Cooper's auger, or like the point of a wooden bucket:[31] most are yellowish or a light brown, with black stripes and little veins: some have black little spots: one will also find completely white ones; they have a thin, light and smooth shell, and can jump out of the water, as if they had been shot from a bow.[32]

XX. *Strombus mangiorum*,[33] is a coarse Needle, about the length of a finger, rough outside, and deeply furrowed, steel-green and lusterless, with a broad lip on its mouth; it sojourns in swampy places, that have firm ground and stones on the bottom, near the roots of *Mangium caseolare*, and on stones that are strewn about there; it is not particularly beautiful, but is reckoned a Curiosity because of its shape, and sought after as food by the Natives, like the foregoing *Sipot aijer.*

The Strombi, *called* Needles *by the Author, are known to all Dutch Collectors as* Pens, *and each has its own name. The one by the letter* A. *on plate* XXX., *is called the* Thick Tyger pen. *The one near the letter* B. *the* thin Tyger pen. *The one by* C. *the* wrapped Pen. *The one by the letter* D. *the* white spotted Pen. *By the letter* E. *the* notched Pen. *Those by the letters* F., G., H., *and* I., *are called* Needle pens, *because they are so thin and so sharp. The one by* K. *the* Snout pen. *By* L. *the* Knob pen. *The one by* M. *the* single Drum screw; *of which there is another larger kind, which is called the* double Drum screw. *The one by the letter* N. *is the* thorny Snout pen. *The one by the letter* O. *is called the* West-Indian Papal Crown, *probably because it first came from there. The one by* P. *is a* Snail pen. *The one by* Q. *a* bastard West-Indian Papal Crown, *most likely for the same reason as the foregoing. The one shown by the letter* R. *is another kind of* Snail pen. *The one by* S. *is the* stippled Auger: *of which there are other kinds; such as the* white, *the* striped, *the* flamed, *etc. And the one by the letter* T. *is reckoned among the* Band whelks. *There are many other kinds of these* Pens, *but since the most important ones are shown here, those will have to suffice.*

CHAPTER 21
Volutae. Bia Tsjintsjing.

The eighth genus include those which we call *Volutae;*[1] in Dutch, *Whirlpools;*[2] in Malay, *Bia Tsjintsjing,*[3] and *Krang lanke;* these are all kinds of *Volutae,* from which one can make rings, although the latter really applies to a particular kind.

Voluta,[4] is an art word, from Architecture, by which is meant the curls, which one sees on Ionian and Corinthian pillars, and these Whelks have been called after them because of their resemblance to this: because they have a flat head, made from many whorls that flow into each other, in the shape of a *linea spiralis,* or snake line: the body is oblong, fashioned from many narrow whorls, rolling over one another, and coming to a point in the back, so that they, when standing on their head, resemble a cone or pyramid. Owing to that form, they have a long, narrow mouth, and the Animal has no lid whatsoever, drawing itself so far inside, that one cannot see any part of it. When they are in the sea, almost all of them are covered with something woolly and with a slimy skin, which can be scraped off easily. Their shapes differ quite a lot from each other, as will become clear from the following kinds.

I. *Cymbium,*[5] Crowned Scoop,[6] or Crown Whelk, Mal. *Bia sempe,*[7] called *Wina* by the Natives of the South-Easter Islands.[8] This one has a different form from the subsequent *Volutae,* because if you hold it upright, and look at it from behind, it resembles a Coat of Armour, or an Emperor's Cloak (*paludamentum*),[9] with many teeth on top, which stand in a circle, as if a crown; from the front it resembles an oblong Scoop, with a wide mouth, somewhat rough on the outside, and dark brown, with large whitish spots here and there; it is a dirty white inside, like ivory; the Whorls on one side only cover half of the width, and within it lies a large Animal, that has a hard grayish flesh, bare, and without a lid. The largest ones are 15 to 16 inches long, and 9 inches wide. They do not occur in Amboina, but all the more in the South-Easter Islands, especially on *Kei.* The Natives eat the meat, roasting the entire shell on coals, but they break the innermost whorls of the largest ones, and make bowls and plates from them; profitable Household stuff, because it does not break easily, and when they have finished the meal, these people use them to bail out the water in their boats. They sometimes offer these hollowed-out beakers for sale; but it is difficult to obtain a complete one, or one has to order it explicitly. The Chinese call this whelk, *Ongle,*[10] that is, King's Whelk, and make nice spoons from its inner parts, and one has trouble guessing, what Whelk was used to make them; but they serve him best, who is left-handed.

There is a smaller kind,[11] on *Ceram,* not more than 7 or 8 inches long, light-brown and smoother, with a narrow crown on top, and one can get it sometimes in one piece, because they are too small for making scoops.

II. *Meta Butyri,*[12] Butter Wedge:[13] this is the largest of the *Volutae;* the head is somewhat flat, but rounded at the edges, and with a point that protrudes in the center, so that it cannot stand up, yellow over its entire body, like butter, with black or brown stipples, all in a row, but

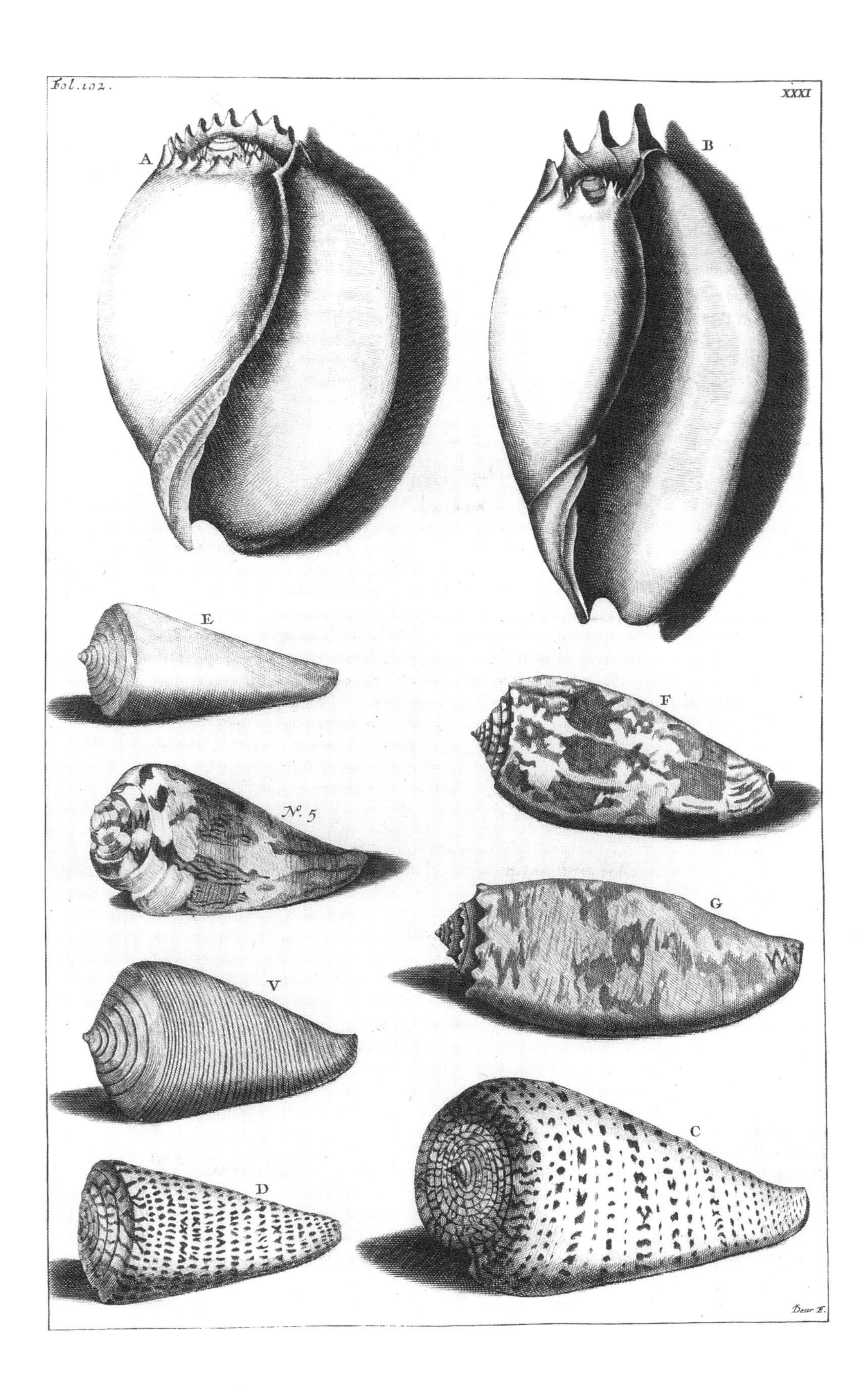
Fol.102.
XXXI
A
B
E
F
N°. 5
G
V
C
D
Deur F.

which [turn into] wide veins or stripes near the head. These are seldom found, particularly those that are sound and have a complete shell; because they often have cracks, or ugly seams.

There is a smaller kind of these, that has the Characters in an orderly fashion, as on the Music Whelk.

III. *Voluta musicalis*,[14] Music Whelk, or Little A. B. C. Books, otherwise Letter Whelks; this has the flattest head of all, so that it can stand up like a Pin,[15] the shell is white, but closely covered with rough, black, and square Characters, all in rows along the whorls, and somewhat resembling Music notes; but the Characters are long on some, and those next to them, are a little shorter, like the letters one sees in the little A. B. C. books of children: the head is marked with similar Characters, and one will find 2 or 3 yellow bands that run across the body, and these are the most beautiful. There is also a kind that has Characters that are very small and superficial, resembling the foregoing Butter Wedge; the mouth is very narrow, so that one can hardly see the Beast, and the outer shell is covered with a thin fleece, which one should immediately scrape off the fresh ones, otherwise it is very difficult to remove. Most often they are full of cracks, so that one will find few smooth ones, and the best ones should not be over three inches; the inner circumvolutions are as thin as Parchment, but as stiff as the tin scrolls on a lantern,[16] rolled tightly together, and without Characters. They also make arm rings from these Whelks in *Landas* on *Borneo*.[17]

IV. *Cereola*.[18] Little Candles, Mal. *Bia Liling:*[19] these are also shaped like a cone, but the head bulges more, so that they cannot stand right well, of a single color, like freshly cut wax; except that in the back by the mouth they have a violet or blackish spot; they have fine ribs across the body, and the fresh ones[20] are also covered with a thick skin, which one should scrape off immediately. A second kind[21] has a broad head, with a short point in the center, and [has] a shorter body, also yellow like wax, and marked with fine white stripes across it; but they do not have the aforementioned violet spot near their mouths.

V. *Voluta Tigerina*,[22] Tygers: these do not have a broad head, but have a long body, and the circles of the head make for several grooves; the body is marked with large chestnut-brown spots, which sometimes are blackish as well, and the background is white, or light red. These spots resemble various shapes, such as Clouds, Animals, or People carrying a large box, or whatever a clever person wants to contemplate therein;[23] they are finely ribbed or striped along the whorls, and covered with a thin fleece which is difficult to remove on the dried ones. They occur on *Hitu*'s coast.

VI. *Nubeculae*.[24] Little Clouds, are also oblong, like the foregoing, but somewhat knobby at the head, as if it were some kind belonging to the Scoops described before, also with a wide mouth, and a thin shell; the biggest kind is brown, with dirty white specks; the other kind is a purplish brown, with blue and white specks, that run together in troops, as if they were clouds, but more beautiful and rarer than the first ones: others call the foregoing Tygers.

VII. *Vespertilio*,[25] Bat; Mal. *Bia morsego*[26] and *Bia buduri*,[27] Ambonese *Ruluton*. This seems at first to be a *Murex*, because the edges of the circumvolutions are covered with sharp knobs, like the thorns of a Rosebush, but the Animal that lives within, proves that it belongs to the *Volutae*, because they all lack lids: the head bulges some as well, like a *Turbo*, and one will also see thorns on it; the shell is a pale white on the outside, sometimes a light red, marked with black spots and tabbied,[28] like the wings of a bat; the mouth is wide, filled with a hard gristly flesh,[29] handsomely painted on the outside, with yellow-green and black stripes.

The second kind[30] is oblong, a light-red background, with a dark-brown tabbying; instead of thorns it has blunt knobs on its head. The third kind[31] is also oblong, and with blunt knobs, the color of smoke, painted with dark-brown stripes instead of the tabbying,

Fol. 104.
XXXII.
II
P
Q
R
S
N
L
M
N°. 1
H
I
K
O
T
Deur F.

and they are somewhat entangled, like thread. The first are quite common on all beaches, although one will find few that are sound, for most are torn, and with dead spots. One thinks that the pale ones are the most beautiful, those that have wide, black tabbying as a background, and high lofty knobs. The Ambonese like to find them for food, because they like that gristly meat, although it can be bitter: The Butonese place them under the heads of their children when they are asleep, so they will be freed from nightly screams and fears, calling them Little Dream Whelks in their own language; in Mal. *Bia mimpi;*[32] others also call them *Bia baduri*, that is Thorny Whelks, a name also given to the *Tribulus*, mentioned before. Some Ambonese also call them *Makiijn horun;* and one often finds, suspended from the rocks where they live, a clump of white, angular eggs, transparent like ice, and clinging together by their narrow necks, not unlike the eggs of the Seacat;[33] one thinks that they are the eggs of these Whelks, but they do not serve for generation, since all *Melicera*[34] of Whelks are nothing more than a superfluity of food.[35] They are called *Cantaruga* on Buton because, as mentioned above, they engender dreams, and let the children sleep peacefully, if one puts them under their pillows.

VIII. *Harpa*,[36] the Harp, is also a wide-mouthed *Voluta*, perhaps the most beautiful of its genus, decorated on the outside with broad protruding ribs, which end on top in pointed thorns, and which resemble the strings of a harp, and the little thorns are along the entire curl; the ribs are the color of flesh, the spots between are browner, marked and decorated with white church windows; they are black near the mouth on the belly; the Animal has a great deal of hard and gristly flesh, nicely painted light brown and yellow, with little stars on top. They have a piece of flesh in front, that is so large, that it does not quite fit in the shell, and they can let go of it and throw it away; but what grows from that, is not known; after all, one finds enough of them that do not have this piece, and if one rips this piece away one will find some white kernels under it, as if they were eggs. They are soon cleaned, but one has trouble getting the meat out, because when one cooks them, and lets them rot, they get dead spots, to wit, wherever the dead blood touches, wherefore one should cut the meat out as far as one can, and let the ants eat the rest. The Ambonese call it, *Tattahul*, (which also means *Sibor*)[37] and the meat is held to be hurtful.

The second kind[38] is smaller, but handsomer than the foregoing, because they have a more beautiful painting, as if with flowers and red spots; the ribs are striped with black stripes, and they have sharp little teeth at the mouth, something which the foregoing does not have; one calls these Noble Harps.

The third kind[39] is small, and oblong, of which only the ribs shine, and the spaces between are a deathly gray, marked with small windows. One finds the first and the third kind all over Ambon: but the second one occurs in the Lias Islands.[40] Collectors give various names to these Whelks, according to their liking, because these are the foremost of Curiosities; the Malay call them *Bia sarassa*,[41] after the beautifully flowered garments of the same name; we also call them *Amourets*,[42] about Love; others call them *Bia basaghi*, or *basigi*, in Malay, that is to say, the angular Whelk. One will catch most of them in the month of May when the rainy Mousson begins; then one will also find the *Mola*,[43] described before, that has fallen off the Animal, shaped like a heart, bossed or round on the outside, and marked with little gold stars or flowers, on the bottom flat, whitish, with purple drops, like measles, where it was attached to the other flesh, which is white at first, but which slowly changes, garnering flowers and little stripes, and the *Mola* is always of a harder flesh than the rest of the Animal.

IX. *Voluta marmorata*,[44] in Dutch Marble Whelks, in Malay *Bia Tsjintsjing*, that is, Ring Whelks, from the following use; these are cone shaped, on the circumvolution's edges jagged or toothed, and with a short point in the center of the head, speckled all over the body, with large white spots under the black background, just like that kind of marble, which is called *Leucosticton*,[45] which gives them a handsome appearance; the mouth is narrow, where the flesh of the Animal lies bare, except for the uppermost corner, where it has a small nail, painted with yellow and black stripes, and by the snout they stick out a narrow tongue, edged in yellow or a light red: They are enwrapped in a thin little fleece, which clings to it, and which has to be scraped off first: Its *Melicera* is a painted little clump of thick thread, or entangled sail twine, white, red, and gristly, good to eat, which goes for the Animal as well; the most beautiful and the most numerous will be found in the *Uliasser* Islands, a few on *Hitu*, and on little *Ceram*. They are eagerly sought after for making rings, which are worn on the fingers not only by the Native women, but by our women as well; this is accomplished with great difficulty, and almost without any tools, because they grind the head on a rough stone, until one can see all the hollows between the curls on the inside, then they break off the hinder part of the body with stones, or saw it off with a thin file, and grind what remains until it becomes a small Ring; and they cannot make more than two small Rings from each Whelk; they are a clean white, like the whitest bone, smooth and shiny, like ivory; because the black color does not go into the white substance of the shell, and is ground out of it: Some leave these Rings smooth; others carve notches in it or flowers; and some know how to carve it so prettily, that there remains a raised little Bevil, with a small black spot on it, as if it were a formal ring set with a stone. One can make these Rings also from the aforementioned third kind of *Voluta musicalis*, also from some other *Volutas*, and Shells; but one holds these *Marmorata* for the best; to do this one should take such Whelks, that have come recently from the Sea, nor should they have been too long in the ground; because such Rings acquire a deathly pallor, without any luster, like those that are worn by unhealthy people: Nor should they be handled roughly, because they break easily, particularly those that are made of Shells. They also make other sumptuous things from these Rings, plating them with gold in such a way, that one can see both, the gold and the Whelk; sometimes they apply a groove in the center of the Ring, and place a little gold hoop in it, or some other ring of a black substance, such as Tortoise shell, etc.

X. *Voluta pennata*,[46] these are oblong like a Roll, the head is not flat, but it protrudes like that of a *Turbo*, with a small red point on top, having two forms, yellow and brown: The yellow kind are called Partridges (*attagenes*)[47] or Gold Cloth, because their entire body is painted with yellow feathers, that have black edges, almost like the feathers of that Bird: The second kind[48] is somewhat smaller, and narrower, straight like a Roll, painted brown and white like feathers, and is called Silver Cloth; both have a narrow mouth, and can stick out a little tongue, that is white, edged with red, and in it is a small bone, or thorn, which will hurt you, if stung by it: The third or brown kind, is different, larger, also painted brown and white, but the feathers are not arranged as orderly; they have a deathly color, and are finely ribbed along the whorls: Although the *Attagenata* are caught and eaten daily, they are not innocent of poison, which was experienced by a slave woman on *Banda*, who knew that she had only held this little Whelk in her hand, which she had picked up out of the Sea, while they were pulling in a Seine net; and while she was walking to the beach, she felt a slight itching in her hand, which gradually crept up her arm and through her entire body; and so she died from it instantaneously.

XI. *Voluta maculosa*,[49] flecked Little Cats, these too are cone shaped, with a sharp point on top, and somewhat sharp on the sides, with broad spots on its body, which are mostly bright yellow, with black or lead-gray speckles, like some Cats, and furthermore strewn with fine, little points, like sand, that are mostly arrayed in rows: Instead of spots, some have black stripes on the upper edge of the whorls, and the head is speckled with similar points; some have so few spots, that they are mostly white or pale yellow, and are considered inferior: One will also find grayish ones, and of different colors; but all are covered or marked at their heads with those black little stipples.

XII. *Voluta cinerea*,[50] in Dutch Cinderellas,[51] are similar to the foregoing, but rounder on the sides of the head, with a sharp point on top, and a dark ashen gray over its entire body, nonetheless smooth and shiny, without any painting, except for a little black spot here and there: they are rare, and are seldom found in one piece, most often they have a crack, as if they were broken off once, and then grew back together again.

XIII. *Voluta spectrorum*,[52] Little Ghosts, shaped like the foregoing but their color is like the Butter Wedge, of which they are a smaller sort, to wit, the color of a yellow yolk, with some thin and hacked Characters, that look quite like the ghosts, which the Map makers paint in the big and dreadful desert called Lob,[53] to the west of *Sina*.

XIV. *Voluta maculosa granulata*,[54] Granulated Little Cats, are a small sort of Cats, with large spots, their bodies covered with spiny grains, with two or three different shapes.

XV. *Voluta filis cincta*,[55] is a kind of Whirlpool, with a roundish head, brown, and wrapped with blackish threads, like Sail Twine; and is rarely found.

XVI. *Voluta filosa*,[56] is a blunt Whirlpool, with a roundish head, and blunt sides; the body is marked with russet, crooked veins, as if it were brown Araken thread, and furthermore several wide bands of mixed colors.

XVII. *Volutae fasciatae*,[57] are of various shapes, but they have one mark in common, they have a broad band around the center of their body: The first kind[58] has a broad head, and is yellow-green or deathly green all over its body, nearly like a green cheese, with a broad white band around its body, that has some black stipples running through it, and they are violet and black around the mouth: The second kind[59] is brown, with a white band, and, on the lower side of the body, grainy: The third kind[60] is the most beautiful and the rarest, scarcely an inch long, cone shaped, with a flat head, with a rough point in the center; the body is brown, above and below, with various speckles, and in the center a white band, which is again divided by a row of speckles, very smooth and shiny like polished Specklewood;[61] these are called Pin Cushions, or *Bia bantal*.[62] There are various other but smaller kinds, and their differences are mostly in terms of colors.

XVIII. *Voluta arenata*,[63] Little Sand Whelks, or Flyspecks, have two different shapes: The first[64] and biggest one has large and black speckles over its entire body, like Flyspecks: the other kind[65] is smaller, notched at the head, and covered with very fine black little dots, as if it were black sand, which are so close together in some places, that they make for black smutches.

XIX. *Musica rusticorum*,[66] Peasant Music, is a short thick Whirlpool, nary an inch long, with a light red body, with large black square droplets, that are in rows, as on the Music Whelk; [it has a] coarse and worn-down shell, because one will always find them on pebbly beaches.

XX. Gray Monks, or Old Wives,[67] is an inferior kind of Whirlpool, narrow of head, somewhat bellied of body, a vile gray, with some wrinkles: There is a smaller sort[68] of this, scarcely a thumbnail long, like a little roll in the center, also somewhat bellied, white, and covered with fine black grains, which make it rough.

XXI. *Terebellum granulatum*,[69] Rough or Granulated Cooper's Auger,[70] which has the shape of the same, since it is narrow and oblong like a roll, but it belongs to the Whirlpools,

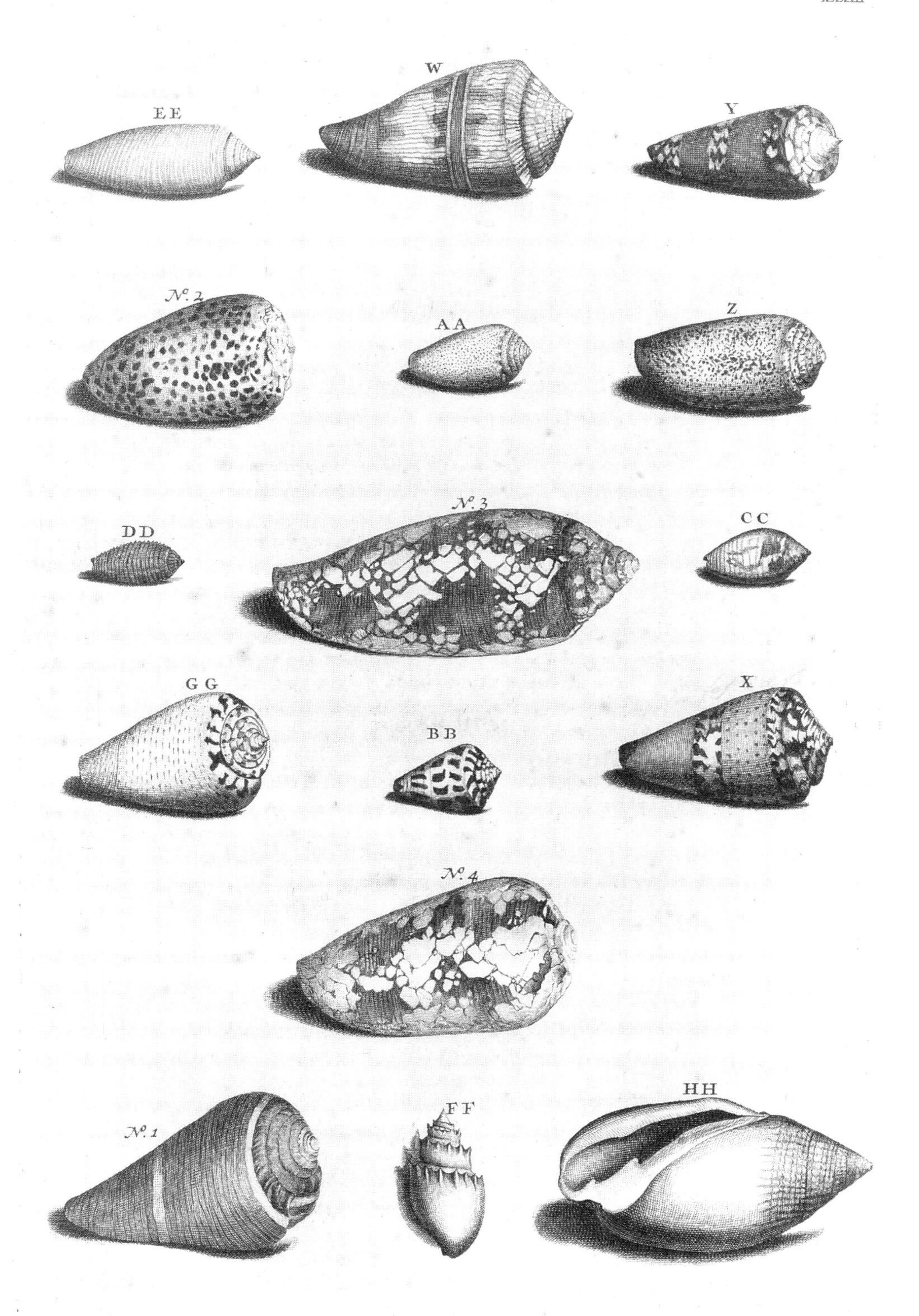

P. d. Sluter f.

according to its head; it is granulated over its entire body, and furrowed as well, and ribbed; its color is light brown, or russet, sometimes with spots as well, like the Little Cats, one finds this kind nowhere as plentiful as on the small Island of *Nussatello.*

XXII. *Voluta fluviatilis,*[71] in Ambonese *Laholun* and *Laborun,* and *Papeijte,* that is *Amarula,*[72] has a mixed shape of a *Voluta* and a Papal Crown,[73] it has a thin shell, is a dirty gray, with thin furrows along the whorls, and with weak little thorns on its sides, sometimes with dull and stiff ones as well: they occur in the mouths of deep rivers. The Natives eat them, although they have a somewhat bitter taste. Others count them among the Papal Crowns.

XXIII. Small Butter Wedge[74] from *Buru,* is not larger than a joint of one's thumb, roundish of head, an ashen-gray body, with many bands drawn athwart, which are made up of pieces and stripes.

XXIV. Midas Ears,[75] is also a Mire Whelk, with a long, narrow mouth, its body a black-brown, they occur in saltish mire, and there are big and small ones. I describe them more fully in the following chapter XXV, the XIIth kind.

Melicera or *Favago,* is like an Ovary, or Spinning[76] (*salivatio*) of some Whelks, which they toss out at certain times of the Year as if it were a superfluous food, usually at the change of Moussons, as in May, being the beginning of the rain-Mousson; and in October, at the beginning of the dry Mousson, or Ambonese summer: it is not of a uniform shape, and I will only describe the two most important ones. The first is the excrement, or the Ovary of this Chapter's VIIth Kind, to wit, *Vespertilio,* or *Bia morsego;* this is quite handsome, and has the shape of a small Bunch of grapes, or white berries, the size of gooseberries, or less so; transparent like crystal, sort of hard, with two or three ridges, otherwise they would be round, all with narrow necks, suspended from one another: If one opens them, or presses them to pieces, a whey-like slime comes out, but if placed in the Sun, they dry up yellow, and then the only thing left on it, is the dry little skin, that looks like the peel of peas: One finds them dangling from the flesh of the aforementioned Whelk, and sometimes sticking to small stones, looking like a Bunch of grapes. They are very like the Eggs of the *Sepia,* but rounder, and have two black little dots, which become the eyes of the Animal. The second kind is a discarded Spinning or little Clump of white and thin little strings, belonging to the IXth Sort *Marmorata,* as well as the *Cochlea Olearia,* resembling the Chinese *Laxat,*[77] after it has been cooked; people keep this stuff, and they call it the Eggs of the Whelks that were mentioned, but it does not do anything for procreation, and is just a superfluous casting-off, at least as far as is known until now. The last kind is sometimes white, sometimes a light red, like gristle, and, when cooked, good to eat, *Plin. Lib. 32. Cap II.*[78] calls such Whelks, which cast out this Spinning, *Melicembales,* which one should correct to, *Meliceribales,* or *Meliceribali,* and ascribes them to the *Buccina* and *Purpurae;* but he compares its *Melicerum,* to the peels of *Ciceren,* which corresponds to the first kinds.

The Voluta, *and all the forms which the Writer included are (with a few exceptions) known to us by the name of* Toots,[79] *and each one has its own particular name as well, among them are the most valuable Whelks, whereof we will mention something in their place. Those that are indicated* A. *and* B. *on Plate* XXXI., *are placed by us among the* Scoops, *and not among the* Toots, *because they are very large and have wide mouths: They have a nozzle in their crown, like a nipple, which is why they are called* Kroontepelbakken.[80] *There are several kinds of these, namely single* Tepelbakken; *also those without nipples, and are called, according to their color, either* Agate Bakken, or White, *also* spotted, *and* striped Bakken. *We reject the name* Crown Whelk [Kroonhoorn], *because it is given to another kind, which is greatly esteemed by the Collectors; and which shall be pointed out hereafter. The* Toot *depicted by the letter* C., *is known to us as* the yellow Tyger, *because it has a yellow background with black spots. And the one by the letter* D., *[is called]* white Tyger's Toot, *because its background is pure white. There are other kinds like these, of which the most common ones have yellow bands. The one shown at the letter* E. *is called the* Mennonite Toot,[81] *because it is without bands or spots, but has a beautiful luster, if it is cleaned properly. The one shown by the letter* F., *is called the* Cloud Whelk, *of which there are also several kinds. The one by the letter* G., *is called the* Agate Kroonbak. *The ones shown by the letters* H. *and* L., *on plate* XXXII., *are called by us* wild Music Whelks, *because some of this kind (and I know at least twenty of them) resemble the true* Music Whelk, *although all of them differ in color and shape. The one by the letter* K. *is called the* gray Chrysanth.[82] *By the* L. *the* Pyed Chrysanth. *And by the letter* M. *the* little Chrysanth; *but the name of* Little Harp *is also beginning to be used these days. There is also a white kind, and one that is rose-red, pyed, and more finely painted, but these are very rare, and only three are known to exist, as far as I know. The one shown by the letter* N. *is called the* Heart Whelk *by us, because its white spots, on a black background, look like Little Human Hearts. We add another kind to these, one that is quite uncommon; it is shown on the same plate, by* No. 1. *added by Mr.* Vincent, *Dr.* D'Aquet *has a similar one, but otherwise I only know of two more; it is called the* Heart Whelk *with bands.*[83] *Those that are shown by the letters* O. *and* P. *are called* Lace[84] Whelks *by us, while some call them* Gold Cloths, *but unjustly so, because the* Gold Cloths *are much browner and have a brighter color. There are many kinds of these* Lace Whelks, *with bands, with white backgrounds, with blue stripes, and more. Those shown by the letters* Q. R. S. *are* Agate Toots, *whereof there are many beautiful kinds, and are held in much esteem: but these here are very common, and are named by the Author. The one by* Q. *[is the]* spotted little Cat. *By* R. Cinderella. *And the one by* S. the Little Ghost, *a name that should be adopted, because we do not have particular names for these kinds. The one by the letter* T. *is an outstanding little* Toot, *and, as far as I know, not to be found among the Collectors here. It deserves a grander name than* granulated little Cat; *but I will not rename it. The one shown by the letter* V. [on plate XXXI] *is called* Oak Leaves: *of which, on plate* XXXIII., *we show a second sort, by* No. 1.[85] *that one has a white band through its crown, and a similar one through the center; except for the present one, there is no other known at this time. Those shown by the letters* W. X. Y. *on the same plate, are called* banded Olive Toots *by us; of that last one, drawn by the letter* Y. *are many kinds, which are considered worthwhile if they are beautiful all around; the Author calls the one by the letter* W., Arakan's thread: *by the letter* X. Green Cheese: *and the one by* Y. Lace cushion: *to which we add one other, shown on plate* XXXI. *with* No. 5., *and is called the* great Oliveband Toot[86] *by us, but it is very rare. The one that is shown on plate* XXXIII. *by the letter* Z. *is known to the Author, as well as to us, by the name of* Flyspeck. *But the one by the letter* AA. *is a* Gnatspeck, *because it has particularly fine stipples; we add one here (but a far rarer kind), shown by* No. 2 *and called a* Fleaspeck.[87] *The one by the letters* BB. *called* Peasant Music *by the Author, is called* spotted Toots *by us, or* spotted Little Cats. *Those by the letters* CC. *and* DD. *are, according to the Author,* Old Wives: *of which* DD. *is granulated. The one by* EE. *is the* granulated Cooper's Auger. *Then by the letters* FF. *is a* river Papal Crown, *which also does not belong among the* Toots. *The one by* GG. *is known as the* Tabby Cat Toot, *because of its fundamental color, whereon it is marked with black and white little dots. These* Toots *are very rare: Doctor* D'Acquet, *and Mr.* Vincent *are the only ones who own one, and except for those, I only know of two more. The one by the letters* HH. *the Author calls* Midas Ear, *because of its shape; but it does not belong among the* Toots: *we have felt it necessary to add a true* Brown Net;[88] *see it shown on the same plate by* No. 3. *of which there are more kinds, but the best one is the yellow* Net Whelk, *which occurs not as often as the foregoing: the one shown by* No. 4. *is the* Brown Net Toot,[89] *and resides in the Cabinet of the artful* Bronkhorst,[90] *distinguished Miniature painter in* Hoorn, *of which only two others are known.*

APPENDIX

Thus far we followed the Author, but since the most estimable Whelks are among the aforementioned Toots *(as was said before), we could not neglect to show to the Collectors, the rarest ones, which the Author could not have known, and which are seldom found, all of them shown on plate* XXXIV. *The one shown at the letter* A. *gets precedence, and is called the* Orange Admiral;[91] *it has a pure white background, and has two beautiful bright Orange bands all around it; furthermore the entire body is nimbly banded, with white and black sharply cutting spots; its crown, with the same color and bands, is very beautiful.*[92] *Of all that comes from the Sea, especially the* Toots, *this is the most beautiful Whelk I know. The Honorable Mr.* D'Aquet, *Mayor of Delft, who is often mentioned here, is the only one who owns one, and no other is known anywhere else. The one shown by the letter* B. *is called the* Supreme Admiral,[93] *since it was, before the other became known, the foremost Whelk. Not counting the one shown here, Mr.* Van der Burg[94] *has a similar one in his excellent Cabinet: its color is a yellowish brown, with flamed stripes that are browner, and that are intertwined, with white spots like hearts; [it] has four yellow bands around it, with fine white little spots, and it is toothed or topped all around its crown. Now there follows the* Admiral,[95] *with the same color and spots as the foregoing, but it has one band less: it is shown by the letter* C. *For the one that is shown here, they once offered 500 guilders (to no avail), wherefrom one can tell, what its value was, although today they are no more known than they were then; it was then in* Zwammerdam's[96] *Cabinet. Now there follows yet another kind, also an* Admiral,[97] *shown by the letter* D., *but differs somewhat from the foregoing. Mr.* Ovens,[98] *painter, possessed this one, but after his death, it was sent, along with his Cabinet, to England, where it was sold. Then there is the one shown by the letter* E., *called the* West-Indian Admiral,[99] *which is also a very fine piece, but not as estimable as the foregoing. Now there comes the* Vice Admiral,[100] *shown by the letter* F., *which is quite an uncommon Whelk, of a bright brown, spotted white, has around its center a white band, with fine brown veins, and is quite pointed. There are various other kinds, that are called the same, but this is the true one, and very rare. Follows the* Guinea Toot,[101] *shown by the letter* G., *a Whelk that was once of the second rank, and was greatly esteemed, a value it can still reasonably expect to command even these days; its color is* gridelin[102] *or pale purple, and has black and white bands with clear spots, which can be both wide and narrow. Now follows the* Crown Whelk,[103] *shown by the letter* H., *I know of six kinds, which are all valuable. Miss* Oortmans, *and the distinguished Collector* Cattenburg,[104] *own the best ones. The one by the letter* I.[105] *is by no means the least: Secretary* Blaauw,[106] *has one in his outstanding Cabinet, and it is second to none. These aforementioned Whelks, or* Toots, *are all first-rate, and if one wants to have a distinguished Cabinet, one should especially try to own any of these, although they are hard to come by: This is as far as the* Toots *are concerned; but there are distinguished ones among the other kinds as well, and which have been shown at their proper place. Now there follows the* Toots, *which are called* Agate Toots, *of which there are many kinds as well, that are valuable to Collectors. The one that stands out, is shown by the letter* K.[107] *on the same plate; and the same for the one by the letter* L. *called the* striped Agate Toot.[108] *The one by the letter* M. *also belongs to the same, but others call it the* Lion Rampant Toot,[109] *because of its spots, which look like small Lions; because of its color it is also called* Tortoise Toot *at times, but then the spots have to be larger; hence far from being the most uncommon or most excellent of* Toots.

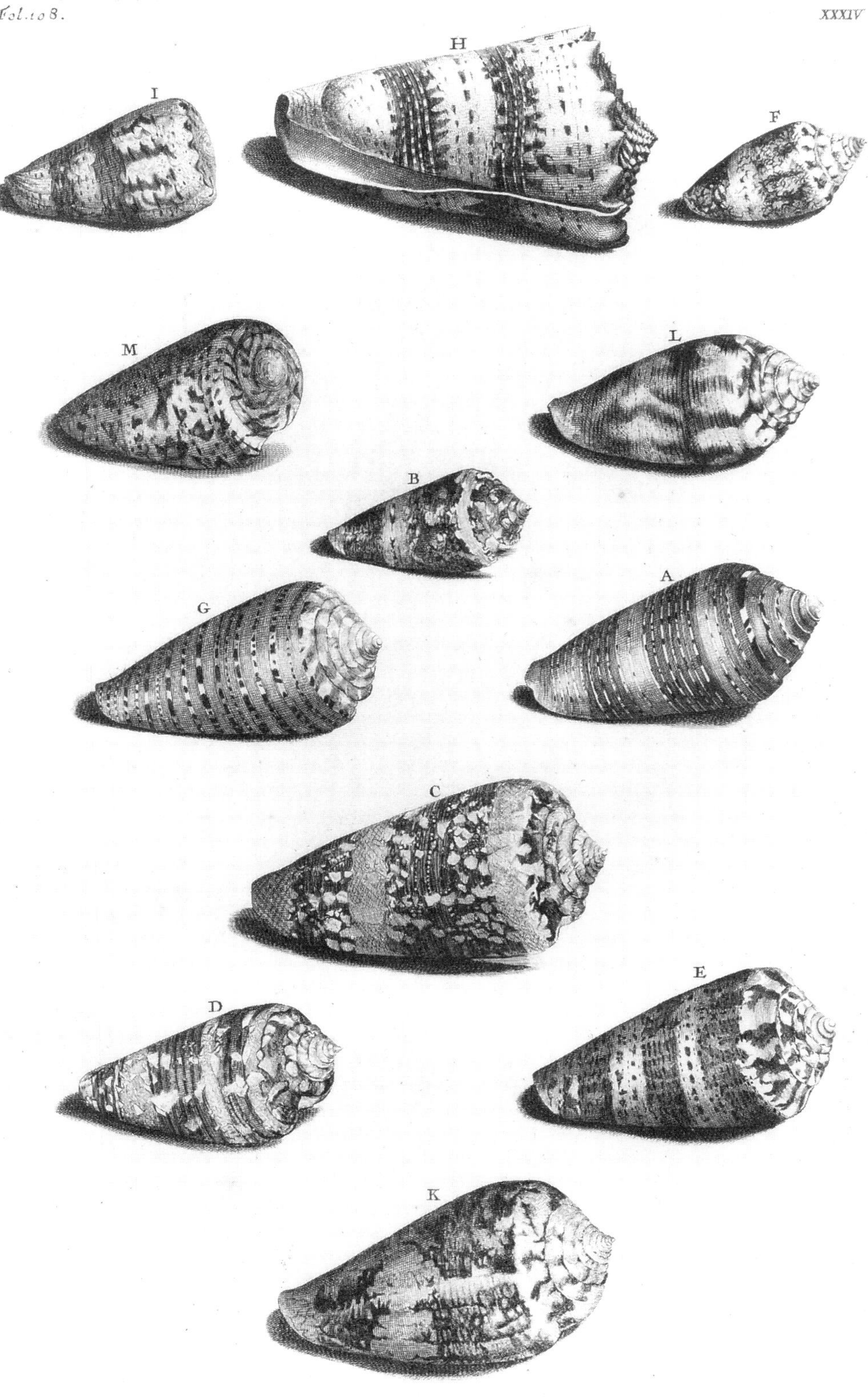

[illegible] Later, fe.

CHAPTER 22

Alatae. Amb. Tatallan.

The ninth Main Genus comprises those,[1] which we call *Alatae*,[2] that is, the winged ones, because the one lip always widens; in Ambonese *Tatallan* and *Talatallan*, after the resemblance to the outspread wings of the *Tallan* bird, which we call the Frigate Bird;[3] they have the mixed shape of a *Buccinum* and a *Voluta;* because the head comes to a point like a *Buccinum;* but the body is oblong, like a *Voluta*, with a long narrow mouth: A special mark of this genus, is that they have a long little bone near their mouth, the color and substance of *Onyx Marina*, sharply sawed on the outside, pointed at the bottom, and on top fastened to a piece of hard flesh, which resembles a little hand, with which this Animal not only accomplishes its movement, pushing itself from place to place, but with which it also fences quite vehemently, as if it were a sword,[4] thrusting everything away, that it encounters; especially a certain kind, called the Pointers, that are true Swordsmen.

The *Alatae* are of two kinds; because with some the protruding lip is divided into several branches; in others, the same is even and smooth, and there are, in order, the following kinds.

I. *Harpago*,[5] Boathook, or Devil's Claw, is the biggest of this genus, the length of a hand or a span, if one includes the branches; the body is that of a flat *Voluta*, the head of which ends in a pointed *Turbo;* the back is obliquely ribbed, covered with bosses, and variegated with brown, or blackish spots; but they are a light rose color near the mouth; they have 6 long branches on the edge, which are hollow inside and come to a point, and of which the rear one, which is around the *turbo*, is the longest and straightest; the two next to it, which proceed from it on either side, are curved towards the foregoing: the other three are shorter, and as crooked as a hook, so that one could suspend the Whelk from it; of which the two in front, or by the snout, are turned away from one another, like Buffalo horns and right next to it the shell has a wide notch; the frontal flap of the Animal, which one will see in its mouth, is soft, thin, of a green and white variegation, and this has an excrescence, that extends to all the branches. One divides them into the male and female animal; of which the male has narrow and thick branches, most of which are vaulted, in the old ones dense and not hollow; the spots are brown or russet rather than black, and they are more blunt and open like a channel; but the female has a thin shell, and is heavily speckled. There is also a particular kind, with a broad head, and a body that suddenly comes to a point, without branches or protruding lip, called *Stumpies*,[6] with a thin, sawed, and as if protruding edge at the mouth, also speckled black over its body, so that one could consider it an imperfect *Harpago.* I am of the opinion, that these Whelks were called *Pentadactyli* by *Plinius*,[7] that is Five Fingers, because they, perhaps in the Mediterranean, have no more than 5 branches. One will find the most beautiful ones on the Banda Islands, there are also many on *Bonoa*, and *Manipa;*[8] but those have a lot of sea grit on them, which is difficult to get off, and which often gnawed tiny little holes through the shell. The Animal is sought after for food on *Banda* and the South-Easter Islands, where they roast the Whelk, turned over, on coals, whereafter they smash the shell.

II. *Cornuta.*[9] *Heptadactyli* by *Plinius.*[10] In Dutch Crabs. In Malay *Bia Cattam.*[11] In Ambonese really *Tatallan.* In Bandanese *Sipe cornuti.* This one is smaller than the foregoing, commonly the length of a hand, its tip divided into 5 branches, or 7, if one counts the ones of the head and the tail, of different lengths, because the two near the *turbo*, or the head, are the longest, and those on the side, the shortest: They have two or three high and narrow bosses on their backs, and on top of that they are ribbed sideways; some are a drab yel-

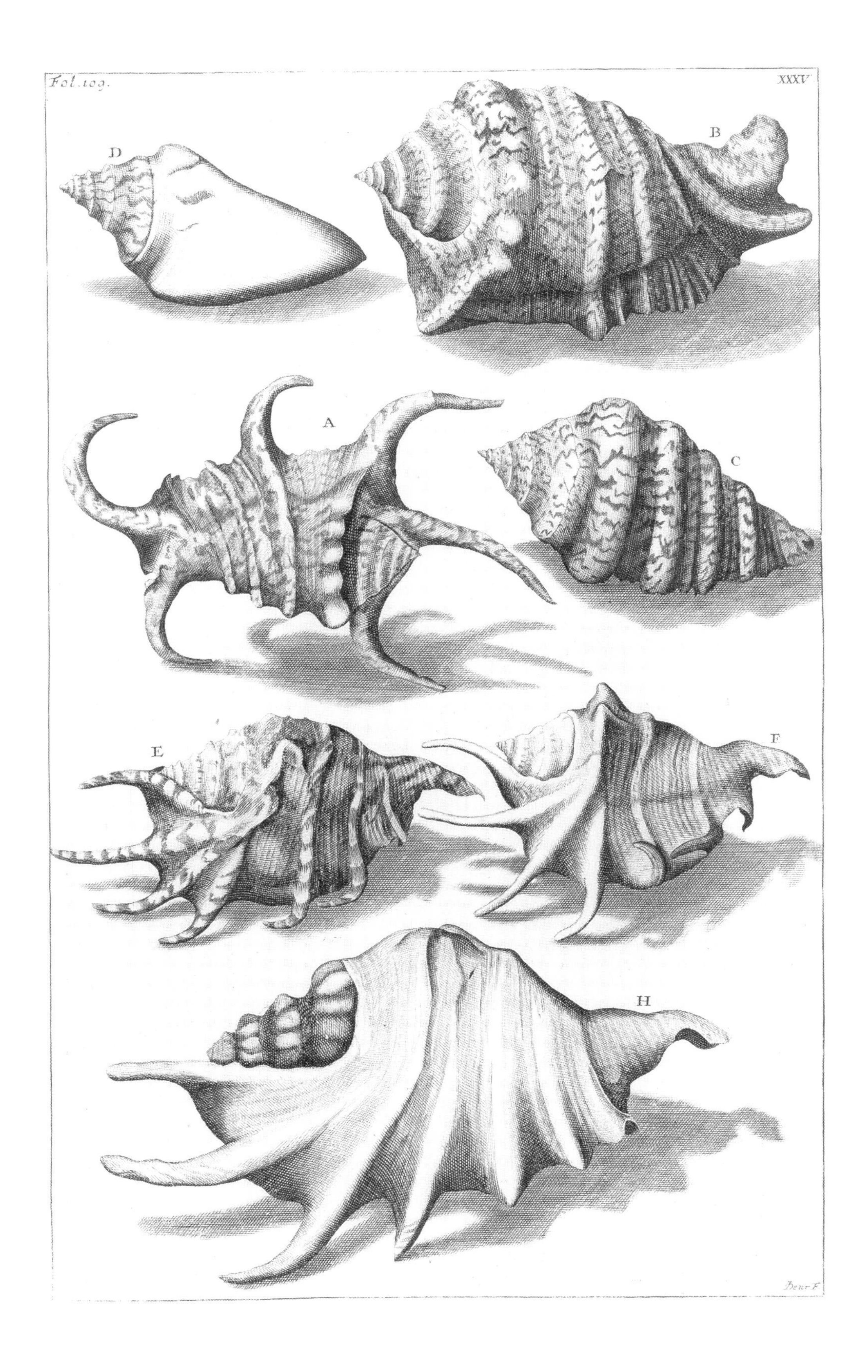
Fol. 109.
XXXV
D
B
A
C
E
F
H
Deur F.

low, without speckles, and with little sheen, but these have the longest branches, which also curve upwards: Others have russet and black spots on their backs, their branches are shorter, some are yellow near the mouth, some violet; the Animal inside is divided into flaps, as said above, but near the snout it sticks out a long finger, round and stiff, with a small cleft in its forehead, which is the Animal's mouth: close behind it, one will see two short little Horns, with hard eyes on them, like Crabs do.

The female has shorter and wider branches, partially hollow below, like the little pipes of a lamp, and usually with a thinner shell: Of these there are also two kinds; one an ordinary, drab yellow; the other with black spots: They too have their Stumpies, like the foregoing, which make for a particular sort, without branches, with a sawed or broken-off edge, of a yellowish color. These too are sought after for food, and are commonly found on all beaches; though one will find only few, that are not defiled with Sea grit. *Jonston Histor. Pisc.*[12] considers these Whelks to be *Murices*, and calls them *Coracoides*,[13] because the branches resemble the beaks of Ravens.

III. *Cornuta decumana*,[14] is a heavy, lumpish Whelk, easily ten times heavier than the foregoing one, because of its thick shell, which also has 7 short branches on the protruding lip, including the tail; the back is not bossed very much, but is so much covered with Sea grit, that one would think it were a rock. It is rarely found, perhaps only on *Bonoa*.

IV. *Cornuta millepeda*.[15] In Dutch Millipede, this one is flatter than the foregoing, drab on the back, densely speckled, with russet stripes; the lip is divided into 10, sometimes into 11 short branches, all of them curved backwards, of which the snout and the head are the longest; it has several knobbles on its back, and the lips are handsomely painted with black and red stripes.

V. *Cornuta nodosa*.[16] In Dutch Podagra, or the Scorpion; this is the smallest one of the genus, also divided into many branches, with a long tail, that is curved like that of a Scorpion: all the branches, as well as the back, are knobbled, just like the fingers of those who have had *Podagra* for a long time; it is ribbed near its mouth, painted black and red, though the protruding ribs are white. It is seldom found, mostly in the Liasser Islands.[17]

VI. *Alata lata*,[18] Wide-lip, hence quite unlike the *Alatae* with branched lips; now follow the ones without branches, with only a simple lip, and among them Wide-lip is the largest, because it spreads out quite wide, with a thick and angular lip, that is sometimes hollow inside; its body is a dark yellow, with a few russet spots, on the inside smooth and red, seldom found; some people call this the true *Tallan*.

VII. *Epidromis*,[19] a Mizzen, the lip of this one is the broadest, and its *turbo* or head sticks out, like a *Buccinum*, and right next to it the lip is at its widest and biggest; if one holds the tip down, they look quite like a Mizzen; they are flat on the back, yolk-yellow, mostly of a uniform color, some have a few russet stripes.

VIII. *Epidromis gibbosa*.[20] Little Lumps, are like the foregoing, but smaller, lumpier and with a thicker shell, a brighter yellow, and usually with crooked russet stripes, like threads: both have a sword, with which they fence; one sees a lot of them in the Bay of Ambon.

There is another *Epidromis*,[21] longer than the foregoing, with narrower wings, which is called the narrow or rolled-up Mizzen; but these are rarely found, and are reckoned to be a separate kind.

IX. *Epidromis minima*,[22] Little Mizzens, are scarcely a finger joint long, with a thick shell, brown on its back, a beautiful yellow in the front part of its mouth; they are found on Hitu's coast, on stony beaches. The Malay call these 3 kinds *Bia Leijer*.

X. *Lentignosa*,[23] Freckles, are wide, with a thick lip, their bodies covered with lumps and

Fol. 110.
XXXVI.
G
N°. 7
N°. 6
I
K
L
M
N
O
P

knobbles, speckled black, russet, and green, but of a dull color, yet inside they are smooth, a yellowish incarnadine, and with little dents below the outer knobbles. *Bia Taijlala*[24] in Malay, because its speckles look like freckles.

XI. *Pugiles*,[25] Fencers, in Malay *Bia t' unjockan*,[26] that is Pointers, because they have, besides the tip in front, a small finger, with which they seem to point to something: these are not wide either, but have a thick lip; they have a noticeable *turbo* or pointed head in front, and in back a tail that is arched up high, like an Ambonese, who is standing on one foot while dancing the *tjakalele:*[27] There are two kinds; the first kind is gray, very ribbed, and covered with knobs; the edge of the mouth is thick, but it has a sharper side; they are light red inside, with white veins; the Pointer, or small finger, is almost as long as its body, and is usually somewhat curved: The second kind[28] is brown, more even and smoother, with white little speckles on its back, and with a few knobbles on top; its lip is also thick, but is rounder than the foregoing, a beautiful red inside: the sword of these Animals is toothed, and more to their liking, than any of the former ones; the same is affixed to a little hand, and they are so skillful with it that, if 3 or 4 are in a saucer that is filled with other Whelks, they will get to fencing and clearing away until few others are left; and there is not one person strong enough to tear this little sword from one that is freshly caught: They are common on all kinds of beaches in the Ambonese Islands, and, like the other ones, are eaten by the Natives, though they have the drawback that if one eats too many of them, one will break out in a sweat that smells like that of a billy goat.

XII. *Luhuana*,[29] in Dutch Luhunese Whelks, these resemble an oblong *Buccinum*, about ½ a finger long, though they have the normal mark of the *Alatae*, to wit, a round notch in the lip around the snout, with a thick shell, even on the back, whitish, with broad drab spots, which make for some broad bands, without any particular luster; inside the mouth they have as a special mark, that they are a beautiful red at the lip, for as far as one can see inside; on the side that is over against it, they are black, with green and yellow edges: These are good Fencers as well, and are often eaten. One will find them often on all kinds of beaches, but nowhere as much as in the *Luku*[30] region, which gave them their name, and in the same entire bight of *Kaijbobo:* they are not esteemed because there are so many of them.

XIII. *Canarium*, in Malay *Bia Canarij*,[31] so called after the shape of a peeled Kanari, having the form of the foregoing, but usually smaller, with 4 or 5 different kinds: The First[32] and largest one is the length of a small finger, plain and with an even shell; some are dark gray or the color of smoke; some have yellow bands against a white background: One may also find, though very rarely, ones that are entirely white. The fourth kind[33] is no more than the length of a finger joint, with lumps on the top, and with a raised lip at its mouth, gray, with blackish or brown speckles: The fifth kind[34] is even smaller, though it is the most beautifully decorated, because, against a white background, one will see light-red and greenish spots and dots, which resemble some flowers; others have bands and stripes; and others have dots that are lined up. These too are Fencers, and when the rightful Animal has abandoned it, a *Cancellus* will grow in it, or a *Cuman;*[35] and these differ from the other *Cumans*, in that they do not have a large pincer, wherewith it can close the door of its house, as the *Cumans* of the Winkles can.

XIV. *Canarium latum*,[36] differs from the foregoing, because it widens, almost like a Mizzen; its body is a dark gray, with yellow and white speckles, and one will sometimes find oblong ones among them, which look like Mizzens that have been rolled up.

XV. *Samaar*;[37] belongs to the Canaris as well, oblong, round, with a pointed tip, with folds on its circumvolution.

Fol. 112.
XXXVII.
Nº. 4
Nº. 5
V
Nº. 2
R
Y
Nº. 1
Q
S
Nº. 3
T
X
W
Deur F.

All the Collectors in the Netherlands call the Alatae, Crabs; *(with a few excepted) which are called* Flap whelks. *There are many kinds of these* Crabs, *and known to us as Males and Females. The males are those, that have their branches close together, and closed; the Females, those that have the branches flatter and open like channels. The Author provided us with several kinds, with names, which we will pass by. The first is shown on plate* XXXV., *by the letter* A. *which he calls the* Boat-hook, *or the* Devil's Claw: *Another one of this kind, is shown by the letter* B. *which we consider the Female of the foregoing: The Author calls another kind, by the letter* C., Stumpy; *of which a second one is shown by letter* D. *The* 2d *kind, indicated by the letter* E., *is called the* spotted Crab; *and the one by the letter* F. *the* yellow Crab, *because it is yellowish, white, and has no spots. Its Female is shown on plate* XXXVI. *by the letter* G. *The one by the letter* H. *is the* 3d *kind, and can be seen on plate* XXXV., *of which one can only say, that it would be serviceable for making fountains or grottoes; because its lumpishness keeps it out of Curiosity Cabinets. But the Author is wise to show, all the different kinds Nature is capable of bringing forth. Then there follows the* 4th *kind, shown on plate* XXXVI. *by the letter* I. *which the Author calls* Millipede: *This one, as well as the following, depicted by the letter* K. *are rare; the latter we call the* Podegra crayfish, *perhaps because of the knobs that it has in its branches, while we had the Female depicted on plate* XXXVII. *by* No. 1.[38] *Now come the kinds which we call* Flap whelks, *because of their big wings or flaps that are not split, and which protrude next to their mouth: although the Author puts them with the foregoing. But we will follow the author who calls the following one the* 6th *kind, and which is shown on plate* XXXVI. *by the letter* L. *and which we call the big* Flap whelk. *The* 7th *kind, by the letter* M., *is called the* Mizzen *by the Author, but we call it the* Little Dove. *His Honor calls the* 8th *kind, by the letter* N. *the* Little Lump; *of which a* 2d kind, *by the letter* O. *is called the* rolled-up Mizzen. *The* 9th *kind, by the letter* P. *is a smaller variety of the foregoing. The* 10th *kind, is shown on plate* XXXVII. *by the letter* Q. *called* Freckles *by the Author, but* Frog *by us. The* 11th *kind, indicated by the letter* R. *the* Fencer, *or the* Pointer, *we call the* knobbled Flap whelk. *The* 12th *kind, by the letter* S. *is the* Luhune Whelk. *The* 13th *kind, shown by* T. *is the* bossed Canari; *and another one of these by the letter* V. *is the* banded Canari; *and a* 3rd *one by* W. *is the* flowered Canari. *The* 14th *kind, by the letter* X. *is the* broad Canari. *The* 15th *kind, by the letter* Y. *a* Samaar. *All of these, as they are named by the Author; we will add some essential pieces, which belong with these, and which are not well known. On plate* XXXVII., *by* No. 2.[39] *one will find the showpiece of the* Flap whelks, *to wit the* Orange Flag, *so called, for its beautiful orange, and white bands, and which truly carries the standard of all the* Flap whelks; *only this one, belonging to Mr.* Vincent, *is known. I added to this the* spotted Flap whelk *from Lapland, belonging to Mr.* De Jong, *who got it from Lapland, and of which no other is known, besides this one, indicated by* No. 3.[40] *It is a yellowish white, with black speckles and spots.* No. 4.[41] *is another kind, which I received from the West Indies, but it has little color. The one shown by* No. 5.[42] *is also counted among the* Flap Whelks; *there are many kinds of these, and one will scarcely find two, that resemble one another, which is why the Collectors call them* Ruffs.[43] *The previous plate* XXXVI., No. 6.[44] *shows a small* banded Flap whelk; *and a similar one by* No. 7.,[45] *though it is a remarkable one, due to its winding spindle or tip, which is why I valued it myself.*

CHAPTER 23

Porcellana Major. Bia Belalo.

The tenth Main genus comprises the *Porcellanae*,[1] a special kind of Whelks, which differ a great deal in shape from all of the foregoing, because these do not have any *Turbo*, or protruding head, at all; their backs are roundly bossed, somewhat flat on the belly,

Fol. 116.
XXXVIII.
E
A
Q
M
L
I
D
K
P
O
C
B
N
H
G
F
Deur F.

along which one will find a long slit or narrow mouth, longer than the entire Whelk, the lips of which curve inwards on either side, but in such a way, that the one on the right is not rolled or wound inwardly, but is notched or has blunt saw-teeth; the side opposite from this, is also a little jagged near the mouth, full of ribs or furrows, and then curves inwardly, rolled up with many twists and turns; the slit's two ends are the same; but one can see on the top one some traces[2] of the twists, and on some of these a little point or *turbo:* Furthermore, all these kinds are smooth by nature, and as shiny as a mirror, as soon as they come from the sea: Their common name in Latin is *Concha venerea;*[3] in *Ennius, Matriculus; Pliny Lib. 9. Cap. 25.*[4] scarcely distinguishes them from the *Mutiano-Murex*, but there is nothing sharp or spiny about any of these kinds, that would fit a *Murex*. One calls them today *Porcellanae*, in imitation of the Greeks, who call them *Charinae: Apud utorsque nomen acceperunt à similitudine pudendi muliebris, quod Graeci Chaeron, Latini porcum & porculum vocant, cujus aliquam similitudinem refert hujus Conchaerima;*[5] some of the Dutch call them Klipkoussen [Clack dishes][6] for the same reason; but one can also call them with a more proper name, *Porcellanae;* in Malay *Bia* or *Sipo bilalo*,[7] that is to say, Lick Whelks, because one can use them to lick, that is, to smooth, linen, paper, and the like; in Ambonese, in Leytimor, *Huri;* on Hitu, *Aulilu*. There are the following kinds:

I. *Porcellania guttata*,[8] in Dutch *Klipkoussen*, in Malay properly *Bia bilalo* and *Krang krontsjong*.[9] This one is the largest and most beautiful of this kind, about the size of a small fist, very round, with a smooth back, that is heavily speckled with black drops, shot through with smaller brown and yellow ones; and there is a golden yellow stripe along its back, but not on all of them; the more uniform the black drops, the rarer it is; as soon as they come out of the sea they will shine like a mirror; the belly is not very flat, but even, so that they can lie on it, very white and shiny; the only thing one will see of the Animal, is a thin flap, speckled almost the same way as the shell, to wit, with black, brown, and yellowish drops, whereon one will find little white grains, and it sticks out a short tongue down below, and attached to that are two short little horns with eyes: The one which is held to be the Female, is thin and has a light shell, gains nearly its entire size, before it curves the one lip, which is sharp and thin like parchment; the shell is beautifully painted with black, blue, and yellow; and the more blue it has, the rarer it is: One finds them on beaches that have white sand, with large rocks, that lie bare in it; they mostly stay hidden in the sand, since anything that sticks out above the sand becomes rough and dead, but when there is a new and full Moon, they come out of the sand at night, and cling to the rocks: One tries very hard to get the Animal out, so that the shell will retain its luster; because if one buries them, or let them lie under the Sky, they will get a deathly pallor, which shines through beneath the outer smooth shells, so that it seems that this Animal, while it is dying, will rob the shell of its luster; the surest way, therefore, is to put the ones that have just been caught briefly into hot water, so that they die, and then remove the meat with hooks, as much as possible, and put the rest with the shell in some shady spot, where neither rain nor sun can reach it, and let the Ants eat the remainder: One should not put them in fresh water, as long as there is some meat in them, because they will pale if they stay just one night in it, nor should the dead blood of one ever touch the other, because that will spoil them as well; one should give them to drink every second or third year, that is, place them in salt water for half a day, rinse them with fresh water, and then dry them in the Sun again; tobacco boxes and spoons are made from them, but these Natives have no use for them, except that some poor Folks, pressed by hunger, roast them on coals, and eat them; which affects them so badly at times, that some have to pay with their lives, while others are barely saved in the nick of time: Those people do not remember the basic rule, as laid down by the Natives, that any Seawhelk that is smooth and shiny, or any one that has only red speckles, is not fit to be eaten, but that the rough and spiny ones are always much better. But if someone carelessly did eat of it, one shall give him immediately a great deal of sugar water, or some thick syrup, to prevent strangulation, so that the throat cannot contract; and thereafter administer as an antidote black *Calbakaar*[10] (described in Book XII of our *Herbarius*) rubbed in some water, which will eject that hurtful food by means of vomiting; this has to be done quickly, in order to get the remedy into the stomach, before the throat contracts, because it is of no use otherwise; as happened in the year 1664, to a Woman in the *Passo Baguala*,[11] who died after eating this Animal, because the *Calbakaar* came too late, and the throat was already closed.

II. *Porcellana montosa*,[12] is somewhat smaller than the foregoing, and its luster is duller, its shell is a pale white, whereon are many angles and hills with russet stripes, as if one were seeing a distant range of round hills, the way the Island of *Ascension*[13] is painted: We call it the *Cape*, meaning the *Cape de bon Esperance;*[14] but it is rarely found, and is well-nigh unknown in Amboina.

III. *Concha testudinaria*,[15] Karet, is more elongated than the foregoing *Cape;* with beautiful spots on the back, as one will see on a Turtle shell, or karet, furthermore as smooth as a mirror; it was once wrongly called *Cape falso*, but is now called *Karet*, and is considered to be one of the rarest.

IV. *Argus*,[16] this is an elongated *Porcellana*, with a brown shell, usually nicely painted, with white eyes, each eye is fashioned from 2 or 3 rings, the way one paints the giant *Argus:* These too are seldom found, because they sojourn deep down in the sea, and crawl in weels[17] sometimes, or are tossed up by earthquakes, as happened in the year 1674, at the hook of *Ciel*,[18] on little *Ceram;* but most of them have a dull color; and the shiny ones are very rare.

V. Snakeheads, large;[19] these have a flat belly, and have a narrow edge all around the belly, the width of a finger; above the edges they are black or dark brown all around, shiny like a mirror; the back is speckled with tawny[20] spots.

VI. Snakeheads, small;[21] are of the same shape and color, but smaller, no longer than the joint of a finger, also very smooth and shiny: They are common on shingle beaches, such as those of *Laricque* and *Nussatello*. One finds here a rare kind, bluish on the back, with a yellow vein, which are called blue Snakeheads.

VII. White Jambus,[22] are similar to the foregoing in shape and size, but entirely white, like unto a certain kind of white, wild, Water Jambu; the back has a protruding edge across it, and one will note a pale white little grain at each end, set in a circle, like a mounted pearl, and near it is a light-red spot, which pales over time; the Animal is white as well, thin and transparent, with black little stipples, and is seldom found.

VIII. *Talpa*,[23] Mole, this one is elongated, round as if it were a roll, black or dark brown on the belly, and on the sides; the back has two pale bands, with a russet one between them: The natives often use them to smooth Pissang[24] leaves, which they use to make their little tobacco rolls,[25] and have the color and shape of a Mole.

IX. *Carneolo*,[26] in Malay *Bia daging*, is elongated like the foregoing as well, but rounder at the ends, where the former ones have a little hole; uniformly light red, or the color of flesh.

X. *Porcellana salita*,[27] Salt Grains, these are short, round *Porcellanae*, about 2 inches long, gray-brown on the back, with white drops, which stick out somewhat, as if grains of salt had been strewn over it; they are reddish on the belly, but they mostly grow a pale yellow; the Female has a thin shell, with few grains that do not stick out. They are commonly found on all beaches, and therefore slighted.

XI. *Porcellana Litterata sive Arabica*,[28] Letter Whelks or Arabian Whelks, while some consider them Music Whelks: They have a flat belly, with a thick edge all round, which sticks out a little; blackish on the belly, and with black drops on the sides; the back is marked with small tawny or russet stripes, and stipples, quite like Arabian script; but others make it out to be a *tablature*[29] with Musical notes; one can also see some of our Letters in them. A second kind is light of shell and color, and instead of script, it has many entangled stripes, as if it were Landscapes; see the odd[30] figure of a Musical Whelk in *Rochefort Hist. Antill. Lib. I. Cap. 19. Art. 9.*

XII. *Porcellana lentiginosa*,[31] Cockroaches, are easily the length of a finger joint, with a high boss; their backs are covered with russet, black and bright-yellow speckles, with some light-blue ones mixed in: One would think, that these are the Young of the first kind, but they do not grow, and they can be distinguished by some protruding ribs that run from one end to the other: The Females have a lighter shell, without ribs on the belly, uniformly covered with russet speckles, and so light, that one can blow them away.

NB. When in the telling of these Whelks we speak of the Male and the Female, one should not assume that there is indeed such a distinction as to gender, but rather that, as is customary, we take the Female to be the one that is the lightest and smoothest.

XIII. *Variolae*,[32] Measles, are once again somewhat elongated, and of two kinds: The first and largest one[33] has large black drops on the sides, like a certain kind of small pox, which the Malay call, *Lute lute bessi*, that is, Iron Pox: The other kind[34] is somewhat smaller, and has purple drops on the sides, similar to the measles (*morbilli*)[35] both are intricately speckled on their backs, with russet and tawny dots; but the last one is the most beautiful.

XIV. *Ovum*,[36] in Malay *Bia* or *Sipot Saloacco*, these have the shape of a Duck's egg, or a little longer; the mouth's tip is longer than the entire Whelk, and notched; the entire shell is pure white, smooth as a mirror and shiny, except for the thick lip, which sometimes is yellowish or a dirty white; the inside of the shell is violet, and the exterior of the Animal is completely black, and it also melts away in a black ink. The Alphorese, or the wild mountain men of *Ceram*, greatly esteem these white eggs, which are found most often on Ceram's beaches; none of them may wear these Whelks around the neck, or from a lock of hair, except for their Champions and those who have gotten some heads of their enemies: After they have been smashed to pieces, and then ground on a stone until they get round, elongated or other shapes, they are used to inlay their long shields, called *Saloacco*, which makes them quite handsome, because these bits and pieces shine, as if they had a white glaze, and they stand out very handsomely on those black shields, which are edged with red and yellow, and thus these [shells] were called after them. One will encounter another kind, that has a thin shell, with a single lip, sharp, not curled over, which are held to be the Females, as was mentioned previously about the first Kind. A third kind remains small, is somewhat knobbly or grainy, and has a thick shell.

These Whelks gave that precious Chinese earthenware the name of Porcelain, because they used to believe that it was made from the same material; or, and which is more believable, that those Porcelain dishes corresponded in whiteness and smoothness to these Whelks: But whether our present Porcelain pottery corresponds to what the Ancients called *Vasa Myrrhina*,[37] is being examined by Scholars; the learned *Julius Scaliger Exercit. 92*,[38] thinks it likely that Porcelain pottery got its name from the *Concha porcellana;* and *P. Bellonius*[39] in his *Observ. Lib. 2. Cap. 71.* does indeed call Porcelain ware *Murrhina*, as is common usage, but doubts that they represent the true *Murrhina* of the Ancients, asserting, more plausibly, that the true *Murrhina* were made from the Precious stones *Sardonyx*, *Onyx*, *Jasper*,[40] and the like: this is confirmed by *Georg. Agric. de Re Fossil. lib. 6.*[41] and *Boetius de Boot,*[42] in his *Gemmarium Lib. 2. Cap. 85.* but for the right history of Porcelain, see the next Book.[43]

In order to catch these Whelks, take a piece of *Caju sonti,*[44] or *Perlarius primus,* put it in sea water, where it is not more than 4 feet deep, and let it rot there, and these Whelks will gather around it: The pulverized shell is also used for poultices, which are smeared on the bodies of those, who are swollen and dropsical; the Animal itself is not fit to eat, since it causes one to choke which is often followed by death. *Phil. Bonannus Part. 3. Fig. 252.*[45] calls it *Ovum marinum.*

The Porcellanae majores *are all called* Klipkoussen *by us, and called worse by Sailors; but one could perfectly well call them* Rock Whelks (Kliphoorns), *which I do all the time. There are many kinds, of which only a few have specific names: these are but common Whelks, because they are so plentiful, except for some, which will be singled out. The first kind, depicted on plate* XXXVIII. *by the letter* A. *is quite common; but there is a second kind, which is of a lighter material, and thinner shells, and also rarer than the foregoing. The 2d kind, shown near letter* B. *is also unusual. But the 3d kind, by the letter* C. *is far more valuable: it is called the* Turtle Whelk, *because of its brown spots which glow and are transparent; it is the most excellent of all these kinds, and especially the one, which is kept in the cabinet of Secret.* Blaauw.[46] *The 4th kind, by the letter* D. *is called the* Double Argus *by us, because of its spots, since each one of them is surrounded by a round little ring: one used to pay a great deal for this kind, and it is still quite rare, which is why it is still valuable. The 5th kind, by the letter* E. *is too plentiful to be of value. The 6th kind, by the letter* F. *is like the foregoing, and was called the* Snakehead *by the author: there is a second kind of these, indicated by the letter* G. *The 7th kind, by the letter* H. white Jambus, *though we call it* Arched Back. *The 8th kind, by the letter* I. *we call like the author,* Moles. *The 9th kind, by the letter* K. *is quite common. The 10th kind, by the letter* L. *are* Grains of Salt. *The 11th kind, by the letter* M. Arabian Letters. *The 12th kind, by the letter* N. Cockroach. *The 13th kind, by the letter* O., Red Measles; *of which another kind, is shown by the letter* P. *The 14th kind, depicted by the letter* Q., *is called the* white Porcelain Whelk *by us; there is a 2d kind of this one, much thinner and of a lighter material, inside as white as outside; but it is quite uncommon.*

CHAPTER 24

Porcellanae Minores. Bia Tsjonka.

The small *Porcellanae* comprise the eleventh Main Genus, and there are many kinds, which differ even in shape from each other, but since they all share the common mark of the *Porcellanae,* we will come to know them in this chapter in no particular order. The Malay commonly call them *Condaga,* and *Bia tsjonka,*[1] because they are often used in the game called *Tsjonka,* in which one counts many small things into certain holes, which have been hollowed out of a thick plank: The most important among these are the ones we call *Thoracia,* that is, Breastplates,[2] because they look like the breastplate of an armor: called in general *Cauri* or *Caudi,* and have the following four kinds.

I. *Thoracium oculatum,*[3] White Eyes, about as long as a finger joint, with a broad and notched side, that has a wide blackish spot on it like a burned mark; the back is a drab gray, with many little round white eyes; the mouth is very jagged, and a light red on the inside: Some are as wide as a breastplate, and they are the handsomest: Some are oblong, do not have a protruding edge, and dainty eyes. They are also called Burn spots.

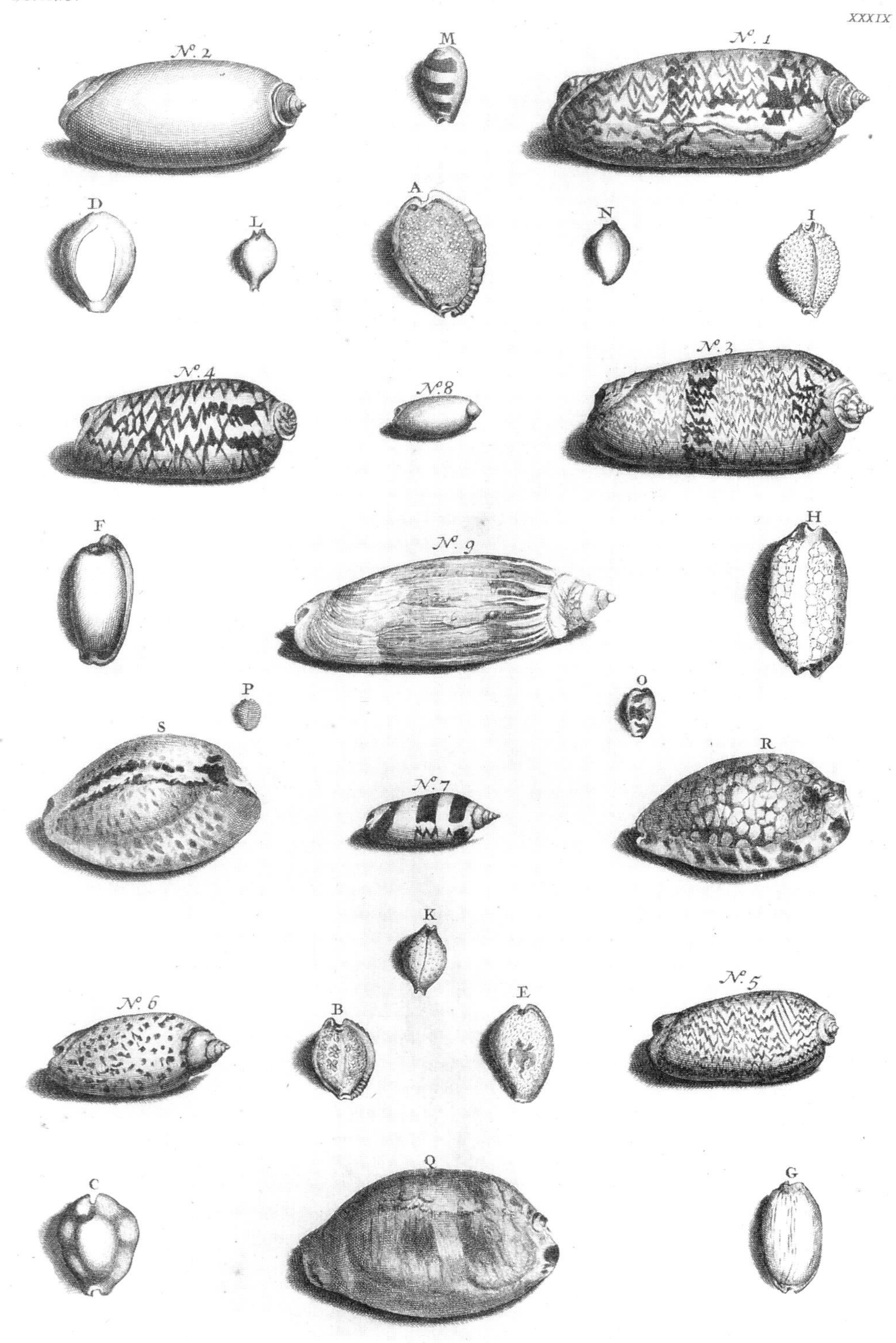

XXXIX
N°. 2
M
N°. 1
D
L
A
N
I
N°. 4
N°. 8
N°. 3
F
N°. 9
H
P
O
S
N°. 7
R
K
N°. 6
B
E
N°. 5
Q
C
G

II. *Thoracium stellatum,*[4] Little Stars, are somewhat smaller than the foregoing, a light brown or russet on the belly and sides, with a little purple towards the ends; the back is a pale white and decorated with small russet or brown stars, like the stone *Astroites.*[5]

III. *Thoracium vulgare, sive Cauricium,*[6] common yellow *Cauris:* these have the true shape of a breastplate, nicely adorned with small bosses, very smooth and a pale yellow, sometimes with a bright yellow circle on the back, white on the belly, and notched near the mouth, usually the size of a finger nail, or the nail of the thumb; but the smaller they are, the nicer those little bosses appear: These represent that famous money, which they haul away by the shipload from the Maldive Islands, and bring to *Bengale* and *Siam,* where it serves the function of coins; but there are other kinds of small *Porcellanae* among them, which are called *Condagae.*

What the Arabs call *Wadaat,*[7] are small white Whelks, belonging to the genus *Concha Venerea,* which they use in those parts to adorn the necks of their dogs, and such like, and which I can only judge to be *Cauris.*

In his second chapter *Franciscus Pyrard* writes the following about *Caudi* or *Cauri:* the *Maldives* knows other riches, which are small little Whelks, as thick as a little finger, completely white, some smooth and shiny, which they fish for twice a month, three days after the new and three days after the full Moon, and are not found at other times; it is the Womenfolk that look for them on the seashore, and they sometimes go up to their waist into the water: These Whelks are carried hence in such great quantities to all parts of the Indies, that I have sometimes seen 30 or 40 ships loaded full with them. Those from *Bengale* think they are so valuable, that they are ordinary money there, although those Nations have enough Gold, Silver and other Metals; and that, which is to be marveled at, the Kings and other great Lords, have warehouses built, in order to store them, and consider them part of their treasure. The Merchants only get them in order to bring them to *Bengale,* since they are not found elsewhere. In *Cambaja,* and in other places in the Indies, they plate the most beautiful ones with silver and gold, and consider them a rare Curiosity, indeed, as if they were Precious stones.[8]

IV. *Thoracium quartum,*[9] plain *Cauris,* are smaller than the foregoing, without bosses, with smooth sides, its back bluish or the color of lead, with a bright yellow ring around it, as if one were looking at a Turquoise in a setting, but the blue color pales over time and turns a pale gray: They are common in *Amboina;* but the true *Caures* are not: The Chinese put the blue-backed *Cauris* of the fourth kind in Lemon juice overnight, until they melt, then drink it down against passing painful or chalky water, perhaps the way we use the eyes of Crayfish.

V. *Casuaris* Eggs,[10] which are oblong, with round and smooth sides, usually the length of a finger joint, with pale green, brown and russet speckles, like the eggs of the *Casuaris* bird; sometimes of a uniform color; other times with a broad burn mark on their backs.

VI. *Dracaena,*[11] Little Dragon heads, have a pale gray or pallid shell, with two round bands athwart, with a large burn mark in between, which resembles an Island at times, or a flying dragon: These are rarely found.

VII. Bluebacks,[12] are oblong, with a white protruding edge or frame, with several colors on one side of its back, with tawny and blackish stripes, which look like some small Landscapes;[13] they sometimes have 4 black little eyes on both ends, which are rare; sometimes the outer skin has been rubbed off by sand, and then they usually are purple on their backs: They are also called Frames.

VIII. *Isabella,*[14] these are also oblong and round, like a roll, without an edge on the sides; both ends are an Orange yellow, and blunt; the back is a pale red with a few black stripes: One will also find ones that are entirely white.

IX. The back of the small *Argus*,[15] is pale brown, and it is densely covered with large white eyes that are connected to each other by narrow edges: These are not found very often.

X. *Nussatellana granulata*,[16] Rice grains, are no longer than a finger nail, round and bossed, covered everywhere with protruding little kernels, which are separated with a groove across the back, ribbed on the belly: The most beautiful ones are entirely white, though sometimes there might be a little purple in the background: The plainer ones veer towards a gray or paler color, and have finer grains: A third kind, is as large as a finger joint, without grains, its background color a steely blue: some are a reddish gray, and so shiny, that they seem enamelled.[17] This small fry is nowhere as abundant as on the most northern island of the three Brothers, or *Nussatello*,[18] which are three small, but high Islands off the west hook of greater *Amboina*, where these, as well as the three foregoing kinds, and the following ones, occur in abundance, wherever there is a beach of fine white sand: One finds them as well, although fewer, on *Hitu*'s coast, but those are never as white, as the ones from *Nussatello.*

XI. *Globuli*, Little Buttons, these are smaller, rounder and higher bossed than the foregoing ones, with a protruding little corner on either end, and if this has been worn down, they resemble the small buttons of a Doublet: One kind[19] is covered with little grains; some are a pale yellow; some are white; Another kind[20] is somewhat smaller, entirely smooth, except that it has a very delicate groove on its back; also yellowish at times; other times white.

XII. *Aselli*,[21] Little Donkeys, are white, but have three wide black bands sideways across their backs, as if they were three sacks that had been put on the back of a Donkey, or even better, as if two white sacks had been put on a black Donkey.

XIII. Small Pearls,[22] have the shape of the *Bia Saloacco, No.* 14, as described in the previous chapter; but not larger than a finger's width, and they stay that small, and are entirely white, so that one could wear them in one's ears as if small pearls.

XIV. *Ursulae*,[23] Little Bears, are small stuff, mostly white, with a large grayish spot on their backs, which one could make out to be a Bear or some such animal.

XV. *Pediculus*,[24] the Louse, are the very smallest of all these genera, usually not much larger than a thick fat louse; some like Wall lice: They are entirely white, with ribs and furrows athwart, as if one were looking at the ribs of a louse; but all the ribs end up in a groove that goes the length of the back. One will find a large kind, the size of an average hazelnut, ribbed and furrowed in the same manner, but not as white as the small one.

The Porcellanae minores, *are the smaller kinds of the foregoing: The first one is shown on plate* XXXIX., *by the letter* A., *and these are called* White eyes. *The* 2d *kind, by the letter* B., *are* Little Stars. *The* 3d *kind, by the letter* C. *are* Cauris, *though the Germans call them* Snake heads, *and use them often to adorn the harnesses of their horses. The* 4th *kind, by the letter* D., *is a* blue Cauris. *We lack a picture of the* 5th *kind. The* 6th. *kind, by the letter* E., *is the* Little Dragon's Head. *The* 7th *kind, by the letter* F. *is called* Blueback. *The* 8th *kind, by the letter* G. *is* Isabella. *The* 9th *kind, by the letter* H., *is the* little Argus. *The* 10th *kind, by the letter* I. *is the* Rice grain. *The* 11th *kind, by* K., *is the* grainy Button: *A second kind, by the letter* L. *shows the* smooth Button. *The* 12th *kind, by the letter* M. *of which there is another kind, with Orange spots. The* 13th *kind by the letter* N. *The* 14th *by the letter* O. *The* 15th *is shown by the letter* P.; *thus far the Author, but we added a few unusual ones. The one by the letter* Q.[25] *is a beautiful* cloudy Achate Rockwhelk. *The one shown by the letter* R.[26] *is called the* white-spotted Achate. *The one by the letter* S.[27] *is very unusual, sent to us from Cartagena, which is why we call it the* Cartagenian Rockwhelk: *These are only known to be in the Curiosity Chambers of Secretary* Blaauw, *and Mr.* Feitemaas.

CHAPTER 25
Cylindri. Rolls.

The twelfth and final genus of the Univalve whorled Whelks comprises the *Cylindri*, or rolls, so called because of their oblong shape, which is like a rolled-up piece of paper, or linen: All of these have a thick shell, smooth, shiny, and with a short tutel in front; owing to their Cylindrical shape they are so much alike, that touch cannot distinguish among them,[1] and difference is mostly expressed in terms of colors and decorations, as follows:

I. *Cylinder porphyreticus*,[2] has dark-gray or blackish speckles on the outside, like that kind of Marble, which is called Porphyry stone, and it has a black band right across the center; they are bright yellow inside.

II. *Cylinder niger*;[3] Satin Rolls, these are as black as satin, and as smooth as a mirror, without any other colors mixed in, if they be the true ones; the baser ones veer towards chestnut brown near the sides; the side is edged with one or two protruding ribs, but the truly black ones are usually not edged. This kind is almost never found, except on *Honimoa*, in the small *Tjouw* bay,[4] and are considered the rarest of this genus.

III. *Cylinder tertius*,[5] has a blunt tutel, usually an olive color, with small black speckles, that look like Islands: Others have several bands sideways, just as in the Agate stone: Some have a uniform color, to wit, light brown, and Isabella.

IV. *Cylinder quartus*,[6] has a tutel that is even more blunt, and which almost lies in a dent; the background color is also olive, while it is the smoothest of all of them, with black spots and stripes, all in a row, quite resembling a train or pomp of men, who, wrapped in long cloaks, are following a corpse, which is why they are normally called *Sepulturae*,[7] or Prince's funeral.

V. *Cylinder quintus*,[8] though this one is greenish, and speckled with yolk yellow and some purplish-blue spots, they are considered the most common and basest.

VI. *Cylinder sextus*.[9] Gray Monks, these are densely speckled on the outside with dark-gray and blackish dots, like the first kind; and those that have negligible spots, are almost entirely gray; its tutels stick out a little more, and one kind has a mouth that is violet.[10]

VII. *Cylinder septimus*,[11] Camelots, these have blackish watering or blackish waves, like camelot.[12]

VIII. *Cylinder octavus*,[13] Blue drops, these differ somewhat from the usual shape, in that they have a noticeable tutel, and a protruding rib, that runs across the back at an angle, a dirty white or yellowish, with some purple or bluish drops on them: One kind remains smaller, with a blunt tutel, and has black drops.

IX. *Cylinder nonus*,[14] are the Large Glimmers, with a pointed tutel, smooth as a mirror; some are striped; some have dark-green and blackish speckles, and these are common: One will also find ones that are entirely white, but rarely.

X. *Cylinder decimus*,[15] are the Little Glimmers, these do not become larger than a finger nail, speckled like the foregoing: One kind is as pointed as a *Buccinum:* A smaller kind is so rare, speckled with stripes and small spots, that one would hold them to be Jasper stones,[16] especially if they have been set in a ring in such a way, that one sees only the back; but one needs to have hundreds of them, in order to find some of these.

XI. *Cylinder*;[17] Little Agates, also are not much longer than a finger nail, its tutel is blunt, a light red or purplish, with veins like an Agate. All the foregoing eleven kinds of cylinders have a rather hard white meat, without any lid, and as the common rule says not fit to eat: All that is smooth and shiny is not meant to be eaten. Their Fatherland is the *Liasser* Islands,[18] especially *Honimoa*,[19] the big Reef near *Guli guli* up to *Keffing* off *Ceram*'s eastern corner,[20] together with the north side of the Bay of Ambon.

XII. *Cylinder lutarius*,[21] Mud Rolls, this is a particular kind, a combination of *Buccinum* and Roll, 4 inches long, and more than 2 fingers wide, the color of soil, with an Orange mouth, and a thick lip: One will find them in the Marshy Sagu Forests on *Ceram*, or similar Swampy rivers, and are therefore not to be counted among the Sea whelks; they are good to eat, are therefore sought after, and have a sweet meat and juice, like the aforementioned *Sipot aijer:* One could also count this kind among the *Volutae*, and they are commonly called Midas Ears: There is a smaller kind; but both have the same want, which is that their tips are broken off. In Chapter XXI, no. XX, they were counted among the *Volutae*.

The Cylindri, Rolls, *whereof there are many kinds, are called* Dates *by us, because they rather look like the pits of these fruits. The Author provided us with several kinds, which are all depicted on plate* XXXIX. *The one by* No. 1. *is surely the largest one; although there are even bigger ones of this genus. We call the one by* No. 2. *the* black Date, *because of its beautiful black color.* No. 3. *is a* motley Agate Date. No. 4. *is called the* Prince's Funeral *by the Author. We lack pictures of the* 5th *and* 6th *kinds, and owing to the great variety of this genus, it is difficult to guess which kind the Author had in mind here. The* 7th *kind is shown by* No. 5. *The* 8th *by* No. 6. *The* 9th *by* No. 7. *The* 10th *by* No. 8., *which can all keep the names the Author bestowed on them, because we do not have other ones. There are no pictures of the last* 3 *kinds that were described; but we add in its stead an exceedingly rare one, indicated by* No. 9.[22]

CHAPTER 26

Conchae Univalviae.

After the twelve genera of Univalve, twisted Whelks, follow the Univalves that are not twisted or turned, divided into two genera. The first one dealt with in this chapter, includes those which have a shell on only one side, while the other side is naked and clings to the rocks; of which there are the following kinds.

I. *Lopas* or *Lepas, sive patella*,[1] a Little Lamp, or Little Dish; in Malay, *Bia sabla;* has a shell with a raised boss, like a swelling, or a flat Pyramid, and clings to rocks; coarse on the outside, gray, and lusterless, inside a dirty Mother-of-Pearl color: The biggest kind[2] is mostly smooth on the outside, or with some dark bands; the mouth is the size of a rixdollar, but oval: The second kind[3] is as small as a Bloody ulcer; sometimes deeply ribbed, not much bigger than a stiver; sometimes with corners that stick out, and somewhat flat, so that it resembles a small Turtle: The third kind[4] is well-nigh flat, with a round circumvolution, on the outside, slightly speckled with black dots, inside the color of silver.

They are commonly called *Patellae*, that is, Little Dishes, in Portugal *Lambroa*,[5] that is, Little Lamps, and they cling to the rocks very tightly so that they have to be pried off with sharp chisels, not without doing some damage to the shell: One cannot tell if they have moved, because they make a bare spot, where they are: All are good to eat, especially those one finds on the Sea rocks in Portugal, and which are generally larger than the Ambonese ones: One turns the shell over and puts it on hot coals, roasting them in their own juice, and then one gets the meat out with a pointed piece of wood: One can also simply cook them in water, like beans. The River-*Lopas*[6] is called *Ahaân*, and is the size of a fingernail, and it

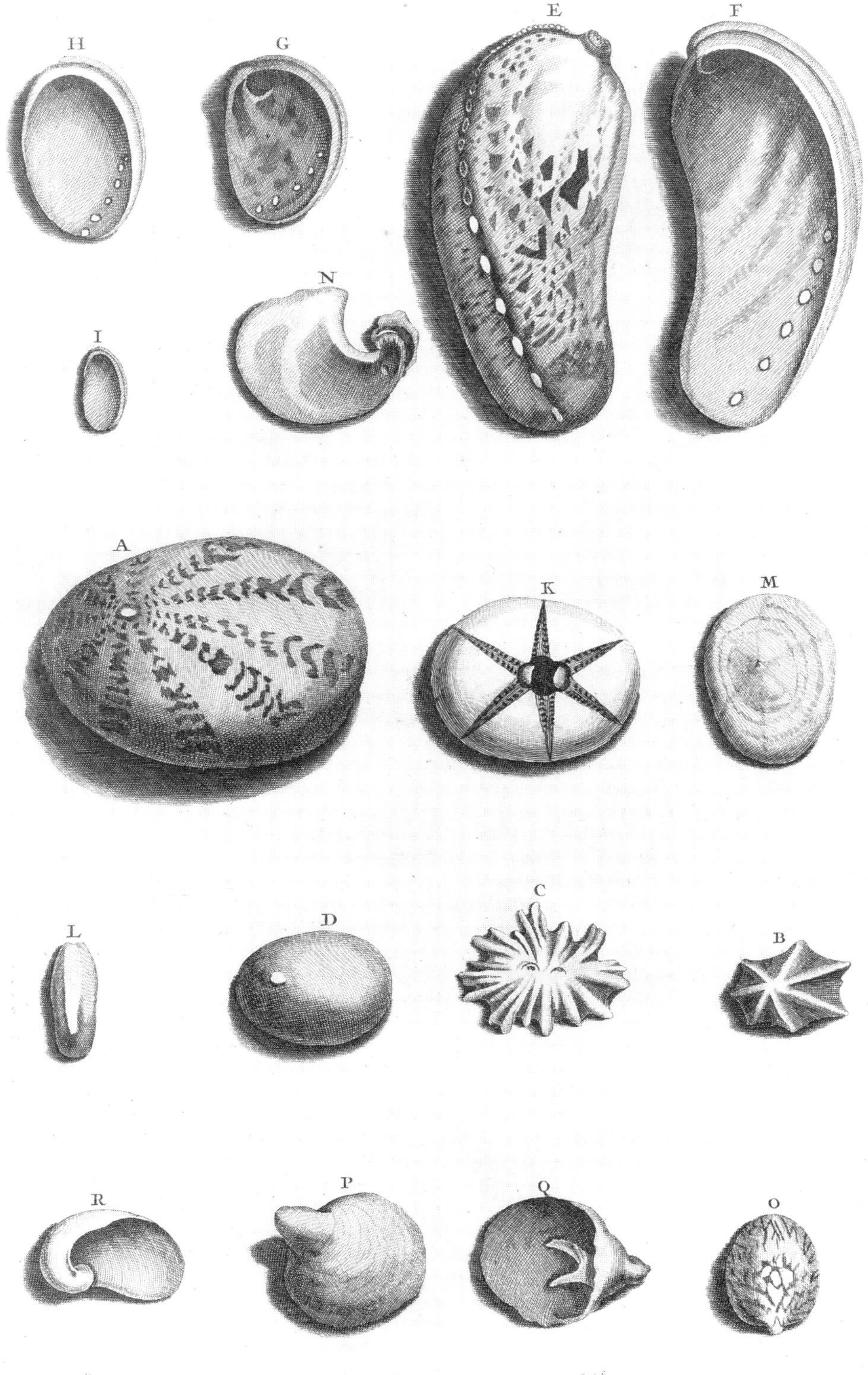

J: Lamsveldt: fec.

has a thin and white little bone in its flesh, the shape of a small penknife. Another kind is thinner and flatter, and is called *Bessi mattal*,[7] and has a thin shell; its meat is edible, but it does not have the aforementioned little bone or penknife.

II. *Auris marina;*[8] in Malay, *Telinga maloli* and *Bia sacatsjo;* in Ambonese, *Hovileij;* this is shaped like an oblong ear, to wit, on one side with a raised edge, which in one corner ends in a curl; along that side the curl has many small holes, of which the ones in front go all the way through, but the ones in back do not; it is coarse on the outside, and lusterless as well; on the inside a beautiful color of silver, or like Mother-of-Pearl; it has a lot of meat, and it clings naked to the rocks, like the *Patellae*, but they are not as good to eat; the largest and handsomest kind is easily the length of a finger, smooth on the outside and speckled with dark-green and blackish dots like a snake: The second and most common kind[9] is shorter and wider, covered on the outside with weals, white and chalky: The third kind[10] is even wider, with a thin shell, smooth on the outside, but slightly ribbed: The fourth kind[11] is very small, scarcely the length of a finger joint, smooth on the outside and with some speckles, without the little holes; one will find them on small stones along the beach.

III. *Balani*,[12] Acorns, Weals, in Malay, *Gindi laut*, and opening Tulips, have the shape of a burst tumor, they grow together in bunches, big and small, and can be found on ships in clumps, also on some large rocks, glued to them with a flat bottom, so that they are closed off at the bottom; from there they rise like a blunt Pyramid, with uneven sides, ribbed or furrowed on the outside, red with some gray mixed in; they have an opening on top, which some have compared to a Tulip that is opening up; the sides of the mouth are sharp, and as if burst; inside one will see a contrivance of two flat and jagged little bones, slightly curved, like the beak of a Parrot, which close together like teeth, and that are attached to the flesh with thin little skins; but every tooth is split in two again: When they open these teeth, one will find a bunch of 12 short and curved little plumes peeping out, of which the two in the center are the biggest ones, and which become gradually smaller on either side; on the inside as hairy as the legs on some crabs; the Animals use these as a way to get food from the sea, and after pulling the same back in again, it closes its teeth; the innermost flesh, when raw, is quite slimy, without a sandbag, or *Papaver*, because it seems to live on nothing but water: once cooked, it becomes white and somewhat hard, but with an excellent taste, like the white meat or fat of the best crabs, with a tart and peppery sauce: This pertains mostly to the big Knobs, that grow on ships and shallops;[13] because those on the rocks do not taste as good: If one chisels them off the wood in such a way, that they are not damaged on the bottom, and puts them in salt water, one will see how they raise the aforementioned beak outside the mouth, deploying the beards or little plumes, and lick off the slime and moss that sticks to the sides of the shell with the rough kernels that are on the beards, which must undoubtedly be its food; but as soon as one moves them, they will pull those beards and beaks back inside again: Note here once again how careful Nature is, because she fastens moss and slime onto the outside of these Knobs, so that this immovable Animal can live from it. The cooked ones are considered good eating by the Chinese; they are also removed raw, and pickled like *Balatsan*,[14] but it has to be kept for at least half a year: Inside, the little house has a protruding edge around the center. One will find them in clumps the size of one's fist, or as large as one's head, fastened onto the ships; the biggest one in the center, and the small ones around it. One will find two or three smaller kinds, which are not ribbed on the outside, but which are an ugly gray, coarse, and full of holes; the mouth opens up into three or four stiff points, which can hurt you quite nastily, if one carelessly steps on them: One will find the latter both on rocks and on turtles. The Chinese take the biggest clumps, place them be-

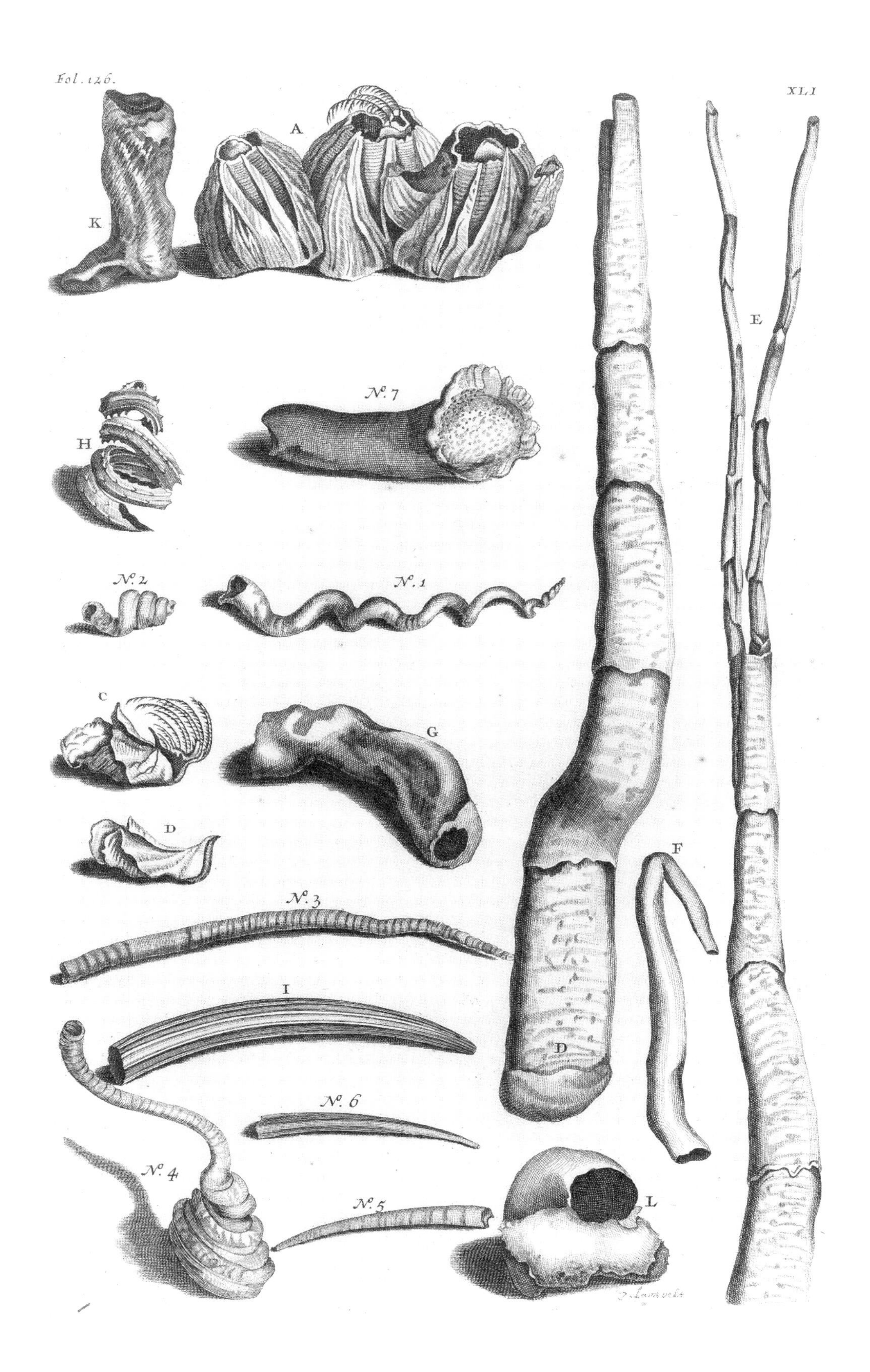

Fol. 126.
XLI
A
K
E
H
N°. 7
N°. 2
N°. 1
C
G
D
F
N°. 3
I
D
N°. 6
N°. 4
N°. 5
L

fore their House gods, and put small candles in them, like a candelabra: One also sees such knobs, on other Whelks, like drooping warts, closed on top and open on the bottom, where they have a sharp little bone, with which they can bore through the shell they are sitting on, and make a round hole in it, as if art had drilled it. They are called *Gindi batu,*[15] or *Gindi laut,* in Malay, because they resemble the spout of a porcelain jug; in Chinese, *Tsjip.*

IV. *Verruca testudinaria,*[16] in Malay *Kutu totruga,*[17] that is Turtle Louse. This too is a *Balanus,* which one will only see on large Turtles, not raised, but flat, as long as the joint of a thumb, half the thickness of a finger high, and roundish, put together with 6 bones of unequal size, which hang together with seams and grooves; whitish and smooth: they are flat on the bottom, wherewith they cling to the shell of the Turtle, and half of the underside is the thickness of the shell, spongy, and full of holes; they are open in the center, and put their naked flesh against the Turtle: They are also open on top, and there they have two teeth joined together, like the first *Balanus,* with still other smaller bones in between, all of which move in the flesh: In the center is a slimy Animal; with the lower, spongy part, they cling onto the Turtles, and do not change their place: One will not find them on all [Turtles], but mostly on those that live on the Turtle Islands *Lussapinju,*[18] 25 miles south of Amboina: On Ambon one will find a 2d kind of the first sort, to wit, with gray or raised Warts.

V. *Opercula callorum,*[19] these are the Little Lids of certain pieces of flesh, which we call *Callus,* being a kind of *Limax Marina,* which occur on sandy and muddy beaches below the sand, bearing this lid on its back like a nail, of which I note three kinds: The first[20] is an oblong little bone, the length of a finger joint, shaped like a small oval Ambonese shield, slightly hollow on the bottom, and bossed on top: The other one[21] is larger and rounder, like a small saucer, with sharp edges, a dirty white, somewhat like Mother-of-Pearl: The third one[22] is a small white bone with a little corner that sticks out, near a curve that is shaped like a small pork shank; the first two lie on the back of the Animal, peeping out a little; but the third one, is embedded in the flesh, and is almost entirely hidden, and this flesh has a cloying smell but tastes good, when cooked. It is difficult to find, because it stays deep down in the sand: One will find them on the beach in front of *Victoria* Castle.

VI. Nutshells,[23] these might be considered shells as far as one half of them is concerned, because they have a little edge on one corner, as if it were the hasp of a shell, though this is not the case; for they cling with their open side to the rocks; they have designs on their backs, like tiny chicken feathers.

VII. Knit caps[24] or Fish hoods, are dirty gray warts, with a tutel that hangs down a little, like the knit sea-caps of Sailors, or the hoods of Fishwives; inside they have a round little bone, like half a pipe; and they cling to the rocks with their open side.

VIII. Milkbowls,[25] this is a small flat *Patella,* smooth, and as white as milk inside.

The Author now leaves the genus, known to us as Whelks, *and goes on to the next one, to wit, the* Univalves, *known to us as* Rock clingers, *because the fish always clings to stones, or rocks, and is covered and protected on the outside with a* Shell, *or* Shield. *There are many kinds, of which some are pictured on plate* XL. *His Honor calls the one by the letter* A. *a* little Lamp, *or* little Saucer; *also the ones by the letters* B., C., and D. *The* 2d *kind, by the letter* E., Sea ears *according to the Author, is called the* long Mother-of-Pearl Shell *by us: Its outside is shown by the letter* E. *and the inside by* F. *The ones by* G., H., and I., *are also called* Mother-of-Pearl Shells. *The* 3d *kind, is depicted on plate* XLI. *by the letter* A., *and which we call the* Pox; *they grow on the bottom of ships, and hamper its sailing, also on rocks, shells, stone, and wood; they have all kinds of shapes and sizes, but this is the most important one. That which is shown by the letter* C., *is the Fish, that lives in the same, with its beard stretched out; and by the letter* D. *is the same, but all closed up. The* 4th *kind, is depicted on plate* XL. *by the letter* K. *is a* Turtlepox; *these are very rare. The* 5th *kind is shown by the letter* L. *of which we have very large ones: Of this we show a 2d and a 3d kind, depicted by the letters* M. *and* N. *The* 6th *kind, is shown by the letter* O. *The* 7th *by the letter* P., *on the outside, and by* Q., *on the inside. We call this one the* Fool's Cap: *A Shell or Whelk which was once so rare, that one was wont to say, when one did not have one of these, you are not quite the fool yet;*[26] *it is still a worthy item, if it is large enough. The* 8th *kind, by the letter* R. *is called the* Milkbowl *by the Author; but we have always called it the* white Earshell: *They are very rare, and, for as far as I know, only three are known; there is yet another kind of this one.*

CHAPTER 27

Solen. Cappang.

The second sort of Univalve Shells consists of long pipes of which some are straight and some are crooked, open on both ends; but in such a way: that the one end is always narrow, like a tip that has broken off, where the shell is the thickest, and inside one sees the hollowness divided into two by a partition, which does not go too deeply, however, and this is where the Animal has its beginning, as if it were the root of it: the other end is always larger and with a thinner shell, because the Animal grows its additions there;[1] so that the pipe keeps on becoming wider, and I found three kinds of these.

I. *Solen arenarius*,[2] Sandpipe, in Malay *Cappang bezaar*;[3] in the Uliasser Islands, *Hatu Aatu*, in Dutch Cowgut; some call them erroneously Triton Whelks or Sea Trumpets: These greatly resemble the back end of the large Cowgut, which we call the roots, being a finger thick in diameter, but the shell is thickest there, and divided inside with a partition into two or three parts, and this [partition] does not go deeper than the length of a finger, though in some it is almost entirely grown together with a chalky *substance:* In front they are 2 or 3 fingers wide in diameter, generally 2 and 3 feet long, a few are straight, but most of them are curved and crooked, the shell has at least the thickness of an oaten straw, sometimes that of a quill, white outside, with circles on it sideways, with cracks here and there, that do not go all the way, because the Beast began its accretion there; they are always the purest and whitest around the mouth, but they are dirty around the tail or root, blackish and overgrown; inside lies a slimy animal, which becomes somewhat hard when it is cooked, and is fit to eat, tasting like the best of Mussels; in front it has 2 small bones that close against one another like a Miter, not attached to the shell, but to the flesh, and which are its teeth; with which it makes a path to continue its shell: This large kind of *Cappang* is not found in many places, but it is on *Ceram* in the bight of *Amaheij*,[4] or the bay of *Kaijeli*, on *Buru*,[5] where it grows around the roots of the *Mangi-mangi* trees,[6] in a soil of sand and small stones; which is why it has to change shape and twist while it grows, such as the stones and roots permit, which it meets while it is growing, and which is the reason one finds so few straight ones: Our Ambonese do not make much of a fuss about them, indeed, they would not even look for them, if we did not sometimes want to keep some as Curiosities; but they will tell you, that they were once much sought after by the Malay and by Strangers, when these were still permitted to come to this region to trade, not only to eat the Animal, which they highly praised, as a means of strengthening their manly powers, but also to carry the shell or pipe with them, since they used it for various medicines; but they did not reveal this to the Ambonese: During times of War, when we sometimes captured enemy ships, we found, among their household goods, that they kept pieces of these pipes, along with other coral and sea plants, whereof the Inhabitants of the Xula Islands, who seem to know the most about them, disclosed to us the following: This shell, along with other things from the sea, to wit, three or four kinds of *Calbahaar*, and the subsequent *denticuli Elephantis*, is a proven Antidote against all sorts of ingested poison, and it will kill its powers, and which can then be expelled from the body by means of other emetics; furthermore, it can also be used against particular enchantments or villanies, which they love to inflict on each other, taking someone's manly powers away; which is why they always have these things ready, when they travel far from home; if one breaks the shell into pieces, one will find that it is made up of several crusts that lie one on top of the other, and they are smooth and shiny like slivers of *Bezoar*;[7] although the outer one is quite wrinkled, coarse and lusterless: If one rubs it, it will give off a swampy odor, quite like the *Mangi-Mangi*, near which they grow.

If one examines them more closely one will find, that the *Solenes* grow as follows: The widest and thinnest part is on the bottom, and is closed off with a thin shell, which breaks very easily, from which one can conclude that they grow downwards: The thickest and narrowest part is on top where it is divided in two by a partition: A long pipe, easily two spans long, rises up from each opening, and sticks out above the ground a little, and they can pull it in and out whenever they wish: A small piece of flesh hangs from every pipe, through which the Animal sucks up its food, but if one gets too close, it will pull it in, and spew up water as high as a fathom: These pipes are very thin and brittle, break off easily, but will grow once again, while on top they are somewhat separated from one another: They are difficult to dig up and keep these pipes attached; the pipes themselves are covered with a chalky and brittle crust, that falls off very easily. One finds them in *Amaheij* in a miry swamp, between *Mangi-Mangi* trees, wherein one will sink up to one's knees, but the bottom is firm ground made of small stones. The people of *Amaheij* blow on these Shells when they want the people to come to church and the children to school, wherefore one could call them School Trumpets.

II. *Solen lignorum*,[8] in Dutch Bore-worm, in Malay *Cappang* and *Utor*, in Ambonese *Ahet*. These resemble twisted chicken guts, wherefore some Malay also call them *Purrut aijam*,[9] they are as thick as a Tobacco pipe; some are thicker; some are thinner, but the shell itself is never thicker than double parchment, also made from several little rolls, on the outside a beautiful white with fine little rings, furnished with an Animal like the foregoing: These grow on the rotting wood of ships as well as of trees that float at Sea; especially on *Mangi-Mangi* wood, wherein they multiply so much, that one will find entire trees that are filled with these Pipes inside, all entangled with rare curls and curves, but most are empty, because the Beast rots, if it cannot keep on adding to its shell: It is a hurtful pest to vessels made from native wood, because this Worm will bore right through them, so that they will sink, particularly if they have not been diligently daubed with lime, and some oil, instead of tar: The biggest ones become as thick as a finger; of which people will save pieces for medicine, like the foregoing.

III. *Solen anguinus*,[10] in Malay *Cappang* or *Bia ular*,[11] these are such Pipes, that are no more than the thickness of a quill, twisted all together with many rare curls, like a snake; white on the outside, somewhat angular, and covered with grains; inside with a similar slimy Animal, that has a toothed little Miter[12] in the front part of its mouth: They do not grow on wood, nor in the ground, but on such rocks, that are very much dented, and which have protruding knobbles, around which these little Snakes will coil themselves, their mouths sucking on the rocks: Some other very tiny Pipes belong to these, which have grown fast onto the rocks, or perhaps partially loose, also with serpentine curves, but no thicker than an oaten straw. The first one is rarely found, and is considered one of the rarest of Curiosities.

IV. *Denticuli Elephantis*,[13] Elephant Tusks, in Malay *Tando laut*,[14] these are small *Solenes*, easily the length of one's middle finger, below scarcely as wide as one's little finger, coming to a point on top, but in such a way, that there always will be an opening, and slightly curved like the horn of a Goat, or somewhat straighter, like Elephant's tusks; it has protruding ribs on the outside, green, whitish at the tip: A second kind[15] is longer, lanker, not as green, rather a pale Spanish green:[16] The very smallest ones are no longer than a joint of one's little finger, such as those found on Persian beaches: All of them have a slimy Animal inside, and if you put it away, the flesh will melt and ooze away: There are no teeth in the flesh, but it has a round little mouth, with a continual little vein on the same, without any hardness, wherein one could stick a little twig; with this little mouth, they suck up mud and fine sand; placed in water, they will not stick it out. One will find them on flat beaches, that are firm, and where there is a calm Sea, as is the case in the Bay of Ambon; the widest half sticks in the sand, while the tip sticks out at an angle: The Natives do not look for them for food, because of their small size, but their rare shape makes them Curiosities.

V. One will find yet another kind of *Solenes*,[17] mostly on and in Coral rocks: of which some are straight, like pieces of gut, or sausages; with a thick shell, regularly ribbed, brownish, others are thicker, smoother, and with a rounder shell, but twisted, like pieces of a kinkgut (*colum*);[18] open on one end, and clinging to the rocks with the other.

The Elephant Tusks will also be found stuck in the sand at an angle, with the tip down, and attached to it a little vein like a small leaf, with a wide mouth just above the sand, that has sharp edges which will hurt the feet; those that lie there with their tip up, have been turned over by the beating waves, and are mostly empty. One will often find them at the mouth of the *Waijnitu* river, and on Cape Gallows: one will not find a sharp little bone in their flesh, and sometimes it seems that two of them are stuck inside one another.

Now follows the Solen, *which we call* Sea Pipes, Sea Snakes, *and which we will hereafter name according to their shapes. The Author says that these are the second kind of Univalves, although this is a branch of the* Whelks, *while those of the former Chapter, were a branch of the* Shellfish. *The first kind of these is pictured on plate* XLI. *by the letter* D., *it is the lower part of a* Sandpipe, *which we call an* Oxgut; *and the upper part is by the letter* E. *The* 2d *kind, is by the letters* F. *and* G. *The* 3d *kind, by the letter* H. *The* 4th *kind, by* I., *and which we also call an* Elephant tusk. *The* 5th *kind, by the letter* K., *and another one by the letter* L. *Thus far the Author: We add hereby some pieces which are very unusual. The one indicated by* No. 1.[19] *is a* little Whelk-snake, *its beauty resides in the twisted little tail, which must be very sharp.* No. 2.[20] *is a similar one, but shorter and squat.* No. 3.[21] *is a* Whelk Snake *that has been stretched out. And* No. 4.[22] *is a coiled one that grew on a Snail.* Nos. 5.[23] *and* 6.[24] *are both kinds of* Elephant Teeth. *And* No. 7.[25] *is the so-called* Venus Sheath; *a very rare piece, that is owned by very few Collectors.*

CHAPTER 28

Chama Aspera. Bia Garu.

We have discussed up till now the univalve Whelks, both twisted and not twisted; now follow the Bivalves;[1] which we really call *Conchae* in Latin, and Shells in Dutch, while in Malay they keep the general name of *Bia*, in High Malay *Krang*, in Ambonese *Kima* and *Ima*. We divide the Bivalves into six genera, all the time including the River types with their similar Sea types; and these are: 1. *Chama aspera*, 2. *Chama Laevis*, 3. *Pecten*, 4. *Tellinae*, 5. *Musculi*, 6. *Ostrea*.[2] We call *Chamae* all those Shells which, lying bare on the ground, for

N
H
A
E
I
D
G
O
F
K
L
M
C
B

the most part gape, and could, therefore, be called Gapers. *Pliny* already divided them into two genera, to wit, *Chametrachea*, and *Camelaea*, to which we give each a particular Chapter.

Chametrachea or *Chama aspera*, are those that have a rough shell, either through ribs that stick out, or because of scales and nails; and those are:

I. *Chama squammata*,[3] Nail Shells, wrongly called by some Clack Dishes [*Klipkoussen*]. In Malay *Bia garu*,[4] that is Scratchers, because one could scratch someone badly with them, or because they scratch the hands, when one handles them. In Ambonese *Maêka*, in Ternate *Kima:* others hold that *Kima* is proper Malay, on Buton, *Morabo*, on Banda, *Maniwoc*, in Macassar, *Alibo*.

These become the largest of all Shellfish; because one finds them so big, that 6 or 8 people have more than enough to carry just one: but since they always keep to the deep; and since one will find around beaches a smaller sort, which do not exceed a hand, we will divide them into *Chama decumanae* or *Pelagiae*, and *Littorales*.[5] Describing the *Littoralis* first, these are one hand long, and less shaped like a normal shell, to wit, oblong,[6] divided into Female and Male:[7] The Female is the most common, its shell is divided into 4 or 5 round ridges that jut out, making deep furrows between both, with sharp edges that close tightly together, except for one side, where they have a narrow opening, with edges that are jagged, so that this shell goes up and down like waves, both inside and outside; on the back are large curved scales, very similar to human nails, round and sharp in front, but most of them, especially the old ones, are broken off and damaged, and the more perfect these nails, the better a Curiosity they make: These small kinds are a dirty white on the outside, without luster, inside a yellow-white like ivory; but with the old ones, the outer shell is very much overgrown with moss, lime, and other sea grit, even with other shells, mussels, and coral trees, so that one would hold it to be a rock, rather than a shell: The Beast that lives within, is dreadful to behold, because if one looks upon one that is gaping, one sees nothing but a taut skin, full of black, white, yellow, and lead-colored veins, painted like a snake's skin; the entire shell is somewhat triangular; of which the front, where the Beast opens up, is the longest side; the second one, on the Beast's left side, when it is in front of us, is the most closed, and which we call the back; the third one, on the Beast's right side, is its belly, with the jagged opening that was described before, and this is so narrow in the Female, that one can hardly get a knife in it: These two narrow sides form a blunt corner, where the two shells close together by means of a *Gynglymum*,[8] to wit, the little corners that protrude on one shell lock into the hollows of the other, while they are furthermore connected with a thick little skin, that lies against it on the outside: In the open Beast one will also see two holes, towards the two aforementioned narrow sides; the one that is to the left side, is small, and almost closed, through which the Animal voids its excess liquid; on the right side it has a larger rounder hole, from which dangles a long brush, made of coarse and tough threads, which one calls the beard; wherewith they cling to the rocks so they will not be tossed around, but they also anchor themselves with these beards when on a sandy soil, being mixed up at their tips with small stones and sea grit, but it is not likely, that they use these to draw up any food; in the center of the shell, but closer to the opening, is a thick pillar, that is the size of an arm in the larger ones, and in the smaller ones the thickness of a finger, being the *tendo*[9] that is made of a tough, wiry flesh, fastened to both shells, with which the Animal pulls them together, and keeps them locked so tight, that there is no force that can open them; there is a hard kind of flesh around this pillar, like a round disk, which is called a *Spondylum*[10] or vertebra, and which is the best for eating; another kind of whitish flesh is attached to the same, with a large yellow lump, like the yolk in an egg, which is the Animal's fat; beneath that is a sack filled with a black slime, with sand and small stones mixed therein: The Male is more oblong, divided into 9 or 10 ridges, and the scales are closer together, but are shorter than those of the Female, and the opening on the one side is much larger: otherwise it has the same shape.

The *Pelagia*,[11] 3 to 4 and 5 feet long, is divided in the same manner as the one before; the scales are quite two knives thick, most are dull and broken off on the outside; and they are so much overgrown on top, that one can hardly clean them; the bottom ones are always finer and smoother; the shell is normally the thickness of a hand across: One finds some that are more than ½ a foot thick, wherefrom one can easily guess how heavy this shell is: If one smashes the shell, one will find that it is made up of several layers, which undoubtedly grow every so often one on top of the other, which one can surmise from the sproutings on the outside, in such a way that the youngest layers always keep growing under the older ones, and which spreads out to the front, which causes the youngest layer also to be the foremost one, and its edges are so sharp that they cut like a knife: Which is the reason one should take care when handling these shells, as long as the Beast is inside, if one does not want to get hurt. In the Moluccan and Papuan Islands, where one will find the largest ones, they experienced in our Shallops,[12] that when the Sailors dropped anchor, and the anchor rope happened to come into this gaping shell, it would, while pinching together, snap it off as if it were cut: Someone would be in danger of losing a hand, if one wanted to touch the gaping Beast, unless he puts something between the shell, in order to prevent it from closing: The Fishermen haul them up as follows: A Diver manages to put a rope around it like a noose, which all hands then hoist up, then they try to get a knife or parring[13] through the opening on the sides, in order to cut through the *tendo* or pillar, wherein lies all the strength of the Beast, and then the shell will gape of itself, and cannot close anymore; and this is the same way one frees any Animal or Person that is held by this shell.

As was mentioned, they gape almost always when on the ground, especially in order to catch the little fishes that come in multitudes to swim and play therein, until all of them together are suddenly locked in there, and come to serve as the Beast's food: This lumpish Beast always has a little Comrade with it, which is its Guard, being a kind of little Shrimp, described before in the first Book under the name of *Pinnoteres*, which pinches its flesh, when it sees that there is a great deal of prey in its house, whereupon the shell snaps shut, and one believes that the Beast cannot live, if that little Pinna Guard happens to be away from it, because the Beast cannot see, and cannot be on guard for robbers.

There is still another kind, thinner and flatter than the former *Littoralis*, having more the shape of a circle, with some trifling scales, though they be long;[14] some are a light yellow; others have a light-red shell, also other variations; but the best and most beautiful to keep are not longer than a finger, pure white and full of scales or nails: our Ambonese do not normally eat them, but other people certainly do: The Inhabitants of *Bonoa*[15] and the Papuas are very fond of it, and I have seen some of them eat the raw meat with great relish, and particularly the yellow fat: The *Badjos* (being people who constantly roam the seas, and get their livelihood from fishing) catch the biggest ones, take the meat away, and smoke it like smoked meat, called *Dendeng bia*, which they then sell in *Makkassar* and *Bima;* a tasty dish for those with iron teeth and thick tongues; worse than dried Seacats, but it is most often dished up at teatime[16] to munch on. In the *Tendo* or band or in the surrounding *Spondylo* one will sometimes find a few beautiful little stones, very like the stone *Calapites* or *Calappus*, to wit like Alabaster; some are of a clear white; others a dirty or yellow-white; some also have a pearly sheen, and are at one corner partially transparent like Agate: one can distinguish them from the *Calappus* stone in that the *Calappus* stone is often the shape of an egg, like a

Lizard egg, usually with a little dark corner, which is the root, wherewith it was attached to the *Calappus* husk; or in the shape of a Lentil, such as one finds in an apple. But these Shell stones, which we call *Chamites*, or in Malay *Mestica bia garu*,[17] are uneven, angular, and for the most part yellowish. One does not find them very often on Ambon, but more often in Maccassar or among the Papuas, I have even found small ones that were partially transparent in smoked *Dendeng;* those that are not thicker than a pea, are the purest and whitest, but those that are as large as a marble, are angular and a dirty white; those who earn their livelihood with fishes and shells, like to carry these stones on them: Our Ambonese are somewhat superstitious, for when they have to bring these large shells across the sea in their boats, they say that they will encounter wind and storms: But once in a while they will bring one of the largest ones, and put it in the yards next to their houses, and let chickens and other tame birds drink from it, which they consider very good for those birds so they will stay healthy, and that no harm will ever come to their eyes; but then the ivory smoothness within begins to pale, and becomes gray; one must also grind down the sharp edges a little, so that no one will be hurt by them: I also had some, that had an excrescence on the inside of the lower shell, very angular, and jagged, almost like a cox's comb, in *substance*[18] like the *Chamites.*

One will find yet another *Chamites* in the Papuan and Ceramsian Shells, long like a small finger, having a round little ball at the bottom, and pointed in front; the little ball is dark and reddish, but with a pearly sheen; the little finger itself is a pure white; at times it resembles a piece of a finger, that was broken off and scaly, but it will always look somewhat pearly; one could call this *Chamites digitalis.*[19] Still another use for the shell; one sometimes finds pieces of broken shell on the beach, cleansed of its outer rough bark by the [sea's] scouring, smooth as ivory, and so hard that one would think they were flintstones; I found some among them that were so petrified, that one could strike a spark with them, like true flintstones; one also keeps them for some use in medicine; to that end one takes the biggest pieces and knocks off the outermost rough bark, the rest one rubs on a stone with water into a poultice, which can be smeared around big bloody ulcers in order to lessen the burning and the pain, similarly for festering eyes: And if someone would want to use the same both inside or outside the body when there is a lack of *Calbakaar*,[20] or white coral, he would not go far astray in my opinion, although they are harder to rub, because of their rougher *substance*, than a common stone.

One also finds, though rarely, among the Inhabitants of *Amboina* and *Ceram*, some round little shields, about 7 inches wide, and half a finger on the sides, but an inch thick in the middle, somewhat bellied on the outside, on the inside slightly hollow, and there it has in the center a boss which is like the nave (*modiolus*)[21] of a wheel, and in the center of the same there is a round hole, through which hardly a little finger would fit: It is of a *substance* as hard as a stone, a dirty white like worn ivory, with many whitish veins that run with many turns and twists, indicating that it is a material that has many layers that have grown one on top of the other; it cannot be a natural stone, since according to my usual test, it did not spark at night, as do other clear flintstones; but if put in sour Lemon juice it will cook, which stones do not do; from which I deduce, that it must have been made from some large Shells, to wit, the *Chama Pelagia* just described; the entire width of the shield, as demonstrated by the course of the veins, has been made from the thickness of these Shells, wherefrom one can deduce that they must be of some incredible thickness in some places; if one holds them up to the light of day, they turn out to be partially transparent around the edges, but one more so than the other. The Inhabitants do not know who has made them, nor from what, only declaring that in the old days, a few were brought into these Lands and were sold for a high price. They are used for putting onto the hilts of their swords, to protect the hand, whereas otherwise the Ambonese have a round table plate of wood at the same spot, and they pour lead on the hilt so they can get the necessary heft for their fist. But the aforementioned swordguards can only be used by their heroes or champions, whom they call *Brani*,[22] believing, that no one has the courage to slice or strike someone who has such a Shield on his sword; but from the time that the Europeans became the owners of these Lands, they no longer care to wear these little Shields as marks of their ancient courage, for fear of drawing too many eyes to themselves; wherefore they now sometimes offer the same to us: Later I was informed with more certainty, that the little Shields were brought into these Parts by the Javanese from *Bantam*, no doubt convincing the Natives, that these things were of great value, and therefore were sold to these People very dearly. Now these Bantamers made them from the aforementioned *Chama decumana*, which occur very thick and heavy in the mouth of the *Sunda* strait, particularly around *Pulo taboang*, that is, Bandit Island, whereby they might mean Princes Island;[23] and also across from there, in the bight of *Lampin*,[24] where there is a deep sea where one also can see at times the Tree, whereon grows the *Calappa Laut*, the top of which one can discern just below the water, but I cannot fathom the reason that no one dares to dive for it; I also understand that they are made in the *Molukkos;* and are still worn to this day by the Alphorese on *Gelolo*.[25]

The Inhabitants call these little Shields *Mossa bula bulang*,[26] that is, Little Shield, shaped like a full moon, and therefore *Clypeolus Lunae* in Latin: If one wants to find the best *Bia garu* for eating, then one should take those that are only a foot or 1-½ long: Such do grow at times in the hollows or holes of the broad rocks called *batu tikar*;[27] and are lodged therein in such a way, that they are not to be taken whole from there, unless one were first to break the edges of the rock into pieces; they are appended to it with a beard, which is why the Malay call this the Beast's *Ackar* or root; and since one will always see the lower side clean, and the upper one overgrown, I must conclude from this, that they never change their place, at least not those that are rooted to the rock. On *Lussapinju*[28] one will find a special kind, that has the mixed shape of a *Bia garu* and *Cururong*,[29] with a shell that does go up and down like the waves of the sea, with short and trifling scales, but it is flat on the left side, like the Cururong, and a dirty gray on the outside; one will find the aforementioned *Mestica* in it as well as in the *Bia garu*, the size of vetch, peas and beans, but in the biggest, one will find only one, as happens in the Pearl shells as well; at first most of them are a clear white and shiny, but they yellow over time, which is a common fault of the *Chamites*.

I still must convey something strange about the *Chamites*, though I cannot be sure whether it was a *Chamites* from a *Bia garu*, or from another Shell, though eyewitnesses said it was the truth: They were fishermen who held these little stones in the highest esteem, and they declared that twice they saw, how a large, angular *Chamites*, kept especially in a little box, over the course of time brought forth another little stone of the same substance and color, and one could see the little hole where the stone had been: I would guess that this happened because the little stone did not form a *continuum corpus* with the bigger one, though it had grown onto it, and that the natural glue had loosened after a while; and though the owner said that the little Mother stone had not diminished in size or weight after birthing, I would still hold that he did not examine the same very carefully. I have discovered myself that, while keeping 6 stones like that, including one that was very angular, in a little box, and examining them again after half a year, I found 7, but no one could show [me], where the seventh had been: On the other hand, I have kept an angular *Chamites* separately for several years now in a small shell, and noted down the proper weight when I first got it, and which I examine and view every year, but do not find that its weight either increases or decreases; nor will it bring forth anything, the more so since one will not find a seam anywhere around it, wherefrom a little piece could have fallen; or it could be, that the owner's disbelief prevents this: After all, all sorts of *Chamites* are now said to be capable of giving birth, and thuswise they may well be the Ancient *Gemonides* and *Peanthides*, mentioned by *Plinius Lib. 37. Cap. 10.*;[30] for the same see [our] Cap. 2., concerning the *Nautillo crasso*.[31] And this is why some Chinese are very fond of these stones, because they bring good fortune to the house, and increase their goods: A certain Chinese who lived here, owned such a *Chamites*, which had birthed, and which he had redeemed from the Moors at great cost, who persuaded him, that he should perfume the stone every Friday with *Benjoin*,[32] whereafter this ordinary poor fellow[33] became greatly enriched (though one should know, that he was a diligent sawer of wood); but his house collapsed in the year 1674, during the great Earthquake, and, having lost the stone, he became poor thereafter: Those who have the time, may consider, how it came about that the Inhabitants of *Ternate* and *Manado* call these Shells *Kemas*, just like what the ancient Greeks called them, to wit, *Chemae*, which the Romans turned into *Chema*, whilst *Chemae* was still used by the old Romans, as is shown by a little poem by *Apulejus* in his *Apologia* taken from *Ennius*.[34]

> Apriclum piscem scito primum esse Tarenti
> Surrentis chemas. Glaucum cumas apud: at quid
> Scarum praeterii cerebrum Jovis poene supremi.

One usually divided the *Chama*, as was said, into *Aspera* and *Laevis*. *Chama aspera* was once held to be a bad sap or food, which was also called *Ostrea:* In Macedonia *Corycos*, *Krios*[35] in Athens. *Chamaeleae* or *Chama laevis* had a better taste, which is why some called them *Pelorides*,[36] and *Basilicae*,[37] that is Kingly, because of their size: But *Plinius Lib. 32. Cap. 11.*[38] tells of 4 kinds translated into Dutch with these words: The kinds of *Chamae* are these, (in the printed books one reads wrongly *Cancrorum genera pro chamarum genera*) *Chamaetrachea*, *Chamaeleos*, *Chamae pilorides*, in various shapes differing from the others in shape and roundness; *Chamae glycymerides*, which are longer than the *Pelorides;* then he adds as a fifth kind the *Colycea* or *Coraphia* (others read *Copycia* and *Coryphaea*). The commentators note as to this kind, that *Chamoetrachea* has the hardest shell: *Chamaelos* the whitest and most delicate: *Pelorides*, called *Conchae-peloriades* by *Athenaeus*, have a sweet taste, and are larger than *Pelorides*, although these were said to have gotten the name from *Pelorios*, that is, big, but it is perhaps better to derive it from *Peloro*, a promontory on Sicily: Others get *Corycia* out of *Colycia* or *Corophya*, which is more likely. *Athenaeus* adds a sixth, which he calls *Chamaenigram*, *Concham molaenam* and *Melaenitem*.

II. *Chama aspera & obtusa*,[39] has a rounder and thicker shell, without raised ridges or billows, and the same is covered with close, though short blunt scales, gray on the outside, with a wide mouth on one side, big enough to stick a thumb in, although the entire shell is not more than a hand wide; yellowish on the sides, common on Ceram's northern coast at *Assahudi* and *Hennetello*.

III. *Chama Striata*.[40] In Malay *Bia Corurong*.[41] In Dutch Little Horse Feet, this one is strangely made, to wit, as if one had cut one corner off a half moon, which made the same side flat, and, if one looks at it from the front, it has the look of a heart, with many ribs above one another going in a half circle, and this way making many little hearts, of which the smallest one is in the back by the lid, representing a transparent drawing; and because of this shape the shell becomes somewhat triangular: The shell is very coarse and prickly all over, covered with small nails like blunt thorns, not only divided into several ridges or billows, but every ridge and valley is divided again into other ribs and furrows, nonetheless locked so tightly together everywhere, that you cannot get a thin knife between it: The animal within is similar to the earlier one, and its skin is also painted most dreadfully. I never saw them larger than a span, with most nails worn down already, but those that are the size of an egg are the most beautiful, full of sharp little nails, and marked with brown spots, and one sees the same kind of brown on the inner sides, on the most transparent side; although it is tightly closed, it does have a thin beard hanging out, with which they cling to the rocks. In the *tendo* or band, one will also find the *Chamites*, but mostly in those shells that occur in *Lussapinju*, differing slightly from the foregoing; for this *Chamites* is as large as an entire hazelnut or marble; the other ones are small like vetch and pinheads: One will find two or three of the largest ones near the head of the band, completely angular and grainy, as if they had been made from a host of little stones, nonetheless hard and clinging tightly together; some are a whitish yellow; some a light purple: The rest of the small stuff lies below that, and there is so much of it in some, that the entire *tendo* seems to have been put together with little stones; none of them very special, nor very beautiful.

IV. *Testae*,[42] Small Sherds, are small round little shells with thick and flat shells, they are the size of a dubbeltje and a sesje,[43] covered with large kernels on the outside, a dirty white with blackish or russet spots on the side: if one sees the single shells lying on the beach, one would think they were thick sherds of a broken dish; the biggest ones have such a thick shell that one could mistake them for a stone; the smaller ones have clearer marks like furrows, dirty white and of a russet color; both are common in Ambon Bay. On the white Sandbeach of *Mamalo*[44] one will find them a clear white, as if they were a small clump of cooked rice, and all are good to eat.

V. *Testae pectinata*,[45] Little Wild Sherds, are also round as a schelling, with a thick shell, with furrows and ridges like the *pectines*, but much coarser, dirty white, and with blackish spots and marks.

VI. *Cartissae*, Little Hearts, *Bia hati*,[46] these are the rarest and most beautiful of all *Chamae*, with a thin shell, truly heart-shaped, with sharp edges, of which some are sometimes toothed, others not; on the underside rather flat, on top with an elevated belly, which comes sharply to a point, and has many half circles or ribs on either side, which become smaller in a transparent way; across this belly is the opening, where the shell comes apart, unlike other shells, yet they hang together in the little cleft of the heart, and each piece separately resembles a half moon. They have three different shapes: The most common ones are bellied both on top and on the bottom, but the least on the bottom, with toothed edges, and have a whitish yellow color: The second kind is whiter, not toothed on the bottom or the sides, and some have raised edges:[47] The third kind[48] is somewhat hollow on the bottom, but very bossed on top, most of them are not toothed, and have reddish dots. The Animal inside is mostly slime, blackish, and, if one puts it in fresh water for a night it will easily fall out. The best and the most are found on *Nussalaut;*[49] a few on Hitu as well. One puts them for two days in fresh water, until the Beast has fallen out, whereafter one should bleach them for half a month in rain and sun, which makes them pure and white, and finally tie them together with a little thread. They are counted among the most excellent of Curiosities.

VII. *Quadrans*, the Little Quadrant[50] has a shape as if one had cut a small cheese, with narrow sides and bellied in the center, into four pieces, so that the side that had been cut off is flat, and shows an elongated heart, and its edges are slightly toothed, usually the color of a mouse or a dark gray, and somewhat striped. One finds them rarely, and then mostly on Amboina's outer islands.

We said that pieces of *Bia Garu, Bia cattam*,[51] and Giant's Ears[52] are changed into stones over a length of time, and are ground so smooth by the sea, that one holds them to be flint-stones or other rare stones: one can distinguish these as follows: If one strikes two white and partially clear flint-stones against one another at night, they will fire brightly and often, but two pieces of Petrified Shell do not, or very little, but the pieces of the Petrified Mountain Shells, as described in the next chapter, do fire a little.

The *tendo* or tendon of this shell is also called *Toncat*[53] and *Toncatnja* in Malay, that is, a pillar. The Badjos and Maccassarese eat all the white and firm flesh from this shell, throwing the white or yellow fat away, because it makes you a little befuddled, as does the black blood.

One tells many wondrous things about a giant *Bia garu*, which was supposedly seen in a lake on the Island of *Timor Laut*, which opened up at night and gave off a clear light or a shining which one could see from a great distance: Another eyewitness from *Hitu* told me, that, around the Islands of *Kei*, he had seen an unusually large *Bia Garu* gaping wide, and in it was something clear like a precious stone, which those Folks held to be its *Mestica*, but no one dared to dive for it, since it was in a place, where there was a strong current.

It is understood of *Bia Garu*, and which one can find written in several Authors, that in the year 1645, on Princes Island, that is in the mouth of the Sunda straits, they found an Oyster (Shell) that was 7 ells in circumference, or 2 and ¼ ells in diameter: Similarly, that one finds Oysters on *Java* that weigh 300 pounds.

In February of the year 1681, in the *Lembe*[54] straits, in the North-East corner of *Celebes*, did Governor *Robbertus Padbrugge*[55] haul two of these Oysters on board, of which the one was eight feet and two inches; the other 6 feet and 5 inches in circumference: They called them *Kemas* there, similar as was said, to the Greek *Chemae*, being of the sort whereafter the village and the mountain are called *Kemas*, since they often find big ones there; the Mountain farmers of Northern Celebes make arm rings from them, knowing how to drill quite nicely right through these hard shells with a piece of porcelain tied to a stick, after which they polish it with Bamboo, and make it smooth and even. This shell, hauled over with pulleys, was lying on the ship's deck, and gaped as if it were dead, wherefore a Sailor pushed an iron lever in it, in order to keep the same open, but the shell pinched close, and held the lever so fast that it bent. They are usually fastened to the rocks, so that one has to break them off with levers: The Maccassarese hold that the best part of the shell is what they call *Subang*, wherewith they understand not only the *Tendo*, but also the *Spondylum* or the tough (*callosa*) flesh, which curves around it like a vertebra.

Here the Writer proceeds to the Bivalves, which we call Shells,[56] *and always with the non-Dutch word* Doublets, Shells, *of which each one has a particular name; which we will now put in their proper places to the best of our knowledge, leaving the Writer's names, where we lack them. The first kind, is shown on plate* XLIII., *the letter* A., *and we call it the* Nail doublet, *also the* Nail Shell: *This has also a* 2d *kind, shown on the same plate by the letter* B. *of which there are more kinds: They differ in shape and color; of which the red ones are the most unusual. The Writer's* 3d *kind, is shown under the letter* C. *but is called by us the* Perspective Doublet; *because it has on its flat side some stripes, that grow faint, and which, curving, display a heart, as can be seen in this depiction. The* 4th *kind by the letter* D. *There is no illustration for the* 5th *kind. But the* 6th *kind, indicated by the letter* E. *is what we call a* Little Venus Heart: *there are several kinds of these as well but they are valuable when they have red dots.* Jr. Oortmans[57] *and Mr.* Vincent[58] *both possess one, which is the color of a lemon; these are considered the most valuable. The* 7th *kind is shown by the letter* F. *and is a very rare* doublet Shell; *I can recall having seen only one, to wit in the Cabinet of the erstwhile Dr.* 's Gravezande *in* Delft.[59] *It is a different kind of* Venus Shell *with hair, which we will show later on the last plate of the Shells.*

CHAPTER 29

Chama Montana Sive Noachina. Father Noah Shells.

One will find enormous bits and pieces of the large *Bia garu* or *Chama decumana*[1] not only in the Ambonese mountains, but also in the neighboring islands, and in the *Molukkos*,[2] about which there have been many disputes about how they got there; wherefore I deem it necessary to describe the same at greater length: I. Its shapes: II. The place, where they are found: III. The wrong reasons for how they got there: IV. The true or most probable reason.

I. As far as their shape is concerned, one may note, that formerly they did not differ at all from the ones which are daily hauled from the sea, but now they are so much overgrown over a long period of time, that people think they are rocks: but, if one looks at them more closely, one can easily determine from the course of the ridges or waves, that they are shells: A few are still almost whole, except that the edges are mostly broken off, and some have entire pieces missing: Others are split down the middle, so that one cannot find the pieces together anymore; some lie bare on the ground, or are covered with just a little earth or some growth, because one will not find them deep down under ground; some are grown fast onto the rocks; some also stick a ways completely down into them: They lie tossed there at all sorts of angles, completely flat, standing up, jumbled together, as if they were cast abroad like seed: All are rough on the outside, mossy, and, on the side that is bare, covered with a sharp stony *Substance*, indeed, entire pieces of sharp flint which are called Lace, are so much fastened to it, that one can hardly break it off; they are a beautiful white inside, solid and dense, as some white marble can be; but one can easily detect the various layers there, like other sea shells have, and though one might find pieces, which are halfway transparent, they will not fire much at night, the way common flint-stones will, though they might give off a few sparks when struck against one another, and give off a noticeable flinty smell, to prove, that they already have pretty much adopted a stony nature: if one strikes them, they sound like porcelain. One will find some, of which just one half will give 4 to 6 men more than enough to carry; others are smaller, and not more than a foot long, and there are also many pieces the size of a head.

II. One will find them the most in Ambonese Territory, also, as I have been told, in the *Molukkos*, and perhaps in other places as well; The largest ones, and the most, I found in the mountains on Hitu, just behind *Hitulamma*, where the *Tomu* village used to be, and where one will still find many of its old walls that were fashioned from stones. The place is very rocky, and so sharp, that one can hardly put one's foot down; this kind of rock is called Lace:[3] I have never been fortunate enough, to find the two halves of these shells still on top of each other, for they are strewn about there; some also, as said before, are stuck to the rocks; others are right in the road which one has to descend, and which the people on their way down look upon as nothing but rocks, and this is also true for the rest of these mountains. I have also found them on the beach of that selfsame coast, driven so far into the rocks, that it was impossible to get the shells out without smashing the rocks, which proves, that the same rocks must once have been soft, while the shells themselves had been scoured white by the sea. There is a hill on *Leytimor*, near the little River *Weynitu*, about a musket shot from the beach, and which the common people have named after me for quite some time now, because there and in the surrounding mountains as well, one will find many of these shells, which are so overgrown, and thorny and black, that they were all thought to be rocks at first, but these are only 1 or 1-½ foot long; the thorny flint, that is stuck to them, can be broken off with much effort, so that one can properly mark the *substance* of the Shell, which is full of holes and broken corners. But it is the island of *Bonoa* that has the sharpest, most barbed and the most jagged mountains in all of the Ambon region, and one will find these Shells there that are very large and thick, many stuck into the rocks, full of holes and completely petrified.

III. Now comes the question, how they got up there in those mountains, which has produced many opinions and much dialogue: Many of our Nation judge one of the following two things to be true; they are either the natural fruit of those rocks, and that they were engendered by that soil, where they are found, like other metals, minerals and stones; and as proof they produce such stones as *Aëtites*,[4] *Gelodes*,[5] *Ombria &c.*,[6] each one with its own particular design, which one will also find lying bare on the ground, or on other stones; or, if one does not want to accept this, then people must have brought them up there some time, in order to eat the meat from them: That this is not very likely either way, I shall prove by way of the following reasons: Firstly I say, that they cannot be a natural or ordinary production of the mountains: 1. Because one finds them with so many different shapes and conditions, such as whole ones, two on top of one another, halves, broken pieces, now lying flat, at other times at an angle or upright, as if they had been sowed there by some mighty hand; some are on top, or a little below the loose soil; some are stuck into, and some stuck onto rocks; here one will see a big one, there a small one, yet all are of the same age, that is, they must have been there for the same amount of time, because they are overgrown and petrified in the same way. Now it is known that all such things, that grow in the ground and out of rocks, are rooted in similar ways, or at least have their own specific *Matrices*,[7] as one can tell about most precious stones: Because the proof of the *Aëtites* stone has nothing to do with this, that is, that one will always find it on the banks of certain Rivers with the same shape and configuration, or in the fields: no one, with even the slightest knowledge of minerals, will be that simple to think that the grains of gold and other metal, which one will find in some Rivers, grew there, instead of coming down from the mountains, where these Rivers arose, and were carried along on their way down: The *Aëtites* and *Gelodes* stones do grow in the ground, like other stones; but they are uncovered, either by Rivers, or by rain, and are then carried off, and strewn here and there, along the banks or in the fields: The *Ombria* stone, which one will find bare in the fields, is generally believed, to have been tossed there by some great tempest: One cannot say any of these things about our Mountain Shells. 2. One will find that the *substance* and shape of these Mountain Shells correspond entirely to those which are hauled from the sea every day, as was said before, and no matter how hard one strikes it with a good Steel, as long as one does not touch the flint, one will not get the tiniest spark from it, or, at the very least, with great difficulty, although they give off a flinty smell when struck; on the other hand, they will boil in sour Lemon juice, although much less so than Sea shells, something that no stone grown in the mountains will do. 3. The flinty crust might be attached to it as firmly as it wants, it does not, however, make for a *continuum corpus* with the Shell, but one can knock it off with hammers and chisels, and then one will usually find something earthy[8] between the flint and the shell. It is even more absurd to think that people brought them up there, since what could possibly have possessed anyone to carry such heavy beasts, if not monsters, all the way from the beach into those high mountains, where it is usually so sharp, that one can barely stand up straight, nor is it likely that people ever lived up there, and all this just to get the meat from them, which one can remove with little trouble on the beach, which was mentioned before about the *Badjos*,[9] who bring the meat from these beasts to Makkassar for sale: even if we suppose that people in ancient times were nearly giants, I would still believe, that their skin was as soft as ours, and that they, therefore, would have avoided those jagged places as much as we do, and would not have stuck the same into the rocks; wherefore I deem this not worthy to be refuted with much reasoning, but laugh at it along with the Natives: For surely, the Native Peoples feel quite injured, when their ancestors are held to be such fools, that they would begin a heavy task like that for such a trifling reason, and also since one never finds such petrified Shells near inhabited places, except for a few, which they have put near their houses for their fowls to drink from, which are always kept clean.

IV. Now since they did not grow there in those mountains, and since people did not bring them there, one cannot think of any other reason than that a great flood carried them up there, which, as we know from Holy Scripture, happened only once, namely in the days of Noah, when all mountains were under water, and when they, and other Sea animals, helped along by the surging waters, crawled up there, and lived in the forests, the way *Ovid* sings about *Deucalion*'s flood, *Lib. Metamorph. I. vers.* 299 & 300.

> Modo qua graciles gramen carpsere capellae
> Nunc ibi deformes ponunt sua corpora phocae.[10]

One might ask why all those Sea creatures did not return to their old home in the sea, when the water receded again? To which I answer: That one can see every day, that all the Sea creatures on flat and smooth beaches, are washed back to the deep when the water recedes, but when these same beaches have hollow rocks and holes, all sorts of small fishes, crabs, shells, &c. stay behind, which otherwise would undoubtedly have been off to their old dwelling, if they had understood that, while they were playing in those holes, the water would run away from them: and so it is also quite believable, that most of these Shells got themselves back to the Sea, where it was smooth and slippery, but what were the other ones to do, that had wound up on those rough and jagged places, and might even have been covered by fallen trees, those surely could not have saved themselves so readily, even more so since the receding of the flood happened much more suddenly than with our normal ebb, which will at most fall 10 feet in 6 hours, except for a few places on earth: If one estimates that the highest mountains are 1-¼ German miles, or 5,000 paces, that is, 25,000 feet high upon a *perpendicul*, and that the waters of the flood must have fallen that much in 125 days, then they must have lowered every day 200 feet without fail, that is, lowered 50 feet in 6 hours, which is easily 5 times quicker than our normal ebb.

And so there is no doubt that God the Creator left behind remnants and marks of the general flood in various places on earth, foreseeing, that in later days there would arise conceited people, who would try to harm the truth of the Holy Histories in this regard as well; namely those who would maintain and protect the *Praeadamites*,[11] wanting us to believe, that the flood did not cover the entire earth, because if so, their invented *Praeadamite* folks would have drowned, but that there was a tall mountain of water, which only covered Palestine, Syria, Armenia, and Araby, and the neighboring lands, namely only where the descendants of Adam lived: And so we will find that in Europe, to wit in France and Italy, they have dug many Sea shells out of the ground in various places: A reliable witness relates that he saw a natural marble rock in the French region of *Avergne*,[12] that had all sorts of jumbled Sea shells and Whelks as if incorporated, as if it were a *massa*, that art had contrived. The bird mountain in Westerwald[13] has the reputation, that they found all sorts of petrified bird bones there, especially when they dug wells in the same mountain.

The Chinese have told me that people sometimes find large anchor cables in the ground in the mountains of Fockien Province,[14] and that they are made of hairy stuff, such as we make in this Country from the *Gumato*[15] tree, being a stuff that is well-nigh imperishable in the earth, and which they hold to be remnants of the great flood, which occurred during the days of their King U,[16] otherwise known as Jao, and which in terms of chronology, does not differ much from Noah's flood: In these Eastern Islands we have the aforementioned Noah's shells: In the West Indies, according to what the Spanish write, they have found some very old vessels and unusual household goods when they were searching for gold in the mountains of *Peru;* whereby we note in passing that, even though the Flood did not change the general shape of the world, many new mountains were tossed up, wherewith these Sea creatures were covered, and in time were petrified, and together with the loam, which contained a stony sap, turned into stone over time: and so these Mountain Shells are indeed the oldest unquestionable remnants of antiquity, and have been in those mountains now for 4,000 Years, their shape testifying to the truth of that incredible antiquity: This corresponds to the common feeling of the Natives, who received these traditions from their forebears; the Moorish Priests in particular, will clearly tell you, that these are remnants of *Nabbi Noch* (as they call Noah): I would not have believed someone else's story all that easily either, if I had not seen the places myself where, considering their shape and situation, I could readily conclude, that they had not grown there, nor had people carried them hither: One half of one of these was presented to Mr. *Jacob Hustard*,[17] former governor of Ambon, in the year 1663, and 6 men had trouble carrying it; he had the intention of bringing it to Europe, and honor some Academy with it, but since the same Gentleman stayed in the Indies I have no idea what happened to it. Another, smaller one, was sent to the Grand Duke of *Tuscany* in 1682, along with other Curiosities of mine.

Someone might suppose that, since these Lands are very susceptible to earthquakes, over the course of time, other violent upheavals, besides the general Flood, were caused by earthquakes, and that many new mountains, that were not there before, arose, and that these Shells were carried upwards along with them: I would never deny writings, that indicate [the existence of] such mountains on earth, but this cannot be said about these Lands, or one should immediately conclude, that all the Islands and Mountains, where these Shells are found, rose out of the Sea along with all their coverings, which would be an absurd thing to say, because one will find them inland on such Mountains and large Islands, which have always been there from the beginning of Creation.

The Author did not provide a picture of Chama Montana, *or* Father Noah's Shells; *because the majority are only big rough bits and pieces of Shells, which, as his Honor mentions, are mostly found singly: I seem to remember having seen three of them, several years ago, in the East-Indies House;*[18] *they were perfect* Doublets, *being uncommonly large, each* Doublet *weighing at least 300 pounds: to give Collectors some idea, one can see their shapes on plate* XLII. *by the letters* A. *and* B.; *it is a sort somewhere between these two, but it resembles more the one by letter* B. *There can be no doubt about the Author's opinion of how and when this petrification happened, and why they are called Father* Noah's Shells: *We could point to many kinds, which were found aplenty everywhere, and of which we possess quite a few; but since mention of petrified things is made again in Chapters* 65., 66. and 84. *in Part three, we will show there several pictures, of those that belong to Mr.* Vincent, *as well as those that rest with us. Those who might still doubt that living and soulless things can become petrified, read Kircherus Subterranean World*[19] *in part two of the eighth book, where he speaks at length about the same and about stone saps: How these are found deep down in the earth, as well as inside mountains, can be consulted in the notes to Dutch Antiquities,*[20] *and particularly, in an eminent little book called* The World's Beginning and End, *written by the English Theologian and Naturalist* de Ray,[21] *which was printed in Dutch in octavo.*

CHAPTER 30
Chama Laevis.

The smooth *Chamae* or Gapers are much smaller than the foregoing coarse ones; they are the size of an ordinary shell, mostly round, and with a thick shell; in this they differ from the *Tellinae*,[1] because those have a thin shell, and are oblong: And the *Chamae* also lie, either bare on the ground, or not very deep in the mud, while one has to dig the *Tellinae* out from under stones or out of the Sand; but in this Chapter we will also include some Shellfish, which waver between *Chamae* and *Tellinae*, so that one could put them with either. There are the following kinds.

I. *Chama Laevis*,[2] properly called, or smooth Gapers, are roundish, or round with three sides; in the back by the hasp, they have a blunt corner, but the front part is round, with a thick shell, entirely smooth and even; some are a pale yellow on the outside, or more pallid; some are a steely green and brownish, but all of them are black on one side; the meat is white and has, of all of them, the sweetest taste, which is why it would not be unreasonable to think these were *Pliny*'s *Glycymerides;*[3] one will find them in pure sand, or that which has some fine mud in it, wherefore their color matches the ground; They are not often found in *Amboina*, and then mostly near the village of Suli;[4] but more so on *Ceram*'s northern coast, where there are long sea beaches, as well as in the South-Easter Islands, especially on *Tenember:*[5] Each of them has a little crab as its guard, the size of a fingernail, with a square shield, belonging to the kind which we called *Cursores* or *Hippi*[6] in the previous book, and which, so it seems, lives inside there, until it has grown large, and can live on its own outside the Shell: The ones one finds in these lands, are not more than 2 or 3 inches wide, but in *Japan* and *China* they are wider than a hand; The Japanese gild or do them over with silver on the inside, and then paint some little trees or figures on them, so that they can be used as small boxes, but they also use these for a game, perhaps to guess, which figure someone would get, the way one does otherwise with cards; because they are so much like each other on the outside, that one cannot know or tell, what was painted inside.

We call them *Chama laevis Glycimeris Indica.*[7] In Malay, *Bia Lebber* and *Bia Tenember.* In Makkassarese, *Tudo.* In Dutch, Smooth Gapers; but they call the large ones Japanese Shells.

II. *Chama Lutaria* and *Coaxens.*[8] In Dutch, Croakers. In Malay, *Bia Codock.* These have the same shape, as the foregoing, more than a hand wide, and bossed, but not as smooth on the outside, steel-green or the color of mud, lusterless, and somewhat coarse because flaps of green fleece cover them, and which will tear in places. One will find them in muddy places, mostly at the mouths of large Rivers, where they are discovered when the water recedes, and they croak, like frogs, when they open and close their shell, so that one can hear them from afar. They are also good to eat, if one puts them first in fresh water for ½ a day, so they will spew forth most of the sand; the largest ones are on *Buro.* Some will have a *Mestica* or little white stone, like the aforementioned *Chamites;* sometimes beautifully round, a pure white, and shiny; at other times it is a dirty white and angular, and they should be counted among the *Paeantides*,[9] and will in time shed a little stone, as if they were giving birth, but this happens rarely.

III. *Chama virgata.*[10] In Malay, *Bia Baguala*, has the same shape, but has a thinner shell, steel-green on the outside, with dark-yellow stripes or rays, which are thrust together in the back by the hasp, violet on the inside. One will find these in the tranquil *Ambonese* bight,[11] near the village of *Baguala*, but they are not much sought after for food, because they taste like mud.

IV. *Chama optica*,[12] Little Perspectives, are roundish and bossed little Shells, about 2 inches wide, with a thick shell, smooth on the outside, and decorated with blackish designs, which represent hills, little houses and peaks, in such a way, that the ones nearest to the edge are the largest and blackest ones; if one follows the other ones towards the back, one will see that they gradually become smaller and bluer, and fainter, just as if one saw a drawing of a distant Landscape or one in perspective. A baser variety has circles or little ribs the length of it, all of them parallel to the edge; the decorations are brown, and all in a jumble. On *Bouton* there is even a third kind, not wider than 2 fingers across, and graven with faint combs like the Jacob Shells,[13] without circles, mostly chestnut brown, and with some white decorations. One will find an even smaller kind, smooth on the outside, that has dark-brown decorations that look like certain tents with tiny flags on top, like a Turkish army in the field.

V. *Chama circinata*.[14] In Malay, *Bia matta doa*, that is, Shells with 2 eyes, this one is somewhat flatter than the foregoing, covered with circles and ribs, which run parallel with the edges, with dark-green or smoky, sometimes blackish speckles and spots as well, all confounded and mixed up with white,[15] at times resembling small towers or houses: the Animal has a white, tasty meat, wherefore this is the most common Shellfish to be used as daily food, particularly with *Papeda*, wherewith they make the sauces. One has to dig them up from moist places, and from sand that is mixed with tiny stones, but they are buried quite shallowly, and betray where they are with two small holes, which one can see in the sand; those that have been dug up have a little pipe on either side,[16] which the Natives call eyes, and through which they squirt water, often hitting the man's eyes who dug them up, even through the aforementioned holes, while they are still stuck in the sand. They are commonly found on all beaches, that is, where there are no great waves.

VI. *Chama litterata oblonga*.[17] In Dutch, Letter Shells. In Malay *Bia letter*;[18] this one is 4 or 5 inches long, 3 wide, with an average shell, also with parallel circles on the outside, with a light-russet or pale-brown foundation, with black stripes or marks, which make for the letter W, and the more these are crowded together, the better the shell: The ones that are called *Bia malam*, that is Night Shells,[19] do not have letters, but are black, with wide, light-red and whitish rays, that narrow towards the back, resembling a painting that depicts a dark night that is pierced by a light shining through: The third kind[20] is smaller, with a thinner shell, grayish with some small black dots, which do not make a picture, hence not of very good quality: The fourth kind[21] is oblong and narrow, with a thick shell, without circles, has brown, flowing tabbying.

VII. *Chama litterata rotunda*,[22] is entirely flat, yet with a thick shell, almost entirely round, full of circles decorated with black letters (M) and (W), which do not have as many branches as the *Oblonga* but rather consist of thin stripes: They are not often found, and therefore rare.

VIII. *Chama pectinata*[23] is flat, with a thick shell, divided by shallow furrows into few but wide combs, that run from the back to the sides. One will not find them on *Ambon;* but rather on *Ceram*'s northern coast *Bolela*.

IX. *Chama Scobinata*[24] is also round, flat, and white like the foregoing; sometimes with faint combs and furrows; sometimes without them, but covered over its entire body with little scales, or little scale-like squares. They are seldom found, and unknown on *Amboina*.

K
E
F
D
C
B
A
I
H
G

X. *Favus;*[25] Wafer Iron, has a thick shell, white, and is marked in such a way with circles and oblique ribs that have sharp edges, that they make square little chambers, as one also sees in Wafer Irons.

XI. *Lingua Tigerina,*[26] Tiger's Tongue, does not differ from the foregoing except that the edges of the circles are sharper, not white, but a dirty russet, blackish at the edges, of various sorts, but of no value, because they do not have nice colors.

XII. *Chama granosa,*[27] is not different from the Tiger's Tongue, except that, instead of squares, it has protruding, rather sharp little kernels; both kinds have some that have very thick and wide lips: There is also the wild Tiger's Tongue, which has neither squares nor kernels, but which is covered with scales like the *Scobinata,* of a smoky color and is not handsome in any way.

XIII. *Remies,*[28] is a flat *Chama,* covered with protruding circles, with a thick and white shell, without any designs, of which the largest kind is 3 fingers wide, but the common ones are the size of a thumbnail, and grow in crowds in white sand, especially on small, deserted islands: They are good to eat, and are pickled shell and all. There is also a small kind, which is of a dark-gray color, but otherwise also covered with rings. Others are somewhat thicker, and striped like the *Pectunculi,*[29] with a smutchy color, and a few black spots: All kinds occur abundantly on white sea beaches, so that, if one pokes around a little in the sand with one's hands, one will find them in heaps; both have a tasty meat, although it is little and not meet for a hungry stomach; they are cooked in water, then taken out of the shell, and a sauce is poured over them, that is made with butter, vinegar or lemon juice, and some pepper and salt. They occur often on lonesome Islands, like Dove Island,[30] one can also sow them, since all you have to do is spread them out on the beach when the tide is coming in, and as soon as the water covers them, they will crawl into the sand, and multiply in very short time; their name in Malay is *Remis.* Some other kinds belong to these, the ones called Sherds,[31] which are granulated on the shell, some are white, others liver colored: The marvelous thing about these little Shellfish is that they grow and live in the sand, and yet there will be little sand in them when they are cooked, as long as you rinse them thoroughly beforehand. One can keep them for several days, if one puts them in a pot or crate with sand but without water, which is why they are also permitted on the tables of the Notables. One will also find such *Remies,* that are as large as a schelling, a dirty white, and have a plain appearance.

XIV. *Tude baija,*[32] is also a *Chama,* covered with circles, that occurs in *Makkassar,* with a thick shell, smoky and brownish, with broad white rays between, round mostly, and bossed, as wide as a small hand: It has 2 smaller sorts, similar to our *Bia Matta doa,* but more bossed, and more blackish brown than the foregoing, and with dirty white rays.

XV. Letter Shells[33] from the Xula Islands, are also like No. VI, but smaller, not more than a finger joint long, and with nicer designs. They occur in the Xulanese Islands,[34] and one could divide them into two kinds, if one paid careful attention to the shape of the letters.

XVI. *Remies gargadja,*[35] is a white *Remies* shell, edged or toothed on one side.

Of the Chama laevis, *some are smooth, others are furrowed, striped, spotted or jagged. The Author's first one is depicted on plate* XLIII. *by the letter* G., *and it is smooth and spotted; but we call it the* Japanese Game doublet, *because it is often used there for gaming: they paint it on the inside with pictures, plants and other ornaments, and then mix them up in two piles, whereafter they pair them again, and he who has done his pile first, wins the game. The* 2d *kind is on the same plate by the letter* H. *and is called the* Croaker *by the Author. The* 3d *kind, is depicted by the letter* I. *and is rarer than the foregoing. The* 4th *kind, by the letter* K. *is called the* Greek A doublet *by us, be-*

cause all its marks look like that: there are many kinds of these, but the finer and sharper those little stripes, the more valuable the shell. The 5th *kind is shown on plate* XLII. *by the letter* A. *The* 6th *kind by the letter* B. *which we call* Bow doublets, *or* Japanese mats; *of which there are many kinds, coarse or finely striped, speckled, spotted, clouded and others. The* 7th *kind, shown by the letter* C. *also has various kinds, and are called* bastard Bow doublets. *The* 8th *kind, by the letter* D. *if one grinds away and polishes the furrows and ribs, one will often find them to be a beautiful red, or white or a soft yellow. The* 9th *kind, is shown by the letter* E. *The* 10th *kind, by the letter* F. *which we call the* Lip doublet, *because their finely toothed wide edges fit tightly together. The* 11th *is another kind of the foregoing, shown by the letter* G. *The* 12th *is yet another one by the letter* H. *The* 13th *kind, by the letter* I., *of which there is another one by the letter* O. *The* 14th *kind, by the letter* K. *The* 15th, *by the letter* L., *and another one by the letter* M. *which we call* tour de Bra. *The* 16th *kind, by the letter* N. *we call the* toothed Venus doublet.

CHAPTER 31

Pectines & Pectunculi.

The third genus of Bivalves is made up of the *Pectines*[1] or flat St. Jacob's Shells,[2] and the *Pectunculi*,[3] which are bossed, as follows:

I. *Pecten primus sive vulgaris*,[4] the common St. Jacob's shells, in Dutch Pyed Cloak, is flat, 3 and 4 fingers wide, with two ears near the hasp in back, of which one will always be the larger one, wherefore they resemble a cloak that was spread out; It has many folds, separated by deep furrows, like a comb, which gave it its Latin name; the folds are obliquely notched and rough, a dirty white or pallid with some black spots. The Malay call them *Bia Sissir*;[5] that is, Comb Shells, or *Bia Terbang*,[6] that is Flying Shell, because at times they jump out of the water, as if they were flying, but not very far.[7] Today's Greeks call them *Chtenia*, which comes from Ancient Greek *Ctenes*.[8] The rarest ones are those that have the back flaps almost identical.

II. *Pecten secundus*,[9] is smaller and more bossed than the former, also with two, but uneven ears, and with rough furrows, gray on the widest part, whitish towards the back, with many black spots and speckles, which is why these are also called Pyed Cloaks; they are white inside, and the color of purple around the edges: All *Pectines* lack the hollowed-out hasp or *Ginglymum*[10] in back that the *Chamae* have, but the ears lie flat on top of one another, and are fastened together by means of a hard and black band that is like a thick thread, and which, when dry, breaks off, and causes the shells to fall away from each other, but otherwise they close tightly all around, and do not have an opening, except for a slight one near the ears. They are not abundant on *Ambon*, and one will find them the most on those beaches that have more stones than sand.

III. *Pecten tenuis*,[11] has a thin shell, without raised folds, and with short ears, divided into four kinds: The first one is a chestnut brown, with short and superficial scales on the folds: The second one[12] remains small, gray, rough from small scales, and with small black spots, common on the very edge of stony beaches: The third one[13] is entirely white, also with superficial scales: The fourth one[14] is [the color of] coral or red lead; some are also lemon yellow, with deeper folds, and roughly jagged on the folds; but they are seldom found, and are therefore considered great Curiosities.

IV. *Radula*,[15] a Rasp, this one is shaped as if one divided a Jacob's Shell in two, to wit, straight on one side, and round on the other, deeply furrowed or toothed, furthermore scaly

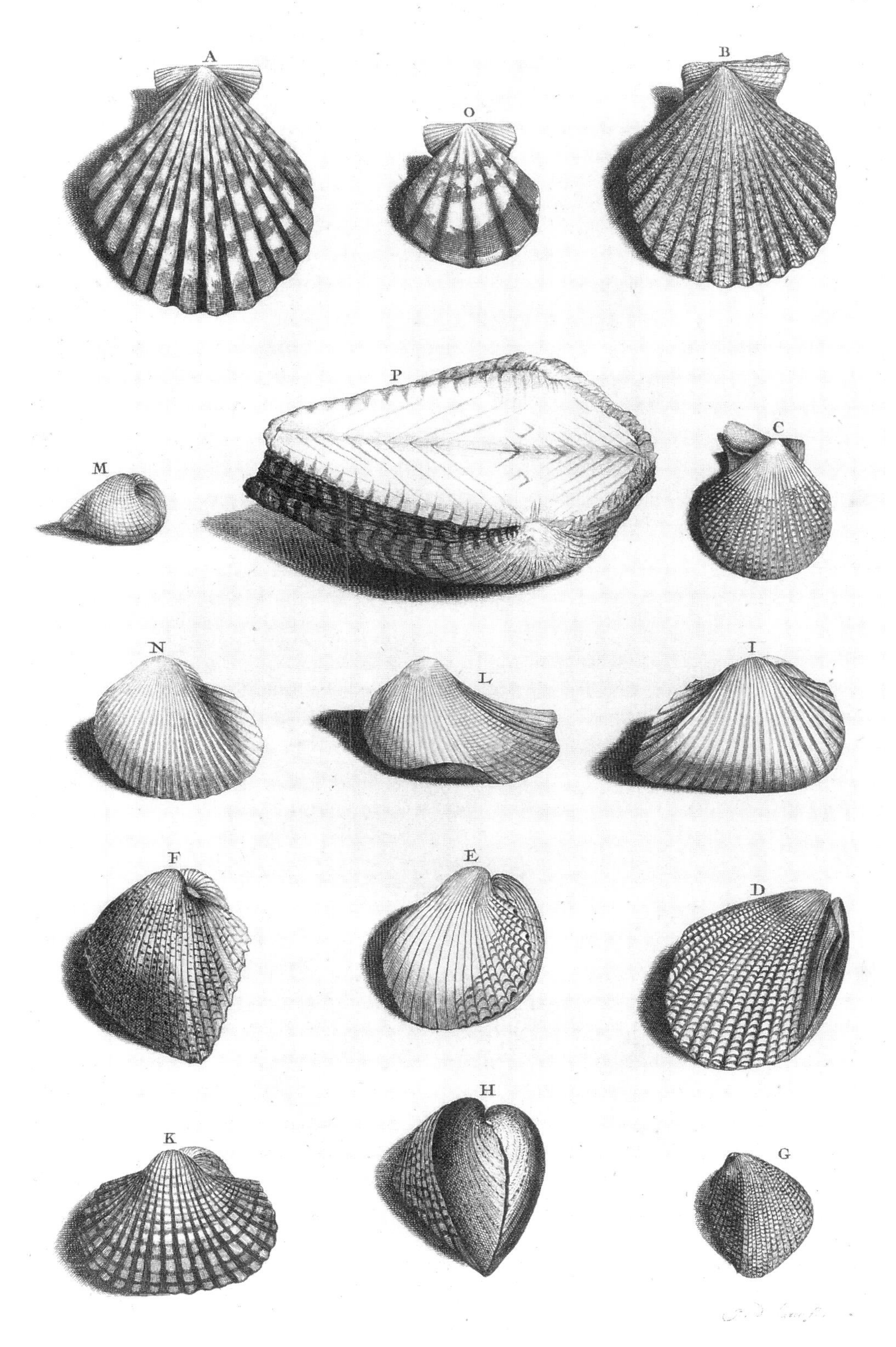
A
O
B
P
M
C
N
L
I
F
E
D
H
K
G

like a Rasp, entirely white, and without any designs. A smaller kind has finer combs, and very small scales, like a smooth file, also white; both fall easily apart when the little black band has broken off.

V. *Pectunculus vulgaris.*[16] In Malay, *Bia cucuran,*[17] because of its jagged edges, which resemble the toothed piece of steel, which they use to scrape out Coconuts: in Chinese *Ham;*[18] this one is toothed like the foregoing, but the folds are rounder, more even, not jagged, though a little rough, and the furrows between are narrow: The shell is roundish and hollow or bossed, without ears, but with jagged edges, that close tightly, and the back of the shells hang together with the same kind of black little band which, when broken, makes the shells no longer stick to each other; their color is a grayish or dirty white, with a few black speckles, which grow as big as three fingers across, they are a lemon yellow in the center of the shell, and noticeably notched on the edges. They have a lot of meat, and stay below the sand, but at low tide they come to the top, open their shells, and gape for a soft wind. Their meat is rather tough in these regions, and not pleasing as a food, but in Portugal, and along the Spanish coasts, where they are called *Brigigoins,*[19] and are no wider than 2 fingers across, they are considered a dainty dish, cooked in its own juice with a Brazilian pepper pod,[20] and which the rich and noble people also find to their taste, *ut ajunt, ad excitandam Venerem.*[21]

VI. *Fragum,* a Strawberry. In Malay *Cucuran mera,*[22] is also edged, bossed, and hollow, somewhat flat on one side, but in such a way, that in the center the edges protrude a little, as if a half moon had been cut in two; the folds are closely covered with small brown-red thongs or scales, which make the entire shell rough, and therefore resembling a Strawberry, normally at least two fingers long, and of the same thickness: These and all other *Pectunculi,* which are round and hollow, have a hasp in back, with which they close, and behind that a little skin, which keeps the shell together. They are common in Ambon Bay, and are not deeply buried in the sand. Some among these have some folds, are quite bare, without any thongs: while on *Nussapinjo*[23] the entire flat side lacks thongs or notches.

VII. *Fragum album,*[24] white Strawberries, these are scarcely an inch, but of the same shape as the foregoing, except that the flat side has sharper edges; the folds are closely covered with fine little thongs, a pale or whitish yellow, and are rarely found. There is a larger sort[25] as well, that has one side entirely flat and smooth; the folds finely covered with yellow little thongs, which look much more like scales; one will find them in the inner cove[26] of Ambon Bay.

VIII. *Pecten virgineus.*[27] In Malay, *Bia anadara,*[28] is a thick-shelled, toothed Shell, with a protruding corner on one side, which makes it crooked, as if one chopped one horn of a half moon with a diagonal cut; the back of the head has two round bosses, where the shells lie on top of each other without a hasp, hanging together on the outside with a little skin, and one will see some fine notches in the shell; the folds are flat, separated by narrow furrows, wherefore the edges are not very notched either, a pale white, without any marking, but right out of the sea they have a dark-gray or soil-colored woolly covering; hairy on the sides, which can be easily rubbed off with sand; the meat is rather hard,[29] and has a pointed little hand the color of red lead, with which it can do a great deal of violence, discharging a light-red juice, which the Natives compare to a Virgin's first flower,[30] whence the name. They are common on all of Ambon's beaches, where there is some mud in the sand, and they are much sought after for food, although they are hard to digest.

Ctenites is a white, round little stone, the size of a pea, or smaller; sometimes shining like a pearl, or crowned with a little white sun, and which one will find in this and in the former *Pectines,* though rarely, and is considered a Curiosity. The Natives wear it for luck when gathering mussels and shellfish. There is another *Ctenites*, being a hard dark-gray stone in the shape of a *Pectunculus*, about which one can read in the next Book, under Stones.[31]

IX. *Pecten granosus*,[32] has the shape of the foregoing, but with a rounder and thinner shell; the folds stick out quite a lot, and are covered with rough kernels, which make the shells somewhat thorny; this one is unknown on *Ambon*, but is found on *Ceram*'s northern coast, and on *Celebes* near *Makkassar.*

X. *Pecten saxatillis.*[33] In Malay, *Bia batu*,[34] is more oval and narrower than the *Bia anadara;* behind it has a flat back, with 2 bossed heads over it, the shells hanging loosely together with a little skin; it is dark and combed in an orderly fashion, and has an opening on the front side, as if a piece were broken off; it is a little hairy by the protruding corner: One will find those that are so narrow and oblong, that they resemble a small baker's trough (*mactra*),[35] and the shells are usually separated from one another and strewn all over the beach, otherwise they cling to rocks with a stony growth via the aforementioned hole, as if they had clamped onto it; wherefore they do not change their place at all.

XI. Small Buttocks,[36] are no bigger than a thumbnail, with a protruding little corner, bossed, and with round buttocks behind, a brownish gray and divided into little squares by folds and stripes, and therefore coarse: Others[37] are flatter, and have 4 or 6 sharp scales on one side, also divided into little squares: Both of these are common in the Bay of Ambon.

XII. *Pecten bullatus.*[38] In Malay *Bia filoos*, has a very thin shell, so that one can easily press them to pieces; on the outside a light russet color or yellowish, with a few speckles, dark and slightly combed, and the shells are loosely attached to each other with a little skin; on the inside it is a light red, very hollow, and puffed up like a certain pan-fried dainty in the shape of a bubble, called *filoos*[39] in this country: The *Pectines plani* are able to fly, or rather jump from the water, because of the little hand, which is a quite hard (*callosa*) piece of flesh, with which they hit the ground, and therefore fly out of the water, with their hindermost or sharpest side first: So *Jonstonius*,[40] and other Writers, have no reason to doubt this flying because this Shell cannot spread its shells like wings, which is not necessary at all when they shoot forth like that.

XII. *Bia Sabandar*,[41] or small Pyed Cloak, is like an ordinary Jacob's Shell, with a few, round ridges, that are tabbied in black, and spotted like the fish called *Ikan Sabandar.*[42]

We call the Pectines & Pectunculi *for the most part* Pyed Cloaks; *of which we will point out some. The first kind, is shown on plate* XLIV., *by the letter* A. *The* 2d *kind, by the letter* B., *these* Pyed Cloaks *are valuable, especially if they have sharp little nails, and are a bright red with sharp white flames, but the ones that have black spots on a pure white foundation are considered the best. The* 3d *kind, by the letter* C., *are called* bastard Pyed Cloaks, *also* West-Indian Pyed Cloaks, *because they often do come from there: there are many kinds of these, with different colors and markings; some are a beautiful bright red; others orange, purple, with white clouds or striped, also speckled. The* 4th *kind, indicated by the letter* D., *is the whiter the better, and is called* Ice Doublets *because of their watery transparency. The* 5th *kind, by the letter* E. *The* 6th *kind, by the letter* F., *which we call* red Strawberry Doublet. *The* 7th *kind, is shown by the letter* G. *and is called the* white *or the* yellow Strawberry Doublet. *The one by* H. *is the* double Venus heart, *because it shows a distinct little heart on either side; these were once greatly valued, and one, that had red speckles or spots, was sold for* 60 *Ducatons, but it was then the only one; now they are slightly more common, but still valuable. The* 8th *kind, by the letter* I. *is called by us a* Bastard Ark. *The* 9th *kind, by the letter* K. *The* 10th *kind, by the letter* L. *The* 11th *kind, by the letter* M. *The* 12th *kind, by the letter* N. *All of these are various kinds of* Bastard Arks; *while we added the true* Noah's Ark, *by the letter* P.[43] *The* 13th *kind, depicted by the letter* O. *belongs with the* Pyed Cloaks, *and is called the* Clouded One; *these are very rare, and are seldom seen.*

CHAPTER 32

Amusium. Wind-Rose Shell.

Although the rare *Amusium*[1] Shell or Compass Shell, seems to be related to the *Pectines*, it deserves, because of its rarity, a separate Chapter.

It resembles a flat Jacob's Shell, fashioned from two thin, round, and mostly flat shells, its size about a hand across;[2] but unlike all other *Pectines*, it does not have protruding ribs on either shell, for they are even and smooth: The most wondrous thing about it is, that the bossed side is entirely white, while the flat shell, which is only slightly bossed, has the color of dark liver, and is covered with green stripes, which come together in back by the hasp, and which spread towards the circumvolution, just like the stripes on a Wind Rose, which gave it its name; on the inside, however, both shells have some thin, protruding ribs, which do not run all the way to the back, but which vanish around the center, and which one can also notice from the outside, if one holds them against the light a little; and so it seems that this Shell is made from two different shells, and one would not believe it, if one did not see them fastened onto each other; they are stuck to each other with a little band, back by the hasp, near the two protruding corners, just like the other *Pectines.* Its Inhabitant has a soft, yellowish flesh, almost like that of the *Pinna:*[3] The second wondrous thing is that it is found so rarely, and that one cannot point to any place on the beach or in the sand, where these may dwell, except for those which happen to stray onto the sand of flat beaches: Only two places are known, up till now, where they are caught, to wit, the north-western corner of *Xula mangoli*,[4] and the beaches from *Waro* to *Hote* on *Ceram*'s northern coast. They float in the sea near the beaches, usually with the white or bossed shell on top, but they are so agile, that they will turn up first the white, then the brown shell, and they also float obliquely; and if they want to go to ground, they pull themselves together, and cut through the water like an arrow: One catches them with nets along with other fish, gaping while they float, and they recoil when one approaches them. One will find them during few months of the year, and then only in the East or South-Eastern *Mousson*, which makes for a lee shore in the aforementioned places. The first ones were brought from *Hote* to *Amboina* in the year 1666; thereafter one did not see them for 20 years, until a few were brought in from *Xula mangoli*, in the year 1685: after that time a greater number of them were brought from *Waro.* Because of their rare shape and because they are so difficult to find, these shells have always remained one of the rarest of Curiosities.

The Natives do not have their own name for them, except that some call them *Bia terbang*, that is Flying Shell, because they float on the water as if they were flying; but it is better to give that name to all the Jacob's Shells, which jump as if they were flying. We Dutch gave them the more fitting name of Wind Rose Shells, after the stripes, which one can see on the flat shell, and that might also be the reason why they are called *Amusium* in Latin. This shell is also described, as a masterpiece of the sea, in the Shell Book by *Philippus Bonanus*,[5] and is depicted in the Third Part, the 354th Figure.

These Shells also became known in Batavia in the year 1696, when fishermen brought them from the Islands.[6]

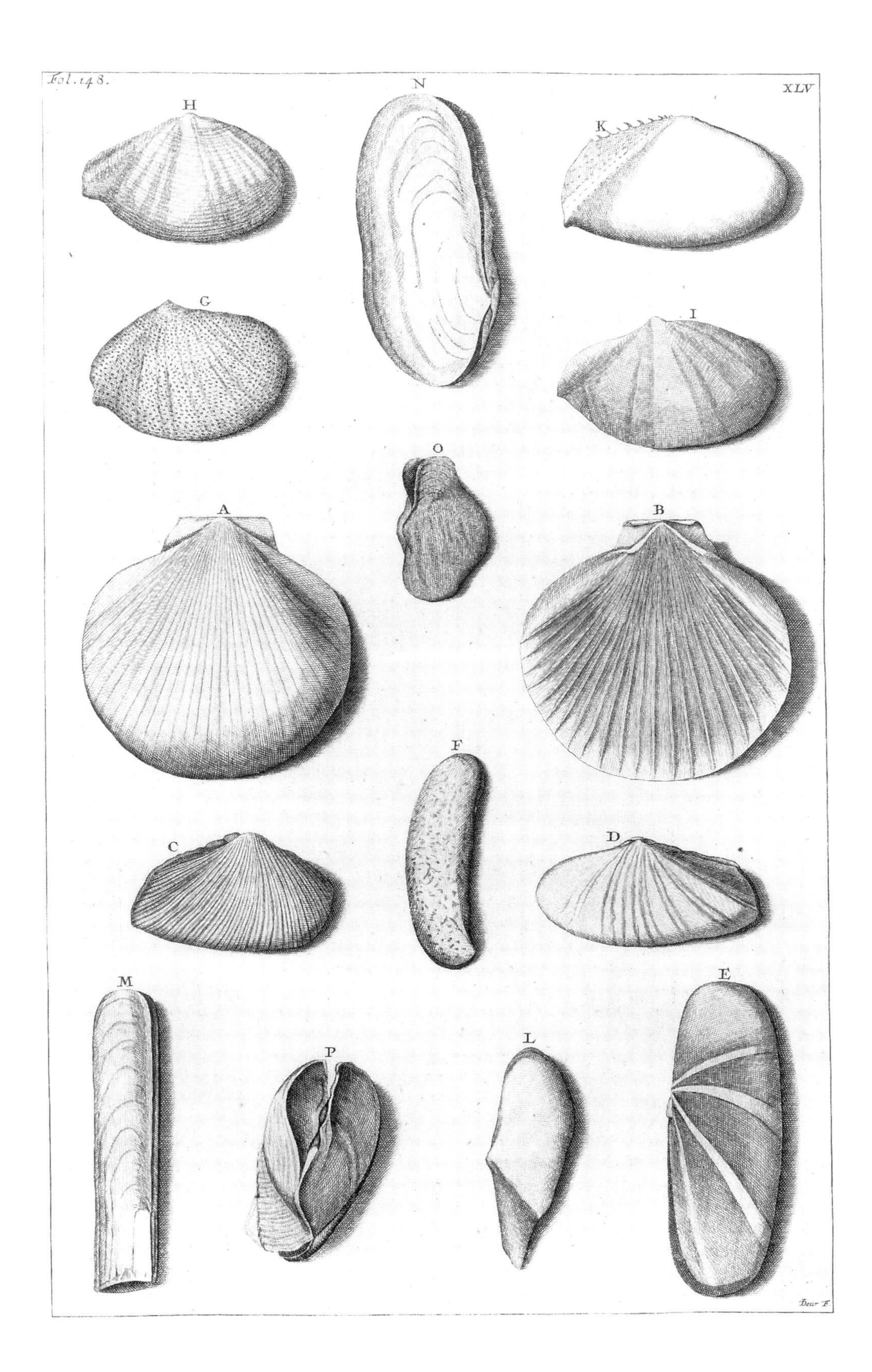
Fol. 148.
N
XLV
H
K
G
I
O
A
B
F
C
D
M
E
P
L

In the year 1698 I obtained them from *Bima* in the Sapi *Straits*,[7] and they had the same shape and characteristic as those from *Ceram*, but those from *Bima* were larger and more beautiful, and were found there in the seine nets as well.

The Amusium *is called the* Moon Doublet *by us, yet also, as does the Author,* Compass Shell, *see the same depicted on plate* XLV., *that by the letter* A. *is the outside, and by the letter* B., *is the same seen from inside.*

CHAPTER 33
Tellinae.

We include among the *Tellinae*[1] all those that have a thin shell and are oblong, either smooth or striped, with the following kinds.

I. *Tellina arenosa.*[2] In Malay, *Bia passir*;[3] is the length of a finger, and the width of 2 fingers across, with a rather thick shell, on the outside striped like the *Pectines*, but the ribs are coarser and narrower, they run somewhat crookedly, and are covered with thorny kernels; some are a light gray or whitish; others are a mixture of blue and gray; but most are reddish, including the inner part of the sides: The Animal has a white flesh; on one corner are 2 hollow little pipes, edged with a red fringe, through which it sucks up water, and then spews it out again with force; there is a hole in the flesh near the other corner, through which it expels filth from the black *papaver*,[4] which is full of sand; the shells close together *per ginglymum*,[5] and behind that is a thick, strong skin, which holds the two shells together, and when the Beast is gone, this same skin pulls so strongly backwards, that one has to use force in order to press them together again; the hasp and the head are not in the center of its back, but nearer to one corner, and the two aforementioned spouts are by the longest corner, and the exit for the filth is by the round corner; closer to the front is a little hand[6] of a rather hard flesh, which peeps outside the shell of the living Animal, and with this it feels the ground, and moves itself up and down, because all *Tellinae* are about a hand long, and sometimes a foot deep in the hard shingle bottom:[7] The 2 water spouts end presently in the stomach or sand sack, to be distinguished from the *papavar*, which is near the round corner: and all *Tellinae* have an oblong shell and two *tendinus* or tendons; one by the water spout, in the longest corner; the other by the round corner, because one was not sufficient to close that rebellious shell: In the center is a small lump of white flesh shaped like the yolk of an egg, which is what is used for *Bocassan*, but the Chinese leave that black *papaver* with it. All the *Tellinae*, be they on a muddy bottom, or on a shingled one, stand upright, at best buried a foot down into the ground, but when the water rises, they rise as well, until they stand only half a foot deep, and can be dug up quite easily. One will find them on the outer edge, where the sea will come, in coarse sand mixed with stones, having a round little hole to mark their place, and which one can see, if one scrapes the sand away a little which the aforementioned little pipes accumulated when sucking in and blowing out water; one might also find a small pearl in there, yellowish or purple, depending on the color of the shell, and this is a *Tellinites:* For yet another stone by that Name see the next Book, under Stones.[8]

These Shellfish are often dug up for making *Bocassan*, and are therefore called by many *Bia bocassan;* but since they are very sandy, the following, blue Shellfish, are thought to be better for that; and the present ones are only used when the others are lacking, and then one needs to keep them in sea water for a week in order to purge them of the sand, during which time one will see them constantly spew forth water, so that it is well nigh impossible not to get hit in the eye when you are standing next to them.

II. *Tellina saxatillis*,[9] *Bia batu*, is a thick-shelled coarse Shell, with its back clinging to the rocks, not much striped, dirty and mossy, hairy on its sides, and those hairs are stuck to the shell, so that one cannot rub them off; they do not close well in front, which is why they are always partially gaping; they are not much sought after. These do not change their places, unlike the *Pecten saxatilis* in the previous chapter.

III. *Tellina Gari*.[10] In Malay, *Bia bocassan*. In Ambonese, *Blastor* and *Blastol*, is a finger long, and the width of a finger across, not striped, but with some circles or tears along the sides; the rest is smooth, blackish or dark blue, with a few pale white and dark rays at the corners, does not close very well, and is in front bent inwards a little; furthermore, it is hollow, hiding a great deal of meat, spewing forth water from its little pipes, like the foregoing: The meat is white and soft, the *papaver* is not filled with sand, which is why they are considered the best ones for making *Bocassan*. One digs for them on soft beaches, where the sea is calm, and where the bottom is either black muddy sand or silty sand.

Bocasan or *Bocassan*, is that famed Ambonese *Garum*, that probably resembles the *Garum* of the ancient Romans the most, and which is used when at table to arouse the appetite. On *Amboina* they make it from two kinds of Shellfish, to wit, from the first and the third kinds of this Chapter, and sometimes also from the subsequent *Tellina picta*, but the third kind is the best and the most common: One should put the Shellfish for several days in sea water, so that they are thoroughly cleansed of sand, which the third kind accomplishes in about one or two days, and, when used in two different ways, will make two different kinds of *Bocassan*, white, and black: The white *Bocassan*, which our Nation prefers is made as follows: one opens the raw Shellfish, removes the white meat, and cuts the *papaver* away, after it is thoroughly washed and cleansed, it should stay in salt for 8 days, and the whiter one gets this meat, the more notable the *Bocassan;* this pickled meat consists mostly of the little thongs and skins, and it is put in a good *Towak*-vinegar[11] mixing it with diced roots of *Galanga major* or *Lanquas*,[12] white ginger, and pods of *Siliquastrum*[13] or peppers, also, for those who can bear it, some garlic; the little pots that one keeps this in, should have a narrow mouth, whereafter one pours some Olive oyl on it, and ties it tightly, because *Bocassan* will not tolerate air, and if prepared in this way it can be kept for a year: Others, instead of red peppers, take black pepper and the bark of *Culit Lawan*,[14] cut very small; if one wants to partake of it, one should always take just a little for the table, and tightly tie the remainder. It is an excellent and pleasant sauce, fit to eat with all sorts of food, especially with roast meat, arousing one's appetite, while it makes all kinds of food appealing, which is why it is transported from *Amboina* all over *India:* The black *Bocassan*, is the one the Chinese and Malay prefer, and which resembles the Roman *Garum* more, nor does it set one's teeth on edge as much as the white one does, because it does not have any vinegar, and is prepared as follows: the Shellfish, after they have been in water for a decent amount of time, until one guesses, that they are thoroughly cleansed of sand, are either left whole, and pickled shell and all, or, if one wants to eat it soon, one opens the shell and removes the white meat along with the black fat that hangs from it, which is yellow in other Shells, and is called the egg, and one throws away only the black Sand sack or *Papaver;* if one leaves this meat in brine for eight days, it will turn a brownish black, and one takes from it as much as one needs for one meal, then one pours some sour Lemon juice over it with sliced ginger and Spanish peppers, and that is how it is prepared; the taste equals that of the Roman *Garum*, which they make from small Fish guts, which they do in this country as well.

If one wants to send the white *Bocassan* overseas, and one does not have Olive oyl to pour over it, one can cover the little pots with the leaves of the *Capraria* or *Caju Cambin* tree,[15] which will preserve it from spoiling, something that the *Culit Lawan* will do as well; some people prefer white pepper to black, because they like that aroma better.

IV. *Tellina Violacea,*[16] is larger but thinner than the foregoing, 4 and 5 inches long, 1 wide, shaped like a sheath, and has such a thin shell, that one can easily press it to pieces, gaping a little at both ends, of a light violet-blue color, with whitish wide rays, with a smooth and flat shell; some are roundish; some are flatter, and 2 fingers across, straight at the corners, as if something had been cut off; this one is rarely found, and is considered a Curiosity; it does not always have its purplish rays in the center of the back, but for the most part gathered in one corner: One will find them on the eastern shore of *Passo Baguala* and *Xulij,*[17] standing upright in fine sand, where water is at its lowest level possible, which happens yearly by a full Moon in *November;* one can tell where it is by a small hole in the sand, that looks like a keyhole: One will not find them every year; hence they are just as rare as the following *Bia pissou,* which is why they are called the Sunbeam of *Baguala,* but the Malay regard them as being a *Bia pissou.*

V. *Tellina cultriformis.*[18] In Malay, *Bia pissou,*[19] also has a thin shell, and is oblong, gaping a little at both ends, slightly bellied in front, and sharp, with both corners bent backwards like a Saber, or like Chinese Sabers;[20] it has light-brown and russet speckles on the shell. One finds it rarely around the *Weijnitu* river, sticking upright in the sand, like the subsequent *Ungues.*[21] While one digs for it, it will spew water from the upper mouth, which gapes the most, whereafter it sinks deeper into the sand, which is why one should dig it out quickly, taking care that its sharp edges do not cut one's hand; since they are rare, they are considered a Curiosity.

VI. *Tellina picta.*[22] In Malay, *Kappija,*[23] is a little Shell with a common shape, with few stripes, grainy, and painted with small landscapes and turrets; some are a light gray on the outside, and white inside; others are russet on the outside, and red inside, but both have markings that are dark gray or blackish, and a joint of a thumb long: They live on hard stony beaches, where they are scraped out with great difficulty, near the edge where the highest tide will reach, sinking towards the deep when the water recedes, and coming up again when the same rises again, which can be said about all *Bocassan* Shells. They occur the most on the beach near *Rumatiga,* which has a shingle bottom; they have a soft meat, and little sand, and are therefore also fit to be used for *Bocassan,* when they are pickled shell and all.

VII. *Lingua felis,*[24] Cat tongues, are broad and flat Shells, truly round on one side, on the other with a narrow corner, covered with many fine little scales, and therefore as rough as the tongue of a Cat: They are white, with reddish rays that come together at the back by the head; others are a plain white, with crooked and sharp ribs, and with small squares without rays; the first kind occurs in fine sand, and is deemed a Curiosity; the others occur under stones, and are not beautiful.

VIII. *Tellina virgata,*[25] Sunbeams, in Malay, *Bia matta harij,*[26] are barely as long as one's finger, two fingers wide, and more, also round at one corner, while sharp at the other one, like a Westphalian ham, covered with circles or ribs, running parallel with the edges, somewhat sharp to the touch, a pale yellow, with reddish rays, coming together at the back by the head, resembling the rays of the setting Sun, which one sees in the evening, when it paints the water,[27] followed by rain: Some of these rays are broad, some narrow, while some Shells have none or merely some traces; they do not have much meat, because they are so flat, and one will usually find them empty on the beach, lying there gaping, but with the shells still attached, which happens, when they come above the sand when the water recedes, and, suddenly touched by the hot sun, become so powerless, that they cannot keep the shell together, because the dryness makes the hindermost little skin shrink towards the outside, which makes them prey to little crabs. One finds them in abundance in the bight of Ambon, near *Waijnitu* and on the *Uliassers*, on flat and hard beaches: They are not sought after for food. One also will find those that, though rarely, have a red shell, with white or yellow rays. Still a third kind, is entirely white, and of a plain color. A fourth kind occurs on eastern *Ceram*, and is the biggest of the *Tellinae*, four inches long, two fingers wide, smooth and brown over the entire body, without rays, and these are rare.

IX. *Tellina laevis*[28] is shaped like the foregoing, just as wide and thin, entirely smooth, without ribs or paintings; some are as white as ivory; some are light yellow, with a red spot or small cross at the back by the hasp; some also have a few dark rays; which are white; these are rare.

X. *Folium.*[29] In Malay, *Bia Lida,*[30] this one is entirely flat and thin, does not have corners that stick out, like the hams did, but is round at both ends, or straight at one end, as if cut off, and sharply toothed there, like a tongue or a leaf, of a bright yellow color, with and without rays: they are seldom found, and therefore considered a rarity; one should handle them gently, nor dry them too long in the sun, because that will fade them.

XI. *Petasunculus,*[31] Little Ham, has the shape of a Banquet Ham, to wit, with one round corner, the other narrowing, a finger long, easily one wide, with a very thin shell, uniformly red: some are incarnadine, and these are broader and rounder: Others are a light yellow, like dirty ivory, and are bigger and plainer: There are also small ones, almost without a corner, and round, also red.

XII. *Petasunculus striatus,*[32] is a small, oblong Shell, ribbed lengthwise, white or a light gray.

XIII. *Vulsella,*[33] Beard Nippers, is an oblong Shell; thick at one corner, and there [the two halves are] attached to each other with a hasp; at the other [corner] it is thin, flat, and round, so that it can open and close; like a pair of tongs or Chinese Beard Nippers, the biggest ones are 3 or 4 inches long, pallid on the outside, and rough, without any beauty, and looks rather like the bill of a Duck: Others are smaller and nicer, and look like a Pincer; they are the length of a small finger, somewhat curved like a Saber, of a dark gray or the color of soil. One will find them all clumped together in groups of 70 or 80, clinging together under a cover of coarse moss, and which one has to soak in water, in order to loosen them.

The Tellinae, *this* 1st *kind is shown on plate* XLV. *by the letter* C. *There is no picture of the* 2d *kind. But the* 3d *one is depicted by the letter* D. *The* 4th *kind, by the letter* E., *which we call the* purple Sunbeam, *or the* large Tour de Bra. *The* 5th *kind, by the letter* F. *is a* Pod doublet, *and is also called the* Polish Knife. *The* 6th *kind lacks a picture. But the* 7th *kind is shown by the letter* G. *and is called by us* Sea tongue, *because of its sharp little scales, also* Shagreen doublet:[34] *these will either be coarse or fine; the coarse ones have brown spots; and the fine ones have red rays on a white foundation: Which is shown here, as the* 8th *kind, by the letter* H. *The* 9th *kind, by the letter* I. *is smooth, and belongs to the* Rose doublets; *of which there are yellow, red, white, flamed, rayed and striped ones. The* 10th *kind, by the letter* K., *is another kind of* Venus shell, *described before. The* 11th *kind, by the letter* L., *is called the* Little Ham *by the Author; though we also count it among the* Rose doublets. *The* 12th *kind we have to ignore since we lack a picture. But the* 13th *kind, is shown on plate* XLVI., *by the letter* A. *which the Author calls* Beard Nipper.

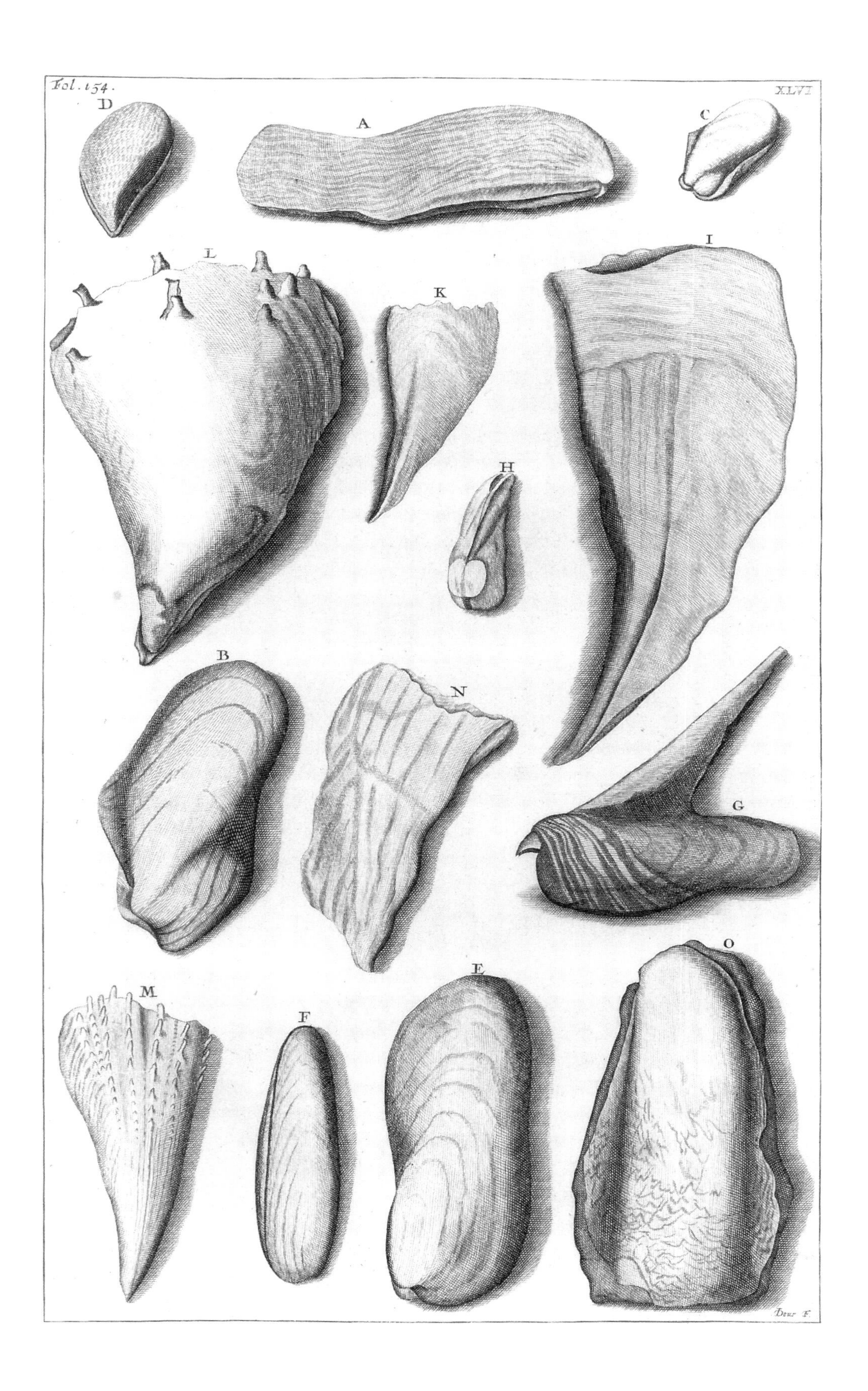
Vol. 154.
XLVI
D
A
C
L
I
K
H
B
N
G
O
M
F
E
Deur F.

CHAPTER 34
Solenes Bivalvy.

We described the *Solenes solidi*, those which consisted of only one pipe, in Chapter XXVII; now come the Bivalves, making the fifth genus of Bivalves which are oblong, their two shells resembling a sheath or pipe, of which the following kinds exist:

I. *Solenes bivalvy, Ungues, Dactyli, vulgo, Vagini*,[1] in Dutch Organ pipes, in Malay *Bia butu* and *Bia saron*, also *Krang sissij* and *Bia kuku*,[2] because its shell appears to have been assembled from nails or scales. They are 4 and 5 inches long, a finger wide, with a thin shell, roundish and flat, on one side somewhat joined by a little skin, on the other side open, but not gaping wide, because immediately behind it stretches a little skin: Both ends are open, of which the top one has straight, and the bottom one has rounded edges: The shell's color is a light brown, some with white spots and rays, even and smooth, or with tiny cracks sideways, indicating that it grows downwards; the flesh is hard and sinewy, with a little piece sticking out on top, the length of a thumb joint, shaped like a *Membrum virile*, with a black foreskin that it can pull out and in as it pleases, and it spews forth water: They stand upright in a bottom of pure sand, always at the edge of the lowest water mark, where the shore begins to fall away: One will know where they are by a small round hole in the sand, where one should dig them up carefully by scooping them up from below, because once they are touched they withdraw more deeply into the bottom. The Chinese pickle and eat them, although their meat is very hard and indigestible. They must be better and softer in the Mediterranean, especially in the *Archipelagus;*[3] because I understand that the Greeks and Italians esteem them highly there. Today's Greeks call them *Sulinees;*[4] the Italians *Lanquetti;*[5] but in *Jonstonus Lib. de Aquatilibus*,[6] I found the Venetian name of *Cappa longa;* in Bononia,[7] *Pische canella;* and it may well be *Rondoletius' Concha longa;*[8] although the Mediterranean *Solenes* differs somewhat from ours. One should not scold *Aristotle*, when he writes, that the *Solenes* is closed on both sides;[9] because, indeed, the loose side does not open very far, being, as was mentioned before, closed from the inside with a little skin.

II. *Rostrum anatus*.[10] Ducks' Bills, are also a kind of *Solenes bivalvij*, perhaps the Females, because Pliny divides them into Males and Females; although these are also of two kinds: The first or largest Duck's bill has the thicker shell, is wider, ribbed along the edges, as if full of cracks, not brown, but an ashen gray, on top with a wide-gaping mouth, of which the lips are slightly curled outwards; the lowest corner closes a bit more, and it stands upright, about a hand deep into the sand, having a black and sinewy end sticking out on top, wrinkled, pulling in and out, and makes two little holes in the sand on top, wherein one could stick a small finger, and wherefrom the Animal spews water; the inner meat is worthless, full of black blood and sand, like a skin and unfit as food, except for a small part, which is pickled: The second kind[11] is shorter, and has such a thin shell, that it is nearly transparent, bellied in the center, widely gaping on top, and with the same kind of flesh; the shell is a dirty white or grayish color, and as rough as Shark's skin, the meat is good for nothing: One will find them in coarse sand, with a bottom that is muddy or marshy.

A third kind[12] is the length of a small finger and as wide, with a shell that has oblique stripes or ribs on it. The *Musculus arenarius*,[13] *Asussen passir* or *Bia pissou*, also belongs to the *Solenes bivalvij*. The *Asussen passir* also sticks its flesh out above the sand, spreading the same all around as if it were a rose, but which immediately goes down, if one touches it, and the mussel withdraws very deeply into the sand, which makes it very difficult to dig out entirely: One puts them in hot water for just a moment, until they begin to gape; and one eats the meat half raw, because it is soft, with a good taste, without sand, white, and better than many Oysters, particularly the lower part, for the upper part is a little hard, divided into two oblong lips; of which the top one has the wrinkled little skin, that sticks out of the sand.

III. *Bia catsjo.*[14] In Ambonese, *Hua ilij* or *Huilij.* These are small Shells about the length of a thumb's joint, white, or bluish mixed with green, its shape is like that of a small shovel with a flat ridge in the center of every shell; below it hangs a tough white sinew, as if they were Longnecks,[15] that was going into the sand; and, when they are touched, they sink downwards: They stand upright but not very deep in the sand, and are marked with a small tear, as if it had been stabbed with a knife; One can dig them up on flat beaches, close together in large numbers, and are often eaten with *Papeda.* They are called *Bia Catsjo* after the shape of the holes in a *Pinang* basin, wherein one can put a *Catsjo,* that is, a *Pinang* implement, which is what they resemble.[16]

IV. Sea brushes,[17] can be found, standing upright in the sand on the beach near *Waijnitu,* being a finger long, the thickness of an oaten straw, whitish, and partially transparent like glass, looking very like the quill of a bird's feather, and is divided on top into many little strings, like a fine brush: One will also find some other stuff, somewhat thicker, like thin guts, flexible and tough, 1 or one-and-a-half feet long, sandy on the outside, filled with water inside, which are dug up like worms:[18] A third kind is pipes, the thickness of one's little finger, a finger long, rough on the outside, smooth on the inside, filled with a slimy water, and so brittle, that they seem to have been made from sand only, for they will soon break, if one presses on them.

In order to catch all sorts of *Solenes,* when one first sees the little holes that they are in, one should push down from the top a broom straw or the stiff central sinew of a *Calappus leaf,*[19] push it down as far as one can, and quickly dig to all sides down there; because, pricked by that sharp switch, they forget to sink deeper.

As already mentioned, one can also put the fifth kind of the foregoing Chapter, *Tellina Cultriformis* or *Bia Pisso,* among the *Solenes,* because it sticks in the sand in the same fashion, nor does it differ much in shape. The shell is oblong, in such a way, that the sides run parallel, furthermore thin and breakable, marked on the outside with fine ribs and furrows, which run parallel with the edges in a half circle. It is always pure, and can be easily polished, wherefore it is accepted as a Curiosity, the color is a yellow brown.

The Solenes Bivalvij, *of which some belong to the* Mussels; *the first kind is indicated on plate* XLV. *by the letter* M. *and which the Author calls* Organ-pipes, *but which we call* Gutter-doublets, *because, being either whole or half, they resemble complete or half gutters. The* 2d *kind, by the letter* N., *the Author calls the* large Duck's beak; *of which another kind is shown, by the letter* O. *and is called the* small Duck's beak, *but we call them the* eternal Gapers, *because they can never close. The other kind, by the letter* P., *is like the aforementioned big one. Of the remainder we did not receive any pictures from the Author.*

CHAPTER 35
Musculi. Mussels.

The same genus of Shells include the *Musculi* or Mussels, and the *Mituli*, of which we have noted the following kinds.

I. *Musculus vulgaris major.*[1] In Malay, *Asusseng.*[2] This is the common Mussel with the familiar shape, but in this country it veers more towards brown than blue, hanging from the rocks or from a piece of wood by a beard, giving away a great deal in taste to those one finds in our Western Sea.

II. *Mitulus anatarius.*[3] In Malay *Asusseng Bebec*, that is, Duck Mussels, in Ambonese, *Jhul*, on Hitu, *Lulat.* These are little Mussels, scarcely the length of a finger joint, half a finger wide, pale brown, and with a slightly wrinkled shell: They stick together in large numbers in a muddy bottom, making a complete crust, common in the Bight of Ambon, and on *Huconalo*,[4] and in other still bays: They are often dug up as fodder for the Ducks, the people either roll a stone over the crust they dug up, or break it by hand, and the Ducks will eat them eagerly: Pigs also like to eat them, especially the wild ones, the ones the Hunters are after.

III. *Mitulus saxatilis*[5] is even smaller, scarcely the length of a finger joint, but wider, and shaped like an ear, wrinkled or grainy on the outside, and covered with hairs around the edges, and it has a green reflection: They hang from hard rocks, that lie flat, and on other hard, flat beaches: They are not fit to eat, except for Pigs and Ducks.

IV. *Musculus arenarius.*[6] In Malay, *Asusseng passir.* This one is entirely hidden in the sand, though it stands upright, like the *Solenes*, shaped like a Mussel, though the sides run more parallel, and the head below is rounder, with a thin and brittle shell, divided into circles, along which it will break, if one presses on them; they are uniformly gray, going towards a pallid color, and clean: They do not have a beard, like the Mussels, but follow the nature of the *Pholates.*[7] The meat is white, soft, and good to eat. One finds them, but rarely, in hard and fine sand on flat beaches, where one can detect its place by a round little hole, and at low tide this same hole shows a little rose of reddish flesh, the size of a schelling, and in the shape of a *Boletus*,[8] which the Animal sticks out on top in order to bask in the sun.

Myrites[9] is a small stone that has the shape of the foregoing Mussels, though it is no bigger than a thumbnail, brown or violet on top, round on the bottom, whitish with a pearly reflection, and partially transparent, which one might find in these Mussels; because of their handsome shape and rarity, those Mussels are considered Curiosities.

V. *Pholas.*[10] In Malay, *Asusseng batu.*[11] In Dutch, Stone sheath. This is an oblong, black Mussel, that has the length and width of a finger, though usually smaller, sometimes also as long as a hand, with some circles by the upper end, which comes to a sharp and flat point, blunt and whitish at the bottom: One will not find this one bare anywhere, but always embedded in large Coral stones (*Lib 12 Herbar. Ambon. Saxumcalcarium*,[12] also called Cat heads). They reside in a small hollow in these Coral stones, and this little hole fits the Mussel so perfectly, that it seems as if art had hewn it, and it stands there with its tip always erect or slanting a little, and it always has a very small hole that runs from there, right through the stone, and through which it sucks up water and spews it out again. All around the Shell one will find a mealy substance like a mush, and the less of this there is, the more beautiful, blacker, and smoother the Mussel will be; but those that have a lot of mush around them, will be rough and grainy, a dull brown, and not handsome: The meat is slimy, but will harden when cooked, not very good to eat, because they taste like rocks: One can only obtain them when the Lime kilns are in use, when they smash those Cat heads to pieces, to burn lime, and one sees with amazement how those Shells lie hidden in the stone, without any sign thereof on the outside; and yet they have to suck their food through those little holes; and so one will have to open 3 or 4 stones, before one will find one, that contains these; the small ones have small, and the big ones big Mussels: There are some East-Indian Philosophers who surmise, that these Mussels are there first, and that thereafter invisible little Water animals grow the stone around them, together conveying that stony material hither, and then fastening it there layer by layer, the way the Bees make their houses.[13] The Cat heads grow like other Sea plants, which we described in the aforementioned *Herbarium;* which is why I included them among the Water plants; now whether such little Animals really make this happen, is still not known to me, but I cannot agree that they grow around these Mussels; the reason for this is, that one will not find these Mussels bare, nor in a promiscuous heap among those stones, but always standing either upright or at an angle, the animal undoubtedly possessing the powers and qualities, to moulder away or grind up the surrounding soft stone while it grows, so it can enlarge its little room, and expand its shell, which one can tell from the circles which the shell has at its thinnest end, where the growth takes place.

VI. *Avicula.*[14] In Malay, *Asusseng burong.*[15] This is a black Mussel, in the shape of a small Bird or little Swallow, which is pointing its two wings upwards, but if one spreads them, it resembles a flying little Bird with a long tail, usually a finger long; they hang with their heads down and their tails up, suspended from sticks, which have been in the sea for a long time,[16] also from the wales and rudders of Shallops by means of the beard, which is near the head. One will also find them growing on the little sea trees or *Akkarbaar,*[17] particularly the black kind, which are no more than the length of a finger; some are black; some red, both with long tails, and they fall off easily; the meat is like that of other Mussels.

VII. *Pholas lignorum*[18] resembles the top joint of a finger, to wit, blunt in front, pointed in back, rather flat, and with a round little hole, opening up into two parts, gray on the outside, ugly, and with a very brittle shell; they grow on the rotting posts that stand in salt water, like the *Solenes integri.*

Pholas' nature corresponds to what *Matthiolus*[19] says about certain Winkles or Snails, which are found living in certain rocks near the castle *Duin*[20] in the Adriatic Sea, where one smashes them into large pieces with iron hammers, and then into smaller pieces, wherein they then find these Snails,[21] good to eat, with a nice taste: They also find similar snails in a rock near the Tartar City called *Tarku;*[22] one also will find various Shellfish in the highest *Alpes* beyond *Verona. Aldrovandus*[23] *Lib. 3. de Exanguibus* also confirms the foregoing: On the shore near *Ancona,*[24] they haul stones of 50 and more pounds from the sea which, after they have been smashed with hammers, deliver small, lively and very tasty little fishes, *Idem Aldrovandus.*

The Musculi, Mussels, *is for us the* 1st *kind, depicted on plate* XLVI. *by the letter* B. *The* 2d *kind, by the letter* C. *The* 3d *kind, by the letter* D., *this one is very rare, and is called the* Duck Mussel. *The* 4th *kind, is shown by the letter* E. *The* 5th *kind, by the letter* F. *is a* Stone Mussel, *because the same grows in the Sea Coral stones, which they smash to pieces; and get therefrom. The* 6th *kind, by the letter* G. *is called the* winged little Bird, *because, if one takes it apart, and places it together, it looks like a flying* Bird. *The* 7th *kind, can be seen by the letter* H.

CHAPTER 36
Pinna. Bia Mantsjado.

The seventh genus of Shells belongs to the *Pinnae*,[1] in Malay, *Bia mantsjado*,[2] after the shape of a small Indian ax, in Ambonese *Kima Omin*, in Dutch Holster[3] Shells. They have a particular shape, different from all other Shells since they are triangular, long, flat, wide on top, and gaping, below sharp and coming to a point, like a holster; they are divided into the following kinds.

I. *Pinna prima sive oblonga*,[4] is one or 1-½ foot long, and more than four inches wide, becoming pointed below, standing upright, with half of it sticking into the muddy bottom; the one long side is closed; the other is loose, and opens about the width of 1 finger, and one will see there on the bottom a beard of black-green threads, about a joint long, narrow [where it is attached] to the Animal and wide in front, which is called *byssum*, and with which the Animal clings to stones and sand; the upper side does not close very well, but is always gaping on top, and is slightly round, the shell there being very thin and as sharp as a knife, so that one can get badly hurt, if one carelessly steps on it or when diving for it: The outer shell is black, or mixed with some brown, and the part that sticks out above the ground, is covered with sea grit or chalk; the rest is clean, covered with short and superficial scales, that stand in rows, and which are well-nigh gone on the old ones; the lower point is white, while inside the upper half is also black, but the lower part is the color of silver: The Animal has a large *Callum*,[5] and a large greasy clump that is either bright yellow, or the color of red lead, which is called the egg, also broad blackish flaps, throughout the Shell, covering the flesh no further than the silver-colored spot, and below that lies a large sand sack filled with black blood and sand.

The young *Pinna*, not even half a foot long, has a white shell, that is very thin and transparent, outside ribbed or striped lengthwise, and covered with sharp scales, gradually assuming a brown color, and the topmost and narrowest side is not round, but straight, as thin and brittle as glass: all *Pinnae* have a pinna guard, called *Pinnophylax* and *Pinnoteres*, which is a small shrimp, being ½ finger long, the color of ice with white dots, and nearly transparent, with thin and very sharp pincers, which always lives in the *Pinna*, for as long as it lives, but when the *Pinna* begins to die, it will walk away from it: Only one shrimp lives in each *Pinna*, but the same seems to form its eggs (which one will find under its tail) in there, and brings forth its young there as well, though they need to move afterwards, and seek other *Pinnae:* Its office is to pinch the *Pinna*, so that it will pull itself together, when there is prey in the Shell, or when there is some danger afoot, which we said before in Chapter XXVIII about the *Chama Squammata*.[6] They do not grow in the open sea, but rather in still bays, where there is a muddy bottom; peeping out with the upper, round part, but if there is a hard sandy bottom, half of it protrudes bare above the ground, together in great numbers, with the sharp side always up, so that one cannot put a foot between them; they are 4 to 5 feet under water: One will find them in the inner bay of Ambon Bay, in the entire marshy bight from *Hennetello* to *Cauwa*, on Ceram's north-western corner, and in the bight from *Tanuno* and *Kaijbobo*.[7] They are used for food, though one should throw the black *Papaver* away, because it will make you somewhat befuddled and dizzy, nor is the rest of any special taste: They must be better in the Greek sea,[8] especially in the *Propontis*,[9] or in the river of *Constantinopolis*,[10] where one still calls them *Pinna*, and where they are supposed to get as large as 2 feet.[11]

One also finds pearls inside at times, just as is written about the *Pinnae* of the Mediterranean, but they are small, round, purple, or a light violet, and lose their luster over time:

I believe, however, that they are whiter and more beautiful in such *Pinnae*, which are not black on the outside, but brownish, and grow in a hard bottom, and in clearer water, while most of the Ambonese ones thrive in muddy water; but [the water] must be somewhat clear, if one wants to dive for them, because some are more than a fathom down; and one should also take care that one does not injure oneself on the sharp edges: For the *Pinnae* see *Plinius Lib. IX. cap. 42.* and *Cicero Lib. 2 de natura Deorum.*[12]

II. *Pinna lata.*[13] *Plinius, Perna.*[14] This one is thought to be the female of the *Pinna*, and is somewhat shorter than the foregoing, but much wider, and with a thicker shell, very much like a small Westphalian ham: I had one that was 16 inches high, and 1 foot wide, slightly hollow on the open side, with a small curve, where the *byssus* is; the shell is also black or the color of soil, covered with rows of sharp, narrow scales, which it maintains in old age: One will find them by themselves, separated from the others, because they need a firmer bottom, and freer sea; but I never found any fresh-water rivers with *Pinnae*, as other writers have reported.

III. *Pinna alba*,[15] is also held to be the female, and is much smaller than the foregoing, it has a white or light yellow shell, thin, partially transparent like glass, not quite triangular, but somewhat bossed in the center, as if bent together; it is smooth on the outside, but slightly pleated. The largest ones are the length of a hand; others are 3 and 4 inches long, and so crooked in the center, as if they had been folded together. They do not stick in the ground, but cling by their beards to stones; they are seldom found.

The Pinna, *called* Holsters *by the Author, are called* Ham doublets *by us, because of their color and shape, also* Sticking doublets, *because the same are most often found sticking out of a soft bottom. The* 1st *kind is shown on plate* XLVI. *by the letter* I. *I had one of these* Ham doublets *that was easily two feet long: Another kind of this is depicted on the same plate, by the letter* K. *but it is much smaller than the foregoing. The* 2d *kind by the letter* L. *of which another kind is shown by the letter* M. *The* 3d *kind is shown by the letter* N.

CHAPTER 37

Ostreum. Oysters. Tiram.

The Oysters make up the eighth genus of Shells, with a variety of shapes, except that they all agree in having a shell that is on the outside rough and scaly, while inside they are silvery or like Mother-of-Pearl, and they do not change their places: We will not deal at length with Oysters here, since they are known well enough, and only note some kinds that are found here in *Amboina*.

I. *Ostreum radicum sive lignorum.*[1] In Dutch, Pole oysters. In Malay, *Tiram besaar* and *Tiram akkar.*[2] These are the largest ones, oblong, the length of a hand or less, of no certain shape, but tortuous with twisted edges, the same sometimes jagged or pleated; they are blackish on the outside, but silvery inside, they twist themselves around the roots of trees with their lower and thickest shell, and grow on them, wherefore their shape has to suit itself to the place, where they hang. The largest and best ones are found on *Manipa, Buro, Kelang, Bonoa*,[3] and on flat beaches, that have many *Mangi-Mangi* trees,[4] for they cling to their roots, and one often has to chop off a piece of the root as well; one could consider them *Pliny*'s *Tridacna*,[5] while today's Greeks call them *Chaederopada*,[6] so called after the resemblance to a donkey's foot. Some countries and beaches have such a luxurious growth of these Oysters, that the Ships that anchor there, haul up their anchors and find the shank

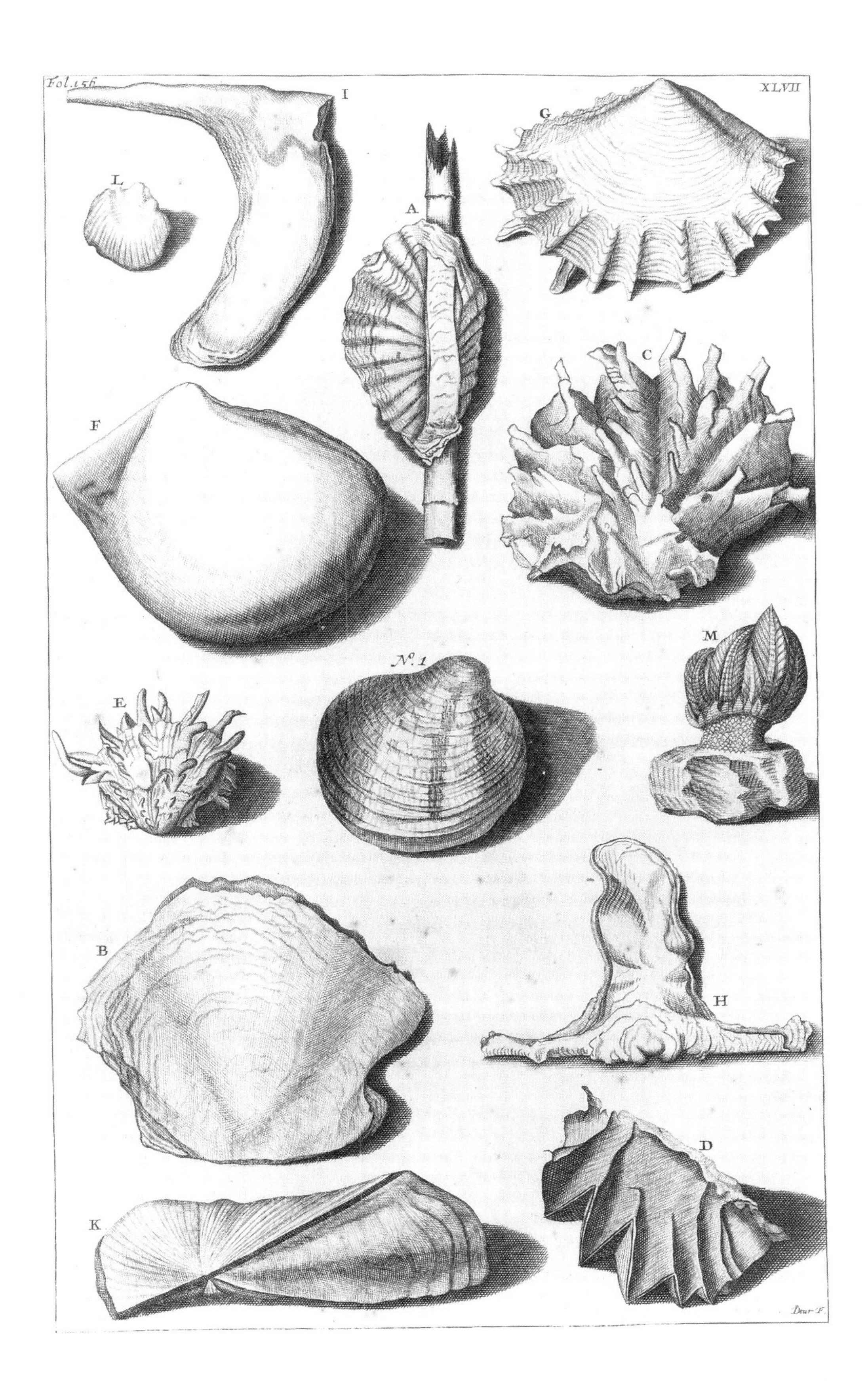
Fol. 156
XLVII
I
L
A
G
C
F
M
N°. 1
E
B
H
D
K

covered with beautiful large Oysters, and the keel of the ship as well, which happens particularly in *Siam;* where certain Mussels also grow on the keels, and they are oblong, gray or a light brown, shaped like the aforementioned *Musculus arenarius;*[7] only rarely does one find little white stones inside, the size of a pea, or vetch; some are entirely white; others have a pearly reflection on one side, and these can be considered *Chamites.*[8]

II. *Ostreum Cratium.*[9] *Sery* oysters.[10] These are smaller, about a finger long, with pleated and jagged edges, gray and scaly on the outside; some are russet, clean, and with a raised ridge on the upper shell; with the lower one they grasp the little sticks or reeds of the *Serys,* that have been in the water around 1 half year; they do this with many short little feet or arms, which encompass the edges of the sticks, in the same manner, as one can see with the roots of *Polypodium:*[11] They grow on these reeds in clumps, on top of each other, so that one has to cut off the sticks as well, but the most beautiful ones are those that grow by themselves: One will also find them on the roots of all kinds of *mangium fruticans,*[12] but those are small, sharp, and very jagged; one will find a great number of them on the *Passo* of *Baguala,* on either side of the Isthmus.[13]

III. *Ostreum Saxatile,*[14] these are small, with many corners, and all sorts of shapes; some are like a little dish, with one side clinging to stone; some are round like the Mother-of-Pearl shells, and these stand upright: They grow on all sorts of stones; even those that are no bigger than a nut, also on shards, and pieces of bricks, &c. they cover the stones in such a way, that one will not recognize them from above. They have a good taste, but it takes some trouble to knock them off and clean them: One will find them a great deal on the North side of the Bay of Ambon.

IV. *Ostreum placentiforme sive Ephippium.*[15] In Dutch, English Saddles[16] and Saddle Shells. In Malay, *Tiram lebber.*[17] In Ambonese, *Ea* and *ija.* In Butonese, *Calepinda.* In the Xulase Islands, *Calvinde,* they occur on *Toccuve*[18] and contain pearls there. These are thin, and flat as pancakes, but slightly bent, in the manner of the flat English Saddles; some are 6 and 7 inches in *Diameter,* with a shape quite like that of a Saddle; the shell is as thick as a quill, scaly, as if it were made solely of skins. It is a blackish gray on the outside, but a clear Mother-of-Pearl on the inside with the colors of the rainbow, especially the round spot, where the *Tendo* is, where one also might see at times some kernels like pearls stuck to the shell: It does not have much meat, merely some flaps, but it tastes good. One will find them on top of each other on *Bonoa,* in still bays where there is a hard coral bottom, with some veins of fresh water coming in. Another and thinner kind will be found in the Bay of Ambon, which resemble pancakes more, and which have scales that can be pulled apart more easily, with a smaller circumference than the foregoing, on the outside it is a russet gray, inside like Mother-of-Pearl, but mixed with a great deal of red, like red copper, hence not as beautiful as the foregoing. One will find the large ones, which are also called *Calepinda,* in the Straits of *Bouton,* and around *Pangesane,*[19] where the people use its Mother-of-Pearl as an inlay for the carved prows of their boats: This kind is also called *Calepinda* on the east coast of *Celebes;* where they have a smaller variety of this, that is flat, and thin, such as one finds on *Baguala:* The people from *Tambocco* string these shells together and hang them on the prows of their boats, after they have been on a raid, and had good luck obtaining severed heads; and their champions do the same by hanging them on their houses, so that, when the wind blows, people will be warned by their clattering, that here lives a brave man, who will not be trifled with; so that these Shells are like the Trumpets of *Fame.*

All these kinds produce pearls,[20] but one will find few of them in the Ambonese shells, and they are small, angular and yellowish; more beautiful ones will be found on the Xula

Islands, and on the Papuan Island *Messoal*,[21] where there are beautiful pearls, which quite resemble the true Ceylonese ones: One will find these Oysters 3 and 4 fathoms deep; some are flat, some lie obliquely like slates, and they are stabbed with a harpoon and lifted up. They are called *Giwan* on *Balij*, where they cut beautiful pieces from them and use them for inlay on their saddles and bridles.

When discussing the *Chama squammata*, I noted the story which the Inhabitants of *Keij* tell,[22] that one can see a large Shell on a rock several fathoms deep near one of their Islands, which gives off a great light, but if one dives for it, it vanishes, or, perhaps, pulls itself together: If this is not a *Bia garu*, then it must undoubtedly be a large Saddle Shell, for I have often seen such a gaping shell myself, lying under water, casting forth a beautiful Mother-of-Pearl reflection.

V. *Ostreum plicatum majus*[23] is an Oyster with a thick shell, fashioned from raised pleats, their ridges covered with long scales that are rolled up like nails, entirely rough and mossy, with a slime that hangs on the outside and that will cause an itching, while the skin is like that of other Coral stones; the inside is white, and black around the edges; One will find them on *Manipa*, but not much, and not on the roots of trees but under stones. One kind occurs on *Huconalo*[24] in the Bay of Ambon, several fathoms deep, and between stones; this one is also pleated, but without nails or horns, though it has small holes: It is so rough and unsightly, that one would take it to be a rock, but inside it is a beautiful white like alabaster.

VI. *Ostreum Plicatum minus*[25] is barely the width and size of one's hand, fashioned from few, though large pleats, without protruding nails or horns, but with small holes and squares scooped out, gray and clean: These pleated Oysters are rarely found, and make for a great Curiosity.

VII. *Ostreum Echinatum*,[26] in Malay *Bia tsjuppo*.[27] This one has two hollow shells, dark brown, and is covered all over with spines and branches, like an *Echinus marinus* or Sea Apple, of uneven length; some are wide, some are round; they are difficult to take hold of, except behind the head, where the spines are shortest. Another kind is short, and as deep as a bowl or teacup, with the other half lying on top as if a flat lid, covered with short and superficial thorns, all somewhat jagged on the sides, and some are as red as coral[28] on the outside: There are others, of a plainer shape, with short and rather broad scales, with a small lid on top, a whitish gray, chalk-like, and therefore not beautiful. All *Tsjuppi* have a wide hole in back, like a coral growth, with which they cling to the rocks: The most beautiful ones are Chestnut brown,[29] not very hollow, full of spines, and without sea grit; and one will sometimes find two or three of them with their buttocks grown together: They will be found on hard russet flintstones, on *Hative* beach, and many do also have a little crab as their guardian: The meat is seldom or never eaten, because it will cause a faint oppression or dizziness.

VIII. *Ostreum echinatum maximum & monstrosum*,[30] this appears to be the King of all *Tsjuppi;* of which I saw only one, that had been pulled up with a net before *Porto* in the Uliassers: it was uncommonly large, and had a misshapen appearance, to wit, the Shell itself was the length of a hand, covered all over with long spines; some were the length of a finger, some shorter, some flat, and as sharp as a lancet, some round as thorns, white and gray, all turned to the front, so that one could not lay hold of it, except, and that barely, behind the head: The whole Shell sounded like fine porcelain, and no Native had ever seen the likes of it: It was sent to the Grand Duke of *Tuscany* in the year 1682.

IX. True Mother-of-Pearl Shells[31] are not found on these Islands, but there is a kind that does resemble them, although it has no pearls; we call them little Table dishes, or small silver platters.[32] In Malay, *Telinga andjin*.[33] In Ambonese, *Asso telinay*, that is Dog Ears: They are mostly as round as a small table dish, behind and on one side with an ear, like the *Pectines*, but blunter and wider, covered on the outside with long scales, with the longest around the edges, flexible, and when dried in the sun, crumbly as parchment that was wet and then dried up again: When these scales and nails have been scoured off, the shell turns out to be somewhat smooth, and speckled with white spots like a snakeskin: The inner sides are black, the rest white and silvery, the meat like that of other Oysters, though white and harder. There is another kind, a russet gray, with a thin shell, which tears easily when put in the sun, and is of no consequence.

In the Molukkos, the Alphorese of Hallemahera[34] have certain little shields on their swords which were made from Shells, thick and ground round, that cover their hands: They call these little shields *Ngnotsjo*, which they seem to fashion from various Shells; the most common ones are made from a large, wide Shell, like those in this, the IXth kind, being easily a span wide, somewhat less in length, scarcely the thickness of a finger, fashioned from three layers of crusts, of which the outermost one is coarse and rough; the central one is like that of the *Bia garu*, but more shiny; and the innermost one is true Mother-of-Pearl: One will see on the inside a large spot full of wrinkles, where the hasp used to be, and that is where the pearls grow, but not in all of them: at first it is attached to the Shell with a narrow little neck, which is licked so long by the Animal, that it becomes loose, and that is how the pearl gets its pear shape: Whether they occur in the sea around the Country of *Hallemehera* and *Gelolo* I do not know, because the ones the Alphorese possess have been buried in the ground for a long time, and were, as is said, brought to *Ternate* in the year 1661 by a great Chinese *Wanckan*,[35] that had loaded them at *Soloc*[36] (a region on *Borneo*'s north side), and was going to take them to China, to be used as inlay in trunks and chests; but since the men on the aforementioned junk were considered *Coxinga*'s[37] Spies, ours did seize it and declared it a prize, and so these Shells were scattered on *Ternate* in various places in great number, and might have been searched for and kept by the Alphorese, who dared to demand a Rixdollar a piece; they generally call them *Mutiara* there, that is, Mother-of-Pearl, and they say, that they should be buried for some time, because the fresh ones are too crumbly to work with: The King of the aforementioned land of *Solok* has a certain flat beach there, where the aforementioned Mother-of-Pearl Shells occur; and where, as I am to understand, not only the Shells, but also the fishes, that can be found around there, contain pearls: This beach is rigorously guarded, so that no one can fish there, nay, they cannot even sail towards Land; but when it pleases the King, he has the Shells hauled up, which he then sells to foreign Merchants.

Similar Shells occur in the Gulf of Persia, especially on the side of Arabia, around the Islands of *Bahareijn*,[38] where there once used to be a fabled oyster bank, and they are brought for sale on the Persian beach opposite *Congo*,[39] mostly used to roast raw Oysters inside them, for they can be used for a long time before the outer, rough bark has been burned off and then they still have good Mother-of-Pearl. They also bring such from the West Indies to Holland, where they also use the largest ones for roasting Oysters: The finer ones are sawed into small pieces and used as inlay by cabinetmakers and gunmakers:[40] one often will find deep holes and dents in the two outer barks, wherein certain worms live that are nearly as thick as one's little finger, and one will find dead ones even in the Shells that were shipped, and the ones from *Ternate* have these dents as well: If these Shells have pearls in them, they came from such coasts and beaches that are entirely barren and dry, without any fresh water, and where the ground is brinier than the sea itself: That is why one will not find any Pearls in the Ambonese Table dishes, of which I had some, that were also a span wide, but barely half a finger thick, and inside with a much paler Mother-of-Pearl than the foregoing. These occur a great deal in the Bay of Ambon, but are no wider than 4 inches, have a thin shell, and are very scaly: One will find larger ones in the *Talega* or inner bay of the neck of little *Ceram*, situated between *Laala* and *Tanunu*, where sweet springs flow in, also in the Xulase Islands, but all without pearls.

X. *Ostreum divisum.*[41] In Dutch, Mallet.[42] In Malay, *Bia kris.*[43] This too is a rare kind of Oyster, shaped like half a cross or like the letter T, but the two upper horns are narrow, thick, and of unequal length; the long leg does not always go straight, but is often curved to one side, while it also is undulated like the sea. It is black, flaky, and scaly: The Malay compare them to a *Kris;* we to a Mallet: The Meat is like that of other Oysters, and good to eat; because of its rare shape, it is considered a Curiosity. One finds them around *Assahudi*, and *Henetello* on little *Ceram*, also in Ambon Bay on the *Leytimor* side.

XI. *Ιστγνόμον.*[44] In Dutch, Carpenter's Square. This is a Mussel-shaped Oyster, flat and oblong like a tongue, with a horn on one side in the shape of a carpenter's square, or like the blade of a *Kris;* thick in back, thin in front, and scaly: One also finds another kind, completely irregular, not truly black, but brownish.

XII. *Ostreum tortuosum,*[45] this is a rare Oyster, from the Papuan Island of *Messoal*, oblong and as if twisted, with a protruding little corner at one end, on the outside gray, striped and clean, its crooked shape cannot be compared to anything, while it is unknown in these Islands; except that one will find a few on *Ceram*'s northern coast by *Hote:* The shell has really 3 sides, and its ribs run the way the sides have been twisted.

XIII. *Ostreum electrinum:*[46] Amber Oyster, [this] is a small Oyster, scarcely a finger long, of which only the upper part will show, the other half remains under water clinging to rocks: The upper half is hollow like a Nutshell, but crooked, thin, of a yellow color inside, yellow on the outside as well and pearly, partially transparent like dark amber, with a beautiful reflection when the sun shines on it. Up till now they have been found only around the Northeast corner of *Buro*, near the small bay called *Waylila*, and *Karamat* mountain, but the place where they grow is hidden from man's eyes, and only the upper part is tossed onto the beach: The *Buro* Moors make holes in them, and string them together in such a way, that two and two are opposite each other, like Cat ears, and which they then wear on their Turbans, delighting in the way the golden yellow luster glitters in their eyes. One will find them on *Xula taljabo*[47] as well, and I have obtained whole ones from there, the lower part being a thin lid, with which they cling to the rocks, but so brittle, that they can hardly be handled without breaking. They have also been found on *Amboina*, by the *Waynitu* River, tossed onto the beach, in January of the year 1686.

XIV. *Mitella,*[48] Little Miters, is a kind of *Balanus*, shaped like a flat Miter: fashioned from 4 or 5 little bones, which look like the claws of a bird, notched and ribbed sideways, a dirty yellow and greenish: All these small bones stand upright, with their tips up, and close the Miter, but they can open them on top, where the Beast sticks something out, with which it licks food from the shell, being the slime that the sea has left there: On the lower or widest part of the Miter one will see other small bones like teeth, which in time will bring forth other Miters; wherefore one will see 4 and 5 together in one clump: Each one has a short neck, that is a scaly skin, with which they are fastened to the rocks. One will find them on tall and sheer Sea rocks, against which the highest tide just manages to splash, but they are never under water, [there are] a large number on *Nusatello* and on the rock near *Laricque* called Sugerloaf:[49] The Natives look for them to make *Papeda*[50] sauces, only for the taste, for there is little meat in them; one can hardly keep them together for a long time, because after a long rainy season they will fall apart.

XV. Longnecks,[51] is also an Oyster growth, in the shape of a tall pointed Miter, fashioned from 6 little white bones; of which the two narrowest ones make up the edge of the neck, corresponding to the *Clavicula* of animals; the two broadest ones are the cheeks or sides; it also has two narrower ones on its belly: They have a long neck in back, made of a tough skin, which shrinks when dry, and with this they cling to ships, though they do not bore into the wood; this Mussel opens up a little at the top: On the side of the belly (because they are closed on the back) protrude curved little brown plumes, like a cock's tail, or like what was said of the *Balanus;*[52] these little plumes serve the Animal for sucking food from the slime that hangs from the shell on the outside. They grow on boats, neither below, nor above water, but where the boat is just washed by the water, and they open up when they are covered with water; but if they are not touched by water, they will dry up and die. Suchlike Oysters or Mussels are thought to be the same, wherefrom one believes that the Barnacle geese[53] or Scottish Brent Geese are generated, of which *Isaac Lydius*[54] in his happy hour of death writes on page 88 that he has seen this in Mr. *Colvius*, and whereof he writes as follows: *The ones that the oft quoted Mr. Colvius has, are like little Mussels, that hang from seaweed, which is stuck onto a piece of wood, wherefrom one has thought, that birds do grow, because one will find a substance inside those Mussel shells (which are of a white color, which is why they are called Oysters by others), that looks very like the feathers of the tail, or wing of a small bird, that is fastened onto the edges of the Oyster or Mussel with a thin little skin, which I noted myself after I opened one; and on the side where the Mussel is stuck to the seaweed, one will see another substance, which appears to be the body of some creature; but one will not find there the true shape of a little bird, nor the beak, nor its head:*[55] It has been sufficiently discovered during the journeys of the Dutch to the North, that the Wild Geese have been found brooding on their eggs in the Northern regions, such as Greenland and *Nova Zembla*, and from there they fly to the Southlands every year.

XVI. *Ostreum Pelagium*,[56] is a kind of pleated Oyster, with a thick shell, but without scales, that grows on Sea rocks at a reasonable depth, attached to them with a large hole, and therefore difficult to remove. People think that they taste better than the big Pole Oysters, which one will find on the roots of trees.

Ostreum, Oysters, *and other kinds that belong to this: The Author has shown us several of these; among which many are quite rare, and which we will point out as follows. The first is depicted on plate* XLVI. *by the letter* O. *and is called the* Pole Oyster. *The* 2d *kind, on plate* XLVII. *by the letter* A. *we call the* Laurel Leaf, *because of its shape. There is no picture of the* 3d *kind. But the* 4th *one is indicated by the letter* B. *which we call* Polish Saddles, *sometimes* English Saddles *as well: One will find many kinds; but the largest I have ever encountered, is in the possession of Doctor* d'Aquet. *The* 5th *kind, depicted by the letter* C. *we call the* branched Coxcomb; *but it is very rare. The* 6th *kind by the letter* D. *is the* Coxcomb; *of which there are also doubles. The* 7th *kind, shown by the letter* E. *we call the* Lazarus Clapper;[57] *because they close tightly together with their crooked hooks, and will not fall apart, even if one clacks with them as one does with a* Lazarus Clapper: *There are many kinds of these, which differ from each other in color and shape; for there are* penned, nailed, scaled, white, gray, pyed, brown *with* white nails, *beautiful* orange, *also* lemon yellow ones; *but the most outstanding of them all is shown on plate* XLVIII., *by* No. 1.,[58] *which is a beautiful blood-red, and is the only one known. The one depicted on the same plate by* No. 2.,[59] *is a grayish white, flamed with red spots, whereof Doctor* d'Aquet *owns the largest and most beautiful one known to us. These are excellent pieces, which seldom occur, which is why I have added them: Mr.* Griffet,[60] *in Rotterdam, has an outstanding piece, to wit a Flintstone on which various of these shells have grown: One will also see one like that in the Cabinet of the* Feitemaas *Gentlemen in Amsterdam: To this kind also belongs*

Fol. 153.
XLVIII
N°. 8
N°. 7
N°. 5
N°. 2
N°. 1
N°. 10
N°. 6
N°. 3
N°. 11
N°. 4
N°. 9

the one indicated on the same plate as No. 3.[61] *an outstanding piece, called the* Rock doublet; *which is owned by Secretary* Blaauw, *and few other ones are known. Now we return to the Author again, who names an* 8th *kind, but did not provide a picture. The* 9th *kind, is shown on plate* XLVII. *by the letter* F. *and it is known to us as the* Mother-of-Pearl Shell: *I saw one of these in Dr.* d'Aquet's *Cabinet, about a foot*[62] *across, a very unusual piece; whereof a second kind is shown by the letter* G. *The* 10th *kind, by the letter* H. *is called a* Polish Hammer, *also an* Indian Kris, *because of its shape, and formerly a* Cross Doublet; *they were once greatly esteemed: I have been informed that a certain Gentleman once paid one hundred ducats for one, in order to honor the then Pope with it as a miracle (because the same represented a true cross, whereon one could imagine the semblance of a body); but its rarity is no longer, since one now knows, that it is a genus. The* 11th *kind, is shown by the letter* I. *called a* Carpenter's Square *by the Author, but* Venus-sheath Doublet *by us; these too are very rare. The* 12th *kind, by the letter* K., *the* twisted Oysters, *are called the* curved Noah Arks *by us. The* 13th *kind by the letter* L. *and the* 14th *by the letter* M. *called the* Miter *by the Author, because of its shape. The* 15th *and* 16th *kinds lack pictures: wherefore we fill the gap with a very unusual shell, indicated by* No. 1.[63] *Having followed the Author thus far, we thought it necessary to have a few others, very unusual kinds, which should have been put in before at various places, follow hereafter; since we only obtained them after the foregoing had been printed. In Chapter* XXVIII on *plate* XLII. *by the letter* F. *is shown a kind of* Venus Shell; *see* No. 4.[64] *on plate* XLVIII. *for another one: This one is the true* Venus Shell with Hair. *And by* No. 5.[65] *the wrinkled* Old-Wife Shell, *which is very uncommon. And by* No. 6.[66] *the* ribbed Venus Doublet, *a very rare piece, because it is seldom paired, and is always damaged. To the* Pyed Cloaks *described in Chapter* XXXI, *belong the* Coral Doublets, *shown on the aforementioned plate by* No. 7.,[67] *where the same is depicted on its one side, ribbed and furrowed; but its other side is shown by* No. 8. *where it has some knobbles on its ribs, resembling coral; that is how most are: Secretary* Blaauw *owns an unusually large one without knobbles, the like of which I have never encountered anywhere else; It is a reddish brown mixed with grayish white: But Miss* Oortmans *owns one, which is a beautiful Lemon yellow, with white flames; and which is the only one known among collectors. One should add to the* Nail Shells, *as described in Chapter* XXVIII, *the one that is shown on the same plate by* No. 9.[68] *and which is outstanding and very rare: To which we should add another one, which we have never seen doubled, like this one, shown by* No. 10.[69] *and which is called the* double Fool's Cap Shell. *But* No. 11.[70] *is the picture of a Shell that need not give way to any other; a piece of perfect beauty, both as to shape, color, clear design, residing in the outstanding cabinet of the* Feitemaas *brothers; wherewith the descriptions and the images of the Shells are concluded.*

APPENDIX

Since several Whelks *came to us after their descriptions had been printed and since they are difficult to get, if not impossible to see, I deemed it necessary to add them here, and to show some on plate* XLIX. *which the Author had passed by, perhaps, because His Honor did not encounter them, or perhaps because they do not occur in those regions: They belong for the most part to that kind, which the Author described in Chapter* XIX, *under* Buccinum, *and where he shows one, called the* Wentletrap, *which we said was just a subsidiary kind in our Commentary; we now depict the true one on plate* XLIX. *by the letter* A.:[71] *It only has a nice white color, but its composition is wondrous, and cannot be imagined as such, the way it is in fact; its circle is loose,*[72] *and diminishes in thickness from the head to the tail, as is the case with most* Whelks; *for it is only joined together by oblique bands, that run parallel across it, but in a diminishing fashion: the center is hollow, so that one can see through it from the head down to the tail; which I encountered in only one other, though inferior, kind of* Whelk: *Owners of such as these are the* Arch Duke of Tuscany &c. *and Mr.* La Faillie,[73] *Chief Officer of the city of Delft, and*

Fol. 263.
XLIX
I
A
H
D
C
B
E
F
M
G
K
L
J. d later fe.

besides them only one other is known, which is somewhere in England (and came from the Cabinet of the painter Ovens).[74] *Its value can be deduced from the fact that the deceased* Volkaartsz[75] *refused an offer of 500 guilders for the same. The one shown by the letter* B.[76] *cannot be matched; and which we, because of its little white round eyes against the brown bands, have called the* Banded Argus eyes. *Those depicted by* C.,[77] D.[78] *and* E.[79] *are various kinds, all white with chestnut-brown spots, also very uncommon, especially the one by the letter* E. *of which also very few are known. The one depicted by the letter* F.[80] *is called the* wrap-around Whelk, *because it has a brown thread on a gray-white background, parallel from top to bottom, as if art had wrapped it around. And the one shown by the letter* G.[81] *the* grooved Spotted Whelk. *The one by the letter* H.[82] *is the* banded Agate Whelk: *of which there are several kinds; some brown; others with lighter spots. the one shown by the letter* I.[83] *is an unusual kind of* Oyl-Cake Whelk, *but of a grayish white color. The one by the letter* K.[84] *is the* banded Knobble Whelk. *The one by the letter* L.,[85] *is another* branched Swiss pants; *of which one was shown on the previous plate* XXIV. *by the letter* A.: *Others call it the* Spiked Mace, *after a certain implement of war;*[86] *because of its sharp points: Older collectors know it by the name of* pied Branch Whelk. *And the last one by the letter* M.[87] *is called the* French Whelk; *but I do not know the reason why. We could add many more here, which the Author did not mention, but since the book is called the* Ambonese Curiosity Cabinet, *we would do the Author wrong, if we turned it into a general description of Whelks and Shellfish; since this was not the intention of his Honor: And yet we could not neglect to add a few of the most important ones, so they could serve to enlighten the ignorant.*

CHAPTER 38

Division of the Sea Snails and Shellfish from Pliny.

In Book IX, Chapter 33, *Pliny* took it upon himself to divide everything that he included under the common name of *Concha*, and which were known in his day to be in the Mediterranean Sea into 33 variations;[1] I will order those here and show, how the same corresponds to the Whelks and Shells described in this Book, as follows:

1. *Concha plana*, are all the flat and smooth Shells, like some *Pectines*, and flat Oysters.
2. *Concava*, those that are bossed and hollow, like the *Pectunculi*, and *Ostreum echinatum.*
3. *Longa*, oblong, like the *Solenes.*
4. *Lunata*, half-round like a half moon. *Chama Squammata.*
5. *In orbem circumacta*, truly round like a disk, such as a kind of *Chama litterata*, and the Table Dishes.
6. *Dimidio orbe caesa*, as if a piece has been cut off a half moon, like the *Chama striata, quadrata*, and *Pecten virgineus.*
7. *In dorsum elata*, bossed, Tiger's Tongues, and *Pectunculus granosus*, all *Cassides* among the Whelks.
8. *Laevis*, smooth like *Chama laevis, Porcellanae.*
9. *Rugata, Chama aspera.*
10. *Denticulata*, those that have toothed or notched edges, like all the *Chama aspera* kinds.
11. *Striata*, which have ribs that stretch from the head to the edges, *Chama striata* and *Pectines.*
12. *Vertice muricatim intorto*, which end in a tutel, and have thorns, like *Pseudopurpura, Turbo caudatus, & Voluta aculeata.*
13. *Margine in mucronem emisso*, which have edges with branches and horns like *Pentadactyli* and *Heptadactyli.*

14. *Margine foris effuso*, that have edges which are curved outward, like the *Cassides majores.*
15. *Margine intus replicato*, with edges that curve inwards, like all the *Porcellanae.*
16. *Distinctione virgulata*, which have designs of small stripes or marks, like the Letter Whelks and Letter Shells.
17. *Distinctione crinita*, painted with fine stripes like hairs, such as fine Letter Shells and Arabian Whelks.
18. *Dist. crispa*, that have entangled designs, like the Arabian Letter whelks.
19. *Undata canaliculatim*, which have large folds, that go up and down like Sea waves. *Concha squammata.*
20. *Undata pectinatim*, which have shells that are striped like combs; all of the *Pectines*, like the so-called Strawberries.
21. *Undata imbricatim*, with shells that are notched on the folds, in the manner in which rooftiles are laid.
22. *Cancellatim reticulata*, which are twisted square-like, the Tiger's Tongues and Wafer Irons among the *Chamae.*
23. *Expanse in obliquum*, which have a protruding corner, *Pecten virginius* and *Tellina.*
24. *Expensa in rectum*, those which are spread out in front, like the Oysters, and the *Bia kris.*
25. *Densata*, short and squat, like the Knobbles.
26. *Porrecta*, long and stretched out, like the Spindles and Babylonian Towers.
27. *Sinuata*, curved. *Bia kris.*
28. *Brevi nodo ligata*, which are short and round, like the Winkles.
29. *Toto latere connexa*, those that do not open on one side, *Solenes*, *Tellinae*, and *Pinnae.*
30. *Ad plausum aperta*, those that open up when you knock on them.
31. *Ad buccinam recurva*, so crooked, that they can serve as a Trumpet, a Triton's Horn.
32. *Navigans & Velificans, Venera sive Nerita. Nautilus.*
33. *Saliens ex aquis.* & *seipsam Carinans. Pectines.*

CHAPTER 39

How One Should Gather and Clean Shells.

Our Compatriots and Friends in the Fatherland are commonly of the opinion that we find the Whelks and Curiosities on the beach, or haul them from the Sea, just as clean and pure as when we send them to those folks, and that consequently it requires little more than picking them up. It is almost not worth refuting this misconception at length; except that it is necessary to purge us, who live in *India*, from the suspicion that it is impolite or unreasonable of us, when we sometimes cannot fulfill our friends' demands nor answer to their cleanliness: Wherefore I need to point out, how much time and effort it takes, if one wants to put a set of Curiosities together, since I ascribe to myself a reasonable knowledge about such things, though no fame, for I have devoted a great deal of time to it in my idle hours. Hence I declare that, over a period of 28 years,[1] I gathered together a Cabinet of about 360 items, existing only of Whelks and Shellfish, that can be found in the Ambonese region, and that there were many kinds among them, whereof I possessed only one item, and that all of it was sent to the Grand Duke of Tuscany, His Highness *Cosmus* the III, in the year 1682, at the insistence of several friends, to whom I was obliged.[2]

Besides the length of time, that is required for this, I will also indicate all the trouble and tedium one has to endure, in order to clean them, and make them presentable: I will gladly confess, that I have seen them just as clean with other Collectors, though far fewer kinds; and if one does not shirk the pains it will take, and devote oneself to the task at hand, having greater opportunity than I did, he will undoubtedly be able to collect and clean as many as I did, and in much less time as well.

1. One should know first, that all the Whelks, which one finds here and there on the beach, are for the most part dead, broken, or torn: We call such Whelks dead, not because the Animal within has rotted or died, but because they have lost their natural luster and color; for everything that one wants to keep as a Curiosity, should have been taken freshly from the Sea, and with the living Animal still inside.

2. One does not find all kinds of Whelks on every beach, but each one has its particular beach and Island, where they should be looked for, just as was noted for every kind in this Book; but there are some, that do not have a particular beach, but which one will happen to find in the open sea, such as the *Nautilus major* and *minor.*

3. Looking for Whelks takes place mostly at night, with both a new, and with a full moon: It must be low tide, or rising tide, when the Whelks show themselves out of the sand: With dark nights one should do this with torches, so that one can see in front of oneself.

4. The best time is the first two rain months, which on *Ambon* are *May* and *June* into *July;* but if it rains too long, they will hide themselves in the sand again, which they do in the hot months as well; but they appear when there is a full moon.

5. Flat sandy beaches have mostly Shellfish and little *Turbines,*[3] but other Whelks, especially the branched ones, should be looked for among the rocks, but only where the shore is not steep, because one cannot search where it is steep.

6. Searching on rocky beaches is just as much trouble and effort as on the flat, sandy beaches, because there one must be constantly afeared of the large Sea Killer, the *Kaiman,*[4] and if one has to go through some marshy holes, then one can easily step on the sharp Sea Hedgehog or Sea Apples, or on the poisonous *Ican Swangi*[5] fish. Though it is true that the *Kaiman* does not come onto rocky beaches, one can still easily hurt one's feet on the sharp coral stone, and such wounds can easily become malignant from the slivers of stone and the burning sea slime that hangs on the rocks everywhere; and one should also take care of one's hands and feet because of the *Echinus sedosus,*[6] called *Bulu babi* in Malay, because its long darts, that are like fine needles, will stick in one's skin at even the slightest touch, and cause a great deal of pain, as was told in the foregoing Book.

7. Now after one has gathered some, one should put them for two days in a tub of seawater, so that one can divide them at leisure.

8. One should be generally warned that all smooth and shiny Whelks, should not be put in fresh water, rain, nor in the sun, as long as the animal is still inside; because otherwise they will lose their luster, will change color, and become dead, and no skill can redress that; nor should one place them on top of one another, so that the dead blood of one will touch the other, because this will spoil them.

9. On the other hand, all coarse, ribbed, and branched Whelks, that do not have a natural sheen, can endure all this much better, they can even be cooked in seawater, after one has left them just long enough in hot water, for the Animal to die.

10. One should do one's best with all whelks to get as much flesh out as possible, so that the rest can rot out of there more quickly; with some one can do this by cutting it out, one puts them aside and waits until the animal has crawled outside a ways, then one quickly gets a sharp knife behind it and cuts it, the way one does with *Auris gigantum,*[7] *Murex ramosus,*[8]

Buccina Tritonis,[9] *Harpa*,[10] &., while one should put salt, vinegar or some other sharp stuff on other ones, without having it touch the shell too much, because the quicker the animal dies, the less luster the shell is likely to lose.

11. But the *Porcellanae*,[11] *Volutae*,[12] & *Strombi*,[13] all of which have a narrow mouth, cannot be dealt with in said manner; one should put these in a shadowy spot, safe from sun and rain; where the black ants can eat it clean, placing it always with the mouth down, so that the dead blood can run out of it, which cannot be done without considerable trouble and stench.

12. Others spread the Whelks on a *parparre* or grate made from slats, and build a smoky fire under it, so that the Animal will shrivel and die from the smoke, where they will lie for at least 2 or 3 weeks; but it is not advisable to bury them in sand, because this will render many Whelks dull.

13. When the Animal has dried up or rotted away, then one should put them first in fresh water for a night, to wit, the smooth ones, that have a sheen, because those should be rinsed off early the next day, and, if required, rubbed with a piece of coarse linen and fine sand; but the coarse and branched ones should be left several weeks in the rain and sun, so that the sea grit and other mossiness will loosen, and which can then be easily rubbed or brushed off.

14. Sea grit, *Salsilago marina*,[14] we call that chalky and hard substance, both white and red, which attaches itself to the shell, sometimes with small grains, sometimes in big splotches. One can scour it off smooth Whelks with some fine sand or coarse linen, but with the branched ones one has to scrape it off with a penknife, and that which will not come off, one should sprinkle with aqua fortis, and then scrape or scour it again.

15. Everyone should be warned, not to believe the story, that one can clean all Whelks that are defiled with sea grit, by placing them in vinegar, or spoiled rice, because this means certain death for all smooth Whelks.

16. But if one wants to make cups out of the *Nautilus major*, or if one wants to get the Mother-of-Pearl out of the Giant's Ear, then one should put these to soak in vinegar, spoiled rice, or grape leaves, in order to get the outer skin off, as was said before in the Chapter on the *Nautilus*,[15] and one must be careful and remove the Whelks several times, and grind off as much as one can from the corners where their crusts are thickest; whereafter, dried off, one should put them back in the vinegar, taking care that no greasy or sweaty hands get into the vinegar.

17. The Volutae,[16] Crabs, and Pope's Crowns,[17] are covered with a skin, when they first come out of the sea: some have a thick, woolly skin, which can always be scoured off; but others have a thin little skin, that is firmly attached to it, and difficult to remove when it is dry, wherefore it is necessary that one scrapes this skin off first as soon as it comes out of the water; but the little skin, that covers the *Pseudo-purpura*[18] should be left on, because it will not cause any disfigurement.

18. The brushes which one uses, are made from *Gomuto*,[19] being the black hairs of the *Sagueer* tree, and are better for this work than pig's bristles; one should make various brushes, rough and fine ones, thick and thin ones.

19. While one is drying the cleaned Whelks, one has to guard them at night against the *Cumans*[20] or little Crabs, because during the night these thieves steal the Whelks they consider fitting for their bodies, leaving you with their old and broken houses: But these thieves must sometimes pay with their lives, to wit, when they crawl into a house that is branched or has a tail, with which they get stuck in this or that place; if one wants to chase them out of the shell, one should hold a glowing coal or burning match at the back, and heat up their

ass, so that they will stick their head out, which should then be grabbed quickly and pulled out; otherwise if they are Whelks that do not have much luster, one can put them in fresh water for a night, cover them up, until they have choked, because some of them hang on so tightly to their shells, perhaps having grown in there, that they rather let themselves be burned, than to relinquish their houses.

20. *Porcellanae*,[21] *Strombi*,[22] Harps,[23] *Cylindri*,[24] are clean and smooth by nature, and do not need to be cleaned, but those that come from the sea dead, cannot be restored, and should be thrown away.

21. Nor are those Whelks worthy of being cleaned, which are covered with the white, chalky sea grit that has eaten holes in the shell.

22. Whelks that have edges or branches that are broken or torn, can be mended by smoothing them with a coarse file, or by rubbing them against a coarse stone.

23. One should not use too much aqua fortis on smooth Whelks, except for the *Buccina Tritonis*,[25] which one can rub with a weak solution of aqua fortis, as one does here in the Indies, because that will enhance its colors.

24. One should rub the cleaned and dried Whelks with a rough piece of cloth for a long time, but not too hard, until they become warm, because this will make them smooth, and enhance their colors.

25. All Whelks that are like Mother-of-Pearl, and which have a coarse shell on the outside, should be soaked overnight in a lukewarm, but strong solution of lye, which will bring out the colors and spots quite nicely.

26. With the bivalve Shells one does not need to take so much trouble, one puts them out to dry until they gape, and then one should cut the *Tendo* shrewdly into pieces with a sharp knife, which will prevent them from closing their Shells, whereafter one can remove the meat from them at leisure.

27. Nail Shells,[26] Jacob's Shells,[27] and all others which are rough on the outside should, after the meat has been removed, be put in fresh water, or in the rain, until the filth that sticks to them has been soaked off, whereafter one should vigorously brush them: The broken Nails should be filed until round with a file or stone.

28. The Hearts[28] and white Strawberries[29] should be soaked in fresh water for a day and a night, so that one can shake the meat out of them, holding onto the shell in back with one's finger, so that the shell will not break; thereafter one should bleach it in the rain and sun, and, finally, brush them gently.

29. Some Whelks and Bivalves are covered on the outside with short bristles or hairs, like the dried pears,[30] *Auris hirsuta*,[31] *Buccinum pilosum*,[32] and *Tellina saxatalis:*[33] One soaks these with their bristles, if the latter are still attached to them, and are simply placed in the rain, and gently brushed.

30. Some Bivalves with thin shells, especially the *Pinne*,[34] and the Swallows,[35] should not be dried in the sun at all, but in the wind, since the sun will crack their edges.

31. The clean Whelks which one wants to dispatch, should not be packed in cotton or similar warm stuff, because the heat will pale their colors, instead one should use fine, thin shavings from the joiner, as well as Chinese or other soft paper: The thorny items, or those that have fine nails, but no exceptional colors, can be kept in cotton: Those one keeps for oneself in one's house, can be laid bare on top of one another.

32. Some Collectors are in the habit of giving the Curiosities to drink every second or third year, which takes place, if one puts them in seawater for a day and a night, then rinse

them with fresh water and dry them again, because it seems that this livens up the colors, as if they had been in their element again; in any case, if it does not help much, it cannot hurt either.

33. Some Whelks, especially the *Nautilus major*[36] and *Buccina Tritonis*,[37] will by nature begin to sweat when they lie in rooms during the long rainy months; drops of water will appear on them, and one should wipe these off very carefully, so that they will not spoil the other things around them.

END OF BOOK II

THE AMBONESE
CURIOSITY CABINET

The Third Book,
Dealing with

MINERALS,
STONES,

And other rare things.

by

Georg. Everhard. Rumphius

The Third Book
Of the
MINERALS,
STONES,
And other rare things.

CHAPTER 1
How They Falsify Gold in These Countries.

Although in olden days gold was taken out of the Indies by the shipload (but the ships were never filled), we will find that today it is no longer as plentiful, nor will one stumble over clumps of it, the way one did over the stones of *Jerusalem* in *Solomon*'s day; because these days people have been asking for a long time already, where one might find a gold mountain. Today *Sumatra* only takes the lead, followed by *Borneo;* one will find a little on *Timor*, there is some dreaming about *Celebes*, as well as *Nova Guinea*, while *Mindanao* and the Philippine Islands supply a little gold dust:[1] We will not include the Old Indies, *China* or *Japan* within the compass of our Water Indies:[2] But no matter there be little, everyone wants to have some of it in his house, for no family counts itself happy where this House God is not present, and this is the reason why the small amount, that one can find today, is thinned, stretched, beaten, and falsified, so that it will at least resemble gold. I will not bring up all the ways this can be done, but only touch a little upon how it is falsified in these Islands, to serve as a warning to our Countrymen, who might be so fortunate as to have to deal with it. One does not have to fear any *Alchemical* tricks from these dull Natives, nor are they privy to the art of gilding or silvering; but they do have another cunning way with which they are able to deceive the experts in metals: It happens as follows; they beat the gold into thin little plates, and then dress silver plates with it in such a way that the entire piece of work is thought to be of pure gold, one will not know this by touch, or one has to scrape so much of it away that one gets to the silver, something that has not often been suspected: Nor can one make sure by cutting a piece off the sides in order to see what holds fast to it, because after cutting or chopping it, the gold lies in the gash in such a way that it shows yellow all the way through: If you toss it in the fire, and melt it down to a clump, you will get some good silver from it: So the nearest thing to do is to learn the shape of the wrought pieces, of which I have come to know two kinds.

The first and most common kind is, what one is wont to call Ambonese gold Snakes, fashioned from joints, like unto the bones from the spine of a Snake, thin plates linked together, and provided with two Snake heads at either end; one can discover these somewhat from their weight; because they are much lighter than massive gold, nor are they as cold to the touch as is true gold.

The second kind is a similar Snake, but not made from joints, but it is a continuous chain, the wire of which is braided crosswise[3] in the manner of the spine from the fish *Serdijn*[4] or Herring; some of these have been made from pure silver; and are painted yellow on the outside, but one can discover that easily by touch: Both kinds were once often made on *Amboina* by the goldsmiths who live in *Ibamahu*, and can be readily found among the common people; but such Snakes that are made of wire are not or are seldom falsified, and can therefore be taken to be what they are, as is proven by touch. One has to believe that the golden Creese-handles[5] have only a gold plate on the surface, while the rest inside is filled up with some kind of resin, as are the heads of the aforementioned Snakes. If one deals in gold in the South-Eastern Islands, one will come across large dishes and table plates, which seem to be of gold, but which are as thin as spangle gold, and a bad alloy; yet they are better than the massive pieces they will show you, since these are mixed with brass, which gives a pernicious mixture which is not fit to be worked in any way.

CHAPTER 2

Water Test of Gold and Silver.

History will tell you that *Archimedes*, builder for King *Hieronis* in *Syracuse*,[1] was the inventor of this art, but it has not been used very often, partly because one needs large clumps of gold and silver to do this, and also because the water will belly, swell, and make quite a curve outside and above the vat, when one puts the gold in it, namely if one is dealing with small pieces, so I wanted to add my own heureka here, and show how one can test small parcels of gold and silver, keeping in mind, that the pieces of gold and silver one wants to test, must weigh at least 2 ounces, and be melted down to small bars or to some other dense shape: This requires certain preparations,[2] from which one can derive some basic rules, followed by the work itself, to wit.

First Preparation.

Take some of the finest gold that can be had, nothing below 23 carats, fine silver and fine copper, all melted down into bars, flat on one side, then you take some soft wax,[3] cover the aforementioned little bars with it after they have been smeared with oyl, so that they go in and out smoothly, trim the little forms at the top, so that their flat ends are all even, let the wax grow cold and hard, and then remove the bars again: They should have been weighed in advance, for which I used an Apothecary weight of 1 ounce, which contains 8 dragmes or quints or 480 graines: Every dragme contains 2 scruples or 60 graines: Each scruple contains 20 graines; and every graine has the weight of a white pepper corn or of 2 grains of peeled rice; an ordinary pound of silver is 6 dragmes heavier than 16 ounces of Apothecary weight.[4]

Second Preparation.

Fill the aforementioned little forms with water, as far as it will go, pour the same into a cup or little bowl, carefully weigh it, and note each one's weight particularly: To start with gold for example. I have a small bar of fine gold, of more than 23 carats, weighing 3 ounces less 25 graines or almost 71 scruples, or 1415 graines, but since five graines of gold casts out so little water, I take it to be an even 71 scruples: After pouring the water in the aforementioned little box of wax it is found to weigh 90 graines, so according to the rule of three,[5] 71 scruples equals 90 graines of water; what is the equivalent of 24 scruples or an ounce? Comes to 30-½ graines.

First Rule.

Since what follows shows that the water difference from carat to carat of an ounce of gold, amounts to more than a graine of water, I can come up with the first basic rule, that is, that an ounce of fine gold of 24 carats, which one is wont to call Cupel gold,[6] casts out 29 graines of water, hence I could make the following table:

Carat.	Water.	Carat.	Water.
24	29 graines	18	36
23	30	17	37
22	31	16	38
21	32	15	39
20	34	14	40
19	35	13	42

All gold that contains more silver than the 12th carat would warrant is no longer called gold, but gold-enriched silver.

Third Preparation.

A small bar of Silver from Siam weighs 3 ounces 2 quints 2 scruples, or 80 scruples; its wax form will first hold water to the amount of 3 dragmes 5 graines, or 185 graines: So according to the rule, 80 scruples equals 185 graines of water; so what equals 24 scruples? 55-½ graines.

Second Rule.

One ounce of Siam silver casts out 55-½ graines of water.

Third Rule.

Divide the aforementioned water weight of gold of 29 graines by 24 parts, which are carats or, if one is dealing with ounces, scruples; so 1-1/5 part of a graine of water comes the closest to equaling 1 scruple, the same for the silver weight of 56 graines also divided into 24; so one gets close to 2-1/3 graines for 1 scruple, subtract the first from the second, and you have 1-2/5 graines left, which we call the water difference from carat to carat, and since all our gold must have a mixture of a 24th part of a scruple of silver, it follows that if its alloy were to be one carat lower, the water difference I just described will indicate the difference in carats.

Fourth Preparation.

A small bar of Japanese copper, which is considered the purest, weighs 7 ounces 1 dragme and 2 scruples; it casts out 7 dragmes 25 graines of water, or 445 graines; according to the rule, one gets nearly 61 graines for 1 ounce.

Example I.

You will be able to tell whether your gold or silver, which you want to put to this water test, contains either silver or copper, as follows; silvery gold is a pale yellow and soft to the touch; the coppery kind is reddish and hard to the touch. I have a bar of ordinary gold, which I find to be enriched with silver, weighing 12 dragmes and 45 graines, or 745 graines, its wax form holds 1 dragme or 60 graines of water, so I say that the aformentioned 12 dragmes and 45 graines is 38 scruples, and that 38 scruples is equal to 60 graines of water; so what is equal to 24 scruples, or 1 ounce? 38 graines.

Now in order to know the alloy of this bar, you say that 24 scruples or one ounce of fine gold equals 29 graines of water, so what equals 38 scruples? 46 scruples: subtract from this the aforementioned 38 graines of water, and you are left with 8 graines of water difference, which is 8 carats, subtract that from 24 carats, which leaves 16 carats, which is the alloy of the bar, as you will also find in the above table; from which follows, that this bar contains 16 parts gold and 8 parts silver, that is, a third of silver.

Conclusion.

This further proves that this test will be somewhat uncertain, if the bars you want to test, do not weigh at least two ounces; the more you take, the more certain the issue; and one should also be very careful that, when you make the little forms, the wax is applied thoroughly, and that, when one shakes the bars to empty them, you do not press into the form.

Example II.

A gold Ducaton from Frankfurt was suspected of having some silver inside, and a golden sheen on top, because it seemed too light for its size, although it seemed Cupan gold[7] to the touch, and, if one scraped it, it sounded hollow, it weighed an even 9 dragmes, or 21-½ *dobbeltjes;*[8] after applying wax in the aforementioned fashion, and cutting the same down the middle, and then weighing the water that could go inside, it was found to weigh 50 graines; according to the above Rules one should get nearly 33 graines of water for 9 dragmes of fine gold, which, when subtracted from 50 leaves a water difference of 17 graines; which shows, that it was more than 15 carats less than fine gold; but since it tested as 21 carats, it follows that it had to contain a piece of silver, though this will not tell you how much.

CHAPTER 3

How Gold and Silver Will Test with Things Other Than a Touchstone.

Since I have noticed that the color of gold and silver shows up differently with things other than a touchstone,[1] I wanted to note here some of those experiences.

From *Makassar* I received a *Mestica Pinang*, that was an old *Pinang* which had changed into a dark-brown stone, like the one described later on:[2] Both gold and silver retained their color if rubbed on its side; but when rubbed on the flat head, where normally the *matta* or eye would be, fine gold showed its true yellow color; while everything that was mixed showed up as russet, as if it had been mixed with copper; but the silver remained white.

The metals maintain their colors with the *Sangites* or the *Mestica* of *Caju Sanga.*[3]

With *Mestica parrang,*[4] fine gold becomes a light yellow, thereafter, the more mixed the gold the more russet the color: the silver is white, though there might be some red in it.

When you use pieces of *Bia garu,*[5] all gold turns the color of lead, but when one turns it to the light, it will show up yellowish; on the round part of *Bia garu*, silver turns the color of lead, but when moved back and forth, it becomes yellowish.

With *Calbahaar puti,*[6] gold becomes the color of lead, inclined towards the light, fine gold becomes yellow, mixed gold becomes dark yellow, as if there were lead in it.

With white Flintstones the metals maintain their color, but they do not stick out, nor do they adhere.

Smooth Flints or Flintlock stones[7] will not test very well, though the new ones from Bima are better, even if they are transparent.

Bones will not test at all, including the teeth of large fishes.

The Ambonese green and petrified *Amianthus*[8] will test, but the real one will not.

A whet stone made from Makassarese Cofassus wood[9] will show fine gold as yellow and mixed gold as russet, but when held towards one and away from the light it will gradually darken.

The black kernels of the Saguweir tree[10] will test as well as black touchstones; though one will sometimes find one that is partly petrified, and which therefore will not test due to its smoothness.

CHAPTER 4
Suassa. What Kind of Metal It Is.

S*uassa*[1] is a mingling of various metals, similar to what is called *Electrum antiquorum*,[2] and is held in great esteem by all Eastern Nations, even by the Portuguese; the Natives prefer to wear it as rings, in preference to gold. There are two kinds: *Nativum* and *Artificiale*, though both differ from what is described by European Authors. I will therefore set down here the way it is found in the Indies, that is to say in our Water Indies.[3]

Electrum nativum or the natural *Suassa*, which one also calls *Tambagga Suassa*, is only found on the island of *Timor*, in its Eastern province of *Ade*,[4] and if one just looks at it, it seems like red copper, except that it is denser and heavier than ordinary copper; [it is] mixed with a little gold which, when one cleaves it in half, one can see glittering inside, but in such a way that the greatest amount is copper, and could therefore be considered a copper rich in gold. One will find it there in massive clumps, both large and small, of which some are sometimes so large, that eight men are needed to lift just one piece, while I personally was told by *Radja Salomon*,[5] the erstwhile King of *Ade*, that there was a piece in front of his house that he used for a chair, and that it required 100 men to carry it up the Mountain, but the same was taken from him by the Portuguese in 1665, causing a great deal of strife, with the result that in the end he had to flee his Country. The Mountain where it was found, lies more than a mile from the beach, south of *Ade*, is easily known by its shape and is called by us Copper Mountain. The Natives say that two large, white buffaloes kept house on this mountain, and that they were of a very wild nature, ate stones, and that they cast out the aforementioned gold-enriched copper in their excrements, and that one of them was still alive. They said it was completely forbidden to damage that mountain, and were quite satisfied with the small clumps that were washed down by the river that had its origin up there; indeed, they would rather fight than injure that mountain in order to dig mines in it. They once thought this Metal to be valuable, but it has been learned that they merely used it for fish hooks, until they were persuaded otherwise by the Makkassarese and Portuguese.

And to be sure, this metal should not be considered *Electrum*, but a copper enriched with gold, since a fifth of it is not gold, and whatever is fashioned from it grows black just like other copper.

Electrum artificiale is much more beautiful and precious than the aforementioned Natural one, and is therefore used more widely: It does resemble red copper, but with a far clearer sheen, also whiter, with a yellow reflection that, after it has been polished, is very much like a smoldering coal. It is made in several ways, of which the following is the most common; one takes equal parts of fine gold of 23 or at least 22 carats, and fine red copper, like the Japanese kind, and melts it together; if one uses more gold, then the *Suassa* will become a darker yellow and is not as beautiful: Others take equal parts of gold and copper, a quarter of fine silver, and two *dubbeltjes'* weight of fine steel, then they smelt first the gold and copper together, the silver thereafter, and add the steel last; but they must know some other clever tricks, which they keep hidden from us, because when our goldsmiths wanted to counterfeit this mixture, they usually spoiled the entire clump, so that it is best to stick to the first method.

Once this *Suassa* is polished it should always stay clean and shiny, even if salt water or vinegar gets on it, indeed, even aqua fortis and *Spiritus salis armoniaci*,[6] for anything that turns black or green, is not true *Suassa*. As said before, it is commonly used by all the peoples of the East Indies, especially the Malay, who make plain, smooth rings from it to wear on their fingers, also lime boxes for their *Pinang* basins,[7] plates to dress the sheaths of their Creeses as well as the underside of the hilt, and the great Kings also make drinking cups and beakers from it: They ascribe powers to it, which one ought to consider superstition; nonetheless, this is so rooted with these people, that one cannot dissuade them, basing their boast on all kinds of experience. For surely, even if only half of it were true, one could easily prefer it to gold, and it is therefore all the more remarkable, that those powers are not lodged separately in either the gold, or the copper, but derive only from commingling: I have seen people, who had burned their mouths with very tart lime, cured by sticking a *Suassa* ring in their mouth; that is why they like to make lime boxes from it, which renders all lime harmless: Any poison that comes in the vicinity of this metal causes it to blanch, and to assume strange colors, and makes it gnash;[8] superstition will add that the wearer of this metal will remain free of various misfortunes and, if something were to hit him or an illness overwhelm him, this would make the ring burst on his hand. Which is why the Malay and the Makassarese like to wear the same on their Creeses, and other weapons, thinking they will be lucky in war, as long as they do not encounter the Dutch, because they believe that even their Devils cannot last against them.

Its worth, when bought, is at least half the weight of gold, but if one wants to get it from them, one usually has to pay more: Many of our notables[9] are wont to dress the Rottangs,[10] which they hold in their hands, with *Suassa*, because it is not as costly as pure gold, has a finer sheen, never turns black, and does not defile one's hands, as silver does.

Bad *Suassa* comes about when ones takes young gold of 17 or 18 carats, that is gold that has at least ¼ part silver in it; and though this, mixed with red copper, will give a *Suassa* that becomes very white, it will not shine after polishing, while soon after one touches it, it will turn blackish or a dark brown, and stain the fingers black, though every time one rubs it it will turn white again: This was discovered by chance through the unfaithfulness of the goldsmiths, who used young instead of old gold; but these black rings still have their use, if one is in mourning.

The *Electrum Antiquorum* which the Ancients describe, to wit, a mixture of 4 parts gold with 1 part silver, is no longer known by that name, because the Malay call such a mixture *Maas Mouda*,[11] that is, young gold, because of its young (that is pale) color.

Black *Suassa* comes from *Tonquin*[12] and *Japon*, being a kind of red copper, which is always black on the outside, but when it is rubbed, it turns the color of copper, while its mixtures are still unknown to me; it is often used for the hilts of swords, for the metalwork around Rottings, and for the buttons of Doublets that are inlaid or gilded with little gold flowers, since gold easily wears off from long use: some say that it is a mixture of red copper and iron; others that it is the heart of red copper: The Chinese call such copper *Lyukung tang*,[13] that is, Thunder copper, and say that this is copper's best and oldest kernel, which turns black on the outside, but not that it was struck by thunder.

Aurichalcum sive Orichalcum Antiquorum[14] does not seem to have been our yellow copper or brass, which is made by people from common copper that has been colored with *Cadmia* or *Chalmey;* but it is a copper rich in gold, either obtained naturally from the Mountains, just as we said about the Timor kind before, or fashioned by art, just like our *Suassa* from the East Indies, because the Persian Kings seem to have made their costly drinking cups from this red *Suassa*, and they called them *Batiahas*, and these were shaped like a pumpkin cut in half, wherefrom we have retained in Malay the word *Bateca*,[15] which now means a Watermelon. This also seems to have been what the two precious vessels were made of in Ezra Cap. 8, and which the Kings of Persia honored in the Temple in Jerusalem: The text of Ezra in the VIIIth Chapter, verse 27, is as follows: *And two vessels of fine copper, precious as gold.*[16] In Hebrew this is נחשתמצהב, which is better translated as shiny copper, which is the way the Greeks translated it, χαλκῷ στίλβοντος,[17] this word στίλβοντος derives from στίλβω, and indicates an attribute that is part of true *Electrum*, and not yellow, as some Translators would have it.

חשמכ Chasmal in Ezekiel in the first Chapter, verse 27,[18] was also just like *Suassa*, which the Greeks translate as ἤλεκτρον *Electrum*,[19] as was proven succinctly by the learned *Sam. Bochardus*[20] in his addition to his *Hiroizoicum*, wherein he derives the word חשמכ Chasmal from two Chaldean words נחש Nechasch and מככ Melal, raw gold, the way it comes out of the mines, and these correspond entirely to the words *Aurichalcum*[21] and *Chalcochruson:*[22] and he thinks the same, and with good reason, of the word χαλκολίβανον, *Chalcolibanon* (whereof one reads in the Revelation of Joh. I:15 and II:18),[23] which is the same thing as *Chasmal* or *Electrum metallicum*, that is, beautifully polished *Suassa*, to which we assigned before the color of a smoldering coal or molten copper, when it is aglow: We therefore conclude, that the true *Electrum metallicum* is a mixture of pure gold and copper, which, according to *Pliny* and the Ancients has the attribute of betraying with its rainbow colors whatever poison was ingested,[24] and which a mixture of just gold and silver will not do at all.

The Javanese make their *Suassa* as follows: They take fine gold that has the weight of a real, a Rixdollar of silver, 3 sesjes *Brongo sari*[25] or fine yellow copper, red copper of each a quarter, lead of at least a *dubbeltje*, smelt all this together, though they do not add the lead until everything is molten, then they pour it into small bars and hit them softly with a hammer, but as soon as they see that it cracks, they put it back into the fire again, and then let it cool off by itself, which they have to do often, until it has become as thin and wide as they want; they can get it as thin as the plate with which they dress their creese sheaths, and it can be soldered as well: This *Suassa* is more valuable to the Javanese than mere gold, and their Emperor, the *Sussuhunan*,[26] wears it more often than gold: They also make lime boxes from it, which takes the tartness away, which the *Suassa* rings do as well, if one holds them in one's mouth.

CHAPTER 5

How Some Metals Are Affected by Spiritus Salis.

Spiritus salis communis[1] has the following effect on metals: With fine gold it only cleans it: It does the same thing with inferior gold.

It blackens on silver, veering towards blue a little, like ink.

True *Suassa* remains clear, and does not change at all, which is a sure test of its silver.

Bad or blackish *Suassa*, when it has first been polished clean, will soon be blackened by the *Spiritus*.

It becomes a bluish black on *Ceraunia metallica* or Thunder Shovel,[2] and dries like salt, while it turns a true blue in the dents of the same shovel.

A bar of another kind of Thunder Shovel does not turn blue, but becomes rather a musk gray, with some purple in it.

Electrum nativum from *Timor* turns red on flat copper, but it turns a dark blue in the dents on the side: Japanese copper first turns dark brown, but dries with a yellowish color.

Spiritus salis Armoniaci does the following things: It purifies all sorts of gold, but it will in time dry up blue on young gold.

Silver turns a dirty blue; true *Suassa* does not change at all, which is also a true test.

Bad *Suassa* turns a black mixed with blue: Thunder Shovels become a dirty yellow, because there is gold in them.

Timorese copper becomes reddish, with some green in it on the sides. Japanese copper turns a dark purple, with some yellow in it.

More precise tests with *Spiritus salis communis:* It turns black on the silver of a rixdollar,[3] and on the third day the spot is reddish, while it dries up blue on the sides: With Japanese copper the spot remains a beautiful, shiny red, and a Spanish green on the sides on the third day: It stays black and dries quickly on Tonkinese black *Suassa*, which is used to make buttons.

CHAPTER 6

Rare Iron in the Indies.

Our Water Indies does not have much iron, and this is not so much because Nature fails to produce it, but rather because of the ignorance of the Natives, who do not know how to extract it from various iron stones.

Crimata,[1] a small Island, west of *Borneo*, has many Blacksmiths who do know how to extract it from the stones: That is where Crimatese axes come from, which are traded all over our East.

There is also a great deal of iron on *Celebes*' East coast, in the district of *Tambocco*,[2] and this is used for making the Tamboccan swords, known throughout these Islands, though of a bad alloy.

A better variety comes from the neighboring province towards the West, around the lake called *Tommadano*,[3] and these swords are far better and worth more than six from Tambocco; because they damascene[4] repeatedly, they know how to work the iron in such a way, that it almost becomes steel, the water from the aforementioned lake helping a great deal: One can recognize these swords by the fact that lengthwise, on their backs, they have many curved veins, this being a sign that they have been damascened often. I will not expand on all the changes iron is subject to in the East Indies, to wit, that this durable metal, which constrains the entire world, is so corruptible here because of the rust, which plagues it constantly, so that, when it stays in the open air, it will waste away in a few short years, and this is caused by the warm, moist, and penetrating air of this country, but I deem that sufficiently known: I will only comment on a few rare kinds, whereof one tells fabulous as well as likely things, the way the Natives owned up to them.

Ghiry[5] is a Mountain and open City[6] on *Java*'s East coast, behind the town called *Grisek*,[7] and the *Penimbaan*[8] lives there, like a Patriarch or a Priestly King who, despite his hypocrisy and false miracles, is esteemed as a holy person by all the People, even by the great Emperor or *Susuhunan:* Among other deceits, this false holyman gives to strangers, and to all who come to visit him, certain iron rings and bracelets, of which the rings are solid and the bracelets hollow, filled with a holy soil that was dug up in the place where he lives; some rings are thin and completely hollow inside, so that they can float on water, which the ignorant consider a miracle: These will never rust, and they are good against bites and stings of venomous animals, equally for lime that burns one's mouth, as was related before.

Upon closer examination, this all proves to be superstitions and Moorish[9] deceits, because that Priest makes them from rusty nails which he pulls from his Temple, saying that this iron is the so-called *Bessi keling*,[10] because it was brought to him from the coast of *Coro-*

mandel, and it will only acquire these powers after his persuasions; it is true that they stay clean as long as one wears them on one's fingers, but if they are put aside, they will rust like any other iron; this superstitious deceit is not known to the common man, so that our Chinese let their women wear them; indeed, even the simplest Christians consider them valuable, because they believe that Nature's powers are in that iron. But they will not so easily let their women wear them, reserving them for the men, and they wear them on their forefinger or thumb, wherewith they grasp their creeses or other weapons, so they will be lucky in war. I trust that they use the same against their Countrymen, because they do not work against the Dutch, as that Moorish Priest found out to his detriment when, in the year 1680, on the 25th of April, he and his kindred of 50 armed men fell upon our people, who, as charged by *Susuhunan Amancurat*, were to storm that holy mountain, because the *Tenimbahan* was secretly scheming with the rebel *Trunajaja*, who did not want to pay him any *Homage:* The Priest and his men did boldly fall upon our people, killing a Dutch captain and 15 Soldiers, who, inflamed by liquor, did approach him somewhat carelessly, but he was presently hit in the knee by our muskets, so that he was carried back up his mountain again, and was stabbed there in his bed by a *Madurese*, as were his two oldest sons the next day by *Amancurat* himself, on the spot where they had slain our Dutchmen; and so that holy family was completely exterminated to the wonderment of entire *Java*, since *Amancurat*'s father had attempted to storm this mountain at Ghiry many times before, but always in vain.[11]

The aforementioned Priest was also wont to give strangers a certain white soil, dug from that same holy place, resembling nothing so much as dirty Crystal or Alabaster, and to which he also ascribed the aforementioned powers, and moreover, that if it were rubbed with water and swallowed by nursing women, it would increase their milk: Concerning another rare iron, see below, Chapter XXXII on *Dentritis metallica*.

The Japanese prepare iron as follows for making their cutlasses; forged into flat bars, they bury it in a marshy place for several years, until it is completely rusted, whereafter they forge it again, and put it back in the swamp for another 8 to 10 years, until all the bad iron has been wasted by the rust; what remains is pure steel, and from this they make their cutlasses: our Ambonese blacksmiths also know how to make the finest *Parrangs*[12] from old and completely rusty hoops; for the sword edge they use pieces of broken Chinese pans which are called *Tatsjos*,[13] instead of steel: the Singalese[14] also take the rustiest hoops from our people, and forge them into costly barrels, layering the same back and forth and turning it until it is a round bar, which they then drill out: The pocket pistol that belonged to Mr. *Rijcklof van Goens*[15] had such a barrel, 17 inches long.

CHAPTER 7
About the Metal Gans.

In *Pegu* and *Siam* there is a common metal, called *Gans*,[1] which appears to be a commingling of copper and tin; it is red like pale copper and is used for money, but it is forbidden to be exported: They smelt it, and make pots from it; perhaps the large piece, which one can see today in *Baly*, is made from this metal, it has the shape of a massive wheel which has a piece of the axle still in it, turned bluish or black on the outside, and which the Natives say has fallen out of the sky. Others give the following account: Near the city *Pedijng*[2] they show the spot, where this uncommonly large piece of metal lies in the same place where it fell, having two large *Martavans*[3] on either side: The wheel is about 4 feet in diameter, and the axle is about the same length, all of one piece, and now turned bluish. The people from *Baly* truly believe that this is a wheel of the Moon's wagon, which once shone so brightly, that it illuminated the night, but a certain Villain pissed against it one time, because the light bothered him, since it prevented his nightly thefts, and so it has been rusty and dark from that time on: But the King of *Baly* never had the heart to remove the piece from its place, or to chop a part off it; but lets the same rest there as remembrance; if it is true, that man did not fashion it, just as one cannot give a reason whereto such a shapeless lump could possible serve, then it is probable, that it was produced by thunder and tossed down there; we will say more about this later on, that thunder stones of various metals have fallen down, and that thunder can also knock off large stones, which can be seen in the City of *Grave*,[4] where they show a very large stone in the Chancel of a certain church, which was thrown there by thunder.

CHAPTER 8

Ceraunia. Thunder Stone. Gighi Gontur.

I believe that one will easily find as many Thunder stones,[1] but many more different ones, in the Indies as in *Europe;* I will note such as I have found and examined in the Indies, dividing them into two main groups, to wit, 1. *Ceraunium lapidem*[2] or the real Thunder stone, and 2. *Cerauniam metallicam*, to be dealt with in the next Chapter.

True Thunder stones have various shapes and colors, but as I have noticed, and came to understand from other eyewitnesses as well; all are wont to have the shape of an Instrument that can strike or wound, such as a hammer, axe, chisel, arrow, gouge, &c. In *Europe* they sometimes have a hole, but I have not yet seen any like that in the Indies: they usually are black in the Indies, like ordinary touchstones, but smoother and shinier, some are steel green, others are a russet mixed with green, a few are white and partially transparent like dark crystals; and they are as hard as any stone might be, and can be used to test all metals.

The name in Latin is *Ceraunius lapis:* In Pliny, *Ceraunia* and *Boetylus:*[3] in Malay, *Gighi gontur*[4] and *Batu gontur:* in Chinese, *Luykhy*,[5] that is, Thunder tooth, because the Chinese and Indians[6] believe that the Thunder has a large head like a Bull, and that Thunder stones are its teeth, but the subsequent metallic ones are supposed to be its Jaw teeth,[7] which he spews forth when he is angry, since they suppose that thundering is nothing but the roar of that Bull: Some are so bold as to say that they saw with their own eyes the shape of such a Bull in a Thunder cloud, when it opened up; but others compare it to a Baker's oven: I came across the following shapes of Thunder stones, and had some drawn, as will be shown here as well.

1. Is a Thunder stone in the shape of a flat gouge, 4 inches long, 2 fingers wide, but somewhat narrower at the tip, where it is square and as if chopped off, with blunt edges on the sides, thickest in the center, marked here and there as if splinters had broken off, which could have happened when it drove into the wood with great force, while one could also conjecture from these same notches that the stone was fashioned from many slivers lying on top of one another, while curved stripes ran all around the body, as if the stone had been joined there; one side had yet another vein, which seemed to be the color of lead, and which tested as iron, when it could be touched with a touchstone; and vinegar turns russet, when it is rubbed on this vein, from which one could conclude, that there has to be some material like iron in this stone, even more so because it makes a hollow sound, when one hits its sides; its color was usually a dark green, almost like green cheese, and not transparent at

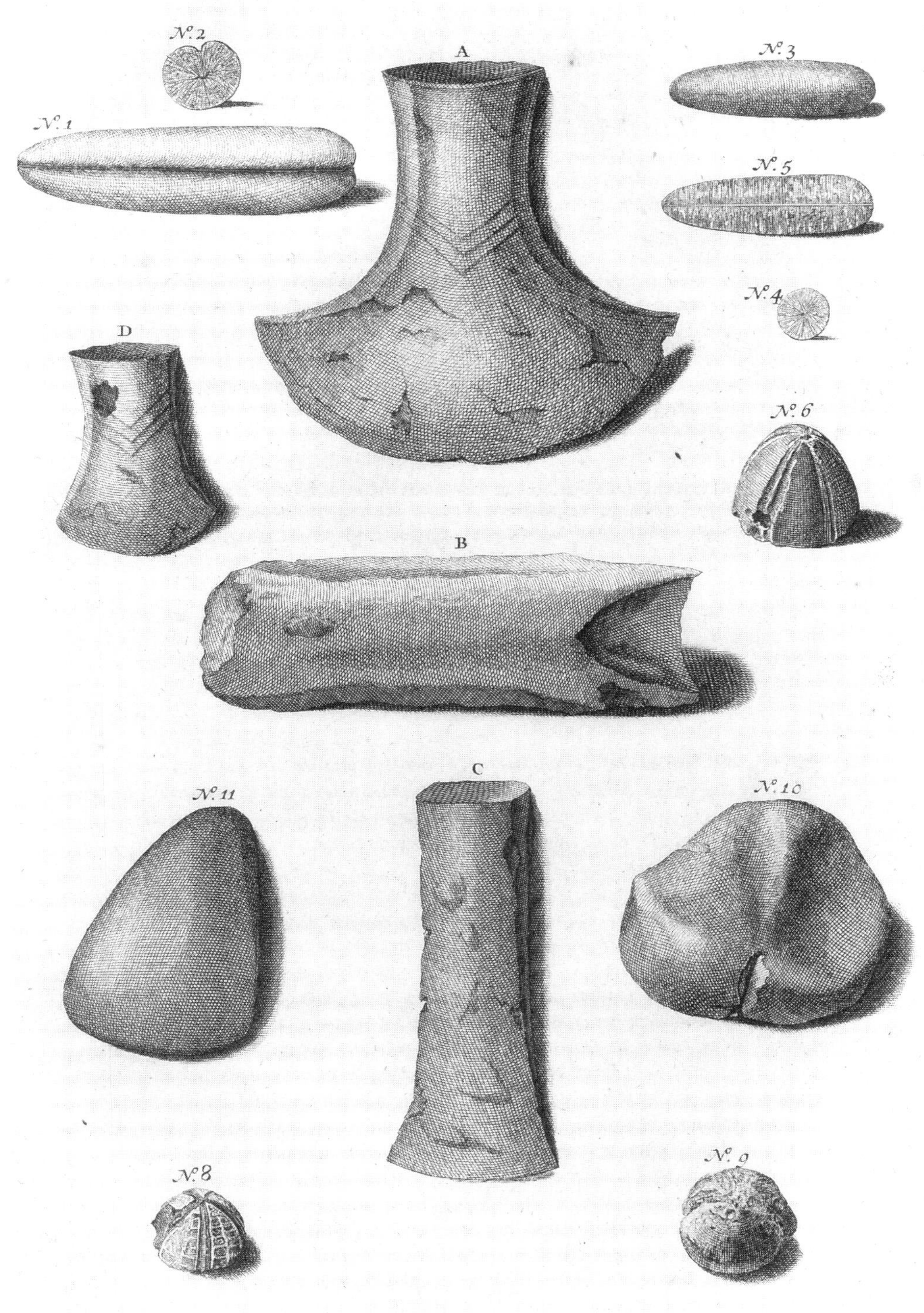

N°. 2
A
N°. 3
N°. 1
N°. 5
N°. 4
D
N°. 6
B
C
N°. 11
N°. 10
N°. 9
N°. 8

all, and so hard, that even if one hits it with a steel[8] one barely gets a spark. All metals that are rubbed against it, test as well as the best touchstone, and if one whets iron on it, it does not diminish the stone at all, as other whetstones will, but the iron becomes sharp; wherefore one can sharpen all kinds of weapons on it; in terms of weight, it is not as heavy as an ordinary flintstone, nor as cold to the touch: A piece of linen that had been sewn around it in such a way that it was smooth all over, remained undamaged after it was put on burning coals for the count of one hundred, even though the stone became very hot, and the linen finally began to smell a little burned: The Natives consider this the sure test of a Thunder stone; but I experienced that all hard and smooth stones will do this. Thunder had driven it into a thick ironwood tree, that had been cloven from top to bottom, and in its heart this stone was stuck, on the Island of *Gape* near little *Bangay*[9] in the year 1677 in January, which caused me to write the following Anniversary couplet.[10]

> When *Gape*'s lord bowed to *Kabonna*'s might,
> Thunder drove me into this ebony heart without a fight.

Because at the same time that the real King of *Gape* had to flee from this *Cabonda* [who was] from *Ternate*, who drove him out of power, said King wanted to fell this tree, not knowing it had been torn by thunder, and use the wood to make *Nadjos*[11] for his Corre Corres, and thus it happened, that this stone was found in the presence of some Dutch citizens, who only got it from the King with great trouble and after much pleading, because he declared, that he would rather lose a slave than this stone, as will be told below.

2. This one is shaped like a crooked little finger, somewhat wider at the back, scarcely 2 inches long, completely smooth and as black as a touchstone, while it also tests all metals, and when it was sewn into a piece of linen, it withstood the same test as the former one.

3. Was found on *Buru* around *Tomahau*, struck by thunder into a *Pangi* tree,[12] shaped like an old, peeled *Pinang*,[13] to wit like a blunt kail,[14] with a dent on the bottom, also very black and smooth, but in the center of that dent, which on the *Pinang* is called the eye, it sounded like Bell metal, which makes one surmise that it contained copper.

An *Imam* or Priest from *Tomahau*, who had found this stone and had worn it in a belt around his body, assured me, that he was once hit by a musketball during the Madjerasensian war,[15] precisely on the spot where he wore the stone, and he fell, but neither he nor the cloth that was wrapped around the stone, was harmed in any way: He also said that if he held it in his hand, the stone sweated so much, that his entire hand became wet.

4. From someone else's story: Another thunder stone found on *Buru*, was driven like a small wedge through a tree into the ground; this one was stone on one side, and iron on the other.

5. A similar stone, seen on *Amboina*, was like a common egg, perhaps a bit more oval, also black and smooth, and could be used for testing.

6. I have come to understand from credible eyewitnesses; that one can see an unusually large thunder stone in the city of *Grave*,[16] to wit, in the Chancel of a local church, where it was bricked in at the spot in the church where thunder had thrown it, to wit on the floor, with the thickest part in the ground; it has the shape of a Sugar loaf, but it is at least a foot and a half thick at its biggest part, has a dark color, mixed with brown and green: when it is rainy weather, this one sweats many drops, which cling to it.

7. Another thunder stone struck into a knee of a ship near Ceylon; it too had the shape of a wedge or small chisel, and had the length of one's middle finger; it was white and well-nigh transparent like dark crystal or like clear flints, but also very hard.

8. In *Goslar*[17] in Saxony, there was a thunder stone that was entirely like a wedge, the length of one's hand, somewhat narrower in back, its color both white and gray, mixed together like alabaster, but very hard: its owner had set half of it in silver, and hung it from a rope, and people noticed that it always moved when there was a thunderstorm; but one time, when the master of the house was not home, and there was a heavy thunderstorm, it smashed (so people guessed) so hard against the wall, wherefrom it was suspended, that it broke off as far as where the silver commenced; One might conjecture that it broke at that spot, because it was loosely joined there, or that it happened because of the Antipathy which this Martial stone might have for silver.

9. The thunder stones from Tambocco are mostly as black as touchstones, but smoother; always shaped like an instrument for striking.

10. Another thunder stone had the shape of an ordinary cook's axe, though at the cutting edge it was also somewhat curved like a flat gouge, 7 inches long, 3 inches wide on the cutting edge, but 2 at the upper end, so that it was also square, like the one under No. 1., also with rounded edges on the sides, bellied in the center with a few notches, as if slivers had broken off: The color cannot really be described, but the closest thing one can compare it to is salted meat, which is a little green and blue on the sides: The entire body had dark-green stipples which made it seem greenish: It had many curved veins that ran sideways, and of which the largest ones went all around, others were short and broken off, of a whiter color than the other substance, furthermore hard, smooth and shiny like marble, with something of a reflection; it was found after the thunderstorm had passed; its owner from *Makkassar* only sold it to me for 10 Rixdollars for lack of money, because he otherwise would not have parted with it for 20.

11. In his Letters, part one, Letter 73, page 205, *Zeilerus*[18] tells us that a large Whale was stranded in *Pomerania*, and that they found a thunder stone in its body.

12. In *Devonshire*, England, in the Year 1622, an enormous thunder stone fell into *Robbert Pier*'s field, it was 3-½ feet long, 2-½ wide, and the thickness of 2 and a half feet; it resembled a pebble in terms of hardness and color, it was smashed into many pieces, and everybody took some so they could show it as something miraculous: It came down with such force, that it struck at least an ell deep into the ground; as soon as the formidable stone was laid there, the thunder ceased.

I only cited these Examples among many, because in my opinion they are sufficient proof that the thunder stone is not engendered by the earth, wherein it is found, as if lightning created it that very instant when it strikes the ground, because it makes the sand melt as if it were glass, but that the opposite is more likely, and that is, that it is engendered in the clouds from substances that are metallic and earthy, which are drawn up from the earth along with vapors, and then are concentrated by lightning's great force and then fashioned into a stone: The reason that it always has the shape of a striking Instrument, must be ascribed to the hidden quality of a Martial spirit: Because *secretiori Philosophia & Magia naturali*[19] teaches us, that all Natural bodies are generated, after particular concepts have incorporated the Astral flow, which is the basis of natural signatures; because that it is not begotten from sand in the ground is sufficiently proven by such stones, that are found in unsuitable material (for instance growing in dry wood, whale flesh, &c.).

About the use of Thunder stones I will only relate what I have learned from the Natives, not disputing whether any of it is superstition: First of all, all thunder stones can be used as Touchstones, because metal clings better to them than to the common ones. The Moors whet their creeses on it, until they have the right cutting edge, believing that in this way they can pierce any kind of hardened skin: The Indians use it the most for *Cabbal*,[20] which is also why they esteem it so highly, for that is when they harden [their skins] and make themselves invulnerable, and to be bold and brave in war; which is quite necessary for those who want to storm a stronghold, being the sort of thing the Natives desire the most. A certain *Sengadje* from *Waytina*, who ruled *Xula Mangoli*,[21] found a remarkable thunder stone in a place that had recently been struck by thunder, and I never heard of or saw any other like it; it looked like a thick bracelet, through which you could stick 3 fingers, the ring itself was about a finger thick, round, smooth and black as jet, as if it were black glass; he would not surrender it for money or persuasion, but always wore it on his navel when he went to war, and intimated that he had never been wounded; now if this is not due to any Astral faculty, and everything is held to be a superstition, though I do not say this, then these tricks serve at least to strengthen the imagination of its wearer, making him bolder, so he will attack with more courage; nor should it be unbeknownst to a Christian, that he cannot attribute his triumphs to any creature, not to mention a lifeless stone, but only to a steadfast trust in the Lord; but I should also mention, that this custom is quite ancient, since *Pliny Lib. 37. Cap. 9.* quotes *Sotacus* that there are two kinds of thunder stones, a black and a red one, which resembles an axe; the black ones were supposedly used to conquer cities, and defeat fleets, and they were called *Baetylus;*[22] but the oblong ones are called *Ceraunia;* he presently adds: That there are other *Ceraunia*, that are very rare and seldom found, and which are highly esteemed by the Parthian *Magi*, and which are only found in places that were struck by thunder;[23] there is no doubt that the Parthian *Ceraunia* is the same as the one described by *Sotacus*.

When placed in water and then drunk down, and also when one is washed with this, it will extinguish raging fevers, especially for those who walk around naked in cold rainy weather or in a thunderstorm, and catch themselves a fever that way.

Some Cattle herders and Farmers in the Fatherland place the thunder stones in water, and make the animals drink from it when they are subject to some illness: And so I also saw how in *Europe*, Women who had inflamed and festering breasts, rubbed the same with these stones, in order to extinguish the inflammation.

13. I saw a piece of a Thunder stone, black and smooth, almost like a Touchstone, around *Baguala*, struck by Thunder against a hard rock on land, which cracked it, and close to it they found this piece; it is not certain whether the other half went into the stone or not.

14. A *Patty*[24] from *Tamilau* kept a Thunder stone, that was square and smooth, and like a black Touchstone at the largest end, while at the other it was like red copper, but he refused to show it to any of our people.

15. A certain Native from *Nau* on *Hitu*'s coast, found a similarly mixed Thunder stone in a rock by the little Sugar loaf on *Larique*'s beach,[25] that was struck by Thunder before his eyes; the *Patty* from *Tamilau* found one the same way, this was round, the size of a Pederero ball,[26] one half like a black Touchstone, the other half like red copper; after it was put in water, and then used to wash inflamed eyes, it cured the same: But when I wanted to borrow the same in the year 1670, they kept it hidden, pretending they had lost it;[27] they usually place the Thunder stones in water, and drink from it, while they also wash themselves with it, in order to extinguish any of the heat that burns inside the body; this healing is much more agreeable than with the subsequent copper Shovels: No one should consider it strange, that a stone, engendered by the hottest of Elements, is capable of extinguishing another fire, because the lesser must always give way to the superior, as we can see when the sun can extinguish the light of a candle, or that lime, smeared on parts that were burned,

draws out the heat: In our Herbal one will find examples, of small nettles drawing out the heat of other leaves.[28]

In the year 1690, thunder smashed the top-mast of a ship that was sailing around the Island of *St. Paulo* in the Pacific,[29] and brought it down; and since they still smelled the stench of sulphur after three days, they searched the ship, until they found the Thunder stone behind a crate, because it still stank, but the smell gradually disappeared: It was about 4 inches long, round and blunt in back, pointed in front, a bluish gray. The skipper took it with him to Holland.

The Author provides us with a tale of Thunder stones, *about which many have doubted that they existed; but these recorded experiences will remove many objections, the more so, if one considers the reasons given here, which, as I feel, have always existed and which concur with the Author's opinion, concluding*, that when fine earthly and metallic particles, which are mixed with water particles by means of the rarefaction of air and therefore carried aloft, are met by the great force of irresistible lightning, this causes them to melt in its rapid passage, and constrains them to a clump according to the size of the materials, which are contained within the same fire, and through the force of which they are pounded into a hard material, obtaining such shapes and characteristics, as is permitted by the constraint of the particles and the mixture of materials: *Thus far I am of one mind with the Author; but another way seems just as likely to me, because I have seen many which through smelting appeared to me to have uneven sides, were rougher, and seemed more earthy, which I imagine happened as follows:* To wit, that when lightning, in air that is purified of earthly matter, meets wood, stone or metallic particles on earth, it causes them to melt and, because of their isolation, causes them to be constrained to rougher and more uneven clumps. *This notion might have taken place as well, although the Author rejects the same; wherefore we believe, that the regular shapes are engendered in the air, and the irregular clumps in the ground:* Mr. Rumphius *provided us with several pictures of those he described, but not for all of them; we will, therefore, as we did in the previous chapters, show some of the ones we own, to make up for the difference. The Author's first kind is depicted on plate* L. *by the letter* A. *and the* 2d *kind by the letter* B. *We add one, shown by* No. 1. *that is about a finger long and thick, becoming somewhat thinner toward the end; it has a gray color, and a furrow or notch that runs down the center for the length of it, wrinkled on both sides and somewhat pleated. Its material is very hard, but breaking one in two, I found it to be brittle and glassy, with rays that shot out from a central point, as is shown here by* No. 2. No. 3. *is another, smaller kind, of a glowing, transparent brown, but much firmer and harder than the foregoing: We broke one of them across, and found that all its shiny rays ran outward from a white center, as can be seen in* No. 4., *and when I clove another lengthwise, I discovered that the same also ran outwards from the aforementioned center, as can be seen by* No. 5. *One can easily conclude from this, that they were formed by a sudden melting and congealing. The one shown by* No. 6. *is of a different shape, namely flat on the bottom, while towards the top it looks like a Sugar loaf, but with a thicker and blunter shape, with five ribs that run separately between two furrows or seams, starting from the top, and ending in the center of the lower flat part; it is shown by* No. 7., *another kind of this is shown by* No. 8. *which is somewhat flatter than the foregoing, and appears to have a net, whiter than the background, between its ribs and all around: It has the color and hardness of a Fire stone. The one shown by* No. 9. *is yet another kind, somewhat flatter and not as regular as the foregoing, also like a Fire stone. I have other kinds of this, but since they differ little, we will leave it at this, and will show just one more of this kind, which we believe was formed by Lightning in the earth; one can see it by* No. 10. *It is uneven, bubbly and flattened, and, since it was still weak, it burst in two places, one of which is clearly seen here; It too is as hard as fire stone, but it is covered with a shiny crust the color of Chestnuts: We could give more examples, but since this plate is also supposed to show the so-called Thunder shovels, we will end it herewith.*

CHAPTER 9

Ceraunia Metallica. Thunder Shovel.

As far as I am concerned it is now sufficiently known that stones are engendered by the Clouds, which is accepted by most learned as well as common Folk; but that metallic bodies are created by the Thunder clouds is, as far as I know, still little known in *Europe*, and will probably be difficult to accept there, especially when one hears about and sees the wondrous shape of this kind of thunder stones, which resemble our domestic Instruments so naturally. I must confess, that at first I could barely be persuaded when they showed them to me, even more so when our Work masters undertook to copy such thunder stones, but they could not achieve it, however, yet I assure the Reader that I did not act carelessly or gullibly, but first carefully questioned such eyewitnesses, who seemed creditable to me: Nor do I perceive any difficulty with the notion that vapors draw metallic substances from the mountains into the clouds, just like earthy and stony substances, and then are smelted to stone by lightning, particularly when one considers, that the bloody and pyed rain, which one sees sometimes, can have no other origin than with such earthy vapors: But the big problem lies in the question how this kind of thunder stones, which I call Thunder Shovels because of their shape, acquired the amazing shape of our domestic Instruments; but I confess that I would rather be astounded by the unfathomable powers of nature, than to lapse into some kind of error because of too punctilious a scrutiny; the closest to what I hold to be true can be found in the *Philosophia secreta Veterum sive Magia naturali*,[1] which teaches, that all of mankind's arts and sciences, and the tools that go with them, and which serve Man in this temporal life, were originally an Astral influence, depending on whether it was favorable or unfavorable; wherefrom one may conclude, that the Astral smith who informed Mankind how to make hammers, chisels, and other tools for the smithy, could also make such tools in the Thunder fire that he commands, without the aid of human hands, just as one can see that many plants have the shape of household goods or weapons of war; so that every star shapes the things of this earth after the sketch which God ascribed to it at the first Creation: I will now present to the Reader some of these metallic Thunder Stones one after the other, which were shown to me or about which I was informed, leaving everyone to his own conclusions, while recommending a closer examination. I was presented with a Thunder Shovel, that very nicely resembled a small shovel, that was roundishly square, with a piece of a handle in it, it was 5 fingers wide at the cutting edge, 3 fingers high at the edges, and the handle was another two fingers long, hollow on the inside; at first glance it seemed a dirty black like rusty copper, and also sounded like copper that one takes hold of with greasy and sweaty palms, but once it was scoured it acquired the color of ordinary red copper, but somewhat paler and veering towards yellow, with some black spots or dents here and there, which could not be cleaned, and which appeared to be of another substance: It was fairly sharp near the cutting edge, and harder than common copper, while 3 or 4 bulging veins ran on either side of the handle towards the Shovel's sides; on one side quite uniform, as if made by human hands; as said, the handle was hollow, but that hollowness ran somewhat crookedly on one side into the Shovel, nor was the handle round, but pressed down on either side with a blunt edge, while it also was irregular at the top of the handle hole.

In the year 1679, the King of *Tambocco*[2] sent several ambassadors to *Ambon* in order to settle some affairs, and they brought this Shovel with them, and I managed to get it from them after a lot of coaxing and a generous payment; and they swore on their manhood, that this Shovel had been found 3 years earlier in the neighborhood of *Tomadana*,[3] a Village on Celebes' East coast, where one will find a large Lake and a large number of iron huts, so that nearly the entire Village consists of blacksmiths; now on a certain night in the same year of 1676, there was a severe Thunderstorm, and the *Radja* of that place went to see the next morning where the thunder had struck, and discovered some trees that had been smashed to pieces, and near it a hole filled with muddy water, which one had not seen there before, and wherein he rummaged around with his feet until he found this Shovel; Two years later those from *Tambocco* took it away from him, when they waged war on him, and since I was well acquainted with the Radja of *Tambocco*, and had bothered him for years for some of the Thunder stones, he was finally moved to part with one: When the aforementioned Thunder shovel arrived here, and was shown to our Work masters, they could only accept it as the work of human hands, convinced mostly by the four furrows or veins which were mentioned before, because they ran so uniformly, plus the hollowness of the handle: They also examined it more closely, and found it to be of an uneven substance, to wit, harder on one side than on the other, and in the dents as coarse and black as rusty iron: They also examined the sides of the handle, which was not soldered, but could not find any indication thereof, hence concluded that it must have been cast, though this too did not seem possible because the hollowness ran so crookedly: The men who brought it from *Tambocco* laughed at them, and asked why anyone in their country would bring himself to cast or forge such a useless tool, which was not fit for either digging or fighting, moreover in a region that had iron in abundance, and asked if they could not tell that it would take a great deal of trouble and a lot of hard work for a Human being to make such a useless and irregular tool, especially the aforementioned protruding veins, which did not contribute to the perfection or beauty of the tool at all, since the tool itself was so irregular, and full of dents and curves: Furthermore, neither they nor their forebears had any recollection that such misshapen tools were ever made there from either copper or iron; now in order to test the real substance of that Thunder shovel, since I had often heard that the Thunder blows most of the metal away, making it unfit to forge, or too crumbly, I caused a small strip to be sawed off the top of the handle, and forged it, which worked the second time after we added some *Borax*,[4] and it became a square little bar, 4 fingers long and of the thickness of a quill: I also had them make another small bar of common red copper of the same length; and when we compared them we first noticed some difference in color, the Thunder shovel being paler and yellower than the ordinary one, but they did not differ much in weight, except that the Thunder shovel was about a stiver heavier, but the greatest difference was in terms of sound, for this was much clearer and brighter with the Thunder shovel, than with the ordinary one, and it came close to sounding like gold, wherefrom we concluded that it must contain some gold, since it also did not turn as black as the ordinary kind when we rubbed it with *Spirites Salis*.[5]

One of my servants, who was born in *Lubo* or *Tulubo*, a Province in the large curve of southern Celebes, told me, that there was a river in that same Land, that flowed with a red water, and that it came down from a mountain, where they had found iron and copper; he said he had traveled along this river with his Uncle, and when they came to a spot where they knew Thunder had struck, they found a sidelong rent or tear in the ground, and they dug down into it, and found a Thunder Shovel, nearly akin to the one just mentioned, and his Uncle had them smelt it, without adding anything to it, and then made rings from it, which he wore on his forefinger in order to be lucky in warfare.

An Ambonese citizen, who was born in *Makkassar*, tells us that he saw various thunder stones in his Country, that were of a metallic substance, and differing much from the above-mentioned shovel, to wit, the substance sometimes resembled bell metal or a similar metal, which are used to cast those pieces, others were like iron, while some were a mixture of iron and copper: They did not have the shape of a shovel, but rather that of a stone wedge, though it was split into two ears at the top: They were very crumbly, and could not be used for casting, unless one mixed it with at least half its weight in gold.

This is contradicted by a story told by the son of a Chinese who lives in *Makkassar*, who said that he had seen other thunder shovels that had the same shape and substance as mine; for instance, thunder had once struck a large Waringin tree, that used to be in front of the King's house on *Bontuala*,[6] where one suspected that the former Kings had buried their treasures: *Radja Palacca* had them search the spot, after the tree had been felled, and found part of the treasure as well as such a Shovel, which he showed to the Dutch President, in order to convince him that such things are struck off by thunder: The same *Radja* wanted to have it melted so he could use it to band his creese, but it would not melt into a clump, no matter what one did: He had seen other ones, some red, others yellow, with various shapes, and [said] that most of the thunder stones that fell on *Makkassar* were either metallic or mixed with metal, just like the Shovel he had seen on *Java*, near the city *Demack*,[7] struck off there by thunder; [he] Said *Radja Palacca*, who is a bold Warrior, and whose courage has been sufficiently proven in the Javanese war,[8] always wears, as I have been told, such a small thunder shovel on his body, when he goes to War, and he never seems to have been hurt, though he made many bold attacks on the Enemy in full view of our Dutch officers, attacking the Enemy partially clothed, only armed with a small shield, a saber and a short spear:[9] I heard the same thing from the Tamboquese Ambassadors concerning their war with *Buton*, but since that would give the impression of beating one's own drum, I was not permitted to refer to it here.

I also received a small clump of metal, that weighed about an ounce, black on the outside, dirty and full of dents, reddish like copper on the inside, being part of a piece struck down by thunder on *Totoli*, which lies in the North-West corner of *Celebes;* I had this smelted with *Borax* as well, but it turned out totally unfit to be worked, even though it was three times recast, for it burst with every hammer blow; but I had a jagged ring made from the same piece, though with the addition of half its weight in gold.

The aforementioned Shovel was sent in the year 1682, along with other Curiosities of mine, to *Cosmus*, the third Grand Duke of *Tuscany*.

The Thunder stone from *Bangay* is by the letter B., and is the same as the one mentioned by the letter B. in the previous Chapter.

2. In the previously mentioned year 1679 I acquired the second little Shovel from the same Tamboquese Ambassadors: This one was somewhat smaller than the first one, and was not square in front, but rather round with a sharp edge, and so hard, that one could barely scrape off anything: Its substance was an even paler yellow, also with various spots and dents, which could not be scoured smooth, from which one can observe, that it was a mixed metal, and here too two curved and raised veins ran on either side of the handle towards the sides, like two sickles; the handle was rounded and flat, with some veins, hollow inside,

and it had a piece of wood therein, which the sellers could not tell me how got there; they suspected, however, that one of their Kings must have had a handle made for it, so he could carry it, which is quite likely, because when we removed the little piece it turned out to be good *Ciaten* wood, carved with a knife to fit the hollowness: The sellers told me, that the thunder had driven it through the head of an Ox and into the ground, and it weighed around 8 ounces, with a width of about a hand, partially round like a half moon: The handle also had two tutels or buttons at both sharp edges.

3. The third Shovel I obtained in the year 1683, also from *Tambocco*, for the price of 6 Rixdollars. This one was bigger and heavier than the two others just described, otherwise of the same shape, rounded in front with similar raised veins that started near the handle, and which ran towards the sides: Its substance differed somewhat, because it seemed more like bell metal than copper, and it had lost some corners at the sides, while on the flat side were some blackish spots, and it smelled strongly of copper; it was found near *Bonsora*, 4 miles south of *Tambocco*, in a hard and stony ground, buried quite deep by the thunder: This Shovel was sent in the same year, with other Curiosities, to Mr. *Christiaan Mentzelius*,[10] Court Physician and Counselor of the Elector of Brandenburg in *Berlin*.

4. This was the smallest Shovel of them all, only 2 fingers wide and 2 finger joints long, with a natural hole on one side of the handle, which was very flat and hollow, with sharp edges on the sides, with only one vein or rib on either side: The front was round and flat, quite sharp, but with a corner missing: The substance was mixed with copper (which tested as high as gold) and with bits of iron. It was found by the Alphorese behind *Tambocco*, struck by thunder into the palmite[11] of a Calappus tree, that had its entire crown singed by lightning: Three owners had worn it on their creeses, which made it as smooth and greasy, as if it had been varnished.

About that monstrous piece of copper or bell metal, which has the shape of a wheel, which a thunderstorm tossed down on *Baly*, see above, Chapter VII, which deals with a metal called *Gans*.

The previous Chapter stated that all these Heathens and Moorish Natives ascribe to thunder the shape of a Bull's or a Horse's head, and that, for instance, these Thunder shovels are its jaw teeth, which stick with their stems or hollow parts into the jawbone, and which the thunder spits out when he is very angry: But a better conjecture might be that lightning is accompanied by a strong wind, which propels these shovels from behind, and in this fashion produces this hollowness; meanwhile the surrounding lightning fire holds that material together on the outside, thus forming a cut, like the ones one can see on some spent bullets, that have a dent in the back, when they hit some pure soil. As far as use is concerned one can say practically the same thing as we said above, and that is, that they wear the Thunder stones on their body, or fashion rings from them which they then wear on their forefinger, when they go to war, particularly if they have to storm something: They are so infatuated by these things, that one can only obtain them after much cajoling and a great deal of money, or by means of open warfare; we particularly laid our hands on many of them after our glorious victory over the great Makkassarese army in the year 1667 on *Buton:*[12] For there is no other Nation that deals so much in all the tricks of war, as do those from *Makkassar* as well as the other Inhabitants of *Celebes;* wherefore the common man is not permitted to have these things on him or to hide them, but he has to surrender them to his Chiefs or *Radjas;* even if they were to commit a lethal crime, they would often save their lives by worshiping a Thunder Stone: a man who wears such a stone on his body is not only, according to their belief, courageous and *Cabbal*[13] or invulnerable, but even when small pieces of such a Shovel are hidden in the haft of a creese, and the weapon has been whetted on a Thunder Stone, it will pierce through any kind of firmness that magic can produce: Similarly, when one mixes some small pieces of thunder copper with the lead that is used for making bullets, the same will penetrate any weapon or firmness. They use these Shovels as Medicine the same way that they do the Thunder Stones, putting them in water, then rubbing them in the water, whereafter they give it to someone to drink, and wash the body of someone who has a hot fever in order to quench all the internal fire, but I would not care for this water because of the nasty copper taste, and in that case I would rather use the Stones.

Georgius Agricola 5. de ortu subterran: refert haec ex Avicenna: Avicenna verò inquit, in Persia decidunt, cum coruscat, corpora aenea & similia sagittis hamatis, quae in fornacibus non liquescunt, sed eorum humor in fumum resolvitur, terra autem residua fit cinis. Decidit quoque prope Lurgeam ferri massa quinquaginta librarum, quae prae duritie frangi non quivit; cujus pars ad Regem Torali est missa, is verò ut enses cuderentur jussit; sed illa nec frangi nec cudi potuit. Adhaec Lydiatus de fontibus Cap. 6. refert, in Hispania massam lapideam venis metaillicis infertam, è nubibus delapsam fuisse.[14]

These words translated into Dutch, mean the following: *Georg. Agric. de ort. subterr.* relates the following from *Avicenna:* When there is thunder and lightning in Persia, copper bodies fall down, that are like barbed arrows, which will not smelt in any kind of smelting oven, yet their moistness vanishes in smoke, and the remaining earth turns to ashes. In *Lurgea* there also fell a large piece of iron, that weighed 50 pounds, which could not be broken because of its hardness; and when a piece of it was sent to the King of *Torali*, he ordered, that they should make swords from it, but due to its hardness it could be neither smashed nor forged. *Lydiatus de Fontibus* says in the VIth Chapter, that in *Hispania* a stone clump fell from the clouds, which was filled with metallic veins.

If one rubs Aquafortis on a *Ceraunia Metallica*, and leaves it on for the space of a day and night, it will show up as follows: After 6 hours most of its body turned slightly green; the next day it veered towards Spanish green, though there were two kinds of spots mixed in; black and brown ones: The black ones are supposed to come from an earthy substance, and the brown ones from something resembling iron: In the center were some spots without color, which must be gold. Rubbed at the same time on common copper, it becomes dark green; and blackish on one side.

The Author presents us here with the Thunder Shovels, *also called* Thunder Chisels; *they consist mostly of metallic materials; of which I saw many different kinds, of similar color and shape, in Captain* Krytsmar's Cabinet;[15] *but since the same was scattered mostly among foreign Collectors, I was unable to locate any, except for the one shown on plate* L. *by* No. 11. *It is black as jet, shiny and, as far as I can tell from the outside, does not contain any metal: This one was also worn around the neck of a superstitious Indian, and to that end there was a hole pierced through it at the top, and he gave up his life in order to save it. Mr.* Rumphius *provides us with only two pictures of this, of which the first is indicated on the same plate by the letter* C. *and the other by the letter* D.

CHAPTER 10

Auripigmentum Indicum. Malay: Atal.[1]

One can find on *Java* a kind of *Auripigmentum pallidum*[2] which our Painters call *Opermont* and *Realgar*, being yellow little lumps like stone but lacking the golden sparkles in it such as true *Auripigmentum* has. Its taste is not astringent but almost tasteless, or perhaps veering a little to *Vitriolum*. One sometimes finds little red pieces among them which must be *Sandaracha Graecorum;* the Malay and Javanese call it *Atal*, the Chinese *Tsjio uijn*, to wit, yellow Stone.

The Javanese and Chinese do not regard it as a poison, for they not only use it with water or oyl in order to be able to paint yellow, but also will administer it without fear to a body, but in small quantities, wherein they are not careful enough in my opinion. After all, it was observed in the year 1660, in *Batavia*, that it was given to a Woman who became mad from it, and climbed up the walls like a cat, the quantity having probably been too large. Otherwise they administer it in small amounts, mixed with *Dju djambu*,[3] to be taken internally once every three months, saying that the same serves to protect the skin from all impurities, and makes for a smooth skin, which the Women strive for. One finds it for sale in the markets of Javanese trade towns, as well as on *Bali* and in *China*.

On *Bali* it is also used to color yellow a kind of linen, called *Krinsing;* this is a kind of cloth sprinkled with gold-yellow, red and white, etc.; but when this gold-yellow has been smeared on it, one has to hang the linen in smoke, because it is this that gives the color its permanence. The Chinese painters use this stuff to paint a gold-yellow on paper.

CHAPTER 11
Tsjerondsjung.

In the *Volcano* or burning Mountain on *Banda*[1] are several caves and overhanging rocks like small cellars or vaults, except for the *Crater* or erupted mouth on the West side of that mountain. Within the *Crater* and vaults one finds various mixtures of sulphur and alum which, I trust, are to be found in all sulphur mines; and among them a kind of *Alumen plumosum*.[2] I do not mean *Amianthus*[3] which bears these days that name unjustly in Apothecaries, but a true substance like alum, white as lime that is slightly damp, with a sour taste and a marked astringency, which the true *Alumen plumosum* must have. The Natives call it *Tsjerondsjong* or *Tsjerondsjung*, and there are two kinds: The first, the whitest and best, one finds hanging from the top of the caves in blunt cones, as one sees in Europe in cellars which are open to the naked sky and much rained on, and with long cones hanging down of a sulphurous substance: This *Tsjerondsjung* is not that hard but rather brittle, so that one is not able to bring the entire cone away: At first it is as pure white as lime, but in time it pales to a grayish color, and becomes almost never properly dry, and has in the mouth a piercingly sharp sour taste, and tart: The lesser kind is also as white as is the dry lime one finds on the ground of the aforementioned caves, and is especially plentiful on the ground of the large *Crater*, near the edges of that hellish hole which constantly produces a warm vapor and a subterranean noise: this kind is far more impure than the previous one, mixed with earth and little crumbs of sulphur, wherefore it also burns more or less when proximate to fire. Other small clumps are white, dry, and light, are less sour, and do not burn, so that they seem to be a burnt-out substance. For the best *Tsjerondsjung* one should look not in the *Crater*, but in the Caves which are on the sides of the mountains, because the rainwater that seeps through there forms these cones. Around the year 1660 this was a profitable merchandise on *Java*, since the Emperor of *Mataram*[4] sent for it from *Banda*, costing for a pound of the pure kind 3 Rixdollars and ¾ of a Rixdollar for the impure sort: For a long time it was not known what the Javanese used it for; some wanted to make us believe that it served to slow down a man's power during the pastimes of Venus or stay it for a long time; but this is against the nature of the Mineral, which is more flowing and expulsive than retardant. Afterwards we have learned with certainty that the Moorish Women use it for themselves, in order to make their nature pure and dry by means of that same expulsive power, and so becoming more pleasing to their Men; such a trick is necessary in such Nations where many Women are kept by one man, and each one seeks to make herself the most comely; it is taken in small quantities, to wit, the size of a pea, with *Djudjambu*,[5] which is made from seeds and flowers that have an astringent quality, and then they should not take it more often than once a week; for one has experienced that when it is used too often, it causes a perpetual flow, because the power of this Mineral is much more expulsive than contractive. Afterwards it came into disfavor again when the Bandanese imported the bad and dry kind into *Java* rather than the best one, perhaps caused by the ignorance of both buyer and seller; because even among the Bandanese are few who know how to find the right kind and how to distinguish it from the dry. They have brought me the same *Tsjerondsjong*[6] from *Nila* and *Damme*[7] where are rich sulphur mines: white and dry as lime, some of it like flour and some like clumps the size of a fist, almost tasteless or with a little tartness, nor did it burn, although there were yellow lumps in it; because when it contains a great deal of yellow it will burn so powerfully though slowly, that a little lump the size of a bean, when lit, will burn more than a quarter hour, making a hole in the thin board it is tossed on; there are some of our nation who do not know this Mineral right well, unjustly holding it for *Arsenicum album*,[8] but the majority and the most ignorant hold it for saltpeter,[9] not knowing that nature never mingles sulphur and saltpeter in a sulphurous or burning mountain, because this would make a mountain explode like a Grenade, because it is certain that those eternal fires are caused by mixing sulphur and alum; the sulphur serving for burning and the alum for retarding it.

CHAPTER 12

Tana Bima. Badaki Java.

This too is a soil like alum, somewhat drier and harder than the foregoing *Tsjerondsjong*, but also crumbly, resembling coarse sulphur, to wit, a pale yellow mixed with a dirty white, although it is not found in any sulphur mountains, nor does it burn when placed on the fire, but it sizzles and cooks; its taste is sour and it is astringent like ordinary alum. In Malay it is called, *Tana Bima*[1] and *Badaki Java;*[2] on *Bima*, *Batungontsjo:*[3] One will find it on the island *Sombawa*[4] on Mount *Sarri*, which lies between *Sape* and *Bima*, where one will also find the best flints: It is kept in earthenware pots, and they hang it for some time in smoke; it is difficult to preserve, since it will melt into a thick Syrup when it rains or if it is in a damp place, though it can be dried out again.

It is most often used to paint one's teeth black, since this is considered becoming by the Malay and the Javanese; they take a little lump as big as a pea, chew it with *Tiripinang*,[5] which makes the lips and teeth black, but it can easily be removed from the lips, though it immediately sticks to the teeth: It does not bother the Natives if they swallow some of the sap when they are chewing it; without having some *Siripinang* in one's mouth it will not turn black at all: One also adds it to the *Zappan* wood or *Lolang*,[6] if one wants to paint something black with *Indigo*, because it provides for a bright black color, and renders it steadfast as well; so that this soil can never be used to paint something black unless some red is added to it: And this is also the reason why it is merchandise, for it is sold in the markets of *Java* and *Bima*.

CHAPTER 13

Some Kinds of Marga and Argilla. Batu Puan.

The Ambonese Islands are rich in all kinds of *Marga* and *Argilla*, though as far as we know, they do not possess any medicinal virtues,[1] and the following are the most important:

The first, some kind of *Terra Sigillata*,[2] is only found in small pieces, the size of a walnut, drier and lighter than true *Terra sigillata*, though it has the same taste, it sticks to the lips, but melts in your mouth like other fine earth, of a light-gray color, with some veering more towards yellow, others to white; they also differ in hardness, for some are as weak as *Bolus*,[3] while others are as hard as half-baked loam, exhibiting a nature that is between stone and *Marga*.

In Malay these are called *Napal*,[4] and *Batu Puang*[5] or *Poang;* and it is called *Puan* because they tie it with *Siripinang* to bridal garlands, of which the Malay and Chinese make a fine show on their wedding tables: in Ambonese, *Hatu mina:* In the Uliassers, *Hatu menal*, that is, Fat Stone, because it turns out to be greasy when it is eaten. Some would like to derive *Batu poan* from Ambonese, and not Malay, since most Ambonese call it *Hatupuan*, and *Hatu puanno*, which comes from *Puan* or *Appuan*, that is, to smelt, to grind, since it is a stone, which melts and grinds up easily in one's mouth.

One will find them in nearly all the Ambonese Islands, though one is better than the other: The one from the Island *Honimoa*,[6] in the mountains there by *Ithawacca*, *Ulat* and *Ouw*, is considered the best kind, being light gray, and resembles *Bolus:* Others prefer the ones from *Waccasihu*,[7] that is on great *Amboina;* these are harder and yellower than the foregoing, sticking firmly to the lips, and which is dug behind the same Village from an oven where the hill juts out; One will find small and large lumps in the small river *Waypya* in the region around old *Hative*,[8] and they are harder and whiter than the foregoing, so that one has to break it off while biting into it, and it is also considered to be among the best: The other kind, which one will find in Hitu country, such as in old *Eli* for instance, or *Senalo* and *Pelissa*, is considered not as good, and is often mixed with sand. One digs for this earth 2 or 3 feet below the common dirt, that is ordinary black, or red, or gray mountain soil; one has to look for places, where a hill juts out steeply, and where they then make holes like ovens, where one can extract the pure *Batu poan*, without any other earth mixed in, but rainwater will stay in those open holes, and this makes the *Batu poan* impure.

None of this *Batu poan* is used as medicine, but is eaten by the handful; the Natives relish it, especially their women, who, when they are with child, are seized by strange appetites: One never eats it fresh, the way it comes out of a Mountain or River, but it is put in a new unglazed pot, and hung in smoke for about a month, wherefrom it gets a smoky taste (called *Wangi*), almost like fresh rice, which is pleasing to the Natives, and then they sell it in the market: The Women really cannot tell you, why they like to eat it so much, except that they have a taste for it, and persuade each other, that they will get a pale color from it, and so in their way want to appear white, I think like unbleached linen: Experience teaches that mother and child do whiten somewhat because of it, but those who eat it daily and eat too much of it, become bloated, get an enlarged spleen, and finally a schirrhus liver or shortness of breath (*Tehatu*), with a foul stomach, that causes *Cacohymia:*[9] But one considers it healthy for those people who have a *sour* stomach and the bloody flux in order to cleanse the hurt intestines, and to salve and grease them, so that the sharp vapors have no hold on them, and at the same time [it] gently binds the bowels.

N°. 6
N°. 5
N°. 9
N°. 7
N°. 10
N°. 2
N°. 8
N°. 1
N°. 11
N°. 3
N°. 12
N°. 4
N°. 13

The color of *Marga Ihana* greatly resembles that of wet ashes, heavier and greasier than *batu poan*, and with an earthy taste: One will find this kind all over the region of old *Iha* or *Hatuana*, so that it appears that the entire country is named after it, though it is mostly found on the banks of the rivers there, especially by *Waysalee;* if one digs 2 or 3 feet deep into this *Marga*, one gets to the true *batu poan*, which one can easily distinguish from it by color and substance. One will find an entire rock of it near the small river *Eijer Guru Guru Kitsjil*, where it falls down from up high, and which is where the long boats from the ships get their drinking water, and is situated on the North side of the Bay of Ambon; this little river has a stone-like sap in it, which leaves a stony crust on wood and branches, that fall into it, but if one lets the water stand, until this substance has sunk to the bottom, then it is good to drink: The Ambonese call this *Marga*, *Hatu suhu*, that is *batu lompor*[10] or mud stone, because, if one rubs it in water it immediately becomes mud; in the Uliassers *Hatu kullul.* One uses this *Marga* in the Uliassers to cut forms, wherein they pour molten tin, copper and lead.

Terra Nussalaviensis resembles *Terra Sigillata* more closely, in that it is white, veering somewhat towards gray, as if there was some light blue in it, as greasy as soap, though it does not stick to one's mouth, and so brittle, that any kind of handling will make it break into small pieces, nonetheless each little piece is firm in its own right, and greasy: The taste is remarkably close to our *Terra Sigillata*, and it is so greasy, that it is difficult to mix with water, like an *Axungia*,[11] or *Medulla Saxorum.*[12] One digs for it on *Nussalaut*[13] where the mountains jut out, and also in the nearby hills behind the villages called *Sila* and *Titaway.* That which has been freshly dug, is put in a *Baly*[14] with water, then stirred with a stick, until the yellow and red clay has separated; because one digs for it 2 and 3 feet deep, below that kind of soil, and since the Natives are rather careless, that hole gets plugged up again with that soil after every digging, and gets mixed up with it, because the vein or course is otherwise pure: After it has been washed, put in pots, and hung in smoke, it is eaten by the Women of that Island like any other *Batu Poan*, and it does not bind the bowels as much, and might resemble the true *Terra Sigillata* a great deal; but the Natives prefer the first or dry kind, because this Nussalaut earth is too greasy, and sticks to one's teeth. In the aforementioned hills one will find yet another kind, whiter, with larger clumps, less brittle, which can be considered a white *Bolus.* It is smoked and eaten like the foregoing.

The *Batu poan* from the Island of *Oma*[15] is not as good as that of the Uliassers, because it contains a lot of yellow clay, mixed with sand and small stones, and smells somewhat like sulphur.

Bolus ruber[16] has two colors, dark red and light red, one will find small lumps of it on top of the *Batu poan* in the mountains of *Ulat* and *Ouw*,[17] and sometimes mixed with it as well; it is not eaten, but is in demand by the Potters, who gather the purest lumps, rub it with water, and then smear it on the outside of their new pots, before they put them in the fire, so they will turn red. They call it *Haca caul* and *Tiaul* there: One finds this *Bolus* abundantly on *Leytimor* on the red mountain, where the same is steep and broken down: The same goes for the red and white *Bolus* near the Hitu village of *Hausihol.*

Terra Aurifabrorum,[18] called *Umepyal* in the Uliassers, is a heavy and greasy *Marga*, consisting of many small lumps, a pale yellow with a silvery reflection, brittle, does not stick to the lips, and always with some fine sand in it: One digs for it in flat ground, near *Pya* and *Kullur*, places on the North side of *Uliassar*;[19] near Cape *Umeputi.*[20] *Iha*'s Goldsmiths use this earth to make forms and crucibles, mixing it with pulverized Javanese Gargoulettes:[21] They also use it to fill the hollow parts of their finished gold pieces, such as the heads of the gold snakes and the creese hafts, which other Goldsmiths fill with *Ambalo*, that is *Gummi lacka;*[22]

one will find such earth also by the hook of *Nussanive* in Portuguese Bay,[23] where the Highway passes, and where the road becomes so sticky when it rains, that one has trouble lifting one's feet.

Both brown and yellow *Ochra*[24] is dug in Hitu, on a mountain, where the old village *Pelissa* used to be, near the place called *Amahutetto*, 2 and 3 feet deep; the brown kind resembles dark *Lacca*, but becomes ugly when it pales, turning a bluish violet on the outside: It is entirely mixed with yellow, so that one will rarely find big lumps of one color; one will also find there red, black, and white earth that veers more towards *Bolus* than *Ochre;* diggers also encounter a blood-red sap there, which is undoubtedly a kind of *Bolus* that melted, but these superstitious people considered it a bad omen for the diggers: The brown and the yellow kind are smoked and eaten like other *Batu poan*. There was once[25] a hamlet, called *Lissaloho* near said village, and it being their turn to build their Town hall, they found the two aforementioned *Okeres* while they were digging, and it was called *Lissaloho* Earth after it: In the same mountains were once the villages *Senalo* and *Eli*, where one also dug *Batu poan*, to wit, at *Senalo* by the origin of *Waccahuli* river, a white, fine, and dense earth; but the one from *Eli* is grayish, mixed with pieces like iron rust and fine sand, hence unfit. One will also find red and white *Bolus* near *Mamalo* in the *Hausihol* mountains, it is fine, and in big clumps, and the red ones can be used for oil paints; just like the ones from the red mountain on Leytimor.

Argilla is a white, greasy and very clay-like earth, found near the wellspring of the river *Way yla*, which flows into the sea by *Hukonalo:* The place is called *Pannat*, about an hour going East-south-east from aforementioned *Pelissa;* it is a low and flat place, where they find this *Argilla*, which hardly ever dries, wherefore one cannot pass there because of the tough clay. It is not fit to eat, nor to make pots, because it contains too many small gray stones: It is called *Pannat*, after the place where it is found.

When it has rained for a long time, some flintstones will emerge in this clayey clearing, the kind that are called *Androdamantas*,[26] heavy and of a hard substance, the color of yellow copper. The lumps are the size of dice, big and small ones; some are square, some are irregular triangles, as if a *Pentagon*, or five-sided body, were hidden in a *Cubus*, with some of its corners sticking out here and there; one might think that, due to its weight and color, it might contain some metal, but it cannot be smelted, as we will tell at greater length later on: But a Moor wants to tell me, that he saw them obtain small lumps of Amurassean gold from such white clay on *Timor*'s southern coast: one will find this white *Argilla* also at the foot of the red mountain, towards *Hative kitsjil*, on the beach, but it is salty and much mixed with sand, grayish, and it does not dry very well, though there is no doubt that a whiter and drier kind could be found, if one dug deeper into that mountain: It is a true *Terra Cimolia*[27] or Fuller's earth,[28] which the Natives themselves use instead of soap, to wash their clothes.

They dig big pieces of *Marga* from *Nussanive*, around the most western corner of *Leytimor*, called *Weyhuki*, which does resemble a *Marga*, but it is much drier and harder, the color of mice with a yellow reflection, as firm as liver, and crunchy if one bites a piece off it: this too is held to be a good *Batu poan*, and does not need to be smoked for three days, and if it is still not hard enough by then, it is not considered *Batu poan*, but is held to be a *Marga* called *Hatu patta*, that is crumbly Stone, on *Leytimor:* One will find similar red, yellow, and gray lumps in all the rivers of *Leytimor*, especially in the Waytommo and the Olyphant, and they are smoked and eaten as *Batu poan*, although they are hard when you bite into them; these clay-like clumps become reasonably hard stones over time, covered on the outside with knobbles and dents, but can be distinguished from others, in that they are as slippery as soap when you touch them, are easy to grind, and cut with a knife, showing inside all kinds of spots, veins, and clouds; see about them in the subsequent chapter on Ambonese *Marmor.*[29]

The village *Labo* is in the eastern part of *Bangay* or *Gapi,*[30] called *Bulu,* and near it is a river wherein one will find small lumps and stones, that make for a good pigment with which to paint red, yellow, and white.

Ampo is a reddish earth on *Java,* wherefrom they make the gargoulettes in *Cheribon,* a pale red like a half-baked stone; pieces of that earth, or special little cakes of it, baked for that particular purpose, are sold in the markets, and are eaten by the Women, like the *Batu poan* on Ambon.

Mr. Rumphius *described various earthy substances in this Chapter, while His Honor also mentions some stones that grow in the same; and since pictures are lacking, and since we possess many kinds of the same, we felt compelled to add some that belong to these. On plate* LI. No. 1. *is one, that is shaped like a perfectly square die, and which is often found in Switzerland by the River baths, and also, but smaller, in the Lei Mountains. The one shown by* No. 2. *has Five sides, as it was described: while we add a very unusual one, indicated by* No. 3. *with more irregular sides, and which is fastened to a rock. By* No. 4. *one will find an entirely different sort: all of them like wide arrowheads, which, all jumbled up, are still stuck in its* Argilla, *or mother earth.*

CHAPTER 14

Some Stones That Can Be Used for Painting.

1. *Comalo*[1] is a Stone that can be used to paint.
2. *Antsjor* is a green Stone, that also can be used to paint; both come from *Suratte.*
3. *Suligi* is a white Stone, from *Atching,*[2] which, when rubbed with other spices and then ingested, is a good remedy against stabs in one's legs.

CHAPTER 15

Hinghong. Chinese Mineral.

Hinghong[1] is a red and grayish Mineral or Stone, which is found in *China* in such places where the Peacock has made its nest three years in a row, lodged about the width of a hand under the sand or soil, and manifests itself by certain rents which one finds there in the ground. It has two substances: one like iron rust, the other somewhat redder.

The Chinese call it *Hinghong,* but in *Siam* it is called *Jaklak,* that is: Medicine against Ringworm.[2] The Peacocks put it next to their eggs in order to keep them safe from snakes. When rubbed with a weak *Arak* while also drinking as much as one can hold, it cures the bites of all venomous snakes, including those that live in the sea and which cause a person to swell up; crushed and mixed with *Sajor Cancong*[3] and applied as a poultice it cures legs suffering from dropsy. The light or reddish pieces when rubbed with water, paint an Orange color, and the other brown ones look like snuff.

CHAPTER 16

Black Sand.

Black Sand[1] is found on *Amboina*, near the village *Hitulamma*,[2] not all year round, but when the North- or North-West wind is blowing hard, it is tossed by the sea onto the beach; since that Sand is plentiful on *Java*, especially on the beaches around *Grisec*,[3] those from *Hitulamma*, whose *Radja* and *Tanahitumessing* clan[4] are of Javanese origin, want you to believe, that this black Sand was stuck to the keel of their Junk, wherewith their ancestors sailed from *Java*, and landed in this place, whereafter they stayed and multiplied; yet it does not appear unless there is a West wind blowing, and that is when the foreign merchants arrive: One finds it in greater abundance on *Batschian*,[5] and on *Halmahera*, the West side of the big Island *Gelolo:*[6] After this Province[7] suffered the great quake of 1674, it also manifested itself near the red mountain,[8] East of *Victoria* Castle, but it does not appear every year. I have found that Ambonese Sand is larger than Moluccan sand, and easily a fourth part heavier; also that black Ambonese Sand, when compared with the fine white kind, is heavier as well: In the year 1678 Mr. *Robbertus Padbrugge*,[9] at the time Governor of *Ternate*, noted that a Loadstone attracted the black sand from *Gelolo* quite well, shaping it into little hairs, just like the file dust of steel, from which one can determine, that it must be rich in iron: The Loadstone also attracts the Ambonese kind, but much less so than that from Ternate. Since that time I have also noticed, that the file dust of iron or steel, that has been attracted by a loadstone, will, when one holds it to a candle, melt together immediately, and form short little hairs; the black sand has otherwise no use, except to be strewn on wet writ by our Scriveners, wherefore our Secretaries and Commercial Agents obtain their yearly supply of it from *Hitulamma:* but before using it, one should first boil it in fresh water, and then dry it in the sun.

Mr. Calf,[10] *the Sardammer who has traveled far and wide, and who is a punctilious collector of excellent, both natural as well as artificial items, presented me with a small box of it, brought by his Worship from* Puretta, *which lies over against* Genua, *which has the same attributes as those mentioned here, to wit, that a Magnet attracts it, just like the file dust of steel and iron: It has the exact same shape as sand; and it is grainy, hard, and shiny, but as black as pitch.*

CHAPTER 17

Cinis Sampaensis. Abu Muae.

There is a City by the name of *Colong*, in a region called *Tsjamsiaa*,[1] which we call *Tsjampaa*, where one will find a place ½ a day's journey from that city going inland, that is the size of our Castle's inner courtyard here on *Amboina*,[2] and this is where in the hottest part of the Summer, a murky water, like mud, bubbles up, as if it were at a boil, or the way porridge bubbles: As soon as the hot Sun shines down on this water, it dries up, and changes into ashes, whitish, veering somewhat towards yellow, or a dirty white: The place is surrounded by a wall, and is guarded by an Officer and 50 Soldiers, to ensure that no one carries any of that ash away; because it must be brought to the King of that Country, who is the only one who can deal in it, being merchandise desired throughout India, and costs in *Tsjampaa* 10 taels[3] of silver for each Picol,[4] that is 32 Rixdollars: The ash also comes down some distance outside the wall, where it becomes dirty, because mixed with sand and soil, and everyone can take of it as they please.

That which falls closer to the wall is better quality, but one durst not take that either, unless one has permission from the Farmer, but the innermost ash, particularly that which springs up in the center of the enclosed square, is the whitest and finest, without any sand, in clumps the size of a fist, though they do easily fall apart, when they are rubbed; its weight is like that of other ashes: It does not spring up at other times of the year, when the square remains unguarded.

The people from *Tsjampaa* call it *Catlu.* The Chinese *Tsjamsia soa,* that is Ashes from Tsjampaa. In Malay, *Abu muae,*[5] that is, Ashes that well up, the way good rice does while it is cooking; some people also call it *Abu maleijo.*

It is mostly used for anything, for which one normally would use soap, for instance washing clothes and linens; also to whiten the hands: One cooks the oval rice, called *Bras bule,*[6] in its lye,[7] [which] thereby acquires a pleasant smell, and is then used to make all sorts of cookies: it is particularly used for making *Casomba* paint,[8] instead of the ashes which otherwise would come from *Durian* rinds[9] and *Calappus* sprouts, and one can do more with just one spoonful of these ashes, than with a big handful of the other ones; the paint will also be much clearer because of it; but there are others who say that if linen is often washed in it, and when the lye has been made too strong, it will burn and wear out much quicker.

CHAPTER 18

Various Flints. Batu Api.

There are several places in the Ambonese Islands where they find good Flints, not only fit to be used with a tinderbox, but also to be used as Flintlock stones: They have three colors; black or the color of horn, which are the best ones; then gray or liver colored; while the brown or reddish ones are the least: Nor are the best ones any larger than a fist, because the big pieces are of a mixed substance and crumbly, chalky on the outside, gray and coarse, like other flintstones, but heavier, they can also be angular, smooth and shiny: The best ones occur on Pig's Island,[1] on its southern shore, in the small bay of *Sonahola* is a little dry river that runs along a hill, and one will find them there among other stones: There are several rocks of this flintstone on the beach, but they are full of cracks and crumbly: One will also find them on Ceram's North coast, near *Caloway,* and on the southern coast near *Guli Guli* from the river *Wattalomi* up to *Assan* and *Dawan,* where one will find fine, massive pieces in the sea, that are the size of a head: The same thing can be said for the North-west corner of *Buro,* to wit, *Balatetto,* and on the southern coast near *Karike:* They also have them in Makkassar, and I am to understand that they have entire rocks of this stone, and some are said to be so fine and so beautiful, that one mistakes them for Agates; and to be sure, there is such a great resemblance between Agate and Flint, that it is conceivable that the outside, or coarsest part of the Agate is nothing but Flint, just as many precious stones are wrapped in a rough mother, called *Matrix Gemmarum* in the *Gemmariis:*[2] At least, they have sent me small pieces of rough Agate from *Suratte,*[3] of which one would think that most were Flints. The only differences being, that Flints will test more or less, when one rubs them with gold, while this is not true at all for Agate, while the latter also has many broad veins or bands, and is partially transparent, while Flint is not: The best East Indian ones I have seen up till now, came from *Bima,* from the *Sarri* mountains;[4] I refer the reader to the previous chapter on *Tana Bima.*[5] They have three colors, red, the color of horn, and white, and if they had had veins, one would have thought they were Agates: One can cleave them as easily as the ones from the Fatherland, and thus produce excellent sparks even when one strikes them lightly against one another or against a steel:[6] They also are so smooth, especially the white kind, that one cannot use them for testing; it is a known fact, that in Europe they make Flint-lock stones from Italian Agate,[7] which are used in pistols, and which provide little, though red and penetrating fire. Our Ambonese Flintstones have the fault that they are difficult to cleave or sliver as one can do with the European ones, or the fault lies with our ignorance of how to deal with them. The people from Makkassar teach us, that one should first cook entire clumps in water for a day, and then they will cleave better, which is done with a small hammer, after one has placed the stone on the flat of one's hand or on a hard pillow, and hit it at the corners, so that it is struck at an angle, which is the safe way of doing this: They do give off much fire, but burst easily, while they also become dull much more easily than the European ones, but one can make do with them.

My Servants found in the center of a reddish Flint, that had come from the southern coast of *Buro,* another white flintstone, the size and shape of a pigeon egg, but more oval and a dirty white, just as one sometimes finds in European ones; I deem this to be the *Enorchis* from *Plin. Lib. 37. Cap. 10.*[8] For which see also the *Gemmarium Boëti lib. 2. Cap. 202.*[9] There are good flints in other places on *Bima,* but they are not as fine, and become dead in time, that is, they no longer give off any fire.

A small Stone similar to that pigeon egg is shown on plate LI. *by* No. 5. *The one we own is a pure white, somewhat transparent and hard; and it has on its longest end a small round growth like a little pearl; it is wrinkled all around the joint, as if it were a strangulated wart.*

CHAPTER 19

Androdamas. Maas Urong.

Maas Urong[1] is a beautiful and rare *Pyrites* or Flint, very heavy, hard, and dense, like unto yellow copper, so that one could mistake it for little lumps of it, but it does not test well on stone, but when struck with a steel it produces sparks like any other flint: One will find it in large and small pieces, all very angular and irregular, yet in such a way, that all the angles follow a certain order, because though most are roundish, they still have corners sticking out, called by the *Geometrae, anguli solidi,* that is, solid angles,[2] consisting of 3 flat sides joined together in an angle, and according to the determination of the angles they seem to be parts of a figure called by Artists *Dodeca-etron,*[3] which is made up of 12 faces, and each face has 5 stripes like a *pentagonus:* One might say that there are many such *Dodeca-etra* in a *Corpus* or figure, but if one smashes the stone, one will find it solid inside, and not made up of parts at all, which one will notice about Crystals as well, which are on the outside both regular as well as irregularly six-sided, yet solid inside: Other pieces are square like dice; others are quite irregular, and angular, yet with some *Dodeca-etra* sticking out here and there. I call this stone *Androdamas* after *Pliny,*[4] and *Pyrites solidus,* in Malay *Maas Urong.* The biggest and finest pieces on *Java* are found near the City of *Mataram;* Also on *Celebes, Borneo* and *Succadana* which, when polished, are the width of a hand and shine like a metal mirror: On *Amboina* and *Ceram* one will only find small lumps the size of a pea or like the ball of a pistol,[5] and which show the aforementioned *Dodeca-etra* at their best: They grow in hard and russet flintstones, which one will find on the banks of certain rivers; that is to say near *Hila*, *Larique* and old *Hative*, in the river *Waypia*, which, as was mentioned before, also delivers the best *Batu poan*. One needs to smash these russet stones, and one will find these little clumps stuck so firmly in there, that one has to chop them out; but one will find that only one among ten contains one of these: Nor are they always solid clumps, for often they are very small pieces, like coarse sand, and kernels of yellow copper, dispersed throughout the body of the stone. One will find reasonably large *Pyrites* like that in European marble stones, especially in the black or blue ones, which the Stone cutters call Knots, and *Centrum marmoris* in Latin, and which they do not like to encounter while they are working; because when they chop it out, which they must, and it stays stuck in there, it will leave a small hole in the stone, which can often ruin an entire tablet; but when such a *Pyrites* is at the corner of a stone, they are gladly cut out, polished, and used in Firelocks, and are then called Italian flints: Other pieces, which occur on the surface of a marble table, have to be patiently cut even by the workmen, and then polished with the rest of the table, which makes it rare and distinctive.

On the coast of Finland, in that part of the East Sea, which is called the Finnish bottom,[6] one will find some dreadful rocks in that sea, called the Finnish Skerries,[7] from which they hew blue marble or blewstone,[8] wherein one will find a great number of these flints, somewhat larger than a musket ball,[9] of a paler color than the ones from the East Indies, as if yellow copper were mixed with silver; which, when polished, shone like a mirror, but which did not test on the stone at all; while on the other hand the solid Ambonese ones do test a little on a stone, but the streak is soon gone, when one rubs it with one's hand.

In the previous Chapter XIII, I described the place called *Pannat*, which consists of only soft white *Argilla*, and where one will find many of these *Pyrites* when it rains, though they are not as regular as the ones which are chopped out of the russet stones of the river near *Larique*.

One will also find large clumps of this stone on *Timor*, but they are very crumbly, and seem to have been fashioned from many different pieces, and are therefore not suitable for either polishing or Flints.

There is a kind of *Marcasita aerosa* or Copper stone that also belongs here, which one will find in various places in these Eastern parts; it is crumbly, coarse and sharp, angular, with glimmers and glitters like yellow copper, mixed with gray mountain stone, although it is clearly different, since it is not only crumbly, and one cannot strike a spark with it, but it also has a sharp corrosive salt, like *Salmiac*,[10] which it exudes, if it is left in the open for a while.

One finds it often on the eastern part of the Island *Messoal*,[11] near the beach, where it is marshy besides: Also on Ceram's Northern coast near *Hote;* on *Nussalaut*, around the warm water near *Sila*, on *Timor*, and in other places.

The use of the coppery flint is not properly understood by the Natives, but many Europeans dote upon it, particularly the first kind or *Pyrytes solidus*, because of its weight and coppery color, and they want to force some metal out of it by means of fire, though there is none in it, so that it remains a bare stone. The Javanese, Malay, and Makkassarese polish the nicest pieces until a flat tablet, either round or square, and affix them to simple rings or to their Turbans; because they believe, that it makes the wearer courageous and a conqueror; There is a mountain called *Giry*[12] behind *Grisec*,[13] where the *Penimbaam*[14] lives, who is revered by all Javanese as a Holy Teacher or Pope; there are several graves of his forebears and relatives on the top of that mountain, and these graves have large pieces of this polished stone on them, which give off a wondrous sheen when the rising sun falls upon them, which the ignorant consider to be a miracle. The second or crumbly kind, is more likely to produce metal, although not much has come to pass thus far: We found that out here on *Ambon* in the year 1666, with a piece brought from *Messoal;* we plied it with a hot fire in crucibles, but it brought forth nothing but a sulphurous stench, leaving a small clump behind on the bottom of the crucible, which looked like zinc,[15] but if one tapped it, it fell apart; but it is thought to be certain, that the Papuas on Messoal extract that stuff from it, with which they make the copper basins called *Gongs*, which we call *gomme*,[16] after adding a little copper, but they have to dig for the stones rather deep and under water, because there they are denser and harder.

That I deem the first kind to be *Pliny*'s *Androdamas*, is due to his words, which will be found in *Lib. 37. Capt. 10.*, where he says: *Androdamas* has a silvery sheen like a Diamond, square and always like dice: The *Magi* suspect, that it got its name, because it constrains people's tempestuousness and wrath:[17] The Authors do not explain whether *Argyrodamus*[18] is the same stone or not; *Clarissimus Salmasus*[19] notes at this place on p. 564 of *Solinus:*[20] *Pliny* says, that it has the powers and hardness of a Diamond. And a little later he adds: *Androdamas* is surely a kind of *Haematites*, that is a Bloodstone, of excellent hardness and weight, but with the color of iron; so that he concludes that the ancients knew no other *Androdamas* than this one, and that *Pliny* compares it to a Diamond, because it is so hard, and because it constrains people:

It is my sense, however, that *Androdamas* is not a kind of *Haematites*,[21] but a *Pyrites*, and that it got its name not from being like a Diamond, but from constraining people, as I already said before about the Javanese; *Pliny* does grant it a silvery sheen, but one should understand that to be a yellowish silver, just like polished *Maas Urong:* True, it mostly has the shape of dice, but I said that here and there it has the five-cornered sides of a *Dodecaetron:* I will describe another stone hereafter, which I hold to be the real *Argyrodamas:* If you still want another stone from *Pliny*, which corresponds to our five-angled or multi-angled *Maas Urong*, I present to you in the same chapter the *Hexicontalithos*, which shows many colors in one body, and which is found in the *Troglodytica regione:*[22] Since it is difficult to believe that a small stone can display 60 colors, it is likely that *Pliny* was speaking of angles: I guess that *Bellonius*[23] also wrote about this stone in *Lib. 2. Observ. Cap. 67.*, saying that he saw in *Tauris*, a city near the red sea, such stones in reasonable amounts, which a Greek monk preferred to think was a *Lapis Arabicus;* it was round or globular, the hardness of a *Pyrites*, composed of many square little pieces, in the manner of an *Androdamas;* it would have been better if he had said it was an *Androdamas* itself, because if the *Arabicus Lapis* is the same as *Arabic. Plin. loc. cit.*[24] then it must be so much like ivory, that it could be thought to be such, if it were not so hard: One holds it to be the kind of Marble that *Orpheus* in his stone poem calls *Lapis Barbarus.*[25]

On Java *Maas Urong* is called *Crapo*, where they get big pieces as wide as one's hand from the region called Mattaram, then they polish it smooth, and set it in silver or copper, and they use it also for striking sparks.

This entire Chapter is a disquisition on a kind of material which we call Marchasita: *It is only apparently a metal, because it has the color and shape of Copper ore; of which they found entire mountains in Moscovy, wherefrom they obtained with great difficulty some silver, but very little; and if it is put in fire, it evaporates, leaving a horribly stinking smoke, and very little* Caput mortuum, *or* Skull,[26] *wherefrom they can obtain the silver: It is found in mines in* Hungary, Bohemia, Saxony, *and other places, but in clumps that have strange shapes, and those same miners consider it burned sulphur, and (as they say) it came from sulphurous waters, which trickle down from the minerals, carrying along particles of what it went through and when it comes to rest in small holes, a constraining force makes it a solid, and gives it such shapes, as the minerals permit, each one different according to either salt, or sulphurous particles, with which they are mixed; if by chance some material is thrown into this water, or happens to get into it, it will be permeated with it and consumed, and retain these particles, which will then evoke the shapes of what was thrown in; but rougher, since a crust will grow over the same; wherefore we thought it a good idea to show some here on plate* LI. *The one shown by* No. 6. *is a rough clump, but it has a beautiful glittering color.* No. 7. *has the shape of a Goat's hoof, while it has on top a knob or clump of different material than that of the little claw.* No. 8. *is a piece, that has a branch running through its center, that stops at the ends, and which we suspect of being a root, which changes into* Marchasita, *and the surrounding crust has grown through the same material.* No. 9. *is a clump, with more of a yellow color, also cloudy; if it is smashed, one will see that all of them have rays that shoot out towards the edges from a central point.* No. 10. *is an entirely round Ball, somewhat cloudy on the outside, but inside with rays coming forth from a central point.* No. 11. *is a similar ball, but on the outside covered with sharp protruding arrowheads.* No. 12. *is an oval Flint, partially encrusted with* Marchasita. No. 13. *is a large, rough clump, in the shape of a copper door knocker: We could add many more kinds, such as Whelks, Shells, and other shapes, which are all enwrapped in this material, or entirely transformed, but we think this sufficient.*

CHAPTER 20
Argyrodamas. Batu Gula.

On the beach and in the mountains on the South-West side of Buton,[1] one will find a kind of white Stone that is the size of a head or a fist, a dirty white on the outside, irregular, full of cracks and as hard as a common Flint, but if one hits it, it falls to pieces, and one can divide these into as many small pieces as one wants, and some will have the shape of dice, others of flat tablets, and these too can be slivered; they are white, transparent like crystal, but on the surface they give off a reflection as if they had been silvered, or as if it were a silvery foam: They are of a soft substance, do not spark, nor do they light up at night, if one strikes them together, as a Crystal will or a Flint. I never saw such Stones again, nor found anything about them in any Author, but I think that these may be *Pliny*'s *Argyrodamas*, to wit a stone that resembles a Diamond, and has a silvery reflection, about which see *Lib. 37. Cap. 10.*[2] This stone is called *Batu gula* in Malay, since it quite resembles the pieces of sugar candy, called *Saccarum cantium* in Latin, and not *candidum*, since *cantium* is a new Greek word, meaning angular sugar:[3] One could hold it to be a *Lapis specularis*,[4] except that it cannot slake to lime or plaster,[5] while it is also heavier and harder than any *Lapis specaluris*. Here and there in the mountains where it is found, there flows a small vein of black *bitumen*,[6] which, when it reaches the sea, becomes hard, and which is sold to strangers as black Amber,[7] and this might be the reason why this stone does not acquire a hard and dense substance, or why it does not become a proper Stone. In the year 1666, they brought me several baskets full of these stones from *Buton*, to which I paid no attention at the time, because I could not see what those from *Buton* declared, to wit, that it had all kinds of beautiful colors, such as silver, pearly, and rainbow, whereas it looked to me just like a somewhat uncommon Stone, wherefore I threw them away; but when 10 years later I happened to smash a large piece of it, I discovered that it was much more beautiful and harder inside, than I had thought, but by then the other pieces were gone. Those from *Buton* refused to bring more of them, not so much because the King I had known was no longer alive, but rather because they did not particularly care to disclose these stones to our Nation, since, as I came to understand, they use the same in place of balls to fire from their cannon, because they break into 100 pieces, as if one were firing langrel:[8]

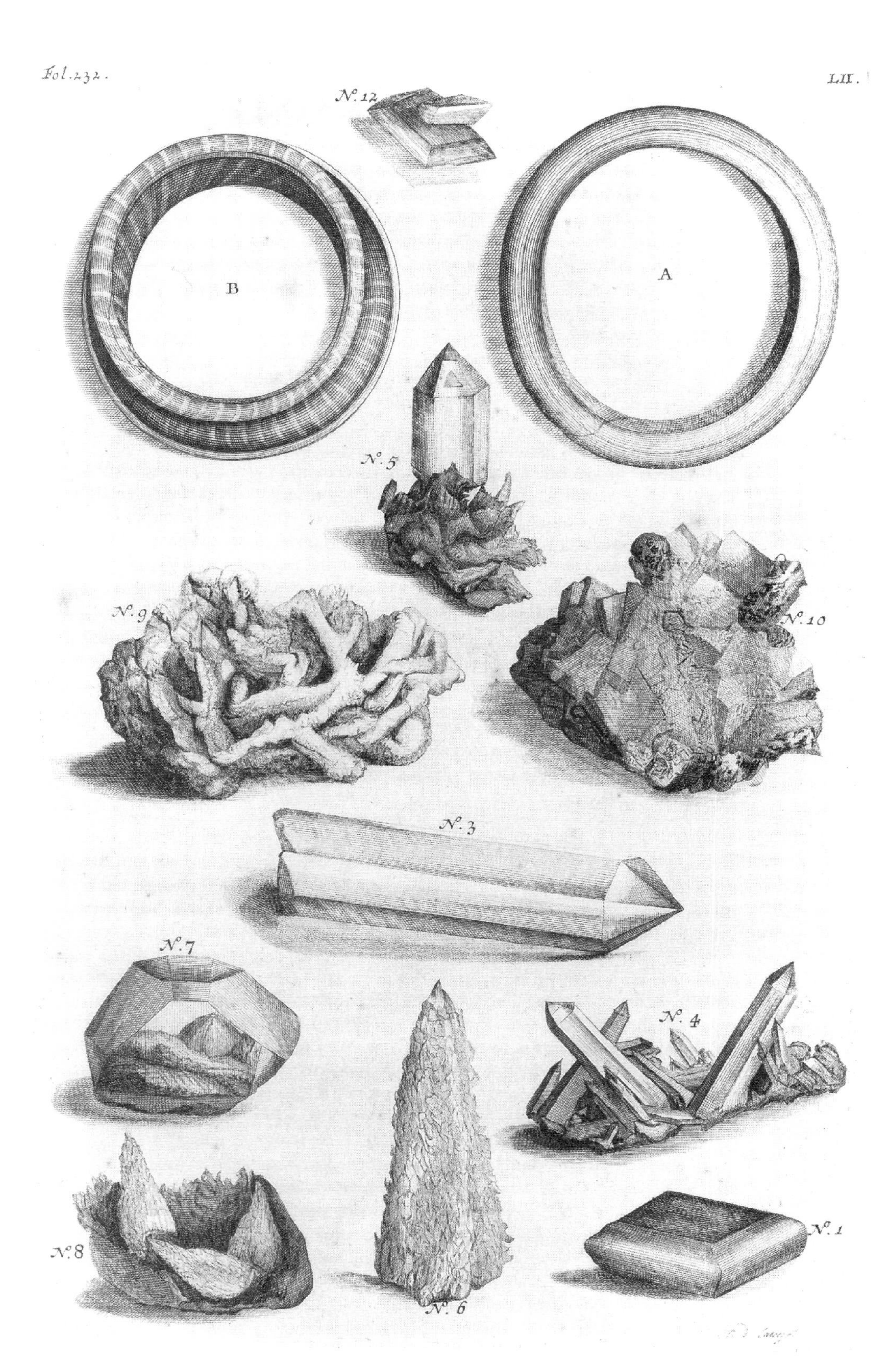
N°.12
B
A
N°.5
N°.9
N°.10
N°.3
N°.7
N°.4
N°.8
N°.6
N°.1

One will also find this kind of *Batu gula* in *Makkassar,* about a day's journey beyond *Marus,* in very high and cold mountains, which one has to traverse, if one wants to go to the land of the Buginese, no one can endure there at the top because of the great cold, indeed, one will not see any birds or other animals up there; about halfway up these mountains, are large caves, wherein one will see large pieces and cones hanging down, attached to narrow necks, in all kinds of shapes, hideous to behold; and these do fall down over time, they are white on the outside and as transparent as crystal, though without the latter's shine, except for some silvery reflections here and there, not dense, but put together from many oval little pieces, of which most have four sides, some five and six sides, but quite irregularly, and therefore easily separated by means of just a light tap, and one can remove them and put them back in again: Most have the thickness of a quill, a few that of a finger or a thumb: The Makkassarese take home the pieces that have fallen down, and polish the largest ones until they become flat tablets or squares, which they then use to adorn their creeses; the polishing is easy to do with water and a rough stone, because they are soft and exude a white sap; they do not fire at night, nor do they smell like Flints, just like the aforementioned *Batu gula,* otherwise one would hold them to be Crystals. In *Makkassar* they are called *Batu bakilat,*[9] that is shiny stones.

Saccharites Ambonicus, or Ambonese *Batu gula,* differs from the two preceding ones in shape, but not in substance: the Ambonese kind also occurs in large clumps, which are dull and chalky on the outside, but directly beneath it lies the *Saccharites,* put together from innumerable little pieces, as clear as crystal, but much softer; These pieces are quite irregular in size and shape; those in the first layer are long and irregular, so that one can scarcely find a piece that is six-sided and like a Crystal; Beneath these are other, bigger pieces, sometimes in the same layers, sometimes in the ones directly below that; the layers being separated by a chalky crust: These pieces are more beautiful and more like Crystal, they have one, two, or several smooth sides, but no regular shape, and the other sides are striped, as if they had been put together with small pieces, like unto *Amianthus:* All these small lumps, both small and big ones, no matter how irregular the angles might be, lie so close together, that there is no empty space between them, yet one needs only a slight tap to knock even the largest clumps apart, and one can remove the pieces with one's fingers; since few or none of these pieces resemble dice, they cannot really be true *Argyrodamas,* a name that befits the aforementioned Butonese kind the best, although the Ambonese *Batu gula* is much clearer and more beautiful than the Butonese, which is of a darker substance, with a silvery reflection here and there, and can be separated into flat tablets or dice, which are the characteristics *Pliny* ascribes to his *Argyrodamas:* One will find the best ones here on *Leytimor* in the valley of the *Waytommo,* at the foot of the mountains, and close to the river's edge, sticking in other rocks in large clumps, the coarse ones and the ones that are similar to coral are of an impure, mixed substance, just as the true or six-sided Crystals have another russet or gray stone for their Mother or foot, that is of the same substance. Nor do true Crystals grow that close together, although the pieces have their feet close together. And all Crystals grow outside the body of the stone they are growing from, as if they were the fruit or buds of the selfsame stone, the *Argydrodamas,* both the Butonese and the Ambonese kind, is covered on the outside with the aforementioned crust. In the river close to the red mountain, one will find a baser kind of *Batu gula,* put together mostly with long pieces, like a coarse *Amianthus,* without tablets or surfaces: besides the shape, *Batu gula* differs from crystal in that *batu gula* always has a dull shine, in that it can be smashed with just a slight tap, in that if they are struck against one another at night, they do not spark, nor does it have a flinty smell when rubbed, all things a crystal must do: One could compare this Ambonese *Batu gula* with that imperfect kind of bastard crystal, which is called *Gletscher*[10] in Switzerland, though it does not seem to have its origin in a stony sap, which is sweated out by the cold, moist rocks, but seems to come from some obsolete ice.

We have no picture of the Argyrodamas, *as described by Mr.* Rumphius, *because he did not send us any, perhaps because they are shapeless things to His Honor, just like the ones we have, which are also coarse clumps, not worthy to be depicted: The miners here call this stuff* Crystal talc,[11] *sometimes* Ice crystal[12] *as well; of which there are several kinds, and the most excellent ones come from England, and in particular from Cornwall: They have a beautiful silvery shine, white, and mostly transparent; Its shape is always that of a crooked square, flat, and at the edges coming to a sharp ridge, as if they had been polished; they do grow that way naturally, however, which is proven by the fact that often another one will stick out in the center, or on the side, that has the same shape. See on plate* LII. *a large example by* No. 1. *and another by* No. 2., *which has half of another growing out of it: The substance and characteristics correspond with the description by Mr.* Rumphius.

CHAPTER 21

Crystallus Ambonica.

On Ambon and on the surrounding islands one will find a kind of plain Crystal[1] that grows on coarse rocks in the same manner as the true kind: One finds few pieces that resemble the right kind, and they are small, not more than the thickness of a quill; the other ones are usually full of cracks, scurfed,[2] of a dirty color and uneven; some are nothing more than dark pebbles on one side, a dirty white or yellowish, and on the other half transparent like Crystal: they all have 6 sides, just like a Mathematical *corpus Chrisma*[3] or right-angled *Cylinder,* and they end on top in a point that has as many sides as a Diamond, though it is often broken off: The largest ones I saw on Amboina, were the length and the thickness of a finger: they grow like a fruit on rough and coarse mountain stones, always a host of them together, not straight up, but at an angle and entangled, usually being one or two large ones, surrounded by others that become smaller and smaller; one will also often find a crystal-like crust on the surface of the stone, which serves as their foot: their natural place is on large, gray rocks, that have small holes in them, both above the ground as well as under it, even those that are under water; the largest ones will be found in high mountains on such cliffs that overhang, where there is the source of a river, and one will find these holes covered with small crystals on the inside, a joint of a finger long, the thickness of a quill or an oaten pipe, all of them with six sides and pointed, the sides are all equally wide, but irregular, so that at times a whole side has grown onto it; which gives the stone five sides: One will find these small pieces scattered in rivers, sometimes in fields as well, when one is digging, though they surely did not grow there, but were washed down from the mountains; for it appears that these little stones stray from their mother rock either because they are ripe for it, or by means of a windstorm, downpours, or in some other way, and are washed away by the rivers; and so, for instance, the River *Waytommo* conveys many of these little stones, and if one digs 1 or 2 feet down into its steep banks, one will find whole layers of coarse, russet, and gray stones, which are covered with these little Crystals, especially in the little holes, also in broken stones that are hollow inside, so that the big rocks have big, and the small stones have small Crystals: In the country around old *Hative* are some rivers, where one can see big rocks some 6 to 8 feet below the surface, and there are fine, big, pure crystals on top of them. There is a beach near *Lima*, a village on the Hitu coast, that is covered with large, gray stones, both on land and in the water, and their holes are full of little crystals, pure ones, though small: There was a certain soldier in that Redoubt's Garrison, who still had the small chisel, pliers, and little hammer, that a Soul Merchant from Amsterdam had given him, so he could snip diamonds and pearls straight off those rocks, because, as the Soul Merchant had told him, he would find them on rocks all over the Indies; That blunt Jeweler had chopped off several pieces already, rock and all, which he, upon his return to Holland, was going to shove in the Soul Merchant's face, as a token of gratitude for having been deceived like that; but an early death prevented this evil intention, and I am only adding this here to uphold the veracity of those honest Merchants of Souls.[4]

Another example of how these crystals grow under water; there is a mountain called *Tolocco* on *Bima*,[5] which is their refuge,[6] about a day's journey from the King's fort on the beach, which is flat and wide on top, covered with seed land and fruit trees; behind the Village, on a higher hill, is a hole with fresh water, where the King goes to perform *Batappa*,[7] and where a *Djing* or Devil appears to him in the water with large buffalo horns on his forehead, and who points out these stones to him on rocks in shallow water.

Batappa is a Godless relique of their Heathendom, which the Moors perform against their law, and, therefore, in secret; when they desire something from a *Djing*, that is, *Daemon* (which they distinguish from Satan or the Devil), or when they want to learn a trick, or want riches, or how to be lucky and invulnerable in warfare, how to rob, steal, or commit thievery, gamble, or love, etc., they go to such distant places and high mountains, stay for a while, day and night, and bring some offerings to the *Djing*, firmly resolved not to be scared by its appearance nor to let themselves be chased off, and so the *Djing* finally gives them a small piece of wood or a little stone, which they are supposed to wear in order to get the things they prayed for, and so they even think that they are Religious in their fashion, and this is why they call these crystals *Batu Djing*.

One will also find beautiful, large crystals in the goldmines on *Sumatra*, more than a finger thick and the length of a joint, with pure and smooth sides, with a perfect point, as if they had been sharpened, but the edges are irregular, some lie loose on the ground, others are stuck to pieces of gray and white stone, and it is likely that the loose ones were also attached like that, for I do not deem it possible, that any crystal would grow loose and without a foot in the earth; one does find some fine pieces here on *Amboina*, several feet below the ground, when one is digging wells, but one can readily perceive, that they were broken off from something: It is noteworthy, that the entire plain, that has *Victoria* Castle and the surrounding town on it, is heaped-up soil, partially done by people, but for the most part the result of rivers overflowing their banks; the proof of this is, that when one digs a well, some 8, 9, or 10 feet deep into the sandy soil, one will find many things, like *Coconut* husks and Kanary nuts, which did not grow there.

The Ambonese crystals are scorned because they are so small and scurfed, but the Javanese take the finest pieces, and know how to grind them down to tablets and points, and how to put them in rings, so that they look like Diamonds, though they can be distinguished from them at a glance; that is why they would rather put some painted foil[8] under them, thereby faking Emeralds and Rubies, and one sees them often on their creeses: The Natives also use them for medicine, which requires a great deal of faith, for they rub them on a smooth touchstone with water, howbeit nothing comes off, yet they give the same to drink to extinguish the fire of fevers and other hot sicknesses, for this purpose choosing those that grow under water, and this is the reason that, every so often, *Radja Bima* has to send a host of these Crystals to the King of *Makkassar.*

There is another kind of Crystal on *Ambon,* which one will find in big pieces in the mountains, to wit, the size of an egg or walnut, not six-sided but irregular with many corners, like a piece broken off a larger clump, otherwise reasonably transparent and clear: It is still not known how these grow, because one will find them mostly in the high cold mountains, where few people go, and since the Native is not curious about it. I have found them stuck among the roots of large trees, undoubtedly washed there by the rain.

One will find such small pieces on *Nussatello,*[9] [they are] small, very clear, and veer towards yellow, like some yellow Diamonds, but also totally irregular, and as if broken off, having such sharp corners, that they can cut glass to some degree; some small six-sided crystals can do the same, and these are found on *Nassalauw*[10] by the Marcasites:[11] The Sea Eagle, called *Kiappa* in Ambonese, seems to delight in these little stones, because on *Manipa,*[12] a nest of such a sea Eagle, that was up in a tall *Casuaris* tree, was once blown down by a wind, and in it one found four or five quite large six-sided Crystals, of which the largest one was the length and thickness of a finger, surrounded by smaller ones, and all of them attached to a piece of gray rock, which the Eagle had found in the mountains, and perhaps thinking it was prey because it shone so brightly, carried it to its nest; but since those on *Manipa* say, that they often find such stones in those nests, one should conclude, that the Eagle must have some use for it, wherefore they gave it the name of Eagle Stone: About other stones found in such nests, see the subsequent Chapter on *Aetites.*[13]

For the difference between a true Crystal and *Batu gula*[14] or *Argyrodamas,* see the addition to the previous Chapter.

There are no illustrations of these Crystals either, wherefore we thought it necessary to add a few rare ones here: The ordinary crystal, mentioned in the beginning of this Chapter, is depicted on plate LII. *by* No. 3. *being six-sided:* the Little Mountain *shown by* No. 4. *is like all the smaller kinds, that are strangely thrown together in a huddle; this one has grown like that.* No. 5. *is another Little Rock, that from one point grew a six-sided piece, that comes to a point with flat sides, as if it had been ground, having three wide and three somewhat smaller bands or sides.* No. 6. *is a piece of a pyramid, also six-sided all the way to the top: and all those sides are covered with countless sharp six-sided Crystals, which all come to a point; they are as wondrously fashioned, as I have ever encountered.* No. 7. *is a piece of clear Crystal, and inside it one can see an entire Landscape with hills and dales, all of it covered as if with a green moss.* No. 8. *is a small brown* Little Mountain, *that has some of them lying on it that look like Pyramids that have fallen down, and the same are very white and knobbly like* Shagreen leather.[15] *The rock that is depicted by* No. 9. *has, on a brown background, some white candied*[16] *floes or flat pieces, all helter-skelter.* No. 10. *is a* Little Mountain *of yellow Topaz, resembling Crystals that are bestrewn with pure white kernels, and overgrown with* Marchasita *or* Little Mountains *that have a metallic luster.*

CHAPTER 22

Silices Crystallizantes. Batu Dammar.

There is yet another kind of small Crystal on these Ambonese Islands, besides the ones just described, which one cannot deem to be more than clear little Pebbles, but since some among them have such a clear reflection that they come close to resembling precious Stones, which they seem to provoke, I wanted to honor them with their own particular Chapter. I call them *Silices crystallizantes;*[1] the Malay *Batu Dammar,*[2] because they greatly resemble the clear Ambonese *Dammar batu.* Some of them are partially transparent, somewhat round on one side, the rest irregularly angular, sometimes clear and sometimes transparent like ice; some have a blue sheen, like the white of a cooked egg; others, when you tilt them, have a yellow reverberation inside, like a small light that darts from one spot to the next;[3] some so much resemble clear pieces of *Gummi tragant,*[4] that one would easily hold them to be such, if their weight and hardness did not prove otherwise: One finds all of them in Ambon Bay, on its north side, on stony beaches, near *Hative:* On *Nussatello, Locki,*[5] and on the south side of *Buro* one will find other little white Pebbles, the size of a hazelnut, some round, others flat, even and dull on the outside, inside transparent and of a uniform substance, some are yellowish, others bluish, so that one could think they were rough Diamonds, and especially if the ones from *Nussatello* were polished, they would not yield to Amersfoort Diamonds,[6] but that has not been proven as yet. Similar little Stones, though slightly larger, were given to me as true Mountain Crystals, which come from the coast of *Coromandel,* but I cannot accept that, since I hold it to be a substantial characteristic of crystals that they are six-sided by nature.

CHAPTER 23

Vasa Porcellanica.

It is not my intention to describe Porcelain in detail, since the same can be sufficiently found in other Authors, especially in the precise histories of the Chinese Ambassage as described by *Johan Nieuhof;*[1] I will only note here some things, concerning that precious Pottery, which, to my knowledge, were not mentioned by other Authors or, if they did, obscurely: First I will render my judgment on where the name Porcelain comes from, to wit from a certain kind of whelk, the ones called *Concha Veneris* and *Murex Mutiani* in Latin, and described earlier in Book II Chapter XXIII: These Whelks are today called *Porcellanae* by the French and Italians;[2] from this (I say) the Chinese Pottery got its name, and not originally in *Sina,* but as it seems in Spain, and more properly in *Majorca,* where one makes fine and white earthenware which got this name because of its smoothness, which it shares with those Whelks: Or one should say, that the Chinese Pottery is so called, because one has persuaded the world, that it is made from smooth seawhelks, smashed, mixed with the white of eggs, and buried I do not know how many years in the ground: It does not have the same name here in the Indies, among the Malay, because it is generally called *Manko,* but that really refers to deep Bowls: *Pingan* are all sorts of Dishes: *Piring,* Table plates: *Tsjauwan,* Cups: *Mamolo,* large Vessels with a neck, which are put on Cupboards for show.[3] I also feel that *Pliny* and the Ancients called this Pottery *Myrrhina Vasa,*[4] which was very expensive at that time, and was only to be found in royal treasuries: But it seems, since the world today is as if crammed full with it, that nature wanted to take this excellent Pottery away from us by means of the Tartars, Haters of all that is beautiful, who chased away the old Chinese masters, and destroyed the practice thereof:[5] In order to fill this want, there appeared some years ago several masters in the Province of *Quantung* near the city *Tikoa;* but since they did not have the right earth, they took white marble or more often alabaster, which allows itself to be easily pounded, and made this dust[6] with water into a dough, which was then shaped into small Dishes and Bowls. And though these are a clear white and are shiny, they do not have the blue reflection of true Porcelain, and are sold for a lower price: And from *Japan* there now comes also a surfeit of Porcelain to provoke the Chinese, which the native Chinese can easily detect because of the *characters* and figures, but our Nation makes the distinction, that the Japanese [porcelain] never has the clear blue that one sees in the Chinese [kind], while the Japanese kind has three rough little knobs on the underside of the bottom as well; the greatest difference is that true Chinese Porcelain does not crack when one puts hot broth in it, while the Japanese does so very easily, unless one heats it first: One can amend this deficiency however, if one places the Japanese stuff in a large kettle, pours water over it, then boils it thoroughly, and afterwards let it cool on its own.

What one considers the best Porcelain is the kind that is painted inside with the flower *Serune*, which is the *Matricaria Sinensis*, for which see my *Herbarium lib. 8.*[7] and wherefore one calls such Dishes *Pingan Serune.* For a more exact description of how Porcelain is made, see *Nieuhof* p. 9.[8]

One has found that today's Porcelain, made under the Tartar government, and called Crackle ware,[9] once it is brought to Holland, cannot endure the cold of winter, and will burst into pieces, which was never heard about the old ware.

I shall add here the description of Porcelain by Thomas Brown, in his Book II (*concerning common errors*) in Chapter V:[10] which, skipping the false opinion or error, begins as follows: But *Gonzales de Mendoza*, a man employed into China from Philip the second, King of Spain, upon inquiry and ocular experience, delivered a way different from all those that came before; For inquiring into the artifice thereof, he found they were made of chalky earth, which, beaten and steeped in water, affordeth a cream of fatness on the top, and a gross subsidence at the bottom: From the top, he says, they make the finest dishes, the coarser ones from the bottom stuff, which, after they have been shaped, they paint, and they do not bury them in the ground for 100 years.

And thereafter is in all volumes the story of Linschoten, *a diligent inquirer, in his Oriental Navigations. Later confirmation may be had from* Alvarez *the Jesuit, who lived long in those parts, in his relations of* China: *that porcelain vessels were made but in one town of the province of* Chiamsi; *that the earth was brought out of other provinces, but, for the advantage of water, which makes them more polite and perspicuous, they were only made in this; that they were wrought and fashioned like those of other countries, whereof some were tinted blue, some red, others yellow, of which color only they were presented unto the king. The latest account hereof may be found in the voyage of the Dutch ambassador, sent from Batavia unto the emperor of* China, *printed in French, 1665; which plainly informeth that the earth, whereof porcelain dishes are made, is brought from the mountains of* Hoang, *and being formed into square loaves, is brought by water, and marked with the emperor's seal; that the earth itself is very lean, fine, and shining like sand; and that it is prepared and fashioned after the same manner which the Italians observe in the fine earthen vessels of* Faventia *or* Fuenca; *that they are so reserved concerning that artifice, that it is only revealed from father unto son; that they are painted with* indigo, *baked in a fire for fifteen days together, and with very dry and not smoking wood: which when the author had seen, he could hardly refrain from laughter at the common opinion above rejected by us.*

Now if any inquire, why, being so commonly made, and in so short a time, they are become so scarce, or not at all to be had; the answer is given by these last relators, that under great penalties it is forbidden to carry the first sort out of the country. And of those surely the properties must be verified, which by Scaliger *and others are ascribed unto china dishes: that they admit no poison, that they strike fire, that they will grow hot no higher than the liquor in them ariseth. For such as pass amongst us, and under the name of the finest, will only strike fire, but not discover aconite, mercury, or arsenic; but may be useful in dysenteries and fluxes beyond the other.*

CHAPTER 24

Pingan Batu. Gory. Poison Dishes.

Where these are concerned, the people from *Europe* have been persuaded, that the finest Porcelain that is called Crackle ware,[1] should have the property of cracking or crackling, if one puts a poison on it: But when examining the Chinese here about this, they cannot point to any Porcelain of the usual kind, no matter how fine it might be, that has this property; but there is a kind that does, though it differs a great deal from the usual stuff, called *Pingan Batu* by the Malay.

The usual *Pingan Batu* is a heavy, large dish, as if it had been made from stone, level, on the bottom a finger or more thick, half as thick on the sides that have curved rims,[2] smooth, pale green, but one will veer more towards blue, another towards gray; most are uniform of color, though some have flower figures that peep through from under the glaze:[3] One will also find ones that have many round and raised backs on the outside, which were once highly prized by the people of Ternate, and called *Gorange Mangati,* that is Shark's liver, either after the shape of the backs or because of the color of raw Shark's liver. All the other kinds of *Pingan Batu* they call *Suggi:* This kind must be smooth, whole and massive, but if it gets too many little cracks in it, they consider it dead and good for nothing: This pottery was once made in *Sina*, but it had no value there, meant only for the common people, and was indeed a base kind of pottery, and was called *Tschopoa* in Chinese, that is, stone dishes: Today's Tartar government[4] does not export it, and perhaps it is not made anymore, which is why one now values the old kind that can still be found; especially the kind that has white flowers painted on it. The Ambonese, and nearly all the inhabitants of the Moluccan world,[5] used to, and still do, bury these stone dishes in the ground, as they do with all kinds of fine Porcelain, so they will not easily lose it when enemies attack them, or when their straw houses burn down, and they do not do this near their houses, but in the forest gardens, where they will plant a little *Codiho* or *Terminalis* bush,[6] in order to mark the spot: Others bury it in caves or under overhanging rocks, particularly the Alphorese or wild mountainmen,[7] first wrapping the same with a large amount of *Gumut* and *Rottangs*,[8] and they do not dig up this buried treasure until they have a great feast or want to sacrifice to their Devils; they are so secretive about that buried stuff that they will not reveal it even to their children: With the result that, when they die, this buried treasure remains unknown, until it is discovered by accident, or because of an earthquake, or from a heavy downpour or a landslide; And in this manner our Hunters have found some beautiful Dishes in the mountains. Similar Dishes have been brought in from *Japan* since then, some smooth and plain, some painted, containing more gray than green, in beauty not giving way to those from China, except that they too have that fault[9] of easily breaking or cracking when one puts hot food in them or places them on coals, but which are improved in the same fashion as we described in the previous Chapter: These are now common, and one can buy them for 1 Rixdollar a piece, or 1-½ for a painted one, which one sometimes can trade to the Alphorese for a nice profit.

Rare or costly *Pingan Batu* are also flowered and flat dishes, somewhat smaller than the foregoing, to wit, like a common food plate with wide edges[10] that are not curved, nor of the same color or substance: Most of them are thick, not transparent with that wondrous reflection that leaves one unable to describe its true color. The base of the glaze appears to be yellow, on top of that a light purple: The bottom is light blue or gray blue surrounded by a circle of the same color that does not reflect, here and there with brown spots or speckles, and if one beholds the dish in a shadowy spot, it seems green: The outside is a gray blue, and if one carefully examines the inside, one will discover it full of very fine cracks, which one cannot feel, however, except for the thin little ribs on the sides, some are browner, some blue, some green; one will also find some that are partially transparent near the edges and which are held to be the most costly, because one will give 70, nay, even 100 Rixdollars for each one of them: The Kings of *Makkassar* had two of these, but one can get the ordinary ones for 20 Rixdollars; the reason that these dishes are so costly, comes from a general notion, with which all Natives are quite taken, and one cannot dissuade them from it, although one will not find many that can withstand the test; the test consists of their ability to betray poison, to wit, if one puts some liquid poisoned food in it, the broth has to boil or cast up bubbles, the dish has to crackle and show cracks, or at least change color: If one puts cooked rice in it, it will stay good for three days, and not become sour, though it will dry up; A Chinese *Kyalura* or Receiver[11] from *Grisek* assured me to have tried this; if one rubs lime on the outermost bottom, where the dish is not glazed, the lime will lose its sharpness, so that one can hold it without injury in the mouth: What these things might be, I cannot say for certain, whilst luck has not permitted me up till now to obtain such a dish, which can endure the aforementioned test, except for an ordinary one, of which I discovered the following: we dissolved a sublimate[12] and poured it in the dish, the place, where the water was, acquired a different, that is, lighter color, but we did not notice any crackling nor any cracks, and when we took the water away again, the dish resumed its former color: I had lime rubbed on the undermost bottom, and after a short time the onlookers tasted it, and holding the same in their mouths they said they had not experienced any sharpness; I did it as well, and tasted the natural taste of lime, but since the onlookers said they had not felt anything, I said the same thing as well: I hope not to have belittled this costly pottery, and prefer to think that I did not have the best kind, and recommend the results to others, who have the opportunity thereto, because in these Eastern regions they are so rarely found, that one hardly hears about either one or the other, and those who possess them will hardly show them. Nor can I say anything for certain about its origin, but the usual opinion is that formerly they came from *Sina*, and that the Natives of these Islands buried them here and there: Asking the Chinese about this, they imparted to me the following:

The renowned Chinese Admiral *Sampo*[13] is said to have been the first inventor and work master of this pottery, but they do not agree on where he made it, or from where he got the material: Some say, that he made it when he was still in *Sina*, and when the rough Dishes were still on a bench, before they were put into the furnace, a large bird, according to their description either the *Geruda*[14] or a Griffin, flew over the workplace, and the whole house shook from it; and it is from this that the Dishes got their cracks, which they hold to be a sign of authenticity.

Others contend, that the aforementioned *Sampo* made these Dishes on the Island of *Condor*;[15] that lies near the Country of *Cambodja*, and which is nothing more than a tall mountain, when *Humvuus*,[16] the first Chinese Emperor, after the expulsion of the Tartars, despatched [Sampo] to foreign lands, in order to search for and bring back to his Emperor whatever was most precious in each Country; be that as it may, it is certain, that one will find the dishes mostly in the Countries *Sampo* visited, and where he lived for a while: Among these they mention primarily, *Borneo*, *Makkassar*, *Baly*, *Java*, and *Malacca*, where one will still find the Chinese well near the small mountain [called] *Bukit Sina:*[17] And it is because of him that the Chinese still call these Dishes Sampopoa, and they assign them the following attribute, besides the ones already mentioned, to wit, that they will turn salt water that has been poured into them into fresh and potable water; but the common *Pingan Batu* they call *Tsju Pijan*. Those from Ternate call all rare Porcelain ware *Picca Radja*, that is, Royal Household Goods.

Poison dishes or *Gori* are the same as or similar to these Dishes, though most veer towards the green ones with similar reflection, some in the shape of Dishes, some like Table plates, and others like small bowls: They are bought in *Pegu*, where they come from *Ava*,[18] and are transported to the Old Indies,[19] where they are held in great esteem, so that the Guzatts[20] once paid 150 to 200 Rixdollars for these Dishes, because, they said, they would not tolerate any poison, but since avarice introduced quantity, and since the greed of the people from Pegu made them imitate and fake them, they came to be despised, and, as I am led to believe, they remain in the Warehouse of the Company in *Golconda*,[21] because the Chinese will not buy them, since their conceitedness makes them discount anything that did not come from *Sina:* and it is also not certain if the maker, *Sampo*, knew it himself, or if he persuaded people, that these dishes would not tolerate poison, or if these same Dishes received some qualities from the earth, wherein they were buried for so many years: Because I imagine that one will find Pottery in various places, buried by the Ancients, that was not made by *Sampo;* for example, in the mountains behind *Cheribon* on *Java* they find now and then brown or dark-red bowls and Saucers, also little mugs that are a dirty white with blue flowers and mostly cracked, which the Chinese think were made by *Sampo*, and that passes the test of making lime sweet, but it is more likely that they, especially the brown and red ones, were made in *Cheribon* itself (where they still make red gargoulettes[22] and other suchlike Pottery), and have been for some time in the ground, wherefrom they may have gotten their power to make things fresh again, and also in time might have lost it again.

On *Buton*[23] they call these Saucers *Manco Paso*, because they have on the bottom 1 to 3 black stripes like large nails; the King of *Buton* has one that is an *Asta* or short ell[24] wide, light green, thick, heavy, covered with cracks, yet smooth, and which, as he boasts, can turn salt water into fresh.

On *Belitton* and *Crimata*[25] one digs up not only these Dishes, but also large antique Vessels that are 1 and 3 and 4 feet high, glazed on the outside, yellow or a yellowish green, marked with Snakes, Dragons and suchlike creatures, on top having a wide mouth, wherein one can stick one's head, and are called *Tadsjos;* one surmises that they used to bury the ashes of their dead in them: one does not attribute to them the power to withstand poison, yet they are priced not only because they are rare, but also because one can do good business with them with the Alphorese on *Borneo*, who will gladly give a slave for each one, so they can put the heads of their enemies therein; because it is their fashion that no one may marry a woman or build a new house until he has gotten himself the head of an enemy, which they then proceed to keep among their treasures.[26]

The *Pingan Batu* of the first kind must have a clear sound, and be without cracks, but the second or precious ones have a dull sound, and the cracks therein are praised, not those that came about through banging or smashing, but those that show beneath the glaze, are very fine and are baked into it: And so there appears as well to be a difference between the green Poison dishes and Gorys from *Pegu* which, according to the Chinese, were once made in *Sina* near the city of *Jautscheu* in the Province *Kiangsi* or *Kangsay*,[27] but no longer for many hundreds of years, and which were once brought into these Indies by *Sampo:* They must be lighter than ordinary Porcelain, and make a dull sound, which they have acquired from their age and by having been buried under the ground for such a long time: they are considered quite valuable in *Sina*, not because of any rare powers, such as the Malay ascribe to them, but only because they are ancient Masterpieces, which no one nowadays can make anymore, wherefore these are bought up by the Chinese in the Indies, and transported back to *Sina*.

The story of the Chinese Admiral *Sampo*, and these Dishes, called *Pingan Batu*, which he brought to these parts, as we are told by our own Ambonese Chinese, is as follows:

These Dishes, called *Pingan Batu* in Malay, and *Sampo* in Chinese, were ordinary Dishes in China some 350 years ago, and even though the Chinese Emperors sometimes changed Porcelain as they pleased, both in terms of material as well as shape, these aforementioned Dishes obtained their value as follows: When 324 years ago, or, as *Pater Martinus*[28] states, in the Year of Our Lord 1368 *Hungvu* or *Humvu*[29] chased away the Tartars, and thus became the first Chinese Emperor from the *Tsju* lineage, which *Pater Martinus* calls *Taiminga*, and who ruled for 37 years, he left his realm after his death to his Son's Son *Kembun*, both residing in *Nankin*, but this *Kembun* was driven out by his Uncle *Englok*, *Humvu*'s fourth Son, in the third year of his rule, perhaps the one *P. Martinus* calls *Taichangus:* This *Englok* brought the imperial chair from *Nankin* to *Pakin*, and there among his most trusted friends, who had helped him to become ruler, was an *Ong-Sampo*, Counselor and Steward of his entire house, a very learned, wise, and powerful man, experienced in the natural arts: One time the Emperor asked this *Sampo*, where the expelled *Kempun* might be hiding, and whether he was alive or dead, and if there was any chance of laying hands on him, and *Sampo* answered: I can tell from the course of the stars that he is still alive, he hides in distant places, but you will not be able to find him or lay hands on him; the Emperor became troubled and anxious from the thought that *Sampo* could well know where *Kembun* was keeping himself, and might speak to him in secret: So he admonished *Sampo* that if he, being a wise man who was experienced in all the natural arts, were going to betake himself on a sea voyage and look for *Kembun* in every Country, and at the same time bring back something rare for him, he would supply him with ships, men and victuals: *Sampo*, trying to rid himself of the Emperor, by going on long sea voyages, although at the time the Chinese had not occupied themselves for several hundred years with great sea voyages and had well-nigh forgotten the art of navigation, accepted such, assuring the Emperor, that he knew what to do about everything, and had personally written a new Navigation book for his Countrymen; he then equipped a Fleet of about 18 ships that were well outfitted in the 13th year of *Englok*'s reign in the year of Our Lord 1421,[30] when *Sampo* commenced his first voyage: The Fleet was commanded by three Chiefmen,[31] all 3 by the name of *Sampo* and Royal Factors, of whom the first and greatest was this *Ong-Sampo*, Admiral of the Fleet, considered not only a wise and learned man, but also a holy one, one who was a practiced miracle man, so this one took with him, among other things, for his service and household, and not at all as merchandise, all kinds of Porcelain dishes, bowls, and vessels, as was usual in *Sina* at the time,[32] all of which, be it from *Sampo*'s holiness, or by means of some rare art, which the Chinese will not talk about, acquired that power and those attributes, such as that the drinking water in the vessels never stank, that the cooked rice and other food that was put on plates remained unspoiled for many days, that all medicines that were put in the bowls doubled their healing powers, even that the water drunk from them could slay the hot burning of fever; but no one ever said, that the poison, called *Pisun* in Chinese, *Warangan* in Malay, which we call *Arsenicum*, when put into them, would make the dishes crack and change color, which, according to them, does happen with ordinary Porcelain, but that on the contrary, *Sampo*'s dishes should not show any changes, and still rob the poison of its power: And it is this kind of Porcelain that *Sampo* sometimes gave as a gift to the greats of the Countries of *Tunkin*, *Cautschi*, *Cambodja* and *Siam*, finally arriving at *Schor*;[33] where there were so many powerful sorcerers that they hauled several ships up the mountains for him, and finally came to *Malacca*, where he lost some ships owing to storms and lightning, and was thereby forced to rest there and make some new ships: The remnants and monuments of this in *Malacca*, are the small Mountain *Bukit Sina*, and the Chinese Well, that is the only one with good drinking water:[34] He returned to his Emperor after five years, bringing with him beautiful fine white linen, *Sanhosu*[35] or blood coral trees, and *Calambac*[36] or Paradise wood; but *Englok* was discontented that he had lost so many ships and men, and therefore kept him in *Pakin*, although *Sampo* persevered in trying to get new ships and people, and wanted to make a second voyage, which, however, did not occur until the second year of the reign of *Englok*'s Son's Son *Suan Tic*, or 28 years after his first voyage, around the year of Our Lord 1450 when *Sampo* set sail for the second time,[37] visiting even more Countries than before, such as *Borneo*, *Baly* and *Java*, where one is still shown his monuments everywhere, and where he once again left behind those dishes and vessels, which one, though rarely, still finds today, after some have lain buried for more than 100 years according to the ways of these Eastern peoples, who usually bury their most valuable household goods: Finally he came to *Siam* and, as it seems, was not inclined to return to his Fatherland, and this is where he died, after he endured a long contest with a Siamese Sorcerer, to see who could perform the greatest miracles; and it has been reported that there still is a large Palace in *Siam*, that *Sampo* built in 3 days and nights, while the Siamese could erect merely a *Tâ* or *Pyramid* in the same amount of time.

The precious and true dishes, which are called *Poa Sampo*, are as follows: The biggest among them are more than an ell wide, others smaller and deeper, though most are of the same shape as the newly imported *Gori* or Poison dishes, coarse and thick of substance, but not heavy, also producing a dull sound; as to color, some are a dirty white, others yellow, pale green, blue, dark brown, and some also black, but never red, all with fine barely noticeable cracks in the outside glaze, having on the bottom a brown circle of about 2 fingers' width, where they are only a little or not glazed at all; from the shards one can tell that the inner earth is neither white nor as hard as stone, as is the case with the common ones, but the color of wood, and one can scratch it with one's nail, as you can do with the inside of a brick: These same dishes are these days highly esteemed even in *Sina*, so that the Chinese buy them for 40 and 50 Rixdollars on *Java* and in other places and bring them to *Sina*, where they sell them for 100 and 150 Rixdollars a piece.

They do not know the *Gory*, brought to *Amboina* in the year 1684, nor where they were made, just as they do not have any marks for the *Poa Sampo* or *Sampo*'s Dishes; but they do know that they are carefully imitated in *Japan*, and those are bought here for 1 and 2 Rixdollars, although they do not show any of the aforementioned distinguishing marks.

CHAPTER 25

Mamacur or Macur.

I will now relate how an insignificant thing can be counted among the most precious treasures, simply because people considered it such and because they imagined it was of great value. A thick, lumpish Armlet, which our Nation considers to be of glass, and which is indeed a mere product of human hands, made of glass, *Amaus*,[1] and similar clear little stones, of such a width, that one can just get a proper hand through it; on the inside always even, coarse, barely an inch wide, while some have a ridge on the outside, others are somewhat triangular like a roof, but with a rounded ridge, while they always seem to have been made from two kinds of substances, because the inner side or bottom is coarse and dark, while the top is smooth and partially transparent like thick glass; it is generally called *Mamacur* and *Macur:*[2] The Moors *Mamacul:* On Ceram *Mamaur:* Those from East Ceram call it *Dittir:* In the South-Eastern Islands *Ditti;* while they are called *Sely* on *Lety* and *Moa.* I have seen three kinds, which differed considerably in terms of substance and price: The first one, considered the best, and highly esteemed, is a true glass-green with a round ridge, partially transparent like the thickest of English bottles,[3] heavier than normal glass; if one holds them against the light, one is supposed to see some drifting clouds in them, which change when one inclines them, and which the Alphorese consider to be snakes and dragons.

The second kind, which some consider the best, is of a bright blue, also partially transparent, sometimes with some purple in it, with drifting clouds inside or grains of sand: Both of these always have a rounded ridge, no color on the outside, and increase in value depending on how many different colors, clouds and water[4] they show inside.

The third kind is much plainer, though in the South-Eastern Islands, as far as *Timor*, it is more highly regarded than on *Ceram:* These are triangular with a rounded ridge, of green, blue and brown glass, much pitted, and filled with yellow and red lac work,[5] which in time drops out; these spots are so ordered, that there is a red one between two yellow ones, all about an inch wide, running obliquely along the top.

The Natives argue quite seriously, that these are not stones made by man but that they are natural, that they come from either the sea or the mountains, and our Nation's greed is such, that we let them keep on thinking that, since it is one of the most famous pieces of merchandise for these Natives, so that one can get a slave for just a plain one, but one can get 5, 10, or more slaves for one that they consider beautifully watered or clouded; in fact, they will even go to war with one another over one of these: Now where these Armlets come from, is not really known, most likely, the Portuguese brought the same to these Islands and persuaded the people, that these were precious Stones; for it is certain, that today's Priests in *Bengale*, and on the *Coromandel* coast, give such Armlets to those who sacrifice in their Temples; and they surely make them from some coarse glass: Yet one cannot really tell, from what or how they make these things, and how they get that watered effect: It is surprising that the Javanese, who have become quite shrewd from dealing with Europeans for a long time, still buy up the true green *Mamacurs*, and present them to their King as if a great honor.

Now we will relate, what *Ceram*'s Alphorese do with these Armlets; the common man is not allowed to possess one, at least not openly, and it must be a great *Radja*, who will have one, and he will consider himself quite wealthy if he owns these three things, a stone Plate or *Pingan Batu*,[6] a *Mamola* or stone Pot, which is made of white Porcelain and adorned with figures, branched leaves, and dragons, and such a *Mamakur.*

The first two things, as was mentioned in the preceding Chapter, are tightly wound with cotton, gomuto, and rottang, so that one can spend an hour, trying to cut one loose, and they are buried in the ground: The *Mamakur* is also wrapped in cotton or linen, but it is hung from a high beam in the nook of a house, because, as they say, it will not be locked in any trunk: They take it down with the new moon, and sacrifice a chicken to it; they even dip it into the blood of a chicken, saying that this will really bring out the watered effect; when they go to war or on raids, they consult it, and want to foresee good or bad luck in it; it is likely, that the Lying Spirit[7] has something to do with it, and by means of this *κατοπτρομαντια*[8] it shows its servants something they can abide by.

When *Herman van Speult*[9] was Governor during the time when the first of the Ambonese wars was beginning,[10] we overran *Lissabatte* on Ceram's northern coast, and our *Gnatahudi*[11] *Paulo Gomas* found such a *Mamakur* hanging from a beam in the Chief's house: It was blue, veering towards green, like the glass of the foot of fine Wine rummers, watered, also like clouds, and with four colors inside.

Radja Saulau, the most powerful Alphorese on *Ceram*, used to have a glass-green *Mamakur*, and he was constantly fighting about it with his neighbors, so that now one then the other had it; it finally came into our hands, and we sent it to the Fatherland soon after, and had similar Armlets made from glass, and we showed them along with the true ones to those from *Ceram* in order to sell them, but the people from *Ceram* quickly picked their old and true Armlets, and did not want the others, although we could hardly see any difference, perceiving grains of sand but no clouds; so if one desired to make peace with them, one had to return the true Armlet to them: I think that one should not make these Armlets from ordinary glass, but from transparent *Amaus*, just like the stone *Paternosters*,[12] which look like Amber.

President *Simon Cos*,[13] later the Governor of *Amboina*, had to remove such a *Mamakur* in the year 1655, from the village of *Noccohay* on Ceram's northern coast, because the neighboring village was going to war over it, but the Captain *Hulong*, who was in charge of that Region, and who otherwise was well disposed towards us, became greatly displeased, saying that such an Armlet was worth 100 Slaves, indeed, it was worth an entire Village; but they never got it again, because it was lost: As was said before, they enamel these Armlets in the South-Eastern Islands with coarse red and yellow enamel, which falls out very easily, but one will usually get a slave for one; but they consider the ones which have a uniform blue or green color, wherein ply, according to their taste, many clouds, infinitely better: The people from *Seru Matta* and *Baber* have been accused of making fake *Macurs*, but they can easily distinguish them from the true ones; how they do this is not known to me, since I do not know if the aforementioned Islanders are versed in the art of making glass, except those from *Ceram Laut*,[14] who diligently look for pieces of broken green bottles, and then smelt them in a piece of hollowed-out charcoal, and make such beautiful Stones from this, that one would consider them to be Topazes or Chrysoprases,[15] but a truly knowledgeable person can easily distinguish them from the natural ones, if they are taken out of the Armlet, because one quickly becomes aware of the glassiness at the edges; nor do they spark at night: From the same Islands I received a *Makur* that on the bottom was easily an inch wide, on the inside very uneven and hilly, to look at it one would say it was black, but if one held it against the light, it became partially transparent and a dark blue, and it had yellow and brown spots on it that were crooked and all commingled, of which the hardest ones seem to have been enameled, but the protruding ones were filled with lac. It did not spark at night, except on the thick sides of the bottom, where it was struck with an Ambonese crystal and then it sparked a little, as all glassy stones will; an ordinary one is worth 7 Rixdollars, a good one 15 or 16; if it is entirely blue with brownish blue or purple clouds in it: Some Malay and Makkassarese contend that they bought these *Makurs* in *Atchin*[16] and brought them to the aforementioned Islands, which is quite likely, since the Portuguese do not frequent these Islands, or very little, while on *Timor*, where they live, *Makurs* are of no consequence: The real grass-green *Makurs* are not found anymore these days, and if one were to come to light, the Javanese and Malay would buy it, encase it in gold, and cover it with other stones, since they are still of the opinion, that it is a natural Stone.

I have been informed that the glassblowers on the coast of *Coromandel*, make them in the following way; they collect the pieces and clods of coarse glass that has all sorts of colors, and throw them into a pot, and let it stand for several days in a kiln; then they fashion a clump of it, that is green, and other colors of glass according to whether the stuff had been clear or dirty, as the case may be; but if they try to make it on purpose, they do not succeed, and it is therefore quite likely that they make these *Mamakurs* from such a clump, although they themselves only wear Armlets of pure glass.

Mr. Rumphius *provided us with two depictions of the Armlets here described, which are on plate* LII., *together with* His Honor's *remarks; to wit, the one by the letter* A. *was of a dark-green and bluish glass; on the inside smooth and very dark like stone; one could also tell, that the upper Armlet had been fastened together at the corner. Letter* B. *was an Armlet of pure and clear glass, with a raised ridge, which was hollow beneath, with white stripes running obliquely across the circle, as if they had been enameled on it: The first one, by the letter* A., *was more esteemed by the* Papuas, *who said that it had been brought to the* Moluccos *by the Portuguese: The second one, by* B., *they considered made by the Chinese.*

CHAPTER 26

Armilla Magica & Coticula Musae.[1]

Inspired by the aforementioned *Mamakur*, I have to tell about a similar Armlet, but of an entirely different substance, and a black stone as well, about which the following was passed on to me: In the year 1668, when I was still living in *Hila*,[2] the following two things were offered to me for sale by a Moor from the village *Mosappel*, who, as he told me, was forced to do so because they no longer brought any luck to his family.

The first item was a stone Armlet that was so large, that one could easily get it over one's hand, triangular, flat on the inside, the width of a small finger, coming to a blunt ridge on top, dull, without any luster, seemingly made by human hands, its substance was like hard and black Slate, with silver sparkles here and there, like *Antimonium;*[3] I saw such large rocks on *Huamohel*, especially on the beach near *Erang:* One could easily scrape these with a knife until one got a gray powder; the seller's Uncle had obtained it as follows: While he was working in the forest on the hook of *Mamoa*, cutting some vines for his weel, he heard something fall down from the sky through the branches into the underbrush, and when he looked for it he found this Ring, which at first had been small, but which gradually grew to the size [it had when he showed it to me]; he believed that a *Djing* or *Daemon* must have favored him, and had destined him to have this Ring for his use; and he always wore it on his hand, when he went to war, especially if he were to storm a place, since, as Captain, he often executed such an order and according to his nephew, no one thinks he was ever wounded in his life; or lost a drop of blood; and when he had reached a ripe old age, he died poor and without a male heir, leaving a daughter behind, who married a stranger and was left by him, [and this man] left this Armlet and the subsequent stone as a rare treasure to his friends; but his aforementioned nephew, not willing to endure the luck just mentioned, whereby he would come to know poverty and loneliness in his old age, sold these two things to me, assuring me, and I completely agreed with him, that they would no longer have any powers with me, which they say of all Curiosities which one has not found oneself or that were not given as a gift, but were bought with money.[4]

The second item was a Stone, by the looks of it a black Touchstone once, smooth, rounded and slightly flat, about the size of a round plum, as it is delineated[5] in the Ring that was just described: The man who had discovered the ring had found it [the stone] stuck in the trunk of a *Pissang* tree,[6] not very far above the ground, in that same spot on *Mamoa*, and whilst he had found it in such an unusual place, he honored this one as well and kept it for the use just described, wearing the same in his belt when he went to war. I could make nothing more of it than a black Touchstone, although it was a little too greasy and smooth to test gold and silver well; it might have lost its testing power,[7] because it had been so long in the moist *Pissang* trunk, or because it had been smeared with musk and other grease, as they do with those stones which they keep for superstitious reasons: My guess is that it is a Thunder stone[8] because one will not find Touchstones in the place where this *Pissang* tree grew; but since the trunk was whole and unblemished, it cannot be that either; from which follows, that this little Stone happened to have been thrown there, and was enclosed by the trunk of the growing *Pissang* tree, as happens more often, when ordinary Stones are found in the lower part of the trunk near the root, and which undoubtedly had been lying there before the trunk began to grow.

In the year 1684, they brought me a large, flat and coarse Stone, that had been found at the bottom of a *Callappus* trunk, and which might have been put there by the planter on top of the Coconut as a lid, when he planted the nut: Both of these are more extensively discussed in the next Chapter.

CHAPTER 27
Mutu Labatta.[1]

Although this Chapter can also be found in Book Twelve of our *Herbarium;* I still wanted to include it here, because it should be with the strange Stones; but what is remarkable, is that, though it occurs mostly on *Timor,*[2] *Solor,*[3] and neighboring Islands, where the Portuguese and Dutch have traded and lived now for many years, its origin and source are still completely unknown to our Europeans, while the Natives who wear it are so uncertain and diverse in what they report, that one does not know, what one should write about it: I have taken great trouble to find out its origin and nature, with the natural inhabitants of those previously mentioned Islands, as well as with my friends, who are in Command there, and who were also charged by certain Gentlemen to find out some answers, but it was all for nought, as I will disclose to the reader in the following statements.

Jacob Wykersloot,[4] who was the Chief of *Timor,* and who resided at *Cupan,*[5] sent me the following missive in the year 1680: One cannot rightly find out what *Mutu Labatta* is, or where it is found, because the Natives from the East side of *Timor* think it comes from the Western part of the Island, and those from the West think it comes from the East, while those from *Rotty*[6] are of the opinion, that it comes from *Sawo;*[7] but when one asks the older

Rotinese about it, they will tell you that they got the *Mutu Labatta* from an Island, that drifted past *Rotty* and *Sawo*, and which can no longer be found, and this is only a guess or a fancy of these Nations, so that no one really knows where it comes from, nor can they explain the powers of the same; but we know from the Murderer *Talo*'s Brother, that the *Mutu Labatta* was obtained on *Sabo*,[8] about a musket shot from the *Village Timor*, from a deep hole on a small mountain, almost as long as the length of a man, the way one sees it now, with holes and as if polished, but, since they all died shortly thereafter, the people from *Timor* and *Talo* himself filled up the hole with soil that was mixed with stones.

Radja Salomon,[9] the exiled King of *Ade mantuttu*, on the eastern corner of *Timor*, who was here on *Amboina* in the aforementioned year, and who is a cautious man with a great deal of experience, could only tell me that the *Mutu Labatta* is commonly worn in his Country as well as in the Islands further to the East, up to *Tenimber* and *Timor Laut*, but the Natives can only say about its origin, that it comes from the West coast: At the end of the same year, he also sent me a small string of beads, which he said were made of the true Mutu Labatta, being 1-¾ feet long, with flat kernels strung tightly together, big ones in the center, and then gradually becoming smaller towards the ends, of a different shape from even the smallest coral stones, but of an orange red, weighing 7 *dubbeltjes*, and estimated by him to be worth 1-½ tail[10] of gold or 15 rixdollars.

When I inquired of several Natives from *Timor* and Rotty what kind of growth this was, they told me so many different things, that one would think they had agreed to hide from the Dutch what *Mutu Labatta* is and where it occurs; because I refuse to believe, that so many Natives, who all wear it quite commonly, do not know where such a thing comes from: Those from Rotty say, that it occurs on Timor, in the mountainous region near *Suuneba*, usually called *Sonnebay*, around those places where gold is found: One will find it in the mountains as small and angular kernels which, when held in the fire, melt together, or, as others say, do not melt, but only become hot and soft, so that one can easily tap them into small flat slices with a hammer, and then pierce them with a hot piece of wire: Those who stick to the smelting say that one should string many kernels on a wire, and hold it over a fire, until they have melted together, and then rub it with another iron until they are smooth; and when this has cooled, one will have a long, small pipe, which one can then break into the pieces one desires, grind these flat at their ends, and then string them on threads: There is one kind that is a dyed yellow, striped on the outside like coral, drilled through inside and black, which seems to be pieces of tobacco pipes that were used a long time ago: now if you ask the *Timorese*, where this grows, they will all deny that *Mutu Labatta* grows in their Country, and direct you to *Sawo*, perhaps it occurs in both places.

So if we skip its place of origin, we will describe it the way we found it: Being kernels of different shapes, most like thin, little slices, some like small cheeses, others like pieces of pipes, in substance resembling pale or yellow coral the most, the majority a reddish yellow, some more russet, others paler, so one would think they were coral that had been worn for a long time already, if they had had that regular shape, and it comes most likely from the mountains, and is therefore a mineral, occurring in both small pieces and long branches, which will be naturally hollow inside; now how fire smelts and forms them, we defer to a closer examination: At least, I did not succeed, because when I held the little pieces of Mutu Labatta over a fire, they became red and glowed, but they did not become soft at all, and when forged, it did not flow.

It is divided into two kinds, and the biggest kernels, which resemble pale coral the most, are called *Ua-Boa* or *Waboa* by the Natives of the aforementioned islands; they put only a few of these on a string, and they are costly, with some kernels the size of hazelnuts, and their *Orangcay*[11] only wear three of them on a string around their neck: The second kind they call *Tzeda*, which are small kernels, more yellow than russet, and much lower in price; besides these two true kinds, there is also a bastard *Mutu Labatta*, which they fashion from ordinary little stones, and even seems more like a baked earth, and they mix these among the true ones in order to make the pile look bigger; but others say, that they receive this bastard coral from the *Kelings*[12] on the *Coromandel* coast, who bake them in their Country from a particular kind of earth; other strands that are worn by the common man and by women, are 3 feet long, and have little lead Pyramids hanging from the end, as if they were tassels; these strands have a lot of rubbish mixed in, of which not more than a fourth is real, mostly 2 pairs of large kernels in the center; The rest is a mixture of Crystal, Amber, black and colored Glass coral, as well as some copper kernels, which decrease in size away from the center, but all of them have a particular number, because they make up these mixtures, so that each one can recognize his own strand.

To us and to the Indians it is only known by the Malay name of *Mutu Labatta* or *Muttu Labatte*, which seems to mean a stone Pearl, though this name is unknown to the Timorese and Rotinese, who only know the previously mentioned *Ua-Boa* and *Tzeda*.

In the aforementioned Islands of *Timor*, *Solor*, and the surrounding ones, up to *Tenimber*, there is nothing more valuable, except for gold, than this *Mutu Labatta*, which usually is paid for with half the weight of ordinary Timor gold; there is not an *Orangkay* or his Wife, who does not wear some kernels of this *Ua-Boa* around their necks: Nor is there anything better with which to buy slaves in these Islands than *Mutu Labatta*, since one can get a slave for two strands of the aforementioned length.

The Honorable Company took a great deal of trouble to find the real mine of the true *Mutu Labatta*, on the Island of Greater *Sawo*, which lies West-South-West of *Timor*, but it lost a ship doing this, whereafter they did not go on, since there was some treachery there, as was intimated in Mr. *Wykersloot*'s story; since there are such a large number of *Orangkays* on *Timor* and *Rotty*, it is reported that one village had 100 *Orangkays*, though there were only 109 subjects, and yet all the *Orangkays* had to wear some of this *Mutu Labatta*, so one can conclude that there must be a lot of fake stuff, with which those that have to trade in those places, must be thoroughly familiar: After these dirty people have worn them for a long time, these Beads will turn very pale and black inside, and they will clean them by rubbing them softly in a watery lye and then dry them: I tried the same thing with some small pieces, which were striped and like pipes on the outside, and found that they would indeed become clean, though they turned pale, whereafter they were grayer than they were before, so that it must be a particular kind of lye. It otherwise has no known Medicinal powers, so that there can be no doubt that the only reason they are so valuable is partly due to the fancy and whimsy of the Natives, who probably wear them for luck and health, and partly because it is found so little and so rarely; be that as it may, it is certain that the Natives will not tell us where they find it, being the envious Nation that they are, just as they also keep other valuable products and metals hidden from us.

CHAPTER 28

Mare Album. White Water. Ayer Puti.

Twice a year the sea around the Banda Islands turns white and shines brightly not by day but at night, and gives off light in such a way, that one can hardly distinguish the sky from the water, so that there is some danger in going to sea in small vessels at that time, because one will not be aware of the approaching heave and fall of the waves, since they have the same color as the sky: The water has its normal color during the day, and one cannot detect any difference, neither in smell nor in taste, except that the Boats that are on this water rot more easily than at any other time; as soon as the Sun has set and the darkness of night has commenced, one will see the water of the Sea as white and clear as some rotted Wood, just like the slime of seacats and herring in our parts, and the further out one goes, the whiter the Sea will appear; this is why on some charts the Sea between *Keram* and the familiar South-Land,[1] is called the *Milk-Sea*, usually white Water, in Latin *Mare Album*, although one should really call it *Mare Noctilucum*, in Malay *Ayer Puti.*[2]

Its season is midway through the East or rain *Mousson*, being the months *June*, *July* and *August*, and it is divided into little and big white Water.

The little or first white Water begins with the new or dark Moon in *June*, which paints the sea slightly white, but does not give it any clear light, and it gradually lessens, but before it can entirely fade, the big white Water arrives with the new Moon in *August* and lasts to *September*, when the entire sea around Banda, as far as the eye can see, shines white and bright, the further at sea the whiter it gets, with only a little normal water here and there near the beaches; yet it does not keep to a particular time, because it can appear earlier in one year, and later in another; indeed, in some years it is not seen at all, or, at least, very little, for the most part it is the clearest when the South-East wind is blowing hard, and makes for showery and rainy weather, and this seems to affect the entire sea: The most dangerous thing is when the sea, especially at night, rises with a heavy swell, even though there is not a breath of wind in the sky, an indication that some vapors must come up from the bottom and impregnate the waves: The white Water comes early and in abundance after the South-East wind has come through early and strong, and it is at its peak when the wind veers towards the south, bringing a lot of dark rainy weather: For I noticed in *Banda* in the year 1663,[3] when the wind blew easterly, even at the end of *August*, when I sailed from *Banda* to *Amboina*, that there was hardly any white Water, except for a lot of fiery sparks, with which the sea seemed to be filled, just like the sea I saw west of *Amboina*, in the month of *December* of the year 1668, when the North-East wind blew, without rain; on the other hand, when the south or south-East wind starts late, one will also have the white Water late, and in some years it can last all the way into *October*.

One will see this white Water as far as our people from *Banda* have sailed to the south-Easter or southern Islands, to wit from *Arou* and *Key* to *Tenimber* and *Timor laut*, and from there up to *Timor* in the West, in the North close to the southern coast of *Keram*, but it does not appear north of the *Liasser* Islands and *Amboina*, but passes the same in the south, so that one can see it clearly from the mountains on *Leytimor;*[4] what its origin is, is still not known, but one generally thinks that it originates in the large bay, formed by the familiar and neighboring South-Land and the long neck of *Nova Guinea;* because our Bandanese seafarers note that it is the biggest and makes the heaviest swell around *Tenimber:*[5] it has happened as well that some coming from *Key* in the year 1664, to the West of *Banda*, some 30 miles from *Key*, found no white Water, but when they drew near to *Banda* they found the

Sea completely white, wherefrom one can conclude once more that it must come from the South-East or south-south-East: It does not mingle with other seawater, which makes for a rare show at night that often strikes terror in ignorant seamen, the way it did with a certain ship, that sailed in *September* from *Amboina* to *Batavia*, and when it had gotten just beyond the straits of *Amboina*, it suddenly went from black into white Water, and they thought that they had run upon a sand shelf; but since they could not sound any bottom, and seeing the sea white, as far as the eye could see to the south, they finally recalled this event, yet they were so oppressed by the Milk-sea, that they hurried full speed back to the black water: Now when this white water has remained within the aforementioned boundaries until the beginning of *September*, it will gradually begin to drift to the West, to wit, its entire *massa* will go along the southern Islands, *Ombo*, *Luhu*, *Bala*, *Ende*, and *Bima*,[6] stretching out 4 or 5 miles south of *Amboina:* From there it starts to break up into large bands, which will go past *Buton*, and, if they do not encounter a West wind, go even further, beyond *Saleyer* and *Makkassar*, until they eventually dissipate and mingle with normal seawater.

The white Water does no notable harm, except for what we said above about the rotting of vessels and the heavy swell, but in *Banda*, where it is at its height, it does cause a scarcity of fish around that time, not only because the fish, since they shy away from the brightness, flee to the black Water, but because even the few that remain, are difficult to catch, because they see the lines and boats much too clearly, and are scared off by the least movement: But then again, it can also cause a surfeit of fish, when the white Water covers the deep and forces the black or normal Water to the beaches, which makes the fish go there as well: When it is over or is going away, one can see the Sea cleanse itself, tossing all kinds of foulness onto the beach; the Sea itself is filled with stinging Sea Gellies, which we call Mizzens,[7] just as in the aforementioned year 1663 I sailed for an entire day through such mizzens from *Pulo Ay* to *Amboina*.

At the end of the white Water in the year 1669, the sea threw such filthy and nasty stuff onto the beach along the entire south-Eastern corner of *Neira*,[8] as well as across from it between *Ranan* and *Wahan*, that no one had ever seen the likes of this before, at least not anyone who was 60 or 70 years old: This excrement looked like a tough slime or *Papeda*, red as blood, that had a nasty sulphurous stench when the Sun hit it: All the fish and crabs that were in the holes where the receding water had stayed behind, died if they had been touched by it, even the harmful land crabs; this stuff was as red as blood during the day, and the beach as if covered with a bloody sheet, but at night it was fiery and sparkled with little stars: In *September* of the same year and the entire next year of 1670 the air was exceedingly dry in both *Banda* and *Amboina*, and during that time there was first a severe outbreak of smallpox among both the old and the young, as bad as the plague, followed by raging fevers and a terrible bloody flux.

The little white Water did not occur on *Banda* in the year 1673, and the big one came late in *Augustus*, because the South-Eastern wind came through late that time.

Our people encountered it in the year 1681 between *Timor*'s Eastern corner and *Kisser*,[9] and the distant sea rose with such a fiery swell, that it seemed as if a ship were sailing ahead of them that had a large lantern on its stern.

What might be the origin of this white Water that lights up at night, I have not been able to determine up till now,[10] and there is certainly enough material here to exercise anyone's mind: I discovered neither information nor curiosity among the Bandanese, because they do not want to waste their time on such trifles, as they call this, and pay no attention to it because it happens every year: in fact, there is no one who wants to observe at what time and with what kind of weather and wind this white Water comes every year, whether it is more or less, earlier or later.

Many people say it is caused by innumerable little animals, that light up at night and paint the sea similar to so many other fishes and stuff in the sea, that light up at night, for example *Sardines*, *Ikan Moor*, and *Gatzje*,[11] or the red fishes that are like the Stone Bream, their innards in particular; especially the Spanish Seacat, *Sepia*, in Portuguese *Siba*, does this, but more so in Portugal than in the East Indies: I am not satisfied with this opinion, since it is most unlikely if not impossible, that so many little animals can all suddenly surface at once, and give a great sea the same color, without any differently colored water in it, nor has anyone ever seen those animals, although one can scoop up the water with a clean vat, and let it stand until daylight, when it is clear again and has the color of ordinary water: Even more so, how could those animals stay together without mixing with that other water, since the white Water is separated from the black as if by a line, and flows away towards the West in large fields, bands and twists: Of which the following is proof: In the year 1679, a ship, sailing from *Banda* around the middle of *September*, going westward towards *Batavia*, when the white Water was over in *Banda*, met the same next to *Amboina*, and sailed for two nights through it until it was halfway past *Buru*, where a western current of ordinary water pressed against it, so that it could not go any further to the west, the more so because it was becalmed: Others suspect, more correctly in my opinion, that it is a thin substance in the shape of a vapor that issues from the bottom of the sea, and which mixes with that water, because it cannot be a thick substance, since the water is not colored during the day; one supposes it to be a sulphurous vapor, that has impregnated the entire bay, which is demonstrated by the many sulphur mountains, which one finds on those Islands, such as *Banda*, *Nila*, *Teeuw*, *Damme*, all the way to *Lery*, and as we know from *Jacob Lemaire*'s journey, there are many large ones on *Nova Guinea*.[12] The second proof consists of the stinking red excrements, the fiery mizzens, as we told before: The *Alchemists* could be of service here, and tell us whether a *Spiritus sulphureus* (sulphurous vapor) when mixed with *Aqua salsa* (salt water) can produce that nocturnal illumination, and the same thing could be done by our *Collegium Naturae Curiosorum*,[13] established in the Holy Roman Empire, which has contrived a *Phosporum liquidum* (which gives off a clear light), of which they sent me a small vial, containing a yellowish water, which, when daubed on something at night did not only produce a clear light, but also showed very fine vapors and small flames, lasting nearly an entire night; but one should note here, that there is still a difference between those well-known things that give off light at night, such as the aforementioned ones, or fish guts, centipedes, the poisonous Campernoyle, and even the aforementioned *Phosphorum liquidum*, which do all fire at night like starlight, and our white Water, which does not really fire, but only gives off a white sheen like snow or milk.

Then there are others, who guess that, during the East *Mousson*, when it rains hard, some kind of material is carried down the big rivers of the South-land and *Nova Guinea*, and infects the entire bay; but since this is not based on experience, we will not elaborate on it any further.

Cusparus Scotus[14] in *appendice ad Physicam curiosam cap. 3.* quotes various things from *Thomas Bartholinus*,[15] *Fortunius Licetus*,[16] *Olaus Wormius*,[17] and others, that give off an unusual light at night, and to which one paid no attention before, such as the freshly slaughtered meat of oxen and sheep in *Montpelieers*[18] in the year 1641 around the skin-like and fatty parts, the head and skin of a *Scorpius marinus*, certain small oysters in a clay-like substance hiding in the sea near *Ancona*, a noble lady in Italy, whose body displayed fiery stripes and rays, a Monk, whose head hair sparked when it was stroked; in England they have also noticed for some years now, that all the freshly slaughtered meat in the halls gives off light at night around the parts that are like skin; the new Philosophers ascribe the cause of this to some tiny, only recently known little animals, which are engendered in the air when the plague is about, and cause it, because they say, this spectacle is never seen, unless there is a great deal of dying: But all these arguments are not sufficient to account for the night-lighting nature of our white water; because in truth one would need an *Apollo*, in order to explain this to us, just as *Bartholinus* wished for his night-lit meat.

It says in *Herbert*'s[19] travel book on p. 16 that they found the sea at four degrees South and North of *Majottos*[20] as white as snow for at least 10 miles, and not because of foam or wind, but when there was a great calm: they understood this to happen every year around the 19th of *September*.

CHAPTER 29

How the Ambonese Goldsmiths Clean Gold, and How They Give It Color.

They take a Chain, or something else made of Gold, that is besmirched and dirty from having been worn too long, put it on glowing coals, until all the dirt and lint has burned off, and the Gold has turned black; to clean this you take a potsherd, put equal amounts of finely ground saltpeter, alum and ordinary salt on it, put the Chain in and pour water over it, so that it is just covered, boil it all together, until most of the water is gone and a thick Syrup is left on the Chain, though one should turn it every so often; when you see that, take it out, wash it in clean water or with some Lemon juice, rub it vigorously, brush it with a small brush, and the Gold will be clean and clear: One normally uses a small kind of Lemon, to wit Limon Maas, which we called *Limonellus aureus* in our *Herbarium;*[1] but *Limon nipis*[2] can also be used.

In order to give the cleaned Gold the color, which the Malay call *Suppo*,[3] you take another pot, rub some sulphur on the bottom, pour a Quartern of water in it, along with some *Tamarind* and a *sesje*[4] of coarse salt, cook it up, dip the Chain in it, and finally rinse it with fresh water: One should dip young Gold only 2 or three times, while for the old you need to do this 5 or six times, until it is properly red.

Gold is also a remedy against *Lapper Garam*,[5] which is a red rash with a painful itching and a biting moisture; Take some Gold, put it in water, and wash the aforementioned illness 2 or three times with it; the Natives say that it is efficacious, though one cannot provide a reason for it, since, in our opinion, Gold cannot dissolve in water.

CHAPTER 30

Myrrha Mineralis Mor.

Mor[1] flows like a thick moisture or honey from rocks on *Crimata*,[2] that are near the shore, whereafter it becomes as thick as porridge, so that one can handle it; it has a slight salty taste, but not bitter, somewhat sandy, it does not occur every year, but sometimes in the dry Mousson, when it flows like thick honey: Mixed with *Ramak Dagin*,[3] usually with some *Djudjambu*[4] as well, it is used as a remedy against looseness of the bowels and stomach cramps.

I call it *Myrrha Mineralis*, not so much because it resembles true *Myrrha*, but because our Indians call it *Mor*, that is *Myrrha;* because it has neither the bitterness nor the pleasant smell of *Myrrha*, just a salty taste without any sharpness, wherefore I deem it to be the perspiration of the briny rocks: A similar substance, but a little whiter, can also be found on *Banda*, on the Southern side of the high land called *Lontor*, where the rocks jut out of the beach, but the Natives do not use it: I also described this *Mor* in the Ambonese Herbal, Book III, in the chapter on *Dammar Selan*, under *Ramak Daging*,[5] since it can be deemed either a *Resina* or a *Mineral.*

CHAPTER 31

Touchstone from the Coast.

These Stones are encased in red copper, black, without a sheen, and jagged, one strikes gold on it quite hard, and then presses it on wax that has been mixed with coals; but if a lot of gold is sticking to the wax, the same is melted.

On the Malabar coast they use a Malabarese Amaril,[1] called *Tsjanitalla*, which is a stone that occurs in large rivers up to *Malabar* and *Goa*, and they make Whetstones from them in the following manner: The Natives take this Stone, pound it very fine, add raw *Gummi Lacca*,[2] just as it comes off the sticks, melt the same and add the ground *Tsjanicalla*, until it becomes a thick plaster, which can be malaxed;[3] then one puts it around a wooden Disk, about a span across, and this clump will glue onto its sides and becomes hard; then one smooths the upper edge, which is about a finger wide, and it can be used to whet all kinds of tools; to wit, one sticks an axle through the Disk that has its two ends resting on two small pillars, and which is then turned with a small piece of rope, just as the Turners[4] do, and one holds the metal against the disk.

CHAPTER 32

Dendritis Metallica.

I heard it said by the Javanese that they sometimes find small pieces of iron in an unknown kind of wood, which they gather in the forest for firewood, and that they hold them in great esteem and carry them on their bodies, when they go to war: Another eyewitness from *Makkassar* told me that he found a small lump of metal in a piece of red and knotty wood (which he had picked up in the forest in order to make a sheath for his creese from it, and which he thought was red *Lingoo*);[1] It was half iron and half copper or, rather, a mixture of both: One can think of many ways that such things get into wood, but none of them seem very probable to me; since, as I said earlier, these metallic substances are broken off by Thunder, it might well be, that the said pieces are *Ceraunia Metallicae:*[2] There does remain the problem, however, of how such pieces of wood can remain whole despite the Thunderbolts, because people told me that the aforementioned pieces were discovered in solid wood; but I believe that the finders were not very precise when they considered all the circumstances, for just as nature can obtain a stony sap from plants, which produce *Mesticae*, it similarly can draw up a metallic substance via the suckers of subterranean roots, which subsequently congeals into a clump inside the wood: This is the Metal that *Julius Scaliger* describes in *Exercit.* 187.,[3] calling it *Metrosideros*,[4] but I am of the opinion that this name properly belongs to ironwood; the Javanese use this to make themselves invulnerable, adding 16 other *Mesticae*, which they tie so tightly to their bodies, that the skin and fat sometimes grow right over them; one of the most important among those 16 is the *Cochlites rarus*,[5] which resembles a small Winkle, and which changes into a white and partially transparent little stone.

CHAPTER 33
Stone Files on Java.

Javanese Smiths do not use Files when they make their creeses, but rather small stone sticks that look like rolls, and which they make in the following manner: they take ground-up Porcelain, melt some *Dammar Selan*[1] or some other hard *Dammar*, stir the said Porcelain in it, until it becomes a thick clump, which they then fashion into rolls by hand, being the length of a hand and the thickness of two fingers: They twirl these rolls very firmly on the creese by means of a string which is wrapped around them (like the Turners do), and in this way they remove the iron as well as with the best of Files.

CHAPTER 34
Stone Bullets and Stone Fingers.

The stone Bullets are a kind of *Goedes*,[1] the Fingers are *Belemnitae* or *Dactyli Jdaei*;[2] both are found in the Xulassian Islands,[3] wherefore we will explain these places a little more: *Mangoli* is the largest of the three Xulassian Islands, but it has only two villages, *Mangoli* and *Weytina*, that are both on the southern side, otherwise it is wild and uninhabited: One sails 3 hours from *Mangoli* until the small bay of *Gorangoli*, which has an anchorage, with three small Islands in front of it; another 6 hours sailing from there, brings one to the small bay *Buja*, where are three small Rivers, and two noticeable mountains two miles Inland, all with the same name; if one sails from *Buja* another three hours in a westerly direction, one gets to the extreme Western point called *Batu Malula* or *Batu Lacki Lacki*, from there one crosses over to *Taljabo:* There is a Rock there in the shape of a Man, and when the Xulassians sail past it they usually throw some fruits at it as an offering, so they will not encounter any disaster there, because it makes for a narrow and very dangerous Strait between *Mangoli* and *Taljabo*, and there is a Maelstrom near this Rock: At its narrowest, the Strait is no wider than a musket shot, but soon it widens out again on both sides; in the center of this Strait of *Taljabo* is Cape *Langoy*, which has a small bay.

On the beach of *Buja* bay, at low tide, one will find certain stone Bullets that are very smooth and roundish; some are as round as if they had been cast, but those are few, of a blue-black color, like inferior Touchstones, which they also resemble in terms of substance and hardness, but not in weight and ability to test, [and they also resemble] the Metals, solid inside and difficult to smash; the smallest ones are like a Pistol bullet, and then they gradually get bigger, until they become like the one, two and 3 pounders of iron [cannon] balls;[4] some are grayish, a few are coarse and russet like iron stone, some are dark red as well, the crooked and angular ones are the most common: The Xulassians look for the roundest ones and use them as Musket bullets and munitions for Pedereroes;[5] one has used them to shoot through a plank, and the Bullets did not break: If one smashes them and rubs them on a stone, they smell sulphurous like spoiled gunpowder; inside is a uniform black substance, and on the outside they seem as if enwrapped in a stone skin, from which one can observe that they did not get this roundness from being rolled around by the sea, but that it is their own nature, like other kinds of *Geodes* in *Europe:* One only finds them on the beach and not in rivers, whether they also might be on *Buja*'s mountains is not known; because the Natives will not go there, neither alone nor with our people, fearing they will be killed or murdered by the Devil that lives in those mountains, and who would smite them with the stones just described; they tell the following story about it: Around 80 years ago there were some populous villages in those mountains, that waged war with each other for a very long time, until finally the Devil procured these Bullets for them, and one is supposed to find other weapons on that aforementioned mountain; one hears at different times on the same mount, especially in the month of *October*, when the North or rain *Mousson* begins, diverse battles with Cannon and Muskets, which one can sometimes hear at *Fatumatta:* In the aforementioned Straits, called *Seranna*, are yet other rocks under water, and the current once drove a Portuguese ship onto them, and they killed the crew and robbed the ship.

Similar stone Bullets, black and very shiny, are also found on the holy mountain *Basagi* on *Baly*,[6] up there on the same plain where the *Pygmaei*[7] or Little Mountain men live, some are white, some are cockroaches as to their skin and hair,[8] and have a pointed hump on their back, so that they always have to lie on their side, they live in small houses, surrounded by a temple;[9] the latter is on an even stone floor, as if from a hard Blew stone,[10] that is so smooth that one can hardly stand on it, and in front of its door stand 4 iron levers, that grow straight out of the earth; all around this stone floor lie all kinds of these round Bullets, which those from *Baly* are allowed to pick up and take with them as a remembrance; such being as it is, this shows that these Bullets were not polished round by the rolling of the waves or by the rain, as some would have it, because they lie [there] level and still; I add *Jerusalem*'s graveyard as further proof, because it is strewn with stone *Ciceres.*[11]

In the small *Langoy* bay on *Taljabo*, is a swampy brackish river, at the mouth of which one will find stone Fingers, shaped entirely like *Belemnites* or *Dactyli Jdaei*, as they are depicted in the *Gemmarium* by *Boëtius;*[12] but these stone fingers are usually broken, and of a weak substance, so that one can cut and scrape them, of a dark substance, mostly a yellowish gray, inside with a heart like a wooden twig, and with many rays on the outside:[13] The whole ones look like the tip of a blunt arrow, all of them with a notch or furrow on one side, along which they will break if one hits them: some are yellowish like *Dammer Selam*,[14] others are russet and partially transparent like dark amber, but these are rarely found, and the Natives keep them to make an *Adjimat:*[15] Rubbed together they smell nasty, or like mud, like *Mangi Mangi:*[16] Nothing is known of their origin; afterwards it was found that the Sea washes them on shore more often in places other than *Taljabo*, at least on the East and North side of the same Island, as well as on *Kelang:*[17] One will find another rare Stone in those same places, the size of a schelling,[18] flat on the bottom and regular and striped on top like the little Whelk called *Umbilicus;*[19] some are brown, which are the hardest ones, some gray, and so soft, that one can scrape them, some sit on a hard pebble: I had one that was found on *Kelang*, a small hand wide, crooked and twisted like a *Carina Nautili*[20] on one side: One will find pieces of the stone Fingers that are like Coral with a hole in the center, and among them are some as thin as a knife, which one could consider *Muttu Labatte*[21] from Timor, if they were of that color.

CHAPTER 35
Ambra Grysea. Ambar.

From the stories told by various Writers about the origin and attributes of Amber, I will only repeat those which seem necessary to me in order to arrive at my own conclusions. This Gum, like unto a Spice, which is counted these days among the most precious items in the world, and which has in terms of weight already risen to the value of gold, is, as far as I know, known among all Nations only by the name of *Ambra*, or *Ambar*;[1] which it is also called by our Malay, except for the Inhabitants of the South-Eastern Islands[2] and *Timor*,[3] who call it, unworthily so, *Ijan-taij*,[4] that is, fish dung: Following the usual way, we will divide it into 3 kinds, to wit, gray or *Ambar grys*, white, and black, which differ a great deal in substance and powers; This chapter shall address the first kind, which one is wont to call *Ambra* and *Ambra Grysea*, because of its gray color. Some call it, because of its excellence and cost, *Ambra Chrysea*, that is, golden Ambar: We mean by that such *Ambar*, that is besmirched or blackish on the outside, but gray on the inside like dried cowdung, tough and never diminished between the teeth, no matter how long one chews on it, of light material, its own odor being pleasant but weak, smelling somewhat of the Sea, or, really, like *Unguis Odoratus;*[5] but once dissolved in a warm liquid, it has a much stronger odor; placed on a hot tin of iron, or, as other people are wont to, of gold, it melts entirely, without smoke, but on coals it gives off a smoke that goes straight up, but it is not much and has a meager column.

One will find various opinions on the origin of Ambar, according to the differing judgments of a number of Authors: The oldest opinion, and the one which well-nigh all these Natives hold, is that it derives from the Whale, but not from all of them, but from a special kind, which the Arabs call *Azel:*[6] Among many others, I will relate here the strange story about Ambar's origin from *Ferdinandus Lopes de Castagneta*,[7] who in the fourth book about the achievements of the Portuguese in the East Indies *cap. 35.* writes, that the choicest *Ambar* occurs in the Maldivian Islands,[8] where there are many large birds, called *Anacangrispasqui* which, after feeding on all kinds of spice-like herbs, cast their dung on the cliffs near the sea, which is then fine and true Ambar, which they call *Pona Ambar*, that is golden Ambar, although it be white, but is the costliest, because it is found only rarely and with much trouble: This dung, hanging from the cliffs in large crusts, is in the course of time beaten off by rain and wind and, falling into the sea in large pieces, floats there back and forth, until it is washed up and tossed on shore, being of a gray color, and is called *Coambar*, that is, Water Ambar, because it has rolled around for a long time in the water and has lost a great deal of its virtue, and this is the second kind: The third and basest [kind] is the black one, called *Mani-Ambar*, that is, Fish Ambar, because it has been swallowed by Whales and other large Fishes and, being indigestible in their stomachs, is spit out again, and has, therefore, lost almost all its virtue and powers; thus far *Castagneta:* Leaving this story for what it is worth, I will allow, that one finds *Ambar* among such bird dung, but it seems suspicious to me, that *Garzias Lib. I. cap. 1* makes no mention of these birds, howbeit he was in his time an exact recorder of East Indian things; and at least *François Pierard*[9] should have known about it, because after the year 1602, he lived for many years in the Maldivian Islands, yet [he] states about Ambar only that the sea casts large quantities of it onto those Islands, and that the Islanders know nothing more about its origin than that it comes from the sea.

Carolus Clusius writes in his annotations to the aforementioned Chapter, that he spoke in Frankfurt with a certain Burgundian, *Servatius Marel*, who had traveled in many Countries, dealing in spices, gems and pearls: This one had assured him that Ambar was nothing more than the superfluous gathering in the stomach of the true Whale, and he did not mean this to include the *Orca*, the *Physeter*, nor several other large and toothed Fishes: To wit, the true Whale, which does not have teeth and has a narrow throat, swallows only little fishes, particularly Seacats[10] and *Polypi;* they stay in his stomach for a long time, and make for a slimy gathering, which it cannot readily digest, and finally, when the whole lot is bothering it too much, it will spit this out again: That first spewing is base and foul Ambar, but that which stays a long time inside and is cooked well, becomes good Ambar; now whether this happens once every year or several times, is uncertain, but this is for sure, that when the Whale has emptied its stomach and is then caught, one will not find Ambar in it: Some time later it has some raw Ambar again, but the best [kind] is that which has been with it for a long time: This is the reason why one will sometimes find the beaks of the aforementioned Seacats along with Ambar, which are erroneously thought to be bird beaks: To find the best Ambar one chooses the one that is gray inside, and when a small piece is put on a hot tin of iron, it should completely melt until it is an oyl, and gives off a pleasant smell; thus far *Servatius Marel. Julius Scaliger Exercit. 104.*,[11] after he has quoted many examples from the books of the Mauritanians, to the effect that Ambar grows in Whales, and that the Whale itself is called *Ambar* in the language of *Fez* and *Marocco*, declares his opinion, that this seems entirely unlikely to him, while he has seen so many Whales opened up in the bight of *Biskay*, and heard, though not understood from anybody, that they never found the least bit of Ambar in those Whales; whereafter he declares his opinion, to wit: That Ambar grows on the bottom of the sea in the manner of a *Fungus* or Campernoyle,[12] and comes to float to the top after the waves have washed it loose; He was moved to give this opinion after he saw some balls of Ambar that were covered with a little skin just like a *Fungus*, and also with bits or scales, like an old *Fungus:* He writes in the same place that when Foxes find Ambar, they will eagerly gulp it down, and then eject it again by way of their Excrements, but the same is merely foul and corrupted Ambar; a story which will serve its purpose hereafter: He has the same three kinds of bird dung that *Castagneta* mentioned, but with different names, calling those Islands *Palandura* (perhaps corrupted from *Maldivas*) and the three kinds *Povambar*, *Puambar*, and *Pinambar*,[13] but the aforementioned names have more in common with the Portuguese text.

Radja Salomon Speelman,[14] banished King of *Ademanduta*, a trustworthy man who was an eyewitness to this tale, has declared to me in all sincerity that in his day, around the year 1665, a dead Whale washed up on shore in his Country, near *Baturou*, which he called *Iju-Ambar*, 15 Fadoms[15] long, whereof the head was one fathom long and came to a point, but the forehead was round: There was a fin on its head, as tall as a man, and the length of 5 fadoms along its back, which it could put down; after cutting it open, they found the belly to be the size of a room, and entirely filled with ambar, whereof that in front and near the throat was white and watery, the center part was yellow and grayish; the back part black and soft as tar, which it gets rid of with its excrement, but it spits out the first two mentioned: Those from Timor did not know it and used it as a tar for their prahus, until they were made the wiser by the Macassarese, who scraped it off their boats again, bought it from them, and took it to their Country: Other Whales were stranded on *Timor* some years before, and these had short and thick teeth, some a foot long, some shorter, like those from *Leti*[16] that will be mentioned in *histor. animalium;* but he had forgotten if any other of those fishes had teeth as well.

The Javanese think that Ambra is the excrement or dung of the great Bird *Geruda*,[17] which is said to live in the Tree called *Paos Singi*, which grows in the South Sea, and its excrement is swallowed by the whale, which, finding the Ambar indigestible, spews it forth again.

Now I will dish up something new, which might perhaps not have been heard by the world before, conveyed by a *Hubert Hugo*, some years ago in charge of the Island of *Mauritius*, who wrote in a letter to his Lordship General *Joan Maatsuyker*,[18] of 14 *December* 1671, the following.[19]

Concerning the differences among the scholars as to Ambar-grys, from where it is supposed to derive, or as to what animal or plant produced it, one can take notice, but one cannot make a determination; for what I discovered from diligent inquiry and from experience is that it is not the foam or dung of a whale nor a kind of *bitumen*[20] or glue, but that it sprouts from the root of a tree (the name of which is unknown to me), and this tree, no matter how far it grows inland, will always shoot its roots towards the sea, seeking to get rid of its own greasy gum, which will course downwards but not up, in the warmth of the sea, otherwise it would be choked or burned by its own great greasiness; wherefore the same sends it on down so it can disburden itself with the help of the warmth of the sea: When the root of this tree discharges its greasy gum into the sea, the same is so tough, that one cannot easily break it off the root, unless it becomes such a clump that it, too heavy for the root's little vein, will break off all by itself from the heat and movement of the sea, and so will float in the sea, and the current, where it is wont to go, will cast it on shore because of its greasy lightness; wherefore I made it my task to observe the movements of the current for an entire year, but owing to a lack of vessels, which would be fit for this, I could not work out anything, which is why I am obliged, with the next vessel newly fashioned by the carpenter, to spend most of my time on this in order to serve the Honorable Company in the best way: And once this is found and applied, I would judge that by planting these same trees on the beach, or in those places where the current ebbs, that searchers cannot miss finding it, or should not, once it was cast on the beach, and would richly increase the Honorable Company; although I opine that, if the trees are grown further inland, they will have fatter gum, than those on the beach, that is why one ambar is held to be better than another. We sent Your Honor 7/8 pound with the small ship (and it was found by a soldier *Joan Westphalen* from *Straalsont*,[21] who was also going to Batavia, in order to sail to the Fatherland, because he had served his time and, at his request, had been dismissed from Fort Good Hope), in addition to 2 pieces which I found myself; which, I hope, will please Yr. Worship, and be an honor to me.

The kinds [called] *Ambar-noir*[22] and *Succinum*,[23] are found more often than the other ones, to wit more precious, of which I sent Yr. Worship as much as I could lay hold of; but [which] in my opinion [are], of little value: But since the letter of the Honorable Commandeur *Jacob Borghorst* to search for everything that smells and has a taste, and to send it to His Worship, I did not want to be negligent in this; thus far *Hugo*.

If the foregoing were true, then I am amazed that this tree would only be on the Island of *Mauritius*, since Ambar is found not only there, but along the entire East coast of *Africa*, on *Madagaskar* and the *Maldivas*, also here in the South-Eastern Islands, indeed, in a host of places all over the world, such as in the West Indies or on the beaches of *Marokko*, and that this weird Ambar tree has remained unknown for such a long time: And since it also grows in the Countries just mentioned, especially on the East coast of *Africa*, and *Madagaskar*, [it is strange that] the subtle Arabs, who have searched the same Coast for centuries, did not know of it sooner: And so one will soon say that all kinds of fragrant Gums, which resemble Ambar somewhat and drip from trees, are like unto Ambar, and so I can show you such trees in the Province of Amboina in my Ambonese *Herbarius*, in the chapters on *Nanarium minimum* and the *Buronese Rowayl*;[24] They sent me from *Manipa*[25] a certain soil which smells quite naturally of Ambar, but closer scrutiny revealed that this same earth had been dug up under the aforementioned tree, and was undoubtedly imbued with its sap.

And how could the aforementioned tree exist, when considered along with what I was told by another person, who had been a long time on the Island of *Mauritius*, that Ambar grows in the stomachs of wild pigs; and why should I not believe him? since he saw it with his own eyes, pulled out of the stomachs of said pigs, just as *Hugo* saw it hanging from the roots of his tree; verily, if one wants to believe, that Ambar grows in the bellies of so many Land and Water Animals, wherein one can find it sometimes by chance, then those who live in the West Indies can also believe, that it grows in their foxes, about which one can read in the French histories *Antillarum cap.* 20. written by Monsr. *Rochefort;*[26] there it is said, firstly, that much good Ambar-gris is found on the Coast of *Florida Tabago*[27] and other Caribbean Islands: And one reads shortly thereafter, that in those Countries where one will find the most Ambar, one will find many foxes, which patrol the beaches at night, where it is most often tossed, and finding the same immediately swallow it, but since they find it indigestible, get rid of it in their excrement, thereby losing a great deal of its virtue, and this is called by them *Ambar des renards* or Fox Ambar, and is not useful either as medicine or as incense, and is therefore not brought into France.

While it appears from the foregoing examples, that Ambar is found inside so many different animals which after finding it indigestible for their nature, expel it unconsumed, it is highly unlikely that it grows within those same animals: So I will hold to the opinion of many scholars who assert that *Ambar*, as well as all Amber [*bernsteen*], is a soft Glue, that is sweated out of the bottom of the Sea, and which, rising to the surface, gradually hardens due to the coldness and brininess of the water; this soft substance has a strong odor at first, but it does not stink as the French history contends, but, like all glue-like saps, such as Naphtha[28] and the like, attracts animals as if it were bait, which is why the Whale and other animals swallow it, along with all kinds of Seabirds, as long as it floats on the sea: once tossed onto the beach, it becomes the prey of wild Pigs, Foxes, perhaps even a Cayman, as the French history testifies of the West Indian Crocodile. Our Ambonese fishermen have experienced this when, seeing something floating around in the Sea in these parts, and seeing the birds peck at it and many fishes surrounding it, would chase them away, and find a small black lump like soft pitch, full of pits and holes, which they thought had been done by the birds and fishes; when they brought it to me, I discovered the outside to be black Ambar, rather soft, but after a few months as hard as other pitch; yet when one handled it and warmed it in one's hands, it would become somewhat soft again. In its very center was hidden a small piece of good Ambar-gris weighing about ½ ounce: And so the Inhabitants of the South-Eastern Islands informed me, that most of it is found in the same way surrounded by a black and soft substance, which, when still fresh, rather stinks, perhaps because it comes first from a Whale; but on coals it smells naturally like *Bitumen* or Naphtha. In the year 1682, a similar black piece, easily a fathom wide, was cast onto the beach at *Luhu*[29] by the East wind, where it melted from the sun and was mixed with sand and other filth; the Chinese and some Natives pried some small pieces from it and made a good haul: The black and soft substance resembling tar, being entirely mixed up with sand and quite dirty, was distilled here in a simple fashion; and from it they derived a brown oyl, on the one hand smelling strongly like *Petroleum*, but always retaining a smell of the sea, as one will notice in the *Unguis odoratus*, wherefore I once more conclude that Ambar is a glue-like substance, and it is also proof, that Ambar cannot grow in the Whale, because one finds such large pieces of it, that they are compared to small Islands, and these could never come out of the narrow throat of a whale; see, among others, *Linschoten* Cap. . . . [*sic*][30] about a piece found near Cape *Comoryn*, weighing 30 Centeners: But one considers it Whale Ambar, when one finds the beaks of seacats in both the black and the gray Ambar, as well as small shells and bones, which could only have gotten there in its stomach; although I now determine Ambar to be a *Bitumen* or Glue, I do not want to consider it impossible, that the same *Bitumen*, attracted by several roots of large trees, is changed into a Resin that resembles Ambar somewhat, as long as there remains a clear distinction between the true Ambar and such Resin, just as it is certain that one and the same greasiness of Prussian earth, once it gets into the sea, is changed into Amber [*bernsteen*], but whilst it is drawn up by Pine trees, it changes somewhat and becomes Resin: And in this way the aforementioned tale about the Ambar tree on Mauritius can become probable; the more so since the same *Hugo* writes that other glue-like kinds occur on *Mauritius*, such as black Ambar, and *Succinum* or Amber [*bernsteen*], as long as he did not think that some resin that had fallen from the trees and was hardened in the sea, was Amber [*bernsteen*], just as people in these Eastern Islands do, who hold that all kinds of resin, which they call *Dammar*;[31] and which is found on the beaches, is really Amber [*bernsteen*].

In order to select and test the true Ambar-gris, one should first note its appearance and color: It is true what *Rochefort* writes in his French history, to wit that the original Ambar must come in pieces that incline to roundness, not perfectly so, but pressed or rounded as if bossed, dirty on the outside, besmirched, sometimes a kind of black, sometimes an admixture of grays and russets, and as if they were surrounded with a fleece, not unlike the *Tubera*, which the Italians call *Tartufoli* and the French *Truffles;*[32] which made many think that Ambar was a *Tuber* or *Fungus Marinus;* the same Author says that it should, on the inside, be gray, dry and light, or really, as if ashes had been mixed with wax, in such a way that one can distinguish the ashes and the wax separately: But it is my opinion that these distinctions do not always hold, because the Natives are wont to adulterate the Ambar with wax, dry earth, and some inferior Ambar, giving the balls the same shape and color on the outside as true Ambar has: In order to distinguish the false stuff one must learn from experience how much a ball of true Ambar weighs, of the size one has encountered, because the fake one is noticeably heavier than true Ambar; the outside does not matter too much, since, for instance, our Eastern ambar is usually found to be black ambar, which one must scrape off, until one can get to the harder [stuff], which is the true amber, yet it keeps on the outside a blackish or dark gray color, must be gray inside, not uniformly so, but mixed with small white spots: Our people compare it to dried cowshit; and if it has some sand, small stones, or shells, then this does not make the ambar of a worse quality, but it lowers the price: If the seller does not permit you to break the balls into pieces, though you would like to know if it contains sand, pebbles, and the beaks of Seacats, take a long needle which you have heated, thrust it into them, and you will become aware whether the needle encounters anything that is hard, whereafter you can talk about the price, which in *Amboina* and *Banda* is 12 to 15 Rixdollars for an ounce, being the price of ordinary gold: Many have considered it a proof, if one obtains some oyl from it with the hot needle, but this will not hold, because the false [kind] does the same thing: One can be more certain, if one puts a piece on a hot iron or knife, because it should melt like wax and burn completely away or go entirely up in smoke, without leaving anything behind.

One cannot describe the smell, which one must have experienced, because ambar has a particular smell, which cannot be counterfeited: We called it a Sea smell before, and the closest to it is the smell of *Unguis odoratus.* The people from *Keram* have taught me yet another kind of proof, but it does not strike me as true, which is that if one gives Ambar to a hen or a rooster, the same will die, which would present a wonder because ambar does no other animal any harm: if one puts ambar in one's mouth it does not, as one says, diminish, or at least not as easily as other gum-like things: but I have also found this proof to be uncertain, because I have had good Ambar that snapped off if you bit into it, like glass or wax, that had become hard from having endured a great deal of cold, which could be ground between one's teeth, so that one could swallow it: and there was yet another lump of fine *Ambar* which also could be ground to pieces between one's teeth, but it was easier to reduce it to a lump, which, however, did gradually lessen through lengthy chewing: But if the tough black Ambar is put in one's mouth, it will not diminish, even if one chews it for hours on end; so that one should not rely on this test: If, after putting it on coals, Ambar gives off a smell of fish or a glue-like smell, like stonecoal or black ambar, or if it resembles such a smell even a little bit, then you have bad Ambar; because the Malay hold that it should have a pleasant smell of flowers, wherefore they call the best kind *Ambar Bonga*, that is Flower Ambar: If one holds one corner of Ambar in the fire, so that it melts a little, it should stick so much to your fingers that you cannot rub it off, and the melted piece should cool to a gray-blue color, while the false kind cannot do either.

As to the powers and attributes of Ambar I will refer to the European books, saying here only that ambar does not mix with water, oyl or any other liquid, nor does it let itself be ground to dust, though if one scrapes it fine enough and mixes it with some dry powder, it will mix with other things: One can also undo it somewhat in the finest spirit of wine that has been warmed over a coal fire: Its best-known power is that it strengthens the brains or nerves, that it refreshes the vital spirit and, especially, which makes it so costly, that it increases the powers of manhood, an art that is particularly esteemed in the Indies, and also for making pomanders[33] and fragrant *Paternosters*,[34] which one wears on one's underclothes or on one's skin, or one can lock it into hollow or pierced goldware so one can smell of it: For this one cannot use solely Ambar, but one should mix it with black ambar, which grants it its hard consistency (*consistens*), also Benjoin, *Muscus* and *Unguli odoratus*, all of it washed in the finest rosewater, and then made into a lump with *Styrax liquida* or *Rasmalla*.

As to the places where one will find *Ambar*, they are scattered all over the world: The History of the *Anthill.*: divides it as to places in the West (*Occidental*) or in the East (*Oriental*): He calls *Occidental* that [Amber] which is found in the West Indies, *Florida*, and the Islands of the *Caribe*, which he esteems above all others, and since he understands *Oriental* to be [the kind] that is found on the coast of *Barbaria*,[35] *Marocco* and *Guinea*, he is right to some extent, whilst the same is also black: But if one understands *Oriental* to include the East Indies, than one cannot agree with what he says, since the entire East coast of *Africa*,[36] particularly from *Sophala* to *Melinde*, if not as far as the Red Sea, the Islands of *Madagascar*, *Mauritius*, and the *Maldivas*, provide as precious an amber-gris as his West Indies, although the black [kind] is found everywhere: I believe that, up till now, people have not sufficiently noticed that the good ambar gris, is often found wrapped in the black kind, and one should, therefore, always look at the black to determine if it does not conceal some hard little lumps of good ambar; after all, we have found such quite often in these parts; It is also found in the South-Eastern Islands, particularly around *Tenember*;[37] but our Merchants have made the Islanders so cunning, that they will counterfeit the same with wax, of which one becomes aware if the lumps are too yellow from being handled, and, when placed in warm water, they become a little soft, something true ambar does not do.

Three small pieces of ambar-gris, called *Pissang*, *Luhu*, *Baber*, tested on a hot tin, resulted in the following: All three did not have a good smell, *Luhu* in particular smelled like burned salt fish, and left a foulness on the tin, as did its black Habit: The *Pissang* and *Baber* smelled better, turned entirely into smoke without leaving anything behind, including the black Coat of the *Pissang*, which remained rather soft; so that there is no ambar that gives off a pleasant smell when it is placed on coals or a hot tin, whilst *Luhu* and its Habit (which, being mixed with a great deal of sand, had turned into stone) had the best odor when raw, quite similar to *Unguis odoratus:* The *Pissang* and its Habit do not diminish between one's teeth, as was the case with the piece from Ceylon, but *Luhu* and *Baber* did diminish and decrease: With *Ambar Pissang* I refer to the above-named small piece of ambar found at sea: With *Luhu* I refer to the large piece that, as mentioned, was found on the beach at *Luhu:* and with *Baber* that which came from the South-Eastern Islands.

In the year 1693, the King of *Tidore*[38] received from his Papuan fishermen an uncommonly large piece of Ambra de gris, weighing 194 pounds, for which the Hon. Company offered 11000 Rixdollars; it had the shape of a turtle, though the head was broken off, which caused him to imprison some of his subjects, but he could not obtain it.

I should add here the story which Mr. Dr. *Andreas Cleyer*,[39] former Chief in *Japan*, communicated to me in several letters, concerning an Ambar fish, which is said to be present in *Japan*, of which the first notice is given below, extracted from his letter of the 18th of February 1695.

The Fish, from which the Ambar de gris derives, is called in Japanese, *Hay ang kie*,[40] and is in shape very similar to the smallest kind of Whale, differing only in that the Whale (North Cape Whale) has no teeth, but a fishbony mouth, while this Fish had a mouth full of teeth, among which two big ones stuck out, as they do with a wild boar, or you can see on a Walrus.

In another letter from the previous year, 1694, his Honor wrote about the Ambar fish as follows: In order to be of service and inform you of the Amber de gris, which in Japan is often found in abundance in various places, principally in the Liquesche Islands,[41] which are under Japanese government, and also in other places, that can be found in *Japan;* it comes from a fish that is very similar to a North Cape Whale, in terms of shape and size, except that it is different in the following, to wit, that this one has two protruding white teeth in the front of its mouth which descend downwards, and has a tail that is not cleft, which was told to me by a Japanese interpreter, who himself was a whaler for many years; adding further that there were certain signs from which they could conclude whether those fishes had a quantity of Amber in their bellies: When these fishes were caught sometimes, these fishermen were obliged to bring such fishes ashore, and there, in the presence of some envoys which their Lord had placed over them, to open them up at the peril of their lives, and then the biggest, most massive pieces were taken from them for the Lord, while the remaining small pieces, which were slimily thin, were left for the fishermen, along with the body, from which they boiled the Whale oyl: And as far as it concerns Yr. Worship, those big pieces of Ambar, which were found in the belly of the fishes, of which Yr. Worship writes: that they could not be swallowed because Whales have a very narrow throat, in that I am of one mind with Yr. Worship, since the substance of the Ambar is at first soft and liquid, and is hardened in the sea, as those from *Japan* confirmed, and that the fish, powerless to spit out the aforementioned stuff that they swallowed, of necessity has to become sick, and must finally die from it or choke; and those large pieces, after the fish has perished or rotted, are then tossed onto the beach by the waves and are often obtained in this manner.

Description
Of the Piece of
GRAY AMBER
Which the
Amsterdam Chamber received from the East Indies
Weighing 182 pounds; together with a brief discourse concerning its origin and powers.

My Lord,[1]

You know that Amsterdam's East Indies Chamber caused several plates of a piece of Gray Amber to be cut and printed. I had the good fortune of seeing this piece, which, because of its uncommon size, was most unusual. A Pound holds 16 ounces, and the normal price for Gray Amber is from 30 to 40 guilders per ounce, one will find that it is worth around 116400 guilders,[2] and that, consequently, this precious lump must surely have been quite some treasure for whoever it was who found it, supposing he knew its value.

Even as the pictures of this piece have been scattered far and wide, and have now fallen into the hands of all those exact Scrutinizers of nature's mysteries, I do not doubt that they will have various opinions concerning the origin, nature, qualities and attributes of this kind of Amber, in order, wherever possible, to search out the source as well the manner in which it has grown together.

If I could be assured that among these Scholars one would take the trouble of providing us with some information about the same, and would unfold his thoughts concerning this material, then I would not so readily bring forth [my opinion], nor be so quick as to propose my doubts concerning same, but only after I had examined his judgments.

But whilst these Gentlemen, being burdened with weightier activities, are not likely to share their thoughts with us any time soon, and since it would be a pity if curious lovers of learning[3] would be deprived even longer, I will be the first to break the ice, in order that, by doing so, it will spur them on to enrich the world with their findings and thoughts about this piece. And so I send you herewith, My Lord, what I have devised about it, and since I know that thou dost favor me, I am entirely convinced that thou wilt take the trouble of inquiring after my views in this matter, and wilt thereafter unfold thine [thoughts] pertaining to my doubts and conjectures. But before I do present them to You, it might not be unwarranted to lay before you the opinions of various Authors as to the origin of Gray Amber.

Matthiolus[4] made some remarks in his explanations of *Dioscorides;* but since these are neither as abundant, nor as expansive as the reasons given by *Justus Klotius*[5] in his Philosophical Treatise on England, I will pass them by at this point.

I did not have the good fortune to get hold of the latter of the two Writers, no matter how much trouble I took, but according to what I read about it in the Philosophical Treatises on England, he draws up eighteen notions about Gray Amber, and particularly examines that [opinion] which holds that this Gray Amber derives from the Excrement of a certain Bird, which is found on the Island of Madagascar, and which is called *Aschibobuch* by its Inhabitants, and the Portuguese [writer], *Odoardus Barbosa*,[6] has given us a description of this Bird, other Writers confirm that it is as large as a Duck, with a large head, and with uncommonly beautiful feathers. I have looked for descriptions of this Bird in *Aldrovandus*,[7] and *Willughei*,[8] the latter an Englishman, but could not find any, nor is any mention made of a bird with this name. The same *Justus* says, that this kind of bird is not only found on

Het stuk Amber der E. Oostindische Maatschappye zoo als het zich in de lengte vertoont.

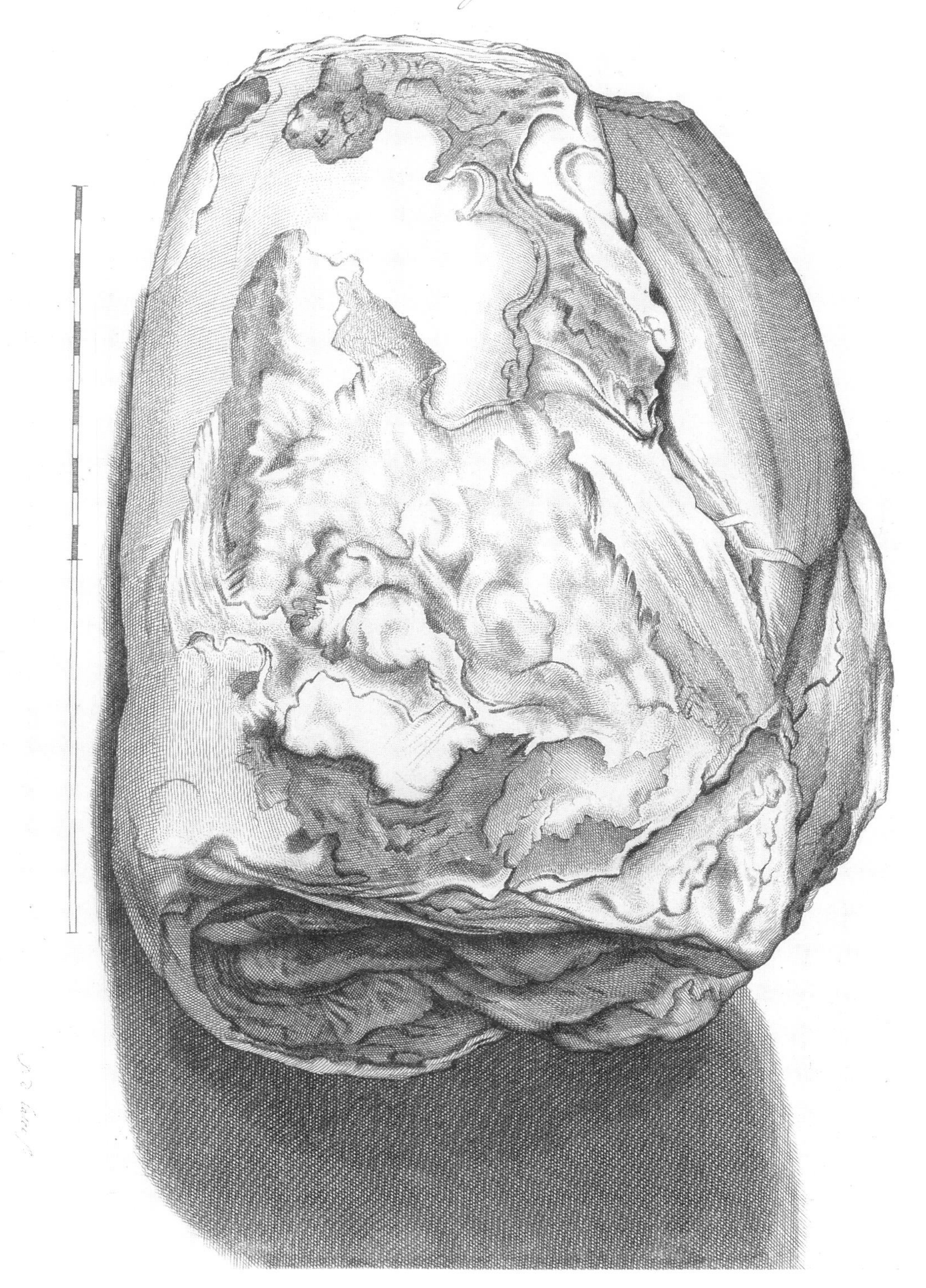

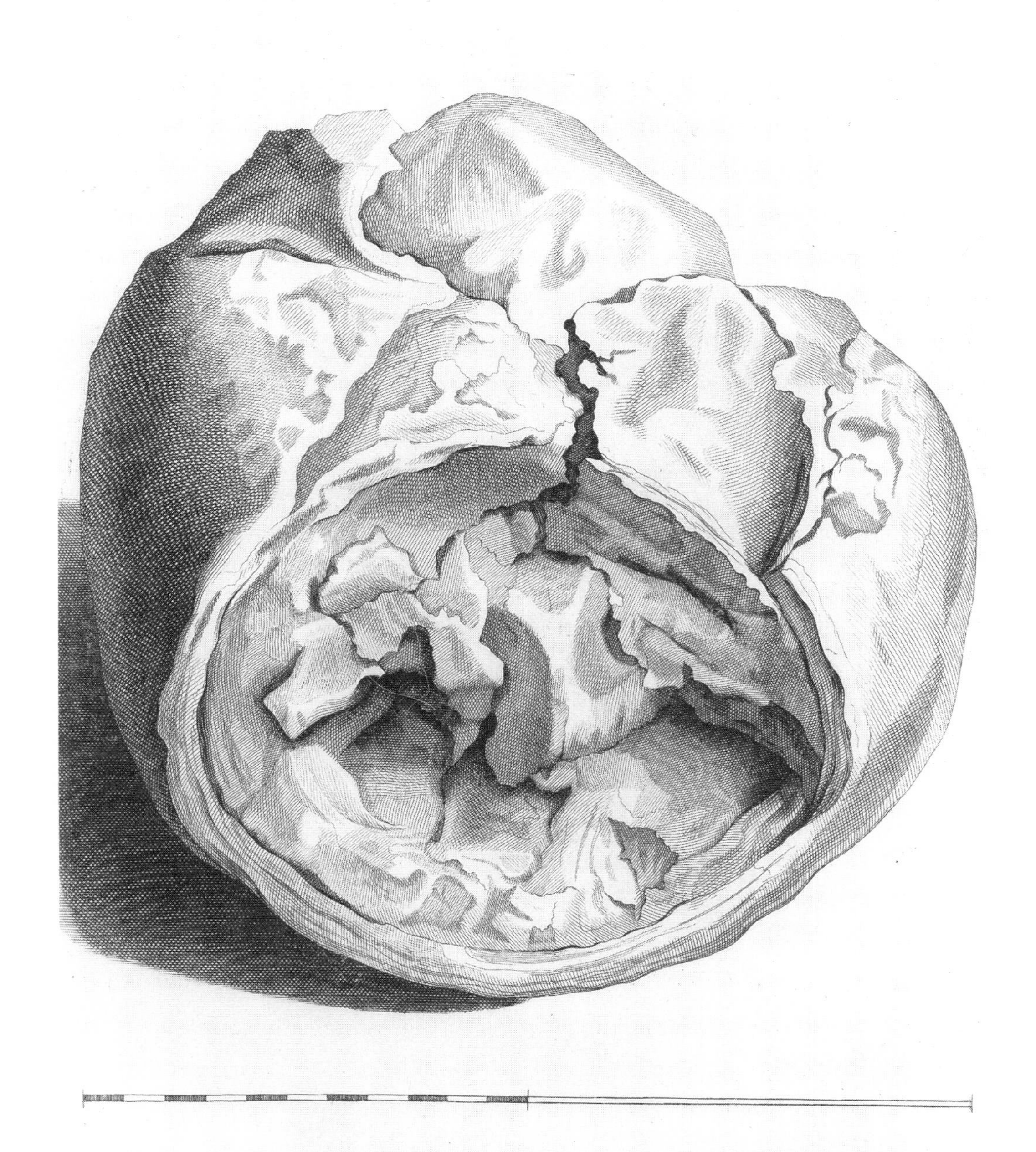

Het stuk Amber der E. Oostindische Maatschappye op zyne breette te zien.

the Island of Madagascar; but that there are also a large number of them in the Maldivian Islands: That they assemble, like unto the Cranes, and that they usually stand around on tall rocks on the beach, from where their excrements fall into the water. This same Writer, after he has discussed this notion, and has expressed his thoughts, goes on to state one, which he deems to be more likely, to wit, that Gray Amber is the excrement of a certain Whale, and which we will consider hereafter.

Now since this lump, as to its value, is worth more than gold, and whereas until now no one has been able to discover whence it came, there have been many persons who, every so often, tried to determine what constitutes its fundamental aspect, of which one is generally so ignorant. We can read about the same in a book, that bears the title *Miscellanea Natura Curiosorum*,[9] and was printed in Holland in the year 1630 [*sic*], wherein *Honnius Vollquadus*[10] tells us of his observation, pretending it was one of 300, concerning a piece of Gray Amber, that weighed five hundred eighty-four and a half ounces. This piece of amber that had been bought in Venice in the year 1613, was deemed to be an uncommonly fine [and] amazing piece, although it cannot touch the one whereof I write here; because it, as we already mentioned, weighed two thousand nine hundred and twelve ounces.

The Writer of the book I mentioned, to wit *Miscellanea Natura & c.*, notes by this example of *Honnius Vollquadus*, that *Gargius* of *l'Orto* tells us in his History of Simples, and other Writers after him, who might have copied him, that one will find at sea not only large pieces of amber, but entire Islands of it. The Writer who makes this Comment in his book *Miscellanea Natura & c.*, decries this information as ridiculous, while adding that if guesses are permitted here, these stories are based on tales told by those who, after sailing across a piece of ocean, where Gray Ambar was lying on the bottom, could smell the odor of amber everywhere, which moved them to believe, and thereafter to spread the word, that it was an underwater Island of amber which gave off this smell. In order to confirm his opinion, this same Writer, relates what *Joannes Fabri*[11] has put forth about *Gregorius* of *Bolmar*'s[12] account, to wit, that one had found a piece of amber of 100 pounds in this part of the sea, and which *Gargius* also divulged, that in the year 1555, around Cape *Komoryn*, a piece of 300 pounds was found, and was sold as if it were loam.

Joannes Ugoliscotus[13] tells us, that around the same Cape *Komoryn* another piece of 1500 pounds was sold, and *Monardus* writes, that 100 pounds of amber was taken out of the body of a Whale. Our Writer, who makes these remarks in his book *Miscellanea Natura*, adds, that what the same *Gargius* has written about the stories of others, comes closer to the truth; to wit, that one had found a piece of Gray Amber, the length of a man, and another one the length of 30 handpalms, and 18 wide. Moreover, the same Writer notes that *Montanus*,[14] during his mission to Japan, accomplished in the year 1659, mentions that the King of Sassumen[15] possessed a piece of Gray amber of 130 pounds, valued at 14000 Tailis,[16] a certain kind of Japanese money.

According to what the Dutch tell, they threw away, in the year 1666, near the green Cape, in the Jambes stream, a piece of Amber that weighed 80 pounds, and *Gargius* confesses, after all these various stories, that the biggest piece of ambar that he ever saw was only 15 pounds.

In the cited book,[17] called *Miscellanea Natura*, I discovered that this amber is not a gum from a tree, that has fallen into the sea by chance; that it, also, is hardly the excrement of any bird, nor a honeycomb, but that it is some slime from the bottom of the sea, that, just like seapitch, grows and congeals; and what ties in with this observation is that a certain merchant from Batavia, by the name of *Gabriel Nakke*, declared that he, as an eyewitness, saw the Natives of the South-Easter Islands obtain amber with pitch from the bottom of the sea, and that one will find this amber on land on the beaches of the Island Mauritius, of Madagaskar, and of the South-Easter Islands, and that pigs are very fond of it and will greedily devour it. In the second picture of the oft-mentioned book, by the 21st Observation made by *Andries Kleijer,*[18] one will find two Whales depicted, of the smallest kind, captured in the Islands of Japan, with a great deal of Gray Amber in their bellies; and this tale has led to the counterfeit story that Amber may well come from the slimy moistness, that grows in the stomach of the Whale, and which is caused by indigestible foodstuffs, and which is thereafter vomited forth: or according to some others, who mean to say that it is the dung, or the seed of these Whales. The people who feel, along with Avicenna and Serapion,[19] that Amber brings death to animals who eat it, are not one whit better.

All these opinions, which I have quoted, are to be found in the general Dictionary of the French language, which was made and published by Abbot *Furetiere.*[20] I will put them once again in order here, one after the other, so we can understand this piece even better.

[1.] This Writer first proposes, based on others, that the gray amber is a kind of Gum, which gives off a pleasant smell, and is sweet to the taste. He adds that this Gum, whereof the Naturalists have not as yet discovered the basic nature, is found on the seashore.

2. That some maintain that amber is the dung of some birds, which can be found on the Island of Madagaskar, beyond the furthest regions of the Eastern Archipelago.

3. That some others contend that it is the excrement or seed of the Whale; and *Joost Globius*[21] says, in his history of gray Amber, that this precious stuff derives from the intestines of a certain Whale, called *Tromp;* since it has a snout[22] on its head that has teeth and that is a foot long and the thickness of a fist, and that the Amber, otherwise called *Spermace,*[23] is found in the head of this Whale.

4. That others hold that this Amber is a kind of seafoam, or a kind of Loam, which comes from deep down in the earth, or from the rocks over which it is hauled.

5. Then there are those who believe that a certain fish, called *Azel,* is very fond of amber, and that it constantly looks for this amber, but that, as soon as it has eaten of it, dies; whereafter the fishermen, who see it float on the water, catch it, in order to remove the Amber, which it has swallowed, from its belly.

6. And that others believe that gray Amber is put together from small drops of honey, which, in time, are cooked again by the sun, and then fall from the rocks and cliffs into the sea, where the movement of the salt and the waves completes the cooking, and so delivers this boble-body[24] wherever it is found. But this notion is not very likely; since one has found pieces of amber that weighed two hundred pounds.

This gray amber, as Furetier notes, is the color of marble, veering towards gray. It is often found lighter or darker, with white stripes, and sometimes a little yellowish. Sometimes one also finds bird beaks in it, or bloodless animals, and other things, which were mixed with it while it was still a liquid Loam. I have a few pieces of it that have five or six little bird beaks, and the like, in it.

The best gray amber is found on the island of Mauritius, in Africa.

This amber is usually found after a thunderstorm, and the pigs, who can smell it from far away, do not neglect to walk up to it and to swallow it down greedily. *Izaak Vigni,* a French traveler, reports, that in a certain region they found gray amber in such profusion, that one could have loaded a thousand ships with it, and that he received a piece there, on the spot, which he sold for 1200 pounds sterling, that is 14300 Dutch guilders. But since up till now this place has not been found, no matter that one has looked where this region was located, foot by foot, for six weeks without stopping, I am convinced that what this Writer says is a fiction.

Besides these opinions, which I have gathered here, there is still what others think, to wit that this precious stuff is nothing but seaslime that acquires the durability of gray Amber, and which, after the sun has dried it, is melted on a fire, in order to cause its true power and odor to appear, and is furthermore mixed with other simples.

I will add to these various opinions about the origin of gray Amber, which *Furetiere* collected, the ones by *Garcias du Jardin*, the Annotations of *Antony Colyn*,[25] and the notions of Mr. *Nikolaas Monard*,[26] physician in Sevile: And this is what *Garcias du Jardin* notes about it: the Amber which the Latins call *Ambarum*, and the Arabs *Ambar*, is, as far as I can guess, known as such under the same name, with little, or no variation.

The Writers who have written about this stuff, are most varied in their opinions about the generation or origin of amber.

Some ascertain that amber is the roe or seed of a Whale; others that it is the excrement of a certain sea animal, and others that it is seafoam. To tell the truth, none of these opinions is based on any kind of sense, since, in the places where one finds the most Whales, and where the constant movement of the waves also stirs up a great deal of foam, no gray Amber is found. Then there are those who suppose that Amber, like Loam, comes from certain subterranean grooves or pipes in the sea; and it is this opinion that appeared the best one to many Writers, and the closest to the truth.

Avecenna in the second book, the 63d chap. and *Serapion* in his herbal the 196th chap. say that amber grows on the seacliffs the way the Campernoyles grow on trees, and that occasionally storms toss this amber along with gravel onto the beach. Among all the things *Avicenna* quotes, it is this notion that is most probable. Whenever the wind blows mostly from the East, they find a great deal of Amber at Sofolan, and on the Islands of Komora, Emgora, Mazamgiquen, all over their beaches; it is tossed there from the Maldivian Islands which lie to the East. But when, on the contrary, the Wind blows from the West, then Amber is found in profusion on those Islands, called the Maldives, from a corrupt word, because one should call them *Naledives*, since *Nale*, in the Malabar tongue, means *four*, and *Diva*, *Island*. Therefore, they should be called *Naledives*, as if one were saying Four Islands, as we call the Islands *Angedivian*, which lie twelve miles from Goa, where we traffic in East Indian trade, because these Islands are five in number, and lie next to the other ones; because *Ange* means *five* in the language of these people. Although it is not pertinent here, I did want to say a word about this in passing, concerning *Malediva*.

These two Writers, quoted by us at the same place, add, that the Amber is devoured by a Fish, called *Azel*, which suddenly dies after it has swallowed it. That the Inhabitants of the Country, seeing it float on the waves, drag it onto the beach with steel hooks, and after disemboweling haul the amber from it, which is of little value, except for that which was stuck to the backbone, which will become quite excellent in due time. But I feel that this opinion is wrong, since, what is at most the truth, is that the animals do not seek any other prey than that which corresponds to their nature, and that when they eat something hurtful, this only happens because they have been deceived by the appearance of something else, with which the detrimental is mixed, the way Rats are caught by eating *Arsenicum*, or Rats' bane, mixed up with something that is tasty to them. Hence it is not very likely that the Fish, called *Azel*, would prey on Amber, if it is true that, when he eats of it, he will suddenly die. And furthermore, since Amber is one of those things that strengthens the heart, this Fish must be very poisonous indeed, if it dies after swallowing such an excellent and powerful medicine.

Averroes, in the fifth book of his *Coll.* the 56th chapt., tells us that a certain kind of Camphor is found, which grows in seagrooves, and which floats thereafter on top of the water, of which the finest and most excellent is that which the Arabs call *Aschap*.[27]

It is not necessary to gather here many reasons, in order to show, that such an opinion is far removed from the truth, and unworthy of such a great Philosopher; First of all, because he says that Camphor grows in the Sea; secondly, because he makes of this same Camphor, which is cold and dry to the third degree, a kind of Amber, which he, however, puts at hot and dry to the second degree. We will here repeat some words, which can be found in *Serapion* and *Avicenna. Serapion*, in his Herbal, the 196th chapt., confirms, that a great deal of Amber is transported from the Land of *Zinga*, that is *Sofola*, since *Zinga* or *Zanga* means in the Persian and Arabic tongue, what *Zwart* [black] means in Dutch; and because this entire coast of Ethiopia is inhabited by Negroes or blacks, so does *Serapion* call it *Zinga*. And similarly, *Avicenna*, in his second book, the 63d chapt., gives Amber the name of *Almendeli*, as if one were saying of *Melinde*, and those of *Selachristicum*, perhaps from the name for *Zeilan* (*Cylon*), one of the most renowned Islands in the East. *Lacuna*, in the first Book of his elucidations of *Dioscorides*, the 20th chap. thinks that it is a city, wherein he is mistaken, since it is perceived as an Island with many cities.[28] And there you have everything that the Arabian Writers mention about it; and as far as the Greeks are concerned there is not one, excepting *Aëtius*, who makes mention of it.

Now take note of what my feelings are about this piece. Just as according to the nature of Countries, the soil is sometimes as red as the Armenian *Bolus*,[29] or sometimes white as chalk, and sometimes also blackish, so it is also likely that one will find Islands, or Regions, that are of an Amber color, or of a shape, wherefrom the soil is light or spongy, or, like a Campernoyle, with small holes. The large amount that is found of it, shows as clear as the light of day that this is true, for one will even note pieces that are often as big as a man, 90 handpalms long, and 22 wide. Others are convinced that they have seen an Island made of pure Amber, and though people have searched for it diligently since then, it has never been found.

In the year 1555 they found a piece of Amber around Cape Komoryn, over against the Maldive Islands, that weighed three thousand pounds; but since the man who found it thought it was pitch, or a kind of loam, he sold it for a very low price.

The largest piece I ever saw, did not weigh more than 15 pounds, but those who sail to Ethiopia, to traffic in trade, assure us, that they have seen much bigger ones; because this entire coast, down from *Sofala*, up to *Brana*, has an abundance of Amber. It is also sometimes found, but rarely, in the countries of *Timor* and *Brazil*.[30]

I also noted, that in the year 1530, according to what the Writers said, a large piece was found in a harbor in the Portuguese sea, called *Setubal*.

Sometimes they find pieces of Ambar, wherein one sees something like the beaks of birds, and it is possible that these little animals nestle in it. And one will also find at times that these pieces of Amber contain the shells of Sea oysters, which probably stuck to it, when the Amber floated up against it.[31] Further, the best Amber is said to be the one that is the purest, and which comes the closest to being white, that is, when it is as gray as ashes; or also when it has either white, and then again ash-gray veins, and which is also light, and wherefrom, after a needle has pierced it, an oyly liquid will drip. The black amber on the other hand, is not held in great esteem, and is usually of little value; and when *Serapion*, in his book that was quoted before, does not approve of the white kind, it is because he is speaking of the one that is adulterated with plaster soil.

In passing we will here mention that *Nanardus*,[32] in his refutations, contends, in his selection *Of Stones*, in the first one, that Ambar is a newfangled thing, of which he makes short shrift. But a little further on, speaking of the mixing of *Diambra*, as if he had forgotten what he had just said, he praises it in this mixture as if the sky were the limit, because he is aware, so he says, that Amber, is an excellent simple, which he uses quite often, he assures us, with women, as well as with very old people.

Amber is highly esteemed by the rich and wealthy Indians, who often use it instead of medicines, and also put it in their normal food. The price of amber is higher or lower according to whether the pieces are big or small; because the greater the pieces, the higher the price, just like precious stones, the size of which determines their price.

But there is no other place where Amber is more esteemed, nor higher in price, than in China. Because several Portuguese, who brought some of it to that realm, sold a kat, a weight of about 20 ounces, for up to 1050 Rixdollars.

Now we will give you *Antony Colyn*'s annotations.

Ferdinand Lopes de Castagueda, who wrote a history of Portugal, wherein he tells of the heroic deeds of the Portuguese in the East, assures us, that they find excellent Amber in the Maldive Islands; but owing to his origin, he is of a totally different opinion than most of the Writers who have written about it. He says that these Maldivian Islands bring forth many savory herbs, which are eaten by certain large birds, called *Anacangrispasqui* by the Islanders. These birds, which are found in great numbers on the rocks along the coast, shit three kinds of Amber: The first one, which is white, remains stuck to the rocks, as soon as it has been excreted; this amber is held to be the best, and the Inhabitants of the Islands call it *Sonombar*, that is, golden amber; this one costs more than the other two kinds, because it is not found very often. The other two kinds of amber are the color black or the color of gray ashes, and not worth as much. The Writer adds, that these pieces of amber, ripped away by the force of the storm winds; and having been tossed into the sea, after a while are thrown onto the beaches again; and that this kind of ambar is called *l'Oamber*, that is, being entirely permeated with sea water. The third kind of amber, which is blackish, is called *Hansamber*, as if one said Fishamber; and this Amber came about, because it was swallowed, and was then spat out again, by Whales or other Fishes, which could not digest it; this is the least of them, since it has lost all its powers.

The precise lovers of learning in our time, who traffic in trade, and who have traveled in strange Countries, affirm that Amber is nothing else than a superfluity, which accumulates over time in the stomach of a true Whale. One should know that true Whales do not have teeth, and therefore have to swallow the fish whole, and the more those fishes are light and soft, the more there will grow of necessity a slime or glue-like substance which, over time, overburdens the stomach of these Whales, so that they are required, either all year long, or at certain times, to vomit it forth again. This stuff, which has been kept such a long time in the stomach, and which is subsequently thrown out, is what one calls the true Amber, and floats on the water. Furthermore, one should know that true Amber should be the color of ashes, and if one puts it on a hot knife, it will melt like oyl, and gives off a pleasant smell.

If one wants to read more about Amber, you only need to read what *Julius Scaliger* wrote in Latin in his book of Subtilties directed at *Cardanus*, discourse 104, 10th part, where he expands on it greatly.

There can be little doubt, that there is not a great overabundance of amber along the coast of Ethiopia, since one discovers it quite rarely along the entire beach. Such is proven to be the case with *Garcias du Jardin*'s authority. *Avicenna* also mentioned it in his Latin work, but more fully in what he wrote about it in Arabic. We will include this passage here, translated by a contemporary Writer.

"As to what some say, that Amber is Seafoam, or the Dung of a certain animal, this is quite absurd. As far as I am concerned, I heard from a trustworthy man that, while he was at sea in his youth, he came to a certain region, nigh to the sea, called *Bachach* by the people who lived there, and that he, after he came unto the beach with some others, found various pieces or lumps of amber, of different colors: That he kept the amber he found, and that he asked the Inhabitants where it came from, and that all they could respond was that they saw it regularly, with occasional lapses of time in between."

In Sevil, one of the most renowned Trade towns not only in Grenada, but in all of Spain, they bring a certain oyl to market, which comes from America. This oyl is reddish, and one ascribes to it a wondrous power against defects of the womb. It is called oyl or *Liquidambar*, and has almost the same scent as Styrax.

This oyl is drawn from a certain tree, called Ocosolt, as the Writer of the Mexican history, recounts as follows: *Among the trees of Mexica, there is one called Ocosolt, which is very tall and beautiful, having leaves like ivy,*[33] *the Liquid of this tree, which the Inhabitants call Liquidambar, is very good for all kinds of wounds.*[34] Mixed with the tree's bark, after it has been pulverized, it brings forth a very lovely and powerful scent.

After these remarks and annotations, we should also consider what Mr. *Nikolaas Monsard*,[35] Physician from Sevil, feels about this. He begins his observation with Florida, a Province in New Spain, from where we get the gray Amber these days, which one will find tossed onto the beach, from Kanavazal,[36] to the Cape of St. Helena: whereof he says this.

One will find various opinions concerning the origin of gray Amber. But what is certain, is that it is a kind of loam, which flows from fountains in the depth of the sea, and which, after it comes into the air, hardens, like many other things, which are weak and soft in the water of the sea, but once removed from it, become hard, like coral, and the yellow Amber.

Among the Greeks one will find that only *Simon Lethi*, and *Etius*,[37] make mention of it, of whom the first confirms that Amber, like loam, flows from wellsprings, and that the one swallowed by fishes is the worst.

This knocks down the notion of those people who say that amber is the roe, or seed of the Whale, because it is sometimes found in the stomach of Whales, which they, thinking it is their own food, swallow.

It is a true case that, in my time, a Whale was caught around the Canarian Islands, called the Happy Ones, and in its innards they found one hundred pounds of amber. Thereafter they killed a multitude of Whales and their young, but did not find any amber at all.

Those who come from Florida, say, that there are many Whales in that sea, but that, although many of them were caught or killed, along with their young, one did not find any amber in them. The Americans know how to catch these little Whales with a marvelous dexterity, as follows.

One of these Americans takes a strong rope, makes a noose of it, then he goes and lies in front of the Whale in a little boat, where it is wont to go with its young, and when he has reached one, he jumps on top of it, and throws the noose around its muzzle. As soon as the little Whale feels this, he dives with the American to the bottom, and he grabs hold of him with both arms around the neck. Now one should mark, that these Americans are great divers and good swimmers, so that they can stay a long time under water. But the young whale, wanting to take a breath, is forced to come up again, and then the American sees his chance, and stabs him by hand with a sharp wooden wedge, or pole, in the nose or vent holes, through which it breathes so that it stays stuck in it, without the little Whale being capable of getting rid of it. And after giving it a lot of rope, he climbs back into his little sloop, and waits in it until the fish, which cannot breathe anymore, has choked; and pulls it ashore. Surely a neat but very dangerous way of fishing. And these Americans are also so powerful and well prepared[38] that one of them can kill a Kaiman, a kind of Lizard, or Crocodile, that is at least thirty feet long and the cruelest of all sea animals.

And then there are those who maintain that gray Amber comes from a certain fragrant fruit, that grows close to the seashore, and ripens in the months of April or May, and which, when it has fallen down, is swallowed by the Whales, as if the fruit, that tends to be the food, could bring forth something else besides flesh and blood.

Although in the first chapter of the book by *Garcias du Jardin*, various things are brought to the fore, concerning the origin and description of gray Amber, we will nevertheless, to satisfy the curiosity of the Reader, cite several particular opinions concerning it, and then knit the final conclusion.

There be some, like *Serapion*, who propose, that gray Amber occurs on the bottom of the sea, or on trees, or rocks, that stand in it in the same fashion as a Campernoyle, or Toadstool in the earth, and that it is ripped loose, during thunder and lightning, by the force of the waves, and tossed onto the beach. And *Scaliger* assures us as well that in the Pyrenees mountains, and in the country of Rouergue,[39] one will find fragrant Campernoyles.

Garcias du Jardin's feelings seem to come closer to the truth. That Writer contends, that gray amber is a Loam, gray earth, or some soil of another color. It is less than one hundred years ago that they found a piece that weighed 100 pounds between Bajonne and Lappreton.[40] The sea threw also a piece of 35 pounds on the neighboring coast of Buch, and afterwards another one of eleven pounds and a half on the beach of Maren.

Eduard Barbossai says in his book on the Indies, that the Inhabitants of the Palandurian Island, in the Indian Sea, think that amber is the excrement of certain large birds, which gather at night on the rocks near the sea. That the dung of these birds is purified by the air and the sun, and that the sea tosses it back up again during a thunderstorm; to which he adds, that it is not harder to comprehend that a Bird shits amber, than that another Animal generates musk and civet.

Simon Lethi asserts, that Amber comes from certain fountains or wells of a gray, savory loam, be they in the sea, or near the sea. He greatly praises the reddish and gray Amber, which is found on Ceylon in the Indies, equal to what one will find in a certain harbor called Sycheon, and he thinks that the black kind is the worst. This opinion has been followed by *Falopius*,[41] our Writer, *Agrikola*,[42] *Garcias*, and a few others.

In his description of Prussia, *Erasmus Stella*[43] says that one knows from experience, that Amber flows from the mud of certain mountains, and is hatched by the power of the sun; that, when it falls on the shrubs at the foot of these mountains, it becomes hard, and that when the sea is at high tide, it is dragged away, and then tossed back up again onto neighboring shores. This Writer adds, that he has seen Amber taken from the soil as weak as wax, which, after it had been soaked in seawater, became hard.

So regarding the hardness or firmness of gray Amber, this can be given to it by the sea, because the force of the waves, whereon it floats, salts and contracts it. Or this gray Amber can harden in the air, like coral, of which it is said that, *in mari herba, si in aërem transferatur, in lapidis firmitatem solidatur;* that is, that it is a herb in the sea, but when one brings it into the air, it becomes as hard as stone. Just as the Asphaltian loam, which, when tossed on the beach, *vapore terrae & vi solis inarescit, ita ut securibus dissindatur,* that is, becomes so hard from the vapors of the earth, that one needs axes to break it. And also like unto the Pissasphaltus of *Dioskorides*, which floats in streams, and which, thrown by the wind onto the shore, becomes as hard as yellow Amber, which one holds to be a russet loam, contrary to the error of the Ancients, who thought that it was a sap, or liquid, that came from trees, which stood close to the sea.

The Writer of the *upright Merchant*, or the *general discourse on Simples*, says in the second part of his book, the XXVIth Chapter, that gray Amber is nothing but a compression of honeycombs, which fall from the rocks into the sea, or which are washed away by the waves, be it by the force of the wind or by something else. He adds that these waxcombs full of honey, while floating at sea, are either weakened by virtue of the seawater or by the power of the sunbeams and will float on the water.

This Writer, in order to confirm his proposal, says, that Mr. *Morkonys*, general Lieutenant of Lion, on the 71st page of his Travels, assures us, that he has heard in England, that gray Amber is the wax and honey which bees generate on the big rocks, which are on the shore of the Indian sea, and that these honeycombs, roasted by the sun, become loose and fall into the sea, which, by its movements, turns it into Amber. He adds that one has found in the center of a large piece of Amber, that had not quite matured yet, after it had been broken open, both the comb of wax and the honey. And in order to strengthen his opinion, he assures us that, if one melts the gray Amber with the spirit of wine or Tartar[44] one finally will be left with a stuff that is like honey.

Renanus Frančiskus,[45] Preacher to the King of France, demonstrates in his Proofs of the miracles of nature, that he is not at all of this opinion; for after he has quoted all the opinions, and has said, that it comes neither from the Fish *Azel*, in the belly of which, as *Serapion* says, it is perfected, after it has been swallowed by that Fish; nor from certain trees, from which flows a glue-like liquid which, when it falls into the sea, becomes hard, and is then tossed onto the beach by the waves; nor from the excrement of the Whale that, far from changing into a substance that has such a nice smell, according to those who traffic along the coasts, where the Whales sojourn, and who catch them, gives off an unbearable stench; nor from the excrement of certain birds, that live on top of the rocks and cliffs; he acknowledges openly, that one does not know what it is, and that no one has yet discovered the cause of such a precious substance. But he seems to lean towards the notion, that it is a Loam which, transported through subterranean pipes on the bottom of the sea, hardens, and is perfected in various ways: This notion is far more demonstrable than that of the Writer of the *general discourse on Simples*, who takes Amber to be a combination of wax and honeycombs. For besides the fact that amber is found in places, where there are no beehives; and that such a great abundance of this precious stuff, cannot come from the wax and honey that falls from some rocks into the sea, this becomes all the more apparent because the Inhabitants of the Maldive Islands, where a plenitude of amber washes up on shore, have never been able to determine wherefrom it came, nor did anyone else who lives along those shores, where this gray amber is found, which one surely would have discovered one way or another, if it originated from honeycombs. And in this regard, the Maldive Islanders, according to *Pirard*'s story in his Travels,[46] answer those who ask them for the origin of the gray Amber, that they do not know where this precious stuff comes from; that they did know that it originated in the sea, but whether it came from the bottom, or from the surface of the water, or from some rocks, or from somewhere else, no one had been able to inform them with any certainty.

Now having quoted the diverse opinions of the Authors, concerning the origin of gray amber, it is only left to me to present my own conjectures, which I will only do in order to rouse one of our scholars, so that they will convey their thoughts to the careful inquirers, concerning what they may have discovered about it.

In order to reach this twofold mark, I deem it necessary that one observes the form or shape of all the lumps of amber, which are brought over to us from time to time, in order to know whether it is a Drop that has fallen off, which one can perceive in the [piece] in the East Indies House.

If those who have seen it, took care to examine it closely, they must have surely perceived, and which is also noticeable in our two prints, that this piece of Amber has the shape of a large Drop. That the side, drawn as A, was fastened to something, which one can readily remark from the impressions, which are clearly visible in it; and that it is on the other side somewhat pointed, which indicates that one should hold it to be an Ickle-drop,[47] which brings me close to the opinion of those who hold that it cannot be anything else, than a Loam, that came from below the earth or the rocks.

The first reason that persuaded me to believe, that gray amber derives from Loam, is that, according to all of our Philosophers, there is a smell of sulphur, so that, according to whether there is more or less of this substance in a body, the smell of the same will also be stronger or weaker.

This being determined, I go on to a second reason, to wit, that sulphur, being a kind of earthen fat, or an oyly liquid, will mix more readily with Loam, than with any other substance. Now, since this Loam is constantly scorched in an eternal fire at the center of the earth, which renders it liquid and makes it boil, it travels through the veins of this same earth, and spreads, like unto the blood in the veins of the body, through various tubes, or grooves, which usually pass under hollow mountains, and is spewed forth by them from time to time. The Loamstreams, which came from the mountain *Gibel*[48] and *Vesuvius*, are incontrovertible proof of what I say, and one observed at the time, that the further these boiling floods had gone astray from their wellspring, [the more] they thickened owing to the cold air; so that in the end their liquid congealed as hard as stone, and they ceased flowing.

My third remark is, that some of these grooves could well end up inside the rocks, just as I do not doubt that they reach that far, and that there are hollows with salt water under these cliffs, which *Kircherus* calls *Hydraphilaci*,[49] the burning substance that constantly flows inside these rocks, this water, which is not pure, but like oyl, and, therefore, filled with sulphur, thickens. Meanwhile, since it cannot congeal into a Loam, because of the salt that is mixed in it, it acquires another consistency wherefrom comes the gray Amber, which, because of a perpetual boiling, as Chemistry confirms, acquires this pleasant smell, wherefore it is so highly esteemed.

One can make the objection here, to wit, that if there were salt in the gray amber, one would discern it in the taste, but one finds the opposite, since, as we said before, it has a very sweet taste. To which I answer with the chemical Philosophers, that salts become bitter from long boiling, and that they finally lose all their bitterness, and become sweet.

Since everything I proposed has now been confirmed, I suppose that there are small holes or vent holes in these rocks, through which the fire is lit, it brings the Loam to a boil, which then, from the force and from the fearful heat, flows through these openings, and since it is a thick and heavy substance, it soon congeals, when it comes into the coolness of the air, and becomes, like other liquids, an Ickle-drop which, since the cold air coagulated it, has many cracks, and which will burst open from the shining of a great heat, as can be seen in the piece, that is on display here.

There is no gainsaying, that heat will make bodies split open. One only needs to behold the earth in summer, after a long drought, in order to be convinced of this truth. One can then see that portions, shrunk from the great heat, and from lack of moisture, are full of cracks and splits. The same can also be true of some bodies by means of a great cold, to wit, when the moisture that is in them, freezes, as was proven by the famous high School of Florence, about crystalline, and other metal tuns of various sorts. These tuns, filled with water, and closed off in such a way that there could not be any exhalation, broke into pieces as soon as the water was frozen. So that, if one finds several cracks in the piece which concerns us here, as in the Print, which was made of it, one can conclude from this, that it was filled with moisture, and that its surface was hard, so that, this moisture having been congealed by the cold, it could have come forth from these cracks.

But just as I do not judge this lump to have a crust, or hard shell; and being convinced myself, that it does not have one at all, I am not able to embrace this notion.

I will add to my Remarks here, the thought of the Paracelsus of our time, to wit *Pieter Jan Fabri*, who says of this Amber in his *Panchianitus*, IVth Book, the 49th Chapter:

The gray Amber is a thick and coarse material, that comes from the sea, and, like the Campernoyles, grows in the rocks off the fat, moist, and loamish substance of water, that is sucked into the stones and rocks of crags; and in its little holes the water boils, and ferments, and its smallest particles evaporate.

And this Author adds, that one should know that the earth, because of its central fire, locked up within, casts up vapors everywhere, even from the bottom of the sea, where the coarsest substance brings up the salt, which also fashions the stones of the rocks that are found in the midst of the sea: And Amber is made from the thickest stuff, that comes from the small holes in these stones, which, after the manner of sulphur, is gray and black, and that same sulphur, when mixed with seawater, renders a very nice odor. And so one can say, that gray Amber is a glue, or a thick substance from the water of the sea which, like wax, melts by the fire, and stiffens from the cold.

According to this same Writer, gray Amber also occurs in sand, since the slimy seawater, that is drawn into it, also thickens in the heat of the sun; he adds, that no Sailor ever saw this Amber float on the water, of which there are three kinds; the one gray, the other the color of earth, and the third black, but that the gray Amber is the best of all, not only because of its smell, but also that it is an excellent remedy to strengthen the heart, which cheers all lively spirits, and that it has the power of a balm.

We conclude our remarks, by saying that, of all the various opinions I have collected, none is better proven than the one which posits that it is a kind of Loam, or that it derives from it; note that, as I just proved, the wells and fountains of the same can be found at the bottom of the sea, in caves of burning Loam, that can be found there as well as under mountains. That there is such a subterranean fire at the bottom of these great depths, is easily accounted for by the new Islands, which one sees engendered every so often from Pumice,[50] which can only be a casting forth, or a vapor caused by fire.

I could augment my remarks about this substance, but since I did not take up my pen for any other reason than to get some of our Scholars going, so that they would tell us what they have discovered about this piece, and about the nature and origin of Amber, I deem it not advisable to amplify my contemplations any further. I submit them, My Lord, to your judgment, begging you to believe that I will always be ready to follow reason and show the utmost regard for the judgment of a person such as you, who has more insight than I do.

I remain,

MY LORD,

Your truly obliged Servant,

N.N.

CHAPTER 36

Ambra Nigra. Ambar Itan.

Black Ambar, usually called *Ambar de Noort,* from a corruption of the French *Ambar de noir,* is usually thought of as the excrement of a whale, because, when such fishes are opened up, it is usually found in the back end of the stomach, but, as I stated in the preceding Chapter, true Ambar is often found wrapped in this black kind; the whale must cast it out by means of the mouth as well as the excrement; it might well be that the whale changes or increases it, yet I absolutely do not believe, that either the black or gray Ambar is produced by the whale, but rather that both are a glue-like or oyl-like moisture that derives from the bottom of the sea and which, when it floats on top of the water, is ingested by all kinds of large fishes, and perhaps is changed somewhat in their stomachs.

They sell two kinds of substances in these Eastern regions that are known as black Ambar, and which, though they share a common nature, are markedly different: The first one is the true black Ambar, which is found in the sea: The other is the Stone Pitch from Buton, called *Ambar Batu.*

The real black Ambar, which one will find with and without Ambergris, is that which the Whale casts out the most, and of which large pieces float in the sea; it derives from the bottom of the same around the South-Easter Islands, and [drifts] from there to *Nova Guinea* and the southland,[1] where one will find many Whales: and it is not unreasonable to guess, that the *Ambar Fountain* is in that large Bay, since it is usually tossed up after a big storm in the East *Mouson;* at first it is soft like melted pitch, but it hardens over time, breaks off into short pieces, and is a grayish black like an extinguished coal, that has been buried a long time in the ground, but when it is fresh, it has a nasty fishy smell; once dry and put on hot coals, it almost smells like *Bitumen*[2] or stone coals, though it will always retain a certain smell of the sea: I have noted two kinds.

I would guess that the first kind is spit out by the Whale with and without Ambar-Gris, wherefore I call it the selfsame's covering; it never becomes truly hard, while it always retains some of its fishy smell, especially if put on coals; one will in fact find some that has a bad odor like shit, while both will go up in smoke on a hot plate: That this kind derives from the Whale I base on the fact that in the year 1677, they brought me a piece of Amber-Gris, the shape and size of a *Pissang,*[3] which looked like good Amber de Gris, wrapped entirely in a black, sticky substance, which in time became hard, being true black Ambar which also retained the smell of Ambar: Many Seacat beaks were stuck in this black covering, and they must have gotten there in the Whale's stomach, since it swallows such soft fishes for its food, and it could well be that the tough slime from those fishes helps in the production or increase of black ambar: It was a clump the size of a Butonese Chest, floating at Sea by *Manipa,* with many fishes and birds gathered 'round.

The second kind seems never to have been in a Whale, it dries much harder and breaks like glass, the broken pieces also shine, and it is quite pitted, but it does not contain the aforementioned beaks; put on live coals it smells like Earth oil,[4] without a fishy smell, and therefore more pleasing than the foregoing, while it slowly dissipates as smoke if put on a hot plate.

One will encounter reasonable amounts of both kinds all over this *Archipelagus* between *Celebes* and *Nova Guinea,* especially in the Papuan Islands, where one will find the hardest kind which also smells the worst, and is often adulterated with some *Dammar;* for it will happen at times, that black *Canari Dammar*,[5] which came by river down to the sea where it hardened, is sold for black Ambar, just as other, lighter, harder and yellow *Dammar*, after it has hardened in Seawater, acquires the shape of yellow Amber [*bernsteen*].[6]

Ambar Batu or Butonese black Ambar is not a fruit of the sea at all, but a kind of Stone pitch, *Bitumen durum*[7] or *Pisasphaltum*, that has sweated from certain rocks on Land, and which at first is soft like tar, then hard as stone, and once it gets into the water it is a shiny black like *Gagates*[8] or Stone coals; put on coals it has the same smell; the treacherous Butonese (to put it briefly) like to pass this off as black ambar, and dare to sell it as such, telling strangers, that it came from the sea: After much effort and digging I came to know that, like other *Bitumen*, it sweats from the earth and rocks in the mountains of Buton's southern regions, sometimes as thin as oil from rocks, other times thicker, like a soft tar, shaped like a horn or turnip from the soil, and so soft that it almost sticks to one's hands, though it will harden in a few weeks: the place is called *Cotawo*, being a mountain between the city of *Buton*[9] and *Culutsjutsju*, but not many people come there, because the place is inhabited by ghosts and vermin,[10] though this is not very likely; because I know from experience that snakes do not like to stay in places where *Bitumen* is burned, particularly the Butonese kind, which is why I saw some Natives who fume their gardens with it: A river wells up in the aforesaid mountain, and the soft *Bitumen* from the rocks seeps into the water, and the river carries it along and into the large bay of *Culutsjutsju*, where it comes into the sea, and hardens and does not become soft again when it is handled; while still soft and adrift in the sea, it will have fishes and birds circling around it just as they do with other Ambar, especially Sharks, and *Tsjukalan* (which is a small kind of *Thynnus*).[11] This is why the Butonese believe, or at least try to persuade other people, that it was spewed forth by a Whale: Others, who know that it comes from Land, say or feign, that there lives in the aforementioned place in the mountains, a big Snake, called *Bokulawa* or *Bonkulava*, being the largest kind of *Boa*, which the Malay call *Ular petola;*[12] this Snake is supposed to spit out the said Ambar, when it comes out of the water in order to drink, and which is then carried into the Sea by the river: Now hear what else happens to this *Bokulawa*, after it has grown very old and very large, acquiring the size and the length of a *Calappus* tree; once it is that big its heavy body prevents it from moving, so that it stays mostly in one place, feeding itself with whatever happens to pass by, and it gathers so much dirt and dust on its body from lying so still for that long, that all kinds of bushes and shrubs grow on it, and when all food is gone, nature will stir it anon, and remind it, that its time has come to quit the land, and the *Bokulawa* will bestir itself with a bold intention and take its leave of the land, but not without frightening the spectators, because it will break through trees and bushes with a dire noise, breaking everything down that comes in its way, resembling nothing so much as a big tree that is dragged down the mountain, and it continues straight down to the sea, wherein it will start to swim, because it has been changed into a long and thin Whale, with scales and teeth, though retaining its old nature of being able to spew forth black Ambar.

The people from the Island of *Binonco*, which lies south-West of *Buton* often see such Fishes and Snakes around their Islands, where one will find the preceding black Ambar: a certain man from Makkassar, who sometimes speaks the truth, assured me that he had seen such a beshrubbed Snake, the thickness of a tree, around *Turatta* in the Land of *Makkassar*, coming down the mountain, on its way to the sea, and making such a noise, that the neighboring people fled before it, but whether it turned into a fish he could not tell: My guess is that this Fish, is really the long and toothed *Pristis* Whale,[13] which attains a great size in the Indian Ocean: In my time such a fish was cast upon the big reef near *Lacker*, one of the south-Easter Islands, more than 30 common fathoms long, or 150 feet, not very thick, though it had scales, and a mouth full of teeth, all of different sizes, like those of Snakes; because the biggest ones were the length of an upright hand or a span, four fingers wide, a little curved on top with a blunt tip, solid and a dirty yellow, but inside it was white and like ivory, though much harder: I own one like it due to the kindness of Mr. *Qualbergen*,[14] at the time Governor of *Banda;* others are shorter, though as wide and blunt on top.

To return to black Ambar, I suspect that there must be places on *Buton* where *Bitumen* sweats forth, because on the West side of *Buton*, on the beach of *Waloba*, others say *Waccocco*, one will find gray-black stones on the shore, which are thought to be ordinary stones, but when they are examined more closely, one discovers that it is a clump of *Bitumen*, mixed with coarse sand, and small, black and dark-green little stones, which can be used as touch-stones, and all of it will harden to stone over time: If one holds the black kind in a fire that corner will become somewhat soft, brittle, though it will hardly burn and will smell strongly of Stone coals; the gray kind is too old and hard, nor does it melt well in fire, but turns brittle and smells like the previous one: The people from *Buton* call it *Batu Ambar*, that is Ambar stone, and it is not esteemed by them nor do they use it. It appears that the West wind tosses this soft Ambar on the beach near *Tomahus*, being the western corner of *Buro*,[15] at least 40 miles from the *Dwaal*-Bay or bight of *Culutsjutsju*, where pieces of pure black and Ambar not mixed with sand often drift onto the shore, though some were so hard and gray, that they were thought to be pure stones.

Black Ambar is usually of little value and is scarcely used: The first kind of Whale Ambar is most often used to make fragrant Beads[16] or Paternosters, as was said in the preceding Chapter; The second kind or Butonese *Ambar batu*, is melted down by the people of *Culutsjutsju* and poured into dishes, obtaining big cakes that are clean, black, and harder than tar; they shine if one breaks them, and they are sold to strange merchants for whatever they can get from them: The Malay and the people from Makkassar will buy it at a lower price, because they use a little of it mixed in with their *Dupa;* for it will grant other sweet incense a manly fortitude, that is, it makes them long-lasting and powerful, but one must know the exact amount, because if one uses too much, it will spoil the rest of the incense: One uses the fresh kind to scent fishlines, nets, and seris[17] in order to lure fish. In the year 1669, the people from *Buro* had a cake of black Ambar, which they used to fume the smallpox (which was very virulent at the time), and drive away the painful itching and scratching: Ambar stones are only good to fume gardens, but I found that it was better to use this with incense than pure stone pitch, since it acquired more of an Ambar smell after it was washed in the sea: The King of *Buton* had a water trough made in his town on the mountain, using the stone pitch that had been softened with oil in place of cement, because it becomes hard again in water: The people from *Buton* also mix it with wax, oil, and various odiferous resins, molding it into gray-black balls, which they sell as Amber, and one should watch out for this deceit, because it is a dirty and base stuff, that remains soft for a long time, and smells like *Calappus* oyl, though in time it will become as hard as stone.

Ambar Gunong[18] is another black substance that those from *Buton* try to sell us as Ambar, saying that it grows in the mountains, where it is found not very deep beneath the surface around the roots of certain trees: I consider it the *resina* (resin) of a Tree that is unknown to me, because it resembles *Gummi Elemi*[19] a great deal, but without any smell; parts of it have the color of horn, others are a pale yellow or russet; burned on coals it smells like fat, but not unpleasantly so: One will also find stubble, dry twigs and leaves in it, and some corners look like thick green glass, held in fire it burns like any other *resina* (resin). Those deceitful

Butonese, who give a false name to anything they find in their Country, tell strangers that it is a precious commodity, a kind of Ambar that comes from the Mountains: they use it with savory salves and with other incense, when it would be better to use plain *Dammar Selan:*[20] The aforementioned black Ambar stones from *Buro* are properly found around *Foggileko* or old *Foggi*, but they are neither known nor appreciated by the Natives.

CHAPTER 37

Ambra Alba. Sperma Ceti. Ambar Puti.

The third kind of *Ambar* is white, though I do not mean to infer that this is the white Bird called *Ambar* by *Castagneda*, as we told in the preceding Chapter XXXV, since if it were such, it would have to be included with Ambergris; but rather that white, flaky stuff, usually called *Sperma Ceti:* Our East Indian *Sperma Ceti* has a different shape, but it is substantially the same as the European one, which floats at sea in huge swathes, yellowish or a dirty white, and is preserved in white slivers. The East Indian kind, on the other hand, is found packed together with big and small pieces, some the size of a fist, others as big as a head or larger, mostly a clean white that veers slightly towards yellow, it flakes easily, particularly between one's teeth, is greasy to the touch, though it does not contaminate the fingers at all, nor does it leave them greasy, it is tasteless or it might have a faint sea odor that smells a little like Ambar, but on hot coals it smells like melted sewet: The big pieces are usually flat, with many dents and corners, when they are tossed onto the beach; covered with sand on the outside; inside one finds, just as with ambar, the beaks of seacats, and pieces of Shells and Oysters; some are no more than 2 fingers thick, and when held against the light they are clear and partially transparent like white Amber [*bernsteen*], but those are rarely found, though I sent one weighing 5 pounds to *Florenço:* other pieces have a black stuff inside that looks like mire, often spread throughout the entire clump, and which is thought to be a black mud: But I am of the opinion that this too gets there in the whale's stomach, just like the beaks of seacats and the Shells; though I was once of the opinion that *Sperma Ceti* is a marine grease, and that it had something in common with Ambar's origin, I now think differently because of what the Natives told me, to wit, that it is not only what a Whale spat out, but also something that it produced itself: This is not only proven by the previously mentioned things which are found mixed in with it, but also by the greasy and flaky substance as well as the sewety smell, all of which are markedly different from all the foregoing kinds of Ambar, making it quite clear that it must be a fishy substance. Our Europeans maintain it is the seed or essence of the Whale, wherefore it received the name of *Sperma Ceti* or Whale Spawn; but the flaky stuff they scoop up in the North Sea, cannot be the same thing as that in the East Indies, because it is so much different from the nature of other animals' seed, and is found in large clumps, with the aforementioned things mixed in, but what it will ultimately turn out to be, will only result from closer experience and more careful examination.

We normally call it in Latin *Sperma Ceti*[1] in order to distinguish it from the European flaky kind: Also *Ambra alba*, in Malay *Ambar puti*,[2] since no whiter Ambar can be found than this: Some Malay call it *Ican punja monta*,[3] that is what a fish spits out, a name that rightly belongs to the Sealard that comes next; otherwise also known as *Bene gadjamina*,[4] that is, Whale seed, but I trust that our Nation persuaded them to say that.

There are years when one will find reasonable amounts of it in the Ambonese Islands, especially on *Buro* and *Manipa* where it is tossed on the beach in some places after the end of the East *Mousson*, but it had better not lay there too long, or the pigs will gobble it up: I have also noticed that it is found more often in *Amboina* and the *Moluccos*, as well as along the East coast of *Celebes*, than in the South-Easter Islands.

What they call the best kind is white or slightly yellowish, flakes easily and does not have the aforementioned black filthiness; the greasy kind does not flake as easily, and especially the kind that the Buronese melt and sell in pieces of Bamboo is much worse, since it is mixed with Sealard, either by the Whale or by the people, so that it has a sewety smell.

We use it here the same way and as a remedy for the same ailments as *Sperma Ceti* in the Fatherland, namely, for the pain caused by Gravel in the reins, for diseases of the lungs and spitting up pus, for blood clotted inside and for expelling it: Experience has taught, that it works quite differently for different people, since some people who are afflicted with Gravel in the Reins did expel the stones so forcefully that it was followed by bleeding, while with others it had but little effect. It is difficult to take because it congeals so easily in one's mouth, even if it is melted with some liquor, which is why one does better to take a small piece as big as a hazelnut and chew on it, then swallow it, which is not too difficult since it does not have a nasty taste or smell, follow this up with some warm liquor, and then let it all melt in the stomach: Otherwise it will easily melt in water or any other kind of liquor, and when it is cold again, it becomes a *Massa* once more, though it will readily mix with heated oils; one can purify it somewhat in this manner, when one melts it in hot water, which causes most of the grossest filthiness to settle on the bottom: But this melted stuff acquires another shape, nor does it stay as white, and becomes more like sewet, wherefore I deem it wiser, that one keeps the pieces the way they came from the Sea, even if there is some black stuff in them, which will not harm anyone, as long as they are cleaned of the outermost sand and filth; one can deprive them of their brininess, if one puts them in fresh water several times, drying them each time in the sun; some also know how to bleach them white, after first melting the smaller pieces into thin and flat cakes, which they then bleach with seawater in the sun until they turn white and become flaky again, whereafter they fashion it into balls and clumps.

One will also find it in the bay of *Lampon* in *Sumatra*, but if it is brought into the village, they will forbid young Women even to look at it or handle it, saying that they will get the whites[5] from it, I believe from ingesting it, wherein they might be unfair to this stuff: Some Chinese were trying to tell me, that it is some kind of Gum that sweated out of old and cloven *Casuaris* trees, a notion that does not require much refutation, since the lumps are too big, nor is there any Gum that will melt like fat in water, or that has this kind of smell. *Mathiolus*[6] and others have sufficiently proven that it is not the ἅλος ἄνθος[7] of *Dioscorides* and *Pliny*, and I would agree with them; but that it is a greasiness or *Bitumen*, that was sweated forth by either the sea or the ground itself, does have (as said before) its own problems: After all, the smell of trane oil it gives off when kindled on hot coals, as well as the fact that it melts so easily in water and will mix with oil should argue that it is not *Bitumen*, but a form of fat that was produced by some Animal. If one wants to put the pieces in a Vat in order to keep them longer, and in such a way that they will not melt into one lump, one should first wrap them in dry grass, and then preserve them in some sawdust: One can also keep them in a stone pot in cool places, but one should never put them on linen or paper, because those will rot from the saltiness, if it stays therein for a long time, even though this is also the means of preservation; because every 10 years one should take *Sperma Ceti* and other kinds of Ambar and, when it has become too old and too dry, one should grant it a night in seawater, and then dry it in the sun, since this will renew its powers. It appears from the French Histories of the *Anthill.*, that this stuff also occurs in the West Indian sea, and that they call it *Sapo marinus* there.[8]

I will now proffer several means of proof, which might support my own contention that *Sperma Ceti* has its origin in certain kinds of Whales.

First of all, the reader can find in the first Volume *ad Annum* 1670 *Obs.* 136. of the *Ephemeridibus Natures Curiosorum*[9] that a certain person kept as a secret for a long time, though it finally came to light, that he had caught many Whales in the Bay of *Biscaija*, and that he had been able to prepare the real *Sperma Ceti* from the brain of the same: for he took the brain of the Fish, put it in a large pot, that had a small hole at the bottom with a stopple in it; he let this stand in a warm place for some time, whereafter he opened up that bottom hole, and let the oyl or trane oyl out, which left the pure *Sperma Ceti* in the pot: They call this kind of whale there the Male, in Latin *Orca*, and it has many teeth, with some that weigh almost a pound. In the same *Obs.* are some Examples, that the English who live in the *Barmudas* Islands, did not only get the *Sperma Ceti* from the brain but also from the stomach of a toothed whale, which they also thought of as the Male.

Furthermore, we have had such experiences here in Amboina since the year 1670, when we took the brains out of the heads of dead Whales, most of whom were toothed, and kept the same for some time in Bamboos and pots; after several months, though mostly during the cold weather or rainy season, we found a pure white and flaky *Sperma Ceti* on the sides of those vats, and if we made a hole at the bottom of the vat, then the oyl or trane oyl would run out: But this kind of *Sperma Ceti* always has a taste and smell that is far more like trane oyl than the previously mentioned pieces, which are found in the sea. For this toothed Whale or *Orca* see my Animal Book.[10] The English in *Bermuda* have learned, that one can get the fresh *Sperma Ceti* out of the Whale with *Omphacinum*,[11] that is, verjuice, but it must be washed and strained in order to remove the smell of trane oyl.

I think it meet here to add a detailed story about *Sperma Ceti* from *Thomas Brouwn lib.* 3. *part.* 1. *cap.* 26.[12] "What *spermaceti* is, men might justly doubt, since the learned Hofmannus, in his work of thirty years, saith plainly, *Nescio quid sit*, that is, I don't know. And therefore need not wonder at the variety of opinions; while some conceived it to be *flos maris;* and many, a bituminous substance floating upon the sea.

"That it was not the spawn of the whale, according to vulgar conceit or nominal appellation, philosophers have always doubted, not easily conceiving the seminal humour of animals should be inflammable or of a floating nature.

"That it proceedeth from the whale, beside the relation of Clusius and other learned observers, was indubitably determined, not many years since, by a *spermaceti* whale, cast on our coast of Norfolk; which to lead on to further enquiry, we cannot omit to inform. It contained no less than sixty feet in length, the head somewhat peculiar, with a large prominency over the mouth; teeth only in the lower jaw, received into fleshy sockets in the upper. The weight of the largest about two pounds; no gristly substances in the mouth, commonly called whale-bones; only two short fins seated forwardly on the back; the eyes but small; the pizzle large and prominent. A lesser whale of this kind, above twenty years ago, was cast upon the same shore.

"The description of this whale seems omitted by Gesner, Rondeleius, and the first editions of Aldrovandus; but described in the Latin impression of Pareus, in the *Exoticks* of Clusius, and the *Natural History* of Nirembergius; but more amply in the icons and figures of Johnstonus.

"Mariners (who are not the best nomenclators) called it a *jubartas*, or rather *gibbartas*. Of the same appellation we meet with one in Rondeletius, called by the French, *gibbar*, from its round and gibbous back. The name, *gibbarta*, we find also given unto one kind of Greenland whales; but this of ours seemed not to answer the whale of that denomination, but was more agreeable unto the *trumpo* or *spermaceti* whale, according to the account of our Greenland describers in Purchas; and maketh the third among the eight remarkable whales of that coast.

"Out of the head of this whale, having been dead divers days and under putrefaction, flowed streams of oil and *spermaceti*, which was carefully taken up and preserved by the coasters. But upon breaking up, the magazine of *spermaceti* was found in the head, lying in folds and courses, in the bigness of goose-eggs, encompassed with large flaky substances, as large as a man's head, in the form of honeycombs, very white and full of oil.

"Some resemblance or trace hereof there seems to be in the *physiter* or *capidolio* of Rondeletius; while he delivers, that a fatness, more liquid than oil, runs from the brain of that animal; which being out, the relicks are like the scales of *Sardinos* pressed into a mass; which melting with heat, are again concreted by cold. And this many conceive to have been the fish which swallowed Jonas; although, for the largeness of the mouth, and frequency in those seas, it may possibly be the *lamia*.

"Some part of the *spermaceti* found on the shore was pure, and needed little depuration; a great part mixed with fœtid oil, needing good preparation, and frequent expression, to bring it to a flaky consistency. And not only the head, but other parts contained it. For the carnous parts being roasted the oil dropped out, an axungious and thicker part subsiding; the oil itself contained also much in it, and still after many years some is obtained from it.

"Greenland enquirers seldom meet with a whale of this kind; and therefore it is but a contingent commodity, not reparable from any other. It flameth white and candent like camphor, but dissolveth not in *aqua fortis* like it. Some lumps containing about two ounces, kept ever since in water, afford a fresh and flosculous smell. Well prepared and separated from the oil, it is of a substance unlikely to decay, and may outlast the oil required in the composition of Matthiolus.

"Of the large quantity of oil, what first came forth by expression from the *spermaceti* grew very white and clear, like that of almonds or *ben*. What came by decoction was red. It was found to spend much in the vessels which contained it; it freezeth or coagulateth quickly with cold, and the newer soonest. It seems different from the oil of any other animal, and very much frustrated the expectation of our soap-boilers, as not incorporating or mingling with their lyes. But it mixeth well with painting colours, though hardly drieth at all. Combers of wool made use hereof, and country people for cuts, aches, and hard tumours. It may prove of good medical use, and serve for a ground in compounded oils and balsams. Distilled, it affords a strong oil, with a quick and piercing water. Upon evaporation it gives a balsam, which is better performed with turpentine distilled with *spermaceti*.

"[Had the abominable scent permitted, enquiry had been made into that strange composure of the head, and hillock of flesh about it. Since the workmen affirmed they met with *spermaceti* before they came to the bone, and the head yet preserved seems to confirm the same. The sphincters inserving unto the *fistula* or spout, might have been examined, since they are so notably contrived in other cetaceous animals; as also the *larynx* or throttle, whether answerable unto that of dolphins and porpoises in the strange composure and figure which it maketh. What figure the stomach maintained in this animal of one jaw of teeth, since in porpoises, which abound in both, the ventricle is trebly divided, and since in that formerly taken nothing was found but weeds and a *loligo*. The heart, lungs, and kidneys, had not escaped, wherein are remarkable differences from animals of the land: likewise what humour the bladder contained, but especially the seminal parts, which might have determined the difference of that humour from this which beareth its name.]"

CHAPTER 38

Lardum Marinum. Sealard. Ikan Punja Monta.

A different filthy and greasy stuff would like to pretend to be a kind of Ambar, and we call it *Lardum marinum*, that is, Sealard or Sea sewet, in Malay, *Ican punja monta.*[1] It is undoubtedly something that a Whale spat out, to be found in big and little lumps, a pale yellow, dirty like old and rancid bacon, and it makes your hands filthy, inside it is quite hairy and stringy, so that it is difficult to pull apart, though it can be cut, it has a greasy, trane-oyl smell and taste, which becomes even stronger and more unpleasant when put on coals, being like trane oyl or old fat, so that it can fill an entire room with its nasty smell, which will last for quite a while: At first it is whitish or a dirty yellow, but it becomes grayish and black on the outside within a few years, and most of the grease dries up in time, so that almost nothing but coarse fibers remain; melted on fire and strained through a cloth it turns into a thick sewet.

It first became known here in the year 1664, when a large piece floated up on *Manipa*'s shore, and it lay there melting in the sun and no one noticed it, until a hunter with his dogs came upon it, and seeing that his dogs ate it, he picked it up: Around that time they also found a piece on *Nussatello*, and while it was not known to the common man, they persuaded each other, that it was a kind of Ambar, and was sold therefore at quite a high price: They brought it to *Java*, where they laughed at these merchants, and they learned from the Javanese and Malay that it was some disgusting stuff that a Whale had spat out: But they would take it off their hands at a low price, because it could be used by those who live from fishing; because if the same is heated, and smeared on the door of the Seris and the mouth of the weels or Robbers,[2] it attracts fish, and this (as far as I know) is the only real use it has; but there are others who do not like to have this stuff on their Seris, that is if they have sore legs, because the sores will get worse in Seawater that has been infected by this stuff. One might hazard the guess, that the Sealard's tough fibers have their origin in the tough and stringy flesh of the Seacats, which is hard to digest, and perhaps causes this mixture in its stomach.

CHAPTER 39

The Malay Names of Some Precious Stones.

Adamas,[1] a Diamond, *Intam* in Malay: The Malay have three kinds of Diamond: 1. *Intam Bun*,[2] that is, the water or dark Diamond, with dark waters;[3] this one is considered the healthiest to wear: 2. *Intam Api*, or fire Diamond, with a reddish water, while its body is slightly yellow; this one heats the blood too much, and is considered unlucky: 3. *Intam Puti* or white Diamond, with clear and white waters; these are the most beautiful to wear, because they sparkle from a distance. All three occur in *Succadana* and *Landas*.[4] The Chinese call the Diamond *Suanu*,[5] and it is not highly prized in Sina; only some humble people buy them in order to cut glass or for engraving.

Canjang[6] is a bastard Diamond, and its waters remain on the bottom and do not shoot forth, these might be our *Clabbeken:*[7] *Clabbeken* occur in Sina, they are very hard and difficult to cut, and are called *Tsjutsju*[8] there, that is Water stone, because they lack the Diamond's fire.

It is not surprising, that Diamonds are not as hard these days, the way they were when the Ancients wrote about them, if it is true that in the Golkondalian mines[9] Diamonds grow back again within a few years, at the place where they were dug up before. Today's covetousness and pomp wants them in such profusion, that every Tradesman is capable of wearing one, so that there is no time for the stone to become old, while before it might have had 1 or 2 thousand years to rest in its mine: For we find the earliest mention of Diamonds in *Moses*, who was the first to place the Diamond, under the name of *Jahalon*,[10] that is, Hammer stone, in the holy breastplate; when the world had existed 2500 years already: but it is not known if *Tubalkain*,[11] the work master in copper and iron, knew already before the Flood how to cut Diamonds.

Rubinus,[12] a Ruby, of which the biggest one, which we consider a Carbuncle, is called *Gomala*[13] by the Malay, while the ordinary ones are known as *Manidam* and *Permatta mera;*[14] because *Permatta* commonly refers to a Precious stone, as if one were saying *per meij matta*, that is, something that is beautiful to the eyes: Nor do the Malay make any distinction between true Rubies, which are as red as fire; or *Spinelles*,[15] which are a bright red; *Lychnis*,[16] which is a yellowish red, and *Balace*, which is light red, or whitish: Some of these are so white, that people think they are crystals, but when one tilts them they show some reddishness, which is quite becoming, and this is the *Eristalis*, or better the *Erytrylla*[17] of *Pliny Lib. 37. Cap.* 10. All the varieties of Rubies are called *Sia Liudsi*[18] in Chinese, that is, kernels of Pomegranate, and they dig them up from a great depth in the ground; the women have them on their hair ornaments, but the men do not wear them.

Smaragdus,[19] in French Esmaraude, in Malay *Permatta Idjou;*[20] they are not often found in these parts: Emeralds come from the Chinese Province called *Sussuwan*,[21] where they are found at the source of certain rivers, where they spring forth from the rocks: One finds them there in pieces the size of a hand, but there are only a few of them in *Sina* and they are very costly.

Saphyre; *Nila* in Arabic, and in Malay *Nilam* and *Permatta biru*.[22]

Granate;[23] in Malay *Bidji de Lima*, rose colored, its shape and color quite resembling the kernels from a ripe Pome-granate, wherefore I would guess that it cannot be *Pliny*'s *Kargedonius*[24] or the *Grenater* of the Ancients, since those are supposed to be bright red and fiery, and since the common Granate properly belongs with the Balaces: They do not get their name from [the word] Palace, as if it were the house or mother of the Ruby, but rather from the peoples called *Balukes* or *Balotsjes* who live to the west of the *Indus* river, where one will find a large number of these stones:[25] I have seen Natives use them to clear up cloudy vision[26] and cool inflamed eyes; one rubs it on a hard whetstone with some water, just until something comes off, and this is then applied to the eyes.

Amethistus[27] is known for its purple color; it is hard, smooth, shiny, but not very transparent, blunt on the sides, and greasy in appearance; the best ones are considered those that have the color of Amorel wine:[28] One will also find some, which have such a light color, that they look like purple crystal, sometimes they even have violet veins inside; they are not set in rings because they are too plain, but are used instead in Bracelets and Necklaces.

Sardius,[29] a *Carnelian*, of which many are found in the Red Sea near *Mocca*, is held to be an Agate by the common man, and is also called red and yellow Agate; because some are as reddish as raw meat, or as water wherein bloody meat has been washed, and it is not transparent at all, which is the case with a true Carnelian; others have an Orange color, are particularly transparent, with broad veins, which sometimes look like images; others can be whitish like another kind of Agate: All have blunt sides and have a greasy appearance: Then there are some which are such a fiery red, that one would hold them to be yellow Rubies, though they always have some dark spots or veins at the side.

CHAPTER 40

Cat's-Eyes. In Malay, Matta Cutsjing.

Cat's-eyes, called *Olhos de Gathos* by the Portuguese,[1] are a kind of *Opalus*,[2] that is particularly distinguished by the fact that they must show a white beam that traverses their bodies, if one holds a certain corner against the light, and they may be dark, partially, or completely transparent, and be whatever color they wish, but if they lack this beam, they are not Cat's-eyes; but one should note, that if one does not turn the correct corner to the light, one will not see the beam or only faintly; the beam is also more radiant by candlelight than during the day. The most common ones have a grayish-yellow color, some veer towards green, just like the eyes of Cats; some do not, some are partially transparent, and their beams veer from white to yellow. One should not grind them to have facets or edges, but they should be left round and simply polished. The most ordinary ones are used for Bracelets and *Paternosters*, the ones that are partially transparent are set in rings, but they have to be flat on the bottom, and not set too deeply into the bezel.

The kind that is the most beautiful comes close to resembling an *Opal*, and is called the Elementary Stone, *Gemma Elementorum*, by the German Jewelers, it is as large as a pea, transparent like a dark crystal, and one can see the colors of the four Elements play within; because the transparent body with the white milkbeam represents the sky, a small yellow flame at the side represents fire; a small blue spot, water; and the other dark corner represents the black earth: They are more fiery after sunset than during the day: They are soft and tender stones, which should be handled delicately; they are found on *Ceylon*, in *Cambaija*, and in *Pegu*.[3]

They are held in greater esteem in the Indies than in Europe and they also cost more here, because the Javanese have a common notion or superstition, that the wearer's goods cannot diminish but only steadily increase;[4] this might be because there are many people who envy those who are doing well, and the former will close their eyes so they will not see another's good fortune. They also believe that Women, who wear this stone, will be more pleasing to their menfolk: Like the Amethyst, it is also a Dreamstone, causing pleasant dreams, and yet it does not want to be worn by any dreamer: The Ancients called such a stone *Paederos*,[5] Orphan in Dutch, because it deserves to be loved just like a handsome little orphan boy.[6]

I also noted of the Elementary Stone that, when set in a ring, wherefrom it protrudes a little like a pea, it will show the images and colors of anything, that has a repercussion of light, especially if one is standing under a roof; for instance, I once sat in an *Oranbay* (being a covered vessel that respectable people sail in)[7] and had such a ring on my finger while we were sailing along a shore, and I could see in it the green trees that stood on shore, also the movement of the boatmen behind me, as well as the faces of those who were sitting next to me. And if this Elementary Stone durst not aspire to be a true *Opalus*, then I would think it is at least Pliny's *Astrobolos*[8] in *Plinius Lib. 37. Cap. 10.* which he compares to a fish eye, and that catches the glitter of the stars.

There is another rare Cat's-eye, but this derives from a wild Cat, and is discussed in its place among the *Mesticae*.

Those we call Cat's-eyes are divided by the Chinese into three different stones: The first is the ordinary or dark Cat's-eye, with a white beam, this one is called *Njaun kang bac* in Chinese, or *mita Cutsjin*, and is worthless if it is not clear: 2. The best and costliest is the

Elementary Stone, which has all sorts of colors that are variable, which is why it may well be the true *Opalus;* because of their excellence these are called *Tsju* and *Po-tsju,*[9] that is, Noble stone, and are found near the rivers of *Sussawan*, hidden in the sand: 3. The third is the most common one, which comes from *Ceilon*, it is pierced and made into beads; it is almost not transparent at all, a dirty white on one side, yellowish on the other, with a mother-of-pearl beam, which cannot be seen if it is tilted away from the light; it does not come from the mountains, but is a *Mestica* from the *Chama Striata* or *Bia Corurong*, which was described earlier in Book II;[10] they occur in number and large sizes in the Bay of the Province *Quang say*, around the city of *Kengtsjuhu*,[11] which is why this Cat's-eye is called *Hamtsju*,[12] that is Shell pearl, but they are ordinary and not worth very much.

Mestica Bulang[13] seems also to be a kind of Cat's-eye or Elementary Stone, but it is rarely found and I have never seen it; according to what others say, a woman from *Makkassar* had one once, which was flat and round on top, protruding at the bottom as if it had a stem, clear and partially transparent; during the day one saw a light in it, which grew larger at night, and the same became larger or smaller, according to whether the moon waxed or waned: The woman who owned it, wanted to use it to make prophesies, but the common man thought, that she had a familiar spirit (*Spiritus familiaris*). *Pliny* in *Lib. 37. Cap. 10.* describes such a stone under the name of *Selenites*, that is, Moonstone, to be distinguished from *Selenites lapis*, which is Muscovian glass. I saw another stone, brought from East Ceram, that had the size and shape of a musket ball, greenish like bottle glass, with a small wandering, yellow light in its center, which the Italians call *Girasole*,[14] and the Latins *Asteras*.

Now since they have often tried to tell me, that the Ceilonese Cat's-eyes come from the Sea or from some kind of Sea animal, I charged a certain Surgeon, who was going to make a journey to *Ceilon*, that he should diligently inquire about this, and when he returned, he informed me, that he asked about it, to wit, about the same stones which we in Dutch call Cat's-eyes, in Portuguese *Olhos de Gathos*, in the Malabar language, *Ponekenbeijdudi*, and that the only thing he could find out from the Natives was that the Sea washes them up on Ceilon's shores: But I can hardly take this tale seriously about all Cat's-eyes, at least not where it concerns the clear and transparent ones, called Elementary Stones before, but it might be true about the other, dark ones, which might perhaps grow in some shells or are fashioned from the selfsame shell: And I particularly do not think this to be the case with a stone, which we hold to be a kind of Cat's-eye as well, and which the Chinese wear on top of their Tartar Caps, and whereof the Chinese confirm, that it is obtained from a Seashell: it is rounded and the size of a musket ball, pierced, the color of dirty ivory, but on either side it has a spot like an eye, with a pearly reflection; and so one should also distinguish the aforementioned Elementary stone from another, fabricated one that is a clever kind of glassware, partially transparent, with blue and yellow flowing nicely into each other, especially if they are angular, wherefrom they make earrings and Paternosters, being noticeably lighter than the natural stones.

But I did see myself pierced kernels of the Ceilonese stones, although they are of a more common quality.

And again, my Son-in-law, Mr. *Wybrand Laurentius*, has informed me at my request, that the Cat's-eyes on Ceilon are found in the big rivers,[15] which might very well carry them down to the beach at times.

CHAPTER 41

Onyx. Joc.

For us Europeans, the Diamond is the King of Precious Stones, but whether that has been the case among Eastern Nations, is highly doubtful: After all, the Holy Writ did not honor the Diamond as such; it was rather *Onix*,[1] called *Schoham* in Hebrew, which includes the *Sardonix*,[2] that had the honor of being called the first, while its birthplace was said to be earthly Paradise, *Genes. Cap. II.*[3] in the holy breastplate, *Exod. Cap. XXVIII.* Nor was the Diamond assigned to any first names or to a royal lineage, as Levi and Juda were, but instead it was delegated to one of the least, Benjamin or Gad. And so one will also find in the Histories of the Chinese, that they paid no attention to Diamonds, though they did to *Onyx*, which makes me conclude that this is the oldest stone in the world and one that was the first to be known. In common Chinese it is called *Jok*, Father *Martinus*[4] says *Yu;*[5] it must be as white as a human nail, though it must also have some blue-black and red, be somewhat less than transparent, yet it must have a beautiful luster. China's Imperial Signet[6] is made of it, and all other subjects are forbidden to make their Signets of *Jok* on pain of decapitation, but they can use *Satsjoo*, which is a kind of white marble mixed with a light flesh color. It is rarely found and with much effort, and only in the heart of large and hard rocks; wherefore it is the custom in China that, when someone knows where there is such a rock, whereof he is sure from certain indications that it contains *Onyx*, the Emperor will order this rock to be broken open; and if he finds *Jok*, he will honor the man who pointed it out, as well as his entire family, with many gifts; but if it is not found, the Emperor will have one of his feet cut off, because he was so rash as to point out something that caused so much fruitless effort. Yet in olden days there was someone called *Banho*, who showed such a rock to two Emperors, losing a foot each time, because after much effort they could not find any *Jok;* but he insisted that his proposal was right, and he indicated the same rock to the third Emperor, Grandson of the first one, who had the rock removed down to the ground, and thoroughly examined, and they finally did find the *Jok;* then he richly rewarded the poor man, and all of his family, and caused the Imperial Signet to be made from this *Jok*, and this was handed down from generation to generation until such time, when this same lineage was eradicated through a revolt, that had been incited by three Kings called *Sam Cocsi;* but to ensure that this Signet would not fall into the hands of the enemy, they tied it to a woman, and threw her down a well; not long after a certain man saw that something in that well was giving off light, so he pulled it up, together with the body which, through the stone's power, was still unspoiled. This man's son, now being in possession of the Imperial Signet, became later one of the 3 Kings or *Sam Cocsi*,[7] who ruled China together around 314 years after Christ's birth. But this Signet was later lost again, when the last Emperor of the Chinese, expelled by the Tartars in the year of our Lord 1270, wanted to flee by sea to *Coinam*, and somewhere between *Heijnam* and *Braselles*,[8] he and his fleet suffered from a terrible thirst and (so they say), out of despair, he threw this stone into the Sea. One will still find a spot there that has whitish water, which is supposed to be fresh and potable, and which our Seamen have also seen though they did not taste it: I do not present these Chinese tales here, as if I considered them true to the letter, but only to point out what great esteem *Onyx* is held in by the Chinese, whilst we consider it an insignificant stone.

The aforesaid *Jok* or *Onyx*, which bore the Imperial escutcheon, was said by others to have been lost around the year of our Lord 1618, when the Tartars became Masters of *Sina* for the second time,[9] and that the aforesaid Signet was secretly stolen by a Tartar and supposedly brought to Tartary, where it was completely lost: It was a piece longer than an upright hand and as thick as a hand.

CHAPTER 42
Achates. Widury.

Red Achates are common in the Indies, white ones as well, with and without figures; also in Coromandel, but [there are] few black ones: The white ones, that do not have any figures, are partially transparent, like the white of a hard-boiled egg; *Pliny* calls these, *Leuc. Achates:*[1] The white or yellowish ones that are adorned with trees, are called *Dendra Achates*,[2] while we call them Tree Achates: They are very common and can be found in large numbers around the city called *Suratta*[3] in the Hindostani mountains. The Malay and Javanese call the white ones, both with and without figures, *Widury*,[4] some wrongly *Belour*, a name that really belongs to a Beryl, just as the Chinese still call all red Achates *Belo*,[5] and consider the white one Ambar [*bernsteen*].[6] Our Nation has similar misconceptions, such as saying that the black Achate is the same as black Ambar [*bernsteen*] or *Gagates*,[7] though they are easily told apart, because black Achate is much heavier, harder, and shines more, as is proper for a stone, while *Gagates*, called *Aljo bidjo* in Portuguese and which is used for making Bracelets, is light, and when held in a fire will burn somewhat with the strong smell of *Bitumen*. One can distinguish the same way between white Achate, and white Ambar [*bernsteen*], except that they have some specks like salt or sand inside. I once saw the hilt of a sword that belonged to a great Lord, which was entirely made from white Achate, and the entire piece was of one color, except that on one side there was the shape of a fly as black as jet, which made the entire piece quite unusual. It is well known that one will find all sorts of figures in white Achate, but in this country one will find them mostly adorned with little trees, which mostly resemble the shape of heather from the Fatherland (*Erica*),[8] of a black color, which stands out quite handsomely on a white stone; sometimes they are many entangled twigs, at other times [they resemble] proper trees, that cluster so naturally on a little hill as if they had been painted there, whilst other twigs lie hidden inside the body [of the stone]: One may also recognize flowers in them, or clouds, and birds, but that requires very sharp eyes. Among the yellow ones are those that have drops,[9] eyes, and other images of people and animals, too numerous to count. I am curious what reasons and causes the new Philosophy[10] would care to propose for those figures, for it teaches, that most of these figures come about from an accidental mixture of substance; yet an Achate surely teaches us, that it is not an accident, but nature's certain and sure design, and that the same is a knowing spirit, that knows how to use any substance in order to fashion a body, that will have the same figures, as the Creator ascribed to every creature's particular nature in the beginning, while its particular destiny and disposition depends on the stars, for its *Zopyra*[11] or sparks are scattered here and there all over the sublunar world, and attract and secure the beams or influences of the stars: These *Zopyra* or Seminal Sparks copy in all plants, that have a seminal force in them, precisely those figures, that were imprinted on them in the beginning: But while making Precious stones in the mountains, nature had more freedom to imprint various figures in the same, deriving or copying not only natural objects, but also such works and paintings, which have been contrived by humanity, so that in this respect one could easily compare the nature of such mountains to a pregnant woman, who sometimes imprints figures on the fruit of her womb,

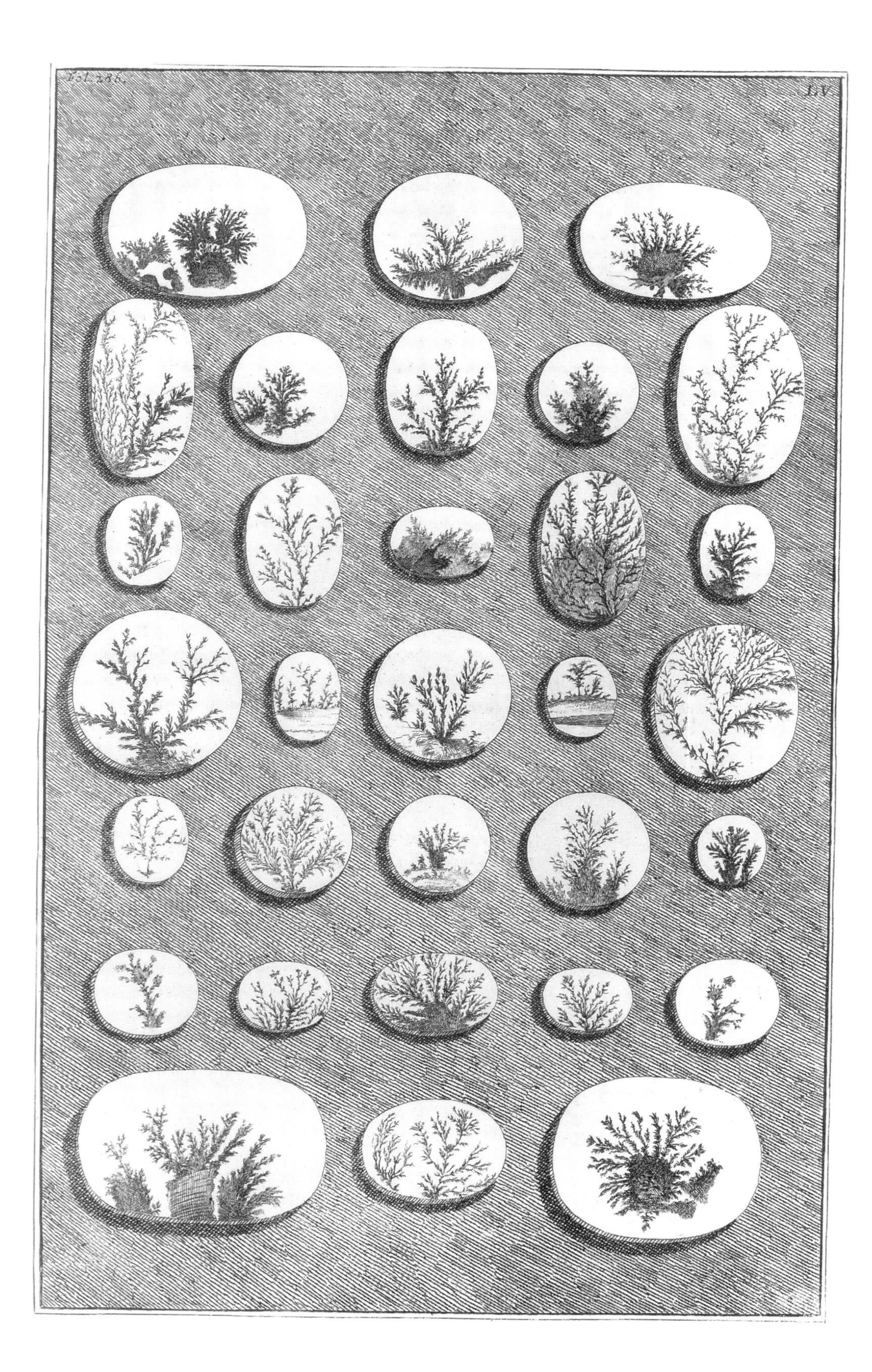
Tol. 286.
LV

as she conceives of them in her thoughts: To achieve this it is not necessary that a mountain, pregnant with Achates, should have seen the same figures which it impresses on the Achate, at a particular time, and with the same size and thickness, the way Jacob's sheep did with the painted sticks,[12] or as some suggest, that the Achate Mountain's pregnant nature derived or copied the shape of moss or heather from the moss, which grew on an opposite mountain; for it will derive such a figure from absent things as well, as long as they can be seen in nature, such as animals, domestic tools and paintings, which cannot be found on a facing mountain: This game that nature plays is so miraculous and varied, that man's judgment and wit cannot possible comprehend how this is accomplished; for we will find nature playing more freely in the Achate than in any other creature, nor will it let itself be captured by rules; for example, we stated in the Chapter on Thunder stones,[13] that the Martial spirit, the natural presiding power of thunder, engenders the same stone in the shape of a tool that can strike and hurt: And the heavenly directors do the same for plants in such a way, that one can know their natures to a certain degree; for the Jovial spirit renders its herbs soft and woolly; Mars renders his fiery and thorny; the Sun and Venus make for beautiful flowers, and so on: But what can one say about the Achates, which show such a great variety of shapes, to wit, people, animals, eyes, hands, trees, heather, moss, horsemen, carts, stirrups, saddles, reins, and so on: And these can hardly be gathered under one heading, or one should assign them all to *Jack Pudding Mercurius:*[14] so let us rather be amazed at the possibilities which nature affords, and praise the wisdom of the Creator, and refrain from trying to be too wise or from examining things which are beyond our understanding, and be content, if we can only guess at them.

As far as the power of this Stone is concerned, I have found, that it is far more appreciated by the Malay and the Javanese than by our own Nation, especially the Tree Achates, and all those which nature has adorned with some figure: They ascribe it the power of preserving its wearer from being bitten or stung by a venomous or harmful animal; and I will assert having had some experience of this: It was recently put to work on a slave, who had been stung in his hand by the fish called Stingray, *Pastinaca marina*,[15] which caused his entire arm to swell, accompanied by much pain and fire, but he was cured by a Tree Achate that had first been put in water and cooled, and then tied to the wound.

Another Achate, adorned with red drops and eyes, removes the fire from lime, so that one can hold it in one's mouth without harm, and which has been often accomplished here by a certain goldsmith, who owns such a Stone, and who, in my presence, smeared large clumps of lime on the Stone, and held it in his mouth without injury.

The Moors make their *Tasibée*[16] or Prayer beads, from the red and veined Achates, which the common man wrongly calls *Batu Ceylon*, and these Prayer beads must always be 99 on one string, while the hundredth and largest one remains in front all by itself.

The Author did not provide us with any illustrations of these Tree stones, but Mr. Feitemaas *conveyed some to us from his rich collection of these stones, and they are depicted on plate* LV. *They are shown at [slightly less than] their normal size, though most differ greatly in color and have wondrously sharp figures. I also add a few other kinds of Tree stones. See plate* LVI. *by the letters* G. *and* H. *This is similar to the olive stone, with black speckled little trees: Also another kind, shown on the same plate by the letter* I. *These are found inside coarse rocks, after one has split them: They are also black against a grayish background, but are often all entangled, going hither and yon, even upside down; yet no matter how one splits the stone, one will always discover entire Landscapes inside: They can neither be polished nor ground; I have tried it myself more than once, but I always lost the little trees. The piece that is shown here has the largest trees I have ever seen.*

CHAPTER 43

Ophtalmius.

A*chates*, *Onyx*, and *Sardius* are three stones, that have a lot in common,[1] and they sometimes make for a mixed version that is a fourth stone, and which could be relegated to any of them: The *Ophtalmius*[2] has such a nature, but comes closest to an Achate, or perhaps slightly darker, with depictions of the eyes of various animals, which gave it its name: For instance a Fisheye Achate[3] has an outer edge that is the color of horn, followed by a ring of a darker color, and in the center a white eyeball, just like those people who have pearls of the eyes.[4]

CHAPTER 44

Cepites.

This is also a kind of Achate with various veins, stripes and colors, in such a medley, that they resemble the shape of a garden or a Fortress;[1] for example: I own such a stone, the size of a hen's egg, but it is hollowed out somewhat on the bottom, like the Cabeletten which are put on pistols;[2] it is the color of horn at the edge of the uppermost or bossed side, and it is partially transparent like Caret;[3] it also has 6 white veins, of which the outermost and the innermost are the widest with four stripes between them, all running parallel, and showing the picture of an irregular pentagonal castle; the innermost courtyard is somewhat bluish and has in its center an oval encircled by two brown rings, and inside the color of the other yard,[4] one can see a house; its wall has a stripe running across it, which forms a cross and a candlestick, and on one corner one will see the shape of an eye that is also surrounded by 6 fine white circles, which spread towards the corners of the castle; the lower or hollow side is bluish, wherein one will see many brown stripes, also crooked and coming together in circles; it is so smooth that one can see one's reflection in it, and it fires at night as well as other Achates: It was found in a river in *Coromandel:* I do not know the Native name for it, wherefore I called it *Cepites*, that is Garden stone, as mentioned by *Pliny*, *Libr.* 37. *Cap.* 10.[5]

Once again we added these Illustrations, which were also put at our disposal by the aforementioned Mr. Feitema.[6] *From the many offered to us, I chose those few that were necessary for this Chapter; See plate* LVI. *The one shown by the letter* A. *is a hexagonal fortress, surrounded by its walls and moats; it is a brightly transparent red: The one by the letter* B. *is of a transparent orange color, though the center island is white, edged by a red-speckled beach; but it is so nobly and delicately drawn, that it is impossible to show the same in copper. The one by the letter* C. *has a white background, but is not transparent; the figure is of a gray purple, but edged with white stripes. The outer edge of the one shown by the letter* D. *is as transparent as glass; the fort is a purplish brown and is of a very clear color in between. The one depicted by* E. *has a transparent, brown outline, with three stripes thereafter of different colors, resembling a dike, surrounded by a transparent but icy water, that has a large island in its center, with many smaller ones around it; the latter are blue, purple, and reddish: This is the strangest and most exceptional piece of its kind I have ever seen. The last of these is shown by the letter* F. *being white and bluish, but hardly transparent. Thus far we showed some which belonged to this Chapter and also in order to show, nature's strange play in these stones, yet we cannot pass up the chance to show a few similar ones on the same plate; by No.* 1. *is a kind of*

Fol. 188.
LVI
N.º 1
N.º 4
N.º 3
N.º 2
A
B
C
D
E
F
G
H
I

Egyptian marble or achate pied and of various colors, which shows the shape of a praying Pope, both in form and color, as is depicted here, though the sky has more spots in it: I found it among many different kinds of stones, which had been broken off a wall of an old temple outside Rome. No. 2. is a German Achate stone, whereof one can see a subterranean rock with its exact color, and right through it some mountains in the distance, and up in the arch two holes, which permit a radiant beam of light to enter: Near it we show two outstanding pieces of so-called Nature stone from Ferrara, by Nos. 3. and 4., which are easily the finest ones among the stones in my possession; they come mostly from Italy and are used on the front[7] of Cabinets: Those who have never seen them will not believe them; one will see in them Deserts, mountains, rivers, ruins, cities, cloudy skies, and other rare sights, and no matter how thin one saws them or polishes them, they will always retain their figures, but with some small alterations, after the colored dust has been removed.

CHAPTER 45

Testing Precious Stones.

I happened upon a sure test, based on my own experience, of how to tell apart all sorts of Precious stones, even if they are only a little transparent, from the fake ones or those made from glass, a feat I have not encountered in any Author,[1] and wherefore I deduced this basic rule: All natural Precious stones, that are found in the mountains, and which are either entirely or partially transparent, should fire at night or when held in darkness, that is, they should display a clear radiance, or at least a spark, when they are struck with another, similar stone. Glass does not do this, nor any *Mestica* from any animal or plant, that is not transparent, nor any dark or soft stone, even if it is natural;[2] yet this spark or shine varies, more in one and less in another, harder to get out of one than out of another, wherefore I note the following experiences.

The Diamond, Ruby and Saphyre, fire little or with great difficulty, to wit, the Diamond has a small but sharply lightening spark, though I do not mean that this is a projecting spark, but a small light or sheen, which can be seen inside the stone: These three are hard and smooth stones and should be struck with their equals, quite hard on such corners, where they would be least damaged. One should not strike pointed Diamonds together, for they will scratch one another: but one can obtain fire from those that were ground flat along the *Fusetre*,[3] if one rubs them vigorously together, not with the point but with the edge at the side or laveur.[4] The ruby's spark will be red and the saphyre's blue. I did not dare to test the Emerald or the Elementary stone,[5] because they are too soft.

Achate, Crystal and all Crystalline stones that are white and partially transparent, and that are found in great number in Amboina, light up the most, so that nearly the entire stone is aglow, after they have been struck either with their own kind or with other stones; indeed, Crystals luminate so easily, that they will fire if one merely lets them fall against one another or even if one strokes them very gently. The Achate also produces ignitible sparks, so that they can be used as flints for wheellocks and flintlocks; which is why I said before, that the Achate and the flint have a great deal in common, so that it seems that the mother of Achate is nothing else but a partial flintstone, at least those that I received from *Surat*.

All *Mestica*, be they from animals or plants, and which are not transparent, will not fire, so if you get a stone like that, no matter how beautiful or how much it might look like a real stone, and it will not fire, you must consider it a *Mestica*, even though the clear stones, which were sold to me as *Mestica Ular* or Snake stones,[6] will fire, but not as much as a Crystal.

Pieces of glass that are struck with a clear flint do fire, but very little, and if you observe very carefully, you will see that the sparks come more from the flint than from the glass; so one has to learn from experience how to tell the fire of glass from that of natural stones.

The *Cochlites* or *Cochlea saxa*[7] does not fire, at least not on the flat sides, even though it is partially transparent and as hard as Flint.

All blue and gray Touchstones[8] fire somewhat if struck hard enough, but the black ones do not fire at all.

All petrified Coral stones, and white *Kalbahaar*,[9] that is striped, fire a little; but the white *Kalbahaar* will not do so at all.

The white *Enorchus* is a russet Stone found on *Buro;*[10] it does not fire.

CHAPTER 46
Mestica in General.

In Malay they call *Mestica* or *Mostica*[1] any little stone that one finds in a plant, in wood or in other stones, in a way that is not nature's wont, and that was produced by the same, not from a corruption that would afflict an animal, but purely from an excess of good food, which has attracted a stone like sap[2] from the external food. Hence we do not include such stones, which happen to get inside animals or plants or which produce any disease in the same, such as those that are found in the stomach, *Urine*, gall bladder, kidneys, gut, etc., it does not even include the *Bezoar*,[3] the pig stone,[4] and particularly not any stone an animal might swallow, as the Kaiman and deer do: But a *Mestica* must be a hard little stone, smooth, seldom or never transparent, produced in places where nature would not have engendered it herself, such as in flesh, brains, fat, and on the bones of animals, in pure wood and in the fruits of certain plants. All the Indians have a far greater esteem for these little stones than does our nation, although they do not have any particular beauty, but they ascribe them many hidden powers, which is mostly a superstition and a fancy. They generally ascribe them the power to make its wearer lucky, be it in warfare or in trade or in any other kind of traffic, according to the animal or plant it was taken from,[5] wherefore they try to gather as many as they can and wear them under a belt on their body, in such a way that the stones are pressed so closely to the naked flesh, that sometimes the skin grows right over them; this is particularly done by those who want to make themselves bold and invulnerable, which the Malay call Cabbal,[6] and they may have learned such tricks from the Jews, who promise that their *Cabala* will make someone strong as well as other wondrous things. It is true that the common people tell of many strange events, that they have seen people, who could not be killed with any kind of weapon, until one or more of those little stones had been cut out of their bodies, where the same had been pushed in; our people have confronted such cabalized[7] people in war, and these things have been said by honest officers, so that I do not care to contradict the same: But every healthy Christian knows full well, that such powers cannot be produced naturally by a stone or a piece of wood, but only by the devilry of the children of ignorance (be they Moors or Nominal Christians): Therefore, when we proceed to write such things about some *Mesticae*, it does not mean that one should believe them, for it was only done to show what the Natives say about them, and why such stones of which we disapprove are so valuable to this nation; for all of it exists in the imagination and whimsy of certain people.

We Europeans should not mock the Natives, because we too have been infected at times by the same disease: For how else did the Bezoar come to be esteemed so highly in the past and is now so much despised; while on the other hand, the ugly Pig stone now fetches such a high price, that people will give a lot of good gold for it if it is big, because people have convinced themselves that it has powers far greater than it has in reality.

In order to distinguish *Mestica* from other stones, which have the same shape and color, they have contrived many little tests, though not all are to be trusted; they propose as a general test, for example, that if a *Mestica* is put in sharp vinegar or acid Lemon juice, it should boil, that is, it should raise little bubbles all around it, which is mostly true, but I do not think that the test is a good one, because it will spoil the best stones, and cause them to lose their shine: It were far better to learn exactly the shape, color, stripes, and especially the place, where any *Mestica* was found that had any attribute at all, just as we will do in detail for each one. One cannot test them by trying to get them to fire at night, since no dark *Mestica* will do so, except for a few and some that are partially transparent, as will be pointed out at the proper place.

I have no intention of describing all the *Mesticae*, but only the ones I have encountered myself, first those that grow in people and animals, followed by those that are found in woody plants and in other stones: While it should be specifically mentioned, that nowhere else are *Mesticae* found in such abundance as in the land of *Makkassar*, in the entire island of *Celebes*, as well as its neighboring Islands, being without a doubt a particular attribute of those regions, where animals and plants attract more stone sap from their food than in any other place. They also ascribe the *Mesticae* the general attribute, that they will only be lucky for the one who found them, or the person who received one as a gift, but not for someone who buys them: and the Moors also desire their priests to exorcize them first, and then cense them every Friday with some good incense.

We also count among *Mesticae*, some stones that are not found in animals, wood, or plants in the aforesaid manner, but are unexpectedly found in other stones that are of another substance.

CHAPTER 47

Encardia Humana. Muttiara Manusia.

Beginning with *Mestica*, that grow in animals, we will first present the Human stone, *Encardia humana*,[1] in Malay *Muttiara* or *Mestica manusia*;[2] of which I saw only two: The first one grew, according to the seller, in the heart of a person, a Javanese priest, who had been killed by the people of *Tomboco*[3] for a reason I do not know, and they found this stone in the very tip of his heart: It was the size and shape of a peeled *Pinang*,[4] that is, like the blunt, lowest point of a heart; rather flat on the upper side; its color black; blacker on the flat side; veering towards gray at the tip, while it had very small holes all around as if it had been pricked by pins: Around the top one could see many clear stripes that snaked all around it, but they remained tortuous and never formed circles: Furthermore, it was very hard, black, and always colder than other stones; if it did not have the aforesaid holes and stripes, which diminished its smoothness and black color, one would hold it to be a touchstone, because it could also test metals. The Natives say that it is lucky when worn in battle, and that it makes people brave and clever, while it also discloses what other People's intentions are.

Another Human stone, though it is not known what part of the body it came from, is smaller, the size of an Amarelle cherry,[5] colder than the previous one, of a dark red that is also somewhat russet, covered with noticeable holes, rough on the outside and grainy without stripes, but with bosses, as if it had lost some pieces. It was sent to me as a Human stone from *Makkasser*, but I could not ask the owner, how and in what person it had been found.

Pliny Lib. 37. Cap. 10. has three kinds of stones, which he calls *Encardia* and *Cardisce:* The 1st. one shows a black heart, delineated against a surface that has another color; The 2d. a green heart in the same manner: The 3d. one showed a black heart, but it was white everywhere else.[6] All three may well be varieties of *Achate* or *Onyx;* but we wanted to give our *Mestica Manusia* the same name, since in Greek *Encardia* means a stone that was found in a heart, while we used the name *Cardisce* before in Book II, Chapter XXVIII for a heart-shaped little shell.[7]

Nota bene: Stones that are found in the human bladder, kidneys, or gall, should not be considered *Mesticae.*

CHAPTER 48
Steatites.

S*teatites*[1] is a white stone similar to Albast[2] or the Calappus stone,[3] but without a firm shape, resembling a small clump of fat, as *Pliny* also says about his *Steatites:* It is found in the fat of some strong and fat Indians, in their head, neck or hip, where it grows without causing any pain or hindrance to the person: it is so hard and smooth, that one can hardly scrape it, not transparent at all, made up of one, sometimes two or more little clumps, of which I saw only one. The Indian Kings know of certain signs or rather guesses, which make them suspect that there is such a stone in someone, whereafter they will kill such a person for minor reasons, simply to get that stone from their body, that is how little value they put on a human life. They wear them in order to become clever and smart.

CHAPTER 49
Tigrites. Tiger Stone. Mestica Matsjang.[1]

They found the Tiger stone on *Sumatra* in the head of a Tiger about where the brains were, the size of a round plum, not perfectly round, but slightly angular, with some tiny holes here and there as if there had been sand in it that had fallen out, as hard as stone, of a grayish color but covered with russet spots, which resembled some figures: One could consider them the heads of people, horses, and other animals, which this Murderer might well have devoured when he was alive, and these images had been impressed on his brains and that stone by his imagination. A certain Gentleman, from whom I obtained this stone and who had seen it removed from the head of the Tiger, stated that it had been a very large and furious Tiger, that was nearly impossible to kill despite numerous slashes and wounds: The Indians immediately wanted to hide the stone so they could use it for their wonted superstition, as was mentioned before in Chapter XLIV, but his authority and dignity prevailed, and they surrendered it to him: It was sent in the year 1682, to his Highness the arch Duke of *Toscany.*

A similar Stone was once found in the brain of a *Renoster*[2] on *Java*, which was described in the works of *Jacobus Bontius* Lib. . . . Cap.[3] called a *Mestica abbadac*[4] in Malay. The Malay consider this very valuable, since they take some of it, after rubbing it with water, against a changeable fever,[5] so they can expel it through sweating. And they also found two *Mestica* in the brain of an Elephant on *Jambi*,[6] in the year 1656, of which one was white and the other black: It had been killed by a Porcupine, which had shot its darts in [the elephant's] eyes when the latter tried to trample it to death.

Another Renoster stone was found in the year 1672, in *Batavia*, and it was as long as the joint of one's little finger, shaped like an egg, the pointed end stuck into the skull, while the biggest part protruded from the brains, very hard and of a gray color: According to what other eyewitnesses said, one could see the images on the aforesaid Tiger stone much better if they were examined at night by the light of a candle.

CHAPTER 50

Aprites. Mestica Babi.

Aprites is a stone the size of a Pedrero ball,[1] it weighs only a pound, is nearly round with a small hole at (the end) of one side, white as ivory, without spots, not transparent, somewhat harder than alabaster, but it could be scraped, and when one licked[2] it, it was discovered to have a saltish taste almost like bacon. It was found in the head and brains of a wild boar, many years ago on *Luhu*, which was once the capital of little *Ceram:*[3] It was a small kind of wild Pig, with firm flesh and tough,[4] for as the man who found it explained, they could not kill or even wound this Pig, until it was pierced from behind with a sharp iron stave; these Pigs normally roam the beach at night, where they clean up on small mussels, and one can hear them cracking the shells, which is why our Hunters call them Beachcombers: The aforementioned stone had been inherited or given 3 or 4 times already within the same family, with the notion that it had the power to cause many dreams (which is no wonder, since it comes from a sleepy beast) especially if one has an important goal in mind, then this Stone was supposed to reveal the good or bad outcome by means of a dream: but its last owner gave it to me during the following occasion: some trespasses had put him in chains, and he was delivered to me this way when I took command of the coast of *Hitu* in 1660; he greatly revered the stone, kept it carefully in cotton, censed it every Friday, and placed it under his head every night in order to find out if and when he would be freed from his chains; but it would not reveal anything to him, which was no wonder, since it came from a dumb beast, that spends its life more in freedom than in prisons: So he became angry at the stone and gave it to me, saying that he had noticed that the stone would no longer serve him or his family, the way it had done in the past: I told him do not think it strange that this stone will not serve you, a Moor, since your forebears were heathens who suffered pigs gladly, just as we Christians do: I will now show you that the stone will have good powers when it is with me, and knowing full well, that he had been enchained for some trifles, I removed his chains; so each one of us went home quite content, I with the stone and he with his freedom. The stone was sent in the year 1682 to the oft-mentioned arch Duke of *Tuscany*, without showing any change during the 22 years it was with me, except that I sometimes washed it with soap, when it was yellowing: The aforesaid owner had inherited it from his grandfather, the former Orangkay[5] of *Saluki*, and anyone who wore it on his body when he went to war, was lucky in combat. When I got it, the stone had some spots, like freckles, which disappeared if they were washed in soapy water, and when it was scraped with a knife, it shone inside like Candy sugar, but it would not fire at night. The aforesaid owner also had a small, angular stone, that had been found in the skeleton of a dead snake, but it had fallen into pieces all by itself, before he had been put in chains, and afterwards he tossed the pieces away because he was despondent, and even more so when the *Genius* of the previously described pig stone, which customarily appeared in [his] dreams in the shape of a man, no longer did and no longer revealed a way out, wherefore he rid himself of that one as well. One of the guests at my second wedding[6] found a small stone in the head of a tame pig, right below the eye, the size of a pea, not quite round, first white, but then gradually turning gray; this one was not as hard as the previous one.

CHAPTER 51

Pilae Porcorum. Pigs' Balls.

Since we started talking about Pigs, we should mention a number of other things, which are also rare; one often finds 2, 3 or 4 balls in the stomachs of the aforesaid small wild Pigs[1] on *Huhamohel*,[2] and they are about the size of one's fist or a little smaller, some are square, while others are as round as a cannonball, they are gray on the outside like those gray hats, some are also brown, rough like felt, and one can clearly see the little hairs on them, light but of a tough substance; if one cuts them in half, one will see nothing but such hairs, hard in the center, reddish yet made entirely of hairs, so dense and compacted that the whole ball seems to be one piece of felt. They do not have any particular taste and if the little hairs touch someone's skin it will itch a little: Up till now the Hunters have been unable to tell me where these balls come from, for they do not know of any fruits in the forest nor anything else the Pigs eat, that has such hairs; it is also amazing that one will sometimes find 4 such large balls in the stomach of such a small Pig, nor did the Hunters ever experience that the meat of such a Pig tasted any the worse for it. On the other hand, one has found such balls in the stomachs of some cows, which were smoother on the outside than the former ones, a blackish green and spongier, also hairy inside but not as much like felt, but as if baked together with slime: but these cows had been ill for a long time, languishing and pining away until they died. Such balls are also described by *Doctor Piso Lib. 5. Cap. 25. Hist. Brasil.*[3] and were found in Brazilian cows, and he calls them *Tophos Pilosos Jumentorum:*[4] Those cows keep going for a long time, coughing and languishing until they die, like goats that have a Bezoar; these same balls are completely hairy, enwrapped on the outside with a blackish, hard and smooth skin: He thinks that they happen when the cows feel an itch on their skin, and then they bite it to get rid of it, and while doing so they pull out some hairs and swallow them, whereafter these remain in the stomach and cannot be digested and are then wrapped in the aforementioned slimy skin. If this is the case with Brazilian cows, I have no notion how this compares to the wild Pigs, because their balls are a light gray on the outside and coarse, while inside they seem to be made of little russet hairs, while all Pigs here have black hair.

They use the Brazilian *Tophus*, after it has been ground and ingested, against a persistent *dysentheria*, applied on the outside like hare's hair,[5] it will stop a nosebleed, which *Piso* says has happened. The Ambonese Pig stones are burned, and given to the Hunting dogs to smell so they will become eager to pursue game.

Near the city of *Tuncham*[6] in the Province *Xantung*,[7] and particularly in the region of the fourth capital *Cingchieu*,[8] which is about 20 miles to the East of *Tuncham*, they found, as the Interpreter told us, saying that he had heard it from the Inhabitants, a stone in the stomach of cows, which the Chinese call *Nieuhoang*, which in Dutch means something like the yellow of cows, since *Nieu* is a cow, and *hoang* means something like yellow;[9] this stone can have various sizes but is often not much smaller than a goose egg; on the outside it seems to consist of a soft stony substance, except that it usually has a yellow color, wherefore it is much softer and tenderer than a Bezoar, which is why some would hold it for some kind of

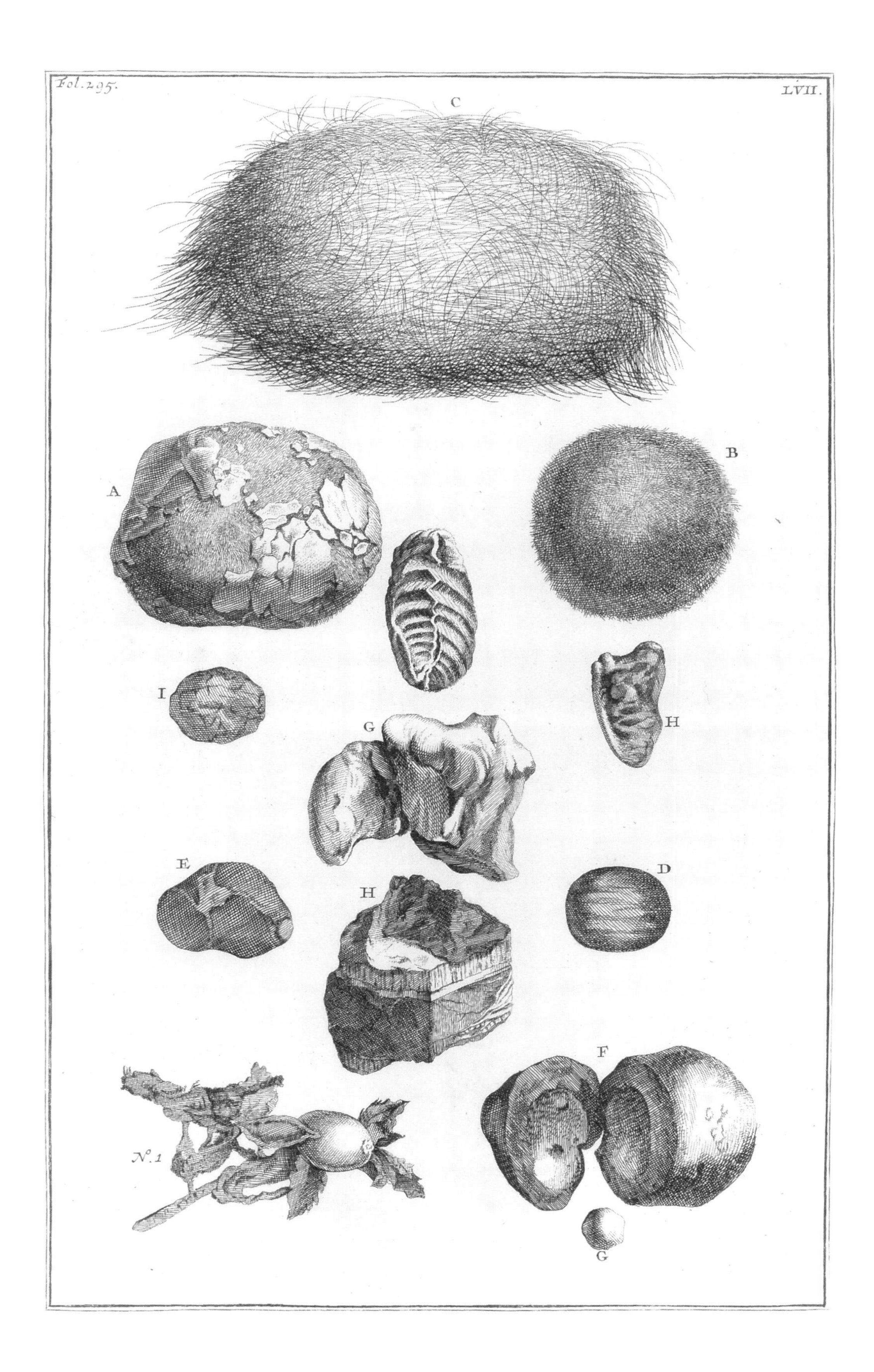
C
A
B
I
G
H
E
H
D
F
N.1
G

Bezoar; The Chinese Physicians praise these stones to the sky and will take great pains to get hold of them. The Chinese write that this stone has a very cold nature, and that it possesses the special power to cure and expel all Catarrhs: If, according to what the Chinese say, one throws some of this stone's dust into boiling water, it will immediately stop boiling, and if one sprinkles some cold water on it, it will produce a smoke-like vapor.

They found such a ball in a cow on *Honimoa*, the size of a Chinese apple,[10] on the outside as black as pitch, filled inside with coarse and thorny hairs, most of them *Gomutu:*[11] since these cows like to eat the young and tender leaves of the Saqueir tree, and also the *Ela* or bran of Sagu flour,[12] I would guess that these balls were undoubtedly formed from swallowing many *Gomutu* hairs, which then got stuck in the stomach and rolled up into balls, which were daily enwrapped with new slime.

They found in an old Polish Billygoat in Cracow a stone, that was as large as a walnut, round and somewhat flattened, as hard as a Stone with some holes, its color was a mixture of gray and white, as heavy as an ordinary stone; it could be scraped quite easily with a knife, producing coarse sand, put in water it produced bubbles all around.

Here we lack once again any pictures from the Author, but Doct. d'Aquet *sent us some from his treasure: See the same on plate* LVII. *by the letter* A. *This one was sent to His Honor from Amboina, and is wrapped in a chestnut-brown crust, crumbly and scaly, inside with fine hair, that is densely packed together. The one by the letter* B. *is another kind, but without a crust and with much spongier hairs: His Honor calls it* Pila marina.[13] *Then there is a third one by the letter* C. *which, as His Honor told me, was cut from the stomach of a tame pig right here in this country: One will also find these often in the stomachs of cows; and I own a few of those, but they have different colors and shapes, some have fine hairs and are densely compacted, others are clothed in a smooth crust: They recently sent me one from Copenhagen,*[14] *that had long hairs, the color of gray rabbit's fur, yet it was very dense and hard inside.*

CHAPTER 52

Hystricites. Pedra de Porco.

The name of Pig stone comprises various [stones] which differ from one another in both substance and power, nor do they come from the same animal: All such medicinal stones, which can be ground and ingested, are not counted among the *Mesticas*, but are really called *Culiga*[1] by the Malay. The true Pig stone derives from the gall of the Porcupine, called *Hystrix* in Latin, and is therefore called in Latin, *Hystricites*, in Portuguese, *Pedra de Porco d'Espinha*, in Malay, *Culiga Landa:*[2] This one is small, no larger than a musket ball, russet, smooth like Spanish soap,[3] particularly if one wets it, at the slightest touch conveying its bitterness to the tongue, it is rarely found and very costly, for the value of this stone has climbed so high these days, that it is sold for 60 to 100 Rixdollars a piece: Others have written sufficiently about its powers; the common test is that, when placed in water, it will turn bitter instantaneously, ingested, it expels sweat and thereby extinguishes the fire of a fever. The Chinese call it *Ho-tischoo*, that is wild Pig stone,[4] perhaps also including the Porcupine, as we do. They also have another test, when they place the stone in their hand, hold it tight, until it becomes warm, then they suck on their arm and if they taste any bitterness, they consider it a genuine stone. We did the same thing after them, but it seems that our tongues are too thick, or this test may not always be trustworthy. The best ones are found around *Malakka*, and across from it in the city called *Bancalis*[5] on *Sumatra;* one will also find

Porcupines on the *Coromandel* coast, on *Ceilon*, and in other places, but they do not have a stone, or if they do they are not the best ones.

The common Pig stone is also called *Pedra de Porco*, in Malay, *Mestica de Scho*, though *Mestica babi*[6] is better, and is said to grow in the stomach of wild Pigs. They occur in *Sucatana* and in the *Tampas* region,[7] where the wild mountain men bring them down to the people who live on the coast, and which they say they got from some wild Pigs. These are larger than the former, lighter and not as russet, but a pale red like a half-baked brick, not as smooth, while they must be in water for at least 6 hours before they become bitter; they often have a felt-like growth, like the *Pilae Porcorum* of the previous Chapter, wherefore we first thought, that we had found the true *Pedra de Porco*, but when we could not taste any bitterness from it, we decided it was inferior. I have seen 6 of those stones, sold by the Portuguese to our *Resident* on *Timor*, as true *Pedras de Porco*, which might well be the case, because I think they came from wild Pigs that occur on the island called *Ende:*[8] They were the size of a walnut; some with even less substance, and of the previously mentioned color: After they had been in water for 6 hours, they gave the water the natural taste of Snakewood,[9] wherefrom one might conclude that the Pigs eat the roots of Snakewood, and that the Snakewood that grows on *Ende* as well, has the same power as the Pig stone; but one will also find Porcupines on the same Island, but with smaller bodies than those from Makassar, and with short quills, resembling Hedgehogs rather than Pigs.

Lapilli è Felle Apri.[10]

One will also find small, angular stones in the gall bladder of wild Pigs, like ciceres, bright yellow or dark ochre, some are rather hard, others crumbly, with a taste as bitter as gall, but not very strong; they have not been examined as to what powers they may have: I found 7 pieces in one bladder, all of them easily as large as a cicer,[11] angular, smooth, and crumbling between one's teeth. They found 3 small stones, the size of pistol balls, in another wild pig on *Buro*, and these were also angular and brittle, as if put together from many small pieces: The first 7 stones were found in the gall bladder of a skinny wild Pig or wild Boar, on *Hitu* in the year 1667, the shape and size of small angular beans; they could easily be scraped and ground into powder, its color just like the dry yellow gum of *Gutta Cambodja.*[12]

Pedra de Porco from Larentuque.

This *Pedra de Porco* is the size of a *Lemon nipis*[13] or smaller, and is brought by the Portuguese from *Solor Larentuque*,[14] so they say, and obtained from some wild Pigs; they are very brittle little balls, irregularly round, scaly here and there with pieces missing, on the outside a mixture of russet and gray, inside also gray and so tender, that one can pulverize them to a fine meal with one's fingernails, which is why they are ground very gently on a stone and then taken against fever. The taste is slightly spicy at first, veering towards the medicinal roots *Lussa Radja* and *Raiz Baffado*,[15] but finally noticeably tasting like snakewood, so that one would say, that the aforesaid Pigs must have eaten a large number of those roots.

The *Pedra de Porco* that came from *Malakka*, the size of an average plum, weighed 10 Johor maces[16] or 12 Japanese maces, amounting to 120 Condoryns,[17] worth 364 Spanish reals.[18] There is a third kind of *Pedra de Porco*, baser than the other ones, resembling fine felt or those balls, that come from the stomachs of certain cows: This one is brittle as well and should be ground very gently, whereafter one takes just a spoonful, the bitterness is slight.

Some Experimental Powers of Pedra de Porco Quoted from a Certain Writer's Written Work (MS.) and Presented Here as Follows:

1. It is useful against stones,[19] and I cured a certain Franciscan friar with it.
2. Against Bloody flux, and I cured many with it.
3. Against Colick, as I found out quite often.
4. Against all diseases caused by winds.
5. Against cholera or mordexin,[20] which arrives with much vomiting and looseness of the bowels.
6. It is more potent against venom than a *Bezoar.*
7. Against Pleurisy, a deadly disease.
8. Against all kinds of stomach pain.
9. It is a strong medicine against fevers.
10. Against all kinds of palpitations.
11. Against falling sickness.[21]
12. Against any wind in the body.
13. Useful for expelling and killing worms.
14. Against the innermost *Apostemata.*[22]

These aforesaid powers I have experienced many times, not to mention several others, which were described by other Authors.

Put this Stone in some water, and if the illness is severe, one should scrape a little from the substance and then the same water should be drunk; the price is sometimes 200 Ducats, especially when it is large, and the lighter it is, the better. Was signed *Albertus Polonus.*

CHAPTER 53

Lapis è Cervo.[1]

After our Hunters caught a Deer on *Combello,*[2] they discovered in its stomach a white Stone the size and shape of a pigeon egg, so hard and smooth, that we thought it was a flintstone, and we still do because it fires at night like any other flintstone: My guess is that the Deer swallowed it while drinking: It might have been angular at first or irregular, like other stones, so it was most likely shaped like this in the stomach; I know that such things happen from small pieces of white *Kalbahaar;*[3] which I gave to a Duck once, and when I killed it the next day I found that the pieces of *Kalbahaar* had become smaller and rounder, howbeit its hardness did not differ much from the flintstone's.

CHAPTER 54

Oculus Cati Feri.[1] *Mestica Cutsjing.*[2]

I have encountered two kinds of *Mestica* in cats, and they were found, according to what their owners said, one in the eyeball of a tame and the other in the eyeball of a wild Cat: The latter, which was found on *Buton* in the eyeball of a wild Cat, had been set in a ring and was the size of a pea, and, so it seemed, ground flat and round, it was not transparent, a dark russet on one side, and in the center a white spot like a pearl, which showed up larger and clearer by candlelight at night than during the day: The Butonese said that in their country a wild Cat was larger than a tame one, that it was cruel and shunned people, that it was seldom seen in the high mountains, but it would at times come down to a Village, dig up graves and devour recently buried corpses; their eyes light up at night like fire: It was not a civet cat nor the animal called *Lau*,[3] both kinds of wild cats known on Amboina, wherefore I would guess that it must be some kind of *Hyaena; Bellonius* said that the civet cat belonged to them, while *Pliny lib. 30. c. II. & lib. 37. cap. 10.* also attests that small stones had been found in the eyeball of the African *Hyaena;* and which he calls *Hyaeniae*, of two colors.[4]

The first stone was found in *Makassar* in the eyeball of an old tame House cat, also the size of a pea, somewhat rounded but with a crooked notch across it as if the stone had shrunk when it dried out, it produced a white and partially transparent beam, as shiny as the beam in other stone Cat's-eyes, dividing the one half that was yellowish from the other, which came closer to being brown, but one could feel the irregularity better with one's nail than detect it by looking, and the beam was also lighter and larger by candlelight than during the day; and since the aforesaid notch made a curve, like a hook, one could see that the stone was partially transparent, with a light within, especially if looked at by the light of a candle. They could not tell me to what purpose the people from Buton and Makkassar wore these stones; one of the owners was a Chinese, and they are always eager to make a profit, which is why they might be in need of a cat's nature. I call them *Aelurites*[5] in Latin, that is Cat stone, and must add that, when the aforementioned pearl beam is turned towards you, it looks like the number 9, while it had yet another little hook towards the other side, as if one had joined 6 and 9 together: It also boiled in Saguer vinegar, but it did not fire at night, as the stone Cat's-eyes will.

CHAPTER 55

Bezoar. Culiga Kaka.

The *Bezoar* should not be considered a *Mestica* but a *Culiga*, which is described at length by others:[1] The ancient and still well known *Bezoar*[2] is divided into Oriental and Occidental ones, and both grow in a particular fleshy little purse, many together, in the stomachs of certain wild Goats, called *Pazan* in Persian. Some would like to derive the name *Bezoar* from a Persian word *Belzahar*, which is supposed to mean Lord of the poison:[3] The Occidental ones come from *Peru:*[4] The Oriental ones come from *Persia*, *Curassou*,[5] *Arabia*, *Malabar*, and the Islands *Ilhas de Vaccas*[6] which are near *Jasanapatam*, but those from *Curassou* are considered the best, and therefore called *Hagjar Curassou*[7] by the Arabians, see the writings of *Philippus Baldeus*[8] and others. We will describe here a new kind of *Bezoar* which, to my knowledge is still unknown to European writers; it is called in Malay, *Culiga kaka* or *Culiga kees*,[9] with a partially Dutch word; in Portuguese, *Pedra* or *Culiga d'Buzio;*[10] in Chinese, *Gautsjo;*[11] all of these mean Monkey stone: It does not differ much in color and shape from the ancient ones, for most are also the color of oyl, others are green-yellow, some are brownish, usually the size of a hazelnut, others are oblong like a piece of a finger, all made of husks layered one on top of the other, with a little hollow inside, wherein one will find something like chaff, and which is held to be the best part of the stone; one brings them from *Succadana* and *Tambas*,[12] places on *Borneo*, where the people living on the coast obtain them from the mountain people; according to what the latter say, these stones are supposed to grow in a certain kind of Monkey or Meerkat, because they could not properly tell us what was the real shape of the animal, wherein they grew, only agreeing that it was a particular kind of Monkey, called *Kaka* by the Malay; some say that they are big Monkeys without tails, like the ones we call Baboons; others say that they are small Monkeys with long tails, called *Mothien* [13] on *Borneo;* then there are those who say that they grow in the stomachs of those Monkeys, many together; others, being the majority, say that one finds them as follows: The mountain people set out at a particular time and shoot these Monkeys with blunt arrows or with the darts, which they blow from a pipe, and which wound the animal but do not kill it; now this Monkey has the habit that, when it gets a hole in its body, it will enlarge it by scratching the same, whereafter it goes looking for Medicinal herbs, which it chews in its mouth, and then uses to stopper the holes, and the skin grows over them and from this in time come these stones, to wit from the chewed leaves and the animal's blood;[14] after several years have passed they go back to the same place in the mountains, where they wounded the Monkey before, kill it with sharp arrows, and feel all over its body, until they find a lump, cut it open and discover the *Bezoar* within:[15] I will not vouch for the truth of this, leaving it to further examination: In any case, the Chinese and the other inhabitants of *Borneo* laugh at our people, because they want to believe that the *Bezoars*, which are for sale in *Bantem*,[16] and which are the only ones known in these parts, come from goats, but they are just Monkey stones.

The *Bezoar* enjoyed far greater esteem in former times; today it has greatly dropped in price, partly because one did not find the great powers which had been ascribed to it, and partly because it is so often falsified.

The test of a real one used to be, that one took a thread, that had been drawn through the sap of some poisonous herb, and pulled it with a needle through the leg of a dog, and left the thread there; when the dog began to show that it was dying, one gave it some scraped *Bezoar* in water, and if the dog recovered from it then the stone was true, if not, then it was rejected. The Malay test it as follows: They smear a thin layer of lime on a piece of white linen and then rub the *Bezoar* on it; now if it immediately colors the linen yellow, then it is a good one, but if it colors slowly, or only a little, or a dark yellow, then they consider it fabricated:[17] But since this test is not all that reliable, I would rather stick to the following: Press a heated nail against the *Bezoar*; if it is good, then flakes will come off, if it is fake, then the hot nail will go right through and the stone will melt; or break off some bits and put them on a tin, the true ones will fall apart into tiny pieces and vanish without smoke, the fake ones melt with smoke and a resin-like smell, because these are made from some kind of resin: If the true ones are broken into pieces, they will have a hollow space that will have something like chaff inside, similar to a little dried leaf: The fake ones are dense, and contain something that is easily recognized as having been put there by people. The Chinese have a more tolerable way of falsifying, making a large one out of many small ones in the following manner: They grind the small ones to a powder and make a fine dough out of it with some water, then they smear it around a stone, first the thickness of a knife, leaving the same to dry each time before they smear another layer on it, and so on, until all of the dough has been used, so that the stone will become flaky as well. This fabricated *Bezoar* is difficult to distinguish from the natural one, while it also has almost all of its powers, since it was made from the same material, since they make a large one out of many small ones, because the larger one is worth more: One can detect them however, because, if one removes the uppermost skin from the natural one, the one below it is always smoother and shinier, which is not the case with the fabricated one.

They cost these days a great deal less than they used to, once a Carat of the largest kind was sold for 8 stivers: 1 Carat of the medium kind for 4 and 6 stivers; and the smallest ones for

2 and 3 stivers: Today one can buy a large stone in *Bantem*, which weighs more than 1 ounce for 8 to 10 Rixdollars: and one can buy the small and common kind for 4 or 5 Rixdollars per ounce. One can buy them in *Succadana* and *Banjar massing*,[18] places in *Borneo*, for even less.

According to the Chinese doctors' prescription, one should keep this stone tightly covered, so that its fine particles will not evaporate, which is why they make a special little box from white wood for every large stone, wherein they keep the stone wrapped in cotton and securely closed, which is the way they are sold:[19] They also insist that when one partakes of the *Bezoar*, it should not touch the teeth because it will spoil them, wherefore they pour it into the patient's mouth with a leaf or small pipe. The least *dosis* is 10 grains, but one can take up to 30 grains, because one cannot take too much of it, especially when one needs to sweat. They use it for any kind of poison, burning fevers, illnesses that were caused by the Melancholic humor, such as scabies, early leprosy;[20] and in order to keep one's youth fresh and healthy, they are wont to purge themselves twice a year in March and September, when the Sun reaches the Equator, then they take 10 grains of *Bezoar* with rosewater every day for 5 days, wherewith they hope to sustain youth.[21] All the aforementioned prescriptions of the Chinese only apply to the new *Bezoar* or Monkey stone and not to the ones that derive from Goats, and which are unknown to them.[22] Our people take *Bezoar* these days in greater amounts than used to be the case, for they will divide an ordinary stone that is slightly larger than a hazelnut into only 3 or 4 parts, taking each time one whole part.

The Chinese *Zeuquius*[23] says in the Little Chinese Handbook, that the *Bezoar* comes from the animal called *Monjet* (otherwise *Mothien*), which he only describes as being a Monkey with a tail in the Province *Queicheu*,[24] which he calls *Quitschiu*,[25] which is full of fragrant flowers and herbs; they catch the Monkeys with snares, slaughter them and feel them all over, for lumps, wherein they will find the *Gautscho*, which is tested with lime as mentioned above, and they do not want them dense inside.[26] In *Sina* they are used mostly for women in childbed, to ease the after-pains, and it is given to them with a mixture of cooked *Cassomba*[27] and *Hoijchjee*, that is dragon's blood in small cakes, in order to purify and strengthen. Men ingest it with weak *Arak*,[28] which is their wine. The Bezoar *extracto ex animali sagitta transfixo, quod fovebat partem sagittae in sou corpore*, that is, about the *Bezoar* that was taken from an animal that had been pierced with an arrow, and which still had part of the arrow in its body. *Vide Miscell. Curios. ad Annum 1670. obs. 115.*[29]

CHAPTER 56

Enorchis Silicis.[1] *Mestica Batu Api.*[2]

In the year 1672, on the south side of *Buro*, near the beach of *Waysamma*, where there are many flintstones, my Servants found in the hollow center of a rock, which they wanted to split to benefit their flint-and-tinder box, another white stone, shaped like the hollowness, the enclosed one was also very much like the hollow and the outer stone:[3] It had the size and shape of a pigeon's egg, a dirty white, slightly softer than an ordinary flintstone, though it fired at night when it was struck against another stone. One could consider it to be a kind of *Enorchis*, about which [see] *Pliny Lib. 37. cap. 10.*[4]

In the year 1667, when *Cornelius Speelman* waged war on *Makassar*,[5] a man from Buton found there a similar egg-shaped little stone in the center of a Stone, not white but bluish and gave the same to the aforesaid Admiral. It would not be wrong to consider these stones *Techolidas*[6] or Jew stones, which are always found in the heart of marble or other similar large stones, for it does not seem very likely what a Surgeon told me he had seen in *Surat*,[7] to wit that they found Jew stones in the fruits of a Tree, and he showed [me] a handful which were quite natural and similar to those we have in our Country. I am of the opinion that all such locked-in stones were made first, and thereafter enwrapped by marble or another Stone, which were a liquid material at first.

CHAPTER 57

Ophitis Selonica.[1] *Ceylonese Snake Stone.*

We have now arrived at the Snake stone, about which many true but also amusing things are told, and I will first relate what *Philippus Baldeus*[2] writes about the common Snake stone that is known all over the Indies: He divides them into natural and fabricated ones, both according to what he says, are the size of a *dubbeltje* or a half schelling, flat and round, blackish, and with a white eye in the center. The natural one comes from a Water snake which, when one suspends it by its tail over a pot of water in such a way, that it can just reach and lick the water, will spit out the stone into the pot after a few days, whereafter [the stone] will shortly cause all the water to dry up; it is a valuable physick for dropsy, for if one places one on the belly of a person who is suffering from dropsy, it will draw forth all the water: Others say, that it comes from the most venomous of snakes, called in Portuguese *Cobra de Cabelo*,[3] and in Latin *Serpens pilosus*,[4] but whether all of this is true, and whether this applies more likely to the fabricated Snake stones, I leave to the scrutiny of others, who live in those countries.

The fabricated Snake stone is better known on *Ceylon*, the Coast of *Coromandel* and *Mallabar*, where the *Bramines*[5] contrive it, as *Baldeus* writes, from living parts of the *Cobra Capelo*, such as the head, heart, liver, and teeth, with a goodly portion of *Terra Sigillata;*[6] Others assure me that there is a fair portion of dried cow dung in it, which the *Bramines* consider to be a great Physick since it comes from a beast, which they consider to be holy: Howbeit, the *Bramines* guard this art so closely, that they will not teach it for money or entreaty to anyone, indeed, it is said that there was only one family among them, which knew this art, and that they have now all died out, hence one suspects that all the stones which are made today, lack their proper power: Be that as it may, they are good stones against stings and bites of all kinds of venomous animals, as long as they are genuine; the true test is, if they yield a beam or produce some bubbles after they have been put in a glass of water, or when you press them to the roof of your mouth and they stick so tenaciously that one can hardly remove them: Some are flat and round, thickest in the center, with sharp edges all around like a lentil, with a white eye in the center, which sometimes is only on one side, and which has a black spot in the center, sometimes the white is mostly on both sides; but this one is considered more ordinary: Some are oblong and not bigger than a nail or a broad bean, most often black and shiny with a small white spot on one side.

One places them only on a sting or bite from a venomous animal, after one has first pricked the wound a little, so that there is some blood, and then the stone should stick so tightly to it, that one cannot pull it off, but after it has done its work, it will fall off by itself, just like a Leech: It can happen sometimes that it falls to pieces from sucking too strongly, in which case one should be ready to place another stone on the wound, the stone that fell off by itself should be placed in cow's milk or the milk of a nursing woman, wherein it will discharge its poison, so that the milk will turn blue from it: Whereafter one washes and dries it, and keeps it in a tight little box between cotton; for I have noticed that this stone likes to be closely confined, otherwise it will lose its powers, like the one I have had for several years, the size of half a schelling, a grayish white on one side, which was fine before, but when it was put on a spider's bite[7] in my hand, it would neither adhere nor draw.

I will add here a description of this stone's sympathetic powers, as it was made known to me by Doctor *Andreas Cleijer*,[8] also to be found in the *Ephemerid. German. ad annum 83. Observat. 7*.[9] A certain boy among Mr. *Cleijer*'s servants, was going to clean a large snake, so he could furnish [his master] with a *Scheleton*,[10] but he hurt one of his fingers accidentally on one of the aforesaid snake's teeth, although all the flesh had been removed, and it caused his entire arm to swell, followed by an infection, fever and dizziness; they placed the Ceylonese Snake stone on the sting, which drew the poison out and made the arm thinner; then they were at a loss where to get some milk, so they could wash the stone in it; but a nursing woman who happened to be present, put the stone in a cup, held her breasts above it, and squirted sufficient milk on the same, which turned bluish and the Stone was rendered pure again; but this warm milk that had been squirted on the poisoned stone had such an effect on the breast, that it became completely inflamed, and one had great trouble curing it: For this Snake stone see *China illustrata* p. 81.[11]

I will also add here the story told me by a certain Surgeon, by the name of Mr. *Christiaan Gyrarts*,[12] who wrote me in his letter as follows: I wanted to buy some Snake stones from a Snake charmer, but only after they had been tested, which he did not want to do at first, but when I promised him a Rixdollar, he allowed himself to be bitten in his arm by a young toothed Snake around the artery in his wrist; one could see immediately how the blood drew upwards, so that the arm began to swell: He placed one of my Stones, which I wanted to buy, on it, and the blood was drawn downwards again, and flowed from it at first sluggishly, then profusely, until the Stone fell off; then he simply took some cloves that he had chewed with a *Pinang* and put them on the wound with a *Daun baru* or *folho de Pouw de St. Maria*,[13] and after I had bought them from him, he began to tell and teach me, that if one does not put those Stones after use in sweet milk or, when that is wanting, place them in salt water, in order to draw the venom from them, then they would be spoiled and would never be able to help again, and when I asked him, what kind of substance it really was, he did not want to say anything, but after I had satisfied his demand for a coat, he told me, that certain sea trees are found around the Nicobar Islands,[14] but at a great depth, which are pulled up by means of certain small iron anchors, whereafter they are polished and then sold without anything else being added to them, except for some superstition, that is not worth writing about. I showed him some *Kalbahaar Itam*,[15] which he declared, upon seeing, to be the same, though it was supposed to have a skin; this looked quite like it, except that mine was a shiny black, which he had not noticed with the other, but he assured me, that it was the same kind of material: But the latter does not seem possible to me, and I believe, that the aforesaid Mr. *Christiaan* was cheated by that juggler.

CHAPTER 58

Ophitis Vera.[1] *Mestica Ular*.[2]

There are other Snake stones in the Indies besides the Ceylonese Snakestone, and they tell strange as well as true things about them, so that I do not know what I should hold to be true: Since they are not often found and seldom seen, one cannot describe them in detail but has to make do with whatever is said about them by the common people: They are not of one but of several substances, colors, and shapes, and derive from various snakes, so I will present them one after the other, since various kinds came into my hands.

On *Java* is a Snake called *Ular Maas*[3] in Javanese and *Kim Soa*[4] in Chinese, that is Gold snake, speckled with red, yellow and black, its belly yellow and shining like gold, hence its name; it has many teeth, and has its head adorned with a red comb; it moves with its head raised, assaulting pigs and other animals, and when it wants to fight it puffs up its throat and spits out its venom which, when it hits a person, makes him swell up as if with dropsy and suffer a languishing death.

In the year 1679, I was shown the *Ophitis* or *Mestica* of this Snake the size of a chick-pea, set in a silver ring because, as was said, it cannot be set in gold because it would then lose its powers; it is white mixed with blue, partially transparent, and at the top one saw a little sun and within many clear rays which stirred when it was moved, it draws all poison from sting wounds, and when placed in vinegar with salt it will boil and move a little; it was mostly round except, where it was set in the ring, at the bottom somewhat flat and with a small hole not very deep into the stone wherewith it seemed to have been stuck onto a bone in the [snake's] head; the collet[5] was open at the bottom so that one could remove the stone and put it back again: After I took it out I put it in wine vinegar, but it did not want to boil though it did so in the sap of a *Lemon nipis*,[6] even more in strong Sagueir vinegar,[7] casting up many bubbles, though it did not move, and once it was removed was clearer than before: In the evening it was struck against Crystal or a clear Flintstone but did not spark, and both of these are characteristics of a *Mestica*. N.B. My conjecture is that the aforesaid Snake was a *Cobra Cabelo*, which is also found on *Java*, where the Stone came from.

The *Ular Sawa*, otherwise *Tsjinde* or *Ular Petola;*[8] in Chinese, *Backyong*,[9] in the Macassar language, *Tomma Labu*,[10] is speckled with brown, red, and black, with a flat head whereon is a circle and therein is a cross, together making the shape of an eye, and this gave it its name in Chinese, but in the Macassar language it got its name from moving with its head raised up, it does no harm to people and is therefore raised in the boats and houses of the Chinese and Balinese, who honor it as a house god because it brings them good luck; however, it likes to suck out eggs and does it so cleverly that one will almost fail to notice the hole it makes: The Chinese like to get the true Snake stone from this one which they suppose it carries in its head; but in such a way that it can lay the stone aside when it wants to eat or drink, at which time one must wait on it craftily and steal the stone, because those taken after the snake has been beaten to death do not light up at night as they normally would; if the snake drops it by chance into the water it cannot get it again, and that is another way of getting a stone: I would guess that it is this Stone that *Baldeus* speaks of, which he calls the natural one, which was mentioned in the foregoing chapter; but it should be mentioned here that *Ular Petola* is to the Malay another Snake than the one here in *Ambon*, to wit the *Coulouber boa* or House snake which, after first helping itself to eggs and hens, becomes as fat as a thigh, whereafter it attacks animals and people: But the aforesaid crowned Snake, honored by the Chinese, is no longer than 1 and ½ fathoms[11] and wider than a leg, with a round head, and with its body painted as wondrously as the *Petolae* cloths; it does no harm to anyone and is content with the food the Chinese serve it; which must be a great delight to Satan, for in this manner a large part of the world pays much homage to his image.

Leaving the shape of the Snake (which provides the precious *Ophitis*) to later scrutiny, I will go on to describe the second *Ophitis* which was presented to me as such. It was brought and sent to me by *Radja Tomboucco* who lives on the eastern coast of Celebes, having been obtained from a Snake which was described to me as being so hideous that the true *Basiliscus*[12] could not be painted uglier, because it is supposed to have two feet, a head, and a comb like a cock, so poisonous that its very breath can kill a person, and it lives nowhere else than in distant mountains where the poisonous dart tree[13] grows, whereunder it sojourns. This *Basiliscus* can, therefore, not be caught alive, and this one had been killed from afar with arrows, and in its head this Stone was found; it was the size of a *dubbeltje*, on one side flat, on the other bossed like a half cone though somewhat flatter, clearly transparent like a dark Crystal, though one half was darker than the other, and the sharp sides were like glass. It did not show the Attribute of lighting up at night, but when struck against a Crystal or a clear Flintstone it did fire at night,and it did not want to boil in vinegar or lemon juice; when this was mentioned to the seller, N.B. that this is not Characteristic of a *Mestica*, he replied that the night-lighting Attribute does not remain with the stone if the animal is shot to death, but only with those snakes that are captured alive, and there are few who have seen them, although all the Indians speak a great deal about them: Furthermore, all *Mesticae* which are clear and hard, would not boil in vinegar or lemon juice, and this also has its reason, because a solid and clear Stone has no little holes [*Pori*] for the vinegar to penetrate: I have no other experiences of this stone, and the seller was also unable to tell me anything.

The third Ophitis was brought to me by a Dutch merchant from the large Island of *Mindanau*,[14] which the Natives call *Magindano*, and which he had bought there from the Natives for a piece of linen from Guiny, in a remote village[15] in the South-East corner, and they had found it in the head of a Snake, and brought it to him clandestinely because their King desired all such stones for himself: It was of the size and natural form of a broad bean, with a protruding and glassy edge all around the back, much clearer than the one described before, and so much like a Crystal or Crystalline glass that one would easily mistake it for one: The seller would not have found much Faith with me if I had not read in the old writ that the true *Dracontia*, for that is what one should hold these two stones to be, if they truly came from snakes, should have the same color as a Crystal, and for a shape that of a bean. It did not light by night nor did it boil in vinegar or lemon juice, but it did fire once at night when struck against another Crystal, like the one described previously, with which it also shared the glassy edge; and since they came from such diverse nations, though the same story was told about them, I would believe it provisionally that they had come from some Snakes.

The fourth *Ophitis* was brought to me from the South-Easter Islands[16] which lie below *Banda*, and was also found in the head of a Snake, though the bearer did not know how to describe it: Its appearance was like the ones mentioned before, the size of a chickpea, partially rounded like a musket ball cut in half, at the bottom flat and bluish, without a hole and, when it was brought to me, of a gray or dirty white on top, but after 6 years it markedly inclined towards yellow, with a rent crossing it as if it were burst, without the little sun and hardly distinguishable from the *Chamites* which comes from the *Bia garu:*[17] It boiled in Lemon juice, but did not fire at night.

On *Jambi*[18] is a large Woodsnake or the common Ular Petola, in whose head was found a white *Ophitis*, oblong, easily the length of half a finger, hard, smooth and shiny: the Malay hang them from their *Creeses*[19] or sew them into a cloth and hang them from the necks of their children to save them from witchcraft: In time it too will yellow, and then they put it in Lemon juice where it boils and moves, and soon they take it out again and soak it in the water in which they wash rice, also rub it a little with rice, and it becomes pure and clear again, and is kept in cotton.

Still other Snakes are mentioned which would have a clear Stone, to wit, 1. an *Iste Liong*[20] in Chinese, that is, *Nagga* or *Draco*, which the Chinese hold to be the largest Snake and which is supposed to be found on the Island of *Contung*, which might be *Pulo Condor*, where the Chinese Admiral *Sampo*[21] sojourned for a spell, and where he looked for this stone for his Emperor; which he also did in the city called *Burneo*,[22] where the land is supposed to have plenty of such Snakes: But the Chinese say that he, during his entire journey, could obtain only one that lit up at night, and brought it to his Emperor. 2. *Tambo Sisi* is the biggest Snake of all, clad with hard scales and comes with a great noise from the mountains down to the water. 3. *Terrebelau*, is the Snake on Celebes, that drops its *Mestica* in water and cannot retrieve it.

One shall test a *Mestica Ular* by soaking it in water, and giving the same to drink to a Woman in labor, and rub the Stone over her belly, which shall bring her an easy delivery.

A Surgeon who had been for several years in *Ligoor*, declared to me in the year 1687 that he had seen the Carbuncle of a Snake with a certain Ruler, who, when he was still a child, had once been left in the forest by his mother, who had suspended him in a cloth from two branches of a tree, then, according to what his parents said, a large Snake had come to him and dropped a certain Stone on his body, and this Snake was afterwards always fed by his parents; this Stone was the size of an old peeled *Pinang*, oval shaped, a transparent fiery yellow, veering towards red, and at night so bright that it illumed a room: the Viceroy and Sovereign of *Ligoor*[23] took this Stone away from him, when he was imprisoned, and sent it to the King of *Siam*.

CHAPTER 59

Stones from Various Other Venomous Animals.

Scolopendrites,[1] in Malay, *Mestica kacki sariboes*,[2] in the Chinese Province *Succhuem*,[3] or as the Chinese say *Soutsjouam*, they find large Millipedes, which the Natives raise in their pillows, which are hollow inside like a small box: They carry these pillows with them when they travel through the country, and sleep with them at night so they will be delivered from some bad Snakes, called *sese*, which have a head like the face of a man, and which crawl inside houses at night and attack people; but they will not come near these Millipedes, and if they were to do so, they release these big Millipedes, which will attack those Snakes and destroy them, whereafter they calmly catch the Millipedes again and put them back in their houses: These Millipedes are no longer than a span, and but 3 or 4 fingers wide, called *Jacan*[4] in Chinese; they have a *Mestica*, that is partially transparent and which shines at night, and that cures the bites of venomous creatures, while it protects the wearer from the aforementioned Snakes.

There are also large Millipedes on *Java*, easily as long as a hand or a span, an inch wide, and they too have a night-firing *Mestica*, indeed, in *Batavia* they have often seen a large Millipede in the graveyard's ossuary, at night, alight at its head or mouth, as if afire; but once the bones were removed, it could not be found: It is thought that that light derives from its *Mestica*, though others, who could show *Mesticae* from other Millipedes, were unable to notice any night-firing quality; perhaps it was taken from a Millipede that had been beaten to death.

Salamandrites, *Mestica Tocke*, come from a kind of Salamander which the Malay call *Tocke*, and we *Gecko*,[5] and which occur on *Sumatra*, *Java* and *Celebes;* this one also has a Stone in its head, white, not transparent, resembling the Calappus stone somewhat, they hang it from their cresses, and if it gets dirty or yellowish, they clean it with Lemon juice and rice water, as was described in the previous Chapter.

CHAPTER 60

Aëtites Peregrinus. Mestica Kiappa.

I have the following *Mistica* from Birds, *Aëtites peregrinus;*[1] those were two stones that were found once in *Hulong*, a Village inland from Little *Keram:*[2] A Sea eagle *Haliaëtos,*[3] called *Kiappa* in Ambonese, had a nest in a tree there, and did much mischief to the Natives by stealing their fowl, and he was so bold, that he did not fear people; so the Men of that village decided to declare war on it, and took to the field with bows and arrows, because 2 or 3 men did not dare to attack it, and the first one who climbed its tree, came down as good as lifeless: But luck was with them, and they were able to kill their enemy with arrows, whereafter they threw the nest down, which was large enough for an Ambonese to live in. The dead Eagle had a knob on its head, where other Eagles are flat, and when it was opened they found in it a round stone the size of a musket bullet; it was light brown, with small holes all over, as if it had been pricked with pins, and a noticeable hole on one side, where it might have been attached to the head with a little vein; also dark and hard as a stone: The other stone they found in the nest, the size of a predero bullet[4] or small ball, somewhat round though angular, white like ivory, smooth, and also as hard as stone, with 7 small holes in only one place: Both boiled in Lemon juice but did not fire at night. The first owner wore it to be safe in warfare, and he attributed to [the stone], that the Dutch had never been able to capture or wound him, when he fought them in the year 1654, yet that silly man had seen for himself that the bird had not been immune to arrows, even though it wore that stone in its head.

We did not receive any illustrations of these Eagle stones either, but in order to please the fanciers, we have depicted two from among many different kinds, that were at hand; see plate LVII. *The one by the letter* E. *has many corners, but it is smooth and the color of chestnuts. The one indicated with the letter* F. *is bigger, but rougher, which we cut in two, and found it hollow inside, though it contained another, whitish smaller stone, indicated by the letter* G.

CHAPTER 61

Lapillus Motacilla.[1] *Mestica Baycole.*

When the Country of *Huamohel* or Little *Keram* still had many Villages and people: to wit before the year 1651,[2] it happened that a certain Inhabitant of *Lycidi*[3] was fishing in the Sea near a decayed Seri,[4] whereon a Wagtail had fastened its nest, the way these birds are wont to do; for they always make their nests on such poles that stick out of the sea: now when the Native was busy fishing, this bird began to bother him, round about his head, flying in and out of its nest, such as a worried and curious Bird will do, so he decided to catch it; he smeared the nest with birdlime made from the milk of the *Soccom* tree,[5] and soon the bird got stuck in it, and hung there, so that the fisherman could grab it; but when he grabbed the bird he found a nice little stone in its nest, wherefore he salved the bird with oyl, since it had presented him with this curiosity, thus removing the birdlime from its feathers, and let it fly, but took the little stone [home] with him; the same was rather flat, an oval, shaped like a human face, to wit, wide on top and coming to a point at the bottom, like a chin, an inch long and the width of a finger, hard and shiny like marble, a pale brown, but whitish towards the chin; the front side which looked like the face, showed many small white circles, among them two large ones which could be the eyes, the eyeball being a dark-white spot covered with fine little red veins, surrounded by a brown *Iris* or circle, and that one [surrounded] again by a white circle: These two eyes with a hole below them did very much resemble a face or rather a skull: One could see many small white circles on the forehead that were smaller than the former ones, comprising a small brown spot, crossing each other on the right side; it had more of such little white circles on the back, but all entangled and without any order, also with some little holes: The Lycidian never showed it to anyone until the day he died, although he confessed to owning it, and he imagined that it was due to this stone, that he was happier and richer than his fellow citizens: I obtained it from his heirs after much pleading and for a good price.

Another Wagtail stone I did not see but it was described to me in a story told to me by a trustworthy *Orankay:*[6] This stone had been found by a Native from *Caytetto,*[7] when he was walking on the beach, a Wagtail was bothering him so much by flying around his head, that he could not get rid of it no matter how much noise he made; he finally started waving his handkerchief and happened to hit that bothersome bird, which dropped a small stone in front of his feet. Which also indicates the restless nature of this bird: The Indian picked up the stone and found it to be somewhat like the egg of this bird, brownish at one end, and black on the other, white in the center with a bright sheen. They told of many wonders about this stone, but their faith stays with them.

The man who found it put it in a beaker half filled with sweet Saguier, whereupon the Saguier began to boil; and did not cease until the beaker was full: Mixed in with the rice which he gave to his chickens, he found that not a single chicken would eat one grain as long as the stone was nearby: It made the wearer tough, quick and indefatigable while marching or walking, which he experienced himself when he was pursued by Dutch soldiers during the Hitu war around the year 1642,[8] not mentioning that there was probably half a mile distance between them. It was censed every Friday with *Benjoin* or *Dupa*, and kept in a cloth with *Muscus Zibeth*[9] or some other stinking incense, which gave it a smell like all the other *Mesticae*, which I obtained from the Natives; which shows what idolatry and superstitions they perpetrate with such stones, and why they part with them with more grief than they would with diamonds or rubies. The finder, who had gotten it from the bird itself, mentioned something else; when he came home with the stone, he was met by a small man, completely unknown to him, of quite short stature, who instructed him in the powers of this stone; which was that when they went off to war in order to conquer an enemy stronghold, they should place this stone in some Saguier, and all should drink from it, and they would be fortunate during their campaign and be safe from the enemy's weapons, providing he kept and censed the stone as mentioned above: We read that the idolatrous Israelites did cense the nets and yarn in the same manner when they hoped for a fine catch of fish. *Habacuc. Cap. 1. verse 16.*[10]

When the first stone was struck against an Achate it fired at night, though little, and I never tested it in Lemon juice and Vinegar, since I was worried that it would spoil the beautiful color; but when it was put in a beaker half filled with Saguier, the same filled up immediately, that is, so you will fully understand this, when my servant poured more into it, and in no other way.

Plin. Lib. 37. Cap. 10. also mentioned a Wagtail stone, which he calls *Chlorites*, which grows in the stomach of the Wagtail along with the Bird, and it is as green as grass, a color that does not belong to the one described up above.[11] The *Magi* wanted to put it in an iron ring so as to produce some wondrous powers after their fashion; from which you can tell that it was already an ancient superstition with this Stone, to ascribe more powers to it than nature can bestow: Though I would never deny nor reject the influence of the stars, in that a stone will be more powerful in one metal than in another, although *Pliny* seems to mock this in his fashion.

CHAPTER 62
Stones from Various Other Birds.

On *Buton* one will find a Bird that is a mixture of a wild Hen and a wild Dove, it walks on the ground like a Partridge, but its beak and voice are like those of a wild Dove, with black feathers but with a greenish sheen on its wings. One will find a horn-like little purse in its crop, which usually contains a little white Stone like a Pebble, but softer, the size of a small bean, and can be scraped with a knife; the whiteness does not last long, but within a year it turns a grayish yellow with a small brown spot on top: The Butonese bring these Partridges to market for sale and most of them have this stone.

Another Bird was brought to me from *Taliabo*,[1] it did not differ much from the former, except that the feathers were a dirty gray, nor did it have many of them, while they would fall out if only handled a little, so that it was nearly bald, one seldom hears its voice, perhaps it sighs a little towards daybreak, like a wild Dove: It also had such a horn-like purse in its crop, with a white Stone in it, no different in hardness and color from a Pebble, flat, angular, veering a little towards blue, nor did it lose its whiteness as the other one did. The Bird was brought to me under the name of *Ajan Taljabo*,[2] which is one of the Soulanese Islands, its flesh was tough, and the muscles of its thighs and chest had so many little bones in it, that they seemed fishbones, but the aforementioned Butonese Partridge has a much better taste and meat.

These two little Stones did fire at night when struck against Crystal or a clear Pebble, wherefrom I would guess, that they must be natural Flints, which the Birds swallow and which then get stuck in their crops, but since they always have a particular lodging, it is believable, that they perform a particular service for this Bird.

At the Cape they found a Stone in the yolk of an Ostrich, the size of a pigeon egg, white and shaped like a Calappus stone,[3] laced with tiny blue veins all over; but the shape of a shining sun was much clearer on it than on a Callappus stone. It can be seen at Mr. *Qualbergen*'s.[4]

CHAPTER 63
Sepites. Mestica Sontong.

I obtained the following Stones from fishes; there grows an oblong little Stone in the fat of the Sea cat *Sepia*,[1] called *Sontong* in Malay, it is barely a finger joint long, the thickness of a quill at one end, whereafter it grows narrower, coming to a blunt point, resembling a peeled kernel of the *Gnemon*[2] fruit, but smooth and even, a beautiful white at first and partially transparent, but the narrower end becomes a dirty yellow or like dirt: It was brought to me as the *Mestica* of the aforesaid *Gnemon*, but since then it has become known where they come from, which was when our Nation found such a stone in the fat or gland of a Sea cat.

Almost all fishes have two holes in the *Cranium*, at the back of their head near the brains, towards the bottom, which makes for the roof of their mouths, and these holes are closed off with two small bones, which can be removed; they are not alike in all fishes, but most are small and oblong, with two sharp ends in the shape of a little boat: One will find the biggest and most beautiful ones in certain fishes, called for that reason *Capalla batoë*,[3] that is Stonehead, not unlike a small Carp, but with such a hard head that one can barely split it; in the aforesaid place in the brains of every head, one will find two small stones, which I call *Lapides lithocephali;*[4] they are very much like our Carp stones,[5] and the same size: the width is that of a fingernail, triangular, some oblong with a sharp corner and two other blunt ones, all shaped like a little boat, to wit; bossed on the bottom, even, smooth, though some have a furrow as if two stones had been connected: hollow and wrinkled on top where it lies against the brain, sometimes with a growth: Their color is whitish like the other bones of the fish, some are partially transparent and bluish like white Achate, but they become yellowish over time. The Natives have nothing to say about their use, but I think, that one can use them in all kinds of ways, like our Carp stones, to wit, to clear the Reins of Gravel and sand.

Lacking any pictures, we added here several kinds that are quite rare, and which are depicted on plate LVII. *By the letter* G. *is a Stone from the head of a* Sea cow.[6] *By the letter* H. *a similar one from the head of a* Sea wolf:[7] *And in addition the one shown by the letter* I. *which, as I was informed, came from the head of a Bull, and which was presented to me as something rare: As far as the Stones are concerned from Carps and other fishes, those are so plentiful and common, that they do not warrant an illustration.*

CHAPTER 64

Chamites. Mestica Bia Garu.

In Chapter XXVIII of Book II, which described the *Chama aspera*, I stated that one can get *Mesticae* out of various kinds of *Chama*, this I will now enumerate.[1]

Mestica bia garu comes from the large Shell *Chama* or Gapers, which we call Nail Shells, when they are larger than one foot, but only one in 10 will have a little Stone, which is always in the *tendo*[2] or in the surrounding *Spondylo:*[3] They do not have the same shape, but I put them all in two genera: The first one is oblong like a small finger, in the shape of an Alchemist's fiole,[4] to wit at the bottom round and almost bossed and then gradually coming to a blunt tip, some are straight, some are slightly crooked like a *Retorte;*[5] its color is white, the little globe is reddish or yellow and dark at the bottom; a transparent white at the narrow end, like white Achate, usually it also has a pearly sheen to it: They do not fire at night, but when rubbed they smell like white *Kalbahaar*[6] and will boil in Lemon juice. One will find small pieces of it, which are easily a finger thick at the bottom, and barely a finger joint long, also somewhat crooked like a *Retorte*, a dirty yellow below, torn and broken off, where it was fastened to the shell, because not all are loose, and the loose ones are round and smooth, according to how they were shaped by the animal's leaking.[7] They are smooth towards the tip, a pure white and as hard as a Pebble.

The second kind of *Chamites* is smaller, and lies loose in the flesh usually around the *tendo;* some are oblong and flat like a Lizard's egg, very much like the Calappus stone, so that they are mistaken for it, but they can be distinguished as follows: The oblong Calappus stone, which also resembles a Lizard's egg and which is also a pure white, usually has a little crown at its thickest end that looks like a tooth that has fallen out, wherewith it was fastened to the shell before, and with a sheen or small sun at the narrowest end.[8] The *Chamites* has none of these characteristics, although it does have a pearly reflection, while it shows some small cracks, where it shines the most. Others are irregularly round, with protruding corners and with holes, plain white or yellowish like ivory and of various sizes; some are [the size of a] pea, some of a die, and they become rather yellow over time.

The third *Chamites* comes from the *Chama Striata* or *Bia Corurong*, particularly from a large kind, which one will find on the *Lussapinjos* Islands,[9] and which has the mixed form of a *Squammata* or Nail Shell, and *Striata:* I found various stones like these, some as large as a pea and truly round, some larger and in the shape of a lentil, all so beautifully smooth and white, that one would think they were pale pearls, and they kept their color: these are in the hasp or glandular flesh, and are formed like that by the leaking. Others are no bigger than a *Katjang*,[10] some are baked together in a little clump, all of them as hard as stone, of different colors. The biggest ones are white and smooth, some are yellowish, but the ones in clumps are a pale red, and purple, all as hard as stone, the *tendo* has so many of these little stones, that it seems it was made of them, and the biggest pieces are near the ends. The Ambonese *Bia Corurongs* do not have them or rarely, and the grainy clumps will not boil in Lemon juice, a sign that they are made of a hard and dense substance.

The fourth *Chamites* comes from the smooth *Chama*, particularly those we call Croakers, or *Bia Codoc*[11] in Malay: These are a plain white, but will gradually turn yellow or grayish, some are as round as a pea, some angular as if some little ones had been attached to them, and which do sometimes fall off: This is the reason the *Chamites*, be they from this one or from the previous *Bia Garu* which is angular, sometimes produce a small stone, which makes the common people believe, that the stone can give birth and produce young. A certain Chinese woman from *Amboina* found such a *Chamites* in a Croaker on *Buro*, which was white, angular like a small die, and after she had kept it for several years, and happened to see it again one time, she found a little stone with it, which she said was the child of the larger one; but when the larger one was carefully examined, one could not know where the small one had been: She would not sell the stone for any money, convinced that her possessions would increase, the way the stone had done. All the Stones mentioned I call *Camites* in Latin, after the example of *Pliny*, who describes many Stones like that in *Lib. 37. Cap. 10 and 11*,[12] naming them after the animals and plants wherein they might have grown or with which they share a certain resemblance, and which have remained unknown to our Jewelers for a long time or were thought to be fabulous, but which now are slowly coming to the fore. And what is said about the Stones giving birth is not a new but an old notion, for *Pliny* says the same thing about his *Gemonites* and *Pheantites*, that they also give birth at their appointed time. They are called *Mestica Bia garu* in Malay; or just *Mestica bia*, because they are mostly of the same substance and color; as to which Shells they come from, see above, about the *Chama aspera* in Book II, Chapter XXVIII.

Their only use is to be kept as Curiosities, while the most beautiful ones are set in silver rings; Natives wear them in order to have good luck while fishing, or when gathering Mussels and Shellfish and similar food from the sea: The aforesaid Chinese woman endowed herself with some cleverness because of this Stone, so that, while she is walking on the beach, she can tell from the little holes in the sand what kind of Shellfish and Mussels are hidden there, and to be sure, she knows how to find rare Mussels, which no one else can find. One pays a Rixdollar a piece for the finest ones, which are round and white, but the others are not worth a quarter. The Natives test them like other *Mesticae* in Tuak, Vinegar, or Lemon juice, which I disapprove of, because it spoils their sheen; some of the most beautiful ones, which are hard, stony and clear, also cannot do this or only a little, and yet they are real ones, because no stone, that is hard and of a dense substance, can do such a thing: be it a *Mestica* or not.

The fifth *Chamites* is considered part of the Cat's-eyes, and were up till now mistakenly considered a natural Mountain stone, yet they grow in a kind of *Chama* or Nail Shell: known to the Chinese, because the nails are so sharp, that someone, who hurts himself with one of them, will feel much pain, as if he had been burned by fire. They occur often in the bight of *Cantong* and on the East side of *Ceilon*, wherefore people confuse them with natural Cat's-eyes, which also occur on *Ceilon*. Their Stone or *Mestica* is round, the size of a gray pea, and after they pierce them with holes and string them like beads, they make bracelets from them. A similar Cat's-eye, but larger, was described before in Chapter XL.

CHAPTER 65

Tellinites. Ctenites. Bia Batu.

One will find Stones that have the shape of various Shells, but one is not sure if they grow inside the Shells or were formed outside from a lump of earth: For one will find them outside the Shell, and even sometimes in places where they normally are not found, of which I have noted the following two kinds.

Tellinites is formed like one of the *Tellinae* mentioned previously in Book II, Chapter XXXIII, particularly the sixth kind;[1] they are oblong, while some are a little rounder or triangularly round, like the Shell *Bia matta doa:*[2] They are ugly Stones as if some dirty, dark yellow, russet soil had petrified, with two buttocks behind and a slash, otherwise even, without a stripe, reasonably hard, yet easily scraped with a knife. One will find them on *Java* on the beach near *Remban* and *Lassam*.[3] I know no other use for them than that the Chinese like to have them along on their ships: One will find these *Tellinitis* also around *Grisek*[4] by *Dudunam* in a marshy sand, one does not know if they were tossed up by the sea: Some will carry them on their bodies, and they believe that [these stones] will protect the wearer against the witchcraft of women, which is when they use love potions or *Ubat Guna*.[5]

Ctenites[6] is a hard dark-gray stone like a Flint, shaped like *Bia anadara*,[7] a kind of *Pectines* or bossed *Jacob*'s Shell,[8] with a similar edge and stripes or furrows, but few, and they do not run as they do on natural shells, [and they are] so hard that one can barely scrape them. One will find them on the smaller Islands, on maps called the *Uos*, east of the largest Papuan Island of *Messoal*.[9] One could include them with the kind of stones called *Ombria*[10] or Mother stones, which are strewn about on the beaches of those Islands. The Papuas look for the roundest ones and use them instead of bullets to shoot from their Pederoes, although they can find ones that are even better suited for this, and also occur on *Messoal* itself, and which one could easily hold to be true iron balls of 1 pound each, and they truly are like iron.

Some other petrified Shells also belong to these, shaped like the Stone sheath,[11] and some smooth *Chamae*, but coarse, dark and of an earthy substance, which one will find in some rocks on the beach, which stay above water, and if one looks at them carefully one can only conclude that they are the flesh of the same animals, which were turned to stone by the intruding mud; because here and there one will see some remnants of a shell on them. In the year 1692 we found many of these in the rocks whereon they were building the Fortress *Duurstede* on *Uliasser* Island.[12]

CHAPTER 66

Cochlites. Cochlea Saxea. Mestica Bia.

I deem the Stones discussed in this chapter to have been natural Snails, which in some fashion were changed into hard stones, of which I note 3 kinds.

The first was a white and partially transparent Stone, the size of a Gun bullet,[1] with the shape of a Winkle, called *Bia papeda* in Book II, Chapter VI; because it has the same whorls, while at the back it has the dark remnant of a shell; the rest is dense, filled out and as hard as a Flint, and one would have thought it to be such, if it had fired at night and had a flinty smell: It was found on the beach of *Mamalo*,[2] and I saw only one in my day, and that was sent to the Arch Duke of *Toscana*. One might say that the entire Whelk had been turned to stone, but upon closer re-examination one will find that this could not be the case, since one cannot find anything else between the whorls of the shell, except a little on top, so that the entire beast must have changed to stone.

The second *Cochlites* is also like a Winkle, but in this case the crawling animal hung outside the shell, closely resembling the shape of a complete Snail, but wrinkled, damaged, with corners broken off in some places: Its little house was like the first one, flint-like and partially transparent, but the animal was a dark stone, a dirty white or grayish. One finds them rarely, perhaps here and there on the beaches of *Leytimor*, and I only saw 3 of them, of which the most beautiful one was sent to the aforesaid arch Duke. They will boil in Lemon juice, which purifies them, and they do not fire at night, nor do they have a flinty smell: I do not yet know how these animals changed into stone, except that it is the feeling of some Chinese, to wit, that these animals were crawling on the open beach, and were accidentally hit by lightning, and were suddenly petrified; if this were true, then this would happen with other kinds as well; but no one has experienced this: It is more likely, that a kind of small and white Winkle (which has four to 6 times more flesh, than its little house can store, wherefore it is called *Issi Palesou* in Ambonese,[3] that is, Much Flesh or More Flesh) once it has acquired the measure of growth and size which nature intended for it, it stiffens from its own stone sap and changes into stone: Although this opinion also has problems, for one can say, that in that case this stone ought to maintain the proper shape of the previous animal, which is not what one sees: To which one can answer, that the animals can hardly maintain their previous shape during petrification, since they shrink, are pounded and break in different places, just as one can never really tell with these stone Snails where the head is. The Malay say that they are found in other Countries as well, but rarely, and they wear them, along with other *Mesticae*, around their body, especially the first kind, in order to be invulnerable in war.

Cochlea Saxea is the third kind of petrified Snails, but they do not have the shape of any Snail's house, only showing a piece of this animal's flesh, usually a finger long and barely the thickness of one's little finger, usually crooked and bossed: One might consider them partially transparent Pebbles, but their shape proves sufficiently that they come from a snail, since the round part is the whitest, usually partially transparent and bluish like Achate, sometimes smooth, sometimes wrinkled, with a white heart in the center like a vein; the wings are of a dark, flinty substance; I cannot yet tell from what kind of Whelk they might come, because one will find pieces lying bare here and there on the beach of *Leytimor*, and they are not alike. They boil quite vigorously in Lemon juice, and do not fire at night, except a little on the flint-like wings, if one strikes them quite hard; rubbed together they smell like other pieces of Shells; their use is unknown.

K
D
C
I
B
A
E
F
H
G

The Author treated the petrified Whelks and Shells quite learnedly in Chapter XXIX *of the previous Book* II: *Also, how and why these are found in Stones, in mountains, even in the impenetrable depth beneath the earth. His reasons are undeniable; that is, that they got there, when the Lord God, in order to destroy the first world, inundated everything with the general Flood; wherefore everything that had been created, People, Birds, and Land animals (outside of what the Ark contained) was killed and destroyed; Carrying the Fishes, Whelks, Shells and other things as well, across the entire Earth by means of the Water's fury. And this same fury caused many of them to be forced deep into the Mountains and under the earth; saturated with Stone sap, they petrified, though retaining their proper shape: And they are found daily almost everywhere, be it in mountains or soils, and were therefore called* Noah's Shells *by the Author; although later floods also contributed to this.*

The gainsayers want them to be products of the earth and of stones. But if one inquires in this matter with a measure of reason, it is indubitably impossible, to approve of that notion; even more so, because most of the ones that are found in the mountains and in the ground, correspond in shape with those which are brought forth by the sea; sometimes entirely a shell, other times half or entirely petrified; and one will also find, that when they are opened completely or split right through, the inner material, to wit the fish, had also changed into stone; and the bark or shell will be of an entirely different material than when it enveloped the core: One can even notice a difference between the hard or firm parts of the fish and the slimy or soft ones; which are also harder or softer, and the outer shell against that, again has a different hardness from the inner parts. And so one will also experience that the Whelks and Shells from the Stone mountains, are usually harder and stonier, than those dug from the earth, which are much softer, more calcified and have aged more: It is also observable that when they come from deep within the mountains or earth, they will be more beautiful and more perfect to the eye. Clear proof, that the air particles did not exert as much force on the deeper ones, as would be the case with those nearer to the surface, and also, that some that were brought to light from deep places, still retained most of their colors; on the contrary, those which were closer to the air, were robbed thereof, except for those that were touched by Sulphur, Saltpeter, Arsenic, or other sharp substances; wherefrom one may again notice, that Whelks, Shells or other things were washed down into the Earth or Mountains in one piece, but they did not grow there. We have seen various kinds that were dug from the earth by the Honorable N. Witzen,[4] Burgomaster of Amsterdam, *who, when the well was dug for the Old Men's home,*[5] *and they had gone through many layers, found at a depth of 160 feet sand mixed with whelks and shells, and kept some, along with other rarities; and the colors were still perfect, and they were all kinds that (and this is even more noteworthy) are no longer found on our beaches. Which makes one think that they were carried along by a powerful storm from other regions and brought over here; or that the flooding and then receding of the water, left so much earth and mud behind, that this entire Genus was suffocated and swallowed up by the earth; and one also begins to think that this Land, was for the most part water or sea, and that it gradually increased, by means of more than one flood (which can be perceived in the various grounds or earth crusts, which were discovered while digging, before they reached the deep Sea floor, where the Banks of Shells were found) until it became what it is today, I must add something similar, also observed by* His Honor, *when he, as Member of the Provincial Executive, was in the field with the Army: Around Tongeren,*[6] *which was once thought, though obscurely, to have been a coastal town, it pleased* His Honor *to try to discover something about that; he was informed there, that a place thereabouts, called the sea dike, had indeed been such a dike; given the fact that they first discovered a layer of good heavy clay,* His Honor *continued, making them dig several feet deeper, where they found a soil that was strewn with many Shells and Whelks.* His Honor *put four kinds at my disposal; which I, to please the Collectors, have traced in copper, and can be seen on plate* LVIII. *By the letter* A. *being an* Oyster shell: *Also by the letter* B. *another kind of* Oyster: *The one seen by the letter* C., *is a* Seashell: *And by the* letter D. one that is known to us as Little Fan Whelks, but which today are no longer present on our beaches. It is known about Oysters, that they come from the British coast: And the other ones still reach us every day from the West Indies. Herewith a third example: Doctor Magnus Bromelius[7] from Gottenburg, who is an eminent inquirer after natural curiosities, discovered many such petrified items during his travels in France, and other places, and brought them along and showed me some twenty kinds, which His Honor had all discovered in Champagne, and dug up near a certain village Chameri outside Rheims, in the mountain vineyard la Montagne de Rheims, and in other places, with an uncommonly large one among them, being a Pen whelk that was somewhat calcified at one end, but otherwise very beautiful and pure, as were all the other kinds; indeed, many were as perfect as if they had just been taken from the sea. Now it is well known, how far away this region is from the sea; just as we have many kinds (though all petrified) that were discovered in the mountains of Tuscany, some 50 to 60 fathoms deep, though they were clearly all Seawhelks.

First, because most of these are found in the sea, and not only resemble each other in outline, but also in their inner parts: But with this difference, that most of those that had been in the ground, had lost their Fish parts and had turned to soil; one sees it clearly in the hollows that have the same shape as the animal that was once in there; while on the other hand, the animals in the Stone mountains are for the most part petrified in their own crusts, as a result of the Stone sap that came in from outside.

For another, we found while conducting a test, by grinding them on a whetstone for Rubies, that the dust of the Shell-like ones (providing they had not perished) was the same as for those that had come from the sea.

Thirdly, it also appears that the petrified Fish parts are always softer, than the part of the Shell.

If one says, that the Land, fresh-water rivers and brooks, also have Snails, Oysters or Mussels (compassed with Whelks and Shells) I will gladly concede the point, but one will also find them much lighter and with a thinner shell, both in this Country and in the East and West Indies; to which we add, that those kinds are not found in the sea: For most Seashells and Whelks, have a heavier, denser, and thicker bark and material; in order to withstand the greater violence of a wild element. Furthermore, the thin-shelled ones are found in those places at sea, where they are not subject to a hard bottom or to the waves dashing against the rocks. We deem it appropriate to be permitted to say this much, and would add even more, if such were possible. Those who want to be further enlightened on this matter, read the learned work by Mr. J. de Ray, Chapter IV of his The Beginning and End of the World, translated from English into Dutch, and printed in Rotterdam by Barent Bos.[8] Also Dutch Antiquities, of about 100 years, compiled and newly presented, with learned annotations, by R. Verstegen, printed in Amsterdam by van Rooijen.[9] And as far as Whelks and Shells showing up in the most inaccessible of stone mountains, this is easily understood, if Mr. T. Burnet's contemplation of the holy Globe[10] has as much influence on the reader as it had on me. Read where he deals with cave-ins, floods, and how Mountains arose. That book was also translated from English, and printed in Amsterdam, in the year 1696, by Van Dalen. Many of the aforementioned petrified Whelks and Shells that were dug out of the ground, or found in the mountains, are kept and displayed by Collectors. And those who excel in this, besides the aforesaid Doctor Ruysch,[11] are Mr. Vincent[12] and others. We will show only a few from our vast collection, for the benefit of the Collectors. See plate LVIII. By the letter E. is a rock-hard stone covered with both Whelks and Shells, found by Mr. Kreitsmaar,[13] Captain of the King's Guards, outside Brussels, while digging during the construction of a redoubt: The one shown by the letter F. is a similar kind, but compassed by different Whelks, and sent to me from the West Indies; Another mountain clump by the letter G.: And by the letter H. a complete double Oyster, completely petrified both inside and out, and found in the mountains of Tuscany: by the letter I. a hard, petrified Whelk; and a similar petrified double shell by the letter K. On plate LIX., by the letter A. one will see a Rock with various Shells, among them a large, ribbed, bright-blue one. By the letter

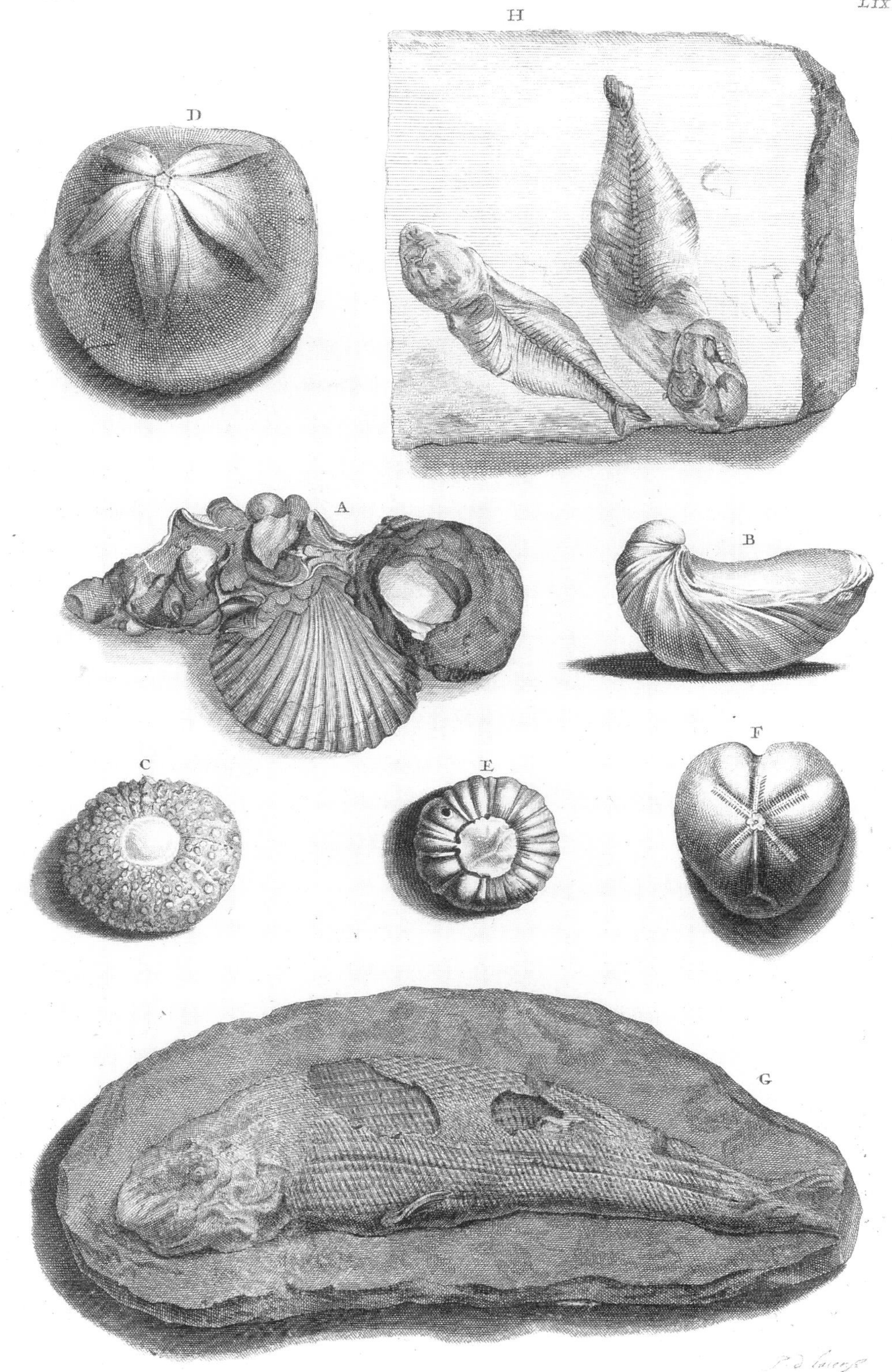
H
D
A
B
C
E
F
G

B. *on the same plate, a petrified Shell, which we call the single* Fool's cap, *of which I also saw double ones that were petrified. One plate* LX. *by the letter* A. *is a* Penwhelk, *called the* Drumscrew, *completely petrified both outside and inside: A similar one by the letter* B., *still in its rock: A double Shell by the letter* C., *and they find many of these in Hungary. By the letter* D. *is a Whelk, called* Cornu Ammonis, *about which there is doubt whether it is a plant or a whelk; but see the pros and cons by the aforesaid* J. Ray. *The one by the letter* E. *(which still has its complete color) is as hard as flint: And the one by the letter* F. *is known to us by the name of* Toot: *By the letter* G. *one will see a black little* Snail's whelk, *with a faint white band: And by the letter* H. *one that we know as the* Posthorn: *What is shown by the letter* I. *is a topped little whelk, which is black and shiny; they all came from Stone mountains, and are petrified both inside and out. These are followed, from among the many that are available, by some that were dug from the earth and are completely petrified, the one depicted by the letter* K. *is a double* Jacob's Shell, *but very pure and beautiful, still containing its fish, but more softly petrified than the foregoing. Such is also the one shown by* L. *and by the letter* M. *a ribbed* Moon doublet, *or* Compass Shell. *By the letter* N. *an oblong Shell with sloping ribs that come together: With these we conclude; but we will write more about other Stone growths in Chapter* LXXXIV.

CHAPTER 67

Myites. Mestica Assousing.

Pliny Lib. 37. Cap. 11. mentions the Stone *Myites*, but says little more than that it got its name from the shape of a mouse; but my *Myites* is named after a Mussel, which in Greek has its name in common with the mouse, it is found in a kind of Mussel, that has a thin shell, and remains deep down in the sand, wherefore it was called *Musculus arenarius* or Sandmussel before in Book II, Chapter XXXV:[1] One finds this little stone very rarely, and it could be considered a little Mussel, if it were not as hard as stone, white at the top and partially transparent, and on the bottom and also the thickest end, purple and bluish, nary the length of a finger joint. I have only seen one of them, and it should be considered more of a rare accident than a normal creature of nature.

To conclude the *Mestica* from Sea animals, I add here the Pearl, which can be found in some Shells and Mussels from the Ambonese Sea, but they are small and inconsiderable and need no particular description. The best ones are found in the flat Saddle shell,[2] which resembles true Pearls the most, though it is yellowish and angular, like the ones that are found in the Bight of Ambon: There are larger and more beautiful ones on *Xula Mangoli*,[3] around the village *Weytina*, also on *Buton*, where this Shell is called *Calepinda*. One will find rounder ones in the Holster Shell or *Pinna*,[4] but they are purple or brown, and pale with time.

The Author did not supply us with an illustration of this Mussel stone, *but since we own one, we show it on* plate LX., No. 1.

CHAPTER 68

Clappites. Klapsteen. Mestica Calappa.

The *Calappus* stone was partially described in *Lib. I. Cap. III.* of our Ambonese Herbal in the History of the *Calappus* tree,[1] but will get here a complete description in its own right; it belongs to those Stones, which *Pliny Lib. 37. Cap. II.* calls *Dendritis;*[2] but since

A
N°. 1
N
K
I
B
H
G
F
N°. 2
N°. 1
C
N°. 3
M
D
E
L

that includes many kinds, which one will find in these Eastern Islands in almost any kind of fruit or tree, I gave it a particular name, formulated in the Plinian manner: It is to be believed, that these little Stones were formed by Stone sap, which is drawn from the earth up into the trees and fruits, where it is concentrated and wherefrom its most precious part a *Gemma* or rare little Stone is made, which as said many times before, is called *Mestica* or *Mostica* by the Malay. The best known of those that derive from trees and fruits, and the most common one, is the *Calappus* stone, in Latin *Calapites*, in Malay *Mestica Calappa*. It is a white little Stone, that looks so much like a white flint or alabaster, that one would hold it to be such, but the *Calappus* stone is softer, strikes no fire, does not have a flinty smell, and has the shape and size of a pigeon's or some other bird's heart, sometimes like a Lizard's egg as well, thicker on one end and with a dark little crown like a tooth that has fallen out, which is the root, wherewith it had been attached to the shell or *Tampurong:* Yet some lack this little crown, which indicates, that they were already loosened from the shell and were floating on water, and then they resemble a Lizard's egg more, they are the clearest at the other and narrower end, which comes to a point that is like a blunt Kayle,[3] and on top of that is a shiny spot like a radiant little sun, which will reveal itself if one holds it against daylight, and if they lack this little sun they are considered dead.

The *Calappus* stone is sometimes truly round, like a large pea, sometimes the shape of a lentil; as large as a flat cherry, surrounded by a narrow edge: They grow in the little apple, which one will find in the old *Calappus* nut, and if they fall off, they will also float on water. Both are pure white, to wit, the first kind is white as milk, the second one is a little bluish white, and both are sometimes partially transparent at the edges, although they are usually dark; The round ones also have the little white sun on the clearest and most elevated side, and one will see very tiny and fine cracks in both of them, that are not deep and which do not make these Stones crooked in any sense: For since the Stone grows in a perpetually moist substance such as the inner cavity of the *Calappus* nut, it stands to reason that the Stone will show cracks, when it comes into the dry air, which also happens to the *Umbilicus Marinus*,[4] when it is freshly cut from its animal. I learned from an eyewitness how they grow in the *Calappus* nuts; he was an Ensign, who had been sent to the South-Easter Islands in 1672, where among other things, he opened a *Calappus* nut, that had a kernel that was mature but not yet hard, and he noticed a red little spot on it, protruding a little from the other flesh, which contained two white flat little Stones, which were still soft and attached to the shell: The Natives said; if that *Calappus* nut could have remained on the tree for two more months the little Stones would have been ripe and hard.

One does not find them in the Ambonese regions, though they open up thousands of old nuts, in order to cook the oil from them, but they bring large numbers from Ceram's northern coast which the Natives say grew in their *Calappus* nuts, but I doubt it, since no one ever came forward who had found any there himself. Most are found on *Celebes* and *Makkasar*, the country of the *Bugis*, *Cajeeli*, and on *Buton;*[5] it seems that those countries have the characteristic of sending more stone sap to their fruits and trees, then happens elsewhere, wherefore those places produce the most *Mesticas*.

One usually tests these Stones with sharp Vinegar or Lemon juice, if one pours a little in one's hand and places the *Calappus* stone in it, it should result in immediate boiling or globules bubbling up all around it, and those that do not, are considered useless or dead; but I have several times rejected this test, since the Stone will lose its shine therefrom and get a deathly pallor, even if it is immediately washed in water. All soft, porous and dark stones do this, since the Vinegar or Lemon juice that presses into the little holes [*Pori*], forces the air out, which causes the bubbles, and this happens in *Europe* as well with *Lapis Victorialis* or *Astroites*[6] which also moves in Vinegar. The second test is the following: if one puts it on a mat and strews some rice or *Pady*[7] around it, then no chicken will dare to eat a single grain, as long as the Stone remains there; but this did not work for me, for if I had not removed the Stone, the chickens would have swallowed it along with the rice, howbeit I thought to have a true one. The third test, which I heard from a Malabar *Empiricus*,[8] is even more unbelievable; if one ties the Stone to the trunk of a *Calappus* tree, all the nuts are supposed to fall down, but I declined to attempt this.

Pliny mentions in the aforementioned place that one of the characteristics of his *Dendritis* is, that when one buries it at the roots of the tree you want to fell, your axe will not blunt; which one might try out on the subsequent *Dendritis*. The oblong *Calappus* stone bears also a great resemblance to a *Chamites*[9] of the first kind, though it can be distinguished by the fact that the *Chamites* usually has a pearly reflection but lacks the little sun, while it also boils less and more slowly in Lemon juice: The *Calappus* stone also becomes smudged, a dirty white or a deathly color, which one needs to clean as follows; one puts it for half a day in the sap of a young *Calappus*, and rubs it with its own *liplap*,[10] or it should be washed and rubbed in water that has been used to rinse the rice one is about to cook.

The *Calappus* stone is considered one of the most important *Mesticae* to carry, in order to have good health and good fortune in any kind of enterprise, and for many other things which the Natives ascribe to it owing to their superstition and fantasy, such as having good fortune in trading, for putting in gardens, and to keep one healthy and invulnerable in war; though the latter is most unlikely, for what has peaceful *Jupiter*, to whom this tree belongs, to do with the War God *Mars?* Those who use it as a remedy for fevers, have better reasons, putting it in some water and then drinking from it, in order to extinguish the burning fever; also after it has been rubbed with some water on a fine stone and then applied to the eyes it will heal their burning. The roundest and most beautiful ones are set in silver rings, because it will not tolerate gold. The largest ones are hung on the creeses and closed in silver plain-rings. The Malabars also make earrings from them, which their women wear in their ears, but set in gold, though they will never stay as white as they do in silver. One usually pays a Rix-dollar a piece for them, though the round ones and those with more beautiful rays cost more.

A certain maid came to me in February of 1691, after she had peeled a *Calappus* nut that had grown on *Bahuala:*[11] She found a *Calappus* stone on the shell or *Tampurong* and under the outer husk, and it was not in the eye but on one side of the *Tampurong* in a shallow little hole: It looked like other *Calappus* stones in color and substance, but it had a different shape; for it was as large as a *dubbeltje* and shaped like a flat heart, flat on the bottom, bossed on top with a small hole, wherein one could see something like a fiber from the husk, and a small sliver had fallen off one corner, thereby uncovering a tiny brush. The maid had placed it in vinegar, and it boiled and moved, but it had also lost some of its shine.

CHAPTER 69

Dendrites Calapparia.

A true kind of *Dendritis* from a *Calappus* tree was found on *Ceilon* in the wood of its trunk, which shortly before had been struck down by thunder, and on that same day 2 Dutch officers and their slaves passed by; the two slaves went over to it in order to get the palm cabbage[1] from the crown, which is when they found this little Stone in the wood of the trunk, and it was quite obvious that it must have grown there, because it was closely encased by that wood; they gave it to their Master the Captain, who later honored me with it in *Amboina.* This Stone was round or a little bossed, the size of a small cherry, hard and smooth like a pebble, not transparent, with a yolk-yellow color, and all around it were little white eyes or circles, with a yellow space[2] inside, some large, some small, as if they had been painted; the topmost eye was the largest one, which had another dark ring inside like the Iris of the eye; some of the other eyes were entangled, some were also entirely white, one will see a few of those little circles in a kind of *Lapis Victorialis* or *Astroites:*[3] There was a little white spot on one side, of a deathly pallor, where one suspects it was hit by lightning a little: The Captain said to me, that the Singalese had told him, such Stones were often found in *Calappus* wood; but they could not produce one, nor indicate one, even though he was the *Dessave,*[4] that is, the governor, who ruled over them: His guess was that they did have such stones, but kept them hidden from him, because that Nation as well as other Indians consider these rare *Mesticae* very valuable, wearing them for good luck, especially in times of war, which I might accept to some degree, since it was able to protect this tree against the might of thunder; but the proverb says: (*the lesser must yield to the greater*).[5] I never saw nor heard of another one quite like it in these Eastern regions; it was set in a ring in the year of 1682 and sent with other Curiosities of mine to the Arch Duke of *Tuscany,* under the name *Dentritis Calapparia:* I did not want to test it in Vinegar or Lemon juice in order not to spoil its sheen, but it did fire at night when struck against a Crystal or Achate, though only a little, as all other hard *Mesticae* will do, that are either entirely or partially transparent.

CHAPTER 70

Pinangitis. Mestica Pinang.

The *Pinang* tree[1] produces two kinds of Stones, which differ much from one another. The first one is the common *Mestica Pinang;* it is a small stone, slightly larger than a green *Catjan,*[2] with the shape of a blunt little cone, pure white, with a radiant little sun on top, nor does it fade as easily as the *Calappus* stone. One will find it in old *Pinang* nuts at the top of the kernel, the nut, which is eaten, is in that little hole, which is called the eye, and wherefrom comes a new sprout: But one should know, that one will hardly find one among a thousand nuts, and then mostly in *Makassar* and *Buton.* They are set in rings because of their beauty and worn like the *Calappus* stones, they are particularly appreciated by those who do the planting and selling of *Pinang.*

The second *Pinang* stone is completely different, and is nothing more than an old *Pinang* grain that has turned into a hard flint, which it resembles the most in shape and size, to wit, at the bottom somewhat flat and otherwise coming to a round point topped by a wrinkly little crown, dark brown with russet spots, very hard, cold, and it seems moist to the touch; it is very rarely found in *Makkassar:* It does not boil in Lemon juice; one can use it for a Touchstone, but it proves gold very unevenly for it will assume the natural color of gold all around the edges and maintain it for a long time, but it will show gold as pale and unsightly on the flat bottom side, as well as on the little crown.

Add here the *Mestica bras,*[3] being an oblong and white little Stone, which one will find at times in *Makkassar* instead of a rice grain, still shut inside its natural husk, and which those that plant the rice, will gladly wear on their person.

Scorodites,[4] that is, *Mestica bawang puti*[5] is a white stone, not transparent, except a little at the thickest end, shaped like a clove of garlic, but twice as large, found in Makkassar in a garlic bulb, just as one will sometimes find garlic bulbs, that consist entirely of one clove: which one is wont to call Male Garlic; it does not fire at night and when rubbed it does not smell at all like garlic.

CHAPTER 71

Dendritis Arborea. Mestica Caju.

This *Dendritis* was found in *Amboina* in the following manner: There was once a certain Chinese, at first a poor and insignificant person, who got his livelihood from stoking Arak,[1] which required a great deal of firewood, and of the best kind, now it happened one time that one of his slaves was splitting wood, and his axe hit something hard inside the wood, which jumped out and fell at the feet of the master who, when he picked it up, was holding a beautiful little Stone of an unusual shape, which, when it was fitted back into the wood, showed that it had indeed been there and, consequently, had grown there, otherwise they would have charged the woodcutter, because he had hit a stone. They did not pay attention at the time as to what kind of wood it had been, but when further examined later on, the servant maintained that it had been *Casuaris* wood,[2] which is what the Chinese use the most for stoking Arak, because it burns hot and leaves solid coals. This Stone was long and round like a finger joint, and surely as thick, and if it had been found elsewhere, it would have been taken for an Achate the color of horn; the center of the Stone showed a milky way made up of 7 narrow little veins, just as one will sometimes see an *Iris secundaria*[3] with a true rainbow; the upper half of the Stone was the color of horn, but of a lighter color below the middle, just like a true white Achate, and the lower corner was the widest and somewhat rough, when held to the light it was everywhere partially transparent, while at night it fired just like an Achate, although a little less, also similar in hardness and shine. The aforementioned Chinese, by the name of *Nonko*, wore it in his belt from that day on, and since he changed from a poor person to a rich one within a few years, he, being a Heathen, ascribed his luck to this Stone; which he rarely showed, nor would he part with it for any amount of money, although I offered to pay its weight in good gold, which would have amounted to around 10 Rixdollars. I had to be satisfied, that he loaned it to me for an hour so I could describe it and draw it, though he never left my room during that time. Later, in the year 1693, the Stone was lost after his death, and none of his heirs knew what had happened to it; and if he wore it in order to be invulnerable, as is the common notion, then it failed in his case, since the tip of his nose was cut off during a quarrel.

Such Stones that grow in wood are usually called *Mestica Caju*[4] by the Malay, and are so rarely found, that I have seen only this one, although it is said that the Natives might have hidden some somewhere. They are found in *Makkassar* in the knobs and burls of *Cofassu* wood[5] and *Camuneng batu*,[6] which I have not seen.

In the year 1674, a Chinese carpenter found a Stone in a *Caju sicki*[7] tree, the size of a *Bonga Manoor*'s bud,[8] whitish on the outside and partially transparent like Achate, but without those veins of the one above, and inside it was a blacker color, almost dark: The Chinese who had found it, did not want to part with it; for they believe that the finder of such a Stone will have great luck, but they consider it powerless if it is sold.

Add two little Stones, which are the color of liver, dark and angular, found on *Camarien*[9] in the fruit of a large *Casuaris* tree, and its fruits were four times the size of normal ones.

CHAPTER 72

Nancites. Mestica Nanka.

The kernels that are found in the flesh of the *Nanka* fruit or *Soursack*,[1] will also turn to stone at times, for instance in *Makkassar*, where these stones are smaller than the kernels themselves, but have the same shape, to wit, like an oblong little egg, hard, smooth, a pale yellow or at least whiter than the kernel: They are seldom found, and though the fruit has well over 100 kernels, only one will turn into a Stone, which are also of different sizes, but usually a finger joint long, some are oblong and round, some are pressed flat a little: The Makkassarese encase them in four little silver bands, and dangle them from their creeses; although this insipid fruit has no warlike characteristics. It can be used a little to test metals, but falsely, and when proved [with a Touchstone] it does not cling;[2] silver reveals itself as slate,[3] while gold will also turn pale and wan; at the lower or thickest end it can be rough sometimes as if a small piece had been bitten off. One will also find *Mesticae* in various other kinds of trees and plants, which I have not seen yet, to wit in the pods of the *Tamarind*, in the fruits of *Tsjeremy* or *Boa Malacca*,[4] in the genicles[5] of the *Sulassi* herb,[6] and others.

Mestica Clompan[7] has the shape of a *Bidji Clompan*, but it is white like a *Calapites*.

CHAPTER 73

Sangites. Mestica Sanga.

How hurtful a milktree[1] the *Sanga* is, can be read in the Amb. *Herbal Lib. 3 Cap. . . .*[2] [*sic*]; from [the tree] they make the Chinese varnish called *tsjad;*[3] it bears a fruit similar to the Ambonese *Gajang*,[4] but much smaller: This fruit is filled with a black-brown milk, which becomes as hard as stone, but it has a dry white kernel inside, which sometimes changes into a Stone; it usually keeps the shape of the fruit, with similar protruding veins, but smaller than the fruit; it is very hard and as heavy as a flint, cold, chestnut brown, dark and proves reasonably well and will show the true colors of metals: some are smaller, full of corners and holes, and brown as well. A particular eyewitness told me that he found a fruit on this tree, of which already half had turned to stone, wherefore he tied a string to its stem so he would recognize the fruit again, but when he returned to the tree about a month later, the fruit had fallen down, and was lying under the tree completely turned to stone. Someone else broke off a partially petrified fruit like that, and put it in a *tomtom*,[5] and it too turned completely to stone after considerable time had passed: Wherefrom one can tell that this tree must have plenty of powerful stone sap, which is also why that Chinese varnish becomes so hard, as one can see on lacquered woodwork. The trees grow in *Makkassar* around the edges of rivers, and I obtained two such Stones.

CHAPTER 74

Parrangites. Mestica Gondu.

This *Mestica* derives from the *Parrang* fruit,[1] otherwise known as *Faba marina* and described in *Lib. 7. Cap. . . .* [*sic*];[2] it has the shape and almost the same size as well of a bean from the same fruit, and was changed into a hard stone, flat and round, with bulges in the center on both sides, blackish at the edges, veering towards blue, so that it seems that the entire bean, which is a dark chestnut-brown, has turned into such a stone, although the Makkassarese *Parrang* bean is much larger than the Ambonese one, but the stone is the size of the Ambonese one, and though this fruit was plentiful in *Amboina*, I never encountered anyone, who had ever found such a Stone in one; it is rare even in *Makassar*, where everything changes to stone, and I managed to obtain one from there that had been set in red copper, in such a way that it could be removed, which they wore on their bodies when they went to war, perhaps because the entire fruit has the shape of a curved sword, which is how it got its name. It too can be used as a touchstone, but it does not show the colors truly, to wit, good gold veers towards a pale yellow, and base gold to russet, while silver turns white, though some red shimmers through.

CHAPTER 75

Manorites. Mestica Manoor.

Not even flowers are spared from petrification in the Country of *Makkassar*, and this is even more wondrous than with the foregoing woods and fruits, because those are far more capable of receiving stone sap, than a fragile flower or bud, yet *Medusa* will not spare them.

And so a *Manorites*, or little Stone from the bud of a *Bonga Manoor*,[1] was brought to me from *Makkassar*, where it had been found in the garden of a Dutch family, where the housewife, who wanted towards evening to gather the buds of this flower from her garden,[2] picked a bud along with others which she only knew as coming from a *Manora* flower, but since it came loose and fell in her hand and seemed to be heavier than a natural flower, she looked at it more closely, and found it to be a little white stone, as if a *Manoor* flower had been carved from alabaster, but one could not tell how the leaves were divided: She called all of her husband's company as witnesses, and indicated the stem to which the Stone had been affixed; and they took the Stone with them and laid it carelessly on the table, where some *Boerepons*[3] happened to have been spilled, a drink made from sour Lemon juice, and when this juice touched the stone it bit off a corner in a very short time. It became mine in the same year, when it was still a pure white, but after eight years it began to lose its pure whiteness and became grayish, but when scraped with a knife, it became white again, and afterwards a Moorish Priest told me, that he had also found such stones on *Buro*, on a *Manora* stock,[4] shaped like a bud, but the stem is always shorter than on the real flower.

CHAPTER 76

Stones That Happen to Come from Certain Trees, or Fruits.

Some Stones, sometimes wrongly included with *Mesticae*, get accidentally in several woods or fruits and were caught inside during growth, and they can be recognized, because they have the substance and color of natural and familiar stones; though there are some among them which are doubtful: I will give some examples here, and then order the others accordingly. During the time I was in *Hitu*, they found in *Mamoa* (a place between *Hila* and *Hitulamma*, where they have many gardens), in the trunk of a *Pissang* tree,[1] a round though somewhat flattened black stone, so much like a common Touchstone that I could see no difference, except that it was greasier and that it did not take well to metals, which might have been because it had been such a long time in the moist trunk; the Natives considered it a *Mestica*, and believed that it had grown inside the *Pissang* trunk, they censed it their particular way every Friday, kept it in a linen cloth, and wore it on their bodies in order to have good luck with their gardens. Others surmised that it was a small Thunder stone,[2] which are also black, but as far as I know they are never found in a round shape, while it must also have been a feeble Thunder stone, since it could not even pierce a soft *Pissang* trunk. My guess is that it was a small stone that was accidentally thrown or fell on top of a young *Pissang* plant which, owing to its hasty growth, embraced this Stone and locked it away inside; to which the Natives objected, that no Touchstones were ever found in that place, but this cannot be relied upon, since I myself found other black Stones in that place, which looked very much like Touchstones: For it is hard to believe, that such a hard Stone would grow in the watery substance that is a *Pissang* trunk, although one knows for sure, that in certain kinds of *Pissang* fruits one will find small, dirty-white Stones, which may be considered the true *Pisangites* or *Mestica Pisang*.

The *Armilla magica* or *Armilla Daemonis*[3] belongs here as well, being an Armlet, apparently made by human hands, but the Natives pretend it was found in the following manner:[4] A certain *Tuhua*, an inhabitant of *Mosappel*, a small village that used to be near *Mamoa*, had his gardens in the aforementioned *Mamoa*, where he also found the aforesaid *Pissang* stone in the following manner; he wanted to cut some vines, which were used for weels, in that place, when he heard something fall down with a rustling sound through the branches and leaves of a tall tree which, when he investigated, turned out to be this Armlet, and he fancied that this was granted to him for his use by a kind *Djing* or *Daemon*, and which he therefore, in his fashion, carefully preserved and censed: In the year 1668, which was many years later, his nephew sold it to me along with the aforesaid *Pissang* stone, both had been well kept in cotton, covered with civet,[5] and he acted as if he had been forced to do so: He pretended that his Uncle *Tuhua* had kept both stones until he was very old, carrying the same on his body, when he went to war, and no one had ever known him to have been wounded or lose a drop of blood, even though he led his troops and had witnessed many an *Attaque;* finally reaching a ripe old age, and being childless, except for a daughter, who had been taken away by a stranger, he handed over both Stones to his nephew, recommending them to him as if they were a secret treasure: And since he feared that his Uncle's fate would befall him as well, he was forced to give those Stones away, noting that they were no longer favorable to his family. He also wanted to make me believe, that the ring had been small at first and had grown to quite some size over time; wherefore I drew it on a piece of paper in his presence, with the intention of showing him the same after some years had passed, and that it would be the same size, but we were separated shortly thereafter, and never saw each other again, but the ring remained the same size.

The Armlet was large enough to push easily over a hand, flat on the inside, as wide as one's little finger, triangular on the outside with a blunt back like the shape of a *Mamakur*,[6] the color of lead, here and there shining like *Antimonium*,[7] not smooth but somewhat striped and coarse, as if it had been shaped with a small knife, its substance very like a hard Slate stone or similar rock, of which I have seen entire mountains in the Country of *Huamohel*, especially on the beach near *Erang*. Hence I can only believe that people carved it from such a Stone, but how it got in the tree I will leave for *Tuhua* and his *Djing* to guess, knowing full well that the deceitful spirit is capable of granting or indicating something that was desired by superstitious people, just as he knows how to supply his servants with Fern seed for St. John's eve.[8] I knew full well that their pretense was nonsense, nor did I see anything unusual or rare about those Stones, but I took them off their hands in order to deliver those simple folk of their superstition, while I am also well aware that in war, victory does not come from such paltry Stones, but I think it advisable to get such things out of the hands of the Natives, because it will make them bold from time to time,[9] which often causes them to wage war on us quite easily.

In the year 1683, a *Calappus* tree was felled in a field near *Victoria* Castle, and they found a large flat Stone in it, which was easily recognized as an ordinary Fieldstone; it was lodged in the lowest part of the trunk, a couple of feet above the root: I suspect that the stone was put there when the *Calappus* nut was first planted and then covered up as is customary, whereafter it was forgotten by the planters and encased by the trunk that was shooting up,[10] although this too has its difficulty, for why did not the nature of the *Calappus* tree, which is to strive and grow upwards, cause the Stone to be borne higher, since this tree as well as the true palmtree, have as their nature to press against all kinds of heaviness? I leave this riddle to those who have heard and seen the grass grow,[11] but it is enough for me to know, that it was an ordinary Fieldstone that had been locked in by the fast-growing trunk. These Stones were also mentioned in the previous Chapter XXVI.

CHAPTER 77

Stones Which Happen to Have an Unusual Shape.

Here do not belong such Coral stones and Seatrees which, in accordance with their normal nature, have the shape of some plant or animal, and which were presented in Book XII of the Ambonese *Herbarium*, because they have a *Vegetavite* nature,[1] commingled with the stones: Nor such Land stones which have a particular outer shape that was produced by their inner spirit, like unto the six-sided Crystals, the four- and five-sided *Androdamantes*,[2] and many others. But I judge them to be Stones of an unusual shape, which they did not acquire at the behest of nature, but by chance, and they can be innumerable; so I will select some, deriving the first example from the Southern corner of Africa, which the East-Indiamen have to round on their way to the Indies: North of the Cape *de bon Esperance*, at the latitude of 29 degrees South, are the mountains which the Portuguese call *Osmontes de Pedra*,[3] that is, Stone mountains, and there, nigh the Elephant River, far inland, our people have found large caves and overhanging rocks of the selfsame mountains, and in their vaults they found hanging, like icicles, as one will find in saltpeter cellars, suspended from narrow necks, various shapes of horses, horsemen, ostrich birds, lions, and many more, just as if they had been sweated out of the aforementioned stone, though they be hard and of a stony substance: The selfsame mountains were of red marble. Our people brought several of these little figures along, being the length of one finger, as hard as stone and smooth, and everyone can perceive that these figures were not purposely made by nature, since they have neither a uniform nor a fixed shape, but were laid bare either by being sweated out of the topmost rocks or because the cover of earth fell away; but the first is the most likely, because one also sees in *Amboina*, near the river *Batu gantong*[4] similar overhanging rocks and caves from which are suspended large teats, and others like them have grown up against them from below, representing the mouth of a large dragon, which keeps on growing every year; and which one can ascertain because all the names that were graven in it for several years now, have been outgrown. And so one will also find here and there on the beach or in the mountains large rocks which, because the soil around them has been washed away, are bare now and have the shape of a human being.

Add to this the rare figure of a person sitting on a chair, on a floor of pure stone, which, according to the Maccasarese, grew like that naturally, somewhere on a tall mountain[5] . . . [*sic*] which separates the Macassarese from the Buginese territory in the Straits of *Saleijer*;[6] and to which those People go in secret in order to transact that cursed thing called *Batappa*,[7] when they request something from the *Djing*[8] or Devil, such as luck in war, good fortune in gaining riches, whoring, and other such things; whereupon the Devil gives them an unknown little Stone, or teaches them some outward solemnities, which they are to accomplish.

Add here as well the stone Maria figure found in a cave of a mountain in *Chili*, where no Christians had lived before, when the Spaniards first arrived, although *P. Kircherus*[9] wants to sell this as a Godlike miracle [in] his underground World, Book . . . , Chapter . . . [*sic*].

In the year 1681 a Coral rock was hauled from the sea in Ambon Bay, which after some scraping and cutting by the Natives acquired the shape of a Woman with half a body, which had stood there with the head down and the belly up, divided as a Coralstone is wont into various branches and corners, as well as the head; chest, belly and sharp corners were misshapen; and since this figure was found exactly near the place called *Hative Kitsjil*,[10] where several hundred years ago a Javanese junk landed, of which the *Anachoda*[11] or Skipper with his wife or sister *Teyfilan* wanting to sail towards land, capsized the proa and *Teyfilan* drowned, so that she never saw the light of day again, and they show you in the sea around there a large Coralstone, which was her *Tudong*[12] or large hat: Now when this rock was brought to light, it had to be *Teyfilan*, and the more so because they spread [the story] among the people that, when they were pulling her out, a sighing had come from her: a great Resort of people came there, especially Heathens and Chinese, who are besotted by such figures, made by nature. The Natives had already hidden it in the forest, and would probably have made it into some idol: Because the *Moors* who now live below *Halong*[13] reckon that they descend from the aforementioned skipper; but I cleverly got that figure out of their hands by paying them 1 Rixdollar, and it makes a fine show now in my garden; and afterwards there grew various small plants and flowers from the selfsame body, because I had put seeds for them in the hollow little holes;[14] Shortly thereafter they brought me another figure of a misshapen man, also hauled from the Sea, but since I cared not to give much money for it, they did not bring me anything else.

CHAPTER 78

Melitites. Boatana.

A Slave found this stone in the mountains, in a brown, fat, clay-like soil beneath the root of the tree *Ubat Tuni*, or *Sesoot*,[1] the roots of which are used to make Saguweir bitter. According to him, it had its own particular shape, to wit eight-sided with four wide and four narrow sides, at both ends becoming narrow, in a square little plate, 2 inches long, in the center thick, brownish, and marbled with white, yellow and black spots and veins, reasonably heavy, but easy to scrape, shaped as straight as the small hammer of a Slater; he most likely lied, because one could clearly see that it had been shaped like that due to the scraping and cutting of a knife, and was perhaps found in nearly that shape: Scrapings from it are the dark yellow of Rhubarb; it oozes a white milk from both ends, in the center, it is yellowish, tastes sweet, and smells like *Terra sigillata;*[2] from which I conclude that it must be a *Melitites*[3] or Honeystone, although it may be a little too hard for that; and I found out afterwards that such Stones are found in both large and small lumps in the river *Waytommo:*[4] Some of them are so soft that they can hardly be called a Stone, but should be considered a *Marga indurata,*[5] of which more hereafter. This slave wanted to perform some strange tricks with this Stone; he wanted to make his master, a Dutch Captain *des Armes*, beholden to him by curing him of a long-standing sickness of the lungs; for which he needed water, fetched by night from a Moorish Temple, but after languishing for a long time, the master died: He said that the Stone was good luck for gardens, which makes some sense, because if one carries said Stone along, one will always have a ground fruit:[6] since the fruits, in Malay, are called *Boatana*, that is to say Ground fruit: He had smeared so much civet[7] on it that one could not get rid of it, notwithstanding that easily 10 years after the master's death, I have often put it in water.

Add here the *Galactites* or Milkstone, which is brought from the coast of *Coromandel,*[8] called *Lama de Costa* in Portuguese,[9] it is whitish, very soft and brittle, light and will stick to the lips: It is transported throughout the Indies to serve nursing women, since it increases the milk and cools inflamed eyes: It is sold, cut into small square pieces.

CHAPTER 79

Satschico. Alabastrites & Lichnites Chinensis.[1]

Satschico or *Satscheo*[2] is a kind of Chinese alabaster that has so many colors, that one can hardly deem all the pieces and variations to be stones.

The first and most common one is a yellowish white, in substance and hardness like true alabaster, not uniform in color, because one half of the stone is yellowish or a dirty white, while the other half is a pale red, where it is also just as hard; some are also a mixture of white and light red like wax, and if they are of a uniform color they are used to make signets for the common man because, as we mentioned before, the stone called *Joc* or *Onyx* can only be used by the Emperor:[3] They carve beautifully shaped cups and saucers from the pieces that are a mixture of white and a reddish yellow, which are considered part of the household goods, and which could be considered with good reason to be the *Myrrhina*[4] of the Ancients.

One also finds another stone by the same name in *Sina*, which we consider a kind of *Marmor* or *Jaspis*, but wrongly so in my opinion, since it has neither the hardness nor the shine of either. One will find framed pieces of it as square as a Slate, dark green with all kinds of veins running through each other, depicting some Landscapes, mountains and rivers: cut into thin tablets and framed, the Chinese hang them in their houses instead of paintings.

Boëtius is of the opinion in his *Gemmarium Lib. 2. Cap. 269.*[5] that there should be a difference between *Alabastrites* and *Alabastrum*, writing that the latter is a soft stone which one can easily cut, which does not hold water well, and which can be used as gypsum, as one can do with all alabaster, that is found in Germany, and therefore called *Spath.*[6] *Alabastrites* on the other hand is hard and not so easy to cut, does not soak up moisture, yet is still softer than marble, and therefore fit for making salve boxes. He contends that *Alabastrum* is an incipient stone, which only begins to harden when removed from its clay and then gradually acquires the hardness of *Alabastrites*, which in turn will become marble; but I do not find such a distinction in *Pliny* or the Ancients, who always consider *Alabastrum* a salve box, and *Alabastrites* the stone from which such a box is made;[7] I would therefore propose *Alabastrites* as a general name for all soft and variegated stones, which must yield to marble in hardness and shine. In the same place in *Boëtius* one will find another kind of *Alabastrites* that grows near the city of *Menz*,[8] and that in terms of the variety of colors and the pretty coursing of the veins resembles *Jaspis*, and is therefore considered an imperfect *Jaspis* by *Boëtius*, while one could also include here the aforesaid Chinese marbled Tablet stone, though I trust, that the learned men among the Chinese would distinguish between the marble Tablet stone and the first *Satschico*, since these are markedly different in hardness and designs.

A third kind of *Satschico*, was brought from *Sina* to *Amboina* in the year 1671. I consider it to be *Pliny*'s *Lychnites*,[9] which is supposed to be a kind of Alabaster or white marble, named such, as if one were saying Candle Stone, since it was hewn, perhaps even carved, in hollow mountains by the light of candles: They brought about 40 pieces of this Stone to *Amboina*, shaped like cups, saucers, and small beakers, nicely carved with dragons, flowers and leaves. The Stone itself had a softer substance than those kinds mentioned before, so that one could think it was gypsum; they have corners, also with bright yellow and red veins, some of which were like wide bands, while others ran together into a circle; some cups were mostly blue with purple spots, yellow and russet veins, in short, one could not find two among the 40 that were of the same color: Rubbed on a stone with water, they turned to a thick mush, like *Marga*,[10] and even a little pushing or pressing would cause them to break into pieces. They came from the Province *Quantung*,[11] where there is a mountain around the city of *Saukinhu*,[12] that is hollowed out and has an entrance to the side, where the diggers enter with burning candles, and the door is closed behind them, leaving only a square hole, through which they breathe. These miserable diggers descend to a great depth, where they finally hew this Stone, and, since it is as soft as chalk beneath the earth, they immediately shape and carve it the way they want and the Stone will permit, which is quite easy, for they can even carve the branches of Garlands, the bodies of dragons and other animals, like filigrane work that is almost transparent: They only beget their proper hardness when they are brought into the air and light of day, when they are also scraped smooth a little or polished. They also occur in the Province *Quancy:*[13] They pack them in rice chaff when they want to transport them. They are sold here at 1 Rixdollar apiece, while in *Sina*, as I was told, they cost no more than a *zesje*,[14] and sometimes even less if one buys them from the mountain people first, so that one should be amazed, that people will risk their lives for so little money, and go down that deep into the earth to get this Stone. The kind from which they make the carved cups, appears to have been smeared with some moisture or polished with it, because moisture that is poured into it will not stick to the sides; and it also seemed that there had been some small holes in them, which had been closed with pieces of the same Stone, and affixed there with some glue: The Stone is so soft, that one can scratch it with one's nail, which is why it is not highly esteemed there; but the really ripe *Satschico* cannot be scratched like that, and they use this to make their framed Grindstones, whereon they rub the Chinese ink when they want to write, being hollowed out in the center.[15]

It seems that the Chinese call all kinds of marble *Satschico;* but they only have four Provinces that can deliver it. 1. *Hokien*,[16] around the city of *Hocscheu*,[17] has the most and the best, mixed with yellow, white and light red, sometimes an ashen blue as well, all with black veins that are arranged into a circle or resemble little branches, and therefore called *Siusan* or *Susantchen*, that is, of many colors. 2. That coming from *Quantung*[18] also has many colors, and some of it is softer, like Alabaster, and this is a young or unripe stone, that is used for the carved cups: The Chinese make table tops from both of them, and it serves the great

Lords for floors, also for framed paintings, if they are covered with landscapes. 3. *Huquan*[19] provides yellow and partially transparent marble. 4. *Leautung*[20] has black and partially transparent marble: And all these kinds are known by the name of *Satschico* or *Satscheo.*

CHAPTER 80

Amianthus Ambonicus. Batu Rambu.

One will find on Ambon in *Leytimor*, a kind of *Amianthus*,[1] which our Apothecaries incorrectly call *Alumen plumosum*,[2] specifically in the valleys of the river *way-hau* or *Batou gantong*,[3] which we call the *Alf*, wherein one will find entire rocks of this bastard *Amianthus*, on the outside hard and a grayish black, but scaly, and if one removes the 2 or 3 outermost crusts, one will find the *Amianthus* as gray hairs, that stick to one another, and lie on top of each other with many sea-green scales, and as such it far more resembles *Talcum*,[4] which is what it is regarded as in *Europe;* It splinters easily, both into little hairs and into scales, which makes one's skin itch, the way ground *Amianthus*, or Chervil,[5] is wont to do. But if one grinds it carefully, one can get large pieces off it, which, if they come from the black-green and gray ones, shine beautifully, and they harden when exposed to air: But the Ambonese kind is usually gray, and it has a lot of green, and because of its scales it looks a great deal more like *Talcum.* The aforementioned river *Alf* and its child *Waynitu*[6] expel many of these pieces, about a finger long, and smaller, which one then finds along the entire *Uvitetto* Hook, which we call Gallows Hook,[7] where they lose their greenness after lying in the sun for a long time and turn from dark gray to light gray, and they acquire such a sticky or clinging power that, if they are held to wet lips, they will stick to them and well-nigh take the skin right off when one pulls them away; but if they are kept at home for a while, this attractive power gradually diminishes: One will also notice a little of this stickiness on those pieces, which were just chopped out of the rocks, if one puts them in the sun for several days; furthermore, they will maintain an earthy smell and taste, and if one rubs them with water on a stone, it turns into a white mush. The entire beach of *Waynitu*, all the way past old *Amahussa*,[8] has various rivers; also blackish rocks, which are crumbly and scaly, black-green and the color of lead, also resembling *Talcum*, and some Chinese mountain experts assured me that there was lead in them, which had to be extracted by means of other lead, but this has not been examined up till now. There are rocks like that in the Oliphant River,[9] which also produces other Stones that have green, brown and black marbling, and which one would take to be green marble, if they were not so crumbly. Though one does find some hard and solid pieces, from which they make small mortars and dishes.

I remember having seen similar rocks of bastard *Amianthus* and *Talcum* in Germany, in that part of the mountain *Melibocus*, which is in the Rheingau and is called the heights (*die Höhe*), around the age-old ancestral castle of *Catzenellebogen*, whose decayed walls are mostly made from this stone and which show gray age, and four miles from it is Castle Königstein;[10] but one thinks that all of these are a kind of Slate or Scalestone.

Ambonese *Amianthus* also has a great deal in common with *Lapis Cononorensis*, called in Portuguese *Pedra de Cananoor*,[11] though others think it is *Lapis Armenus*, which is greenish mixed with gray, though green dominates, it has long threads and splinters easily, it is harder than the Ambonese kind, and does not smell as earthy, but more like horn[12] and when rubbed with water, turns the same into a white color. Other pieces are dark green, producing large scales like the Ambonese kind, but they should be a little harder than *Talcum*, otherwise of the same smell and taste. This stone is transported throughout the Indies for medicinal use, and one has to pay in *Batavia* half its weight in silver; if you rub a little of it in water, which should be done gently so that it will not splinter, and then drink it down, it is found to be excellent for expelling smallpox, while it also extinguishes the heat of fevers by sweating, and expels urine[13] with force: Among the Ambonese kind we found such pieces that were very nearly like the Cananorese kind, and were used if the latter was lacking, particularly for cooling, although it is weaker. It is most likely a cold Stone, which is hewn from the rocks in the *Alf*, since it is the coldest river on *Amboina*, and the rocks are always in the water.

Cananoor has another Stone that does not differ much from the foregoing, with long threads, a greenish yellow with white or gray veins, also producing a white milk when it is rubbed with water, and is called *Pedrafria*, that is Cold stone, to distinguish it from the foregoing, and because it is mostly used to cool fever patients down, and against burning of the eyes; among them are pieces that are a blackish green with some yellow in them, without long threads but it does have broad scales, both, according to what the Malabars say, grow in the same mountain, in such a way that *Pedrafria* is the outside, but *Pedra Cananoor* the inside of the rock, which is likely, and which can also be seen with the Ambonese *Amianthus*.

If *Creagus* be not a special Stone, that has the reputation of pulling at flesh, because it sticks to one's lips, and seems to pull the flesh to it, and which other Authors place with kinds of Magnetic stones, then I still do not know any other Stone, which shows these attributes more clearly than our Ambonese *Amianthus*, after it has been for some time on the beach in the sun.

We again added here a picture of a piece of Amianthus *Mineral, shown with the lower part brown, then the white stripe, which runs through it, hard stone, and the uppermost vein, with little stripes slanting upwards, which is the stuff, of which one says is made inflammable linen; it handles and can be picked apart as if it were silk; it is depicted on plate* LVII. *by the letter* H.: *Those who desire to know more about this, read* Kircherus, Subterranean World, *the 2d Chapter in the 8th book.*[14]

CHAPTER 81
Lapis Tanassarensis.

They brought me a Stone from *Malacca*, which one would think was a *Haematites*[1] or Bloodstone at first, with long threads like a piece of moldered wood, though its color was a dark blue, with a few white stripes; it was found in *Tanassary*,[2] in the shape of a scale stone or Slate, and when rubbed with water it turns a light blue; it is undoubtedly wood from a certain tree which, after falling into the rivers there, turned to stone: Rubbed with water, it is taken as a good remedy against the Bloody flux; One finds it for sale in *Malacca*, where it was brought down from *Tanassari;* I have no other knowledge of it, since I only saw a piece about a hand long and 1 inch thick, which was easily cut through the center with a saw, and I kept one half. Compared with a true *Haematites*, one finds that this *Haematites* is heavier and harder, reddish at both ends, but where it is used (as the polishers do when they are burnishing),[3] smooth and dark purple, blue like burnished steel, but the *Tanassarine* is lighter with more and coarser threads striped lengthwise, a bluish gray like Slate, and their saps are also markedly different; which they exude when rubbed with water, and which is red for the *Haematites*, and bluish for the *Tanassarine*.

Although it was said it was for sale in *Malacca*, one should distinguish it from the *Lapis*

Malaccensis that was described by *Garcias Lib. 1. Aromat. Cap.* . . . [*sic*]: for that is merely the *Histrisitis* or Hedgehog stone, for which see Chapter LII.

CHAPTER 82

The Preacher's Stone.

One will find certain stones in the River *Waytommo* behind *Victoria* Castle in *Ambon*, that are the size of an egg or a fist, covered with bumps and holes on the outside, yet smooth, soft to the touch, of a fine substance; they are so soft that they are easily cut or ground, and when they are rubbed they produce a sticky paste, and some pieces are so beautifully veined and spotted that, if they were harder and shinier, one would think they were some precious marble or Serpentine stone;[1] but now one will find, that this Stone is only an incipient *Marga lapidescens:*[2] They do however differ among one another in terms of hardness, some are as hard as a common Stone, others are so soft and crumbly, that one could easily take a bite of it, and when they are polished they often fall into pieces. There is nothing beautiful about it on the outside, being usually covered with a mossy slime, and it will not show its colors and designs until it is polished; it is most often a russet color, gray and the color of liver, but polished ones will show all sorts of shapes, such as clouds and spots, that are green, ashen blue, red and black with many fine veins, wherein one can imagine all sorts of images, but not with all of them, or darkly so:[3] Some are almost always a green color like seaweed or seagrass, which one could easily hold to be *Pliny's Phycites.*[4]

One can distinguish them from other Stones, in that they are smooth as soap on the outside, when one rubs them a little against another Stone: They got their name from a certain Preacher, who was the first to find them in the *Waytommo* and made much of it, while before they lay there ignored, except for the softest pieces; those that are mixed with fine sand and that are soft, they crack and eat like *Batu puan* or *Terra sigillata:*[5] for they think that these Stones are a similar kind, and are indeed nothing more than a variegated *Marga* that gradually petrified: If they were harder and if they did not have that fault, that they crumble while they are being polished, one would be able to make handsome things from them, just like *Marmor Zeplicium*[6] or Serpentine stone, which is not much harder than the hardest ones of these Stones: The aforesaid Preacher knew how to polish them and make handsome angular knife handles and carved tablets from them, although many failed during polishing, and when they do not have handsome designs, spots or clouds, they are not worth the trouble.

CHAPTER 83

Coticula.[1] *Batu Udjy.*[2]

One also finds Touchstones[3] in *Amboina*, black, dark green and brown, of which only the black are true ones: The best are found between the villages of *Laricque* and *Waccasihu*[4] in a stony corner of the beach, and they are very black and smooth, but no bigger than a chicken egg, while most are like marbles: One will find them much bigger on other beaches of *Ambon* and *Keram*, but those are blue-black, and they show up gold paler than it is, which is better for the buyer than for the seller: The green and brown ones are only small pieces, which are seldom found, nor do they touch as well as the black ones. Goldsmiths are in the habit of grinding the large pieces flat like tablets or whetstones, but I found this not so good for the Ambonese ones, because, once robbed of their outer crust, they do not prove as well as before, which is why one should leave them as they naturally are. The distinction that some people make between the lower and upper sides of the stone, where it is hit by the sun and will therefore prove better, is not noticeable with the Ambonese ones, which lie on the beach and are constantly rolled over. One cannot tell the difference between the black thunder stone and the black Touchstone in terms of color and ability to test, except that the thunder stone has the shape of a blunt axe, small chisel, or finger, while its appearance is also smoother and shinier.

The Touchstones from the Coast are coarse black stones without a shine and jagged, encased in red copper, and they strike the gold on them rather hard, then press them on a lump of wax that was heated over coals, and when much gold stays behind on the lumps, they will melt them and in this way someone who tests a lot of gold can get a large amount of gold without much trouble: Upon closer examination it was discovered that this whetstone was made from Malabar Amaril;[5] see the previous Chapter XXXI.

CHAPTER 84

Cancri Lapidescentes.

Because Dr. *Andreas Cleijer*[1] sent me some crabs a number of years ago which had been brought from *Sina* and which, as he wrote me, were alive under water but now changed to stone, I wanted to put them here with the stones; because they are hard, dense stones, that kept the shape of an ordinary Crab, about the width of 2 fingers, except that I particularly noticed that the shield (*thorax*) had a thick growth on either side, like wings, though fastened to the shield, which is where the legs are, of a color like half-baked brick, though the legs are a dark brown, as is the little flap below the belly; They are very like a hard stone, though they are easily pulverized. *P. Martinus* in his Sines. Atlas. p. 138.[2] while describing *Kaochieu*[3] the 8th Capital of *Quantung*, says that they live on the bottom of the sea, between the city just mentioned and the Island *Hainam*,[4] as well as in a certain lake (*lacus*) on the same Island; and they are alive as long as they are under water, but when they come into the air, they are changed into a hard stone, retaining their former shape, although those that are transported always have some legs broken off. He calls them *lapide schentes Cancri:*[5] The Portuguese *Grangejo de Pedra.*

Our Merchants bring them from *Maccou* and *Canton:* I copied the following about their powers from a written (MS.) book.

1. Good for all swelling when ground with vinegar.
2. For bloody flux when ground with water.
3. For any kind of stool when ground with wine, but if that does not help, grind it in water and drink it.
4. If a Fever begins with some swelling, then grind it with water and apply it where it occurs.
5. For pain in the head ground with vinegar, also stroked on the temples.
6. For all shortness of breath or asthma, ground with water.
7. For some tumors that have embedded themselves in the back or shoulder,[6] ground with vinegar and applied thereon.
8. They are generally used to extinguish burning fevers, as P. *Martinus* also writes, ground with water and drunk.

I understood from a story by Mr. *Christiaan Gyrarts* whom I have mentioned before, that between the City *Boacheu* and the Island *Eynam*[7] they catch Seacrabs which, as soon as they are hauled from the water, change to stone and retain their shape; similar Crabs are also caught in a certain Lake on the Island *Eynam.* There are places there, such as the little Town *Gium*, where there is no ebb and flow, but the water flows from the one half Moon to the East, and from the other half Moon to the West; which the Chinese here confirm.

We previously made some remarks in the appendix to Chapter LXVI, *concerning the petrified Whelks and Shells, that are found in rocks or earth; and since this Chapter deals with petrified* Crabs, *we deemed it necessary to add something else, namely various kinds of Sea apples, also fishes and plants whereof we could show a whole series, if we were not constrained by the present project. It is against reason and experience to say that these too are stony growths; we gave our reasons before: Though we only add here the question, where did nature ever produce stones that had joints, shears, legs, backs, bellies, crusts and other characteristics, which all correspond to the same animals that are alive and that get their food from the water and on beaches? One can see that even more clearly with the internal parts, which (as we said before about the petrified Whelks and Shells) are found to be much softer, than the outer crust or horny skin, and that the Stone saps penetrate all the parts, and will petrify some of them entirely; or only the outer ones, by addition, encased by crusts of Stone. This is amply shown by the* Sea shrubs *and* trees, *which are kept in the Cabinets of the Collectors, and which come in all kinds of shapes and colors, and from which one can break the stony crusts, and in so doing restore the Shrubs completely to their former woody or natural state, the way they once grew, before they were encapsuled in stone: But we will no longer concern ourselves with writing about these things separately, for that would require an entire work; but we will continue to show those things which Mr.* Rumphius *has described, and whereof he did not send us any illustrations; therefore, as we did several times before, we relied once again on those people who own such curiosities; and in order to depict these* Ambonese Stone crabs *we were honored by the* Honorable Burgomaster Witzen,[8] *who permitted us to copy them; see plate* LX., *where* No. 1. *shows the top, and* No. 2. *the bottom:* No. 3. *shows another kind which I obtained from the West Indies: We own other ones but will not show them in order not to exceed the limitations of the plate, since more, for instance* Sea apples, *need to be shown, for they, as far as the inner fishparts are concerned, greatly resemble the* Crabs, *see the illustration on plate* LIX. *The one by the letter* C. *is the crust, as hard as stone, but the inner part is much softer and entirely white: This is also the case with the other kind, by the letter* D. *The one shown by the letter* E. *is as hard as flint, browner than the foregoing, and it came from a Mountain stone: The one by the letter* F. *is another kind, called the* Skull, *and is similar to the one by the letter* C. *Then follow Fish stones or petrified Fishes, whereof I encountered several kinds, but I could only lay my hands on these two; one is shown by the letter* G., *this one is encased in a dense black Stone, and the other by* H. *They are two fishes with all their bones, of a russet color, inside a white, but softer Stone than the previous one. I will bypass all the various kinds of Wood, Fruits, and other things that turned to stone; because the great variety would require a whole treatise in its own right; but in conclusion I would like to point out a small piece on plate* LVII. No. 1. *shows a small twig, with leaves and a small fruit, of the* sweet-brier, *that had entirely turned to stone, and which was given to me by* Doctor Scholts[9] *from Rotterdam, a true Lover of all such curiosities, His Honor owns an even larger piece, as well as several twigs and leaves, with the fruits still on them: The one shown here is only a small one.*

CHAPTER 85

Sal Ambonicus. Garam Ambon.

Although the Ambonese can get real and decent salt from the Europeans, they prefer to stick to their old habits of making a coarse and unseethed salt, called *Sassi*, which far more resembles hard, dark, and ash-gray stones than salt, except that it is always moist: They make it as follows.

They look for pieces of wood, that have been floating in seawater for a long time, and which have been entirely stripped of their bark, they split these into long pieces about 2 ells long, pile them into small heaps, and light a fire under them; they constantly sprinkle seawater on them while they are burning, just short of extinguishing the flame, and they keep doing this until they will not burn anymore: This burned wood falls apart into russet and rough Stones, very similar to *Lapis Calaminaris*,[1] whereon one can already see some salt glittering; then they take baskets woven from simple green leaves, put the burned lumps in them, and put them over some hollow vats, once again pouring seawater on them, which drips like a lye into the vats, and they do this until the lumps fall apart into soil and ashes: Then they place many shards or half pots, which are called *Ules* and *Lamam*, in such a way that one can easily stoke a fire under them. Then they pour the aforementioned lye into those shards and cook it until it has turned into a hard stone, which keeps the form of the potsherds like small cakes, whereafter they hang them in smoke. This is their common Salt which they use in all their food, particularly when they eat *Papeda*,[2] because it tastes better with it than our normal salt, although it smells noticeably of lye. One breaks the cakes into pieces the size of an egg, and rubs one on the food that is going to be eaten, until it is sufficiently salted: It is difficult to keep this salt without smoking it, because it will melt and make the place wet; nor is it any good for pickling, because it is not strong enough. This Salt is good for those who suffer from chin cough,[3] which is caused by some thick and sharp phlegm, which gets stuck in one's throat causing a nasty itching, now if one licks this salt, it will thin the phlegm and lessen the itching. The driftwood that is needed for this work is plentiful during the rainy season, when so many trees float in the sea, that they often cover entire beaches. Especially in the year 1664 so many heavy trees covered the beaches of *Ambon*, *Keram*, *Manipa* and *Buro*, which all faced to the East, that one could no longer recognize the beach; all of these were stripped of their bark, robbed of their branches, and made as smooth as if they had been scoured: No one is able to tell up till now where all this wood came from, since it has never been seen before or since; except that one knows for sure how in *Kelkeputi* bay or *Elipa Puti* on *Keram*'s southern coast, a large piece of land that had tall trees on it, sank beneath the water, and that in the same year *Keram*'s large rivers, swollen by tremendous rains, carried many pieces of land away, trees and all:[4] And the saltburners were well satisfied that time, they gathered all the wood, split it, and used it both for the kitchen as well as to make salt. This saltburning is a difficult and miserable job, because there is a tremendous amount of smoke, when one sprinkles seawater on it; although the split wood needs to be dried for some time beforehand in the sun: Since they cannot distinguish between this and the other driftwood, but use it all mixed up just as it comes, one can well imagine, that one kind of salt is better than another, and they must be particularly careful, that they do not use milkwood,[5] which has a hurtful and caustic milk, which produces a salt that is bad for the teeth; just as coral lime, when it has been slaked with this kind of wood, and sometimes used with the *Pinang* chaw, is known to soften the gums.

CHAPTER 86

Lapis Cordialis.

This is a stone made by the Portuguese in *Goa*, but that is now transported throughout the Indies because of its beneficent powers; I do not know of what it is made, except that from its appearance and taste, I would say its most important ingredients are, ground salt, Bezoar, Ambra and Musk. It has the size and shape of a pigeon egg or small chicken egg, seems gilded on the outside, while inside it is dark gray, glistening with small gold dots, soft to rub, and clearly smelling of *Muscus* and *Ambra;* I translated the following [information] about its uses and powers from a written Book (MS.) in Portuguese, that came from *Goa:*

This Stone is the best and most effective Cordial that has been discovered up till now, and there is no other like it: A Bezoar or other Cordial cannot compare to this; and he who examines it, will find that out for himself, nay, will experience even more than what I say or indicate here.

1. The amount one takes of this Stone, should be equal to the weight of 6 barley grains, but a little more or less will not matter.

2. When a person suffering from hot, burning fevers, is seized by a great thirst, he can be given a little of this Stone with water every hour, for not only will it not do any harm or injury, but it will quench the great burning and thirst, and cause the illness's harm not to reach the heart, while at the same time it strengthens and gladdens in a wondrous way.

3. If it were to happen that the sick person desires some wine, then one should give him some with this Stone in it.

4. Now if someone has no fever, but feels sad and melancholy, then he should take some of this Stone with wine, and he will immediately feel relieved.

5. If one is both feverish and melancholy at the same time, then one should take the same with some water, and if the fever is slight but the melancholy persistent, then one should take the same with some watered wine.

6. A Person who is recovering or healthy, yet feels some melancholy, should drink the same as just mentioned, because it will gladden and strengthen all the limbs of the body.

7. Taken as before, it is also effective against all sorts of poison, taken either by mouth or in some other way; even against the bites of the Spectacle Snakes, commonly called *Cobra Capelo.*

8. The same against Adder bites.

9. Also against the stings of Scorpions, and it immediately removes the pain and poison; though one should, after one has drunk some of this Stone, also put some of the powder on the bites.

10. If someone every morning, while still fasting, takes some of this Stone with water or wine, he can be assured that he will be preserved that day from all kinds of Venom, no matter if it should come from bites, or from something taken by mouth.

11. Taken with some water it serves to stop bleeding from the chest or nose; one should snuff up some of this Stone's dust, the way one takes Snuff, and drink the rest.

12. It is also efficacious against tumors, if taken with water.

13. It also preserves one's sight, if used as before.

14. If taken the same way, it will also protect against bad air.

15. Taken with wine it will protect someone against the four-day fever.

16. Used the same way, it gives one a good memory.

17. Used the same way, it will expel the three-day fever.

18. Used as before, it will keep one from getting the bad, contagious Leprosy.

19. Taken with water it will serve against smallpox.

20. Taken with wine or water, it will stir one's appetite, and quicken those who have been weakened by illness.

21. Taken with water it kills stomach worms.

22. Taken as before, it is good against sores that eat your flesh away as well as for a rash.

23. Taken with wine it clears the reins of gravel.

24. The same thing against bites from a mad dog.

25. If taken as before, it will also serve against any venom from arrow wounds or from any other poisonous weapon, and here too one should first give the wounded man something of the Stone to drink, whereafter one should sprinkle some of its dust on the wound.

26. Taken with wine, it is good against excrements, caused by cold.

27. With water, good against excrements caused by heat.

28. Taken with wine it serves to help those who cannot make water.

29. Taken with wine it serves as a laxative.

30. Similarly, taken with wine or water, it is good against the falling sickness or *Apoplexia*, and a person who is liable to get this illness, should ingest some of this Stone once, twice, or several times a day.

All the foregoing with the approval of the Inquisitors and of the representatives in the assembly of St. *Paulus*, the new order of Jesuits in *Goa* 1655.

They are sold in *Goa*, *Cutschyn*, and on *Ceylon* by weight, one such, the size of a duck's egg, fetches 12 *Pagodis* or 24 Rixdollars.

CHAPTER 87

Succinum Terrestre.

According to what the Chinese tell us, one can find *Succinum*[1] far inland, dug from the earth, mostly in the big pine forests in the Province of *Sukuen*[2] and immediate surroundings. The Chinese say that this Amber [*bernsteen*][3] is nothing more than the rosin of the aforementioned pine trees, which, after lying on the ground for around 1000 years, is changed into such a stone; it is darker and redder than the European *Succinum*, and therefore much baser, so that the European kind retains its high price in both *Sina* and *Japan;* the Chinese call it *Houpek;*[4] the Malay *Alambir:* I also understand that there is a *Succinum fossile* from *Pegu.*[5]

The Rosin, called *Dammar Selam*[6] in our herbal, after it has floated for a long time in seawater, comes to resemble Amber [*bernsteen*] so much, that one would think it such, especially when there are clear and russet pieces, coarse and whitish on the outside, clear and transparent on the inside: When one puts it on coals, it is the smell that immediately indicates the difference; *Dammar Selam* also jumps and crackles too much. On the Island of *Mauritius* they also find a certain Rosin on the beach, which those there hold to be Amber [*bernsteen*], but no one has written yet if it comes out of the trees or out of the sea.

About which see above, Chapter 35, about the *Ambra Grysea* in *Hubert Hugo*'s story.

END OF BOOK III

NOTES

INTRODUCTION

1. This is a letter to Christian Mentzel (or Menzel), dated 20 September 1680 and published first, in Latin, in *Miscellanea Curiosa sive Ephemeridum Medico-Physicarum. Academiae Naturae Curiosorum. Decuriae II. Annus Primus, anni 1682* (Norimbergae [Nuremberg]: Wolfgangi Mauritii Endteri, 1683), pp. 50–52, hereafter referred to as *MC*, with date of the proceedings, *not* date of printing. It was also published in German translation two years after Rumphius' death, in Michael Bernhard Valentini, *Natur- und Materialien-Kammer. Auch Ost-Indianische Send-Schreiben und Rapporten* (Frankfurt am Mäyn: Johann David Zunners, 1704), pp. 117–118.

Any bibliography or biography must start with Rouffaer's catalogue raisonné: G. P. Rouffaer and W. C. Muller, "Eerste proeve van een Rumphius-bibliographie," in *Rumphius Gedenkboek, 1702–1902* (Haarlem: Koloniaal Museum, 1902), pp. 165–220, hereafter referred to as "Rouffaer." It provides a scholar with an index of anything written on Rumphius prior to 1902. There was never much known about Rumphius' youth in Germany, but up to the end of the nineteenth century there were not many details about his life in the Indies either. Then a Dutch scholar consulted the Dutch East Indies Company archives; see P. A. Leupe, "Georgius Everardus Rumphius, Ambonsch Natuurkundige der zeventiende eeuw," in *Verhandelingen der Koninklijke Akademie van Wetenschappen*, (Amsterdam: C. G. van der Post, 1871), 12:1–63, hereafter cited as "Leupe." Leupe's article superseded all previous efforts where it concerned Rumphius' life on Ambon. Not much of biographical significance was added after Leupe and Rouffaer except for the following: Conrad Busken Huet, *Het land van Rembrand. Studies over de Noordnederlandse beschaving in de zeventiende eeuw* [1883–1884] (reprint ed. Amsterdam: Agon, 1987), pp. 592–595; M. J. Sirks, *Indisch Natuuronderzoek* (Amsterdam: Amsterdamsche Boek-en Steendrukkerij, 1915); M. Greshoff's article on Rumphius in the *Encyclopaedie van Nederlandsch-Indië* (The Hague: Nijhoff, 1919), 3:630–45; an article in German that emphasizes Rumphius' Germanity and that was published in Tokyo just before the outbreak of the Second World War: F. A. Schöppel, "Rumphius. Vortrag gehalten vor der Zweiggruppe Batavia am 24. Juli 1939," *Nachrichten Deutsche Gesellschaft für Natur- und Völkerkunde Ostasiens*, no. 52 (Tokyo, 1939), pp. 13–18; and, finally, G. Ballintijn, *Rumphius. De blinde ziener van Ambon* (Utrecht: W. de Haan, 1944). Ballintijn's popular biography is primarily useful for its reproductions of both complete and lengthy excerpts from Rumphius' letters. Its drawbacks are an irritating style and breathless suppositions.

Biographies of Rumphius in English are few. One of the earliest, after those mentioned by Rouffaer, is G. Sarton, "Rumphius. Plinius Indicus (1628–1702)," in *Isis. International Review Devoted to the History of Science and Civilization*, 27, no. 74 (August 1937), pp. 242–257. Another is M. J. Sirks, "Rumphius, The Blind Seer of Amboina," in *Science and Scientists in the Netherlands Indies*, ed. Pieter Honig and Frans Verdoorn (New York: Board for the Netherlands Indies, Surinam and Curaçao, 1945), pp. 295–308. This is a translation of Sirks' chapter on Rumphius in his *Indisch Natuuronderzoek*. An excellent outline of Rumphius' life by the editor of the second memorial volume, published more than half a century after the first one, can be found in H. C. D. de Wit, "Georgius Everhardus Rumphius," in *Rumphius Memorial Volume* (hereafter *M*), ed. H. C. D. de Wit (Baarn: Hollandia, 1959), pp. 1–26. De Wit also contributed a biographical sketch to lead off his contribution, "Orchids in Rumphius' Herbarium Amboinense," in *Orchid Biology*, ed. Joseph Arditti (Ithaca, N.Y.: Cornell University Press, 1977), pp. 49–56. Then there is my introduction to *The Poison Tree. Selected Writings of Rumphius on the Natural History of the Indies*, ed. and transl. E. M. Beekman (Amherst: University of

Massachusetts Press, 1981), pp. 1–40. The present effort, with its new information on the first third of his life, supersedes all previous biographies in terms of Rumphius' years in Germany.

2. It is entitled "Georgii Everhardi Rumphii Peregrinatio, sive iter in Brasiliam," is 218 lines long, and was printed inexplicably after the title page of the *sixth* volume of Rumphius' *Herbal*, which was printed in seven volumes between 1741 and 1755: *Het Amboinsche Kruid-boek*, volume 6 (Amsterdam & The Hague: Hermanus Uytwerf et al., 1750).

Life in Europe

3. There exist a number of variant spellings of his name. His last name was also spelled Rumpff, Rumph, and, of course, Rumphius, and was not unusual in the Wetterau region, where it might have referred to a measure of grain (see *Verslag der Rumphius Herdenking* [Amsterdam: J. H. de Bussy, 1903], p. 17) or to a kind of hod used by millers (with thanks to Mr. Steinl). Variations on his first name are Jörg, Jeuriaen, Juriaen, Jurriaan, and George, while his middle name was also spelled Eberhard, Eberhardt, Everhard, and the Latinized Everhardus.

4. The following is based primarily on Otto Renkhoff, *Nassauische Biographie*, 2d. rev. ed. (Wiesbaden: Historische Kommission für Nassau, 1992); and Karl-Heinz Schellenberg, *Braunfelser Chronik* (Braunfels: Magistrat der Stadt Braunfels, 1990).

5. Thousands of books have been written on this terrible conflict. I have made use only of the following texts. The Geoffrey Parker volume has an extensive bibliography. Gottfried Lammert, *Geschichte der Seuchen, Hungers-und Kriegsnoth zur zeit des Dreissigjährigen Krieges* [1890] (reprint ed. Wiesbaden: Dr. Martin Sändig oHG, 1971). This is a very informative chronology, based on contemporary sources, of the horrors of famine and the plague visited upon Germany. Herbert Langer, *Hortus Bellicus der Dreissigjährige Krieg. Eine Kulturgeschichte.* (Leipzig: Edition Leipzig, 1978). Geoffrey Parker, *The Thirty Years' War* (London: Routledge & Kegan Paul, 1984). C. V. Wedgwood, *The Thirty Years War* (New Haven: Yale University Press, 1939). Golo Mann, *Wallenstein* [1971], transl. Charles Kessler (New York: Holt, Rinehart and Winston, 1976). The English translation of Golo Mann's masterpiece lacks the scholarly apparatus, hence has no bibliography to speak of, and is sometimes unreadable. It is an important book, and I agree wholeheartedly with Mann when he states in the preface to the English version: "Marx, I am convinced, was in error. There is no single secret gist, nor clue to, historical phenomena. Words and deeds show what men are and how events take shape. Narration and elucidation synchronize." This position was anticipated by Plutarch and Samuel Johnson.

6. There are many editions of Grimmelshausen. I refer to a facsimile of the first edition: *Grimmelshausens Simplicissimus Teutsch*, ed. J. H. Scholte (Tübingen: Max Niemeyer Verlag, 1954). I quote from an American translation which I emended when necessary: Johan Jakob Christoffel von Grimmelshausen, *Simplicius Simplicissimus*, transl. George Schulz-Behrend (Indianapolis: Bobbs-Merrill, 1965).

7. Those are the very last words of the chapter (in the Scholte ed., p. 269): "Biß du mit deinen Beweißthumen fertig bist / so bin ich vielleicht wo der Pfeffer wächst."

8. From ch. 19 to ch. 34 in book one and from ch. 1 through ch. 14 in book two.

9. R. Wille, *Hanau im dreissigjährigen Kriege* (Hanau: G. M. Alberti, 1886), p. 465; see also p. 467.

10. "Als ich dieses hörte / sahe ich ferner stillschweigend zu / wie man Speiß und Trank muthwillig verderbte / unangesehen der arme Lazarus / den man damit hätte laben können / in Gestalt vieler 100. vertriebener Wetterauer/ denen der Hunger zu den Augen herauß guckte / vor unsern Thüren verschmachtete / weil naut im Schanck war" (Scholte ed., p. 84).

11. ". . . deßwegen hielte er auch das versprochen Quartier sehr ehrlich und auff Holländisch / deren Gebrauch ist / ihren gefangenen Spanischen Feinden von dem jenigen / was der Gürtel beschleust / nichts zu nemmen . . ." (Scholte ed., p. 249).

12. Mann, *Wallenstein*, p. 33.

13. In ch. 1 of book five (Schulz-Behrend transl., p. 265).

14. P. 268 in the Scholte ed. (my translation; the American translation is wrong).

15. Historians argue about the losses, cold comfort to the dead. Wedgwood estimates a loss of one-third of the population (p. 516), Parkers settles for "about 15 to 20 percent" (p. 211), and Langer proposes losses by region, stating that the region of Rumphius' youth sustained a population loss of between 33 and 66 percent.

16. Emended version of Schulz-Behrend, p. 31; in the Scholte ed., p. 44. The entire allegory can be found in chs. 15–18 of book one.

17. Rumphius identifies August Rumpf as his father in the letter to Mentzel, referred to in n. 1. For his dates, see below.

18. Information provided in a letter to me by Arno W. Fitzler, dated Braunfels, May 1996.

19. Fürst zu Solms-Braunfels'ches Archiv, Braunfels [hereafter FSBA], A.23.6, II.57.4. I am indebted to Gerhard Steinl in Hungen for information about these early years.

20. FSBA, A.96.5, I.197.a. This important passage, transcribed for me by Mr. Steinl, reads: "Bitte deroent Eu(wer) G(naden) ganz un(er)thenig umb gottes willen sie wolten ja nicht allein meiner Sond(er)n meiner haußfrawen und deren leibesfrucht zu deren entbündtnus sie uber 7 oder 8 woch(en) uffs lengst nicht mehr hat, zu gnaden erbarmen." This translates approximately as: "For that reason I humbly beg Your Grace, to be merciful not only to me, but also to my wife and her [unborn] child, for she has no more than 7 or 8 weeks before it will be born."

21. See R. Wille, *Hanau im dreissigjährigen Kriege* (Hanau: G. M. Alberti, 1886), 1:8.

22. The document was printed in Eugen Rieß, *250 Jahre Evangelisch-Reformierte Kirche Wölfersheim* (Wölfersheim: Kirchengemeinde Wölfersheim, 1991), pp. 63–64. As Mr. Steinl pointed out, the printed text has two crucial mistakes. On p. 63, third line from the bottom, "*Vatter* Johan Oth Keller" should read "*Vetter* Johan Otth Keller," and in the second line from the bottom, "*sin tochterman*" should read "*ein tochterman.*"

23. "Tochterman" in the original. Another piece of evidence which proves that the Kellers were his mother's family is part of a sentence from a report Rumphius wrote sometime between 1684 and 1688 at the request of Anthonio Hurdt. This wonderful text is otherwise not pertinent here but, the fragment reads: "'t welk men in Europa ligtelijk kan vernemen bij onse residenten in Muskou met name mijn Neef, den Heer Baron JOHAN KELLER . . ." ("about which one can easily inquire in Europe from our residents in Moscow namely my Cousin, Baron JOHAN KELLER . . ."). The manuscript (not, of course, in Rumphius' handwriting) is in the Royal Library in The Hague: KB, Collectio Cuperus. no. 16. 42 C 14. The passage in question is on folio 24. With thanks to Wim Buijze. It was printed once as: "Antwoord en Rapport op eenige pointen uit name van zeker heer in 't Vaderland, voorgesteld door den Edelen Heer Anthony Hurt, Director-Generaal over Nederlandsch Indië en beantwoord door GEORGIUS EVERARDUS RUMPHIUS, oud Koopman en Raadspersoon in Amboina," in *Tijdschrift voor Staathuishoudkunde en Statistiek*, Tweede Serie, Eerste Deel (1856), 13:137. Keller was Holland's "*resident*" in Moscow from 1677 to 1698.

24. People in Kleve spoke Dutch well into the twentieth century. Culturally it also looked to Holland: students from Kleve preferred to study at Leiden University, for example. See Georg Cornelissen, *Das Niederländische im Preußischen Gelderland und Seine Ablösung durch das Deutsche* (Bonn: Rheinisches Archiv, 1986), pp. 40–41, esp. 41 n. 17. With thanks to Hans Heestermans.

25. Wölfersheimer Kirchenbuch 1651, p. 323. With thanks to Gerhard Steinl.

26. Taufbuch der reformierten Kirche Hanau 1645, p. 8. See Ingrid Krupp, *Das Renaissanceschloß Hadamar* (Wiesbaden: Historischen Kommission für Nassau, 1986), p. 118.

27. "Reformed" is *Reformiert* in German and was at the time another term for Calvinism. Calvinists were called Huguenots in France and Presbyterians in England and Scotland. Like the Lutherans, Reformed Church members concentrated on God's word in the Bible, but to insure total attention, they banned all decorative elements from their churches in order to minimize interference with the study of the Holy Writ. A visual equivalent to the spirit of orthodox Protestantism can be found in the wondrous paintings by Pieter Saenredam (1597–1665) of church interiors.

28. Ernst J. Zimmermann, *Hanau. Stadt und Land, Kulturgeschichte einer Fränkisch-Wetterauischen Stadt* (Hanau, 1917), p. 745; his birth year of 1639 is from Heeres, "Rumphius' Levensloop," in *G*, p. 1. Heeres quotes a source, Mr. Begeer, from Hanau, who reports that the woman we know to have been August Rumpf's wife, "Anne Elizabeth," gave birth to children in 1639 and 1645. 1645 was the birth year of the daughter, Anna Catherina; hence 1639 must have been the year of the younger son's birth. For his name, see n. 73 below.

29. Rieß, *250 Jahre*, pp. 128–29, 97. He might even have been taught by a maternal cousin, Johann Otho Keller; see Rieß, p. 129.

30. *Hanauer Geschichtsblätter*, Neue Folge, nos. 3–4 (Hanau, 1919), p. 168. We know that August taught the count's sons "geometric drawing" and architectural drawing from a note dated 17 January 1633 (FSBA, A.23.6, II.57.4).

31. Rieß, *250 Jahre*, p. 80.

32. Mann, *Wallenstein*, p. 200.

33. Schellenberg, *Braunfelser Chronik*, p. 38.

34. Ibid., p. 42.

35. In a letter from the Indies, dated 20 August 1663, printed in Leupe. See n. 1.

36. Krupp, *Hadamar*, p. 118.

37. Heinrich Heusohn, "Eine Denkmünze auf den Naturforscher Georg Eberhard Rumphius," *Frankfurter Münzzeitung*, 3, no. 26 (February 1903), p. 392.

38. See, e.g., Krupp, *Hadamar*, p. 188; K. P. Decker, of the archives of the Kulturgut Fürst zu

Ysenburg und Büdingen, in a letter to me dated 28 July 1995; Arno Fitzler in the letter cited in n. 18. For Rumpf's work on his son's school, see *Geschichte des Staatlichen Gymnasiums zu Hanau* (Hanau, 1925), p. 49.

39. Helmut Winter, *Festschrift zur 375-Jahr-Feier der Hohen Landesschule Hanau (1607–1982)* (Hanau, 1982), p. 15.

40. Karl Dielmann, *Hanau am Main* (Hanau: Kuwe-Verlag, 1971), p. 17.

41. She was the third daughter of William of Orange and his third wife, Charlotte of Bourbon (1546–1582). She was given the name for political reasons and was joined by her three siblings, named Brabantia (after Brabant), Flandrina (Flanders), and Antwerpiana!

42. Winter, *Festschrift zur 375-Jahr-Feier*, p. 21; see also pp. 17–21. Herborn was reserved for the aristocracy. Rumphius never attended the school. See *Die Matrikel der Hohen Schule und des Paedagogiums zu Herborn*, eds. Gottfried Zedler and Hans Sommer (Wiesbaden: J. F. Bergmann, 1908).

43. For Rumphius' autobiographical poem, in Latin, of 218 lines, see n. 2.

44. FSBA, A.26.8, III.163d. With thanks to Mr. Arno Fitzler, who provided me with the few known facts about Ernst Casimir.

45. Recorded in the diary of Count Wilhelm II, FSBA, A.27.5, III.200f/g.

46. In the original: "Littore, non dubium, nos deposuisset eodem, quo socios reliquae, tertia nostra ratis." This literally means: "Our third ship was going to deposit us on that same shore, not far from where it had left our allies [or comrades] before."

47. A good general history of this company is Henk den Heijer, *De geschiedenis van de WIC* (Zutphen: Walburg Pers, 1994).

48. Ibid., p. 50. Excellent contemporary accounts are: Johannes de Laet, *Historie ofte Iaerlyck Verhael van de West-Indische Compagnie* (Leiden, 1644); "Het dagboek van Hendrik Haecxs, Lid van den Hoogen Raad van Brazilië (1645–1654)," ed. S. P. L'Honoré Naber, in *Bijdragen en Mededelingen van het Historisch Genootschap*, vol. 46 (Utrecht: Kemink & Zoon, 1925); Caspar Barlaeus, *Rerum per octennium in Brasilia gestarum historia* [Amsterdam, 1647], transl. and ed. by S. P. L'Honoré Naber (The Hague: Martinus Nijhoff, 1923); Joan Nieuhof, *Gedenkwaardige Zee en Landreize door de voornaamste landschappen van West en Oost-Indien* (Amsterdam: de weduwe van Iacob van Meurs, 1682). Another good history is P. M. Netscher, *Les Hollandais au Brésil. Notice Historique sur Les Pays-Bas et Le Brésil au XVIIe Siècle* (La Haye: Belifante Frères, 1853). In English there is C. R. Boxer, *The Dutch in Brazil, 1624–1654* (Oxford: Clarendon Press, 1957).

49. Heijer, *WIC*, 50; "Het dagboek van Hendrik Haecxs," p. 137.

50. Pierre Moreau, *Histoire des derniers troubles du Brésil entre les Hollandais et les Portugais depuis l'an 1644 jusques en 1648* (Paris: Chez Augustin Courbe, 1651), p. 103. With thanks to Wim Buijze.

51. Six months was an extraordinarily long time. Several years later it took Rumphius six months to get to Indonesia! Moreau, p. 105.

52. Moreau nicely calls these miseries "les inconstances outrageuses de la mer," p. 105.

53. Moreau states, "la plus fascheuse saison de l'année pour nauiger," p. 105.

54. Ibid., p. 112.

55. Ibid., p. 116. Moreau writes: "Cap de Fineterre, le long des costes de Portugal, puis dix à douze lieuës vis à vis la ville de Lisbonne, & en apres proche les grandes & hautes roches qui paroissent en mer, & qu'on appelèe les coches [roches] de Barrolles." The latter rocks are groups of islands, called Barlengas, very close to the Portuguese coast. One *lieu* was the equivalent of 7.4 kilometers; hence they sailed in view of Lisbon for about 74–88 kilometers or between 45 and 54 miles.

56. In the present text he makes reference to a volume he could only have read in Portuguese; see n. 7 of ch. 35 in book 3. And in ch. 19 of book 2, he speaks of a certain shell (described under no. XXV) that had a particular use in Portugal, with the tone of personal observation. One could say the same thing of the shells discussed under no. V in ch. 31 of book 2.

57. See n. 1.

58. Valentini, *Natur-und Materialien-Kammer*, pp. 52–53.

59. Portugal had six provinces at the time: Algarve, Alemtejo, Entre Douro e Minho, Tras os Montes, Estremadura, and Beyra. During this time Portugal and Spain did not engage in any major battles. Neither nation could afford large military operations; hence both had to be satisfied with patrols, raids, and skirmishes. This would explain why Rumphius never mentions a major engagement. See H. Schäfer, *Geschichte von Portugal*, 5 vols. (Hamburg: Friedrich Perthes, 1852), 4:548, 556–557.

60. In the original: "De zelfde manier van heiningen, dog van een ander gewas, heb ik gezien in 't Zuiderdeel van Portugal, dat men *Algarbia* noemt, en verstont van de Portugezen, datze het geleert hadden van de Brasilianen." *Het Amboinsche Kruid-boek*, 6:89. "Algarbia" is Algarve, what the Moors called "El-Gharb," a region once famous for its orchards, now for high-rise hotels and tourism.

61. Ibid.

62. Bk. 11, ch. 55 of *Kruid-boek*, 6:179–80. "Minius" is the old Latin name for the river commonly known as the Minho. The Minho River forms Portugal's northernmost boundary with Spain, and Minho is also the popular name for this northern province, officially called Entre Douro e Minho. Hence this was clearly a border region.

63. Valentini, *Natur-und Materialien-Kammer*, p. 53. The original passage reads: "Dieses aber weiß ich wohl, das ich das rechte *Sium* in Portugal habe wachsen sehen, und zwar allezeit an feuchten wässerichten Oerthern, also, daß das Würßelgen selbst im Wasser gestanden. Sie nenneten es allda *Peroxil de Agoa*, das ist, Wasser-Petersilien, und ist in solchem Werth gehalten, daß ein jedweder Plaß, da es grünet, seinen eigenen Besißer hat, und also nicht einem jeden frey stehet solches abzupflücken; wie wolen wir als Soldaten dasselbige zuweilen zu Salat und Mues abgeknippet haben, ohnerachtet der Schleudersteine, die wir öffters an die Köpffe bekommen haben."

64. Christel Lentz, *Das Idsteiner Schloß. Beiträge zu 300 Jahren Bau-und Kulturgeschichte* (Idstein: Schulz-Kirchner Verlag, 1994), p. 52.

65. See Renkhoff, *Nassauische Biographie*, p. 556. Renkhoff states that the count hired "flämischer Wallonen," but that is impossible because they don't exist. He means either Protestant Walloons or Flemish workmen.

66. Christel Lentz, "Der Lustgarten des Grafen Johannes von Nassau-Idstein und sein Blumenbuch," in *Hessische Heimat*, 45, no. 3 (1995), p. 84. I am indebted to Christel Lentz for relevant information concerning Rumphius' stay in Idstein.

67. Grimm's *Deutsches Wörterbuch* defines *Bauschreiber* as "*rechnungsführer bei bauten.*"

68. See Hessisches Hauptstaatsarchiv Wiesbaden (hereafter HHStAW), Abteilung 133, Stad Idstein, 11 w, p. 3.

69. See Christel Lentz, "Über die Idsteiner Vergangenheit des Naturforschers Georgius Everhardus Rumphius" in *Heimatbuch für den Rheingau-Taunuskreis* (1991), 42:76–77.

70. Rumphius' stay in Idstein provides us with the only known address of his youth in Germany. He stayed in what was called "Gasthaus zum Hirschen," now 1 Obergasse. We know this from a bill, dated 16 April 1651, wherein Rumphius mentions his landlord, Gottfried Meynhardt. This man ran the Gasthaus. See HHStAW, Abteilung 131, R 200, no. 64. August also stayed there (HHStAW, Abteilung 133 IX, "Amtsprotokolle," dated 10 October 1653), as well as the "Baumeister's son from Hanau," an engineer by the name of Johann Conrad Rumpf (HHStAW, Abteilung 131, R 38, no. 369, dated 28 June 1660).

71. Lentz, *Das Idsteiner Schloß*, p. 49.

72. The count wrote in February 1652 to Johannes Walter, an artist he employed, that "unser großer Bauschreiber ist nicht mehr bei Uns in Dienste. . . . Hat nicht leiden können, daß man wider die Calvinisten predigt." The letter is in the HHStAW, Abteilung 133, Stadt Idstein, 11 w, p. 3.

73. Letter from August to Count Johannes, HHStAW, Ab. 133, Idstein, 9, p. 65. On August's calling his son back home, see HHStAW, Abteilung 133, Idstein, 9, pp. 63–64.

74. He signed his last bill connected with work in Idstein on 15 November 1651.

75. HHStAW, Abt. 133, Idstein, 9, pp. 67–68.

76. Ibid. My translation makes August's German sound better than it was. His style was poor, ungrammatical, without correct punctuation, and was written in a hand that often cannot be deciphered.

77. Letter from Rumphius to the VOC in 1663; printed in Leupe, pp. 40–43.

78. Krupp, *Hadamar*, p. 118.

79. Entry in *Wallonisches Totenbuch, Hanau*, 1666.

80. He mentions this in the same letter as the one in n. 77; see Leupe, p. 41.

81. Ibid.

82. See *Dutch-Asiatic Shipping in the 17th and 18th Centuries*, ed. J. R. Bruyn et al., 3 vols. (The Hague: Martinus Nijhoff, 1979), 2:110–111.

Life in The Indies

The following is a short bibliography pertaining to this period, with the emphasis on works in English. The standard text on the history of the erstwhile Dutch East Indies is F. W. Stapel et al., *Geschiedenis van Nederlandsch Indië*, 5 vols. (Amsterdam: Joost van den Vondel, 1938–1940), especially vol. 3, for the seventeenth century, written by Stapel. This is still the major source, because it is well written, has a great deal of information based on such original sources as the secret documents of the company's *Corpus Diplomaticum*, and labors to achieve a balanced view of its subject. It also has a fine series of reproductions of contemporary scenes and documents. Another valuable source is the various articles in the encyclopedia of the Indies, *Encyclopaedie van Nederlandsch-Indië*, ed. D. G. Stubbe, 4 vols., 3 supps. (The Hague: Martinus Nijhoff, 1917–1932).

In English one can consult the reprint of E. S. De Klerck, *History of the Netherlands East Indies*, 2 vols. [1938] (reprint ed. Amsterdam: B. M. Israël, 1975); for the present period the relevant pages can be found in vol. 1, ch. 9–11, esp. pp. 255–259 on De Vlamingh and Ambon, pp. 275–277 on Speelman and Macassar. What is in my opinion still the best single-volume history in English, with excellent chapters on the period, is Bernard H. M. Vlekke, *Nusantara. A History of the East Indies Archipelago* (Cambridge, Mass.: Harvard University Press, 1945), ch. 5–8. Then there is the vigorously written *History of Java* by Thomas Stamford Raffles [1830] (reprint. Oxford: Oxford University Press, 1979). One should remember that Raffles, vehemently anti-Dutch, had a personal grudge against his colonial rivals, that he was inordinately ambitious, and that he was foiled in his attempt to achieve lasting historical glory. Hence he presented his own efforts in a glowing light and painted a bleak picture of his Dutch predecessors. Ironically, his politics most resemble those of Daendels, who was far from a hero to nineteenth-century liberals. The same may be said for the *History of the East Indian Archipelago*, in 3 vols. (Edinburgh, 1820), by John Crawfurd, who served under Raffles. It also is a biased account. This is the same author of another work which constitutes for all practical purposes the first encyclopedia of the Indies: *A Descriptive Dictionary of the Indian Islands & Adjacent Countries* (London, 1856). Raffles (and by implication Crawfurd) is still presented as the great liberator in a fine study by C. E. Carrington, *The British Overseas: Exploits of a Nation of Shopkeepers* (Cambridge: Cambridge University Press, 1950), pp. 408–411; yet this same text has a very perceptive evaluation of Kipling, correctly noting that this great writer was "the very opposite of [being] the poet of orthodox, conservative imperialism . . . [but rather] the poet of the frontier-rebel, the filibuster, the buccaneer," pp. 674–675.

Another critical account of the seventeenth century is George Masselman, *The Cradle of Colonialism* (New Haven: Yale University Press, 1963). It is an excellent study, based on sources similar to Stapel's and Vlekke's, and has an extensive bibliography. It deals primarily with the earlier years of the company, ending with the death of Coen. As a counterbalance, one should mention praise for the Dutch system from a most unlikely source: A. R. Wallace, *The Malay Archipelago* [1869] (reprint ed. New York: Dover, 1962). Although Wallace concerns himself primarily with botany and biology, he expresses very strong admiration for the Dutch colonial system as compared with that of his own nation. It also contains a brusque dismissal of Multatuli's *Max Havelaar*, judging it to overstate its case in a disagreeable manner. Fulsome praise of the Dutch regime can be found on pp. 72, 97, 187, 194–197, 222–223. One should add that Wallace's book is useful for readers who do not know Dutch in that it is also a kind of travel companion which describes most of the islands mentioned in the present context. His work is not, however, the result of a quick ramble through the islands, for, altogether, Wallace spent eight years in the archipelago.

Typical of the critical works by Dutchmen in the nineteenth century is J. A. van der Chijs, *De Vestiging van het Nederlandsche Gezag over de Banda-eilanden (1599–1621)* (The Hague: Martinus Nijhoff, 1886). As one can tell from the title, it covers a period before Rumphius' time. It is a useful source, since Chijs based his exposition on official documents. The work deals with the devastation of the Banda Islands, concentrating on Coen's actions there. Chijs' thesis is that the Company "systematically destroyed the Bandanese, in order for its rule to be established and remain in the country of that unfortunate nation" (p. 168). The documents Chijs cites are damaging enough, especially if one reads the amiable introduction to each other of two peoples who were soon at war (pp. 1–13, covering the year 1599). Though earlier—even prior to Multatuli—M. Dassen, *De Nederlanders in de Molukken* (Utrecht: Van Heijninger, 1848), is a work much less useful than that of Chijs because its sources are either paltry or absent. Most of it seems to be based on Valentijn. Dassen deals almost exclusively with the Ambonese wars and De Vlamingh's role in them and wrote what is more a polemical diatribe than a work of scholarship.

A modern history, in English, is Leonard Y. Andaya, *The World of Maluku. Eastern Indonesia in the Early Modern Period* (Honolulu: University of Hawaii Press, 1993). By the same author is a study of one of the main actors in the wars with Makassar: *The Heritage of Arung Palakka: A History of South Sulawesi (Celebes) in the Seventeenth Century* (The Hague: Martinus Nijhoff, 1981). Then there is also H. J. de Graaf's *De geschiedenis van Ambon en de Zuid-Molukken* (Franeker: T. Wever B. V., 1977). A large part of the book is devoted to the Ambonese wars, but an even greater portion is devoted to the missionary efforts of the Dutch. No sources are given, and many passages are perfunctory, for instance, the portrait of Rumphius. De Graaf devotes several pages (pp. 157–163) to Captain Jonker, a full-blood Ambonese who fought for the Company and took part in Speelman's campaign against Macassar. J. A. van der Chijs wrote a biography of this man which is far from objective: *Tijdschrift van het Bataviaasch Genootschap*, vol. 28 (1883), pp. 351–473, and vol. 30 (1885), pp. 1–234. A more sober relation of facts is given by F. De Haan, *Priangan*, 1:228–31. *Priangan* is an essential and inestimable work for the first two centuries of Dutch presence in the archipelago. It contains an immense amount of firsthand material, not only for Java but for most other aspects of the colonial East Indies as well. This huge work is arranged as follows: the first volume presents nearly 500 pages of historical narration of events, followed by nearly 300 pages of additional material in the same volume, and which is then continued in the next three volumes of addenda and annotations, covering altogether nearly 3,000 pages. De Haan is invaluable for providing a necessary balance to the biased view of his nineteenth-century colleagues.

Levinus Bors published in 1663 a sycophantic biography, *Amboinse oorlogen, door Arnold de Vlaming van Oudshoorn als superintendent, over d'Oosterse gewesten oorlogaftig ten eind gebracht* (Delff: Arnold Bon, 1663). Bors was De Vlamingh's private secretary, had firsthand experience of many incidents, and wrote a fine prose. After its first appearance in a journal, a biography of Speelman appeared separately: F. W. Stapels, *Cornelis Janszoon Speelman.*

Rumphius' own history of Ambon is still one of the best, though to be sure, he is partial to the Dutch. It is written in a sober prose, with occasional passages of wit and verve, and is organized around the basic metaphor of illness. His narrative begins in 1599 and ends in 1664; hence for ten of those years Rumphius was himself present in Ambon. It appeared only once in that most valuable journal published by the Ethnological Institute of Leyden: Georgius Everhardus Rumphius, "De Ambonsche Historie," in *Bijdragen tot de Taal-, Land- en Volkenkunde van Nederlandsch-Indië.*

1. Rumphius, *Het Amboinsche Kruidboek* (also spelled *Kruyd-Boek*), 1:1.

2. Fifth page of the unpaginated preface (*voorrede*) of the first volume of the *Herbal*, or *Kruidboek.*

3. F. de Haan, *Oud Batavia*, 2 vols. (Batavia: G. Kolff, 1922), 1:100.

4. This is just one indication how vast Indonesia is; superimposed on a map of the North American continent, the archipelago stretches across the entire width of the continental United States, from the Pacific to the Atlantic.

5. De Haan, *Oud Batavia*, 1:207.

6. See *De avonturen van een VOC-soldaat. Het dagboek van Carolus Van der Haeghe 1699–1705*, ed. Jan Parmentier and Ruurdje Laarhoven (Zutphen: Walburg Pers, 1994), p. 80.

7. De Haan, *Oud Batavia*, 1:145.

8. Soldiers mostly stood guard and saluted superiors; see De Haan, *Oud Batavia*, 1:209.

9. Ibid., 1:124. For Germans in the East Indies, see Roelof van Gelder, *Het Oostindisch avontuur. Duitsers in dienst van de VOC [1600–1800]* (Nijmegen: SUN, 1997); Also Peter Kirsch, *Die Reise nach Batavia. Deutsche Abenteurer in Ostindien 1609 bis 1695* (Hamburg: Kabel, 1994).

10. De Haan, *Oud Batavia*, 1:193.

11. Marville's private letter to Batavia, 28 April 1666. See also Rumphius' own words in his preface to *Amboinsche Kruidboek*, vol. 5. Hereafter "Batavia" is used to refer to the colonial government on Java, and "Gov. & Council" for the governor and council in Ambon. Unless otherwise noted, these documents were all published in Leupe's article.

12. Letter to Mentzel, 20 September 1680, printed in *MC* [see "Works" section], 1683, p. 51.

13. De Haan, *Oud Batavia*, 1:204.

14. Rumphius, *Kruidboek*, 1:3–4.

15. From a letter the missionary sent to Lisbon, quoted in Bernard H. M. Vlekke, *Nusantara. A History of the East Indian Archipelago* (Cambridge, Mass.: Harvard University Press, 1943), 81.

16. Vlekke, *Nusantara*, pp. 72 ff.

17. G. E. Rumphius, *De Ambonsche Historie*, in *Bijdragen tot de Taal-, Land- en Volkenkunde van Nederlandsch-Indië*, zevende volgreeks, tiende deel [1910], p. 232.

18. Rumphius, *Ambonsche Historie*, pp. 198–199.

19. H. J. de Graaf, *De geschiedenis van Ambon en de Zuid-Molukken* (Franeker: Wever, 1977), p. 108.

20. Ibid., p. 120.

21. M. Dassen, *De Nederlanders in de Molukken* (Utrecht: W. H. Van Heijningen, 1848), pp. 124–125.

22. Rumphius, *Ambonsche Historie*, pp. 192–193.

23. F. W. Stapel et al., *Geschiedenis van Nederlandsch Indië*, 5 vols. (Amsterdam: Joost van den Vondel, 1938–1940), 3:341–343.

24. Vlekke, *Nusantara*, p. 151.

25. Rumphius, *Ambonsche Historie*, p. 193.

26. Stapel, *Geschiedenis*, 3:346.

27. Vlekke, *Nusantara*, pp. 149–50.

28. See, for instance, his remarks at the end of his description of the "Macassarese Poison Tree," which is ch. 45 in the third book of his *Herbal*, or *Amboinsche Kruidboek*, 2:267–268.

29. Rumphius himself provides us with this date when he writes: "In Amboina one will find few of these trees, for when I arrived in the year 1654, I only found two of them near Victoria Castle. . . ." This is in his *Herbal*, *Amboinsche Kruidboek*, 1:190.

30. François Valentijn, *Oud en Nieuw Oost-Indiën*, 5 bks. in 8 vols. (Dordrecht & Amsterdam:

Joannes van Braam & Gerard onder de Linden, 1724–1726), 2.1:37 (on sketch); Rumphius, *Amboinsche Kruidboek*, 2:4 (on tradition).

31. Valentijn, *Oud en Nieuw Oost-Indiën*, 2.1:37.

32. On the latter, see, for instance, C. R. Boxer, *Francisco Vieiera de Figueiredo. A Portuguese Merchant-Adventurer in South East Asia, 1624–1667* (The Hague: Nijhoff, 1967).

33. Gov. & Council to Batavia, 26 September 1660.

34. Gov. Hustaerdt to Batavia, 9 September 1657.

35. Pieter van Dam, *Beschrijvinge van de Oostindische Compagnie*, 7 vols. ('s-Gravenhage: Nijhoff, 1927–1954), 1:589.

36. In the same document as in n. 34.

37. See the lists at the end of Valentijn, 2.2:45.

38. He was asked to find a suitable spot for a structure that would replace the small reduit called "Neira." See V. I. van de Wall, *De Nederlandsche Oudheden in de Molukken* ('s-Gravenhage: Nijhoff, 1928), p. 25.

39. Gov. & Council to Batavia, 9 September 1657. In the original: "tot de militaire chargie niet wel gehumeurt" (Leupe, p. 5).

40. Ibid.

41. See the bibliography in the fine short history, Femme S. Gaastra, *De Geschiedenis van de VOC* (Zutphen: Walburg Pers, 1991), pp. 181–188.

42. See E. M. Beekman, *Troubled Pleasures. Dutch Colonial Literature from the East Indies, 1600–1950* (Oxford: Clarendon Press, 1996), pp. 17–79.

43. Jan de Vries and Ad van der Woude, *Nederland 1500–1815. De eerste ronde van moderne economische groei* (Amsterdam: Balans, 1995), p. 454.

44. See J. Heniger, *Hendrik Adriaan Van Reede tot Drakenstein (1636–1691) and Hortus Malabaricus. A Contribution to the History of Dutch Colonial Botany* (Rotterdam: A. A. Balkema, 1986).

45. P. H. van der Brug, *Malaria en malaise. De VOC in Batavia in de achttiende eeuw* (Amsterdam: De Bataafsche Leeuw, 1994), p. 29.

46. Valentijn, *Oud en Nieuw Oost-Indiën*, 2.1:113.

47. Ibid., p. 96.

48. G. E. Rumphius, *Ambonsche Landbeschrijving*, ed. Suntingan Dr. Z. J. Manusama (Jakarta: Arsip Nasional Republik Indonesia, 1983), p. 44. It is now called "Batu Suanggi."

49. Rumphius, *Landbeschrijving*, p. 45.

50. Ibid.

51. Van de Wall, *Nederlandsche Oudheden*, p. 192.

52. Emerson, from a lecture to the Boston Society of Natural History given in 1834: "An everlasting Now reigns in nature that produces on our bushes the selfsame rose which charmed the Roman and the Chaldean."

53. Berlin, *The Hedgehog and the Fox. An Essay on Tolstoy's View of History* [1953] (reprint ed. New York: Simon and Schuster, 1970), 1–2.

54. Bk. 3, ch. 42.

55. For instance, J. P. Lotsy, *G*, p. 56; or Rouffaer, p. 180.

56. Rumphius knew Orta's work primarily through Clusius' epitome, though there are some tantalizing suggestions that he might have read Orta's original text, as Conde de Ficaldo suggested (see Rouffaer, *G*, p. 215). Yet I deem it highly unlikely that a young man living under duress in a strange country somehow manages to obtain a copy of one of the rarest books ever published.

57. See n. 12. The Latin says *statim* and Valentini's German "sobald ich." M. C. Valentini, *Museum Museorum* (Frankfurt am Main: Johann David Zunners, 1704), p. 117 of pt. 3. For more information see entry under 1704 in the next section, entitled "Works."

58. *Liefhebber* in the original; Burman's Latin has *amator.*

59. In the original: "Het zelve werk nu lag zonder ordre, als zynde beschreven, zo als my de planten dagelyks voorquamen. . . ." Second (unnumbered) page of the preface in vol. 1 of *Amboinsche Kruidboek*.

60. In a letter to Jacob de Vicq (in Valentini, *Museum Museorum*, p. 55 of the third pt.) he refers to his manuscripts as "*verwirrete Schriften*." One must be careful with his official pronouncements. While he was wrangling with his superiors about his sabbatical leave in 1669, Rumphius uses a phrase which has often been quoted: "myn ondersoeckinge van Amboinaes gewassen etc. voor myn vertreck te mogen voltooyen, zynde dit wel het meeste oogmerck waerom ick my in India begeven hebbe" (his letter to Jacob Cops, dated 18 July 1669; printed in Leupe, pp. 12–14). This translates as "[I hope] to complete my inquiries into Ambonese plants etc. before my departure, being the most important reason why I went to India." But one should not forget that this is rhetoric, intended to convince a recalcitrant adversary of the long-standing importance of his private enterprise!

61. Rumphius, *Landbeschrijving*, p. 28.

62. François Valentijn (1666–1727), a preacher and author of the important but controversial *Oud en Nieuw Oost-Indiën*, lived on Ambon twice in his life. During his first sojourn, from 1685 to 1694, he came to know Rumphius personally, and after the latter's death he had another tour of duty on Ambon, from 1706 to 1714, when he became intimately acquainted with Rumphius' heirs. During Valentijn's first few months in Ambon, Rumphius was one of his instructors in Malay. In his detailed description of the city of Ambon, Valentijn notes very carefully that Rumphius' burial plot (*graf-thuin*) was on Macassar Street. Elsewhere Valentijn announces that, in 1710, his stepson Gerard Leydekker married the widow of Paulus Augustus Rumphius, the only son of "Georgius Everhard Rumphius, that famed blind Gentleman who wrote the *D'Amboinsche Rariteit-kamer*," and that his youngest stepson, Bartholemeus, married in 1711 Adriana Augustina Rumphius, who was the eldest daughter of Gerard's wife as well as being Rumphius' granddaughter. In the third volume of *Oud en Nieuw Oost-Indiën*, Rumphius is called the author's "Bosom friend," and in the process of describing the island Manipa, Valentijn tells his readers that he is adding "a neat little Map . . . which Rumphius (a Man whose neat annotations about this land of Amboina we have often made use of) did draw very handsomely." At another place Valentijn declines to elaborate on a descriptive list of Ambonese stones, "because that has already been done at length by Mr. Rumphius in his *D'Amboinsche Rariteit-kamer*, and we refer the Reader to the same."

It would thus be inaccurate to maintain that Rumphius was never mentioned by Valentijn. Nor does it seem accidental that the only work by Rumphius that is mentioned by title is also the only work that was printed at the time Valentijn was writing. But nowhere, to my knowledge, does Valentijn ever credit the provenance of large segments of his description of Ambon and things Ambonese. Most of it comes from Rumphius, and Valentijn "reworked" or simply lifted a great deal, including the material (now lost) of Rumphius' three "books" on "land-, air- and sea animals," the physical description of Ambon, Rumphius' history of Ambon and his description of Banda, even information from the *Herbal*, which was still awaiting the VOC's permission to be printed. It is likely that Valentijn pirated Rumphius' Malay dictionary as well. See Beekman, *Troubled Pleasures*, pp. 119–144.

63. Valentijn calls it the "Wallohi," today the Wai Loi. "*Wai*" means "river" in Ambonese.

64. In the original, "de Pagger van 't Opperhooft." A variant spelling of the Malay *pagar*, "*pagger*" referred to a fence, either natural (bamboo) or manufactured (mats). To have one's house fenced in like that was both a protective measure and a status symbol.

65. The well-known mango (*Mangifera indica* [L.]).

66. Valentijn, *Oud en Nieuw Oost-Indiën*, 2.1:100. This is Valentijn's personal experience because, though he plagiarized the *Ambonsche Landbeschrijving*, Rumphius never mentions these visits, which must have been very trying to him.

67. Rumphius, *Landbeschrijving*, p. 28. The fort has been restored. Now known as Benteng Amsterdam and only twenty-eight miles by paved road from the city of Ambon, it can be easily visited today. See W. Buijze, "The VOC-Benteng Amsterdam op het eiland Ambon," *Moesson*, 39, no. 4 (15 October 1994), pp. 26–7. The view of Hila as reproduced in this book on p. lxiv was used as a guide to the restoration.

68. Valentijn, *Oud en Nieuw Oost-Indiën*, 2.1:101.

69. The letter is dated 25 October 1664, from Hanau, and the passage reads in the original: "Ich have diese vergangen Herbstmeß einige Schreiben von meinem Sohn Gorg Eberhard auß Ostien vond er Insel Ambönia auß dem Castell Siella woruff er Commetand ist und geht ihme gar woll ist auch Kauffer und mit H[err] im Landt begert nicht wieder nach Hanaw" (HHStAW Abt. 133, Stadt Idstein, p. 48). With thanks to Christel Lentz, who discovered this passage. These few lines tell us several things. The first, and most intriguing, is that Rumphius corresponded with his father, a fact not known before; second, that the transmission was by ship to Europe and then with merchants to the (still famous) fair in Frankfurt. The first phrase indicates, therefore, that August Rumpf received this letter in October of 1663. This passage also indicates that August was barely literate. His handwriting is nearly illegible. August also had no inkling where his son lived, called Hila "Siella," and assumed that the rather modest compound, despite Valentijn's encomium, was a "castle." On the other hand, and it would be a nice and understandable touch, perhaps Rumphius did a bit of boasting to his father.

70. Gov. & Council to Batavia, 26 September 1660. "Arabic" does not imply that Rumphius knew Arabic but rather that he could read Malay, which in his day was printed in Arabic and not Latin characters.

71. One could include some knowledge of the languages spoken on the various islands in eastern Indonesia.

72. *Dagh-Register gehouden int Casteel Batavia vant passerende daer ter plaetse als over geheel Nederlandts-India. Anno 1663*, ed. J. A. Van der Chijs (The Hague: Nijhoff, 1891), 256.

73. See ch. 28 in bk. 3 in the present volume.

74. The XVII to Batavia, 17 February 1662.

75. Leupe printed this important letter in its entirety as an appendix ("Bijlage I"), on pp. 40–42.

76. Marville to Batavia, 28 January 1666. In the original: "Hij heeft den roem en oock het uyterlyck aensien, te syn een man van goede kennisse en ervarentheyt in dese landen, bysonder op de kust van Hitoe. Van een goet en statigh leven en van een onbillycken maer nedrigen en heusschen ommegangh, en vooral van een conscientieux en oprecht gemoet, niet gierigh ofte inhaligh." Printed by Leupe, p. 8.

77. For instance, Leupe, p. 8; or Heeres, p. 5 n. 4.

78. For some instances, see the protracted dispute with De Jager about sandalwood in Valentini, *Museum Museorum* (third pt.), pp. 3–45, or his letters in German in the same volume, pp. 44–48, 54–55, 58. On Padtbrugge's assessment of Rumphius' character, see Pabbruwe, *Dr. Robertus Padtbrugge*, p. 76; or Rumphius' alleged fight with the same governor, mentioned by Valentijn, 4.2:110. Or his letter dated 18 July 1669 (printed in Leupe, p. 12), when Rumphius goes over Governor Cops' head to air his grievances to Batavia. Nor should one ignore the iconographic evidence of the only portrait we have, drawn by his son from life and reproduced in this volume as the frontispiece.

79. For instance, W. S. S. van Benthem Jutting, in *M*, 192.

80. Resolution of 27 February 1667; see Leupe, p. 10.

81. Rumphius, *Amboinsche Kruidboek*, 5:287.

82. There still is a family by the name of Rumphius Twijsel who descends from one of these daughters. As E. J. C. Boutmy de Katzmann from the Central Bureau for Genealogy in The Hague informed me, she was only known as "N. N. Rumphius, daughter of George Everhardus Rumphius and Susanna N. N." "N. N." meant that her last name was not known.

83. Rumphius' letter to Cops, 18 July 1669; printed by Leupe, pp. 13–14. "General good" is "*gemeyne wesen*" in the original.

84. Ibid.

85. Phrase used by Batavia in its missive to Gov. & Council in Ambon, 14 February 1668.

86. Gov. & Council to Batavia, 9 May 1670.

87. Even in such an unlikely venue as his recommendation on the fortification of Victoria Castle; see Valentijn, 2.2:239.

88. Quotes from Rumphius' proem.

89. *MC (1683)*, p. 51. In the original: "quam ego, plus consuens, totum mundum cum omnibus creaturis subito meo visui subtraxit, unde jam per decennium in *tristibus tenebris sedere cogor* [*sic*]."

90. See ch. 28 in bk. 2 in the present work.

91. Milton, *Samson Agonistes*.

92. In his preface to the *Kruidboek* in the first volume.

93. In the previously mentioned letter to Mentzel.

94. See Rumphius' letter to Cops, 17 May 1670, wherein he mentioned "the little light left in one eye which is my only solace."

95. Preface to the *Kruidboek* in the first vol.

96. Dave Sobel, *Longitude* (New York: Walker and Co., 1995), p. 43.

97. Cops had a bad reputation. This is how Valentijn characterized him (2.2:91): "This Gentleman ruled very harshly when he was Governor, and had a bad name among the inhabitants, because he was very covetous, which was evident wherever he had been; he was also very obstinate, and not obliging at all, wanting everything to turn his way."

98. Batavia to Cops, 29 December 1670.

99. Valentijn, 2.1:125.

100. Van de Wall, *Nederlandsche Oudheden in de Molukken*, pp. 167–168. There is a picture of the house, taken in the 1930s, in Rob Nieuwenhuys, *Komen en blijven. Tempo doeloe—een verzonken wereld. Fotografische documenten uit het oude Indië 1870–1920* (Amsterdam: Querido, 1982), p. 79. But one will never be absolutely sure about the exact location of Rumphius' house because the town was bombed by the allies in the Second World War and the house and most of the street destroyed.

101. See Valentijn, 2.1:128–129, for a contemporary account of how these houses were built.

102. Ibid., p. 130.

103. From Rumphius' own description of the event (see "Works" under 1675): *Waerachtigh Verhael van de Schrickelijke Aerdbevinge* (Batavia, 1675), p. 4.

104. There are conflicting reports about Rumphius' loss. In *Waerachtigh Verhael* (p. 4) he states that his wife and their "youngest daughter" were killed. The daily register of Victoria Castle, from which this passage was taken (*Dagregister* for 17 February 1674; printed by Leupe on pp. 17–18) states the same, but Rumphius himself says in his *Herbal* that "that Author did lose his Wife, two Children, and a Maid, all from his house" (6:195).

105. *Generale Missiven van Gouverneurs-Generaal en Raden aan Heren XVII der Verenigde Oostin-*

dische Compagnie, ed. W. Ph. Coolhaas, 8 vols. [Rijks Geschiedkundige Publicatiën, grote serie 104, 112, 125, 134, 150, 159, 164, 193] ('s-Gravenhage: Nijhoff, 1960–1979), 4:781.

106. Rumphius, *Kruidboek*, 6:195.

107. In the 1680 letter to Mentzel (*MC* [1683], p. 51) he says that blindness forced him to dictate, but in his preface to the *Herbal* he adds a reason which echoes Orta: that ordinary people could profit from it if it were written in Dutch.

108. See Leupe, p. 20.

109. See his letter to De Jager of 14 September 1689, in M. B. Valentini, *Museum Museorum* (third section), pp. 44–45. As Rumphius notes, the disaster also consumed his commentary on Bontius, which he had completed only four years earlier.

110. See ch. 39 in book 3 of the present volume.

111. Rumphius to Isaac de Saint Martin, 9 July 1696; printed in Leupe, pp. 56–57.

112. Batavia to Gov. & Council, 17 August 1680. The original reads: "UEd. zullen 't zelve boeck voor een secreet ende zeer dienstich document van de Secretarye houden ende niet verder laten copieeren."

113. See Heeres, *G*, p. 14.

114. Heeres, *G*, p. 11.

115. See J. P. Lotsy, "Over de in Nederland aanwezige botanische handschriften van Rumphius," in *G*, p. 48.

116. Rumphius to Cleyer, 15 May 1688; printed in Valentini, *Museum Museorum* (third pt.), p. 58. We do not know when Paulus Augustus was born. We know from Rumphius' letter to Cosimo—printed in his English translation, from an earlier Italian translation, by W. S. S. van Benthem Jutting, in *M*, pp. 198–199—that Paulus was "studying in Holland" in 1682. Leupe (p. 61 n. 21) states that Rumphius' son was in Batavia in 1686 and transferred the same year to Ambon. Valentijn doesn't list Paulus as accountant on Ambon until 1687. He became the "head" of Larike in 1690, junior merchant ("*onderkoopman*") at Larike in 1693, promoted to merchant there in 1699, and he remained in Larike until dismissed on 30 April 1705, "due to illness," according to Valentijn. He appears to have drowned in view of Batavia in 1706 (see *Encyclopaedie van Nederlandsch Indië*, 4:503).

117. Ibid.

118. *G* prints one opposite p. 46.

119. Johannes Philippus Sipman—who added a superfluous index to the *Ambonese Curiosity Cabinet*, which I omitted as being redundant—was born in Darmstadt, Germany, in 1666 and died in Batavia in 1725. He made a career in the VOC, first on Ambon (as Valentijn's lists in bk. 2, pt. 2, indicate), from junior merchant to "*secunde*," then governor of Makassar, and finally as member of the Council of the Indies in Batavia. On his being Rumphius' assistant for at least four years, see Rumphius' letter to De St. Martin from 1696, printed in *G*, p. 52.

120. See Leupe, p. 61 n. 22.

121. Letter from Gov. & Council to Batavia, 30 September 1694.

122. Leupe, p. 33.

123. Resolution of the XVII in Amsterdam, 15 September 1702.

124. Printed by Leupe, p. 59. My guess is that Paulus wrote it, because the brief missive is replete with Gallicisms, a stylistic lapse to which Rumphius never fell victim, but which one will see abused in bureaucratic documents of that age.

125. See ch. 3 in book 2 of the present volume. Fehr's own description of the shell can be found, in Latin, in "De Carina Nautili elegantissima," *MC*, pp. 210–211.

126. This charming custom was discontinued in 1870. The original entry for Rumphius in the society's *Matricula*, *Tomas I*, is on p. 98 under the heading "XCVIII. Dn. Georgius Eberhardus Rumph. Curios. Plinius I," and reads: "Natus in Comitatu Solmensi, educatus Hanoviae, ubi parens ejus Augustus Rumphius Architectum egit ante 1666. Ipse verò insatialibi rerum peregrinarum cognoscendarum desiderio flagrans mature patriam reliquit, primò in Portugalliam delatus, deinde post triennium redux, A. 1652. in Indiam Orientalem se contulit, sedenque Ambona vixit ubi aliquot volumina rerum Indicarum naturalium conscripsit, iamque ab anno circiter 1670 ex gutta serena caecus factus est." Hence the "Plinius Indicus" on the title pages of his books is probably Rumphius' own invention, for he ends the autobiographical passage in his letter to Mentzel: "Rumphium illum Indicum" (*MC* [1682], p. 51). He was sometimes also referred to as "Plinius secundus," for instance, by Mentzel in *MC* [1684], p. 70.

127. "Das kaiserliche Privileg der Leopoldina vom 7. August 1687," ed. George Uschmann, in *Acta Historica Leopoldina*, no. 17 (1987), p. 9. Mentzel turns out to be an important figure in Rumphius' life. Christian Mentzel (1622–1701) obtained his degree in medicine from Padua in 1654. Four years later he became the Elector of Brandenburg's court physician and remained in the job for thirty years,

until 1688. He was primarily interested in botany and Chinese studies. He published *Icones arborum, fructuum et herbarum exoticarum* in Leiden, and *Index nominum plantarum universalis* in 1682 (2d. rev. ed. in 1696). During the 1680s he became interested in sinology and published in 1685 a lexicon, *Sylloge minutiarum lexici Latino-Sinico-Characteristici ex Autoribus et Lexicis Chinensis eruta.* In 1696 he published, in Berlin, *Kurze chinensische Chronologie*, and the Berlin Library holds Mentzel's unpublished *Flora Japonica* (in 2 vols.), which he wrote with Cleyer.

128. *Amboinsche Kruidboek*, 2:263–268. Speelman's letter can be found in partial English translation in Yule, *Hobson-Jobson*, pp. 956–957.

129. Heniger, *Van Reede tot Drakenstein*, p. 53.

130. See P. C. Bloys van Treslong Prins, "Origineele bescheiden van en over Georgius Everhardus Rumphius," *Tijdschrift voor Indische Taal,-, Land- en Volkenkunde*, deel LXIX, aflevering 3 & 4 (1930), pp. 426–435. In the first document, dated 25 September 1698, it is stated that "my second wife has been Isabella Ras" (p. 427), and the second document is Ras' last will and testament, drawn up when she was dying, and dated 7 March 1689.

131. Gov. & Council to Batavia, 24 September 1700. Sipman took Rumphius' post as president of the "Collegie des Kleynen Gerigts en Huwelyckszaecken"; see Leupe, p. 29.

132. She died in 1692; see Bloys van Treslong Prins, "Origineele bescheiden," *Tijdschrift*, p. 427 n. 1.

133. Ch. 83 of the "*Auctuarium*"—more conveniently known as the *Herbal*'s vol. 7—entitled "Description of the Serpenticida, or Moncus," 7:69–71. This quote on p. 69. The animal is *Herpestes javanicus* Geoffroy; see A. C. V. Van Bemmel, "Rumphius' Treatises on Mammals," in *M*, 32.

134. Rumphius, *Amboinsche Kruidboek*, 7:70.

135. Van de Wall, *Nederlandsche Oudheden in de Molukken*, p. 169. We know it was made from marble from P. Lesson, *Voyage autour du monde entrepris par ordre du Gouvernement sur la corvette La Coquille*, 4 vols. (Paris: P. Pourrat, 1839), 2:181.

136. C. L. Blume, *Rumphia, sive commentationes botanicae imprimis de plantis Indiae Orientalis, tum penitus incognitis tum que in libris Rheedii. Rumphii, Roxburghii, Wallichii, aliorum, recensentur*, 4 vols. (Lugduni-Batavorum: Prostat Amstelodami, apud C. G. Sulpke.—Bruxelles, apud H. Remy.—Dusseldorfiae, apud Arnz et socios.—Parisiis, apud C. Roret, 1835–1848), 2:11.

137. Rouffaer, *G*, 207.

138. Lesson, *Voyage*, 2:181.

139. Ibid., p. 180.

140. Van de Wall, *Nederlandsche Oudheden*, p. 169. It had been designed by Antoine Payen (1792–1853), a painter and architect, who lived in the Indies from 1817 to 1828. Payen was the first teacher of the famous Javanese painter Raden Saleh. There is a photograph of the original monument opposite p. 16 in *G* and in *M* in the section "Illustrations" after p. 462.

141. See *Moesson*, 40, no. 8 (15 March 1996), p. 6; and no. 11 (15 June 1996), p. 8.

Works (except the Ambonese Curiosity Cabinet*)*

1. See also Rouffaer, *G*, p. 174.

2. Rouffaer, *G*, p. 190.

3. See E. M. Beekman, *Troubled Pleasures*, pp. 119–144.

4. Rouffaer, *G*, p. 195.

5. De Wit, *M*, p. 12.

6. Greshoff's article on Rumphius in *Encyclopaedie van Nederlandsch-Indië*, 3:640–645.

7. See Pliny's "Prefatio," Latin paragraph 18. Another direct reference by Rumphius is when he writes in the opening paragraph of his preface: "So I present you [the reader] with something new, and not with some monster dragged from the wildernesses of Africa." This phrase would have evoked at the time the well-known sentence from Pliny: "unde etiam vulgare Graeciae dictum semper aliquid novi Africam adferre." In Rackham's translation: "This is indeed the origin of the common saying of Greece that Africa is always producing some novelty." The edition of Pliny's *Natural History* used in these remarks and throughout the present volume is from the Loeb Classical Library Bilingual Editions, transl. H. Rackham, W. H. S. Jones, and D. E. Eichholz, 10 vols. (Cambridge, Mass.: Harvard University Press, 1938–1962).

8. Mary Beagon, *Roman Nature. The Thought of Pliny the Elder* (Oxford: Clarendon Press, 1992), p. 13. Beagon's study is particularly relevant because she examines Pliny's *thought* and does not rehash the usual criticisms.

9. From Pliny's "Prefatio," Latin paragraph 13; Loeb, 1:8–9.

10. From Pliny's "Prefatio," Latin paragraph 19; Loeb, 1:12–13. I know that "*vigilia*" is usually translated as "awake," but the trope is also possible in classical Latin and more fitting.

11. See bk. 3, ch. 9, in the present volume, for one instance.

12. For Pliny, see the splendid opening of book 33; Loeb, 9:2–5. For Rumphius, the evidence is throughout his work.

13. For Pliny see book 2, Latin paragraphs 155–156; Loeb, 1:290–3. There are several instances in the present volume for Rumphius: for instance in bk. 1, ch. 16, 17, 19, 36; in bk. 2, ch. 26; in bk. 3, ch. 23. Or the following from his *Herbal* (*Amboinsche Kruidboek*, 6:211) could have been written by Pliny: "For one should not be so quick to accuse nature, that she would produce some precious creatures, and keep them from man's eyes for all eternity, while instead everything was created for him."

14. For Pliny, see the opening section of bk. 14; Loeb, 4:188–191. For Rumphius, see the present volume, for instance, ch. 24 and 25 in bk. 3.

15. Bk. 9, Latin paragraph 4; Loeb, 3:435.

16. In the present volume, see ch. 42 of bk. 1, or ch. 5 in bk. 2.

17. Rumphius, *Amboinsche Kruidboek*, 6:148.

The Ambonese Curiosity Cabinet

1. See the illustration accompanying Rumphius' short note to Mentzel in *MC* [1688], opposite p. 9.

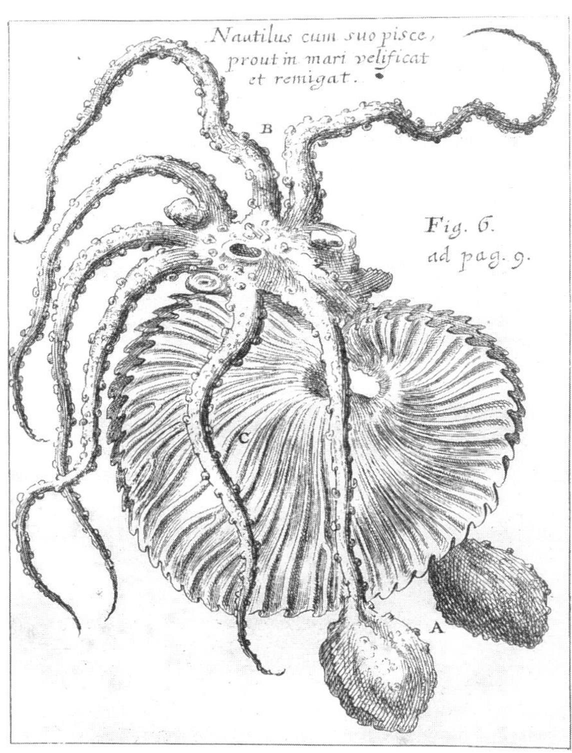

2. Written communication from Hermann L. Strack; see also W. S. S. Van Benthem Jutting, "Rumphius and Malacology," in *M*, pp. 202–205.

3. This was alleged in *Merian, Maria Sibylla, Leningrader Aquarelle*, 2 vols., eds. Ernst Ullmann, Helga Ullmann, et al. (Lucerne: Bucher, 1974), 1:98–106. The same thing is mentioned by Irina N. Lebedeva, "De nalatenschap van Maria Sibylla Merian in Sint-Petersburg," in *Peter de Grote en Holland. Culturele en wetenschappelijke betrekkingen tussen Rusland en Nederland ten tijde van tsaar Peter de Grote* (Bussum: Uitgeverij Thoth, 1996), p. 65. In the Plantage Library of The University of Amsterdam is a copy of the first edition of the *Rariteitkamer* wherein the plates were hand-colored by Merian. In 1997, a second copy was discovered, which I have not seen. For a description, see H. L. Strack and Jeroen Goud, "Rumphius and the 'Amboinsche Rariteitkamer,'" *Vita Marina* 44:1–2 (November 1996), pp. 29–39.

4. Rouffaer, *G*, p. 172; see also Van Benthem Jutting, *M*, pp. 202–203.

5. See Holthuis, in *M*, p. 80.

6. Valentijn mentions this work by Petiver, but he could not have examined it very carefully because he never mentions the fact that the catalogue is based on Rumphius! See Valentijn, *Oud en Nieuw Oost-Indiën*, 3.2:548. James Petiver (?1663–1718) was familiar with the Dutch community of collectors having come to Holland in 1711 as an agent for Sir Hans Sloane.

7. See Rouffaer, *G*, p. 173, and Martens, *G*, p. 111.

8. See Rouffaer, *G*, pp. 188–217; or Van Benthem Jutting, *M*, pp. 186–196.

9. See also n. 1 of the introduction.

10. John Crawfurd, *Descriptive Dictionary of the Indian Islands and Adjacent Countries* [1856] (reprint ed. Kuala Lumpur: Oxford University Press, 1971).

11. Ibid., p. 370.

12. Albert S. Bickmore, *Travels in the East Indian Archipelago* (New York: Appleton, 1869). It was first published in London the previous year. I am quoting from the American edition. Bickmore's book was translated into Dutch and published in Holland in 1873.

13. Ibid., p. 134.

14. Ibid., p. 13.

15. Ibid., p. 141.

16. H. L. Strack, "Results of the Rumphius Biohistorical Expedition to Ambon (1990). Part 1. General Account and List of Stations," *Zoologische Verhandelingen Leiden*, 289 (30 December 1993), 68.

17. Bickmore, *Travels*, 39.

18. Ibid., pp. 251–252.

19. Ibid., p. 250.

20. David Fairchild, *Garden Islands of the Great East. Collecting Seeds from the Philippines and Netherlands India in the Junk "Chêng Ho"* (New York: Scribner's, 1943), p. vii. Sometime in the 1880s Wallace "came to the college in Kansas where my father was President and delivered a lecture on Natural Selection" (ibid.).

21. Fairchild, *Garden Islands*, p. 193. He visited Rumphius' house (p. 195) and had himself and his party photographed with the monument on Rumphius' grave (p. 184d).

22. He was killed because he made a linguistic mistake, the type of error that has been lethal before in human history. See E. D. Merrill, *An Interpretation of Rumphius's Herbarium Amboinense* (Manila: Bureau of Printing, 1917), p. 9. This incident was later used by Maria Dermoût in her incomparable novel, which is suffused with Rumphius, entitled *De tienduizend dingen* (Amsterdam: Querido, 1955) (translated as The ten thousand things, transl. Hans Koningsberger [New York: Simon and Schuster, 1958]).

23. Elmer D. Merrill, *Plant Life of the Pacific World* (New York: Macmillan, 1945), p. 243. Another traveler paid his respects to Rumphius: H. W. Ponder, *In Javanese Waters* (London: Seeleye, 1944), pp. 170, 172, 174–175, and 222. For some inexplicable reason he calls Rumphius "the Blind Butterfly Man" (p. 172). But another British naturalist of renown, Henry O. Forbes, who traveled in Indonesia for five years, never mentions Rumphius, nor does his wife; see Henry O. Forbes, *A Naturalist's Wanderings in the Eastern Archipelago. A Narrative of Travel and Exploration from 1878 to 1883* (New York: Harper, 1885); and Anna Forbes, *Unbeaten Tracks in Islands of the Far East* (reprint ed. Singapore: Oxford University Press, 1987), which was originally published in 1887 with the title *Insulinde.* Another appreciation is S. Peter Dance, *A History of Shell Collecting*, 2d. rev. ed. (Leiden: Brill, 1986), esp. pp. 26–29. Dance bemoans the fact that *D'Amboinsche Rariteitkamer* could not be read because it was written in Dutch (pp. 28–29).

24. Consider the following passage from the last chap. (21) of *The Voyage of the Beagle* (reprint ed. Garden City: Doubleday, 1962), p. 494: "When quietly walking along the shady pathways, and admiring each successive view, I wished to find language to express my ideas. Epithet after epithet was found too weak to convey to those who have not visited the intertropical regions, the sensation of delight which the mind experiences. I have said that the plants in a hothouse fail to communicate a just idea of the vegetation, yet I must recur to it. The land is one great wild, untidy, luxuriant hothouse, made by Nature for herself, but taken possession of by man, who has studded it with gay houses and formal gardens. How great would be the desire in every admirer of nature to behold, if such were possible, the scenery of another planet! yet to every person in Europe, it may be truly said, that at the distance of only a few degrees from his native soil, the glories of another world are opened to him. In my last walk I stopped again and again to gaze on these beauties, and endeavored to fix in my mind for ever, an impression which at the time I knew sooner or later must fail. The form of the orange-tree, the cocoa-nut, the palm, the mango, the tree-fern, the banana, will remain clear and separate; but the thousand beauties which unite these into one perfect scene must fade away; yet they will leave, like a tale heard in childhood, a picture full of indistinct, but most beautiful figures."

25. Darwin's remarks cover several pages of "Notebook C," written in 1838; see *Charles Dar-*

win's Notebooks, 1836–1844, transcribed and edited by Paul H. Barrett et al. (Ithaca: Cornell University Press, 1987), pp. 240–247.

26. Alfred Russel Wallace, *The Malay Archipelago. The Land of the Orang-Utan and the Bird of Paradise. A Narrative of Travel with Studies of Man and Nature* [1869] (reprint ed. New York: Dover, 1962), p. 72. Other passages where Wallace expresses his admiration of the Dutch are on pp. 72–75, 196–197, 222, 271. Wallace even defends the Dutch on Banda and their ferociously defended monopoly in the nutmeg trade: "A small country like Holland cannot afford to keep distant and expensive colonies at a loss; and having possession of a very small island where a valuable produce, *not a necessary of life*, can be obtained at little cost, it is almost the duty of the state to monopolise it. No injury is done thereby to anyone, but a great benefit is conferred on the whole population of Holland and its dependencies, since the produce of the state monopolies save them from the weight of a heavy taxation. Had the Government not kept the nutmeg trade of Banda in its own hands, it is probable that the whole of the islands would long ago have become the property of one or more large capitalists" (p. 221). And see the continuation of this surprising train of thought on p. 222, where he even recommends the extirpation of the spice trees!

27. Ibid., p. 74.

28. Ibid., p. 226.

29. Ibid., p. 224.

30. Ibid., p. 226.

31. Ibid., p. 233.

32. Ibid., p. 224.

33. L. B. Holthuis, "Are there poisonous crabs?" *Crustaceana*, vol. 15, pt. 2 (1968), 217. As is clear from the notes, Prof. Holthuis provided the technical knowledge required for the identifications in the first book.

34. Holthuis, in *M*, p. 63 and passim.

35. Also in ch. 64 of bk. 3.

36. Martens, *M*, p. 112.

37. Ibid., p. 110.

38. Baldomero M. Olivera at the University of Utah has been conducting research in this field. See also L. J. Cruz and J. White, "Clinical Toxicology of *Conus* Snail Stings," in *Clinical Toxicology of Animal Venoms*, eds. J. Meier and J. White (Boca Raton: CRC Press, 1995), pp. 117–127.

39. Strack, "Part 1. General Account," in *Zoologische Verhandelingen Leiden*, 289 (30 December, 1993), p. 69.

40. Martens, *M*, 109.

41. From Chemnitz's "*vorrede*" to *Georg Eberhard Rumphs* . . . (1766). The original reads: "die Anzahl der claßischen Schriftsteller in der Conchyliologie ist . . . ungemein klein."

42. Among many, for instance, William C. Dampier, *A History of Science* (Cambridge: Cambridge University Press, 1942); and H. C. D. de Wit, *Ontwikkelingsgeschiedenis van de biologie*, 2 vols. in 3 pts. (Wageningen: PUDOC, 1982–1989).

43. See the first note of ch. 8 in bk. 3.

44. Wichmann, in *G*, 163.

45. Wichmann's article dates from 1902.

46. The Latin tag from Horace means: "a monument more lasting than bronze." Wichmann, *G*, 164.

47. See De Haan, *Oud Batavia*, 1:66.

48. See, for instance, these lines (234–237) from the second part of Keats' *Lamia:* "Philosophy will clip an Angel's wings, / Conquer all mysteries by rule and line, / Empty the haunted air, and gnomed mine— / Unweave a rainbow, as it erewhile made." Or these lines (25–28) from Wordsworth's *Tables Turned:* "Sweet is the lore which nature brings; / Our meddling intellect / Mis-shapes the beauteous forms of things; / —We murder to dissect."

49. From "The Logic of Totemic Classifications," in: Claude Lévi-Strauss, *The Savage Mind* (Chicago: University of Chicago Press), p. 45. I quote again from this edition (see n. 89), of which the translator is not named, though I needed to revise the text.

50. This is obscure in the original Dutch as well: "Den Raat soude ik hier in 't geheel naar het behagelijk voorschrift bestelt hebben en waar den coopman Rumphius niet tusschen of denselven niet gedacht, dan oft met voordacht daar uijtgelaten is, verwachte ick eerbiedig in het aanstaende te mogen verstaen onderweijlen dat wij hem daar noch in taelen, sijnde in alles wat den landtaart 's-Compagnie's wege tegens is. Te geweldigh met vooroordelen ingenomen en heeft in alles, omdat hem in zijn zinnelijkheid van fremdigheden te verschaffen seer gedienslike zijn meer dan een gemeijne zucht tot deselve, dat zomwijlen vrij wat dwars komt gelijk int opspooren van 's-Compagnie's gronden, en enige andere

saecken niets dan te veel ervaren." VOC archives in the Royal Library in The Hague, no. 1385, folio 52 verso. The report is dated 23 May 1683. Padtbrugge wrote a most peculiar Dutch, and I could only decipher it with the aid of Dr. Hans Heestermans. Fortunately, the crucial words concerning Rumphius' recalcitrance come through loud and clear.

51. Rumphius, *Amboinsche Kruidboek* (bk. 3, ch. 37), 2:241.

52. Ibid., 242. The crucial phrase reads in the original Dutch: ". . .een ervaren Man in de kennisse der planten, die my in dit werk veel geholpen heeft, waarom ik ook behoorlyk agt, zyne gedachtenisse alhier te zetten."

53. ". . . maar mynen Baas, die my dit disteleeren leerde, Iman Reti, Moorsche Priester van Boero"; *Amboinsche Kruidboek*, 2:241. What an enormous difference from Valentijn, who never mentioned a source unless there was a politically correct reason to do so.

54. The translation is by Van Benthem Jutting, in *M*, 198, which I emended. Van Benthem Jutting translated an Italian translation of Rumphius' original letter, which was first published in *Le Collezioni di Giorgio Everardo Rumpf, acquistate dal Granduca Cosimo III de Medici*, ed. Ugolino Martelli (Firenze: Tipografia Luigi Niccolai, 1903), pp. 123–125. The letter and the catalogue, both from 1682, were originally in Latin and reside now in the Sächsische Landesbibliothek in Dresden, *handschriftenabteilung*, B110. The original Latin letter is transcribed on pp. 127–130 of the following article: Rudolph Zaunick, "Georg Eberhard Rumph und Cosimo III von Medici," in *Physis*, 3 (1961), pp. 125–136.

55. The connotation of compulsion is clearer in the original: "op het aanhouden van eenige Vrienden, aan dwelke ik verplicht was" (folio 163). The whole preamble to ch. 39 seems to have been dictated by irritation.

56. Van Benthem Jutting, *M*, 199.

57. For instance, the witty image of "Synodus Marina" in his dedication to D'Acquet in the present volume. Or the following passage in his letter to the retired Governor-General Camphuys, after the latter had written and teased Rumphius with the remark that his slaves had found a number of shells of which several "according to those who pretend to have knowledge of such things . . . cannot be rivaled by those from Amboina or its neighbouring Islands." To which Rumphius replied (in a letter dated 29 June 1695): "It was also news to me that your servants found so many Marine curiosa on the islands of Edam and Alckmaer that they would challenge those of Ambon and the Moluccas; which confirms my old guess and also assures me that it is at least partially true, for I have obtained quite some 40 kinds from various friends which had been gathered on the islands and beaches near Batavia. But if they would dare challenge the rarity and beauteous variety of those from Ambon, then this touches the honor of the Ambonese monarchy that has held sway over Marine curiosa for so many years. I have therefore, by your leave, resolved to put it to the test, and to this end ordered some hundred varieties from among Ambon's champions, packed them together in a little firkin that will come after, in order to challenge those from Batavia, hoping to hear in time the successful outcome of this Battle, but if they lose the sport, they need not return but can remain in captivity."

58. Krzysztof Pomian, *Collectors and Curiosities Paris and Venice, 1500–1800* [1987], transl. Elizabeth Wiles-Portier (Cambridge: Polity, 1990), p. 30. Pomian writes quite cogently: "the semiophores, objects which were of absolutely no use . . . but which are endowed with meaning." On the same page he writes that "usefulness and meaning are mutual exclusive," where I would think one can include economic considerations under the general rubric of "usefulness."

59. Roland Barthes, *Empire of Signs*, transl. Richard Howard [1970] (New York: Hill & Wang, 1982), p. 78.

60. Roelof van Gelder, "De wereld binnen handbereik. Nederlandse kunst-en rariteitenverzamelingen, 1585–1735," in *De wereld binnen handbereik, Nederlandse kunst-en rariteitenverzamelingen, 1585–1735*, eds. Ellinoor Bergvelt and Renée Kistemaker (Zwolle: Waanders Uitgevers, 1992), pp. 29, 31.

61. Ibid., p. 25.

62. Lebedeva, *Peter de Grote en Holland* [see n. 3], p. 18.

63. G. J. Hoogewerff, *De twee reizen van Cosimo de Medici Prins van Toscane door de Nederlanden (1667–1669)* (Amsterdam: Johannes Müller, 1919), p. xxvi.

64. K. van Berkel, "Citaten uit het boek der natuur. Zeventiende-eeuwse Nederlandse naturaliënkabinetten en de ontwikkeling van de natuurwetenschap," in *De wereld binnen handbereik*, p. 179.

65. The crates, totaling nine, which Rumphius sent to Cosimo were transported on VOC ships from Ambon to Batavia and from Java to Europe; see *Generale Missieven*, 4:501–502, 653.

66. Jaap van der Veen, "Met grote moeite en kosten. De totstandkoming van zeventiende-eeuwse verzamelingen," in *De wereld binnen handbereik*, pp. 60–61.

67. Ibid., p. 61.

68. See C. Willemijn Fock, "Kunst en rariteiten in het Hollandse interieur," in ibid., pp. 70–91.

69. *De Wereld binnen handbereik*, p. 28.

70. Ibid., p. 36.

71. Ibid., p. 58.
72. Ibid., p. 295, n. 33.
73. See De Haan, *Priangan*, 1:283.
74. *De Wereld binnen handbereik*, p. 58.
75. H. Engel, "The Life of Albert Seba," in *Svenska Linné-Sällskapets Arsskrift*, (1937), 20:82. This same Engel is also the author of the following useful book: *Hendrik Engel's Alphabetical List of Dutch Zoological Cabinets and Menageries*, ed. Pieter Smit, 2d. rev. ed. (Amsterdam: Rodopi, 1986).
76. Engel, "Albert Seba," 84.
77. Ibid., p. 82.
78. Valentijn, *Oud en Nieuw Oost-Indiën*, 3.2:560.
79. Ibid., p. 561.
80. Ibid., pp. 560–563. Valentijn even lists all the shells they owned; the entire section on shells is on pp. 517–586.
81. Descartes, *Oeuvres et Lettres*, ed. André Bridoux, in Bibliothèque de la Pléiade (Paris: Gallimard, 1952), p. 885. In the original: "tous les effets merveilleux qui s'attribuent à la magie."
82. Ibid., pp. 883–884. In the original: "les simples connaissances qui s'acquièrent sans aucun discours de raison, comme les langues, l'histoire, la géographie, et généralement tout ce qui ne dépend que de l'expérience seule."
83. Ibid., p. 884. The original passage reads: "Pour les sciences, qui ne sont autre chose que les jugements certains que nous appuyons sur quelque connaissance qui précède, les unes se tirent des choses communes et desquelles tout le monde a entendu parler, les autres des expériences rare et étudiées. Et je confesse aussi qu'il serait impossible de discourir en particulier de toutes ces dernières; car il faudrait, premièrement, avoir recherché toutes les herbes et les pierres qui viennent aux Indes, il faudrait avoir vu le Phénix, et bref n'ignorer rien de tout ce qui'il y a de plus étrange en la nature."
84. Pomian, *Collectors and Curiosities*, pp. 53–60.
85. Ibid., p. 234.
86. Ibid., p. 64.
87. Stephen Jay Gould calls the final resting place of curiosity cabinets, the "cabinet museum," a "magic place." See his essay "Cabinet Museums: Alive, Alive, O!," in *Dinosaur in a Haystack. Reflections in Natural History* (New York: Crown, 1997), pp. 238–247; quote on p. 246.
88. Claude Lévi-Strauss, *La pensée sauvage* (Paris: Plon, 1962), p. 52.
89. Ibid., p. 53. This translation is from the edition mentioned in n. 49, but corrected by me after comparing it with the original.
90. Edward O. Wilson, *Biophilia* (Cambridge, Mass.: Harvard University Press, 1984), p. 85.
91. Ibid., p. 10.
92. Ibid., pp. 63–64.
93. Ibid., p. 81.

About This Translation

1. Montaigne, *Essais*, 2 vols. (Paris: Garnier), 1:185–186. The translation is by Donald Frame, *The Complete Essays of Montaigne*, [1957] (reprint ed. Stanford: Stanford University Press, 1985), p. 127.

THE FIRST BOOK OF *THE AMBONESE CURIOSITY CABINET* DEALING WITH SOFT SHELLFISH

Chapter 1. Of the Locusta Marina. Sea Crayfish. Udang Laut.

1. This should be *thorax*, of course, from the Greek for breastplate or cuirass.
2. The last clause is troublesome in Dutch: "*weshalven hy van vooren niet aan te doen is.*"
3. This refers to an incident related by Gaius Suetonius Tranquillus in his third "life" of *The Twelve Caesars*. Tiberius was notoriously cruel and much feared. His favorite retreat was on Capri. When he was in residence there one day "a fisherman suddenly intruded on his solitude by presenting him with an enormous mullet, which he had lugged up the trackless cliffs at the rear of the island. Tiberius was so scared that he ordered his guards to rub the fisherman's face with the mullet. The scales skinned it raw, and the poor fellow shouted in his agony: 'Thank Heaven, I did not bring Caesar that huge crab I also caught!' Tiberius sent for the crab and had it used the same way." Section 60 of chapter 3, Suetonius, *The Twelve Caesars*, transl. Robert Graves (Harmondsworth: Penguin, 1957), p. 139. I had to rearrange the Dutch sentence to make it understandable in English. It was due to Tiberius that the island was called "'Caprineum" ("Goatish") because of his obscene sexual practices, deviations once associated with supposedly lascivious goats.
4. This means "Marine Langouste from the Indies." To avoid confusion, especially for Ameri-

can readers, I shunned the noun "lobster" (Rumphius used *kreeft*), because this immediately evokes images of plates heaped with steaming pieces of *Homarus americanus*, to be dipped in clear, liquid butter. Rumphius does not seem to be familiar with this crustacean, nor did he care particularly for its gastronomic possibilities. To conform to his time, I use "crayfish," a noun with a long history in English, going back to the Middle Ages. It was applied as the general name for all the larger, edible crustacea, and was distinguished from crabs, shrimp, and oysters. Sometimes spelled "crevisse" or "crawfish." J. G. de Man, in his article "Over de Crustacea ('Weeke Schaalvisschen') in Rumphius' 'Rariteitkamer,'" in *G*, pp. 98–104, was one of the first to try to identify Rumphean crustaceans. De Man's efforts were extended and corrected by L. B. Holthuis in "Notes on Pre-Linnean Carcinology (including the study of xiphosura) of the Malay Archipelago," in *M*, pp. 63–125; hereafter referred to as 'Holthuis." I have incorporated Prof. Holthuis's remarks and identifications in the notes to the chapters in this first book. Holthuis identifies this spiny lobster as a langouste of the genus *Panulirus*, probably *Panulirus versicolor* (Latreille). Plate 1 depicts *Panulirus homarus* (L.). I am sure most readers are aware that "L." stands for Linnaeus, and that the name, added in brackets, after the Latin binomial (genus, species) is the author of the species, either the original discoverer or the person who first named the plant or animal correctly in a taxonomic sense.

5. *Udang* means "shrimp" and *laut* is "sea."

6. *Squilla*, also *scilla*, in Latin referred to a small "fish" resembling a lobster, a prawn. Hence this name means something like "a lobster-like fish that goes backwards."

7. With "*seris*" Rumphius is probably referring to large marine fish traps, normally spelled *sero* or *seru*, called *bila* in Celebes. Constructed from bamboo, with *gemutu* (the black threads found between the stems of the leaves of the Aren Palm, similar to horse hair) holding them together, the sero is essentially a series of diminishing triangular traps that fit into one another. A kind of fencing directs the fish to swim towards these chambers, where they go from one to the next until they can no longer turn around and leave.

8. In his description of "seatrees," that is coral, in his *Herbal*, Rumphius equates these fishing traps with "*fuiken*," and says that they are also known as *Bobbers*, *Bubut* in Malay. It is a trap, usually made from wicker or some such material, with compartments that are divided by hoops and that become increasingly more narrow. There is some resemblance, at least in principle, to our New England lobster trap. *Weel* has been used in English since the middle of the thirteenth century, and the following definition from 1725 (in the *OED*) clearly resembles the *fuik*: "Weel . . . made of Osier-twigs, which are supported by Circles or Hoops, that go round, and are ever diminishing. . . . Its Mouth is somewhat Broad, but the other end terminates in a Point: It's so contrived, that when the Fishes are

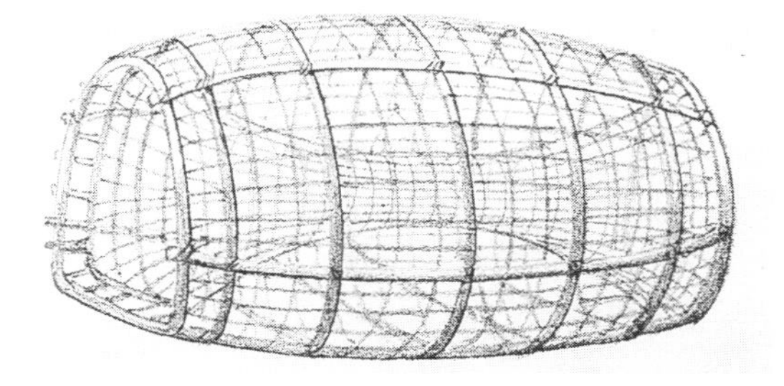

got in, they cannot come out of it again, because of the Osier Twigs, which advance on the inside, to the Place where the Hoops are, and which stop the Passage, leaving but a small opening there."

9. Charles César de Rochefort (1605–?1690) was a seventeenth-century theologian and renowned minister of the Walloon church in Rotterdam. The work Rumphius refers to is *Histoire naturelle et morale des Iles Antilles de l'Amerique. Enrichie de plusieurs belles Figures des raretez les plus considerables qui y sont décrites. Avec un vocabulaire Caraibe* (A Rotterdam, Chez Arnout Leers, 1658). Jean Baptiste Du Tertre (1610–1687) accused Rochefort of having copied the following work of Du Tertre's: *Histoire generale des isles de S. Christophe, de la Gvadeloope, de la Martinique et avtres dans l'Amerique* (published in 1654), and then printed it as his own under the title just given.

Be that as it may, it is an interesting book that is well written. There was a second, revised edition in 1665, also published in Rotterdam, and it appeared in English in 1666 in London under the title *The History of the Caribby-Islands . . . In Two Books, The First Containing the natural; the Second, the Moral History of those Islands.* The translator was John Davies, and it is from his translation I will be quoting throughout these annotations. An integral Dutch translation was published in 1662, also by Arnout Leers in Rotterdam, entitled: *Natuurlyke en Zedelyke Historie van D'Eylanden, de Voor-Eylanden van Amerika, etc.* The translator was H. Dullaart. The title page of the Dutch edition states that the translation was supervised by Rochefort and that many things were added. It also states that Rochefort had been a preacher in America.

Rumphius often refers to Rochefort, yet despite the latter's attractive style and engaging deference to nature's wonders, one can't help noticing how much more precise Rumphius is, how, as it were, his descriptions keep the animal alive in words, while Rochefort usually makes only general comments. For instance, Rochefort's description of what he calls "Homars" is only eight printed lines. What Rumphius is paraphrasing is: "The Inhabitants of the Islands take them in the night time upon the sands, or in the Shallows near the low-water-mark; and with the assistance of a Torch, or Moonlight, they catch them with a little iron fork" (p. 119).

10. I chose this simple spelling on purpose, because what Rumphius used in the original is even odder: "Karibaners." I am presuming he is referring to the inhabitants of those islands, an island group he spells with a "k": "Karibische Eilanden."

11. One of those tireless compilers that belong to the earlier history of natural science, Conrad Gesner (1516–1565) had a difficult youth. When Gesner was still a teenager, his father, a follower of Zwingli, was killed, along with Zwingli, at the Battle of Kappel on 11 October 1531, and young Conrad was left without sufficient financial resources. But he succeeded in his studies in Paris, and in Montpellier, where he attended the lectures of Rondelet (see notes to ch. 22 of book 1). The stamina of people like Gesner is amazing. He traveled a great deal, no sinecure in the sixteenth century, practiced medicine (in fact, while administering to victims of the plague in Basel, he succumbed to the disease himself), and helped various colleagues. Yet before he was thirty he published his huge *Biblioteca universalis*, a catalogue of all past writers in Latin, Greek, and Hebrew, and at the age of twenty-four he was appointed a professor of medicine, physics, and ethics in Switzerland. He was a scholar of Greek, Latin, and Hebrew, a botanist, and the owner of a curiosity cabinet that contained items he had collected during his many travels. Rumphius is alluding to Gesner's *Historia animalium*, published in Zurich in four volumes between 1551 and 1587. More than 4,500 pages long, this huge work represents the beginning of modern zoology. The particular reference is to the third volume, which describes fishes and other aquatic animals: *Historiae Animalium. Liber III. qui est de Piscium & Aquatilium animantium natura* (Zürich: Christoph. Froschover, 1558).

12. This was the physician Hendrik d'Acquet (1632–1706), who became the mayor of Delft. He collected primarily naturalia which he obtained from VOC captains and other personnel. Rare specimens which were not in his collection he often had copied by reputable draughtsmen, while, conversely, he permitted items he possessed to be drawn and published by others. This he did, for instance, with Rumphius' book, which, as was mentioned, he also was instrumental in getting published.

D'Acquet maintained a private garden where he cultivated tropical plants. Like so many other private collections, his was auctioned off after his death and dispersed. D'Acquet also owned a large collection of watercolor images and oils of a number of animals, including many of the ones Rumphius mentions. This collection is now in the Royal Tropical Institute (Koninklijk Instituut voor de Tropen) in Amsterdam.

13. Figure A on Plate 1 is indeed a Spiny Lobster (*Panulirus homarus* [L.]), but B, according to Holthuis, is a freshwater shrimp *Macrobrachium rosenbergii* (De Man), subspecies (subsp.) *dacqueti* (Sunier). The subspecies was called after d'Acquet.

Chapter 2. Of the Ursa Cancer. Udang Laut Leber.

1. A "span" was a measurement calculated as the distance between thumb and little finger of a hand with the fingers stretched out. This is about nine inches.

2. The width of a hand was formerly three inches.

3. The Latin means "Bear Crab" and "Broad Squilla," where *squilla* refers to "any long-tailed crustacean, whether lobster, crayfish or shrimp," as Holthuis has noted. This "Sculptured mitten lobster" is, according to Holthuis, *Parribacus antarcticus* (Lund), and is pictured on Plate II, fig. C. Figure D is *Thenus orientalis* (Lund). The Malay should have been *Udang laut lebar*, a large marine langouste. *Lebar* means "wide."

Chapter 3. Of the Squilla Arenaria. Pincher. Locky.

1. I don't know what a "French" *knipmes* might specifically be, but what was called a "Springknife" was a knife with a blade that could be folded into the haft, what we would call a jackknife. In Spain it was known since the Middle Ages as a *navaja*, and described at the time as a knife with a blade that was attached and between two hafts.

2. The Latin adjective Rumphius adds here, *pinnatus*, means "feathered," and by extension "having fins."

3. Holthuis identifies this as *Lysiosquillina maculata* (Fabricius). Rumphius' Latin means: "Terrestrial Sand Squilla." The Malay means simply "Sea Shrimp."

4. Rumphius calls this creature by contrast "Marine Sand Squilla," and Holthuis identifies this as *Odontodactylus scyllarus* (L.).

5. This specific Rumphian name means "Swan Squilla," because the entire claw was used to make trinkets resembling swans.

6. This is still the same creature: "Swan-carrying" because of what was just mentioned in n. 5, while "Lokki" or "Lokky" was its common Ambonese name.

7. Both methods of fishing use a kind of trap. See nn. 7 and 8 in ch. 1.

8. Reading "betrappen" for "betrakken."

Chapter 4. Of the Squilla Lutaria. The Mudman. Udang Petsje.

1. Holthuis identifies this species—which in Latin literally means "Squilla that Lives in Mud"—as *Thalassina anomala* (Herbst). The animals on Plate V are not the correct ones. Letters K and L are, according to Holthuis, land hermit crabs of the genus *Coenobita*. Letter K shows *Coenobita brevimanus* (Dana), while letter L is *Coenobita* (spec.). From a utilitarian viewpoint, this Mud Lobster has little to recommend itself, as Rumphius noted nearly three centuries ago. See also Holthuis's comments in: *FAO Species Catalogue. Vol. 13. Marine Lobsters of the World. An Annotated and Illustrated Catalogue of Species of Interest to Fisheries Known to Date*, ed. L. B. Holthuis (Rome: Food and Agriculture Organization of the United Nations, 1991), pp. 229–231.

Chapter 5. Of the Cancer Crumenatus. Catattut. Purse Crab.

1. "Canary" refers to the tree commonly known in Malay as *kanari*. "Bread, a hard and nourishing kind called *Bagea*, was made of the seeds of *Canarium Indicum L.*, which were kneaded to a paste with *sago mentah* and then baked. The leaves of *Pandanus bagea Miq.* were used for baking this Kanari bread" (Van Slooten, in *M*, p. 307). Crawfurd, in his *History of the Indian Archipelago* (London, 1820, 1:383), praises the *kanari* (or *kenari* in Javanese) as follows: "Of all the productions of the Archipelago the one which yields the finest edible oil is the *kanari*. This is the large handsome tree, which yields a nut of oblong shape nearly of the size of a walnut. The kernel is as delicate as that of a filbert, and abounds in oil. This is one of the most useful trees of the countries where it grows. The nuts are either smoked and dried for use, or the oil is expressed from them in their recent state. The oil is used for all culinary purposes, and is more palatable and finer than that of the coconut. The kernels, mixed up with a little sago meal, are made into cakes and eaten as bread. The *kanari* is a native of the same country with the sago tree, and is not found to the westward. In Celebes and Java it has been introduced in modern times through the medium of traffic."

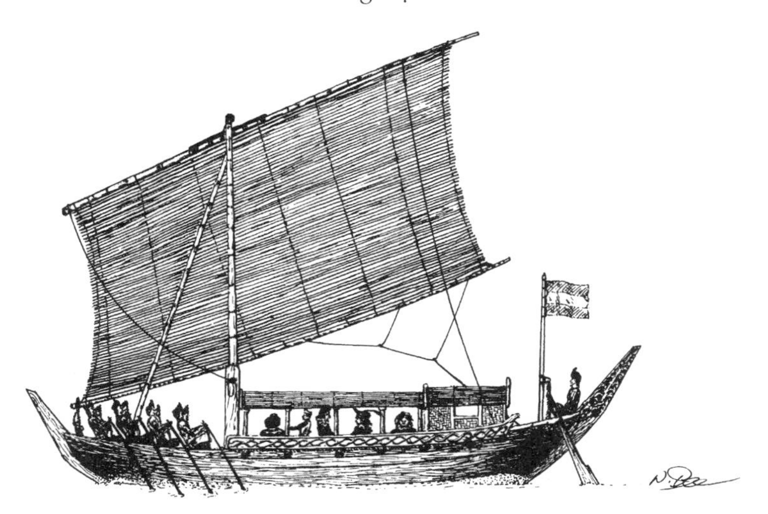

2. This should be "orembaai," a boat particular to the Moluccas. It had a keel, no outriggers, while the prow and stern curved up and over the deck and were usually decorated with carvings. It usually had one mast like a tripod, with one or two *tanja* sails, that is to say, canted rectangular sails. In the absence of wind it could be rowed, sometimes with as many as twenty rowers. The illustration was drawn by Nick Burningham of the Western Australian Maritime Museum; he also provided me with the above information.

3. What Rumphius had in his chapter title is correct: *crumenatus*. This derives from *scrumena* (which contains the same root as *scrotum*), and as *crumena* or *crumina* is Latin for a small money purse which was once worn suspended from one's neck. This is a tropical land hermit-crab, *Birgus latro* (L.), or the robber- or coconut crab.

4. "Cattam" is *ketam*, or Malay for "crab," and "calappa" is *kelapa*, the ubiquitous coconut tree (*Cocus nucifera*), a word that refers to both the nut and the tree.

5. The Uliasser Islands are three islands, frequently mentioned by Rumphius, east of Ambon and south of western Ceram. The names of the islands present some problems. The three are known today as Haruku, Saparua, and Nusa Laut. In Rumphius' day Haruku was also known as "Oma," Saparua was also known as "Honimoa" and "Muar," and Nusa Laut was written as one word: "Nusalaut." Ceram was known as "Seran," and Rumphius' "Mulana" is the small island Pulau Molana, that lies between Haruku and Saparua.

6. "Nusatello" refers to three small islands off the western coast of Ambon's Hitu peninsula, directly south of the Howamohel peninsula of Ceram. They were also known as "The Three Brothers": Lain, Hatala, and Ela.

7. Leytimor is Ambon's southern peninsula, where the city of Ambon is located. Ema was a town about eight miles from the city.

8. "Lussapinje" is a misspelling of "Lucapinho," which referred to a group of small islands alternatively known as the Nusa Penju (also: Nusapinyu) Islands, or Turtle Islands. This small archipelago is far south of Ambon, in the center of the Banda Sea, between Ambon and the Damar Islands. Its neighbors are the Lucipara Islands.

9. The Banda (also called Nutmeg) Islands, which include Pulo Nera (Palmwine Island), Lontar (the name of the palm that provides leaves for writing), Pulo Ai (Water Island), Gunung Api (Fire Mountain, i.e., Volcano), Pulo Run (or Puloron, Chamber Island), Rosingen or Pulo Rozengain, Pulo Kapal (Ship Island), Pulo Pisang (Banana Island), and Pulo Suwanggi (Sorcery Island). Rumphius states that three of these—Pulo Nera, Lontar Besar, and Pulo Ai—were the real centers for these nutmeg trees. Pulo Ai, he feels, grew the best of them and though very small and with little fresh water, gave the appearance of "a continual garden."

The irony seems to be that the natives had little use for the products of these trees but were encouraged in their cultivation by the Chinese and Hindus. The Malay and Javanese were the traders with China and India. The first Europeans to reach them were the Portuguese, led by Antonio d'Abreu, a lieutenant of d'Albuquerque. After nearly a century of their presence, the Dutch displaced the Portuguese in 1609. In a number of ruthless wars they defeated the local Banda peoples and established a commercial monopoly by having the trees grown only on assigned plantations, or what they called "parks" (*perken* in Dutch) and forcing the owners to sell the mace and nutmeg to the VOC at a fixed, but low, price. Rumphius describes the tree, its fruit, and its uses in his *Ambonese Herbal*, bk. 1, chs. 5–7.

10. *Ternate* is an island off the western coast of the larger one called Halmahera or Gilolo. As Crawfurd puts it, Ternate is "the mere pedestal on which stands the active volcano of the same name." Ternate is one of the five islands which were the original Clove Islands or the original Moluccas. The other four—all lying off the western coast of Halmahera—are Tidor, Motir, Makian, and Batjan. These five, along with the ten Nutmeg Islands mentioned before, formed the fabled group known vaguely as the Spice Islands and were the reason why the Portuguese and the Dutch ventured into the eastern archipelago.

11. "Tafuri" might be Tifore Island (also: Tafore), a small island all by itself in the Moluccan Sea, between the northern peninsula of Celebes (Minahasa) and Halmahera. Hiri is a small island off the north coast of Ternate. "Bliau" is most likely "Belau" or Ambelau, the smaller island next to Buru.

12. Rumphius used the word "*dammers*" as well as "*toortsen*." Both words mean "torches," but the first one, better spelled *damar* (*dammar*), is the Malay word for the resin, the resinous torch, and the tree that produces the resin, for instance the genus *Hopea*, once particularly prominent in the Moluccas. Since the distinction is of no great practical difference, I stuck to a single use of "torches."

13. *Rixdollar* is the English spelling for the name of a silver coin current from the sixteenth century on in Holland (*rijcksdaler*, or modern form *rijksdaalder*), Sweden (*riksdaler*), Denmark (*rigsdaler*), Germany and Austria (*reichsthaler*). Particularly known during Holland's commerce with the East. It was a silver coin of fifty stivers (*stuivers*), worth in English money at the time, four shillings and nine pence (Sewel 1691).

14. The *schelling* (cognate with English *shilling*) was the twentieth portion of a "pound," or worth six stivers (*stuivers*). There were a number of these coins and each varied as to exact value.

15. "Siriboppar" is most likely *sirih popar*, or *Ficus leucantatoma* Poir. The roots of this tree were considered an antidote to dizzy spells and the sting of poisonous fishes.

16. "Calbahaar" is probably "Acarbahaar" or "Acarbarium," a coral identified by Bayer (*M*, p. 234) as *Toeplitzella regia* and *Isis hippuris* (*M*, p. 240).

17. For the *Cancelli* see ch. 22 of this book. The link between the large coconut crab (*Birgus latro*) and the small hermit crabs is clear during the initial stages of its life. The larvae are released at sea and after some maturing, pick a shell to live in just like their smaller counterpart, the hermit crabs. They emerge on the shoreline and change shells when their size increases. After reaching a certain size, however, they stop living in shells and move to burrows. They forage at night, because the tropical sun of daytime hours would dry them out. On tropical atoll islands, the coconut crabs are usually the largest animal, easily attaining a weight of more than seven pounds and "measuring three feet from leg tip to leg tip." They feed mainly on coconuts and seem to practice cannibalism. If captured by islanders, they have to be kept separated, because the larger specimen will invariably kill and eat the smaller ones. See the article (with superb photographs) by Gene S. Helfman, "Coconut Crabs and Cannibalism," in *Natural History* 88, no. 9 (November 1979), pp. 77–83.

18. For Rochefort see n. 9 in ch. 1. The "Painted Crabs" as described by Rochefort do not resemble the coconut crab at all but are true land crabs, *Gecarcinus ruricola* (L.). In the English translation, this is indeed the third section but of chapter 22 of bk. 1 (pp. 140–142), where it is said that they use their claws, or "mordants" (!), to defend themselves, "and they ever and anon strike them one against another to frighten their enemies" (p. 140). Rumphius also quotes correctly that when these crabs come down from the mountains in May, they do so in huge numbers, looking "like an Army marching in rank and file [and] they never break their ranks" (p. 141), but coconut crabs don't do this, nor do they migrate in great number to the shore. Darwin encountered this remarkable animal in the Cocos or Keeling Islands in 1836 and described it as follows:

"I have before alluded to a crab which lives on cocoa-nuts: it is very common on all parts of the dry land, and grows to a monstrous size: it is closely allied or identical with the Birgos latro. The front pair of legs terminate in very strong and heavy pincers, and the last pair are fitted with others weaker and much narrower. It would at first be thought quite impossible for a crab to open a strong cocoa-nut covered with the husk; but Mr. Liesk assures me that he has repeatedly seen this effected. The crab begins by tearing the husk, fibre by fibre, and always from that end under which the three eye-holes are situated; when this is completed, the crab commences hammering with its heavy claws on one of the eye-holes till an opening is made. Then turning round its body, by the aid of its posterior and narrow pair of pincers, it extracts the white albuminous substance. I think this is as curious a case of instinct as ever I heard of, and likewise of adaptation in structure between two objects apparently so remote from each other in the scheme of nature, as a crab and a cocoa-nut tree. The Birgos is diurnal in its habits; but every night is said to pay a visit to the sea, no doubt for the purpose of moistening its branchiæ. The young are likewise hatched, and live for some time, on the coast. These crabs inhabit deep burrows, which they hollow out beneath the roots of trees; and where they accumulate surprising quantities of the picked fibres of the cocoa-nut husk, on which they rest as on a bed. The Malays sometimes take advantage of this, and collect the fibrous mass to use as junk. These crabs are very good to eat; moreover, under the tail of the larger ones there is a great mass of fat, which, when melted, sometimes yields as much as a quart bottle full of limpid oil. It has been stated by some authors that the Birgos crawls up the cocoa-nut trees for the purpose of stealing the nuts: I very much doubt the possibility of this; but with the Pandanus the task would be very much easier. I was told by Mr. Liesk that on these islands the Birgos lives only on the nuts which have fallen to the ground.

Captain Moresby informs me that this crab inhabits the Chagos and Seychelle groups, but not the neighbouring Maldiva archipelago. It formerly abounded at Mauritius, but only a few small ones are now found there. In the Pacific, this species, or one with closely allied habits, is said to inhabit a single coral island, north of the Society group. To show the wonderful strength of the front pair of pincers, I may mention, that Captain Moresby confined one in a strong tin-box, which had held biscuits, the lid being secured with wire; but the crab turned down the edges and escaped. In turning down the edges, it actually pinched many small holes quite through the tin!" (Charles Darwin, *The Voyage of the Beagle*, ed. Leonard Engel [Garden City, N.Y.: Doubleday & Co., 1962], pp. 461–462).

Chapter 6. Of the Cancer Saxatilis. Cattem Batu.

1. "*Saxatilis*" means in Latin "that which dwells among rocks"; hence Rumphius' name means "Crab That Dwells among Rocks." Holthuis states that this crab is *Scylla serrata* (Forsskål). The Malay means "Stone Crab."

2. These are mangroves.

3. *Calappus* is the coconut. I have maintained this usage throughout, with a few rare exceptions. No one in Indonesia at the time (or later, for that matter) used the word "coconut" but used either *calappus* or *klapper.*

4. D'Acquet's well-meaning contributions are, once again, off the mark. None of the creatures displayed in Plate V match Rumphius' descriptions. Letter M is *Lophozozymus pictor* (Fabricius).

Chapter 7. Cancer Marinus. Cattam Aijam.

1. The Latin means "Smooth Sea Crab," where "laevis" should have been *levis.* Holthuis says this is possibly *Charybdis feriata* (L.), while the crab displayed on Plate VI by the letter N. is *Carpilius convexus* (Forsskål). Holthuis notes that Rumphius' description of edible crabs (and whether they are dangerous or not) on pp. 9–26 of *The Ambonese Curiosity Cabinet* is the first European discussion of this subject for tropical crabs. His observation here that a crab may prove to be harmful after it has eaten noxious food is testimony to Rumphius' precise observations and insight. See L. B. Holthuis, "Are There Poisonous Crabs?" *Crustaceana* 15:2 (1968), pp. 215–222.

2. This Latin name means "Furrowed Sea Crab," which Holthuis cannot identify with certainty, offering: *Charybdis* aff. *acutifrons* (De Man). The crab depicted by the letter O on Plate VI is the ordinary European shore crab, *Carcinus moenas* (L.).

3. De Wit identifies "Arbor excoecans" as *Excoecaria agallocha* L. This is a small tree infamous for its milky sap, which was once said to induce blindness. It was used as a poison to catch fish.

4. "Cataputia" is, as Holthuis suggests, most likely *Cajuputih* or *Melaleuca leucodendron* L. The fruits are indeed between 3 and 4 millimeters wide and less than 3 millimeters thick, hence can easily be called "kernels."

5. The crab that Schijnvoet assigned to the letter P, *Charybdis feriata* (L.), was identified by De Man as *Goniosoma cruciferum* (Fabricius). Another synonym is *Cancer cruciata* (Herbst). It inhabits the Indian Ocean, Indonesia, and the seas of China, Japan, and Australia. It has a spot on its back that resembles a cross, and De Man conjectures that it might have been this crab that inspired the legend that St. Francis Xavier lost a cross into a raging sea when he tried to calm it, but a crab brought the implement back to him the next day on Ceram. The legend has been incorporated in the dramatic statue that stands today in the city of Ambon, in front of the Catholic cathedral, depicting the missionary brandishing a cross in his outstretched hand, a Bible in the other, while at his feet a crab is holding a crucifix in its claws. St. Francis Xavier (1506–1552), the "Apostle of the Indies," included Ambon in his missionary efforts for the Portuguese in Asia. The Jesuit visited Ambon in 1546 from February to June, converting a goodly number of people to the Catholic faith. See H. J. de Graaf, *De geschiedenis van Ambon en de Zuid-Molukken* (Franeker: Wever, 1977), pp. 35–39.

Chapter 8. Of the Pagurus Reidjungan.

1. For the original's "*zegerynleer,*" once more commonly spelled in Dutch as *segrijn* (also chagrin), and spelled "shagreen" in English. A Persian word, it referred to a tough, coarse leather with a grainy surface, usually made from the skin of donkeys, horses, and, later, rays and sharks. Once commonly used in the Orient for making slippers.

2. A term from antiquity for genera of crustacea. Both De Man (p. 103) and Holthuis (p. 112) state that Rumphius is really talking about two species. The first one (and the one depicted on

Plate VII, letter R) is *Portunus pelagicus* (L.) and the second one, with the three large red spots, is *Portunus sanguinolentus* (Herbst).

3. "*Cattam bulan*," or *ketam bulan*, is simply "Moon Crab." "*Reidjungan*" is the Malay word *radjungan*, an Indonesian name for large crabs of the genus *Portunus*, which were a favorite commodity on the fishmarkets. See H. C. Delman and J. G. De Man, "On the 'Radjungans' of the Bay of Batavia," *Treubia* 6 (1925), pp. 308–323.

4. The "*scheer-vogel*" (better "*Schaarvogel*") is the frigate bird, also known as "man-o'-war bird," a rapacious fowl that goes by the lovely Latin name of *Fregata ariel* (Gould). Rumphius also mentions it in his *Herbal*, bk. 7, ch. 24 (p. 45).

Chapter 9. Of the Cancer Lunaris. Cattam Bulan.

1. The Latin and the Malay mean the same thing: "Moon Crab." *Cattam Bulan Trang* would mean "Moonlight Crab," because *terang bulan* means "moon light," from *bulan* (moon) and *terang* (shining). Holthuis calls this delectable crab *Ashtoret lunaris* (Forsskål).

2. By "*Dammar*" Rumphius means a resin torch.

Chapter 10. Of the Cancer Caninus. Cattam Andjin.

1. The Latin and the Malay mean exactly the same thing. Holthuis offers that this is a *Cardisoma* sp.

2. This refers to the *kanari* or *kenari* nut, which is the size of a walnut, and is the fruit of the *Canarium Indicum* L. tree. See n. 1 in ch. 5.

3. Holthuis identifies this crab as *Ocypode ceratophthalmus* (Pallas).

4. These nimble crabs were called "Hippeus" [ἱππεύς] by Aristotle, which means "riders" or "horsemen," and described as follows: "On the Phoenician coast there are some that they call horsemen, because they run so fast that it is difficult to catch them, and when opened, they are empty, because they have no pastures" (*Historia Animalium*, bk. 4, ch. 2, section 2). I am quoting from a nineteenth-century translation: *Aristotle's History of Animals in Ten Books*, transl. Richard Cresswell (London: Bell & Sons, 1883), p. 77. This is indeed the same genus, because as Holthuis points out, Aristotle is describing *Ocypode cursor* (L.) which occurs in the eastern Mediterranean and on West African coasts.

5. This is not clear, and might be due to the spelling. Rumphius is not referring to *hanze* or *hanse*, a variation of *hansa*, a league or guild, as, for instance, the German Hansa League. This might be a reference to *hamsa* or *hamzah*, however, the unvoiced glottal occlusive in Arabic. "Hamza" is the orthographical sign *alif*, or the first letter of the Arabic alphabet. It roughly corresponds to our "h." This has nothing to do with linguistics, of course, but more with being the *leader* of the alphabet. Hamza is also the name of the paternal uncle of the Prophet, a man of both historical and legendary distinction.

6. That is, cockatoo.

7. "Lussapinju" is a misspelling of "Lucapinho," which referred to a group of small islands alternatively known as the Nusa Penju (also: Nusapinyu) Islands, or Turtle Islands. This small archipelago is far south of Ambon, in the center of the Banda Sea, between Ambon and the Damar Islands. Its neighbors are the Lucipara Islands.

8. "Pandang shrubs" most likely refers to the *pandanus* plants, of the family Pandanaceae, that grow on the coast and in swamps, but also in the mountains. At the coast it often has air roots. The wood, leaves, fruits, and flowers all have a use.

Chapter 11. Of the Cancer Raniformis.

1. This "Crab in the Form of a Frog," as the Latin translates, is identified by Holthuis as *Ranina ranina* (L.).

2. The Malay should be spelled *ketam kodok* and also means "Frog Crab."

Chapter 12. Of the Cancer Terrestris Tenui Testa. Cattam Darat.

1. The Latin means: "Of the Land Crab with the Thin Shell." The Malay means "Dry-land Crab," since *darat* means "dry land" as opposed to the coast.

2. This "Red-eyed Crab" was tentatively identified by Holthuis as *Eriphia sebana* (Shaw & Nodder).

3. This should be "*Cancer Noxius*" and is considered by Holthuis the same species as the above. For this animal see ch. 16.

4. These ingredients are: scrapings of black coral, roots of the *Ficus leucantatoma* (Poir.) tree, *pissang swangi* (a short, thick banana, good for frying), *lalang* grass (a grass better known as *alang-alang*, of which the young roots were used for medicinal purposes), and the *pinang* or areca nut, used when chewing betel (*Piper betle*). The banana seems not to fit here. Rumphius describes it in ch. 2 of

book 8 of his *Herbal;* no. 10 is the "*pissang swangi*" banana, which is not suitable for eating raw, but very good if fried (p. 132). It obviously resembles the *plátanos* from Latin America. In the next ch. 3, Rumphius says why it is a medicinal ingredient. The root of this banana variety, after it is rubbed with water and then drunk, was considered an antidote against bad food (p. 136). Nor should the word "*swangi*" be confusing. *Suangi* most often means "magician" or "casting a spell," but Rumphius says that in his day it meant not only "wizards and witches" but also "wild, uninhabited, dangerous places, rocks, and islands," and, finally, fruits or plants that looked rough or coarse, or had a "resistant nature" (*wederspannigen aard;* book 7, ch. 7, p. 17).

5. See ch. 28 of this book.

6. Holthuis: *Pilumnus vespertilio* (Fabricius). Rumphius' Latin means the "Shaggy Crab," while the Malay means the same thing, with *bulu* referring to being hairy on one's body, not the head.

Chapter 13. Of the Cancer Vocans. Cattam Pangel.

1. The Latin means "The Gesturing Crab" and is identified as *Uca vocans* (L.) by Holthuis. The Malay most likely refers to *panggil* (*memanggil*), which means "to summon" and in a loose sense, "to call," hence in Malay the "Summoning Crab."

2. Pliny (in book 9, but paragraph 97 in the Loeb ed.) does mention the "Hippa" (not the "Mippi," of course), stating only that it is very swift. This is *Ocypode cursor* (L.).

3. See n. 4 in ch. 10.

4. "Bellonius" is Pierre Belon (1517–1564), physician and French naturalist, author of various works on fishes and birds, who also traveled extensively in Greece, Asia Minor, Egypt, Arabia, and Palestine between 1546 and 1549. It is most likely that Rumphius is referring to Belon's book concerning these journeys: *Les observations de plusieurs singularités et choses mémorables trouvées en Grèce, Asie, etc.*, first published in Paris in 1553, and a larger edition in 1555, printed in Antwerp.

5. "Syrtibus" is really Syrtes or Syrtis, better known as the Gulf of Sidra, shallow waters off the coast of Libya. Probably to protect their trade, the Phoenicians circulated stories that the Syrtic Sea was dangerous.

Chapter 14. Of the Cancer Spinosus. Cattam Baduri.

1. This first variety of the "Thorny Crab" is, according to Holthuis, *Micippa cristata* (L.).

2. The Malay also means "Thorny Crab"; *duri* means "thorn" and is the radical of the noun *durian*.

3. *Pikol* has to do with carrying, and *rompot* (*rumput*) refers to weeds.

4. The second variety is, according to Holthuis, *Parthenope longimanus* (L.). The Latin "Longimanus" referred to someone having a long forearm, and was the epithet of Artaxerxes, king of Persia.

5. Sewel calls an "ell" also a "cubit," which was an ancient measurement equivalent to the length of one's forearm, or between eighteen and twenty-two inches long.

6. "Cateam" should be "Cattam," the Malay word for crab.

7. This third kind is said by Holthuis to be a species of *Micippa*.

8. This rare crab, depicted on Plate IX, by itself, is, according to Holthuis, *Daldorfia horrida* (L.).

9. In the original, "*kraalboortjes*." Holthuis, correctly it seems, suggests that this is a misprint for "*kraalboompjes*," Rumphius' noun for "coral." The coral mentioned here appears to belong to *Acropora* (*M*, p. 251).

Chapter 15. Of the Cancer Floridus. Cattam Bonga.

1. Holthuis identifies this as *Atergatis floridus* (L.). The Malay means the same thing as Rumphius' Latin: "Flower Crab."

2. Modern science has proven this observation to be correct. The toxicity of crabs has now been studied rather carefully, and it has been found that crabs with black pincers are indeed often toxic, with these Xanthidae as a prominent example. See Holthuis, "Are There Poisonous Crabs?" *Crustaceana*, pp. 216–217.

Chapter 16. Of the Cancer Noxius. Cattam Pamali.

1. See next chapter.

2. "On that spot," taking "*tersteede*" to be "*ter plekke*."

3. Holthuis guesses that this "noxious crab" is *Eriphia sebana* (Shaw & Nodder). A nice touch is "infaustus": this is then the "Crab That Deals Misfortune." "Cattam Pamali" is probably *ketam pemali* or "Taboo Crab," while "Cattam Bisa" means "Venomous Crab."

Chapter 17. Of the Cancer Ruber. Cattam Salissa.

1. Holthuis (personal communication) explains this curious description of the *Carpilius'* carapace as follows: "The dorsal part of the shield covers the upper part of the body completely and at the lateral margins folds under; the ventral part reaches to the implantation of the legs. The part closest to the legs is half-moon shaped and separated from the rest of the shield by a distinct groove. The half-moon shaped part, called the pterygostomian region, of dry shields sometimes breaks off along the groove, making the lower surface of the shield incomplete, but the dorsal surface of the shield remains intact."

2. Perhaps this tree is *Terminalia catappa* L. These are tall trees that usually grow near the coast. The leaves are used for a tanning material, and the seeds are eaten and used in cooking like almonds. See the *Herbal*, bk. 1, ch. 58 (pp. 174–176).

3. "Caybobbo," or Kaibobo, is a village on the coast of what used to be called "Big Ceram," that is, the main part of that big island, not the narrow peninsula Howamohel, or "Little Ceram." Kaibobo is situated across from the small island called Pulau Babi in Piru Bay.

4. Boano is the island northwest of Ceram's Howamohel peninsula.

5. Holthuis states that this "Red Crab" is *Carpilius maculatus* (L.).

6. This sentence is corrupt. Holthuis states that this "Copper Crab" is *Lophozozymus pictor* (Fabricius). Both the Latin and the Malay (*tembaga*) mean "Copper Crab."

7. This is a village on the northern coast of Howamohel, below Boano.

8. Although a very common last name in Holland, this might be Jan de Jong (?-1712), who was an accountant for the VOC. He seems to have had a large collection of shells and marine fauna, as well as porcelain, coins, etc.

Chapter 18. Of the Cancer Nigris Chelis. Cattam Gigi Itam.

1. Assahudi is a village on the northern coast of Howamohel, Ceram's western peninsula.

2. "Nussanive" refers to Nusanive, a town and cape on the southern tip of Leytimor, or Ambon's southern peninsula.

3. *Hic niger est, hunc tu, Romane, caveto* is from the *Satires* by Horace (book I, 4th Satire, line 85) and means: "This is black, and you, Roman, beware of him."

4. The Latin means the "Black-Clawed Crab," while the Malay means indeed "The Crab with the Black Tooth." Holthuis conjectures that this is *Etisus utilis* (Lucas), while calling the smaller animal *Zosimus aeneus* (L.).

Chapter 19. Of the Cancer Lanosus. Cattam Bisa.

1. Rumphius' Latin means "Woolly Crab," in Malay "Venomous Crab." Holthuis identifies it as *Dromia dormia* (L.).

2. This Latin means "Poisonous Crab" as well.

3. Boano (almost always misspelled in this edition) is the island off the northern coast of Howamohel.

4. Serua is a lonesome island in the eastern Banda Sea. Its closest neighbors are the Damar Islands to the south.

5. "Opblazer" in the original, or the Puffer Fish, of the family *Tetraodontidae.* Called *fugu* in Japan, it is the notorious fish which, if not properly cleaned, can kill you.

6. "Basta laut" or "Basta Marina" is discussed in the *Herbal*, book 12, ch. 29. It seems to be a sponge coral, that is said to be red while under water and that turns black when removed from the sea. It smells like the sea, but when burned it smells "like shrimp" (p. 253). Bayer (*M*, p. 241) says it is a true sponge, *Ianthella basta* (Pallas).

Chapter 20. Of the Cancer Calappoides. Cattam Calappa.

1. This means "Calappus Crab" or "Coconut Crab."

2. This is the nominative case of "Calappus," and it means the same thing as given above. Linnaeus kept the name because Holthuis classifies this animal as *Calappa calappa* (L.).

3. The Latin and the Malay mean the same thing: "Moon Crab." See ch. 9.

4. See ch. 23, no. XIV, in book 2.

5. Pierre Belon's name for the crab means "Crab from [the city of] Heraclea."

6. Jan Jonston (1603–1675) was born in Poland of Scottish parents. He studied at a number of European universities, finally obtaining his medical degree from Leiden University in 1634. His interest in the animal world led him to visit various private collectors, including the well-known Paludanus in Enkhuizen. The volume Rumphius refers to was Jonston's most important publication, an encyclo-

pedia of the animal world, entitled *Historiae naturalis*, published in Amsterdam in 1657, and in a Dutch translation in 1660. It seems to have been used quite often by collectors as a reference tool.

7. For Gesner see n. 11 of ch. 1.

Chapter 21. Of the Cancer Perversus. Balancas.

1. Jacob Bontius (1591–1631) was the son of a professor at Leiden University. When he was twelve, he was admitted as a student and obtained his medical degree in 1614. In 1627 he sailed to Java with the fleet of J. P. Coen as its medical authority. He survived the two famous sieges of Batavia by the ruler of the Mataram realm, Sultan Ageng, in 1628 and 1629, though he was ill for four months with malaria, dysentery, and beriberi. His first wife died on the outward journey, his second wife died from cholera in Batavia, and his third wife survived. Bontius wanted to return to Holland and be appointed professor in Leiden, but he never saw Holland again and died in Batavia. What is remarkable is that in less than four years of tropical experience, while burdened with the responsibilities of a medical man in what was still practically an outpost, in addition to being appointed justice of the peace (*advocaat fiscaal*) and to the position of bailiff, Bontius managed to gather and record botanical information. His books were published only after his death. One was published in Leiden in 1642, as *De Medicina Indorum Lib. IV*, of which book 1 contained notes to the work of Garcia ab Orta. But Rumphius is referring to Jacobi Bontii, *Historiae Naturalis & Medicae Indiae Orientalis*, ed. Gulielmi Pisonis, printed with other texts in Gulielmi Pisonis, *De Indiae Utriusque Re Naturali et Medica* (Amstelaedami: Elzevirios, 1658), pp. 83–84.

2. The noun is too general, since *katjang* was the common word for anything resembling a bean. However, since "Javanese" was added, Rumphius might be thinking of *katjang djawai*, *Arachis hypogaea* L., of the *Leguminosae* family, better known to us as peanuts. One has to remember that these were not necessarily a familiar product in Rumphius' day. Also known as *katjang tanah* (earth bean), *katjang tjina* (Chinese bean), or *katjang pendem* (underground bean), they were commercially sold on European markets as "monkey nuts," "Curaçao almonds," or "oil nuts."

3. *Bocassan* was a mussel sauce; see ch. 33 of book 2.

4. For "op de binnen—of voorzyde van Java."

5. I had to amplify slightly. Valentijn said that *Sawan* for the Ambonese was an "evil spirit" ("*quaden geest*") or a spirit of the dead ("*des dooden geest;*" 2:142–143); Van Hoëvell concurs (p. 120).

6. The Latin means what it says. I kept "perverse" for its name in English because the word was in use from the fourteenth century, and had the original double meaning of something not being right (which is the classical meaning as well: askew, turned the wrong way, overturned) as well as something subversive or corrupt. Holthuis classifies this as *Tachypleus gigas* (O. F. Müll.). Americans would simply call it the horseshoe crab. The latter is also called the king crab, to add to the confusion, since for most Americans king crabs are caught in Alaska and come to their attention as a delectable variety of crab meat. Rumphius' crab is not a crab but is related (as he unwittingly intimated) to scorpions, spiders, and other members of the class Arachnida, phylum Arthropoda. This creature is the only surviving member of the order Xiphosura, of the class Merostomata, and the two genera, *Tachypleus* and *Limulus*, are found only (respectively) in eastern Asia and eastern North America (from Maine to Mexico). Therefore, while it is common on these shores, it was relatively unknown in Europe. What Rumphius calls the "tail" is now considered a postanal spine.

7. "Balancas" must be *belangkas*, a general word for crab, while in modern Indonesian, it seems to refer specifically to a horseshoe crab.

8. "Wall-Lice" for *weegluizen* (which could also be spelled *weekluis* at the time), also known as *wandluis* or *bedwants* (*Cimex lectularius* L.). This is an unpleasant biting bug that inhabits wooden walls, wainscotting, and the beds of Rumphius' day that were built into the wall (*bedstede*). They gave off an unpleasant odor when squashed and were a familiar nuisance, because one will encounter all kinds of expressions that include these pests, as well as homegrown control efforts such as a late seventeenth-century instruction to smear one's bed with cow dung, though it had to be fresh and mixed with some water.

9. This is the Latinized form of *Charles de l'Escluse* (1526–1609), a Belgian scholar born in Arras, in northern France. He was educated at the Universities of Ghent, Louvain, and Wittenberg and studied medicine under the famous Rondelet at the University of Montpellier. He became a mentor of the son of Fugger, the wealthy banker. In 1564 he traveled to Spain and Portugal and came across Orta's work. In England (in 1571) Clusius found the work in Spanish on medicinal plants of the New World by Nicolas Monardes, a Spanish physician from Seville, entitled (in Clusius' translation) *De simplicibus medicamentis ex occidentali India delatis quorum in medicina usus est*, and during another stay in England obtained a copy of the work by Christoval Acosta, entitled *Tractado de las drogas, y medicinas de las Indias orientales.* Clusius became prefect of the Imperial Garden in Vienna and later

the director of the botanical gardens of the University in Leyden. He succeeded to the chair held by Dodoens at the latter's death.

What Clusius appears to have done is collect works on botany and natural history such as the three Iberian works just mentioned (da Orta, Monardes, Acosta), abridge them, translate the abbreviated version into Latin, then annotate and correct them. The problem arises—especially in the case of Garcia da Orta—that the original work was sometimes supplanted by the later version, so that one will often find references to Carolus Clusius' work when in fact it was someone else's.

Clusius' first collaborative effort was on Rondelet's *De piscibus marinis libri XVIII*, and he also translated Dodoens' *Cruydeboeck* into French as *Histoire des plantes* (1557). Among his original works are a flora of Spain (1576) and one on Hungary and Austria (1583). Rumphius is referring to Clusius' *Exoticoram libri decem* (1605), a significant work on exotic flora.

10. This is an interesting term that came into English directly from Dutch. It meant several things, having to do generally with the configuration of land and sea, but Rumphius has the general mariner's term in mind, who called *voorland* ("foreland") any land that would loom up in front of the ship's bow, whether one was steering for it or not. By extension it came to mean, in Dutch, one's fate, whatever was lying in store, whatever would happen and one could not avoid. This was incorporated in the saying "Oost-Indië zal uw voorland zijn" or "the East Indies will be your foreland," i.e., your fate, with a negative connotation.

Chapter 22. Of the Cancelli. Cuman.

1. With the *Cuman* or *Cancelli*, Rumphius is describing various hermit crabs (Coenobitidae), a decapod crustacean, which protects its uncovered abdomen by occupying the shell of a univalve mollusk. *Cancelli* is Latin for "Little Crabs," and the Malay word *Cuman* may be *Kuman*, i.e., a maggot. Hermit crabs are plentiful in the tropics, and some species are capable of living for a considerable period on dry land.

2. *Whelk* for *hoorntje*, because Rumphius uses the Dutch term in a general sense, though it is properly restricted to univalve mollusks with shells coiled in a spiral; bivalve mollusks in Dutch are known by the general name of *schelpen*. The whelk is a marine mollusk with a turbinate shell, and the past participle of the verb is used attributively for any shell that is twisted or convoluted.

3. *Periwinkles* for *alikruiken*. Periwinkle is the name for a small sea-snail, not to be confused with the evergreen plant of the same name. The prefix comes from Latin *pina*, via the Greek for a kind of mussel, and "winkle" is allied to "winch." In English it is better to use "winkle" by itself as a kind of sea-snail; it also means a convoluted shell. The Dutch word *alikruik* dates from the seventeenth century and comes from the province of Zeeland. The last syllable, *kruik*, means crock, probably to indicate the spiral shape of the shell.

4. *Turbinatis* refers to the Turbo shells (described in chs. 6–8 of book 2), which are mostly tropical and littoral, the latter meaning that they live adjacent to the shore between the high and low water marks. The name derives from the Latin *turbo*, for "whirlwind" and "spinning top." *Turbo* can loosely be applied to any shell that is spiral in shape and resembles a top. Note that the mollusks mentioned by Rumphius are all related by virtue of their form, including the shell Pliny mentions (see below).

5. See ch. 5 and n. 1 of ch. 4. The link between *Birgus latro* and the small hermit crabs is clear during the initial stages of its life. The larvae of *Birgus latro* are released at sea and, after some maturing, pick a shell to live in, just as their smaller counterpart, the Coenobitidae, do. They emerge on the shoreline and change shells when their size increases. After reaching a certain size, however, they stop living in shells and move to burrows.

6. They are described by Rochefort in his *Histoire des Antilles* in ch. 14 of book 1, under the heading of insects, and are called a kind of snail by Rochefort, who describes their behavior as "to take possession of a shell as they find most convenient for them, within which they fit and accommodate themselves, as Souldiers, who having no settled habitation take up their quarters in other men's houses, according to their necessities, and they then present posture of their affairs" (pp. 78–79).

7. *Rondolet* refers to a famous physician of the sixteenth century, Guillaume Rondelet (1507–1566). He spent most of his life in Montpellier, where he was born, studied, practiced medicine, and taught. In 1537 he took his medical degree at the University of Montpellier and after various other appointments, returned to the university in 1551 and became its chancellor in 1556. His two passions in life were anatomy and ichthyology. As an anatomist, he is said to have "discovered" the *vesiculas seminales*, to have been the prime mover in the construction of the University of Montpellier's first *theatrum anatomicum*, and he was accused of dissecting one of his own children when it died at an early age.

Besides various medical works (on fevers, urinary infections, elephantiasis) his main work is on fishes and other aquatic animals. Gulielmi Rondeletii, *Libri de piscibus marinis in quibus verae piscium effigies expressae sunt; Universae aquatilium historiae pars altera* (Lvgdvni [i.e., Lyons], apud Matthiam

Bonhomme, 1554–1555). His pupil, Laurent Joubert, published a French translation of this in 1558 under the title *Histoire entière des poissons.* It is a large compendium of freshwater and marine zoology which was not restricted to fishes but also included mollusks, marine mammals, and amphibians. The book was somewhat controversial in its time because Rondelet argued for firsthand observation and anatomical investigations—something Rumphius would have endorsed. Nineteen kinds of animals had been named after Rondelet by 1850: eleven fishes, three mollusks, two crustaceans, two sea anemones, and one ophiuroid or Brittle Star.

Rondelet has one other distinction. He met and became friends with Rabelais when both were studying in the medical faculty at the University of Montpellier. He appears in *Gargantua and Pantagruel* under the name "Rondibilis," one of the "experts" who advises Panurge on the burning question of whether he should get married (book 3, chs. 31–33). Although at first glance Rondelet seems an unlikely candidate as a Rabelaisian companion, it becomes clear from the only contemporary biography that Rondelet was a jovial character, very fond of food, drink, music, and dancing. In fact the reputed manner of his death would have been sanctioned by Rabelais, since Rondelet died from "Bloody Flux" (dysentery) after eating too many figs. See the biography by his devoted pupil, Laurent Joubert, "Vita Gulielmi Rondeletti," written in 1568 and published in Joubert's *Operum latinorum* (Lyons, 1582).

8. *Sea wrack* was used to translate Rumphius' phrase *veelderhande ruigte*, which poses several problems. He is referring primarily, it seems to me, to natural detritus cast up by the sea on a shore or beach and which may mingle with littoral life both vegetable and animal. The term "sea-wrack," an older word that came into the language around the middle of the sixteenth century, covers both marine flora cast on shore, seaweed, and litter from ships. The botanical nature of tropical beaches in the Malay archipelago was the subject of a specialized study towards the turn of the century: A. F. W. Schimper, *Die indo-malayische Strandflora* (Jena: Gustav Fischer, 1891). Of interest for the present subject is Schimper's note that where no birds are present—such as on the Cocos Islands—to distribute seeds from the vegetation on the beach, it may be that crabs, and particularly the hermit crabs, perform a similar function, allowing even coral islands to be decked out with some kind of flora. Schimper is, of course, careful to point out that there are many other factors, such as ocean currents and the wind (pp. 75–77). Schimper's scientific study includes Rumphius' *Herbarium Amboinense* in its literature and refers to this work concerning the species *Carapa*, a Mangrove variety. Schimper states that he found two kinds of this tree in Java, contrary to existing sources, and notes that Rumphius had already described the two: *Carapa moluccensis*, or *Martahul latifolia Rumphii*, and *Carapa obovata*, or Rumphius' *Martahul parvifolia* (see pp. 99 and 39). Rumphius' name is also connected with several plants or trees in the same study: e.g., the plant *Cycas Rumphii* in Pegu, a district (also town and river) in Burma (p. 70); a variety of sago-palm, or *Metroxylon Rumphii* (p. 134); a variety of palm-like plants called *Cycadaceae*, which includes the *Rumphii* (Miq.) (p. 101), and a variety of *Bruguiera*, which is an allied genus of the Mangroves, called *Bruguiera Rumphii* (p. 95).

9. *Ikan Swangi* is identified by De Beaufort (*M*, p. 58) as probably a species of *Synanceia*, fishes with poisonous dorsal spines that prefer to lie in crevices of reefs. They are dangerous and said to have caused deaths. *Ikan* is the general Malay term for fish. In the present text Rumphius states that the tails of the *Cuman*, used as a plaster, can cure the dangerous stings of these dorsal spines, and in his *Ambonese Herbal* he asserts that the roots of the Manga-tree (book 4) and the seeds of the Bori tree (*Croton tiglium*, book 6) will do the same.

10. *Faba* is Latin for beans and *Castanea Marina* is Latin for sea chestnut. A more usual name is *Faba Marina*, or sea bean (*Entada scandens*), a shrub of the Mimosaceae family, bearing pods. It is probably these pods, which can be up to eighteen cm long, to which Rumphius is referring.

11. This refers to an extant work of Claudius Aelianus (c. A.D. 170–235) called *De Natura Animalium*, which the *Oxford Classical Dictionary* characterizes as "a collection of excerpts and anecdotes of a paradoxical or moralizing character, concerned chiefly with the animal world." His work enjoyed great popularity with Christian writers. The "carcinium" is mentioned by Aristotle in book 4, ch. 4, section 14.

12. Pliny the Elder (A.D. 24–79) also discusses these shells in his *Naturalis Historia.* For Rumphius' admiration for Pliny, see the introduction. The *Murices* and *Purpurae* are discussed by Pliny in book 9 (sections 52 and 60–64) with dizzying confusion, when *Murices* and *Purpurae* appear to be interchangeable, along with the generic term *concharum*, and in section 61 the whelk—called *buccinum*—is also alleged to be a shellfish which produces the purple dye. Pliny does agree that the best purple came from Tyre and describes the finest shade this dye can produce in the eloquent phrase: *laus ei summa in colore sanguinis concreti, nigricans aspectu idemque suspectu refulgens; unde et Homero purpureus dicitur sanguis:* i.e., "its highest glory consists in the color of congealed blood, black when glanced at but gleaming when held up to the light; this is the origin of Homer's phrase, 'ruby-red blood.' " The reference to Homer is the *Iliad* (17:361).

13. *Pinnotheres* I translated as "Shell-hunters" in keeping with Rumphius' own nomenclature in

the next chapter, wherein he describes the *Pinnoteres*. His reference to this mollusk in Pliny illustrates the unwonted confusion which may arise when spelling rules were not readily adhered to. Rumphius states that Pliny in book 9 of his *Natural History* refers to *Pinnotheres* in section 31, and to *Pinnoteres* in chapter 42. Obviously the only difference is one consonant. It may also be mentioned that the first mollusk (*Pinnotheres*) does not occur in section 31, which deals only with the inflated prices the Romans paid for the mullet fish, and section 42 mentions only the "*maena*" and the lamprey. A small crab, called by Pliny *pinoteres*, is first mentioned in section 51 of book 9 in a listing of varieties of crab; there it is described as being the smallest kind and, therefore, liable to injury. It compensates for this by "stowing itself in empty bivalve shells and shifting into roomier ones as it grows bigger" (Pinoteres vocatur minimus ex omni genere, ideo opportunus injuriae. huic sollertia est inanium ostrearum testis se condere et cum adcreverit migrare in capaciores). This sounds like the *Cuman* Rumphius is describing and is obviously the hermit crab; but Pliny had mentioned earlier, though in the same section, the hermit crab under the name of *paguri*. This *pinnoteres* reappears in section 66, as the companion of the *pinna*, a mollusk Rackham translates as the "fan-mussel," and is not mentioned again. The *pinnoteres*, or pea-crab, also called *pinnophylacem*, or "fan-mussel guardian," by Pliny is, as it were, the messmate of the *pinna*. The story was that tiny fish (*minutis piscibus*) filled up its body cavity, and, when full, the *pinnoteres* signals to shut the opening. The little partner shared in the spoils.

There is no "*pinnotheres*" in Pliny, hence, they can not be "distinguished" from *pinnoteres*, because they are, it seems to me, one and the same. However, the *Cuman* Rumphius describes is still the hermit crab and has nothing to do with the *pinnoteres*, because the hermit crab lives in *univalve* shells, while the *pinnoteres* lives *with* the mollusk in its *bivalve* housing. The reference, once corrected, really seems to belong to Rumphius' next chapter (23) which describes the *pinnoteres* in terms similar to those of Pliny in section 66 of book 9, and in fact Rumphius gives it as expressive a name as Pliny did by calling it the *deurwagter*, or "doorkeeper."

14. This means "the little crabs which come from the warty turbo shells." This seems to prolong the confusion just explained, for it appears to echo the *pinnoteres* shells. Yet Rumphius clearly is dealing with a creature like the hermit crab because it lives in *a hoorntje* which, as was previously explained, refers to a univalvular mollusc or, as it was translated, a winkle.

15. The *Mancenilie Tree* seems at first a tree from lore and legend, but if one glosses the spelling imaginatively, one comes to the realization that it refers to a tree that in English is called the *manchineel*, a member of the Euphorbiaceae family. Its name derives from the Spanish *Manzanilla*, and the particular one referred to has the scientific nomenclature *Hippomane mancinella*, which contains in its Greek name the hazard of this tree, since *hippomane* comes from "horse" and "madness." It is fitting that Rumphius quotes from a history of the Antilles, because Europeans first encountered the tree in the Caribbean. The manchineel secretes a milky and viscous sap, which on contact proves to be caustic and poisonous. The toxin is lodged in the leaves and in the wood, as well as in its fruit, which looks like small, hard green apples. This sap is indeed "pernicious milk." In a popular book on poisonous plants, a case is cited of a botanist in Florida who dripped some of the sap on his hand, with the result that his entire arm turned ulcerous and was paralyzed for several days. People have died from eating the fruit of what the Spaniards called "the tree of death." The popular text is Edward R. Ricciuti, *The Devil's Garden: Facts and Folklore of Perilous Plants* (New York: Walker and Company, 1978), pp. 102–104.

The "oyl" is mentioned by Rochefort (see n. 9 of ch. 1) in his description of the "Souldiers" (in the English translation ch. 14 of book 1, p. 80), while the "Mancenilie Tree," called "Manchenillos" by Rochefort, is mentioned in connection with what Rochefort called "Painted Crabs," which Rumphius mistakenly thought to be the "Purse Crab" (ch. 5).

Chapter 23. Of the Pinnoteres, or Pinna Guard.

1. Strack guesses that these are commensal shrimp (*Anchistus miersi*). See also n. 13 to the previous chapter. Holthuis notes that there are too many varieties of commensal shrimp to be sure which ones Rumphius is talking about. Commensal shrimp belong to the subfamily *Pontoniinae*, and the small commensal crabs are *Pinnotheridae*. Ancient authors, such as Aristotle, used the name "*Pinnotheres*" for both shrimp and crabs, while modern nomenclature reserves it for the crabs.

2. The *Tricdana*, or Rumphius' "Chama Squamata," are described in ch. 28 (no. 1) of book 2; the *Pinna* is described in ch. 36 of book 2.

3. This underwrites what I said in n. 13 of the previous chapter, because Pliny mentions that aquatic animals have senses, in section 67 of book 9, referring to the previous section 66, which discusses the "*pina*" and the "*pinoteres.*"

4. Discussed as number 6 in ch. 30 in book 2.

5. This is impossible; it should probably be 1683.

6. See ch. 25 of this book.

Chapter 24. Of the Cancer Barbatus.

1. See ch. 10 of this book.

2. The Elephant (Oliphant) River was the Dutch name for the same river that was known as "Way Batu Gadjah." *Gadjah* (or *Gajah*) means "elephant" in Malay. The river runs through what was once a pleasant park south of the city, through the city of Ambon itself, and disembogues by what is now a dock.

3. With "Alphorese" Rumphius is referring to the Alfurs, the inhabitants of interior Ceram. *Alfuras* (or a variety of other spellings, such as *Alfoursi* or *Harafuras*) was the name the Portuguese gave to the indigenous peoples who lived in the interior and mountains of the Moluccan Islands and who had not been subdued by them, somewhat equivalent to the Spanish use of *Indios bravos* for all the native Indians of the Americas. Rumphius always uses the term for wild "Mountain People," but one has to remember that they did not live on Ambon or the Uliasser Islands (i.e., Haruku or Oma, Saparua, Nusalaut) but inhabited the interior of Buru and Ceram. Even by the last quarter of the nineteenth century, they were still described as "savages," only clad in a *tjidako* (loincloth) to cover their genitals. On the large island of Ceram, above Ambon, and equal to the size of Halmahera in square miles, the Alfuras were divided into two tribes: the *Patasiwa* (or *Uli siwas*) and *Patalima* (*Uli limas*), with the first in the western, and the second in the eastern, part of that island. Their religion appears to have been a form of animism—a belief that trees, stones, forests, and mountains were populated by spirits, often the souls of their dead (*nitu nitu*), recalling the religion of the Bataks. The Alfuras were headhunters; the number of heads determined a man's status in his community and was indicated by black rings on the belt of his *tjidako* (E. W. A. Laufer, "Schets van de Residentie Amboina," *Bijdragen tot de Taal-, Land- en Volkenkunde van Nederlandsch-Indië*, 3:53–78; and Hoëvell, *Ambon*, pp. 147–59).

4. "*Cattam Gigi Bulu*" is Rumphius' own invention and means something like "crab with hairy teeth." "Cattam" is most likely *ketam*, which means "crab"; *gigi* is Malay for "teeth," and *bulu* stands for "bristle." The Latin means "Bearded Crab." This is *Pseudograpsus setosus* (Fabricius). Indeed, only the males have these tufts.

5. For "Wawo," see the last chapter (44) of this book.

6. What is meant is the no. 2 illustration on plate X.

Chapter 25. Of the Cancelli Anatum. Cattam Bebec.

1. These crustaceans are of the genus *Leucosia*. The Latin means "duck crabs," as does the Malay, since "cattam" is *Ketam*, meaning "crab," and *bebec* (*bebek*) means "duck." The first kind is classified by Holthuis as *Leucosia anatum* (Herbst).

2. The second kind is *Leucosia craniolaris* (L.), according to Holthuis.

3. See ch. 13 of this book.

4. Holthuis states this is *Myra fugax* (Fabricius).

5. This is *Mictyris longicarpus* (Latr.), according to Holthuis.

Chapter 26. Of the Foetus Cancrorum, or Bloody and Fiery Sea-Red Sheets.

1. The sun comes into *Taurus* around April 20 or as Milton put it in *Paradise Lost:* "In springtime, when the Sun with Taurus rides."

2. Holthuis (*M*, p. 98) informs us that the first kind of "small Crabs," presumably the *Foetus Cancrorum*, are "a mixture of Decapod larvae and Mysidacea." The "other kind of small Shrimp," from which they make "Bolatsjang," are of the order Mysidacea of the class Crustacea. Holthuis tells us that these Mysidacea are made into *trasi*. Rumphius calls this red-brown paste "*bolatsjang*," also spelled *belatjan*, a fish paste still used today. *Trasi* and *belatjan* are the same thing.

Trasi is a condiment thought to be essential to *rijsttafel*. The variety Rumphius describes is *trasi udang*, the red-brown paste made from the little shrimp (*udang rebon*). A lesser kind, of a purple-gray color, is made from shrimp larvae, or *udang djambret*, about the size of mosquito larvae. This is the kind Rumphius mentions. The superior *trasi udang* is made from the little shrimps, about two centimeters long, mostly caught off the northern coast of Java. They can be cooked before preparation, but they are also often dried uncooked in the sun after they have been salted. Then they are mashed to a pulp (called *brabon* in Javanese), dried once again, then mashed some more, and finally rolled into little cylinders which are left to ferment. The result can be kept for a month, without refrigeration, and will not rot. The nets Rumphius refers to as "stretched-out pieces of cloth" are called *bandet* in Javanese. These are nets of roughly woven material, over 200 feet long and some 20 feet wide. They are spread their entire length, and the shorter ends are slowly brought to meet each other, enclosing the shoal of shrimp.

3. *Laricque* is an alternate spelling for Larike, a town on the southwestern coast of Hitu on the island of Ambon, where Rumphius resided as "junior merchant" (*onderkoopman*) from 1657 to 1660.

4. *Grissek* is the name of a Javanese harbor near Surabaya, situated north of that city opposite

the most western point of the neighboring island of Madura. Its name was also spelled *Gresik*, *Grisee*, and *Grissee*. The last was the common one adopted by Dutch cartographers. Java's largest river, the Solo or Bengawan River, disembogues in the district around the town, also called Grissee. It was in Grissee that the Muslims, who overthrew Java's native religion, first settled, and the town contains the tomb of one of these religious pioneers; it is, therefore, somewhat of a religious shrine. The town was also the first to be visited by a European, the Portuguese discoverer of the Moluccas, Antonio d'Abreu, who stopped there in 1511. It has been noteworthy for its production of salt, copper implements, a type of cloth woven there, called *songkets* and made from gold or silver thread, as well as sea fishing.

The history of its name is rather interesting. The older Javanese spelling, *Gresik*, means small pebbles or crackling sand. In *Krama*—or ceremonial Javanese spoken by the courtiers—its name was *Tandes*, which means "hard ground." These must be considered synonymous, because *Krama* substituted totally different words or formations for any one word (including pronouns or prepositions) that had become familiar, i.e., vulgar. However, in Malay a word almost homophonic, *tandas*, means "cesspool." When the Chinese arrived in the Grissee district around 1400, they heard the *Krama* form of the town's name—i.e., *Tandes*—and not knowing this most peculiar form of Javanese, assumed it was the Malay word *tandas*. They translated this literally into Chinese: *ts'e*, which means "shithouse," then added their word for village, *ts'un*, and called Grissee *Ts'e-ts'un* or "shithouse village."

Rouffaer even conjectures that a somewhat analogous meaning might be found in Javanese, since Javanese *Tandes* can also mean "the mouth or delta of a river" and a related word, *nandes*, "to empty, all the way to the bottom." Since the Solo River, or the Kali Bengawan (*kali* is "river" and *bengawan* means "the [most important] stream"), does indeed flow into the sea at Grissee, it may well be that *tandes* was a figurative name for Grissee or, as Rouffaer puts it politely, *Cloaca Maxima* or the "main sewerpipe." See the article by G. P. Rouffaer, "De Chineesche Naam Ts'e-Ts'un voor Gresik," in *Bijdragen tot de Taal-, Land- en Volkenkunde van Nederlandsch-Indië*, series 7, part 5 (1906), pp. 178–79.

5. Sebalt de Weert (1567–1603) was a Dutch navigator and vice-admiral during the years when the Dutch attempted to dislodge the Portuguese in various parts of the world. He was born in Antwerp, lived as a boy in Germany, and was educated in Antwerp. He was captain in 1598, part of the fleet of Mahu and De Cordes. He discovered in 1598 the northern islands of the Falkland group and experienced much adversity trying to best the Straits of Magellan. De Weert was part of Olivier van Noort's fleet (which was finally reduced to one ship), which accomplished the first Dutch circumnavigation of the globe. He was vice-admiral under Warwijck and sailed to Atjeh to secure the pepper trade. He had been in Ceylon in 1602 and returned there hoping to take the cinnamon trade away from the Portuguese. De Weert was killed in Ceylon during a feast given by the Maharajah of Kandy, Dom João, in 1604. The date "12 January" is wrong; it should be 10 March 1599. See: *De reis van Mahu en de Cordes door de Straat van Magalhães naar Zuid-Amerika en Japan 1598–1600*, 2 vols., ed. F. C. Wieder ('s-Gravenhage: Nijhoff, 1923–1924), 1:181–182. Rumphius must have read it in the following contemporary narration: *Wijdtloopigh Verhaal van tgene de vijf schepen (die int jaer 1598 tot Rotterdam toegherust werden, om door de straet Magellana haren handel te dryven) wedervaren is . . . meest beschreven door M. Barent Iansz. Cirurgijn* (Amsterdam: bij Zacharias Heijns, 1600). Holthuis notes that what De Weert saw was lobster krill. These are tiny Galatheide crabs belonging to *Munida gregaria* (Fabricius) and *Munida subrogosa* (White). When very young (called "Grimothea"), they can be seen in such large numbers that they do indeed turn the surface of the sea a bright red over large areas.

6. *Rio de Platas*, more commonly known as Rio de la Plata, is the large estuary with the capital of Uruguay, Montevideo, on its northern shore and the capital of Argentina, Buenos Aires, on its southern shore. Sebalt de Weert, coming up the eastern coast of South America, might have "rounded" several points around the Rio de la Plata, such as Cape San Antonio or Cape Corrientes. *Penedos de St. Paulo* is a group of rocks in the Atlantic Ocean at some distance off the coast of Brazil, near the equator.

7. What follows is a direct quote from Padtbrugge's travel journal which he kept on board the *Sparrendam*. He was sailing to the Cape of Good Hope in December of 1670. The original is in the National Archives in The Hague: VOC 1280, fol. 442 recto and verso, and folio 443 recto.

8. An *escharam* is, properly speaking, neither a scar nor a scab, but a "slough" (noun), i.e., a mass of dead tissue formed on the surface of a wound, particularly such a slough caused by caustics.

9. *Manado* is a city near the most northern tip of Celebes, on the Celebes Sea, in Minahasa.

10. *The Indian Ocean between Malabar and Persia* is, properly speaking, the Arabian Sea. Malabar is the Coast of Malabar, the southwestern coast of India, south of Goa, down to Cape Comorin. The coast on the other, southeastern, side of India's point is the equally fabled Coromandel Coast.

11. *Mocha* is a city in what is now Yemen. It is situated southward, on the northern shore of the Red Sea, down where it narrows into the straits called Bab el Mandab and turns into the Gulf of Aden. The name of the city is, of course, best known as a brand of coffee which was grown in Yemen and shipped from Mocha.

Chapter 27. Of the Pediculus Marinus. Fotok. Sea Louse.

1. De Man and Holthuis classify this little decapod as *Notopus dorsipes* (L.).

Chapter 28. Of the Eschinus Marinus Esculentus. Sea Apple. Seruakki.

For identifications of these echinoderms I used primarily H. Engel's article "The Echinoderms of Rumphius," in *M*, pp. 209–223; and "Die Mollusken (Conchylien) und die übrigen wirbellosen Thiere in Rumpf's Rariteitkamer," by E. von Martens, in *G*, pp. 109–136.

1. Aristotle's divisions of the Sea Urchin (= *Echinus Marinus;* the title should not have the "s") mean in translation: (1) "common edible sea urchin," (2) "flat sea urchin"—"spatagus" is related to *spatangius*, which, coming from Greek, is not further identified other than as a "kind of sea urchin," (3) "bristly sea urchin," (4) "echinometra," a term also used by Pliny, which means "mother urchin," and (5) means "a genus [of sea urchins] that live in the sea, with long and very hard spines." For a good translation of Aristotle's text see d'Arcy Wentworth Thompson's translation of the *Historia Animalium* (1910; reprint 1940, 1956) in *The Works of Aristotle Translated into English*, eds. J. A. Smith and W. D. Ross, 12 vols. (Oxford: Clarendon Press, 1908–52), 4:530.

Rumphius mentions Dioscorides after Aristotle but, interestingly enough, not his favorite, Pliny.

2. Rumphius' own, succinct labels are: (1) "Common Sea Urchin," (2) "Echinometra with fingers," (3) "Bristly Echinometra," (4) "Edible Sea Urchin," (5) "Flat Sea Urchin."

3. In the original "*zenuwe*," which, in Rumphius' day, did not mean a "nerve" but something like a tendon.

4. This is a direct reference and correction of Pliny, who, in book 9, paragraph 100, did say that.

5. The Latin word for "crop."

6. The Latin means "Edible Sea Urchin," and "Manuhulu" is Ambonese for "chicken feathers." This name was given to the fifth variety of sea-star in the subsequent chapter 34, and to the "Medusa Head" of chapter 35 as well.

7. Both Martens and Engel agree that this is *Tripneustus gratilla* (L.). They both agree as well that letters B and C on plate XIII are not the animal that Rumphius described. The animal depicted by A, B, and C is the edible sea urchin of the Northeast Atlantic, *Echinus esculentus* L.

8. This "Stony Sea Urchin" is *Echinometra mathaei* (Blainville). Aristotle has *bryttus*, not "brisius."

9. This "Black Sea Urchin" is, as Martens and Engel agree, *Echinothrix diadema* (L.). Not illustrated.

10. Ruma Tiga is a village on the southern coast of Hitu, over against Hative on Leytimor.

11. Neither Martens nor Engel can identify this animal, which was baptized by Rumphius with the expressive name "Polyphemus' Eyes." Polyphemus was the cyclops, one of the one-eyed giants, who imprisoned Odysseus and his crew. The first word of the title of this section should be "*Oculus*," of course.

12. Pliny repeats the same thing in book 9, paragraph 100.

13. "Sajor songa," according to Heyne (p. 1437), is the herb *Wedelia biflora.* Its favorite habitat is on the beach and along the banks of rivers. It smells like a mixture of licorice, oil, and tar, and the marrow of its roots is said to lessen the pain from the sting of the *ikan suanggi*, or Scorpion Fish (family Scorpaenidae), and relieves one's eyes when they're smarting from the tropical sun. The leaves of this plant were eaten, but not frequently, because they have a strong taste and are quite diuretic; hence they were most often cooked together with fish or turtle. Rumphius called it "Seruneum aquatile" in vol. 5 (p. 423) of his *Herbal.* De Clercq states (no. 1342) that "sajor pacu" (*sajur paku*) is a kind of fern, *Diplazium esculentum* (Swartz), of which the young and tender leaves were eaten as a vegetable, either raw or cooked. See K. Heyne, *De Nuttige Planten van Nederlandsch Indië*, 2d. rev. ed., 3 vols. (Batavia: Departement van Landbouw, Nijverheid & Handel in Nederlandsch-Indië, 1927); and F. S. A. De Clercq, *Nieuw Plantkundig Woordenboek voor Nederlandsch Indië*, 2d. rev. ed. (Amsterdam: J. H. de Bussy, 1927).

14. Engel states that what is depicted on plate XIII, by the letter A is "a rather bad representation of Aristotle's lantern" (p. 212). The simile comes from Aristotle (*Historia Animalium*, IV, paragraph v), and in Engel's usage it refers to what Webster's dictionary calls "the protrusible 5-sided masticatory apparatus of a sea urchin, each side being made up of a tooth with its supporting ossicles and the muscles that activate it." Aristotle writes: "In respect of its beginning and end the body of the urchin is continuous, though in respect of its superficial appearance it is not continuous, but similar to a lantern lacking its surrounding skin," *Historia Animalium*, 3 vols., transl. A. L. Peck, Loeb Classical Library (Cambridge, Mass.: Harvard University Press, 1965–1970), 2:49. There has been much debate about this passage. Rondolet was the first (1544) to say that this referred to the sea urchin's

mouth, but Scaliger in 1619 argued that it meant the entire animal. Jacob Theodor Klein (1685–1759) was the first to call it *laterna Aristotelis* in his *Naturalis Dispositio Echinodermatum* (1734); see Peck's comments on pp. 49 and 352 in the Loeb edition.

Chapter 29. Of the Echinometra Digitata Prema. Djari Laut.

1. "Echinometra" is a term from Aristotle. Pliny described this creature in book 9 (paragraph 100). The name derives ultimately from the Greek for "hedgehog." Both Engel and Von Martens agree that this is probably *Heterocentrotus mammillatus* (L.).

2. "*Loblonga*" is a misprint, and should be "*oblonga.*" The Latin phrase means something like "Fingered Sea Urchin; First Class: Oblong." "*Digitata*" means "having fingers."

3. "Tenuous" for *ydel;* somewhat confusing here because of the gem simile, but I think Rumphius was referring to the consistency of the "warts," hence "tenuous" in the sense of "meager of substance."

4. Taking "*blinkende*" to be a misprint for "*klinkende.*"

5. Where the "*spadel*" in the original text should have been "*spatel.*" *Spatula* dates in English from the sixteenth century.

6. "To couch" for "*vellen,*" in the old sense from the Middle Ages, in Malory for instance, of lowering a pike or lance to the position of attack.

7. This means "Sea Fingers"; the Malay means the same, since "*djari,*" or *jari,* means "finger." Note that the title of this chapter has a misprint and should read: "Digitata Prima."

8. *Babbelstokje* might be as Holthuis suggested, a misprint for *sabbelstokje,* which literally translates as "a little stick to suck on." Surprisingly, English did not have a word for this device, a pacifier of sorts, in the seventeenth century. Hence I translated with the neutral substantive "teether."

9. "Oculi canceri" means "crab's eyes." These are oval calcium concretions described by T. H. Huxley in *The Crayfish. An Introduction to the Study of Zoology* (London: Kegan Paul, 1880) as follows: "two lenticular calcareous masses, which are known as 'crabs'-eyes,' or *gastroliths,* and were, in old times, valued in medicine as sovereign remedies for all sorts of disorders. These bodies are smooth and flattened, or concave, on the side which is turned towards the cavity of the stomach; while the opposite side, being convex and rough with irregular prominences, is something like a 'brain-stone' coral" (pp. 29–30).

10. "Xulasse Islands" refers to the Sula Islands in the Moluccan Sea between Celebes and Buru. Mangole was the most significant island at the time. See also ch. 34 of book 3.

11. These are all remedies against poison or infections. For "Solenes" see ch. 27 of book 2; "denticuli Elephantis" is described under no. 4 of that chapter, and the "Sandpipe" under no. 1.

12. "*Notunda*" should be "*rotunda*"; hence this means something like: "Fingered Sea-Urchins; Second Class: Round." There is no consensus about the particular species, but it is definitely a sea urchin of the family Cidaridae. Engel (*M,* pp. 213–214) suggests *Cidaris cidaris* (L.) and notes that the figure was not provided by Rumphius but added by Schijnvoet.

13. *Tholus* referred to a dome in Latin, a cupola or rotunda.

14. "*Cidaris*" was Latin for a diadem or tiara, and "Mauri," though specifically referring to a denizen of Mauretania, came, by extension, to mean "Moor."

15. "*Sutura oranii*" must be a misprint for "*sutura cranii*" or "skull seam." *Cranium* is Medieval Latin.

16. The first edition has "*Zeegriffen,*" where "*griffen*" is most likely the plural of *grif,* which means the same thing as *griffel,* or "slate pencil." English did not have such a specific word. The noun "pencil" was used by Shakespeare for a painter's brush, and the notion of a lead pencil dates from half a century *after* Rumphius' death. The only compromise I could come up with was "style" which, in its oldest meaning, referred to an instrument used to incise letters in a wax table and was, by extension, so used on other materials.

17. "Cholera" for "*bort.*" This is not the disease of epidemics, as the *WNT* points out. Rumphius himself called it "cholera," specifically "cholericus fluxus" in ch. 26 of book 6 in his *Herbal,* which discusses the *upas bidji.* The Dutch term *bort* and the original meaning of cholera referred to a disorder that caused bilious diarrhea, vomiting, cramps, and stomachaches. It was particularly likely to strike at the end of summer and the beginning of autumn; hence it was known as "Summer Cholera," "Bilious Cholera," "European Cholera," "British" or "English Cholera." Another term for it, often found in early texts, is "cholerine" or "cholera nostrus." The same conclusion is arrived at in a study of cholera in the Indies; see J. Semmelink, *Geschiedenis der cholera in Oost-Indië vóór 1817* (Utrecht: C. H. E. Breijer, 1885), pp. 1–3, 123–125.

18. "*Syringa*" must definitely refer to a "syringe." It is a medieval Latin word, and the instrument was larger and blunter than the modern needles.

19. The OED provides this English phrase for the Dutch "*wervel van een spil*" in a 1648 definition of the Dutch phrase. A whirl or whorl was the flywheel or pulley of a spindle.

20. "Burst ones," because I am reading "*uitgeberste*" for the text's "*uitgeverste.*"

21. For "*Opblazer*," a fish that is covered with spines and which can distend its body with air, probably of the family *Tetradontidae.* It is very poisonous.

Chapter 30. Of the Echinometra Setosa. Bulu Babi.

1. In the original *handkaasje.* This looks like a Dutch word, but it is not. It is a Germanism. The original German is *Handkäse,* which literally means the same thing: a cheese made by hand. It was, and still is, a common food in Hesse. It is a little cheese made from cow's milk, about three inches in diameter, an inch and a half thick, that looks like a "stone" used in shuffleboard. A common food in Hesse, it is not known in the Netherlands.

2. "*Setosa*" or "*Setosus*" from *saetosus,* which means bristly. Engel identifies these as *Diadema setosum* (Leske). There is a better depiction of this animal on plate XIV, letter B, added by Schijnvoet.

3. *Bulu* is Malay for bristles; hence *bulu babi* means pig's bristles.

4. For "*besterve.*"

5. "Crusta Accarbary" most likely refers to the "crust" of a particular kind of coral, which Rumphius calls "gray Accarbaar" (or "Acabahar"); see chs. 11–14 in book 12 of his *Herbal.*

6. Both Martens and Engel suggest this might be *Astropyga radiata* (Leske).

Chapter 31. Of the Echinus Sulcatus. Skulls.

1. Rumphius' name means "Furrowed Sea Urchin." Neither Martens nor Engel is sure, though both are in agreement that the animals depicted at C and D on plate XIV are not from Ambon, and might not even be from Indonesia. Engel (*M,* p. 215) proposes that letter C is the West-Indian *Clypeaster rosaceus* (L.).

2. *Amianthus* is a fine form of asbestos. See ch. 80 of book 3.

3. This should be the Weynitu (Wainitu) River, which flows to the west of Ambon City.

4. Nos. 1, 2, and 3 are, according to Engel (*M,* pp. 215–216) *Metalia spatagus* (L.), *Schizaster lacunosus* (L.), and *Echinolampas ovata* (Leske).

Chapter 32. Of the Echinus Planus. Pancakes and Sea Reales.

1. The first variety of these "Flat Sea Urchins," as Rumphius calls them, is classified by Martens and Engel as *Arachnoides placenta* (L.). It is depicted on plate XIV by the letter G, not E, as Schynvoet would have it. We would call them sand dollars.

2. A *real* was a small silver coin minted in Spain but used all over the world in Rumphius' day. The British, in the early seventeenth century, called it the "Spanish sixpence." It might be, however, that he was thinking of the larger *real* of eight, better known as a "piece of eight," which the Portuguese had first put into circulation in the Indies and which the VOC also used during the earlier part of the seventeenth century. The subject of money, the VOC's real god, is extremely confusing and complicated in terms of coinage, fiscal policy, and regulatory attempts. It also does little honor to the VOC, whose position seems to have been, quite literally, to make money on its coinage.

3. Engel calls this *Echinodiscus auritus* (Leske).

4. Engel informs us that these "young ones" belong to the species *Echinodiscus bisperforatus* (Leske).

5. This river flows past the city of Ambon in the west.

6. Engel states that this species is *Astriclypeus manni* (Verrill), which up till now was only found in Japan.

7. Both Martens and Engel classify this animal as *Laganum laganum* (Leske), and say that it is shown by the letter E on plate XIV.

8. In the original, a "*Ruiters schildje.*"

9. Engel: *Peronella orbicularis.*

10. The "Schelling" was a Dutch coin that was the twentieth part of a "*pond*" ("pound"); hence its value depended on the value of the "pound," and this could vary widely. Rumphius might also have had the silver coin in mind, which was worth six stivers (*stuivers*).

11. Reading "*regen*" for "*regten.*"

12. "Small stones" probably refers, according to Martens, to coral that grows around a mollusk. Schynvoet says that one is depicted by the letter H on plate XIV, but this is not the case. It is rather, according to Engel, a crustacean of the subclass *Cirripedia.* Holthuis identifies this species of barnacles that live on whales as *Cetopirus complanatus* (Mörch).

13. Where I assume that "*voorhanzels*" is "*voorhangsels,*" an old, but precise, term for a curtain, or for things that might make up a curtain.

14. Taking "*knoopen*" (which means "buttons" in Dutch) to be *knoppen*, which means "buds." One has to remember that Rumphius considered coral a plant.

15. Engel states that this "Sunbeamed Pancake" is *Rotula orbiculus* (L.) from West Africa.

Chapter 33. Of the Limax Marina. Sea Snail.

1. These seem to be large naked mollusks, of the genus *Tethys*. The Latin means precisely the same thing as the Dutch.

2. Strack: *Acanthopleura spinosa* (Bruguière).

3. "Lames" were overlapping metal plates used in the coat of mail armor.

4. "Stone-sticker," to keep the alliteration of Rumphius' "*Klipklever*." For *Lopas*, see ch. 26 of book 2.

5. This is a strange spelling of "Sial," the cape at the southernmost tip of Howamohel or "Little Ceram." The name means "Cape Bad Luck." Valentijn also states that "the dry cape" was another name for Cape Sial—see his *Oud en Nieuw Oost Indiën*, vol. 2, part 1, pp. 35–36—but he does not give a positive reason for the popular name. The only thing he offers is that this piece of coast was so rocky and covered with stones that they seemed like rice kernels. That the place was called "dry" most likely resulted from the fact that there was no water to be had. Valentijn reports the story that the local potentate of Sial was called the "Sagueir-king in olden times because, due to lack of water, he was wont to wash himself in Sagueir (a liquer that flows from the Saguier tree)" (p. 36).

6. This "warty" mollusk is, according to Van Benthem Jutting (*M*, p. 195), *Oncidium verrucosum* (L.).

7. These are sow-bugs, *Oniscus asellus* (L.), of the suborder Oniscidea. "Esel" (German for donkey or ass) and "*Asellus*" (little donkeys in Italian) were old names for these animals.

8. "*Tylos*" is *tilos*, Greek for "knob" or "knot."

9. This probably is supposed to be *callosus*, hard-skinned or calloused.

10. Engel identifies this species as *Dolabella scapula* (Martyn).

11. "*Panegeijers*." I wonder whether with this strange noun, Rumphius has "rowers" in mind. "*Pangayen*" (in various spellings), a variation of *pangaaien*, which in modern Dutch is *pagaaien*, also meaning "to row" or "paddle," from the Malay word *pengayuh*. The implication is that these "rowers" or "paddlers" were a different people.

Chapter 34. Of the Stella Marina. Bintang Laut.

1. I translated Rumphius' Latin name as "Sea Stars," a locution once just as common as "Starfish" is now.

2. This "usual kind" of Asteroidea is identified by Martens and Engel as *Linckia laevigata* (L.). Both experts also note that this animal is depicted on plate XV by the letter E, not A.

3. In Malay it means the same thing: *bintang* = star, and *laut* = sea.

4. Neither Martens nor Engel commits himself to a positive identification of this "Stella Marina Minor."

5. Cape "Martyn Fonso" is now Cape Martafons, near Rumahtiga on Hitu.

6. The Latin means "Sea-star with Fifteen Rays." This was identified by Engel, after Martens, as *Acanthaster planci* (L.).

7. Also spelled "Larike" now, a town on the west coast of Hitu, Ambon's northern and largest peninsula.

8. The "Fourth Sea-Star" is, according to Martens and Engel, *Protoreaster nodosus* (L.). It is depicted by the letter A of plate XV, and was provided by Rumphius himself.

9. "Pasty" is the older English word for *pastel*, or meat pie. I did not want to use the noun "pie" because that would be confusing. From the thirteenth century on, a "pasty" referred to a hollow pastry shell, filled with a ragout of meat or fish, and then baked in an oven, which is what Rumphius has in mind.

10. This means: "Fourth Sea-star and the Bread and Meat of the Sea."

11. This means "Fifth Sea-star" coupled with the Greek word (*scolopendra*) for a centipede. This was not identified; Martens only says that it must be an Ophiuroid.

12. This is from ch. 20 of Rochefort's book. Rumphius is more accurate, but Rochefort's conceit is lovelier. He considers the sea the mirror image of the sky; hence it also should have stars: "To consider narrowly al the rarities to be seen in the Sea, it might be said, that of whatever is excellent in the Heavens there is a certain resemblance in the Sea, which is as it were the other's looking-glass. Hence it comes, that there are Stars to be seen in it. . . ." Rochefort then goes on to describe these starfish in a wonderful passage that contains Rumphius' reference. "If these Sea-Stars may not enter into any competition with those of the Heavens, as to magnitude and light, they exceed them in this, that they are animate, and that their motion is not forc'd, and that they are not fix'd nor confin'd to

the same place: For the fish, which hath taken up its abode in this starry mansion, moves which way it pleases on the azure plains of the waters while the weather is calm, but as soon as it foresees any tempest, out of fear to be forc'd to the Land, which is not fit to entertain Stars, it casts out two little anchors out of its body, whereby it is so firmly fastened to the Rocks, that all the violent agitations of the incens'd waves cannot force it thence" (pp. 126–127).

13. This is Howamohel, or "Little Ceram," the narrow peninsula extending from Ceram's west coast.

14. Could "Condong" be "Gondan," or *Ficus variegata* Blume? The fruits of this tall tree are eaten with *sambal* (the ubiquitous condiment of hot peppers) in *rudjak*, while the shoots are eaten raw as *lalab* or cooked in *sajur*. "Tamaryn" is probably tamarind, *asem* in Javanese, *asam* in Malay; the *Tamarindus indica* (L.) tree. It is a favorite ingredient in the Indonesian kitchen, and is said to make meat more digestible and to remove the smell of fish. *Asem* is also used for various medicinal concoctions.

15. I wonder whether "Santang Calappa" is meant to be *santen*, or coconut milk. This is not the fluid or "water" that is inside the nut and which people drain in Indonesia, as well as anywhere else where coconut groves abound, to quench a tropical thirst. *Santen* is the milky fluid that one gets if you mix and grind grated coconut with water. It is a standby in the Indonesian kitchen, where it is used in all kinds of dishes.

Chapter 35. Of the Caput Medusa. Bulu Aijam.

1. Martens is uncertain, and Engel echoes him, stating that "it is impossible to identify the species of *Euryalae*, since Amboina seems to be the centre of distribution of this group of Ophiurids . . ." (*M*, p. 220).

2. The Latin name for cabbage, a large genus of the family Cruciferae.

3. Both Martens and Engel strike out on this one as well.

4. "*Orang Lamma*" is an interesting use for an old person, an elder. Literally, it means "people of long ago," "the ancients."

5. For *sluitkool*. "To pome" was a wonderful verb, borrowed from French and now in disuse, meaning "to form a close compact head or heart, as a cabbage [or] lettuce."

6. "Medusa Head."

7. This is also called the resurrection plant, because it rolls up when it is dry—what Rumphius has in mind here for his little joke—and expands when you moisten it.

8. Peter the Great visited Holland during his European tour of 1697–1698.

Chapter 36. Of the Sagitta Marina. Sasappo Laut.

1. These are anthozoans, and belong to the Pennatulacea, an order of the Alcyomaria, which include such species as the sea pens, and which usually have the end of the axis embedded in the sand or mud on the bottom of the sea. Bayer (*M*, p. 242) identifies the first kind as *Virgularia juncea* (Pallas). See also ch. 34 in book 2.

2. This means "intestinal worm."

3. The antecedents are unclear, but the noun "Standards" (same word in Dutch) referred to lances or poles, not necessarily to flags. The Makassarese spear or lance was called *poke* and is pictured here after a nineteenth-century drawing by the ethnographer Matthes.

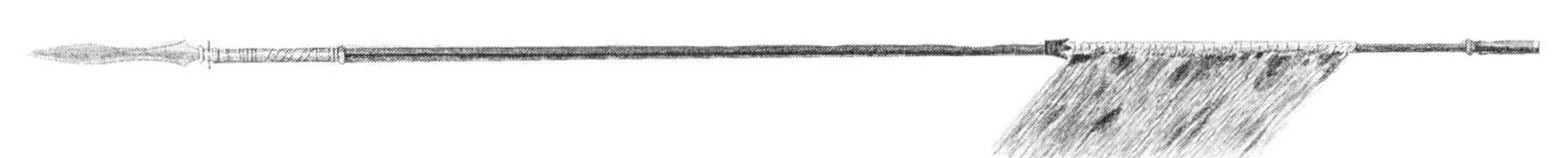

4. Bayer (*M*, p. 243) states that the second kind is *Pteroeides grandis* (Pallas).

5. *Sagitta* in Latin means "arrow," "shaft," or "bolt"; so "Sea Arrows." We know them as sea pens.

6. Once again, what seems the preferred antidote to any kind of poisoning: the outer layer of black coral, here reduced to ashes.

Chapter 37. Of the Phallus Marinus. Buto Kling.

1. This spelling is wrong; it should be "Zoophyta" or what was once known as "zoophytes," i.e., plants that had qualities of animals. "Plantanimalia" describes the same concept.

2. This is a reference to his *Herbal;* in book 12, Rumphius describes coral, which he thought was a plant.

3. For "*Kartouw.*" This was an early type of light cannon, with a short barrel, made from either leather or bronze, in different calibers, throwing a shot from about two to eighteen pounds.

4. Described by Rumphius in book 6 of his *Herbal*, identified by De Wit as *Enhalus acoroides* (L.).

5. The Latin is self-explanatory, but neither Martens nor Engel can identify these creatures. The only thing they are sure of is that these are Holothurians. One should be warned that what Rumphius calls "Holothuriae" four chapters later are not species of Holothurioidea, but Coelenterata.

6. *Buto(h)* or *butu* means "penis" in Malay, and *kling* is most likely *keling*. *Kling* was a Malay expression that referred to anyone from India, but particularly a Tamil. The word derives, according to Yule, from "Kalinga," the name of a former realm on India's eastern coast on the Bay of Bengal, between the mouth of the river Kistna in the south and that of the river Mahanadi in the north. This region was also once known as the "Northern Circars." Also known as Telinga, this Telugu coast appears to have been the region which at an early date traded with the Indonesian archipelago and also sent emigrants to the islands. From an early date, therefore, the Malay people held the name "Kalinga" to be synonymous with India. Subsequently, most of the Indian emigrants were Tamils, who came from a region further south down the coast (below Madras).

7. For the Dutch "*Kafferskullen*." Essentially this means the same thing as the Malay, since *kul* is a dialect slang word for "prick," while in older usage it was a slang word for "testicle." It is difficult to say whether Rumphius had a specific meaning in mind with "*kaffer*." The word was in common use to refer to anyone who was not white, but it was also used in its original Arabic sense of "infidel," i.e., anyone who did not believe in Islam. In the old Indies the word referred specifically to a native court officer, an official who was as much despised as a revenue agent paying a visit to a still in the Kentucky mountains.

Chapter 38. Of the Anguis Marini. Ular Laut.

1. These, according to Martens (p. 133), are the real sea-snakes, or Hydrophidae.

2. Rumphius is most likely thinking of the Arafura Sea.

3. *Onin* is a region in western New Guinea (now Irian Jaya) on a peninsula that juts out into the Ceram Sea below what is normally referred to as New Guinea's "bird's head," opposite Ceram. In later colonial times it was better known as Fakfak. Rumphius and Valentijn called this peninsula *Sergile*, which seems to be a Portuguese word for "mountain range." Even before Rumphius' day, Onin's coast was inhabited by enterprising people who traded with Ceram and with merchants from the Moluccas in slaves and *massooi* (an oil manufactured from the bark of the tree *Massaoia aromatica* Becc.).

Rumphius' passage is somewhat confusing. The "great bay" north of "Wezel's Island" (now called Pulau Adi) would have to be Triton Bay. This cannot be "Ryklof van Goen's Bay," because J. H. F. Sollewijn Gelpke (see below) has convincingly identified that with Sabakor Bay, northwest of Triton Bay. Rumphius equates "New Guinea" with "Onyn Cubicy." The latter word is a corruption of the word "Coveay," or "Cabiay" (also: Cubiay, Covejay, or Coveghay), which, in turn, is probably a corruption of the local name "Kufiai," and would correspond to the southern region of what might be called the "wattle" of New Guinea's "bird's head."

The fact that Rumphius mentions that these waters had been visited by "our people" in 1675 indicates that he was referring to a journey by Johannes Keyts in 1678 (*not* 1675). Keyts' report was kept secret by the VOC, and Rumphius did not read it either. He was, however, familiar with what members of Keyts' crew and local merchants had told him. See J. H. F. Sollewijn Gelpke, "Johannes Keyts. In 1678 de Eerste Europese Bezoeker van de Argunibaai in *Nova Guinea*," in *Bijdragen tot de Taal-, Land- en Volkenkunde*, deel 153, 3e aflevering (1997), pp. 381–395.

Except for some coastal regions, such as Onin, little was known about the second largest island in the world. Well into the seventeenth century, it was thought to be part of Australia. The first Dutch reconnaissance was in 1606. The Spanish mariner Ortiz de Retes called the island "Nueva Guinea" in 1545, thinking that, because of the appearance of the Papuas, he was dealing with an African race. Dutch colonial dominance was not established until 1828.

4. Rycklof van Goens (1619–1682) came to the Indies with his parents when he was barely ten years old. He started working for the VOC when still in his teens and enjoyed a rapid and illustrious career, being appointed to just about every position in the colonial hierarchy, including the highest office, that of governor-general, in 1678, when he succeeded Maetsuycker. He held this office for almost three years before he could retire to Holland, where he died in Amsterdam, barely three months after his arrival. Van Goens was a skilled diplomat as well as military leader. His embassies to the Susuhanan of Mataram between 1648 and 1654 became well known, and his military campaigns in Ceylon and on India's Coromandel Coast between 1657 and 1663 greatly contributed to the VOC's power in that region. The "Rycklof van Goens Bay" is now known as Teluk Sebakor.

5. "Cecilia" refers to the blind snake—*Caecilia* (L.)—and was so used in Rumphius' lifetime.

6. "Acorus Marinus" is identified by De Wit (p. 412) as *Enhalus acoroides* (L.).

7. This means merely "sea-snake," since *anguis* is the Latin synonym for the more familiar

coluber and *serpens*. Martens conjectures that this animal might be of the family Homalopsidae, although I believe these are considered poisonous.

8. The Malay means the exact same thing as the Latin since *ular* is "snake" and *laut* is "sea."

Chapter 39. Of the Tethyis.

1. Neither this word nor the title exists in classical Latin, nor would this have anything to do with Tethys, Oceanus' wife. The fact that Rumphius mentions Greek and translates it as "teats" indicates he was thinking of the Greek *titthos* (τιτθός), a noun which in classical Greek referred specifically to the nipple of a woman's breast. Martens says that these are "undoubtedly" Ascidiacea, an order of Tunicates (*Tunicata*).

2. This reference is to the Jesuit scholar Juan Eusebio Nieremberg (1595–1658). He was born in Madrid of German parents and only entered the Jesuit order at his father's insistence. He was a gifted intellect, author of a great number of books in both Latin and Spanish, and the first professor of "Historiae naturalis" at Madrid. The book Rumphius refers to is: *Historia naturae maxime peregrinae libri XVI*, published in Antwerp in 1635.

3. The Latin phrases mean: "he [i.e., Nieremberg] says that these are fleshy masses" and "[they are called] little glands by Athenaeus."

Chapter 40. Of the Sanguis Belille. Dara Belilli.

1. That is, Rumphius classes these creatures with the ones described in the previous ch. 39.

2. Perhaps Rumphius means the seas around Solor, including what is presently known as the Sawu Sea. Solor is one of the lesser Sunda Islands, southeast of the eastern portion of Flores, and was usually referred to in tandem with its two neighboring islands, Adonara and Lomblem, which are most likely Rumphius' "surrounding Islands." The Dutch had a fort on Solor in the sixteenth century, but abandoned it in the seventeenth century. All three islands are mountainous, and Adonara had the most significant agriculture.

3. Jacob van Wykersloot (?-1680) was the chief in Timor in 1672, was promoted to senior merchant in 1676, and died in July of 1680.

4. "Larentuque" is Larentuka, the largest town in eastern Flores, situated across from Adonara, on the other side of the Straits of Flores. Larentuka was a dominant political entity.

5. *Radja Salomon Speelman* was a Timorese chief, or "king," as Rumphius has it, who was defeated by the Portuguese in 1677, according to Valentijn (3.2:112, 121), and came to Banda, where he had himself baptized with the name of the VOC's general and governor-general. He returned to Timor in 1680. Rumphius stated that the radja's realm, Ade Mantutu, was in the eastern corner of Timor. See his *Herbal*, book 12, ch. 20, and book 6, ch. 12.

6. For "*als teenen van lym*." *Teen* normally means "toe" in Dutch, but Rumphius has another meaning in mind: thin willow twigs which were once used for weaving baskets and other items. *Osier* is the exact translation, for that too refers to twigs or tough shoots of a willow (*Salix viminalis*) used for weaving. The adjective "osiered" means plaited or twisted like osiers, and since Rumphius uses "*teen*" here in a figurative sense, one arrives at the image given in my translation.

7. Snake-wood was also called "Letter-wood," and Rumphius has even a third name for it: "Speckle-wood." This is most likely the tree or shrub *Strychnos colubrina*, which reputedly was used as an antidote to snakebite.

8. I am not sure of "*belille*." *Darah* means "blood" in Malay (*sanguis* in Latin), but all I can think of for "*belille*" is *berlilit*, which means "to twist" or "wind." Martens (p. 131) thinks that these are also *Tunicata*.

9. *Arak* is a variety of native distilled spirits, between sixty and seventy proof. In Rumphius' day it was a distillate of *sageru* (saguweir), a palm wine made from the gemutu palm, or *tuak*, a wine made from the coconut palm. In the twentieth century it was made from the fermentation of molasses obtained from the sugar factories. The yeast for the fermentation of *arak* came from cooked red rice. Most of the liquor came from Java, and the best *arak* was said to come from Batavia. *Arak* is entirely clear and has a light yellow color and a bittersweet taste and smell. The *arak* industry was controlled by the Chinese.

10. "*Bolatsjang*" or *belatjan* is a reddish fish paste still used today.

11. For "cholera," see n. 17 of ch. 29. "Mordexi" or "Mordexim" was another form of cholera ("colerica passio") from India.

12. "Theriac" is not a plant but a medical antidote known to the ancients. Mithridates is reputedly the originator of theriac. Derived from the Greek *theriake* (whence treacle) it is described in the dictionary as an "antidote; specifically, Theriaca Andromachi, or Venice treacle, which is a compound of sixty-four drugs, prepared, pulverized, and reduced by means of honey to an electuary."

13. Reading "*venyn*" for "*venye.*" There are quite a few misprints in this text.

14. In the "Auctuarium," or vol. 7 of his *Herbal*, ch. 36, Rumphius mentions that the root and the berries of "Lussa radja" were used against poison in general and food poisoning in particular (p. 29).

15. Probably *tahil.* A weight equal to forty-one grams at the end of the sixteenth century in the Indies. Around the same time, it was equal to about three grams less in China.

16. A *kati*, "catty" in English, was equal to 16 *tahils*, about one and one-third of our pounds.

Chapter 41. Of the Pulmo Marinus. Papeda Laut.

1. *Zoophitum* is a term from the seventeenth century, after Aristotle, referring to a class of animals which were once considered intermediate between animals and plants. The term is most often applied to sea anemones, corals, and sponges.

2. Despite the Latin name Rumphius gives, this creature has nothing to do with the order Pulmonata, because that order represents a kind of mollusk of the order Gastropoda, and Rumphius is clearly describing a creature without a shell. "Holothuria" is a generic term, originally found in Pliny. These creatures are now considered a class, Holothurioidea, of the Eleutherozoa which are part of the phylum Echinodermata, and are more commonly known as "sea cucumbers." The description, as well as the Dutch name (*quallen*), makes it clear Rumphius is dealing with a medusa, hence with a group of animals called generically Coelenterata; these include the sea-lungs. Holthuis states that these are Scyphozoa.

3. *Patella Marina* which means a sea-pan in Latin, belongs to a genus of mollusks, including the common limpet. See ch. 26, no. I, in book 2.

4. *Papeda Laut* comes from the Malay words for sea (= *laut*) and *papeda*, a word from the Moluccas referring to a kind of porridge cooked from the meal of the sago palm. One variety of sago palm is called after Rumphius: *Metroxylon Rumphii. Papeda* is, properly speaking, the sago pap, also called *popedah* or *lapia*, and is primarily associated with the Moluccas and Ambon. In Ambon it is eaten both cold and heated. Cold, the *papeda* is rolled as a lump in a *pisang* leaf or dried in the shape of cookies. Warm, it is often eaten with a sauce made from fish (*kuah ikan*), or cooked with hot peppers (*tjili*) or with shellfish.

5. See n. 9 of the previous chapter.

Chapter 42. Holothuria. Mizzens.

1. For *Holothuria*, see n. 2 of the previous chapter.

2. Holthuis informs me that this "Mizzen" is *Physalia physalis* (L.) of the Physaliidae family, better known to us as the "Portuguese man-of-war." The Latin means "sea nettle."

3. *Aquafort[is]*, a word coming into English at the beginning of the seventeenth century, stands for the early scientific name for nitric acid; *aqua regis* is a mixture of nitric and hydrochloric acids and gets its name from being able to dissolve the "noble" metals such as gold and platinum. These two strong mineral acids were discovered in Europe towards the end of the thirteenth century.

4. Rumphius may have found inspiration for his expressive nomenclature in Pliny's *Natural History*, especially book 9. Pliny compares several creatures to ships, or sails, or boats. The "nautilus" mimics a fast galley by raising its "sail," i.e., a "thin membrane," between its "arms." It rids itself of "bilge water" like a ship, uses its other "arms" as oars, and its tail as a rudder (p. xlvii); the "*nauplius*" also resembles "a ship under sail: it is a shell with a keel like a boat, and a curved stern and beaked bow," and it uses "the curve of the shell" as a sail (p. xlix); the Venus shell "sails like a ship" (*navigant*), and the scallop "uses its own shell as boat" (p. lii). Rumphius' term *Urticas Marinas* might also have been derived from Pliny, who calls jellyfish *urticae* (book 9, p. lxviii), that is to say "nettles," comments on their stinging, and describes them with the expressive phrase: *carnosae frondis his natura, et carne vescuntur* ("They have the nature of a fleshy leaf and feed on flesh").

5. The word "beard" for the tentacles of this Coelenterate is an early term.

6. The sentence narrating the periodicity of the *mizzens* in the Banda Sea is problematic. I translated *Oostmorissen* as "East monsoon," since *morissen* should be "Mousson" from the Portuguese *monção.* The East Monsoon corresponds roughly to our summer.

7. *White water* is a direct translation of *'t witte water.* This may also be translated as "milk sea," or *mare album*, but I kept "white water" because it is a literal translation of the Malay phrase *ajer putih.* This is a phenomenon which occurs in the summer months during the East Monsoon, around the Banda islands. The sea turns the color of milk, or as Rumphius put it, "a white glimmer like snow or milk" (in book 3, ch. 28). It has also been characterized as if "a fog hung over the surface" (*Encyclopaedie van Nederlandsch-Indië*, vol. 1, p. 32, in the article on the "Banda eilanden"). It seems likely that this phenomenon is caused by huge numbers of minute marine animals. Martens (p. 136) suggests that these are *Polycystines*, belonging to the group of *Radiolarians*, a class of marine protozoans. It is clear from some of his letters and from other texts in this book that Rumphius did not allow the possibility of

tiny animals but sought for a solution of the "white water" in terms of the sulphur which was plentiful in the mountains on the islands in the Banda Sea. Valentijn mentions that this marine phenomenon is at its weakest in June and July but very pronounced in August. See *Oud en Nieuw Oost-Indiën*, 2:137–138. This same "milk sea" is the topic of one of the many manuscripts Droogstoppel finds in the pack of papers Sjaalman leaves with him. See the list in the fourth chapter of Multatuli's *Max Havelaar.*

8. *Shallop* to translate *Chaloep*, a spelling Rumphius modeled on the French word *chaloupe*, although, at least in English, it was most likely derived from the Dutch word *sloep*. I am assuming that this refers to the larger variety, with one or more masts and carrying fore-and-aft, or lug, sails, since Rumphius and the governor were out for a considerable time.

9. *Pulo Ay* in *Banda* refers to a small island in the Banda islands. *Pulo Ai* (*pulau* is Malay for island; *kepulauan* means archipelago or island-world, and *ai* or *air* means "water") was conquered by the Dutch in 1616; they built a fort there called "Revenge." The island rises straight up from the sea and probably consists of coral lime.

10. This was Simon Cos (?-1664). He came to the Indies in 1639. In 1648 Cos was the resident of Hitu, and the governor of the Moluccas from 1656 to 1662. He became Ambon's governor in June of 1662 and died in office two years later, 4 February 1664. Rouffaer (*G*, p. 213) quotes the VOC's "Daily Register," or *Dagh Register*, that Cos had been ordered to make an inspection tour of the Banda Islands, and that Rumphius went along as "*mathematicus*," as the *Dagh Register* of 1663 (p. 256) called him.

11. *Epidromides Marinae* is another expressive name. The *Oxford Latin Dictionary* lists a meaning after Isidorus: "aft-sail"; in other words "Marine Aft-Sails," and the mizzen sail is an aft sail.

12. This Latin translates as: "If the Ambonese people were skilled in the Latin tongue, I believe they would have given them this name, because of their ability to burn."

13. For "*Sajor songa*," see n. 13 of ch. 28.

Chapter 43. Urtica Marina. Culat Laut.

1. "Potto marina" (should be *pota marina*) is now called *polmone di mare* or *polmone marino*, and refers to the common jellyfish.

2. This means "Sea-Nettle." The animal described is a sea anemone, of the order Actiniaria.

3. *Kulat* is mushroom or fungus in Malay; hence this is "Sea Mushroom."

4. I kept this rare usage to directly echo Rumphius' use of the same word. *Campernoyle* is a corrupt form of the medieval Latin *campinolius*, which, in its turn, is an obscure formation derived from *campus*, or "field." It translates as toadstool or mushroom. The *OED* provides the following quotation from the early sixteenth century: "Campernoyles that some men calyth tode stoles."

5. This for "*lekkere monden.*" Related to "lecherous," *lickerish* once meant "pleasing to the palate," "tempting," and also referred to an individual who was fond of delicious food.

6. See ch. 20, n. 6.

Chapter 44. Vermiculi Marini. Wawo.

1. For "*of veel meer Schepzels van Wormen.*"

2. For "*getweernde zyde.*" *Tweernen* was a verb that referred, among other things, to strengthening a thread or a yarn by twisting two (or more) threads together; hence the notion of one-ply, two-ply. Rumphius is clearly only interested in the thinness of the thread, so I used "silken Floss," because that indicated the fine filaments of silk.

3. "Liasser Islands" are the Uliasser Islands. "Latuhaloij" were really two villages, called today Latu and Hualoy, in southern Ceram. "Molukkos" seems strange usage here, but in Rumphius' day this noun only referred to the Spice Islands off the western coast of Halmahera (or Gilolo), from Ternate to Batjan, and *not* to all the islands between Celebes and New Guinea.

4. "Marine Worms."

5. We have an interesting quibble about nomenclature here. Van Hoëvell stated in 1875 that "*wawo*" is incorrect, and that it should be called "*laor.*" He was referring to Valentijn. It is clear from Rumphius that nearly two centuries earlier, "wawo" was used around the city of Ambon (which, after all, fronts the sea), that the northern peninsula, Hitu, called it "*melatten*" and the southern peninsula (excluding the city of Ambon), known as Leytimor, called the worms "*laur*," i.e., Van Hoëvell's "laor." It is a graphic illustration that even on such a small island, only fifty-eight square miles, there is an astonishing and bewildering linguistic diversity, particularly if it concerns something that everybody wants. Van Hoëvell also makes it clear that in his day, the end of the nineteenth century, everything took place exactly the way Rumphius described it. The entire population comes down to the shore at night, armed with countless torches (*lobeh*), to catch the animals. This event is called "*timbah laor* [i.e., *wawo*]," and provides an enchanting sight in March and April. Even the "*bakasam*" was prepared pretty much as Rumphius described it (pp. 214–215). The Dutch ornithologist Kees Heij witnessed the same event in 1997. Calling these worms "laor," he notes: "One of the nicest Ambonese traditions takes

place only in March, on the second and third day after the full moon, between seven and nine at night. During those two hours the sea around the old coral reefs begins to resemble a noodle soup, due to a multitude of worms that emerge from the reefs. At these particular spots, the coast is illuminated by hundreds of torches and kerosene lamps. Entire families collect these worms with nets and sieves. The worms are typically fatter the second night. People in the know contend that the best tasting ones are the worms collected during the first night. The worms, which can be nearly twelve inches long and 0.06 inches thick, are either salted, or dried, or cooked. Mixed with a variety of spices, they make a tasty dish that can be kept for a long time. And if they caught a lot of these worms, they will dry them and press them into cakes" (Kees Heij, "Bericht uit de Molukken," *Moesson*, 42:8 [February 1998], p. 23).

These worms are described in detail by R. Horst in an article in *G*, (pp. 105–108). He states that they are of the family *Eunicidae*, genus *Lysidice*, and that he gave this particular species the name *Lysidice oele* (p. 106), after the word "*oele*," which is what these creatures are called on Banda. The annual phenomenon and festive gathering can be compared to the appearance of "*palolo*" or "*bololo*" in the lagoons of the Samoan, Fidji, Tonga, and Gilbert Islands.

6. The first, and best, kind is called "Little Wawo," and the second, is "Large Wawo."

7. For "*Sajor songa*," see n. 13 of ch. 28.

8. Rumphius seems to imply here that the sauce made of *wawo* is something different from *bocassan* (*bakasam*), while not much later he seems to say that the *wawo* sauce *is* "bocassan" or, at least, a "bocassan" sauce. This is confusing, because elsewhere "*bocassan*" is described as a dish made from mussels. Valentijn, for instance (*Oud en Nieuw Oost-Indiën*, 2:158–159) says that the mussels were kept in salt for five to six days and then carefully washed. A sauce was prepared from boiling vinegar together with green ginger, peppers, white pepper kernels, and nutmeg, and was then left to cool. After it had cooled, the mussels were added and the *bocassan* was stored in large bottles, with olive oil on top to keep it from spoiling.

9. De Wit in his "Checklist" identifies this "Filix esculenta" as *Athyrium esculentum* (Retz.).

10. In book 6, ch. 76, of his *Herbal*, Rumphius calls this shrub "Lignum aquatile," also "Aywayl," or "Waterhout" (Waterwood), stating that the fresh leaves were eaten with *wawo*, and since "wawo" was known as *Laur* in Ambonese, this plant was also called "Daun Laur" (*daun* = leaf).

11. "Sop" exists in English as a noun meaning a "piece of bread" but also as the liquid in which the bread is dipped. The *OED* begrudgingly intimates that this word, like so many others, came directly into English from Dutch.

12. "Lanquas" or *langkuwas* is *Alpinia galanga*, an herb with a long stem and clusters of flowers. The white variety was used for cooking. Valentijn (*Oud en Nieuw Oost-Indiën*, 2:157, 158) says that it was very much like ginger.

13. See ch. 85 in book 3.

14. *Gaba-gaba* refers to the main nerves of the leaves of the sago palm (*Metroxylon Rumphii*). Because they are very durable, these nerves were used to make the walls, floors and attic spaces of native dwellings.

THE SECOND BOOK OF *THE AMBONESE CURIOSITY CABINET* DEALING WITH HARD SHELLFISH

Chapter 1. About Hard Shellfish in General.

1. "Things which remain the same and share the same nature: shells whether they are twisted or broken." Perhaps "*anfractuosa*" should have been "*aufractuosa*."

2. *Bia* is the general Malay word for "shell," and *kerang* is in the Moluccas the general word for "coral."

3. This peculiar word refers to the animal's liver. In a nineteenth-century translation of *Aristotle's History of Animals* (transl. Richard Cresswell [London: George Bell, 1883], p. 83) I found the information that this term derives from Scaliger, while Aristotle called it "mecon": μήκων (*mecon*) in Greek has as its primary meaning "poppy" and μηκώνιον (*meconion*) the "juice of the poppy" or "opium." However, the secondary meaning of *mecon*, after Aristotle, is "the liver of testaceous animals," and the secondary meaning of *meconion* is the "discharge from the bowels of new-born children," now known to any pediatrician as the newborn's *meconium*. From Rumphius' description of the contents of the "little sack" (which he calls the "Sand sack" in ch. 6), saying it contains "a greenish or blackish mud," one can tell that the secondary meaning of *mecon* and *meconion* predominates. The Latin noun *papaver*, however, *only* refers to "poppy"; hence I would suggest that Scaliger simply translated the primary meaning of *mecon*, ignoring the fact that the Latin does not have the physiological connotations.

4. Rumphius' addition of the Latin *callosa* (properly *callosus*) to the Dutch word *weerig* makes

it clear he is talking about something that has the substance of a callous, and it was so used in the sixteenth and seventeenth centuries in Dutch: *eeltig*.

5. "Through the addition of new material."

6. "Circumvolution" for *omloop*, an expressive but difficult substantive to translate. Circumvolution, which refers to "a rolling, whirling, or turning round an axis," is, in English, as old as the fifteenth century.

7. "By extending the entire body of the shell."

8. "Porcellanae" are *Cypraea* (see ch. 23), a genus of gastropod mollusks that includes the "Cauris," as described in ch. 24.

9. *Lamellae*, the plural of the Latin noun for a "thin plate," now refers to the plates that form the gills of bivalve mollusks.

10. "*Spondylo*" from Greek *sphondulos*, which originally meant a vertebra, a joint, but also came to refer to a particular oyster. The first denotation might have been the inspiration for this word in the meaning of the hinge of bivalve mollusks. It seems that Rumphius has this in mind here, because he is referring to a part of the animal's anatomy and not to the family of spiny oysters (*Spondylidae*).

Chapter 2. Nautilus Major Sive Crassus. Bia Papeda.

1. The Dutch *veelvoet*, a correct rendering of "*polypus*," is tersely eloquent, but does not come across as well in English. When Rumphius uses it again I have retained the more familiar "polypus."

2. Rumphius is thinking of "horns of Ammon"—Ammon being the Egyptian deity whom the Greeks identified with Zeus and who was depicted with ram's horns—which are ammonites, or fossil shells of an extinct group of Mollusca. They had an external shell in the shape of a spiral and were divided inside into chambers, just like the nautilus. See ch. 4.

3. Rumphius probably got the second word from Pliny, but "*intergerinum*" should be *intergerivus*, meaning "something that is placed between," such as the wax walls in a beehive. The first word has to do with anything that pertains to walls; hence the phrase means something like "between walls."

4. A bird of the genus *Cacatua* that is generally white except for its crest, which has different colors. The *Cacatua moluccensis* on Ambon and Ceram has a crest that is a pinkish red.

5. The original Dutch *oogappel* can mean iris, pupil, or eyeball.

6. As Schynvoet's remark in his addendum makes clear, this drawing did not survive.

7. *Vochtigheit* in the original. Used attributively or figuratively, "liquid" had an early usage, but usage meaning a "liquid substance" in English commenced *after* Rumphius' death.

8. Sewel has "scale" for the original text's *schilfer*, but this does not fit here.

9. The *schelling* was a silver coin, worth six stivers, about twenty-five millimeters across.

10. This is the famous *Nautilus pompilius* L., so often confused with the animal described in the next chapter, the *Argonauta*. These two were most famously mixed in Oliver Wendell Holmes' poem, "The Chambered Nautilus." It is a lovely poem that starts quite rightly: "This is the ship of pearl . . . ," but then he lets it immediately "sail" "the unshadowed main," and praises its inclination to exchange outworn dwellings, something a hermit crab might do. The message of the final stanza is well worth considering and very American: "Build thee more stately mansions, O my soul, / As the swift seasons roll! / Leave thy low-vaulted past!" But it is not a lesson one can draw from Rumphius' *Bia papeda*, which slowly propels itself through the water at great depths in search of crustaceans.

Martens notes (p. 112) that Rumphius was the first to describe the living animal, and that it remained the only description of it until 1832, when Richard Owen published his famous "Memoir" on the nautilus.

Rumphius Latin phrase means: "The Great Nautilus or the Thick One."

Identifications of the animals in book 2 are based on the following text: E. von Martens, "Die Mollusken (Conchylien) und die übrigen wirbellosen Thiere in Rumpf's Rariteitkamer," in *Rumphius Gedenkboek*, pp. 109–136; hereafter referred to as Martens. Martens' work is dated now, and H. L. Strack completely revised all identifications and provided a current nomenclature. In some cases Strack disagrees with von Martens' identifications, and the German scientist also neglected to identify several species that were not illustrated. Some species defy positive identification. A number of identifications provided in this annotated translation of *D'Amboinsche Rariteitkamer* will have to be revised in the years to come.

11. *Papeda* is the Moluccan name for a type of porridge cooked from sago flour, extracted from the sago palm.

12. The so-called Thousand Islands—actually about 600—are a group of small islands in the Java Sea, northwest of the Bay of Batavia (Jakarta), and which are now a favorite resort area for residents of Indonesia's capital. Now known as Pulau Seribu, they were liked as recreational islands during the days of the VOC as well, while the countless uninhabited islands, with their pristine beaches, enticed

people to collect shells and, these days, to dive among the coral reefs. Only a handful were inhabited, mostly by Buginese and Mandarese from southern Celebes, and they traded in the obvious products: coconuts, copra, fish, mother-of-pearl, and tripang (a Holothurian known as the "sea cucumber").

13. Many of the beautiful "nautilus cups" that were produced in Holland in the seventeenth century used the shells from Ambonese waters. Most were set on top of a stem or foot that depicted some kind of mythological marine deity or animal, which in turn provided the base for a frame of silver or gold that held the shell itself. The latter was either left untouched or engraved with scenes that one will sometimes find in paintings as well. Dutch shell engravers were in much demand at the time, men such as Cornelis Bellekin and Dirck van Rijswijck. The method which Rumphius described was pretty much the procedure that became standard.

14. *Balanus* is a barnacle (Cirripedia).

15. Pliny's confusing passage, which describes the *Argonauta*, is in book 9, ch. 47, paragraph 88 in Latin. See also n. 23 of the next chapter.

16. The Latin means "A Pearly Shell."

17. Robertus Constantinus (?–1605), one of Scaliger's pupils, was from Normandy. He was a physician, linguist, botanist, and historian.

18. This is modern Greek, a phrase of which the second word is unclear. The first is "*thalamoi*," which means "chambers," while the last word is the possessive genitive "*podiou*," meaning "of the foot." Perhaps the second word is the Latin *tectum*, which can mean "room," which is related to the classical Greek τέγος. Hence the phrase could mean "chambered rooms of the foot."

19. This is the Cardanus whom Scaliger attacked in his *Exercitationes*. Jerome Cardan (1501–1576) was born in Milan and remained as ardent a Milanese as Stendhal. A brilliant intellectual, Cardan, who acquired the degree of *Doctor Medicinae* in 1525 from the University of Padua, was a strange and eccentric man, cursed with wanton children. He professed to be an expert in dream analysis, believed in astrology (he wrote a book on it), and suffered from gout (he also wrote a book on podagra). He was a prolific writer, and his oeuvre includes commentaries on Hippocrates and on Galen.

20. The "glands."

21. Valentijn (*Oud en Nieuw Oost-Indiën*, 2:158–159) identifies *Bocassan* as a dish prepared from mussels. The mussels were kept in salt for five to six days and then carefully washed. A sauce was prepared from boiling vinegar together with green ginger, peppers, white pepper kernels, and nutmeg and was then left to cool. After it had cooled, the mussels were added and the *bocassan* was stored in large bottles, with olive oil on top to keep it from spoiling. It is described in detail by Rumphius in this book's ch. 33 (section III).

22. I wonder if Rumphius, who was fascinated by such lore and who was almost obsessed with stone-like concretions called "*mestica*" (see book 3), had "*paeanitis*" in mind—mentioned by Pliny in book 37, ch. 46, in Latin, paragraph 180—a stone also known as "*gaeanis*," or the "earth stone," which is said, in Eichholz's translation, "to become pregnant and to give birth to another stone, and so is thought to relieve labour pains."

23. This refers to Rochefort's *Histoire Naturelle et Morale des Iles de l'Amerique;* see n. 9 of the first ch. of book 1. Rochefort calls it "burgau" and his brief, lyrical description is as follows: "The *Burgau*, which is of the figure of a Snail, being uncas'd out of the outermost coat, presents to the eye a silver shell intermixt with spots of bright black, a lively green, and so perfect and shining a grey, that no Enameller could come neer it with all the assistances of his art. As soon as the fish which had been lodg'd within this precious little Mansion hath been disseiz'd thereof, there is immediately seen a magnificent entry beset with pearls, and afterwards several rich appartments so clear, so neat, and enamell'd all over with so bright a silver-colour, that there cannot in matter of shell anything be imagin'd more beautiful" (from the contemporary English translation, published in 1666, pp. 120–121).

24. This refers to Buonanni's main work, *Ricreazione dell'occhio a della mente nell' osservazione della chiocciole*, published in 1681, Latin edition in 1684.

Chapter 3. Nautilus Tenuis. Ruma Gorita.

1. The second word of *Nautilus Tenuis* is Latin for "thin," or "fine," and is a synonym for "*gracilis*." *Ruma gorita* would mean "the house of the octopus." *Rumah* means "house" and the other word, more commonly spelled *gurita*, refers to a *small* octopus or squid. Here we see again how accurate Rumphius was, because this is not the fabled "Chambered Nautilus," but the female's egg case of a relative of the octopus, the *Argonauta*. The confusion arises because it is also known as the "Paper Nautilus" owing to the egg case's thin shell. This case or "shell" does not have "chambers." The male does not have a shell at all, and both sexes have only eight arms (octopod), whereas the "Chambered Nautilus," which Rumphius described in the previous text, has about ninety tentacles. These animals move by a kind of jet propulsion, forcing water through a fleshy funnel. The male's third arm on the left is used for copulation. The animals embrace each other with their full complement of arms,

and the male inserts this third arm into the female's mantle cavity (palladium) and leaves it there, alive and ready for procreation. This sex-arm, so to speak, is longer and thicker than the other arms, and contains a capsule that holds the spermatozoa. Martens (in *G*, p. 112) identifies the larger variety Rumphius describes as *Argonauta argo* (L.) and the smaller one as *Argonauta hians* (Lightfoot). Both Martens and Benthem Jutting note the fact that Rumphius was remarkably accurate in the description of the present animal, only succumbing to the myth of the animal's "sail." The *Pompilum* (*Pompilus*), mentioned in connection with Pliny, is, properly speaking, the previous "Nautilus Major."

2. *Span* and *finger* are length measurements often used in the seventeenth century. A "span" referred to the distance between the tips of the thumb and little finger, while stretching them, usually an average of nine inches. A "finger," as a length measurement, referred to the length of the index finger, and as width it was the equivalent of four barley corns laid side to side.

3. *Polypus* ("many-footed"), the ancient Greek term (in Aristotle, for instance) for octopus or squid. It is also the accepted term for a pendulous tumor in the nose or intestines.

4. "*Bolitoena*" (Bolitaena), a Greek term, properly *Bolítaina*, or *Bolbítion*, which referred to a small cuttlefish. Aristotle's passage that includes this name is rather confusing; see *The History of Animals* (*Historia Animalium*), book 4, ch. 1, sections 15–16.

5. Here *beards* refers to the "arms" of the octopus-like creature.

6. *Seacat* (*zeekat*), once so used in English as well as a noun denoting the decapod cuttlefish, *Sepia officinalis* (L.), which is a common European cuttlefish. Cuttlefishes are invertebrate animals, related to the octopus, of the class Cephalopoda and family Sepiidae. "Cephalopoda" means "having legs on your head." It has an internal "shell" known as the "cuttle bone," five pairs of arms around its mouth, its "ink" is the brown pigment known as "sepia," and it usually prefers shallow coastal waters. The etymology of "cuttle" is uncertain.

7. *Tingang* or *tinan*, an outmoded large proa from western Java, steered by two oars, and used to haul larger cargoes than an ordinary proa, hence something like a lighter.

8. Rumphius' observation that the nautilus grabs hold of leaves or pieces of driftwood is accurate, and he seems to have been the first observer to note this habit; cf. Martens, *G*, p. 112.

9. This is my translation for "*visfuiken.*" *Fuik* always presents a problem, because, although used in fishing, it is not a net. It is a trap, usually made from wicker or some such material, with compartments that are divided by hoops and which become increasingly more narrow. There is some resemblance, at least in principle, to our New England lobster trap. *Weel* has been used in English since the middle of the thirteenth century, and the following definition from 1725 (in the *OED*) clearly resembles the *fuik:* "Weel . . . made of Osier-twigs, which are supported by Circles or Hoops, that go round, and are ever diminishing. . . . Its Mouth is somewhat Broad, but the other end terminates in a Point: It's so contrived, that when the Fishes are got in, they cannot come out of it again, because of the Osier Twigs, which advance on the inside to the Place where the Hoops are, and which stop the Passage, leaving but a small opening there."

10. *Kakatu*, from the Malay *kakatua*, for parrot. This wonderful bird belongs to the genus *Cacatua*, of the family of parrots called Psittacidae. About a third of all known varieties live in the eastern part of the Indonesian archipelago, i.e., the Moluccas and, especially, New Guinea. The *kakatua* is white with a colored crest—the kind from Ambon and Ceram (*Cacatua moluccensis* [Gmelin]) has a crest the color of salmon—and has the characteristic curved upper beak to which Rumphius is referring. On Ceram and Buru a larger variety was once native: *Eclectus roratus* (Statius, Müller). The male is green and the female bird is red. The related parakeets are also plentiful in this part of the archipelago, especially the large, fan-tailed parakeets (*Aprosmictus*), which are green or blue on their backs and have red bellies. A specific kind on Ceram and Ambon is *Alisterus amboinensis* L. Another related bird that is plentiful in those parts is the lori or Luri, the short-tailed *Lorius* of the family Trichoglossidae. The common lori on Ambon and Ceram is the *Lorius domicella* L., which is black on top of its head, underwing feathers are blue, chest yellow, and the rest red. Cockatoos are often kept in captivity by the local population. There is a thriving business in all these birds which the native population steal from the nests when still chicks and then sell to itinerant traders. Ludeking, in his *Schets van de Residentie Amboina* (p. 53), mentions the Ambonese believed that if loris or parrots screeched at night, it meant that the devil was absconding with a corpse; and if they screeched during the day, and a person was about to commit a burglary, their cries indicated that people were coming.

11. *Carina* is Latin for a "ship's keel" and, by extension, a metonymy for a boat or ship.

12. A *doekhuif* was, even in Rumphius' day, an obsolete word that referred to a woman's white head covering (*huif*), made of woven cloth (*doek*), that had many starched and ironed pleats. In Sewel's *New Dictionary English and Dutch*, which was published in Amsterdam in 1691, the word "*Doek-huyf* " is said to be "out of date" and is there translated as "Linen-coif." *Huif* remained in Dutch usage as the cloth cover over a wagon, for instance, the cloth cover of our "prairie schooners."

13. *Lego Lego* seems to have been a dance with the participants in a circle. Valentijn (2:163) asserts

that they also sang to one another. He also informs us that a "*lego*" (he uses the word as a proper noun) could last for days and that on Ambon the dance started to the right. He noted as well that the sexes did not dance together but always separately. The *Encyclopaedie van Nederlandsch-Indië* (1:572) states that on the neighboring island of Buru the men danced a *lego-lego:* they stood in a circle and, to the rhythm of the music, advanced first the right foot and then the left, slowly, and in a repetitive motion.

14. *Quartern* for "*mutsje*," a liquid measure roughly equivalent to a deciliter.

15. *Rixdollar* is the English spelling for the name of a silver coin current from the sixteenth to the middle of the nineteenth century in Holland (*rijcksdaler*, or modern form *rijksdaalder*), Sweden (*riksdaler*), Denmark (*rigsdaler*), Germany and Austria (*reichsthaler*). Particularly known during Holland's commerce with the East, it was a silver coin of fifty stivers (*stuivers*), worth in English money, at the time, four shillings and nine pence (Sewel 1691).

16. I do not know what "*Zeeusche kan*" refers to. A *kan* was roughly the equivalent of one liter, and there were 2 *pinten* in a *kan*. What a *pint* was varied greatly. For instance, a *pint* of wine was the equivalent of three-fifths of a liter, a *pint* of beer was half a liter, and a *pint* of milk nine-tenths of a liter. The British "pint" is half a quart, or one-eighth of a gallon, and a liter is close to two British pints; hence one could say that a *kan* was a bit more than two of our pints. A *mutsje* was one-tenth of a *kan*, and a "quartern" is less than a quarter of a pint.

17. The sea eagle *Haliaeetus leucogaster* is called *elang laut* in Malay and *bahak* in Javanese. It is found all over the archipelago, especially in the eastern regions. It is white, with a dark-gray back, wings, and upper half of the tail. Its legs are bald. Rumphius is, once again, very accurate, here in terms of the subsequent figurative comparison with the Palladium statue, because this eagle dive-bombs from up high down towards the sea, where it grabs a fish with its claws. Birds of prey are generally known as *lang*, *helang*, *elang*, or *ulong-ulong*. Hence *elang laut* literally means "bird of prey of the sea (= *laut*).

18. The *Palladium* was, in Greek mythology, a statue of Minerva, or Pallas Athena, that fell from the sky or was sent down from heaven by Zeus to Dardanus, the founder of Troy. It was kept in Troy as a tutelary deity, and as long as the Palladium statue was in Troy, the city could not be taken. For that reason, Odysseus and Diomede disguised themselves, slipped into the city, and stole the statue. Helen was an accessory to the crime because, though she recognized Odysseus, she did not raise the alarm.

Many ancient cities had such protective images known as "Palladia," including Rome. It was written that Aeneas rescued the Palladium from burning Troy—clearly at odds with the previous tale—and brought it to Rome, where it was placed in the *penus Vestae*, the innermost part or sanctuary of a temple of Vesta (the goddess of the household) and was presumed to have saved Rome from the Gauls in 390 B.C.

19. The reason Rumphius sent the shell to Vienna is the following. He corresponded with various German scholars such as Mentzel, Schröck, and Fehr. In 1680, Mentzel and Cleyer (for Cleyer, see n. 38 of ch. 35 in book 3) proposed Rumphius as a likely member of the Academy, which was situated in Schweinfurt. He was approved, and he entered the academy as its ninety-eighth member in 1681, hence *not* 1683. At such a time a successful candidate was given a special name that, in some way, was appropriate to his professional life. Rumphius was called *Plinius*, as he proudly displayed on the title page of the present work. Similarly, the man who was the academy's president in 1683, Dr. Fehr, was called the "Argonaut," after the crew of the ship, *Argos*. Its epic journey was the subject of Apollonius of Rhodes' *Argonautica*. Since the shell itself resembles an ancient ship it seemed fit to send it as a present to his learned colleague in Schweinfurt. Is it truly fortuitous that this particular kind of shellfish became known after Linnaeus as *Argonauta?* Linnaeus became a member of the same academy.

The Latin sentence means: "Johan Michael Fehr, Doctor of Philosophy and Medicine, Physician in Schweinfurt, and Elected President of the Academia Naturae Curiosorum as established by order of the Holy Roman Empire, under the name of Argonaut." Fehr (1620–1688) died while he was president of the "Academia." Rumphius' mentor, Christian Mentzel, or "Christianus Mentzelius," was a "Doctor of Medicine and the Court Physician and Counsellor" of the Kurfürst of Brandenburg (the later Kingdom of Prussia), who, at that time, was Friedrich Wilhelm, known as the "Great Elector." The Kurfürst, born in 1620, died in 1688, despite the academy's divine cognomen of "Apollo."

Christian Mentzel (1622–1701) studied medicine in Frankfurt and Padua, and obtained his degree from the Italian university. He was well versed in Chinese and Chinese history. He wrote, for instance, a grammar of Chinese, and a *Lexicon Sinicum*, in ten volumes, which remained in manuscript, and *Sylloge Minutiarum Lexici Latino-Sinico Characteristici*, in four volumes, which was published in *Ephemerides*.

He was also an ardent botanist. In 1695 he deposited into the Royal Library in Berlin a text he had written in collaboration with Cleyer, entitled *Flora Japanica*, which contained 1,360 figures of plants drawn from nature. Earlier, Mentzel had published the botanical *Index nominum plantarum universalis cum pugillo plantarum rariorum* (Berlin, 1682).

20. *Suidas*, a Greek lexicon from the tenth century A.D., in alphabetical order that was a combi-

nation of a dictionary and an encyclopedia. It quoted a large number of ancient authors and sources and provided a great deal of knowledge about ancient life and history.

21. *Athenaeus* was a physician under Emperor Claudius (A.D. 41–54). He founded the school of the Pneumatists, adding *pneuma* (breath or spirit) to the four basic elements. His system, based on Aristotle, was admired by Galen.

22. *Road snail* for "*wech slekke*," which, because Rumphius provides the Latin alternative *Limax*, can be identified as the land slug. This description follows Aristotle quite faithfully; it is in ch. 1, however, and Rumphius quotes most of section 16.

23. Pliny's well-known description of the nautilus incites Rumphius to a rare criticism. He calls his scholarly avatar's prose "rather obscure" (*wat donker*). This is often the case with Pliny, and particularly in the passage concerning the nautilus. Rumphius' Dutch translation of Pliny's Latin, which I then translated into English, is more comprehensible than either the original or, for that matter, Rackham's translation, which is as follows: "But among outstanding marvels is the creature called the nautilus, and by others the pilot-fish. Lying on its back it comes to the surface of the sea, gradually raising itself up in such a way that by sending out all the water through a tube it so to speak unloads itself of bilge and sails easily. Afterwards it twists back its two foremost arms and spreads out between them a marvelously thin membrane, and with this serving as a sail in the breeze while it uses its other arms underneath it as oars, it steers itself with its tail between them as a rudder. So it proceeds across the deep mimicking the likeness of a fast cutter, if any alarm interrupts its voyage submerging itself by sucking in water." This is section 47 of book 9, in the original Latin, paragraph 47; in the Loeb ed., vol. 3:221. Rackham's effort is too modern. For instance, the ship with which Pliny compares the nautilus is called a *Liburna*, which was a fast Liburnian galley, and not a cutter. To my knowledge the word "cutter" was never used for a galley, no matter how swift or slow it could go. Mulianus' description is some improvement, ditto for Bellonius. It is perhaps fruitful to compare the writing of such "authorities" to that of Rumphius in order to appreciate the clarity of his prose.

24. *Mutianus* was Gaius Licinius Mucianus, a Roman Consul of the first century, who was appointed governor of Syria by Nero. Later he aided Vespasian and remained that emperor's chief adviser. He wrote a book of geographical *mirabilia*, or "marvels," which Pliny used extensively in his *Natural History*.

25. *Head* for *snuyt*, which was an older and specialized term for the prow. *Snuit* means "snout" in modern Dutch.

26. *Omen* for "*voorspook*," the wonderful old Dutch noun which could be translated as "anticipatory haunting."

27. *Jonston* was the Polish physician Jan Jonston (1603–1675), who studied medicine at Leiden and Franeker. His interest in the animal world led him to visit various private collectors, including the well-known Paludanus in Enkhuizen. The volume Rumphius refers to was Jonston's most important publication, an encyclopedia of the animal world, entitled *Historia naturalis*, published in Amsterdam in 1657, and in a Dutch translation in 1660. It seems to have been used quite often by collectors as a reference tool.

28. *Thalamis Toukdapodius* should be *thalamis t'oktapodos* in modern Greek and means "of the octopus in the chamber."

29. Born in Caen, Normandy, Robertus Constantinus (Robert Constantine; ?–1605) was a physician, linguist, botanist and historian. He lived in Scaliger's house for a while and after the older man's death, Constantinus published part of Scaliger's commentaries on Theophrastus. Rumphius is referring to *Lexicon graecolatinvm Rob. Constantini. Secunda hac editione* (Geneva: Eustathii Vignon & Iacobus Stoer, 1592). There was also a 1637 edition.

30. *Pectunculus*, Latin for a very young or very small scallop; for instance, Pliny, book 32: original Latin, paragraph 70. See ch. 35.

31. *Bellonius'* passage is troublesome. For instance, the animal's procedure in order to rise to the surface is not clear, and another problem was *platvoeten*, which, literally, means "flat feet." Since nothing is associated with it, the term is difficult to pin down. I am assuming that Belon was using Galen's word *leiopodes* (from Greek *leio* and *pous*, which means "smooth footed") and that Rumphius knew that in Dutch it usually referred to birds, hence, specifically, to the webs between their toes.

32. Jacob Bontius (1591–1631) was the son of a professor at Leiden University. When he was twelve, he was admitted as a student and obtained his medical degree in 1614. In 1627 he sailed to Java with the fleet of J. P. Coen as its medical officer. He survived the two famous sieges of Batavia by the ruler of the Mataram realm, Sultan Ageng, in 1628 and 1629, though he was ill for four months with malaria, dysentery, and beriberi. His first wife died on the outward journey, his second wife died from cholera in Batavia, and his third wife survived. Bontius wanted to return to Holland and be appointed professor in Leiden, but he never saw Holland again and died in Batavia. What is remarkable is that in less than four years of tropical experience, while burdened with the responsibilities of a medical man

in what was still practically an outpost, in addition to being appointed justice of the peace (*advocaat fiscaal*) and to the position of bailiff, Bontius managed to gather and record botanical information. His books were published only after his death. Four were published in Leiden in 1642, as the four-volume *De Medicina Indorum Lib. IV*, of which book 1 consists of notes to the work of Garcia ab Orta. There is one other edition of note. In a work comprising fourteen volumes by Gulielmus Piso and printed by the famed Elsevier brothers, there are six books which are the work of Bontius: four of these are the aforementioned *De Medicina Indorum*, and the other two are Bontius' unfinished *Historia Animalium* and *Historia Plantarum*. Most likely it is the *Historia Animalium* to which Rumphius is referring.

33. *Acetabulis* was Latin for the suckers or cavities in the arms of polypi, the socket of the hipbone, or the cup of flowers.

34. This is a troublesome phrase: "*verwisseld tegens malkander staande*," I translated as (positioned) "alternately across from one another," taking *verwisselen* to mean "alternating."

35. See ch. 42 of book 1.

36. An "*Amsterdamsche houtvoet*" was the equivalent of 11.14 inches.

37. Cosimo de Medici III (1642–1723) was the Grand Duke of Florence. Rumphius was urged to send a large collection of tropical shells (360 species) and other curiosities to the Grand Duke in 1682, an event he refers to several times in his *Ambonese Curiosity Cabinet*. I have a feeling it was imposed upon him, and I do not think that Rumphius liked it. Cosimo III, a nefarious individual and ruler, presided over the demise of the Medici dynasty in Tuscany. His offspring were childless and his dukedom was transferred first to Spain; then, after Don Carlos' conquest of Naples, the succession was once again transferred, this time to Francesco II, Duke of Lorraine, husband of Maria Theresa of Austria.

38. Jan Volckertsz (1578–1651) was a merchant in textiles who left a considerable collection of shells, corals, etc. to his two sons.

39. *Savry* refers either to the etcher Salomon Savery (1594–1665), who also worked in England, or to his son, Jacob Savery III (1617–1666), who was an etcher and publisher in Amsterdam and Dordrecht.

40. Most likely Johan Bernard de la Faille (1672–1727), who held a high position in the city of Delft. His collection was impressive and was said to have "possessed the most famous shell collection of his day" (Engel, p. 85).

41. *Lister* was not the famous surgeon Joseph Lister, but the less illustrious contemporary of Rumphius, Martin Lister (1638–1712), a physician and naturalist. Educated at Cambridge, he practiced medicine in York until 1683, and then moved to London. Two works by Lister are of significance in the present context, both on shells and shellfish: *Historiae Conchyliorum* (1685–1692) and *Conchyliorum Bivalvium* (1696).

42. *Gesnerus* was Konrad von Gesner; see n. 11 of ch. 1 of book 1.

43. *Weals* for *striemen*. The original meaning of "weal" was a ridge raised on a person's body from the stroke of a whip or a rod.

44. "D'Aquet" was the physician Hendrik d'Acquet (1632–1706), who became the mayor of Delft. He collected primarily naturalia, which he obtained from VOC captains and other personnel. Rare specimens which were not in his collection he often had copied by reputable draughtsmen, while, conversely, he permitted items he possessed to be drawn and published by others. This he did, for instance, with Rumphius' book, which, as was mentioned, he also was instrumental in getting published. D'Acquet maintained a private garden where he cultivated tropical plants. Like so many other private collections, his was auctioned off after his death and dispersed.

45. The fact that Schynvoet uses a plural indicates that he was referring to the father Feitama and his two sons. Sybrant Feitama (1620–1701) was from Haarlem. A nephew of the famous cartographer Joan Blaeu, he became a very wealthy merchant and apothecary who lived on the Damrak in Amsterdam. He was a poet and published two collections of poetry, and was otherwise known for his collection of stuffed, exotic animals and a large collection of drawings by contemporary Dutch artists. His son, Isaac Feitama (1666–1709), increased his father's collections, as did *his* son, named after his grandfather, Sybrant Feitama (1694–1758). The last scion was a frail individual who devoted more time to culture than business.

46. Juffrouw *Oortmans* indicates Petronella De La Court (1624–1707), widow of Adam Oortmans. She took over running the brewery "The Swan" ("De Zwaan") on the Singel after her husband's death and, being a woman of wealth, established a large collection of shells, precious stones, art, and other things, including a celebrated doll's house.

47. In the original "*waterlandsche boerinne kap*"; this referred to a bonnet worn by a farmer's wife in Waterland, a region, north of Amsterdam, in the province of North Holland.

Chapter 4. Cornu Ammonis. Little Posthorn.

1. Amon is, of course, the Egyptian god who became prominent during the eighteenth dynasty. He was originally a Theban deity whose name meant "the hidden one" and whose animal was the ram with curved horns.

2. The name Rumphius gives to this cephalopod refers to an extinct group of Mollusca, belonging to the class Cephalopoda, which includes the octopus, squid, as well as the extinct Ammonites. These fossil shells show that the erstwhile animals had an external shell that was coiled in a spiral and that had many chambers inside, hence the resemblance to the nautilus. They could attain a great size. The animal in the text, however, is *Spirula spirula* (L.), a genus of small cephalopods that occur in tropic seas at great depths and only rarely are found on the surface or on beaches.

3. "*Tay*" is *tahi*, the Malay word for "shit," similar in its derisive connotations as our noun. The noun also means the remains of something, dregs, dirt of any kind, and is featured in a host of wonderful expressions such as *tahi bintang* (shooting stars), i.e., star lees; *tahi lalat* (mole or freckle), i.e., fly excrement, or flyspeck; and *tahi ular sawa*, which means the excrement of the python, which was once highly prized for medicinal purposes.

Chapter 5. Carina Holothuriorum. Sea Gellies' Boats.

1. For "*by 't uitgaan van den Oogst.*"

2. This is Sewel's splendid rendering of "*weêrschyn.*"

3. Misprinted in the text as "*Holotuhri,*" which I silently corrected. Strack: *Janthina janthina* (L.). Their popular name is very apt: Purple Sailor.

4. *Nerita* probably refers to some mediterranean Janthina species. Rumphius connects it to the classical Greek form *neō* (νέω), which means "I swim." The name occurs in Aristotle (book 4, ch. 1, sections 16–17 ff.).

Chapter 6. Cochlea Lunaris Major. Matta Bulan Besaar, or Matta Lembu.

1. For *uytspatten* in the text, which once had a subsidiary meaning of something that grew out of something, esp. sores and ulcers.

2. "Spanish-green" was a copper oxide green, vitriol; in older usage, as here, also synonymous with "vert-de-gris," or, in English spelling, verdigris.

3. "*Tsjankor*" is most likely the large shell variously called "*chank*," "*chunk*," "*chanco*," "*chanquo*," "*chianko*," even "*xanxus*." Called *sankha* in Sanskrit, this large shell, said to be the size of a "Man's Arm above the Elbow," is *Turbinella pyrum* (L.). They were, states Yule, "highly prized by the Hindus and used by them for offering libations, as a horn to blow at the temples, and for cutting into armlets and other ornaments (*Hobson-Jobson*, pp. 184–185)."

4. These are shells of the genus *Turbo* (L.) Lam., according to Martens, who identifies this particular shell as *Turbo marmoratus* L. Rumphius' Latin phrase means "Large Moon Shell."

5. Island southeast of the southern peninsula of Celebes, once of some importance, though dominated by either Ternate or Gowa, and thereafter by the VOC.

6. Rumphius' use of *kantoor* does not mean office, but is more directly related to its original meaning in French, i.e., *comptoir*, English *counter*, that is, the table where the merchant counted out his money. Hence, by extension, the older meaning of "trunk" (which I used here) and sometimes also as a synonym for "cabinet" (which I also used).

7. "Sleeker" for the Dutch "*licksteen.*" The latter referred in older usage to a smooth stone that was used to rub something, like the leather of fine kit gloves. *Likken* here in the sense of to make smooth, even, or shiny; to polish.

8. For "Oculi cancri," see n. 9 of ch. 29 in book 1.

9. This means "Small Papeda Shell." Strack: *Turbo (Lunella) cinereus* Born.

10. Valentijn said that *Sawan* for the Ambonese was an "evil spirit" ("*quaden geest*") or a spirit of the dead ("*des dooden geest*," 2:142–143); Van Hoëvell concurs (p. 120).

Chapter 7. Cochlea Lunaris Minor. Bia Matta Bulan.

1. This animal, called "Black Furrowed Snail" by Rumphius in Latin, is identified by Martens as *Turbo setosus* Gmelin.

2. "*Krang sussu.*" The first word is *kerang*, the general word for coral or shell, while "*sussu*," better *susu*, is the Malay word for "*breast*," among other things. *Mata susu* literally means "eye of the breast" and was the Malay equivalent of our "nipple," but only for females.

3. The "Pethola Snail" kept its Malay name in modern nomenclature; Martens identifies this as *Turbo petholatus* L., also called "Tapestry Turban." "*Pethola*" (properly *petola*) refers to a woven pattern of a particular cloth. See n. 7 of ch. 58 in book 3. The snake that is said to have a similar design

is the *Python reticulatus.* One of them gave Wallace some trouble when he was on Ambon in December of 1857; see Alfred Russell Wallace, *The Malay Archipelago* (1869) (reprint ed., New York: Dover, 1962), p. 228.

4. This is the animal's operculum, called "Moon's eye" in Malay. The spelling should be: *mata bulan.* The plural indicator "s" does not exist in Malay; one repeats the root word. See also the next chapter.

5. Martens states that Reeve described this variety as *Turbo variabilis,* but Strack is not sure of its identity.

6. This "Rough Lunar Snail" is, according to Martens, *Turbo chrysostomus* L., and is called "Glowing Oven" by Rumphius in Dutch.

7. Strack thinks this is probably *Turbo bruneus* (Röding).

8. "Spanish-green" is the verdigris color. See n. 2 in ch. 6 in this book.

9. See n. 3 of this book's first chapter.

10. What Rumphius calls the "Smallest Lunar Snail" is difficult to place. Strack thinks it probably is a juvenile form of one of the larger species.

11. Strack: *Turbo argyrostoma* L. Neither this one nor the above is pictured.

12. This means "Moon's Eye Shells."

Chapter 8. Umbilicus Marinus. Matta Bulan.

1. This is the "Giant's Ear" of ch. 6, or *Turbo marmoratus* L. All items in this chapter are opercula of *Turbo* species.

2. "*Liksteen*" in the original; see n. 8 of ch. 6.

3. Operculum of the *Turbo petholatus* L., according to Martens. The Latin means: "Black Sea Navel."

4. This "Grainy Navel" might be, according to Martens, the operculum of *Turbo chrysostomus,* while Strack would also include *Turbo setosus* Gmelin, but, since Schijnvoet seems to have provided the pictures, it is difficult to match the illustrations with the text.

5. Rumphius adds the Latin to indicate that he means "porous." Martens corroborates this experiment.

6. *Hordeolus* was Latin for a sty. One should note Rumphius' melancholy interest in remedies for afflictions of the eye.

7. Candia was the largest city on Crete, and was under Venetian rule until, after a siege of twenty-two years, the Turks conquered the city in 1669.

8. I am not entirely sure what Rumphius has in mind here. In the original: "*wanneer ze den zelven op 't kuiltje van 't hart leggen.*"

Chapter 9. Cochlea Laciniata. Little Ruffs and Spurs.

1. This "Lappet Snail" is, according to Strack, *Angaria delphinus* (L.).

2. Whenever Rumphius uses the phrase "sea grit," he is referring to coralline algae.

3. Ruffs for "Lobbetjes." These collars probably date from the sixteenth century. They were loosely pleated, and not like the rigidly starched disks of the seventeenth century.

4. *Calcar* is the noun in classical Latin for "spur," either on a horseman's boot, or on the leg of a game cock. Strack: *Astralium calcar* (L.).

5. "Red mountain" refers to the Batu Merah hill northeast of Ambon City. It is called that because it consists of red clay, rich in iron oxide.

6. This refers to Levinus Vincent (1658–1727), a merchant and textile dealer who lived in Amsterdam but moved in 1705 to Haarlem. He had acquaintances in England and was elected a member of the Royal Society in 1715. His curiosity cabinet seems to have been well known, containing the usual items, such as shells, corals, and insects.

7. Martens states that the shell depicted by the letter K is *Xenophora solaris* (L.).

Chapter 10. Trochus. Bia Cucussan.

1. *Trochus* was in Roman times a trundling hoop for children. Linnaeus kept the name for this genus.

2. The Latin means the "First Trochus or the spotted Trochus" and is identified by Martens as *Trochus niloticus* L.

3. The "Second Trochus" is, says Martens, *Trochus maculatus* L.

4. A *kukusan* is a conical basket used on Java in steaming rice. It was placed in the neck of a copper kettle (*dandang*) with boiling water, and the steam would permeate the rice in the *kukusan.*

5. This phrase means the "Third Trochus or Papuan Trochus, also the long-lived Trochus." Strack says this is *Tectarius pagodus* (L.).

6. Strack found the smaller species, *Tectarius tectumpersicum* (L.), precisely where Rumphius said it was.

7. Manipa and Kelang are islands between Buru and Ceram, both west of Ceram.

8. "Messoal" is the rugged island Misool, west of New Guinea. It was populated and more prepossessing in the days of the VOC, when it could muster fairly large fleets for raiding purposes.

9. "Nussanive's steep hook" is Tandjong Nusaniwe, a cape on the southern tip of Ambon's Leitimur peninsula.

10. This common belief was already in Pliny, book 8, ch. 54, where one also will find the notion that female bears lick their shapeless cubs into viable offspring. Pliny does not say that they suck their paws for six months.

11. I assume this is a misprint: "*heeft een zoodanige* Trochus *na een geheel jaar op* Sluitens *noch geleeft*." I suspect that "*op* Sluitens" must be "*opgesloten*."

12. The "Fourth Trochus" is, according to Strack, *Monodonta labio* (L.).

13. *Labeosus* meant "having large lips" or to be "blubber-lipped."

14. The second time it is printed more correctly. *Quisquiliae* comes from *quisquilia*, which is Latin for refuse, off-scourings, rubbish, dregs, and, by extension, trifles.

15. "Riffraff" for the wonderful Dutch expression *schorri morri*, usually written as one word, a phrase that derives from the Persian *surmur*, which meant people of ill repute, the scum or dregs of society.

16. "Begyne's Turds" for *Bagyne drollen*. I was going to call these more succinctly "Nun's Turds," but this would not be correct because "*beguijnes*" (*begijnen*) were a *lay* sisterhood of nominal nuns in the Low Countries. The order was founded in the twelfth century by a priest from Louvain reputedly called "Lambert dit le Bègue." They lived in their own kind of nunneries, but they could leave when they pleased, for instance, in order to get married. They had a negative reputation and were said to be lazy, wanton, and spoiled. The term "*begyne*" or "*beguine*" came into English in the fifteenth century with Caxton, and was more often than not used negatively. For instance; this entry by Thynne, which the *OED* gives as a definition: "But this woorde 'Begyn' sholde in his owne nature rightlye haue ben expounded, 'supersticious or hipocriticall wemenne.' "

17. Schijnvoet's figures 1–11 are different species belonging to the family Trochidae, and are difficult to identify from the brief text. No. 12, however, is, according to Strack, *Telescopium telescopium* (L.), a species belonging to the family Potamididae, which also occurs on Ambon.

Chapter 11. Cochlea Valvatae. Bia Tsjonckil.

1. The Latin means something like "Folding-door snails" since *valvae* were the "leaves" of a folding door. The Malay means "Pick Shells" since "*tsjongkil*," now *chungkil*, is the Malay word for, as Wilkinson puts it, "prising up with a pointed instrument," or "to pick." Hence *chungkil gigi* is a toothpick, *ubat chungkil* is vaccine (literally "picking medicine"), and *kelapa chungkil* is copra.

2. This means the "Folding Doors or the Half-moons."

3. This means "opening at their own will."

4. This "First Smooth Folding Door or the Yolk" is, according to Kabat, *Natica stellata* Hedley. All the shells in this and the next chapter belong to the families Naticidae and Neritidae.

5. This "Pallid Yolk" is not illustrated, but according to Kabat it is most probably *Naticarius orientalis* (Gmelin).

6. The "Compressed Yolk" is identified by Kabat as probably *Polinices albumen* (L.).

7. The first species, according to Kabat, is *Naticarius onca* (Röding), while the smaller one might be *Naticarius alapapilionis* (Gmelin).

8. Kabat: the "Fifth Folding Door" is *Natica vitellus* (L.).

9. The "Sixth Folding Door" is, according to Kabat, *Natica fasciata* (Röding).

10. The "Seventh Folding Door" is also called the "Whitish Folding Door" by Rumphius. Kabat says it is *Polinices mammilla* (L.).

11. Kabat: *Mammilla sebae* (Récluz).

12. Kabat: *Polinices aurantia* (Röding).

13. Kabat says this "Eighth Folding Door or the Thin One" is *Tanea undulata* (Röding).

14. According to Kabat, this "Ninth or Gothic Folding Door" must be *Tectonatica bougei* (Sowerby).

15. See number VIII of ch. 21 in this second book.

16. The "Tenth Folding Door" is also called the "River Folding Door" or the "Reddish Folding Door," and is identified by Martens as *Neritina pulligera* (L.). He also notes that Rumphius' description is accurate, except that the "young" are really the egg capsules.

17. "*Bia mattacou*" is a puzzle. Rumphius says that this means "red eye," but, though *matta*

means "eye," there is no "red" in this phrase, since that is *merah*. There is the possibility of a sly joke, since "*matakao*" can mean a chancre sore, gonorrhoea, or syphilis. On the other hand, De Clercq, in *Maleisch der Molukken*, states that *matakau* was an Ambonese word for magic objects which were used to ward off theft.

18. "*Mangi mangi*" are mangrove trees, belonging to the family Rhizophoraceae, perhaps a *Bruguiera*.

19. These "Pans" or "Dishes" are described in ch. 26 of this book.

20. Martens identifies these shells which Schynvoet added as: no. 1 is a land snail from Cuba, *Helix picta* Born (= Polynita picta [Born]), and the second one belongs to the Neritidae: *Nerita polita* (L.).

Chapter 12. Valvata Striata. Bia Tsjonkil.

1. The Latin means "First Furrowed Folding Door or the Alpine One." This, says Martens, is *Nerita polita* (L.).

2. Although Martens sees little difference between the first and third varieties, he states that the second kind might differ slightly, and calls it *Nerita rumphi* (Récluz).

3. This has to do for the phrase "*de derde heeft spitze zwarte bergen, die men daarom spitsbergjes noemt.*" The reason for Rumphius' punning choice is lost in English. We know about the group of islands to the east of Greenland, but the pun does not translate. "Spits" means sharp, pointed, a sharp point, or the pointed summit of a mountain or hill. This group of islands is now known as the Svalbard Islands and is under Norwegian jurisdiction.

4. The "Second Folding Door or the Banded One" is also a *Nerita polita* L. says Martens.

5. Pulau Run is one of the smaller islands in the Banda group. Martens notes that Linnaeus kept Rumphius' name but applied it as "peloronta" to a Caribbean species of *Nerita*, which still seems to bear this name of an (incorrectly spelled) Indonesian island!

6. The Latin means: "The Third Folding Door or the Wavy One." Martens says that this common shell is called *Nerita chamaeleon* L., retaining Rumphius' other name for it, "Camelot" (see below).

7. This has nothing to do with King Arthur's fabled domain but refers to a weave that dates back to the Middle Ages and that was once said to have been made from camel's hair, hence the French alternate spelling—the word comes from French—*chamelot*, in Dutch also *kamelot*. Whether camel's hair was ever used is uncertain; in Rumphius' day it was mostly made from angora wool and was known in English as "camlet," but most often it seems to have referred to a half silk. Rumphius has a secondary meaning in mind here: watered camelot (from the French *camelot ondé*), a watered fabric which once seems to have been quite common, so that "camelot" became synonymous for a while with "watered." "To camlet" exists as a verb (as *kameloten* does in Dutch), meaning to produce watered camlet, or "to mark with wavy veins." This is what Rumphius has in mind.

8. See n. 1 of the previous chapter.

9. "Pickers" for "Peutertjes" which does not refer to little kids but to the verb *peuteren*, which means "to pick at something," for instance, to pick one's teeth is *peuteren*.

10. "Outumurij" refers to Hutumuri, on the southeastern coast of the Leitimur peninsula.

11. The "Grainy Folding Door" is *Neritopsis radula* L.

12. This second kind might be, according to Martens, *Nerita planospira* Anton. Strack, however, considers this doubtful.

13. The "Wakkat" is a kind of mangrove tree, most likely *Sonneratia acida* (L.).

14. The "Black Furrowed Folding Door" is *Nerita exuvia* L., according to Martens.

15. Martens says that the "White Furrowed Folding Door" is *Nerita plicata* L.

16. The "Flattened Folding Door" is *Nerita albicilla* L. or *Nerita planospira* (Anton), according to Strack.

17. These little "*valvatae*," or the genus' riffraff (*quisquilia*) could be small species of *Nerita* or *Neritina*, but cannot be identified with certainty.

18. The "Thorny Folding Door" is the freshwater snail *Clithon corona* (L.), found in local rivers.

19. Strack: *Nerita textilis* Gmelin.

20. Martens gives *Nerita undata* L. as a possibility.

21. Nos. 5 and 6 are both *Nerita chamaeleon* L., according to Martens.

22. Perhaps *Nerita polita* L.

23. *Nerita albicilla* L.

Chapter 13. Cassides Tuberosa. Bia Cabesette.

1. I chose "cask" and not "helmet" for the original *Stormhoed* in order to avoid any confusion with the modern cover of military heads, and also to conform with Rumphius' usage in the next three chapters. A *stormhoed* (or *helmhoed*), as the *WNT* points out, was a simple helmet without a visor, chin

strap, or neck piece, though it usually had an edge. It was once made of leather. This headpiece was called a "cask" in Shakespeare's day, after the French *casque*. It was also spelled "casket" once, but this would have unfortunate overtones in American usage. Our modern notion of a helmet is also that of a simple headpiece, but if one looks at Rumphius' illustrations of these remarkable shells, one must admit they are anything but simple. They remind one of the fanciful helmets in such Dürer prints as "Small Horse" (1505) and "Large Horse" (1505) or Leonardo da Vinci's "Study of a Warrior in Profile" in the British Museum. Martens notes that *Cassis* has survived in modern nomenclature. This chapter and ch. 15 describe species of the family Cassidae, though with species of other families mixed in. Chs. 14 and 16 mostly describe species of the family Muricidae.

2. Rumphius' names mean, in order: "Lumpy Casks," "Warty Casks," "Smooth Casks," and "Murices."

3. With *voluta*, Rumphius is referring to shells now classified under Conus, and which he describes in ch. 21.

4. This "First, Lumpy Cask or the Horned One" is, according to Martens, *Cassis cornuta* L.

5. "*Aculaetus*" means "prickly."

6. Metal helmets were originally not known among Indonesian peoples; hence this word, "*cabesette*," is a borrowing from Portuguese *cabeça* ("head"), because Spanish was not important in the archipelago. This diminutive form is an older variation of modern Portuguese *capacete*, which refers to "helmet" in general.

7. "*Krang*" ("*kerang*") is "shell" and *buku* is the Malay word for a knot, a joint, knuckle, kernel of something: hence the "Knotty Shell."

8. This is not clear. Rumphius' spelling of this "snail" is close to the original Greek: ἀπορραις. This is glossed as a kind of murex, not a snail. The only connection with "dripping" is that this rather uncommon noun in classical Greek is said to derive from αἱμόρροις, which, used in the plural, refers to "veins that are liable to discharge blood"; this was ancient usage for "hemorrhoids," or "piles." But it is true that Aristotle in his *Historia animalium* (4.4.34) uses this word to indicate a kind of shellfish. Strack notes that modern nomenclature uses the name *Aporrhais* for a mainly European genus of rather small marine snails.

9. This becomes confusing. I translated the sequence exactly as in the original, which does not solve anything. The trouble is that both "Oma" and "Haruku" are the names of two villages on one of the Uliasser Islands (the one closest to Ambon), called Oma in Rumphius' day but also Haruku (which is now the official name). Hence I am assuming that in the sentence, "Oma" is the island, and "Haruko" (= Haruku) is the village.

10. See ch. 39, under the first item.

11. Strack: *Cypraeacassis rufa* (L.).

12. Martens states that this "Feathered Cask" is an immature stage of the foregoing, and does not belong, as Schynvoet states in his note, to the "Voluta."

13. This was marbled paper, usually with brown, violet, or black mixed in, used for binding books.

14. This "Prickly Cask" is, according to Strack, *Volema myristica* Röding.

15. See the upcoming ch. 17.

16. "Tukabessi Islands" refers to the Tukang Besi archipelago south of the island Buton, which is situated off the easternmost "leg" of Celebes.

17. "Samat(e)ra" is an old form of "Sumatra," and was often indicated as such on sixteenth-century maps.

18. Martens states that Schynvoet's no. 1 is a juvenile form of *Cassis cornuta* (L.).

19. No. 2 is *Cassis flammea* (L.), from the Caribbean region.

20. No. 3, says Strack, is *Cypraeacassis testiculus senegalica* (Gmelin), from West Africa.

21. No. 4 is *Malea pomum* (L.).

Chapter 14. Cassides Verrucosae. Knobbles.

1. *Pimpeltje* refers to a drinking glass, made from a type of glass now often referred to as hobnail glass. In Dutch "*pimpel*" does not have the teenage affliction as its primary meaning; rather, it was used to indicate a small bird, the blue titmouse (*Parus caeruleus* L.), a carrot, or a liquid measure, and had something to do with drinking in general. These knobbed glasses are no longer known, but from contemporary descriptions I would guess that they were the size of our shot glasses.

Rumphius often uses "*knobbel*" when establishing a similarity between the shell's excrescences and the bossed glass. I therefore selected that noun for *pimpeltjes* in English. "Knobble" was part of the English language since the early fifteenth century, and, as even the *OED* admits, came directly from Dutch. The dictionary defines it as a "small knob," which was another reason for adopting it, since Rumphius employs the Dutch diminutive. The word is also a verb and has an adjectival form.

Martens notes that this chapter is a grab bag of genera and families, rather than exclusively devoted to the family Cassidae.

2. *Verrucosus* meant "warty" in classical Latin, from *verruca*, which has "wart" as its secondary meaning. That is what "verrucose" or varicose veins, that unsightly affliction of old age, means: veins that display wart-like elevations or bumps. Hence the first part of Rumphius' phrase means the "First Warty Cask." What comes after "or" (*sive*) is another problem. "Ceramica" has nothing to do with pottery but refers to Ambon's neighboring island of Ceram. Linnaeus kept the name, although the snail can be found in many other locations. The whole phrase means: "The First Warty Cask or the One from Ceram." Strack identifies this as *Vasum ceramicum* (L.).

3. This is the operculum of mollusks, and is discussed at length in ch. 17.

4. This "Second Warty Cask" is *Vasum turbinellum* (L.) and the other shells mentioned in this section are probably various kinds of the same species or different stages in its growth.

5. This is an old name for what is now Pulau Pombo, which does mean "Dove Island" in Malay. It is a small coral reef off the east coast (near the top of the peninsula) of Hitu.

6. This must be "Oma," the older name for Haruku, the most westward of the Uliasser Islands.

7. The Malay called these "Papua Shells."

8. According to Houart, these "Banda Knobbles" are *Drupa rubusidaeus* (Röding).

9. Although the illustration by letter D on plate XXIV is not very distinctive, Houart tentatively identified this species as *Thais muricina* (Blainville).

10. Houart identifies this as *Drupa ricinus* (L.). The second species mentioned in the text (but not illustrated) could be *Drupa lobata* (Blainville).

11. Martens finds Rumphius' "Hairy Ears" a more fitting name than Linnaeus' *Tritonium (Persona) anus* L. or, today, *Distorsio anus* (L.).

12. See the subsequent ch. 19.

13. These appropriately named shells belong to the genus Bufonaria. The species depicted by Rumphius is probably *Bufonaria rana* (L.). "Ranella" (should be *ranula*) is the diminutive of *rana*, classical Latin for "frog."

14. *Codoc* (*kodok*) means "frog" in Malay; hence this means "Frog Shell." The bivalve species described in ch. 30 under no. II, is also called "Bia kodok" in Malay and is still called such today.

15. Strack states this is probably *Cymatium muricinum* (Röding), a species known from the Indian and Pacific Oceans.

16. Although I would prefer this to be the wonderful mammal from South Africa (*Suricata tetradactyla*), which was often kept as a pet, this most likely refers to monkeys with long tails.

17. Strack identifies this as *Cymatium pileare* (L.).

18. This, according to Strack, is *Vasum capitellum* (L.), a Caribbean species.

19. The only thing I can offer for "*Oorreliezen*" is *oorijzers*. This untranslatable word referred to a kind of clasp, ring, or cap, made of gold or silver, that fitted over a woman's head. This adornment was then covered with a cap of fine linen.

20. "Bed-ticking" for "*Beddeteiken*," which was a coarse cloth (*tijk*), the thickness of damask, and decorated with blue stripes. It was the common fabric used to cover mattresses and pillows, though it was also used to make clothes.

21. The shell depicted is, according to Strack, from the West Indies, and is *Melongena corona* (Gmelin). Schynvoet's no. 3 is a thornless variety.

22. *Volema myristica* (Röding), according to Strack.

23. Houart states that no. 5 is probably *Mancinella alouina* (Röding).

24. No. 6 is too poorly depicted to be identified with certainty, but Houart offers that it might be *Thais aculeata* (Röding) or *Thais bituberculata* (Lamarck).

Chapter 15. Cassides Laeves Sive Cinereae. Gray Casques.

1. I used "Casque" instead of "Cask" since Rumphius also uses a different word, i.e., *Kasketten*, from the previous *stormhoeden*.

2. Strack says this "Smooth Casque the Color of Ashes" is *Phalium glaucum* (L.).

3. This is a variety of *Phalium glaucum*, as Martens suggested, or it might refer to specimens of *Phalium fimbria* (Gmelin).

4. *Alga coralloides* is a species of algae belonging to *Rhodophyta*, identified by F. S. Collins in 1917 as *Gracilaria lichenoides* (L. ex Turner) Harv. See J. S. Zaneveld, "An Identification of the Algae Mentioned by Rumphius" in *M*, p. 279.

5. Ruma Tiga is below Cape Martafons, on Hitu's southern coast, which represents the northern shore of Ambon Bay.

6. "*Bewang*" should be *bawang*, the general Malay word for "bulb." Specifically it referred to

species of *Allium*, in the present context, for instance, *bawang puti* (or "white bulb"), which means "garlic."

7. In his *Ambonsche Landbeschrijving*, Rumphius says that "Hucconalo," or "Hoeconalo," is another name for "Ruma Tiga," but not entirely. He states that it is a "small village, situated across from the big country [*sic*] a mile north from the castle, not far from Cape Martin Fonso which we call Melis Hook, where the bight becomes so narrow that one can easily swim across it, and where towards the back, it becomes a wide bay again. Hoeconalo once resided under Wackal . . . and was divided in two. The first and real Hoeconalo was a good mile up in the mountains behind Batavi kitsjil, slightly to the west of today's Roema tiga. . . . The other part was called Pari, [and was] 1½ hours away from the beach up in the mountains. . . ." G. E. Rumphius, *Ambonsche Landbeschrijving*, ed. Z. J. Manusama (Jakarta: Arsip Nasional Republik Indonesia, 1983), pp. 71–72.

8. *Areola* was a Latin noun which, as its secondary meaning, referred to a small garden bed. Strack: *Phalium areola* (L.).

9. The "Bellies" are *Phalium bisulcatum* (Schubert & Wagner).

10. This "Striped Fringe" is *Casmaria ponderosa* (Gmelin), according to Strack.

11. "Smooth Fringed Casques" are *Casmaria erinaceus* (L.).

12. Martens thinks this smaller variety might be *Nassarius coronatus* (Bruguière).

13. See ch. 46 and later, in book 3, esp. ch. 55.

14. Taking "*gevlakte*" to be "*gevlekte.*"

15. Strack: probably *Phalium bandatum* (Perry).

16. *Phalium flammiferum* (Röding).

17. *Cassis tessellata* (Gmelin), from West Africa.

18. Probably *Phalium exaratum* (Reeve).

19. Probably *Phalium saburon* (Bruguière), from the Mediterranean and West Africa.

20. Nos. 6–9 are specimens of either *Casmaria erinaceus* (L.) or *Casmaria ponderosa* (Gmelin). Since we lack good descriptions, they cannot be identified from the drawings.

Chapter 16. Murices. Bia Unam.

1. Houart identifies this "Branched Murex" as *Chicoreus ramosus* (L.). This is one of the names which Linnaeus transferred literally from Rumphius.

2. *Bia unam* is a general Malay phrase for a kind of edible shellfish.

3. "Resin" is *harpuys* in the original, a word that does not exist in English. It is a Scandinavian word that entered the French and German languages via Dutch. It originally referred to a kind of resin, left over after a substance like turpentine had been tapped from pine trees. It was also known as "yellow pitch," and used on ships to protect masts and spars, if not the entire hull. Although common usage in New England, I did not use "pitch" because it would create confusion with the tar-like substance which is black. *Harpuys*, on the contrary, has a yellow color.

4. See nn. 11 and 12 of the next chapter.

5. Houart supposes this smaller variety to be *Hexaplex cichoreum* (Gmelin).

6. Now Cape (= Tandjung) Martafons on the Hitu peninsula.

7. An archipelago east of the Kei Islands, southeast of New Guinea, in the Arafura Sea.

8. The Tukang Besi Islands are located southeast of the island of Buton, which lies below the most eastern "leg" of Celebes.

9. This "Stony Murex" is, according to Strack, *Cymatium lotorium* (L.).

10. Houart states that without proper illustrations it is impossible to say if these forms belong to different species or if all, or part of them, are variations of *Chicoreus brunneus* (Link.).

11. According to Houart this is *Chicoreus brunneus* (Link.). Rumphius' second name for this shell is the rather unusual word "*munk-yzer*" which was properly spelled *Mink-yzer.* In English this nasty contraption was called a "caltrop," which, in a general sense was a snare, but which specifically (as the *OED* describes it) was "an iron ball armed with four sharp prongs or spikes, placed like the angles of a tetrahedron, so that when thrown on the ground it has always one spike projecting upwards: used to obstruct the advance of cavalry, etc."

12. Houart: *Homalocantha scorpio* (L.).

13. This is *Cymatium pyrum* (L.).

14. *Haustellum* referred in Latin to an object that scooped up water. Linnaeus kept the name, since Houart identifies this as *Haustellum haustellum* (L.).

15. The Latin noun *tribulus* referred to several things. It was a military device used against enemy cavalry, as was the caltrop mentioned before. It was also, according to Lewis and Short, a kind of thistle, and, finally, it also was the name for a water chestnut that had a triangular and prickly fruit. Houart identifies this species as *Murex tribulus* L. Another instance of direct borrowing by Linnaeus.

16. The Malay means "Lemon-thorn Shell." This probably refers to *Citrus hystrix* D.C. because this has thorny branches. The true lemon, *Citrus medica* L., has branches without thorns.

17. This spectacular shell is, according to Houart, *Murex Pecten* (Lightfoot). Schijnvoet provided an illustration on plate XXVI, no. 3; see n. 28.

18. For *zegen*, a large, narrow net that was cast over an area of shallow water, and then slowly pulled together, trapping the fish; a seine net.

19. "*Bia sissia*" must probably be *bia sisir*, since *sisir* is Malay for "comb."

20. Houart: *Chicoreus axicornis* (Lamarck).

21. This is unclear in Dutch. The lovely noun "*Brandaris*" is unique to Dutch, despite its ostensible Latin ending. It generally meant a lighthouse, particularly the one on the island of Terschelling (one is still there today), and had little to do with "branden," the Dutch verb for "to burn," or, at best, only in an extended sense.

22. Houart: no. 2 is *Hexaplex cichoreum* (Gmelin).

23. A snipe (*snip* in Dutch) is a wading bird (genus *Capella*), with a very long and slender bill, in order to probe the mud for food.

24. Sybrand Feitema, and his two sons, were avid collectors. See n. 45 of ch. 3 of this second book.

25. Blue gastropods, as Strack notes, are extremely rare in nature and certainly not part of this species. These specimens were most likely painted so they would fetch a higher price.

26. Houart: *Bolinus brandaris* (L.), from the Mediterranean.

27. Houart: *Bolinus cornutus* (L.), from West Africa.

28. Houart: *Murex pecten* (Lightfoot).

29. See n. 44 of ch. 3.

30. See n. 46 of ch. 3.

31. Probably Harmanus van der Burgh (1636–1708), a merchant and apothecary originally from Delft, and of considerable wealth. He lived in Amsterdam on the Nieuwezijds Voorburgwal, and was married three times. He collected shells and corals, as well as insects, precious stones, coins, paintings, and graphic work.

32. Probably Jan de Jong (?-1712), who lived on the Prinsengracht (a prominent address) and was an accountant for the VOC. Collected pretty much what everybody else did, especially shells and marine items.

Chapter 17. Unguis Odoratus. Sche Chelet. Unam.

1. *Unguis* is Latin for "nail" or "hoof," *onyx* in Greek, and *odoratus* means "fragrant." Such a "nail" refers to the operculum of mollusks, which is a plate that develops at the dorsal portion of the foot and serves to close off the shell when the animal has retracted. The Latin word *perculum* means "cover." Onycha were indeed popular with the Chinese. Needham reproduces a Chinese text on the "perfumed onycha mollusc" from the early thirteenth century and states that the perfume itself was called *chia hsiang* or *chia chien*, of which the latter is most likely Rumphius' "sche-chelet." See Needham, *Science and Civilization in China*, vol. 5, part II, p. 138 (n. c), and p. 139 (fig. 1319).

2. See n. 3 of ch. 16.

3. *Turbines* from the Latin *turbo* (that which spins or twirls) refer to several species of marine snails, described by Rumphius in ch. 19 of this book.

4. A *buccinum* was a Roman trumpet, shaped like the letter C, and means "crooked horn." See ch. 19 of the present book.

5. Thymiamata is plural of the Greek *thumíama*, which means "that which is burnt as incense" or "fragrant stuffs for fumigation." It is the word used in the New Testament for incense, for instance in Revelations 5:8 or 18:13. Usually the incense was in a spoon that was held over a pan filled with burning coals. The smoke from the coals mingled with the vapor of the incense, and together they were said to symbolize prayer.

6. *Aloe* or *Aloe vera* of the *Liliaceae*, is a plant well known for its medicinal purposes. It is not native to the Indies and used to be imported from Bombay and sold on markets in Java as *djadam*. See his *Herbal*, book 2, ch. 11 (pp. 29–33) on "Aloës—en Paradys Hout" or "Agallochum. Calambac."

7. *Dioscorides* was a Greek physician who lived in the first century A.D. and served in the Roman army. He is most famous for his herbal, *Materia medica*, in five books, which described medicinal properties of some 600 plants as well as nearly 1,000 drugs. Although far from reliable, this work remained a standard text for several centuries.

8. *Castoreum* (or *castor*) is what used to be called "beaver oil," i.e., a substance derived from the beaver's perineal glands. It was used as a fixative by perfumers.

9. This is the operculum of *Chicoreus ramosus* (L.); see ch. 16, no. I.

10. *Onin* is a region in western New Guinea (now Irian Jaya) on a peninsula that juts out into

the Ceram Sea below what is normally referred as the New Guinea's "bird's head," opposite Ceram. In later colonial times it was better known as Fakfak. Rumphius and Valentijn called this peninsula *Sergile*, which seems to be a Portuguese word for "mountain range." Even before Rumphius' day, Onin's coast was inhabited by enterprising people who traded with Ceram and with merchants from the Moluccas. The town of Rumbati had a prominent slave market when Rumphius was alive.

11. Except for some coastal regions, such as Onin, little was known about the second largest island in the world. Well into the seventeenth century, it was thought to be part of Australia. The first Dutch reconnaissance was in 1606. The Spanish mariner Ortiz de Retes called the island "Nueva Guinea" in 1545, thinking that, because of the appearance of the Papuas, he was dealing with an African race. Dutch colonial hegemony was not established until 1828.

12. *Byza.* As the end of this chapter makes clear, Rumphius is referring to *Byzacium*, a region in northern Africa which Pliny described in book 5:4 (in the original Latin, paragraph 25; Rumphius' citation is erroneous). This region, in what is now the modern nation of Tunis, was in Roman times considered part of the "Province of Africa," from the Gulf of Hammamet to the Gulf of Gabes. It was either of Carthaginian or Phoenician origin. Rumphius is copying Pliny fairly closely, except that "Adrumetum" was the chief city Hadrumetum, now known as Sousse (on the Mediterranean coast).

13. *Astangi*, variant of *istanggi*, commonly *setanggi*, means "incense" in West Java. The word *dupa* also generally means "incense," but its several varieties could include any mixture of fragrant seeds, fragrant woodchips, aromatic herbs, or roots. The simplest way of using *dupa* was to burn the seeds of the musk mallow (*Hibiscus abelmoschus*) to scent clothing. To obtain a mixture that would also be firm, the dark balsam of the tree *Sindora galedupa* was used; its second word became synonymous with *dupa* as a general term for a mixture of various ingredients that produced incense. Rumphius considered "*galedupa*" the base for all perfumes (see Van Slooten, *M*, pp. 302–304).

14. "Mother sickness" for *moeder ziekte.* This was also known in Dutch as *miltvuur* or *miltbrand*, an inflammation of the spleen, i.e., anthrax. The latter was known in English as splenetic fever and is a serious cattle disease. Humans can be infected by cattle and contract a carbuncular variation. "*Anthrax*" in Greek means "carbuncle" or "coal."

15. Martens: the operculum of *Fasciolaria trapezium;* see ch. 19, no. V.

16. Strack: the operculum of *Hexaplex cichoreum* (Gmelin).

17. This, as previously noted, is *Chicoreus ramosus* (L.).

18. *Aru Islands:* an archipelago of about eighty-five islands east of the Kai (Kei) Islands, southeast of the Onin peninsula on New Guinea, in the Arafura Sea. The main source of income used to derive from pearl fisheries, fishing for *tripang*, or sea cucumbers (Holothuria), and sharks, and hunting birds of paradise.

19. *Tuccabessii* most likely refers to the small group of Tukang Besi islands, southeast of the island Buton, which lies below the most eastern "leg" of Celebes.

20. Strack: operculum of *Charonia tritonis* (L.). See ch. 19, no. II.

21. Strack: operculum of *Syrinx aruanus* (L.). See ch. 19, no. I.

22. Consists of the opercula of two species; see nn. 23 and 24.

23. Operculum of *Haustellum haustellum* (L.); see ch. 16, no. VI.

24. "Murex Tribulus" is precisely that: *Murex tribulus* (L.). See ch. 16, no. VII.

25. Probably the operculum of *Thais muricina* (Blainville). See ch. 16, no. IV.

26. "Knobbles" for *Pimpeltjes;* see the previous ch. 14.

27. Operculum of *Chicoreus brunneus* (Link); see ch. 16, no. III.

28. Operculum of *Pila ampullacea* (L.). See ch. 18, no. XV.

29. *Macassar* was and still is the chief city of Celebes (now Sulawesi) on the "foot" of the Western "leg" of that island. Historically the city and nation were very important as the center of power in the Moluccas and Celebes, and as the driving force of opposition to Dutch colonial rule.

30. *Marus* or *Maros*, a town about forty miles north of Macassar which was once the residence of the native chief. East of Maros is a famous waterfall.

31. *Nardus* (or *nardos*) is a plant (*Nardostachys jatamansi*) which produced the root called spikenard—once a costly aromatic ingredient for ointments and oils—which only grew at the highest elevations in Nepal. Pliny discusses nard at length in book 12:26. Spikenard is of the Valerian order. There is a bewildering array of plants associated with spikenard; hence the confusion Rumphius describes is not surprising. *Malabathrum* (or *Malobathrum*) was often confused with *nardi folia*, but in fact it is the leaf of cinnamon (*Cinnamomum tamala*) from India. Pliny discusses malobathrum in book 12:59.

32. "Nicolas Mirepsicus" is Nikolaos Myrepsos, a thirteenth-century physician from Byzantium. He was the physician of Johannes III, the Byzantine emperor who ruled from 1222 to 1252.

33. Bitumen was a kind of asphalt, used in ancient times as a cement; see the passage in ch. 36 of book 3 about "black amber."

34. Pliny does indeed mention that "there is a substance called ostraceum, called by some onyx."

But this is in book 32:46 (or paragraph 134 in the Latin) and not what Rumphius gives. *Ostraceum* or *ostracium* comes from the Greek *ostrakion*, which is the diminutive of *ostrakon*, which means an earthen vessel (*testa* in Latin), also a potsherd or tile (as Rumphius states), and the hard shell of snails, mussels, and tortoises.

35. Schechelet (or *schecheleth*) is indeed one of the four ingredients (stacte, onycha, galbanum, and frankincense) of the holy incense as legislated in Exodus 30:34. Schecheleth is "onycha" or "onyx," and is said to have come from the operculum (Rumphius' "lid" or "cover") of a mollusk of the genus *Strombus*. Rumphius is quite correct when he disbelieves Bochartus because *bdellium* (*bedolach* in Hebrew) is a sweet-smelling, yellow resin of the *Balsamodendron mukul* tree, which once grew in northern India, Afghanistan, and Beluchistan. It is mentioned as a precious commodity in Genesis 2:12, and it was used, contrary to what Rumphius says, as incense. What he calls *Caryophyllata* in the same critical passage refers to the dried flower buds of the clove tree (*Eugenia caryophyllata*). Similar to Rumphius' "High-German Jews" (below, in text), cloves are called "*kruidnagels*" in Dutch, which literally means "spice nails."

36. The second word of *Unam Carbou* is most likely *kerbau*, the water buffalo.

37. Alvin P. Cohen (personal communication) suggests this might be *lo-pi*, meaning "snail shutter." It is impossible to conjecture what "*Hiole*" might be.

38. Samuel Bochart (1599–1667) was a French theologian and Hebrew scholar. He was a priest in Caen from 1625 and wrote several major works. One work dealt with biblical geography—*Geographiae sacrae*, in 2 vols., 1646—and another, which is the one Rumphius is referring to, on biblical zoology and ethnology, *Hierozoicon . . . Sacrae Scripturae*, published in 1663.

39. *Galenus* (?130–?201) was the most influential physician of antiquity. A student of Greek medicine, he codified Greek medical knowledge on the basis of Aristotle. Arab scholars translated his work into Arabic, and only in the Middle Ages was Galen translated into Latin. Until the second half of the sixteenth century, his authority was supreme, and his influence extended into the next century. In 162 Galen became the court physician of Marcus Aurelius, also of that emperor's infamous son Commodus. The corrupt political atmosphere drove him away from Rome, and he returned in 168 to Pergamon.

40. This fabled town in the *Arabian Nights*, during the reign of Harun-al-Rashid, is better known today as Basra, Iraq's third largest city. It was situated at the time some ninety miles from the Persian Gulf, and was only accessible by ships via a man-made canal. Suffering a decline, it was captured by the Turks in 1668. In 1690, during Rumphius' lifetime, the city was stricken by plague and some 80,000 people died.

41. This strange word hides, as Holthuis informed me, the name of an Arab scholar, a compiler from the thirteenth century, by the name of Muhammed el Kazwini, who died in 1283. I had the notion that one might be able to track the author down because Rumphius' text says that he published a "treatise on water animals," which was indeed the case. Rumphius probably saw the name in Bochart, who often quotes El Kazwini in his *Hierozoikon*. The first part of El Kazwini's *Cosmography*, the only one printed, it seems, is a disquisition on water animals. See J. Victor Carus, *Geschichte der Zoologie bis auf Joh. Müller und Charl. Darwin* (München: R. Oldenbourg, 1872), pp. 166–169.

42. Avicenna (978–1036) was a Persian physician and natural historian, whose full name was Abu Ali al Hussain ibn Abdallah ibn Sina el Bochara. The last word refers to the city where Avicenna was the court physician of the ruling sultan. His works were standard literature well into the seventeenth century, of which his *Canon Medicinae*, a huge compilation of medical knowledge from ancient sources, proved to be the most important. Avicenna recognized the importance of Aristotle and copied him (and Dioscorides) in detail, accompanied by commentaries. The Aristotelian passages in Avicenna were translated from Arabic into Latin in the early thirteenth century by Michael Scot, and were a principal mover in the rediscovery and promulgation of Aristotle's works and thought.

43. I doubt that Rumphius is referring to Mecca, Islam's holy city. Most likely this is meant to be Mocha (Mokha) in Yemen. Situated in a desert, it had as its chief export coffee. Water was a precious commodity and was brought in from Mosa, some twenty miles away in the mountains. It was a prosperous trading port where both the Dutch and the British maintained "factories," or trading posts.

44. Judda (Jidda or Jeddah) was a port city on the Arabian coast of the Red Sea. Well fortified, it was used by the Turks as a port for a fleet of galleys that patrolled the Red Sea, forcing foreign vessels to dock at Judda first. Judda is the principal port for pilgrims to Mecca, some forty-six miles inland. Originally settled by Persian merchants, it became an important trade center during the fifteenth century, a commercial link between Egypt and India.

45. Jemen or Yemen is equated here by Rumphius with *Arabia Felix* ("Happy Araby"), following the tradition initiated by Ptolemy, although the ancient geographers included the entire Arabian peninsula under that name. Mokha, Aden, and Hadeda (now Hodeida) were its most important har-

bors. Beyond the arid coastal strip lie a great desert and rugged mountain ranges. The Romans divided the general landmass they called Arabia into Arabia Petraea (after the city Petra), Arabia Deserta (which speaks for itself), and Arabia Felix. The latter was roughly the area that had the Red Sea as one boundary. The "happy" appellation, according to Pliny (5:12, or paragraph 65 in the original), came from its perfumes and wealth, since this region procured the most frankincense for trade with the Mediterranean world. That "perfume" was translated into "felix" is made clear by Pliny in book 12:41, where he criticizes the use of such expensive fragrances, especially at funerals. It seems that the perfumes were first used to cover up the stench of death, whereafter it became a habit of the rich. Pliny alleges that Nero burned more than one year's output during the funeral of his wife Poppaea. Pliny thought such offerings to the gods a waste of money.

46. "Bacharyn" is Bahrein, the small, but strategically important, archipelago in the Gulf of Bahrein, off the Hasa coast of Arabia and north of the Qatar Peninsula, in the Gulf of Persia, south of Kuwait. Bahrein was once famous for its pearl fisheries, and it was also a great exporter of dates (as was Basra). Of signal importance is the fact that these islands have an abundant freshwater supply, which derives from submarine sources. The Portuguese held sway there for about a century after 1507, then the Persians took over, to be followed by Arabs from the mainland.

47. Abandan, which Rumphius states is "near the Tigris," is in the general region of Iraq and Iran that includes Basra. It is an island in the delta of the waterway called Shatt-al-Arab, which disembogues into the Persian Gulf east of Kuwait. It was once the southernmost settlement of Mesopotamia. The reason Rumphius mentions the Tigris River is its proximity. The Shatt-al-Arab River was formed by the confluence of the Euphrates and the Tigris, and is now over one hundred miles long. The problem with the Euphrates, Tigris, and Shatt-al-Arab rivers is the constant threat of silting up.

48. This is a misprint for "Unguis."

49. According to Al Cohen this is most likely *tz'u-erh lo*, or the "thornlike snail."

50. Quantung might refer to the province Kwangtung in southern China or the city known in the West as Canton (Kwangchow). Canton was most likely the port called "Cattigara" by ancient geographers and was, from the sixteenth century on, China's principal port that dealt with European traders. The Portuguese, Dutch, Spanish, and British vied with one another to gain a monopoly of the coveted trade in Chinese tea, silk, and rhubarb. The Dutch established a "factory" in 1762. In 1757, the Imperial Court declared Canton the only port that was permitted to deal with Europeans. In the late eighteenth century, Britain increased its import of opium in China, via Canton, a policy that resulted ultimately in the Opium War in 1841–1842. Britain came out on top and acquired the island of Hong Kong, only eighty miles from Canton. Hong Kong quickly became Canton's most important rival.

51. No. 3 is probably the operculum of a large *Murex* species, according to Strack, while no. 4 might possibly be one of a *Charonia*.

52. Nos. 5 and 6 are the opercula of *Natica stellata* (Hedley), and *Natica vitellus* (L.); see ch. 11, nos. I and V. Strack notes that Rumphius *did* send two figures, together with the shells to which they belong. Schijnvoet for some reason decided to use them for this chapter, although neither operculum can be used for incense.

Chapter 18. Cochlea Globosae.

1. The snails described in this chapter under the general rubric of "Spherical Snails" are, as Martens indicates, a mixed bag of marine, freshwater and land snails.

2. The "Striped Snails or the Oil Snails" are *Tonna tessellata* (Lamarck).

3. *Minjac* (*minjak* or *minyak*) is "oil," "grease," or "ointment," in Malay, hence "Oil Shells."

4. *Favus* means "honeycomb" in Latin.

5. As mentioned before, this word refers to a kind of tumor, and was so called because the yellow discharge had the color of honey (*mel* or *mellis*). Both "Melicera" and "Favus" refer to the animal's egg capsule. Both terms are found in Pliny.

6. This is a stony coral; various species of the genus *Acropora*.

7. This does not make much sense in modern Greek.

8. The "Other Striped Snail" is *Malea pomum* (L.).

9. This "Winged Snail" is *Tonna perdix* (L.).

10. *Culit* (*Kulit*) is skin and *bawang* is onion in Malay.

11. This "Other Winged Snail" is *Tonna cepa* (Röding).

12. According to Houart, this "Broad Snail" is *Purpura persica* (L.).

13. *Rapa* (*rapum*) is a turnip, and is identified as *Rapa rapa* (L.).

14. *Bulla* means "bubble" in classical Latin, particularly a "water bubble," and as Martens indicates, Linnaeus kept the name for his genus *Bulla*.

15. The first species is *Bulla ampulla* (L.).

16. The second species is *Atys naucum* (L.).

17. The third species is *Hydatina physis* (L.). Specimens with black and red bands are *Hydatina aplustre* (L.).

18. Martens even informs us what is meant by "the Prince's Flags." The Prince of Orange used to fly a special flag which was essentially double the national flag, consisting of six bands (instead of three) of red, white, blue, and red, white, blue.

19. "*Imbrium*," or "*imbricum*," comes from the adjective *imbricus* which means "rainy" or "rain bearing." This "Rain Snail" is *Pythia scarabeus* (L.). Martens notes that these snails are only noticeable during and immediately after rain.

20. *Ribut* is Malay for a strong wind or storm (like the storm once called "Sumatra," which, as Wilkinson notes, was a strong westerly squall). Hence these snails are called in Malay "Storm Snails."

21. Since Rumphius' figure depicts a shell with dark spots, this species must be *Ficus subintermedia* (d'Orbigny).

22. Strack states that the "Navel" is *Architectonica perspectiva* (L.) and notes that the first of the two smaller species could be *Architectonica laevigata* (Lamarck). Rüdiger Bieler of the Field Museum in Chicago identifies the two small snails, the size of a fingernail, as follows. The first, grey one, he says, "could have been a small specimen of *Architectonica laevigata* (Lamarck, 1816) or of *A. perdix* (Hinds, 1844) that had lost its luster. Equally likely, Rumphius was referring to a member of the architectonicid genus *Heliacus*, such as *Heliacus variegatus* (Gmelin, 1791)." The second, light-brown species, he thinks, was *Psilaxis radiatus* (Röding, 1798).

23. *Arcularius* was an unusual noun that referred to someone who made small boxes, like jewelry boxes (*arcula*). According to Kool & Strack this is *Nassarius arcularius* (L.).

24. Kool and Strack consider this to be *Nassarius pullus* (L.).

25. The *totombo*, commonly woven from *pandanus* leaves, usually came from southeastern Ceram.

26. The Goram (Gorham) or Gorong Islands is a small archipelago southeast of Ceram; its main islands were called Goram, Suruaki or Pulu Pandjang, and Manawoka. These islands were primarily known for fine timber.

27. The illustration depicts a land snail, *Planispira zonaria* (L.). Although Rumphius describes two forms that should live in seawater, he was probably referring to species of the same land-snail genus that were found on the beach. The third species is *Chloritis unguiculastra* Von Martens. See also n. 38.

28. "Gamaron," also "Gamron," "Gamrun," or "Gombroon," was the ancient name for a town on the Persian Gulf now called Bandar 'abbas.

29. These "Land Snails" are *Nanina citrina* (L.).

30. *Lutarius* meant "living in mud" (= *lutum*). The larger species is certainly *Pila ampullacea* (L.), while the smaller species with the more pointed tip could be *Pila polita* (Deshayes).

31. See ch. 11, nos. 1–3, of this book.

32. "Cotihomera" is most likely Valentijn's "Codiho plant" (*Oud en Nieuw Oost-Indiën*, p. 231); *Codiho merah* or *Codiaeum variegatum* (L.) Blume. It was a favorite garden plant because its beautifully variegated and colored leaves were used for decoration at parties. Backer notes that the scientific name is a Latinization of the local name for the plant in Ternate: *kodiho* or "codino." See C. A. Backer, *Verklarend Woordenboek van Wetenschappelijke Plantennamen* (Batavia & Groningen: P. Noordhoff, 1936). Prof. Holthuis provided me with this information.

33. Tombuko (Tombuku) is on the eastern coast of Celebes' central "leg." Better known as Bungku or Tobungku, it is a region divided into northern and southern parts by the Gulf of Tolo or Tomaiki. In Rumphius' day the northern region in particular was inhabited by people who had not succumbed to Islam, and who were famed for their locally forged swords.

34. "Pomatias" is Greek, derived from πῶμα which means a "lid" or "cover." The noun existed in ancient Greek. Rumphius found it in Dioscorides' *Materia Medica* where it is said to refer to a snail which in winter covers the opening of its shell with a lid.

35. Spelled "Coetschien" in the original, this harbor on the coast of Malabar was also known as Cochien, Coechin, Cochijn, Cuchin, and, best known today, as Cochin.

36. A variety of smaller species belonging to different families.

37. Probably *Pyrene punctata* (Brugière).

38. Probably one or more species of the land-snail family Cyclophoridae. As Martens already noted, the snail depicted as fig. R. is misplaced. This is *Chloritis ungulina* (L.) and is most likely the "Elephant's snout" under item XIII.

39. This is the Mediterranean species *Galeodea echinophora* (L.).

Chapter 19. Buccinum. Bia Trompet.

1. *Tuit* has a variety of meanings in Dutch, from the spout of a teapot to a nipple, but all meanings share the sense of something pointed that sticks out. It can refer to details in architecture, tools,

caps, hats, etc. Since "snout" won't do here and words such as "tip," "protuberance," "top," and the like would be confusing in the present context, I used the rare substantive "tutel." Listed in the *OED*, it is clearly part of the linguistic cluster related to the Frisian *tute*, our present English verb *toot*, and the Dutch noun *tuit*.

2. This ancient usage of *Buccinum* refers to *Charonia lampas* (L.), a large marine snail in the Mediterranean and adjacent seas. This chapter discusses species which are more or less similar in form to that species.

3. The Malay means "Trumpet Shell." *Trompet* is a borrowing from Dutch.

4. This is *Syrinx aruanus* (L.). Named after the island Aru, which is really an archipelago of five large and some eighty small islands in the Arafura Sea, south of the "neck" of New Guinea's "bird's head" and east of the Kei Islands. In Rumphius' day the inhabitants lived primarily from pearl fishing, shooting birds of paradise, and tending coconut trees. Until the end of the eighteenth century, the VOC granted Banda a mercantile monopoly with these islands.

5. "*Blatta*" refers to the operculum of Mollusks, as described by Rumphius in ch. 17 of book 2.

6. This is *Charonia tritonis* (L.).

7. Tritons, the mermen of Greek mythology, were usually depicted as blowing on conches and were rather indistinct creatures. Triton himself was reputedly the son of Neptune and the nymph Salacia, and would, at his father's command, blow on his shell to either incite or calm the sea. Using these shells as trumpets is well known from the Pacific. One will find it in Mexico as well. The people who originally populated the present state of Colima on Mexico's Pacific coast—now generically known as los Colimenses—also used conches for instruments after they cut the tip off the apex of the shells. These shells were important enough to be depicted as earthenware objects. The conch was also an instrument for the Aztecs; in fact, the "simplest aerophone from pre-Cortesian Mexico is the shell horn. It is a natural horn with only a blow hole. The shell horn was a ritual instrument used exclusively by rulers and priests. It is an attribute of the creator-god Quetzalcoatl (the Feathered Serpent) and his priests, and was worn on a necklace. According to legend, Quetzalcoatl created man by blowing on the holy shell horn, symbol of fertility and spiritual rebirth, of the Lord of the Kingdom of the Dead." Elisabeth den Otter, *Pre-Columbian Musical Instruments*, Bulletins of the Royal Tropical Institute (Amsterdam: KIT/Tropenmuseum, 1994), p. 21.

8. "Paternosters" were rosary beads.

9. "Turkish paper" was a marbled paper that was used for binding books. This mottled paper could have spots in brown, black, ocher, even red and orange.

10. "Bright as daylight" for "*schoonlicht*," an obscure usage for *klaarlicht*, meaning as clear as in broad daylight.

11. The Italian priest Martinus Martini (1614–1661). Rumphius is most likely referring to the supplement, entitled "De bello tartarico historia," added to Martini's once-famous atlas, *Novus atlas sinensis* (n.p., 1655).

12. This does not wash. As an onomatopoeic verb, *kinken* refers to a bright sound, such as one used to hear in a smithy, not a soft rustling. The word *kink* in *kinkhoorn* means a "twist," "turn" or "coil" in something, and exists in English with the same meaning. Hence the name derives from its shape, not from a purported sound.

13. Since there used to be confusion between the Liu-Kiu Islands and Formosa, this could refer to either.

14. In the original "*marrelpriem*," but should have been *marlpriem*. This was an iron tool, shaped like a thick needle, used to splice rope on sailing ships. Rumphius quotes correctly, because Rochefort clearly states: "If a little hole be made at the small end of these Cornets, they become a kind of musical Instrument which makes a sharp and piercing sound, and forc'd through the windings of the shell, may be heard at as great a distance as the smallest kind of Trumpet might be: But there is a great secret in the sounding of it" (p. 123 of the English translation).

15. "Buccina tuberosa" is *Tutufa bubo* (L.).

16. "Buccina tuberosa rufa" is *Tutufa rubeta* (L.). "*Tuberosa*" derives from a rare Latin usage, *tuberosus*, an adjective that means "full of lumps or bumps," while "*rufa*" (from *rufo*) refers to a reddish color.

17. "Tombucco" is a region subsequently known as Bungku, or Tobungku. It is on the eastern coast of Celebes, across from the Banggai Islands and divided by Tomori Bay into a northern and a southern Bungku. It was ruled by Ternate for a long time. The so-called heathens, or non-Islamic population, were indeed quite martial. The finest product of the Bungku region, for instance, was swords made from local iron ore and forged by local smiths. See also ch. 6 in book 3.

18. Assuming that "Gengber" is a misprint for *Gember*.

19. The phrase "*witgeknoopte*" is not clear. I can only guess that the participial form might be "*geknobbelde*," and translated it as such.

20. This is *Fasciolaria trapezium* (L.).

21. This would be an operculum which, when burned as incense, would smell like musk. See ch. 17 of this book.

22. *Fusus* means a "spindle," the instrument on which the Fates spin our destinies. Given the description of the color, this must be *Fusinus colus* (L.).

23. This must be *Fasciolaria filamentosa* (Röding), although the figure shows a rather sculptured specimen. The rarer species that lives at a greater depth must be *Fusinus longissimus* (Gmelin).

24. The Latin has *pilosus* ("hairy," "shaggy") in it; hence the phrase means something like "the first hairy buccina, or the thick one" and can be identified as *Cymatium pileare* (L.).

25. Martens surmises that this has nothing to do with gender, but that it is the juvenile form of the foregoing. Shells that were similar but lighter were often called "female," though this was not taken literally. See also ch. 22 in the present book.

26. "*Mitra*" was originally the headband or turban worn by Asians, just as the word "tiara," which is what Rumphius is alluding to, was a crown worn by Persian kings. The pope's crown is a triple decker, but Rumphius does not use the noun "tiara." On the other hand, his anti-Catholic bias is clearly evident. I also think that he had the well-known rumor in mind that some of the papal rings, for instance those worn by the Borgias, concealed a poisoned needle which, when a person kissed the ring, would prick and kill. The name stuck, since today it is still called *Mitra papalis* (L.). Martens stated that the papal connection was older than Rumphius.

27. The second species is *Mitra cardinalis* (Gmelin).

28. The third species is *Mitra stictica* (Link).

29. *Mitra mitra* (L.).

30. Possibly *Mitra eremitarum* (Röding).

31. This is *Turris babylonia* (L.). The comparison with painted representations of the Tower of Babel is intriguing. The most famous instances are the two paintings by Pieter Breugel, one of which is now in Vienna and the other at the Boymans in Rotterdam. It is unlikely, however, that Rumphius saw these paintings, but the image was widely circulated in graphic form; hence there is a more likely provenance. Robertus Padtbrugge, who was governor of Ambon from 1683 to 1687, had an older brother, Dionysius Padtbrugge (1629–1683), who was a graphic artist. Dionysius produced an etching of the Tower of Babel (after the French artist Nicolas Cochin) that had a religious poem by his brother—the subsequent governor of Ambon and, therefore, Rumphius' superior—engraved on it. See: H. J. Pabbruwe, *Dr. Robertus Padtbrugge (Parijs 1637–Amersfoort 1703), dienaar van de Verenigde Oost-Indische Compagnie en zijn Familie* (Kloosterzande: Duerinck, 1995), p. 12; hereafter referred to as "Pabbruwe." One will find a reproduction of the print in F. W. H. Hollstein, *Dutch and Flemish Etchings Engravings and Woodcuts ca 1450–1700*, vol. XV (Amsterdam: Menno Hertzberger, n.d. [?1963]), p. 77.

32. The description matches several species of the family Turridae, but is too short to permit a positive identification.

33. Probably *Lophiotoma polytropa* (Helbling).

34. The Latin means something like "Flat Granulated Buccina." It is *Gyrineum gyrinum* (L.).

35. Rumphius' Latin name means "Round Granulated Buccina." According to Kool and Strack, this species is *Nassarius papillosus* (L.).

36. The Latin means the "Prickly Buccina," from *aculeatus* (furnished with stings or prickles). It is *Phos senticosus* (L.).

37. The Latin phrase means the "Wavy Buccina" (from *unda* = wave), and Linnaeus kept this with the name *Cantharus undosus* (L.).

38. *Cantharus fumosus* (Dillwyn).

39. *Linatella succincta* (L.).

40. This "Lined Buccina" is, according to Kool and Strack, *Nassarius glans* (L.).

41. This must be one of the smaller species of the genus *Colubraria.*

42. *Colubraria nitidula* (Sowerby). Letter Q seems to depict this second species and not the first.

43. *Vexillum vulpecula* (L.).

44. "*Bidji*" or *biji* means kernel or pit in Malay; hence this shell is called "the shell (that resembles) the kernels of the gnemon tree." *Gnetum gnemon* (L.) is called *bagu* or *bogor* in Javanese, and *malinjo* in Malay, and was particularly cultivated on Java. Its bark (*bagu*) was used to make a kind of rope, its fruit was eaten as a vegetable or used for making a kind of *krupuk* (*krupuk malinjo*), while its flowers, buds, and tender leaves were also consumed as a vegetable.

45. *Vexillum vulpecula* (L.), a species that is very variable.

46. This is *Vexillum plicarium* (L.).

47. Probably *Vexillum rugosum* (Gmelin).

48. "Oilstones," which has a similar usage in English, were, as the *OED* points out, smooth and

finely grained whetstones used for honing iron implements, and the surface of which was lubricated with oil.

49. Rumphius' phrase means "Little Tower [that is] Girdled With Thread." Martens rightly suggested that the figured species is *Cancilla granatina* (Lamarck), but the description refers more to *Cancilla filaris* (L.) or *Cancilla praestantissima* (Röding).

50. This is *Vexillum sanguisugum* (L.).

51. Martens did not hazard a guess. This species is not shown, and the description is very brief. Strack can suggest only that it is one of the many species of the family Turridae.

52. This "Ladder Buccina" is not the famous Wentletrap (*Epitonium scalare* [L.]) but possibly *Epitonium perplexum* (Pease) or *Epitonium pyramidale* (Sowerby).

53. The figured specimen is *Vexillum exasperatum* (Gmelin). The Portuguese species referred to by the general noun *Bozios* definitely refers to a different species.

54. Now spelled *búzio*, this Portuguese noun is the normal word for any whelk or univalve mollusk. The stringing of these shells into necklaces was still done in the Azores in recent times. Maldivian cowries were also called *búzio;* see n. 8 of ch. 24.

55. This "Leafy Buccina" is the snail *Littoraria scabra* (L.).

56. De Wit identifies this tree as *Aegiceras corniculatum* L.

57. This is *Turbinella pyrum* (L.).

58. From *meliceris*, originally a Greek word, for a kind of tumor, wart, or wen. It refers here to the animal's egg capsule.

59. This word does not refer to a temple but to a coin which was current in the East Indies towards the end of the sixteenth century. It was minted in India, in the region once called Canara (on the coast of Malabar, south of Goa). It was a gold coin that depicted Visnu in his fourth incarnation as a man-lion (*Nara singa*) and was worth four guilders and four stivers.

60. Most likely Harmanus van der Burgh (?1636–1708), a rich merchant and apothecary in Amsterdam, who had a collection of shells, insects, gems, porcelain, antiquities, etc.

61. See n. 88 of ch. 21. "*Topschoon*" makes even less sense here, nor will the conjecture from the *WNT* do, because this clearly must have something to do with color. My only feeble guess is: very beautiful?

62. The brothers Feitema—Sybrant Feitema (1620–1701) and Isaac Feitema (1666–1709)—were mentioned before. Like van der Burgh, they were wealthy merchants and apothecaries, and collected similar things.

63. This refers to the flag of the States-General, and I assume that Schynvoet is not thinking of the normal tricolor but the elaborate flag of the beginning of the Republic, which had a yellow field, with a red lion on it that in its right claw held a blue sword, and seven blue arrows (or darts) in its left paw. I capitulated to "*pen*" and simply translated it as "shell," because *pen* in Dutch can mean a feather, a writing tool, or a peg.

64. If this is Jan de Jong (?-1712), then he came, like the Feitamaas, from Haarlem, but lived in Amsterdam. He was an accountant for the VOC, and collected the same kind of things, especially, so it seems, shells and marine varia.

Chapter 20. Strombus. Needles. Sipot.

1. These "Strombi" (a term from Pliny for a kind of spiral snail) are the modern genera *Cerithium*, *Turritella*, *Terebra*, and other related ones.

2. "*Pennen*" has various meanings in Dutch: the writing tool (with all its derivative metonymies), feather (wing), plectrum, any pointed object, pin, peg, wooden nail, a knitting needle, and so on. Given the shape of these shells, I opted for 'pegs.'

3. "*Sipot*" or "*sipat*," now *siput*, means "shell" in Malay.

4. The "First Strombus or the Awl" is *Terebra maculata* (L.).

5. The "Second Strombus" is *Terebra subulata* (L.).

6. The first one mentioned of the "Third Strombus," and the one depicted by letter C is *Terebra chlorata* (Lamarck).

7. *Terebra felina* (Dillwyn).

8. Probably *Terebra dimidiata* (L.).

9. The "Fourth Strombus" is *Terebra guttata* (Röding).

10. Though not pictured, the description of the color indicates *Terebra dimidiata* (L.).

11. The "Toothed Strombus" is *Terebra crenulata* (L.).

12. The "Lussapinjo Islands" are the Lucipara Islands, a lonesome archipelago of four islands in the Banda Sea. Part of a large coral reef, these islands were not inhabited.

13. The "Seventh Strombus" is probably *Terebra cingulifera* (Lamarck).

14. The "Eighth Strombus or the Lancer" is *Hastula lanceata* (L.).

15. The "Ninth Strombus or the Grainy One" is, as far as the figured specimen is concerned, *Terebra anilis* (Röding) or *Terebra cumingii* Deshayes.

16. "*Chalybus*" should probably be "*chalybeius*" from *chalybe*, "steel," a Latin noun straight from Greek. Hence this would be: the "Steel Strombus," which is *Hastula strigilata* (L.) or the very similar species *Hastula acumen* (Deshayes).

17. This "White, Tailed Strombus" is *Rhinoclavis vertagus* (L.).

18. *Rhinoclavis fasciata* (Bruguière).

19. *Rhinoclavis aspera* (L.).

20. This "Grainy, Tailed Strombus" is *Cerithidea cingulata* (Gmelin).

21. This "Drum Strombus" is *Turritella terebra* (L.). The Greek "*Tympanotonos*" makes less sense, since it means "drum sound."

22. This "Strombus Full of Lumps" is *Pseudovertagus aluco* (L.).

23. This "Strombus Full of Corners" is *Cerithium nodulosum* (Bruguière).

24. This "River Strombus" is a freshwater snail, *Melanoides torulosa* (Bruguière).

25. "*Sipot aijer*" means "water snail."

26. The adjective *paluster* means "swampy" or "marshy"; hence this is the "Marshy Strombus," which is *Terebralia palustris* (L.).

27. Should be Buru, i.e. the island west of Ambon.

28. This "Smooth Marshy Strombus" is *Faunus ater* (L.).

29. "*Terrebellum*" derives from *terebra* or *terebrum*, a device used for boring, hence an auger. This is *Terebellum terebellum* (L.).

30. See ch. 29. The "Rolls" are a group of shells now known as the family Olividae.

31. I am not sure about this. The original has "*teil*" which, even in Rumphius' day referred to a bowl, saucer, or tub of earthenware or wood. This does not sound logical, but there was something that could be construed as a pointed wooden bucket. In the region Rumphius came from men used a wooden container that came to a point. It was carried on one's back and used particularly for cement, mortar, or stones. Rumphius' father featured such a device in his seal and signet ring. It was called *Bütte* at the time, *Kübel* in modern German, something like a hod (or "hot" prior to the sixteenth century) in English.

32. This phenomenon has been verified; see Martens, *G*, 120.

33. This "Strombus" is named after the tree it frequents: "Mangium caseolare" which, according to De Wit, in *M*, p. 442, is either *Sonneratia alba* (J. Smith) or *Sonneratia caseolare* (L.). This shell today is called *Terebralia sulcata* (Born).

Chapter 21. Volutae. Bia Tsjintsjing.

1. With "Volutae," Rumphius refers to the families Volutidae (plate XXXI, letters A and B, and on plate XXXII, letters H and I), Harpidae (letters K, L, and M on plate XXXII), as well as numerous species of the Conidae family. Because of their variety and beauty, the Conidae has always been much sought after by collectors.

2. *Wellen* is a deceptive term. One would immediately assume that it means a "well" from which one draws water, but that clearly would have little in common with a shell. There was, however, an older usage, meaning "draaikolk," or "whirlpool," which makes figurative sense.

3. "*Tsjintsjing*," would also once have been spelled *tjintjin*, now *cincin*, or *chinchin*, and refers to a finger ring. Hence the Malayan name for these shells is "Ring Shells."

4. The volutes were the spiral scrolls attached to the capital of an Ionic pillar, or below the corners of the abacus on a Corinthian column.

5. This species is *Melo aethiopica* (L.). "*Cymbium*" was the Latin word for a small drinking vessel, a cup or bowl.

6. "Scoop" for *bak*. The Dutch noun has different meanings, most of which would not fit here. Rumphius' description implies that these shells were also used to bail water out of boats, while Strack notes they are still popularly known as "Bailer Shells." But "bail" as an English substantive has too many modern connotations, hence my choice.

7. "*Sempe*" might conceivably be *simpai*, which is Malay for a flexible band or hoop.

8. Rumphius' "South-Easter Islands" was a generic name for various groups of islands southeast of Banda. They included the Tanimbar Islands, Kei Islands, Timorlaut Islands, Watabela Islands, and the Goram Islands. The Kei (or Kai) Islands were probably the most significant for Rumphius.

9. A *paludamentum* was a particular cloak which indicated the rank of general in the Roman army.

10. Al Cohen suggests that this might be an attempt at transcribing *wang-luo; wang* meaning "king," "kingly," or "royal," and *luo* or *lo* referring to a variety of shellfish with spiral shells, including

the Conus shells. See also B. E. Read, "Chinese Materia Medica, Turtle and Shellfish Drugs," *Peking Natural History Bulletin*, 12:2 (December 1937), p. 70 (item 233).

11. This is probably *Melo amphora* (Lightfoot).

12. The Latin means "butter cone." This is *Conus betulinus* (L.).

13. For "*boterwegge*," which seems to have been a word to indicate a wedge-shaped clump of butter, later a slice of white bread with butter on it, from which came the Dutch word "*boterham*."

14. The Latin is self-explanatory for these shells: *Conus litteratus* L.

15. I chose to translate *kegel* here as a bowling pin, since the context seemed to warrant such a realistic interpretation, and since Rumphius used it elsewhere, in his *Herbal* as well, in this sense. After this I went back to "cone." *Kegel* can mean both "cone" and "[bowling] pin."

16. The original Dutch phrase—"*doch zoo stijf als de hoorne blikken aan de lantaernen*"—posed some problems. Suffice it to say that *blik*, which now usually refers to "tin," can be a synonym for *blad*, that is to say, a "plate" or "leaf" of some material, in this case horn. For many centuries the sides of lanterns were made from horn, witness the false folk etymology "lanthorn." This is what Rumphius seems to be referring to: what are called "lantern leaves" in English, sheets of horn for lanterns, as the *OED* defines it.

17. There is a region called "Landak" in the western part of Borneo.

18. From *cereolus*, which is in Pliny, meaning "the color of wax," and which is *Conus virgo* L.

19. This means "Candle Shells," since *lilin* (*liling*) refers to a wax candle. *Lilin* can also mean just wax.

20. He means the ones that have just been caught.

21. *Conus quercinus* (Lightfoot).

22. These "tiger volutes" are *Conus striatus* L.

23. This is only an approximation of "*naar dat iemand schrander is in 't bespiegelen*."

24. From *nebula*, a mist, a vapor, by extension a cloud; these are *Conus geographus* L. The other one, more beautiful and rarer, is *Conus tulipa* (L.).

25. This means "bat" in Latin, from *vesper*, i.e., evening. Linnaeus kept Rumphius' name, and it is now known as *Cymbiola vespertilio* (L.).

26. *Morsego* means "bat" in Portuguese; hence their name in Malay is "Bat Shells."

27. This should probably be "*baduri*," a word that contains the noun *duri*, which means "thorn," also a sting or a quill (of a porcupine). Hence: "Thorny Shells." From this derives the name for that extraordinary fruit *durian*, which means "thorny fruit."

28. "Tabbied" and "tabbying" for the Dutch *wateren*. We do have that verb as well, but it inevitably invites confusion. "Tabby" however does not have that problem and is more precise. The *OED* provides as the principal meaning of "tabby" a general term "for a silk tafeta, originally striped, but afterwards applied also to silks of uniform colour waved or watered." To give cloth that wavy appearance, a machine, called a calender, was used. Basically a set of heavy rollers, the machine inspired the phrase "calendered paper." And, yes, the brindled gray or tawny cat comes from the original textile term.

29. "Gristle" for "*knarsbeenig vleesch*." *Knarsbeen* was an older synonym for *kraakbeen*, which means "cartilage" or the older English form, "gristle."

30. This is also *Cymbiola vespertilio* (L.).

31. This third kind is another form of that very variable species *Cymbiola vespertilio* (L.).

32. *Mimpi* is "dream," *bermimpi* is "to dream"; hence, "Dream Shells."

33. The "Seacat" was the squid or cuttlefish.

34. "Melicera" from *meliceris*, which means a wen or wart, from the Greek *melikeris*, which means "dark wax."

35. The original has "*overvloed van nering*," which is the definition for the previously mentioned wart or melicera. The old meaning of *nering* is "food." They are the egg cases of the bat snail, *Cymbiola vespertilio* (L.).

36. This is *Harpa ventricosa* Lamarck.

37. This might be a misprint for "*sipot*" (*siput*), a generic Malay term for shellfish.

38. This is *Harpa harpa* L.

39. This smaller species is *Harpa amouretta* Röding. It would gladden Maria Dermoût's heart that this particular Rumphian name has been preserved.

40. The "Liassersche Islands" are the Uliasser Islands, i.e., Haruku, Saparua, and Nusa Laut.

41. "Sarassa" is most likely the Javanese word *serasah*, which referred to a cotton print fabric that used to come from Coromandel.

42. Usually spelled "*amourette*," it was a word used in Dutch for a love affair, a dalliance, what the French call an *amour passager*.

43. This had various meanings in Latin: millstone, molar tooth, or a jawbone, as well as a very

particular meaning for grains mixed with salt and strewn over a sacrificial animal. Perhaps Rumphius had "millstone" in mind?

44. *Conus marmoreus* L.

45. I do not know what kind of stone this is, but it means "white spots." It is not in Pliny, to my knowledge.

46. This means "winged volute" and is *Conus textile* L. It is illustrated on plate XXXII as letter P, not O.

47. Lewis and Short identify this *attagena* as a kind of meadow bird, a "heath-cock," i.e., the gamebird called black grouse.

48. This one and the third brown species could be *Conus aulicus* (L.) or *Conus omaria* Hwass. The one depicted as letter O (and *not* P) is certainly *Conus omaria.*

49. "Spotted volutes" or *Conus augur* Lightfoot.

50. "*Cinerea*" from *cinerarius*, "pertaining to ashes," now *Conus cinereus* Hwass.

51. This is not a usage derived from Perrault's fairy tale, because the collection of stories did not appear until 1697, less than five years before Rumphius' death. But the *WNT* lists a usage that predates Perrault by a quarter of a century, arguing that "Asschepoester" (*Cendrillon* in French) was a name in vogue *before* the celebrated *Contes.*

52. These "ghosts volutes"—from *spectrum*, Latin for an image or an apparition—are called *Conus spectrum* L.

53. Rumphius most likely derived this from Kircher (see n. 9 of ch. 77 in book 3). In his *China Illustrata*, published in Amsterdam in 1667, he mentions a "Desertum Lop" which is in "Tartaria Magna" (p. 56); later he states that this is located between "Peim and Ciarciam" and between "Kamul and Tarfan" (p. 108). My page references are to the contemporary Dutch translation: Athanasius Kircherus, *Tooneel van China, Door veel, zo Geestelijke als Wereltlijke, Geheugteekenen, Verscheide Vertoningen van de Natuur en Kunst, en Blijken van veel andere Gedenkwaerdige dingen, Geopent en Verheerlijkt*, transl. J. H. Glazemaker (Amsterdam: Johannes Janssonius van Waesberge, 1668).

54. These "grainy, spotty volutes" are *Conus coccineus* Gmelin.

55. The Latin means something like "volutes that are girded with thread," *Conus figulinus* (L.).

56. This "threaded volute" is *Conus miles* L.

57. "Swathed" or "banded" volutae.

58. The first kind is *Conus capitaneus* L., and is shown on plate XXXIII, letter X.

59. This second kind is probably *Conus vexillum* Gmelin.

60. The third and rarest kind is *Conus generalis* L., and is depicted on plate XXXIII, letter Y.

61. This is Snakewood (*Strychnos muricata* Kostel.).

62. *Bantal* is Malay for cushion, hence "Cushion Shells."

63. From *harena*, i.e., "sand"; thus, "sand volutes."

64. These are *Conus stercusmuscarum* L.

65. And these *Conus arenatus* Hwass.

66. Its Latin name is the same as the Dutch and is the ubiquitous *Conus ebraeus* L.

67. No. XX is *Conus catus* Hwass.

68. The smaller species might be *Conus proximus* Sowerby II, but certainly not *Conus glans* (Hwass.), as Martens suggested.

69. "*Terebellum*" might be a misspelled form of a variation of the Latin verb *terebro* (to bore or perforate), or the noun *terebra* (*terebrum*), which refers to an instrument for boring. This species is *Conus nussatella* L., after the "Three Brothers Islands" or "Nussatello," now more commonly known as Pulau Tiga, off Ambon's west coast.

70. *Kuipersboor* in Dutch.

71. "*Fluviatilis*" from *fluviaticus*, which means "belonging to a river."

72. This means "little bitter one," and the name seems to have stuck, since this could be the species *Thiara amarula* L.

73. This is Rumphius' *Mitra papalis*, which he described as the tenth kind of "Buccina" in ch. 19 of this book.

74. *Conus glaucus* L.

75. The larger species and the one illustrated is *Ellobium aurismidae* L., while the smaller one that is not shown is *Ellobium aurisjudae* (L.)

76. "Spinning" for *spinzel*, that which is produced by the process of operating a spinning wheel. The confusion lies in the Latin word *salivatio*, which Rumphius adds as if a clarification. But *salivatio*, a Medieval Latin term, can only mean "salivation."

77. Is "*Laxat*" perhaps the Chinese freshwater snail *loshi?*

78. I have not been able to find this reference. None of the passages on chickpeas (*cicer*) mention the *melicerus*, and the passages that mention *melicerus* do not refer to the skin of chickpeas.

79. The *WNT* uses this particular passage in order to illustrate the obscure and obsolete usage

of *tooten* for the shells of members of the Conidae family. It persisted into the eighteenth and nineteenth centuries.

80. I translated *bak* as "scoop," *tepel* is "nipple" in Dutch, and *kroon* is "crown"; hence this becomes something like: "Crowned Nipple Scoops[!]"

81. This is a special Dutch usage, derived from "mennist" or a follower of Menno Simonsz, i.e., a Mennonite, which, by extension, could also refer to something sober in appearance, unadorned, yet still quietly beautiful.

82. Is "*Chrisant*" the same as "*chrysant?*" If so, the latter is short for the chrysanthemum flower.

83. *Conus bandanus* (Hwass.), a species closely related to *Conus marmoreus* (L.).

84. "Lace" for "*speldewerk*," better "*speldenwerk*," which referred to lace made with the aid of pins; in English bobbin lace. Once again—this must be one of the most quoted texts in the *WNT*—Schynvoet's passage is quoted for a specific usage pertaining to a shell. One can get a good idea of how this kind of lace was produced in the wonderful painting by Caspar Netscher entitled *The Lacemaker*. The painting is now in the Wallace Collection.

85. This is the same species, *Conus figulinus* (L.) as shown on plate XXXI, letter V.

86. This is *Conus namocanus* (Hwass.).

87. This *Conus pulicarius* Hwass. is shown on plate XXXIII, no. 2.

88. This is *Conus aulicus* L.

89. Probably *Conus pennaceus* Born.

90. This painter was Johannes Bronkhorst (1648–1727). He was thirteen when his father died, and his mother apprenticed him to a pastry cook in Haarlem. This profession became his livelihood for the rest of his life, but when he moved to Hoorn he began to paint a vast number of watercolors, mostly of birds and insects. He owned a cabinet of shells. With thanks to Prof. Ton Broos.

91. *Conus aurisiacus* L., still a rather rare species.

92. This is an attempt to render "*topschoon*," a locution the *WNT* does not know what to do with either. It cites this same passage by Schynvoet. They conjecture the meaning "smooth at the top," but it might just mean "very beautiful."

93. *Conus ammiralis* L.

94. Most likely Hermanus van den Burgh (?1636–1708), an apothecary and collector in Amsterdam.

95. *Conus ammiralis* L.

96. This probably does not refer to the remarkable Jan Swammerdam (1637–1680), but to his father: Jan Jacobsz. Swammerdam (1606–1678), an apothecary in Amsterdam, famous for his impressive curiosity cabinet.

97. *Conus ammiralis* L.

98. This is Jurriaen Ovens (1623–1678), a German painter who was Rembrandt's pupil. He lived in Amsterdam between 1640 and 1663, and was a collector of shells, among other things.

99. This is probably *Conus pulcher* Lightfoot, which occurs in southwest Africa.

100. *Conus acuminatus* Hwass. which inhabits the Red Sea.

101. *Conus genuanus* L., from West Africa.

102. "Gridelin" for virtually the same word in Dutch, *gridelyn*, both from the French phrase *gris de lin* or "gray of flax." This is a gray with a violet sheen to it.

103. *Conus imperialis* L.

104. Dirck van Cattenburgh (1616–1704) was a merchant residing in Amsterdam. He made money but did not seem to amass it. He was acquainted with Rembrandt for many years, perhaps because, besides being a wine merchant, Cattenburgh was also an art dealer.

105. Another form of *Conus imperialis* L.

106. Most likely Pieter Blaeu (1637–1706), son of the famous cartographer. Besides running the family business, he also became the municipal secretary of Amsterdam. A very rich man, he collected a variety of things.

107. *Conus ermineus* Born, from the Caribbean and West Africa.

108. *Conus achatinus* Gmelin.

109. Probably the West Indian species *Conus spurius* Gmelin.

Chapter 22. Alatae. Amb. Tatallan.

1. All species described in this chapter belong to the family Strombidea.

2. Poetic usage in classical Latin for "winged," derived from *ala*, "wing."

3. Identified by Van Bemmel (*M*, p. 44) as *Fregata ariel* (Gould), and described by Rumphius in his *Herbal* in book 7, p. 45.

4. This is the operculum of this snail, like the "Blatta" in ch. 17, though it was not used for incense.

5. In classical Latin *harpago*, as a noun, referred to a hook with which one could tear things

down; a grappling hook, for instance. Clearly related to the verb *harpago,* which meant "to rob" or "plunder." Strack: *Lambis chiragra* (L.).

6. For "*Stomyjes*," a spelling that is impossible in Dutch, and which should be "*Stompjes.*" As Martens already noted, this is not a different variety but, like the so-called females, a younger stage of the snail. See also ch. 23, no. XII.

7. The Pliny reference does not mean much. In book 32, section LI, Pliny lists the names of 144 species from the sea—a strange mixture of all sorts of creatures. The "pentadactyli" are merely listed in section 147.

8. Boano and Manipa are islands off the eastern coast of Ceram (off the eastern coast of its Howamohel peninsula). Manipa is just about halfway between Ceram and Buru, while Boano is further north, above Kelang island.

9. These "Horns" are *Lambis lambis* (L.).

10. I have not found any mention of this "seven-fingered" snail in Pliny.

11. "*Bia Cattam*" should be "*Bia ketam.*" *Ketam* is Malay for a crab; hence these shells are called "Crab Shells."

12. For Jonston, see n. 27 of ch. 3 in this book.

13. There is a fairly rare usage *coracicus:* "of or belonging to a raven."

14. "*Decumana*" refers to *decumani* or *decimanus,* which, in a metonymic sense, meant "large" or "immense." Its main meaning was the tenth part of something, a tithe. According to Strack, this must be *Lambis truncata sebae* (Kiener).

15. Strack: *Lambis millepeda* (L.).

16. From *nodosus,* "full of knots." Strack: *Lambis scorpius scorpius* (L.). With "Podagra" Rumphius clearly has arthritis in mind, but in classical Latin it referred to gout in one's feet.

17. The "Liasser Islands" are the Uliasser Islands, southwest of Ambon.

18. The Latin means something like "Broadly Winged," and the species is, according to Strack, *Strombus latissimus* (L.).

19. "*Epidromis*" meant an aft sail, Rumphius is more specific: *bezaan,* was the triangular aft sail, copied from Mediterranean tradition, called a mizzen in English. Linnaeus kept Rumphius' name: *Strombus epidromis* L.

20. This "Hump-backed aft sail" is, according to Strack, *Strombus canarium* L.

21. Strack: *Strombus vittatus vittatus* L.

22. Strack: *Strombus minimus* L.

23. *Lentiginosus* means "full of freckles" from *lentigo,* a lentil-shaped spot. Strack: *Strombus lentiginosus* L.

24. This should be *tahi lalat. Tahi* means dregs, lees, or plain "shit"; *lalat* is the Malay word for the common housefly, hence "freckles" were literally called "fly specks" in Malay.

25. The Latin means "boxer" (pugilist), of course, not a swordsman. Strack: *Strombus aurisdianae* L.

26. This strange spelling hides the word *tunjokkan,* which means "to point out," "to show."

27. For "*sagelilt*" in the original, which is even more peculiar than the foregoing. I unscrambled this to mean *tjakalele* (*cakalele* now), the once-famous war dance of the Moluccan peoples. Hence the original attempted to conjugate *tjakalele* as a Dutch verb! The illustration depicts such a dancer. From a contemporary painting (*De Vismarkt te Bativia* [Fish market in Batavia]) by Andries Beeckman:

28. Strack: *Strombus bulla* (Röding).
29. Strack: *Strombus luhuanus* L.
30. This should be "Luhu," a region on Howamohel or Little Ceram, the peninsula of the large island of Ceram that points at Ambon from the north. Kaijbobo or Kaibobo is a town on a point of Ceram proper or "Big Ceram" (*Seram besar*), east of Howamohel or "Little Ceram," and near the small island called Pulu Babi.
31. The Latin and the Malay after the fruits, which look like almonds, of the kenari tree (*Canarium*) of which De Wit lists twelve varieties mentioned by Rumphius in his *Herbal.*
32. Strack: *Strombus gibberulus* L.
33. Strack: *Strombus labiatus labiatus* (Röding).
34. Strack: *Strombus mutabilis mutabilis* Swainson.
35. See ch. 32 of book 1.
36. Strack: this "Broad Canarium" is *Strombus marginatus succinctus* L.
37. I do not know what this "*Samaar*" or "*Samar*" might be. The only remote possibility is a kind of wood from a Moluccan tree. Rumphius calls it "Metrosideros molucca mas" (*Herbal*, 3:25) and says that it is called "*Samar*" on Ambon. Heyne (*Nuttige Planten*, p. 1139) lists it as *Homalium foetidum* Benth. and says that this tall tree has a wood that turns red when it is freshly cut, gradually turns brown, and becomes black if it is put in seawater. This must be *Strombus dentatus* L., although the characteristic denticles on the outer lip do not show very well.
38. Strack agrees with Martens that this is an immature *Lambis chiragra* (L.) rather than *Lambis scorpius* (L.).
39. This is *Harpulina arausiaca* (Lightfoot), a species only found in Sri Lanka.
40. This *Harpulina lapponica* (L.) is found in Sri Lanka and southern India.
41. This belongs to the family Volutidae, but the illustration is not clear enough to be more specific.
42. Strack: *Strombus gallus* L., a species from the West Indies.
43. This most likely refers, as Strack suggests, to the bird species *Philomachus pugnax* (L.), known as Ruffs in English. The males were famous for their variegated plumage. As an eighteenth-century commentator put it: "The males or Ruffs assume such variety of colors in several parts of their plumage, that it is scarce possible to see two alike" (Pennant, in *British Zoology*, 1768).
44. Difficult to identify. Strack offers that it might be a juvenile specimen of a species of the Turbinellidae family.
45. A juvenile specimen of *Turbinella pyrum* (L.).

Chapter 23. Porcellana Major. Bia Belalo.

1. Rumphius' "*Porcellanae*" are snails belonging to the families Cypraeidae (Cowries) and Ovulidae.
2. Taking "*voetstap*" not to mean "footstep."
3. The Latin means "Venus Shell."
4. The reference to Pliny does reveal some confusion, which Rumphius, wisely, ignores. The passage is indeed in book 9, but in the Loeb ed. it is the second half of chapter 41, paragraph 80. Mucianus was one of Pliny's informants. Pliny does describe a shellfish that he calls a "Murex," and then cites the strange tale that an infestation of these animals clung to a ship that was carrying some noble youths who were to be castrated. But the weight of these creatures halted the ship "in full sail" and the youths were saved, which is the reason why these shellfish are consecrated to Venus.
5. The Latin means: "Both of them received the name from the likeness to a woman's genitals, which the Greeks call Chaeron, and the Romans pig [*porcum*] and piglet [*porculum*], these also resemble Conchaerima." The final phrase is not entirely clear. The other problem is presented by the supposed Greek words "*Charinae*," "*Chaeron*," and "*Conchaerima*," It is, once again, the spelling of the Greek that leaves one bewildered. "Charinae" should probably be "*choirioi*," from "*choirion*" [χοιρίον] which means a piglet or porker. "*Chaeron*" most likely should be "*choiros*" [χοῖρος], which also means a young pig or a porker, a word which in ancient Greek, according to Liddell and Scott, was "like *porcus* in Varro . . . of the *pudenda muliebria*, often in Comic poets, who are always punning on the word and its compounds." Rumphius then seems to make an unwarranted connection between these misspelled Greek words and "*conchaerima*," as if "*con*" were somehow a prefix of "*chaerima.*"
6. The Dutch vulgarism *klipkoussen* is more obscure. *Klip* means "rock" or a fissure in a rock. *Kous* in modern Dutch means "stocking," but it did have a vulgar meaning in Rumphius' day as a euphemism for the female genitals, and, by extension, it also meant "woman." To come up with some sort of translation for this otherwise incomprehensible locution, I used the Shakespearean "clack dish." This was a beggar's bowl for alms, and was used figuratively for the female pudendum in *Measure for Measure* (III, ii, 127–129), where it is said of the Duke: "Not the duke? Yes, your beggar of fifty, and his use was to put a ducat in her clack-dish; the duke had crotchets in him."

7. *Bilalo* is a Makassarese word for a shell that was used to smooth things.

8. *Guttata* is Latin for drops of any kind of liquid and, by extension, for spots on stones, animals, anything. These are *Cypraea tigris* L.

9. "*Krang*" is *kerang*, Malay for coral or shell; *krontsjong* is another puzzle. Perhaps this is meant to be *kerontjong*, which was not the popular Indo music "*krontjong*" in Rumphius' day, but would refer to a tinkling sound made by little bells on an anklet, by extension, a jingling sound. Rumphius would be indicating the decoration of bells, of course, not the sound.

10. This is black coral, ground up, and mixed with other ingredients for medicinal purposes.

11. This is the isthmus that connects Ambon's two peninsulae, called Hitu and Leytimor. It was known both as "Paso" or "Beguala" and was only 400 feet wide.

12. "*Montosa*" probably for *montuosus*, i.e., "mountainous." This is *Cypraea mappa* L.

13. Ascension Island was discovered by the Portuguese navigator João da Nova on Ascension Day (the Holy Thursday forty days after Easter) in 1501. The island, midway between Africa and Brazil, is a peak that rises half a mile out of the ocean.

14. Cape of Good Hope.

15. "*Testudinaria*" from *testudo*, Latin for turtle, also tortoise shell, hence meaning "like a turtle." *Karet* is Javanese for "tortoise shell." Just as rare on Ambon as the foregoing, this is *Cypraea testudinaria* L.

16. Argus, of course, was the giant with a hundred eyes (he never slept with more than two closed) who was appointed by Juno to watch over Io. The reference to the painted eyes is not clear. These are also rare shells, called *Cypraea argus* L.

17. See book 1, ch. I, nn. 7 and 8.

18. "*Ciel*" refers to Tandjong (= cape or hook) Sial, a notorious cape at the southwestern end of Ceram's peninsula, here called "little Ceram" by Rumphius, but also known as Hoamohel. In *Ambon, en meer bepaaldelijk De Oeliasers* (1875), Van Hoëvell notes (p. 16) that the cape was feared (its name means Cape Bad Luck) for its strong current, and that one felt earthquake tremors all year round.

19. These are *Cypraea mauritiana* L.

20. "*Taneit*" in the original. This is now an obsolete word for a brownish yellow, hence my translation "tawny."

21. This is *Cypraea caputserpentis* L. The rare blue variety is *Cypraea onyx* L.

22. "*Jamboezen*" refers to *jambu* or, in older spelling, *djambu*. This refers in general to the fruits of a number of trees and shrubs of the genus *Eugenia*, and specifically to *Eugenia acquea* or *djambu ajer*; i.e., the "water jambu." This variety can indeed have white fruits, though jambus are usually red. Rumphius presents these trees in his *Herbal* in chapter 32 of book 1, where he says that "the Arabs and Persians" call them "Indian apples." He finds them superior to durians, calling them "the most excellent fruit of the Indies" because they "delight sight as well as smell and taste." The Jambu tree, he says, resembles an apple tree, and the fruit a "red, round, elongated apple or even better a peach, but rounder on both ends, and not as even outside, as a peach." Strack calls this *Calpurnus verrucosus* (L.).

23. *Talpa* is Latin for a "mole," and this is now called *Cypraea talpa* L.

24. This should be "*pisang*," the *Musa* fruit, better known as the banana: described by Rumphius in chs. 1–4 of book VIII of his *Herbal*.

25. This sounds like a variety of native cigarettes. Cigarettes were not known in Europe until the middle of the nineteenth century. Indonesian clove cigarettes, *kretek*, are now also known in the West. Rumphius notes in his herbal that in his day (second half of the seventeenth century), the Malay already mixed cloves and tobacco for smoking purposes; see *Herbal*, book 2, ch. 1 (on cloves or "caryophyllum"), p. 8. They used dried banana leaves to make these "*Bonckos*," small rolls five to six inches long, which, the way Rumphius puts it, were "placed in one's mouth, lit in front, and smoked in that fashion" (p. 136). The same banana leaves, if they had not been torn too much, were smoothed with one of these Porcellana shells. This activity was known as "*Bilalo*," hence the shells were called "*Bia Bilalo*" (see n. 7 of this chapter). See the *Herbal*, book 8, ch. 2 (pp. 130–136).

26. From *caro*, Latin for flesh. *Daging* means flesh or meat in Malay. These are *Cypraea carneola* L.

27. *Cypraea vitellus* L.

28. *Cypraea arabica* L.

29. "*Tablature*" referred to an interesting method of musical notation once used for the lute and other stringed instruments. The lines of the stave were meant to indicate the instrument's strings, whereon letters were placed to indicate where they were to be "stopped" by the fingers.

30. *Kluchtig* means primarily "comical" or "funny," but I do not think that this applies here. The illustration being what it may, Rochefort has a wonderful passage on this shell which he did not see himself in its proper environment. He says that a Mr. Montel saw such a "Musick-Shell" "that had five Lines, a Key and Notes, which made good Musick: Some person had added the Letter, which it seems Nature had forgotten, and caus'd it to be sung, and the Musick was not undelightful. This might af-

ford the ingenious many excellent reflections: They might say among other things, that if according to the opinion of *Pythagoras* the Heavens have their Harmony, the sweetness whereof cannot be heard by reason of the noise made upon Earth; if the Air resound with the melody of an infinite number of Birds who sing their several parts there; and if Men have invented a kind of Musick, after their way, which by the Ears recreates the Heart; it were but just that the Sea, which is not always toss'd and troubled, should have within its territories certain Musicians to celebrate, by a Musick particular to them, the praises of their Sovereign Maker. The Poets might adde, that these natural tablatures are the same which the *Syrens* had in their hands, when they had their melodious Consorts; and that being perceiv'd by some eye which came to disturbe their recreations, they let them fall into the water, where they have been carefully kept ever since" (p. 125 of the English translation).

31. The Latin locution derives from *lentigo*, which referred to freckles or lentil-shaped spots. Their scientific name is *Cypraea lynx* L.,

32. This Medieval Latin term was more commonly used for smallpox.

33. Letter P depicts the first and largest kind: *Cypraea caurica* L.

34. Letter O shows the other, smaller kind: *Cypraea chinensis* Gmelin.

35. Most likely a misprint for *morbidus*.

36. This is *Ovula ovum* L. What a "*Saloacco*" is, is explained a little later, but I think that these shields were not exclusively Alfuran. Rumphius states in his *Herbal* (book 5, ch. 27, p. 176) that a particular wood was used to make these "Ambonese long-shields or Salowackos." This was wood from the "Shield Tree" ("*Schildt-Boom*") or "*Clypearia*" known in Malay as "*Caju Salowacko.*" This might be *Albizzia moluccana* Miq.; see De Clercq, *Nieuw Plantkundig Woordenboek voor Nederlandsch Indië*, p. 10. The illustration is from Valentijn, *Oud en Nieuw Oost-Indiën.*

37. Pliny discusses "*Vasa Myrrhina*" especially in book 37, chs. 7 and 8. But they clearly were not made of porcelain.

38. Julius Caesar Scaliger (1484–1558), during his lifetime considered the greatest scholar in Europe. He began his public life as a soldier. After rheumatic gout stopped that career, he devoted himself to his studies and to writing Latin verse. He attacked Erasmus, wrote a Latin grammar, published a flood of Latin verse, and wrote commentaries on classical texts. Rumphius is referring to the last, particularly *Exercitationes* (published in 1557), which was Scaliger's best-known work. The *Exercitationes* are commentaries on the *De subtilitate* by Jerome Cardan, and display Scaliger's encyclopedic knowledge.

39. P. Bellonius was the French zoologist Pierre Belon (1517–1564), who was particularly known for his books on fish (1551–1552) and evergreens (1553). Belon was an ardent Catholic who fought against the Huguenots. He deliberately wrote in French and traveled a great deal in the Near East, as well as in Italy and England. He published accounts of these travels. He was murdered in Paris while he was conducting botanical research in the Bois de Boulogne. I am not sure that Belon wrote any "Observations," but he did produce a *Commentarium* on Dioscorides.

40. These three minerals are described by Pliny in book 37, and often overlap. Onyx and sar-

donyx are both a chalcedony, and jasper is a quartz. D. E. Eichholz, the translator and editor of volume 10 of the Loeb ed. of Pliny's *Natural History*, states (p. 178) that *murrina* was most likely fluor-spar.

41. Behind this abbreviation hides Georgius Agricola (1494–1555), the well-known metallurgist from Germany. The work referred to is probably *De natura fossilium*.

42. Anselmus Boëtius de Boodt (?1550–1632) was a Flemish physician from Brugge, a city where he was also appointed magistrate. He became the court physician of Emperor Rudolf II, the troubled and controversial Habsburg monarch who was emperor of Germany from 1576 to 1612, king of Hungary from 1576 to 1608, and king of Bohemia from 1576 to 1611. An eccentric individual, Rudolf was known for his interest in the occult, astrology, and curiosities. He was also a patron of the arts. Over a number of years he was forced to relinquish power to his brothers, and by 1611 he even had to abdicate as king of Bohemia. De Boodt returned to Brugge in 1612, where he continued his own interest in curiosities, in art (he was a fine watercolorist), and in natural history. He became famous with the book on gemstones: Anselmi Boetii de Boodt (Brugensis Belgai, Rudolphi Secvndi, Imperatoris Romanorum, Personae Medici), *Gemmarum et Lapidum Historia, Qua non solum ortis, natura, vis & precium, sed etiam modus quo exiis, olea, salia, tincturae, essentiae, arcana, & magisteria arte chymica confici possint, ostenditur* (Hanoviae: Typis Wechelianis apud Claudium Marnium & heredes Ioannis Aubrii, 1609). It was later reprinted in Leiden in an augmented edition, with notes and illustrations, edited by Adrianus Toll: Anselmus Boetius de Boot [*sic*] (Brugensis, Rudolphi II. Imperatoris Medicus), *Gemmarvm et Lapidum Historia* (Lvgdvni Batavorum: Joannis Maire, 1636).

43. This would be ch. 23 of book 3.

44. De Wit, in *M.* (p. 449), identifies *Perlarius primus* (*Caju sonti*) as *Pipturus paniculatus*.

45. The ubiquitous "Phil. Bonannus" refers to Filip Buonanni (1638–1725) and his once-famous book published in Latin in 1684 as *Recreatio mentis et oculi de testaceis*.

46. This is most likely Pieter Blaeu (1637–1706), son of the famous cartographer Joan Blaeu, who followed his father into the business. He also became Amsterdam's *stadssecretaris* (municipal secretary), a title often used before his name. A very wealthy man, Blaeu owned large collections of curiosities from nature, antiques, and artwork.

Chapter 24. Porcellanae Minores. Bia Tsjonka.

1. "*Tsjonka*," also spelled *chongkak*, was a Malay word for the cowry (*Cypraea*) shell. *Chongkak* lent its name to a once-popular Malay game called *main chongkak*, which used cowries as counters (*buah chongkak*) on a board shaped like a boat (*papan chongkak*). Called *main dakon* on Java, it was a game once played in the Middle East and the West Indies as well.

2. A "*thorax*" does indeed, in classical Latin, have breastplate, cuirass, or doublet for its secondary meaning (primary is "breast" or "chest"), but the Dutch is confusing. "*Borststuk*" was not the common noun for breastplate; Rumphius also connects it to "*wapen*," but in that word's secondary meaning of "armor," while its primary, and far more common, meaning was "weapon." I also have some trouble comparing a small shell to a cuirass, especially since he, while describing no. I, clearly states that this small shell is "as wide as a breastplate" (*zommige zyn breed als een borststuk*).

3. This "Breastplate Furnished with Eyes" is *Cypraea erosa* L.

4. This "Breastplate Furnished with Stars" is *Cypraea helvola* L.

5. From Pliny, book 37, paragraph 133 (Loeb ed., 10:272). Pliny says little about this "star stone" except that Zoroaster claimed it had magical properties.

6. This "Common Breastplate or Cauris" is the cowry used for money, a fact reflected in Linnaeus' nomenclature, since this is *Cypraea moneta* L.

7. Martens thinks that these are *Marginella monilis* (L.), but Rumphius is probably correct. The Germans, according to Schynvoet, used to adorn their horses with true *Cypraea moneta* L.

8. The word "cowry" (*kauri* in Hindustani) comes originally from Sanskrit (*kaparda*). Use of these shells as currency was ancient and included China, India, Southeast Asia, and Africa. In China the cowry (*pei-tzu*) was still currency in the seventh century, and even longer in the northern part of Indochina. Edward H. Schafer, *The Vermilion Bird. T'ang Images of the South* (1967; reprint ed., Berkeley: University of California Press, 1985), p. 208. Towards the end of the thirteenth century, Marco Polo stated that in the Mongol province of Yun-nan, the people used for money "certain white porcelain shells that are found in the sea, such as are sometimes put on dogs' collars; and 80 of these porcelain shells pass for a single weight of silver, equivalent to two Venice groats, i.e. 24 piccoli. Also eight such weights of silver count equal to one such weight of gold." (*The Travels of Marco Polo.* The Yule-Cordier edition, 2 vols., 1903 and 1920 [reprint ed., New York: Dover, 1993], 2:66).

The shell was a basic monetary unit in Bengal for a long time. Yule notes that one region still paid its revenue in cowries in 1801, and in Siam *bee-a* or cowries were still in use in 1865. As Yule's sources indicate, during Rumphius' time the shells came indeed primarily from the Maldives, which at the time were even known as the Cowrie Islands. The fourteenth-century Arab traveler Ibn Batuta called them *al-wada*, and they were *wada* in Egypt during the fifteenth century (Yule, *Hobson-Jobson*, pp. 269–271). This corresponds to Rumphius' "*Wadaat.*"

Rumphius' passage is a fairly close quote from *Voyage de François Pyrard de Laual. Contenant sa Navigation aux Indies Orientales, Maldiues, Moluques, Bresil & Etc.*, 2 vols., 3rd rev. ed. (Paris: Chez Samvel Thibovst, 1619). This curious book was published in abridged form into German (during the eighteenth century) and English (Purchas, for instance), but the only complete translation was in Portuguese, published in Goa in 1858 and 1862, until Albert Gray translated it in its entirety into English towards the end of the nineteenth century: *The Voyage of François Pyrard of Laval to the East Indies, the Maldives, the Moluccas and Brazil*, translated into English from the third French edition of 1619 and edited, with notes, by Albert Gray, 2 vols. (London: Hakluyt Society, 1887). I will be quoting from this translation; Rumphius' passage is from 1:236–240.

François Pyrard (?-?1621) took part in a French attempt to find an alternate route to the Indies in 1601. Two French ships, the *Corbin* and the *Croissant*, met the superior fleets commanded by the well-known Dutch captains Heemskerk and Spilbergen, but otherwise they had little luck and were handicapped by slack discipline and indecision. After a lengthy stay at Madagascar, the French ships attempted to cross the Indian Ocean in the summer of 1602. In July the *Corbin* foundered in the Maldives. Only forty men made it to shore, and those who survived were captives of the Maldivians for a number of years. Pyrard stayed in the islands for four years, was treated well, and traveled the entire archipelago, a privilege accorded him because he took the trouble to learn the Maldivian language. Pyrard and only three other companions were freed by a Bengali invasion force, which took them to Chittagong. A couple of months later they arrived in Calicut but, after an eight-month stay, they were kidnapped by a Portuguese band outside the city and taken to Cochin as prisoners. Pyrard's account (1:429–433) of his stay in the "Tronco" prison in Cochin is chilling; to my knowledge the Dutch never had anything resembling such a place. He was released, and nearly ten years after he left France, he returned to Laval in 1611.

Pyrard published the first edition of his book in 1611, a much revised version in 1615, and the most complete one, which contained a Maldive vocabulary, in 1619. As is true of so many other travelers in the seventeenth century, Pyrard's is a strange story, but his book will always remain a unique firsthand account of what, for a long time, were islands unknown to the Western world.

The Arabs, Portuguese, Dutch, and especially the Bengali came to the Maldives to buy cowries, which were called *boli* in the Maldives, while the Dutch and English copied "cowry" from Hindi *kauri*. The Portuguese simply called them *buzios*. A surprising fact is that the Bengali were not the only people to hoard cowries in large quantities. These shells were also much prized as currency on the west coast of Africa, particularly in Ouida and Guinea on that infamous gold and slave coast. During the eighteenth century the VOC traded in slaves and in cowries to buy the slaves. The English and the French purchased cowries from the Dutch in Amsterdam. A rare account of this kind of transaction was published in the middle of the eighteenth century by an anonymous Dutchman. He had left Holland in 1747, "desirous of visiting several parts of the East Indies," but he never got that far. After his ship rounded the Cape, it had to be abandoned during a storm. After five days at sea, the survivors reached the island of "St. Maurice" (Mauritius?), where they were forced to stay for six months until a passing ship took some of them, including the narrator, to Ceylon. He stayed there for two years, "knowing that the Island would furnish the best of Entertainment for my Curiosity." Because he was there that long, the Dutchman discovered that his countrymen traded with the Maldives for cowries (p. 19). He goes on to state that the shells were a "great Currency on the coast of *Africa*, particularly *Guinea*, where the Negroes value them as much as Gold and Silver, and call them *Bougies*" (p. 20), and that "the Company" supplied Europe with these shells. He then briefly elaborates:

"Formerly Twelve Thousand Weight of these *Cowries* would purchase a Cargo of five or six Hundred Negroes; but those lucrative Times are now no more; and the Negroes now set such a Value on their Countrymen, that there is no such thing as having a Cargo under twelve or fourteen Tuns of *Cowries*.

"As payments in this Kind of Specie are attended with some Intricacy, the Negroes, though so simple as to sell one another for Shells, have contrived a Kind of Copper Vessel, holding exactly a hundred and eight Pounds, which is a great Dispatch to Business. However, the *Maldives* must not be thought the only Place which affords these Shells; they are also found in the *Philippine* Islands, but they don't come up to the *Maldavian*, either in Colour or Clearness. The chief *European* Market for these Shells is at *Amsterdam*, where are spacious Warehouses of them, the *French* and *English* Merchants buying them up to send to *Africa*."

This curious document of only twenty-three pages was translated into English and published in London in 1754, as *A Voyage to the Island of Ceylon & Etc.*, "Written by a Dutch Gentleman." My quotes are from this translation.

9. This "Fourth Breastplate" was also used as money, and is *Cypraea annulus* L.

10. *Cypraea cylindrica* (Born).
11. Probably *Cypraea errones* L.
12. This is *Cypraea teres* Gmelin; specimens with "four black eyes" probably belong to the species *Cypraea felina* (Gmelin).
13. "Landscape" is one of those words that demonstrate the considerable influence of Dutch on the English language. *Landschap* is a pure Dutch word that came into English during the seventeenth century as a technical term in painting. Ben Jonson spells it the Dutch way, "*Landtschap*," in his *Masque Blackness* from 1605.
14. *Cypraea isabella* L.
15. *Cypraea scurra* Gmelin.
16. The "Kernels from Nussatello" (which are represented by the figured specimen) are *Cypraea nucleus* L. The plainer ones could be *Cypraea staphylea* (L.), while the third species could be *Cypraea limacina* (Lamarck).
17. This is the first time Rumphius uses the verb "to enamel" ("*geëmalieert*"), which is the common word for us. Up till now he always used "*brand-schilderen*," which argues for a prose style which, like many of his contemporaries in Holland, preferred to use Dutch original compounds over foreign, imported vocabulary.
18. These are the three small islands (now called Pulau Tiga) off Ambon's southwestern coast of the Hitu peninsula. They are called Ela, Hatala, and Lain (or Nussatello).
19. *Cypraea cicercula* L.
20. *Cypraea globulus* L.
21. Rumphius' Latin and Dutch names mean the same thing, and the Latin was adopted by Linnaeus, who named this species: *Cypraea asellus* L. See also n. 7 of ch. 33 in book 1.
22. This could be the rather scarce species *Cypraea eburnea* Barnes or possibly *Calpurnus lacteus* Lamarck.
23. This is difficult to identify. It could be *Cypraea kieneri* Hidalgo, *Cypraea hirundo* L., *Cypraea ursellus* (Gmelin), or even smaller specimens of *Cypraea stolida* (L.).
24. This is *Trivia oryza* (Lamarck). The larger kind probably is *Cypraea childreni* (Gray).
25. This is an immature specimen, probably of *Cypraea tigris* L.
26. Although Schijnvoet says that this species is from the Caribbean, it is probably *Cypraea grayana* Schilder or *Cypraea histrio* Gmelin.
27. Schijnvoet says that this species is from Cartagena; hence it must be the well-known West Indian species *Cypraea mus* L.

Chapter 25. Cylindri. Rolls.

1. This phrase indicates that Rumphius relied mostly on touch when he examined these shells.
2. This translates as the "Porphyry Cylinder." Strack: *Oliva miniacea* Röding.
3. This "Black Cylinder" is *Oliva vidua* Röding.
4. "Honimoa" is Saparua. Tiouw Bay to the west of the island's capital also called Saparua.
5. This "Third Cylinder" is *Oliva reticulata* Röding.
6. This "Fourth Cylinder" is probably *Oliva vidua* Röding, a very variable species.
7. This means "Obsequies." Depicted is the kind of seventeenth-century funeral procession Rumphius had in mind.

8. The "Fifth Cylinder" is *Oliva tricolor* Lamarck.
9. The "Sixth Cylinder" might be *Oliva reticulata* Röding.

10. The kind with the violet mouth must be *Oliva caerulea* Röding.

11. This "Seventh Cylinder" is possibly *Oliva elegans* Lamarck, but there are other species with similar color patterns.

12. "Camelot" was a kind of cloth. See n. 6 of ch. 12 in this book.

13. The "Eighth Cylinder" is *Oliva annulata* Gmelin. The smaller kind could be *Oliva tessellata* Lamarck.

14. The "Ninth Cylinder" is *Oliva oliva* L.

15. The illustrated specimen is probably *Oliva carneola* Gmelin.

16. Jasper (*jespis* in Latin) is an opaque quartz that is most often green.

17. Might be *Oliva carneola* Gmelin as well.

18. The Uliasser Islands.

19. "Honimoa" is Saparua.

20. There are many coral reefs off the southeastern coast of Ceram. Guli Guli is a town almost on the eastern tip of the island, with Kefing, a small island, not too far to the east of that point.

21. These are the "Midas Ears" Rumphius first mentioned in Chapter 21, no. 24: *Ellobium aurismidae* (L.) and the smaller one is *Ellobium aurisjudae* (L.).

22. This must be *Agaronia acuminata* (Lamarck), a West African species.

Chapter 26. Conchae Univalviae.

1. *Lopas* and *lepas* were ancient names, straight from Greek, for a snail that clung to rocks. In English better known as limpets. *Patella* referred to a little dish, pan, or plate. "*Sabla*" is probably *sebelah*, which means one of a pair, a side, a portion, even "flat."

2. This is *Cellana testudinaria* (L.).

3. *Patelloida saccharina* (L.), as shown by letter B. The species shown as letter C is *Siphonaria laciniosa* (L.).

4. The illustration by D on plate XL is not clear enough, but it might be a young specimen of *Cellana testudinaria* (L.).

5. This gets confusing. The forty-volume *Grande Enciclopedia Portuguesa e Brasileira* lists a decapod crustacean with the name "Lambro," which they say is the genus *Lambrus* Leach, and is not all that common. Rumphius, however, is talking of limpets or *Patella*. These were commonly eaten (still are) in Portugal, and particularly the Azores, and are/were called *lapa* (plural: *lapas*), which is the gastropod mollusk the limpet. How the text arrived at "*Lambroa*" is a mystery to me, except that the translation, i.e., "Little Lamps," might hold a clue. The word in Portuguese would be *lámpara* (as in Spanish).

6. *Septaria porcellana* (L.). This same animal reoccurs under no. VI.

7. I do not know what "*bessi mattal*" might mean. Perhaps the first word is meant to be *besi*, which means "iron."

8. This means "Sea Ear." The Malay equivalents are not readily comprehensible, except that *telinga* means "ear" as well. The largest kind here is *Haliotis asinina* L. "*Sacatsjo*" is most likely *sekacha*, which, according to Wilkinson, is a shrub (*Clerodendrum deflexum*).

9. *Haliotis varia* L.

10. *Haliotis glabra* Gmelin.

11. This is probably *Gena planulata* Lamarck or *Gena varia* Adams.

12. The creatures discussed under nos. III and IV belong properly to book 1, because these are marine crustaceans, popularly known as barnacles, mostly the large family Balanidae, of the order Cirripedia. *Balanus* is classical Latin for "acorn" and, by extension, the common noun for the glans penis. Holthuis, in *M*, p. 69, identifies these barnacles as *Balanus tintinnabulum* L.

13. This refers to a large sloop with one or more masts.

14. "*Balatsan*" (also spelled "*bolatsjang*" by Rumphius), now known as *belacan* (a fish paste), was explained by Rumphius in book 1, ch. 26, as a brown paste made from crushed and pickled little shrimp.

15. "*Gindi*" is not a form normally found in Malay, but I am wondering if Rumphius had *gindi* in mind (the Anglo-Indian spelling was *gindy*), which is a word in Telugu, referring to what Yule describes as "a vessel resembling a coffee-pot without a handle." Yule also notes that this word would be *kindi* in the Malay language, while Wilkinson gives *kendi*. Hence the Malay phrase would translate as "stone gindy" and "sea gindy."

16. "*Verruca*" meant an excrescence on precious stones or a wart on the human body; the second element probably from *testudineatus*, an adjective that means "arched" or "vaulted." The latter might also refer to *testudineus*, an adjective meaning "pertaining to a turtle," from *testudo*, which meant both a turtle and an arch or vault. Holthuis identifies this as *Chelonibia testudinaria* L. and is depicted on plate XL, letter K.

17. *Kutu* means "louse" or "flea" in Malay, and "*totruga*" must be the Spanish and Portuguese "*tortuga*," or "turtle."

18. "*Lassapinjo*" is *Nusa pinjo*, an island twenty-five miles south of Ambon.

19. The Latin has a misprint, which in a chapter of much confusion, is not surprising. It should read: "*opercula callosum*," or "hardened lids." From *callus*, which referred to the hardened, thick skin or flesh of animals or plants.

20. This is *Scutus unguis* (L.).

21. This is *Umbraculum umbraculum* (Lightfoot).

22. This is the inner shell of *Dolabella auricularia* (Lightfoot), which Rumphius described before in book 1, ch. 33, no. II, under the name "*Limax marina verrucosa.*"

23. *Septaria porcellana* (L.). This is the "River *Lopas*" under no. I in this chapter.

24. This presented another problem. The original has "*Orlamjes*," which are further described as "*gelyk de Orlamsche mitjes der Matroozen.*" Now the old noun *oorlam* in Dutch has a drink of gin as its main meaning, though it could also mean an old Indies hand, i.e., a veteran of the tropics, or a veteran sailor. The index to the superb volumes of the "Linschoten Vereniging" (the Dutch equivalent of the Hakluyt Society) thankfully confirmed my guess: *oorlam* could also refer to a sailor's knit cap (*Tresoor der Zee-en Landreizen. Beredeneerd register op de werken der Linschoten-Vereniging*, ed. C. G. M. van Romburgh and C. E. Warnsinck-Delprat ['-Gravenhage: Martinus Nijhoff, 1957]). The latter is the only thing that makes sense in this context when considering the illustrations by P and Q on plate XL. These are *Cheilea equestris* (L.).

25. According to Kabat this is *Sinum haliotoideum* (L.).

26. This translates: "*noch niet in den kap te zyn.*"

Chapter 27. Solen. Cappang.

1. The original text uses several times the phrase: "*dewyl het Dier aldaar zyn aangroeijen heeft.*"

2. "*Solen*" is a noun supplied by Pliny, who merely lists them among other mollusks in book 32 (paragraph 151) but who mentions in books 10 (paragraph 192) and 11 (paragraph 139) that these clams can hear. Although not related to the species discussed in this chapter, they are more familiar to us as razor clams (Solenidae family). The "Sand Solen" is *Kuphus polythalamia* (L.), a bivalve species of the family Teredinidae.

3. *Kapang* is the Malay word for this knife-shaped mollusk that attacks wood in water. *Besar* means "large."

4. The Bight of "Amaheij" (Amahai) is, say, northeast of the Uliasser island Saparua, towards the middle section of Ceram.

5. "Kaijeli" (Kayeli) Bay is on the eastern coast of Buru, the largest island to the west of Ceram.

6. *Mangi-mangi* trees are mangroves.

7. See ch. 55 in book 3.

8. This "Wood-gathering Solen" is one or more of the species within the family Teredinidae. Letter F just shows an empty tube, and letter G seems too thick and heavy to belong to the Teredinidae; it is most likely a species of the gasteropod family Vermetidae.

9. *Ayam* means "fowl," specifically "chicken." "*Purrut*" is most likely *perut*, a word that refers to the stomach itself and to its contents, also the womb. In this context, for instance, *perut uda* is the large intestine, and *tali perut* (literally, "stomach rope") is a phrase for "viscera."

10. This "Snaky Solen" is the gasteropod species *Siliquaria anguina* (L.).

11. The Malay means the same thing: "Snake Shell."

12. The "toothed little Miter" is the mollusk's operculum.

13. This first kind of "Elephant Tusks" is *Dentalium elephantium* L.

14. *Laut* means "sea," and "*tando*" most likely should be *tandok*, the common Malay noun for "horn"; hence the Malay called these shells "Sea Horns."

15. The second kind is *Dentalium aprinum* (L.).

16. "Spanish green" is verdigris.

17. This is probably *Dendropoma maxima* (Sowerby) or *Serpulorbis grandis* Gray.

18. "*Kinkgut*" for the expressive "*konkeldarm.*" *Colum* is the same as *colon*, or the large intestine.

19. Martens: *Vermetus (Vermicularia) lumbricalis* L.

20. Martens: *Vermetus semisurrectus* Bivona, from the Mediterranean.

21. Martens: *Vermetus (Thylacodes) protensus* Gmelin.

22. Martens: *Thylacodes rumphi* Mörch.

23. *Dentalium entale* (L.), a European species.

24. Possibly *Dentalium octangulatum* Donovan.

25. This is the rather bizarre bivalve *Brechites penis* (L.).

Chapter 28. Chama Aspera. Bia Garu.

1. The noun was in use by the last quarter of the seventeenth century. It originated with Aristotle (*History of Animals*, book 4, ch. 4, section 2).

2. These invertebrate animals are Bivalvia, also called Pelecypoda or the Lamellibranchia, a class of the phylum Mollusca. One could also call them bivalve mollusks, and they include animals with such confusing, because interchangeable, popular names as oysters, clams, mussels, and cockles. They all share certain characteristics such as a generally sedentary mode of life, a shell consisting of two valves that can lock tight together and which are connected with a ligament. They range in size from minuscule to gigantic, and live in both freshwater and the sea, although most species live in the sea. Rumphius' divisions do not correspond to modern nomenclature, of course, but some names are still used. *Chama* is *chema*, from the Greek for a "gaping mussel"; the first kind could be translated as "rough mussel" (from *asper* = rough), and the second "smooth mussel" (from *levis* [not "*laevis*"] = smooth, used as the opposite of *asper*). *Pecten* (from the word for "comb") referred to a sea scallop, *tellinae* (from the Greek *telline*) for a kind of shellfish, *musculi* for the "sea mussel" and *ostrea* for oysters. These terms are still in use: the Pecinoidea, Chamoidea, Tellinoidea, and Ostreoidea are superfamilies within the class Bivalvia. *Musculus* is still used for a genus of small mussel-like bivalves.

3. This means literally "scaly mussel," from *squama*, for the scale of a fish or snake. For "Clack Dishes," see ch. 23 of this book, n. 6.

4. *Garu* can mean a "harrow" or "rake," so one could translate Rumphius' name for this mollusk as "Harrow Shell."

5. "*Chama decumanae*" means "large mussel," deriving from a metonymy from *decima* or *decuma*, which referred to a tenth part, a tithe; the larger animals are then classified by Rumphius as "belonging to the sea" (from *pelagius*), while the smaller variety are "coast dwellers" (*littorales*). These belong to the family Tridacnidae. Pliny provides the funny anecdote that the name *tridacna*, from the Greek *tris* for "three" and *daknō* "I bite," came about because someone jested that he would like to have the oysters that reputedly came from the "Indian Sea" because they were so big that they needed three bites to be consumed (book 32, section 21, paragraph 63).

6. "Oblong" for "*uit den ronden langwerpig.*"

7. The "female" is *Tridacna squamosa* Lamarck, and the "male" *Tridacna maxima* (Röding).

8. "*Gynglymum*" is strange usage. Perhaps Rumphius derived this from the Greek *γάγγλιον*, a tumor on a tendon or sinew, *ganglia* in Latin. If this surmise is correct, then Rumphius should have spelled it "*ganglium.*" In any case, he is clearly not using it in the modern sense of a nerve center. This refers to the articulating hinge teeth.

9. From the Latin verb *tendo*, "to stretch," hence our noun "*tendon.*"

10. *Spondylus* actually referred to both a vertebra or joint of the spine, and the muscle of an oyster or other bivalve.

11. This is the real *Tridacna gigas* (L.) or giant clam.

12. *Shallop* to translate *Chaloep*, a spelling Rumphius modeled on the French word *chaloupe*, although, at least in English, it was most likely derived from the Dutch word *sloep*. I am assuming that this refers to the larger variety, with one or more masts and carrying fore-and-aft or lug sails.

13. "Parring" is the *parang*, or Indonesian machete.

14. This is probably *Tridacna crocea* (Lamarck).

15. "*Bonoa*" should be the island called Boano, which lies to the west of Ceram's Howamohel peninsula. It is separated from Ceram by the Straits of Boano. The soil cannot be tilled; hence the inhospitable environment forced the population to work elsewhere. They were notorious pirates from the very beginning of the Dutch presence in those waters, and they constantly made a great deal of trouble for the VOC.

16. The Dutch "*theeback*" is a dated metonymy for drinking tea. It comes from *bak* for a tray, on which were placed the tea service and the goodies that came with it. The *WNT* gives a quote that states that the usual time was between 3:00 and 4:00 in the afternoon. It seems that the food was not sweet (as it would be today) but more like hearty *tapas!*

17. Rumphius described these *chamites*, or the mollusk mesticas, in ch. 64 of book 3. Martens says that these are calcareous concretions inside the hinge ligament of the large *Tridacna*, usually only in the older ones and even so only in one out of ten. The Malay means "*mestika* of the harrow shell," and was printed as one word, which must be wrong.

18. Rumphius prints this word as if it were a foreign one, but it was already in use in the Middle Ages.

19. This translates as "Finger Mestika."

20. Rumphius spelled this as "*Calbahaar*" in his chapter on coral in his *Herbal* (ch. 1 of book 12). He distinguished red, white, and black coral.

21. Rumphius took this term from Pliny. Strangely enough the English equivalent "hub" is much later usage. The *OED* states that it did not enter normal use until the nineteenth century! Before that the noun used was "nave."

22. *Brani* or *berani* means brave, to dare, to venture something. These champions were called, in Dutch, *voorvechters* in Rumphius' day, and they did indeed challenge and fight the enemy in front of the main body of fighting men, like the "champions," such as Samson, in the Old Testament.

23. This island is what the Dutch called *Prinsen-eiland*, in Malay *Pulau Panaitan*. Between it and Sumatra, in the Sunda Straits, lies Krakatau Island.

24. This is Lampong Bay, a large bay that cuts into southern Sumatra. The "*Calappa laut*" is described by Rumphius in his *Herbal*, book 12, ch.8.

25. "Gelolo" is Gilolo, which is the other name for the northern Moluccan island of Halmaheira. This island, shaped like Celebes, lies to the east of Celebes, and the Alfurs ("Alphorese") used to live in its northern region.

26. The first word must be Ambonese, while the second element is supposed to refer to *bulan*, Malay for "moon." The Latin works; this would mean a little shield shaped like the moon. "Clypeolus" should be *clipeolum*, which derives from *clipeus*, which referred to the round, brass shield of a Roman soldier.

27. *Batu* is "stone," and *tikar* referred to a mat, including one for sleeping.

28. "Lussapinju" is "Lucapinho," a small group of islands also known as Nusa Penju or Nusapinyu, otherwise the Turtle Islands. This small archipelago is in the center of the Banda Sea, between Ambon and the Damar Islands.

29. "*Cururong*" is the "*Chama striata*" or "*Bia Corurong*," described below under III. See also nn. 40 and 41.

30. I believe Rumphius is referring to the phrase in Pliny (book 37, section 26, paragraph 180 in the Latin) "*Paeanitides*, quas quidam *gaenidas* vocant" (emphasis added), which are said (in Eichholz's translation) "to become pregnant and to give birth to another stone."

31. "*Nautilus crasso*" means "thick nautilus"; Rumphius is referring to the second chapter of this book.

32. Benzoin is, properly speaking, the resin of the *Styrax benzoin* tree, which was once cultivated almost exclusively in the Batak country in Sumatra and in the Brunei region of northern Borneo. In Malay and Javanese it was known as *manjan* or *minjan*, a term obviously related to what Rumphius also calls *minjac*. In English it was called "gum Benjamin." Benzoin was gathered from the tree when it was about six years old, the resin oozing from a transverse cut just below the lower branches.

The word itself is a corruption of the Arabic *luban jawi*, which means "incense of Java." This term does not refer to the dominant island of the East Indies but was Arabic for Sumatra. The *Styrax benzoin* tree does not grow in Java. Arabic *luban jawi* was corrupted by the Portuguese to *bensawi* and *benjoin* and was thus incorporated into other languages. The ancient Chinese imported benzoin resin from the Malay archipelago under the name of *nan-si hian*, saying they obtained it in a country there called *Po-se*. See Berthold Laufer, *Sino-Iranica*, pp. 464–467 and 468–487 (on Po-se as a Malay region).

33. My rendering of the substantive *kalis*, an interesting word in Dutch which originally meant a "vagabond," and came to mean, by extension, a poor soul. The word comes originally from the language of the Gypsies: *kales*, a plural of *kalo*, meaning "black," which they used when referring to themselves.

34. This is a case of wishful thinking. Apuleius does quote some lines from Ennius (the "father" of Roman poetry) in his *Apologia* (section 39). They derive from the surviving eleven lines from Ennius' disquisition on edible seafish, entitled *Hedyphagetica* (Good Things to Eat). The lines pertaining to Rumphius' quote are: "apriculum piscem scito primum esse Tarenti, / Surrenti tu elopem fac emas, glaucumque aput Cumas. / quid scarum praeterii cerebrum Iouis paene supremi." This can be literally translated as: "Know that the best fish is the *apriculus* at Tarentum, make sure that you buy *elops* at Sorrento, and a *glaucus* at Cumae. I almost forgot the *scarus*, which is as delicious as Jove's brain." (*Apriculus*, *elops*, *glaucus*, and *scarus* are kinds of fishes which have not been definitively identified.) I used the Butler edition in a German reprint: *Apulei Apologia*, ed. H. E. Butler and A. S. Owen (orig. 1914; reprint ed. Hildesheim: Olms Verlag, 1967).

Rumphius, who might have been quoting from memory, got the second line wrong. There is no "*chemas*" in the original; hence the word was *not* in the poetry of this "old Roman" called Ennius. By supplying it incorrectly, Rumphius comes up with some lines that make less sense than the original: "Know that the *apriculus* is the best fish of Tarentum, [and that] the clams are from Sorrento. The *glaucus* is from Cumae; but why did I almost forget the *scarus* [as delicious as] Jupiter's brain?" The mistake is one thing, but what is more interesting is to think of this blind man, in the eastern reaches of the archipelago, either having access only to a battered copy or quoting from memory some scarce extant lines from an obscure Roman poet! As I have mentioned elsewhere, Dutch colonialism was not an exercise in barbarism, at least in the seventeenth century.

35. "*Krios*" is yet another word for "mussel."

36. This gets to be monotonous: "*Pelorides*" are "big mussels;" from *pelōr*, which means "monstrously big."

37. From *basileus*, Greek for "king," hence also "big mussels."

38. The subsequent cascade of names does not warrant annotation. Modern nomenclature is different, and even the contemporary references are confusing. Rumphius did get most of it from Pliny, book 32, but chapter 53 (paragraph 147 of the Latin): it reads "cancrorum genera, chemae striatae, chemae leves, chemae peloridum generis, varietate distantes et rotunditate, chemae glycymarides, quae sunt maiores quam pelorides, coluthia sive coryphia, concharum genera. . . ." One can tell that most of the names Rumphius gives are present in Latin. The Latin phrase he adds, of which the first part is in the above quoted passage, reads: "the genus of crabs [should be] the genus of clams." "*Chamae glycymerides*" stands for "sweet clams."

39. "Rough and Dulled Clams" must be *Tridacna maxima* (Röding).

40. "Furrowed clams" are *Hippopus hippopus* (L.). In other words, Linnaeus used Rumphius' Dutch name.

41. The Malay means "Cabin Clams," since the second word contains the element for an enclosed space, *kurang*, for instance a cabin on a ship.

42. From *testa* for "potsherds," but also the shells of clams. It is *Gafrarium tumidum* Röding.

43. This was a small coin, worth very little.

44. Mamala is a village on the coast of northeastern Hitu, east of Hitulama. It is known today for ritual combats at the end of Ramadan, when participants fight each other with brooms made from the fiber of coconuts (*sapulidi*).

45. "Comb Sherds" says the Latin. This could be *Gafrarium pectinatum* (L.).

46. The problem is with "cartissae," which should be "*cardissae.*" My surmise is that it derives from the Greek *kardia* for "heart." *Hati* is the Malay word for "heart." Voskuil identifies the figured specimen (letter E) as *Corculum cardissa* (L.).

47. "Raised edges" for "*aan sommige opgeworpen.*" Martens specifies these as *Cardium roseum* Chemnitz.

48. *Carsium junoniae* Lamarck, according to Martens.

49. Nusa Laut is one of three islands near Ambon, which together are called the Uliasser Islands.

50. Rumphius did not call this shell thus because of the navigational instrument, but strictly from *quadrans*, Latin for "a fourth." This is *Hecuba scortum* (L.).

51. "*Bia cattam*" is *bia ketam*, i.e., "Crab Shells."

52. For "Giant's Ears," see ch. 6 of this second book.

53. "*Toncat*" is *tongkat*, Malay for a strut, something that props up, a crutch.

54. The Lembe(h) Straits are indeed off the northernmost tip of Minahasa, on Celebes' topmost, long peninsula.

55. Robert Padtbrugge was born in Paris in either 1637 or 1638. Obtained his degree as a physician from Leiden University in 1663, and thereafter started his career in the VOC. Was in Persia, Ceylon, and was governor of the Moluccas from 1676 to 1680. He was governor of Ambon from 1682 to 1687, when both Rumphius and Valentijn were living there.

56. The Dutch has "*schulpen.*"

57. This must be Petronella Oortmans-de la Court (1624–1707), the woman who assembled an impressive collection of curiosities in Amsterdam.

58. Most likely Levinus Vincent (1658–1727), who had one of the most admired collections of his time. He became a member of the Royal Society in London in 1715 and published a catalogue of his acquisitions between 1706 and 1715.

59. "Dr. 's Gravezande" is more correctly Cornelis Gravensande (1631–1691), a physician and municipal official in Delft. He lectured on anatomy and medical science and owned a collection of shells. *Hendrik Engel's Alphabetical List* made a mistake owing to similar names. It lists this Gravensande as the man under no. 565 (p. 101). That is impossible, since that man, Willem Jacob 's-Gravenzande, died in 1742, and Schynvoet clearly states (p. 133) that he is referring to "*wylen* [i.e., deceased] *de* Hr. Dr. 's-Gravezande *tot* Delft."

Chapter 29. Chama Montana Sive Noachina. Father Noah Shells.

1. "*Chama*" is *chema*, which referred to a large, gaping mussel, while *decumana* meant "huge" (see n. 5 of ch. 28). The title means something like "Mountain Chema or Noah Chema."

2. Although today we might consider Ambon part of the Moluccas, in Rumphius' day this was

not the case. The word comes from "Molucco," which was another term for Ternate, and the original "Moluccas" were, in fact, just the string of small islands off the western coast of Halmaheira, from Ternate to, and including, Batjan.

3. As Valentijn describes it (2, part 1, p. 97), the village of Hitu Lama is situated "at the foot of very sharp mountains, which begin just behind this village, and very near it, on the short road that leads to the other side of that region, or to Three-Houses [this is Rumahtiga], there is a place called the Pincushion, from the sharpness of these rocks, and which one had to cross on horseback or on foot, and many a horse's cropper broke here due to the steep incline, which later was removed at the order of the Merchant *Joannes Moris* [he came after Rumphius], while he was Head of the Hitu Coast. It took a great deal of trouble to explode all those rocks by means of holes they had bored and filled with gunpowder; nor had the natives, who are lazy by nature, any desire to do this, but he [i.e., Moris] made them understand that no one was going to profit more from this than they, since they used the road constantly, and with great difficulty, when they had to carry Chiefs, Ministers, and other people, at the risk of their lives, on those rocks, which were as sharp as needles, with a litter on their shoulders and a person therein, whilst otherwise (and which time bore out) they could use it with the greatest of ease and without any danger. This convinced them to begin the job, which was completed very shortly, for the general benefit of all."

4. This term, meaning "Eagle Stones," derives from Pliny. Rumphius discusses such stones in several places in the third book, for instance, ch. 60.

5. This should be "geodes" and is discussed in ch. 34, for instance, of the third book.

6. The *ombria*, or "Rain Stone," also derives from Pliny (book 37, paragraph 176), and was said by the Roman to fall out of the sky.

7. Rumphius uses *matrix* in the original sense of a place of propagation, origin, or cause. In classical Latin this word was never applied to women.

8. In the original "*aardig*," which is a common adjective meaning "nice" and so on, and as an adverb it can indicate degree ("fairly well," etc.). This does not fit here. It turns out that in the seventeenth century there was a rare usage that took the radical (*aard*, i.e., "soil") literally, so that the present usage means "of the earth"!

9. The *orang badjo* were the ubiquitous sea gypsies who frequented the coasts of Borneo and Celebes.

10. These two lines from Ovid's *Metamorphoses* (1:299-300) mean, in Frank Justus Miller's straightforward translation: "And where but now [i.e., before] the slender goats had browsed, the ugly sea-calves [i.e., seals] rested."

11. This refers to one of many religious disputes of that time. The notion of human beings having been on earth prior to Adam—based on a quibble in Paul's *Epistle to the Romans* (12-14) where he said that there was sin in the world before the law had been given to Adam—clashed with orthodoxy. A seventeenth-century sect was founded on the beliefs expressed in a book, entitled *Prae-Adamitae*, published in 1655, by Isaac La Peyrère. It was said at the time that La Peyrère was the second husband of Robertus Padtbrugge's mother. Valentijn, at least, reports this, and also gossips that Padtbrugge, who was then the governor of Ambon, adhered to these views. He goes on to tell that shortly before he arrived on Ambon (which was 1 May 1686), Rumphius and Padtbrugge had been engaged in a violent argument about this matter "on Soya mountain," and that thereafter they could no longer remain friends. A descendant of Padtbrugge contends, however, that La Peyrère was not the governor's stepfather, but that another Frenchman, Elias Poirier, was. For Valentijn, see Stellwagen's edition of just his descriptions of his voyages, extracted from his larger work: *Van en naar Indië. Valentijns 1ste en 2de Uit- en Thuisreis*, ed. A. W. Stellwagen ('s-Gravenhage: Stemberg, 1881), p. 69. For Padtbrugge, see H. J. Pabbruwe, *Dr. Robertus Padtbrugge (Parijs 1637-Amersfoort 1703), dienaar van de Verenigde Oost-Indische Compagnie, en zijn familie* (Kloosterzande: Duerinck, 1995), p. 11.

12. This must be, of course, Auvergne, a region in southern France, south of the Loire, and west of Lyon.

13. The Westerwald is part of the Rheinische Schiefergebirge, an upland plateau through which the Rhine cuts a deep valley, between Coblenz and Bingen. This is the same general area that Rumphius remembers in ch. 80 in book 3.

14. What used to be known as Fu-kien province on China's eastern coast, across from, say, Taiwan.

15. "*Gumuto*" is *gemuto*, the fiber, resembling horsehair, of the sugar palm (*Arenga saccharifera*), once used to make rope and thatch.

16. My guess is that Rumphius has Yü in mind, one of the three famous rulers (Yao, Shun, Yü) who presumably came after the Yellow Emperor (Huang Ti) and who were the first kings mentioned in the *Su Ching*, according to legend China's oldest extant written work, and who were revered by Confucius as models. Yü was the monarch who supposedly found a way to drain the waters left be-

hind by a great flood. This legendary king is also supposed to have founded the first ruling dynasty, the Hsia dynasty, presumably in the second millennium B.C. I would guess that the "Jao" with whom Rumphius equates Yü is probably the aforementioned Yao.

17. Jacob Hustaerdt was the sixteenth governor of Ambon, from 1658 to 1662. Thereafter he became governor of Coromandel, then of Ceylon, member of the Council of the Indies, and died in 1665 in Batavia.

18. This was the headquarters of the VOC, on no. 24 on the Oude Hoogstraat. A complex of buildings, the compound expanded between 1603 and 1643. Towards the end of the nineteenth century it became neglected, part of which was used by the internal revenue services.

19. Athanasius Kircher (1601–1680); the work mentioned is *Mundus subterraneus*, published in two volumes in Amsterdam in 1678. See n. 9 of ch. 77 in book 3.

20. In the original "Nederlantsche Outheeden." This is *Antiquitates Belgicae, of Nederlandsche Oudtheden* (Amsterdam: Jacob Royen, 1700); I am citing from the fifth impression of 1733. What Schynvoet is referring to, here as well as in ch. 66 (n. 8) in Book 3, is most likely pp. 19–27, which discusses the presence of sedimentary shells. The book was written, according to the preface, for it was otherwise published anonymously, by R. Verstegen. This was Richard Verstegen (1550–1640), a Flemish archaeologist, satirist, and polemicist, who studied at Oxford, removed himself to France because of his Catholic faith, and proceeded to write against the British queen from Paris. He was an ardent adherent of the Counter Reformation and wrote many satirical verses aimed at the Protestants.

21. Schynvoet had most likely the Englishman John Ray (1628–1705) in mind. Ray gained fame as both a naturalist and a preacher. He lectured in mathematics, Greek, and "humanity" in Trinity College, and was ordained at the end of 1660. He had to leave the college owing to a religious matter and was for the rest of his life dependent upon the generosity of his pupil, Francis Willughby. Those two traveled on the Continent between 1663 and 1666, collecting floral and faunal specimens. Ray's fame as a naturalist is lodged in his *Historia generalis plantarum* (1685), and his *Catalogus plantarum Angliae* (1670) seems to have been the basis for later floras of English plants. His immediate fame and lasting reputation derive from two theological disquisitions, however; the once well-known *The Wisdom of God manifested in the Works of the Creation* (1691)—which argues God's perfection in nature—and the *Miscellaneous Discourses concerning the Dissolution and Changes of the World* (1692). Schynvoet is referring to the last book, which contained three essays (based on earlier sermons), "The Primitive Chaos and Creation of the World," "The General Deluge, its Causes and Effects," and "The Dissolution of the World and Future Conflagrations."

Chapter 30. Chama Laevis.

1. These are discussed in ch. 33.

2. "*Laevis*" or *levis*, which means "smooth." The one described is *Meretrix meretrix* (L.), while the one shown on plate XLII, letter G, could be *Meretrix lusoria* (Röding), which is the mussel shell that is used for the Japanese game.

3. Rumphius read his Pliny carefully but to little avail, because this is once again a reference to Pliny's simple list of names in book 32 (paragraph 147), a listing that names items but does not further discuss them. "Glycymarides" means something like "Sweetness of the Sea."

4. The Leitimur peninsula pinches off the main island of Ambon at the appropriately named Paso, a narrow strip of land that separates Ambon Bay from Baguala Bay. Suli is a village on the northern shore of Baguala Bay, hence part of northeastern Hitu.

5. What is meant by "Tenember" is Tanimbar, Tenimbar, or Timor Laut Islands, about sixty-six of them southeast of the Banda Islands, and considered today the southernmost of the Moluccas. In Rumphius' day, these islands were populated by warlike peoples, who hunted for human heads and were said to be cannibals. The Dutch came to the Tanimbars in 1645 but found little reason to stay or fight the local warriors. They tried again in the nineteenth century. The Tanimbars are better known to botanists, for this is the only place where the Larat orchids grow.

6. I presume he is referring to the "*Hippi*" in ch. 10 of the first book. See also n. 4 of that chapter.

7. This translates something like "Sweet Indian Smooth Chama."

8. This "Mud-living Chama or the Croaker" is, according to Martens, *Cyrena coaxans* (Gmelin). But Strack, who collected many near Laha in 1990, says it is *Polymesoda coaxans* (Gmelin), and he also notes that the native population still use the Malay name for it: *Bia kodok* or "Frog Shell" (Strack, p. 57). They make a croaking noise when they open or close the valves of their shells and water or air suddenly rushes in.

9. This should be "*Paeanitis*" (which means "Apollo Stone," since *paean* was an appellation of Apollo as the healing deity), also called by Pliny "*gaeanis*" or the "earth stone" and of which he says that it "can become pregnant and give birth to another stone, and so is thought to relieve labour pains" (Loeb ed., vol. 10:311). It was considered a precious stone but is now unknown.

10. This "Striped Chama" is probably *Mactra grandis* Gmelin.

11. This is Baguala Bay (see n. 4), not Ambon Bay.

12. This is *Lioconcha castrensis* (L.). Rumphius' nice image of an army camp is kept in this name, since *castrensis* is an adjective meaning "pertaining to the camp." The baser variety is probably *Circe scripta* (L.). The third kind from Bouton is more difficult to identify; it might be *Glycymeris reevei* (Mayer). The smaller kind could be *Lioconcha ornata* (Lamarck).

13. Discussed in the next ch., no. 31.

14. This "Circled Chama" is *Anomalocardia producta* Kuroda & Habe.

15. Assuming that "Verwert" is *verward.*

16. These are the animal's siphons, with which it sucks in water in order to obtain oxygen and food particles after it has been filtered through the gills.

17. The "Oblong Letter Chama" (*littera* means a letter of the alphabet) is *Tapes litteratus* (L.).

18. The Malay means the same thing as the Latin, because in Dutch *letter* is cognate with the English noun. There are other examples of purely Dutch words that went unchanged into Malay. Most of them pertain to technical terms, or to things that didn't exist previously in the Archipelago.

19. These superb shells, of which the Dutch name translates the Malay exactly, are *Tapes litteratus* (L.), which Chemnitz named *nocturna.*

20. This is another form of *Tapes litteratus* (L.)

21. This must be *Paphia textile* (Gmelin). For "tabbying," see n. 18 of ch. 21.

22. This "Round Letter Chama" is *Circe scripta* (L.).

23. This "Combed Chama" is *Codakia punctata* (L.).

24. *Scobina* is Latin for "rasp." This "Rasp Chama" is *Scutarcopagia scobinata* (L.), but the shell depicted as letter E on plate 43 does not provide a good likeness of this species.

25. *Favus* means "honeycomb" in Latin. This must be *Fimbria fimbriata* (L.), but the drawing by letter F on plate 42 looks more like *Fimbria souverbii* (Reeve), a species from the southwestern Pacific.

26. The "Tiger's Tongue" is *Periglypta purpura* (L.).

27. This species, and not the former one, was named *Codakia tigerina* by Linnaeus.

28. "*Remies*" or *remis* is a general Malay name for small bivalve shellfish. This is one of many names that have stuck, for the modern scientific name is *Tellina remies* L.

29. See next chapter.

30. This small island off the east coast of Ambon was called such (*Pulu Pombo*), because of a large population of black and white doves.

31. See ch. 28, under nos. IV and V. The little shell depicted by letter O on plate 43 defies identification. According to Martens, it could be a species of *Corbicula*, which is a freshwater mussel.

32. This could be *Katelysia hiantina* (Lamarck). "*Tude baya*" is Macassarese for a shell that is covered with what look like little hairs. *Tude*, in the same language, means a shell or a vagina.

33. Letter L could be *Sunetta contempta* Smith, while figure M could be *Sunetta truncata* (Deshayes). Both illustrations are on plate 43.

34. The Sula Islands (Mangole and Taliabu) are to the east of Celebes in the Moluccan Sea.

35. This is *Tellina gargadia* (L.).

Chapter 31. Pectines & Pectunculi.

1. These are the scallops, although this chapter discusses other shells as well. Among various other meanings, *pecten* was Latin for "comb."

2. "St. Jacob's Shell" is a reference to the famous scallop symbol of St. James (Santiago), the patron saint of pilgrims, associated with the Spanish town of Santiago de Compostela. The scallop *Pecten jacobaeus* became a widely known symbol of pilgrimage—Christian pilgrimage, that is—hence of no consequence in an Islamic country.

3. "*Pectunculi*" are really small or immature *pecten.*

4. Martens states that this "First or Common Pecten" is ubiquitous in the Indian Ocean. Dijkstra identifies it as *Decatopecten radula* (L.).

5. *Sissir* (*Sisir*) can mean a comb, a harrow, a rake, a bunch of bananas (in Dutch, *kam*, which also means "comb"), or any toothed instrument.

6. *Terbang* means "to fly."

7. Scallops can indeed jump out of the water, something observed as early as Aristotle (book 4, ch. 4, section 4) and Pliny. In book 9, Pliny states (paragraph 103) that the "scallop leaps and soars out of the water" and that it makes a "whizzing sound" (from *strideo;* book 11, paragraph 267) when they do.

8. The "modern Greek" *chtenia* means "combs," but the word which Rumphius asserts is classical Greek (it should at least have an "h," i.e., "*Chtenes*") does not exist.

9. Dijkstra: *Gloripallium pallium* (L.).

10. For this strange usage, see n. 8 of ch. 28.

11. This "Thin Pecten" is, according to Dijkstra, *Laevichlamys squamosus* (Gmelin).

12. Dijkstra thinks this might be *Mimachlamys albolineata* (Sowerby).

13. Dijkstra: *Mimachlamys lentiginosa* (Reeve).

14. Dijkstra thinks this might be *Mimachlamys senatoria* (Gmelin). "Red lead" is a red oxide of lead, also known as "minium," once used as a pigment and later as a base paint. The original had "*minie*," but it is now more commonly spelled *menie*. In the older usage, this was a bright red color; more modern fabrications are a bright orange.

15. A *radula* was a scraping iron of the Romans. This is *Lima lima vulgaris* (Link).

16. Martens agrees with Rumphius that this "Common Pectunculus" is easily found everywhere in the Indian Ocean; he calls it *Cardium rugosum* (Lamarck), which is *Trachycardium flavum* (L.). But it could also be another species within the genus *Trachycardium*.

17. This "Scraper Shell" is correctly translated, because *cucur* (*kukur*) means "to rasp" or to scrape a surface with a rough implement.

18. Al Cohen conjectures that this is probably *han*, a shell which B. E. Read, in his article "Turtle and Shellfish Drugs," lists as no. 229, and glosses as an Arcidae, though Rumphius' animal is a member of the Cardiidae.

19. "*Brigigoins*" is most likely the Portuguese word *berbigão* (plural *berbigões*), a name for a local Iberian cockle variety, *Cerastoderma edule* (L.).

20. In the original "*houwe*," which should be *hauw*, the Latin *siliqua*, although the etymology of *hauw* or *houw* is uncertain. This means the pod or husk of a pulse.

21. The Latin means ("*ajunt*" more commonly *aiunt*) "to arouse Venus, as they say."

22. *Mera* is "red" in Malay. This is *Fragum unedo* (L.).

23. "*Nussapinjo*" is Nusa Penyu, a small archipelago in the center of the Banda Sea.

24. This "White Strawberry" is *Fragum fragum* (L.).

25. *Lunulicardia hemicardia* (L.).

26. "Cove" for the original *sak* (or *zak*), a rare meaning of which is a sack-like bay which is removed from the sea and which is for the most part surrounded by land; an inner bay.

27. This "Virginal Pecten" is *Anadara antiquata* (L.).

28. "*Anadara*" is most likely *anak dara*, which is the same as *perawan*, meaning maiden, young girl, virgin, unmarried young woman. The word *anadara* is still in use.

29. See below under no. XII, where it is said that this protrusion permits the scallop to jump. Rumphius' "little hand" is the animal's foot with which bivalves bury themselves.

30. Rumphius is referring, of course, to menstrual blood. Sewel also gives "Women's flowers" for menstrual period.

31. See ch. 65 in book 3.

32. The "Pecten Full of Grains" is *Anadara granosa* (L.).

33. This "Pecten That Lives among Rocks" is *Arca ventricosa* (Lamarck).

34. The Malay means "Rock Shells."

35. "*Mactra*" is a Roman spelling of the Greek noun for "trough."

36. *Anomalocardia squamosa* (L.).

37. For example, *Timoclea marica* (L.).

38. This "Rounded Pecten" is *Fulvia aperta* (Bruguière).

39. This word is pronounced almost identically to "phyllo," as in "phyllo pastry," the thin, flaky pastry shell used in such delectables as baklava and spanokopita. Rumphius' brief description seems not to exclude the possibility of a similarity.

40. This is J. Jonston, *Historia naturalis*, published in Amsterdam in 1657; Dutch edition in 1660.

41. Dijkstra: *Decatopecten plica* (L.).

42. "*Ikan Sabandar*" is a fish of the Siganidae family: *Siganus oramin* (Schneider). See W. H. Schuster and Rustami Djajadiredja, *Local Names of Indonesian Fishes* (Bandung: W. van Hoeve, 1952), p. 134.

43. This shell from the West Indies might be *Arca noae* (L.), a species from the Mediterranean.

Chapter 32. Amusium. Wind-Rose Shell.

1. "*Amusium*," usually spelled *amussium*, was the Latin word, according to Lewis and Short, for "a horizontal wheel for denoting the direction of the wind." This is *Amusium pleuronectes* (L.), known to collectors as the "Asia Moon Scallop."

2. I am not entirely sure about this. In the original: "*in den grootte van een dwarshanden meer.*"

3. See ch. 36.

4. Mangole is the easternmost of the Sula Islands.

5. This is Filippo Buonanni (1638–1725), a pupil of Kircher, also a Jesuit, who wrote a celebrated work on shells entitled: *Ricreazione dell'ochhio e della mente nell' osservazione delle chiocciole*, published in

Rome in 1681. The Latin version was also published in Rome, in 1684, and was called: *Recreatio mentis et oculi in observatione animalium testaceorum.*

6. Martens (p. 127, in *G*) conjectures that these are the so-called Thousand Islands in the Bay of Batavia. See n. 12 of ch. 2.

7. "Bima" refers to the eastern part of Sumbawa (one of the Lesser Sunda Islands), near the Sape Straits.

Chapter 33. Tellinae.

1. An old term for these shells, already in Aristotle, also to be found in Pliny, and which include the popularly known "sunset shells."

2. This "Tellina That is Full of Sand" is, according to Martens, *Asaphis violascens* (Forsskål).

3. The Malay means the same as the Latin, since *pasir* means "sand" or "sandy."

4. For *papaver*, see n. 3 in ch. 1 of this second book.

5. See n. 8 in ch. 28.

6. This "little hand" is the foot of these bottom-dwelling mollusks.

7. For "*zingelgrond*," which should be *singelgrond*. The English "shingle" is more than likely cognate with this Frisian noun.

8. Ch. 65 of book 3.

9. The Latin and Malay convey the same sense: this "Tellina That Dwells among Rocks" is a species of *Barbatia*, perhaps *Barbatia fusca* (Bruguière).

10. The second word in the Latin phrase derives from *garum* or *garon*, a word that refers to a rich fish sauce of which the ancient Romans were fond and which was made from various smaller variety of fishes, including the Atlantic mackerel. Rumphius found it in Pliny, who, in book 31 (paragraph 43), calls it a "choice liquor" and says that the sauce consists "of the guts of fish and the other parts that would otherwise be considered refuse; these are soaked in salt, so that garum is really liquor from the putrefaction of these matters." He goes on to tell us that *garum* was an expensive luxury: they paid 1,000 sesterces for twelve pints of the "scomber fish," most likely the mackerel: "Scarcely any other liquid except unguents has come to be more highly valued, bringing fame even to the nations that make it." *Garum* was a good medicine for burn wounds, according to Pliny, but only if the word were not mentioned, and for dog bites, ulcerous wounds, and crocodile bites (see pp. 434–439 in the Loeb ed., vol. 8).

The Indonesian mollusks which are used for the Moluccan version of *garum* are probably *Psammotaea elongata* (Lamarck).

11. "*Towak*" is *tuak*, "toddy" in English, made from the coconut palm. The sap was collected after cutting off the palm blossom, left in a bamboo cup to ferment, producing a liquor of about 5 percent alcohol. In the Moluccas this sap was not drunk as a refreshment but used as vinegar, while on Java or in the Malay States it was a regular drink.

12. A kind of ginger root which De Wit (*M*, p. 433) identifies as *Languas galanga* (L.).

13. This is some kind of peppergrass or pepperwort, mentioned by Pliny (book 19, paragraph 187, and elsewhere).

14. De Wit identifies this as cinnamon: *Cinnamomum culilawan* Bl. (*M*, p. 427).

15. De Wit identifies this as: *Garuga floribunda* Dec. (*M*, p. 422).

16. This "Violet Tellina" is *Siliqua radiata* (L.).

17. Most likely Suli, a village on the northern shore of Baguala Bay, hence on eastern Hitu.

18. The Latin should have been *culterformis*, i.e., in the shape of a knife (in Roman times a vintner's knife, butcher's knife, cook's knife, or a razor). From *culter* derives the expression "being under the knife" (*sub cultro*), i.e., being in extreme danger or peril. This is *Ensiculus cultellus* (L.).

19. *Pisau* means "knife" in Malay, hence "Knife Shells."

20. "*Zeepmes*" in the original. Generally speaking, a *zeepmes* was an ironic term for a short, curved saber, and a "Chinese *zeepmes*" specifically referred to a machete-type cutlass with a long shaft.

21. Rumphius is referring to no. I of the next chapter (34).

22. "*Picta*" can mean "decorated" or "painted"; hence the Latin means "Painted Tellina." This is not shown and is therefore difficult to identify. It probably is one of the smaller species of the genus *Gari*. The red one could be *Gari occidens* (Gmelin).

23. I do not know what "*kappija*" might mean in Malay. There is the word *kapis* (*kekapis*), which refers to a scallop shell, or perhaps *kopiah*, the ubiquitous headdress that resembles a fez without a tassel.

24. Martens: *Tellina linguafelis* L. The plain white one could be *Tellina palatum* (Iredale).

25. Linnaeus seems to have kept Rumphius' name of "Striped Tellina," for this is *Tellina virgata* L.

26. The Malay means "Sun Shell," since *mata hari*, which literally means "eye of the day," is "sun" in Malay.

27. In the original: "*als hy water trekt.*" I am not sure of this, though among the myriad uses of *trekken*, there was an old usage that meant "to paint."

28. This "Smooth Tellina" is *Tellina chloroleuca* Lamarck.

29. This "Leaf" is *Tellina foliacea* L.

30. *Lida* means "tongue" in Malay, hence the "Tongue Shell."

31. This is *Tellina rostrata* L. The larger, yellow one is probably *Tellina perna* (Spengler), while the little red one could be *Tellina minuta* (Lischke).

The Latin diminutive is a rare word that, according to Lewis and Short, referred to a small leg of pork, from *petaso*, which specifically referred to a pork shoulder. Lewis and Short specify that this was not the ham, which was called *perna*. Yet there is some connection here, for Pliny, in his catalogue in book 32, takes the time to describe a shell called *perna* (at the end of paragraph 154), of which he says (in Jones' translation) that "they stand like pigs' hams fixed bolt upright in the sand; and, gaping not less than a foot wide where there is broad enough space, they lie in wait for food. They have, all around the edges of the shells, teeth set thick like those of a comb; inside is a large fleshy muscle" (Loeb. ed., vol. 8, p. 561).

32. There is no picture and only this one-sentence description. Could be any of the small species of the family Tellinidae.

33. *Vulsella vulsella* (L.). "*Vulsella*" or *volsella* was the Latin word for a pair of tweezers with which one removed hair. I did not use "tweezers" because that was very recent usage in Rumphius' day. Expressive in their own right, the words "nippers" and "pincers" were far more common.

34. "Shagreen," which Rumphius has mentioned before, was an untanned, coarse, grainy leather, made from the skin of a shark, seal, horse, or donkey.

Chapter 34. Solenes Bivalvy.

1. The Latin is fairly transparent: "Bivalve Solenes, Nails, Fingers, and, Vulgarly, Vaginas." It is probably *Solen truncata* Wood. Linnaeus referred to Rumphius' illustration for his description of *Solen vagina* L., and I have noticed throughout that Carolus Linnaeus was more likely to keep sexual names than other scientists. "*Solenes*" are popularly better known as "razor clams."

2. The Malay phrases mean: the "Prick Shell" (*bubu* or *butoh* is vulgar usage for penis); "Sheath Shell" (taking "*saron*" to be *sarong*, used in its original meaning of "sheath"); "Scaly Cockle" (*Krang* [*kerang*] is "cockle" and "*sissij*" might be *sisek*, or "scale"); and "Nail Shell" (*kuku* can mean "claw," "talon," and fingernail).

3. The "Archipelagus" must be Greece.

4. This would be romanized today as "solenes" (pipe or tube), hence the same name these mollusks had and still have.

5. The four Italian phrases for these shells all refer to the same animal. For instance "*Cappa longa*" is *cappalunga*, which is the same as *cannello*, or *cannella*, both of which are diminutives of *canna*. But whatever the various local names, they all refer to *Solen marginatus* (Pulteny), an edible mollusk.

6. This is a reference to the section on marine animals in Jonston's *Historia Naturalis* (1656).

7. "Bononia" is Philippus Bonanni (also Buonanni) (1638–1725), mentioned before.

8. Guillaume Rondolet (1507–1566); the work referred to is *Libri de piscibus marinus* . . . (1554–1555). See n. 7 of ch. 22 of book 1.

9. In *History of Animals*, book 4, ch. 4, section 2.

10. *Rostrum* here in its original meaning of the bill or beak of a bird. *Anas* is Latin for "duck." This "Duck's Bill," letter N on plate XLV, is *Lutraria australis* Reeve.

11. Shown on plate XLV as letter O: *Laternula anatina* (L.).

12. This is *Spengleria plicatilis* (Deshayes) and is shown as letter P on plate XLV.

13. See the next ch. 35, no. IV.

14. This is a brachiopod, according to Martens, either *Lingula anatina* (Bruguière) or *Lingula murphiana* (Reeve).

15. See no. XV of the upcoming chapter 37.

16. This is a confusing passage that pertains to the common ritual of betel chewing, better known in Indonesia as *sirih*, or lime chewing. One takes a leaf of the *sirih* plant (*Piper betle*), spreads some *chunam*, or lime, from calcined shells, adds a small piece of *gambir*, which are little balls of solidified juice cooked from the leaves of *Uncaria gambir*, and finally a small piece of the pinang nut (a nut of the Areca palm). Then the leaf is folded together, and the chew is placed behind the teeth. These ingredients are kept in a *bekas pinang*, or sirih box, which contains small vessels of various shapes, fitted to hold the diverse ingredients. Besides the leaf, the *chunum*, *gambir*, and *pinang* nut, the box also

holds a special tool for cutting the pinang nut, called *kachip;* Rumphius is referring to the shape of the hollowed-out place that holds this pincer or slicer. All this is not clearly conveyed in the original: "Men noemt haar *Bia* Catsje naar de gedaante van de kuiltjes in de Pinang-beeken, daar in men een *Catsjo,* dat is, een beet *Pinang* leggen kan, naar welke zy gelijken" (p. 150). I am assuming that "*beeken*" is a misprint for "*bekken,*" translated as a "basin" at the time. Perhaps *beet,* which normally means "a bite" or a "morsel," refers here to the small chisel?

17. This is not a mollusk either but a sea pen. Sea pens are a group of marine animals belonging to the order Pennatulacea, one of the orders, highly modified, in the subclass Octocorallia or Octocorals. They usually have an elongated axial polyp with secondary polyps along the side of the stalk. Most sea pens have a brittle, horny, central skeleton, the so-called pen.

18. Taking "*sormen*" to be a misprint for *wormen.*

19. A leaf of the coconut palm.

Chapter 35. Musculi. Mussels.

1. The "Ordinary Large Mussel" is *Modiolus philippinarum* Hanley.

2. De Clercq, in *Het Maleisch der Molukken,* states that *asusseng* (*asuseng*) is Ambonese for a kind of edible snail that occurs particularly on the beach near Galala.

3. *Modiolus (Brachydontes) subramosa* Hanley. "*Mitulus*" is a term for a mussel, found in Pliny. Even today the common edible mussel is called *Mytilus edulis* (L.). The second part of Rumphius' name means "relating to ducks," and the Malay means the same thing.

4. "Huconalo" (also "Hockonalo" and "Hukunalo") seems to have been another name for Rumah Tiga. See Rumphius' *Ambonsche Landbeschrijving,* p. 59.

5. *Septifer bilocularis* (L.).

6. Martens thinks this "Sandy Mussel" is *Modiolus vagina* Lamarck.

7. See no. V.

8. A mushroom; a term derived from Pliny (book 22, paragraph 92 ff.).

9. See ch. 67 in book 3.

10. *Lithophaga teres* (Philippi) and *Leiosolenus obesus* (Philippi). "*Pholas*" comes from the Greek φωλάς, from a phrase that means "lurking in a hole." It was used of the hibernating bear in its cave, and as a metaphor for a courtesan. Specifically it referred to a kind of mollusk that bore holes in stone.

11. The Malay means "Stone Asusseng."

12. Rumphius' *Herbal:* vol. 6, book 12, ch. 27. Martens says this refers to masses of dead coral, or Nullipores.

13. It is interesting to note that these "East-Indian Philosophers" had a fundamentally sound notion of how coral reefs grow.

14. The Latin noun means "Little Bird." Probably *Pteria crocea* (Lamarck), but it could also be *Pteria avicula* (Holton).

15. The Malay means "Bird Asusseng."

16. Unclear in the original: ". . . *stokken, en dus houd men die lang in zee staan, alsmede aan de berghouten.*"

17. Black corals (Ceriantipatheria).

18. Martens: *Martesia striata* (L.).

19. "Matthiolus" was Pierandrea Mattioli (1501–1577), a physician who, at the age of twenty, became Pope Leo the Tenth's personal physician. Later he was doctor for Emperor Ferdinand in Prague and Emperor Maximilian II. He returned to Trente in Tirol, where he died of the plague. He became famous with an encyclopedic work, a compilation mostly of plants, entitled *Commentarii in Sex Libros Pedacii Dioscoridis,* published in 1544.

20. This must be "Duino," the castle high above the Adriatic (one can see both Trieste and Venice from it), which once gave shelter to Dante in the fourteenth century and to Rilke in the twentieth. As the guest of Princess Marie von Thurn und Taxis before the First World War, Rilke wrote portions of his magnificent *Duino Elegies* at this castle.

21. These are *Lithophaga lithophaga* (L.), a Mediterranean species closely related to the Pholas as described by Rumphius. Aldrovandus' species is the same. It was much sought after for food.

22. H. L. Strack suggests that Tarku might be a city near Derbent on the Caspian Sea.

23. Ulysse Aldrovandi (1522–1605) was, like Mattioli, a former student of Rondolet. Suspected of heresy, Aldrovandi was brought as a prisoner to Rome, where he was ordered to limit his studies to antique statuary. A zoologist and botanist, Aldrovandi wrote an enormous number of texts, which were mostly compilations of existing knowledge, characterized by a modern commentator as "display windows of erudition." The work Rumphius refers to deals with "bloodless animals."

24. Italian port on the Adriatic, below Rimini, eastwards from Florence.

Chapter 36. Pinna. Bia Mantsjado.

1. The name *Pinna* was given to these shells by the ancients, spelled either "*pina*" or "*pinna*," and is preserved in modern nomenclature.

2. "*Mantsjado*" is most likely *mentjadu* (after the Portuguese "*machado*"), an Ambonese word for a particular axe.

3. Given the illustrations on plate XLVI, Rumphius' name of "Holster Shells" is very fitting if one thinks of the holsters that were strapped to saddles in the seventeenth century to hold horse pistols. The noun itself came into English from Dutch.

4. This "First or Oblong Pinna" is *Pinna muricata* L. Letter I on plate XLVI is an adult and K a juvenile. The specimen that was over a foot long probably belongs to the species *Pinna bicolor* (Gmelin).

5. Or *callus*, a hardened, thick skin.

6. Rumphius mentioned it not only in ch. 28 of this second book, but also in ch. 23 of the first book. The description of the commensal shrimp here matches that of ch. 23 in book 1, and both derive from Pliny, book 9, paragraph 142 in Latin.

7. All these are villages on Ceram.

8. The Aegean.

9. The Sea of Marmora.

10. I cannot be sure what Rumphius means by "river," but perhaps he meant the Bosporus, because the Golden Horn is really Istanbul's harbor. The Bosporus is not a river, but one does not find any *Pinna* in rivers.

11. The two-foot-long *Pinna* referred to here is the so-called Noble Pen (*Pinna nobilis* L.). This bivalve's byssus was used as a cloth material! It had a golden color and was used to make clothing: "The byssal tufts were washed in soap and water, dried in the shade, combed and finally carded. A pound of byssus would produce only three ounces of high-grade threads." This "sea silk" was used to make stockings, gloves, and collars, and was said to be the legendary material of the Golden Fleece. See R. Tucker Abbott, *Kingdom of the Seashell* (New York: Crown, 1972), pp. 184–185.

12. The Pliny reference is correct, except that the Loeb edition has it in ch. 41, and Pliny says that these "pinnae" can contain pearls. Cicero (*De Natura Deorum*, book 2, section 48) does not mention pearls at all but cites the "*pina*" as an example of a relationship between it and the little commensal shrimp that is beneficial to both, an association that Pliny had mentioned in book 9, chapter 46. Marcus Tullius Cicero, *De Natura Deorum & Academica*, transl. H. Rackham, The Loeb Classical Library (Cambridge, Mass.: Harvard University Press, 1961), p. 240.

13. This "Broad Pinna," letter L on plate 46, is *Atrina vexillum* (Born), the one by letter M is probably *Atrina pectinata* (L.).

14. Rumphius took this name from a passage in Pliny (book 32, paragraph 154) which I mentioned before, in n. 30 of ch. 33. The Roman epitomizer clearly stated that the "*perna*" was a "*pinna*": ". . . appelantur et pernae concharum generis." He also compares them to "pigs' heads fixed bolt upright in the sand" (Jones' translation in the Loeb ed.; vol. 8, p. 561).

15. *Streptopinna saccata* (L.).

Chapter 37. Ostreum. Oysters. Tiram.

1. The Latin means the "Rooted or Wooden Oyster." *Sacostrea cucullata* (Born).

2. *Tiram* is the Malay word for edible oysters that adhere to mangrove roots and fishing stakes. "*Besaer*" is *besar* or "large," and "*akkar*" is *akar* or "root." Hence: "Big Oyster" and "Root Oyster."

3. This series of islands are all in the vicinity of Ambon. The largest island is Buru ("Buro") to the west of Ambon; then going east from Buru, one first encounters Manipa, then Kelang, and Boano is northwest of Kelang, or north of Ceram's Howamohel Peninsula (or "Little Ceram").

4. Mangroves.

5. In his *Natural History*, book 32, paragraph 63, where it is said that these are the oysters from the "Indian Sea" that are very large. See also n. 5 of ch. 38 in this second book.

6. M. Philippides informs me that this is a rather unusual word to know for someone not a native speaker. The first part of the compound is *gaidaro* (γάιδαρο), which is a word for "donkey" that probably derived from Turkish. The last element should have been *pous* (πους), genitive of *podos* or "foot."

7. No. II of ch. 35.

8. See ch. 28.

9. This "Wickerwork Oyster" is *Dendrostrea frons* (L.).

10. The Scottish zoologist Henry O. Forbes saw *seros* in Ambon in the late nineteenth century. Calling them "fish-maises," he went on to describe them as "a double line of close bamboo palisades, reaching above the level of the water, enclosed a lane, which extended shorewards from its seaward

entrance a little way beyond low-water mark, and doubling back terminated in deep water in a circular well, where the fish that had entered during high tide, and whose escape had been prevented by the ebb, were enclosed and captured from a trap door in a little platform erected over it." Henry O. Forbes, *A Naturalist's Wanderings in the Eastern Archipelago. A Narrative of Travel and Exploration from 1878 to 1883* (New York: Harper & Brothers, 1885), p. 289.

11. De Wit (*M*, p. 451) identifies this as *Drynaria sparsisora* (Desv.) Moore.

12. This, according to De Wit (*M*, p. 443), is *Aegiceras corniculatum* (L.) Blanco.

13. This "*Passo*" is the narrow isthmus that pinches off Leitimur from Hitu, and separates the Bay of Ambon from Baguala Bay.

14. Since there is no picture, this species, belonging to the family Ostreidae, cannot be identified.

15. *Placuna ephippium* Philipsson. Rumphius' name has some problems. *Ephippium* was a rare word, from Greek, that referred to a "horse cloth" or caparison, according to Lewis and Short. They compare it to a *clitellae*, which was a pack saddle, especially for asses. "*Placentiforme*" is more of a problem, for it depends on whether Rumphius had "late," or Medieval, Latin or classical Latin in mind here. If the latter, he is referring to a cake, since *placenta* meant no more than that in Rome. A *placentarius* was a pastry cook.

16. "English saddles" were distinguished from Continental ones by having no horn and lower pommels and cantles. The seat was also broader and flatter. As mentioned in the introduction, Rumphius was acquainted with them because his stable in Hila was equipped with British tackle.

17. "*Lebber*" is most likely *lebar*, Malay for "wide," for something that is relatively broad.

18. The "Xula Islands" are the Sula Islands, but not one island can be said to come close to being called "Toccuve." Perhaps Taliabu, one of the three largest Sula islands, although in Rumphius' day, the VOC dealt mostly with Mangoli.

19. Perhaps "Pangkadjene," a region on the western "leg" of Celebes, on this peninsula's east coast, north of Makassar.

20. Rumphius does not say much about pearls, because real pearls were not found in his general region. Pliny, however, used pearls as his prime example of human greed in a fascinating passage in book 9 (paragraphs 107–123). He states that most came from the Indian Ocean, especially Ceylon, something Rumphius repeats. Pliny insists that a pearl is entirely the product of water and weather and that it tries very hard to avoid capture. Hence it has a shell with sharp edges, and lives among rough rocks in deep water where "sea-dogs," that is, sharks, patrol. Oysters have remarkable leaders that are "marvellously skilful in taking precautions" (paragraph 111), but even so, nothing can "protect them against women's ears" (*nectamen aures feminarum arcentur;* paragraph 111; Loeb ed., vol. 3, p. 237). Pliny goes on to detail the incredible sums of money that were paid for pearls by the Roman upper classes and imperial consorts. But the largest ones in the world were said to be two pearls owned by Cleopatra, who, as a wager, dissolved one, worth ten million sesterces, in a cup of vinegar, proving she could spend that sum on a single banquet. It is said that Cleopatra's second pearl was taken to Rome after her capture, and was large enough to be cut in two and turned into a pair of earrings for the statue of Venus in the Pantheon (ch. 58; paragraphs 117–123).

21. The island called Misool is to the west of New Guinea's "bird's head" and belongs geographically to that large island rather than Ceram or the Moluccas. The Aru Islands, off the coast of New Guinea in the Alfura Sea, were the main suppliers of a lesser grade of pearls but of excellent mother-of-pearl.

22. See ch. 28.

23. This "Great Pleated Oyster" is *Hyotissa hyotis* (L.).

24. The name of the village "Huconalo" was also spelled "Hockonalo" or "Hukunalo." See n. 24 of the previous ch. 35.

25. This "Small Pleated Oyster" is *Lopha cristagalli* (L.).

26. The figured specimen of the "Prickly Oyster" is probably *Spondylus sinensis* (Schreibers.) The other ones are also species of Spondylus but cannot be definitely identified.

27. De Clercq in *Het Maleisch der Molukken* notes that *tjipu* is Bandanese for a kind of shell. He lists this shell as *tjipu bil.*

28. This Spondylus species cannot be identified.

29. Another Spondylus species.

30. This "Large and Monstrous Prickly Oyster" must be *Spondylus regius* (L.) or *Spondylus imperialis* (Chenu.).

31. *Pinctada margaritifera* (L.). Letter F is a specimen that has been polished on the outside, while letter G is a specimen in its natural state.

32. "*Piering*" in the original. An unusual word, no longer in use, for a small porcelain platter. It derived from the Malay *piring*, which indicated a shallow-rimmed plate made of some kind of metal.

The word occurs in seventeenth-century inventories of porcelain goods, and there were rice, sugar, and tea *pierings*.

33. *Telinga* in Malay means "ear" and *anjin* is "dog."

34. Halmaheira (also called Gilolo) is the island, shaped like a miniature version of Celebes, to the east of the original Moluccan Islands.

35. Also spelled "*Wang-kangh*" or "*wankan*," this once referred to a Chinese junk. See n. 13 of ch. 24, book 3.

36. I cannot find a Solok in northern Borneo, but there is a place by that name in Sumatra.

37. "Coxinga," also "Koksinga" or "Koxinga," was a Chinese leader who conquered Formosa (Taiwan) in 1662, captured a Dutch fort, and expelled the Europeans. He also attacked Dutch trading posts in Cambodia in 1667.

38. This is Bahrein Island, also a small archipelago by that name, in the Persian Gulf, just above the Qatar Peninsula, which was indeed the gulf center for pearl fisheries.

39. This is Kung, a city on the Persian coast between Taub and Farur, where the VOC had a trading post.

40. In the original, "*bussenmakerswerk*." A *bussenmaker* made guns in the seventeenth century (artillery in the Middle Ages) and was synonymous with *bussengieter*, or someone who worked in a foundry where they made weapons. I highly doubt that they made cannon inlaid with shells or mother-of-pearl, but the practice was common in the manufacture of firearms and handguns.

41. This "Divided Oyster," popularly known as the Hammerhead Oyster, is *Malleus malleus* (L.).

42. In the original, "*Meshamer*," which might be "*metshamer*"—from *metsen*, a variation of *metzelen*, which means "to lay bricks"—or a hammer-like tool used by a bricklayer. My "mallet" is a compromise.

43. The Malay compare this shell to their well-known dagger, the *kris*.

44. Martens identifies this shell as *Isognomon isognomon* (L.), hence Linnaeus kept the word, which in all likelihood was contrived by Rumphius and combined from *iso* (ἰσο; the tau is not possible) and gnōmōn (γνώμων; the omicrons should be omegas). The latter had as its principal meaning an "interpreter," but its secondary meaning was the indicator on a sundial that by the position of its shadow indicated the time of day. It furthermore could mean a "carpenter's square" (*norma* in Latin) as well as a figure in geometry. Hence Rumphius has both his "carpenter's square" and something that might mean "likeness indicator."

45. This "Tortuous Oyster" remains that way, for it is identified as *Trisidos tortuosa* (L.).

46. This "Amber Oyster" could be *Anomia sol* (Reeve). In English, popularly known as jingles.

47. This is Taliabu of the Sula Islands.

48. *Mitella* is the diminutive of *mitra*, Latin for turban. Holthuis identifies this as *Mitella mitella* (L.).

49. Strack (p. 38) says that this rock near Larike is now called *Batu Suangi*, or "Haunted Rock."

50. *Papeda* is a porridge made from sago meal, a bland affair that certainly asked for seasoning.

51. These are Goose barnacles (see ch. 26), one or more species of the genus *Lepas*.

52. See no. III of ch. 26.

53. In the original "*kranten*," which means "newspapers" in Dutch. However, "*branten*" once existed as the plural of *brant*, normally spelled *brand*, a shortening of *brandgans*, which is the barnacle goose (*Branta leucopsis*). These birds visit Dutch shores in winter, but they do not breed or nest there.

54. He was part of a family of Dutch clergymen, established by Martinus Lydius (1539–1601), a moderate Calvinist. He had two sons, both of whom became ministers, called Balthasar and Johannes. Balthasar, the more prominent of the two, had two sons in turn, Jacobus and Isaac. Jacobus Lydius (?1610–1679) is meant here, not Isaac. Jacobus was a Protestant theologian, minister, and archaeologist. The book in question is: Jacobus Lydius, *Vrolycke Uren, ofte Der wysen Vermaeck* (Dordrecht: Hendrik van Esch, 1650).

55. This refers to the long-standing myth that the wild black geese seen in Scotland were produced from goose barnacles, the reason being that no one had seen their nests or their eggs. The barnacle goose (*Branta leucopsis*) breeds in Spitsbergen, Greenland, and northwestern Siberia, while a different species, the brent goose (*Branta bernicla*), has a circumpolar breeding range. Hence, Rumphius was right.

56. This rather generically named "Sea Oyster" cannot be identified.

57. "*Lazarus-klap*" in the original. Sewel has for this, under "Lazarus," "a Clapper where withall the lepers go a begging in Holland."

58. *Spondylus sinensis* Schreibers.

59. Possibly *Spondylus americanus* Hermann, a Caribbean species.

60. Engel (p. 101) mentions that Griffet was an apothecary in Rotterdam, who in 1724 was eighty-six years old.

61. *Chama lazarus* (L.).
62. A "*Hout voet*" was the equivalent of 11.50 inches.
63. Martens thinks this might be *Venus puerpera* (L.), but it could also be another species within the large family Veneridae.
64. This is the West Indian species *Pitar dione* (L.).
65. This is *Chione paphia* (L.), a species present from the West Indies to Brazil.
66. The West African *Cardium costatum* (L.).
67. Nos. 7 and 8 are both *Lyropecten nodosa* (L.), to be found from the southeastern U.S. to Brazil.
68. This West Indian shell is *Trachycardium isocardium* (L.).
69. Martens thinks that this might be the core of a fossilized mussel.
70. This might be the European *Acanthocardia tuberculatum* (L.).
71. *Epitonium scalaris* (L.), the most desirable collectors' item in the seventeenth and eighteenth centuries.
72. The original: "*zyn slakke kring is los.*"
73. See n. 40 of ch.3 of this book.
74. Jurriaen Ovens (1623–1678), a German painter who was one of Rembrandt's pupils. He lived in Amsterdam between 1640 and 1663, and collected curiosities.
75. This refers to a collector named Volkertsz, Volckertsz, Volkaartsz, Volckers, or Volckersen (1576–1651), who lived in Amsterdam. His cabinet of curiosities was inherited by his sons.
76. *Argobuccinum pustulosum* (Lightfoot), a species occurring in South Africa but also in Chile, down to Tierra del Fuego.
77. *Babylonia areolata* (Link).
78. *Babylonia spirata* (L.).
79. *Perrona nifat* (Bruguière), from West Africa.
80. *Latirus infundibulum* Gmelin, a species found from the West Indies to Brazil.
81. *Colubraria muricata* (Lightfoot).
82. *Fasciolaria tulipa* (L.), from the West Indies and from Florida to Brazil.
83. *Ranella olearium* (L.), from the Mediterranean, Africa, Australasia, and Bermuda.
84. *Pleuroploca trapezium* (L.).
85. *Vasum ceramicum* (L.).
86. "*Morgenster*" in the original, a medieval weapon, which basically resembled a club with spikes on top. The term came originally out of German (*Morgenstern*), and was said to be a joke, which was based on comparing this murderous weapon to Venus or the Morning Star. But the term stuck and became standard in the sixteenth century. The English name for this type of weapon was "mace."
87. A juvenile specimen of one of the larger *Strombus* species, probably *Strombus gigas* L., from the West Indies.

Chapter 38. Division of the Sea Snails and Shellfish from Pliny.

1. The number is correct, but the chapter, at least in the Loeb ed. (vol. 3), is 52 (Latin paragraphs 102–103). I will provide Rackham's translation of the Latin phrases (on pp. 231 and 233 in the third volume of the Loeb ed.), followed by the chapter number in Rumphius' book if he is speaking of some shell in general, and the specific item if so indicated.

1. "Flat shell-fish"; chs. 31 and 37.
2. "Hollow"; ch. 31 (no. V) and ch. 37 (no. VII).
3. "Long"; ch. 28.
4. "Crescent-shaped"; ch. 27 (no. I).
5. "Circular"; ch. 30 (no. VII), ch. 37 (no. IX).
6. "Semicircular"; ch. 28 (nos. III and VII), and ch. 31 (no. VIII).
7. "Humped"; ch. 30 (no. XI); ch. 31 (no. IX); chs. 13–15.
8. "Smooth"; ch. 30; chs. 23–24.
9. "Wrinkled"; ch. 28.
10. "Serrated"; ch. 28.
11. "Furrowed"; ch. 28 (no. III); ch. 31.
12. "With the crest bent into the shape of a purple [seasnail]"; ch. 19.
13. "The edge projecting into a sharp point"; ch. 22 (no. I and no. II).
14. "The edge spread outwards"; ch. 13.
15. "The edge folded inwards"; chs. 23–24.
16. "Picked out with stripes"; ch. 23 (no. XI); ch. 30 (no. VI).
17. "Picked out with flowing locks"; ch. 30 (no. VI); ch. 23 (no. XI).
18. "Picked out with curls"; ch. 23 (no. XI).

19. "Parted in little channels"; all "scaly shells."
20. "Like the teeth of a comb"; ch. 31 (nos. VI and VII).
21. "Corrugated like tiles."
22. "Reticulated into lattice work"; ch. 30 (nos. XI and X).
23. "Spread out slant-wise"; ch. 31 (no. VIII); ch. 33.
24. "Spread out straight"; ch. 37 (no. XI).
25. "Close-packed"; ch. 14 (no. IV).
26. Rackham has "diffused" for this, but Rumphius' description is more accurate; ch. 19 (nos. VI–VII); ch. 19 (no. XII).
27. "Curled"; ch. 37 (no. XI).
28. "Tied up in a short knot"; winkles.
29. "Linked up all down the side"; chs. 27, 33, 36.
30. "Opened so as to shut with a snap."
31. "Curved so as to make a trumpet"; ch. 19 (no. II).
32. "Navigating & Sailing, the Venus Shell or the Nerita"; Rumphius added the latter, although it is Pliny's term (not in this passage, however) for a sea mussel that resembles the nautilus; ch. 3. This entry, and the next one, were reduced from complete sentences in Pliny. This one reads: *navigant ex his Veneriae, praebentesque concavam sui partem et aurae opponentes per summa aequorum velificant.*
33. "Leaps out of the water & [uses] itself as a boat"; ch. 31. Pliny: *saliunt pectines et extra volitant, seque et ipsi carinant.*

Chapter 39. How One Should Gather and Clean Shells.

1. If we subtract twenty-eight years from 1682, we arrive at 1654, a very early date to have started collecting. One should remember that Rumphius arrived in Java in June of 1653, and that he went to the Moluccas in the fall of that year. In 1654, Rumphius was still a soldier.

2. Cosimo III (1642–1723) visited Holland twice between 1667 and 1669. He became particularly acquainted with Pieter Blaeu (1637–1706), who had taken over the famous printing and cartography business from his father, Johannes Blaeu, in 1662. It was Pieter Blaeu who arranged for the duke to see private curiosity cabinets, including Jan Swammerdam's. Rumphius was in contact with Blaeu; he sent botanical samples to him for Padtbrugge, who was the governor of Ambon from 1683 to 1687. I think that it was primarily at Pieter Blaeu's insistence that Rumphius parted with his beloved collection.

3. See ch. 19.

4. "*Kaiman*" was a common Dutch usage, but should properly be applied only to the West Indian cayman. Rumphius is speaking of a crocodile (*Crocodilus porosus* Schneider) that used to be found all over Southeast Asia from India to the Fidji Islands. It can grow up to thirty feet, resides in both salt and fresh water—with the shore as its favorite location—and is by choice a night animal. It has always attacked humans, hence Rumphius' fear was totally justified.

5. De Beaufort (*M*, p. 58) states that this fish is a species of *Synanceia*, possibly the Stone Fish (*Synanceia horrida* [L.]), which is indeed dangerous.

6. This should be "*Echinus setosus*;" see ch. 30 in book 1.

7. For this "Large Ear," see ch. 26.

8. Ch. 16.

9. Ch. 19.

10. Ch. 23.

11. Chs. 23–24.

12. Ch. 21.

13. Ch. 20.

14. "*Salsilago marina*" is redundant. The latter word means, of course, "of the sea," while the first one, also spelled *salsugo*, meant "saltiness" or "brackishness," and by extension "salt water," or "brine." This has little to do with "grit" (*gruis*). This "grit" is coralline algae.

15. Ch. 2.

16. Ch. 21.

17. Ch. 19, no. X.

18. Ch. 19, no. V.

19. *Gemutu* (or "*Gomuto*") refers to the black "hairs" between the stems of the sugarpalm leaves. This is Rumphius' *Saguerus*, the Aren palm (*Arenga saccharifera* Labill.), from which saguir wine is made. *Gemutu* looks very much like black horsehair and was traditionally used to make rope, brushes, and floormats.

20. Ch. 22 of book 1.

21. Chs. 23–24.

22. Ch. 20.

23. Ch. 21.
24. Ch. 25.
25. Ch. 19, no. II.
26. Ch. 28.
27. Ch. 31.
28. Ch. 28, no. VI.
29. Ch. 31, no. VII.
30. Ch. 16, no. V.
31. Ch. 14, no. VI.
32. Ch. 19, nos. VIII–IX.
33. Ch. 33, no. II.
34. Ch. 36.
35. Ch. 35, no. VI.
36. Ch. 2.
37. Ch. 19, no. II.

THE THIRD BOOK OF *THE AMBONESE CURIOSITY CABINET* DEALING WITH MINERALS, STONES, AND OTHER RARE THINGS

Chapter 1. How They Falsify Gold in These Countries.

1. I translated "*zandgoud*" as "gold dust." As to the question of gold in the Indies: In ancient times it was supposed that the Indonesian islands were rich in gold. Sumatra was considered a "gold island" and had Mount Ophir, the fabled gold mountain of lore and legend. Arab traders said the same about Nias, the island off the west coast of Sumatra, and the *Ramayana* instructed that Java was rich in ore. This turned out to be fool's gold. Gold was and is found only in Sumatra (in the Benculen area on the island's west coast), southern and western Borneo, and northern Celebes, though in the second half of the twentieth century large gold deposits have been found in New Guinea. Some primitive mining seems to have taken place in those former regions, while in western Borneo the Chinese did a lot of gold washing. The VOC seems to have tried seriously only in Sumatra to obtain the precious ore for commercial purposes. In 1669 they constructed the Salida mine near Padang. In 1670 the VOC brought European miners to Sumatra for the first time. They were mostly Germans from Thüringen, as well as Hungarians. The first load of Sumatran ore sailed to Batavia in 1670, but the whole enterprise never amounted to much. For some reason, the Dutch would not give in to reality. Though they ceased operations several times—the first time was in 1697—they did not abandon the notion of mining gold in the Indies until 1737.

2. The *Water Indies* are defined by Rumphius in his introduction to his *Ambonese Herbal.* It is the part of the East Indies "to the south and east of Malacca, consisting of countless great and small Islands, and where the Malay tongue is generally used." He then makes it clear that the Water Indies start with Sumatra and proceed, via Java and the other great Islands, in a long row eastward. Scaliger called this curve of islands "the eyelids of the world." At the end of the lesser Sunda Islands, it "then curves upward and to the North with the South-Easter Islands, and to the Northwest, through the Moluccas, and ends in the North with the *Philippine* or *Manilhase* Islands. In the most extreme corner of these Water Indies to the East, are the three governments of Amboina, the Moluccas, and Banda; which are closed off to the East and separated from the South Sea by the land, which men call *Nova Guinea.*"

Old Indies was defined by Rumphius in his preface to the *Ambonese Herbal* as *Indiam intra Gangem*, or Hindostan, and *India extra Gangem*, or "Asia east of the Ganges down to the Sea, including Bengal, Aracan, Malacca, Siam, and Cambodia." In other words: India and what used to be conveniently called "Indochina."

3. This is my solution for "*in 't kruis gevlochten*," a troublesome phrase.

4. A "*Serdijn*" is indeed related to the herring, and was more commonly spelled *sardijn.* The name of this fish derives from *sardina*, Latin for the island of Sardinia. The fish itself (*Clupea sardinus*) was always called *sardijn* in Dutch when it was still alive and well. Once it had been caught and processed, the food was known as "*sardientje*," that is to say, sardine.

5. "*Creese*," an old English approximation of *kris* or *keris*, the Malayan dagger.

Chapter 2. Water Test of Gold and Silver.

1. No one would call Archimedes (c. 287–212 B.C.) a builder—Rumphius uses the Dutch *bouwmeester*, which means the same as the German *baumeister* (which is what his father was; see introduction), which referred to someone who was a kind of superintendent of building works, not an architect

necessarily, but rather a mathematician and inventor. After studying in Alexandria, he returned to his native Syracuse to pursue mathematical studies. He considered his various inventions beneath the dignity of a pure scientist, though they made him famous. The one Rumphius refers to is, of course, the famous water proof of the gold content of Hieron's crown. Hieron, the king of Syracuse, was wondering whether a custom-made gold crown actually contained a portion of silver. He asked his friend to devise a test. As Thomas Heath puts it, Archimedes was puzzled, until one day "as he was stepping into a bath and observed the water running over, it occurred to him that the excess of bulk occasioned by the introduction of alloy could be measured by putting the crown and equal weights of gold and silver separately into a vessel of water, and noting the differences of overflow" (*Encyclopaedia Britannica*, 14th ed.). Archimedes is supposed to have been so happy at his solution that he ran home naked shouting *εὕρηκα, εὕρηκα*, which means "I have found it, I have found it." Note that Rumphius gets closer to the original Greek by spelling the famous exclamation with an "h." The above procedure is clearly what Rumphius had in mind for his own experiments.

2. For *toestel*, which now generally means an apparatus of some kind, but which once also referred to preparations for something.

3. Spelled "*wax*" in the original! Although now commonly spelled *was* in modern Dutch, "*wax*" was once an alternative spelling.

4. Except for "quints" all these weights were once used in England as well. A contemporary table states that a graine was the smallest weight in apothecaries, that twenty graines made one scruple, three scruples made one dragme (or drachme; from the Greek for the coin and the weight), eight dragmes made an ounce, and sixteen ounces a pound. One can tell that the weights are exactly like Rumphius'. He gives the weight of one white peppercorn or two grains of peeled rice as the equivalent of a "graine," while the *Encyclopaedie van Nederlandsch-Indië* (under "*maten en gewichten*") gives 0.000065 kilogram as the official equivalent in twentieth-century colonial times, and the *OED* states a "graine" equaled 0.007 of a pound.

5. This mathematical rule exists in English as well. The *OED* defines it as "a method of finding a fourth number from three given numbers, of which the first is in the same proportion to the second as the third is to the unknown fourth."

6. In the original "*Capelle-gout.*" The word was later spelled as "*kapel*" or "*kapelle*," but it had nothing to do with "chapel." It derives from the French fifteenth-century word *coupelle*, which comes from medieval Latin *cupella* (diminutive of *cupa*, or "cask") which referred to a small vessel once used to refine precious metals. The *OED* describes this as "a small flat circular porous vessel, with a shallow depression in the middle, made of pounded bone-ash pressed into shape by a mould." As a verb, "to cupel" usually referred to acquiring silver from lead in large quantities by the process of "cupellation," or separating silver from argentiferous lead in a cupel. A "cupel" was also the furnace wherein the vessel was placed. As mentioned, this process usually involved silver; hence there was a French expression "*argent de coupelle*" (in Dutch *kapelzilver*), and Rumphius used "*Capelle-gout*" by analogy.

7. "Cupan" might be *kupang*, a common small weight for gold in the Indies and later also, according to Wilkinson, in the Malay States.

8. The only thing I can make out of this is that it means what it says: twenty-one-and-a-half *dimes*. The silver dime (*dubbeltje* or *dobbeltje* in Dutch) was used as both a coin and a weight.

Chapter 3. How Gold and Silver Will Test with Things Other than a Touchstone.

1. A touchstone was a black stone related to flint that was once used to test the purity of gold and silver. This was determined by the mark the metals left on the stone after they had been rubbed on it. It was also known as Basanite or, more commonly, Lydian stone (*Lydius lapis*).

2. See ch. 70 of this book. The "*Pinang*" is the nut of the Areca palm.

3. See ch. 73.

4. See ch. 74.

5. See ch. 64.

6. This is white coral, described by Rumphius in his *Herbal*, book 12, ch. 16.

7. "Flintlock stone" refers to the flint that was fixed in the hammer of a flintlock gun and that, upon striking the battery of the pan, ignited a charge. "Smooth" means that they were artificially made and not natural.

8. See ch. 80.

9. A variety of *Vitex;* see his *Herbal*, vol. 3.

10. The oft-mentioned Aren or Sugar Palm (*Arenga saccharifera*).

Chapter 4. Suassa. What Kind of Metal It Is.

1. Wilkinson provides the interesting translation of "pinchbeck" for *suassa.* "Pinchbeck" was an alloy so named after its inventor, a British watchmaker during the early decades of the eighteenth

century, consisting of five parts of copper, one part of zinc, and was said to resemble gold. *Suassa* was (at least up to the early part of the twentieth century) an alloy of copper with a little gold. Wichmann notes (p. 140) that a natural mixture of copper and gold does not exist, while he also informs us that from the sixteenth century to the twentieth century, the Portuguese asserted that there is a fine-quality copper to be found in what was once Portuguese Timor (p. 141). Subsequent identifications of metals and minerals in this, the third book, are based on a long article, the only one known on this topic in Rumphius, in *G*, by Arthur Wichmann, entitled "Het aandeel van Rumphius in het mineralogisch en geologisch onderzoek van den Indischen Archipel," on pp. 137–164. Cited hereafter only as "Wichmann."

An expert on Indonesian crafts quoted this text by Rumphius in his description of gold- and silversmithing in the archipelago (see J. A. Loeber's article "*Metaalbewerking*" in *Encyclopaedie van Nederlandsch-Indië*). The main point he makes is that *suassa* is an example of the Indonesians' different interpretation of beauty. They do not admire precious metals, as we do, for their purity, but for their appearance; the emphasis is on the *color* of the object. Hence density and purity of the metal are of no great importance, and one will find that a lot of work in precious metals is purposely very thin. We have to keep this emphasis on the *aesthetic* appearance, and not monetary value, in mind when reading Rumphius' text.

2. This means, of course, "electrum of the ancients." I am sure that Rumphius had primarily Pliny in mind, who discusses *electrum* in book 33 of his *Natural History* (Latin paragraphs 80–81). "Electrum" is not to be confused with amber, though in Latin it bears the same name (see the notes to ch. 35 of the present book). Pliny uses it to indicate either a gold ore that contains one-fifth silver (the "natural" electrum) or gold that has been artificially diluted with silver ("artificial" electrum; note that Rumphius uses the same terms and distinctions). The Roman notes that Homer already mentioned electrum in his description of Menelaus' palace in the *Odyssey* (book 4:71 ff.), and that the temple of Athena on Rhodes had a goblet made of electrum dedicated by Helen and said to have had the same dimensions as her breast. Of more importance to the present text is Pliny's comment that electrum shines "more brightly than silver in lamplight" and that it detects the presence of poison (see below).

3. See n. 2 of the first chapter of this book.

4. This large island was visited by the Portuguese in the beginning of the sixteenth century; they established a modest fort around 1520 on Solor, an island off Timor's northern coast. This was conquered by the Dutch captain Schot in 1613, and one might say that from that year dates Portuguese-Dutch rivalry on the island. Timor remained a backwater for the greater part of the colonial era, and the Timorese regularly attacked the strongholds of both European nations. The Ade region was in eastern, that is to say Portuguese, Timor.

5. *Radja Salomon Speelman* was a Timorese chief, or "King" as Rumphius has it, who was defeated by the Portuguese in 1677, according to Valentijn (III.2:112, 121), and came to Banda, where he had himself baptized with the name of the VOC's general and governor-general. He returned to Timor in 1680. Rumphius stated that the *radja*'s realm, Ade Mantutu, was in the eastern corner of Timor. See his *Herbal*, book 12, ch. 20, and book 6, ch. 12.

6. Ammonium chloride. "Armoniac" was a common corruption in medieval Latin, French, and English. The word derives from the name of the Egyptian god Ammon, and this white crystalline salt was called "Ammon's Salt" because it was supposedly first prepared from camel dung near the temple of Jupiter Ammon.

7. The lime mentioned was part of the sirih chaw, and was known as *kapur*. The *kapur* was kept in a *chepek*, a round box with a lid, which was part of the sirih paraphernalia that was kept in a box called *tampat sirih* (on Java: *pakinangan*). The lime in Rumphius' day was slaked from crushed seashells, and as his text indicated, could be quite sharp at times. *Suassa* metal had an effect on it and made the lime milder or sweeter. Sirih chewing was in Rumphius' day a habit also indulged in by the Europeans. De Haan (*Priangan*, 2:367n. 5) remarks that Coen chewed, as did Bontius, Baldaeus, and Valentijn. Women were particularly fond of the habit, and even brought their sirih boxes, plus the requisite maid, to church, where every so often they'd spit in the cuspidor which the attendant would hold for them.

8. Pliny says the same thing about *electrum*. In Rackham's translation: "Natural electrum also has the property of detecting poisons; for semicircles resembling rainbows run over the surface in poisoned goblets and emit a crackling noise like fire, and so advertise the presence of poison in a two-fold manner" (Loeb ed., vol. 9, p. 63).

9. One of those interesting problems which the text, the language, and the society imposes. The original "*onze groote*" means literally "our great ones," but the language and the society resist using "nobles" because they were not important in the Netherlands. Any other grandiloquent form of address—such as "the great," "the mighty," "majestic," "magnificence," or the Spanish *grande*—would sound very odd in a Dutch context.

10. "*Rotan*," or "rattan" in English, referring to the pliable stems of the Calamus palms (*rotan*

in Malay) used for various things, for instance caning furniture, but also as a switch used for beating people or simply to indicate a higher social station. Something like a swagger stick. Rumphius describes *rotan* in his *Herbal*, book 7, ch. 53, pp. 97–101.

11. This proportion is from Pliny. *Mas* means "gold" in Malay, and *muda* means "young."

12. "*Tonquin*" is Tonkin in the former Indochina, now better known as North Vietnam. J. P. Puype mentions that Rumphius' "black Suassa" is *shakudo*, and notes that this passage is "the earliest Dutch reference (I dare say the earliest Western mention as well) of *shakudo*." See J. P. Puype, *The Visser Collection. Arms of the Netherlands in the Collection of H. L. Visser*, vol. 1, in 3 parts (Zwolle: Waanders Publishers, 1996), part 3: 194–196.

13. Al Cohen thinks this is *hung t'ung*, which literally translates as "thunder [sound] copper." *Lei* means "thunder."

14. "Aurichalcum or Orichalcum of the Ancients" was a highly prized copper ore (see Pliny, book 34, Latin paragraph 2), and "*cadmia*" is what Pliny called "*cadmea*" (and which he discussed at length in book 34, ch. 22) and what Rackham identifies as silicate and carbonate of zinc (Loeb ed. vol. 9, p. 201).

15. "*Bateca*" should be *beteka*, or *mendikai*, listed in Wilkinson as the word for watermelon (*Citrullus vulgaris* Schrad.).

16. I am using the King James version.

17. This would translate as "gleaming in copper." Liddell and Scott translate the second word as the adverb meaning "brilliantly," while στίλβω is the verb for "to glitter" or "to gleam," especially in connection with polished or bright surfaces, such as shields.

18. The King James version has "and I saw as the colour of amber," which Moshe Greenberg translates as "I saw the like of *hashmal*," which he does not otherwise identify. See *The Anchor Bible Ezekiel 1–20*, transl., with introduction and commentary by, Moshe Greenberg (Garden City, N.Y.: Doubleday, 1983), p. 38. However, Ezekiel translates the Hebrew *ḥašmal* with *electron*; see *The Anchor Bible Revelation*, transl., with introduction and commentary by, J. Massyngberde Ford (Garden City, N.Y.: Doubleday, 1975), p. 383.

19. One has to remember that "*elektron*" in Greek could refer both to amber and to a compound of gold and silver.

20. "Bochardus" was Samuel Bochart (1599–1667), a French scholar born in Rouen, who moved to Leiden in 1621, but spent most of his life in Caen. The book is *Hierozoicon seu de Animalibus Scripturae*, which was published in Holland in 1692.

21. "*Aurichalcum*" should probably be *orichalcum*, after a Greek word meaning "mountain copper," a yellow copper ore.

22. This is a compound noun of which the first part means "copper" and the second one "gold."

23. In the King James version, 1:15 of *Revelation* reads: "and his feet like unto fine brass." The Greek word used there for "brass" (also translated as "bronze") is the one Rumphius quotes from Bochardus: *chalkolibanos*. Massyngberde Ford (see n. 18) states that no one is sure what it means (although I would note that it contains the Greek for copper, *chalkos* [χαλκός]), though, he adds, that "it must be a metal or alloy like gold ore or fine brass or bronze" (p. 383). Interestingly enough, he notes that the word is a hapax legomenon; i.e., it is used only once in the text.

24. See the quote in n. 8.

25. *Zesjes* were small silver coins worth very little. *Brongo*, in the phrase "*brongo sari*," is the Javanese word *prunggu*, which refers to bronze, while *sari* means the quintessence of something; hence the phrase means "the very best kind of bronze."

26. *Sussuhunan*, often abbreviated as "*sunan*," is the title of the sovereign of Surakarta or Solo, the Javanese city and realm in central Java.

Chapter 5. How Some Metals Are Affected by Spiritus Salis.

1. This translates as "spirit of common salt" and is hydrochloric acid.

2. See the subsequent ch. 9.

3. In the original "*Spaansche mat.*" The subject of colonial money is a large and hopelessly confusing one. There is no need to unravel it in terms of this book; hence only some fundamental aspects must suffice. One should first mention that it is impossible to give stable equivalents for any coin in Rumphius' day. Value fluctuated wildly, both on an official and on a personal level; hence no one can provide a reader with a table that would be accurate in every circumstance. Second, there was a proliferation of different coins, both European and Asian, which confused the picture even more. Third, the inevitable profit mentality of the VOC created unnecessary problems.

If there was such a thing as a standard denomination in Rumphius' day, it would have to be the Spanish real, known to Hollywood as "piece of eight," at the time in the colonies also known as "dollar" or "piaster" (see De Haan, *Priangan*, 3:730–732). The Portuguese introduced this coin, and the

Dutch maintained its existence until the beginning of the eighteenth century, when the *rijksdaalder*, or "rixdollar," became more prominent. In fact, De Haan states that in the Moluccas the two coins were synonymous (2:691) and worth forty-eight stivers. Here we touch on the other problem. Most of the time the individual value of coins was expressed in *stuivers*, "stivers" in English, our "nickels." The *stuiver* value was set by the VOC on its own accord, which meant that it was arbitrarily determined by the prevailing desire (also known as greed) for profit. The argument was that Asians coveted silver in both minted and unminted form; hence the VOC instituted the absurd logic of saying that, since the precious metal was so highly prized, they would lower the silver content in their coins—for instance 1.5 percent less for the colonial as compared with the European real; the latter was supposed to contain 25.306 grams of pure silver—while at the same time increasing the number of *stuivers* a real was worth! This led to the bizarre situation that, although the real was supposed to be worth forty-eight *stuivers* at first, it was soon determined to be worth sixty. *Stuivers* of the first kind were considered the tangible ones, actual coins, and were therefore called "heavy *stuivers*," while the other variety of the inflated price were imaginary, hence known as "light *stuivers*." These two totally different exchange rates for the same coin persisted into the nineteenth century. One can imagine that the confusion was horrendous, and the opportunity for fraud a common incentive.

The Spanish real was a silver coin. Most other coins were also silver: *Leeuwendaalers* (worth presumably forty-eight *stuivers*), *kruisrijksdaalder* or "Cross rixdollar" (about sixty *stuivers*), ducatons (about fifty-three *stuivers*), while after 1678 they added the *rijksdaalder*, or "rixdollar" (sixty "heavy" *stuivers* and seventy-five "light" ones), the "silver rider" (about ninety *stuivers*), *ropijen* (twenty-four "heavy" and thirty "light" *stuivers*), "small silver rider" (with the horseman facing to the left), and so on. The same insanity applied to gold coins, such as the Japanese gold *kobangs* (worth presumably ten rixdollars), and Japanese gold *itzebus* (worth about two and one half rixdollars), and the Dutch gold ducatons (worth about two rixdollars). Many other coins were also legal coin, so that monetary exchange was a true nightmare. Counterfeit coins were everywhere, silver and gold contents varied, values could only be called mercurial, and reform impossible. That is why I do not try to affix a firm value or weight—since coins were used as weights as well—to anything Rumphius might mention. The situation did not significantly improve until 1854. See the extensive article under the title "*Muntwezen*" by G. Vissering in the *Encyclopaedie van Nederlandsch-Indië*, vol. 2.

Chapter 6. Rare Iron in the Indies.

1. "*Crimata*" is presumably Karimata, the main island of the Karimata group, some sixty islands off the southwestern coast of Borneo, below Pontianak, in the Karimata Straits. They were not heavily populated and of little renown, though Wichmann quotes sources that prove that there was a mine and that the local population knew a number of blacksmiths. The "iron," according to Wichmann (p. 143), was limonite ("*bruinijzererts*"), which is a naturally occurring hydrous ferric oxide.

2. "*Tambocco*" is Bungku or Tobungku or Tombuku, a region on the east coast of Celebes (on the eastern of the two "legs," not the long northern one), divided by Tomori Bay into north and south Bungku. It was indeed once praised for the manufacture of its swords, which were forged in the village called Toèpé. Wichmann notes that this is the first mention in recorded writing of the presence of a special ore in this region. See also Albert C. Kruijt, "Het Ijzer in Midden-Celebes," in *Bijdragen tot de Taal-, Land- en Volkenkunde van Nederlandsch-Indië*, zesde volgreeks—negende deel (deel LIII der geheele reeks), 1901, pp. 148–160. Kruijt quotes Valentijn but does not mention Rumphius.

3. "*Tommadano* Lake" is Matano Lake in southern Bungku. It is the deepest lake in the archipelago (1,934 feet), and it also is one of the most elevated. The water temperature is more than 80 degrees, and the water is very clear.

4. The original has "*wellen*," which is a Germanism. This "wavy" (what the German means essentially) or "watered" pattern on the blade was highly prized. That this process was completely unknown in Europe is proven by the fact that most Western European languages made a verb out of the city in the Orient where they first encountered this process. "Damascene" as a verb and a noun was derived from Damascus and in usage in the sixteenth century, though the name of the city, Damascus, was known already in the Middle Ages.

Wichmann makes the important point that Rumphius' brief but accurate description of these swords contains the first mention of the *pamor* technique of producing wavy lines on a blade, also known as "flames." It was an effect much prized in Java, but Rumphius is the first to note its equally important presence in the eastern archipelago. There *pamor* refers both to pattern and the metal. *Pamor* appears to have been a white iron that was easily worked. As Rumphius correctly noted, the smiths made these blades from alternating layers of iron and *pamor*. The local technique gave these swords durability as well as resilience, so that the weapon would not break and yet could cleave through thick material; see Wichmann, pp. 143–146. For a lengthy discussion of *pamor* motifs on krisses and how they are achieved, see J. Engel, *Geschiedenis en algemeen overzicht van de Indonesische wapensmeedkunst* (Amsterdam: Samurai, 1980), esp. pp. 28–56.

5. *Ghiry* (Giri) is the name of a 409-foot-high hill, half an hour to the southwest of Grisee (south of Surabaya), where one will find the holy grave of Raden Paku or Sunan Giri, a pupil of Raden Rachmat, one of the original Muhamedan proselytizers who spread the doctrine of Islam throughout Java.

6. "Open" here means not walled in.

7. A city below Surabaya. See n. 4 of ch. 26 in book 1.

8. This should be *panembahan*, a royal title.

9. Rumphius generally means Muhamedans when he uses "Moors" or "Moorish."

10. *Besi* means "iron," and *keling* or *kling* referred to someone from southern India.

11. This truncated story, which is not entirely correct, refers to an important episode in the seventeenth century, part of the VOC's larger scheme to gain control of Mataram, Bantam, and Makassar, as well as insuring certain trade monopolies. The VOC was literally and figuratively stuck between the powerful kingdom of Bantam in west Java and the ancient power of Mataram in central Java. In the eastern archipelago, where it got spices, its main opponent was Makassar. Governor-General Maetsuycker disdained force and tried to settle differences and gain advantages by means of diplomacy, but the VOC soon found itself obliged to use force in order to intervene on behalf of Mataram, an intervention that was solely meant to increase the VOC's hold over that ancient kingdom.

The reason this once-glorious realm was in serious decline was the rapacity and debauchery of its sultan, Amangku-Rat, son of the more illustrious Sultan Ageng. We need not detail Amangku-Rat's egregious behavior; suffice it to say that he killed a great number of people, led a thoroughly debauched life, and held sway with a reign of arbitrary terror which did not exclude his immediate family. Although the Javanese population was never quick to rise against its sovereigns, there was a great deal of dissatisfaction. A man named Truna Djaja (Rumphius' "Trunajaja") took advantage of this. Ironically, he had been brought up in the sultan's *kraton*, or royal compound, in Plèrèd (central Java), even though he was born in Madura, the diminutive version of Java that hovers in the space cut out of the northeastern coast (above Surabaya). Truna Djaja declared himself Mataram's implacable enemy and set himself the task of becoming the supreme ruler of all Java. He found military support in a large number of Makassarese warriors who had quitted their homeland when Speelman defeated that realm in 1669. They lived by pillaging coastal areas on Java, but in 1671 they offered their allegiance to the Sultan of Bantam, who was the implacable foe of the VOC, thereby hoping to undermine Mataram's power. As was so often the case in those days, problems about a woman created bad blood between the sultan from Bantam and the Makassarese leaders, and the latter allied themselves to Truna Djaja and proceeded, with his aid, to burn and pillage eastern Java in 1675. They were joined in the same year by a third ally, the priest-king or Panembahan of Giri, the same one whose inferior rings and bracelets solicited Rumphius' ire. The priest-kings of Giri were traditional enemies of Mataram; hence they were eager to support Truna Djaja, while Bantam soon followed suit because the opportunity to undermine both Mataram and the VOC was too good an opportunity for the sultan to miss.

In 1676 Truna Djaja attacked Mataram. The time was well chosen because the court was in disarray. The Susuhunan (the preferred title of the Mataram sovereigns) was paying little attention to government, his four sons were warring among themselves, with the succession to the throne as the prize, corruption was rife, and the people were ready to disobey. The Makassarese were devastating Mataram's coastal holdings in east Java, and the VOC was still watching and waiting. The success of the Makassarese prompted the Europeans to intervene militarily, but it was not enough. The badly led Mataramese forces were defeated by Truna Djaja. Now Batavia knew it had to act. It therefore sent Speelman, the successful general from the wars with Makassar, to try and save Mataram. This proved to be easier said than done. In 1677 Mataram fell to the forces of Truna Djaja; the Susuhunan fled and died in exile. His eldest son, Anom, succeeded his father as Susuhunan Amangku-Rat II, the "Amancurat" in Rumphius' account. The new Susuhunan depended completely on the Dutch to overcome his many enemies, who, because he was in league with the Europeans, accused him of being unfaithful to Islam, and incited his opponents to a religious war.

Suffice it to say that many more complications arose until in 1678, the Dutch sent a force inland, hoping to defeat Truna Djaja once and for all. The expedition set out from the coast in September, and this represents the first major attempt by Europeans to penetrate Java's heartland. The former governor of Ambon, Antonie Hurdt, was given command, and Hurdt, together with Tak, St. Martin, the legendary Buginese leader Aru Palakka, and the equally legendary Captain Jonker with his Ambonese found themselves in a difficult and hard-fought campaign which only ended with Truna Djaja's defeat in November of 1678, when the capital, Kediri, fell to the Europeans and their allies.

Truna Djaja managed to escape, however, and once again found support from the Makassarese, who only after very severe fighting were finally destroyed in February of 1679. Truna Djaja was not captured until Captain Jonker cut off all his retreats from a mountain outpost, forcing the Madurese to surrender in December of 1679. Brought before Amangku-Rat II, the Susuhunan killed Truna Djaja himself in January 1680. The destruction of the priest-king of Giri by Amangku-Rat II happened in the same year. Again with the help of VOC troops, the Susuhunan managed to defeat the Panembahan but only after a desperate fight on his holy mountain. As Rumphius reports, the Panembahan was wounded and captured by Amangku-Rat II, who had him and the twenty-five members of his family exterminated. As Veth notes, echoing Rumphius, "now the house of Mataram had finally been able to revenge itself on the priest-kings of Giri, who had never acknowledged its authority" (p. 57). Mataram had been rescued, but this was a Pyrrhic victory since it was now entirely dominated by the VOC. Bantam was to follow shortly, and by the end of the century the company was in firm control of the former independent states of Java.

The above is based on P. J. Veth, *Java, geographisch, ethnologisch, historisch*, 3 vols. (Haarlem: Bohn, 1898), 2:1–60; and Bernard H. M. Vlekke, *Nusantara. A History of the East Indian Archipelago* (Cambridge, Mass.: Harvard University Press, 1945), pp. 154–161.

12. The *Parrang* (*parang*) is an Indonesian type of machete.

13. "*Tatsjos*" must refer to *tatju*, which De Clercq (*Maleisch der Molukken*) says was a shallow iron pan.

14. The Singalese were the people from Ceylon.

15. That Rumphius mentions Rycklof van Goens (1619–1682) here makes sense because this illustrious colonial figure from the seventeenth century was the governor-general (a position he held from 1678 to 1681) who ordered Antonie Hurdt into Java's interior in order to restore the Susuhunan to his throne, making Mataram something like a vassal state. Van Goens came to the Indies with his parents in 1628 and was given employment by the VOC while still in his early teens. He had a distinguished career in Indonesia as well as in India and Ceylon. He became known for his diplomatic skills, displayed during various embassies to Sumatra, Mataram, and Ceylon, but his military career was equally distinguished. He captured a Portuguese and a French fleet and dislodged the Portuguese from Ceylon and the Coromandel coast. After he had held the highest colonial office, he requested leave to return to Holland. He did so in 1682 and died three months after arrival. Van Goens' "pocket pistol" seems too long to me and, indeed, as Mr. Puype informed me, optimum length was nine inches.

Chapter 7. About the Metal Gans.

1. "*Gans*" is "ganza," a name, as Yule has it, "given by old travellers to the metal which in former days constituted the inferior currency of Pegu." Some people said it was lead; others that it was an alloy of copper and lead. Pegu was the name of a city and kingdom on the Burma coast, in the delta of the Irawadi.

2. This is Pejeng, a Balinese region famed for its antiquities, inland, northeast from Bali's capital, Den Pasar. Rumphius' "wheel" is the famous Pejeng "Moon," the most prized kettledrum from the Early Metal (Bronze) Age. "The kettledrums (*nedara*) of Southeast Asia are strictly speaking not drums (membranophones) at all, since there is no membrane over their opening. They are percussion instruments in the same category as bells, cymbals, [or] gongs. . . . All of these are 'idiophones,' made of one kind of material, and which when hit 'sound all of their own.' The material used is usually metal." (Kemper, p. 23). "Any Southeast Asian kettledrum consists of a hollow body or mantle which is open at the bottom and closed with a metal plaque or tympana at the top. According to the drum's height and width, it might be compared with either a metal cup, or a metal beaker, nicely profiled, turned upside down" (p. 24).

The Pejeng drum mentioned by Rumphius is the largest one known on Bali. It is just over six feet tall, with a diameter of over five feet. By Rumphius' day it was already a "mysterious object, having no connection whatever with Man's history . . . [and] no apparent connection with latter-day Bali, nor for that matter with extraneous early civilization in that part of the world. If the Pejeng drum was made on the island, using earlier technological methods and combining earlier decorative motifs, the casters did this in a remarkable way which shows them as a gifted and spirited people. The 'Moon' represents a culminating point in a technical and artistic development not to be equaled, as far as we know, anywhere in the same field" (p. 39). Bernet Kempers describes the Pejeng "Moon" at length in his archaeological guide to Bali: A. J. Bernet Kempers, *Monumental Bali. Introduction to Balinese Archaeology and Guide to the Monuments* (The Hague: Van Goor, n.d. [1977?]), pp. 23–29. The Balinese are serious in their veneration of their antiquities; hence few pictures exist. Bernet Kempers prints a rare drawing of the drum (p. 37). He also notes that the legend was mentioned "for the first time in the archaeological record" by Rumphius in this text, printed in 1705. He also notes quite sensibly that Rumphius' fanciful notion of its having fallen from the sky is "an explanation as satisfactory as anything else which could have been suggested" (p. 39), a more balanced and charitable point of view than De Haan's irascible condemnation of Rumphius' "monstrous fantasies," which, according to De Haan, prove that Rumphius was "not broad-minded." We would rather say today that it was Rumphius' generous mind which could entertain all sorts of peculiar lore that stamps him as a humanist. See De Haan's *Priangan*, 2:456–457 n. 9. To countermand De Haan's enlightened criticism, one might also mention Bernet Kempers' interesting assertion that this little text by Rumphius "marked the first appearance in world literature of any Indonesian antiquity" (p. 93). One should add here that Rumphius could not have seen it firsthand because he never left Ambon (except for a few excursions to neighboring islands), and that he most likely, as Bernet Kempers conjectures, got the information from Hendrik Leydekker, "one of the few people at that time who had visited interior Bali" (p. 93). This is confirmed by G. P. Rouffaer in his article "Een paar aanvullingen over Bronzen Keteltrommen in Ned.-Indië," *Bijdragen tot de Taal-, Land- en Volkenkunde van Nederlandsch-Indië*, zesde volgreeks—zevende deel (deel LI der geheele reeks), 1900, pp. 284–307. Henrik Leydekker sailed several times to Bali, where he was the guest of the Balinese king; Leydekker was Valentijn's predecessor, see *Oud en Nieuw Oost-Indië*, III, part 2, p. 257. Hence Rumphius heard it *told* by Leydekker and did not copy a written account, so that one can safely say that his text is the earliest printed European account of this important artifact.

3. "*Martavans*," also known as "martabans," were large earthenware, glazed vessels in the shape of vases. Their name derives from the town of Martaba, in Pegu, the same region that used *ganza* as currency. They were so large that, as Valentijn stated, it required two persons to carry an empty one.

4. Grave is a town in North Brabant. See also the next chapter.

Chapter 8. Ceraunia. Thunder Stone. Gighi Gontur.

1. The text deals with prehistoric tools and weapons, and, I suspect, some meteorites. Rumphius' theory of their provenance is nonsensical, of course, but it was shared by just about everybody at the time (see below). One should note, however, that Rumphius is greatly intrigued that these objects look like weapons. Most are axes; hence he was on the right track. Rumphius was the first European to write about these prehistoric tools from Indonesia. See Robert von Heine-Geldern, "Prehistoric research in the Netherlands Indies," in *Science and Scientists in the Netherlands Indies*, eds. Pieter Honig and Frans Verdoorn (New York: Board of the Netherlands Indies, Surinam and Curaçao, 1945), p. 129.

2. The term comes from Pliny, see below, but the Latin noun, *ceraunius*, derives from the Greek κεραυνός, which means "thunderbolt."

3. This is in book 37, ch. 51 (Latin paragraphs 134–135). In Eichholz's translation: "Among the bright colorless stones there is also the one called 'ceraunia' ('thunder-stone') which catches the glitter of the stars and, although in itself it is like rock-crystal, has a brilliant blue sheen. It is found in Carmania. Zenothemis admits that it is colourless, but describes it as 'containing a twinkling star.'" "*Boetylus*" should be "*baetulis*" and was presumably a different kind of stone (see below).

4. *Gigi* means "tooth" in Malay, and *guntur* means "thunder"; the other Malay phrase means "thunder stone."

5. Al Cohen conjectures this might be *lei-ya* and notes that B. E. Read, in his *Compendium of Minerals and Stones*, no. 114, gives *lei-shih* as the Chinese name for meteorites.

6. I doubt that both "Indian" and Chinese regarded thunder as a bull. This was true of the Hindus, but the Chinese considered it a dragon. I should mention here that when Rumphius uses "Indians," he generally uses it properly: i.e., people who live in India. Needham (*Science and Civilisation in China*, vol. 3, p. 480) mentions the dragon but also that the ancient Chinese believed that lightning (*tien*) and thunder (*lie*) were "the result of a clash between the two deepest physical forces which they could imagine, the Yin and the Yang," and notes that this conception was already partly comparable to the principles of positive and negative. In Vedic myths, Indras, the god of thunder, was a bull, and his horns were the thunderbolts. These thunderbolts could bellow, i.e., thunder that follows lightning. See Angelo de Gubernatis, *Zoological Mythology*, 2 vols. [1872] (reprint ed., Detroit: Singing Tree Press, 1968), 1:7–41.

7. "Jaw teeth" for "molars."

8. Rumphius uses this word in the sense of a piece of steel specifically designed to use with a flint to strike a spark. In a flintlock or pistol it was the piece of steel that was struck by the "cock" that held the flint.

9. "*Gape*" is most likely Gapi, in Rumphius' day another name for the Banggai Islands, off the east coast of central Celebes.

10. This is what is known as verses written to commemorate an event by incorporating the date in Roman numerals. In the original text the capitalized letters were to add up to MDCLXXVII or 1677: "ALs *Gapes* opperheer boog Voor *Kabonna* stoVt, / Voer Ik Door DonDers kraCht In 't hart Van 't yser-hoVt." It works, if one disregards the "A," adds up the three "D"s (each one representing 500), plus the "C," equals 1600. "L" is 50, five "V"s equal 25, plus two "I"s, equal 2. My translation could not keep this, of course.

11. These are the actual outriggers of an outrigger proa. In Dutch these frames are called, quite appropriately, *vlerken*, or "wings"; Valentijn spelled this same word "Njadjos," stating quite correctly that they keep the vessel in balance by touching the surface of the sea; *Oud en Nieuw Oost-Indiën* 2:182.

12. This is the *Pangium edule* tree, particularly famous for its seeds, which were used as food, oil, and a salt substitute, while the leaves were used to daze fish so they were easier to catch. The reason the latter worked—and also the reason that preparation of the seeds had to be done with care—is that the pangi, pangei, or pangium tree contains Prussic acid.

13. This is the areca nut (*Areca catechu* L.) used in the sirih chaw.

14. In the sense of a bowling pin, since I doubt that Rumphius had the geometric "cone" in mind. In Dutch both are conveyed by the same word.

15. This refers to the series of military campaigns the VOC waged against the Sultanate of Ternate between 1650 and 1656. See the introduction.

16. Grave is a Dutch city in the province of North Brabant, situated on the banks of the river Maas.

17. Goslar is a town in Lower Saxony, at the northern edge of the Harz Mountains. Heine describes it at length early on in his *Die Harzreise*, complaining that its pavement was as bumpy as "Berlin hexameters."

18. Perhaps this "Zeilerus" is Martin Zeiller (1589–1661), a German scholar. He lost the use of one eye in his youth but was reputed to have used the other better than most men with normal vision. He studied in Ulm and Wittenberg then traveled around the world between 1612 and 1630 as *Hofmeister* in the retinue of wealthier gentlemen. This traveling whetted his appetite for geography, and he published what was once a celebrated thirty-volume series of *Topographia* between 1642 and 1655. He settled in Ulm at forty-one. Rumphius might be referring to Zeiller's *Epistolische Schatß-Kammer*, published in 2 vols. in Ulm in 1683, or to *606 Episteln oder Send-Schreiben von allerhand Politischen Sachen*, published in 2 vols. in Marburg in 1656.

19. This is the passage, mentioning "secret philosophy" and "natural magic," that provoked De Haan's ire (*Priangan*, 2:456–457, n. 9), causing him to mock Rumphius for a number of misconceptions. What De Haan finds a "hare-brained notion" was once widely held all over the world. Schafer quotes Geoffrey Grigson and C. H. Gibbs-Smith's *Things: A Volume of Objects Devised by Man's Genius Which Are the Measure of His Civilization* (New York, 1957), wherein they state that "the Greeks and Romans did not identify the actual stone implements of early men. The ones chipped to shape escaped their notice; the ground and polished stone axes they took for thunderbolts fallen from heaven, a superstition found in every part of the world" (quoted by Edward H. Schafer in his *The Vermillion Bird. T'ang Images of the South*, p. 153). De Haan assumes a particularly uncharitable tone that one will find often noted in Dutch scholarship and criticism. He was writing in the early decades of the twentieth century, and could not see that his erudition was not common fare more than two centuries earlier. He also seems to have forgotten—even though he himself suffered the same problem in Batavia—that scholarly texts were rare or nonexistent in seventeenth-century Indies. Rumphius' greatest authority was Pliny, not a contemporary, and I suspect that Rumphius was also susceptible to the intriguing lore of the islands.

20. Wilkinson does indeed gloss *Cabbal* (*kebal*) as a kind of invulnerability obtained by magic art (*ilmu kebal*). This may refer to a hardening of the skin (*keban kulit*) until it resembles the rough rind of the jackfruit or the skin of a toad; or to making one's skin so slippery that weapons will glance off it (*keban minyak*); while there was also *kebal raksa*, which was a method of rubbing quicksilver over one's body and thus obtaining a metallic armor beneath one's skin, the idea being "that quicksilver rushes at once to a wounded spot and so prevents further penetration."

21. This refers to the Sula Islands, particularly the island Mangolé.

22. The reference in book 37, ch. 51, reads: "Sotacus distinguishes also two other varieties of the stone, a black and a red, resembling axeheads. According to him, those among them that are black and round are supernatural objects; and he states that thanks to them cities and fleets are attacked and overcome, their name being 'baetuli,' while the elongated stones are 'cerauniae.'"

23. This reads in translation: "These writers distinguish yet another kind of 'ceraunia' which is quite rare. According to them, the Magi hunt for it zealously because it is found only in a place that has been struck by a thunderbolt." One can tell that Rumphius' final comment here makes no sense, nor do I understand the addition of "Parthian" to the Magi. The Parthians were famous for their skills as horsemen and as warriors but not for gathering information. The Magi were a priestly caste of the Medes in Persia.

24. A *Patty* (*patih*) was an official who worked for both a local potentate and for the VOC.

25. Larike is a village on the southwestern coast of Hitu, and the "Sugar Loaf" is a rock, now called Batu Suangi, in the sea in front of the village. Rumphius was stationed there from 1657 to 1660.

26. This ammunition might well have been a stone since the "*pederero*" or "*pedrero*" was a small

piece of ordnance that could fire anything. Stones and pieces of iron were the most common items used.

27. A sober statement that has a chill all its own. This was the year Rumphius became blind, and it hints at his desperation. The hostile attitude of the population is puzzling.

28. "*Bernnetel*" is a Germanism; *brandnetel* is the common Dutch word. The text in the *Herbal* is in book 10, chs. 33–34, pp. 47–49. The Indonesian nettles discussed are the red and white "Urtica decumana or Daun gattal besaer," identified by De Wit as *Laportea decumana* (Roxb.), and "Urtica (molucca) mortua or Daun gattal matti," said to be *Acalypha boehmerioides* Miq. by De Wit.

29. This might be San Paulo (or San Pablo or San Pedro), one of the Desventuradas Islands in the Pacific, first encountered by Magellan in 1521.

Chapter 9. Ceraunia Metallica. Thunder Shovel.

1. *Philosophia secreta Veterum sive Magia naturali* means "the secret philosophy of the ancients and natural magic" and refers, I would assume, to a whole body of work on alchemy and magic. Despite De Haan's fulminations on this subject (cf. n. 19 of the previous ch. 8), this was a perfectly respectable branch of learning before and during Rumphius' lifetime. In fact, Zedler's *Lexicon*, published more than a quarter of a century *after* Rumphius' death, devotes considerable space to "magic" (vol. 19, pp. 287–303), and this in a work that prides itself to be, as the title indicates, *Grosses vollständiges Universal Lexikon Aller wissenschaften und Künste, Welche bisher durch menschlichen Verstand und Witz erfunden und erbessert worden* (The Great and Complete Universal Lexicon of *All the Sciences and Arts*, which, up till now, have been discovered and improved upon by Human Intellect and Wit [italics added]). Zedler treats the subject seriously, stating that "natural magic" (*magia naturalis*) is the "skill to disclose natural, as well as hidden powers and causes, of rare and unusual activities" (p. 301), emphasizing that this inquiry is not merely speculative but includes "purely practical things." Natural magic is then divided into "permitted" and "forbidden" magic, of which the first, or *philosophica*, is further divided into "curious" and "essential" magic. This kind of inquiry is based on what Martius in his *Magia naturali* calls "Philosophia corpusculari," which amounts to a kind of atomism, which resembles Rumphius' text. Basically, Martius contends that his atoms are "motes," tiny "corpuscles" that contain both spirit and matter, and which can be controlled and manipulated at man's discretion. Such manipulation or interference, which would now be called experimentation, can be used, for instance, "to increase and improve plants, trees, and fruits" or "to increase, change and enlarge flowers and shrubs." An example is given of a lettuce that grew so fast from seed that one could actually *see* it grow, gaining an entire inch within two hours, while the authority of the Bible is invoked with the example of Jacob obtaining spotted and "ring-streaked" cattle by having the animals drink from water wherein he first put "rods of green poplar, and of the hazel and chestnut tree; and pilled white streaks in them." Then he brought these herds to drink and "the flocks conceived before the rods, and brought forth cattle ringstreaked, speckled and spotted" (Genesis 30:32–43), thereby besting Laban.

Rumphius, therefore, is referring to what was once a perfectly reputable mode of inquiry and experimentation. For once, he disagreed with Pliny, who, in book 30 (ch. 1), attacks magic only to proceed to give a lengthy account of strange cures and ointments. This early form of science had a tradition that included Plato, Pythagoras, Democritus, Empedocles, Philo, and the Hebrew scholars of the Kabbala. Sources which were more contemporaneous with Rumphius were Thomasius (*De crimine magiae*), Johann Nicolaus Martius (*De magia naturali*), Rüdiger (*Physica Divina*), and Vossius. Earlier sources were Albertus Magnus, Raymundus Lullus, Arnoldus Villovanus, Caspar Schottus (*Magia naturali*), and Giambattista della Porta (1539–1615), who was a disciple of Paracelsus (1490–1541), the controversial German physician who preached the doctrine of signatures and his belief that all terrestrial things were infused with an astral spirit.

2. See n. 2 of ch. 6.

3. See n. 3 of ch. 6.

4. Borax is the well-known sodium borate.

5. Hydrochloric acid.

6. Bontuala (Bontoala) was Arung Palakka's stronghold in Goa, southern Celebes, after the defeat of Makassar.

7. Demak is a city near Semarang in north-central Java, and an important center for Islam in Indonesia. It was the ruler of Demak, Raden Patah, who was instrumental in destroying the Hindu realms of Madjapait during the early sixteenth century.

8. The "Javanese War" is the campaign described in n. 11 of ch. 6, and lasted from 1676 to 1679.

9. Arung Palakka (1635–1696) was the legendary king of the Bugis who was an ally of the Dutch in their efforts to defeat Makassar. His involvement precedes the "Javanese War" mentioned above, though he also aided the Dutch on Java because his archenemy, the Makassarese, were on the island as allies of Truna Djaja. What was really involved here was a fight for political supremacy in southern

Celebes, where for decades the Makassar kingdom of Goa had been supreme. This supremacy was eventually challenged by the Bugis kingdom of Bone, which, under the leadership of Arung Palakka, saw its chance to attain hegemony when the VOC decided to dislodge the Makassarese in 1666. In other words, the Indonesian nations were using the Dutch as much as the Europeans thought they were manipulating the Indonesians. This so-called Makassar War lasted from 1666 to 1669, when Speelman and Palakka and Jonker with his Ambonese defeated Makassar at Sombaopu. The Dutch general Speelman was in many ways as remarkable a man as Arung Palakka, but the Dutchman was always afraid that the Bugi leader would die because of his reckless behavior in combat. Arung Palakka was, as Rumphius noted, greatly feared as an *individual* fighting force, always in the vanguard, attacking the enemy on his own, performing feats of daring, but he was more valuable as a leader of his people. There is an excellent study, in English, of Arung Palakka and his role in Celebes' politics: Leonary Y. Andaya, *The Heritage of Arung Palakka. A History of South Sulawesi (Celebes) in the Seventeenth Century* (The Hague: Nijhoff, 1981). See also the introduction.

There is a problem in the text. In the description of Palakka's weapons he is said to have in his hands "*een schildeken, zabel en toraan.*" The first two words are plain enough, but the last one is not. "*Toraan*" must be an alternative spelling for *torana*, which Valentijn (2.1:182) describes as a "*werp-spiets als een korte piek,*" hence a short spear, a dart.

10. See n. 19 of ch. 3 in book 2. Rumphius sent this prehistoric axe to Mentzel in September of 1683, part of a collection of 104 different items. Menzel published a drawing of the axe and an excerpt of Rumphius' letter in *MC, 1685*, pp. 212–213.

11. From the Spanish and Portuguese *palmito*, this referred specifically to the tender shoots of the palm tree which were eaten as a vegetable. Was also known as "native cabbage." Hence this is *not* necessarily a palmeto palm.

12. On the 4th of January 1667, the Goa commander in chief, Karaeng Bontomarannu, surrendered to Speelman and lifted his siege of the island of Butun. Many Bugis came to Arung Palakka's side, and this signaled the beginning of Speelman's ultimately successful campaign to defeat Makassar.

13. See n. 20 of the previous chapter.

14. Rumphius' translation is exact except for the small points (which might well be due to the printer) that it should be "near Lurgea," that "Lydiatus" was an author, who wrote a book on fountains (*de fontibus*), and that "metaillicis" should not have that first "i."

15. "Captain Krytsmar," also spelled "Kreitsmaar" by Schynvoet in the later ch. 66, was the military officer Johan Christian von Kretschmar und Flämischdorf (1650–1693). He was the commander of a regiment of Dutch guards in the service of King William III of Britain, who was also Stadholder of Holland. Kretschmar had an extensive cabinet that contained minerals, fossils, and gems, as well as shells and marine fauna.

Chapter 10. Auripigmentum Indicum. Malay: Atal.

1. This text and those on *Tsjerondsjung* and *Hinghong* have several things in common. All three are, most probably, native minerals. Knowledge of these minerals, what their properties were, and how to use them came more than likely from ancient China. Note that all three texts make reference to China, although it seems that *atal* came from Europe and that *atal bato* was the rough *auripigmentum* that was imported from China. Auripigmentum does not seem to be found in Java, and it appears from Rumphius' description that he might have been writing about an artificial product obtained from mixing sulphur and arsenic. See Wichmann, *G*, p. 143.

The particular minerals of these texts are discussed in Joseph Needham, *Science and Civilization in China*, vol. 5, pt. 2, "Spagyrical Discovery and Invention: Magisteries of Gold and Immortality," *Chemistry and Chemical Technology* (1974).

Needham notes that orpiment and realgar were "prominent . . . in the oldest data we have on Chinese proto-chemistry" and that the two arsenical substances were used to produce "artificial gold and silver" (p. 224). Realgar (arsenic disulphide, As_2S_2) was called *hsiung huang* ("cock yellow"); orpiment (arsenic trisulphide, As_2S_3) had the Chinese name of *tzhu huang* ("hen yellow") (p. 191). Hence in the present context, Rumphius' "Atal" refers to both arsenic sulphides. Their medicinal properties were, until relatively recent times, believed in despite a dangerous potential for poisoning and death. The Chinese believed that realgar and orpiment had aphrodisiac properties, "effectively lifting impotence," as Needham puts it (p. 285), and when combined with cinnabar (mercuric sulphide) and sulphur into an "Elixir of the Four Magical Ingredients," would grant longevity and divert hunger. Various other minerals were also used to alleviate sexual deficiencies (Needham, pp. 285–286).

2. *Auripigmentum* (from Latin *aurum* = gold, and *pigmentum* = coloring matter) is now usually referred to as *orpiment* (what Rumphius calls *opermont*). It is a bright yellow mineral, an arsenic sulphide (As_2S_3), used as a pigment for the paint called King's Yellow. It is also known as yellow arsenic. In Holland's translation of Pliny it is said to resemble "burnished gold." *Realgar*, also known as ruby sulphur,

was formerly called red orpiment and red arsenic. This is also found in mineral form, and was used as a pigment in paint and for medicinal purposes. *Sandara* or *sandarac* is the same as realgar; Pliny (book 34) calls it *sandaracae*, and also reports its wholesome effect on a woman's complexion, and that it can even produce "a clear and melodious voice." When mixed with sulphur, orpiment and realgar have been used for pyrotechnics. Rumphius' mention of its medicinal use by the Javanese and Chinese is not unlikely. It can be taken internally, but in very small doses, for example, to cure certain forms of anemia, while it also has a tonic effect on the nervous system. It has been used against malaria and can be effective against chronic skin affections. In fact, it has been noted that people who live in Tirol and Stiermarken (the latter are known as Styrians) eat arsenic regularly to improve their strength and health and to improve their complexion. Toward the end of the nineteenth century this practice formulated a specific noun: arsenicophagy. According to Rumphius this was apparently practiced by the Javanese as well.

3. *Djambu* refers to the fruit of various trees, such as *djambu bol* (the "Malay apple" or *Eugenia malaccensis*), the *djambu ajer* (*Eugenia aquea*), and *djambu bidji* or *djambu klutuk*, better known as guava.

Chapter 11. Tsjerondsjung.

1. Rumphius is referring to *Gunung Api*, or "Fire Mountain," an always dangerous volcano that erupted many times in colonial times, covering the island in ashes, burying houses under hot rocks, and stirring up the sea at its beaches.

2. *Alum* is a mineral salt, very astringent, used in dyeing and medicine. *Alumen plumosum* should be *Alumen plumbosum* and means "alum with lead."

3. *Amianthus* is a mineral (variety of asbestos) that splits into long white fibres and is incombustible. It used to be called sometimes Earthflax or Salamander's Hair. See also ch. 80.

4. *Mataram* was a realm in middle Java which at one time held sway over the greater part of the island. In 1775 it was divided into the realms of Surakarta and Jokjakarta.

5. *Djudjambu*, *djambu*, or *jambu* refers to various fruits of, in European nomenclature, *Eugenia*, or *Syzgium* in English, part of the *Myrtaceae*, or myrtle family of plants.

6. *Tsjerondsjung* is probably not a Malay word but may be what the ancient Chinese called *tzu-jan t'ung*. Needham indicates that this was somewhat of a generic term that included iron pyrites, arsenical ores, and related forms of chalcopyrite "which [were] so glittering and highly coloured as to be called 'peacock ore'" (Needham, p. 200, n. "g"). The reference to iron pyrites relates it to the subsequent description of "Mestica Ular," while "peacock ore" may well be ascribed to the text in ch. 15, describing a mineral called *hinghong* by Rumphius, which doubtlessly is realgar, an arsenic ore. It is of interest to note that since *tzu-jan t'ung* was also translated as "natural copper" (Needham, p. 200), some of it was imported from "the East Indies, especially from Tan-Mei-Liu, a kingdom in Malaya [that was] tributary to Palembang" (p. 201). One may also note that Rumphius states that this material is easily confused with *Arsenicum album*. This "white arsenic" (arsenious oxide, As_2O_3), called in Chinese *yü shih* or *p'i shih*, was yet another form of arsenic (along with realgar and orpiment) used by the ancient Chinese in their metallurgy (Needham, p. 191).

The role assigned to *Tsjerondsjung* in sexual pleasure is also consistent with the aphrodisiac properties of arsenic mentioned in n. 1 of ch. 10. Needham quotes various sources, ancient and modern, which state that arsenic has most probably an erectile tendency due to its "vaso-dilation" of the blood vessels in the penis (see Needham, pp. 286–291).

It should be mentioned that Wichmann (*G*, pp. 151–152) identifies this *tsjerondsjung* as *Pickeringite*, which is a native magnesium alum.

7. *Nila* and *Damme* are two small islands in the southern Banda Sea, northeast of Timor. Damme, more often spelled *Damar* (also *Dammer* and *Daam*), is the remainder of an old volcano and still has solfataras, i.e., volcanic vents which emit sulphurous and aqueous vapors, encrusting the rim with sulphur deposits. At the beginning of this century, these sulphur deposits were exploited commercially. The same is true of Nila.

8. *Arsenicum album* is white arsenic, a white mineral chemically known as trioxide of arsenic (As_2O_3).

9. *Saltpeter* is a crystalline potassium nitrate. It has a saline taste and was the chief ingredient in gunpowder.

Chapter 12. Tana Bima. Badaki Java.

1. This means "Bima earth" in Malay, and Wichmann (p. 151) contends this is the "misy," mentioned by Pliny in book 34, esp. ch. 31. Rackham says that "misy" was copper pyrites (Loeb ed., vol. 9:214), while Wichmann says it is a ferrous sulphate.

2. Wichmann states in a note on p. 151 that this might mean "copper rust from Java," in that "*badaki*" might be *padakki*, which means "copper rust" but which really refers to a ferrous vitriol.

3. Wichmann glosses this as meaning "sour" or "tart" because the second component of this word should be *ngontjo;* hence this phrase means "sour stone."

4. This is, of course, Sumbawa, the large island between Lombok and Flores.

5. It is spelled correctly the second time: *sirih pinang*, the areca nut used in the sirih chaw.

6. *Zappan* (*Sappan*) wood is a small tree that grows in thickets, its Latin nomenclature being *Caesalpinia sappan* L. Common and widely distributed in the Malay Archipelago, India, and the Philippines. It yields a desirable, red dyeing substance. Crawfurd (p. 376) notes that its proper Malay name is *sapang*, and in Ambon it is called *lolan*. From the names Rumphius gives in various tongues, it would seem that he knew that the present shrub, *Anticholerica* (*Sophora tomentosa* L., according to de Wit, *M*, 415), is closely related to Sappan wood.

Chapter 13. Some Kinds of Marga and Argilla. Batu Puan.

1. *Marga* is "marl," and *argilla* was Latin for a white potter's clay. Rumphius found the notion of medicinal uses in Pliny, book 35, chs. 53–59 (Latin paragraphs 191–202). Marl is a calcareous clay, that was once used to top-dress arable land. Rich in lime, this clay is soft, white, gray, or brownish. Pliny described it as a mulch as well; see book 17, ch. 4.

2. This refers to a particular calcareous clay that was used to make a famous kind of ancient pottery known as *terra sigillata*, or samian, Arretine, or red-slip pottery. It seems to have been among the very few classes of pottery mentioned by ancient authorities; cf. Pliny, book 35, ch. 160. It was red-gloss tableware and one of the most important products of earthenware in Roman times, when it was traded as far as southern India, Africa, and Russia. Cf. D. P. S. Peacock, *Pottery in the Roman World. An Ethnoarchaeological Approach* (London: Longman, 1982), pp. 114–128. See also ch. 13.

3. Known in English as "bole," this is a friable earthy clay, usually reddish because it contains iron oxide, otherwise consisting mostly of hydrous silicates of aluminum.

4. *Napal* is the Malay word for edible earth. Both the *Encyclopaedie van Nederlandsch Oost-Indië* (under the heading "*Aarde*") and Wilkinson say that pregnant women, in particular, ate this kind of marl. Eating earth, otherwise known as geophagy, is not common in the Western world, but, according to the *Encyclopaedie*, was practiced all over the archipelago. The brief article only mentions *Ampo*, as the earth is called on Java, that it was roasted in the form of cakes, and that it was sometimes garnished with coconut oil. But the Moluccas are not mentioned, nor Rumphius' long chapter. Although it is said to lead to constipation, I would hazard the guess that geophagy might well be a peculiar way to obtain calcium in a society where dairy products were scarce. In any event, it does not seem to have been considered a pathology. It is in our society, and when children do this, the problem is called "pica," after the Latin word for "magpie."

5. As Wilkinson puts it, the primary meaning of *Puang* (*puan*) is a "caddy-shaped large betel bowl," i.e., a receptacle (gold if that of a king) that contained the leaves of the sirih plant, the main ingredient for the sirih chaw. It can also be a shortened form for *peremuan*, meaning "lady." Rumphius says that pieces of *batu puan* or "puan stone," otherwise known as *napal*, were used to decorate the table at weddings. This refers to a cultural pun, for Wilkinson notes that *makan napal*, or "to eat *napal*," was a figurative expression for being pregnant; hence the presence of *napal* expresses the wish for a fruitful union. The *pinang*, or areca nut, was also associated with marriage. For instance, an areca nut and sirih accompanied a man's formal marriage proposal (*sirih peminang*), or a young areca nut (*pinang muda*) was symbolic of a perfect match. Hence *pinang* can be said to be a literary but also a social metonymy for courtship and marriage.

6. This is the island known as Saparua, the most eastern of the Uliassers. Itawaka was a village in the north, Ulat in the south, and Tiouw ("Ouw") on the central southern coast.

7. *Waccasihu* (Wakasihu) on the far western coast of Ambon's Hitu peninsula, below Larike.

8. This would be on the southern coast of Hitu, diagonally across the bay from Ambon City. *Way* means "river."

9. From Galen's Greek, *cacochymy* indicates a bad or unhealthy state of a person's humors; this is the original and literal meaning of being "ill-humored." Physicians who subscribed to the doctrine of balancing the four humors in one's body would usually recommend purging for this rather vague condition of ill health.

10. *Batu* means "stone" and *lompor* (*lumpur*) is the Malay word for mud or slime. Malay has various expressions for different kinds of mud; for instance, mud on a road is *bechak*, and so on.

11. In Latin, the word for axle grease.

12. This means the "pith of stones." I do not know what this is.

13. This is the smallest of the Uliasser islands, just south of Saparua.

14. I would guess that "*baly*" should be *baldi*, a word which, coming from Portuguese *balde*, referred to a pail.

15. Oma or Haruku, the first island of the Uliassers, next to Ambon.

16. Wichmann says this "red bolus" is laterite.

17. Ulat and Ouw (Ow) are on Saparua, the Uliasser island that is shaped like a butterfly; these villages would have been low on its eastern "wing."

18. Terra means "earth," and the second word is made up of the word for "gold" (*aurum*) and a genitive plural of *faber*. Now a *faber* was, as Lewis and Short indicate, a worker in wood, stone or metal, that is to say a "forger," "smith," "artificer," etc., *faber* being synonymous with *artifex*. Hence my guess is that this is an unusual construction for "goldsmith" and that this substance is "goldsmith earth," on the analogy of "silversmith earth," which Pliny mentions in book 35, ch. 58 (Latin paragraph 199).

19. Another name for Saparua.

20. This cape is on Saparua's northernmost point, above Kulor.

21. "*Gorgeletten*" in the original. I used the French spelling as being the closest to Rumphius'; in Portuguese *gargoleta*, "gurgulets" in English, in Sinhalese *gurulota*, while the word survived in India and Ceylon as "*goglet*," synonymous with a "water bottle." As Gray puts it in his edition of Pyrard (see n. 8 of ch. 24, book 2) this "was an earthenware vessel with spout, whereby the liquid was poured into the mouth from a distance, to avoid contact."

22. Usually spelled *embalau*, this is a gum-lac or shellac.

23. In the westernmost part of the Leytimor peninsula.

24. This is an earth that is a mixture of hydrated oxide of iron with clay. Varying from light yellow to orange and brown, it is known as limonite.

25. Gerard Demmer, who was governor from 1624 to 1647, ordered all these villages in interior Hitu to be deserted or destroyed and their inhabitants shifted to the coast. This was the result of the war with Kakiali on Ambon's Hitu peninsula. It all concerned the VOC's insistence on maintaining a monopoly in the clove trade and forcing the local population to comply. In 1643, Demmer had just about accomplished depopulating the interior of the peninsula, except for Kakiali's ally, Telukabesi, who had entrenched himself in a fortress in Kapaha in northeastern Hitu. It took until 1646 and three Dutch campaigns to dislodge Telukabesi. The latter escaped but surrendered in August. He was beheaded in September. It spelled the end of Hitu sovereignty, and, despite future uprisings, the VOC remained in control. It is clear from Rumphius' history of Ambon, that he did not agree with Demmer's harsh policies. For those in the know, the present text would have been something of a quiet lament for a world and a way of life that was violently eradicated.

26. See ch. 19.

27. "Cimolian earth" can be found in Pliny, book 35, ch. 57, who lists a truly amazing list of things it can cure, from dysentery to swollen testicles.

28. A clay (hydrous silicate of alumina) once used to clean cloth by beating it or treading on it; this process was expressed by the verb "to full."

29. See ch. 82.

30. Wichmann identifies this as the island Peleng of the Banggai Islands just below the central "arm" of Celebes.

Chapter 14. Some Stones That Can Be Used for Painting.

1. Perhaps this is "*komala*" or "*kemala*," which is the same as *gemala* (or *gumala*), which refers to a luminous bezoar stone. Given Rumphius' fascination with bezoars (see ch. 55), this might be the case, although I do not understand how this can be turned into pigment. I have no idea what the other two stones might be.

2. This is an alternate spelling of "Atjeh" in Dutch or "Achin" in English, the name for the Islamic sultanate in northern Sumatra.

Chapter 15. Hinghong. Chinese Mineral.

1. *Hinghong* is almost certainly what is called in Chinese *Hsiung huang*. This is realgar, a native disulphide of arsenic, and is so identified in B. E. Read and C. Pak, *Chinese Materia Medica. A Compendium of Minerals and Stones* (2nd ed. 1936; rpt. Taipei, 1977), item 49.

As previously mentioned (see notes to *Auripigmentum Indicum*), Needham clearly states that realgar and orpiment were two forms of arsenic used a great deal by the ancient Chinese. He also notes that realgar was carved by the Chinese into "cups of great beauty" (p. 285).

Hinghong's purported efficacy against snake bites reminds one of a curious passage from an eighth-century Chinese pharmaceutical treatise. There it is said that gold can be poisonous. After remarking that "gold comes like red and black gravel," the author continues: "The southerners say that gold develops in places where the teeth of venomous reptiles have dropped out, or where they rested or where their toxic excrement lay about on the stones; and its poison is of the same category as that of orpiment and realgar. If a man has been poisoned by gold, he can be cured by applying snake preparations" (see the entry for the "Golden Snake," Needham, p. 62). Since realgar was often the name given to a mixture of sulphur and orpiment (also found as a native mineral), and since it was once thought to be produced by brimstone, and is also associated with snakes, one has yet another series of connections which link *hinghong* to *tsjerondsjung* and *auripigmentum indicum*, as well as to the text describing *Mestica Ular.*

2. I translated *Coerap* (*kurap*) as "ringworm;" and *zugtige beenen* as "legs suffering from dropsy," a disease characterized by the accumulation of water in the tissues of the body.

3. *Cancong* may be *kangkung* (*Ipomoea reptans*), a creeper which grows in swamps. It has pale red flowers and its leaves, shaped like arrowheads, are eaten as a vegetable and used for medicinal purposes. *Sajor* or *sajur* is the word for "vegetable."

Chapter 16. Black Sand.

1. Wichmann in *Gedenkboek*, p. 149, calls this "titanic iron-ore," which is ilmenite, a mineral of little value that is often part of basic igneous rocks. When such rocks deteriorate, one often obtains this "sand" which is indeed slightly magnetic (though not polar), black, and looks metallic.

2. Hitu-lama and Hitu-messeng together form the village of Hitu on Ambon's northern Hitu Peninsula.

3. For *Grisce* (i.e., Grisee); see n. 4 to ch. 26 of book 1.

4. This clan (Rumphius has "*stam*") is from the Hitu-messeng region (*tanah*, used in somewhat similar fashion to Mexican usage of *pueblo*).

5. "*Batschian*" is the island Batjan (now Bacan), west off the southern peninsula of Halmahera. In the sixteenth and seventeenth centuries it was involved in the constant warfare between Tidore and Ternate sultanates, with the Portuguese and with the VOC. Dutch influence dates from the first decade of the seventeenth century.

6. "*Gelolo*" is Gilolo (Jailolo), a name once synonymous with Halmahera, an island off the eastern coasts of Celebes, and shaped like that larger island.

7. That is: Amboina.

8. This is *Batu Merah*, east of the river Wai Tomo, which flows right next to Fort (or Castle) Victoria.

9. Robert Padtbrugge (1637–1703), an interesting individual, was governor of Ternate 1676–1680, and of Ambon from 1682 to 1687; hence during a time when both Rumphius and Valentijn were living in Ambon.

10. This is Nicolass Calf(f) (1677–1734), a merchant from the town of Zaandam (spelled "Sardam" in the seventeenth century), who traveled a great deal and owned a cabinet of various natural objects and gems.

Chapter 17. Cinis Sampaensis. Abu Muae.

1. Note that this country is variously called "Sampa," "Tsjamsias," and "Tsjampaa," but it is difficult to know for sure whether this is Champa, the name of a kingdom in what was formerly known as the east coast of Indochina, Annam, or Edward Schafer's "Nam-Viet" (discussed in his *Vermilion Bird*). Nor can I find a city by the name of "Colong." Finally, I have no idea what these bubbling ashes might be.

2. This was the fortress in the city of Ambon, first built by the Portuguese in 1580, and then enlarged, repaired, and reconstructed by Dutch governors for the next 200 years. It was known as Victoria Castle.

3. A "tael" or "tail" was a Japanese coin and weight. As a coin it was worth around three-and-a-half guiders; as a weight it equaled one Dutch ounce.

4. Usually spelled *pikul*, this was a weight originally from China, the equivalent of 100 Chinese *kati*, which translated into about 62 kilograms or between 130 and 140 pounds.

5. *Abu* is Malay for "ashes" and *mual* is *bual*, which Wilkinson translates as "bubbling up" or "frothing up."

6. "Bras" is *beras* the specific word for husked rice that has not been boiled yet (after it is cooked it is known as *nasi*). I suspect that the second word is *bual* or *mual*, which means "to expand" or "swell" especially of boiling rice.

7. One must be careful not to think this is the acid that mobsters throw at the faces of their victims. The older sense intended here is an alkalized water obtained by leaching or percolating vegetable ashes in water. In a general sense, "lye" was synonymous with any detergent material used for washing. There was even a lye for washing one's hair, not to mention the fact that urine was used as a detergent, called "Chamber-lye."

8. In Malay normally *kasumba* and in English "safflower," i.e., the dye obtained from the petals of the thistle-like plant *Carthamus tinctorius* L. It is usually a reddish orange obtained from the dried flowers which are known as *kembang kasumba;* it was used specifically as a dye for silk and yarns. As

Van Slooten mentioned (*M*, pp. 310–311), the process of obtaining this dye from the flowers is quite complicated, because the color is not fast to begin with. It may be that Rumphius mentioned this color also because the rind of the durian is used—i.e., the lye from its ashes—to extract the color, which naturally is yellow, from the flower heads.

9. Rumphius discusses the durian in his *Herbal*, book I, ch. 24 (pp. 99–104). He mentions that it is the Chinese who "burn these peels down to an ash that is used for making or preparing the color *Cassomba*. The lye from those ashes is also used to extract other colors from dried herbs and flowers which can be used to color thread and linen" (p. 103).

Chapter 18. Various Flints. Batu Api.

1. This is the island called Pulu Babi in Piru Bay, western Ceram.
2. This translates as "Books on gems."
3. Surat was once an important harbor in the Gulf of Cambay, north of Bombay.
4. Between Sapé and Bima on Sumbawa; see ch. 12.
5. Ch. 12.
6. As the *OED* puts it: "a piece of steel shaped for the purpose of striking fire with a flint [and] in a pistol or firelock, the piece of steel that is struck by the 'cock' carrying the flint."
7. But see the next chapter, where Rumphius says it is some kind of pyrite.
8. I do not know why Rumphius says this because "enorchis" only occurs in a long alphabetical list of lesser stones, book 37, chs. 54–70. It is mentioned in ch. 58, Latin paragraph 159, and all Pliny says is that "enorchis is white, and when it is split up into pieces [it] reproduces exactly the shape of the testicles" (Loeb ed., 10:295).
9. Anselmus Boëtius de Boodt: *Gemmarum et lapidum historia* (1606); see also n. 42 of ch. 23 in book 2.

Chapter 19. Androdamas. Maas Urong.

1. Normally spelled *mas*, this is the Malay word for "gold," and *urong* (*orong*) comes from the Javanese *wurung*, which means "worthless"; together *mas urong* means "iron pyrites." Wichmann says this is marcasite.
2. Rumphius is trying to get across that he is dealing with solid geometry not plane geometry.
3. This is the geometric figure dodecahedron, which is defined as "a solid having 12 plane faces and commonly in two forms." It is clear from the rest of the sentence that Rumphius is thinking of the first form: the pentagonal dodecahedron.
4. In Pliny, book 37, ch. 54 (Latin paragraph 144). Wichmann (n. 1 on p. 142) notes that Pliny's *androdamas* cannot be a pyrite. The name means "tamer of people."
5. This is maximum eleven millimeters, or 0.43 inches.
6. "Finnish bottom" is a partly literal rendition of *Bottonhavet*, the Swedish name for the lower portion of the Gulf of Bothnia between Sweden and Finland.
7. "Skerry" is an Orkney word that derives from Norwegian *skjer*, Swedish *skät*, and Danish *skaer*. It refers to rugged, isolated rocks which are submerged by the sea at high tide or bad weather.
8. For *arduinsteen*, a hard stone mined in Belgium. There was a white variety, but the blue one, called "blue stone" in English, was the most desirable; the gray variety was called "freestone." It was famous for being hard and was often used as a substitute for marble.
9. No more than eighteen millimeters, or 0.71 inches.
10. "*Salmiac*" is a contraction of *sal ammoniacum*. Sal-ammoniac is ammonium chloride. Wichmann, however, states (p. 142) that this is not what Rumphius is referring to, but is properly an iron vitriol.
11. This is the island Misoöl, off the coast of the "Bird's head" or New Guinea, north of Ceram. This rugged island is part of New Guinea in terms of its flora and fauna, not Indonesian. It was more important in the seventeenth century than it is today.
12. See also nn. 5 and 11 of ch. 6. This is the hill, half an hour's travel southwest from Grisee, where one will find the holy grave of Raden Rachmat or Sunan Giri. There are many other graves, the oldest dating from the fourteenth century.
13. For this city, Grisee, see n. 4 of ch. 26 of book 1.
14. This was the priest Puspa Ita.
15. In the original "*spyauter*," which was also spelled "*spiaulter*" and which corresponds to the English spelter and the French *spiautre*. This peculiar word for "zinc" probably came into English from Dutch.
16. This was indeed an old alternative noun for "gong" among the Dutch in the colonial Indies.
17. This is a close paraphrase of the passage in Pliny mentioned before: book 37, ch. 54, in the Loeb ed.

18. "*Argyrodamas*" is mentioned only once by Pliny in that same passage (Loeb ed., vol. 10:282); the name means "silver tamer."

19. *Clarissimus Salmasus* is the Latinized name of Claude Saumaise, better known as Claudius Salmasius (1596–1653). Salmasius was a French scholar, born in Burgundy. He was said to have been a child prodigy who understood Pindar at ten, and could compose his own Latin and Greek verse at the same tender age. Educated at Heidelberg and Paris, he became one of the most famous continental scholars of the seventeenth century, primarily known as a scholarly critic. Salmasius was feared as well as respected, though his arrogance earned him many enemies. It is only natural that many learned institutions invited him to join their faculties, but he turned them all down with the exception of Leiden, where he succeeded Scaliger (see n. 38 of ch. 23 in book 2). He wrote a great number of critical and polemical works, of which one was his detailed attack on Solinus (see below), entitled *Exercitationes Plinianas in Solinum Polyhistora*, published in Paris in 1629.

Salmasius had a most successful life except for two things: his wife and Milton. He married the daughter of another famous scholar at the time, by the name of Johann Mercerus. Anne Mercerus turned out to be a shrew who refused to feed her husband and who took his money from him. Soon she became known as "the other Xantippe." It is said that she hastened his death—Salmasius lies buried in Utrecht—but Milton also played a role. The English poet found out about Salmasius' sorry home life and used it to attack the continental scholar during a famous controversy between 1651 and 1655.

Charles II, in exile in France, had engaged Salmasius to write an elaborate tract in Latin, addressed to the heads of government in Europe, attacking the men in Britain who had committed regicide. Salmasius complied, and his text, entitled *Defensio Regia pro Carolo I*, appeared in 1649. Milton, the republican, was asked to write a defense, which he did in a long Latin reply entitled *Ioannis Miltoni Angli Pro populo anglicano defensio contra Claudii anonymi, alias salmasii, Defensionem regiam*, published in 1651. Also known in English as the *Defense of the English People against Salmasius*, Milton's tract is probably his most vicious polemic. Mocking Salmasius as a "mere grammarian" and a "bore of a pedant" whose writings were "the squallings of his professorial tongue" and who could only boast of "his presumption and his grammar," Milton harangued Salmasius as a foreigner who had no business meddling in British affairs of state, as a bogus scholar, and as a sorry husband. It was said at the time that Milton's personal attacks and intellectual invective caused the French scholar to lose his sinecure at the court of Sweden's Queen Christina, where, as Edward Philips wrote in 1694 in his biography of Milton, Salmasius was "dismissed with so cold and slighting an adieu, that after a faint dying reply, he was glad to have recourse to death, the remedy of evils, and ender of controversies."

Given the present context of Rumphius' life, it is an ironic coincidence that Milton lost his sight during the time he wrote his attack on Salmasius, a piece of work that established his reputation. It seems to me that Rumphius was unaware of this furor because he quotes Salmasius various times in his work, at some length, for instance, concerning cloves in his chapter on that spice in his *Herbal*, book 2, ch. 1 (p. 4).

20. Solinus is Gaius Julius Solinus, a Latin author who wrote shortly after A.D. 200. His main work is *Collectanea Rerum Memorabilium*, also known in Rumphius' day as "Polyhistoria." In this work Solinus describes the known world, in terms of history, origin, products, customs, etc. It turned out that Solinus had copied most of it from Pliny without acknowledgment, a fact that earned him the posthumous nickname of "Pliny's monkey." Solinus' only work was critically demolished by Salmasius; see the above note.

21. This confusing passage is based on ch. 37 of book 36 in Pliny, wherein he introduces "*schistos*" and "*haematites*." The latter are, as the name indicates, "blood stones" or a kind of red iron ore. The trouble is that Pliny then goes on to state in the subsequent chapter 38, on the authority of Sotacus, that there are five kinds of haematites, and "*androdamantas*" is one of them.

22. Not quite in the "same chapter" but in fact ch. 60 of book 37. "*Hexecontalithos*" means "sixty stones in one," and is said to show a multitude of colors. Eichholz translated the "Troglodytae" (Pliny has "Trogodytae"—both are correct) as "Cave-dwellers." These "Troglodytes" were Ethiopian cave dwellers.

23. The French zoologist Pierre Belon; see n. 39 of ch. 23 in book 2.

24. This "Arabian stone" is another mere name in the long list of stones mentioned by Pliny in ch. 54 in book 37; Loeb ed., 10:282. All he says is that it "closely resembles ivory."

25. This is a spurious text, wrongly ascribed to the legendary Thracian poet, who presumably initiated what became known as "Orphism." The poem in question is the *Lithica*, a text dealing with precious stones.

26. Once a legitimate term in chemistry, it referred to the residue left after distilling or melting down minerals. It applied most often to what remained after vitriol had been used.

Chapter 20. *Argyrodamas. Batu Gula.*

1. In ch. 36, Rumphius is more specific: there he calls it "Cotawo," a place situated between the city of Buton and "Culutsjutsju." This is a bay called Kalinsusu.

2. In book 37, but in ch. 54. One has to know that Rumphius assumes "*argyrodamas*" and "*androdamas*" to be the same thing. Pliny, however, did not. Hence Rumphius is simply transferring the characteristics of "*androdamas*"—that it has a silvery sheen and "always resembles small cubes"—to "*argyrodamas.*" Wichmann identifies this mineral as calcite, very helpfully described in a dictionary as "consisting of calcium carbonate crystallized in hexagonal form, cleaving readily into rhombohedrons, and including besides common limestone chalk, marble, dogtooth spar, Iceland spar, stalactites, and stalagmites." One can tell that Rumphius' description is quite accurate.

3. Sugar in our sense was unknown in ancient times; hence "*saccarum*" is medieval Latin. "*Cantium*" does not exist in either classical or modern Greek. "*Candidum*" would have to be related to *candida*, which means "white." Perhaps he had *cubium* in mind? The Malay *batu gula* is assembled from *batu*, the Malay word for "stone," and *gula*, which means "sugar." This phrase means literally, therefore, "stone sugar" or "sugar stone." Later it also came to mean "sugar candy" or lumps or cubes of sugar.

4. Wichmann says that this is gypsum, while Eichholz identifies Pliny's "stone" as selenite, which is a variety of gypsum in crystallized form. Pliny's *specularis lapis* or "mirror stone" is described in book 36, ch. 45. He says that it can be split into plates "as thin as may be wished" and says that the best kind was mined in Spain. He cites the charming practice of strewing flakes of it on the sand of the Circus Maximus's arena during the games, to give an attractively bright and glittery effect. Eichholz notes (n. d., pp. 126–127 of vol. 10) that selenite was often used as window glass into the eighteenth century.

5. In the original sense of becoming hydrated, as in the sentence: "Air-slaked lime has slaked by simple exposure to air."

6. Pliny describes this remarkable material in book 35, ch. 51. The Romans got most of it from Palestine, especially the Dead Sea. It occurs as a liquid, called petroleum, as a liquid solid, also known as tar, and as a solid, called asphalt.

7. Described by Rumphius in ch. 36.

8. What he has in mind is something like "grapeshot." But that term did not come into use until after Rumphius' death, around the middle of the eighteenth century. I also could not use it because, properly speaking, grapeshot referred to small iron balls that were connected. On impact it produced great havoc in a relatively large area. I used "langrel" (also "langrage") because that is what the original word *schroot* implies: case shot loaded with pieces of iron such as nails, bolts, and bars used to damage rigging and sails.

9. "*Batu bakilat*" is most likely *kilat bahu*, which is the Malay expression for "an armlet worn on the upper arm," according to Wilkinson. "*Bakilat*" most likely should be *berkilat*, which means "to flash," or "to be shiny." *Berkilat* derives from *kilat*, the word for "*lightning*," which, as Wilkinson notes, was explained "as the flash of the whip with which the Angel of the Thunders drives the clouds before her."

10. Since Holland had little need for a word like "glacier," Dutch borrowed the term *gletsjer* from Swiss. It has been in that language, according to Jan de Vries (*Nederlands Etymologish Woordenboek*) since 1507, deriving ultimately—via the Italian-Swiss dialect in Tessin, French, and vulgar Latin—from Latin *glacies*, which means "ice."

11. "Talc" for the original *talk*. Wichmann notes (n. 3 on p. 150) that this erroneous identification originated with Charleton and was iterated by Caesalpinus, Salmasius, Wormius, and Boëtius de Boodt—all authorities we have encountered before in this book.

12. In the original, "*glets-krystal.*" The first word (also spelled *glatsch* or *gletsch*) is the German Alpine dialect word for the Latin *glacies*, and it means "ice" or "gletsjer" (see n. 10).

Chapter 21. *Crystallus Ambonica.*

1. This seems to be ordinary quartz crystal.

2. For "*schurft.*" I doubt that Rumphius had the disease in mind but was using it figuratively. The English *scurf* (both noun and verb), cognate with Dutch *schurft*, had once a similar usage.

3. This is not entirely clear because the original has: "zy hebben alle 6 zyden, gelyk het Mathematische *corpus Chrisma* of eenhoekige *Cylinder*" (p. 231). I have a feeling that if one substitutes an indefinite article for the text's definite article, the meaning of the simile would be clearer, and I did so accordingly. The other problem is the final three words of the phrase. I am assuming that "*eenhoekig,*" not a normal word, means a "right angle," so that it translates as a "right-angled Cylinder."

As W. Buijze informed me, the simile is most likely based on the sacred monogram of Christ's name in Greek (called a "chrismon"): an I (iota) and an X (chi) combined within a circle. *Chrisma* is from the Greek *chrism* for "ointment," hence it means oil. If one looks at an interference figure of a uniaxial crystal, one will see a pattern that resembles this sacred monogram, one that is often included in the religious art of churches and chapels, for instance, in the mosaics in Ravenna. If one has this information, one can appreciate that Rumphius was quite specific for a time when crystallography was in its infancy. Up to the seventeenth century Europe thought that quartz was formed by freezing. The first significant examination was performed by the Danes Nicolaus Steno (1669) and Erasmus Bartholinus (also in 1669), while the great Dutch scientists Christiaan Huygens (in his *Traité de la lumière*, 1690) and Antoni van Leeuwenhoek (in his examinations under a microscope in 1695) made significant contributions. The foundation for a real science of crystals was not established until the last quarter of the eighteenth century.

4. Here is published evidence that this practice, about which one often reads in various personal testimonies, was in fact true. Gullibility is a historical constant, but one also wonders whether Rumphius himself encountered such a pitch. One should remember that he was once deceived by a compatriot.

5. "Bima" was a region on eastern Sumbawa, and the western part of Flores (this region was also known as Manggarai), as well as all the islands between Sumbawa and Flores. The first Dutch contact with this powerful region was in 1605, when it had to succumb to the redoubtable Speelman, since Bima was an ally of Gowa.

6. For "*vlugtberg.*"

7. As the subsequent passage makes clear, Rumphius is talking about *bertapa*, which constitutes an individual's withdrawal from the world to live a life of austerity. The notion of *tapa* originally had religious significance, and referred to an act of asceticism, when one mortified the flesh to obtain spiritual enlightenment. It was originally practiced by Hindu recluses in India. Niels Mulder described this religious practice as a form of asceticism "which may consist of fasting, praying, sexual abstinence, meditation, keeping awake through the night, *kungkum* (sitting for hours immersed in rivers during the night at auspicious places), or retreating to the mountains and into caves. The purpose of *tapa* is purification to reach *samadi*, which is a state of mind that can be described as a world-detached concentration in which one is open to receive divine guidance and ultimately the revelation of life, of origin and destiny" (Niels Mulder, *Mysticism & Everyday Life in Contemporary Java*, p. 23). This exalted purpose is alien to Rumphius. However, as is so often the case in Indonesia, everything has a dual purpose. The same practice of *tapa* is often portrayed as a way of obtaining magical powers, a sojourning with lesser forces such as spirits, ghosts, devils, or angels. See also ch. 77.

8. This is the very thin metal foil that once was put behind mirrors and on the backs of gems in order to increase the light's reflection.

9. Nusa telo (or *telu; telu* is Javanese for "three") is one of three small islands below Ambon's Hitu peninsula. Known as "The Three Brothers," these islands were called Lain, Hatala, and Ela.

10. This is Nusa Laut, the smallest of the Uliasser Islands, east of Ambon.

11. Marcasite (or the latter variant "Marchasita") is an ancient name for pyrites or crystallized forms of iron pyrites. In Rumphius' day this term was applied to a bewildering number of ores, minerals, and stones. They were once worn for decorative purposes.

12. A small island in Manipa Strait between Buru and Ceram.

13. See ch. 60.

14. *Batu gula* means "sugar stone"; in more modern usage this expression means a "sugar lump." See the previous chapter.

15. For the original's "*zegerynleer,*" once more commonly spelled in Dutch as *segrijn* (also *chagrin*), and spelled "shagreen" in English. A Persian word, it referred to a tough, coarse leather with a grainy surface, usually made from the skin of donkeys, horses, and, later, rays and sharks. Once commonly used in the Orient for making slippers.

16. "*Gecandilisseerde*" comes from the verb *kandiliseeren*, which derives from *kandij*. The verb refers to crystallizing sugar in order to make candy. Also used for the process of making candied fruit.

Chapter 22. Silices Crystallizantes. Batu Dammar.

1. These "crystalline pebbles" are identified as quartz by Wichmann, with some chalcedony mixed in (p. 147). Although the Dutch *keisteen* can mean "flint," I opted here for "pebble" because the Latin *silex* can mean both "pebble" and "flint," but flints do not shine like diamonds.

2. *Dammar* (*damar*) is the resin or gum that is exuded by certain trees, particularly Hopea and Shorea of the family of Dipterocarpaceans. The Moluccan *damar* used to come from a gigantic Hopea tree; the resin would simply flow from the trunk and branches, and be gathered either in pieces that had fallen to the ground or in coagulated chunks that were still sticking to the branches. *Batu* is Malay for stone, and the phrase, *batu damar*, referred to ordinary *damar*, not the better varieties such as *damar mata kuching* (cat's eye *damar*) or white (*damar putih*) or black (*damar hitam*) *damar*.

3. In the original, the verb was *verschieten*, which among other things, can refer to a shooting star, but "to shoot" in English was too ambiguous. Wichmann says that these stones were probably chalcedony.

4. This might be the gum obtained from the tragacanth bush (*Astragalus tragacantha*); it contains tragacanthin, which is a substance soluble in water, and bassorin, a substance that is not soluble in water but rather swells up to form a gel. Pliny mentions the "goat-thorn" rather perfunctorily in book 13, ch. 36.

5. A place on the southern coast of West Ceram.

6. "Amersfoort diamonds" are glossed in the *WNT* with a German quote from 1821, which informs us that these were crystalline pebbles that looked like diamonds, and which were found around the Dutch city of Amersfoort.

Chapter 23. Vasa Porcellanica.

1. Joan Nieuhof (1618–1672) was the steward of the diplomatic mission which the VOC sent from Batavia to China in 1655 and which lasted until 1657. The effort was not successful politically or economically, but Nieuhof's book was. His brother published it in 1665 in Amsterdam as: *Het Gezantschap Der Neêrlandtsche Oost-Indische Compagnie, aan den grooten Tartarischen Cham, Den tegenwoordigen Keizer van China* (The Ambassage of the Dutch East Indies Company to the Great Tartar Khan, the Present Emperor of China). It was translated into Latin in 1668, and into English in 1669. As Jörg (see n. 8) noted: "This published report of which many reprints and translations appeared, had a great influence on the later Chinoiserie mode, since the 150 or so copper engravings, made after drawings done on the spot in China . . . for the first time gave a wide public a visual image of all sorts of aspects of the highly mysterious Celestial Empire" (n. 16 on p. 321). For a discussion of the illustrations, see Leonard Blussé and R. Falkenburg, *Johan Nieuhofs Beelden van een Chinareis 1655–1657* (Middleburg: Stichting VOC Publications, 1987).

2. Rumphius' opinion is essentially correct.

3. Taking "*kassen*" as a misprint for "*kasten*."

4. Not likely. Pliny (book 37, section 708) describes these "myrrhine vessels" as being fashioned from a mineral deposit. Experts have identified it as fluorite, a translucent mineral that comes in various colors. Pliny did say that it came from the Orient, but he identified that as Parthia, not China. It is true that outrageous sums were paid for myrrhine items; Nero, as usual, outdoing everyone by paying 1 million sesterces for a single bowl.

5. This is my conjecture for the misspelling "*werstellingen*." The first edition confirms my interpretation. As to Rumphius' strange remark, one might surmise that he loved Sung and Ming pottery but did not think much of the cultural achievements of the Manchus of the Ch'ing dynasty.

6. A solution for "*gestotene*."

7. De Wit identified this flower as the *Chrysanthemum indicum*.

8. This is incorrect. In the original edition Rumphius refers to, porcelain is discussed on pp. 90–91.

9. Here I must forego my principle of not using words that were not current prior to 1700. The Dutch noun *kraakwerk* is properly "crackle ware" in English, but the *OED* does not have an instance of printed usage until 1867! This is one more proof how far ahead the Dutch were in various aspects of European society. Dutch usage precedes English by nearly two centuries. Despite Wedgwood, the English were quite late in mastering the manufacture of fine china.

The VOC was very interested in trade with China from the very beginning, but this was easier contemplated than done. The trade was dominated by the Portuguese (Macao), and the Dutch made little headway until they established themselves in Formosa and Japan. Coxinga's victory in Formosa once again diminished Dutch chances in China. Towards the end of the seventeenth century, the Dutch no longer tried to establish a mercantile foothold in China proper because the Manchus permitted fleets of junks to sail to the Indies and trade in Batavia. This was far more convenient and cheaper than outfitting their own ships; hence the VOC quickly availed itself of this favorable arrangement. Direct trade with China, including porcelain, was not established until the eighteenth century. For that development, see C. J. A. Jörg, *Porcelain and the Dutch China Trade* (The Hague: Nijhoff, 1982). For Dutch-Sino relations during the seventeenth century, see John E. Wills, *Pepper, Guns and Parleys. The Dutch East India Company and China, 1622–1681* (Cambridge, Mass.: Harvard University Press, 1974).

10. Rumphius is referring to Sir Thomas Browne (1605–1682), the British physician who obtained his medical degree at Leiden, perhaps in 1633. The work in question is one of Browne's lesser known texts: *Pseudodoxia Epidemica; or Enquiries into very many received tenents and commonly presumed truths, which examined prove but Vulgar and Common Errors.* First published in 1646, it was translated into Dutch in 1668, and a German translation appeared in 1680. Rumphius is quoting section no. 7 in

chapter 5 of the second book. Since Rumphius' rendering is nearly verbatim and Schijnvoet's addition entirely so, I simply quoted from Browne's original text.

Chapter 24. Pingan Batu. Gory. Poison Dishes.

1. See n. 8 of the previous chapter.

2. "Curved rims" for *omgewelde kanten.*

3. The word *glaze*, a superior layer on pottery, is also a usage not recorded in English until the nineteenth century. The word was not in Johnson's dictionary, which, by the way, has a large number of entries that are of Dutch origin.

4. Since Rumphius mentions the date 1684 at the end of this text, the ruler he is referring to was a Manchu, K'ang Hsi, who ruled China from 1661 to 1722. He is the subject of the book by Jonathan D. Spence, *Emperor of China. Self-Portrait of K'ang-hsi* (London: Jonathan Cape, 1974).

5. "World" for *kreits*, a Germanism (from *kreitz*). This was poetic usage for a circle (of influence, for instance), sphere, the English word "ken." By extension *kreits* could mean "region" or "world."

6. "Codiho or Terminalis bush." Perhaps the first is *Codiaeum variegatum*, of the *Euphorbiaceae*, also known in the Indies as *Puring* and *Croton*. Rumphius describes "Terminalis" shrubs in his *Herbal*, book 6, chs. 39–40, noting in particular that the leaves were used as an antidote to the stings of the poisonous fish, "Ikan Swangi." Merrill (p. 137) states that "Terminalis" is *Taetsia fruticosa* (Linn.).

7. With "Alphorese" Rumphius is referring to the Alfurs, the inhabitants of interior Ceram. But since he later applies the same term to the Dayaks of Borneo, he is clearly using it in a general sense. The *Alfuras* (or a variety of other spellings, such as *Alfours* or *Harafuras*) was the name the Portuguese gave to the indigenous peoples who lived in the interior and mountains of the Moluccan Islands and who had not been subdued by them, somewhat equivalent to the Spanish use of *Indios bravos* for all the native Indians of the Americas. Rumphius always uses the term for wild "Mountain People," but one has to remember that they did not live on Ambon or the Uliasser Islands (i.e., Haruku or Oma, Saparua, Nusalaut) but inhabited the interior of Buru and Ceram. Even by the last quarter of the nineteenth century, they were still described as savages, only clad in a *tjidako* to cover their genitals. On the large island of Ceram, above Ambon and equal to the size of Halmahera in square miles, the Alfuras were divided into two tribes: the *Patasiwa* (or *Uli siwas*) and *Patalima* (*Uli limas*), with the first in the western, and the second in the eastern, part of that island. Their religion appears to have been a form of animism—a belief that trees, stones, forests, and mountains were populated by spirits, and they also revered the souls of their dead (*nitu nitu*). The Alfuras were headhunters; the number of heads determined a man's status in his community and was indicated by black rings on the belt of his *tjidako* (E. W. A. Laufer, "Schets van de Residentie Amboina," *Bijdragen tot de Taal-, Land- en Volkenkunde van Nederlandsch-Indië*, 3:53–78; and Hoëvell, *Ambon*, pp. 147–59).

8. "*Gumut*" is most likely *gemutu* (or *gomito*). These are the black threads found between the stems of the leaves of the Aren Palm (*Arenga saccharifera*), similar to horse hairs. "*Rottang*" is better known as rotan (also rattan), a climbing palm used to wrap or bind things and make mats, baskets, even furniture. Wilkinson lists sixty varieties.

9. Taking "*gebek*" as a misprint for "*gebrek.*"

10. Reading "*randen*" for "*ranken.*"

11. *Kyalura.* It is not entirely clear what Rumphius has in mind here. The first element of the word is probably *kjai*, which was an honorific. *Lura* referred to a "chief" or "head" and was so used on Java.

12. This "sublimate" was most likely mercuric chloride, a white powder that is a violent poison, and which was once used to kill people surreptitiously. As Ben Jonson put it in *Silent Women:* "Take a little sublimate and goe out of the world, like a rat."

13. *Sampo* refers to the Chinese Admiral *Chêng Ho* (now also Zheng He), who lived in the fifteenth century. Chêng Ho was a prominent eunuch—known in chronicles as the "Three-Jewel Eunuch"—at the court of an Emperor of the Ming Dynasty, who sent him on a long journey into the "Western Ocean" because he believed that his predecessor had fled across the seas. Chêng Ho left China probably in 1405 with 62 ships and with nearly 40,000 men under his command. Altogether Chêng Ho made seven voyages exploring the Indian Ocean and the South Seas. On the first three voyages (1405–1407, 1407–1409, 1409–1411) he visited, among many other countries, Java (called *Chao-Wa* and also *Shê-Pho* in Chinese) and Palembang (*San-Fo-Chi*) and went as far as Ceylon. On the fourth voyage (1413–1415) he went as far as the Persian Gulf, while on the last three (1417–1419, 1421–1422, 1431–1433) he explored the east coast of Africa and even brought back a giraffe to the imperial court.

On all seven voyages these Chinese fleets visited the East Indies, including Java, Sumatra, and Borneo (anchoring at Brunei). The last and "perhaps most dazzling of the Grand Treasure-Ship Fleets . . . left in 1431 and before it returned in 1433 its commanders, with their 27,550 officers and men, had established relations with more than twenty realms and sultanates from Java through the

Nicobar Islands to Mecca in the north and the coast of al-Zanj (East Africa) in the south" (Joseph Needham, *Science and Civilization in China*, vol. 4: *Physics and Physical Technology*, part 3 [Cambridge: Cambridge University Press, 1971], p. 490).

The voyages of Chêng Ho became famous in ancient China and were the subject of several books recording his deeds and accomplishments. A fourteenth-century term for the ships carrying the Chinese ambassadors was "Starry Rafts" (Needham, 4:492). That Rumphius would have heard of this great admiral is not unlikely. Needham puts it as follows: "So great was the fame of Chêng Ho and his companions in South-East Asia that they entered at last . . . into the realms of heroic hagiology. For the admiral was adopted as a tutelary deity by the Chinese communities of the Malayan diaspora, and incense burns to this day in the temple of Sam-Po-Tai-Shan at Malacca" (Needham, 4:494; note that the Temple's name contains the word "Sampo"). Chêng Ho is also revered in Semarang, a large harbor in Java. The city has a temple, called Sam Po Tong or "Sam Po's Grotto," where both Chinese and Javanese pray to his statue. They also have a large celebration in his honor in July. See Nio Joe Lan, "Cheng Ho: de te Semarang als heilige vereerde Chineesche eunuch," in *De Indische Gids* 62, no. 8 (August 1940), pp. 604–615, and no. 9 (September 1940), pp. 656–665.

Chêng Ho, as Needham puts it, may well be dubbed the "Vasco da Gama of China" (4:502). What he and his compatriots tried to accomplish was never a matter of conquest but rather a search for cultural contacts. "The Chinese voyages were essentially an urbane but systematic tour of inspection of the known world" (4:529). In keeping with our particular area of interest, Needham writes that "the Chinese had maritime and commercial relations with the Philippines, Java, Bali, Borneo and Sarawak, and the Moluccas and Timor," not only in Chêng Ho's time during the Ming Dynasty (i.e., the fifteenth century) but as far back as the Thang Dynasty, that is to say, from 618 to 906. This has been ascertained by ceramic wares found in the Indies. There was a particularly busy interchange with Borneo (note that Rumphius mentions Borneo particularly as a place where "Sampo" might have looked for the *Mestica Ular* in ch. 58 of book 3), where they traded ceramics, beads, and metal tools for edible bird's nests, hornbill ivory, and rhinoceros horn. This was true from the Thang Dynasty (which begins in the seventh century) to Chêng Ho's era in the fifteenth century (see Needham, 4:536–537).

It is abundantly clear that the Chinese anticipated European commerce with, for instance, the Spice Islands. The Chinese enjoyed nautical technology, maps, and superior ships. For example, the treasure ships or galleons of Chêng Ho's fleet were 1,500 tons or more, while Vasco da Gama's ships were never over 300 tons. The fleets of the Ming Dynasty were far greater than any fleet of the European nations (Needham, 4:484), and they had the ability and know-how to cross vast bodies of water, certainly as early as the ninth or tenth century. In contrast, the Portuguese first crossed the equator in 1474, first rounded the Cape in 1488 (Bartolomeu Dias), and Vasco da Gama was the first to enter the Indian Ocean in 1498 with the aid of an Arab pilot from Africa who directed him to India. This was half a century after the Ming Dynasty's navy had stopped frequenting the shore of Africa. In 1512 Serrao, a Portuguese captain, finally reached the Spice Islands, and used a junk to explore beyond Celebes. In 1513 Jorge Alvares anchored his trading ship in China. In the seventeenth century the Dutch supplanted the Portuguese. Needham is quite right to remark that "it is indeed an extraordinary historical coincidence that Chinese long-distance navigation from the Far East reached its high-water mark just as the tide of Portuguese exploration from the Far West was beginning its spectacular flow" (4:487).

The importance of ancient China in maritime history is undeniable; what is baffling is China's disappearance from the world's oceans. We can assume that the major development of a Chinese navy began in the tenth century and that its heyday was between 1130 and 1433. Its decline was far more rapid; Needham states that "by the middle of the sixteenth century almost nothing was left of its former grandeur" (4:484). As with other cultural achievements in the world, the death blow was delivered by politicians. The Confucian bureaucracy was firmly landlocked in its policies and decidedly xenophobic in its thinking. From 1435 Chinese policy changed permanently, and the navy was doomed. "Administrative thugs in the service of the Confucian anti-maritime party" went as far as to burn the records of Chêng Ho's travels, probably around 1477, because their contents were judged to be "deceitful exaggeration of bizarre things far removed from the testimony of people's eyes and ears" (4:525). For politicians, no matter what their ethnic background, the old adage is and will always remain true: *plus ça change, plus c'est la même chose.*

Primary emphasis has always been given to facets of the Indonesian archipelago which have little to do with maritime interests. But in an island realm of more than 13,000 islands, matters nautical are obviously of great importance. It is such a nautical item that argues for a link between Indonesia and China. Relief sculptures on the walls of Java's great Buddhist monument, the Borobodur, show a variety of ships, mostly Indonesian. But there is one which shows a striking similarity to a Chinese vessel, including a clearly visible rigid Chinese lug-sail, and Needham remarks that "we may have in this picture the oldest representation of a Chinese sea-going ship in history." The Borobodur was built around 800 (or just before), but Java was well known to the Chinese since the fifth century.

There may have been an interchange of Indonesian and Chinese nautical technology. Half a century before the Borobodur sculptures, a Chinese monk commented on ships from the south. Among those were "great ships of the sea" which were called *po.* One such type of large ship was called a Khun-Lun ship and "many of those who form the crew and technicians of these ships are Ku-Lun people." Needham's gloss on this entry mentions that *Khun-Lun* and *Ku-lun* meant "Malaysian." The binome *Khun-lun,* after first having had other meanings, came to mean both Malay and its people the Malays, because it seems to have been used to designate "Poulo Condor Island transliterating Pulo Kohnaong" (4:459). It may be remembered that Rumphius refers to "the Island *Contung,* which might be *Pulo condor,* where the Chinese Admiral *Sampo* sojourned for a spell." It is quite possible that *Contung* transliterates *Khun-Lun.* Be that as it may, the carving on the Borobodur walls, the *Khun-Lun po* ships (or "Malaysian deep-water ships") and *Khun-Lun* meaning "Malay" certainly argue for an early and significant reciprocity between ancient China and the Malay archipelago (4:457–459).

More evidence may be gleaned from the word "junk," that common term for a Chinese sailing vessel. This is not a Chinese word. It derives from the Portuguese *junca,* which is a corruption of the Malay and Javanese word for a large ship, *ajong,* and was abbreviated to *jong.* The Malay call the junk a *wangkang* and have the name *top* for smaller Chinese boats (Crawfurd, *Dictionary,* pp. 193 and 379). Yule in *Hobson-Jobson* corroborates this, adding that in Javanese the constellation of the Great Bear (Ursus Major) is called *Lintang jong,* which may be literally translated as "the heavenly transverse called the large vessel," and in Malay *Bintang jong* (or *djung*), i.e., "the stars resembling a large vessel (*Hobson-Jobson,* 472–473).

These remarks were based on Joseph Needham's scholarship. The major section on Chêng Ho is to be found in his *Science and Civilization in China,* vol. 4: *Physics and Physical Technology,* part 2: *Civil Engineering and Nautics* (Cambridge: Cambridge University Press, 1971). For sailing charts Chêng Ho may have used, see Needham, vol. 3, *Mathematics and the Sciences of the Heavens and the Earth* (1959), pp. 556–561; for information about a "rutter" Chêng Ho may have known—a "rutter" is a book which provides a navigator with crucial piloting information—see Needham, vol. 4, *Physics and Physical Technology,* part 1, *Physics* (1962), pp. 279–288. In the latter volume Needham records that the earliest mention of spectacles in China derives from a tribute of ten pair of spectacles from the king of Malacca in 1410 to the Chinese Emperor. It was Chêng Ho who brought these exotic articles home (pp. 119–120).

14. This is the Garuda, the mythical bird of Hindu legend, carrier of Vishnu, and still venerated in Java. See n. 16 to ch. 35.

15. This island lies due south of the Mekong Delta in south Vietnam.

16. This strange name presumably refers to Hung Wu, the first ruler of the Ming dynasty. See below, n. 28.

17. This low hill's Malay name means "Chinese Mountain," and it is still known as such today. It is in a suburban area northeast of Malacca's old town, above Egerton Road, and became a Chinese cemetery in the twentieth century. The "Chinese well" Rumphius mentions was still in use in the twentieth century. Known as the "Old Military Well" it kept on producing a very pure water, and was jealously guarded by Malacca's population. The well, also known as Hang Li Poh's Well, was reputedly dug by Chêng Ho's men. Near this same hill one will find a temple dedicated to Chêng-Ho as a deity. See Sarnia Hayes Hoyt, *Old Malacca* (Kuala Lumpur: Oxford University Press, 1993); plate 4 is a picture of Chêng-Ho's temple.

18. Pegu used to be a kingdom in Burma, also known as Martaban, now the Rangoon region. If the rest is correct, then "Ava" is "Awa," the easternmost province of Shikoku in Japan.

19. This is India.

20. "Guzatts," or "*Guzeratters,*" as Linschoten puts it (*Itinerario,* ed. Kern, 1:167–169), were "people from the land of Cambajen." That name refers to the principal city and major port, called Cambay, of the province of Guzerat in India. Also spelled *Gudsjerat,* this is a region on what could be called a peninsula between the Gulf of Cambay and the Gulf of Cutch. The people of Guzerat were Muslims and renowned traders. They traded, according to Linschoten, in cotton, linen, rice, precious stones, and also in silk, indigo, and lac. He makes special note of their superior culture, which manifested itself in their skill in writing and mathematics, which gave them an advantage not only over the other peoples of India but also over the Portuguese. Linschoten furthermore reports that "they will not eat that which has life in it or blood, nor kill ought in the world, no matter how small or needless it may be: for they firmly believe that all things have a soul." They would buy birds from the Christians and set them free, and in Guzerat appeared to have been places like animal shelters or "Guest-houses where one cures all the birds and other beasts . . . as if they were human, and when they are healthy again they let them fly or walk away again, which they consider a work of great caritas." Nor do they eat any vegetable that is red, nor eggs, nor do they drink wine. They'd rather die from hunger or thirst "than partake of the food and drink of Christians or anyone else," and they also had the unusual habit,

at that time, of washing up "before they eat and when they have eased themselves or made water." Physically they resemble "anyone from Europe," and "one will find women among them who, as to the whiteness of their skin as well as their beauty, rival Portuguese women."

The kings of Cambay or Guzerat became fabled for their riches, as one can imagine, and Yule quotes some amusing lines from Butler's *Hudibras* concerning Mahmud Bigara, Sultan of Guzerat: "The Prince of Cambay's daily food / Is asp and basilisk and toad, / Which makes him have so strong a breath, / Each night he stinks a queen to death."

21. Until 1687, an independent kingdom in the Deccan, a huge country that once included almost all of southern India. Golconda was up the eastern coast, with Masulipatam as its principal harbor.

22. In the original "gargaletten." I used the French spelling as being the closest to Rumphius'; in Portuguese *gargoleta*, "gurgulets" in English, in Sinhalese *gurulota*, while the word survived in India and Ceylon as "goglet," synonymous with a "water bottle." As Gray put it in his edition of Pyrard (see n. 8 of ch. 24, book 2), this "was an earthenware vessel with a spout, whereby the liquid was poured into the mouth from a distance, to avoid contact" (1:329). If a person was clumsy doing this, it marked him as a newcomer, *reinol* in Portuguese or *baar* in the East Indies.

23. Buton or Butong is an island, southeast of Celebes.

24. *Asta* was originally a Sanskrit term for the measurement "ell," which was twenty-seven inches long. What a "short ell" might be I do not know.

25. I would surmise that "Belitton" is Billiton, the tin island east of Bangka, and "Crimata" the Karimata Islands, which can be found off the western coast of Borneo (west from the town of Sukadana) in the Karimata Straits.

26. A century and a half later, these jars were still highly prized by the Dayaks (Rumphius' "Alphorese") on Borneo. The fabled "white radjah," James Brooke, testified to their value in 1841: "Some Dyaks, lately from the interior, have brought one of the celebrated Jars; I do not buy it, since it is far too dear as a mere curiosity. It stands three feet high, and is narrow at the top and bottom, with small rings round the mouth, for the purpose of suspension. The colour is light brown, traced faintly with dragons, and its chief merit and proof of antiquity is the perfect smoothness of the bottom. The ware itself appears coarse and glazed, and those in which the dragons are in alto relievo are valued at a hundred reals. They are not held sacred by the Dyaks as objects of worship, or as venerable relics, though none can be manufactured at the present time; but they are collected as a proof of riches, in the same way that the paintings of old masters are in Europe. There can be no shadow of doubt that the manufacture is Chinese, since similarly formed dragons are unknown in any country except China. They are the real grotesque monsters peculiar to that nation, and were probably introduced many centuries ago"—Rodney Mundy, *Narrative of Events in Borneo and Celebes, down to the Occupation of Labuan. From the Journals of James Brooke, Esq., Rajah of Sarawak, and Governor of Labuan. Together with a Narrative of the Operations of H. M. S. Iris*, 2 vols. (London: John Murray, 1848), 1:253–254.

27. He might be referring to the old Sung potteries at Yung-ho Chên near Chi'i-an Fu (is this "Jautscheu"?) in Kiangsi.

28. This must be Martinus Martini (1614–1661), an Italian priest, who studied mathematics under Athanasius Kircher and went to China as a missionary. His main work, commonly known as the *Atlas Sinensis*, was *Novus Atlas Sinensis*, published in Amsterdam in 1655. But the work Rumphius mentions here is Martino Martini, *De bello Tartarico historia, in quâ pacto Tartari hac nostra aetate Sinicum Imperium inuaserint, ac ferè totum occuparint narratur; eorumque mores breviter describuntur*, published in Amsterdam in 1655. It was translated in the same year into Dutch as *Historie van den Tartarschen Oorloch, In dewelcke wert verhaelt, hoe de Tartaren in dese onse eew in't Sineesche Rijck sijn gewallen, ende het selwe gelijck geheel hebben verovert, mitsgaders haere manieren in 't kort werden beschreven*, transl. G. L. S. (Utrecht: Gerard Nieuwenhuysen, 1655). I made use of the Dutch edition, but in both volumes one will encounter the same strange spellings of Chinese names which Rumphius used.

29. There is no intention on my part to enter the trackless wastes of Sinology. So much has changed—if only the nomenclature—since Rumphius' day that nonspecialists gladly leave that realm to more knowledgeable barbarians. Hence the following are no more than educated guesses, inflicted by necessity, that hope to be accurate enough to serve as guides for those who care to pursue these matters at greater length. I am assuming that "*Hungvu*" or "*Humvu*" is the same man as the peculiarly named "Humvuus" (see n. 16), i.e., Emperor Hung Wu, (1327–1398), the first ruler of the Ming Dynasty. He was indeed the Buddhist monk Chu Yüan-chang who expelled the Mongols (these are Rumphius' "Tartars," I presume) in 1368. He established the capital in Nanking. Hung Wu did bypass his own sons and chose his grandson, Zhu Yunwen, to become emperor. Zhu Yunwen ascended to the throne in 1398, but was overthrown by Zhu Di, the fourth son of the dead emperor Hung Wu, in 1402. The eunuch Ma He was one of Zhu Di's trusted generals and was instrumental in the taking of Nanking. The grateful Emperor Zhu Di bestowed the name Zheng (formerly Chêng) on his loyal

commander, hence the name Zhen He, or Chêng Ho. In 1403 Zhu Di ordered the building of a large imperial fleet that would visit all the known ports in the general vicinity of China and the Indian Ocean as well. Chêng Ho became the commander.

One can tell that there are just enough details that match, or that are close, in Rumphius' account to warrant the suggestion that "Kembun" must be Zhu Yunwen. He was rumored to have escaped the sacking of Nanking and the burning of his palace, and was said to have been seen in a variety of places in China, eventually fleeing overseas. His avuncular nemesis, "Englok," must be Zhu Di, and the only suggestion I have for this peculiar cognomen is that Zhu Di assigned the name of "Yongle" to his reign, which lasted from 1402 to 1424. Another corroborating fact is that Zhu Di did move the capital from Nanking to Beijing, that is "Peking" or Rumphius' "Pakin," formally naming the city as his realm's capital in 1420.

The only explanation I have for "Sampo" is that at court, Chêng Ho was known as "San Bao," which means "Three Jewels," a name that might easily be corrupted into the name by which the admiral is still known in the Malay archipelago. For a more popular retelling of Needham's material see Louise Levathes, *When China Ruled the Seas. The Treasure Fleet of the Dragon Throne, 1405–1433* (New York: Simon and Schuster, 1994).

30. This is not quite correct. The first fleet left China in 1405, only the third year of Zhu Di's reign, and counted 317 junks. In total there were seven voyages between 1405 and 1433.

31. Chêng Ho was, in fact, the supreme commander of the imperial argosy.

32. In *When China Ruled the Seas* Louise Levathes writes the following about the porcelain that was taken on board the first voyage: "The imperial porcelain works at Jingdezhen in Jiangxi province increased from twenty kilns at the beginning of the Ming dynasty to fifty-eight in the reign of Xuande (1426–1435), producing for export mainly white porcelain and the bluish or greenish Qingbai, a delicate, thin-walled porcelain. The treasure ships were also certainly laden with Cizhou, a northern Chinese stoneware with painted or incised decoration under a clear glaze; Dehua, a Fujian-made porcelain with a lustrous brownish glaze; and pale green celadons, which were considered to possess magic qualities. Ample quantities of all types of Chinese porcelains have been found from the Philippines to East Africa" (p. 84). Celadon is the green porcelain from the Sung dynasty that was the first Chinese porcelain to be exported to Europe. It was manufactured in the Lung-chüan district in Chekiang province.

33. Things get confused and difficult to untangle from now on. On Chêng Ho's first voyage, the fleet anchored in Campa (Champa; what is now southern Vietnam), Java, Sumatra, Ceylon, and Calicut, on the west coast of India. "*Tunkin*" might be "Tongking" or Tonkin (Cochin China), which is northern Vietnam. If so, then the second name, "*Cautschi*," might be a corruption of "Kiau-chi," the ancient Chinese name for that region. "*Schor*" might be Shehr, once an important port on the Yemen coast. The Chinese called it Shi-ho.

34. See n. 17.

35. *Sanhosu* is undoubtedly *shan-hu-shu*, the coral the Chinese bought from the Arabs in the Red Sea and from the Philippines.

36. *Calambac*, the best kind of aloes wood (eagle wood), came from Champa, the southeastern region of Vietnam.

37. The second voyage started in 1407, and Chêng Ho was not part of it. He died in 1432, at sea, during the seventh voyage.

Chapter 25. Mamacur or Macur.

1. The *WNT* thinks that "*Amaus*" was a general term for "enameled," but I doubt this is what Rumphius had in mind. I think he equated it with "*hammonitrum*" (or "*ammonitrum*"), which is natron mixed with sand, and which he read about in Pliny. In book 36, ch. 66 (Latin ch. 194), Pliny discusses a "new" way to make glass. One takes a newly discovered "white sand" and "wherever it is softest, it is taken to be ground in a mortar and mill. Then it is mixed with three parts of soda, either by weight or by measure, and after being fused is taken in its molten state to other furnaces. There it forms a lump known in Greek as 'sand-soda' (in the original: *hammonitrum*)" (Eichholz transl., Loeb ed., 10:153).

2. Valentijn also describes these arm bracelets in his *Oud en Nieuw Oost-Indiën*, vol. 2, part 1, pp. 73–75. The details are suspiciously similar, and I would guess that Valentijn's description is an unacknowledged paraphrase of Rumphius' original text.

3. In the original "*Engelsche bottels.*" The normal Dutch word for such a glass container is *fles*, but "*bottel*" was once used, most likely borrowed from English (which had been using it since the fourteenth century), and which survived for a while in the word *bottelier*, that is someone who poured drinks in a wine cellar (later synonymous with "steward"). It normally referred to bottles made from very rough glass or stoneware. This is what Rumphius had in mind here.

4. "Water" here used in the sense of giving a variegated or clouded appearance that resembles watered or moiré silk.

5. The use of "*lakwerk*" in the original is problematic. Rumphius does not mean "lacquered," in the sense of varnished, but using lac as a filler. What he is referring to is the once-common, dark-red resin (*gummi resina laccae*) that was the product of a plant louse (*Coccus* or *Carteria lacca*). The animal's excrement would encrust twigs and branches, and if these twigs were sold as such, it was known as stick lac; if the red pigment was separated from the resin it was known as lac-lac, and the resin that was left, usually in kernels, was known as sand or seed lac (granular lac). When these grains were melted and made into cakes, it was known as block lac, and if flattened into thin sheets it was known as shell-lac (shellac). Since it was a gum made by these insects, the way bees make wax, it could also be used as a pliable material which could be shaped and applied. I think that is what he has in mind here.

6. See ch. 24.

7. This is my literal translation of "*leugengeest*," a word also used by Vondel, that seems to have been synonymous with "lie," "lying," "words that lie," or even the devil.

8. A Greek compound noun, assembled from *μαντεία*, which means "divination," and *κατοπτρικός*, which means "in a mirror"; hence divination by means of a mirror, or specular divination.

9. Herman van Speult (?-1625) was governor of Ambon from 1617 to 1625, and a tough autocrat in Coen's mold. He was engaged in skirmishes with Muslim representatives of the Sultan of Ternate on Ambon and, though it is merely a footnote for the Dutch, involved in what became known in England as the "Ambonese Massacre." In 1620 the Dutch Republic and England signed a pact of mutual cooperation. One result was a small British "factory" set up next to Victoria Castle in Ambon. Van Speult made sure the British could find the real estate for a staff of about twenty people, but he never trusted them. In February of 1623 the Dutch were alerted by what they considered the suspicious behavior of some Japanese mercenaries in their employ. The soldiers were interrogated, which in those days meant torture, and the Dutch found out that the British were allegedly planning to attack the castle and dislodge the Dutch. Van Speult immediately took countermeasures. He arrested all the Englishmen and interrogated them, which, once again, meant various degrees of torture, and all the principals confessed to the plot. Van Speult issued swift justice. He condemned the prisoners to death, and beheaded twelve Englishmen, ten Japanese, and one Portuguese in March. This act, by no means a particularly vicious or extreme one in those days, caused an outrage in Britain and was used for propaganda. It has been argued that the Ambonese executions were the cause of the first Anglo-Dutch War (1652–1654). The tragic event is the basis for Dryden's play *Amboyna* (1673), which he wrote on commission. Van Speult left the Indies in 1625 with a return fleet, but was delayed in the Red Sea, and he died in Mocha in 1625.

10. In his "History," Rumphius considers Van Speult's military action against the town of Hutumuri (also spelled Outomoerij) in 1618, the "First Ambonese War." Hutumuri is on the southern coast of Ambon's southern peninsula, called Leitimor (Leytimor). The people there were in league with the VOC's implacable enemy, the Sultan of Ternate, and appeared to have molested anyone who was friendly to the company. Van Speult attacked the fortification in July with, according to Rumphius, 1,000 native troops and 60 Dutchmen. He took the place after five assaults. Van Speult's next adversary was the more formidable Daja, who is described by Rumphius as "a small, skinny little man, but very courageous and a bitter enemy of the Christians" (p. 52), but Van Speult did not live to vanquish him.

11. The man's title was "*ngatahudi*," which indicated the head of a Mardika village, where the "Mardijkers" lived. These were Asian freedmen not born on Ambon, who lived inside the city of Ambon, in their own neighborhood. Rumphius lists this Paulo Gomes ("History," p. 36) as one of the fourteen "*orangkai Cameras*," or Indonesian councillors, who assisted the governor, and who were in fact the heads of the fourteen most important villages on Leitimur. Note the Portuguese names this early in the colonial history of the Moluccas. The head of the Mardijkers was also a prominent member of the *hongi* raids—police actions by water, one might call them—and he sailed at the front of this collection of large native proas. That is why he is mentioned here as part of the attack on a town in Ceram.

12. This was a general term for local beads in the colonial Indies, not necessarily rosaries. To be sure, the word originally referred to a rosary and to the prayer beginning "Our Father." In Latin these first two words are *pater noster*. Rumphius always uses this phrase, written as one word, for beads.

13. Simon Cos (?-1664), the seventeenth governor of Ambon, from 1662 to 1664. When he was appointed to that post, he had already been governor of the Moluccas for six years. It was with Cos that Rumphius sailed to Banda in August of 1662 to inspect the fortifications there. Rumphius notes that "the most important event" of Cos' tenure as governor was building a stone fortress on the neighboring island of Buru (in the Bay of Kajeli) to replace a wooden one. This is a strange item to recommend as the most important achievement in someone's career, but I suspect that Rumphius was the architect for this fortress, and that this is the reason why he singled it out. After all, Cos had directed military campaigns, had been governor of the Moluccas and of Ambon, and commissioner of the same islands. He was even a member of the powerful Council of the Indies. Hence my conjecture that the original layout of what became known as Fort Oostburg on Buru was Rumphius' work.

14. This is a small island off the easternmost tip of Ceram, between Ceram and the Gorong Islands.

15. The chrysoprasus is another stone from Pliny (book 37, chs. 20 and 32). Rumphius says that it is green, and Pliny (in ch. 32) says that it is the color of leeks. Hence it might well be the modern "chrysoprase," which is a green chalcedony.

16. This is Rumphius' own spelling, which is very close to the English spelling, Achin, of what the colonial Dutch spelled Atjeh, the devout Islamic sultanate on the northern tip of Sumatra.

Chapter 26. Armilla Magica & Coticula Musae.

1. The title means: "Magic Arm Bracelet & A Small Touchstone from a Banana Tree." This whole story is repeated in ch. 76.

2. Hila was (and is) a village on the northern coast of Ambon's upper peninsula called Hitu. Hila is now known for the fact that the local population venerate some Portuguese helmets as *pusakas.*

3. The Medieval Latin name of antimony, an elementary body (one of the series nitrogen, phosphorus, arsenic, antimony, bismuth, etc.), of a brittle metallic substance, that has a bright, bluish-white color, and a flaky crystalline texture. Known in ancient times as *stibium.*

4. This would be a correct interpretation of the Indonesian notion of *pusaka,* but I also can't help thinking that this was a subtle dig at the manic collectors in Europe, or even Cosimo III, to whom Rumphius was forced to send a collection of tropical curiosities.

5. This is not clear. The Dutch is: "*of gelyk hy in den voorschreven Ring afgetekent staat.*"

6. Banana tree.

7. Touchstones were a variety of quartz or jasper, very smooth, usually black or a dark color, which were once used to test the quality of gold and silver alloys. They did this by rubbing the latter on the touchstone, and then determined quality from the color of the streak they produced.

8. See ch. 8 of this book.

Chapter 27. Mutu Labatta.

1. Wichmann (p. 137) states that "*mutu labatta*" was also known as *muti salah* or *muti tanah,* which would make more sense. *Muti* would refer to the Sanskrit word *mutia,* which means "pearl" or "shell" (*siput mutia* is "mother of pearl"). On Java it was spelled *muti* or *mute* and, according to Wilkinson, referred especially to beads. *Salah* means "at fault," "amiss" or "fake," so that the phrase would make sense in the given context as "fake pearl-like beads." Given Rumphius' description, the second phrase would also apply since *tanah* means earth. Rumphius' own translation towards the end would argue that "*labatta*" is some form of *batu* ("stone") or mineral. Rouffaer thought these were beads made from coral or sardonyx (p. 430) while the best ones were most likely beads carved from quartz or opal (433). Rouffaer comes to this conclusion (in an otherwise rambling and confusing text) on the basis of what Rumphius wrote about these same beads in his *Herbal* (*Amboinsche Kruidboek* 6:237–239). The two texts are virtually alike except for the last paragraph, wherein Rumphius (in the *Herbal* text) quotes Dapper's book on Africa. Dapper refers to the Gold Coast's *akori* beads, which were fashioned from blue coral, and this reference was enough for Rouffaer to suggest the identification of true *muti-sala* beads as quartz, lesser ones as probably Venetian glass, and the worst ones ceramic beads from India. See G. P. Rouffaer, "Waar kwamen de raadselachtige mutisala's (aggri-kralen) in de Timor-groep oorspronkelijk vandaan?," in *Bijdragen tot de Taal- Land- en Volkenkunde van Nederlandsch-Indië,* vol. 50 (1899), pp. 409–675. In other words, Rumphius' "*mutu labatta*" most likely refers to aggry or aggri beads. They were once widely distributed across the globe, for they have been found in Africa (the Guinea coast, for instance), Europe (England and Germany), Southeast Asia (Indonesia), Micronesia (Palau Islands), and elsewhere. A German ethnographer from the nineteenth century suggested that the aggry beads found on Timor came from Japan and were similar to the *usi-isi* beads. This sounds plausible until one realizes that the VOC traded with Japan and would have been aware of their existence. The same ethnographer suggests that the original source of these beads was Egypt and that the Phoenicians distributed them along their trade routes. See Richard Andree, "Aggri-Perlen," *Zeitschrift für Ethnologie* 17, no. 3 (1885), pp. 110–115.

2. *Timor* is the island, mentioned before in these pages, that comes, as it were, at the end of the string of smaller Sunda Islands, and which was half Portuguese and half Dutch. It seems to me that Rumphius had visited it.

3. *Solor* is a small island south of the extreme eastern shore of Flores.

4. For Jacob Wykersloot, see n. 3 of ch. 40 in book 1.

5. "*Cupan*" is *Kupang,* still the capital of western Timor, and the largest town in this region of Indonesia. Kupang is much closer to northern Australia (Darwin) than to Jakarta; it was the first port Bligh reached after the mutineers removed him from HMS *Bounty.*

6. *Rotty* (Roti) is a small island off the most southwestern tip of Timor. It was known for its large stands of lontar palms; the Rotinese were congenial to the Dutch, serving at times as colonial soldiers.

7. "Sawo," now Savu, is a small island, northwest of Roti, between Roti and Sumba. It is a barren, stony island, also blessed with large stands of lontar palms. *Lontar* palms (*Borassus flabellifer* [L.]) were highly prized. They provided sugar, syrup, and tuwak; the fruits and seeds were eaten, and the leaves were used for various purposes. Perhaps the best-known use was for paper, made from the fibers of the young, whitish leaves. The same fibers were woven into sarongs and also used to make boxes and baskets. In southern Celebes they used the lontar stems to make *songkos*, brimless hats with gold and silver thread interwoven. Savu's *ikat* weavings are among the finest Indonesia has to offer.

8. This must be a misprint for "Sawo" or Savu.

9. For Radja Salomon, see n. 5 of ch. 4.

10. Also spelled "*tael*," "*taes*," or "*tahil*," this Japanese coin was worth three and a half guilders, and was the equivalent as a weight to one ounce.

11. "*Orangcay*" literally means people (*orang*) who are rich or powerful (*kaja* or *kaya*). It was the title of village chiefs or heads of districts.

12. General word for people from southern India, specifically Tamils, and later used as a derogatory term.

Chapter 28. Mare Album. White Water. Ayer Puti.

1. "Het Zuydland" or "Southland," was another name for Terra Australis Incognita, and was for centuries considered a huge continent south of the Indonesian archipelago. For a long time no one bothered to find it, nor would anyone do so by accident because navigators crossing the enormous Pacific took Magellan's northern route. The Spaniards were quite satisfied with the established routes between the Philippines and Acapulco, so they did not venture into the southern Pacific during the sixteenth century. When mariners did begin to go further south during the seventeenth century, they kept on mistaking islands for portions of the "mainland." For instance, on the famous voyage of Quiros and Torres in 1605, the commander thought that the New Hebrides were not islands but part of a continent, hence he called it "Australia del Espiritu Santo." After this discovery the two captains were separated; Quiros got back to Acapulco, but Torres was forced to go further west and discovered that Australia was separate land while sailing between New Guinea and Australia on a northwesterly course. But when he reached Manila, his important information was kept a secret; hence it was not until 1762 that it became known that New Guinea was an island and not part of the mythical "Southland." Torres Strait commemorates this remarkable captain.

During the seventeenth century and during Rumphius' lifetime, it was the Dutch who expanded Western knowledge of Australia. After all, northwestern Australia is not a huge sailing distance from Flores and Timor, across the Timor Sea. In 1605, the Dutch ship *Duyfken* sailed into the Gulf of Carpentaria and the crew went ashore, but the country did not strike them as inviting. Over the next half a century, the Dutch, setting sail from Java, explored the northern and southwestern coasts, all the way down to Cape Leeuwin (Dutch for "She Lion"). By 1665, the Dutch had charts of the west coast of the land they called New Holland. Still during Rumphius' lifetime, the great captain Dampier sighted Australia in 1688 and returned in 1699 to explore the western coast of Australia. So when Rumphius speaks of what is "familiar," he's probably referring to the known western coast of Australia and to Tasman's explorations of the east coast (after doubling Van Diemen's Land).

2. The Dutch, the Latin, and the Malay all mean exactly the same thing, except for "Mare Noctilucum," which, deriving from *noctiluca* (*nox-luceo*), means "Sea That Shines by Night."

3. Rumphius was involved with the construction of redoubts in Banda during 1662 and 1663.

4. Note that these are all islands south and southeast of Banda. The white water seems to be restricted to the Banda Sea, perhaps into the Arafura Sea. It seems that Rumphius considers these two waters constituting one large "bay." The Aru and Kei Islands are more easterly in the Banda Sea, closer to New Guinea, while the Tanimbar Islands and Timorlaut are southerly, toward Australia. The "Liasser" Islands are the Uliasser Islands next to Ambon, while "Leytimor" is Ambon's southern peninsula.

5. Tanimbar Islands.

6. This sequence is not entirely clear. Ombo is a small island in the Haruku Straits between Ambon and Haruku, Luhu is a village on Howamohel peninsula on Ceram, "Bala" makes no sense unless it is a misprint for Bali, Ende is an ancient name for the island Flores, and Bima is on Sumbawa's east coast.

7. See ch. 42 of the first book.

8. This is Banda Neira.

9. "*Kisser*" is Pulau Kisar, right off the easternmost tip of East Timor, and below Wetar.

10. Martens (p. 136) thinks that this magical effect is caused by protozoa which Rumphius, lacking a microscope, could not see, and suggests more specifically that these are *Polycystines*, which belong

to the Radiolarians, a class of marine protozoans. Valentijn also mentions this event (*Oud en Nieuw Oost-Indiën*, 2:137–138), and Multatuli witnessed it during his brief tenure at Ambon (February to July of 1852) and included it as a topic of one of the many manuscripts Droogstoppel finds in the pack of papers Sjaalman leaves with him; see that list in the fourth chapter of Multatuli's *Max Havelaar*. Darwin noted a similar phenomenon while sailing down the east coast of Latin America on the *Beagle* in 1833: "While sailing a little south of the Plata on one very dark night, the sea presented a wonderful and most beautiful spectacle. There was a fresh breeze, and every part of the surface, which during the day is seen as foam, now glowed with a pale light. The vessel drove before her bows two billows of liquid phosphorus, and in her wake she was followed by a milky train. As far as the eye reached, the crest of every wave was bright, and the sky above the horizon, from the reflected glare of these livid flames, was not so utterly obscure as over the vault of the heavens" (p. 163). Darwin's description is in ch. 8 of *The Voyage of the Beagle*. He did not know the real cause either.

11. "*Ikan Moor*" or "*Anniko Moor*" (= Anak Moor) is a "Slender Grouper" or *Anyperodon leucogrammicus*. See *Fishes, Crayfishes, and Crabs. Louis Renard's Natural History of the Rarest Curiosities of the Seas of the Indies*, ed. Theodore W. Pletsch, 2 vols. (Baltimore: Johns Hopkins University Press, 1995), 1:103. And the "*Gatzje*" fish is most likely *ikan gatja* or, according to W. H. Schuster and Rusami Djajadiredja, *Local Common Names of Indonesian Fishes* (p. 45), *Lutjanus fulviflamma* (Forsskål, 1775).

12. This is an interesting detail. Rumphius must have read Le Maire's account of his famous voyage to find a new route to the East Indies. Jacob le Maire (1585–1616) and his pilot, Willem Schouten, performed the feat in the years 1615–1617, besting the terrifying waters along Tierra del Fuego, discovering what is now called the Strait of Le Maire, south of the Straits of Magellan, as well as discovering and rounding—no mean feat—Cape Horn (named after the town of Hoorn in northern Holland). The latter event was a unique achievement because they were sailing on an uncharted ocean and had rounded the very bottom of South America in open waters rather than threading their way through the treacherous passages between the mainland, islands, and capes as Magellan and Drake had done, and Darwin's Captain Fitz Roy was to do in the nineteenth century. Then they survived the enormous trek across the Pacific Ocean to the East Indies, staying far north of Australia. On that final portion of the journey, when they sailed into the Indonesian archipelago, Le Maire and Schouten proceeded along the northern coast of New Guinea. In July of 1616 they saw three or four "Mountains that burned," and they only knew what they were because Schouten had seen the active volcano, Gunung Api, on Banda. They called them "Vulcani Mountains." Engelbrecht identified the place as Manam Island, the first contact with New Guinea (which they thought was part of Australia) after they sailed southwest from the Purdy Islands. See *De Ontdekkingsreis van Jacob le Maire en Willem Cornelisz. Schouten in de Jaren 1615–1617*, 2 vols., eds. W. A. Engelbrecht and P. J. van Herwerden (The Hague: Martinus Nijhoff, 1945), 1:82–83. An intriguing question is how Rumphius came across this information. Le Maire's account was published in Amsterdam in 1622, Schouten's before that in 1618, while the same information was published in 1646 in the second volume of a contemporary history of the VOC (*Begin ende Voortgangh der Vereenighde Nederlantsche Geoctroyeerde Oost-Indische Compagnie*).

13. Rumphius, under the name "Plinius," was a member.

14. This is the Latinized version of Caspar Schott (1608–1666), a German scholar from Würzburg. The text mentioned is *Physica curiosa, sive mirabilia naturae et artis libris XII comprehensa*, published in Herbipoli (which is Würzburg) in 1662.

15. Thomas Bartholinus (1616–1680) was a Danish physician and scholar. He was appointed professor of mathematics and anatomy at the university in Copenhagen, and in later years Christiaan V made him his personal physician. He wrote a number of books, of which the ones on anatomy are probably the most important. It is impossible to pinpoint which text this author quoted from, but given the context, perhaps *De Luce animalium*.

16. Fortunius Licetus (1577–1656) was an Italian from Rapallo (near Genoa). He too was a physician and an enormously learned man, who was something of a celebrity in his lifetime. He wrote and published an astonishing number of books.

17. "Olaus Wormius" was plain Ole Worm (1588–1654), a Dane who also became a physician; he obtained his medical degree in Basel. He taught at the same university as Bartholinus, and was well known as a physician and naturalist. He acquired an enormous collection of natural objects and began to catalogue them in 1642. That catalogue, entitled *Museum Wormianum*—listing his own objects from nature as well as artificial items—was published in Amsterdam in 1645. This is undoubtedly the work that was quoted.

18. This is, of course, Montpellier, the city in southern France, chief town of Languedoc. It was particularly renowned for its medical school, established during the twelfth century, which had such illustrious alumnae as Rondolet and Rabelais. Its university was founded towards the end of the thirteenth century.

19. My guess for "Herbert" the traveler is Sir Thomas Herbert (1606–1682). He was mostly in-

volved with court intrigue and the English Civil War, but he also traveled a great deal, particularly between 1627 and 1629, when he was a member of the retinue of Sir Dodmore Cotton, Britain's ambassador to Persia. Several years later he toured the Continent. His journey to Persia inspired his major work, which had considerable success, entitled *Description of the Persian Monarchy now beinge: the Oriental Indyes, Iles and other parts of the Greater Asia and Africk*, published in 1634. Herbert published a second, enlarged edition under a shorter title, in 1637, and lived to see a third edition published in 1664 and a fourth in 1677.

20. "*Majottos*" is now called Mayotta, one of the four principal islands of the Comoros, a group of islands in the Indian Ocean, north of Madagascar, between that large island and Mozambique on mainland Africa.

Chapter 29. How the Ambonese Goldsmiths Clean Gold, and How They Give it Color.

1. Two misprints. It should be "Herbarium," of course, and the lemon variety is spelled *Limonellus aurarius* (i.e., "golden lemon"), which De Wit (p. 440) identifies as *Citrus aurantium* L. subsp. *amara* Engl. var. *pumila* Heyne.

2. "*Limon nipis*" is a thorny citrus tree with warty fruits, generally known as *Citrus hystrix* D.C. The acid fruits were mostly used for medicinal purposes.

3. "*Suppo*" probably refers to *sepoh*, a word used to indicate a mixture of alum, saltpeter, and other ingredients that was used to darken gold.

4. *Sesje* or *zesje*, I would guess, is an obsolete coin used as a weight. In Holland it was a silver coin worth six stivers, hence worth very little. It became proverbial for a niggardly tip; spelled the Flemish way as *zeske*, it could refer to a tiny, flat copper weight. It derives from the noun "*six*" (*zes*).

5. I do not know what the medical name for this skin disease might be. *Lapar* means "hunger" and *garam* is "salt."

Chapter 30. Myrrha Mineralis Mor.

1. "*Mor*" is probably meant to be *mur*, Arabic for "myrrh."

2. This is the island called Karimata, off the west coast of Borneo.

3. This is most likely "damar daging." The *Agathis alba* Foxw., of the Pinaceae family, is a tall tree that exudes a clear, sticky resin that turns as hard as stone in a few days. They used to find pieces of this hard resin buried in the ground at the foot of old or vanished trees, and this was sold as copal at high prices.

4. *Djudjambu*, *djambu*, and *jambu* all refer to a variety of *Eugenia* fruits, which is part of the *Myrtaceae*, or myrtle, family. See also n. 22 of ch. 23 in book 2.

5. The exact reference is: ch. 8 of book 3 in the second volume of the *Herbal*. Rumphius writes that "*Dammar Selan*" (or "*Dammar Sila*," also "*Dammar Malejo*") is an "ordinary resin" (p. 168) that flows naturally from Ambonese trees in great quantities (p. 169). It looks like ordinary amber. Rumphius discusses "*Ramak Daging*" in the subsequent ch. 9 (pp. 173–174), calling it "*Resina Carneola*" because it looks like red agate, and says that he considers it an "old" *dammar selan* that is "naturally refined" and only flows from rocks on Crimata.

Chapter 31. Touchstone from the Coast.

1. "Amaril" might refer to amarilite, a mineral which today is identified as a hydrous ferric sulfate.

2. This is the red resinous deposit with which the *Coccus* insect encrusts the twigs of some trees, and which was used in the Orient as a scarlet dye.

3. "*Malaxeren*" in the original. This rare verb was once used in English as well, certainly in Rumphius' day. The direct borrowing was from French (*malaxer*), but it ultimately derived from Latin (*malaxare*) and ancient Greek. It means "to soften by kneading or rubbing" and was primarily a medical term. The *OED* gives a quote from 1693 ("as if he did malax a plaister") and several from the eighteenth century, but this one from 1639 makes its use perfectly clear: "Powder all the medicaments severally, then mixe . . . and beate and malaxe them into a mass."

4. Rumphius is not so much thinking of a potter as of a craftsman who once fashioned objects from wood or metal on a lathe or "turner's wheel."

Chapter 32. Dendritis Metallica.

1. Perhaps this is what Rumphius in his *Herbal* (vol. 4, p. 135) called "Lignum aquatile," which De Wit (p. 439) identified as *Oreocnide rubescens* (BL.) Miq.

2. This would translate as "Metallic Thunder Stones." *Ceraunus* could also mean a "meteoric stone," but Rumphius clearly is unaware of the existence of meteors. See also ch. 9.

3. Julius Caesar Scaliger (1484–1558), discussed in n. 38 of ch. 23 in the second book. The work

in question is most likely *Exotericarum exercitationem liber XV. De subtilitate ad Hieronymum Cardanum* (1557).

4. Rumphius describes this tree in his *Herbal*, book 4, ch. 6 (pp. 16–20), borrowing Scaliger's name of "*metrosideros.*" Scaliger seems to have applied this to the "heart" or kernel of the tree, thinking it was truly iron, but Rumphius disagrees and calls the wood of the entire tree "*metrosideros,*" or *Nani* in Ambonese and *Caju Bessi Besar* in Malay, the latter meaning the "big iron tree." He says that it is even harder than ebony but not as "fine" and that it won't polish as well as ebony. The real "iron wood" is the pith or kernel of the trunk, which is the color of rust. The native population used this wood for anchors and rudders because it was considered more durable in salt water than metal. If they used it around their houses, they had to tie it to other wood because one couldn't hammer a nail into it. De Wit (p. 444) identifies it as *Metrosideros vera* Roxb.

5. See ch. 66.

Chapter 33. Stone Files on Java.

1. *Damar* is a generic noun for a variety of resins that used to come mostly from the Moluccas, Borneo, and Sumatra. The resins are harvested from trees of the Dipterocarpaceae family, Hopea and Shorea genera. The "*dammar selan*" which Rumphius (in his *Herbal*) also spells "*Dammara selanica*" is, according to De Wit (*M*, 428) the hard resin from the tree *Shorea selanica* (BL.) BL.

Chapter 34. Stone Bullets and Stone Fingers.

1. This should be "*geodes,*" that is, a stone that has a cavity often lined with crystals or other mineral matter. It was once also called an "earth stone" because the word, after all, derives from the Greek for "earth": γῆ.

2. These are fossils found in rock. The *OED* describes them as "straight, smooth, cylindrical objects, a few inches long, convexly tapering to a sharp point." They were also known as "thunder stones" and are the European variety of the stones Rumphius describes in ch. 8 of this book. These "bullets" and "fingers" are, according to Wichmann (*G*, 158–60), belemnites. These are Mesozoic fossils of extinct cephalopods.

3. "Xulassian Islands" are the Sula Islands, northwest off Buru, between that island and southern Celebes. The largest one is called Taliabu (Rumphius spells this "Taljabo"), to the east of that Mangole, the island Rumphius is primarily talking about, and directly below Mangole is Sanana Island, the most populous. Rumphius is sailing in a westerly direction. They had little to offer the local population, mostly primitive tribes, except for a sprinkling of Muslims on the coasts, and the population tried to find work elsewhere, for instance in the manufacture of kajuputih oil on Buru. For a long time, the Sula Islands were a haven for pirates. The Sula Islands also got embroiled in the strife between Ternate, Tidore, and the VOC. It was essentially subservient to Ternate, and as a result Mangole and Sanana were invaded by Arnold de Vlaming in 1652.

All the islands are extremely mountainous, relieved by plateaux, while Sanana has no plains at all. Mangole, the island Rumphius is chiefly interested in, has tall mountains (3,400 feet) in the eastern section and the Buja (Buya) Mountains in the west. These are about 2,600 feet high and are mentioned by Rumphius. The dangerous strait he describes between Mangole and Taliabu (Taljabo) is most likely Capalulu Strait. Rumphius' second name for the rock, "Batu Lacki Lacki," means the "Male Rock." The description has the feel of personal experience.

4. These sizes would range from 11 millimeters (or 0.43 inches) for the pistol bullet to 5 centimeters (nearly 2 inches) for the one-pounder, 6.3 centimeters (nearly two-and-a-half inches) for the two-pounder, and 7.2 centimeters (nearly 3 inches) for the three-pound cannonball.

5. *Pederero* (there are all sorts of different spellings) for *bas*, most often spelled "pedrero" in English usage. A small piece of ordnance formerly used to fire stones, broken iron, or just about anything except proper ammunition. Though inaccurate, it was a lethal contraption, once defined as "a murthering piece used in warres, to shoot chaine-shot or stones from."

6. In n. 4 on p. 157 of *G*, Wichmann states that Rouffaer thought "*Basagi*" most likely referred to "Besakih." This makes perfectly good sense. Besakih is a large and very ancient temple complex on the slopes of Bali's holiest mountain, Gunung Agung. As Bernet Kempers notes, it was once very difficult to get to, "far away in the mysterious world of mountains and forest, peopled by ghosts and supernatural powers." As already mentioned in various places, stones were readily venerated in Indonesia, and the "4 iron levers" that "grow straight out of the earth" are probably the phallic lingas that are prevalent in Besakih. See A. J. Bernet Kempers, *Monumental Bali* (Den Haag: Van Goor, n.d.), pp. 177–180.

7. I doubt these are pygmies, though tribes of pygmies were discovered in New Guinea during the first two decades of the twentieth century. More likely these were dwarfs, used for religious purposes.

8. These are not mythological creatures but albinos. Valentijn describes these "cockroaches" as follows: "They are nearly as white as a Dutchman, although to someone else it is a hideously faded color, or a deathly pallor, especially when one sees it from up close. They have very yellow, as if singed, hair; many large freckles on their hands and countenance, and have a scaly, rough, and wrinkled skin. By day they are purblind, nay, they are almost half blind, so that their eyes seem to be almost closed, leering constantly, but they can see very well at night. They also have gray, where other natives have black, eyes, and, though also born from black parents, they are despised by their own Nation more than other natives, and also considered hideous by them. I have known a King of Hitu and his brother, who were cockroaches, [though] they had several black Brothers and Sisters, and had themselves black children, and [have seen] others of the female sex, but not many. One will also find such people in the Realm of Lovango, in Africa, and other places."

Valentijn goes on to say that the name for these people derives from an Asian beetle called *Schallebijter* or *Scharrebijter* in Dutch, a large beetle of the species *Carabus* which, when it sheds its skin, is pale and wrinkly, resembling a cockroach. See Valentijn, *Oud en Nieuw Oost-Indiën*, 2:146–147; also A. Hallema, "Kakkerlakken als bijnaam voor Indische Albino's," in *De Indische Gids*, 62, no. 8 (August 1940), pp. 617–621.

9. This is an interpretation of "*rondom met een tempel bezet.*"

10. "Blew," or blue, stones for *arduin*, the hard, bluish-gray limestone from southern Belgium, used in buildings for thresholds, steps, and so on. When this stone is polished, it acquires a dark-blue, almost black luster and is extremely smooth.

11. This whole sentence is a puzzle: "*breng hier by tot verder bewys den akker van* Jerusalem, *die vol steene* Ciceres *is.*" The "*akker van* Jerusalem" is probably "*dodenakker*," or "*graveyard*," but the only meaning I can construe for "Ciceres" is that it is *ciceris*, plural of the Latin noun for the chickpea (*cicer*). This would make sense figuratively, since Rumphius is talking about small round objects. One must remember that the bullets of his day were round and did not have the missile shape of modern cartridges.

12. Anselmus Boëtius de Boodt (1550–1632), was a Flemish lawyer, naturalist, and poet who became a physician at the court of Rudolf II. His best known work was *Gemmarum et lapidum historia*, which was published in Hanover in 1606. This is the book to which the text refers to.

13. Again an approximation for something that is not clear: "*en in de omtrek veele straalen.*" This has to refer to the stone, but makes little sense.

14. One of the resins (*damar* or *dammar*) from certain trees used for a variety of things: from incense to torches.

15. They make *jimat*, a charm or a talisman, from such a stone. The "a" in front of "*adjimat*" for once might not be a misprint but an echo of the Arabic *azimat* (amulet or cabalistic charm), from which the Malay word derives.

16. "*Mangi Mangi*" was a local name for what we know as mangroves. Rumphius describes them in ch. 58 of book 4 of his *Herbal* (pp. 102–105), stating that they could grow as tall as an alder tree, that the leaves resembled those of the laurel, and that they only grew in swamps or on the coast. He particularly noted, as any traveler would, the difficult obstacle presented by their entangled roots. The Dutch used mangrove wood as firewood and in the construction of foundations for fortresses. They called it "lalary wood" and for many years obtained their main supply from Kayeli on the neighboring island of Buru.

17. Kelang Island is off the western coast of Ceram's Howamohel peninsula.

18. A coin worth one-twentieth of a seventeenth-century pound.

19. See ch. 8 of book 2.

20. The nautilus.

21. See ch. 28 of book 3.

Chapter 35. Ambra Grysea. Ambar.

1. *Amber* (*Ambra*, *Ambar*) is the common word for two substances which are easily confused, a muddle of which Rumphius also becomes a victim. His text is really only concerned with a biliary substance discharged from the intestines of the sperm whale (*Physeter catodon*). This rare substance is now known as "ambergris"; Rumphius also spells the last word *grysea* and *grys.* The other substance, known as "amber," is a fossil resin that should yield succinic acid, and which is then the true Baltic amber. The greatest region for amber, mentioned by Rumphius, was in East Prussia, particularly the coastal area known as Samland, above Königsberg (now in Poland). Hence this is a yellow fossil, vegetal substance that is also known in Dutch as *barnsteen* or *bernsteen.* When Rumphius uses the latter substantive I added it in brackets to the normal English word "amber" to minimize confusion.

The two substances are not at all alike. Ambergris, as its name indicates, is a grayish, sometimes even blackish, substance, while amber or barnsteen, is yellow. One possible source of confusion might derive from the fact that both substances float on the surface of the sea and may be found on beaches, sometimes even in the same ocean, such as the Atlantic. The other source of confusion is the noun itself. "Amber," as Rumphius correctly states, comes from Arabic (*anbar,* or *ambar*). This Arabic term, however, properly referred to the whale product, the animal substance. Yet the fossil resin gave us the word "electric," from the Greek *elektron*, because the ancient Greeks recorded the phenomenon of amber possessing an electrical charge as a result of friction.

Chapter 35, however, deals only with ambergris, the intestinal secretion of only one species of whale, the sperm whale (*Physeter catodon*). For practical purposes, we are only concerned with two families of whales: the baleen (*Balaenidae*) whales and the toothed whales, of the suborder *Odontoceti.* Most of the whaling industry was based on the killing of baleen whales, which included the right whale, the Greenland whale, and, in more recent times, the blue whale. The whale Scaliger mentions is a baleen whale, the Atlantic right whale or Biscay whale (*Balaena glacialis*), which has just about been exterminated. Arctic whaling started in the early seventeenth century and was dominated by the Dutch, with Spitsbergen as the central depot. The only toothed whale that was hunted was the sperm whale, of the family *Physeteridae*, especially the gigantic *Physeter catodon*, the kind that secretes the ambergris. The *Orca* Clusius' informant, Servatius Marel, mentions, seems to be a single species (*Orcinus orca*) of the toothed whales. It is also called the killer whale or grampus, and is known for its consumption of seals and dolphins, as well as fish and birds. It has nothing to do with ambergris.

In sum, the only source Rumphius mentions that is close to the truth about ambergris is the Burgundian Servatius Marel, who is quoted by Carolus Clusius. He is right that it comes from the stomach of a whale, and right that it is not the "right whale," by which he means the baleen whale. He is also right that it is not from an Orca, although it does come from a "Physeter"—a term derived from the Greek word for a blowing tube (*phuseter*)—which already in Pliny's time referred to the sperm whale. Hugo and Rumphius are both wrong in their assessments, though the fiction of an arboreal provenance gave Dermoût the wondrous image of the mythical tree at the center of the South Sea (see below, n. 16).

Ambergris was and remains a prized substance. One indication of how rare it was is the recording of the first amount of ambergris brought to England by a British whaler. This took place in 1790! See Edouard A. Stackpole, *Whales and Destiny. The Rivalry between America, France, and Britain for Control of the Southern Whale Fishery, 1785–1825* (Amherst: University of Massachusetts Press, 1972), p. 114.

2. The "South-Eastern Islands" are various small archipelagoes northeast of Timor, in the Alfuran Sea, between, say, the Babar Islands and New Guinea. The Dutch colonial administration included the following islands in this group: the Tanimbar (or Timorlaut) Islands, the Kei (or Kai) Islands, the Watubela (or Matabela) Islands, and the Goram Islands. If one follows these insular groups in this sequence, one notices that they curve up, in a semicircle, to the west of the Aru Islands, towards the southern point of the large Moluccan island of Ceram. All these coral islands were, in Rumphius' day, rather primitive and of little consequence to the VOC. The island called Great Kei, in the Kei Islands, had the unfortunate distinction of being the place where Coen transported the Bandanese, after he decided to depopulate their island after 1621, in order to maintain the company's monopoly in the production of nutmeg.

3. *Timor* is the last large island in the line of volcanic islands, east of Flores, and northwest of Port Darwin in northern Australia. The Portuguese were the first Europeans to come to the island, where their first settlements date from 1520. The first contact with the Dutch was in 1613, when Apolonius Schot (or Scotte) conquered the Portuguese fort "Henricus." Since that time, one will encounter a history of constant opposition by the Timorese people against the Portuguese and the Dutch, culminating in the capture of Fort Concordia in Kupang (the capital of the western half of the island, which used to be under Dutch control) in the first half of the seventeenth century. During the eighteenth century, the Dutch paid more attention to Timor, but it was always considered a barren backwater of little value. The Timorese people—who recount the legend that their ancestors came from Malacca, via southern Celebes and Flores, to the present island, calling it "the East," i.e., *timur*—are not purely Malay and seem to be related to the Papuan peoples further east.

4. *Ijan-taij* must be *ikan tahi; ikan* means "fish" and *tahi* is "shit," and is still used today as a swearword. *Tahi* can also imply the remnants of something, the dregs, sediments, dirt of anything, for instance *tahi chandu*, or opium dross, or "rust," which is *tahi besi* in Malay (*besi* = iron).

5. See ch. 17 in book 2.

6. *Azel:* Rumphius got this name for a whale from Orta, *Coloquios dos simples*, third colloquy; in Clusius' edition, ch. 1.

7. This is Fernando Lopes do Castanheda, who was born in Portugal in the beginning of the sixteenth century and died in 1559. He was the illegitimate son of the first governor of Goa, entered the Dominican order, and lived from 1528 to 1538 in India. Back in Portugal he wrote an important book, *Historia do descobrimento, e conquista da India pelos Portuguezes* (History of the discovery and conquest of India by the Portuguese), published in seven volumes in Coimbra between 1551 and 1561. One wonders if Rumphius read it in the original Portuguese. The only available translation of the *com-*

plete work during Rumphius' lifetime was in Italian, published in Venice in 1578. Otherwise only the *first* volume was translated: published in French in 1553 in Antwerp, and in Castilian in 1554, also in Antwerp. Rumphius refers to the fourth volume. I have found no evidence that he read Italian, and the fact that he lived for three years in Portugal makes it very tempting to conjecture that he knew Portuguese. Castanheda's work covered the Portuguese presence not only in India (Goa), but also in Ethiopia, Arabia, Persia, China, and the Moluccas.

8. This is the name of an archipelago of coral islets, south of India, southwest of Ceylon, and directly below the Laccadiva Islands. The Portuguese tried to establish themselves on the Maldive Islands from 1518 on, but the population, especially of the northern atolls, was not very hospitable to the Europeans and preferred an alliance with the rulers of Ceylon, a policy the people actively pursued from the seventeenth century on. The central islands traded with Malaysia and Arabia, hence the early references to these islands. Skilled seafarers, the people traded in coconuts, copra, tortoise shell and cowrie shells. The reason these islands are invoked here is that ambergris was often found by these islanders, as well as the fabled coco-de-mer. Because of the latter, they were often confused with the Seychelles Islands; see below, under *Geruda*, in n. 17.

9. Normally spelled François Pyrard or Pirard. He was a seventeenth-century Frenchman from Bretagne who traveled in the Indies and was imprisoned in the Maldives. On his return to France, he wrote a description of the Maldives that was published in Paris in 1615. See ch. 24, n. 8, in book 2.

10. *Seacat* (*zeekat*) was, as mentioned before, a Dutch term for squids or cuttlefish. When Rumphius speaks of their "beaks" (*bekken*), he may have had their horny jaws in mind, or even the cuttlebone, which is the internal, calcified shell.

11. Julius Caesar Scaliger (1484–1558) was born in Italy and led an adventurous early life as a soldier, yet when he died, he was a noted and respected physician in France, although he never obtained a degree in medicine. He became particularly notorious for his disputatious character, as is illustrated in his quarrel with Rabelais—who satirized Scaliger's philosophizing in *Gargantua and Pantagruel*, book 5, ch. 9, in the section on "entelechy"—and his attack on Erasmus, which represents his first printed work (1531). His botanical work consists of three annotated editions on botany by the pseudo-Aristotle and Theophrastus: *In libros duos, qui inscribuntur de plantis, Aristotele autore, libri duo* (1556); and *Commentarii et animadversiones in sex libros de causis plantarum Theophrasti* (1566). It seems, though, that Rumphius is referring to Scaliger's *Exotericarum exercitationem liber XV. De subtilitate ad Hieronymum Cardanum* (1557), a long (more than 1,200 pages) attack on the natural philosophy of Cardan which, among many other things, rejects such beliefs as the notion that a swan sings when it is dying, that a bear licks its cub into shape, or that the peacock is ashamed of its ugly legs. It was a scathing attack which, though justified in its ridicule, was needlessly acerbic and moved even Scaliger to write a funeral oration for its victim when he thought Cardan had died. Scaliger was a famous man. As great a figure as Samuel Johnson desired fervently to become "another Scaliger." See Thomas de Maria's 1993 biography, *Samuel Johnson.*

12. *Campernoyle.* I kept this rare usage to directly echo Rumphius' use of the same word. *Campernoyle* is a corrupt form of the medieval Latin *campinolius*, which, in its turn, is an obscure formation derived from *campus*, or "field." It translates as toadstool or mushroom. The *OED* provides the following quotation from the early sixteenth century: "Campernoyles that some men calyth tode stoles."

13. The first kind, which should be spelled *ponabar*, is white ambergris, *puambar* is greyish, and the brown ambergris is, according to Barbosa, *minibar.*

14. *Radja Salomon Speelman* was a Timorese chief, or "king" as Rumphius has it, who was defeated by the Portuguese in 1677, according to Valentijn (3.2: 112, 121), and came to Banda where he had himself baptized with the name of the VOC's general and governor-general. He returned to Timor in 1680. Rumphius stated that the radja's realm, Ade Mantutu, was in the eastern corner of Timor. See his *Herbal*, book 12, ch. 20, and book 6, ch. 12.

15. *Fadom* or *fathom* (in Dutch *vadem*) was a unit of measurement that originally referred to the distance between the tips of a man's middle fingers when he stretches his arms as far as he can. A "Rhenish fathom" was about 1.88 meters, an "Amsterdam fathom" about 1.69 meters, and an "English fathom" about 1.83 meters. Hence it is safe to say that a fathom is roughly six feet in length.

16. *Leti* could refer to the single island Lèti (Letti) or the group of islands called the Lèti Islands, which are part of the Southwest Islands in the Banda Sea. The single island by that name, Pulau Lèti, lies to the east of the northernmost tip of Timor. It is a barren, dry place, with tall, steep coral cliffs on its beaches. The native population prefers the meat of dogs to that of cattle.

17. The Bird *Geruda* (*Garuda*) is the well-known mythical bird that is still popular in Indonesia, particularly in Bali, and is today the emblem of Indonesia's national airlines. This mythical creature has the beak and talons of a bird of prey (it is often thought to be an eagle) and the body of a man, and was originally the vehicle (*vahana*) or mount of Vishnu, the original Vedic sun god, who became in subsequent Hindu myth an important deity, one of the Hindu triad (*trimurti*). Both Vishnu and

Garuda are solar incarnations, and the latter was also known as an implacable foe of snakes. It is noteworthy that Hindu mythological thought closely allied birds and serpents, a notion which modern science has proven to be a fact. The ancient Hindu texts state that Kasyapa, that is to say, "the turtle," a great sage who married the daughters of Daksa (i.e., "dextrous," a son of Brahma, and father of the daughters who became the wives of the gods), had Kadru and Vinata among his wives. The first one became the mother of the Naga serpents, and Vinata became the mother of Garuda and the race of birds. Kadru once made Garuda's mother her slave by means of a ruse, and said she would only free her if her son would steal the ambrosia (*amrta*) from the serpents. Though considered an impossible task, Garuda made himself small, insinuated himself among the monsters and, after blinding them with dust, succeeded in carrying off the sacred food, thereby regaining his mother's liberty. This is in the *Mahabharatam.* A variation in the *Bhagavata Purana* states that Garuda had to steal the ambrosia from the gods and give it to the snakes, which he did, but he put the divine substance on a bed of sharp *kusa* grass, and when the snakes, after releasing Vinata, licked it off the sharp grass blades, they cut their tongues down the middle, and are still so afflicted to this very day.

In Hindu myth, the sun bird is always battling the snakes, hence his epithet of destroyer and devourer of snakes: *Nagantaka* and *Sarparati.* Garuda is also popularly believed to cure victims of snakebite. In another legend, Garuda defeats Kadru's son, the evil serpent Kaliya, with a blow of his left wing. This wing shone like gold, as befits one who is called Suparna, or "He who has beautiful feathers."

A bird's element is the sky and its domicile is the tree; hence many legends and myths have a tree to represent the sky or the world, and the bird which sits among its branches is the sun. In Malay legend, as Rumphius notes, this tree is called *Paos Singi*, a phrase reasonably close to *Pau Jangi*, Skeat's term for the same mythical tree, which is said to grow in a central whirlpool in the sea, a place called the "Navel of the Seas." The mythical explanation of amber, i.e., that it is the droppings of the divine bird, is not that strange, nor that the bird should sit in a tree, but what might seem incomprehensible is a tree that grows on the bottom of the ocean. The connection here is a rare palm tree and its fruits.

For centuries, people on the Maldive Islands (see also n. 8 of ch. 24 in book 2) found on their shores both amber and large floating fruits that resembled coconuts. These nuts were once known as the fabled *coco-de-mer*, *coco-des-Maldives*, or *Double Coco-nuts.* Large fruits, which can weigh up to forty or fifty pounds each, they are covered with a thick fibrous husk, and contain from two to three nuts, each about twenty-eight inches long! These huge nuts are divided lengthwise into two lobes (hence the name "double coco-nuts") and covered by a very hard, black shell. After the seed had germinated, the empty fruits, in their hard black shells, might be found floating on the ocean, most ending up on the Maldive Islands. For a long time no one knew where these large nuts could be coming from, and legends were created to substitute for knowledge. Only much later (after Rumphius) did it become known that these nuts were the product of a very large palm tree, called *Lodoicea sechellarum*, which was only found on one of the Seychelles Islands. This remote archipelago of coral islands in the Indian Ocean lies southwest of the Maldive Islands, and over 600 miles northeast of Madagascar. These were uninhabited islands, though Portuguese charts marked them as early as 1502. The first recorded landfall was made by a British crew in 1609; the second recorded visit dates from 1742. The Seychelles (so called after 1756 in honor of Louis the XV's finance minister!) play a minor role in Dutch colonial history in that the French usurpers of this archipelago, observing that these islands were not subject to hurricanes, ordered spice plantations to be grown in order to wrest the spice monopoly from the Dutch. This was a secret project, and when, in 1778, a ship bearing the British flag was seen off shore, the local authorities ordered the plantations to be burned to the ground. They succeeded admirably, only to discover that the ship was a French slave ship which had raised the Union Jack as a subterfuge, under the mistaken impression that the islands had been occupied by the British. It proved to be an accurate omen because in 1814, the French had to cede the archipelago to the British crown.

Owing to the unknown provenance of these mysterious floating fruits, all sorts of legends were attributed to them and to the place from which they were thought to come. In the earlier sixteenth century, the inner husk was thought to be a potent antidote to poison (a notion repeated by Camões in his *Lusiads;* 136 in the final canto X), which represents one connection with Garuda. Pigafetta also reported that their size could only be exploited by such a huge bird as the Garuda. The Malay, according to Magellan's chronicler, called these fruits *Buapanganghi* and the tree *Campangganghi.* The final element that these two nouns share *ganghi* or *janggi* or (as Rumphius has it in his Herbal) *zangi*—most likely derives from the Arabic *zangi*, meaning "Ethiopian" and, by extension, East Africa. Hence the name Rumphius cites for them in this text, *Paos Singi*, has the same half of the compound in the second word—*singi*, a variation of *zangi*—while *paos* is *pau* according to Skeat, Malay for a fruit, particularly the wild mango. Highly prized and of great worth—Bohemia's Emperor Rudolf II, an ardent collector of curiosities, once offered 4,000 guilders to the family of the Dutch admiral Wolfert Hermansz. for a coco-de-mer he had received as a present from the King of Bantam—they were extolled for their medicinal virtues throughout the East. In India these nuts were known as *Darydi nariyal*, "coconut[s] of the sea," and were mostly obtained from the Maldives. The Maldavians called these windfalls *Tava'karhi*, or the "hard-shelled nut." Rumphius, who believed in their marine origin, thought that there was a real tree (not *Paos Singi*) which produced these fruits. In his *Herbal* he called it the *Calappus marinus* (book 12, ch. 8), or "sea coconut palm," since the first word was the usual noun in the Indies for that common tropical tree, after the Malay *kalapa*.

In the next section, which is the anonymous letter about the lump of amber, one will find the wild conjecture, repeated from dubious sources, that the gray amber comes from the excrement of a bird (echoing Rumphius) which is from Madagascar (not likely, but perhaps a confusion with the Maldives), has beautiful plumage (an echo of the mythical Hindu sunbird), and is as large as a duck! The latter would have offended Garuda, a bird ample enough to transport a god.

Sources for the preceding commentary, besides the two major works by Rumphius, are Yule's *Hobson-Jobson* (under "coco-de-mer"), Jones' checklist *Arabic Loan-Words in Indonesian*, De Gubernatis' *Zoological Mythology* (esp. vol. 2, pp. 94–95, 167–179, and 184), various editions of the Hindu epics and myths, and Skeat, *Malay Magic*, pp. 6–8.

18. Governor-General Joan Maetsuycker (1606–1678) was in some ways an exceptional man. In the seventeenth century there were far more of his ilk than in later years. He was trained as a lawyer and came, most surprisingly, from a Catholic family. Yet the staunchly Protestant company employed him because he was the best legal mind they could find. Valentijn, who didn't like Maetsuycker, presented a negative portrait in his *Oud en Nieuw Oost-Indiën*, 5:297–306. He repeats the damaging rumor that Maetsuycker was a "Jesuit" and called him a "sly and consummate fox" (p. 298). The average tenure of this, the highest office in the Indies, was usually five years, but Maetsuycker was Governor-General for a quarter of a century. He followed the edicts of the company to the letter, and they kept on refusing his requests for resignation because they valued him too highly. Maetsuycker never left Batavia and would not subscribe to a policy of expansion. Yet his hand was finally forced in Celebes and in the Moluccas.

An interesting fact is Maetsuycker's reaction to the Protestant clergy's demand that only the Dutch Reformed Church be allowed in the Indies: "When the Consistory of Batavia pointed out to Governor-General Maetsuycker that the Law of Moses forbade the tolerance of non-Christian religions, he simply answered: 'The laws of the old Jewish republics have no force in the territory of the Dutch East India Company.'" Clergymen were in the service of the company and could be fired and sent back home; hence the church could never exert a great deal of influence on political matters. See B. H. M. Vlekke, *Nusantara. A History of the East Indian Archipelago* (Cambridge, Mass.: Harvard University Press, 1945), p. 132.

19. The letter from Hubert Hugo (a name Nabokov would have liked) is dreadfully written. His life was as strange as his biological notions. Hubert Hugo (?1618–?1678) became an employee of the VOC in 1639 and worked for the company until 1653, mostly in India. He gained rapid promotion and returned to Holland in 1653 as second in command of the retour fleet. He was back in Holland in 1655, but we don't know what he did until in 1661 he overpowered a ship called "The Black Eagle" and proceeded to execute marauding raids in the Red Sea and off the east coast of Africa. His backers were French, and he had a letter of marque from a French nobleman. Hugo was a pirate from 1661 to 1665! He targeted primarily Turkish shipping but did not disdain VOC prizes, wherefore the VOC tried to capture him. They never did, and he disappears again. Then suddenly in 1671 Hugo makes the VOC an offer they couldn't refuse. He persuaded them that he could make money for the company in the slave trade and in amber if they granted him legitimacy. Evidently won over, the VOC appointed him head of their factory in Mauritius, provided him with his own ship, and made his position in Mauritius independent of the government at the Cape of Good Hope, which meant that for all intents and purposes, Hugo was his own boss. He also told them that he had designed a ship much swifter than any owned by the VOC, specifically adapted for the slave trade, but no one ever saw one under sail. His business dealings did not amount to much during his tenure in Mauritius from 1673 to 1676. He resigned at the end of the contractual five years and arrived in Batavia in December of 1677. He presumably died soon thereafter. See F. W. Stapel, "Hubert Hugo. Een zeerover in dienst van de Oostindische Compagnie," *Bijdragen in de Taal- Land- en Volkenkunde* (1930), pp. 615–635.

20. *Bitumen* was a kind of mineral pitch found at the time in Palestine and Babylon. Used as a kind of mortar, it is basically the same material as asphalt, and was also known as "Jew's pitch," or *Bitumen judaicum*.

21. *Straalsont* was a small town in Westfalen, a German region on the other side of Limburg's eastern border.

22. For *Ambar-noir*, or "Black Amber," see the next chapter.

23. *Succinum* is Latin for "amber," and is the root of "succinic," as in "succinic acid," an acid obtained from the dry distillation of amber, formerly known as salt or spirit of amber.

24. Rumphius' "Nanarium Minimum sive Oleosum" is a canary tree which produces a certain resin. It is discussed in the *Herbal*, book 3, ch. 6 (2:162–165). Merrill (p. 303) reduces it to *Canarium oleo-*

sum (Lam.) Engl. In the aforementioned chapter, Rumphius makes the same critical remarks (p. 164) as he does here. "Rowayl" is more complicated. It is another resin-exuding tree, found on Buru, but Rumphius never saw it. It is discussed in a different chapter describing the *Sasuru* tree or "Pseudo-Sandalum Amboinense," which Merrill (p. 406) lists as *Osmoxylon umbelliferum* (Lam.). Rumphius writes on p. 55 that this tree from Buru is similar to the Sasuru tree, but he never observed one in nature (p. 56). I cannot find any modern identification.

25. *Manipa* is a small island west of Ceram (or "Keram," as Rumphius also spells it). It produces a great deal of sago, and it once used to do a good business in cloves, the unopened flowers of the *Caryophyllus* tree. Manipa's crop fell victim to the VOC's monopoly policies, and the population declined and became paupers after the terrors perpetrated in 1651.

26. For *Rochefort* see notes to ch. 1 in book 1. The paraphrase is almost exact (book 1, ch. 20, p. 129 of the English translation), except that Rochefort adds that the amber which the foxes excrete is "only [used] in perfumes." Rochefort considered amber a kind of bitumen, said that it smelled like "rusty bacon," and gives pretty much the same advice as Rumphius does on how to determine whether one is dealing with the real article.

27. There must be a connective or a comma missing in the original text, for Rochefort does not mention this hybrid, but clearly refers to separate islands and regions (p. 127). He insists that the "greatest abundance" of amber is found "on the Coasts of Florida" and states that it was this "rich Commodity" which made the Spaniards build forts there (p. 127).

28. *Naphtha* for *aardolie*, which, literally translated, means "earth oil." This could be called petroleum, and naphtha is also an *oleum petrae*, but Rumphius uses both "aardolie" and "petroleum" separately; hence I wanted to use a different word as well. Naphtha originally applied to a volatile liquid that was a by-product of asphalt and bitumen, and that issued from the earth.

29. *Luhu*, which should be spelled *Luwu*, is a region on the eastern coast (and inland) of the western "leg" of Celebes, fronting the Gulf of Boné. Its capital is Palopo and it is populated with (the dominant) Buginese as well as with Toradjas.

30. The *Linschoten* reference is to ch. 70 of his *Itinerario*. When one reads Linschoten's text on amber, one will recognize a number of things that are echoed by Rumphius. The huge piece of ambergris was found in 1555 near Cape "de Comorijn", i.e., Cape Comorin, the extreme southern tip of India. The lump weighed thirty "centeners," according to Rumphius, which is the same measure as in Linschoten, where it is called the more common "quintal," a word taken over from Portuguese (originally Arabic) meaning a hundredweight, or 100 kilograms, 112 pounds. Linschoten got most of his information from Orta. Rumphius may well have copied this information from Rochefort, who mentions the exact same discovery by Linschoten on p. 128 of the English translation.

31. *Dammar* (*Dammer, Damar, Demmar*) is the Javanese word for liquid storax.

32. *Truffles*, the subterranean fungus, most often the genus *Tuber*, was highly prized in ancient times. Theophrastus (third century B.C.) and Pliny expressed their amazement; Pliny repeating the popular myth that thunder engendered these fungi (see book 19, section 11). Truffles do have a fleece or skin, the hymenium, and have a dark color with lighter marbling. At the time, their manner of existence was still shrouded in mystery, although about a decade after Rumphius' death, Tournefort discovered their spores, the first ascospores to be described. Pliny used the image "this blemish of the earth" (*anne vitium id terrae*) for truffles.

33. *Pomanders* for *reukballen*. The Dutch noun means literally "scent balls." The English "pomander" derives from French *pome* (= apple) and *ambre* (= amber), hence an "amber apple," that is a ball of amber carried on one's person as an aphrodisiac scent or an agent against infection or the plague. The noun came to mean a mixture of various aromatic substances, kneaded into a ball, and carried in a small box or bag either in one's hand or suspended around the neck. By Rumphius' time the pomander was more than likely a hollow ball of gold, silver, or ivory that contained an aromatic mixture of spices and other ingredients. An instance of this usage is Thackeray's phrase from *Henry Esmond:* "The courtier bowed out of the room, leaving an odour of pomander behind him."

Rumphius might well have copied the notion that ambergris increases potency from Linschoten (who uses the euphemism "*oncuyscheyt*"), as well as his admixture of ingredients for a pomander and his mention of the custom of carrying such a fragrant ball inside something else. Both Linschoten and Rumphius recommend benzoin (*benuin*), or "gum Benjamin," from the *Styrax benzoin* tree, musk (*almiscar* or *mosseliaet* in Linschoten), while Rumphius adds the ingredient *unguis odoratus* (see ch. 17 of book 2), rose water and liquid styrax, which is *storax*, a resin from the small tree *Styrax officinalis*, that smells like vanilla. The latter could be substituted by "rasmalla" or "rasamala roots," roots from the *Canarium microcarpus* tree (in Malay: *kaju rasamala* or *kanari minjak*), which, after processing, also smells like vanilla. The ambergris was used as a fixative.

Linschoten mentions that such a mixture of "savoury materials" are made into "handsome Apples and Pears, finished [= *beslaghen*] with Silver and Gold bands, for carrying about and to smell, also in the heft of knives, the handles of Poniards, which have been very neatly made of transparent Silver, . . ." (*Itinerario*, 2:31).

34. *Paternosters.* Named after the first two words of the prayer, "Our Father," it normally referred to beads, particularly the beads of the rosary. The rosary was also known as a "paternoster" because every tenth small bead (which were to remind one to say an Ave Maria prayer) was preceded by a larger one, which was to induce the practitioner to say an "Our Father." By extension, the notion of strings of beads informed the usage of "paternoster" for necklaces of beads worn by exotic races, such as in the Indies. From figurative usage it came also to refer to an antique torture when a twisted rope, studded with knots (hence the name, since it reminded people of the prayer beads), was tied around a victim's forehead and then twisted at the back of the skull by means of two small sticks, at the discretion of the presiding judge. In the plural it was also a slang term for thumbscrews or handcuffs.

35. "*Barbaria*" stands here most likely for the Barbary Coast, the region near Tangiers.

36. The "East coast of Africa" from Sophala to Melinde, stretches along the coasts of Mozambique, Tanzania, and Kenya. Sofala (its usual spelling now) is not that far below the Zambeze River, while Melinde is today's Malindi, in Kenya, just above Mombasa. Linschoten considered this whole coast the "Melinde Coast" because Malindi was the northernmost port Vasco da Gama visited on his way to India. Sofala had gold mines and, according to Linschoten, regularly found ambergris on its shores.

37. The Tenember (Tanimbar) Islands form an archipelago of about sixty-six islands near the confluence of the Banda and Arafura Seas, south of the Kei Islands, and east of the Babar Islands. The main island is called Jamdena (Yamdena). The small population was non-Islamic in Rumphius' day, hunted heads, committed cannibalism, were quite primitive and constantly at war with each other. They wore shells as decorations around their necks and wore rings cut from shells around their upper arms.

38. *Tidore* and Ternate were two small islands off the western coast of the large island Halmahera. The latter looks like a smaller version of Celebes and, separated by the Moluccan Sea, lies to the east of Celebes and north of the "bird's head" of New Guinea. The small islands of Tidore and Ternate were once powerful sultanates in the sixteenth and seventeenth centuries, controlling territories much vaster than the original islands. By the seventeenth century, Ternate, the dominant force in the Moluccas, controlled the northern part of Halmahera, the southern portion of Mindanao (the large and most southern island of the Philippines), the eastern third of the northern peninsula of Celebes, the Sulu Islands, and Ceram, together with the adjacent islands of Ambon, Huruku and Saparua. The Sultan of Tidore ruled over the southern half of Halmahera, the islands of Batjan and Makian, and the "bird's head" of New Guinea, the peninsula which Rumphius and Valentijn called "Sergile," and which had the important region called Onin. The VOC was at war with these sultanates for most of the seventeenth century.

39. The colonial historian F. de Haan found the following information about Andreas Cleyer in Batavia's archives and published them as "Uit oude notarispapieren II. Andreas Cleyer," in *Tijdschrift voor Indische Taal-, Land-, en Volkenkunde*, vol. 46 (1903), pp. 423–468. Cleyer (also spelled Kleyer) was born in Cassel (Hessen), Germany and, like Rumphius, went to the Indies as a soldier in the employ of the VOC. He must have had influential connections from the start because he had an illustrious, though controversial, career. Official documents refer to him as a "Doctor" or "Physician," also as a "Chemist"; in 1665 he proposed to make chemical drugs for the VOC at a 50 percent higher profit than was customary in Europe. In 1667 he was appointed rector of the Latin School in Batavia, a job which he did not relish, and which lasted only six months. In May of 1667 Cleyer acquired a more lucrative position: he was put in charge of the "Medicinal Shop" in Batavia's Castle. Here Cleyer's genius for lining his own pockets—a talent he shared with many VOC officials, including governors-general—knew its first flowering. One will keep reading about demands during his tenure for stricter controls of the way business was conducted in the medical warehouse, yet he held this position for almost ten years. In 1676 Cleyer was appointed the company's apothecary, a position he held until 1682, and eventually he was installed as the VOC's surgeon-general. Evidence of high-level protection is Cleyer's appointments to various official and semi-official bodies (College van Huwelijken, Vaandrig van de compagnie Suppoosten van het Kasteel, member of the Weesmeesteren, etc.) and the presence of high officials at social functions. For instance, Governor-General Maetsuyker attended the christening of Cleyer's daughter Cornelia in 1667, and Governor-General Speelman was Cleyer's son's godfather. In a small society pathologically sensitive to social hierarchy, such events said a great deal.

Cleyer's most surprising appointment, and relevant to the present text, was his tenure as director of the Dutch trading mission in Deshima, Japan, from 1682 to 1683 and 1685 to 1686. One will recall that the Dutch were the only nation permitted to trade with Japan. Japanese laws were strict and rigorously enforced, yet there was an extensive private trade going on that far exceeded official VOC business. Proof that Cleyer was heavily involved in this illicit *kambang* trade can be derived from the fact that he was expelled from Japan as persona non grata in December of 1686, along with eight other Dutch officials. Exile was to be preferred to the fate of Japanese citizens who had been complicit in the forbidden smuggling: they were tried and executed, either by beheading or crucifixion.

The Japanese adventure had been profitable for Cleyer. He acquired a large and expensive house on the Tijgersgracht (Tiger Canal) in Batavia, worth about 50,000 guilders, a substantial sum in those days, and was involved in a number of real estate deals that brought him in contact with his powerful protector, Governor-General Speelman. He was also favored by Speelman's successor because Camphuys had also been the director in Deshima three times, and was an admirer of things Japanese. Furthermore, Cleyer, like Camphuys, was a passionate naturalist and botanist. In 1681, Cleyer sent the apothecary Hendrick Claudius to the Cape of Good Hope to collect plants, seeds, and other botanical specimens. If they could not be shipped, he was to draw them, and he was also instructed to gather "curious stones that might have a medicinal usage," i.e., Rumphius's *mestika.* Cleyer was rich enough to send Claudius at his own expense and powerful enough to do so on a company ship and under company protection.

Andreas Cleyer married a Dutchwoman, Catherina van Rensen, from Middelburg in Zeeland, and fathered three children, who all did well thanks to his influential connections. Cleyer died either in December 1697 or January 1698.

Given that Cleyer was from Germany, was in the health-care business, and a fervent botanist, it seems quite normal that he came in contact with Rumphius. Part of their correspondence has survived. Cleyer sent Rumphius plants from Java, and Rumphius reciprocated with Ambonese specimens and shells. Rumphius gave Cleyer a Chinese treatise on how to gauge someone's blood pressure from the pulse, a text that may well have been the basis for Cleyer's Latin treatise on Chinese medical practices published in 1682. Four years before that publication he was admitted as a member of the "Academia Naturae Curiosorum" in Schweinfurt, Germany (November of 1678), and De Haan states that it was Cleyer who, with the aid of his friend and fellow member, Mentzel, proposed Rumphius for the same honor; see n. 19 of ch. 3 of book 2.

It was reasonable for Rumphius to ask Cleyer for details about whales in Japanese waters, and the two old men exchanged letters on the subject in 1694 and 1695. The dubious nature of the information might be owing to the Japanese not permitting the Dutch to travel. As Cleyer wrote to an acquaintance about his sojourn in Deshima: "What kind of information can you gather during a Japanese journey when all of us, without distinction, are treated as prisoners and are never permitted to leave the house?"

40. As Al Cohen conjectures, this phrase might well be *hai-weng-yü*, a Chinese (not Japanese) locution which means "old-man-of-the-sea fish." The original text's "*hay ang*" would be *hai-weng*, which means "old man of the sea," and "kie" would be *yü*, or "fish." The fact that the "fish" described has teeth would be correct, for the sperm whale, as noted earlier, is a toothed whale, although I do not know what is meant by the two tusks. It is also of impressive size.

41. *Liquesche Islands* might refer to what are now known as the Ryukyu Islands (with Okinawa in the central group), which stretch from Japan's southernmost Kyushu Island to Taiwan (Formosa). Taiwan, or Formosa, was in Rumphius' day known also as "Lequeo pequeno," while "Lequeo major" once referred to the Liu-ch'iu Islands, and Luchu was another name for the Ryukyu Islands.

Description of the Piece of Gray Amber Which the Amsterdam Chamber Received . . .

1. To sign off a letter with "N.N." was once a convention if the author either was not known or preferred to remain anonymous. The letters are the initials of the Latin phrase "*nomen nescio,*" meaning "I don't know the name," or "*non nominandus,*" "not to be named." The author is Nicolas Chevalier, who, at his own expense, published in 1700 in Amsterdam the original French version: *Description de la pièce d'Ambre Gris que la chambre d'Amsterdam a receu des Indes Orientales pesant, 182 livres; Avec un petit traité de son origine et de sa vertu.* See John Landwehr, *VOC. A Bibliography of Publications Relating to the Dutch East India Company, 1602–1800*, ed. Peter van der Krogt (Utrecht: HES Publishers, 1991), p. 595. Nicolas Chevalier (1661–1720) was a Huguenot refugee, born in Sedan (France), who left France after the revocation of the Edict of Nantes. He came to Holland in 1685 and lived in Utrecht and Amsterdam. He died in Amsterdam. Chevalier was a bookseller, printer, and art dealer, as well as an engraver of coins and seals. In the last function he appears to have acquired a bad reputation for producing medals that were the work of other people. See *Werken der Hollandsche Maatschappij van Wetenschappen*, 19:153–174. L. B. Holthuis helped me identify some of Chevalier's obscure sources.

2. To put this amount in perspective, one should know that at the time a mason made 360 guilders a year, a cloth shearer made 270 guilders a year, what was called a "sedentary worker" made about 250 guilders a year, and an able-bodied seaman between 120 and 180 guilders per year. Except for the sailor's wages, these incomes were considered quite good. See C. R. Boxer, "Sedentary Workers and Seafaring Folk in the Dutch Republic," in *Britain and The Netherlands. Papers delivered to the Anglo-*

Dutch Historical Conference 1962, ed. J. S. Bromley and E. H. Kossmann (Groningen: Wolters, 1964), pp. 155–160.

3. It is interesting to note that in English the notion of someone pursuing learning for its own sake or for the love of it, and as a nonprofessional, is a rather late usage. Towards the end of the eighteenth century, the word "amateur" was still positive, indicating someone who was fond of something, but at the beginning of the nineteenth, it was already restrictive and considered an alien passion. The *OED* quotes a definition of "amateur" from 1803 which states that it was "a foreign term introduced and now passing amongst us, to denote a person understanding, and loving, or practicing the polite art of painting, sculpture, or architecture, without any regard to pecuniary advantage." In contrast, the Dutch equivalent, and so used here and throughout Rumphius' work, is *liefhebber*, which dates from the early seventeenth century. The notion of an "amateur" was so unusual in England that the language borrowed *liefhebber* during the seventeenth century because English did not have a comparable noun.

4. This author is better known as Petro Andrea Matthioli de Sienna, or, Latinized, Mateolo Sinense, or Matthiolus (1501–1577). At twenty he became the personal physician of Pope Leo X, and later of the Emperors Ferdinand and Maximilian II. He published a book of commentaries on Dioscorides, entitled *Commentarii in Sex Libros Pedacii Dioscoridis*, in 1544, which became something of a best-seller. It was translated into German, Italian, Dutch, and French.

5. Augerius Clutius, according to Rumphius in his *Herbal* (book 12, ch. 8), published a book on the coco-de-mer in Amsterdam in 1634. More prosaically known as Outgaerts Cluyt (1577–1636), he was the son of Dirk Outgaertszoon Cluyt (Theodorus Augerius Clutius; ?1550–1598) who had been an assistant of Clusius. The younger Cluyt/Clutius studied pharmacology in Montpellier. He traveled a great deal between 1602 and 1608, for instance, to Germany, France, Spain, and even northern Africa, and sent many plants and seeds back to the botanical garden (or *hortus*), in Leiden. He practiced as a physician in Amsterdam between 1634 and 1636, and eventually became the director of Leiden's *hortus*, which had been designed by his father. He wrote a useful treatise on how to prepare and transport trees over long distances: *Mémoire pour indiquer la vraie manière d'emballer, et d'envoyer au loin les arbres, les plants, les graines, etc.* (Amsterdam, 1631). But he was of more interest to Rumphius because of his little book on the coco-de mer, a text Rumphius never saw. This rare little book is called *Opuscula duo singularia; Historia cocci de Maldiva Lusitani, sece nucis medicae Maldivensium; de Hemerobio, sive Ephemero Insecto et Majali Verme* (Amsterdam: Jacobi Charpentier, 1634). To commemorate the service of Cluyt, father and son, to the University of Leiden as well as to botany, Boerhaave named a genus of shrubs and trees (mostly from South Africa) after them: Cluytia.

6. The Portuguese author Duarte Barbosa published a description of the coasts of East Africa and Malabar in 1554. An English translation was published by the Hakluyt Society in 1866. Barbosa joined Pigafetta and Magellan on the latter's attempt to circumnavigate the globe.

7. Aldrovandus was the botanist Ulysse Aldrovandi (1522–1605), director of the botanical garden in Bologna, which he designed and stocked. Thought a heretic, he was taken to Rome, where he studied but could officially only devote himself to the study of classical statuary. He was also a pupil of Rondolet. Chevalier is particularly interested in Aldrovandi's three-volume work on birds: *Ornithologia* (1599–1603).

8. The spelling is remarkably close, for this refers to Francis Willughby (1634–1672), a British ornithologist and ichthyologist who was a friend and collaborator of John Ray (1628–1705), the British pioneer of systematic descriptions of nature. Ray and Willughby toured the Continent together in 1663, brought their collections home to England, and intended to use them for a systematic description of nature's flora and fauna. But Willughby's early death prevented completion, and Ray published his friend's *Ornithologia* in 1676 and in a translated version in 1678. In 1686, Ray published Willughby's *Historia Piscium*. Ray went on to fame with his own *Methodus plantarum nova* (1682) and his three-volume *Historia generalis plantarum* (1686–1704).

9. "*Miscellanea Natura Curiosorum*" were collections of notes on the medical and physical sciences published by the members of the Academiae Naturae Curiosorum, which was established in 1652 in Schweinfurt, Germany. This is a perfect example that Chevalier did not know what he was doing and just flailed around in the dark. This was, in fact, the learned journal published by the organization Rumphius belonged to; see the introduction. It was never published in Holland, nor was it the work of a single author, as he mindlessly keeps insisting. Chevalier also mindlessly copies what he read there or heard about without any comprehension. This is a fine example of the difference between "information" and "knowledge."

10. Behind this name hides Henricus Vollgnad, who published an article on amber in the periodical *Miscellanea curiosa sive ephemeridum medico-physicarum Germanicarum Academiae Naturae Curiosum, Decuria I Annus 3 (1672)*, (published in Leipzig and Frankfurt in 1681), pp. 448–449, entitled "De Ambra Augustana insolentioris ponderis." The date of "1630" which Chevalier gives in his dreadful text is wrong, and should be 1681.

11. Joannes Faber (?1570–?1640) published in 1651 "Novae Hispaniae Animalium N. A. Recchi imagines et nomina J. Fabri expositione," in F. Hernandez's *Rerum medicarum Novae Hispaniae Thesaurus* (published in Rome in 1651).

12. This is the Italian scholar Gregorio de Bolivar, who wrote "Notizie d'alcuni animali del Mondo Nuovo," which was published in *Tozzetti dei progressi delle scienze in Toscana* (vol. 2, pt. 1, pp. 246–258).

13. This strange monicker hides Jan Huyghen van Linschoten, and the reference is to ch. 70 of his *Itinerario.*

14. Arnoldus Montanus (?1625–1683) was more plainly a preacher called van Bergen or van den Berg who, in 1669, published a compilation in Amsterdam entitled *Gedenkwaerdige gesantschappen der Oost Indische Maetschappij in 't Vereenigde Nederland, aan de Kaisaren van Japan* (Memorable embassies of the East Indies Company of the United Netherlands to the Emperors of Japan).

15. I wonder whether "King of Sassumen" refers to Satsuma (Xaxuma), the southwestern region of Japan's southernmost island, Kyushu.

16. "Tailis" is most likely the Chinese word for a measure of weight: "Tael" (also spelled "Tahil"), equivalent to about thirty-eight grams.

17. Taking "*gedagte*" to be a misprint for "*gedaagd.*"

18. This is Andreas Cleyer; see ch. 35, n. 37.

19. This is not the church father but one of two Serapions, the elder and the younger. The writer was most likely referring to the younger Serapio[n] who wrote a treatise on medicines in Arabic. Orta quoted him often and was probably using *Obras de Joao Serapio* (*Jahiah ben Serabi*) published in Venice in 1497 (Markham, p. 493).

20. Antoine Furetière (1619–1688), a French scholar and writer who became an abbot in the diocese of Bourges. The dictionary the author refers to got him into trouble with the Académie Française, which had admitted him in 1662 but expelled him in 1685 because the members accused him of using their exclusive privileges to publish a dictionary of the French language. His *Dictionnaire universel* was published posthumously in Rotterdam in 1690.

21. This is Justus Fidus Klobius who published a pamphlet of seventy-six pages on amber in Wittenberg in 1666, entitled *Ambrae historia.*

22. I used "snout" for *tromp*, knowing full well that it usually meant "trumpet." However, that is too bizarre, even for Chevalier. *Tromp* could also mean a trunk (such as that of an elephant), a snout, or a beak.

23. "*Spermace*" is, of course, spermaceti, the wax found in the head cavity and blubber of the sperm whale. The confusing phrase is from the erstwhile mistaken belief that the substance was the whale's (*cetus*) sperm (*sperma*). It was used in medicine, and the manufacture of candles. It has nothing to do with ambergris.

24. "Boble-body" is my solution for a rather ugly neologism in the Dutch original: "*druppelende lyvigheit.*"

25. This is a French apothecary, Antoine Colin, from Lyon, who published French translations of Orta, Monardes, and Acosta in 1619, most likely translating from Clusius' edition.

26. For "Nikolaas Monard," see n. 35.

27. All of this, and what is to follow, Chevalier took either from Orta himself or, more likely, from Clusius' digest. He even copies the tone of Orta's original censure of Averroës.

28. This learned display was derived from Orta via Clusius.

29. This strange item is not an Armenian ball, but some arcane knowledge lifted from Orta. In his third colloquy he mentions "Armenian *bola*," which Markham glosses as referring to "*bolarmenico*" which, he says, is a red silicate of aluminum; what is also called red ochre or *terra sigilatta* (see ch. 13).

30. This information is also straight from Orta's third colloquy, as well as the item that ambergris was discovered in Setubal, Portugal.

31. Chevalier got this wrong as well. Orta clearly states that pieces of ambergris sometimes came to the Maldives *on* the birds or *in* their beaks, while it might also stick to the shells of mollusks. It is a question of involuntary transportation of the ambergris, not ingestion. But the above notion, incorrectly copied, probably started the idea that there were "beaks of birds" in the whale's intestinal tracts. These "beaks" were most likely the bony jaws or even the cuttlebone of the cuttlefish and squids the animal ate.

32. Again the tone is one of knowledge, yet Chevalier did not read "Nanardus," but copied another statement from Orta, although he did so incorrectly. There was no "Nanardus," but Orta and, therefore, Clusius as well mention a "Manardi." This was Joao Menardo, also known as Johannes Manardus, who published *Epistolarum Medicinalium Libri XX* in 1540 and a book on gems called *Literaria de Gemis.* The writer then tries to make the reader believe that he knows what he is talking about by his rather sharp remark about "Diambra," as if it were another substance. He copied this from

Clusius, who also has "Diambra." This is incorrect, however, because all Orta said was "of amber," i.e., "di Ambra." Both Chevalier and Clusius seem to have misinterpreted this passage in Orta; in any case, Chevalier's "stricture" makes little sense in the original either.

33. Ivy for *both* Dutch words "*Veil*" and "*Eiloof*," since they were once synonyms for the evergreen ivy *Hedera helix*.

34. Chevalier copies this exotic item verbatim from Clusius' addition to his epitome of Orta's third colloquy. "*Liquidambar*" is a balsam from a tree of the *Altingia* genus. Liquid storax was highly prized in Asia as well, where it came from Indonesia. In the latter country it was known as *dammar*.

35. This Spanish physician was Nicolo Monardes (1493–1588), who in his birthplace, Seville, wrote about the natural products from both the East and the West Indies, for instance in a book called *De las cosas . . . de las Indias Occidentales* (1565–1574). Clusius produced a Latin epitome, which he included in his *Exoticorum Libri Decem* from 1605.

36. This is most likely Cape Canaveral, wherefrom we hurtle people into space.

37. The "Etius" Chevalier mentions is almost certainly Aecio, for Orta stated that "this simple" was not mentioned by the Greeks except for "Aecio." As Markham notes in his index, this Aecio came from Amida in Mesopotamia, and became a physician in Constantinople. He was one among a fairly large number of compilers of medical knowledge from ancient and contemporary sources, for instance, his own "Tetrabiblos," published in Bazel in 1533, 1535, and 1542. Clusius calls him "Aëtius."

I used *loam* for "*leem*." The two words are certainly cognate, and the oldest meaning in English is clay, or a clay-like soil, mud, or even the "mortal clay" of one's body. From 1300 on, it was loosely used for earth, ground, or soil. I do not think Chevalier meant to use it as "lime" or "glue," although *loam* is allied to *lime*, but he may have had a viscous deposit in mind.

38. "Powerful" and "well prepared" for "*rauw*" and "*afgerecht*." Both are troublesome words, particularly *rauw*, which normally means "raw," and was once also common for "rude" and "uncivilized." But the *WNT* does print a rare, seventeenth-century usage of "*pittig*" or "*fors*."

39. The "country of Rouergue" was the name for a former province in France, roughly speaking, in the Garonne region.

40. This is in the Navarre region in southwestern France, where Bayonne is a harbor on the Gulf of Biscay.

41. Fallopius was the Latinized name of the Italian anatomist Gabriello Fallopio (1523–1562). A student and colleague of Vesalius, Fallopius seems to have been a diligent and dedicated scholar who revealed a number of things about the human body, such as the sphenoid sinus, the muscles of the forehead and tongue, a number of nerves, the vagina and the ovarian tubes in the human female, the placenta, and so on. Only one of his works was published during his lifetime: *Observationes anatomicae*, published in Venice in 1561; his collected works were published in the same city in 1584.

42. I am assuming that Chevalier is referring to Georg Agricola, or Georg Bauer (1490–1555), the German scholar who has been called the "father of mineralogy." It is difficult to say which work is referred to, though I doubt it was Agricola's historical work or his famous book on metals (1556). Perhaps it was *De natura eorum quae effluunt e terra* (1546) or *De animantibus subterraneis* (1548).

43. Erasmus Stella was a historian and physician who was born in Leipzig and died in 1521. He studied at the University of Leipzig, and studied medicine in Bologna. The writer is probably alluding to Stella's *Interpretamenti gemmarum libellus*, published in 1517.

44. "Tartar" for *wijnsteen*, which was once also known as *Tartarus vini*, the red or white crust that collects on the inside of wine vats. This calcium salt comes now purified as "*cremor tartari*."

45. Most likely, this refers to Erasmus Franciscus (1627–1694), the author of a compendious work entitled *Ost- und West- Indischer wie auch Sinesischer Lust- und Stats-Garten* (Neurenberg, 1668).

46. See n. 8, ch. 24 in book 2.

47. "Ickle-drop" for "*kegeldrop*" because Chevalier is presumably referring to drops of liquid that fall from a cone-like structure. Quite often this was the word for an icicle as well, but since I wanted to avoid the notion of winter, I used the general term then in vogue, i.e., "ickle."

48. Chevalier copied this from Kircher (see n. 9 of ch. 77), who says that Mt. Etna was called "Montis-gibelli" by the Italians (see ch. 9 of book 4 of Kircher's *Mundus subterraneus*). Hence this is supposed to be Mt. Etna.

49. This too is from Kircher's *Mundus subterraneus* (chs. 10–11 in book 2), where he discusses "water caves" from Asia and from Africa.

50. The original has "*pruimsteen*," which must be a misprint for *puimsteen*.

Chapter 36. Ambra Nigra. Ambar Itan.

1. What we know as Australia.
2. In the original Rumphius added the word "*lym*" between brackets after *Bitumen*. *Lijm* nor-

mally meant (and means) "glue," but in his day it was also used to translate the biblical "asphaltum" and "bitumen." I eliminated "*lym*" because the redundancy would be confusing in English.

3. Banana.

4. This former locution for "petroleum" was once current in English as well as in Dutch.

5. Most likely he is referring to resin from the kanari tree (*Canarium indicum* L.).

6. This is the resin from trees that contain succinic acid, known as yellow amber or, in Dutch, *bernsteen*. I left the Dutch in the text between brackets to prevent confusion. See n. 1 of the previous ch. 35.

7. This would translate as "hard bitumen." *Bitumen* was a mineral pitch that the ancients found in a natural condition in the Near East. The Latin *bitumen* corresponds to the Greek ἄσφαλτος, whence comes our "asphalt." It is confusing when Rumphius equates this with "pisasphaltum," a term he found in Pliny (book 24, Latin paragraph 41), where it is spelled closer to Greek as *pissasphaltos*. Pliny says that this was a mixture of pitch (πίσσα or *pissa* in Greek) and bitumen (ἄσφαλτος or *asphalt* in Greek). Presumably, this product was a natural one, but was not hard as stone.

8. Greek for "jet." The term, from Dioscorides, was taken from the name of the town Gagas in Lycia.

9. Buton (Butung) is a good-sized island southeast of Celebes' southeastern peninsula. My surmise is that Rumphius' "Buton" is the present capital, Bau-bau, on the west side of the island, and that "Culutsjutsju" is Kaling-tjussu or Kale-susu, both a harbor and a bay on the east coast. The Bay of Kaling-tjussu is very large and was also known as *Dwaalbaai*.

10. For the original *ongedierte*, which is more general and covers any creature that is considered objectionable. "Vermin" once had the same inclusive meaning; hence one should include here reptiles, and wild beasts, and not restrict it to the noun's modern meaning, which primarily refers to insects.

11. The "*tsjukalan*" fish is most likely a garfish or needlefish, of the family Belonidae. "*Thynnus*" should be "*thunnus*," better known to us as "tuna."

12. This snake might have been a python, perhaps a *Python reticulatus*, also known as *ular sawah* in Malay. In any case, a reptile belonging to the *Boidae* or giant snakes. Valentijn (vol. 3, pt. 1, pp. 285–288) describes these snakes in detail, saying they were not native, and that they grew to "unbelievable lengths" (p. 285). He writes that these snakes normally were eight to twelve feet long, but that they grew much larger in the forests and, especially, in the mountains. Now, a "*voet*," or "foot," was between twenty-five and thirty-two centimeters; hence the smallest was at least six feet. But, says Valentijn, he had seen them from eighteen to twenty-five "*voeten*," while the skeleton of one was thirty-six "*voeten*" (p. 285), hence nearly forty feet! One petola snake that was captured in the mountains in 1706 left a trail in the sand the width of a heavy beam, and another one captured on Run (a small island near Banda) left a trail like that "of a heavy coconut tree" when eight sailors "had trouble" dragging it down the mountain. It had died because it couldn't swallow the rack of a deer (p. 286). Valentijn states that they were a dirty brown, with black and white "eyes" and "yellow spots" here and there, altogether quite resembling "Sarassa and Petola" fabric (p. 286).

13. The "*Pristis* whale" is the type and genus of a small family of cartilaginous fishes, related to rays and skates, known as Pristidae. They are distinguished by an elongated body that ends in a snout and resembles a blade with teeth along each side. Some are known as "sawfishes," and these can attain the length of twenty feet.

14. Cornelis van Quaelbergen (? –1687) came to India's Coromandel coast as "assistant" in 1639, was promoted to junior merchant (*onderkoopman*) in 1644, merchant (*koopman*) in 1647, and senior merchant (*opperkoopman*) in 1650. Van Quaelbergen was in Masulipatam from 1652 on, but was recalled to the colonial capital Batavia in 1657 after he was caught at extensive private trading. Fined 3,000 reals (a hefty amount), he repatriated to Holland. Less than a decade later he was appointed "commander" of the Cape of Good Hope (1666), and once again, he was accused of illegitimate trade activities, again recalled to Batavia, but cleared of all charges. In July of 1672 he was put in charge of a fleet of warships off the coast of India, and in September of 1673 Van Quaelbergen defeated a British fleet off Masulipatam, a feat that was remembered for years to come. Back in Batavia in 1674, Van Quaelbergen was appointed governor of Banda, where he served for two years (from February 1677 to February 1679), at which time Rumphius must have known him. From 1680 to 1684 he was governor of Malacca and a member of the influential Council of the Indies from 1682. He died in Batavia in 1687. Van Quaelbergen is a good example that, though contrary to company policy, enriching oneself was common practice, and it did not prevent one from advancement. Valentijn's personal opinion of Van Quaelbergen (he spells the name "van Quaalberg") was that he was a "testy character" but "a good soldier," who always had problems with servants and colleagues. Interestingly enough, Valentijn mentions that while Van Quaelbergen was governor of Banda, he sponsored "several" climbs of that island's active volcano, Gunung Api. See *Oud en Nieuw Oost-Indiën*, vol. 3, part 2; pp. 91–92).

15. This should be "Buton," a surmise affirmed by the mention of "Dwaalbaai" in the next phrase, because that was another name for "Culutsjutsju Bay," as mentioned in n. 10.

16. Reading "*koraalen*" as a misprint for "*kraalen.*"

17. Seris are fish traps. See n. 10 of ch. 37 in book 2.

18. Malay for "Mountain Amber."

19. I wonder whether Rumphius had the term *Gummi elasticum* in mind, an ancient name for caoutchouc, better known as rubber. In Rumphius' day this was only obtained in the wild, most often from the *Picus elastica* tree, *karet* in Malay. Commercial exploitation did not begin until the last quarter of the nineteenth century.

20. See n. 1 of the previous ch. 33.

Chapter 37. Ambra Alba. Sperma Ceti. Ambar Puti.

1. Usually written as one word, it is a late Latin formation combining *sperma* (sperm) and *ceti*, genitive singular of *cetus* (whale). It is clear from the rest of the passage that the native population had totally different names for whales. There were whales in the archipelago and they might have been hunted. In a short article in *G* (pp. 89–93), Max Weber notes that he was personally cognizant of ten varieties of whales in Indonesia, and provides evidence that the inhabitants of Solor (east of Flores) were accomplished whalers.

2. The Latin "*Ambra alba*" and the Malay "*Ambar puti*" mean the same thing: white amber.

3. *Ikan* means "*fish*," and *punya* indicates ownership or, in *pasar* Malay (i.e., simple market Malay), it could indicate a genitive case which is what it does here. "*Monta*" is most likely *muntah*, which means "vomit"; hence the entire phrase means "fish vomit." One might think this was the proper name for a variety of whale, but it isn't; *ikan paus* is the name for the *Physeter macrocephalus* whale.

4. *Bene* (*Beneh*) means "seed" (also plant seed), and the second word is *gajah mina*, which literally means "fish elephant"; *gajah* (Rumphius' "gadja") means "elephant." The latter indicates that the Indonesian peoples had no specific name for the whale, for *gajah mina* is the Hindu monster *makara*, which was depicted as having the body of a fish and the head of an elephant.

5. "The whites" is an old English term for *leucorrhoea.* This is a fungal vaginitis, with a white discharge, that is caused by the *candida albicans* fungus.

6. Matthiolus (1501–1577) published a commentary on Dioscorides, which is what Rumphius has in mind here. See n. 4 of the interpolated "Description of the Piece of Gray Amber."

7. Properly this lovely Greek term for sea salt or brine is ἁλόν-ανθον (*halos anthos*), which literally means "flower of the sea."

8. The reference is to the oft-quoted *Histoire Naturelle et Morale des Iles Antilles*, by Charles de Rochefort. I did not find this phrase in Rochefort's description of ambergris.

9. These were contributions from members of the Academia Naturae Curiosorum in Schweinfurt, of which Rumphius became a corresponding member in 1681. These publications, one might call them "proceedings" or "acts," were printed—starting in 1670—under the title *Miscellanea Curiosa sive Ephemeridum Medico-Physicarum Germanicarum Academiae Naturae Curiosorum*, published first in Frankfurt and Leipzig. The particular reference is to *MC*, pp. 266–269 (the article Rumphius mentions in his text), and pp. 268–269 (which has the Bermudas reference).

10. Rumphius' "Animal Book" has not survived.

11. A term used by Pliny (book 12, Latin paragraph 130), for the oil of unripe olives or the juice of unripe grapes. "Verjuice" (my translation of the original text's *verjuis*) came into English from French very early in the fourteenth century. This sour juice from green grapes or crab apples was once a common condiment and medicine.

12. The remainder of this chapter is a verbatim quote of chapter 26 of *Pseudodoxia Epidemica, or Enquiries into very many received tenets and commonly presumed truths, which examined prove but Vulgar and Common Errors*, first published in 1646, by Sir Thomas Browne. Rumphius is most likely quoting from the Dutch translation published in 1668, or the German one from 1680 (see also n. 9 of ch. 23 in this book). In the original Dutch there is nothing to indicate after "cap. 26" that the rest of the text is a direct quote. I added the quotation marks and kept Rumphius' translation of "*nescio quid sit*," which is lacking in Browne's original text. Otherwise I simply copied Browne's original English text of the entire chapter, except for the very last paragraph, which Rumphius did not include.

Chapter 38. Lardum Marinum. Sealard. Ikan Punja Monta.

1. This means "fish vomit"; see n. 3 of the previous chapter 37.

2. "Seris," or *seros* (see n. 10 of ch. 37 in book 2), are fish traps, and "Robbers" should be "Bobbers," what in English were once called "weels," another device for trapping fish.

Chapter 39. The Malay Names of Some Precious Stones.

1. The Latin names come from Pliny, book 37 of the *Natural History;* references in notes to this chapter are to vol. 10 of the Loeb ed., transl. by E. Eichholz, pp. 164–333. "*Adamas*" is discussed on pp. 206–209. "*Intam*," more commonly spelled *intan*, is a Kawi form of *hira*, or *hinten* in old Javanese. Wilkinson specifically states that *intan* referred to "small and imperfect Indonesian diamonds" in contrast to the imported *berlian*.

2. *Bun* means "dew" in Malay; hence this is the "dew diamond."

3. A usage already found in Shakespeare, "water" referred to the luster of a diamond. The three highest grades were the "first water" (usually the white diamond), "second water" (when yellowish), and "third water" (when there is even more color in the stone). From this first usage we still have the phrase "of the first water" to indicate the finest quality of something. Dutch has the same usage; hence I kept it. In Valentini's *Museum Museorum* (1704; see section "Works" in introduction) one will come across Herbert de Jager's report of how much rough diamonds cost at the time in the Indies. One has to keep in mind that around 1600 a minimum survival wage per year was eighty guilders (see A. Th. van Deursen, *Mensen van klein vermogen. Het kopergeld van de Gouden Eeuw* [Amsterdam: Bert Bakker, 1992], p. 19). De Jager states that a rough diamond of ten carats fetched fl. 180; nine carats, fl. 160; eight carats, fl. 150; seven carats, fl. 130; six carats, fl. 120; five carats, fl. 110; four carats, fl. 90; three carats, fl. 70; two carats, fl. 40; and one carat, fl. 30 (Valentini, p. 103).

4. Places on the west coast of Borneo (Kalimantan), *Succadana* (Sukadana) is a town diagonally across from the Karimata Islands, and *Landas* (Landak) is a mountainous region north of Pontianak.

5. Needham (vol. 3, section 25, p. 670) says that the common name for diamonds was *chin-kang*, and notes that "the Chinese were not acquainted with cut and polished diamonds until the Portuguese brought some to Macao" (p. 671). Al Cohen plausibly suggests *tsuan-shih*, meaning "drilling stone," another term for diamond in Chinese.

6. I am not sure what this might be, but I wonder whether it could be *kaca*, *kacha*, or *katja* in older spelling, meaning "glass."

7. "*Clabbeken*" (also *klabbeek*, *klabbeker*, etc.) once referred to a fake diamond, and derives from a crystal that was found near Clabbeke, a village near Brussels. It seems to have been a stone that looked a lot like a diamond.

8. This might be *shui-ching* ("water crystal"), a term for rock crystal or quartz (Read, "Minerals and Stones," no. 37).

9. It is confusing to have the "Golkondalian mines" in the same passage which states the ancient belief that diamonds "grew" naturally as if a fruit of the earth. Diamonds (the very word may well be a corruption of the original Greek noun *adamas*, which meant "invincible") were not very important to the ancients—pearls, for instance, were considered far superior—and they did not know where diamonds were mined in India. One such locality was Golconda, but this was not mentioned by Pliny or other ancient sources, nor by Orta, who wrote an important early text on diamonds in the forty-third colloquy of his *Coloquios dos simples e drogas he cousas medicinais da India*, published in Goa in 1563. To my knowledge the first detailed account in a European text of the mines at Golconda was in Tavernier's account of his travels in the Far East and India. He had visited them first in 1641 (also in 1645) and described them in *Les Six Voyages de Jean Baptiste Tavernier, Ecuyer Baron D'Aubonne, Qu'il A Fait en Turquie, en Perse, Et Aux Indes*, published in 2 vols. in Paris in 1676. A second edition of this work was printed in Amsterdam in 1678. A German translation was published in 1681 and a Dutch one in 1682. All these publishing dates are within Rumphius' lifetime. Everything in Tavernier's text that pertained to India was translated and annotated in a late nineteenth-century English edition: *Travels in India by Jean Baptiste Tavernier, Baron of Aubonne*, ed. V. Ball, 2 vols. (London: Macmillan, 1889). My references are to this edition. The mines were in various locations, but they were collectively known as the Golcondian mines because they were under the jurisdiction of the Kingdom of Golconda, a town seven miles from Hyderabad, in Andhra on the Indian subcontinent (Ball identifies Golconda in nn. 1 and 2 on p. 151 of the first vol.; see also his listing of India's diamond mines in appendices 2 and 3, in 2:450–461). Tavernier does not mention the ancient belief in the diamond's growth capacity, either in his description of Golconda (ch. 10 in vol. 1 in the English ed.) or in his account of the mines (chs. 15–17 in vol. 2). The fortress of Golconda was also the treasury of the Nizam of Hyderabad, which probably caused the word "Golconda" to become synonymous with great wealth.

It should be noted as well that Tavernier clearly identifies Borneo as a place where diamonds were found, specifically "Succadan," which is Rumphius' "Succadana." Neither Rumphius nor Tavernier made the common mistake, at the time, of saying that diamonds were found in Malacca. This misconception comes from Orta, who in the aforesaid colloquy states that diamonds were found in a "rock in the strait of Tanjampur, in Malacca." As Ball points out (2:462), the old Portuguese writers meant "Borneo" when they wrote "Malacca," and "Tanjampur" is most likely Tandjong Pura, a place "30 miles up the river Pawán in the northern part of the Mátan District, adjoining Sukadana."

10. The Hebrew *Yahalom* was originally rendered into English as "diamond," though it was more likely onyx. Rumphius derives this from the description in Exodus (28:18) of the high priest's breastplate, with its twelve stones representing the twelve tribes of Israel. His translation of "hammer stone" is feasible since the Hebrew root can mean "to hammer."

11. In Genesis 4:22, Tubal-cain is said to be "an instructor of every artificer in brass and iron," hence the founder of the profession of metalworkers.

12. For many centuries there was a great confusion concerning the names of precious stones. Ruby is a case in point. The original Dutch has *robijn*, a word that was once used in English as well, transformed as "rubine," and meant "ruby." Just as Rumphius does, rubies and carbuncles were often considered the same stone. The true ruby, however, is and was very rare, and the word was more commonly applied to garnets and red spinel, both inferior stones. "Ruby" is not a word in classical Latin. Pliny uses *carbunculi*, a term that seems to have covered rubies, red garnets, and red spinel (see Eichholz's note c, on p. 238 of the Loeb ed.). *Carbunculus* is the diminutive of *carbo*, which usually referred to live or glowing charcoal; hence the name means something like "small pieces of glowing charcoal." Pliny discusses *carbunculi* in book 37, section 25, pp. 238–243 of the Loeb ed., vol. 10.

13. "*Gomala*" might well be *gemala*, which now refers to a stone that is presumed to have magical powers. But Wilkinson, more appropriately, glosses this as "luminous bezoar," or a stone that grew "out of the head of a dragon" and lit its way at night. As just mentioned, rubies were rare, but the association here is with a fiery red stone; in other words, rarity bestowed exceptional powers to something seldom seen. See below, chs. 46–75, for Rumphius' obsession with bezoar stones, *mestika*, and *guliga*, all of which are related to *gemala*.

14. Manidam (*Manikam*) was a generic noun for precious stone, a gem, though its other meaning is germ, quintessence, or embryo. *Permatta* (*Permata*) also means "gem" or "jewel," but Rumphius' explanation sounds like folk etymology to me. The "*matta*" in "*permatta*" is not "eye" necessarily, and he seems to construe that the prefix "*per*" means "for." "*Per meij*" should be *permai*, Malay for "pretty" or "beautiful."

15. A spinel is also fiery red but is not a true ruby. Not in Pliny.

16. "*Lychnis*" is from Pliny, and might have been a red garnet (see note e, on p. 247 of Eichholz's Loeb ed.). The lovely name is from the Greek, and has to do, as Pliny states, with "the kindling of lamps, because at that time it is exceptionally beautiful" (p. 247).

17. For "*Balace*," see n. 25. The stone Rumphius equates it with is in Pliny. In Latin paragraph 160 (p. 294–295 of Loeb ed.) Pliny states that erythallis, "although white, looks red when it is tilted."

18. This probably transcribes *shih-liu tzu*, or "Kernels of Pomegranate," according to Al Cohen. Being the Chinese name for garnet, it is mentioned by Read in his *Chinese Materia Medica*, "Minerals and Stones," no. 35g.

19. "*Smaragdi*," or emeralds, were discussed at length by Pliny (Loeb ed., 10:212–225), but briefly by Rumphius because they are not common in the Indies. Rumphius must have known that Pliny considered emeralds especially soothing to the eyes, relieving eye strain, and having the ability to concentrate one's vision.

20. "*Idjou*" most likely is *hidjau* (*hijau*), Malay for "green." Hence the Malay phrase means no more than "green gem."

21. This is Suchuan Province.

22. In Malay, *batu nilam* is indeed the phrase for "sapphire," and *biru* means "blue" in Malay; hence the second phrase simply means blue gem. Sapphires are not discussed by Pliny.

23. "Granate" is most likely the garnet, of a deep, transparent red color. Not discussed as much by Pliny. The Malay "*Bidji de Lima*" means "a pomegranate kernel," since *bidji* (*biji*) means "kernel" and *delima* is the Malay name for the fruit.

24. "*Carchedonia*" are discussed by Pliny in section 30 of book 37 (p. 249 of the Loeb ed.). Eichholz says these are Carthaginian garnets (note a, on p. 248), and Pliny tells us that they were found in Libya, at night, "when they reflect the moonlight, particularly at full moon" (p. 249).

25. In the original "*Balace*" was *Baleis*, a spelling which permitted Rumphius to play his word game in Dutch with *paleis*, which means "palace." This too might have been derived from Tavernier, who has *balet* for this stone (in ch. 8; in the English ed., 1:382). The name was more commonly spelled *balass*, with many variations (for instance "*balax*," "*balassi*," etc.), including "*ballace*" and "*balace*." I used the last in my version, for it afforded me the exact same wordplay with "palace." The reason why I think "*baleis*" equals "*balass*" is that Rumphius explicitly states that it is "rose colored" and that the stone was named after a people called "Balukes" or "Balotsjes," "who live to the west of the *Indus* river." Yule (in *Hobson-Jonson*, under "*balass*") states that this stone is a "rose-red spinelle" that came

from "the famous mines on the Upper Oxus, in one of the districts subject to Badakhshan." *Balass*, along with all the variations, would be a corruption of this name. Rumphius' information is generally correct, the Oxus River (or Amu-Dar'ya), which separates Afghanistan from Russia, is indeed "west" of the Indus, the large river that flows through Pakistan. Yule also mentions the same wordplay. He cites a British source contemporary with Rumphius (Ovington in 1689) who wrote that the "Balace Ruby is supposed by some to have taken its name from *Palatium*, or Palace."

26. "To clear up cloudy vision" is my rendition of "*donker gezicht te verklaaren*." *Donker* could once mean "blind," but I think it could also mean *troebel*, hence my "cloudy," since nothing, as Rumphius unfortunately knew, would help or improve ("*verklaren*" in the basic sense of "clearing up") blindness.

27. This account of the amethyst is from Pliny (Latin paragraph 121; Loeb ed., pp. 262–267). The peculiar reference to wine has to do with the fact that the name "amethyst" derives from the Greek phrase meaning "not drunken." This is mentioned by Pliny.

28. "Amorellen-wyn" in the original. The main ingredient of this wine was a particular kind of cherry, known as *morel*, also *morelle*, and, in Rumphius' day, also written *amarelle*. These dark-red, almost black cherries were steeped in red or white wine, with cloves and cinnamon.

29. The sard stone is indeed a carnelian as well. Pliny discusses it in Latin paragraph 105 (Loeb ed., pp. 248–251) of book 37.

Chapter 40. Cat Eyes. In Malay, Matta Cutsjing.

1. The Portuguese probably derived from Orta, who mentions "cat's eyes" in the forty-fourth and in the last colloquy, or from Linschoten, ch. 86 (2:67).

2. The opal (*opalus*) is discussed by Pliny in book 37, Latin paragraphs 83–84 (Loeb ed. 10:228–231). It seems to me, however, that Pliny's *asteria* or "star stone" (Latin paragraph 13; Loeb ed., 10:270–272), corresponds more closely to Rumphius' cat's eye, since Pliny states that its distinguishing mark is "that a light is enclosed in it, stored in something resembling the pupil of the eye. This light is transmitted and, as the stone is tilted, is displayed successively in different places, as if capable of locomotion within."

3. Ceylon was traditionally considered the place where the best cat's eyes were found. See Orta in his forty-fourth colloquy.

4. Orta mentions the same superstition in India in the forty-fourth colloquy, p. 360 in Markham's ed.

5. This term is from Pliny (Latin paragraph 129 in book 37; Loeb. ed., pp. 268–271), who says that the "*paederos*" is the finest of white stones. Eichholz (note b on p. 270) also thinks the *paederos* was an opal.

6. Eichholz translates "*paederos*" as "favourite" (p. 271), and Rumphius as "orphan." But it seems to me that the latter's addition—"because it deserves to be loved just like a handsome little orphan boy"—has the more correct allusion since παιδέρως was the synonym for παιδεραστής, which, as Liddell and Scott note, referred to a lover of boys "mostly in a bad sense."

7. "*Oranbay*" refers to a vessel particular to the Moluccas. Normally spelled *orembaai*, it referred to a ceremonial boat, without outriggers, in the center large and strong enough to support drummers (of *tifa* drums) and gong players. They set a pace for the rowers. The stem and stern curved up high and were elaborately decorated. It was rigged like a schooner.

8. Eichholz reads *astolos* for *astrobolos* in other editions (note 4 on p. 272 of Loeb ed.) and translates Pliny's description as "resembles the eye of a fish and sheds brilliant white beams like the sun." Eichholz (note d on p. 272) says that this "astolos" might "possibly [be] cat's-eye quartz." The second part of Rumphius' description, that the stone supposedly "catches the glitter of the stars," is by Pliny ascribed to a stone called *astrion*, whereof he says that the "name is sometimes explained by the fact that the stone, when held up to the stars, is supposed to catch their glitter and reflect it" (Loeb ed., p. 273).

9. "*Tsju*" is most likely *chu*, according to Al Cohen, and "*po-tsju*" is *pao-chu*, both common words for "pearl."

10. This is confusing. The Chinese got their cat's eyes primarily from Ceylon (see Hirth and Rockhill's ed. of *Chau Ju-Kua*, pp. 228–229), and definitely considered them stones. Rumphius says that the "common" Ceylonese ones are not stones but growths (see the subsequent ch. 46), or *mestika*, of a particular shell which he described before in book 2, ch. 28, under number III. To confuse matters even more, Rumphius also discusses (subsequent ch. 54) a *mestika* from the eye of a cat.

11. "Quang Say" is Kuangsi Province and "Kengtsjuhu" is most likely Kuan-chou-fu or Canton.

12. "*Hamtsju*" is *han-chu*, which does indeed mean "shell pearl."

13. Rumphius here continues the confusion. *Bulang* means "moon" in Malay; hence this is not a moonstone but a "moon *mestika*," yet he equates it with Pliny's *selenitis*, or "moonstone" (Latin

paragraph 181; Loeb ed., pp. 310–313). The Roman compiler describes this stone as being "transparent, [and] colourless . . . with a honey-coloured sheen" and "contains a likeness of the moon, and reproduces, if the report is true, the very shape of the moon as it waxes or wanes from day to day" (Eichholz' transl.).

14. A "*Girasole*" is a fire opal, and "*Asteras*" is probably the asteria mentioned in n. 2.

15. Chau Ju-Kua says the same thing, but in reference to Malabar, which he calls Nan-p'i (p. 229 of Hirth and Rockhill's ed.).

Chapter 41. Onyx. Joc.

1. This is correct. What has been translated as "diamond" was, according to most scholars, most likely onyx.

2. Pliny (Latin paragraph 85; Loeb ed., pp. 232–233) assigns sardonyx the second place of honor, after the opal.

3. Genesis 2:12.

4. For Martinus Martini, see n. 28 of ch. 24.

5. I think that Rumphius is writing about jade, not onyx. According to Needham (vol. 3 of *Science and Civilisation in China*, p. 641) the radical *yü* "applied to jade and all kinds of precious stones." Laufer says the same thing; see Berthold Laufer, *Jade. A Study in Chinese Archaeology and Religion* (1912; reprint ed. New York: Dover, 1974), p. 22. The word "jade" was not common currency until late eighteenth century, if not the nineteenth. Before that time jade was known by any number of names ranging from A to Z (see Fischer, 314–318), but "jade" was not one. For an extensive discussion of the bibliographical history of jade see Heinrich Fischer, *Nephrit und Jadeit, nach ihren mineralogischen Eigenschaften sowie nach ihrer urgeschichtlichen und ethnographischen Bedeutung* (Stuttgart: E. Schweizerbart'sche Verlagshandlung, 1880). In Dutch the above is true as well. "Jade" is only found in French-Dutch dictionaries, and is not recorded in common usage until possibly the end of the eighteenth but really not until the nineteenth century. This is a large and complicated subject; Laufer's study provides a wealth of information.

6. These tablets or insignia of imperial power are discussed by Laufer in his second chapter (pp. 80–103).

7. A reference to the "Three Kingdoms," the Wei, Wu, and Shu, which, after the demise of the Han dynasty, fought each other for many years. As Al Cohen notes, two things are confused within that peculiar "*Sam Cocsi*" phrase. It attempts to transcribe the title of a historical *text*, called *San-kuo chih*, commonly translated as "Accounts of the Three Kingdoms," though, Cohen states, a more accurate translation would be "Accounts of the Tri-part Kingdom." The text refers to three states: the Wei Kingdom (220–265), the Shu Kingdom (221–265), and the Wu Kingdom (222–280). By 314, according to Cohen, "these three states had been unified into a very shaky single state called Chin," but that disintegrated two years later, in 316.

8. Khubilai and his Mongols kept on pursuing the Sung court, which, defeated time and again, fled south. In 1276 the empress dowager conceded defeat, but some Sung loyalists fled further south with two imperial children. They enthroned the oldest boy, though they were forced to flee from port to port on China's southern coast. The ten-year-old boy emperor died in 1278, and his little half brother was installed. By this time, the Sung followers had been forced to flee Kuang-chou, called Canton in modern times, and had traveled to the Lei-chou peninsula, then to a small island. As the *Cambridge History of China* puts it: "Yet again the Mongols' persistent attacks compelled them to flee, this time to the island of Yai-shan, across the waters from Kuan-chou. The Mongols countered with a blockade of the island. On 19 March 1279, the Sung fleet attempted to break through the blockade, but during the ensuing battle Lu Hsiu-fu, with the child emperor in his arms, drowned. The last Sung emperor had perished at sea, and the Sung dynasty had at last fallen to the Mongols." *Cambridge History of China*, Vol. 6: *Alien Regimes and Border States, 907–1368*, ed. Herbert Franke and Denis Twitchett (Cambridge: Cambridge University Press, 1994), p. 435. There is plenty of drama in both Rumphius' story and history's version.

The three place names in Rumphius' story are so garbled that it is impossible to identify them with certainty. My guess is that "*Coinam*" might be Champa, the region now known, roughly, as eastern Vietnam, and that "*Heijnam*" is Hainan, the relatively large island off the southern tip of the Luichow (Lei-chou) peninsula. "*Braselles*" might refer to the Paracel Islands in the South China Sea, directly east from Hur in Vietnam, and south of Hainan.

9. Rumphius might be alluding to the Manchus, who overthrew the Ming Dynasty in 1644, and who ruled China for nearly three centuries. Cohen notes, however, that 1644 is the starting date given by traditionalists, but Manchu historians begin the history of the Ch'ing Dynasty (this is the name the Manchus used for their dynasty from 1626) in 1616 "with the first Manchu territorial claims

and establishment of a political state," while Ming loyalists argue that the Ch'ing Dynasty did not start until 1661. The Ch'ing Dynasty ended in 1911; hence, "depending upon one's favorite date of beginning, the Ch'ing Dynasty lasted 296 years, 276 years, 268 years, or 251 years."

Chapter 42. Achates. Widury.

1. In the Loeb edition (10:276) Pliny's term is spelled *leucachates*, i.e., "white agate." Agate is a chalcedony, of different colors, and often striped or banded. Cut into sections, such stripes or bands can suggest various objects in nature, such as an eye (e.g., the Mexican agate), a fort, or moss.

2. Eichholz, in the Loeb edition of Pliny, prints this as one word as well: *dendrachates* or "tree agate." Pliny discusses agates, which he spells *achates*, in Latin paragraphs 139–142.

3. Once an important Portuguese and then British stronghold, Surat is a city north of Bombay, not quite on the coast of the Gulf of Cambay. Until the end of the sixteenth century it was part of the kingdom of Guzerat. For more than half a century Surat was the most important city of the British in Continental India.

4. Horne (*Javanese-English Dictionary*) has a "diamond-like jewel" for *widuri.*

5. "*Belo*" makes no sense.

6. I continue my practice of adding "*bernsteen*" whenever I have to use "amber."

7. See the previous ch. 42. The word is Greek for "jet."

8. Latin, from Greek, for "heather."

9. In the original "*drop*" came at the end of the printed line followed by a hyphen, which makes one read it as part of the first word printed on the next line, i.e., "*oogen*"; hence something called "*drop-oogen.*" But the next to the last paragraph on p. 288 makes it clear that "*drop*" and "*oogen*" are meant to be separate nouns, because the paragraph starts out: "*Een ander Achaat, met roode droppen en oogen geteikent. . . .*"

10. This very personal remark reflects the celebrated intellectual debate of the seventeenth century which gave birth to modern science and to our modern age. It is interesting to note that Sedler's *Universal Lexicon*, which is roughly contemporaneous with Rumphius' life (the particular volume in question was published in 1741) has a special entry on the "new philosophy" (vol. 27, p. 2122), which designates it as both a general term for "contemporary philosophy" (or *philosophia recentioris aevi*) and specifically the philosophy written since the Reformation. The lexicon correctly states that the main attack was on scholasticism, that "the new philosophers" repudiated the medieval Aristotle and rehabilitated pre-Socratic philosophy, while it names as the main proponents Cardanus, Hobbes, Descartes and Leibniz. This is a very important topic—after all, it engendered our age without resolving the questions which still haunt us—unfit for a note, even a lengthy one. I mention it in my introduction as well. Suffice it to add here that, given Rumphius' exhortation, the "new philosophy" attacked the theocentric bias of man's preeminence in the natural world.

11. "*Zopyra*" most likely goes back to the Greek *ζώπυρον*, a term found in both Plato and Aristotle. It literally means a "spark," a "hot coal." In *De Caelo* (book 4:308) Aristotle equates this particular word with movement: in Stocks' translation: "the inquiry into nature is concerned with movement, and these things [i.e., "heavy" and "light"] have in themselves some spark (as it were) of movement. . . ." See *The Works of Aristotle*, 12 vols. (Oxford: Clarendon Press, 1908–1952), 2:307.

12. This refers to the strange tale in Genesis 30:37–42, when Jacob took twigs, peeled strips of bark from them so they would have white streaks, and then put them into the watering troughs of Laban's flocks: "And the flocks conceived before the rods, and brought forth cattle ring-streaked, speckled, and spotted."

13. See ch. 9 of the present book.

14. I presume this is a bit of sarcasm. Mercurius here both in the sense of "mercurial," that is volatile and inconstant, and as one of the five principal elements of life. The alchemists believed quicksilver to be the base or seed of all metals. Hence the use of "*Mercurius*" here is something in the sense of fundamental inconstancy. "Jan Potage" had the same meaning as "*Jack Pudding*" in England: a fool, a buffoon.

15. This is close to modern nomenclature. The stingray, which goes by the general name of *ikan pari* in Malay, is technically *Trygon pastinaca.* The stings from its tail are indeed very painful and dangerous.

16. "*Tasibée*" (*Tasbeh*) was the Arabic word for an Islamic rosary of one hundred beads. The beads were made of carnelian, mother-of-pearl, ebony, olive stone, onyx, or coral.

Chapter 43. Ophtalmius.

1. This is generally true, since all three are varieties of chalcedony. Pliny implied the same thing. He classed onyx and sardonyx together and seems to say that agate is very much like onyx (Latin paragraphs 90–91 in book 37).

2. Kunz states that the Middle Ages believed in such an "eye stone," or *opthalmios* (from the Greek for "eye"), which was more than likely a white opal. Opals were reputed to cure diseases of the eye. G. F. Kunz, *The Curious Lore of Precious Stones* (1913; reprint ed. New York: Dover, 1971), pp. 146, 148.

3. Rumphius is probably thinking of eye agates.

4. This short text is a melancholy example of Rumphius' preoccupation with his blindness. It does not particularly connect to any of the previous texts or the ones that follow, but seems to have been entirely dictated by his misfortune. Because of this text and several other instances, one gets the feeling that he tried all sorts of cures, including a large number of folk and natural remedies. The real importance of this brief passage lies in those final words: "*gelyk die menschen die een paerl op de oogen hebben.*" English once had the same usage as Dutch: a "pearl" (*paerl* in Dutch) in one's eye referred to a "thin white film or opacity growing over the eye," and the *OED* gives "a kind of cataract" as a secondary meaning, and provides a quote that says it is "glaucoma." This is precisely what Rumphius suffered from. The only portrait we have of him, drawn by his son from life (reproduced at the beginning of this book), clearly depicts those "pearls." Valentijn uses the same term in his *Oud en Nieuw Oost-Indiën* (vol. 2, part 1, 252–253), where he says that if one takes the leaves of a certain plant ("Ailoaha"), "washes them in cool valleys" and then put them on the eyes, this will soothe eyes afflicted by "the heat of the Sun," and "even remove" the beginning symptoms of "pearl of the eye" or the "dimness of one's sight." He adds that he saw Chinese blow "pulverized, refined white sugar" onto a patient's eye by means of "a small reed or pipe"!

Chapter 44. Cepites.

1. If one looks at the illustrations, one can see that this does not refer to a view of an actual fortress but to a contemporary blueprint or plan. For instance: this eighteenth-century plan of a fort on Banda.

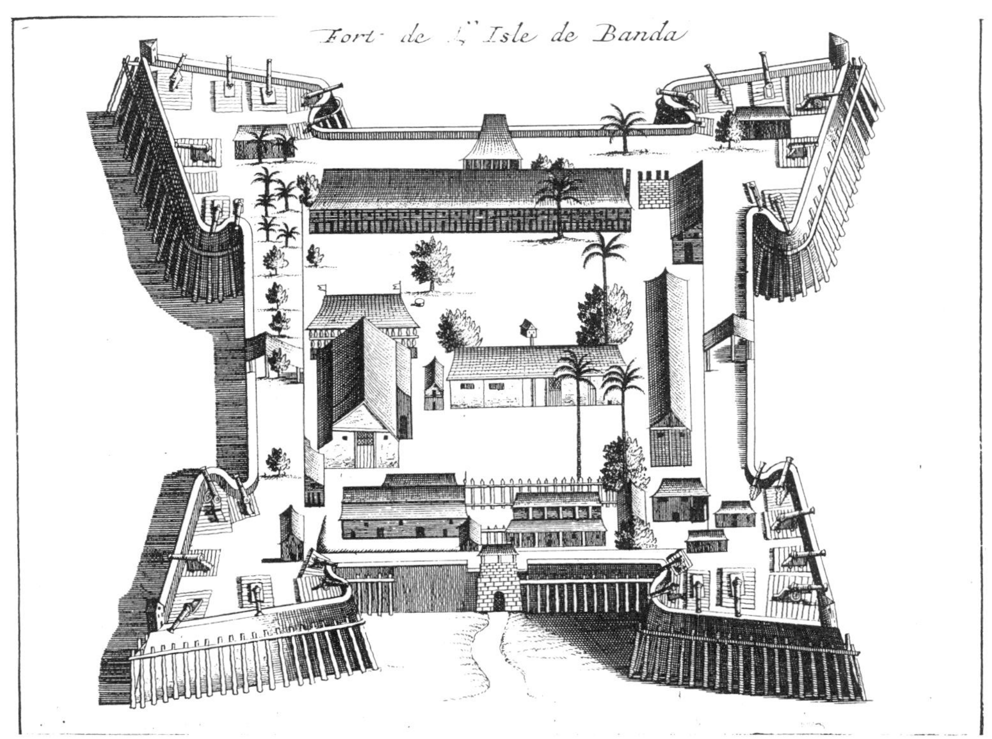

2. "Cabeletten" in the original text is a strange noun. As such it does not exist, but I wonder whether it is related to the Portuguese "*cabeset*" or "*capacete*," which originally meant "helmet." Rumphius uses it himself, spelled as "*cabasette*," in chapter 13 of book 2, as a name for one of his "Helmet Shells." In other words, Rumphius might be referring to some sort of "cap" or "helmet" which was put *on* pistols to protect the powder in the pan of a flintlock weapon. In the tropics gunpowder would quickly get soggy. Mr. Puype informed me that in the seventeenth century, mounted harque-

busiers were equipped with portable guns or muskets which had a flap intended to keep dust and moisture from the priming pan. See C. Beaufort-Spontin, *Harnisch und Waffe Europas: die militärische Autrüstung im 17. Jahrhundert* (München, 1982), pp. 72–73.

3. More often spelled *karet*, this refers to tortoise shell.

4. This is not clear in the original: "*en binnen in de koleur van het ander plein, afbeeldende een woonhuis.*"

5. Spelled *cepitis*, these stones are mentioned by Pliny in book 37, Latin paragraph 152 (Loeb ed., 10:288), but the mention is brief and of little help: "The 'cepitis' also known as 'cepolatitis,' is white, with lines of veins that meet at a single point" (Eichholz's translation, p. 289). The stone is unknown. Rumphius derived "Garden stone" from Pliny's second name, since "*cepo*" most likely comes from κῆπος, Greek for "garden," and "*latitis*" might have come from λατομεω, the Greek verb for "to quarry." Hence this name might well mean something like "Rockgarden stone," as Marios Philippides suggests.

6. For Feitema see n. 45 of ch. 3 in book 2.

7. "Front" for *voorstuk*.

Chapter 45. Testing Precious Stones.

1. This is true of Rumphius' particular test, though it should be said that Pliny also took it upon himself "to explain the methods of detecting false gems, since it is only fitting that even luxury should be protected against deception." He then tells us that one can test whether a gem is genuine by weight, temperature, and by examining the surface. This would be the Latin paragraphs 198–200, or ch. 76 of book 37, Loeb. ed. 10:326–329.

2. The text is corrupt here. The original has "*nog geen donker en weeke steen al is hy natuurlyk dit vuurtje of glans is. . . .*" (p. 290). The first edition has correctly: "*weeke steen al is hij natuurlijk; dit vuurtje of glans. . . .*"

3. "*Fusetre*" might possibly be a misprint for *fusette*, a term in the diamond industry for the area where the separate surfaces of a diamond that has double crystals join; *Naatwas* in Dutch.

4. "Laveur" is a very rare word. Its meaning is not certain, and Rumphius probably found it in ch. 88 of Linschoten's *Itinerario* (2:71). It refers to the direction of a stone's growth, which is called *was* in Dutch, and "wire" in English.

5. This is the *Gemma Elementorum* mentioned in ch. 40 of the present book.

6. See the subsequent ch. 58.

7. See the subsequent ch. 66.

8. See the subsequent ch. 83.

9. "*White Kalbahaar*," or white coral, is described in Rumphius' *Herbal*, book 12, ch. 15, pp. 226–227.

10. See the subsequent ch. 56.

Chapter 46. Mestica in General.

1. Rumphius clearly defines this concretion to distinguish it from the bezoar stone, discussed in the subsequent ch. 55. It was also spelled *mustika* and might derive from the Sanskrit *mastaka*, meaning "head."

2. Rumphius could have borrowed the notion of "stony sap" from Kircher, who in his *Mundus subterraneus* (1:310; see also 2:42–45) defines it as a "boiled moisture that derived from a Stone-like substance, which, when it dries, changes into stone." He then gives the example of such a sap in a plant: "For if any plant draws Stone-making water to itself as nourishment, then it will necessarily follow that the Stone-making water is drawn after it at the same time." Or he might have read about it in the works of Paracelsus (1493–1541), who wrote in German. The Swiss physician used the term *tartarus* for the accumulation of earthy materials in bodies, a process he called *Steinbildung*.

3. See the subsequent ch. 55.

4. See ch. 52.

5. In the original: "*na dat het dier of plante is daar ze van genomen zyn. . . .*"

6. Usually spelled *kebal;* see n. 20 of the previous ch. 8.

7. "*Gecabalizeerd*" in the original, that is, people who have become *kebal*.

Chapter 47. Encardia Humana. Muttiara Manusia.

1. "*Encardia*" is strictly a term from Pliny (see below); the phrase means "human heart stone."

2. *Muttiara* (*Mutiara*) means "pearl," and *manusia* means "human" (also humanity): hence Rumphius calls this growth a "human pearl."

3. See n. 18 of ch. 19 in book 2.

4. The *pinang* is the areca nut, one of the ingredients of preparing a *sirih* chaw.

5. The "Amarelle cherry" is a sour cherry from Germany, where it is also known as *Glaskirsche* (*Prunus cerasus*).

6. These stones are mentioned in book 37, but the Latin paragraph is no. 159, while in the Loeb edition (10:292–294) they form the beginning of chapter 58. Eichholz translates the description of the three stones as follows: the first one "shows the likeness of a heart in high relief on a black ground. Another variety bearing the same name displays the likeness of a heart in green, and a third in black, the rest of the stone being white." Rumphius' second name should be spelled "*cardices*" and is related to the Greek *kardia* ("heart"), whence our adjective "cardiac." To my knowledge, Pliny does not use "*cardices.*"

7. See number VI of ch. 29 in book 2.

Chapter 48. Steatites.

1. Mentioned by Pliny only once in book 37, in Latin paragragh 186, where he states, in Eichholz's translation, that the stone is so called "after the fat found in one animal or another." It would presumably derive from the Greek *stear* (genitive: *steatos*), which referred to hard fat, suet, or tallow. Eichholz says that this might well be soapstone (note b, on p. 315).

2. Might "Albast" be a short form for either "alabaster" or "alabastris?" Rumphius discusses the latter in subsequent ch. 79.

3. This is a "stone" found in a coconut; see subsequent ch. 68.

Chapter 49. Tigrites. Tiger Stone. Mestica Matsjang.

1. This was the word for "tiger" on Java (from *matjan* in Ngoko). The more familiar Malay term is *harimau.*

2. This is an old synonym for "rhinoceros." "*Renoster*" was the usual term used in the East Indies in the seventeenth century and thereafter. See De Haan, *Priangan,* 2:554.

3. The incomplete reference refers to Jacobi Bontii, *Historiae Naturalis & Medicae Indiae Orientalis,* ed. Gulielmi Pisonis, printed with other texts in Gulielmi Pisonis, *De Indiae Utrisque Re Naturali et Medica* (Amstelaedami: Elzevirios, 1658), book 5, ch. 1, pp. 50–52.

4. The Indonesian word for "rhinoceros" is *badak.* The peculiar spelling Rumphius employs was probably copied from Linschoten, who, in ch. 47 of his *Itinerario* (1:206), spells it as "*abada,*" a Portuguese version of the Malay *badak.*

5. In the original "*veranderde koorts.*"

6. "*Jambi*" is Djambi, a city and once an important region on Sumatra's east coast.

Chapter 50. Aprites. Mestica Babi.

1. The original has "*baskogel.*" A *bas* was, as mentioned before, a pedrero, a rather common piece of ordnance during the seventeenth century. A pedrero's ball would be about five centimeters in diameter, or about two inches. *Kogel* means "bullet" in Dutch, but I could not use this since that would give the wrong connotation, because a pedrero fired *stones,* and all sorts of junk, but not what we would consider standard ammunition. "*Aprites*" probably derives from *aprinus,* which means 'belonging to a wild boar"; hence this would translate as "stones belonging to a wild boar."

2. Reading "*likte*" for "*slikte.*" The latter would mean "swallowed" and must be a misprint.

3. "Little Ceram" was the contemporary nickname for Ceram's narrow peninsula that was close to Ambon. Also known as Howamohel. Luhu was one of its towns.

4. I tried but could not keep the expressive terseness of the original: "*hard van vleesch en leven.*"

5. An "orankay," or *orangkaja* (*orangkaya*), which means literally a rich (*kaya*) man (*orang*), was the title of a village chief on Ambon.

6. Rumphius' second wedding took place in either 1690 or 1691.

Chapter 51. Pilae Porcorum. Pigs' Balls.

1. Hair balls in pigs are a possibility, though one wonders, because bristles are so tough that they are not easily removed. Another possibility might be that the wild pigs ate some small hairy animals and that the fur of their victims formed a ball in their stomachs.

2. This is Howamohel, Ceram's narrow peninsula opposite Ambon.

3. This is bibliographically somewhat confusing. Rumphius had the following in mind: *De Medicina Brasiliensi,* by Guilielmi Piso, which has, in "Liber Quartis, qui agit de facultatibus simplicium" (ch. 103, pp. 121–122), a discussion of "De Globulo bovino, illiusque usu." Piso's treatise was printed, together with someone else's book (entitled *Historiae Rerum Naturalium Brasiliae*), in a volume that bears the overall title of *Historia Naturalis Brasiliae* (Lugduni Batavorum: Apud Franciscum Hackium & Amstelodami: Apud Lud. Elzevirium, 1648).

4. This phrase means something like "a bunch of hair from cattle."

5. A contemporary medical text in Dutch does not list this particular use of a hare's hair. It

does say that if this animal's hair were reduced to powder and ingested, it would break up and expel stones (bladder, kidney, etc.). See Johan van Beverwyck, *Schat der Ongesontheydt, ofte Genees-konste van de Sieckten* (Amsterdam: By de Weduw' van J. J. Schipper, 1672), p. 281. The entire chapter 38 concerns itself only with expelling "stones." This use is also not mentioned in the chapter specifically devoted to items that staunch the flow of blood (ch. 38) in Beverwyck's *Heel-konste ofte Derde deel van de Genees-konste* (1671). These texts, with others, were published in a single volume to which I am referring: Johan van Beverwyck, *Wercken der genees-Konste* (Amsterdam: Weduwe J. J. Schipper, 1672).

6. "*Tuncham*" cannot be identified positively, but one suggestion, offered by Cohen, is Tung-ch'ang.

7. "*Xantung*" is most likely Shantung Province.

8. "*Cingchieu*" is probably the city Ch'ing-chou.

9. "*Nieuhoang*" is *Niu-huang*, which Read (in his *Chinese Materia Medica*, "Animal Drugs," no. 337) translates as the "Cow Bezoar."

10. This is an orange.

11. "*Gomutu*," more commonly spelled *gemutu*, are the black hairs between the leaf stems of the Aren palm, called "Saguerus" by Rumphius. *Gemutu* looks very much like horsehair and was once used for making rope. The tree was more important as the producer of a sap that was made into vinegar, a beverage, and the once well-known aren sugar. The collecting process resembled our tapping of sugar maples in early spring.

12. In book 1, ch. 18, of his *Herbal*, Rumphius describes *ela* as "the reddish remains of Sagu meal that looks like coarse bran." He notes on the same page (p. 79) that *ela* was either thrown away or used as pig fodder. On the heaps of *ela* left in the bush, mushrooms would sprout (which he called "Culat Sagu" or "Diamor"), and they would also generate the thick, fat sagu worms which the Ambonese consider a delicacy.

13. "Marine hair ball."

14. "Koppenhaven" in the original.

Chapter 52. Hystricites. Pedra de Porco.

1. Rumphius clearly distinguishes *mestika* from *Culiga* (*guliga*), though this is still confusing since Skeat, as well as others, seem to think they are similar objects: stones found in animals and trees (Skeat, p. 275). Wilkinson says that *guliga* is "a talismanic or curative bezoar" and even glosses it as a "snake stone," while the latter is traditionally the bezoar.

2. "*Landa*" (*Landak*) is porcupine in Malay.

3. "Spanish soap" was a hard white or marbled soap, once made from olive oil in southern Europe. It was used for medicinal purposes or as a toilet soap.

4. Al Cohen offers the suggestion that "*Ho-tischoo*" is an approximation of *Hao-chu*, which means "porcupine," but this does not include the idea of a bezoar stone. Read does list a boar bezoar in the section "Animal Drugs," no. 359.

5. Most likely the city on Sumatra known as Bencoolen.

6. "*Scho*" might be the Chinese *chu*, which means "pig." *Babi* is "pig" or "hog" in Malay.

7. "*Sucatana*" might be the town of Sukadana on the west coast of Borneo in Sukadana Bay, east of the Karimata Islands. "*Tampas*" (*Tampak*) should probably be Landak, a region north of Pontianak.

8. Ende is not an island but a formerly important town in the center of the southern coast of the island of Flores.

9. Also known in Malay as *Bidara laut* or *Bidara pait*, this "Snake Wood" (*kaju ular*) is *Strychnos ligustrina* Bl. This is a small tree with yellow wood and roots, all of which tastes as bitter as the seeds of lemons. The connection with Timor makes sense because, according to Rumphius, it only grew on Timor, Roti, Leti, and neighboring islands (*Kruidboek*, 2:121). It was used medicinally, particularly to cure fever, as a tonic, and to withstand snakebite.

10. This means "Little stones from a pig's skin."

11. Once a common term for "pea."

12. This is the once famous "guttegom," or *gummi-resina gutti* or *gummi cambogia*. It is the dried sap of *Garcinia morella* Desr. tree and was popular in apothecaries. Later this yellow gum was often used in the manufacture of varnishes.

13. A small citrus fruit (*Citrus acida* Roxb.) known in Malay as *djeruk nipis*.

14. This is not one place. "*Solor*" is an island, south off the eastern tip of Flores. "*Larentugue*" (*Larantuka*) is a town on Flores, roughly speaking, opposite of Solor. Larantuka was the administrative capital of the Solor archipelago under Portuguese rule in the sixteenth and early seventeenth centuries.

15. According to De Wit (p. 441), "Lussa radja" is *Brucea javanica* (L.) Merr. One suggestion for the baffling "*Raiz Baffado*" is "*Rais Battata*" or the sweet potato (*Ipomoea batatas*), or it might be a

corruption of "*Raiz de Cobra*," or "snakeroot," known in Malay as *ubi ular*, better known to us as yams (*Dioscorea alata* L.).

16. Yule quotes various sources, who agree that the *mas* or *mace* was a weight in Malacca and Sumatra equal to 1/16th of a tael.

17. "Condoryns" (Condryn) is most likely a variant spelling of "*candareen*" (from the Malay *kanduri*) or "*quandreen*," also "*condrin*," and was the word for the hundredth part of a tael.

18. In the original "*Spaansche matten*," the subsequent name for the Spanish real or, in English, pieces of eight. Worth about the same as a rixdollar. That these "stones" fetched ridiculously high prices can be seen from a sales catalogue from 1727, wherein the value of one of these stones is estimated at 3,000 guilders! See H. Engel, "The Life of Albert Seba," in *Svenska Linné-Sällskapets Årsskrift*, vol. 20 (1937), p. 84.

19. Such as kidney stones, bladder stones, and the like.

20. "Mordexin" is meant to be the Portuguese word *mordexim*, later corrupted by the French into "mort-de-chien," which in turn became corrupted into "mordisheen" and so on. Mordexim referred to a terrible form of cholera, typified by severe vomiting and diarrhea.

21. Epilepsy.

22. Refers to a large abscess that can fester quite deep inside the body.

Chapter 53. Lapis è Cervo.

1. "Stone from a Deer." The "è" means "from."

2. This was Kambelo (or Kambello, also Combella), the place from which cloves were shipped on Ceram's western peninsula of Howamohel.

3. White coral.

Chapter 54. Oculus Cati Feri. Mestica Cutsjing.

1. This means "Eye of a Wild Cat," where *oculus* and *ferus* are legitimate Latin words, but the second word is not, unless this refers to the late Latin word *catus*, of obscure origin.

2. Normally spelled *kuching* (*kucing* now), the second word means "cat."

3. The mysterious "Lau animal" is the Palm Civet, of the genus *Paradoxurus*, specifically, *Paradoxurus hermaphroditicus* (Pallas). See De Wit, *M*, p. 31; also Eleanor M. O. Laurie and J. E. Hill, *List of Land Mammals of New Guinea, Celebes and Adjacent Islands* (London: British Museum, 1954), p. 84. Rumphius' contemporary Valentijn described the animal as follows in his *Oud en Nieuw Oost-Indiën*, vol. 2, part 1, p. 276: "One will find another Animal among the *Ambonese* which does not differ much from a Cuscus . . . One could rightly call it a Forest Cat; but they call it Lauw.

"It is about one-and-a-half foot long, or a little longer, and looks very much like a Civet Cat; but the tail is half its length. Its color is rather dark, also with more colors, sometimes speckled with dark spots.

"It has red eyes, white teeth, and its head is almost like that of a Civet Cat, or like the head of a Fox, the front part near the eyes curves inward a little, just like the top of its muzzle, below, and near the eyes, it has many long hairs, almost brushes, like the Civet Cat, while it also has some white hairs on its forehead. Its eyes look sad, the nose is cloven, and it has round, wide-open ears.

"The body is like that of a Cat; but speckled with brown or black spots.

"The tail is black, thickly covered with hair, and it has a white spot or brush there.

"It does not look friendly at all.

"The front paws of this Animal are also its hands, each one has five toes, with short, broad nails, divided, and which are not hairless (as those of the Cuscus) but covered with tiny little hairs. The toes or claws on the hind legs are just as long, and it uses them while walking, which makes them smooth and black on the bottom, while on top they are covered with hair, just like the front ones.

"It walks slowly; but it is a fast swimmer, and it can be handled like a Cat, it knows how to twist its body in various ways. It usually lives in trees. Its voice is seldom heard; but it sounds like that of a grunting pig."

4. These animals are, of course, not related at all. Pliny was fascinated by the hyena and described it in book 8 (Latin paragraphs 105–107), repeating the lore of the time, including the notion "that this animal alone digs up graves in search of corpses" (Loeb ed., 3:77), something of which the fictitious Indonesian hyena is also guilty. In book 37, Pliny does indeed mention "*hyaeniae*" or "hyena stones" (Latin paragraph 158; in the Loeb ed., 10:300–301), and alleges that people actually attacked the animal in order to obtain these stones from its eyes!

5. There was indeed a rather rare usage from the Greek: *aelurus* meant "cat" or "cat-like." Hence "cat stones."

Chapter 55. Bezoar. Culiga Kaka.

1. Rumphius insists on the distinction between a bezoar and a *mestika* (see his previous ch. 46), though here he equates bezoar and *guliga*, as do most other commentators. The second phrase is troublesome because it is not clear what the antecedent is of "*wiens*." One of the clear differences between a bezoar and a *mestika* seems to be that the bezoar was a calculus (this is the original meaning of this word, i.e., "little stone" from Latin *calx*) or concretion in the stomach of goats (the "old" or ancient ones in Rumphian nomenclature) and monkeys (his "new" kind). A bezoar could not come from plants, but it could be scraped and, when mixed with water, ingested as a medicine. It seems that for Rumphius, a *mestika* was more of a fetish than a medicine, and that it was a concretion that could derive from plants as well as animals. Indonesia, particularly Borneo, was a favorite place for bezoars, at least for the Chinese. The bezoar has had a long history. For the present notes I have used some selected texts (including some classic ones) on the subject which were particularly relevant, though this survey is by no means comprehensive. In chronological order these texts are the following:

Das Steinbuch des Aristoteles, mit literargeschichtlichen Untersuchungen nach der Arabischen Handschrift der Bibliothèque Nationale, ed. Julius Ruska (Heidelberg: Carl Winter, 1912). No known lapidary by Aristotle survived. This edition prints and translates into German a ninth-century manuscript in Arabic, which was ascribed to Aristotle.

Chau Ju-Kua, etc., ed. Friedrich Hirth and W. W. Rockhill (1911). This text has been alluded to before. In these Chinese descriptions of "barbarous peoples" from the twelfth and thirteenth centuries, bezoars are said to come from northern India and Asia Minor.

Garcia da Orta: *Colloquies on the Simples & Drugs of India* (1563). The edition used is the rare annotated English version by Clements Markham (London: Henry Sotheran, 1913).

Joseph de Acosta: *The Natural & Moral History of the Indies* (1590). The edition used was a contemporary English translation, published in 1604.

Jan Huygen van Linschoten: *Itinerario* (1596). The Linschoten edition was used before. Most of his information derived from Orta.

A. Baldaeus: *Naauwkeurige beschrijvinge van Malabar en Choromandel, Derzelver aangrenzende Ryken, En het machtige Eyland Ceylon* (Amsterdam, 1672).

W. W. Skeat, *Malay Magic* (1900). This was also used before.

Henry Yule, *Hobson-Jobson* (1903), often used in these notes. Yule prints quotes from early Portuguese texts as well as later British sources.

H. A. van Hien, *De Javaansche Geestenwereld en de betrekking die tusschen de geesten en de zinnelijke wereld bestaat*, 4 vols. (Semarang & Bandung: Van Dorp & Fortuna, 1896–1913). This is an invaluable work on Javanese magic and spiritualism when still practiced at the beginning of the twentieth century. In 4:30–31, Van Hien discusses *mustikas* (i.e., Rumphius' *mestikas*). Contrary to Rumphius, Van Hien states or implies that the bezoar, *mestika*, and *guliga* are all one and the same thing.

A. Hart Everett, "On the Guliga of Borneo," *Journal of the Straits Branch of the Royal Asiatic Society*, no. 4 (December, 1879), pp. 56–58.

Carl Bock: *The Head-Hunters of Borneo* (1881; reprint ed. Singapore: Oxford University Press, 1985).

J. Kreemer Jr.: *Volksheelkunde in den Indischen Archipel* (The Hague: Bijdragen Taal-Land-en Volkenkunde, 1914). This is a valuable and lucid essay on folk medicine in the Indonesian archipelago.

2. The bezoar was indeed used well into the eighteenth century as an antidote, especially against poison. Samuel Johnson has a substantial entry on this "medical stone" in his *Dictionary*, published in 1755, saying that "at present, it begins to be discarded in the practice of medicine, as of no efficacy at all." So we can safely say that it was still used by the middle of the eighteenth century. Johnson, like Rumphius, divided the bezoars into Oriental and Occidental ones.

3. Rumphius commingles the wrong and the right etymology of the word. The first, wrong one, he derived from Orta, who, in the forty-fifth colloquy of his *Coloquios dos simples e drogas* (published in 1563), said that the word derived from the Persian goat called "*pazam*," although Kreemer does mention that the stone is a gallstone from a Persian goat (*Capra aegagrus* Gm.), information he gleaned from Führer (Kreemer, p. 39). Linschoten, who, after all, copied Orta for many of his descriptions of exotic items, says the same thing (*Itinerario*, ch. 87; Linschoten ed., 2:68–69). Most experts have since then agreed on the fact that the word is a corruption of the Persian *pād-zahr*, which became in Arabic *bādizahr* or *bazahr*, which in turn was bastardized in Portuguese as *besar* or *bazar* (which gave rise to various strange etymologies), from which English derived "bezoar." The Persian really means "Ex-

peller of Poison," as a ninth-century Arabic lapidary states (see Ruska, p. 147). Rumphius is, therefore, quite close, though he probably derived it from Baldaeus.

4. That the "Occidental" bezoars come from Peru is correct. Rumphius most likely found that information in Acosta: *Historia natural y moral de las Indias*, first published in 1590. Linschoten translated Acosta's book, and his translation was published in 1598 and 1624. Baldaeus copied Linschoten. I am referring to and quoting from an English translation from 1604. Acosta mentions something which accords with what Rumphius reports about the monkey bezoars from Borneo. The vicuñas that harbor the Peruvian bezoars graze in a province where "there are many herbs and venomous beasts, which poison the water and the pastures where they eate and drinke, and where they breathe; amiddest which venomous hearbes there is one very well knowne of the Vicuña by a naturall instinct, and of other beasts that ingender the Bezoar stone which eate this hearb, and by means thereof they preserve themselves from the poisoned waters and pastures; and they say that of this hearb the stone is compounded in the stomacke, whence it drawes all the vertue against poyson and other wonderfull effects." Joseph de Acosta, *The Natural & Moral History of the Indies* [English transl. 1604], ed. Clements R. Markham (London: Hakluyt Society, 1880), p. 293.

5. This must be Chorasan or Khorasan, a region in northern Persia, which might well have been the Parthia of ancient times. Yule gives the "Persian province of Lar" as the place of origin, but both Orta and the ninth-century Arabian lapidary agree that the best ones came from Khorasan.

6. This island is mentioned by Orta and Baldaeus. Ilha das Vacas was called Delft Island by the Dutch and is located in Palk Strait, between southern India and northern Ceylon. Orta said that it was "near Cape Cormorin," but that is hardly the case. Cape Cormorin is the southernmost tip of India, while Delft Island is closer to Jaffna, on the northern tip of Ceylon.

7. "*Hagjar*" should be *hadjar*, Arabic for "stone." Baldaeus probably came across this word in Orta (forty-third colloquy), where Linschoten also found it (2:69).

8. Rumphius is referring to Philippus Baldaeus, *Naauwkeurige Beschrijvinge van Malabar en Choromandel* (1672).

9. This is more confusing than it seems at first sight. *Kees* was indeed once a general slang word for "monkey," but in particular, according to the *WNT*, *kees* could also refer to a "meerkat," an Indonesian monkey, or to the baboon of South Africa. South Africa is also the region where one will find the delightful suricate, an animal commonly tamed as a pet and not related to the monkey at all. What Rumphius says later on—that the animal might be a variety of monkey that is large, without a tail, resembling the baboon, or a monkey with a long tail, like the *monjet* ("monthien")—might indicate that he is talking about members of the *Cercopithecidae*, which belong to the Catarrhina. These Macacus monkeys can be found from Thailand to the Philippines, generally gray brown, with a lighter face and underside. The tail is usually as long as its body. This *Macacus fascicularis* Raffl. might well be Rumphius' monkey, in a general sense, because this animal is called *monjèt* in Sundanese and Javanese (which is what Rumphius says later on in the text) and *kera* in Malay. One should not think he was wrong when he says that one of the monkeys from which they obtained the bezoar resembled a baboon. There is a black baboon (*Cynopithecus niger* Desm.) in the archipelago which is considered a transition type between the Macacus and the true baboon from Africa. It is entirely black, has a stumpy tail, and lives mostly in Celebes. The fact that Rumphius mentions Borneo prominently complicates identification. The only monkey among the ones mentioned that is said to live in Borneo as well is the *Macacus speciosus* Cuv., which is dark and has a rudimentary tail. Hence my guess would be: Macacus monkeys of the Cercopithecidae. In the nineteenth century, Carl Bock reports that the Dyaks obtained bezoar stones from the "buhis monkey" or *Semnopithecus cristatus* (see Carl Bock, *The Head-Hunters of Borneo*, p. 206). The *Semnopithecinae* are a subfamily of the *Cercopithecidae* and have long tails. One variety is listed by Wilkinson as *kokah* (*Semnopithecus siamensis*), which is most likely the text's "*kaka*," a spelling that otherwise makes little sense.

10. This should be *Pedra de Bugio*. *Bugio* is indeed Portuguese for a species of macacus monkey.

11. This might well be *hou-tsao*, or "monkey berry," a Chinese term in the same class as *niu-huang* ("cow bezoar") and *kou-pao* ("dog bezoar").

12. "*Succadana*" (Sukadana), a city and region on Borneo's southwestern coast, opposite the Karimata Islands, was an important trading center in the sixteenth century. The Dutch as well as the British had offices there. The Dutch were engaged in a brisk business in diamonds and bezoar stones, which they sold at great profit. This pleasant arrangement existed until 1622, when Mataram attacked Sukadana and defeated it. I wonder whether Rumphius' "*Tambas*" is the region called Sambas, north of Pontianak. In 1609, the VOC signed a treaty with the sovereign of Sambas, a Malay king, in order to obtain the region's diamonds.

13. I do not know what this word might be, but later on Rumphius states that it is synonymous with "*monjet*." *Monjet* was the general Malay word for "monkey."

14. This is essentially what Acosta stated about the Peruvian vicuñas; see n. 4.

15. Corroborated in general outline by A. Hart Everett, in his article "On the Guliga of Borneo," *Journal of the Straits Branch of the Royal Asiatic Society*, no. 4 (December 1879), pp. 56–58. Skeat quoted from the same article in *Malay Magic*, pp. 275–277.

16. "*Bantem*" (Bantam), also spelled Bantèn, was the most westerly region of Java on the Sunda Strait.

17. A. Hart Everett reports substantially the same test in his article mentioned in n. 15, that is to say, at the end of the nineteenth century! The dog test is from Baldaeus, p. 170.

18. This is Banjarmasin, a city in southern Borneo, where the Dutch came early in the seventeenth century to trade in pepper.

19. Here is another example of the remarkable longevity of such practices, for Skeat reports towards the end of the nineteenth century that he was offered such a bezoar stone, "which its owner kept in cotton-wool in a small tin box" (*Malay Magic*, p. 275).

20. It is strange that Rumphius ascribes all these characteristics of bezoars to "Chinese doctors," for most of this is in Orta. One reason might be that he was simply quoting Baldaeus, *Naauwkeurige Beschrijvinge*, p. 169. In Markham's edition on pp. 364–365, one will find his recommendation that one can take "as much as 30 grains for all illness caused by poison or melancholy," while he also recommends it against leprosy and "prickly heat," which might well be Rumphius' "scabies" (*schorstheit*).

21. This futile ritual to preserve one's youth is also from Orta: "All wealthy persons purge themselves twice every year, in March and September, and after purging they take 10 grains every morning for five days, in rose water, and they say that with that it preserves their youth" (Markham's ed., p. 364). Also in Baldaeus, pp. 169–170.

22. Considering what I just stated in the previous two notes, one must wonder about this statement. All this information is not Chinese but from Garcia da Orta's masterpiece, which, at the very least, Rumphius knew from Clusius' edited Latin version, from which follows that this same information pertains to *goat* bezoars, and not monkey bezoars, although the Chinese had been obtaining the latter from Borneo since the eleventh century.

23. Al Cohen thinks that this is not the name of a person but of a famous historical text, *Shih-chi*, that was constantly quoted as an authority on all kinds of things. I suspect that "Little Chinese Handbook" is the same thing. In the original text "Handbook" was "*Landboek*," but I think that the letter "L" is a misprint.

24 and 25. Both words refer to Kuichou Province.

26. I am not sure about this: "*en hy wil ze binnen niet vol hebben.*"

27. This is most likely *kasumba* (*Carthamus tinctorius* L.). Known as "safflower" in Europe, this plant that resembles a thistle was once cultivated for the red dye that was obtained from its petals, while the seeds rendered an oil for lamps. In the Indies, *kasumba* was also used as a medicinal herb.

28. Rumphius most likely means a kind of palm wine, though *arak* could also be a rice brandy and, later, a strong liquor (between sixty and seventy proof) made from molasses, known as *arak api* ("fire *arak*").

29. This refers to the first volume of the "proceedings" of the Academy Rumphius belonged to: *Miscellanea Curiosa sive Ephemeridium Medico-Physicarum Germanicarum Academiae Naturae Curiosorum.* Decuriae I. Annus Primus. Anni 1670. (Francofurti & Lipsiae: Joh. Fritschii & John Fried. Gleditschii, 1684), pp. 234–235.

Chapter 56. Enorchis Silicis. Mestica Batu Api.

1. "Enorchis" is a term from Pliny. The word derives from the Greek for testicle (*orchis*), the same root as in the word "orchid"; see below, n. 4. "Silicis" is the plural of *silex*, which referred to any hard stone found in fields, but particularly flints. Hence the phrase means something like "Testicular Flints."

2. The Malay means "Fire stone *Mestika.*"

3. The original is obscure: "*ook veel meer de holligheit en de uitwendige steen naar den ingeslotenen.*"

4. In Pliny's *Natural History*, book 37, Latin paragraph 159 (Loeb ed., 10:294–295). Eichholz translates: "The 'enorchis' is white, and when it is split up into pieces reproduces exactly the shape of the testicles."

5. Cornelis Speelman (1628–1684) was a celebrated military leader (usually addressed as "Admiral") for the VOC in the seventeenth century. His successes eventually earned him the highest colonial office of governor-general. Among several major achievements was the brilliant campaign against Macassar in 1667. For the Macassar campaigns, as well as a solid biography of this fascinating figure, see F. W. Stapel, *Cornelis Janszoon Speelman* (The Hague: Martinus Nijhoff, 1936).

6. I do not know what to make of this. "*Techolidas*" does not seem to exist, but *Tecolithos* seems very close. The name means "solvent stone" and is described by Pliny (book 37, paragraph 184) as looking "like an olive stone and has no value as a gem, but when sucked breaks up and disperses stone

in the bladder" (Loeb ed., 10:315). This, of course, has nothing to do with "Jew stones," in terms of the names, but Pliny's description corresponds with Kircher's, who calls it *Judaicus Lapis* and says that it is shaped like an olive, was found on Mt. Olive, and was used against strangury and kidney or bladder stones (see book 8, ch. 6, of Kircher's *Mundus subterraneus*).

7. City on the Indian subcontinent's west coast, above Bombay, near the Gulf of Cambay.

Chapter 57. Ophitis Selonica. Ceylonese Snake Stone.

1. This means "Ceylonese Snake Stone." *Ophitis orophites* is yet another term from Pliny (book 36, Latin paragraph 55; Loeb ed., 10:42–44), but the Roman does not use it in the sense Rumphius likes to give it. For Pliny it is a kind of marble, which has markings that resemble snakes. If this kind of marble was worn as an amulet it was said to "relieve headaches and snakebites. . . . But as an antidote to snakebites some praise particularly the variety of serpentine [= ophites] known as 'tepharis' from its ashen color" (Eichholz transl.).

2. The first half of this chapter is verbatim from Philippus Baldaeus, *Naauwkeurige Beschrijvinge van Malabar en Choromandel*, pp. 168–169.

3. "*Cobra de Cabelo*" means the "Snake of the Hood." See n. 6 of the next chapter.

4. Rumphius copied Baldaeus's peculiar Latin name for the cobra. Normally *pilosus* means "hairy," but I do not think there exist anywhere a hirsute cobra. In his *China Illustrata* (Amsterdam, 1667), Kircher mentions the same translation for "Cobra de Cabelos." He gives the following explanation for this peculiar name: "It has in the upper part of its head a certain kind of covering, resembling a flat cap, which rises and turns, according to how it turns and moves" (p. 97). I am quoting from the contemporary Dutch translation: Athanasius Kircherus, *Tooneel van China, Door veel, zo Geestelijke als Wereltlijke, Geheugteekenen, Verscheide Vertoningen van de Natuur en Kunst, en Blijken van veel andere Gedenkwaerdige dingen, Geopent en Verheerlykt*, transl. J. H. Glazemaker (Amsterdam: Johannes Janssonius van Waesberge, 1668). The "Snake stone," or "Piedra de Cabelos," is discussed in part 2, ch. 5, pp. 97–99. The present chapter gives the reader an account of the traditional or "true" snake stone. In the next chapter, Rumphius lists more specifically Indonesian varieties, which are fabricated from a mixture of fable and truth. A good discussion of these stones is given by J. Kreemer in his previously mentioned essay "Volksheelkunde in den Indischen Archipel" (pp. 50–64). He quotes from Rumphius' chapters 57 and 58 at length, while noting that these pages are "not only the most important, but almost the only source for our knowledge of these stones in our Archipelago" (p. 51).

5. Better known today as "Brahmans," these were members of the priestly class, who were specialists in magic, as well as experts in religious rituals.

6. This was a specific calcareous clay once used to make a famous kind of ancient red pottery. See n. 2 of ch. 13 in this third book.

7. Rumphius is specific: he was bitten by a "*houtspin*," which, literally translated, means a "wood spider." I do not know what variety this might be.

8. For Cleyer, see n. 38 of this third book's ch. 35.

9. This refers to the series of publications sponsored by the Academiae Naturae Curiosorum that Rumphius belonged to. The quoted text was published in *MC*, pp. 18–25.

10. A misprint for "skeleton," of course, a word from Greek that came into the European language via Latin (*sceletus*).

11. *China illustrata*, by Athanasius Kircher, was published in Amsterdam in 1667.

12. Also spelled "Christiaen Geraerts," "Christiaen Gerards," or "Christiaen Gieraards," this man, a "Surgeon," was Rumphius' son-in-law. He was the one who found the manuscript of Rumphius' Malay dictionary in a market in Batavia in 1695. See Rouffaer, *G*, p. 168.

13. The Malay is probably what Rumphius spells in his *Herbal* as "*dawan batu*" and which De Wit (p. 428) identifies as the plant *Pometia tomentosa* T. et B. The Portuguese should read: Folha de Pau de St. Maria.

14. Archipelago in the Bay of Bengal, north of Sumatra, nearer the Malay peninsula than to India.

15. The ubiquitous black coral.

Chapter 58. Ophitis Vera. Mestica Ular.

1. The name means "True Snake Stones," and Rumphius proceeds to list and describe six of them, mixing fantasy and reality. One can tell that these stones were used more as talismans or fetishes than for medicinal purposes. Kircher was also fascinated by these stones and featured them rather fancifully on the frontispiece of his book *Magneticum naturae regnum*, published in Rome and Amsterdam in 1667. Kircher describes them at length in section 2, ch. 5, pp. 50–57.

2. *Ular* means "snake" in Malay; hence this simply means "Snake *Mestika*."

3. *Ular Maas* is Malay for "gold" (= *mas*) and "snake" (= *ular*). Rumphius conjectures that the

Ular Mas is the Cobra Cabelo snake on Java. The latter is an old term for the cobra (*Naja tripudians*), derived from Portuguese: cobra (from Latin *colubra* = snake) means "snake," and Rumphius' spelling *cabelo* is a variation of *capello*, Portuguese for "hood." This Portuguese name was usually reserved for the cobra of India and Ceylon (where it is also called the "snake god"); the variety living in the Malay Archipelago is slightly different. Its color is dark brown, almost black, with a blue sheen; the underside of the head is lighter in color. The two sides of the head have a faded red color, and when the hood is spread it displays the characteristic pince-nez, while younger snakes have two light spots with a darker splotch in the middle. This would seem close enough to warrant the description by Rumphius of a snake he—we must hope—never encountered. The "comb" does not tally with the cobra and might have been borrowed from the legendary basilisk (see below). The description matches the behavior of the Malay cobra when threatened: the raising of the body, the hood puffed out, and the spitting.

There is also an Indonesian subspecies of the Indian cobra, that is, spitting cobra: *Naja naja sputatrix*. It attacks big animals only if pursued; its normal diet consists of rodents, birds, frogs, and lizards. It is interesting that Rumphius should mention pigs because, though not the cobra's prey, wild boar are one of its natural enemies. The Sundanese of Java have interesting names for this snake, making distinctions in terms of its age: the younger one is called "spoon snake" (*orai-sinduk*) because they find the figure on its neck to resemble a spoon, and they call the older snakes "pig snake" (*orai-babi*) because its blue-black color resembles the hide of the Chinese swine. In Sumatra it is called the "poison-spitting snake" (*ular bieludakh*) and also the "sun snake" (*ular matahari*). There is a second variety of cobra in the Malay archipelago, the giant or king cobra (*Naja* [*s. Ophiophagus*] *bungarus*), which is the largest poisonous snake in the world. It is seldom seen and feeds on other snakes, both poisonous and nonpoisonous, hence its name, since *ophiophagus* means "snake eater" in Greek.

One will also find that this snake is sometimes called "chapel snake" because, as a writer noted in 1700, it "keep[s] in Chapels or Churches, and sometimes in Houses" (*Hobson-Jobson*, p. 224). Yule's glossary also prints the following quote: "In my walks abroad I generally carry a strong, supple walking cane. . . . Armed with it, you may rout and slaughter the hottest-tempered cobra in Hindustan. Let it rear itself up and spread its spectacled head-gear and bluster as it will, but one tap on the side of its head will bring it to reason" (p. 225).

4. Most likely this should be *chin she*, which was both a real snake and a fabled one. Needham, in *Science and Civilization*, vol. 5, pt. 2, p. 62, identifies this reptile as the *Coronella bella*, which has scales that look like metallic ones of silver or gold; hence, according to the doctrine of signatures, the animal was considered a viable antidote to gold poisoning. This colubrid snake is not venomous, but its defensive reaction resembles that of cobras: it bends its head back, hisses, and strikes.

5. *Collet*, when used as a term in jewelry, means the circle or flange in a ring wherein the stone is set.

6. *Lemon nipis*, in Malay *djeruk nipis* (*Citrus aurantifolia*), is a sour lemon with a "pleasantly sour and scented" juice (*M*, 317).

7. This was the sap from the Aren Palm (*Arenga saccharifera*), which, in the Moluccas, was not made into a wine, but used as vinegar.

8. At first Rumphius identifies this snake with an *Ular Sawa*. There is a snake by that name in the Indies: "Ular Sawa," a python (*Python reticulatus*), that can attain a length of over thirty feet, which turns out to be the same snake as the *Ular pethola*. In the first book, second chapter, of his *Ambonese Herbal*, Rumphius describes a snake by that name that likes to climb the coconut palm. First he states that it is the largest snake in those islands, i.e., the Moluccas, and describes it as "beautifully covered with patches of white and black, with a little yellow mixed in, almost like a silken garment called Pethola." He then goes on: "And one may wonder how it gets up that steep tree, though it might be helped along by two little claws which are on its belly, towards the tail, by the exit of the gut, and which are so well hidden by the scales that one can hardly see them. Nor can one rightly know what it wants up a Coconut tree, for it leaves the fruits untouched; and people are of the opinion, which may well be true, that it goes into such a tall tree so it can bask its limbs in the Sun because they may have stiffened up from the cold of the night, since it has been observed to display its entire body in many curves upon a branch where the Sun shines upon. And the climbers of this tree have to be aware of this, because it has happened that people while climbing up this tree were frightened by this unusual guest and have fallen down from up high. Yet this Ular Pethola is the most harmless of all Snakes. It also climbs other trees, for I have found it in my garden in a Pandang tree, while others have found it in the bunches of fruit of the Musa or Pissang tree [i.e., banana tree], and so well hidden that one could not discern it, nor was it to be driven away even if the entire bunch was hacked to pieces."

An *ular petola* also figures prominently in a legend about one of Ambon's heroes: an Islamic priest, Hasan Soleiman, also known as "Radja bulan" (the "Moon King"). This is a historical figure, though many legends were created about him. For instance, he was singled out as an extraordinary pupil when his forehead shone in the dark when he was asleep, he is said to have been clothed in garments of cloves when he visited Java, and his fortune was based on his possession of a remarkable stone. He obtained it one night when he went to the beach to relieve himself—this was near the town Hila on the northern peninsula of Ambon called Hitu—and saw a strong light on the coast of Ceram (i.e., Cape Sial, the southernmost tip of the Ceramese peninsula called Howamohel). He took a small prahu and rowed himself over. What he discovered was a large petola snake (or "patola," in variant spelling) with its forehead aglow. He killed it and found in its head an extraordinary *mestika.* The stone became the basis for his fortune and fame. G. W. W. C. Baron Van Hoëvell, *Ambon, en meer bepaaldelijk de Oeliassers* (Dordrecht: Blussé en Van Braam, 1875), pp. 99–100. The same legend is told in a drastically curtailed and sanitized version in: H. J. de Graaf, *De geschiedenis van Ambon en de Zuid-Molukken*, p. 103. Van Hoëvell also reports another Ambonese legend that tells of a large dragon or *naga* which lived in that same region around the town of Hila and was said to be so large that its head was under the mountain Wawani and his tail in the plateau called Fransana (Hoëvell, pp. 100–101). Valentijn also mentions this snake in *Oud en Nieuw Oost-Indiën* (3:285–287); for Valentijn's description of other Ambonese reptiles see pp. 285–292 in the same volume.

The word *petola* (or *patola*) refers to several other things: a kind of cloth used as a garment that has a distinctive design, the *kain petola;* the name for a plant with serpentine fruit of the cucumber family, *Luffa petola;* and a shell, *Bia petola,* so named by Rumphius in the present work, book 2, ch. 7, with a variegated design. Rumphius also gave the name *petola* to an orchid—the last plant described in book 10 of his *Herbal.* This *daun petola* (or "*petola* leaf") is said to be *Anoechtochilus reinwardtii* by de Wit (in his chapter "Orchids in Rumphius' Herbarium Amboinense," *Orchid Biology*, pp. 57–59). Rumphius mentions that it is seldom found and that owing to this rarity, as well as the intricate and beautifully colored designs on its leaves, he named it after the equally rare (because expensive) silken cloth, dyed with many colors, called "*petola.*"

9. This might be *pai-ching*, meaning "white eye."

10. This is *tu-mala'bu* in Macassarese, or "the long python."

11. One-and-a-half fathoms is approximately nine feet.

12. The *basilisk* is a legendary creature. Albertus Magnus (*Animalia*, p. 25) states that the basilisk is hatched by the sun from an egg laid by an old cock. What emerges is "a serpent just like a cock in every way except that it has the long tail of a serpent" (*The Book of Secrets of Albertus Magnus*, ed. Michael R. Best and Frank H. Brightman [Oxford: Oxford University Press, 1973], pp. 83–84). It was of course a favorite of *bestiaries*, either under its own name or confused with the cocatrice. T. H. White's superb translation of a twelfth-century bestiary provides the following details: The basilisk was considered a serpent, not to mention "a king of serpents." Its smell as well as its glance could kill a man or animal, so could its hiss, and it could burn people even before biting them. White quotes Sir Thomas Browne's firm belief in the basilisk; its ability to kill with a mere look was interpreted by Browne as a lethal form of airborne poison. Aldrovanus, Clusius, and Avicenna all thought the basilisk to be a serpent and most probably a hooded one. Since the bestiary also relates that the basilisk "can be conquered by a weasel" (which sounds very much like a mongoose), the conjecture is not far fetched that we are dealing with a king cobra (*basileus* is "king" in Greek) or a spitting cobra (*The Bestiary*, ed. and translated T. H. White [1954; reprint ed. New York: Capricorn Books, 1960], pp. 168–70).

We should mention dragons in the present context. White's *Bestiary* calls the dragon or "draco" the largest of serpents, notes that it does not need poison but kills by winding itself around its opponent, and places it in countries "where there is perpetual heat." White points out in a note that the word "dragon" more often than not was simply synonymous with "huge snake," most likely the python. The wondrous acts of some of these beasts, be they real or imagined, cannot all be enumerated, although they constitute the glory of medieval bestiaries. In the case of the dragon there is, for instance, the marvelous detail that it "lassoes the legs [of elephants] with its tail and destroys them by suffocation" (White, pp. 165–167), a passage taken from Pliny's *Natural History* (book 8, section xi).

The third ophitis brought to Rumphius was a "Dracontia." Not only does the word obviously relate to the dragon as a large serpent, but it was also the term commonly used for "snake stone." *Draconites*, according to Albertus Magnus, came "from the Dragon's head. And if the stone be drawn out from him alive, it is good against all poisons, and he that beareth it on his left arm, shall overcome all his enemies" (*The Book of Secrets of Albertus Magnus*, p. 45). Note that its use as a counterpoison is also characteristic of the *ophitis.* Rumphius' ironic statement about the devil echoes the lore of the Middle Ages, for the *Bestiary* mentions that "the Devil, who is the most enormous of reptiles, is like this Dragon" (White, p. 167).

The *Bestiary* includes "Fire Stones" from "a certain oriental mountain." They are said to be male and female: "When these are far apart from one another the fire in them does not catch light. But when by chance the female may have approached the male, at once the flame bursts out, so much so that everything around that mountain blazes." T. H. White notes that these were most probably *pyroboli*, or what Pliny called *pyrites.* The latter word was most often translated as "fire stone" or "flint."

In 1794 Sullivan noted that "heated Bath waters . . . owe their origin to the contact of common water with pyritae, whose composition is iron, sulphur and the vitriolic principle" (*OED*). This quality of striking fire and of providing heat in water is also characteristic of Rumphius' *ophitis.* Certainly there are some similarities between these stones and Pliny's *pyrites.* Perhaps there is also a connection with the locality of the Indies. In the previous description of *Tsjerondsjung* mention was made of sulphur and arsenic. A pyrite called arsenopyrite is often found with gold in gold deposits and is the world's primary source of arsenic.

Pliny explains pyrites in his *Natural History* (book 36, Latin paragraphs 137–138). He mentions that some are called "live stones" (*quos vivos apellamus*), and when struck with another stone, "they give off a spark, and if this is caught on sulphur or else on dry fungi or leaves it produces a flame instantaneously." In book 37, Latin paragraph 158, Pliny mentions and describes "snake stones" under the name of *draconitis* or *dracontias:* "The 'draconitis,' otherwise known as 'dracontias,' the 'snake stone' is obtained from the brains of snakes [*e cerebro fit draconum*], but unless the head is cut off from a live snake, the substance fails to turn into a gem, owing to the spite of the creature as it perceives that it is doomed. Consequently, the beast's head is lopped off while it is asleep. Sotacus, who writes that he saw a gem in the possession of a king, states that those who go in search of it ride in two-horsed chariots, and that when they see the snake they scatter sleeping-drugs [*somni medicamenta*] and so put it to sleep before they cut off its head. According to him, the stone is colorless and transparent, and cannot subsequently be polished or submitted to any other skillfull process." Pliny's *dracontias* seems to have sired the *ophitis*, and I would think that the "old writ" Rumphius mentions might be Pliny's *Natural History.* It is of interest to note, however, that Rumphius' precise description ("that the true Dracontia . . . should have the same color as a Crystal and for a shape that of a bean") is not from Pliny's section on *Dracontias* but from his description of *Alectorias* or "cock stones" (book 37, §liv), which are "in appearance like rock-crystal and in size like beans" (*Alectorias vocant in ventriculis gallinaceorum inventas crystallina specie, magnitudine fabae*). Pliny also catalogues a stone called *ophites* (book 36, §xi), but this one does not come from snake skulls but is found in the earth like, as it were, a proper stone: "From serpentine, the markings of which resemble snakes—hence its name—[*differentia eorum est ab ophite, cum sit illud serpentium maculis simile, unde et nomen accepit*], these stones differ in that their markings are grouped differently. Those of the Augustean curl over like waves so as to form coils, while the Tiberian has scattered grayish-white spots which are not rolled into coils. . . . It has two varieties: one is soft and white, the other hard and dark. When worn as amulets, both are said to relieve headaches and snakebites. Some authorities recommend the white variety as an amulet to be worn by sufferers from delirium or a coma. But as an antidote to snakebites some praise particularly the variety of serpentine known as 'tephrias' from its ashen color."

13. The "Dart Tree" is the Upas Tree (*Antiaris toxicaria*), fabled in legend, and of some poisonous consequence in reality. Its toxic sap was used by native peoples for hunting and warfare. See my translation of Rumphius' description of this tree in *The Poison Tree*, pp. 127–158.

14. *Mindanau* (*Magindano*) is the island of Mindanao, the southernmost of the large Philippine islands, and the first large island north of Celebes.

15. I translated *negorij* as "village" rather than "plantation," as some translators would. *Negorij* or *negerij* comes from the Malay for town: *negri.* In Dutch usage it means either a native village (what the Malay call a *kampong*) or, figuratively, a boring, sleepy town.

16. The *South-Easter Islands* (*Zuid-Ooster Eilanden*), with Timorlaut as the largest, lie east of Timor, south of Ceram, and diagonally south of what may be described as the wattle of New Guinea.

17. There is confusion in the nomenclature because previous to Linnaeus, the generic term *Cheme* or *Chama* referred to bivalve mollusks of a more or less smooth appearance. Rumphius, using the same terminology as Aristotle and Pliny, means in particular *Tridacna*, a genus of bivalve mollusks which included the largest ones such as the Giant Clam (*Tridacna gigas*), which can grow up to five feet in width. In chapter 28 of book 2, Rumphius described what he called *Chama squammata*, or "Bia garu" in Malay, which appears to mean "Nail Shells" (*bia* is Malay for "shell"). It may be added that his account represents the first description of the living mollusk, its habitat, and its brilliant coloring in the literature of conchology. Rumphius gave a particular name—*chamites*—to the *mestika* of these mollusks, and devoted an entire chapter to them in the third book, chapter 64. What *chamites* particularly refers to is a calcareous concretion inside the hinge ligament of this large mollusc. It occurs only in the older ones and even so only in one out of ten. Hence the phrase in question might be circumscribed as follows: "the *mestika* of the Giant Clam, or chalky concretions found in the large mollusks of the Moluccas." See Martens, *G*, pp. 124–125.

18. *Jambii* must be *Djambi*, a large region on the east coast of Sumatra.

19. *Krisses* is the common Malay spelling (and Dutch plural) for this dagger; in early English travel literature one will often find the spellings *creese*, *kreeze*, *crease*, and others. It has typically a wavy blade of serpentine shape. In his book *East Monsoon*, G. E. P. Collins describes a war dance which

a Biran radja performed for him with a kris. He well describes the kris as having a "slender, snaky" or a "sinuate flame-formed blade," which in such a dance can come almost alive and "writhe like a snake." Collins also reports the same local radja's explanation of how the island of Celebes got its former name: "When the Portuguese first came to Makassar, they went ashore and were met by some of the Raja of Goa's men. They asked them the name of the country. . . . Though the white men spoke Malay, which they had learnt in Malacca, the Goans couldn't understand them very well. They had seen the white men's eyes fixed intently on their krises—they were probably on their guard against a sudden attack—and thinking they asked the name of the weapons answered 'Selé Besi, [or] Iron Kris' " (G. E. P. Collins, *East Monsoon* [New York: Scribner's, 1937], pp. 245–257). The local people were probably simply saying "*sulawesi*," which was (and is again) their name for the island, the third largest of Indonesia.

20. This one and the other two large animals (*Iste Liong* [the second word might be *lung*, or "dragon," in Chinese], *Tambo Sisi*, and *Terrebelau*) may well be snakes since, for instance, *naga* in Malay can mean a very large snake. However, it can also mean dragon, and from the few details given, one may wonder whether these creatures are large lizards of the family *Agamidae*. Some of these are quite large, have a threatening display that is imposing (hissing with the mouths open, puffing up crests and neckfolds, whipping long tails around), and live in a habitat that for many of them is restricted to the general region of the Moluccas with which Rumphius was most familiar. The fact that he mentions the island *Contung*, which is now called Pulu Condor and lies off the coast of Cambodia, may provide a possible clue. The Oriental or Chinese water dragon (*Physignathus cocincinus*) inhabits Indochina. Then there is a variety of water lizard Soa-Soa, which is said by R. F. Ellen (*Bulletin of the School of Oriental and African Studies, University of London*, 46:2 (1983), p. 400) to be the monitor lizard *Varanus indicus*.

21. *Sampo* refers to the Chinese Admiral Chêng Ho, who lived in the fifteenth century; see the nn. to ch. 14 of this book.

22. Strange as it may sound, the city of *Burneo* is not readily found. In the text on the *Durian* in his *Herbal*, Rumphius mentions that there is a city which has the same name as the Island of Borneo. In *A Voyage to and from the Island of Borneo*, by Captain Daniel Beeckman (published in London in 1718), we find a statement similar to that of Rumphius: "the Island of *Borneo* (so called from a City of that Name) . . ." (p. 34). Later (p. 45) Beeckman is even more explicit: "There are in this Island four chief Ports of Trade, *viz.* the City of Borneo, situate on the North, in the Latitude of 4 Deg. 20 Min. North." He also has a city by that name on his frontispiece map. From this information it would seem that the city of *Brunei* (Dutch spelling: *Broenei*) is the city of "Burneo" or "Borneo." It is situated on the river of the same name and is the capital of what was once a powerful sultanate. Pigafetta mentions in his account of Magellan's voyage how splendid and imposing he found the court in Brunei of Sultan Bulkeiah (1521). As for the snakes, Borneo has been known for the presence of many of them and was once famed for its bezoar stones.

In the present context it is noteworthy that Beeckman mentions that "the Country abounds with Pepper, the best Dragons-blood Bezoar . . ." (p. 36). "Dragon's-blood" refers to a bright red resin from the palm tree *Calamus draco*. Beeckman's *Voyage* has been reprinted and offers a fascinating account in English of the journey itself as well as of the southern coast of Borneo. Captain Daniel Beeckman, *A Voyage to and from the Island of Borneo* (Folkestone: Dawsons of Pall Mall, 1973).

23. *Ligoor*, or *Ligore*, is another name for the Siamese province of Nakhon Sri Tammarat. It is part of the Malay peninsula, on its eastern coast, down from the Gulf of Siam into the South China Sea. It is below the Isthmus of Kra and is now part of Thailand. The Portuguese were the first to establish relations with Siam, but they were gradually replaced by the Dutch in the seventeenth century. The *Pinang* is the nut of the *Areca cathecu* L., incorrectly called in English the "betel nut."

Chapter 59. Stones from Various Other Venomous Animals.

1. This ancient term for a centipede is from Pliny (for instance book 8, Latin paragraph 104).

2. "*Kacki*" (*Khaki*) means "leg," and *seratus* or *saratus* "one hundred."

3. This is Ssuchuan Province.

4. In his *China Materia Medica*, Bernard E. Read (in the section "Insect Drugs") notes that the more common Chinese name for these ferocious centipedes is *chi chi'ü*. He states that such a centipede "is able to overcome snakes . . . (and when it) sees a big snake it crawls over to it and bites it on the head" (p. 162). A little further on, he notes that "according to the *Yu Yang Tsa Tsu* by *Tuan Ch'eng Shih*, at *Sui Ting Hsien* (Szechuan) the centipedes are so large they can suck snakes and lizards toward them from 3 to 4 feet away and the meat and bones just disappear. . . . Centipedes can overpower dragons, snakes and lizards but they are afraid of toads, slugs and spiders" (p. 164).

5. "*Tocke*" is the domestic lizard *tokek*—so called as an onomatopoeic approximation of the sound it makes—called *gecko* by the Europeans. A smaller version is the *tjitjak* (now *cicak*), *Hemidactylus*, which can be found in every house in the Indies. All these nocturnal animals belong to the *Geckonidae*, and were the subject of several superstitions, such as the notion that they were poisonous. They are

not, they are useful because they eat insects, and I found their cry a comfort at night rather than something to be feared.

Chapter 60. Aëtites Peregrinus. Mestica Kiappa.

1. "*Aëtites*" is a term from Pliny. In book 10, Latin paragraph 12, he says that this eagle stone (*lapis aëtites*) is also called *gagites* and that the eagles "built it into their nests." "The stone in question is big with another inside it, which rattles as if in a jar when you shake it. But only those taken from a nest possess the medicinal power referred to" (Rackham transl.; Loeb ed., 3:301). This is clearly the same thing about which Rumphius is writing.

"*Peregrinus*" in classical Latin meant "strange," "exotic," "foreign," "coming from foreign parts." Hence this name means "Exotic Eagle Stone." Pliny discusses the stones again in book 36 (Latin paragraphs 149–151; Loeb ed., 10:118–121), where he emphasizes that they are only found in pairs (Loeb 10:119). This would corroborate Rumphius' statement.

2. Ceram's narrow peninsula, otherwise known as Howamohel.

3. This is most likely *Haliaëtus leucogaster* (Gmelin), according to Van Bemmel (*M*, p. 44). The text's "*Kiappa*" could also be spelled "Jappa."

4. The shot, usually stones, tossed from a small cannon, usually on board ships, about two inches in diameter.

Chapter 61. Lapillus Motacilla. Mestica Baycole.

1. *Lapillis* is Latin for "little stone," and *motacilla* is Pliny's name for what Eichholz calls the "water-wagtail" (book 37, Latin paragraph 156; Loeb ed., 10:291). The name survived in that the family to which these Old World birds belong are called Motacillidae. The closest we have in the United States is the flycatcher. This is, of course, not the same bird. Van Bemmel (*M*, p. 55) identifies it as *Rhipidura leucophrys melaleuca* (Quoy & Gaimard). "*Baycole*" was the Ambonese name for the wagtail.

2. This is a melancholy statement relating the fact that this oft-mentioned narrow peninsula of Ceram adjacent to Ambon was depopulated after a series of fierce campaigns of the VOC against Madjira, the viceroy of the Sultan of Ternate. 1651 is the year when the fifth "Ambonese war," as Rumphius called it, began, with Arnold de Vlaming in command. De Vlaming defeated Madjira and executed harsh policies to ensure that the VOC's authority would never be questioned again. In 1656 he ordered all villages on Howamohel burned, and the population moved to Hitu on Ambon. The peninsula, also known as "Little Ceram," never recovered.

3. A once powerful village on Howamohel; also spelled Lessidi.

4. Fish trap.

5. Also spelled "Soccus" or "Saccus" by Rumphius, it was identified by Merrill (p. 190) as *Artocarpus communis* Forst. Char. Gen. (1776), or more commonly known as the breadfruit tree.

6. Village chief.

7. "*Caytetto*" might refer to Kaibobo on Ceram.

8. Probably the constant campaigns at the time to defeat the Makassarese and the leader Kakiali.

9. "*Muscus Zibeth*" is musk or civet, from the gland of *Viverra zibetha* L. 1758, or civet cat.

10. The reference is to Habakkuk 1:16, which reads in the King James version: "Therefore they sacrifice unto their net, and burn incense unto their drag."

11. What Pliny wrote (book 37, Latin paragraph 156; Loeb ed., 10:290–291) was: "The 'chloritis,' or 'greenstone,' which is of a grassy colour, is said by the Magi to be found as a congenital growth in the crop of the water-wagtail. They recommend that it should be set in an iron bezel so as to produce certain of their all too familiar miracles" (Eichholz transl.).

Chapter 62. Stones from Various Other Birds.

1. This is the island Taliabu, in the Sula Islands. These islands are east of Celebes' middle, in the Molucca Straits. Prof. Voous identified this bird as the Nicobar Pigeon (*Caloenas nicobarica* L.), a relatively large pigeon that can be found from the Nicobar Islands in the west of the archipelago to the Solomon Islands in the east. The birds eat large fruits that have fallen on the ground. These fruits contain large seeds which rub against one another in the bird's stomach, whereafter they are expelled. The islanders bestow both the bird and the "stones" with supernatural powers. The pigeon needs these "stones" for digestion. They may be real stones or the hard kernels of various fruits. See A. Hoogerwerf, "Verspreiding en voedsel van *Caloenas nicobarica* and *Ducula bicolor*," *Ardea* 55 (1967), pp. 256–257.

2. This is most likely *Ajam Taliabu*, or the "Taliabu Chicken."

3. Rumphius describes this concretion in the coconut, which he calls "Callapites" in the third chapter of his *Herbal*, where he described it as "oval in shape, like a stubby little Kayle [a kind of bowling pin] or lizard's egg, the size of a little bird's heart, sometimes also like a dove heart. It is of a dirty-yellow color at the hinder and widest end, and has a little crown, like a tooth that has fallen out,

with which it was fastened to the tampurong [the coconut's shell]: If one holds it against the light one can see at its foremost and narrowest end a little shiny spot that resembles a radiant little Sun, and if they do not have this they are considered dead or counterfeit." *Herbal*, book 1, ch. 3, p. 22.

4. For Van Quaelbergen, see n. 14 of ch. 36 of this book.

Chapter 63. Sepites. Mestica Sontong.

1. "Sea cat," English from the Dutch *zeekat*, is an old synonym for the cuttlefish. The cuttlefish is what Latin called "sepia" (Pliny: *saepia* in book 9, Latin paragraph 83), an oval cephalopod of the family Sepiidae. Rumphius' name means "Cuttlefish stones." The Malay for "cuttlefish" is normally spelled *sotong*.

2. This is the fruit of the tree *Gnetum gnemon* L., which was once cooked as a vegetable or turned into small flat cakes to be eaten as a kind of *krupuk*.

3. The *Ikan kepala batu*, or "stone-head fish," is *Pelates quadrilineatus* (Bl.) of the family Theraponidae. See W. H. Schuster and Rustami Djajadiredja, *Local Common Names of Indonesian Fishes* (Bandung: W. van Hoeve, 1952), p. 74.

4. This rather redundant name means "Stones Called Headstones."

5. A "Carp stone" refers to a small, oval, semitransparent little bone that is part of the fish's skeleton where the head is attached to the spine. It was once sold in apothecaries as having medicinal powers.

6. "Sea cow" refers most likely to the dugong, the aquatic grazing mammal that, with the manatee, make up the entire order Sirenia.

7. The "Sea wolf" could either be a seal, a Sea elephant, or the wolf fish.

Chapter 64. Chamites. Mestica Bia Garu.

1. These "stones" were described by Martens as calcareous concretions inside the hinge ligament. See ch. 29 in book 2 for identifications of the various mollusks.

2. The ligament.

3. The muscle of the oyster or other bivalve.

4. A phial in the alchemist's laboratory.

5. The retort was a glass vessel that had a long neck, bent downwards, used for distillation.

6. Coral.

7. "Leaking" for *lekken*, which can also mean "to lick." But one needs a tongue for that, and I don't think oysters have tongues. I think that Rumphius' "leaking" is what we would call "secreting."

8. See n. 3 of the previous ch. 62.

9. At one time spelled "Lucapinho," the Nusa Penju or Nusapinyu Islands form a small archipelago in the center of the Banda Sea.

10. Bean or peanut.

11. These are discussed in ch. 30 of book 2.

12. Book 37, Latin paragraph 180. See n. 30 of ch. 28 in book 2.

Chapter 65. Tellinites. Ctenites. Bia Batu.

1. The name for these stones—ultimately derived from the Latin *tellus*, or "earth"—corresponds to the name for the shells as described in ch. 33 in book 2. Number VI there was "*tellina picta*." These are fossil shells.

2. For this shell, see ch. 30 in book 2, no. V.

3. These Javanese towns are most likely Rembang and Lasem on the northern coast of central Java between Mt. Murjo and Mt. Lasem.

4. Better spelled Grisek, this town is above Surabaya, opposite Madura.

5. "*Ubat guna*" means "love potion" as well; from *ubat*, "potion," and *guna*, "magic."

6. Called such in ch. 31 in book 2, under no. VIII.

7 and 8. Refer to the same text under no. VIII in ch. 31 of book 2.

9. "*Messoal*" is Misoöl or Batan Me, an island southwest of New Guinea's bird's head. It is surrounded by a host of tiny islands, of which the largest one is called Los. This is most likely Rumphius' "Uos." This group of small islands was once also known as the "Kanari Islands" because they had many *kanari* trees (*Canarium commune*). The *kanari* nuts were a staple trade commodity.

10. This "rain stone" is from Pliny, book 37, Latin paragraph 176 (Loeb ed., 10:306). These are fossilized sea urchins. See also ch. 29 in book 2.

11. In other words, like the shells discussed in ch. 35 in book 2, under no. V.

12. Fort Duurstede was completed in 1691. It had been built on a hill overlooking a bay on the island Saparua. The governor of Ambon, Nicolas Schaghen, ordered its erection. It was of no great significance during the eighteenth century, but in 1817 it became famous when guerrillas under the

command of Pattimuri sacked the fort and killed all inhabitants except for the resident's eldest child, who somehow escaped the massacre. For a description of this building, see V. I. van de Wall, *De Nederlandsche Oudheden in de Molukken* ('s-Gravenhage: Martinus Nijhoff, 1928), pp. 196–197.

Chapter 66. Cochlites. Cochlea Saxea. Mestica Bia.

1. That is to say, between fifteen and twenty millimeters, or between 0.59 and 0.79 inches. Wichmann (*G*, p. 158) says these are fossilized gastropods (snails, slugs).

2. *Mamalo* (Mamola) is a place about midway between Hila and Hitulama on Ambon's northern peninsula, called Hitu.

3. See ch. 11 in book 2, no. VII.

4. Nicolaas Witsen (1641–1717) came from a rich patrician family in Amsterdam. He lived in that city all of his life and was its burgomaster thirteen times between 1682 and 1705. He studied mathematics, developed into a competent cartographer, and geographer, and became a specialist in shipbuilding. He advised Peter the Great on the last industry, when the Russian czar was in Holland in 1697–1698.

5. The "Old Men's House" is on the Kloveniersburgwal in Amsterdam. Established in 1601, when the first structure was built with funds raised by a public lottery, it survived to become part of the University of Amsterdam today.

6. Tongeren is a town in Belgian Limburg, three miles from Maastricht. It is Belgium's oldest city, dating back to ca. 50 A.D.

7. I think Schynvoet is confusing father and son. The father was Olaus Bromelius (1639–1707), the most important Swedish botanist prior to Linnaeus. He received his medical degree in Leiden in 1673, and practiced in Stockholm and Göteborg. The botanical family of Bromeliaceae was named after him. His son, Magnus, also studied medicine in Leiden, as well as Oxford and Paris, though he obtained his degree from Rheims in 1703. He became the professor of medicine in Uppsala university in 1716, and the era's most prominent anatomist in Stockholm. He published articles on mineralogy and paleontology. He was knighted in 1726 and was thereafter known as Magnus von Bromell.

8. See n. 20 of ch. 29 in book 2.

9. See n. 19 of ch. 29 in book 2.

10. Thomas Burnet (1635–1715) was a controversial British divine from Yorkshire. Schynvoet is referring to one of Burnet's books that was very popular at the time: *Telluris Theoria Sacra, or Sacred Theory of the Earth*, published in London in 1681. Burnet presented the notion that the globe had been crushed like an egg when the deluge occurred, that internal waters erupted, and that the remaining pieces of the earth's shell became mountain ranges. Popular as well as controversial, the book was praised by Johnson in his *Lives of the Poets*. The Dutch edition mentioned by Schynvoet is *Heilige Beshouwinge* [sic] *des Aardkloots, Vervattende in vier Boeken De Oorsprong en Algemene Veranderingen Van onse Wereld, Welke die nu al ondergaan heeft, ofte nog na desen ondergaan zal* (Amsterdam: Daniel van Dalen, 1696). It is clear (book 2, ch. 4, pp. 216–217) that Burnet was against Descartes, while he asserts on p. 9 that only 5,000 years have elapsed since the first creation of the world.

11. Frederik Ruysch (1638–1731) was a celebrated physician and professor of anatomy in Amsterdam. His fame rested on his various anatomical discoveries and on his method of preserving human organs, if not entire corpses, with new techniques of injecting various fluids. His anatomical and zoological collections were celebrated, and were bought by Peter the Great of Russia in 1717 and still exist rather complete in St. Petersburg today; see Anna Radzjoen, "De anatomische collectie van Frederik Ruysch in Sint-Petersburg" in *Peter de Grote en Holland*, ed. Renée Kistemaker et al. (Bussum: Thoth, 1977), pp. 47–53.

12. For Vincent, see n. 5 of ch. 9 in book 2.

13. Johann Christian von Kretschmar und Flämischdorf (1650–1693) was a military officer who became the colonel and commander of the "Second Battalion Guards of his Majesty of Great Britain," that is to say Willem III of Holland. His cabinet was primarily known for its collection of fossils, minerals, gems, and shells.

Chapter 67. Myites. Mestica Assousing.

1. This remains as mysterious as when Rumphius first mentioned it in ch. 35 in book 2. "*Myites*" is not to be found in Pliny. In Latin "mouse" is *mus*, also in Greek. F. S. A. De Clercq states in *Het Maleisch der Molukken* that *asuseng* is an edible snail found most prominently on the beach of Galala (p. 8). Wichmann (*G*, p. 158) says that these are fossilized lamellibranchia (clams, oysters, and the like).

2. See ch. 37 of book 2, no. IV.

3. Mangoli is one of the Sula Islands.

4. See ch. 36 of book 2.

Chapter 68. Clappites. Klapsteen. Mestica Calappa.

1. Rumphius discusses these vegetal calculi at the very beginning of his *Herbal;* hence his fascination with such concretions was there from the outset of his tropical career. It indicates the insecurity of his life when he, at the beginning of the *Herbal*, feels compelled to state "that I intended to describe it in another place among other stones and minerals. But since it is found in the Calappus nut, and since I am worried that the mentioned work will be not completed by me, I deemed it better to include its description here" (*Herbal*, book 1, ch. 3, p. 22). For my translation of Rumphius' description of this stone, repeated here as well as in ch. 64, see n. 3 of the previous ch. 62. Perhaps one should mention again that the "Calappus" is the coconut.

2. Pliny, book 37, Latin paragraph 192; Loeb ed., 10:320–321. It reads in Eichholz' translation: "As for the white 'dendritis' or 'tree stone,' it is said that if it is buried beneath a tree that is being felled the edges of the axes will not be blunted." Rumphius repeats the latter.

3. An Elizabethan bowling pin.

4. See ch. 8 of book 2.

5. With "the country of the *Bugis*" Rumphius is most likely referring to the Bone region along the Gulf of Bone on Celebes' western "leg." "*Cajeeli*" is Kajeli, in Kajeli Bay on Buru. Buton is the relatively large island southeast off Celebes' eastern "leg."

6. Perhaps this "Victory Stone" is chrysophrase; see Kunz, *The Curious Lore of Precious Stones*, p. 67. "*Astroites*" is most likely a misprint for Pliny's *astriotes*, or "star stone," which was highly regarded by Zoroaster (book 37, Latin paragraph 133; Loeb ed., 10:272).

7. *Pady* (*Padi*) is rice.

8. This wonderful term (mentioned derisively by Pliny in book 29, Latin paragraph 6; Loeb ed., 8:184) had a specific meaning in Roman times: it referred to a physician who derived his knowledge exclusively from experience, not theory.

9. See ch. 64 of this book, though Rumphius mentions it specifically with the *second* kind of Chamites.

10. *Liplap* was the *kelambir* or soft white marrow of the young coconut.

11. "*Bahuala*" is most likely Baguala, a town and bay on Ambon's southern peninsula, known as Leitimor.

Chapter 69. Dendrites Calapparia.

1. For the original *palmijt.* Also spelled *palmiet*, this referred in the Indies quite specifically to the tips of leaves and to young leaf buds of the palm tree, which were eaten as a vegetable. One finds this often translated in English with the word "palmetto," but that refers to a tree, not part of a palm.

2. My guess for the original "*plein.*"

3. See n. 6 of the previous chapter.

4. Also spelled *dissave* or *dissava*, it originally meant "governor of a province" on Ceylon. This could have been Isaac de Saint Martin (?1629–1696), who served on Ceylon from at least 1663 to 1671. He was a friend of Van Reede tot Drakenstein and devoted to botany. De Saint Martin was a rich man, an intimate of Camphuys and Chastelein, and like these two officials, Rumphius' friend and protector. He was also known as an expert on the "religion and politics of the native peoples" and seems to have been fluent in Malay. See De Haan, *Priangan*, vol. 1, "Personalia section," pp. 15–21.

5. In the original: "*Als meerder komt, moet minder wyken.*"

Chapter 70. Pinangitis. Mestica Pinang.

1. Rumphius' "Pinang tree" is the Areca palm.

2. "*Catjan*" or *kachang* is a generic noun for "bean."

3. "*Bras*" or *beras* is husked rice that has not been cooked.

4. "*Scorodites*" means "stones from the scordion plant," also spelled *scordotis.* This plant was described by Pliny in book 25, Latin paragraph 63; Loeb ed., 7:182–183. Since Rumphius compares it to garlic, this is most likely water germander (*Teucrium scordium* L.), also known popularly as English treacle or wood garlic.

5. The Malay word "*puti*" or *puteh* means "white," and *bawang* means "bulb"; the Malay phrase "white bulb" refers to "garlic."

Chapter 71. Dendritis Arborea. Mestica Caju.

1. Arak is a native distilled spirit, at the time probably the sap of the date palm. Later the liquor was distilled from sugarcane molasses, and the business seems always to have been dominated by the Chinese.

2. The Casuarina tree had feathered leaves, like a lacy conifer. Rumphius had the beach variety

in mind here, which De Wit identifies as *Casuarina equisetifolia* J. R. et G. Forst (*M*, p. 423). Rumphius discusses this tree in book 4 of his *Herbal*, ch. 50, entitled "Strandt Casuaris Boom" (Beach Casuaris Tree). He mentions the same Chinese man, Noncko, and notes that arak distillers preferred this wood because it burned longer, and that Sea Eagles liked to nest in these trees (p. 88).

3. "*Iris*" was the ancient name for "rainbow," hence this phrase means "secondary rainbow." This phenomenon happens sometimes and manifests itself as a much fainter rainbow outside the primary one, with the order of the colors reversed, i.e., violet on the outside and red on the inside, while in the primary one this is the other way around.

4. "*Caju*" or *kaju* means "wood."

5. Most likely *Vitex cofassus* Reinw. ex BL. (*M*, p. 424).

6. Perhaps *Murraya koenigii* L.?

7. In his *Herbal*, book 4, ch. 17, Rumphius mentions that this tall Sicki tree (or "Sicchius") is used for lumber and that it burns easily (pp. 40–42).

8. This is *Jasminum sambac* L. (De Wit, p. 430), popularly known as Arabian jasmine. Rumphius described it in the fifth volume of his *Herbal*. An English translation can be found in my *Poison Tree*, pp. 184–189. See also the subsequent ch. 75.

9. "*Camarien*" is Kamarian, a coastal town on Ceram's southern coast, almost straight across from Liang on Ambon's northeastern tip.

Chapter 72. Nancites. Mestica Nanka.

1. This is somewhat confusing. "Soursack" (misspelled *soorsatk;* should be *soorsack* or *suirsack*) is what in English is called the jackfruit (*Artocarpus integrifolia* L.). In the first volume of his *Herbal*, Rumphius calls this tree "*Soccus*" or "*Saccus.*" It was a large fruit, the size of a pumpkin, with a sweet pulp, containing many kernels the size of chestnuts.

2. This passage deals with this "stone" as if it were a touchstone to test gold and silver.

3. I translated the original *potloot* as "slate" because, though this noun means "pencil" in modern Dutch, such a writing tool did not exist at the time. The reference is to a slate pencil.

4. "*Boa*" is probably *buah*, or "fruit," and "*Malacca*" might refer to the tall *Phyllanthus emblica* L. tree, the fruits of which were eaten raw or candied.

5. For "*knie.*" Rumphius is most likely referring to the joint in the stalk of a plant. That was called a *geniculum* in Latin, from *genu* or "knee."

6. Also spelled *selaseh*, "*sulassi*" refers to sweet basil (*Ocimum* sp.).

7. This is no doubt *kelumpang*, the *Sterculia foetida* L. tree, famous for its flowers, which smell like excrement. The wood was used for coffins, the leaves for medicinal purposes, and the seeds (which is what Rumphius is referring to) for making a condiment, while their oil was also used to light lamps.

Chapter 73. Sangites. Mestica Sanga.

1. A "milktree" is a tree that produces a sap (*getah* in Malay), in this case a poisonous sap.

2. In his *Herbal*, book 3, ch. 44, Rumphius calls the "Sanga Tree" the "Vernis Boom" or "Varnish Tree" (*Arbor Vernicis*). This is *Gluta renghas* L. (*M*, p. 417). He notes that "estimable Chinese" call this type of varnish "*Cil*" (p. 259), while the common Chinese call it "*Tsjad*" ("*Amrac*" in Malay). In either case, it was used for Chinese lacquer work (p. 260).

3. "*Tsjad*" is *chi'i*, Chinese for "varnish."

4. In book 1, ch. 55, of the *Herbal*, Rumphius notes that the "Gajang" or "Gajanus Tree" produces a red sap that leaves indelible spots. It is now classified as *Inocarpus edulis* Forst. (*M*, p. 433). Both the Ambonese people and the Europeans ate the tree's fruits (pp. 170–171).

5. A "*tomtom*" was a straw trunk; see ch. 10 of book 2.

Chapter 74. Parrangites. Mestica Gondu.

1. "*Parrang* fruit" refers to *katjang parang* (*Canavalia ensiformis* L.), a climbing plant with pods that are curved like a *parang*, the Indonesian machete. It likes coastal plains, hence Rumphius' Latin name, which means "Sea Bean."

2. The plant is described in his *Herbal*, book 7, ch. 4, pp. 5–9.

Chapter 75. Manorites. Mestica Manoor.

1. This is the same jasmine flower, *Jasminum sambac* L., mentioned before in ch. 71. Rumphius described it in his *Herbal*, book 2, ch. 29, pp. 52–55, where he tells an abbreviated version of the following story on p. 54.

2. The reason for the specific time is that the jasmine is a flower that blooms at night, when its scent is at its strongest. Rumphius noted in his *Herbal* that "at Sunset one should pick such buds which are likely to open that same evening, since about an hour after one has picked them they open up on

their own, and then smell most potently." The buds were strung in garlands and worn in a woman's hair, strewn on bed linen ("which is most pleasant on hot days"), and used as a scent: "When these flowers are plucked in the evening and put in fresh water in such a way that they float, and left as such for 6 to 8 hours, though covered, it renders such a strong smell that it would seem to be rosewater wherewith one washes one's face to feel refreshed and to cool burning eyes."

3. This literally translates as "Peasant Punch." *Pons* was a variation of the English "punch." There seems no agreement on the ingredients, but one recipe quoted in the *WNT* seems to make sense: arak, sugar, raisins, and Rumphius would add lemons. This passage is quoted by the *WNT* as an instance of the use of "*boerepons.*"

4. "Stock" for *stok.* This is a venerable word in English. In Middle English it was often spelled like the Dutch *stok.* Rumphius uses it here in the sense of the trunk of a tree, or the thick main stem of a shrub, to distinguish it from the branches. An instance of its early sixteenth-century use in English, in a figurative sense, is: "Of the whiche tree, fayth, hope, & charite, be compared to the stocke, to the barke, & to the sap" (*OED*).

Chapter 76. Stones That Happen to Come from Certain Trees or Fruits.

1. The banana tree.

2. See ch. 8 of the present book 3.

3. These Latin phrases mean "Magic Armlet" and "Demon's Armlet."

4. The same story is told in ch. 26, which should be consulted because some things are clearer in that version.

5. The civet cat is a predator belonging to the *Viverridae.* It is larger than the housecat but smaller than a badger. There are two major varieties, and both are held captive for the secretion. The one normally referred to in Europe is the African civet, *Viverra civetta,* while the Asian one is the *Viverra zibetha.* The latter is called in English "Rasse" after the Javanese name for the animal, *Rasé* (*Viverricula malaccensis*). It is a good climber and feeds on small animals, insects, as well as on fruits and roots. The civet substance, obtained from a pouch near the genital area of the animal, has a strong musky odor and is called *dedes* in Malay. It is used both in Europe and in Asia for the manufacture of perfumery.

6. Described in ch. 25.

7. Antimony; see n. 3 of ch. 26.

8. St. John's Eve refers to the feast day of St. John the Baptist on the twenty-fourth of June. It was a very popular holiday in the Middle Ages and after, and perpetuated many folklore traditions. What we are dealing with is better known as "midsummer night," the eve before the twenty-fourth of June, when midsummer fires were lit in Germany and Austria (particularly in Steiermark, in the Austrian Alps), known in German as *Johannisfeur,* "St. John's Fire" in English. Folklore also said that during that night the ferns would bloom, while in England they spread white linen around the plants hoping to catch their "seed." Not until the nineteenth century was it known that ferns do not have "seeds" but spores. Before that time no one had even *seen* these spores; hence what was believed to be the "seeds" of the ferns were considered mysterious and invisible. By extension, anyone who owned some of these "seeds" would be capable of becoming invisible.

9. Something seems missing in the original: "*waar op zy by tyd en wyl stout worden en hun verlaten . . .*" (p. 327).

10. This was also first mentioned in ch. 26.

11. This expression was in the original "*dewelke gehoort en gezien hebben hoe het gras groeit,*" and has a different connotation in Dutch. Someone "*die het gras kan horen groeien*" ("who can hear the grass grow") is conceited, thinks a lot of himself.

Chapter 77. Stones Which Happen to Have an Unusual Shape.

1. *Vegetavite* is a Latin adverb meaning "with the quality of plants." Rumphius thought that coral was a plant, a notion that persisted into the second half of the eighteenth century, and discussed it, therefore, in his *Herbal.*

2. *Androdamantes* is probably a kind of pyrite and is described in ch. 13 of book 3 (on *Marga* and *Argilla*), as well as by Pliny in book 37, Latin paragraph 144; Loeb ed., 10:280–283.

3. It is difficult to identify Rumphius' "Steenbergen" (Stone Mountains; *Os Montes de Pedra*). The given latitude, 29° south, does not help because on most maps that would be too high up the western coast; the information "North of the Cape" of Good Hope is likewise useless because too vague; and there are any number of "Olifant" or "Elephant" rivers in South Africa. Rumphius is quoting a Portuguese source; hence one would have to surmise that the reference is to a region known before 1650. At that time little more than the peninsula between Tafelbaai and Vals Baai was familiar, the area now known as the Cape Peninsula. And indeed, there is a mountain north of the Cape of Good

Hope, known as Steenberg, or "Stone Mountain," at the time, between the Sandvlei (Zandvallei) and the Valley or Vlakte.

4. The river Batu Gantong (also formerly spelled Wai Batu Gantu) flowed, in Rumphius' day, on the western outskirts of the city of Ambon. It is an intriguing name for a river because *Batu* means "rock" and *gantung* is the root for words that express "hanging," for instance, *pegantugan*, place for hanging, gallows. Hence the name could mean "Hanging Rock."

5. It is almost impossible to locate the seated stone person and the mountain where this is supposed to be. People worshiped a king who had been turned into stone with his entire family and who presumably was calcified in a limestone cave near Maros, but Maros is above Goa (the realm of Macassar) towards the western coast of Celebes' southwestern "leg" and nowhere near the Straits of Salajar. Generally speaking, the various peoples inhabiting Celebes worshiped large stones because they thought these stones were dwellings of Karaëng-lôwé, "the Great Lord." Nevertheless, this was most likely the cave Rumphius had in mind. The cave near Maros was called Bulu-Se-pong, and was supposed to be a stone replica of a king's bamboo palace—his name was Karaëng-Borong—complete with the king and his family, who were petrified because his wife had a dog look for and retrieve a spool she had dropped while weaving. Perhaps the ceremony Rumphius refers to was not quite that sinister. Even in the twentieth century, local people came to the cave and the petrified king, and asked him to intercede for them. They pledged to bring him offerings if their wish was fulfilled. Such a pledge usually was translated into natural products such as rice, meat from a goat or buffalo, water, or palm wine. A priestess would pray to the Karaëng and his family, burn incense before them, and smear them with oil. If everything went well, the whole affair would be concluded with a *selamatan*, or ceremonial dinner. See "Het Animisme bij de volken van den Indische Archipel," part 3, in *De Indische Gids*, 6:2 (1884), pp. 78–80.

6. *Straits of Saleijer* most likely for the Straits of Salajar, between the southernmost tip of Celebes (Sulawesi), in the region called Bonthain and the island Pulau Salajar.

7. *Batappa* is difficult to place, but I wonder whether this was meant to be *bertapa*, the practice of mortifying the body, and other religious practices, performed by ascetics. To the Malay this was less a religious practice than a means of securing magical power.

8. *Djing* (*djin, jin, jihan*), from the Arabic *jinn*, a word which in old Arabic referred to a spirit, a supernatural being, that was not necessarily evil. In the Indies *djin* was often used as a euphemism for *hantu*, which is an evil spirit. Rumphius clearly has the latter in mind.

9. *P. Kircherus* refers to Athanasius Kircher (1601–1680), a German scholar who became a Jesuit in 1618, assumed the post of professor in Würzburg until the Thirty Years War forced him to flee to Avignon in France. The pope summoned him to Rome, where Kircher taught mathematics at the Collegium Romanum, and, later, studied hieroglyphs and archaeology. He wrote a large number of books, among which one on China—*China illustrata* (published in Amsterdam in 1667)—and one on Egyptian hieroglyphs—*Oedipus aegyptiacus* (4 vols., published in Rome between 1652 and 1654)—are perhaps the most significant. Kircher was the first scholar to draw attention to hieroglyphs. Rumphius is referring to *Mundus subterraneus* (published in 2 vols. in Amsterdam in 1664–1665), and one can tell from the incomplete reference that he was citing from memory and did not have the book at hand. The text in question is a tale told by Kircher in his *Mundus subterraneus*, book 8, part 1, ch. 9, which the German Jesuit read in the *History of Chile*, by Alonsius de Valle. It concerns a stone Virgin with Child, made up of variously colored "Jaspis" "according to the rules of Perspective." A black stone depicts the Virgin's hair "spread across the back and shoulders; a white stone forms [her] face turned to one side, in such a way, that only one eye is wondrously depicted. Her Gown is pink up to the waist, the Mantle is Lemon-yellow and blue; in a word, the image attracts the eyes of its beholders most marvelously." This "*mirabilia*" was found by an Indian mother and her child near Araucus, on the coast of the "South-sea, called Inbulia." It goes without saying that this natural wonder, the result of God's providence working in nature, persuaded many Indians to become Christians.

Kircher had a great interest in Oriental languages, and when he was teaching in Rome, he interviewed many of the missionaries returning from China. His compilation of information on China became famous and very influential. A Dutch translation appeared in 1668, a French one in 1670, and a (partial) English translation in 1669. See B. Szczesniak, "Athanasius Kircher's *China illustrata*," in *Osiris. Commentationes de scientiarum et eruditionis historia rationeque*, (Brugge, 1952), 10:385–411.

10. *Hative Kitsjil* means "Little Hative (or Hatiwe)," a village near Hative Besar, or "Great Hative," about seven miles north of the city of Ambon.

11. *Anachoda* (or Nakhoda) is a Persian word for shipmaster. "*Kapitan*" or "*Kapitan kapal*" was used only for European masters.

12. *Tudung* (or tudong) was a large, broad hat, woven from the leaves of the gebang palm. Perhaps this is the same coral rock as the one called today "Batu Capeo," just over three miles from Kota Ambon.

13. *Halong* is a village on the way to Hative, on the Leitimor peninsula, about four miles north of the city of Ambon.

14. This coral rock in the shape of a woman, becomes the "coral woman" ("*koralen vrouw*") in Dermoût's *De tienduizend dingen* (The ten thousand things, 1955); cf. *Werk*, pp. 131–132.

Chapter 78. Melitites. Boatana.

1. Rumphius calls this "Sesoot tree" "*Pharmacum Sagueri*" in his *Herbal* (vol. 2, p. 136). Identified by Heyne (p. 1091) as *Garcinis picrorrhiza* Miq., these tall trees are found only in the mountains of Hitu and Leytimor. The very shallow roots were also known as *obat saguer*, or, in Ambonese, *obat tuni*.

2. This refers to a particular calcareous clay that was used to make a famous kind of ancient pottery known as *terra sigillata*, or samian, Arretine, or red-slip pottery. It seems to have been among the very few classes of pottery mentioned by ancient authorities; see Pliny, book 35, Latin paragraph 160. It was red-gloss tableware and one of the most important products of earthenware in Roman times, when it was traded as far as southern India, Africa, and Russia. See D. P. S. Peacock, *Pottery in the Roman World. An ethnoarchaeological approach* (London: Longman, 1982), pp. 114–128. See also book 3, ch. 13.

3. Pliny, in book 36, paragraph 33, states that the stone "exudes a liquid that is sweet and is like honey." The name for this stone comes from *melitinus*, which means "of honey" or "mixed with honey."

4. *Waytommo* (commonly spelled Wai Tomo) is a river that flowed (and still flows) through the city of Ambon beginning at the northeast corner of Fort Victoria.

5. *Marga indurata*, this refers to a kind of marl, literally "hardened marl," and may be what used to be called "steel marl." Marl (*mergel* in Dutch; Rumphius did not use this word although it was common during his lifetime) is a kind of soil, usually clay, mixed with lime, the latter often from the remains of mollusks. It was highly prized as a fertilizer, in antiquity as well (see Pliny, book 17, paragraph 4).

6. *Ground fruit* for *aardvrucht* (*Erdfrucht* in German). Rumphius' usage is not clear. *Aardvrucht* normally referred to a root (*ubi* in Malay; *knol* or *wortel* in Dutch), or a plant's edible product that grew underground, but it doesn't seem that Rumphius had this in mind. He seems to be using it literally, i.e., a stone as the "fruit" of the soil or ground, because I would guess that "*boatana*" was probably meant to be "*buahtana*," of which the first element *buah*, or "fruit," and "*tana*" is *tanah*, or "soil" or "ground."

7. Rumphius is alluding to the animal's secretion.

At one time the English word "civet" was synonymous with the civet scent, perfume in general, lasciviousness, and foppery. A "civet cat" also meant a heavily perfumed, foppish male, as in "a fine puss-gentlemen that's all perfume" (Cowper), or Pope's "courtly civet-cats." In *As You Like It* (3, 2:63–67) it is noted that "the courtier's hands are perfumed with civet," which is denigrated with the reply: "Civet is of a baser birth than tar, the very uncleanly flux of a cat." Hence the smell of the secretion as well as the location of the pouch near the genital area made both the animal and the scents synonymous with lubricity. In his *Oud en Nieuw Oost-Indiën*, 3:270–272, Valentijn gives a detailed description of the Ambonese "Civet Cat," and the scent derived from it, which, he asserts, is used to induce the menses, to "cleanse" the uterus when it is "cold," and to increase a man's potency. He also mentions that the civet is so clever that it hangs its tail in the sea, where it is mistaken for food by shrimp, which fasten to the hairs. The animal then pulls up its tail, shakes the shrimp loose, and eats them (p. 272).

8. *Coromandel* was the former name given to India's eastern seaboard, between Point Calimere (opposite Jaffna in northern Ceylon) and the mouth of the Kistna River. A hot, low coast, the most important cities associated with it are Madras and Pondicherry.

9. *Lama de Costa* means in Portuguese "Coastal Mud."

Chapter 79. Satschico. Alabastrites & Lichnites Chinensis.

1. This phrase means something like "Chinese Lamp-like Stone"; *lychnis*, which means "giving light" or "shining," from *lychnus*, a lamp. Rumphius got the term from Pliny (Bk. 36, Latin paragraph 14) who mentions a white marble from Paros "which they proceeded to call 'lychnites,' since according to Varo, it was quarried in galleries by the light of oil lamps" (Loeb ed., 10:12–13).

2. Al Cohen thinks that these two words might transcribe *hsüeh-hua kao* (from *hsüeh-hua shih-kao*), Chinese for "alabaster."

3. See ch. 41 in this book.

4. This was probably fluor spar or burned agate and was described by Pliny, book 37, Latin paragraphs 7–8; Loeb ed., 10:176–181.

5. Boëtius de Boot, *Gemmarum et lapidum historia* (1647), chs. 268–270 (pp. 491–494).

6. I do not know what Rumphius has in mind. The German word *spath* (in Dutch: *spaat*) is said to be connected to a flaky substance, or to a mineral that has lozenge-shaped crystals.

7. Rumphius is correct. The noun *alabaster* (plural: *alabastri* or *alabastra*) primarily meant a box for unguents or perfumes, and Pliny says that *alabastrites* is a stone, a kind of onyx marble, that was used for *making* unguent jars (Pliny, book 36, Latin paragraphs 59–62; Loeb ed., 10:46–49).

8. Boëtius de Boot had "Misnia" for "Menz" or "Mens," and *Misnia* was the Latin name for the town of Meissen.

9. In book 36, Latin paragraph 14; Loeb ed., 10:10–13, where Pliny states that lychnite was a stone which, according to Varro, "was quarried in galleries by the light of oil lamps."

10. Marl. See ch. 13 of this book and n. 5 in ch. 78.

11. Kuangtung Province in southern China.

12. Al Cohen thinks this is the city Shao-chou-fu, or in the modern spelling, Shao-kuan.

13. Probably Kuangsi, the province next to (west of) Kuangtung.

14. A "*zesje*" was a small silver coin worth very little, which became proverbial for a trifling amount of money.

15. Schafer calls these inkstones palette stones; they were quite valuable; see *The Vermilion Bird*, p. 155. Al Cohen states that these stones fetch a high price to this very day. The Dutch sinologist (and mystery writer) Robert van Gulik noted that the inkstone or, as he calls it, the "inkslab" was very important to the Chinese. "Ink, inkslab, brush and paper, became at last those extremely refined implements that are known as *wên-fang-ssû-pao*, the 'Four precious things of the library.' With the stick of ink one rubs the inkslab, then the brush is moistened, and the paper lies ready to receive the columns of graceful characters, or the few expressive strokes that evoke, as if by magic, the very essence of a scene from nature. After some time, however, the stick of ink gets rubbed down to a mere stub, and one has to choose a new one. The hairs of the brush, too, become frayed, and generally after the lapse of one year it cannot be used any longer. So the only thing that remains constant, after all, is the ink-stone. It will last for years and years, and even for centuries. No wonder that of the four precious things the inkslab has always been valued most highly by calligraphers and connoisseurs" (p. 14). Van Gulik translated and annotated a treatise on inkstones from the eleventh century: *Mi Fu on Ink-Stones*, transl., with an intro. and notes, by R. H. van Gulik (Peking: Henri Vetch, 1938). Van Gulik's annotation on Mi Fu's discussion of where the best stones come from contains a description of a place which clearly resembles Rumphius' mountain where the poor diggers had to work under such miserable circumstances. The mountain was near Tuan-chou, Kao-yao-hsien, in Kuangtung Province. There were several quarries that had been exploited since ancient times: "Most famous is the so-called *lao-k'ang*, 'old quarry,' which penetrates deep into the rock. This quarry is subdivided into several caves, some of which are permanently under water; therefore the stones from this quarry are also called *shui-yen*, 'water-stone.' . . . Stones cut from the layer that was perpetually submerged were valued most. Inkstones made from stone of this quarry were offered as tribute. Every time the stones had been hewn out, the entrance to the cave was closed and sealed, so that no robbers could steal the precious material" (Van Gulik, *Mi Fu*, p. 35).

16. This is Fukien Province, northeast of Kuangtung.

17. This is Fo-chow City.

18. Kuangtung Province.

19. Al Cohen notes that "*Huquan*" is the Ming Dynasty province Hu-kuang, an area that covered the modern Hupei and Hunan Provinces.

20. This is Liaotung Province in southern Manchuria.

Chapter 80. Amianthus Ambonicus. Batu Rambu.

1. *Amianthus* is a mineral that is a variety of asbestos. It can be teased into fibers that are white and that are capable of being woven into a fabric. The wonder of amianthus was its indestructibility by fire, already noted by Pliny (book 36, paragraph 139), and which inspired its older name of "Salamander's Hair" (salamanders reputedly were immune to fire). It was also known by the expressive name of "Earthflax." Rumphius' kind of amianthus is most likely the variety of mineral serpentine called chrysotile.

2. This is "Feather or Plume Alum." Alum is a white, astringent, mineral salt.

3. A river that flows down to the coast on the west side of the city of Ambon.

4. The *talcum* Rumphius mentions is mineral talc.

5. "*Krevelkruid*" is most likely a misprint for *kervelkruid* (*Chaerophyllum*), a herb used in soups and salads. The related herb, "rough chervil" (*Chaerophyllum temulum*), is a hedgerow plant with hairy stems that can itch.

6. The "Waynitu" is a river to the west of the Oliphant River and after the village Hative, just before Gallows Hook.

7. Gallows Hook, a point to the west of the Oliphant River and the Wainitu River, as far west of the city as one could go in those days before one came to the straits between Gallows Hook and Hook of Mang. The name is self-explanatory, but the place is now called Benteng.

8. There was (and is) a village by that name to the west of Ambon City.

9. The river Way Batu Gajah (Batu Gadjah), which flows to the city from the pleasant valley by the same name, some twenty minutes from shore, where later the governor's residence was located.

10. Rheingau refers to a region in Hessen that is good wine country. The Melibochus mountain (or Malchen) is the highest point, 1,690 feet, of the so-called Bergstrasse, which is an ancient road that skirts the Odenwald between, say, Pfungstadt and Heidelberg. It is a granite mountain that is heavily wooded. The ruins of the Katzenelnbogen castle are some distance northwest of Melibochus, on the right bank of the Rhine, near the famous Loreley (Lurlei) rocks. The castle belonged to the once powerful counts of Katzenelnbogen, who controlled a region between the Rhine and Lahn rivers from the eleventh to the fifteenth centuries. "Die Höhe" is part of the Taunus Mountains, to the east of the Katzenelnbogen ruins. The Königstein castle is some ten miles north of Frankfurt in the Taunus range, and was a stronghold that is first mentioned in history in 1225. One could say that the two ruins are within reach of Hanau, but Melibochus is some distance south of Frankfurt.

11. The different names all refer to stones that came from the city of Cananor or Cannanore (Kannanur), a port on the northern Malabar coast. It was one of Portugal's most important cities on that coast and was mentioned in *The Lusiads* by Camões in canto 10. Located some sixty miles north of Calicut, it had its first fort built by Vasco de Gama in 1505, while in 1656 the Dutch built a larger and stronger fortress which still stands. These stones from Cannanore are mentioned again a little later, now joined by its close relative "Cold Stone" (*Pedra fria*).

12. This is a guess for a phrase that comes after "*en zo aardachtig niet van reuk*"; the puzzling phrase reads "*maar een hoornachtig.*"

13. This is my guess for the strange noun "*Oeryn*" in the original. The familiar noun was used in both Dutch and English from the Middle Ages, although the simple *pis* was far more common in Dutch. "Urine" did have the alternate spelling "uryn" in earlier Dutch, which led me to believe that this was meant here.

14. Athanasius Kircher (1601–1680); the book mentioned is *Mundus subterraneus* (1678). See n. 9 of ch. 77 of the present book.

Chapter 81. Lapis Tanassarensis.

1. Either hematite (a ferric oxide) or red jasper, and mentioned by Pliny in book 37, Latin paragraphs 144–149 and 169; Loeb ed., 10:114–119, 300–304.

2. Later spelled Tenasserim, "Tannassari" was a city and region on the Gulf of Martaban, west of Siam, and is still the name of a Burmese province. Rumphius' name is Malay, a combination of *tanah* (land or country) and *seri* (delight), hence "Land of Delight" or "Earthly Paradise." It is mentioned by Linschoten in ch. 17 of his *Itinerario* (Kern ed., 1:72).

3. The phrase within parentheses reads in the original: "*gelyk by de zwaartvegers by het bruineeren.*"

Chapter 82. The Preacher's Stone.

1. This is serpentinite.
2. A marl that is turning into stone.
3. This is not clear in the original: "*doch in de meeste niet met al, of donker . . .*" (p. 334).
4. Mentioned by Pliny in book 37, Latin paragraph 180; Loeb ed., 10:310. "*Phycites*" means "seaweed stone," from *phycos* (clearly Greek) which meant seaweed.
5. See ch. 13 of the present book 3.
6. I do not know what kind of marble this might be.

Chapter 83. Coticula. Batu Udjy.

1. *Coticula* is Pliny's term for "touchstone," what the Greeks called *basanos*. Pliny describes them in book 33, Latin paragraph 126; Loeb ed., 9:94–95. It is Pliny who mentions that "the part of these [stones] that has been exposed to the sun is better than the part on the ground," an opinion Rumphius refutes.

2. This means "touchstone" in Malay, from *batu*, "stone," and *uji*, "to test."

3. These siliceous stones were usually black or of a dark color, related to flint, and were also known as "basanite" or "Lydian stones." The stone was rubbed on gold and silver alloys to test their purity. This was determined by the color of the smudge or streak that was left on the metal after the stone had rubbed it. "To touch" was commonly used as a verb to indicate this kind of testing. Wichmann (*G*, p. 148) states that there are no true touchstones in the Indonesian archipelago.

4. Larike and Wakasihu are on the lower west coast of Hitu.

5. Perhaps amarilite.

Chapter 84. Cancri Lapidescentes.

1. For Cleijer, see n. 38 of ch. 35.
2. For Martinus and his *Atlas Sinensis*, see n. 28 of ch. 24.
3. This is, most likely, Chiao-chou, a name, as Al Cohen informs me, of a Ming Dynasty prefecture occupying the Tongking region of northern Vietnam.
4. This must be Hainan, across from Kuantung's Luichou Peninsula. Rumphius is talking of a very long stretch of coast.
5. This means "crabs turned to stone," the same meaning as the chapter's title; from *lapidesco*, "to turn to stone," a Plinian term. The Portuguese means the same thing.
6. To render: "*Voor enige gezwellen, die op de schouder of rug veroudert zyn . . .*" (p. 336).
7. Misprints for the same city and island mentioned in nn. 3 and 4 above.
8. This is the well-known burgomaster of Amsterdam, Nicolaas Cornelisz. Witsen (1641–1717). His cabinet was famous.
9. Nothing is known about this collector.

Chapter 85. Sal Ambonicus. Garam Ambon.

1. Wichmann (*G*, p. 148) identifies this as *galmei*, which is calamine.
2. A sago mush.
3. "Chin cough" was the older name for whooping cough. "Hooping cough" and "whooping cough" did not become general usage until after Rumphius' death, i.e., from about the middle of the eighteenth century. Johnson lists them.
4. Valentijn in *Oud en Nieuw Oost-Indiën*, vol. 2, part 1, p. 66, confirms this, saying that the rain monsoon in July 1664 was extremely severe, with earthquakes at the same time. It seems that large pieces of land simply collapsed into the sea, like giant sinkholes.
5. "Milkwood" for *melkhout*. I think Rumphius means trees that produce a treesap (or *melk*, i.e., "milk"), be it beneficial (such as resin) or harmful (various poisonous saps used for arrows). The same usage exists in English, for instance "coconut milk."

Chapter 87. Succinum Terrestre.

1. Usually spelled with one "c," "*sucinum*" is Latin for amber, also called *electrum*. Pliny discusses it at length in book 37, Latin paragraphs 30–54). For more particulars on amber, see n. 1 of ch. 35 in this third book. Rumphius might have found this in *Sinicae historiae decas prima res a gentis origine ad Christum natum*, by Martino Martini, published in Munich in 1658.
2. Perhaps Suchuan, a province in western China, near Tibet.
3. As I did earlier, I added the Dutch "*bernsteen*" whenever English needs to say "amber" to make sure it is not confused with the excretion from the whale.
4. This is *hu-p'o*, or "amber."
5. The areas of southern Burma, comprising the mouths of the Irrawaddy, on the Gulf of Martaban.
6. This tree, the name of which should be spelled "Selan," is discussed in the third book of his *Herbal* (which is the book that discusses resin trees) under the name of "Dammar Selan" or "Dammara Selanica," book 3, ch. 8, pp. 168–170. De Wit identifies this useful tree as *Shorea selanica* Blume (*M*, p. 428).

BIBLIOGRAPHY

PRIMARY WORKS

See pp. lxxvi–xciii of the Introduction.

SECONDARY SOURCES PERTAINING TO RUMPHIUS

Ballintijn, G. *Rumphius. De blinde ziener van Ambon.* Utrecht: W. de Haan, 1944.

Blume, C. L. *Rumphia, sive commentationes botanicae imprimis de plantis Indiae Orientalis, tum penitus incognitis tum quae in libris Rheedii, Rumphii, Roxburghii, Wallichii, aliorum, recensentur.* 4 vols. Lugduni-Batavorum: Prostat Amstelodami, apud C. G. Sulpke; Bruxelles, apud H. Remy; Dusseldorfiae, apud Arnz et socios; Parisiis, apud C. Roret, 1835–1848.

Dagh-Register gehouden int Casteel Batavia vant passerende daer ter plaetse als over geheel Nederlandts-India. Anno 1663. Ed. J. A Van der Chijs. 's Hage: Nijhoff, 1891.

Engel, H. "The Echinoderms of Rumphius," in de Wit, *Rumphius Memorial Volume*, pp. 209–223.

Greshoff, M. "Rumphius," in *Encyclopaedie van Nederlandsch-Indië.* The Hague: Nijhoff, 1919.

———, ed. *Rumphius Gedenkboek, 1702–1902.* Haarlem: Koloniaal Museum, 1902.

Heusohn, Heinrich. "Eine Denkmünze auf den Naturforscher Georg Everhard Rumphius." *Frankfurter Münzzeitung* 1, no. 26 (February 1903).

Hickson, Sydney J. "The Life and Work of Georg Everard Rumphius." *Memoirs and Proceedings of the Manchester Literary and Philosophical Society* 70, no. 2 (1926), pp. 17–28.

Holthuis, L. B. "Notes on Pre-Linnean Carcinology (including the Study of Xiphosura) of the Malay Archipelago," in de Wit, *Rumphius Memorial Volume*, pp. 63–125.

Huet, Conrad Busken. *Het land van Rembrand. Studies over de Noordnederlandse beschaving in de zeventiende eeuw* [1883–1884]. Reprint ed., Amsterdam: Agon, 1987.

Le Collezioni di Giorgio Everardo Rumpf, acquistate dal Granduca Cosimo III de Medici. Ed. Ugolino Martelli. Firenze: Tipografia Luigi Niccolai, 1903.

Lentz, Christel. "Über die Idsteiner Vergangenheit des Naturforschers Georgius Everhardus Rumphius." *Heimatbuch für den Rheingau-Taunuskreis* 42 (1991), pp. 76–82.

Leupe, P. A. "Georgius Everardus Rumphius, Ambonsch Natuurkundige der zeventiende eeuw," in *Verhandelingen der Koninklijke Akademie van Wetenschappen* 12:1–63. Amsterdam: C. G. van der Post, 1871.

Martens, E. von. "Die Mollusken (Conchylien) und die übrigen wirbellosen Thiere im Rumpf's Rariteitkamer," in Greshoff, *Rumphius Gedenkboek*, pp. 109–136.

Merrill, E. D. *An Interpretation of Rumphius's "Herbarium Amboinense."* Manila: Bureau of Printing, 1917.

Pabbruwe, H. J. *Dr. Robertus Padtbrugge (Parijs 1637–Amersfoort 1703), dienaar van de Verenigde Oost-Indische Compagnie en zijn Familie.* Kloosterzande: Deurinck, 1995.

Rouffaer, G. P., and W. C. Muller. "Eerste proeve van een Rumphius-bibliographie," in Greshoff, *Rumphius Gedenkboek*, pp. 165–219.

Sarton, G. "Rumphius. Plinius Indicus (1628–1702)." *Isis. International Review Devoted to the History of Science and Civilization* 27, no. 74 (August 1937), pp. 242–257.

Schöppel, F. A. "Rumphius. Vortrag gehalten vor der Zweiggruppe Batavia am 24. Juli 1939." *Nachrichten Deutsche Gesellschaft für Natur- und Völkerkunde Ostasiens*, no. 52 (Tokyo, 1939), pp. 13–18.

Sirks, M. J. "Rumphius, the Blind Seer of Amboina," in *Science and Scientists in the Netherlands Indies*,

ed. Pieter Honig and Frans Verdoorn, pp. 295–308. New York: Board for the Netherlands Indies, Surinam, and Curaçao, 1945.
———. *Indisch Natuuronderzoek.* Amsterdam: Amsterdamsche Boek-en Steendrukkerij, 1915.
Sollewijn Gelpke, J. H. F. "Johannes Keyts. In 1678 de Eerste Europese Bezoeker van de Argunibaai in *Nova Guinea.*" *Bijdragen tot de Taal-, Land- en Volkenkunde,* deel 153, 3e aflevering (1997), pp. 381–395.
Strack, H. L. *Results of the Rumphius Biohistorical Expedition to Ambon (1990).* Part 1. *General Account and List of Stations.* Monograph series Zoologische Verhandelingen, no. 289. Leiden: Nationaal Natuurhistorisch Museum, 1993.
Strack, H. L., and Jeroen Goud. "Rumphius and the 'Amboinsche Rariteitkamer.'" *Vita Marina* 44, nos. 1–2 (1996), pp. 29–39.
Treslong Prins, P. C. Bloys van. "Origineele bescheiden van en over Georgius Everhardus Rumphius." *Tijdschrift voor Indische Taal-, Land- en Volkenkunde* 69, nos. 3–4 (1930), pp. 426–435.
Valentijn, François. *Oud en Nieuw Oost-Indiën.* 5 books in 8 vols. Dordrecht and Amsterdam: Joannes van Braam & Gerard onder de Linden, 1724–1726.
Valentini, M. C. *Museum Museorum.* Frankfurt am Mäyn: Johann David Zunners, 1704.
Veen, F. R. van. "Georgius Rumphius (1628–1702). De eerste geoloog in Oost-Indië." *Grondboor & Hamer* 51, no. 5 (1997), pp. 94–103.
Verslag der Rumphius-Herdenking van het koloniaal Museum te Haarlem. Amsterdam: J. H. de Bussy, 1903.
Wichmann, Arthur. "Het aandeel van Rumphius in het mineralogisch en geologisch onderzoek van den Indischen Archipel," in Greshoff, *Rumphius Gedenkboek,* pp. 137–164.
Wit, H. C. D. de. "A Checklist to Rumphius's Herbarium Amboinense," in de Wit, *Rumphius Memorial Volume,* pp. 339–460.
———. "Georgius Everhardus Rumphius," in *Rumphius Memorial Volume,* pp. 1–26.
———, ed. *Rumphius Memorial Volume.* Baarn: Hollandia, 1959.
Zaunick, Rudolph. "Eine handschriftliche Mitteilung von G. E. Rumph über die Korallenbauten." *Mitteilungen zur Geschichte der Medizin und Naturwissenschaften* 14, no. 5 (1915), pp. 330–332.
———. "Georg Eberhard Rumph und Cosimo III von Medici." *Physis. Rivista di storia della scienza* 3, no. 2 (1961), pp. 125–136.

OTHER SECONDARY SOURCES

Aa, A. J. van der. *Biographisch Woordenboek der Nederlanden.* 21 vols. Haarlem: J. J. van Brederode, 1852–1878.
Abbott, R. Tucker. *Kingdom of the Seashell.* New York: Crown, 1972.
Abkoude, Johannes, van. *Naamregister van de bekendste en meest in gebruik zynde Nederduitsche Boeken.* 2d. rev. ed. Rotterdam: Gerard Abraham Arrenberg, 1788.
The Anchor Bible, Ezekiel 1–20. Transl., with introduction and commentary by, Moshe Greenberg. Garden City, NY: Doubleday, 1983.
———, *Revelation.* Transl., with introduction and commentary by, J. Massyngberde Ford. Garden City, NY: Doubleday, 1975.
Acosta, Joseph de. *The Natural & Moral History of the Indies* [English transl., 1604]. Ed. Clements R. Markham. London: Hakluyt Society, 1880.
Andaya, Leonard Y. *The Heritage of Arung Palakka. A History of South Sulawesi (Celebes) in the Seventeenth Century.* The Hague: Martinus Nijhoff, 1981.
———. *The World of Maluku. Eastern Indonesia in the Early Modern Period.* Honolulu: University of Hawaii Press, 1993.
Andree, Richard. "Aggri-Perlen." *Zeitschrift für Ethnologie* 17, no. 3 (1885), pp. 110–115.
Aristotle. *History of Animals.* Transl. Richard Cresswell. London: Bell, 1883.
———. *The Works of Aristotle Translated into English.* Ed. J. A. Smith and W. D. Ross. 12 vols. Oxford: Clarendon Press, 1908–1952.
Backer, C. A. *Verklarend Woordenboek van Wetenschappelijke Plantennamen.* Batavia-Groningen: P. Noordhoff, 1936.
Baldaeus, A. *Naauwkeurige beschrijvinge van Malabar en Choromandel. Derzelver aangrenzende Ryken, En het machtige Eyland Ceylon.* Amsterdam: J. J. van Waesberge, 1672.
Barlaeus, Caspar. *Rerum per octennium in Brasilia gestarum historia* [1647]. Transl. and ed. S. P. L'Honoré Naber. The Hague: Martinus Nijhoff, 1923.
Barthes, Roland. *Empires of Signs.* Transl. Richard Howard. New York: Hill & Wang, 1982.
Beagon, Mary. *Roman Nature. The Thought of Pliny the Elder.* Oxford: Clarendon Press, 1992.
Beeckman, Captain Daniel. *A Voyage to and from the Island of Borneo* [1718]. Folkestone: Dawsons of Pall Mall, 1973.

Beekman, E. M. *The Poison Tree. Selected Writings of Rumphius on the Indies.* Amherst: University of Massachusetts Press, 1981.

———. *Fugitive Dreams. An Anthology of Dutch Colonial Literature.* Amherst: University of Massachusetts Press, 1988.

———. *Troubled Pleasures. Dutch Colonial Literature from the East Indies, 1600–1950.* Oxford: Clarendon Press, 1996.

Belon, Pierre. *Les Observations de plusieurs singularitez et choses mémorables trouvées en Grèce, Asie, Judée, Egypte, Arabie et autres pays estranges.* Paris: G. Corrozet, 1553.

———. *Les Observations de plusieurs singularitez et choses mémorables trouvées en Grèce, Asie, Judée, Égypte, Arabie et autres pays estranges.* 2d. enl. ed. Anvers: Christofle Plantin, 1555.

Berlin, Isaiah. *The Hedgehog and the Fox. An Essay on Tolstoy's View of History* [1953]. Reprint ed., New York: Simon and Schuster, 1970.

Bernet Kempers, A. J. *Monumental Bali. Introduction to Balinese Archaeology [and] Guide to the Monuments.* Den Haag: Van Goor, n.d. [?1977].

Best, Michael R., and Frank H. Brightman, eds. *The Book of Secrets of Albertus Magnus.* Oxford: Oxford University Press, 1973.

Beverwyck, Johan van. *Schat der Ongesontheydt, ofte Genees-konste van de Sieckten.* Amsterdam: By de Weduw' van J. J. Schipper, 1672.

Beverwyck, Johan van. *Wercken der Genees-Konste.* Amsterdam: Weduwe J. J. Schipper, 1672.

Bickmore, Albert S. *Travels in the East Indian Archipelago.* New York: Appleton, 1869.

Blussé, Leonard, and R. Falkenburg. *Johan Nieuhofs Beelden van een Chinareis 1655–1657.* Middelburg: Stichting VOC Publications, 1987.

Bochart, Samuel. *Hierozoici, seu De animalibus S. Scripturae compendium duas in partes divisum.* Franequerae: Johannis Byselaar, 1690.

Bock, Carl. *The Head-Hunters of Borneo* [1881]. Reprint ed., Singapore: Oxford University Press, 1985.

Bontii, Jacobi. *Historiae Naturalis & Medicae Indiae Orientalis,* in Gulielmi Pisonis, *De Indiae Utrisque Re Naturali et Medica.* Amstelaedami: Elzevirios, 1658.

Boodt, Anselmus Boëtius de. *Gemmarum et Lapidum Historia. Qua non solum ortis, natura, vis & precium, sed etiam modus quo exiis, olea, salia, tincturae, essentiae, & magisteria arte chymica confici possint, ostenditur.* Hanoviae: Typis Wechelianis apud Claudium Marnium & heredes Ioannis Aubriii, 1609.

Bor, Levinus. *Amboinse oorlogen door Arnold de Vlaming van Oudshoorn, als superintendent over d'Oosterse gewesten oorlogaftig ten eind gebracht.* Delff: Arnold Bond, 1663.

Boxer, C. R. *Francisco Vieiera de Figueiredo. A Portuguese Merchant-Adventurer in South East Asia, 1624–1667.* 's-Gravenhage: Nijhoff, 1967.

———. *The Dutch in Brazil, 1624–1654.* Oxford: Clarendon Press, 1957.

Britain and The Netherlands. Papers Delivered to the Anglo-Dutch Historical Conference, 1962. Ed. J. S. Bromley and E. H. Kossmann. Groningen: Wolters, 1964.

Browne, Sir Thomas. *Pseudodoxia Epidemica; or Enquiries into very many received tenets and commonly presumed truths, which examined prove but Vulgar and Common Errors.* London: E. Dod, 1646. 2d. rev. ed. London: E. Dod and N. Ekins, 1650.

Brug, P. H. van der. *Malaria en malaise. De VOC in Batavia in de achttiende eeuw.* Amsterdam: De Bataafsche Leeuw, 1994.

Bruyn, J. R., et al., eds. *Dutch-Asiatic Shipping in the 17th and 18th Centuries.* 3 vols. The Hague: Martinus Nijhoff, 1979.

Buonanni, Filippo. *Ricreatione dell 'occhio e della mente nell' offeruation' delle chiocciole . . . .* Roma: Varese, 1681.

———. *Recreatio mentis, lb et ocvli in obseruatione animalium testaceorum curiosis naturae . . . .* Roma: Varesii, 1684.

Burnet, Thomas. *Heilige Beshouwinge* [sic] *des Aardkloots, Vervattende in vier Boeken De Oorsprong en Algemene Veranderingen Van onse Wereld, Welke die nu al ondergaan heeft, ofte nog na desen ondergaan zal.* Amsterdam: Daniel van Dalen, 1696.

Carrington, C. E. *The British Overseas. Exploits of a Nation of Shopkeepers.* Cambridge: Cambridge University Press, 1950.

Carus, Victor, J. *Geschichte der Zoologie bis auf Joh. Müller und Charl. Darwin.* München: R. Oldenbourg, 1872.

Cense, A. A. *Makassaars-Nederlands Woordenboek.* 's-Gravenhage: Nijhoff, 1979.

Chau Ju-Kua: His Work on the Chinese and Arab Trade in the Twelfth and Thirteenth Centuries, Entitled "Chuy-fan-chi." Transl. and ed. Friedrich Hirth and W. W. Rockhill. Taipei: Ch'eng-Wen Publishing Company, 1970.

Chijs, J. A. van. *De Vestiging van het Nederlandsche Gezag over de Banda-eilanden, 1599–1621.* Den Haag: Martinus Nijhoff, 1886.

Clausz, Bruno. *Geschichte des Staatlichen Gymnasiums zu Hanau.* Hanau: G. M. Alberti, 1925.
Clutius, Augerius. *Opuscula duo singularia; Historia cocci de Maldiva Lusitani, sece nucis medicae Maldivensium; de Hemerobio, sive Ephemero Insecto et Majali Verme.* Amsterdam: Jacobi Charpentier, 1634.
Collins, G. E. P. *East Monsoon.* New York: Scribner's, 1937.
Commelin, Izaäk. *Begin ende voortgangh, van de Vereenighde Nederlantsche geoctroyeerde Oost-Indische compagnie.* Amsterdam, 1645.
Constantin, Robert. *Lexicon graecolatinvm Rob. Constantini. Secunda hoc editione, partim ipsivs avthoris, partim Francisci Porti & aliorum additionibus plurimùm auctum, tum quanta fieri potuit diligentia recognitum, ita vt iam stvdiosis possit esse graecae linguae thesavrvs.* Genevae: Eustathii Vignon & Iacobus Stoer, 1592.
Coolhaas, W. Ph., ed. *Generale Missiven van Gouverneurs-Generaal en Raden aan de Heren XVII der Verenigde Oostindische Compagnie.* 8 vols. 's-Gravenhage: Martinus Nijhoff, 1960–1979.
Cornelissen, Georg. *Das Niederländische im Preußischen Gelderland und Seine Ablösung durch das Deutsche.* Bonn: Rheinisches Archiv, 1986.
Corsini, Filippo. *De twee reizen van Cosimo de Medici, prins van Toscane door de Nederlanden, 1667–1669.* Amsterdam: J. Muller, 1919.
Crawfurd, John. *History of the East Indian Archipelago.* 3 vols. Edinburgh: Constable, 1820.
———. *Descriptive Dictionary of the Indian Islands and Adjacent Countries* [1856]. Reprint ed., Kuala Lumpur: Oxford University Press, 1971.
Cruz, L. J., and J. White. "Clinical Toxicology of *Conus* Snail Stings," in Meier and White, *Clinical Toxicology of Animal Venoms*, pp. 117–127. Boca Raton: CRC Press, 1995.
Dam, Pieter van. *Beschrijvinge van de Oostindische Compagnie.* 7 vols. 's-Gravenhage: Martinus Nijhoff, 1927–1954.
Dampier, William C. *A History of Science.* Cambridge: Cambridge University Press, 1942.
Dance, S. Peter. *A History of Shell Collecting.* 2d. rev. ed. Leiden: Brill, 1986.
Darwin, Charles. *Charles Darwin's Notebooks, 1836–1844.* Transcr. and ed. Paul H. Barrett et al. Ithaca, NY: Cornell University Press, 1987.
———. *The Voyage of the Beagle.* Ed. Leonard Engel. Garden City, NY: Doubleday and Co., 1962.
Dassen, M. *De Nederlanders in de Molukken.* Utrecht: W. H. Van Heijningen, 1848.
De avonturen van een VOC-soldaat. Het dagboek van Carolus Van der Haeghe, 1699–1705. Ed. Jan Parmentier and Ruurdje Laarhoven. Zutphen: Walburg Pers, 1994.
De Clerck, F. S. A. *Het Maleisch der Molukken.* Batavia: W. Bruining, 1876.
———. *Nieuw Plantkundig Woordenboek voor Nederlandsch Indië.* 2d. rev. ed. Ed. A. Pulle. Amsterdam: J. H. de Bussy, 1927.
De Jonge, Nico, and Toos van Dijk. *Vergeten Eilanden. Kunst & Cultuur van de Zuidoost-Molukken.* Amsterdam: Periplus Editions, 1995.
De Klerck, E. S. *History of the Netherlands East Indies* [1938]. 2 vols. Reprint ed., Amsterdam: B. M. Israël, 1975.
De Ontdekkingsreis van Jacob le Maire en Willem Cornelisz. Schouten in de Jaren 1615–1617. 2 vols. Ed. W. A. Engelbrecht and P. J. van Herwerden. 's-Gravenhage: Martinus Nijhoff, 1945.
De reis van Mahu en de Cordes door de Straat van Magalhães naar Zuid-Amerika en Japan, 1598–1600. 2 vols. Ed. F. C. Wieder. 's-Gravenhage: Martinus Nijhoff, 1923–1924.
De wereld binnen handbereik. Nederlandse kunst-en rariteitenverzamelingen, 1585–1735. Ed. Ellinoor Bergvelt and Renée Kistemaker. Zwolle: Waanders Uitgevers, 1992.
Delman, H. C., and J. G. De Man. "On the 'Radjungans' of the Bay of Batavia." *Treubia* 6 (1925), pp. 308–323.
Dermoût, Maria. *De tienduizend dingen*, in *Verzameld Werk.* Amsterdam: Querido, 1974.
Descartes, René. *Oeuvres et Lettres.* Ed. André Bridoux. Paris: Gallimard, 1952.
Deursen, A. Th. van. *Mensen van klein vermogen. Het kopergeld van de Gouden Eeuw.* Amsterdam: Bert Bakker, 1992.
Dictionary of Scientific Biography. 18 vols. Ed. Charles Coulston Gillispie. New York: Scribner's, 1970–1990.
Die Matrikel der Hohen Schule und des Paedagogiums zu Herborn. Ed. Gottfried Zedler and Hans Sommer. Wiesbaden: J. F. Bergmann, 1908.
Dielmann, Karl. *Hanau am Main.* Hanau: Kuwe-Verlag, 1971.
Encyclopaedie van Nederlandsch-Indië. 4 vols. and 3 supps. Ed. D. G. Stubbe. Den Haag: Martinus Nijhoff, 1917–1932.
Engel, H. "The Life of Albert Seba." *Svenska Linné-Sällskapets Årsskrift* 20 (1937), pp. 76–100.
Engel, J. *Geschiedenis en algemeen overzicht van de Indonesische wapensmeedkunst.* Amsterdam: Samurai, 1980.
Everett, A. Hart. "On the Guliga of Borneo." *Journal of the Straits Branch of the Royal Asiatic Society* 4 (December 1879), pp. 57–58.

Fairchild, David. *Garden Islands of the Great East. Collecting Seeds from the Philippines and Netherlands India in the Junk "Chêng Ho."* New York: Scribner's, 1943.
Fischer, Heinrich. *Nephrit und Jadeit, nach ihren mineralogischen Eigenschaften sowie nach ihrer urgeschichtlichen und ethnographischen Bedeutung.* Stuttgart: E. Schweizerbart'sche Verlagshandlung, 1880.
Forbes, Anna. *Unbeaten Tracks in Islands of the Far East.* Reprint ed., Singapore: Oxford University Press, 1987.
Forbes, Henry O. *A Naturalist's Wanderings in the Eastern Archipelago. A Narrative of Travel and Exploration from 1878 to 1883.* New York: Harper, 1885.
Franke, Herbert, and Denis Twitchett, eds. *Alien Regimes and Border States, 907–1368.* Vol. 6 of *The Cambridge History of China.* Cambridge: Cambridge University Press, 1994.
Gaastra, Femme S. *De Geschiedenis van de VOC.* Zutphen: Walburg Pers, 1991.
Gelder, Roelof van. *Het Oost-Indisch avontuur. Duitsers in dienst van de VOC, 1600–1800.* Nijmegen: SUN, 1997.
———. "De wereld binnen handbereik. Nederlandse kunst-en rariteitenverzamelingen, 1585–1735," in *De wereld binnen handbereik, Nederlandse kunst-en rariteitenverzamelingen, 1585–1735,* ed. Ellinoor Bergvelt and Renée Kistemaker, pp. 15–38. Zwolle: Waanders Uitgevers, 1992.
Gesner, Conrad. *Historiae Animalium. Liber IV. Qui est de Piscium & Aquatilium Animantium Natura.* Zürich: Christoph. Froschover, 1558.
Gould, Stephen Jay. *Dinosaur in a Haystack. Reflections in Natural History.* New York: Crown, 1997.
Graaf, H. J. de. *De geschiedenis van Ambon en de Zuid-Molukken.* Franeker: Wever, 1977.
Grigson, Geoffrey, and C. H. Gibbs-Smith. *Things. A Volume of Objects Devised by Man's Genius Which Are the Measure of His Civilization.* New York: Hawthorn, 1957.
Grimmelshausen, Johan Jakob Christoffel von. *Grimmelshausens Simplicissimus Teutsch.* Ed. J. H. Scholte. Tübingen: Max Niemeyer Verlag, 1954.
———. *Simplicius Simplicissimus.* Transl. George Schulz-Behrend. Indianapolis: Bobbs-Merrill, 1965.
Gubernatis, Angelo de. *Zoological Mythology; or, The Legends of the Animals* [1872]. 2 vols. Reprint ed., Detroit: Singing Tree Press, 1968.
Haan F. de. *Oud Batavia.* 2 vols. Batavia: G. Kolff, 1922.
———. *Priangan. De Preanger-Regentschappen onder het Nederlandsch Bestuur tot 1811.* 4 vols. Batavia: Bataviaasch Genootschap van Kunsten en Wetenschappen, 1910–1912.
———. "Uit oude notarispapieren II. Andreas Cleyer." *Tijdschrift voor Indische Taal-, Land-, en Volkenkunde* 46 (1903), pp. 423–468.
Hallema, A. "Kakkerlakken als bijnaam voor Indische Albino's." *De Indische Gids* 62, no. 8 (August 1940), pp. 617–621.
Hanauer Geschichtsblätter, Neue Folge, nos. 3–4. Hanau, 1919.
Heij, Kees. "Bericht uit de Molukken." *Moesson* 42, no. 8 (February 1998), p. 23.
Heijer, Henk den. *De geschiedenis van de WIC.* Zutphen: Walburg Pers, 1994.
Heinie-Geldern, Robert von. "Prehistoric Research in the Netherlands Indies," in *Science and Scientists in the Netherlands Indies,* ed. Pieter Honig and Frans Verdoorn, pp. 129–167. New York: Board of the Netherlands Indies, Surinam, and Curaçao, 1945.
Hendrik Engel's Alphabetical List of Dutch Zoological Cabinets and Menageries. 2d. rev. ed. Ed. Pieter Smit. Amsterdam: Rodopi, 1986.
Heniger, J. *Hendrik Adriaan Van Reede tot Drakenstein, 1636–1691, and Hortus Malabaricus. (A Contribution to the History of Dutch Colonial Botany).* Rotterdam: A. A. Balkema, 1986.
Herbert, Sir Thomas. *A relation of some years travails, begvnne anno 1626. Into Afrique and the greater Asia, especially the territories of the Persian monarchie, and some parts of the Oriental Indies and iles adiacent.* London: W. Stansby and J. Bloom, 1634.
Hexham, Henry. *A copious English and Nether Dutch Dictionary.* Rotterdam: De Weduwe van Arnold Leers, 1675.
Heyne, K. *De Nuttige planten van Nederlandsch Indië.* 3 vols. 2d. rev. ed. Batavia: Departement van Landbouw, Nijverheid & Handel, 1927.
Hien, H. A. van. *De Javaansche Geestenwereld en de betrekking, die tusschen de geesten en de zinnelijke wereld bestaat.* 4 vols. Semarang & Bandung: Van Dorp & Fortuna, 1896–1913.
Hoëvell, G. W. W. C. Baron van. *Ambon, en meer bepaaldelijk De Oeliasers.* Dordrecht: Blussé en Van Braam, 1875.
Hollstein, F. W. H. *Dutch and Flemish Etchings, Engravings and Woodcuts ca. 1450–1700,* vol. 15. Amsterdam: Menno Hertzberger, n.d.
Holthuis, L. B. *FAO Species Catalogue.* Vol. 14. *Marine Lobsters of the World. An Annotated and Illustrated Catalogue of Species of Interest to Fisheries Known to Date.* Rome: Food and Agriculture Organization of the United Nations, 1991.
———. "Are There Poisonous Crabs?" *Crustaceana* 15 (1968), pp. 215–222.

Honig, Pieter, and Frans Verdoorn, eds. *Science and Scientists in the Netherlands Indies.* New York: Board of the Netherlands Indies, Surinam, and Curaçao, 1945.

l'Honoré Naber, S. P. "Het dagboek van Hendrik Haecxs, lid van de Hoogen Raad van Brazilië, 1645–1654." *Bijdragen en Mededelingen van het Historisch Genootschap* 46 (1925), pp. 126–311.

Hoogewerff, G. J. *De twee reizen van Cosimo de Medici Prins van Toscane door de Nederlanden, 1667–1669.* Amsterdam: Johannes Müller, 1919.

Hoyt, Sarnia Hayes. *Old Malacca.* Kuala Lumpur: Oxford University Press, 1993.

Huxley, T. H. *The Crayfish. An Introduction to the Study of Zoology.* London: Kegan Paul, 1880.

Jonston, Jan. *Historiae naturalis de piscibus et cetis libri v, cum aeneis figuris.* Amstelodami: J. J. fil. Schipperi, 1657.

———. *Naeukeurige beschryving van de natuur der vier-voetige dieren, vissen en bloedloze water-dieren, vogelen, kronkel-dieren, slangen en draken.* Transl. M. Grausius. Amsterdam: J. J. Schipper, 1660.

Jörge, C. J. A. *Porcelain and the Dutch China Trade.* The Hague: Martinus Nijhoff, 1982.

Joubert, Laurent. *Operum latinorum tomus primus . . . Cui subjectus est tomus secundus, nunc primum in lucem proditus.* Lugduni [Lyons]: S. Michaelem, 1582.

Kircher, Athanasius. *China monumentis qva sacris qua profanis nec non variis naturae & artis spectaculis, aliarumque rerum memorabilium argumentis illustrata.* Amstelodami: Joannem Janssonium à Waesberge & Elizeum Weyerstraet, 1667.

———. *Magneticum naturae regnum; sive, Disceptatio physiologica de triplici in natura rerum magnete, juxta triplicem ejusdem naturae gradum digesto: inanimato, animato, sensitivo.* Romae: Typis Ignatii de Lazaris, 1667.

———. *Mundus subterraneus, in XII libros digestus; quo divinum subterrestris mundi opificium, mira ergasteriorum naturae in eo distributio, verbo pant amorphou Protei regnum, universae denique naturae majestas & divitiae summa rerum varietate exponuntur.* Amstelodami: J. Janssonium & E. Weyerstraten, 1665.

———. *Toonneel van China, door veel, zo geestelijke als weereltlijke geheugteekenen, verscheide vertoningen van de natuur en kunst, en blijken van veel andere gedenkwaerdige dingen, geopent en verheerlykt.* Transl. J. H. Glazemaker. Amsterdam: J. J. van Waesberge, 1668.

Kirsch, Peter. *Die Reise nach Batavia. Deutsche Abenteurer in Ostindien 1609 bis 1695.* Hamburg: Kabel, 1994.

Kistemaker, Renée, et al., eds. *Peter de Grote en Holland. Culturele en wetenschappelijke betrekkingen tussen Rusland en Nederland ten tijde van tsaar Peter de Grote.* Bussum: Thoth, 1996.

Kratzsch, Siegfried. "Das kaiserliche Privileg der Leopoldina vom 7. August 1687." *Acta Historica Leopoldina* 17 (1987), pp. 770.

Kreemer, J. Jr. *Volksheelkunde in den Indischen Archipel.* Den Haag: Bijdragen Taal-Land-en Volkenkunde, 1914.

Kruijt, Albert C. "Het Ijzer in Midden-Celebes." *Bijdragen tot de Taal-, Land- en Volkenkunde van Nederlandsch-Indië* 6, no. 9 (1901), pp. 148–160.

Krupp, Ingrid. *Das Renaissanceschloß Hadamar.* Wiesbaden: Historischen Kommission für Nassau, 1986.

Kunz, G. F. *The Curious Lore of Precious Stones* [1913]. Reprint ed., New York: Dover, 1971.

Laet, Johannes de. *Historie ofte Iaerlyck Verhael van de West-Indische Compagnie.* Leiden: Elsevier, 1644.

Lammert, Gottfried. *Geschichte der Seuchen, Hungers-und Kriesgsnoth zur zeit des Dreissigjährigen Krieges* [1890]. Reprint ed., Weisbaden: Dr. Martin Sändig oHG, 1971.

Landwehr, John. *VOC. A Bibliography of Publications Relating to the Dutch East India Company, 1602–1800.* Ed. Peter van der Krogt. Utrecht: HES Publishers, 1991.

Langer, Herbert. *Hortus Bellicus der Dreissigjährige Krieg. Eine Kulturgeschichte.* Leipzig: Edition Leipzig, 1978.

Laufer, Berthold. *Jade. A Study in Chinese Archaeology and Religion* [1912]. Reprint ed., New York: Dover, 1974.

Laufer, E. W. "Schets van de Residentie Amboina." *Bijdragen tot de Taal-, Land- en Volkenkunde van Nederlandsch-Indië* 3 (1900), pp. 53–78.

Laurie, Eleanor M. O., and J. E. Hill. *List of Land Mammals of New Guinea, Celebes and Adjacent Islands.* London: British Museum, 1954.

Lentz, Christel. "Der Lustgarten des Grafen Johannes von Nassau-Idstein und sein Blumenbuch." *Hessische Heimat* 5, no. 3 (1995), pp. 83–93.

———. *Das Idsteiner Schloß. Beiträge zu 300 Jahren Bau-und Kulturgeschichte.* Idstein: Schulz-Kirchner Verlag, 1994.

Lesson, P. *Voyage autour du monde entrepris par ordre du Gouvernement sur la corvette La Coquille.* 4 vols. Paris: P. Pourrat, 1839.

Levathes, Louise. *When China Ruled the Seas. The Treasure Fleet of the Dragon Throne, 1405–1433.* New York: Simon and Schuster, 1994.

Lévi-Strauss, Claude. *The Savage Mind.* Chicago: University of Chicago Press, 1966.
Linschoten, Jan Huygen van. *Itinerario. Voyage ofte Schipvaert van Jan Huygen van Linschoten naer Oost ofte Portugaels Indien, 1579–1592.* 5 vols. Ed. H. Kern et al. Den Haag: Martinus Nijhoff, 1910–1939.
Lopes de Castanheda, Fernão. *Historia do descobrimento & conquista da India pelos Portugueses.* Coimbra: J. de Barreyra, 1551.
Mann, Golo. *Wallenstein* [1971]. Transl. Charles Kessler. New York: Holt, Rinehart, and Winston, 1976.
Martin, K. *Reisen in den Molukken, in Ambon, den Uliassern, Seran (Ceram) und Buru. Eine Schilderung von Land und Leuten.* Leiden: Brill, 1894.
Martini, Martino. *De bello tartarico historia: in qua, quo pacto Tartari hac nostra aetate Sinicum imperium invaserint, ac fere totum occuparint, narratur; eorumq mores breviter describuntur.* Viennae Austriae: M. Cosmerovium, 1654.
———. *Novus atlas sinensis a Martino Martinio . . . descriptus et serenmo.* N.p., 1655.
Masselman, George. *The Cradle of Colonialism.* New Haven: Yale University Press, 1963.
Meier, J., and J. White, eds. *Clinical Toxicology of Animal Venoms.* Boca Raton: CRC Press, 1995.
Merian, Maria Sibylla, Leningrader Aquarelle. 2 vols. Ed. Ernst Ullmann, Helga Ullmann, et al. Lucerne: Bucher, 1974.
Merrill, Elmer D. *Plant Life of the Pacific World.* New York: Macmillan, 1945.
Mi Fu on Ink-Stones. Transl., with an introduction and notes by, R. H. van Gulik. Peking: Henri Vetch, 1938.
Middelnederlandsch Woordenboek. 9 vols. Ed. E. Verwijs and J. Verdam. 's-Gravenhage: Martinus Nijhoff, 1885–1929.
Montaigne. *Essais.* 2 vols. Paris: Garnier, 1965.
Moreau, Pierre. *Histoire des derniers troubles du Brésil entre les Hollandais et les Portugais depuis l'an 1644 jusques en 1648.* Paris: Chez Augustin Courbe, 1651.
Mulder, Niels. *Mysticism and Everday Life in Contemporary Java. Cultural Persistence and Change.* Singapore: Singapore University Press, 1978.
Mundy, Rodney. *Narratives of Events in Borneo and Celebes, Down to the Occupation of Labuan. From the Journals of James Brooke, Esq., Rajah of Sarawak, and Governor of Labuan. Together with a Narrative of the Operations of H. M. S. Iris.* 2 vols. London: John Murray, 1848.
Needham, Joseph. *Science and Civilization in China.* Vol. 3. *Mathematics and the Sciences of the Heavens and the Earth.* Cambridge: Cambridge University Press, 1959
———. *Science and Civilization in China.* Vol. 4. *Physics and Physical Technology.* Part 1. *Physics.* Cambridge: Cambridge University Press, 1962.
———. *Science and Civilization in China.* Vol. 4. *Physics and Physical Technology.* Part 3. *Civil Engineering and Nautics.* Cambridge: Cambridge University Press, 1971.
———. *Science and Civilization in China.* Vol. 5. *Chemistry and Chemical Technology.* Part 2. *Spagyrical Discovery and Invention. Magisteries of Gold and Immortality.* Cambridge: Cambridge University Press, 1974.
Netscher, P. M. *Les Hollandais au Brésil. Notice Historique sur Les Pays-Bas et Le Brésil au XVIIe Siècle.* La Haye: Belinfante Frères, 1853.
Nieremberg, Juan Eusebio. *Historia naturae, maxime peregrinae, libris XVI.* Antwerp: Ex officina Plantiniana Balthasar Moreti, 1635.
Nieuhof, Joan. *Het Gezantschap der Neêrlandtsche Oost-Indische Compagnie, aan den grooten Tartarischen Cham, Den tegenwoordigen Keizer van China.* Amsterdam: J. van Meurs, 1665.
———. *Gedenkwaardige Zee en Landreize door de voornaamste landschappen van West en Oost-Indien.* Amsterdam: De Weduwe van Iacob van Meurs, 1682.
Nieuw Nederlandsch Biografisch Woordenboek. 10 vols. Ed. P. C. Molhuysen, P. J. Blok, K. H. Kossmann et al. Amsterdam: N. Israel, 1911–1974.
Nieuwenhuys, Rob. *Komen en blijven. Tempo doeloe—een verzonken wereld. Fotografische documenten uit het oude Indië, 1870–1920.* Amsterdam: Querido, 1982.
Orta, Garcia da. *Colloquies on the Simples & Drugs of India* [1563]. Transl. Clements Markham. London: Henry Sothern, 1913.
Parker, Geoffrey. *The Thirty Years' War.* London: Routlege and Kegan Paul, 1984.
Peacock, D. P. S. *Pottery in the Roman World. An Ethnoarchaeological Approach.* London: Longman, 1982.
Pisonis, Gulielmi. *De Indiae Utrisque Re Naturali et Medica.* Amstelaedami: Elzevirios, 1658.
———. *Historia Naturalis Brasiliae.* Lugduni Batavorum: Apud Franciscum Hackium & Amstelodami: Apud lud. Elzevirium, 1648.
Pletsch, Theodore W., ed. *Fishes, Crayfishes, and Crabs. Louis Renard's Natural History of the Rarest Curiosities of the Seas of the Indies.* 2 vols. Baltimore: Johns Hopkins University Press, 1995.
Pliny. *Natural History.* 10 vols. Transl. H. Rackham, W. H. S. Jones, and D. E. Eichholz. Cambridge, MA: Harvard University Press, 1938–1962.

Polo, Marco. *The Travels of Marco Polo. The Complete Yule-Cordier Edition* [1903 and 1920]. 2 vols. Reprint ed., New York: Dover, 1993.
Pomian, Krzysztof. *Collectors and Curiosities: Paris and Venice, 1500–1800.* Transl. Elizabeth Wiles-Portier. Cambridge: Polity, 1990.
Ponder, H. W. *In Javanese Waters.* London: Seeleye, 1944.
Puype, J. P. *The Visser Collection. Arms of the Netherlands in the Collection of H. L. Visser.* Vol. 1, in 3 parts. Zwolle: Waanders Publishers, 1996.
Pyrard, François. *Voyage de François Pyrard de Laual. Contenant sa Navigation aux Indies Orientales, Maldiues, Molugues, Bresil & Etc.* 2 vols. 3d. rev. ed. Paris: Chez Samvel Thibovst, 1619.
———. *The Voyage of Pyrard of Laval to the East Indies, the Maldives, the Moluccas and Brazil.* 2 vols. Transl. and ed. Albert Gray. London: Hakluyt Society, 1887.
Radzjoen, Anna. "De anatomische collectie van Frederik Ruysch in Sint-Petersburg," in Kistemaker et al., *Peter de Grote en Holland*, pp. 47–53. Bussum: Thoth, 1977.
Raffles, Thomas Stamford. *The History of Java* [1817]. 2 vols. Reprint ed., Kuala Lumpur: Oxford University Press, 1978.
Ray, John. *The Wisdom of God Manifested in the Works of the Creation: Being the Substance of some Common Places delivered in the Chappel of Trinity College in Cambridge.* London: Samuel Smith, 1691.
———. *Miscellaneous Discourses concerning the Dissolution and Changes of the World; Wherein the primitive Chaos and Creation, the general Deluge, Fountains, Formed Stones, Sea-Shells found in the Earth, Subterraneous Trees, Mountains, Earthquakes, Vulcanoes, the Universal Conflagration and Future State, are largely discussed and examined.* London: Samuel Smith, 1692.
Raychaudhuri, Tapan. *Jan Company in Coromandel, 1605–1690. A Study in the Interrelations of European Commerce and Traditional Economies.* The Hague: Martinus Nijhoff, 1962.
Read, B. E., and C. Pak. *Chinese Materia Medica. A Compendium of Minerals and Stones.* 2d. ed. Taipei: SMC Publishing, 1977.
Renkhoff, Otto. *Nassauische Biographie.* 2d. rev. ed. Wiesbaden: Historische Kommission für Nassau, 1992.
Rieß, Eugen. *250 Jahre Evangelisch-Reformierte Kirche Wölfersheim.* Wölfersheim: Kirchengemeinde Wölfersheim, 1991.
Riedel, Johan Gerard Friedrich. *De Sluik-en Kroesharige Rassen tusschen Selebes en Papua.* 's-Gravenhage: Nijhoff, 1886.
Rochefort, Charles de. *Histoire naturelle et morale des Iles Antilles de l'Amerique. Enrichie de plusieurs belles Figures des raretez les plus considerables qui y sont d'écrites. Avec un vocabulaire Caraibe.* Rotterdam: Chez Arnout Leers, 1658.
———. *Natuurlyke en zedelyke historie van d'eylanden de Vooreylanden van Amerika. Met eenen Caraibaanschen woorden-schat.* Rotterdam: A. Leers, 1662.
———. *The History of the Caribby-Islands.* 2 vols. London: Printed by J. M. for Thomas Dring and John Starkey, 1666.
Romburgh, C. G. M. van, C. E. Warnsinck-Delprat, eds. *Tresoor der Zee-en Landreizen. Beredeneerd register op de werken der Linschoten-Vereniging.* 's-Gravenhage: Martinus Nijhoff, 1957.
Rondeletii, Gulielmi. *Libri de piscibus marinis in quibus verae piscium effigies expressae sunt; Universae aquatilium historiae pars altera.* Lvgdvni: Matthiam Bonhomme, 1554–1555.
Rouffaer, G. P. "Een paar aanvullingen over Bronzen Keteltrommen in Ned.-Indië." *Bijdragen tot de Taal-, Land- en Volkenkunde van Nederlandsch-Indië* 6, no. 7 (1900), pp. 284–307.
———. "Waar kwamen de raadselachtige mutisala's (aggrikralen) in de Timor-groep oorspronkelijk vandaan?" *Bijdragen tot de Taal- Land- en Volkenkunde van Nederlandsch-Indië* 50 (1899), pp. 409–675.
Scaligero, Giulio Cesare. *Exotericarum exercitationum liber quintus decimus: De subtilitate.* Lutetiae: Michaelis Vascosani, 1557.
Schafer, Edward H. *The Vermilion Bird. T'ang Images of the South* [1967]. Reprint ed., Berkeley: University of California Press, 1985.
Schäfer, H. *Geschichte von Portugal.* 5 vols. Hamburg: Friedrich Perthes, 1852.
Schellenberg, Karl-Heinz. *Braunfelser Chronik.* Braunfels: Magistrat der Stadt Braunfels, 1990.
Schimper, A. F. W. *Die indo-malayische Strandflora.* Jena: Gustav Fischer, 1891.
Schott, Caspar. *Physica curiosa, sive, Mirabilia naturae et artis libris XII.* Herbipoli [Würzburg]: Johannis Andreae Endteri & Wolffgangi jun. haeredum, excudebat J. Hertz, 1662.
Schröder, C. A. *Ethnographische Atlas, bevattende afbeeldingen van voorwerpen uit het leven en de huishouding der Makassaren.* Amsterdam, 1859.
Schuster, W. H., and Rustami Djajadiredja. *Local Names of Indonesian Fishes.* Bandung: W. van Hoeve, 1952.
Semmelink, J. *Geschiedenis der cholera in Oost-Indië vóór 1817.* Utrecht: C. H. E. Breijer, 1885.

Skeat, W. W. *Malay Magic. An Introduction to the Folklore and Popular Religion of the Malay Peninsula* [1900]. Reprint ed., London: Frank Cass & Co., 1965.
Sobel, Dave. *Longitude.* New York: Walker and Co., 1995.
Spence, Jonathan D. *Emperor of China. Self-Portrait of K'ang-hsi.* London: Jonathan Cape, 1974.
Stackpole, Edouard A. *Whales and Destiny. The Rivalry between America, France, and Britain for Control of the Southern Whale Fishery, 1785–1825.* Amherst: University of Massachusetts Press, 1972.
Stapel, F. W. *Cornelis Janszoon Speelman.* Den Haag: Martinus Nijhoff, 1936.
———. *Geschiedenis van Nederlandsch Indië.* 5 vols. Amsterdam: Joost van den Vondel, 1938–1940.
———. "Hubert Hugo. Een zeerover in dienst van de Oostindische Compagnie," in *Bijdragen in de Taal-, Land- en Volkenkunde*, pp. 615–635. 1930.
Stapel, F. W. et al., eds. *Geschiedenis van Nederlandsch Indië.* 5 vols. Amsterdam: Joost van den Vondel, 1938–1940.
Suetonius. *The Twelve Caesars.* Transl. Robert Graves. Harmondsworth: Penguin, 1957.
Tavernier, Jean Baptiste. *Les six voyages de Jean Baptiste Tavernier. Ecuyer Baron d'Aubonne, en Turquie, en Perse, et aux Indes, pendant l'espace de quarante ans & par toutes les routes que l'on peut tenir.* Paris: Gervais Clouzier & Claude Barbin, 1676.
———. *Travels in India by Jean Baptiste Tavernier, baron of Aubonne* [1676]. Transl. with a biographical sketch of the author, notes, appendices, etc., by, V. Ball. London: Macmillan, 1889.
Taylor, Paul Michael, and Lorraine V. Aragon. *Beyond the Java Sea. Art of Indonesia's Outer Islands.* Washington, DC: National Museum of Natural History, 1991.
Thieme, Ulrich, Felix Becker, and Hans Vollmer. *Allgemeines Lexicon der bildenden Künstler von der antike bis zur Gegenwart.* 37 vols. Leipzig: Seemann, 1907–1950.
Valentijn, François. *François Valentijn's Description of Ceylon.* Transl. and ed. Sinnappah Arasaratnam. London: The Hakluyt Society, 1978.
Verstegen, Richard. *Antiquitates Belgicae, of Nederlandsche Oudtheden.* Amsterdam: Jacob Royen, 1700.
Veth, P. J. *Java, geographisch, ethnologisch, historisch.* 3 vols. Haarlem: Bohn, 1898.
Vlekke, Bernard H. M. *Nusantara. A History of the East Indian Archipelago.* Cambridge, MA: Harvard University Press, 1943.
Vries, Jan de, and Ad van der Woude. *Nederland, 1500–1815. De eerste ronde van moderne economische groei.* Amsterdam: Balans, 1995.
Wall, V. I. van de. *De Nederlandsche Oudheden in de Molukken.* 's-Gravenhage: Nijhoff, 1928.
Wallace, Alfred Russell. *The Malay Archipelago. The Land of the Orang-Utan and the Bird of Paradise. A Narrative of Travel with Studies of Man and Nature* [1869]. Reprint ed., New York: Dover, 1962.
Wedgwood, C. V. *The Thirty Years War.* New Haven: Yale University Press, 1939.
White, T. H., transl. and ed. *The Bestiary* [1954]. Reprint ed., New York: Capricorn Books, 1960.
Wijdtloopigh Verhaal van tgene de vijf schepen (die int jaer 1598 tot Rotterdam toegherust werden, om door de straet Magellana haren handel te dryven) wedervaren is . . . meest beschreven door M. Barent Iansz. cirurgijn. Amsterdam: Zacharias Heijns, 1600.
Wille, R. *Hanau im dreissigjährigen Kriege.* Hanau: G. M. Alberti, 1886.
Wills, John E. *Pepper, Guns and Parleys. The Dutch East India Company and China, 1622–1681.* Cambridge, MA: Harvard University Press, 1974.
Wilson, Edward O. *Biophilia.* Cambridge, MA: Harvard University Press, 1984.
Wilson, Edward O. *Consilience. The Unity of Knowledge.* New York: Knopf, 1998.
Winter, Helmut. *Festschrift zur 375-Jahr-Feier der Hohen Landesschule Hanau, 1607–1982.* Hanau, 1982.
Wit, H. C. D. de. *Ontwikkelingsgeschiedenis van de biologie.* 2 vols. in 3 parts. Wageningen: PUDOC, 1982–1989.
Zedler, Johann Heinrich. *Grosses vollständiges Universal Lexicon* 64 vols. and 4 supps. Halle & Leipzig, 1732–1754.
Zierler, Martin. *Das vierdte hundert Epistein, oder, Sendschreiben von underschidlichen Politischen, Historischen und andern Materien und Sachen.* Vlm: Johann Ghorlins, 1644.
Zimmerman, Ernst J. *Hanau. Stadt und Land, Kulturgeschichte einer Fränkisch-Wetterauischen Stadt.* Hanau [Selbstverlag], 1917.

INDEX

Species are generally listed by their modern Latin names. Page numbers in boldface refer to illustrations.